IMPORTANT:
HERE IS YOUR REGISTRATION CODE TO ACCESS YOUR PREMIUM McGRAW-HILL ONLINE RESOURCES.

For key premium online resources you need THIS CODE to gain access. Once the code is entered, you will be able to use the Web resources for the length of your course.

If your course is using **WebCT** or **Blackboard**, you'll be able to use this code to access the McGraw-Hill content within your instructor's online course.

Access is provided if you have purchased a new book. If the registration code is missing from this book, the registration screen on our Website, and within your WebCT or Blackboard course, will tell you how to obtain your new code.

Registering for McGraw-Hill Online Resources

To gain access to your McGraw-Hill web resources simply follow the steps below:

1. USE YOUR WEB BROWSER TO GO TO: http://www.mhhe.com/shieress8/
2. CLICK ON **FIRST TIME USER**.
3. ENTER THE REGISTRATION CODE* PRINTED ON THE TEAR-OFF BOOKMARK ON THE RIGHT.
4. AFTER YOU HAVE ENTERED YOUR REGISTRATION CODE, CLICK **REGISTER**.
5. FOLLOW THE INSTRUCTIONS TO SET-UP YOUR PERSONAL UserID AND PASSWORD.
6. WRITE YOUR UserID AND PASSWORD DOWN FOR FUTURE REFERENCE. KEEP IT IN A SAFE PLACE.

TO GAIN ACCESS to the McGraw-Hill content in your instructor's **WebCT** or **Blackboard** course simply log in to the course with the UserID and Password provided by your instructor. Enter the registration code exactly as it appears in the box to the right when prompted by the system. You will only need to use the code the first time you click on McGraw-Hill content.

Thank you, and welcome to your McGraw-Hill online resources!

* YOUR REGISTRATION CODE CAN BE USED ONLY ONCE TO ESTABLISH ACCESS. IT IS NOT TRANSFERABLE.

0-07-235118-7 SHIER/BUTLER/LEWIS: HOLE'S ESSENTIALS OF HUMAN ANATOMY & PHYSIOLOGY, 8E

ONLINE RESOURCES

REGISTRATION CODE

jalaluddin-98964538

Test Yourself

Take a quiz at the *Hole's Essentials of Human Anatomy and Physiology* Online Learning Center to gauge your mastery of chapter content. Each chapter quiz is specially constructed to test your comprehension of key concepts. Feedback on your responses explains why an answer is correct or incorrect. You can even e-mail your quiz results to your professor!

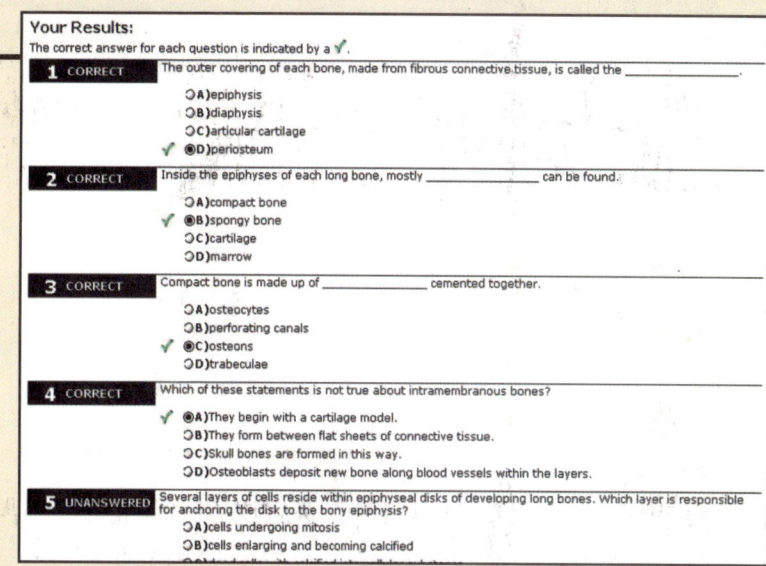

Course Tools

Here you'll find chapter-specific study outlines and a listing of relevant websites, along with links to interactive lab activities. The *Hole's Essentials of Human Anatomy and Physiology* Online Learning Center also features cutting-edge online histology and anatomy atlases plus general study tips and career information.

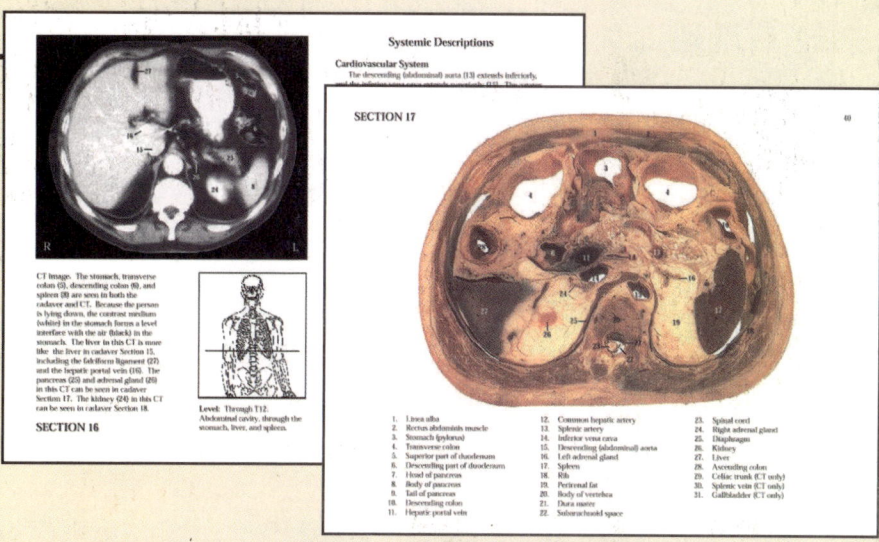

Access to Premium Learning Materials

Hole's Essentials of Human Anatomy and Physiology Online Learning Center is your portal to exclusive interactive study tools like McGraw-Hill's Essential Study Partner and PowerWeb.

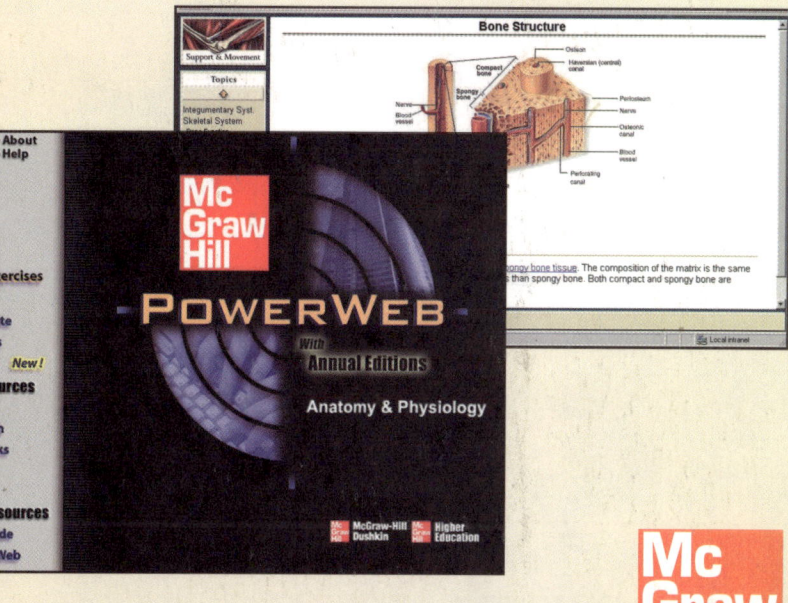

Visit www.mhhe.com/shieress8 today!

EIGHTH EDITION

Hole's Essentials of Human Anatomy and Physiology

DAVID SHIER
Washtenaw Community College

JACKIE BUTLER
Grayson County College

RICKI LEWIS
The University at Albany

Boston   Burr Ridge, IL   Dubuque, IA   Madison, WI   New York   San Francisco   St. Louis
Bangkok   Bogotá   Caracas   Kuala Lumpur   Lisbon   London   Madrid   Mexico City
Milan   Montreal   New Delhi   Santiago   Seoul   Singapore   Sydney   Taipei   Toronto

McGraw-Hill Higher Education

A Division of The **McGraw-Hill** Companies

HOLE'S ESSENTIALS OF HUMAN ANATOMY & PHYSIOLOGY
EIGHTH EDITION

Published by McGraw-Hill, a business unit of The McGraw-Hill Companies, Inc., 1221 Avenue of the Americas, New York, NY 10020. Copyright © 2003, 2000, 1998, 1995, 1992, 1989, 1986, 1983 by The McGraw-Hill Companies, Inc. All rights reserved. No part of this publication may be reproduced or distributed in any form or by any means, or stored in a database or retrieval system, without the prior written consent of The McGraw-Hill Companies, Inc., including, but not limited to, in any network or other electronic storage or transmission, or broadcast for distance learning.

Some ancillaries, including electronic and print components, may not be available to customers outside the United States.

 This book is printed on recycled, acid-free paper containing 10% postconsumer waste.

International   1 2 3 4 5 6 7 8 9 0 VNH/VNH 0 9 8 7 6 5 4 3 2
Domestic   1 2 3 4 5 6 7 8 9 0 VNH/VNH 0 9 8 7 6 5 4 3 2

ISBN 0–07–235118–7
ISBN 0–07–119892–X (ISE)

Publisher: *Martin J. Lange*
Sponsoring editor: *Michelle Watnick*
Senior developmental editor: *Patricia Hesse*
Director of development: *Kristine Tibbetts*
Senior project manager: *Jayne Klein*
Senior production supervisor: *Sandy Ludovissy*
Media project manager: *Sandra M. Schnee*
Senior media technology producer: *Barbara R. Block*
Designer: *K. Wayne Harms*
Cover/interior designer: *Kristyn A. Kalnes*
Cover image: *Doug Pensinger/Allsport*
Senior photo research coordinator: *John C. Leland*
Photo research: *Mary Reeg*
Compositor: *Precision Graphics*
Typeface: *10.5/12 Garamond Regular*
Printer: *Von Hoffmann Press, Inc.*

The credits section for this book begins on page 568 and is considered an extension of the copyright page.

Library of Congress Cataloging-in-Publication Data

Shier, David.
    Hole's essentials of human anatomy & physiology — 8th ed. / David Shier, Jackie Butler, Ricki Lewis.
     p. cm.
    Includes index.
    ISBN 0–07–235118–7
    1. Human physiology. 2. Human anatomy. I. Title: Essentials of human anatomy & physiology. II. Title: Essentials of human anatomy and physiology. III. Shier, David. IV. Butler, Jackie. V. Lewis, Ricki. VI. Title.

QP34.5 .S49   2003
612—dc21   2001044530
                                                                               CIP

INTERNATIONAL EDITION ISBN 0–07–119892–X
Copyright © 2003. Exclusive rights by The McGraw-Hill Companies, Inc., for manufacture and export. This book cannot be re-exported from the country to which it is sold by McGraw-Hill. The International Edition is not available in North America.

www.mhhe.com

Brief Contents

Unit 1
LEVELS OF ORGANIZATION

1. Introduction to Human Anatomy and Physiology 1
2. Chemical Basis of Life 29
3. Cells 47
4. Cellular Metabolism 72
5. Tissues 90

Unit 2
SUPPORT AND MOVEMENT

6. Skin and the Integumentary System 111
7. Skeletal System 126
8. Muscular System 170

Unit 3
INTEGRATION AND COORDINATION

9. Nervous System 205
10. Somatic and Special Senses 251
11. Endocrine System 279

Unit 4
TRANSPORT

12. Blood 305
13. Cardiovascular System 327
14. Lymphatic System and Immunity 365

Unit 5
ABSORPTION AND EXCRETION

15. Digestion and Nutrition 391
16. Respiratory System 433
17. Urinary System 459
18. Water, Electrolyte, and Acid-Base Balance 480

Unit 6
THE HUMAN LIFE CYCLE

19. Reproductive Systems 496
20. Pregnancy, Growth, and Development 526

Contents

Preface vii

UNIT 1
Levels of Organization

CHAPTER 1

Introduction to Human Anatomy and Physiology 1

1.1 Introduction 3
1.2 Anatomy and Physiology 3
1.3 Characteristics of Life 4
1.4 Maintenance of Life 4
1.5 Levels of Organization 7
1.6 Organization of the Human Body 8
1.7 Anatomical Terminology 13

REFERENCE PLATES

The Human Organism 21

CHAPTER 2

Chemical Basis of Life 29

2.1 Introduction 31
2.2 Structure of Matter 31
2.3 Chemical Constituents of Cells 37

CHAPTER 3

Cells 47

3.1 Introduction 49
3.2 Composite Cell 49
3.3 Movements Through Cell Membranes 58
3.4 The Cell Cycle 64

CHAPTER 4

Cellular Metabolism 72

4.1 Introduction 74
4.2 Metabolic Reactions 74
4.3 Control of Metabolic Reactions 75
4.4 Energy for Metabolic Reactions 76
4.5 Metabolic Pathways 79
4.6 Nucleic Acids 79

4.7 Protein Synthesis 81
4.8 DNA Replication 86

CHAPTER 5

Tissues 90

5.1 Introduction 92
5.2 Epithelial Tissues 92
5.3 Connective Tissues 98
5.4 Muscle Tissues 105
5.5 Nervous Tissues 107

UNIT 2
Support and Movement

CHAPTER 6

Skin and the Integumentary System 111

6.1 Introduction 113
6.2 Types of Membranes 113
6.3 Skin and Its Tissues 113
6.4 Accessory Organs of the Skin 117
6.5 Regulation of Body Temperature 120
6.6 Healing of Wounds 120

CHAPTER 7

Skeletal System 126

7.1 Introduction 128
7.2 Bone Structure 128
7.3 Bone Development and Growth 130
7.4 Bone Function 131
7.5 Skeletal Organization 134
7.6 Skull 136
7.7 Vertebral Column 142
7.8 Thoracic Cage 147
7.9 Pectoral Girdle 147
7.10 Upper Limb 148
7.11 Pelvic Girdle 151
7.12 Lower Limb 154
7.13 Joints 156

REFERENCE PLATES

Human Skull 167

CHAPTER 8

Muscular System 170

8.1 Introduction 172
8.2 Structure of a Skeletal Muscle 172
8.3 Skeletal Muscle Contraction 176
8.4 Muscular Responses 180
8.5 Smooth Muscle 184
8.6 Cardiac Muscle 184
8.7 Skeletal Muscle Actions 185
8.8 Major Skeletal Muscles 187

UNIT 3
Integration and Coordination

CHAPTER 9

Nervous System 205

9.1 Introduction 207
9.2 General Functions of the Nervous System 207
9.3 Neuroglial Cells 208
9.4 Neurons 209
9.5 Cell Membrane Potential 212
9.6 Nerve Impulse 215
9.7 The Synapse 216
9.8 Impulse Processing 219
9.9 Types of Nerves 220
9.10 Nerve Pathways 220
9.11 Meninges 222
9.12 Spinal Cord 224
9.13 Brain 225
9.14 Peripheral Nervous System 235
9.15 Autonomic Nervous System 240

CHAPTER 10

Somatic and Special Senses 251

10.1 Introduction 253
10.2 Receptors and Sensations 253
10.3 Somatic Senses 253
10.4 Special Senses 257
10.5 Sense of Smell 257
10.6 Sense of Taste 258

10.7 Sense of Hearing 260
10.8 Sense of Equilibrium 264
10.9 Sense of Sight 267

CHAPTER 11
Endocrine System 279

11.1 Introduction 281
11.2 General Characteristics of the Endocrine System 281
11.3 Hormone Action 281
11.4 Control of Hormonal Secretions 285
11.5 Pituitary Gland 286
11.6 Thyroid Gland 288
11.7 Parathyroid Glands 290
11.8 Adrenal Glands 292
11.9 Pancreas 295
11.10 Other Endocrine Glands 297
11.11 Stress and Health 298

UNIT 4
Transport

CHAPTER 12
Blood 305

12.1 Introduction 307
12.2 Blood and Blood Cells 307
12.3 Blood Plasma 314
12.4 Hemostasis 317
12.5 Blood Groups and Transfusions 320

CHAPTER 13
Cardiovascular System 327

13.1 Introduction 329
13.2 Structure of the Heart 329
13.3 Heart Actions 336
13.4 Blood Vessels 341
13.5 Blood Pressure 346
13.6 Paths of Circulation 351
13.7 Arterial System 351
13.8 Venous System 355

CHAPTER 14
Lymphatic System and Immunity 365

14.1 Introduction 367
14.2 Lymphatic Pathways 367
14.3 Tissue Fluid and Lymph 369
14.4 Lymph Movement 369
14.5 Lymph Nodes 370
14.6 Thymus and Spleen 371
14.7 Body Defenses Against Infection 373
14.8 Nonspecific Defenses 373
14.9 Specific Defenses (Immunity) 375

UNIT 5
Absorption and Excretion

CHAPTER 15
Digestion and Nutrition 391

15.1 Introduction 393
15.2 General Characteristics of the Alimentary Canal 394
15.3 Mouth 396
15.4 Salivary Glands 399
15.5 Pharynx and Esophagus 400
15.6 Stomach 402
15.7 Pancreas 404
15.8 Liver 406
15.9 Small Intestine 411
15.10 Large Intestine 417
15.11 Nutrition and Nutrients 420

CHAPTER 16
Respiratory System 433

16.1 Introduction 435
16.2 Organs of the Respiratory System 435
16.3 Breathing Mechanism 441
16.4 Control of Breathing 447
16.5 Alveolar Gas Exchanges 449
16.6 Gas Transport 451

CHAPTER 17
Urinary System 459

17.1 Introduction 461
17.2 Kidneys 461
17.3 Urine Formation 465
17.4 Urine Elimination 473

CHAPTER 18
Water, Electrolyte, and Acid-Base Balance 480

18.1 Introduction 482
18.2 Distribution of Body Fluids 482
18.3 Water Balance 484
18.4 Electrolyte Balance 487
18.5 Acid-Base Balance 488
18.6 Acid-Base Imbalances 491

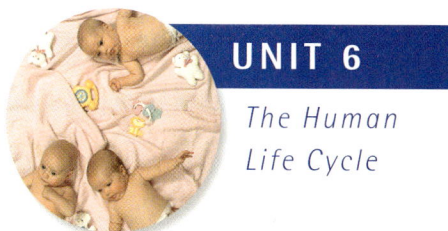

UNIT 6
The Human Life Cycle

CHAPTER 19
Reproductive Systems 496

19.1 Introduction 498
19.2 Organs of the Male Reproductive System 498
19.3 Hormonal Control of Male Reproductive Functions 505
19.4 Organs of the Female Reproductive System 507
19.5 Hormonal Control of Female Reproductive Functions 513
19.6 Mammary Glands 515
19.7 Birth Control 516
19.8 Sexually Transmitted Diseases 520

CHAPTER 20
Pregnancy, Growth, and Development 526

20.1 Introduction 528
20.2 Pregnancy 528
20.3 Prenatal Period 529
20.4 Postnatal Period 544

Appendix 550
Glossary 551
Credits 568
Application Index 570
Subject Index 572

About the Authors

David Shier

David Shier has accumulated twenty-five years of experience teaching anatomy and physiology, primarily to premedical, nursing, dental, and allied health students. He has effectively incorporated his extensive teaching experience into another student-friendly revision of *Hole's Essentials of Human Anatomy and Physiology* and *Hole's Anatomy and Physiology*. David has published numerous papers and abstracts in the areas of renal and cardiovascular physiology, the endocrinology of fluid and electrolyte balance, and hypertension. A faculty member in the Life Science Department at Washtenaw Community College, he is actively involved in a number of projects dealing with assessment, articulation, and the incorporation of technology into instructional design. David holds a Ph.D. in physiology from the University of Michigan.

Jackie Butler

Jackie Butler's professional background includes work at the University of Texas Health Science Center conducting research about the genetics of bilateral retinoblastoma. She later worked at Houston's M. D. Anderson Hospital conducting research on remission in leukemia patients. Now a popular educator at Grayson County College, Jackie teaches microbiology and human anatomy and physiology for health science majors. Her experience and work with students of various educational backgrounds have contributed significantly to another revision of *Hole's Essentials of Human Anatomy and Physiology* and *Hole's Human Anatomy and Physiology*. Jackie Butler received her B.S. and M.S. degrees from Texas A&M University, focusing on microbiology, including courses in immunology and epidemiology.

Ricki Lewis

Ricki Lewis, author of the WCB/McGraw-Hill textbooks *Life* and *Human Genetics*, combines the skills of scientist and journalist. Since earning her Ph.D. in genetics from Indiana University in 1980, she has published more than 3,000 articles in scientific and popular publications. Today Ricki contributes regularly to *The Scientist* and *Biophotonics International*, and has published an essay collection, *Discovery: Windows on the Life Sciences*. She is a genetic counselor for a private medical practice in upstate New York, and is an adjunct professor of biology at the University at Albany and at Miami University. Ricki brings a molecular, cellular, and genetics perspective, with a journalistic flair, to *Hole's Essentials of Human Anatomy and Physiology* and *Hole's Human Anatomy and Physiology*.

Preface

To the Student

Welcome to the eighth edition of *Hole's Essentials of Human Anatomy and Physiology*. We continue our commitment to introduce the structure and function of the human body in an interesting and highly readable manner.

Many of you are planning careers in health care, athletics, science, or education. We understand that you face the challenge of balancing family, work, and academics. This text provides you with many helpful tools that will prepare you for success in the study of human anatomy and physiology.

The Guided Tour to Top Performance, p. ix, highlights the integrated study tools of your text.

- Take the lead with chapter vignettes, objectives, and key terms.
- Attack the chapter content with *Check Your Recall Questions* and *A&P Trivia*.
- Pull concepts together with real-life *Clinical Connections* and *Topics of Interest*.
- Prepare for top performance with *Review Exercises* and *Critical Thinking Questions*.
- Cruise online to a *Learning Center* packed full of activities to complement the text.

Your next step to effective learning begins with a solid study strategy. Many first year students feel overwhelmed by the amount of material in Anatomy and Physiology. Be assured that you can do the work, and you can be successful. Studying anatomy and physiology is like preparing for the Tour de France. Practice, diligence, and perseverance will pay off. Professor Susan Allen of North Harris College, Houston, Texas, offers the following study tips to assist you in preparation for the ride ahead.

1. Go over your notes and handouts everyday.
    - Review material in the first 24 hours after it is presented. You will learn faster and remember longer.
    - Go over your notes at least once a day, seven days a week.
    - Read over all notes taken to date, and read the notes out loud. Seeing, saying, and hearing helps.
    - Tape-record the lectures (after getting permission), and listen to the lectures.
2. Rewrite your notes.
    - Use block letters and an outline form.
    - Put a small amount of material on each page and illustrate facts with drawings. A picture is worth a thousand words.
    - Color code headings.
3. Read each chapter or unit before going to class.
    - Use the SQ3R method when you read: Survey, Question, Read, Recite, Review.
    - Use the chapter outline at the end of each of the chapters in Hole's Essentials text.
    - Answer the questions at the end of the chapter.
    - Pay particular attention to the diagrams and charts.
4. Form study groups.
    - Plan regular times to meet and go over the material.
    - Explain the material to someone else.
    - Talk through a concept to gain a thorough understanding.
    - Make up an exam over the material.
5. Use the Cornell Method of note taking.
    - Organizing the material will cut down on your study time.
    - (For further information on this great method of note taking, refer to the study tips page under Biology on the North Harris College Web Page: http://science.nhmccd.edu/biol
6. Budget your time.
    - Study for short periods of time with breaks in between. Short repeated study sessions are much more effective than one long session.
    - For every hour you are in class, spend two to three hours studying outside of class.
7. Make flashcards of terms and definitions.
    - Make up meaningful acronyms and word combination to help you remember information.
    - Sound out difficult words and practice spelling them.
    - Learn the meanings of the prefixes and suffixes of words. Check out the inside back cover of the text for meanings of these words.

8. Use effective ways to learn terminology.
    - Look at the word.
    - Say the word out loud and repeat the word often during the day.
    - Touch the area on a model or torso, or touch the area on your own body (when possible).
    - Write the word.
    - Color the region represented by the term in an Anatomy Coloring Book.
9. Make models of the chemical structures in chapter 2.
    - Use gumdrops, marshmallows, toothpicks, etc.
    - Look at a diagram and build a model. It will help you learn the material faster and remember it longer.
10. Use additional study aids that are available:
    - *Student Study Guide*
    - *The Dynamic Human* CD-Rom
    - *Essential Study Partner* found in the Online Learning Center
    - Film clips or videos recommended by the professor

The cover image of Lance Armstrong, three-time Tour de France cycling champion, husband, father, and cancer survivor sends a message of encouragement, inspiration, and determination to all. This is your life; don't spend it at the back of the pack. Enjoy the ride; use the study tools provided and sprint to success in anatomy and physiology.

David Shier
Jackie Butler
Ricki Lewis

Guided Tour
to Top Performance

HOLE'S ESSENTIALS OF HUMAN ANATOMY AND PHYSIOLOGY

- Maintains commitment to readability
- Applies concepts to everyday examples
- Emphasizes the interrelatedness and interdependence of organ systems
- Provides you with the right tools to cross the anatomy and physiology finish line

VIGNETTES
take the lead to chapter content. They connect you to many areas of health care including technology, physiology, medical conditions, historical perspectives, and careers.

CHAPTER OBJECTIVES
help you stay on course as you master the information within the narrative. Use them as guides to identify important chapter topics.

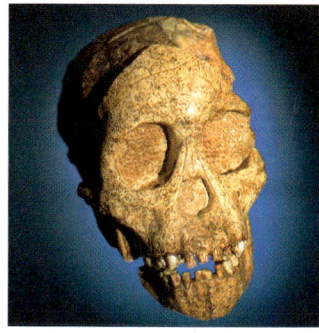

chapter 7
Skeletal System

CLUES FROM SKELETONS PAST. As the hardest and therefore most enduring of human tissues, bone has persisted over time to provide clues to early humans and their forebears. Some glimpses into the past, courtesy of skeletal remains or fossils, include:

7300–6220 B.C. Skulls with circular holes are the earliest evidence of trepanation, a technique used to relieve pressure following a skull fracture or as a spiritual treatment for headache, tumors, or mental illness. A few of the people treated with trepanation were lucky—they survived, as evidenced by new bony growth over the holes made in their skulls. However, most trepanated skulls have gaping, drilled holes, indicating that the "treatment" was lethal.

2.8–2.6 million years ago "Mr. Ples" is the name anthropologists have given to the face and left side of a skull from Sterkfontein, South Africa, which once belonged to a member of *Australopithecus africanus* (see photo), a type of primate that preceded *Homo sapiens*. Using computer modeling to fashion a "virtual endocast" of the entire skull contents, researchers have estimated the cranial capacity of *A. africanus* at 515 cubic centimeters (cc). By comparison, a chimp's cranial capacity averages 370 cc, and a modern human's, 1,350. Expanded cranial capacity correlates to increase in intelligence.

3.5 million years ago Not all evidence of a skeletal system is in the form of preserved bone. On the Serengeti Plain are clues to our ancestors who first began to walk upright, a stance that freed their hands, perhaps making possible the development of tools. This evidence consists of shallow footprints where an animal called *Australopithecus afarensis* once lived. The prints reveal that it had long big toes and arched feet.

Photo:
Australopithecus africanus lived from 2.8 to 2.6 million years ago. Our knowledge of this primate comes from skeletal evidence.

Chapter Objectives
After studying this chapter, you should be able to do the following:

7.1 Introduction
1. List the active tissues in a bone. (p. 128)

7.2 Bone Structure
2. Describe the general structure of a bone, and list the functions of its parts. (p. 128)

7.3 Bone Development and Growth
3. Distinguish between intramembranous and endochondral bones, and explain how such bones develop and grow. (p. 130)

7.4 Bone Function
4. Discuss the major functions of bones. (p. 131)

7.5 Skeletal Organization
5. Distinguish between the axial and appendicular skeletons, and name the major parts of each. (p. 135)

7.6–7.12 Skull—Lower Limb
6. Locate and identify the bones and the major features of the bones that comprise the skull, vertebral column, thoracic cage, pectoral girdle, upper limb, pelvic girdle, and lower limb. (p. 136)

7.13 Joints
7. List three classes of joints, describe their characteristics, and name an example of each. (p. 156)
8. List six types of synovial joints, and describe the actions of each. (p. 157)
9. Explain how skeletal muscles produce movements at joints, and identify several types of joint movements. (p. 159)

Aids to Understanding Words

acetabul- [vinegar cup] *acetabulum*: Depression of the coxa that articulates with the head of the femur.
ax- [axis] *axial* skeleton: Upright portion of the skeleton that supports the head, neck, and trunk.
-blast [budding] osteo*blast*: Cell that will form bone tissue.
carp- [wrist] *carpals*: Wrist bones
-clast [break] osteo*clast*: Cell that breaks down bone tissue.
condyl- [knob] *condyle*: Rounded, bony process.
corac- [a crow's beak] *coracoid* process: Beaklike process of the scapula.
cribr- [sieve] *cribri*form plate: Portion of the ethmoid bone with many small openings.
crist- [crest] *crista* galli: Bony ridge that projects upward into the cranial cavity.
fov- [pit] *fovea* capitis: Pit in the head of a femur.
glen- [joint socket] *glenoid* cavity: Depression in the scapula that articulates with the head of the humerus.
hema- [blood] *hematoma*: Blood clot.
inter- [among, between] *inter*vertebral disc: Structure located between adjacent vertebrae.
intra- [inside] *intra*membranous bone: Bone that forms within sheetlike masses of connective tissue.
meat- [passage] auditory *meatus*: Canal of the temporal bone that leads inward to parts of the ear.
odont- [tooth] *odontoid* process: Toothlike process of the second cervical vertebra.
poie- [make, produce] hemato*poiesis*: Process by which blood cells are formed.

Key Terms

articular cartilage (ar-tik′u-lar kar′tĭ-lij)
bursa (ber′sah)
cartilaginous joint (kar″tĭ-lah′jin-us joint)
compact bone (kom′pakt bōn)
diaphysis (di-af′ĭ-sis)
endochondral bone (en″do-kon′dral bōn)
epiphyseal plate (ep″ĭ-fiz′e-al plat)
epiphysis (e-pif′ĭ-sis)
fibrous joint (fi′brus joint)
hemopoiesis (he″mo-poi-e′sis)
intramembranous bone (in″trah-mem′brah-nus bōn)
lever (lev′er)
marrow (mar′o)
medullary cavity (med′u-lār″e kav′ĭ-te)
meniscus (mĕ-nis′kus)
osteoblast (os′te-o-blast)
osteoclast (os′te-o-klast)
osteocyte (os′te-o-sīt)
periosteum (per″e-os′te-um)
spongy bone (spun′je bōn)
synovial joint (sĭ-no′ve-al joint)

→ **AIDS TO UNDERSTANDING WORDS**
increase your pace in understanding and remembering scientific word meanings. Examine root words, stems, prefixes, suffixes, pronunciations and build a solid A&P vocabulary.

→ **KEY TERMS**
are part of your basic training in building a solid science vocabulary. Phonetic pronunciations and definitions can be found at the beginning of the chapter, within the text, and in the glossary.

Guided Tour

GENETICS CONNECTIONS lead the standings by exploring the molecular underpinnings of familiar as well as not so familiar illnesses. Read about such topics as ion channel disorders, muscular dystrophies, and cystic fibrosis.

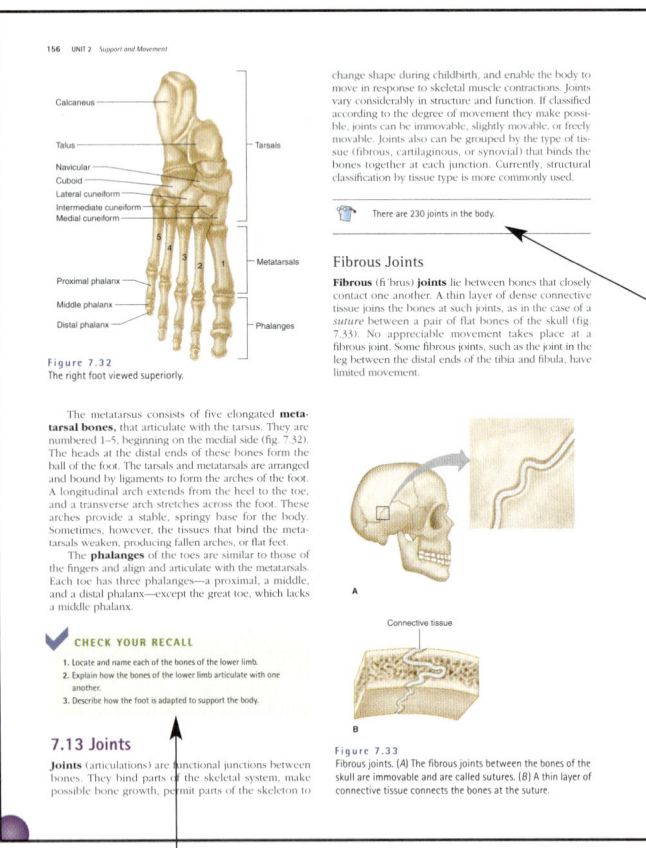

FACTS OF LIFE go all out with amazing bits of anatomy and physiology trivia, adding a touch of wonder to chapter topics.

CHECK YOUR RECALL QUESTIONS attack the material covered in major sections by testing your understanding of key concepts.

ORGANization ILLUSTRATIONS found at the end of selected chapters, conceptually link the highlighted body system to every other system and reinforce the dynamic interplays between groups and organs. These illustrations help you review chapter concepts and reinforce the "big picture" in learning and applying the principles of anatomy and physiology.

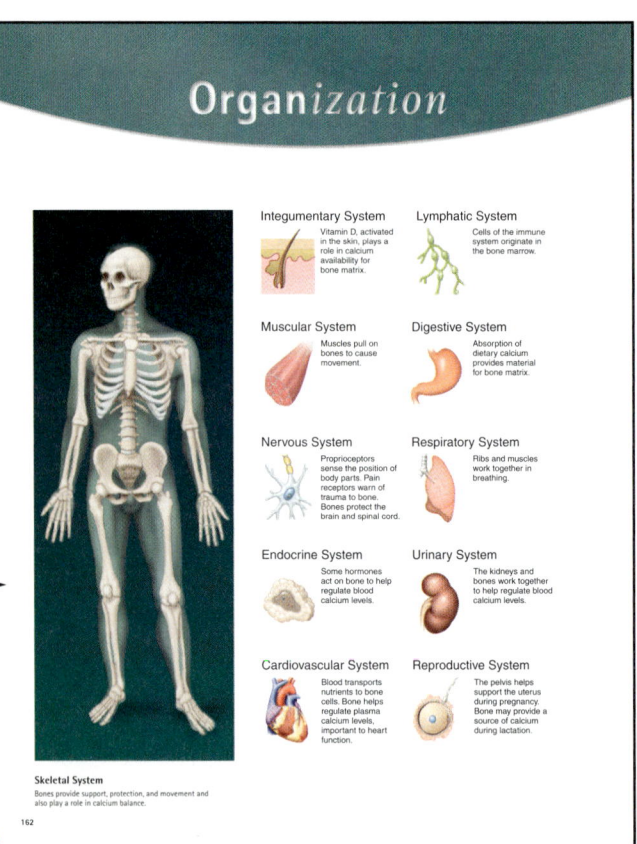

TOPICS OF INTEREST are proven performers in presenting disorders, physiological responses to environmental factors, and other topics of general interest.

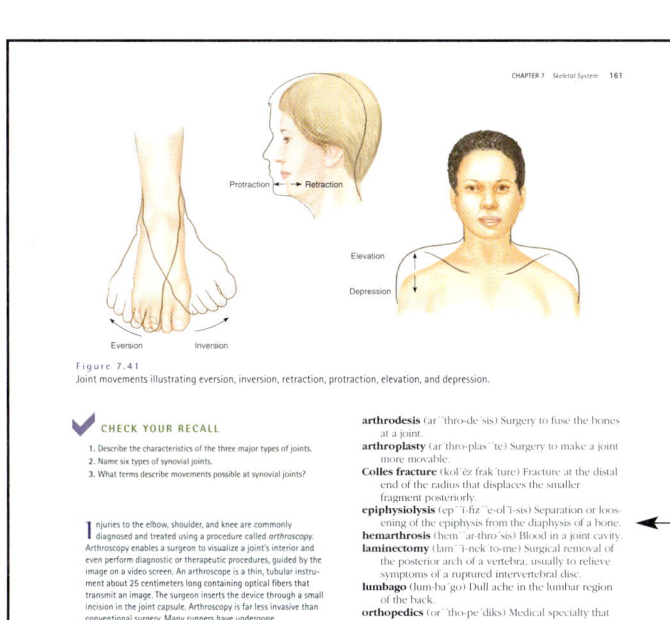

CLINICAL TERMS help you sprint ahead in understanding medical terminology. Lists of related terms often used in clinical situations are found at the end of several chapters.

CLINICAL CONNECTIONS help you go even further by "pulling the chapter concepts together." These short vignettes at the end of the chapter give you a real-life connection to the material covered. Short paragraphs in colored boxes also apply ideas and facts in the narrative to clinical situations.

CHAPTER SUMMARY OUTLINES prepare you for another top performance by helping you review the chapter's main ideas.

xii Guided Tour

REVIEW EXERCISES AND CRITICAL THINKING check your understanding of the chapter's major ideas. Critical thinking questions encourage you to apply information to clinical situations.

REFERENCE PLATES continue to offer vibrant detail of body structures.

---

164 UNIT 2 Support and Movement

  4. Ankle and foot
     a. The ankle and foot consist of the tarsus, metatarsus, and five toes.
     b. Included are the talus that helps form the ankle, six other tarsals, five metatarsals, and fourteen phalanges.

7.13 Joints (p. 156)
Joints can be classified according to the type of tissue that binds the bones together.
  1. Fibrous joints
     a. Bones at fibrous joints are tightly joined by a layer of dense connective tissue.
     b. Little or no movement occurs at a fibrous joint.
  2. Cartilaginous joints
     a. A layer of cartilage joins bones of cartilaginous joints.
     b. Such joints allow limited movement.
  3. Synovial joints
     a. The bones of a synovial joint are covered with hyaline cartilage and held together by a fibrous joint capsule.
     b. The joint capsule consists of an outer layer of ligaments and an inner lining of synovial membrane.
     c. Bursae are located between the skin and underlying bony prominences.
     d. Types of synovial joints include: ball-and-socket, condyloid, gliding, hinge, pivot, and saddle.
  4. Types of joint movements
     a. Muscles fastened on either side of a joint produce movements of synovial joints.
     b. Joint movements include flexion, extension, dorsiflexion, plantar flexion, hyperextension, abduction, adduction, rotation, circumduction, pronation, supination, eversion, inversion, retraction, protraction, elevation, and depression.

REVIEW EXERCISES

Part A
 1. Sketch a typical long bone, and label its epiphyses, diaphysis, medullary cavity, periosteum, and articular cartilages. (p. 128)
 2. Distinguish between spongy and compact bone. (p. 128)
 3. Explain how central canals and perforating canals are related. (p. 129)
 4. Explain how the development of intramembranous bone differs from that of endochondral bone. (p. 130)
 5. Distinguish between osteoblasts and osteocytes. (p. 130)
 6. Explain the function of an epiphyseal plate. (p. 130)
 7. Explain how a bone thickens. (p. 131)
 8. Provide several examples to illustrate how bones support and protect body parts. (p. 131)
 9. Describe a lever. (p. 133)
10. Explain how upper limb movements function as levers. (p. 133)
11. Describe the functions of red and yellow bone marrow. (p. 133)
12. Explain the mechanism that regulates the concentration of blood calcium ions. (p. 134)
    ... axial and appendicular skeletons. (p. 135)
    ... anium and the facial skeleton. (p. 136)
    ... f fontanels. (p. 142)
    ... a. (p. 142)
    ... mong the cervical, thoracic, and lumbar

18. Name the bones that comprise the thoracic cage. (p. 147)
19. List the bones that form the pectoral and pelvic girdles. (p. 147)
20. Name the bones of the upper limb. (p. 148)
21. Name the bones that comprise a coxa. (p. 152)
22. List the bones of the lower limb. (p. 154)
23. Define joint. (p. 156)
24. Describe a fibrous joint, a cartilaginous joint, and a synovial joint. (p. 156)
25. Define bursa. (p. 157)
26. List six types of synovial joints, and name an example of each type. (p. 158)

Part B
Match the parts listed in column I with the bones listed in column II.

I	II
1. Coronoid process	a. Ethmoid bone
2. Cribriform plate	b. Frontal bone
3. Foramen magnum	c. Mandible
4. Mastoid process	d. Maxilla
5. Palatine process	e. Occipital bone
6. Sella turcica	f. Temporal bone
7. Supraorbital foramen	g. Sphenoid bone
8. Temporal process	h. Zygomatic bone
9. Acromion process	i. Femur
10. Deltoid tuberosity	j. Fibula
11. Greater trochanter	k. Humerus
12. Lateral malleolus	l. Radius
13. Medial malleolus	m. Scapula
14. Olecranon process	n. Sternum
15. Radial tuberosity	o. Tibia
16. Xiphoid process	p. Ulna

Part C
Match the movements in column I with the descriptions in column II.

I	II
1. Rotation	a. Turning palm upward
2. Supination	b. Decreasing angle between parts
3. Extension	c. Moving part forward
4. Eversion	d. Moving part around an axis
5. Protraction	e. Turning sole of foot to face laterally
6. Flexion	f. Increasing angle between parts
7. Pronation	g. Lowering a part
8. Abduction	h. Turning palm downward
9. Depression	i. Moving part away from midline

CRITICAL THINKING

1. How does the structure of a bone make it strong yet lightweight?
2. Archaeologists discover skeletal remains of humanlike animals in Ethiopia. Examination of the bones suggests that the remains represent four types of individuals. Two of the skeletons have bone densities that are 30% less than those of the other two skeletons. The skeletons with the lower bone mass also have broader front

---

26 REFERENCE PLATES The Human Organism

PLATE 5
Human female torso with the lungs, heart, and small intestine sectioned.

Labels: Larynx, Trachea, Left subclavian a., Right common carotid a., Right subclavian a., Brachiocephalic a., Arch of aorta, Superior vena cava, Pulmonary a., Pulmonary trunk, Pulmonary v., Right atrium, Left atrium, Right ventricle, Lung, Left ventricle, Lobes of liver, Diaphragm, Spleen, Gallbladder, Stomach, Cystic duct, Duodenum, Transverse colon, Ascending colon, Jejunum (cut), Mesentery, Descending colon, Ileum (cut), Ureter, Cecum, Vermiform appendix, Sigmoid colon, Common iliac a., Rectum, Uterus, Ovary, Tensor fasciae latae m., Uterine tube, Round ligament of uterus, Femoral a., Urinary bladder, Femoral v., Adductor longus m., Great saphenous v., Gracilis m., Rectus femoris m., Vastus lateralis m., Vastus medialis m., Sartorius m.

Preface

To the Instructor

In this eighth edition of *Hole's Essentials of Human Anatomy and Physiology* we continue our commitment to introduce the structure and function of the human body in an interesting and highly readable manner. We have added content only when it can be integrated into the larger concept of homeostasis and maintenance of the internal environment. Indeed, a book at this level is almost a metaphor of the human body itself—nothing unnecessary is retained, and every component contributes to the final purpose.

Given the immensely varied population of most universities and colleges, particularly two-year and community colleges, we have continued to devote top priority to the readability of this text. Our challenge has been to do so while at the same time responding to requests to enhance physiology coverage. We have accomplished this through thoughtful changes in the text and art, carefully designed and implemented pedagogical features, and a wide choice of ancillaries designed to review and augment both in-class and out-of-class student activities.

The level of this text is geared toward students in one-semester courses in anatomy and physiology who are pursuing careers in allied health fields and who have minimal background in physical and biological sciences. The first four chapters cover the chemistry and cell physiology necessary to understand biological processes. Students who have studied this material previously will view it as a welcomed review, but newcomers will not find it intimidating.

General Themes

Commitment to Readability

Even the most basic concepts of human anatomy and physiology can be challenging to the uninitiated, and most of the students in introductory anatomy and physiology courses fall into this category. Students and instructors both are faced with an enormous amount of material to cover in a short period of time. Our approach is to never add unnecessary content, and to maintain readability as our top priority.

Clear Application of Concepts

The content carefully balances structure and function to provide an integrated view of how the human body works. In striking this balance we recognize a clear trend toward a greater emphasis on physiology across the board in the health care professions. All physiological concepts are tied to some level of body structure and organization and presented in a student-accessible way. Numerous practical applications and everyday examples are provided.

Emphasis of the Interrelatedness and Interdependency of Organ Systems

Chapter 1 introduces the concept of the internal environment, along with homeostasis, the mechanism that keeps the internal environment relatively constant. The book reinforces this theme throughout, most strikingly in the end-of-chapter "ORGANization" figures that hammer home the interrelatedness and interdependency of organ systems.

Enhancing the Text with Consistent Illustrations

Detail, clarity, accuracy, and consistency prevail, with frequent use of icons for orientation and to establish a sense of scale. Color is consistent from chapter to chapter—a cell is not blue in one chapter, orange in another.

What's New?

Hole's Essentials of Human Anatomy and Physiology is written with the student in mind. Several new features accompany the eighth edition.

Key Term Pronunciations within the Text

A list of key terms and their phonetic pronunciations at the beginning of each chapter helps build science vocabulary. The key terms are boldfaced, defined within the chapter and followed again by pronunciations when the term is first introduced in the text. These terms are likely to be found in subsequent chapters. The glossary at the end of the book explains phonetic pronunciations.

Chapter Opening Vignettes

Interesting, creative, and thought-provoking vignettes introduce the chapter topics.

Check Your Recall Review Questions

This edition offers review questions at the ends of major sections in each chapter to test understanding of the material just covered.

Design

The revitalized text design injects new life into the study of Anatomy and Physiology. Bright, bold, modern colors are used throughout the feature boxes, tables, and chapter openers making them easy to recognize.

Illustrations

The new art program is designed to support the text and beyond. Labels and legends go only as far as the text itself. However, the detail of the figures is accurate enough to support more detailed discussion if the instructor or the student chooses further investigation.

Facts of Life

These briefs are fun bits of A&P trivia information scattered throughout the text adding a touch of wonder to chapter topics and concepts.

Clinical Connections

These new clinical connections are integrated at the end of several chapters to "pull the concepts together". The short vignettes help students make a vital real-life connection to the chapter material.

Review Exercises and Critical Thinking

Updated end-of-chapter review exercises help the student check understanding of the chapter's major ideas. Critical thinking questions enourage the student to apply information to clinical situations.

Online Learning Center

New OLC activities and resources are available for students and instructors.

Digital Content Manager

This multimedia collection of visual resources allows instructors to utilize artwork from the text in multiple formats to create customized classroom presentations, visually-based tests and quizzes, dynamic course website content, or attractive printed support materials. The digital assets on this cross-platform CD-ROM are grouped by chapter within the following easy-to-use folders.

Active Art Library Key figures from the text are saved in manipulable layers that can be isolated and customized to meet the needs of the lecture environment.

Animations Library Numerous full-color animations of key physiological processes are provided. Harness the visual impact of processes in motion by importing these files into classroom presentations or course websites.

Art Libraries Full-color digital files of all illustrations in the book, plus the same art saved in unlabeled and gray scale versions, can be readily incorporated into lecture presentations, exams, or custom-made classroom materials. These images are also pre-inserted into blank PowerPoint slides for ease of use.

Photo Libraries Digital files of instructionally significant photographs from the text—including cadaver, bone, histology, and surface anatomy images—can be reproduced for multiple classroom uses.

PowerPoint Lectures Ready-made presentations that combine art and lecture notes have been specifically written to cover each of the 20 chapters of the text. Use the PowerPoint lectures as they are, or tailor them to reflect your preferred lecture topics and sequences.

Tables Library Every table appearing in the text is provided in electronic form.

You can quickly preview images and incorporate them into PowerPoint or other presentation programs to create your own multimedia presentations. You can also remove and replace labels to suit your own preferences in terminology or level of detail.

Teaching and Learning Supplements

Online Learning Center (http://www.mhhe.com/shieress8) The OLC offers an extensive array of learning and teaching tools. The site includes quizzes for each chapter, links to websites related to each chapter, supplemental reading lists, clinical applications, interactive activities, art labelling exercises, and case studies. Students can click on a diagram of the human body and get case studies related to the regions they select. Instructor resources at the site include lecture outlines, supplemental reading lists, technology resources, clinical applications, and case studies.

- **Essential Study Partner** The ESP contains 120 animations and more than 800 learning activities to help your students grasp complex concepts. Interactive diagrams and quizzes will make learning stimulating and fun for your students. The Essential Study Partner can be accessed via the Online Learning Center.
- **PowerWeb** is an online supplement that offers access to course-specific current articles referred by content experts, course-specific real-time news, weekly course updates, referred and updated research links, daily news, and the Northernlight.com Special Collection™ of journals and articles.

The **Laboratory Manual for Hole's Essentials of Human Anatomy and Physiology,** 0-07-235120-9, by Terry R. Martin, Kishwaukee College, is designed to accompany the eighth edition of *Hole's Essentials of Human Anatomy and Physiology.*

Student Study Guide, 0-07-243813-4, by Nancy A. Sickels Corbett contains chapter overviews, chapter objectives, focus questions, mastery tests, study activities, and mastery test answers.

The **Instructor's Manual,** 0-07-242523-7, by Michael F. Peters includes supplemental topics and demonstration ideas for your lectures, suggested readings, critical thinking questions, and teaching strategies. The Instructor's Manual is available online through the Instructor Resources of the Online Learning Center.

Computerized Test Bank t/a Hole's Essentials of Human A&P, 0-07-242528-8, is a computerized test generator free upon request to qualified adopters. A test bank of questions contains matching, true/false, and essay questions. The test generator contains the complete test item file on CD-ROM.

McGraw-Hill provides over 400 **Overhead Transparencies,** 0-07-235122-5, of all text line art including fully labeled and unlabeled duplicates of many of them for testing purposes or custom labeling.

The Digital Content Manager, 0-07-242524-5, is a multimedia collection of visual resources that allows instructors to utilize artwork from the text in multiple formats to create customized classroom presentations, visually-based tests and quizzes, dynamic course website content, or attractive printed support materials. The digital assets on this cross-platform CD-ROM are grouped by chapter within easy-to-use folders.

PageOut is McGraw-Hill's exclusive tool for creating your own website for your A&P course. It requires no knowledge of coding. Simply type your course information into the templates provided. PageOut is hosted by McGraw-Hill.

MediaPhys CD-Rom, 0-07-255140-2, combines incredible multimedia and powerful visuals with in-depth textual content. This interactive program provides a friendly and educational environment that allows you to:
- navigate through body systems via detailed graphics, animations, and sound
- explore concepts in a logical order from simple to more complex
- visualize physiological processes and their relationships

Human Anatomy and Physiology Laboratory Manual–Fetal Pig Dissection, 0-07-231199-1, by Terry R. Martin, provides excellent full-color photos of the dissected fetal pig with corresponding labeled art. It includes World Wide Web activities for many chapters.

Web-Based Cat Dissection Review for Human Anatomy and Physiology, 0-07-232157-1, by John Waters, Pennsylvania State University. This online multimedia program contains vivid, high-quality labeled cat dissection photographs. The program helps students easily identify and review the corresponding structures and functions between the cat and the human body.

Dynamic Human Version 2.0, 0-07-235476-3. This set of two interactive CD-ROMs covers each body system and demonstrates clinical concepts, histology, and physiology with animated three-dimensional and other images.

Interactive Histology CD-ROM, 0-07-237308-3, by Bruce Wingerd and Paul Paolini, San Diego State University. This CD contains 135 full-color, high-resolution LM images and 35 SEM images of selected tissue sections typically studied in A&P. Each image has labels that can be clicked on or off, has full explanatory legends, offers views at two magnifications, and has links to study questions. The CD also has a glossary with pronunciation guides.

Life Science Animation CD-ROM, 0-07-234296-X, contains 125 animations of major biological concepts and processes such as the sliding filament mechanism, active transport, genetic transcription and translation, and other topics that may be difficult for students to visualize.

Life Science Animations 3D Videotape, 0-07-290652-9, contains 42 key biological processes that are narrated and animated in vibrant full color with dynamic three-dimensional graphics.

Life Science Animations (LSA) videotape series contains 53 animations on five VHS videocassettes; Chemistry, The Cell, and Energetics, 0-697-25068-7; Cell Division, Heredity, Genetics, Reproduction, and

Development, 0-697-25069-5; Animal Biology No. 1, 0-697-25070-9; Animal Biology No. 2, 0-697-25071-7; and Plant Biology, Evolution, and Ecology, 0-697-26600-1. Another available videotape is Physiological Concepts of Life Science, 0-697-21512-1.

Atlas to Human Anatomy, 0-697-38793-3, by Dennis Strete, McLennan Community College and Christopher H. Creek, takes a systems approach with references to regional anatomy, thereby making it a great complement to your regular course structure, as well as to your laboratory.

Atlas of the Skeletal Muscles, third edition, 0-07-290332-5, by Robert and Judith Stone, Suffolk County Community College, is a guide to the structure and function of human skeletal muscles. The illustrations help students locate muscles and understand their actions.

Laboratory Atlas of Anatomy and Physiology, third edition, 0-07-290755-X, by Eder et al., is a full-color atlas containing histology, human skeletal anatomy, human muscular anatomy, dissections, and reference tables.

Human Anatomy and Physiology Study Cards, 0-07-290818-1, by Kurt Van De Graaff, R. Ward Rhees, and Christopher Creek, is a set of 300 3" × 5" cards with terminology, pronunciation guides, word origins, diagrams, and concise descriptions of anatomical and physiological concepts.

Coloring Guide to Anatomy and Physiology, 0-697-17109-4, by Robert and Judith Stone, Suffolk Community College, emphasizes learning through the process of color association. *The Coloring Guide* provides a thorough review of anatomical and physiological concepts.

Acknowledgments

Any textbook is the result of hard work by a large team. Although we directed the revision, many "behind-the-scenes" people at McGraw-Hill were indispensable to the project. We would like to thank our editorial team of Marty Lange, Kris Tibbetts, and Pat Hesse; our production team, which included Jayne Klein, Sandy Ludovissy, Wayne Harms, John Leland, Sandy Schnee, and Barb Block; and most of all, John Hole, for giving us the opportunity and freedom to continue his classic work. We also thank our wonderfully patient families for their support.

David Shier
Jackie Butler
Ricki Lewis

Reviewers

We would like to acknowledge the valuable contributions of the reviewers for the eighth edition who read either portions or all of the manuscript as it was being prepared, and who provided detailed criticisms, and ideas for improving the narrative and the illustrations. They include the following:

Tammy Atchison, *Pitt Community College*
James Bridger, *Prince George's Community College*
Debra Brossett, *University of Louisiana at Monroe, College of Nursing*
Lu Anne Clark, *Lansing Community College*
Mary P. Harbaugh, *Century College*
Dawn K. Holtzmeier, *Hocking College*
Robyn Jordan, *University of Louisiana at Monroe*
Steve Nunez, *Sauk Valley Community College*
Amy G. Ouchley, *University of Louisiana at Monroe*
Jean Revie, *South Mountain Community College*
Brian Shmaefsky, *Kingwood College*
Kent R. Thomas, *Wichita State University*
Elaine K. Tompary, *William Rainey Harper College*
Debora J. Warshefski, *St. Clair County Community College*
Claudia Williams, *Campbell University*

Thanks also to the following art consultants:

John L. Carr, *University of Louisiana at Monroe*
Lu Anne Clark, *Lansing Community College*
Hermann Nonnenmacher, *Pittsburg State University*
Brian Shmaefsky, *Kingwood College*
Kent R. Thomas, *Wichita State University*

Unit 1
Levels of Organization

chapter 1

Introduction to Human Anatomy and Physiology

THE TALE OF THE MUMMY'S TOE. No one knows her name, but she lived sometime between 1550 and 1300 B.C. in Thebes, a city in ancient Egypt. All that remains are pieces of her skeleton, bound with linen to preserve the general shape of her body in life. Yet, the telltale bones reveal a little of what her life was like.

The shape of the pelvic bones indicate that the person immortalized as the mummy was female. She was 50 to 60 years old when she died, according to the way the bony plates of her skull fit together and the lines of mineral deposition of a particularly well-preserved tooth. Among the preserved bones from the skull, pelvis, upper limbs, and right lower limbs, the part that easily commands the most attention is the right big toe, for it ends in a prosthesis, a man-made replacement for a skeletal part. It was crucial for her balance and locomotion.

The mummy's toe tip is made of wood and painted a dark brown, perhaps to blend in with her skin color. It consists of a long part and two smaller parts that anchor the structure to the rest of the digit. Seven leather strings once attached it to the foot, and it even bears a fake nail. The fact that connective tissue and skin have grown over the prosthesis reveals that her body had accepted the unnatural replacement part. Most amazing, however, is the shape of the prosthesis, which is remarkably like the body part it was intended to replace. Signs of wear indicate that it served its owner well. Unlike other prostheses found with mummies that were placed after death to provide a complete skeleton for burial, this one was clearly used during the person's lifetime.

The old woman with the fake toe is evidence of quite sophisticated medical technology. Modern-day medical sleuths from the departments of pathology and diagnostic radiology at Ludwig-Maximillians University in Munich evaluated the ancient evidence using computerized tomography (CT) scans of the remnants of the natural toe. The researchers detected poor mineral content of the toe, plus calcium deposits in the largest blood vessel, the aorta, suggesting impaired circulation to the feet. Perhaps the mummy in life suffered from type II diabetes mellitus, which can cause poor circulation to the toes. If gangrene had set in, long-ago healers might have amputated the affected portion of the toe, replacing it with a very reasonable facsimile.

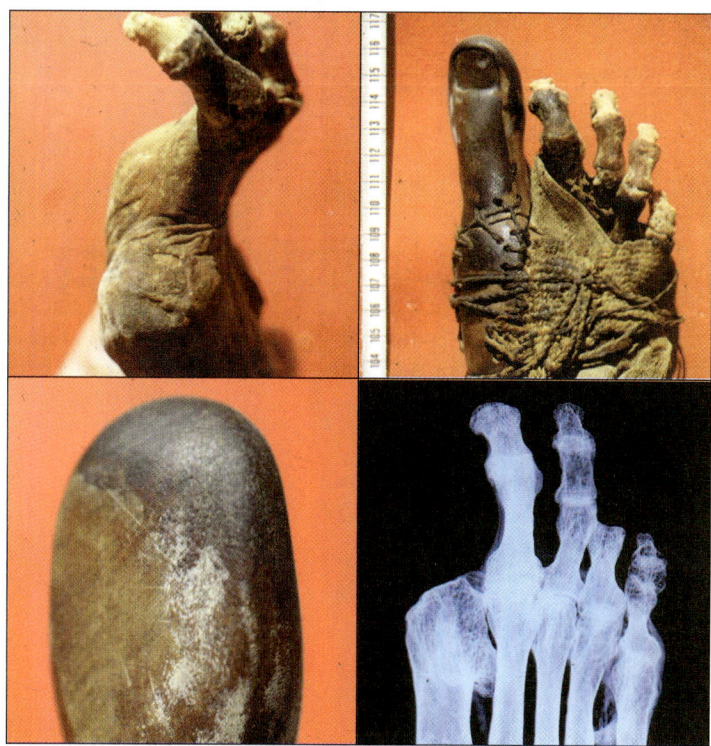

Photo:
A wooden toe on an ancient Egyptian mummy reveals sophisticated knowledge of human anatomy and physiology from long ago.

Chapter Objectives

After studying this chapter, you should be able to do the following:

1.1 Introduction
1. Describe the early studies into the workings of the body. (p. 3)

1.2 Anatomy and Physiology
2. Define *anatomy* and *physiology*, and explain how they are related. (p. 3)

1.3 Characteristics of Life
3. List and describe the major characteristics of life. (p. 4)

1.4 Maintenance of Life
4. List and describe the major requirements of organisms. (p. 4)
5. Define *homeostasis*, and explain its importance to survival. (p. 5)
6. Describe a homeostatic mechanism. (p. 5)

1.5 Levels of Organization
7. Explain biological levels of organization. (p. 7)

1.6 Organization of the Human Body
8. Describe the locations of the major body cavities. (p. 8)
9. List the organs located in each major body cavity. (p. 8)
10. Name the membranes associated with the thoracic and abdominopelvic cavities. (p. 10)
11. Name the major organ systems, and list the organs associated with each. (p. 12)
12. Describe the general functions of each organ system. (p. 12)

1.7 Anatomical Terminology
13. Properly use the terms that describe relative positions, body sections, and body regions. (p. 13)

Aids to Understanding Words

append- [to hang something] *append*icular: Pertaining to the limbs.
cardi- [heart] peri*cardi*um: Membrane that surrounds the heart.
cran- [helmet] *cran*ial: Pertaining to the portion of the skull that surrounds the brain.
dors- [back] *dors*al: Position toward the back.
homeo- [same] *homeo*stasis: Maintenance of a stable internal environment.
-logy [study of] physio*logy*: Study of body functions.
meta- [change] *meta*bolism: Chemical changes that occur within the body.
pariet- [wall] *pariet*al membrane: Membrane that lines the wall of a cavity.
pelv- [basin] *pelv*ic cavity: Basin-shaped cavity enclosed by the pelvic bones.
peri- [around] *peri*cardial membrane: Membrane that surrounds the heart.
pleur- [rib] *pleur*al membrane: Membrane that encloses the lungs and lines the thoracic cavity.
-stasis [standing still] homeo*stasis*: Maintenance of a stable internal environment.
-tomy [cutting] ana*tomy*: Study of structure, which often involves cutting or removing body parts.

Key Terms

abdominopelvic (ab-dom″ĭ-no-pel´vik)
absorption (ab-sorp´shun)
anatomy (ah-nat´o-me)
appendicular (ap″en-dik´u-lar)
assimilation (ah-sim″ĭ-la´shun)
axial (ak´se-al)
circulation (ser-ku-la´shun)
digestion (di-jest´yun)
excretion (ek-skre´shun)
homeostasis (ho″me-ō-sta´sis)
metabolism (mĕ-tab´o-lizm)
negative feedback (neg´ah-tiv fēd´bak)
organelle (or″gan-el´)
organism (or´gah-nizm)
parietal (pah-ri´ĕ-tal)
pericardial (per″ĭ-kar´de-al)
peritoneal (per″ĭ-to-ne´al)
physiology (fiz″e-ol´o-je)
pleural (ploo´ral)
reproduction (re″pro-duk´shun)
respiration (res″pĭ-ra´shun)
thoracic (tho-ras´ik)
visceral (vis´er-al)

The accent marks used in the pronunciation guides are derived from a simplified system of phonetics that is standard in medical usage. The single accent (´) denotes the major stress and identifies the most heavily pronounced syllable in the word. The double accent (″) indicates the secondary stress. A syllable marked with a double accent receives less emphasis than the syllable that carries the main stress, but more emphasis than neighboring unstressed syllables.

1.1 Introduction

Modern medicine began with long-ago observations on the function, and malfunction, of the human body. The study of the human body probably began with our earliest ancestors, who must have been curious about how their bodies worked, as we are today. At first, their interests most likely concerned injuries and illnesses because healthy bodies demand little attention from their owners. Their healers relied heavily on superstitions and notions about magic. However, as healers tried to help the sick, they began to discover useful ways of examining and treating the human body. They observed the effects of injuries, noticed how wounds healed, and examined dead bodies to determine causes of death. They also found that certain herbs and potions could sometimes be used to treat coughs, headaches, and other common problems.

Over time, people began to believe that humans could understand forces that caused natural events. They began observing the world around them more closely, asking questions and seeking answers. This set the stage for the development of modern science.

As techniques for making accurate observations and performing careful experiments evolved, knowledge of the human body expanded rapidly (fig. 1.1). At the same time, early medical providers coined many new terms to name body parts, describe their locations, and explain their functions. These terms, most of which originated from Greek and Latin words, formed the basis for the language of anatomy and physiology. (The names of some modern medical and applied sciences are listed on page 17.)

CHECK YOUR RECALL

1. What factors probably stimulated an early interest in the human body?
2. What kinds of activities helped promote the development of modern science?

1.2 Anatomy and Physiology

Anatomy (ah-nat´o-me) is the branch of science that deals with the structure (morphology) of body parts—their forms and how they are organized. **Physiology** (fiz´´e-ol´o-je), on the other hand, concerns the functions of body parts—what they do and how they do it.

The topics of anatomy and physiology are difficult to separate because the structures of body parts are so closely associated with their functions. Body parts form a well-organized unit—the human organism—and each part functions in the unit's operation. A particular body part's function depends on the way the part is constructed—that is, how its subparts are organized. For example, the organization of parts in the human hand with its long, jointed fingers makes it easy to grasp objects; the hollow chambers of the heart are adapted to pump blood through tubular blood vessels; the shape of the mouth enables it to receive food; and teeth are shaped to break solid foods into small pieces (fig. 1.2).

Anatomy and physiology are ongoing as well as ancient fields. Researchers frequently discover new information about physiology, particularly at the molecular level. Although unusual, new parts of human anatomy are discovered today too. Recently, researchers identified a small piece of connective tissue between the upper part of the spinal cord and a muscle at the back of the head. This connective tissue bridge may be the trigger for pain impulses in certain types of tension headaches.

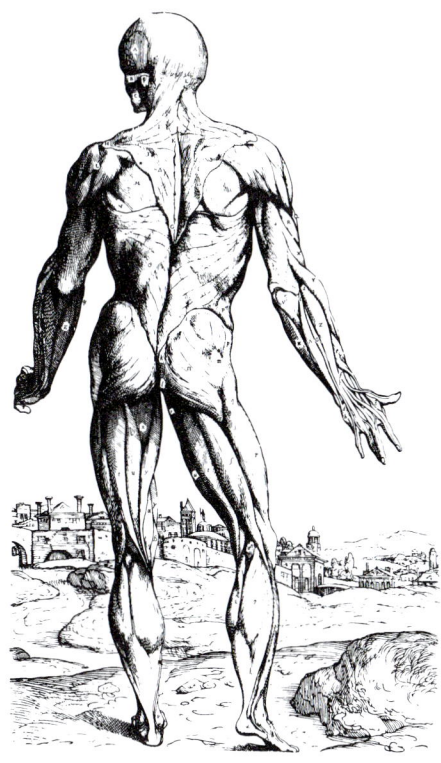

Figure 1.1
The study of the human body has a long history, as this illustration from the second book of *De Humani Corporis Fabrica* by Andreas Vesalius, issued in 1543, illustrates. (Note the similarity to the anatomical position, described later in this chapter.)

 In 2000, a team of international researchers and a private company deciphered the human genome—that is, the biochemical instructions that run the human body. Discovering the activities of our 35,000 or so genes is revealing many new details of physiology.

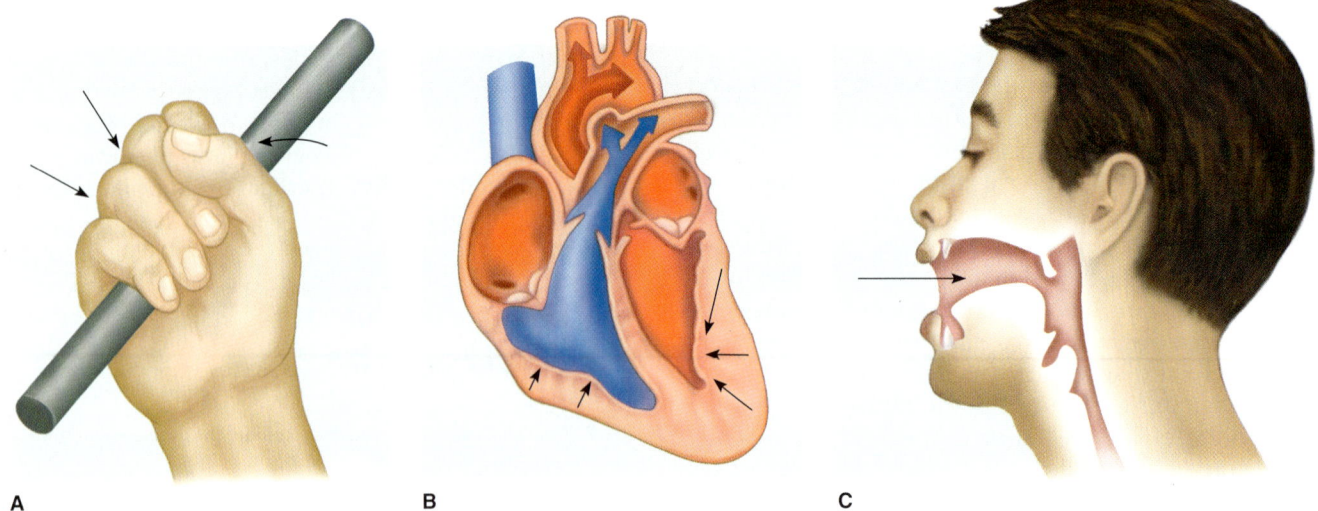

Figure 1.2
The structures of body parts make possible their functions: (A) The hand is adapted for grasping, (B) the heart for pumping blood, and (C) the mouth for receiving food. (Arrows indicate movements associated with these functions.)

CHECK YOUR RECALL

1. Why is it difficult to separate the topics of anatomy and physiology?
2. List several examples that illustrate how the structure of a body part makes possible its function.

1.3 Characteristics of Life

Before beginning a more detailed study of anatomy and physiology, it is helpful to consider some of the traits humans share with other organisms, particularly with other animals. As living organisms, we can move and respond to our surroundings. We start out as small individuals and then grow, eventually able to reproduce. We gain energy by taking in or ingesting food, by breaking it down or digesting it, and by absorbing and assimilating it. The absorbed substances circulate throughout the internal environment of our bodies. We can then, by the process of respiration, use the energy in these nutrients for such vital functions as growth and repair of body parts. Finally, we excrete wastes from the body. The acquisition of food and utilization of its energy, plus excretion, constitute **metabolism** (mĕ-tab´o-lizm). Table 1.1 summarizes the characteristics of life.

CHECK YOUR RECALL

1. What are the characteristics of life?
2. How are the characteristics of life dependent on metabolism?

TABLE 1.1   CHARACTERISTICS OF LIFE

PROCESS	EXAMPLES
Movement	Change in position of the body or of a body part; motion of an internal organ
Responsiveness	Reaction to a change taking place inside or outside the body
Growth	Increase in body size without change in shape
Reproduction	Production of new organisms and new cells
Respiration	Obtaining oxygen, removing carbon dioxide, and releasing energy from foods (Some forms of life do not use oxygen in respiration.)
Digestion	Breakdown of food substances into simpler forms that can be absorbed and used
Absorption	Passage of substances through membranes and into body fluids
Circulation	Movement of substances from place to place in body fluids
Assimilation	Changing of absorbed substances into chemically different forms
Excretion	Removal of wastes produced by metabolic reactions

1.4 Maintenance of Life

The structures and functions of almost all body parts help maintain the life of the organism. The only exceptions are an organism's reproductive structures, which ensure that its species will continue into the future.

Requirements of Organisms

Life requires certain environmental factors, including the following:

1. **Water** is the most abundant chemical in the body. It is required for many metabolic processes and

provides the environment in which most of them take place. Water also transports substances within the organism and is important in regulating body temperature.

2. **Foods** are substances that provide the body with necessary chemicals (nutrients) in addition to water. Some of these chemicals are used as energy sources, others supply raw materials for building new living matter, and still others help regulate vital chemical reactions.

3. **Oxygen** is a gas that makes up about one-fifth of ordinary air. It is used to release energy from food substances. This energy, in turn, drives metabolic processes.

4. **Heat** is a form of energy. It is a product of metabolic reactions, and the degree of heat present partly determines the rate at which these reactions occur. Generally, the more heat, the more rapidly chemical reactions take place. (*Temperature* is a measure of the degree of heat.)

5. **Pressure** is an application of force to something. For example, the force on the outside of the body due to the weight of air above it is called *atmospheric pressure*. In humans, this pressure is important in breathing. Similarly, organisms living under water are subjected to *hydrostatic pressure*—a pressure a liquid exerts—due to the weight of water above them. In humans, heart action produces blood pressure (another form of hydrostatic pressure), which forces blood through blood vessels.

H ealth-care workers repeatedly monitor patients' *vital signs*—observable body functions that reflect essential metabolic activities. Vital signs indicate that a person is alive. Assessment of vital signs includes measuring body temperature and blood pressure and monitoring rates and types of pulse and breathing movements. Absence of vital signs signifies death. A person who has died displays no spontaneous muscular movements, including those of the breathing muscles and beating heart. A dead person does not respond to stimuli, and has no reflexes, such as the knee-jerk reflex and the pupillary reflexes of the eye. Brain waves cease with death, demonstrated by a flat electroencephalogram (EEG), signifying a lack of metabolic activity in the brain.

Although organisms require water, food, oxygen, heat, and pressure, these factors alone are not enough to ensure survival. Both the quantities and the qualities of such factors are also important. For example, the volume of water entering and leaving an organism must be regulated, as must the concentration of oxygen in body fluids. Similarly, survival depends on the quality as well as the quantity of food available—that is, food must supply the correct nutrients in adequate amounts.

Homeostasis

Factors in the external environment may change. If an organism is to survive, however, conditions within the fluid surrounding its body cells, which comprise its **internal environment,** must remain relatively stable. In other words, body parts function only when the concentrations of water, nutrients, and oxygen and the conditions of heat and pressure remain within certain narrow limits. This condition of a stable internal environment is called **homeostasis** (ho˝me-ō-sta´sis).

The body maintains homeostasis through a number of self-regulating control systems, or **homeostatic mechanisms,** that have the following three components in common (fig. 1.3):

Receptors, which provide information about specific conditions (stimuli) in the internal environment.
A **set point,** which tells what a particular value should be (such as body temperature at 98.6°F).
Effectors, which cause responses that alter conditions in the internal environment.

A homeostatic mechanism works as follows. If the receptors measure deviations from the set point, effectors are activated that can return conditions toward normal. As conditions return toward normal, the deviation from the set point progressively lessens, and the effectors are gradually shut down. Such a response is called a **negative feedback** (neg´ah-tiv fēd´bak) mechanism, both because the deviation from the set point is corrected (moves in the opposite or negative direction) and because the correction reduces the action of the effectors. This latter aspect is important because it prevents a correction from going too far.

To better understand this idea of negative feedback, imagine a room equipped with a furnace and an air

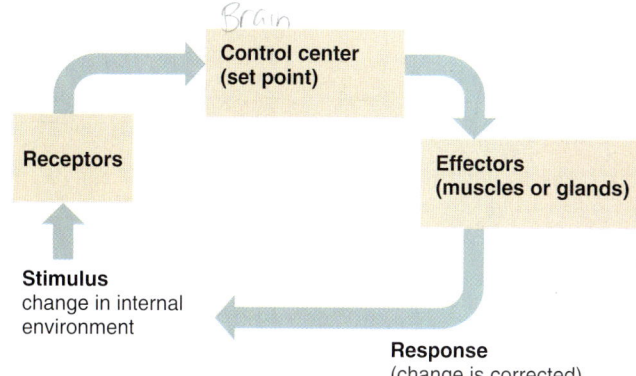

Figure 1.3
A homeostatic mechanism monitors an aspect of the internal environment and corrects any changes.

conditioner (fig. 1.4). Suppose the room temperature is to remain near 20°C (68°F), so the thermostat is adjusted to an operating level, or set point, of 20°C. Because a thermostat senses temperature changes, it will signal the furnace to start and the air conditioner to stop whenever the room temperature drops below the set point. If the temperature rises above the set point, the thermostat will stop the furnace and start the air conditioner. As a result, the room will maintain a relatively constant temperature.

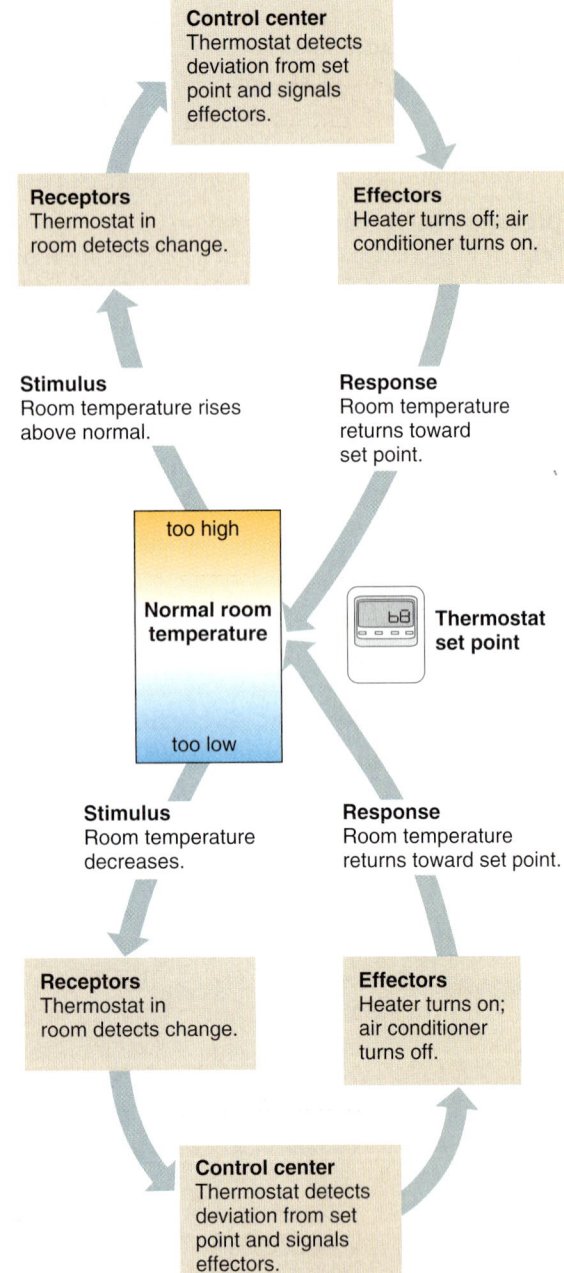

Figure 1.4
A thermostat signals an air conditioner and a furnace to turn on or off to maintain a relatively stable room temperature. This system is an example of a homeostatic mechanism.

A similar homeostatic mechanism regulates body temperature in humans. Temperature receptors are scattered throughout the body. The "thermostat" is a temperature-sensitive region in a temperature control center of the brain. In healthy persons, the set point of the brain's thermostat is at or near 37°C (98.6°F).

If a person is exposed to a cold environment and body temperature begins to drop, the temperature receptors sense this change and the temperature control center triggers heat-generating and heat-conserving activities. For example, small groups of muscles are stimulated to contract involuntarily, an action called *shivering*. Such muscular contractions produce heat, which helps warm the body. At the same time, blood vessels in the skin are signaled to constrict so that less warm blood flows through them. In this way, deeper tissues retain heat that might otherwise be lost.

If a person is becoming overheated, the brain's temperature control center triggers a series of changes that promote loss of body heat. Sweat glands in the skin secrete perspiration, and as this fluid evaporates from the surface, heat is carried away and the skin is cooled. At the same time, the brain center dilates blood vessels in the skin. This action allows the blood carrying heat from deeper tissues to reach the surface, where heat is lost to the outside (fig. 1.5). The brain stimulates an increase in heart rate, which sends a greater volume of blood into surface vessels, and an increase in breathing rate, which allows more heat-carrying air to be expelled from the lungs. Body temperature regulation is discussed further in chapter 6 (p. 120).

Another homeostatic mechanism regulates the blood pressure in the blood vessels (arteries) leading away from the heart. In this instance, pressure-sensitive receptors in the walls of these vessels sense changes in blood pressure and signal a pressure control center of the brain. If blood pressure is above the set point, the brain signals the heart chambers to contract more slowly and with less force. This decreased heart action sends less blood into the blood vessels, decreasing the pressure inside them. If blood pressure falls below the set point, the brain center signals the heart to contract more rapidly and with greater force so that the pressure in the vessels increases. Chapter 13 (p. 348) discusses regulation of blood pressure in more detail.

There are many other examples of homeostatic mechanisms in human physiology. One is the increased respiratory activity that maintains blood levels of oxygen in the internal environment during strenuous exercise. Another is the sensation of thirst created by the nervous system, stimulating water intake when the internal environment has become too concentrated. Negative feedback mechanisms also control hormone secretion (see chapter 11, p. 285).

Homeostatic mechanisms maintain a relatively constant internal environment, yet physiological values may vary slightly in a person from time to time or from

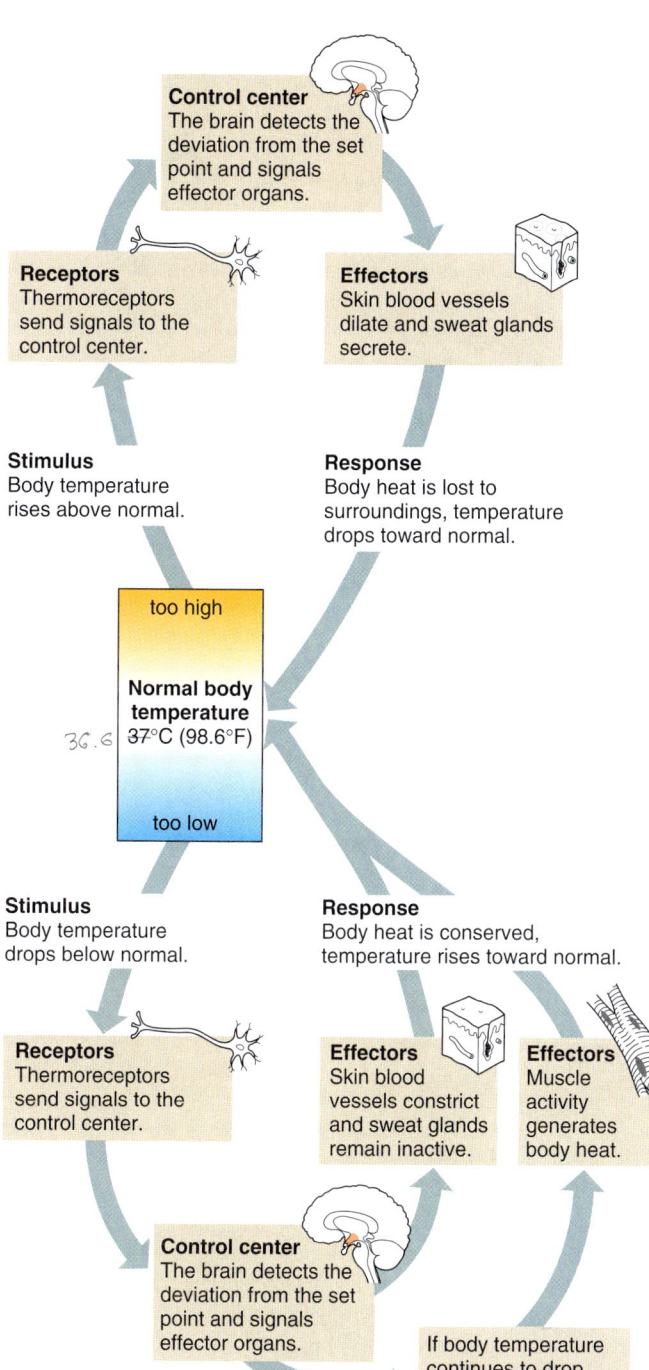

Figure 1.5
A homeostatic mechanism regulates body temperature.

one individual to the next. Therefore, both normal values for an individual and the **normal range** for the general population are clinically important.

CHECK YOUR RECALL

1. What requirements of organisms does the external environment provide?
2. Why is homeostasis important to survival?
3. Describe two homeostatic mechanisms.

1.5 Levels of Organization

Until the invention of magnifying lenses and microscopes about 400 years ago, anatomists were limited in their studies to what they could see with the unaided eye—large parts. With these tools, investigators discovered that larger body structures were made up of smaller parts, which in turn were composed of even smaller ones (fig. 1.6).

Today, scientists recognize that all materials, including those that make up the human body, are composed of chemicals. Chemicals consist of invisible particles called **atoms,** which join to form **molecules.** Small molecules can combine in complex ways to form larger molecules called **macromolecules.**

Within the human and other organisms, the basic unit of structure and function is a **cell,** which is microscopic. Although cells vary in size, shape, and specialized functions, all have certain characteristics in common. For instance, all cells of humans and other complex organisms contain structures called **organelles** (or´´gan-elz´) that carry out specific activities. Organelles are composed of aggregates of macromolecules, such as proteins, carbohydrates, lipids, and nucleic acids.

Cells may be organized into layers or masses that have common functions. Such a group of cells forms a **tissue.** Groups of different tissues that interact form **organs**—complex structures with specialized functions—and groups of organs that function closely together comprise **organ systems.** Organ systems make up an **organism** (or´gah-nizm).

Body parts can be thought of as having different levels of organization, such as the *atomic level, molecular level,* or *cellular level.* Furthermore, body parts vary in complexity from one level to the next. That is, atoms are less complex than molecules, molecules are less complex than organelles, tissues are less complex than organs, and so forth.

Chapters 2–6 discuss these levels of organization in more detail. Chapter 2 (pp. 31–44) describes the atomic and molecular levels. Chapter 3 (pp. 49–58) deals with organelles and cellular structures and functions, and chapter 4 explores cellular metabolism. Chapter 5 describes tissues. Chapter 6 (p. 113) presents membranes (linings) as examples of organs, and the skin and its accessory organs as an example of an organ system. Beginning with chapter 6, the structures and functions of each of the organ systems are described in detail.

CHECK YOUR RECALL

1. How does the human body illustrate levels of organization?
2. What is an organism?
3. How do body parts at different levels of organization vary in complexity?

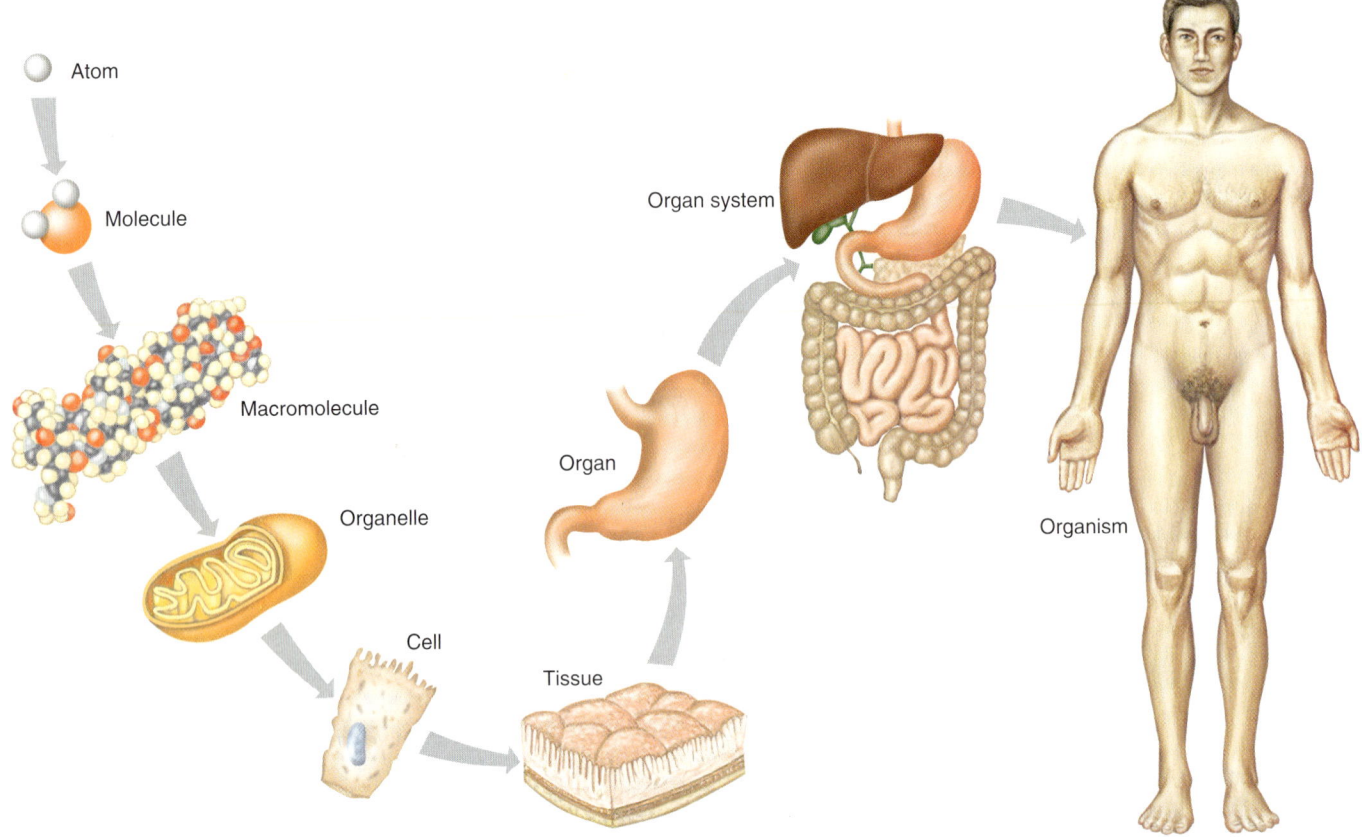

Figure 1.6
A human body is composed of parts within parts, which vary in complexity.

1.6 Organization of the Human Body

The human organism is a complex structure composed of many parts. Its major features include several body cavities, layers of membranes within these cavities, and a variety of organ systems.

Body Cavities

The human organism can be divided into an **axial** (ak´se-al) **portion,** which includes the head, neck, and trunk, and an **appendicular** (ap´´en-dik´u-lar) **portion,** which includes the upper and lower limbs. Within the axial portion are two major cavities: a **dorsal cavity** and a larger **ventral cavity** (fig. 1.7A). The organs within such a cavity are called visceral organs, or **viscera** (vis´er-ah). The dorsal cavity can be subdivided into two parts: the **cranial cavity** within the skull, which houses the brain, and the **vertebral canal,** which contains the spinal cord within sections of the backbone (vertebrae). The ventral cavity consists of a **thoracic** (tho-ras´ik) **cavity** and an **abdominopelvic** (ab-dom´´ĭ-no-pel´vik) **cavity.**

The thoracic cavity is separated from the lower abdominopelvic cavity by a broad, thin muscle called the **diaphragm.** The thoracic cavity wall is composed of skin, skeletal (voluntary) muscles, and various bones.

A region called the **mediastinum** (me´´de-as-ti´num) separates the thoracic cavity into two compartments, which contain the right and left lungs. The remaining thoracic viscera—heart, esophagus, trachea, and thymus gland—are located within the mediastinum (fig. 1.7B).

The abdominopelvic cavity, which includes an upper abdominal portion and a lower pelvic portion, extends from the diaphragm to the floor of the pelvis. Its wall consists primarily of skin, skeletal muscles, and bones. The viscera within the **abdominal cavity** include the stomach, liver, spleen, gallbladder, kidneys, and most of the small and large intestines.

The **pelvic cavity** is the portion of the abdominopelvic cavity enclosed by the hip bones (see chapter 7, p. 152). It contains the terminal portion of the large intestine, the urinary bladder, and the internal reproductive organs.

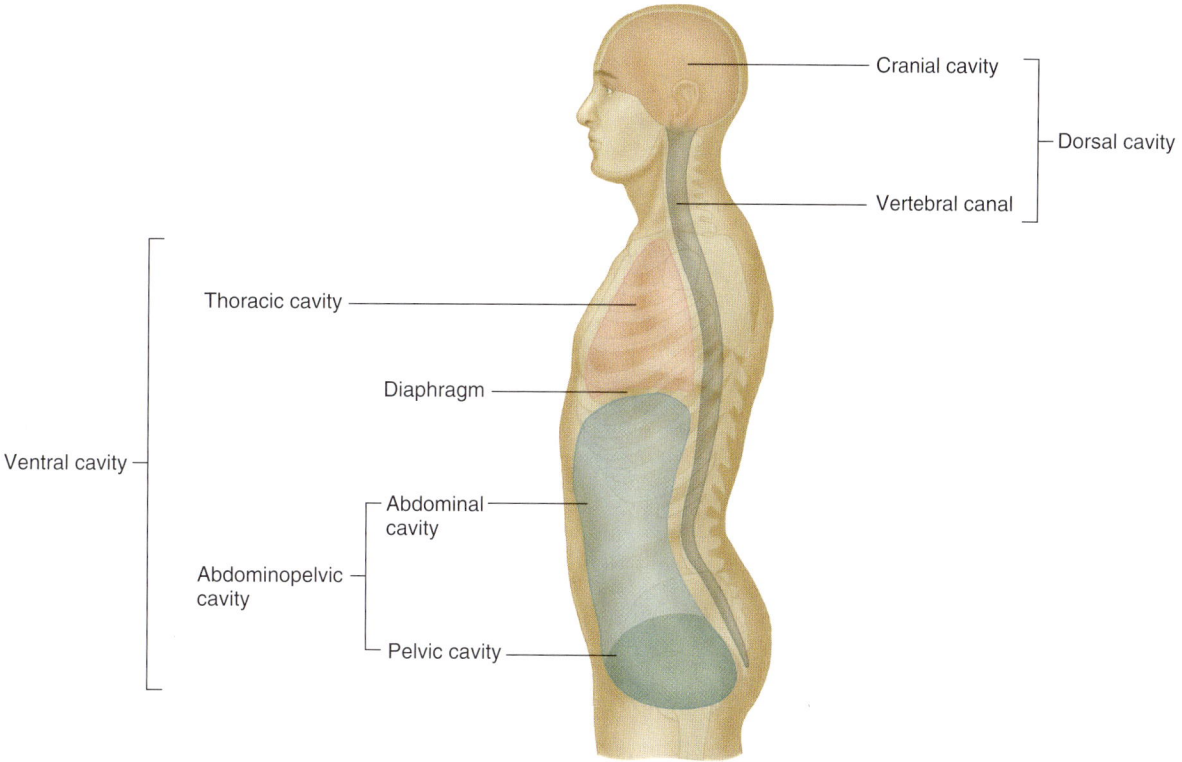

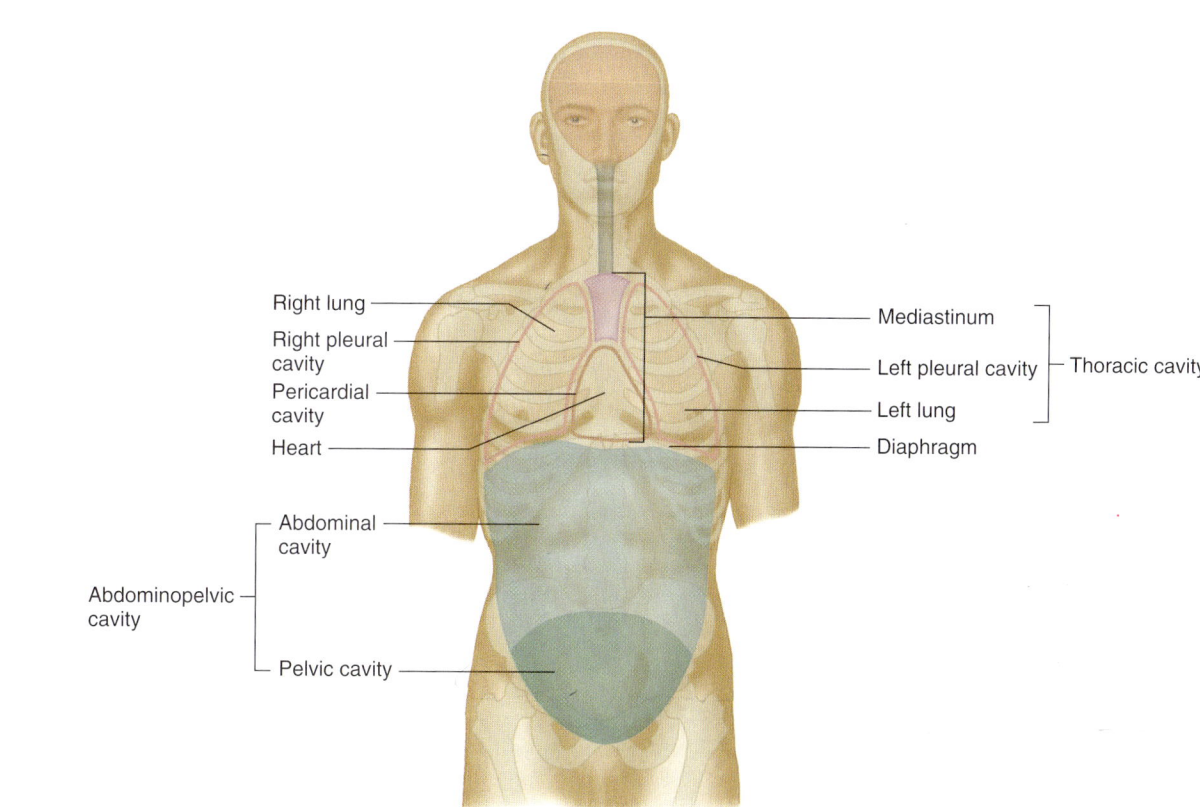

Figure 1.7
Major body cavities. (*A*) Lateral view. (*B*) Coronal view.

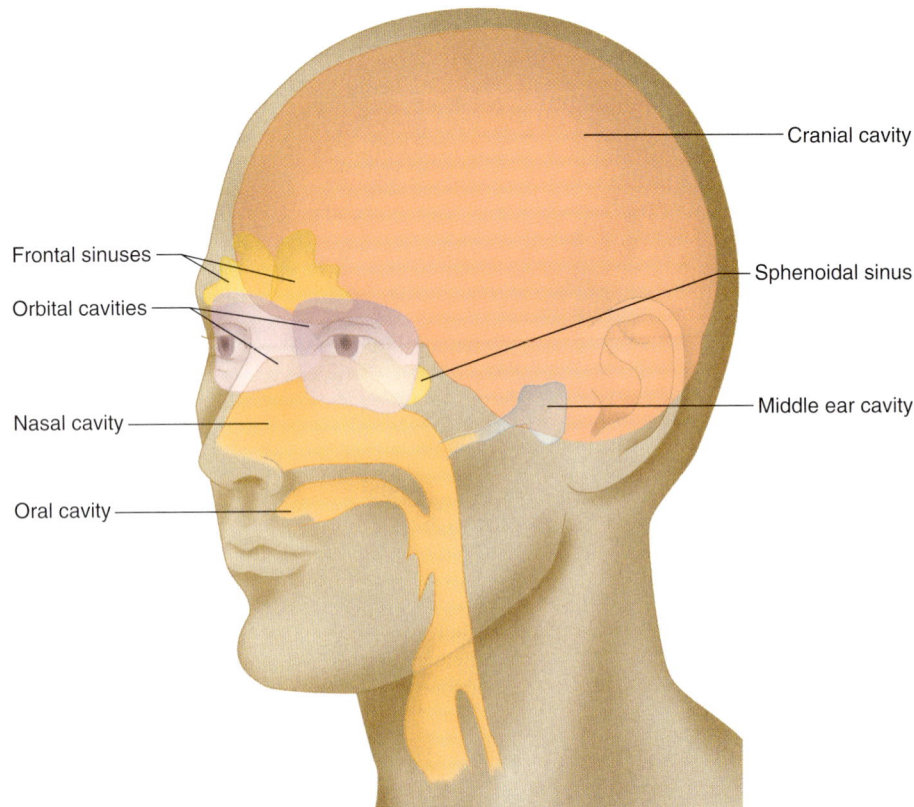

Figure 1.8
The cavities within the head include the cranial, oral, nasal, orbital, and middle ear cavities, as well as several sinuses.

Smaller cavities within the head include (fig. 1.8):

1. **Oral cavity,** containing the teeth and tongue.
2. **Nasal cavity,** located within the nose and divided into right and left portions by a nasal septum. Several air-filled *sinuses* connect to the nasal cavity (see chapter 7, p. 136). These include the frontal and sphenoidal sinuses shown in figure 1.8.
3. **Orbital cavities,** containing the eyes and associated skeletal muscles and nerves.
4. **Middle ear cavities,** containing the middle ear bones.

Thoracic and Abdominopelvic Membranes

The walls of the right and left thoracic compartments, which contain the lungs, are lined with a membrane called the *parietal pleura* (fig. 1.9). A similar membrane, called the *visceral pleura*, covers the lungs themselves. (Note: **Parietal** (pah-ri´ĕ-tal) refers to the membrane attached to the wall of a cavity; **visceral** (vis´er-al) refers to the membrane that is deeper—toward the interior—and covers an internal organ, such as a lung.)

The parietal and visceral **pleural** (ploo´ral) **membranes** are separated by a thin film of watery fluid (serous fluid) that they secrete. While no actual space normally exists between these membranes, the potential space between them is called the *pleural cavity* (see fig. 1.7B).

The heart, which is located in the broadest portion of the mediastinum, is surrounded by **pericardial** (per´´ĭ-kar´de-al) **membranes.** A thin *visceral pericardium* covers the heart's surface and is separated from a thicker *parietal pericardium* by a small volume of fluid. The *pericardial cavity* (see fig. 1.7B) is the potential space between these membranes.

In the abdominopelvic cavity, the lining membranes are called **peritoneal** (per´´ĭ-to-ne´al) **membranes** (fig. 1.10). A *parietal peritoneum* lines the wall, and a *visceral peritoneum* covers each organ in the abdominal cavity. The *peritoneal cavity* is the potential space between these membranes.

 CHECK YOUR RECALL

1. What does *viscera* mean?
2. Which organs occupy the dorsal cavity? The ventral cavity?
3. Name the cavities of the head.
4. Describe the membranes associated with the thoracic and abdominopelvic cavities.

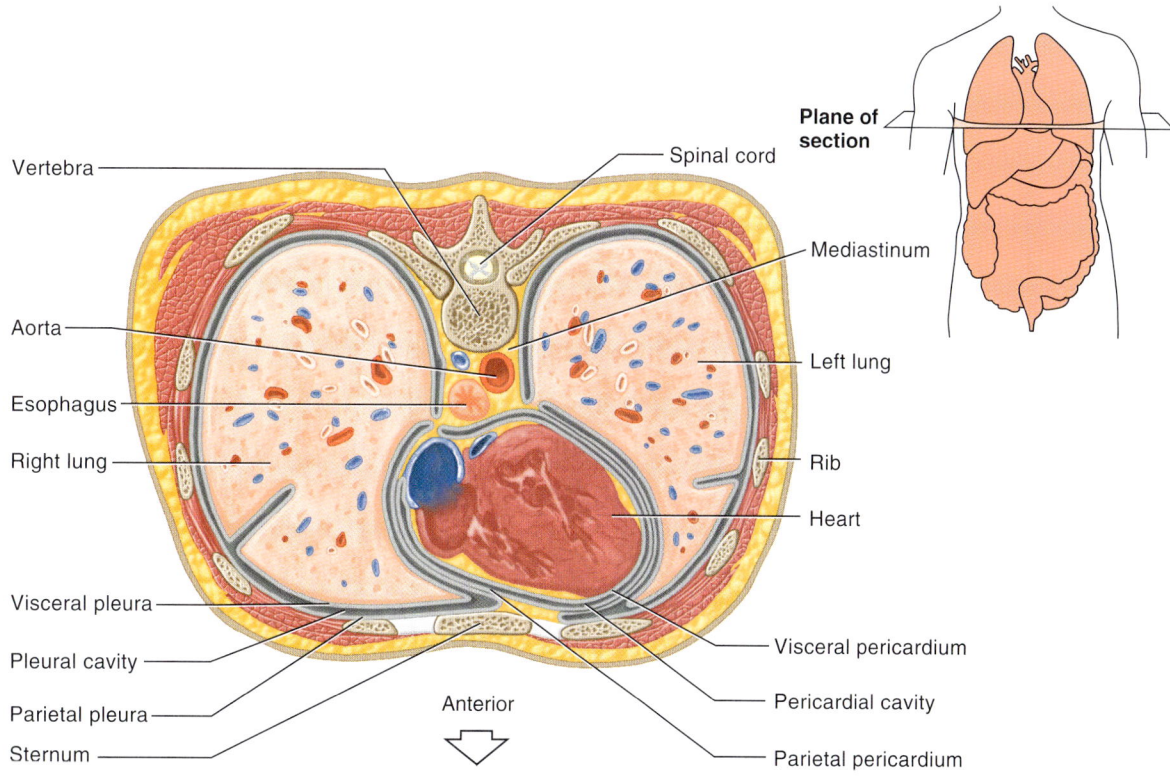

Figure 1.9
A transverse section through the thorax reveals the serous membranes associated with the heart and lungs (*superior view*).

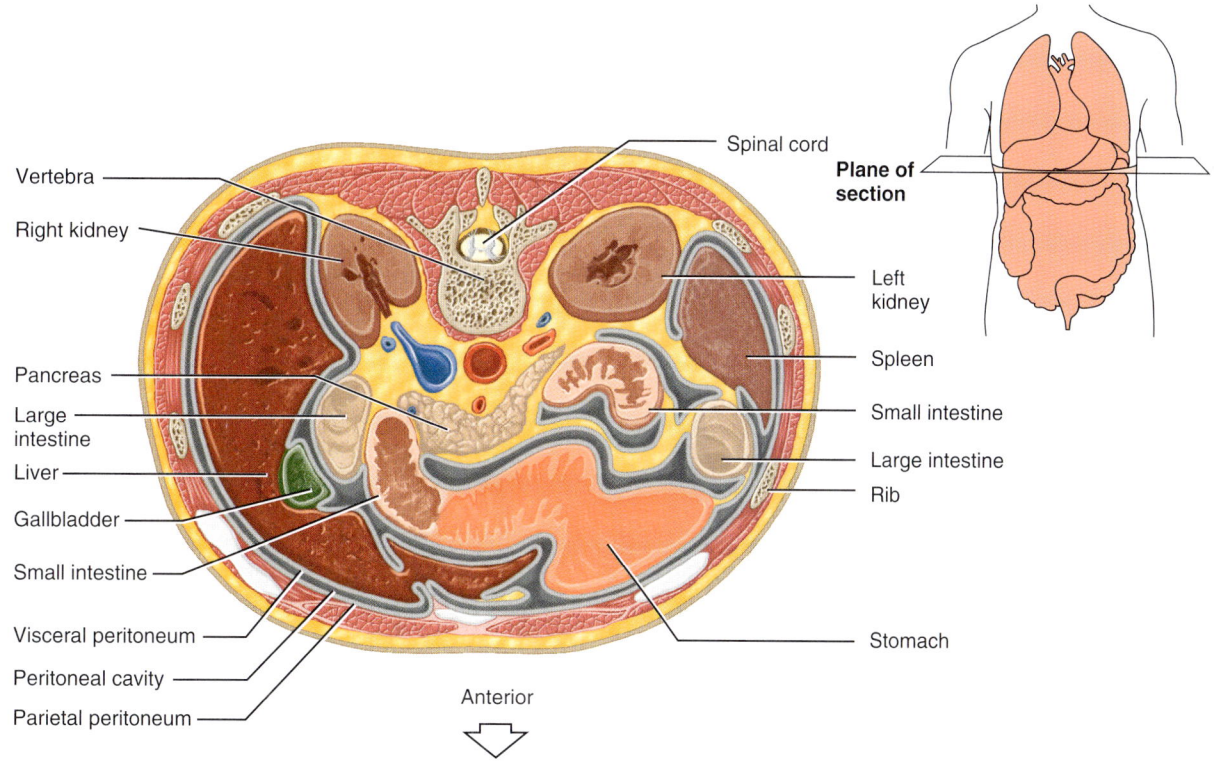

Figure 1.10
Transverse section through the abdomen (*superior view*).

Organ Systems

The human organism consists of several organ systems. Each system includes a set of interrelated organs that work together to provide specialized functions. As you read about each system, you may want to consult the illustrations of the human torso and locate some of the organs listed in the description (see Reference Plates, pp. 21–28).

Body Covering

Organs of the **integumentary** (in-teg-u-men´tar-e) **system** (see chapter 6) include the skin and various accessory organs, such as the hair, nails, sweat glands, and sebaceous glands. These parts protect underlying tissues, help regulate body temperature, house a variety of sensory receptors, and synthesize certain products.

Support and Movement

The organs of the skeletal and muscular systems (see chapters 7 and 8) support and move body parts. The **skeletal** (skel´ĕ-tal) **system** consists of bones, as well as ligaments and cartilages that bind bones together. These parts provide frameworks and protective shields for softer tissues, are attachments for muscles, and act with muscles when body parts move. Tissues within bones also produce blood cells and store inorganic salts.

Muscles are the organs of the **muscular** (mus´ku-lar) **system.** By contracting and pulling their ends closer together, muscles provide forces that cause body movements. They also maintain posture and are the main source of body heat.

Integration and Coordination

For the body to act as a unit, its parts must be integrated and coordinated. The nervous and endocrine systems control and adjust various organ functions, which maintains homeostasis.

The **nervous** (ner´vus) **system** (see chapter 9) consists of the brain, spinal cord, nerves, and sense organs (see chapter 10). Nerve cells within these organs use electrochemical signals called *nerve impulses* to communicate with one another and with muscles and glands. Each impulse produces a relatively short-term effect on its target. Some nerve cells act as specialized sensory receptors that can detect changes inside and outside the body. Other nerve cells receive the impulses transmitted from these sensory receptors and interpret and act on the information received. Still other nerve cells carry impulses from the brain or spinal cord to muscles or glands, stimulating them to contract or to secrete products.

The **endocrine** (en´do-krin) **system** (see chapter 11) includes all the glands that secrete chemical messengers called *hormones*. The hormones, in turn, move away from the glands in body fluids, such as blood or tissue fluid (fluid from the spaces within tissues). A particular hormone affects only a particular group of cells, which are called its *target cells*. The effect of a hormone is to alter the metabolism of the target cells. Compared to nerve impulses, hormonal effects occur over a relatively long time period. Organs of the endocrine system include the pituitary, thyroid, parathyroid, and adrenal glands, as well as the pancreas, ovaries, testes, pineal gland, and thymus gland.

Transport

Two organ systems transport substances throughout the internal environment. The **cardiovascular** (kahr´´de-o-vas´ku-lur) **system** (see chapters 12 and 13) includes the heart, arteries, veins, capillaries, and blood. The heart is a muscular pump that helps force blood through the blood vessels. Blood transports gases, nutrients, hormones, and wastes. It carries oxygen from the lungs and nutrients from the digestive organs to all body cells, where these biochemicals are used in metabolic processes. Blood also transports hormones and carries wastes from body cells to the excretory organs, where the wastes are removed from the blood and released to the outside.

The **lymphatic** (lim-fat´ik) **system** (see chapter 14) is sometimes considered part of the cardiovascular system. It is composed of the lymphatic vessels, lymph fluid, lymph nodes, thymus gland, and spleen. This system transports some of the tissue fluid back to the bloodstream and carries certain fatty substances away from the digestive organs. Cells of the lymphatic system are called lymphocytes, and they defend the body against infections by removing disease-causing microorganisms and viruses from tissue fluid.

Absorption and Excretion

Organs in several systems absorb nutrients and oxygen and excrete various wastes. For example, the organs of the **digestive** (di-jest´tiv) **system** (see chapter 15) receive foods from the outside. Then they break down food molecules into simpler forms that can pass through cell membranes and thus be absorbed. Materials that are not absorbed are transported back to the outside and eliminated. Certain digestive organs also produce hormones and thus function as parts of the endocrine system. The digestive system includes the mouth, tongue, teeth, salivary glands, pharynx, esophagus, stomach, liver, gallbladder, pancreas, small intestine, and large intestine. Chapter 15 (pp. 420–427) also discusses nutrition.

The organs of the **respiratory system** (see chapter 16) move air in and out and exchange gases between the blood and the air. More specifically, oxygen passes from air within the lungs into the blood, and carbon dioxide leaves the blood and enters the air. The nasal cavity, pharynx, larynx, trachea, bronchi, and lungs are parts of this system.

The **urinary** (u″rĭ-ner″e) **system** (see chapter 17) consists of the kidneys, ureters, urinary bladder, and urethra. The kidneys remove wastes from blood and help maintain the body's water and electrolyte balance. The product of these activities is urine. Other portions of the urinary system store urine and transport it outside the body. Chapter 18 discusses the urinary system's role in maintaining water, electrolyte, and acid-base balance.

Reproduction

Reproduction is the process of producing offspring (progeny). Cells reproduce when they divide and give rise to new cells. The **reproductive** (re″pro-duk´tiv) **system** of an organism, however, produces whole new organisms like itself (see chapter 19).

The male reproductive system includes the scrotum, testes, epididymides, vasa deferentia, seminal vesicles, prostate gland, bulbourethral glands, penis, and urethra. These parts produce and maintain sperm cells (spermatozoa). Components of the male reproductive system also transfer sperm cells into the female reproductive tract.

The female reproductive system consists of the ovaries, uterine tubes, uterus, vagina, clitoris, and vulva. These organs produce and maintain female sex cells (the egg cells or ova), receive sperm cells, and transport the sperm cells and egg cells within the female system. The female reproductive system also supports the development of prenatal humans, such as embryos and fetuses, and enables the birth process to proceed.

CHECK YOUR RECALL

1. Name each of the major organ systems, and list the organs of each system.
2. Describe the general functions of each organ system.

1.7 Anatomical Terminology

To communicate effectively with one another, researchers and clinicians have developed a set of precise terms to describe anatomy. Some of these terms concern the relative positions of body parts, others relate to imaginary planes along which cuts may be made, and still others describe body regions.

Use of such terms assumes that the body is in the **anatomical position.** This means that the body is standing erect, face forward, with upper limbs at the sides and the palms forward. Note that the terms right and left refer to the right and left of a body in anatomical position.

Relative Positions

Terms of relative position describe the location of one body part with respect to another. They include the following:

1. **Superior** means that a body part is above another part or is closer to the head. (The thoracic cavity is superior to the abdominopelvic cavity.)
2. **Inferior** means that a body part is below another body part or is toward the feet. (The neck is inferior to the head.)
3. **Anterior** (or *ventral*) means toward the front. (The eyes are anterior to the brain.)
4. **Posterior** (or *dorsal*) is the opposite of anterior; it means toward the back. (The pharynx is posterior to the oral cavity.)
5. **Medial** relates to an imaginary midline dividing the body into equal right and left halves. A body part is medial if it is closer to this line than another part. (The nose is medial to the eyes.)
6. **Lateral** means toward the side with respect to the imaginary midline. (The ears are lateral to the eyes.)
7. **Proximal** describes a body part that is closer to a point of attachment than another body part. (The elbow is proximal to the wrist.)
8. **Distal** is the opposite of proximal. It means that a particular body part is farther from a point of attachment than another body part. (The fingers are distal to the wrist.) or Lateral
9. **Superficial** means situated near the surface. (The epidermis is the superficial layer of the skin.) *Peripheral* also means outward or near the surface. It describes the location of certain blood vessels and nerves. (The nerves that branch from the brain and spinal cord are peripheral nerves.)
10. **Deep** describes parts that are more internal. (The dermis is the deep layer of the skin.)

Body Sections

Observing the relative locations and organization of internal body parts requires cutting or sectioning the

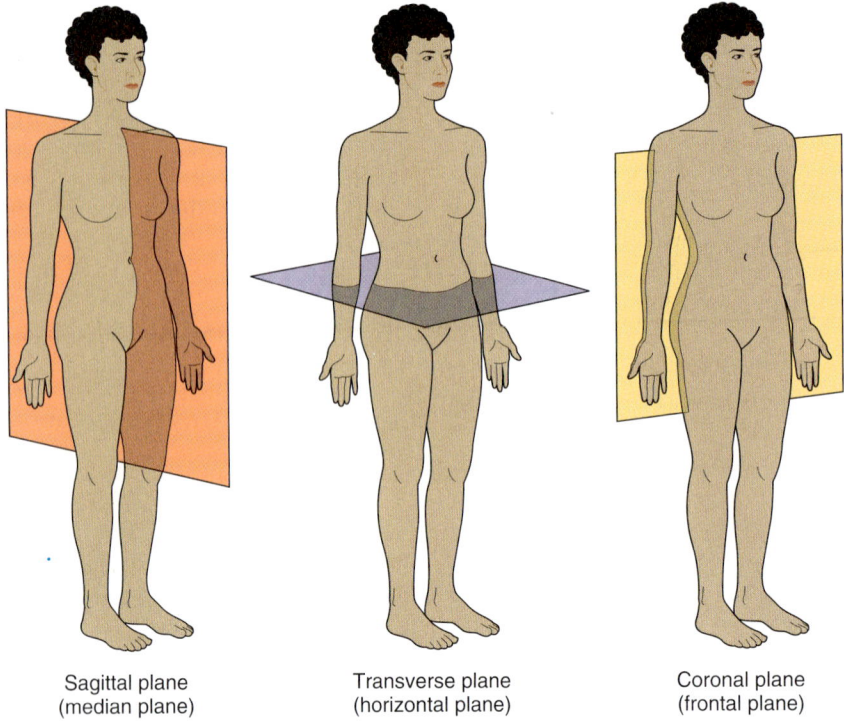

Sagittal plane (median plane)   Transverse plane (horizontal plane)   Coronal plane (frontal plane)

Figure 1.11
Observation of internal parts requires sectioning the body along various planes.

body along various planes (fig. 1.11). The following terms describe such planes and sections:

1. **Sagittal** refers to a lengthwise cut that divides the body into right and left portions. If a sagittal section passes along the midline and divides the body into equal parts, it is called *median* (midsagittal).
2. **Transverse** (or *horizontal*) refers to a cut that divides the body into superior and inferior portions.
3. **Coronal** (or *frontal*) refers to a section that divides the body into anterior and posterior portions.

Sometimes, a cylindrical organ such as a long bone is sectioned. In this case, a cut across the structure is called a *cross section,* an angular cut is an *oblique section,* and a lengthwise cut is a *longitudinal section* (fig. 1.12).

Body Regions

A number of terms designate body regions. The abdominal area, for example, is subdivided into the following nine regions, as figure 1.13A shows:

1. **Epigastric region** refers to the upper middle portion.
2. **Left** and **right hypochondriac regions** lie on each side of the epigastric region.
3. **Umbilical region** refers to the middle portion.
4. **Left** and **right lumbar regions** lie on each side of the umbilical region.
5. **Hypogastric region** refers to the lower middle portion.
6. **Left** and **right iliac regions** (left and right inguinal regions) lie on each side of the hypogastric region.

The abdominal area is also often subdivided into four quadrants, as figure 1.13B shows.

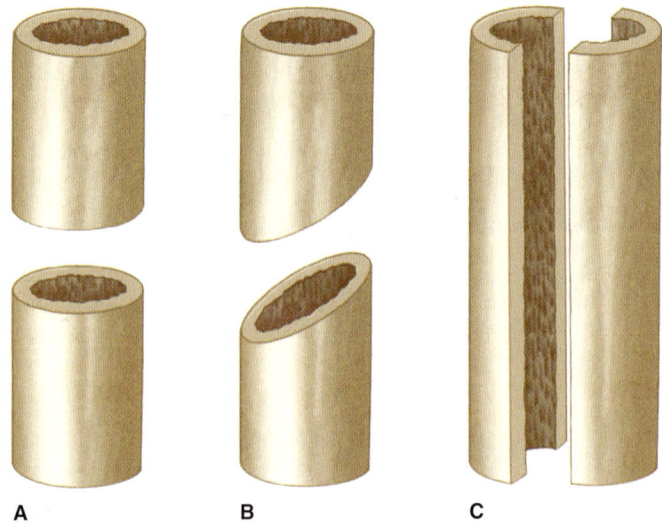

A   B   C

Figure 1.12
Cylindrical parts may be cut in (*A*) cross section, (*B*) oblique section, or (*C*) longitudinal section.

CHAPTER 1 Introduction to Human Anatomy and Physiology

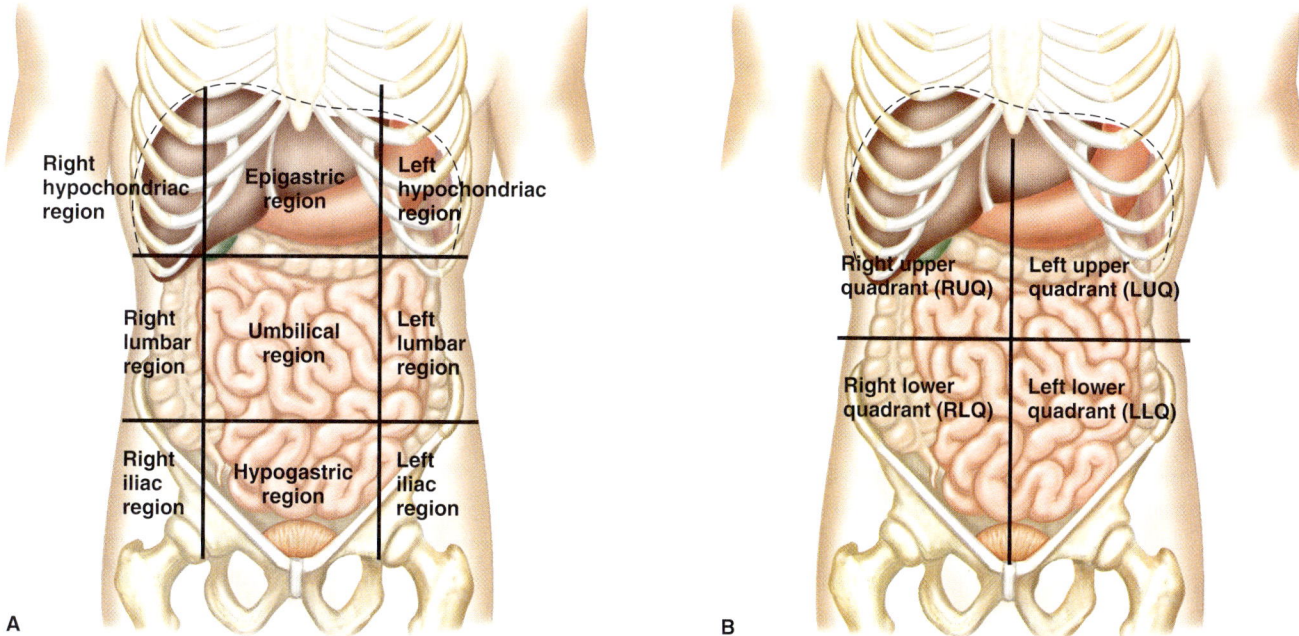

Figure 1.13
There are two common ways to subdivide the abdominal area. (*A*) The abdominal area is subdivided into nine regions. (*B*) The abdominal area may also be subdivided into four quadrants.

Radiologists use a procedure called *computerized tomography*, or CT scanning, to visualize internal organ sections (fig. 1A). In this procedure, an X-ray-emitting device moves around the body region being examined. At the same time, an X-ray detector moves in the opposite direction on the other side. As the devices move, an X-ray beam passes through the body from hundreds of different angles.

Since tissues and organs of varying composition within the body absorb X rays differently, the amount of X ray reaching the detector varies from position to position. A computer records the measurements from the X-ray detector, and combines them mathematically to create a sectional image of the internal body parts on a viewing screen.

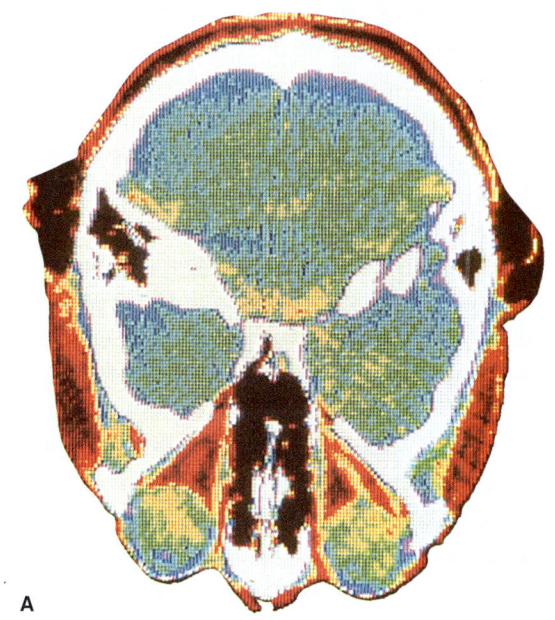

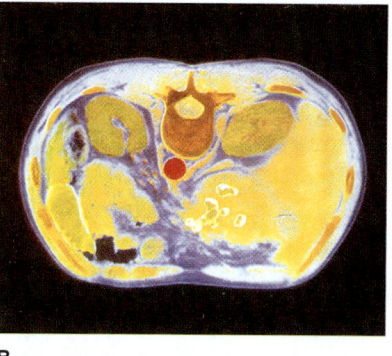

Figure 1A
Falsely colored CT (computerized tomography) scans of the (A) head and (B) abdomen. Note: These are not shown in correct relative size.

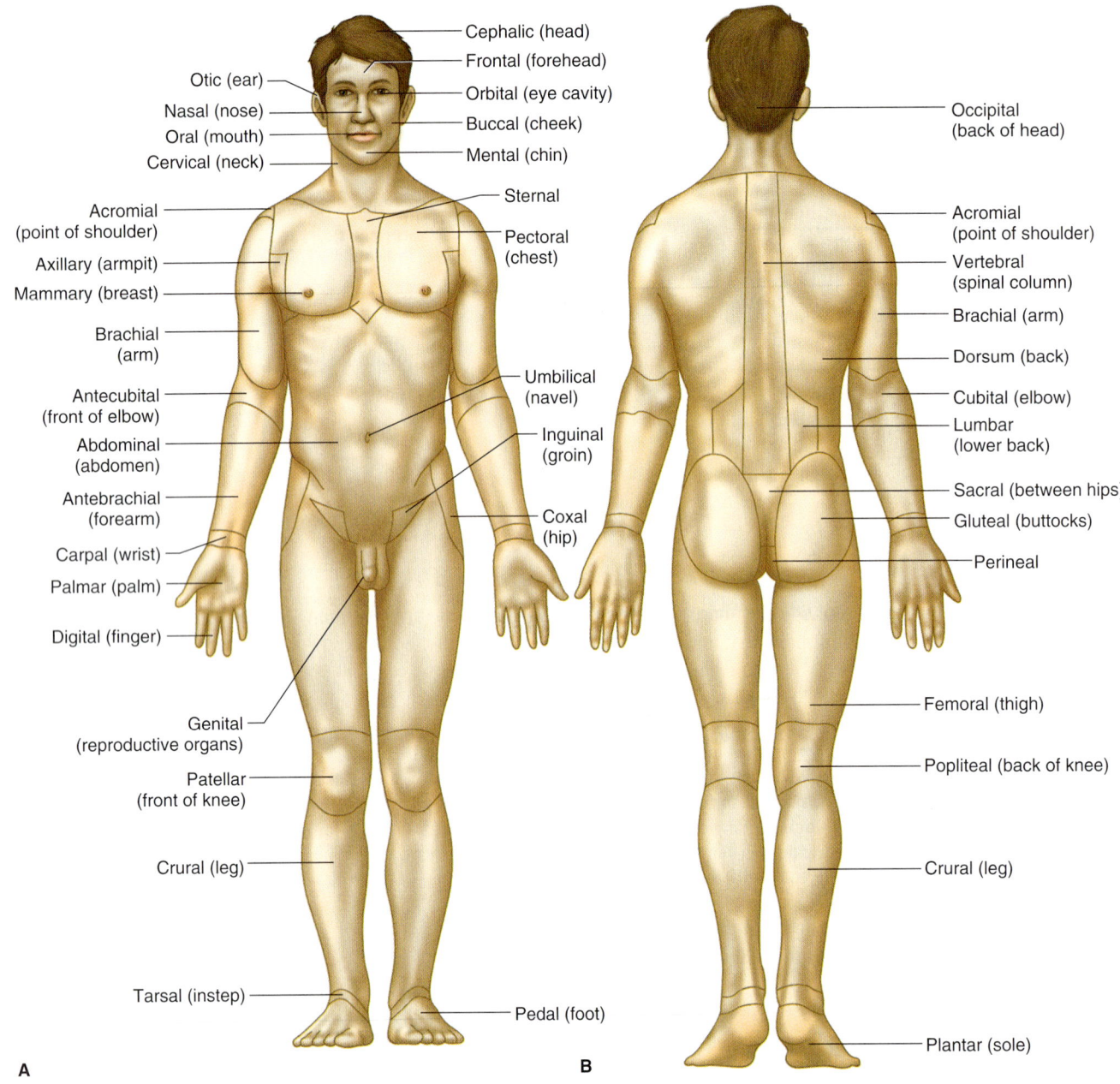

Figure 1.14
Some terms used to describe body regions. (A) Anterior regions. (B) Posterior regions.

The following terms are commonly used to refer to various body regions, some of which are illustrated in figure 1.14:

abdominal (ab-dom´ĭ-nal) The region between the thorax and pelvis.
acromial (ah-kro´me-al) The point of the shoulder.
antebrachial (an´´te-bra´ke-al) The forearm.
antecubital (an´´te-ku´bĭ-tal) The space in front of the elbow.
axillary (ak´sĭ-ler´´e) The armpit.
brachial (bra´ke-al) The arm.
buccal (buk´al) The cheek.
carpal (kar´pal) The wrist.
celiac (se´le-ak) The abdomen.
cephalic (sĕ-fal´ik) The head.
cervical (ser´vĭ-kal) The neck.
costal (kos´tal) The ribs.
coxal (kok´sal) The hip.
crural (kroor´al) The leg.
cubital (ku´bĭ-tal) The elbow.
digital (dij´ĭ-tal) The finger.
dorsal (dor´sal) The back.

femoral (fem´or-al) The thigh.
frontal (frun´tal) The forehead.
genital (jen´ĭ-tal) The reproductive organs.
gluteal (gloo´te-al) The buttocks.
inguinal (ing´gwĭ-nal) The depressed area of the abdominal wall near the thigh (groin).
lumbar (lum´bar) The region of the lower back between the ribs and the pelvis (loin).
mammary (mam´er-e) The breast.
mental (men´tal) The chin.
nasal (na´zal) The nose.
occipital (ok-sip´ĭ-tal) The lower posterior region of the head.
oral (o´ral) The mouth.
orbital (or´bi-tal) The eye cavity.
otic (o´tik) The ear.
palmar (pahl´mar) The palm of the hand.
patellar (pah-tel´ar) The front of the knee.
pectoral (pek´tor-al) The chest.
pedal (ped´al) The foot.
pelvic (pel´vik) The pelvis.
perineal (per´´ĭ-ne´al) The region between the anus and the external reproductive organs (perineum).
plantar (plan´tar) The sole of the foot.
popliteal (pop´´lĭ-te´al) The area behind the knee.
sacral (sa´kral) The posterior region between the hipbones.
sternal (ster´nal) The middle of the thorax, anteriorly.
tarsal (tahr´sal) The instep of the foot.
umbilical (um-bil´ĭ-kal) The navel.
vertebral (ver´te-bral) The spinal column.

CHECK YOUR RECALL

1. Describe the anatomical position.
2. Using the appropriate terms, describe the relative positions of several body parts.
3. Describe the three types of body sections.
4. Name the nine regions of the abdomen.

Some Medical and Applied Sciences

cardiology (kar´´de-ol´o-je) Branch of medical science dealing with the heart and heart diseases.
cytology (si-tol´o-je) Study of the structure, function, and abnormalities of cells.
dermatology (der´´mah-tol´o-je) Study of the skin and its diseases.
endocrinology (en´´do-krĭ-nol´o-je) Study of hormones, hormone-secreting glands, and their diseases.
epidemiology (ep´´ĭ-de´´me-ol´o-je) Study of the factors determining the distribution and frequency of health-related conditions occurring within a defined human population.
gastroenterology (gas´´tro-en´´ter-ol´o-je) Study of the stomach and intestines and their diseases.
geriatrics (jer´´e-at´riks) Branch of medicine dealing with older individuals and their medical problems.
gerontology (jer´´on-tol´o-je) Study of the aging process.
gynecology (gi´´ně-kol´o-je) Study of the female reproductive system and its diseases.
hematology (hēm´´ah-tol´o-je) Study of the blood and blood diseases.
histology (his-tol´o-je) Study of the structure and function of tissues, also called microscopic anatomy.
immunology (im´´u-nol´o-je) Study of the body's resistance to infectious disease.
neonatology (ne´´o-na-tol´o-je) Study of newborns and the treatment of their disorders.
nephrology (ně-frol´o-je) Study of the structure, function, and diseases of the kidneys.
neurology (nu-rol´o-je) Study of the nervous system and its disorders.
obstetrics (ob-stet´riks) Branch of medicine dealing with pregnancy and childbirth.
oncology (ong-kol´o-je) Study of cancers.
ophthalmology (of´´thal-mol´o-je) Study of the eye and eye diseases.
orthopedics (or´´tho-pe´diks) Branch of medicine dealing with the muscular and skeletal systems and their problems.
otolaryngology (o´´to-lar´´in-gol´o-je) Study of the ear, throat, and larynx, and their diseases.
pathology (pah-thol´o-je) Study of structural and functional changes that disease causes.
pediatrics (pe´´de-at´riks) Branch of medicine dealing with children and their diseases.
pharmacology (fahr´´mah-kol´o-je) Study of drugs and their uses in the treatment of disease.
podiatry (po-di´ah-tre) Study of the care and treatment of feet.
psychiatry (si-ki´ah-tre) Branch of medicine dealing with the mind and its disorders.
radiology (ra´´de-ol´o-je) Study of X rays and radioactive substances and their uses in the diagnosis and treatment of diseases.
toxicology (tok´´sĭ-kol´o-je) Study of poisonous substances and their effects upon body parts.
urology (u-rol´o-je) Branch of medicine dealing with the urinary system, apart from the kidneys (nephrology) and the male reproductive system, and their diseases.

SUMMARY OUTLINE

1.1 Introduction (p. 3)
1. Early interest in the human body probably developed as people became concerned about injuries and illnesses.
2. Primitive doctors began to learn how certain herbs and potions affected body functions.
3. The belief that humans could understand forces that caused natural events led to the development of modern science.
4. A set of terms originating from Greek and Latin words is the basis for the language of anatomy and physiology.

1.2 Anatomy and Physiology (p. 3)
1. Anatomy describes the form and organization of body parts.
2. Physiology considers the functions of anatomical parts.
3. The function of an anatomical part depends on the way it is constructed.

1.3 Characteristics of Life (p. 4)
Characteristics of life are traits all organisms share.
1. These characteristics include:
   a. Movement—changing body position or moving internal parts.
   b. Responsiveness—sensing and reacting to internal or external changes.
   c. Growth—increasing size without changing shape.
   d. Reproduction—producing offspring.
   e. Respiration—obtaining oxygen, using oxygen to release energy from foods, and removing gaseous wastes.
   f. Digestion—breaking down food substances into component nutrients that the intestine can absorb.
   g. Absorption—moving substances through membranes and into body fluids.
   h. Circulation—moving substances through the body in body fluids.
   i. Assimilation—changing substances into chemically different forms.
   j. Excretion—removing body wastes.
2. Acquisition and use of energy constitute metabolism.

1.4 Maintenance of Life (p. 4)
The structures and functions of body parts maintain the life of the organism.
1. Requirements of organisms
   a. Water is used in many metabolic processes, provides the environment for metabolic reactions, and transports substances.
   b. Food supplies energy, raw materials for building new living matter, and chemicals necessary in vital reactions.
   c. Oxygen releases energy from food materials. This energy drives metabolic reactions.
   d. Heat is a product of metabolic reactions and helps govern the rates of these reactions.
   e. Pressure is an application of force to something. In humans, atmospheric and hydrostatic pressures help breathing and blood movements, respectively.
2. Homeostasis
   a. If an organism is to survive, the conditions within its body fluids must remain relatively stable.
   b. Maintenance of a stable internal environment is called *homeostasis*.
   c. Homeostatic mechanisms help regulate body temperature and blood pressure.
   d. Homeostatic mechanisms act through negative feedback.

1.5 Levels of Organization (p. 7)
The body is composed of parts with different levels of organization.
1. Matter is composed of atoms.
2. Atoms join to form molecules.
3. Organelles are built of groups of large molecules.
4. Cells, which contain organelles, are the basic units of structure and function that form the body.
5. Cells are organized into tissues.
6. Tissues are organized into organs.
7. Organs that function closely together comprise organ systems.
8. Organ systems constitute the organism.
9. These levels of organization vary in complexity from one level to the next.

1.6 Organization of the Human Body (p. 8)
1. Body cavities
   a. The axial portion of the body contains the dorsal and ventral cavities.
      (1) The dorsal cavity includes the cranial cavity and the vertebral canal.
      (2) The ventral cavity includes the thoracic and abdominopelvic cavities, which are separated by the diaphragm.
   b. The organs within a body cavity are called *viscera*.
   c. The mediastinum separates the thoracic cavity into right and left compartments.
   d. Other body cavities include the oral, nasal, orbital, and middle ear cavities.
2. Thoracic and abdominopelvic membranes
   a. Thoracic membranes
      (1) Pleural membranes line the thoracic cavity (parietal pleura) and cover the lungs (visceral pleura).
      (2) Pericardial membranes surround the heart (parietal pericardium) and cover its surface (visceral pericardium).
      (3) The pleural and pericardial cavities are the potential spaces between the respective parietal and visceral membranes.
   b. Abdominopelvic membranes
      (1) Peritoneal membranes line the abdominopelvic cavity (parietal peritoneum) and cover the organs inside (visceral peritoneum).
      (2) The peritoneal cavity is the potential space between the parietal and visceral membranes.
3. Organ systems
   The human organism consists of several organ systems. Each system includes a set of interrelated organs.
   a. Body covering
      (1) The integumentary system includes the skin, hair, nails, sweat glands, and sebaceous glands.
      (2) It protects underlying tissues, regulates body temperature, houses sensory receptors, and synthesizes various substances.
   b. Support and movement
      (1) Skeletal system
         (a) The skeletal system is composed of bones, as well as cartilages and ligaments that bind bones together.
         (b) It provides a framework, protective shields, and attachments for muscles. It also produces blood cells and stores inorganic salts.
      (2) Muscular system
         (a) The muscular system includes the muscles of the body.
         (b) It moves body parts, maintains posture, and produces body heat.

c. Integration and coordination
      (1) Nervous system
         (a) The nervous system consists of the brain, spinal cord, nerves, and sense organs.
         (b) It receives impulses from sensory parts, interprets these impulses, and acts on them by stimulating muscles or glands to respond.
      (2) Endocrine system
         (a) The endocrine system consists of glands that secrete hormones.
         (b) Hormones help regulate metabolism.
         (c) This system includes the pituitary, thyroid, parathyroid, and adrenal glands, as well as the pancreas, ovaries, testes, pineal gland, and thymus gland.
   d. Transport
      (1) Cardiovascular system
         (a) The cardiovascular system includes the heart, which pumps blood, and the blood vessels, which carry blood to and from body parts.
         (b) Blood transports oxygen, nutrients, hormones, and wastes.
      (2) Lymphatic system
         (a) The lymphatic system is composed of lymphatic vessels, lymph fluid, lymph nodes, thymus gland, and spleen.
         (b) It transports lymph fluid from tissues to the bloodstream, carries certain fatty substances away from the digestive organs, and aids in defending the body against disease-causing agents.
   e. Absorption and excretion
      (1) Digestive system
         (a) The digestive system receives foods, breaks down food molecules into nutrients that can pass through cell membranes, and eliminates materials that are not absorbed.
         (b) It includes the mouth, tongue, teeth, salivary glands, pharynx, esophagus, stomach, liver, gallbladder, pancreas, small intestine, and large intestine.
         (c) Some digestive organs produce hormones.
      (2) Respiratory system
         (a) The respiratory system takes in and sends out air and exchanges gases between the air and blood.
         (b) It includes the nasal cavity, pharynx, larynx, trachea, bronchi, and lungs.
      (3) Urinary system
         (a) The urinary system includes the kidneys, ureters, urinary bladder, and urethra.
         (b) It filters wastes from the blood and helps maintain water, acid-base, and electrolyte balance.
   f. Reproduction
      (1) The reproductive systems produce new organisms.
      (2) The male reproductive system includes the scrotum, testes, epididymides, vasa deferentia, seminal vesicles, prostate gland, bulbourethral glands, urethra, and penis, which produce, maintain, and transport male sex cells.
      (3) The female reproductive system includes the ovaries, uterine tubes, uterus, vagina, clitoris, and vulva, which produce, maintain, and transport female sex cells.

1.7 Anatomical Terminology (p. 13)
Terms with precise meanings help investigators communicate effectively.
1. Relative positions
   These terms describe the location of one part with respect to another part.
2. Body sections
   Body sections are planes along which the body may be cut to observe the relative locations and organizations of internal parts.
3. Body regions
   Special terms designate various body regions.

REVIEW EXERCISES

Part A
1. Briefly describe the early development of knowledge about the human body. (p. 3)
2. Distinguish between anatomy and physiology. (p. 3)
3. Explain the relationship between the form and function of body parts. (p. 3)
4. List and describe ten characteristics of life. (p. 4)
5. Define *metabolism*. (p. 4)
6. List and describe five requirements of organisms. (p. 4)
7. Describe two types of pressure that may act on the outside of an organism. (p. 5)
8. Define *homeostasis*, and explain its importance. (p. 5)
9. Explain the control of body temperature. (p. 6)
10. Describe a homeostatic mechanism that helps regulate blood pressure. (p. 6)
11. List the levels of organization within the human body. (p. 7)
12. Distinguish between the axial and appendicular portions of the body. (p. 8)
13. Distinguish between the dorsal and ventral body cavities, and name the smaller cavities within each. (p. 8)
14. Explain what is meant by *viscera*. (p. 8)
15. Describe the mediastinum and its contents. (p. 8)
16. List the cavities of the head and the contents of each cavity. (p. 10)
17. Distinguish between a parietal and a visceral membrane. (p. 10)
18. Name the major organ systems, and describe the general functions of each. (p. 12)
19. List the major organs that comprise each organ system. (p. 12)

Part B
1. Name the body cavity that houses each of the following organs:
   a. Stomach
   b. Heart
   c. Brain
   d. Liver
   e. Trachea
   f. Rectum
   g. Spinal cord
   h. Esophagus
   i. Spleen
   j. Urinary bladder

2. Write complete sentences using each of the following terms correctly:
   a. Superior
   b. Inferior
   c. Anterior
   d. Posterior
   e. Medial
   f. Lateral
   g. Proximal
   h. Distal
   i. Superficial
   j. Peripheral
   k. Deep
3. Sketch a human body, and use lines to indicate each of the following sections:
   a. Sagittal
   b. Transverse
   c. Coronal
4. Sketch the abdominal area, and indicate the location of each of the following regions:
   a. Epigastric
   b. Umbilical
   c. Hypogastric
   d. Hypochondriac
   e. Lumbar
   f. Iliac
5. Provide the common name for the region to which each of the following terms refers:
   a. Acromial
   b. Antebrachial
   c. Axillary
   d. Buccal
   e. Celiac
   f. Coxal
   g. Crural
   h. Femoral
   i. Genital
   j. Gluteal
   k. Inguinal
   l. Mental
   m. Occipital
   n. Orbital
   o. Otic
   p. Palmar
   q. Pectoral
   r. Pedal
   s. Perineal
   t. Plantar
   u. Popliteal
   v. Sacral
   w. Sternal
   x. Tarsal
   y. Umbilical
   z. Vertebral

CRITICAL THINKING

1. Which characteristics of life does an automobile have? Why is a car not alive?
2. What environmental characteristics would be necessary for a human to survive on another planet?
3. Overweight people who lose weight often find it difficult to keep the weight off because a set point for the body's fat stores changes as the body perceives itself as starving. Explain how this protective mechanism, of great frustration to dieters, might operate.
4. Put the following in order, from smallest to largest: organ, molecule, organelle, atom, organ system, tissue, organism, cell, macromolecule.
5. Why is lung cancer that has spread to the mediastinum very dangerous?
6. You are building an android. Choose a human organ system, and state which materials you would use to model it in the android.
7. In health, body parts interact to maintain homeostasis. Illness can threaten the maintenance of homeostasis, requiring treatment. What treatments might be used to help control a patient's (*a*) body temperature, (*b*) blood oxygen concentration, and (*c*) water content?
8. Suppose two individuals develop benign (noncancerous) tumors that produce symptoms because they occupy space and crowd adjacent organs. If one of these persons has the tumor in the ventral cavity and the other has the tumor in the dorsal cavity, which person would be likely to develop symptoms first? Why?
9. If a patient complained of a "stomachache" and pointed to the umbilical region as the site of discomfort, which organs located in this region might be the source of the pain?
10. How might health-care professionals provide the basic requirements of life to an unconscious patient?

WEB CONNECTIONS

Visit the website for additional study questions and more information about this chapter at:

http://www.mhhe.com/shieress8

Reference Plates

THE HUMAN ORGANISM

The series of illustrations that follows shows the major parts of the human torso. The first plate illustrates the anterior surface and reveals the superficial muscles on one side. Each subsequent plate exposes deeper organs, including those in the thoracic, abdominal, and pelvic cavities.

Chapters 6–19 of this textbook describe the organ systems of the human organism in detail. As you read them, refer to these plates to visualize the locations of various organs and the three-dimensional relationships among them.

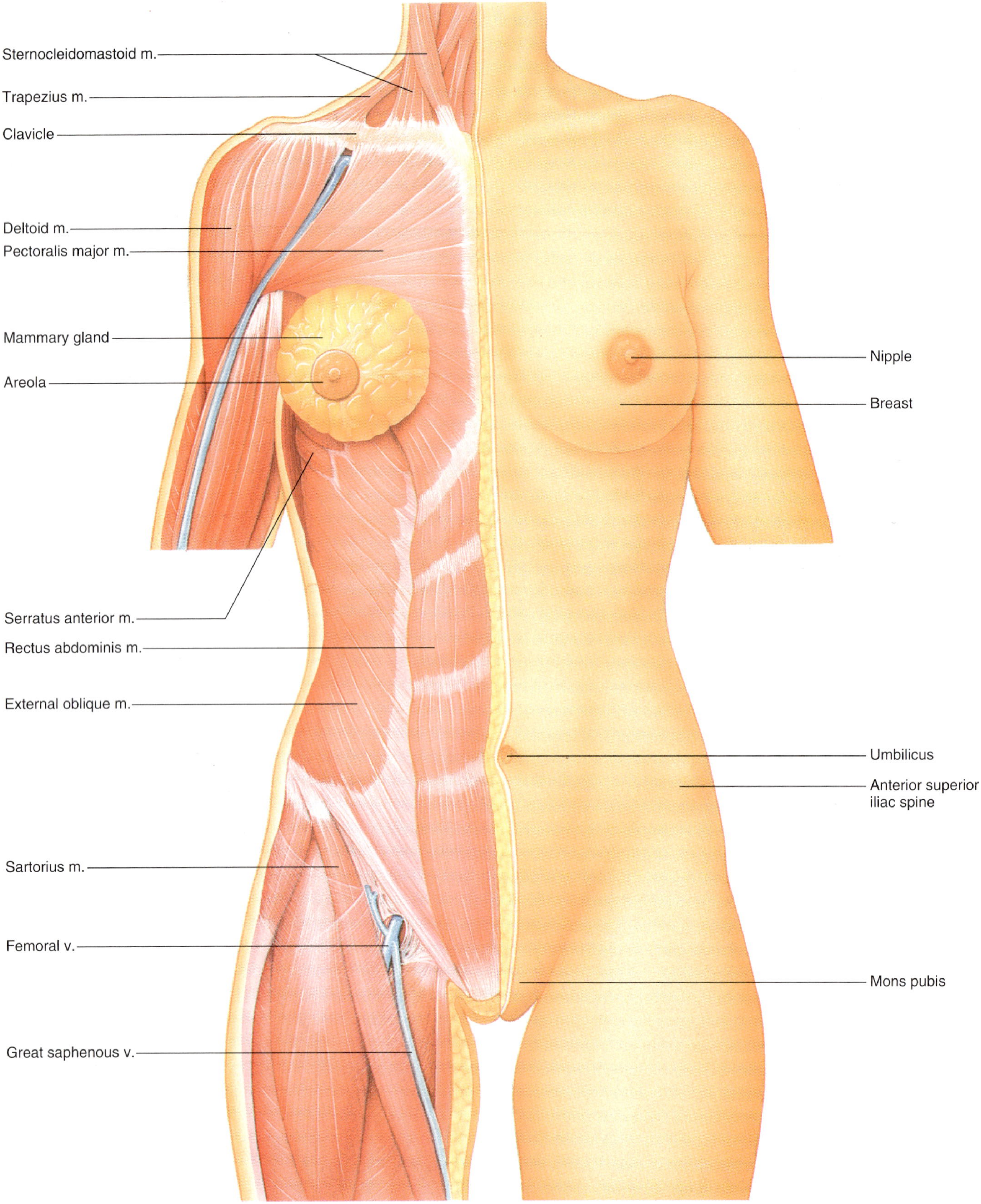

REFERENCE PLATES *The Human Organism*

- Sternocleidomastoid m.
- Trapezius m.
- Clavicle
- Deltoid m.
- Pectoralis major m.
- Mammary gland
- Areola
- Serratus anterior m.
- Rectus abdominis m.
- External oblique m.
- Sartorius m.
- Femoral v.
- Great saphenous v.

- Nipple
- Breast
- Umbilicus
- Anterior superior iliac spine
- Mons pubis

PLATE 1
Human female torso showing the anterior surface on one side and the superficial muscles exposed on the other side. (*m.* stands for *muscle*, and *v.* stands for *vein*.)

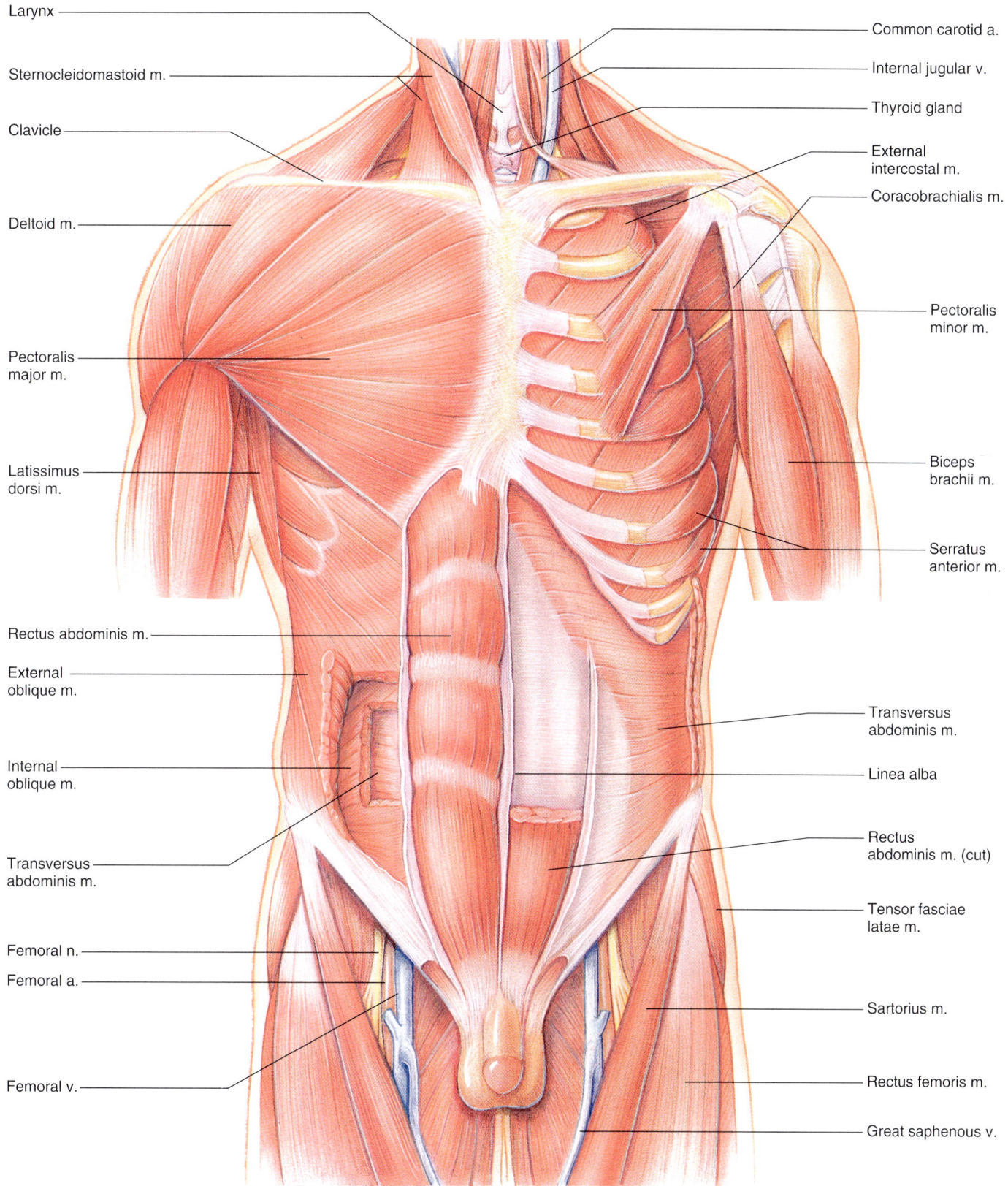

PLATE 2
Human male torso with the deeper muscle layers exposed. (*n.* stands for *nerve*, and *a.* stands for *artery*.)

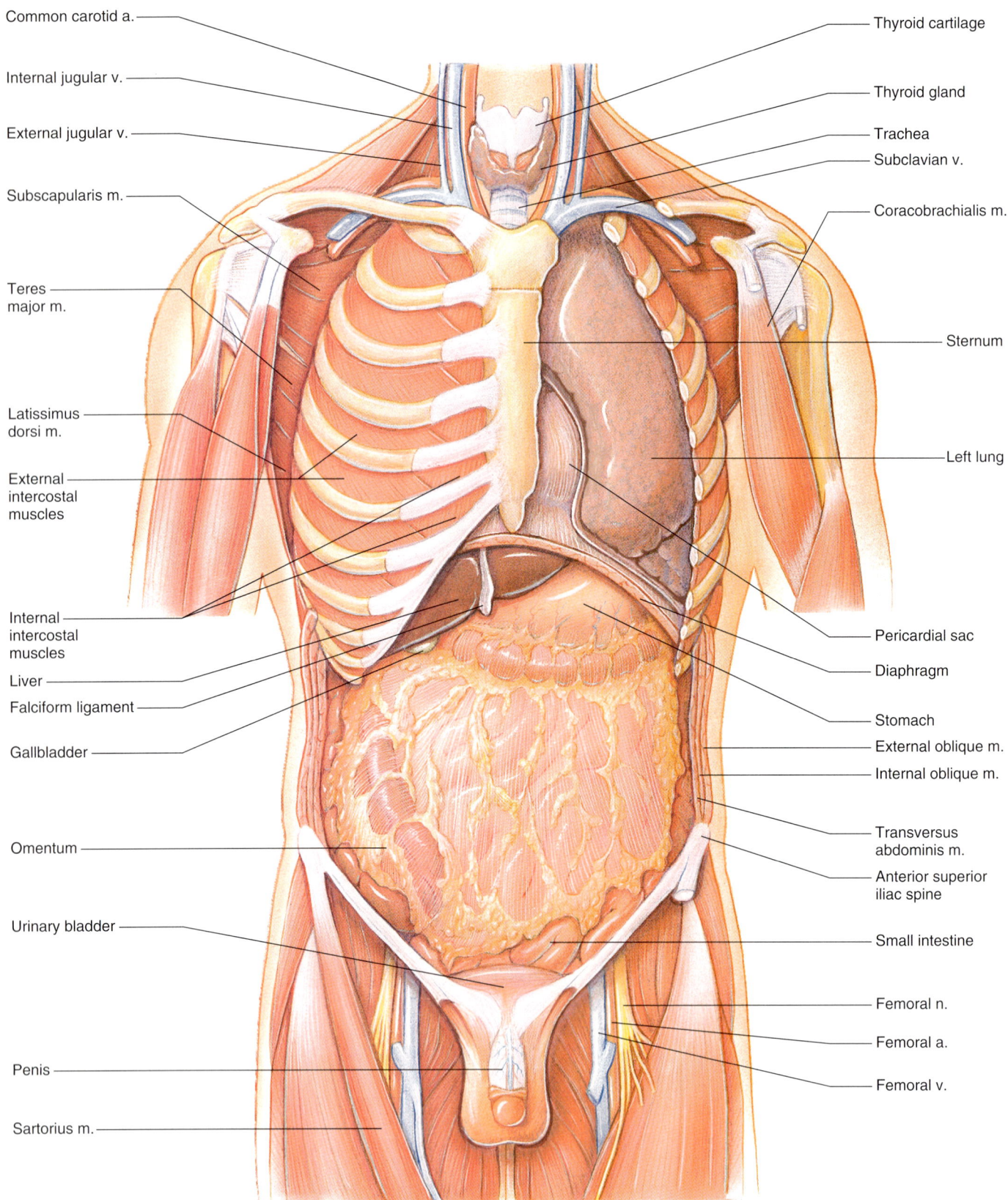

PLATE 3
Human male torso with the deep muscles removed and the abdominal viscera exposed.

REFERENCE PLATES *The Human Organism* 25

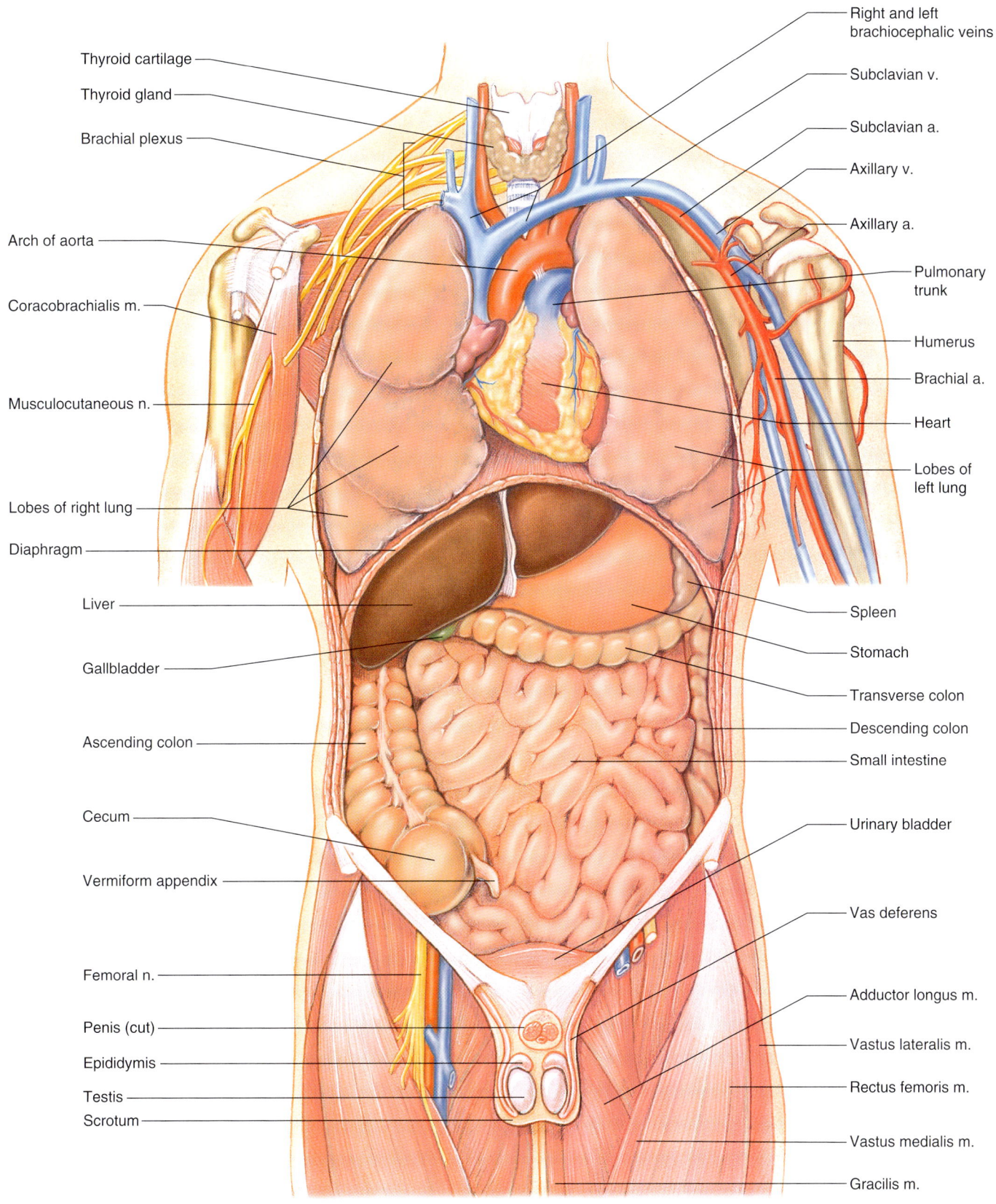

PLATE 4
Human male torso with the thoracic and abdominal viscera exposed.

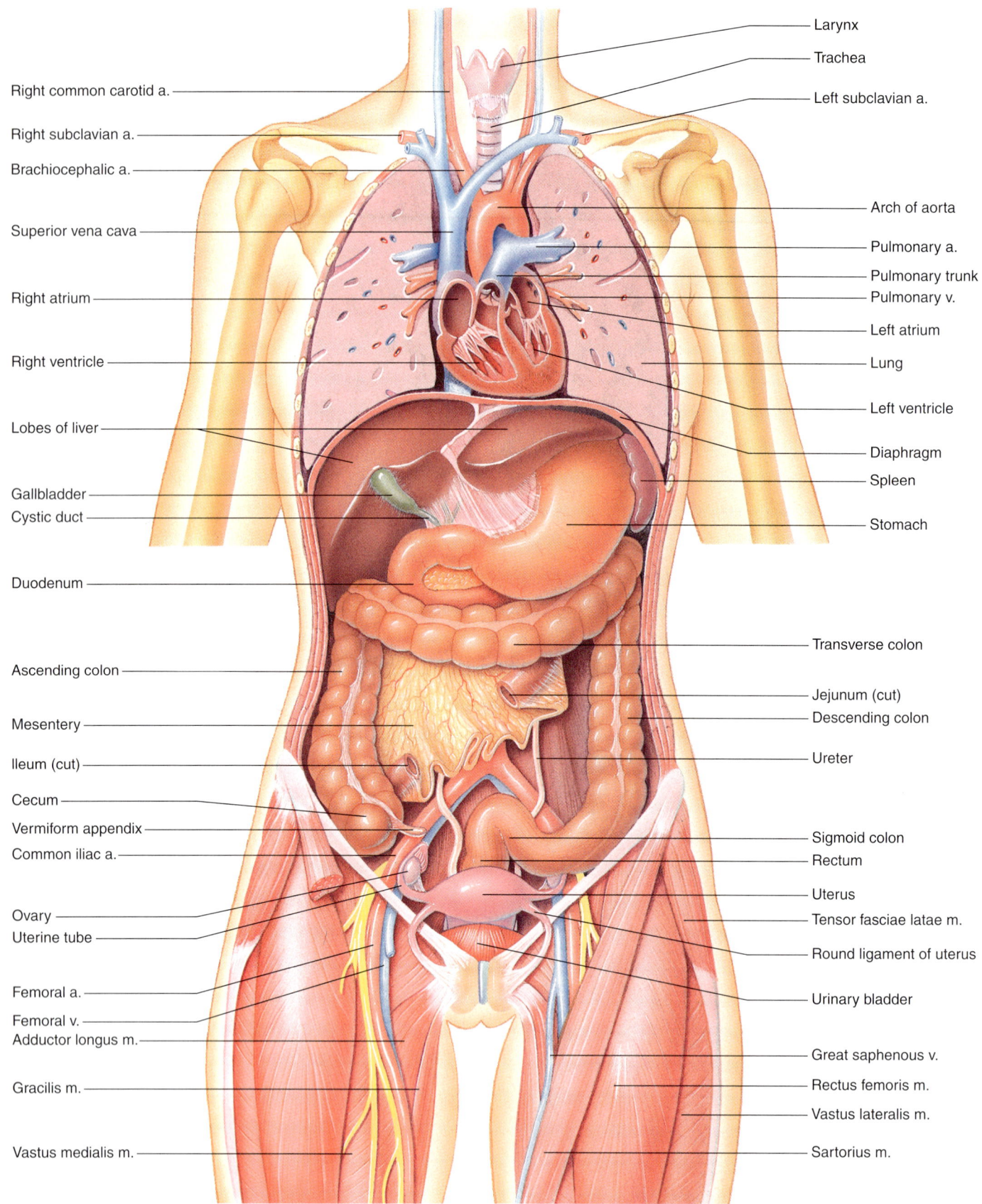

PLATE 5
Human female torso with the lungs, heart, and small intestine sectioned.

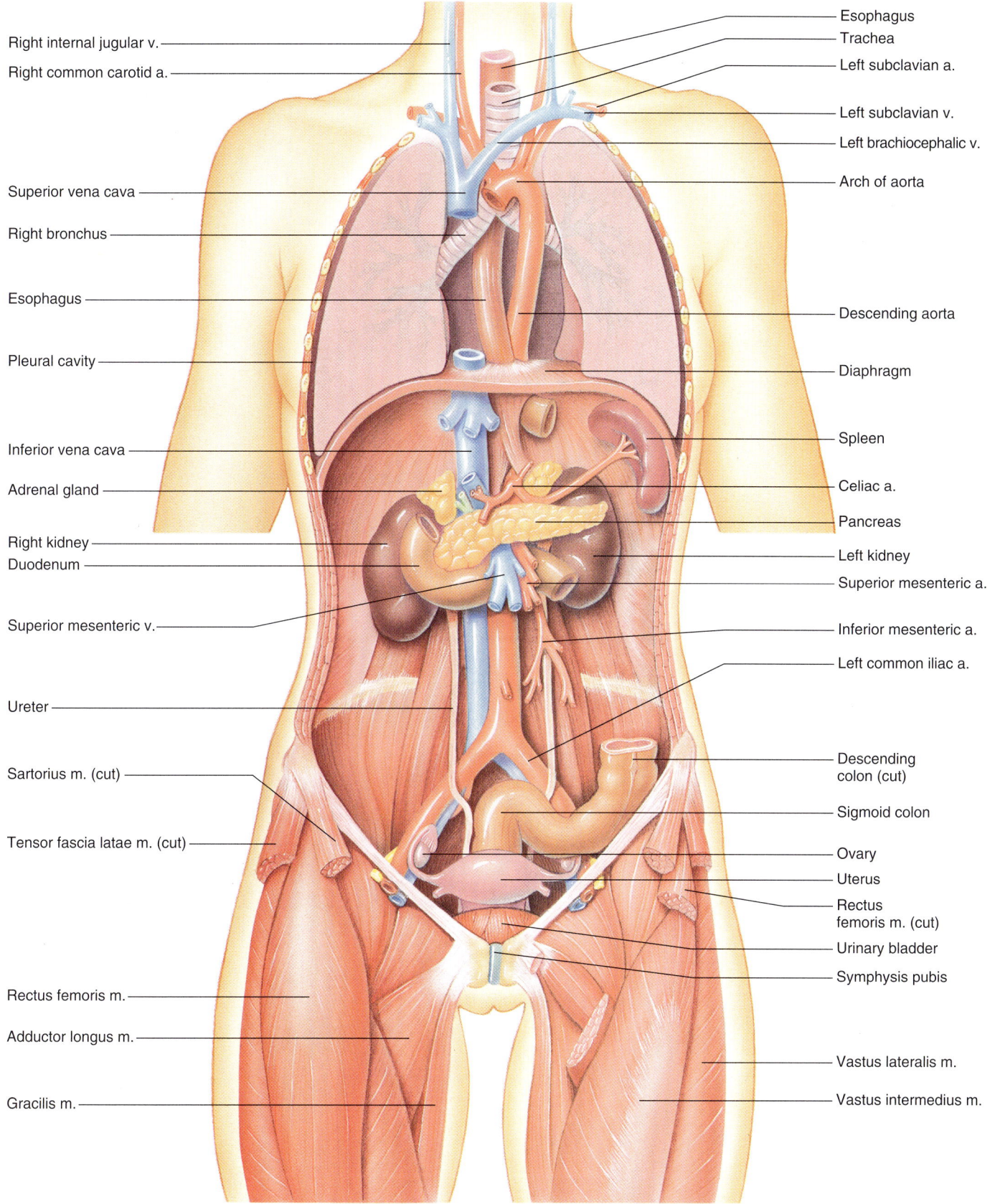

PLATE 6
Human female torso with the heart, stomach, liver, and parts of the intestine and lungs removed.

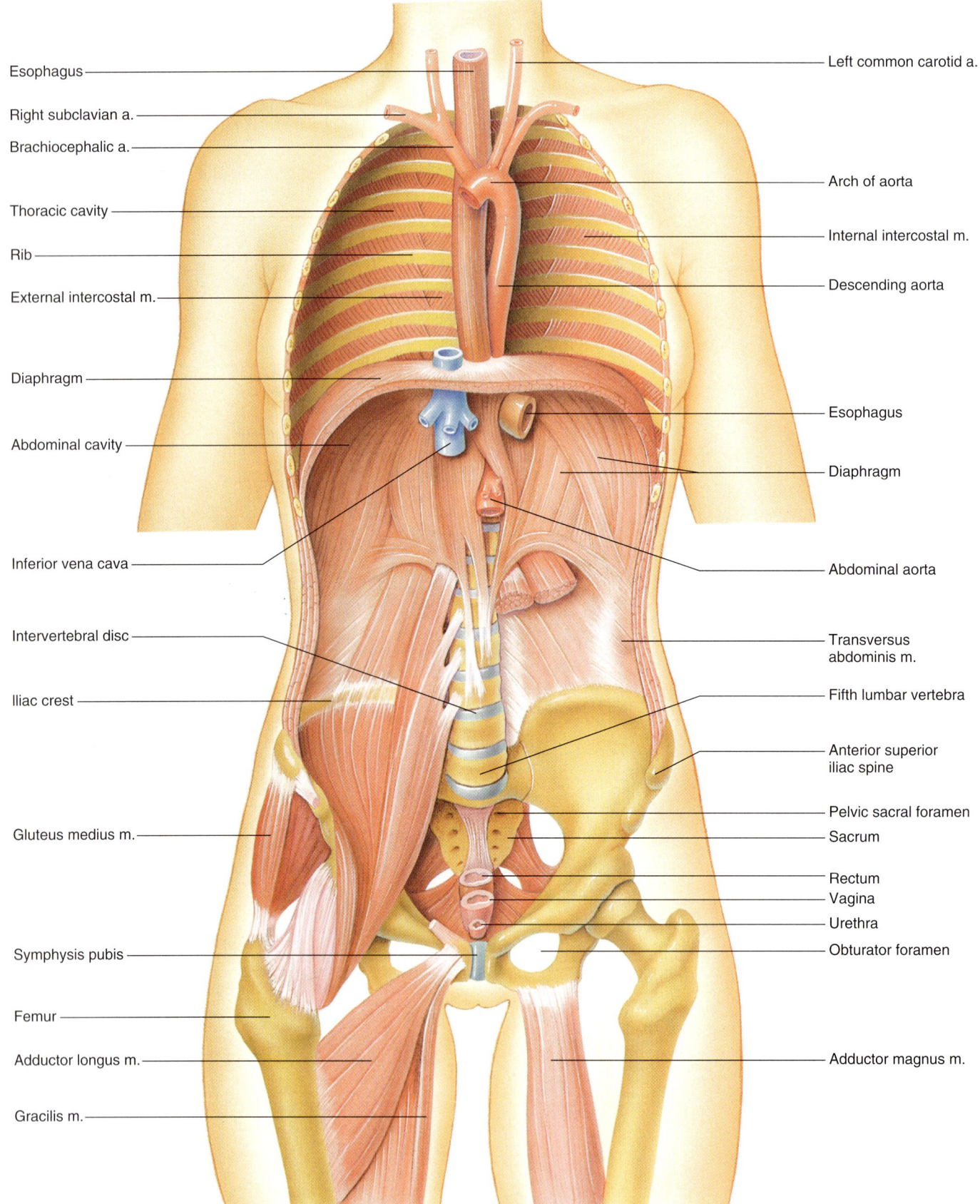

PLATE 7
Human female torso with the thoracic, abdominal, and pelvic viscera removed.

chapter 2

Chemical Basis of Life

TOO LITTLE OR TOO MUCH IRON. A human body is a highly organized, ever-changing collection of chemicals, each present within a certain concentration range. Imbalances of these chemicals of life can have drastic effects on health. Consider iron.

Too little iron in the diet can lead to iron-deficiency anemia. Because of resulting deficits in the oxygen-carrying protein hemoglobin, which has at its heart four iron atoms, the red blood cells that contain the protein are smaller and paler than they should be. On a whole-body level, the person feels profound fatigue, shortness of breath, and rapid pulse, and has pale skin. Fortunately, taking iron supplements can reverse the condition. This form of anemia is the most common nutritional disorder in the United States. Several other types of anemia are inherited.

Hemochromatosis is, in a sense, the opposite of iron-deficiency anemia. The small intestine absorbs too much iron, and over many years, the element is deposited in the heart, liver, and pancreas, slowly destroying them. Hemochromatosis affects five in every 1,000 individuals. Symptoms include fatigue, heart palpitations, joint pain, impotence, weight loss, and increased skin pigmentation. However, many people have the condition without symptoms, learning of it when excess iron shows up on a blood test done for some other reason. Symptoms are less common in women, because they lose some of the iron overload with each menstrual period.

Detecting hemochromatosis is important because the condition can affect major organs without causing symptoms. Like iron-deficiency anemia, treatment is straightforward—in this case, having blood removed every few months.

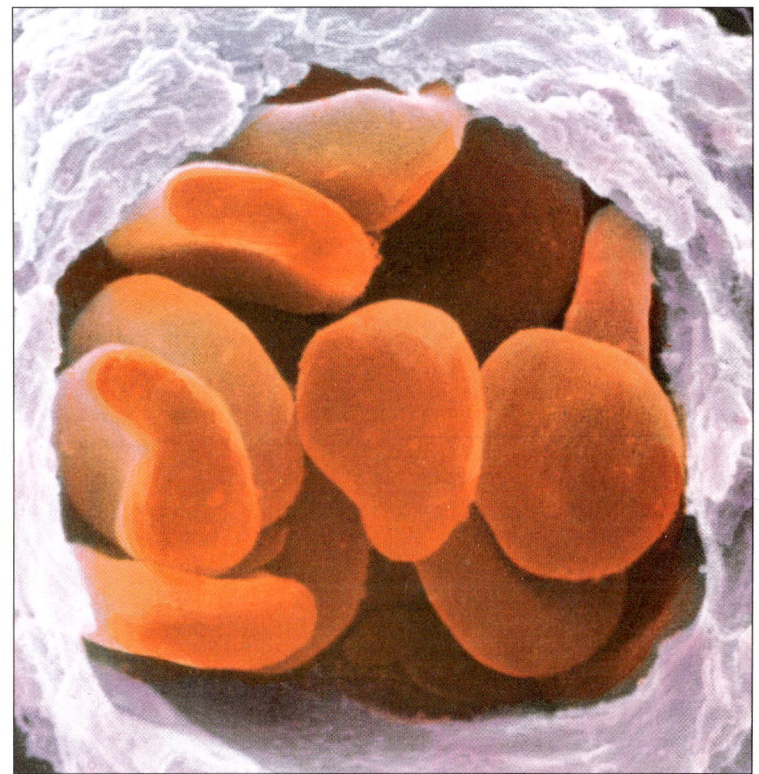

Photo:
Hemoglobin molecules in red blood cells carry oxygen, and include iron. These red blood cells are shown in a cross section of a blood vessel (5,000×).

Chapter Objectives

After studying this chapter, you should be able to do the following:

2.1 Introduction
1. Explain how the study of living material depends on the study of chemistry. (p. 31)

2.2 Structure of Matter
2. Discuss how atomic structure determines how atoms interact. (p. 32)
3. Describe the relationships between atoms and molecules. (p. 35)
4. Explain how molecular and structural formulas symbolize the composition of compounds. (p. 35)
5. Describe three types of chemical reactions. (p. 36)
6. Define pH. (p. 37)

2.3 Chemical Constituents of Cells
7. List the major groups of inorganic chemicals common in cells. (p. 38)
8. Describe the functions of various types of organic chemicals in cells. (p. 39)

Aids to Understanding Words

di- [two] *di*saccharide: Compound whose molecules are composed of two joined saccharide units.

glyc- [sweet] *glyc*ogen: Complex carbohydrate composed of many joined sugar molecules.

lip- [fat] *lip*ids: Group of organic compounds that includes fats.

-lyt [dissolvable] electro*lyt*e: Substance that dissolves in water and releases ions.

mono- [one] *mono*saccharide: Compound whose molecules consist of a single saccharide unit.

poly- [many] *poly*unsaturated: Molecule with many double bonds between its carbon atoms.

sacchar- [sugar] mono*sacchar*ide: Sugar molecule composed of a single saccharide unit.

syn- [together] *syn*thesis: Process by which chemicals join to form new types of chemicals.

Key Terms

atom (at´om)
carbohydrate (kar´´bo-hi´drāt)
decomposition (de´´kom-po-zish´un)
electrolyte (e-lek´tro-līt)
element (el´-e-ment)
inorganic (in´´or-gan´ik)
ion (i´on)
lipid (lip´id)
molecular formula (mo-lek´u-lar for´mu-lah)
molecule (mol´ĕ-kūl)
nucleic acid (nu-kle´ik as´id)
organic (or-gan´ik)
protein (pro´tēn, pro´te-in)
structural formula (struk´cher-ol for´mu-lah)
synthesis (sin´thĕ-sis)

2.1 Introduction

At the cellular level of organization, chemistry, in a sense, becomes biology. A cell's working parts—its organelles—are intricate assemblies of macromolecules. Because the macromolecules that build the cells that build tissues and organs are themselves composed of atoms, the study of anatomy and physiology begins with chemistry.

Chemistry is the branch of science that considers the composition of matter and how this composition changes. Chemistry is essential for understanding anatomy and physiology because body structures and functions result from chemical changes within cells.

2.2 Structure of Matter

Matter is anything that has weight and takes up space. This includes all the solids, liquids, and gases in our surroundings, as well as inside our bodies.

Elements and Atoms

All matter is composed of substances called **elements** (el´-e-mentz). At present, 111 elements are known, although naturally occurring matter on earth includes only 92 of them. Among these elements are such common materials as iron, copper, silver, gold, aluminum, carbon, hydrogen, and oxygen. Although some elements exist in a pure form, they occur more frequently in mixtures or chemical combinations.

Living organisms require about twenty elements. Of these, oxygen, carbon, hydrogen, and nitrogen make up more than 95% (by weight) of the human body (table 2.1). As the table shows, a one- or two-letter symbol represents each element.

Elements are composed of tiny particles called **atoms** (at´omz), which are the smallest complete units of elements. Atoms of an element are similar to each other, but they differ from the atoms that make up other elements. Atoms vary in size, weight, and the ways they interact with each other. Some atoms, for instance, can combine with atoms like themselves or with other kinds of atoms, while other atoms cannot.

Atomic Structure

An atom consists of a central portion, called the **nucleus,** and one or more **electrons** that constantly move around it. The nucleus contains one or more relatively large particles called **protons.** The nucleus also usually contains one or more **neutrons,** which are similar in size to protons.

Electrons, which are extremely small, each carry a single, negative electrical charge (e^-), while protons each carry a single, positive electrical charge ($p+$). Neutrons are uncharged and thus are electrically neutral (n^0) (fig. 2.1).

Because the nucleus contains the protons, it is always positively charged. However, the number of electrons outside the nucleus equals the number of protons. Therefore, a complete atom is electrically uncharged, or neutral.

The atoms of different elements contain different numbers of protons. The number of protons in the

TABLE 2.1 ELEMENTS IN THE HUMAN BODY

MAJOR ELEMENTS	SYMBOL	APPROXIMATE PERCENTAGE OF THE HUMAN BODY (BY WEIGHT)
Oxygen	O	65.0%
Carbon	C	18.5
Hydrogen	H	9.5
Nitrogen	N	3.2
Calcium	Ca	1.5
Phosphorus	P	1.0
Potassium	K	0.4
Sulfur	S	0.3
Chlorine	Cl	0.2
Sodium	Na	0.2
Magnesium	Mg	0.1
		Total 99.9%
TRACE ELEMENTS		
Chromium	Cr	
Cobalt	Co	
Copper	Cu	
Fluorine	F	Together less than 0.1%
Iodine	I	
Iron	Fe	
Manganese	Mn	
Zinc	Zn	

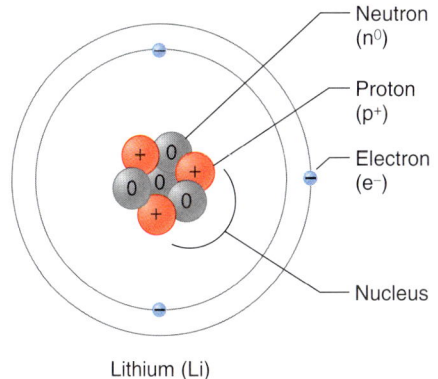

Lithium (Li)

Figure 2.1
In an atom of lithium, three electrons move around a nucleus that consists of three protons and four neutrons.

TABLE 2.2 — ATOMIC STRUCTURE OF ELEMENTS 1 THROUGH 12

ELEMENT	SYMBOL	ATOMIC NUMBER	ATOMIC WEIGHT	PROTONS	NEUTRONS	ELECTRONS IN SHELLS		
						FIRST	SECOND	THIRD
Hydrogen	H	1	1	1	0	1		
Helium	He	2	4	2	2	2 (inert)		
Lithium	Li	3	7	3	4	2	1	
Beryllium	Be	4	9	4	5	2	2	
Boron	B	5	11	5	6	2	3	
Carbon	C	6	12	6	6	2	4	
Nitrogen	N	7	14	7	7	2	5	
Oxygen	O	8	16	8	8	2	6	
Fluorine	F	9	19	9	10	2	7	
Neon	Ne	10	20	10	10	2	8 (inert)	
Sodium	Na	11	23	11	12	2	8	1
Magnesium	Mg	12	24	12	12	2	8	2

atoms of a particular element is called the element's **atomic number.** Hydrogen, for example, whose atoms each contain one proton, has the atomic number 1; carbon, whose atoms each have six protons, has the atomic number 6.

The weight of an atom of an element is due primarily to the protons and neutrons in its nucleus; electrons have very little weight. For this reason, an atom of carbon with six protons and six neutrons weighs about twelve times more than an atom of hydrogen, which has only one proton and no neutrons.

The number of protons plus the number of neutrons in each of an element's atoms approximately equals the element's **atomic weight.** Thus, the atomic weight of hydrogen is 1, and the atomic weight of carbon is 12 (table 2.2).

CHECK YOUR RECALL

1. What is the relationship between matter and the elements?
2. Which elements are most common in the human body?
3. Where are electrons, protons, and neutrons located within an atom?
4. What is the difference between atomic number and atomic weight?

Bonding of Atoms

The chemical behavior of atoms results from interactions among their electrons. When atoms combine with other atoms, they either gain electrons, lose electrons, or share electrons. The electrons of an atom occupy one or more areas of space called shells around the nucleus (table 2.2). For the elements up to atomic num-

ber 18, the maximum number of electrons that each of the first three inner shells can hold is as follows:

First shell (closest to the nucleus)	2 electrons
Second shell	8 electrons
Third shell	8 electrons

More complex atoms may have as many as eighteen electrons in the third shell. Simplified diagrams, such as those in figure 2.2, depict electron locations within the shells of atoms.

The electrons in the outermost shell of an atom determine its chemical behavior. Atoms such as helium, whose outermost electron shells are filled, have stable structures and are chemically inactive, or inert (table 2.2). Atoms such as hydrogen or lithium, whose outermost electron shells are incompletely filled, tend to gain, lose, or share electrons in ways that empty or fill their outer shells. In this way, they achieve stable structures.

Atoms that gain or lose electrons become electrically charged and are called **ions** (i´onz). An atom of

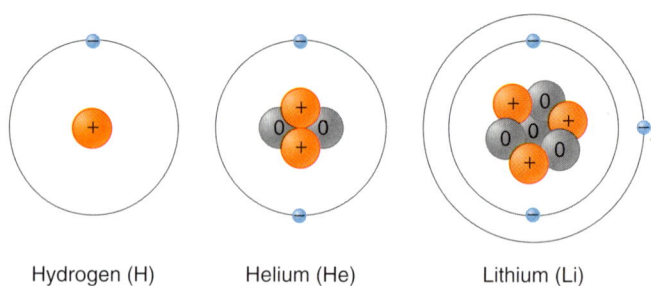

Figure 2.2
The single electron of a hydrogen atom is located in its first shell. The two electrons of a helium atom fill its first shell. Two of the three electrons of a lithium atom are in the first shell, and one is in the second shell.

Topic of Interest

RADIOACTIVE ISOTOPES

All the atoms of a particular element have the same atomic number because they have the same number of protons and electrons. However, the atoms of an element vary in the number of neutrons in their nuclei; thus, they vary in atomic weight. For example, all oxygen atoms have eight protons in their nuclei. Some, however, have eight neutrons (atomic weight 16), others have nine neutrons (atomic weight 17), and still others have ten neutrons (atomic weight 18). Atoms with the same atomic numbers but different atomic weights are called **isotopes** of an element.

How atoms interact reflects their number of electrons. Because the number of electrons in an atom is equal to its number of protons, all the isotopes of a particular element have the same number of electrons and react chemically in the same manner. Therefore, any of the isotopes of oxygen can play the same role in an organism's metabolic reactions.

Isotopes may be stable, or they may have unstable atomic nuclei that decompose, releasing energy or pieces of themselves. Unstable isotopes are called *radioactive*, and the energy or atomic fragments they give off are called *atomic radiations*.

Atomic radiations include three common forms called alpha (α), beta (β), and gamma (γ). Alpha radiation consists of particles from atomic nuclei, each of which includes two protons and two neutrons, that travel relatively slowly and can weakly penetrate matter. Beta radiation consists of much smaller particles (electrons) that travel more rapidly and penetrate matter more deeply. Gamma radiation is similar to X-ray radiation and is the most penetrating of these forms.

Each kind of radioactive isotope produces one or more forms of radiation, and each becomes less radioactive at a particular rate. The time required for an isotope to lose one-half of its radioactivity is called its *half-life*. Thus, the isotope of iodine called iodine-131, which emits one-half of its radiation in 8.1 days, has a half-life of 8.1 days. Half-lives vary greatly. The half-life of phosphorus-32 is 14.3 days; that of cobalt-60 is 5.26 years; and that of radium-226 is 1,620 years.

Because atomic radiation can be detected with special equipment, such as a scintillation counter, radioactive substances are useful in studying life processes (fig. 2A). A radioactive isotope, for example, can be introduced into an organism and then traced as it enters into metabolic activities. For example, the human thyroid gland is unique in using the element iodine in its metabolism. Therefore, radioactive iodine-131 is used to study thyroid functions and to evaluate thyroid disease (fig. 2B). Doctors use thallium-201, which has a half-life of 73.5 hours, to assess heart conditions, and gallium-67, with a half-life of 78 hours, to detect and monitor the progress of certain cancers and inflammatory diseases.

Atomic radiations also can change chemical structures, and in this way, alter vital cellular processes. For this reason, doctors sometimes use radioactive isotopes, such as cobalt-60, to treat cancers. The radiation from the cobalt preferentially kills the rapidly dividing cancer cells.

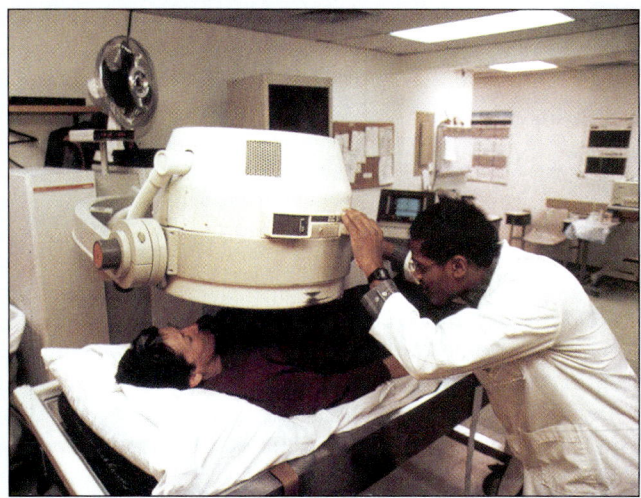

Figure 2A
Scintillation counters detect radioactive isotopes.

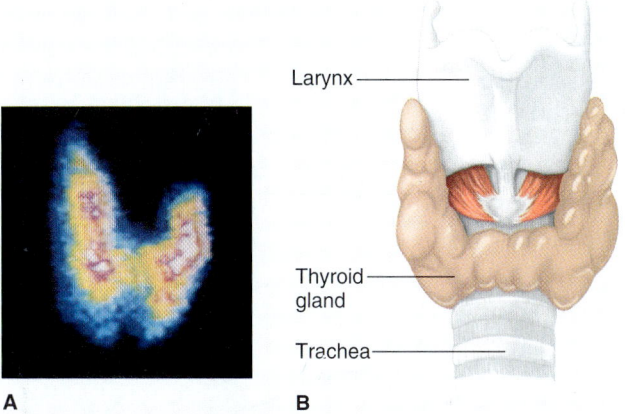

Figure 2B
(A) Scan of the thyroid gland 24 hours after the patient received radioactive iodine. Note how closely the scan in (A) resembles the shape of the thyroid gland, shown in (B).

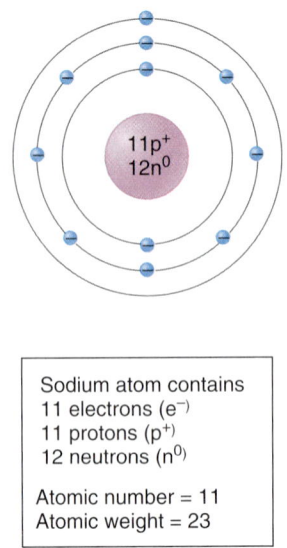

Figure 2.3
Sodium atom.

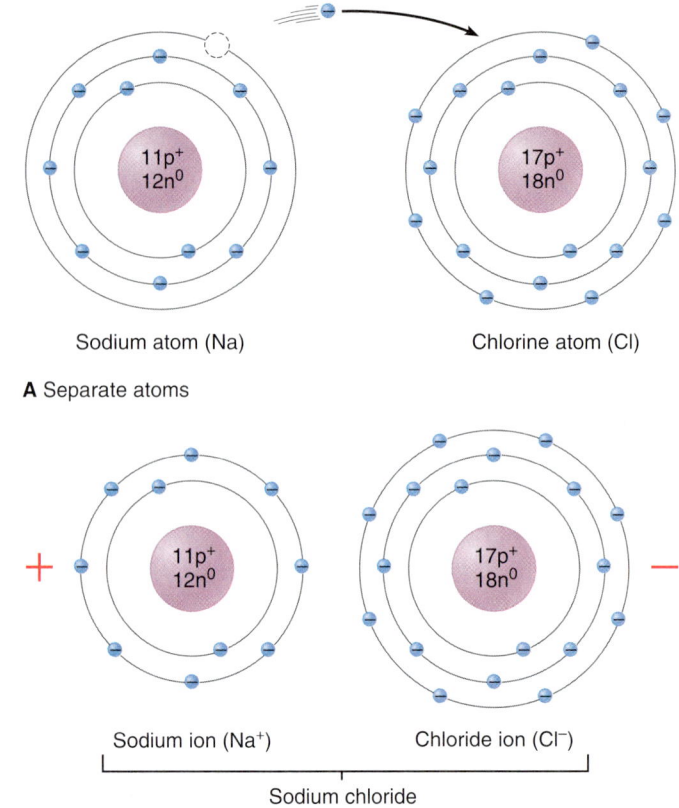

A Separate atoms

B Bonded ions

Figure 2.4
Formation of an ionic bond. (A) If a sodium atom loses an electron to a chlorine atom, the sodium atom becomes a sodium ion (Na⁺), and the chlorine atom becomes a chloride ion (Cl⁻). (B) These oppositely charged particles attract electrically and join by an ionic bond.

sodium, for example, has eleven electrons: two in the first shell, eight in the second shell, and one in the third shell (fig. 2.3). This atom tends to lose the electron from its outer shell, which leaves the second (now the outermost) shell filled and the new form stable (fig. 2.4A). In the process, sodium is left with eleven protons (11⁺) in its nucleus and only ten electrons (10⁻). As a result, the atom develops a net electrical charge of 1⁺ and is called a sodium ion, symbolized Na⁺.

A chlorine atom has seventeen electrons, with two in the first shell, eight in the second shell, and seven in the third shell. An atom of this type tends to accept a single electron, thus filling its outer shell and achieving stability (fig. 2.4A). In the process, the chlorine atom is left with seventeen protons (17⁺) in its nucleus and eighteen electrons (18⁻). As a result, the atom develops a net electrical charge of 1⁻ and is called a chloride ion, symbolized Cl⁻.

Because oppositely charged ions attract, sodium and chlorine atoms that have formed ions may react to form a type of chemical bond called an **ionic bond** (electrovalent bond). Sodium ions (Na⁺) and chloride ions (Cl⁻) uniting in this manner form the compound sodium chloride (NaCl), or table salt (fig. 2.4B). Some ions have more than one electrical charge—for example, Ca^{+2} (or Ca^{++}).

Atoms may also bond by sharing electrons, rather than by exchanging them. A hydrogen atom, for example, has one electron in its first shell but requires two electrons to achieve a stable structure (fig. 2.5). It may fill this shell by combining with another hydrogen atom in such a way that the two atoms share a pair of electrons. The two electrons then encircle the nuclei of both atoms, and each

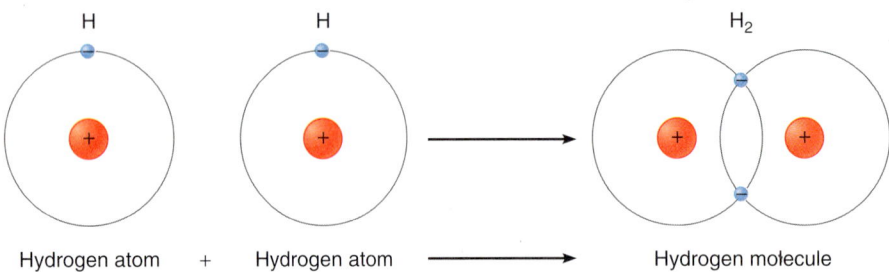

Figure 2.5
A hydrogen molecule forms when two hydrogen atoms share a pair of electrons and join by a covalent bond.

atom achieves a stable form. In this case, the chemical bond between the atoms is called a **covalent bond.**

Carbon atoms, which have two electrons in their first shells and four electrons in their second shells, form covalent bonds when they unite with other atoms. In fact, carbon atoms (and certain other atoms) may bond in such a way that two atoms share one or more pairs of electrons. If one pair of electrons is shared, the resulting bond is called a *single covalent bond;* if two pairs of electrons are shared, the bond is called a *double covalent bond.*

Another type of chemical bond, called a **hydrogen bond,** is a weak electrical attraction between a hydrogen atom (covalently bound to a nitrogen or oxygen atom) and another nitrogen or oxygen atom nearby (see figs. 2.16 and 2.19). Its contribution to protein structure is described later in this chapter (see section 2.3, p. 43).

Molecules and Compounds

When two or more atoms bond, they form a new kind of particle called a **molecule** (mol´ĕ-kūl). If atoms of the same element combine, they produce molecules of that element. Gases of hydrogen, oxygen, and nitrogen consist of such molecules (fig. 2.5).

When atoms of different elements combine, they form molecules called **compounds.** Two atoms of hydrogen, for example, can combine with one atom of oxygen to produce a molecule of the compound water (H_2O) (fig. 2.6). Table sugar (*sucrose*), baking soda, natural gas, beverage alcohol, and most drugs are compounds.

A molecule of a compound always contains definite kinds and numbers of atoms. A molecule of water, for instance, always contains two hydrogen atoms and one oxygen atom. If two hydrogen atoms combine with two oxygen atoms, the compound formed is not water, but hydrogen peroxide. Table 2.3 lists some particles of matter and their characteristics.

TABLE 2.3	SOME PARTICLES OF MATTER
PARTICLE	CHARACTERISTICS
Atom	Smallest particle of an element that has the properties of that element
Electron (e^-)	Extremely small particle; carries a negative electrical charge and is in constant motion around a nucleus of an atom
Proton (p^+)	Relatively large particle; carries a positive electrical charge and is found within a nucleus of an atom
Neutron (n^0)	Relatively large particle; uncharged and thus electrically neutral; found within a nucleus of an atom
Molecule	Particle formed by the chemical union of two or more atoms
Ion	Atom or molecule that is electrically charged because it has gained or lost one or more electrons

CHECK YOUR RECALL

1. What is an ion?
2. Describe two ways that atoms can combine with other atoms.
3. Distinguish between a molecule and a compound.

Formulas

A **molecular formula** (mo-lek´u-lar for´mu-lah) represents the numbers and types of atoms in a molecule. Such a formula consists of the symbols for the elements in the molecule, together with numbers to indicate how many atoms of each element are present. For example, the molecular formula for water is H_2O, which means

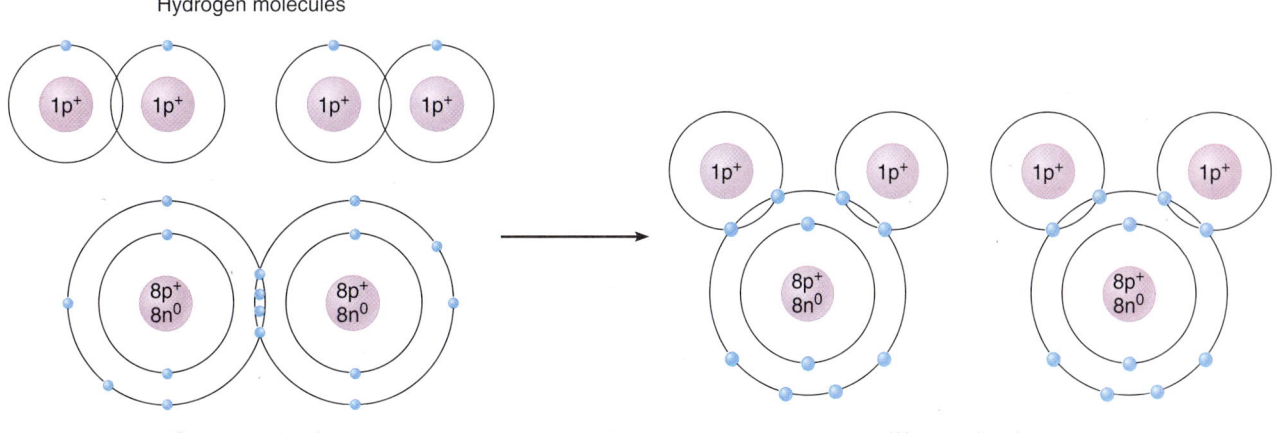

Figure 2.6
Hydrogen molecules can combine with oxygen molecules to form water molecules.

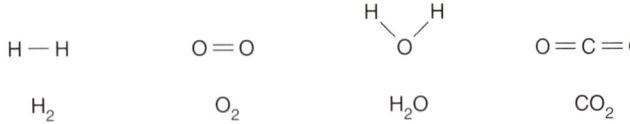

Figure 2.7
Structural and molecular formulas for molecules of hydrogen, oxygen, water, and carbon dioxide. Note the double covalent bonds.

that each water molecule consists of two atoms of hydrogen and one atom of oxygen (fig. 2.7). The molecular formula for a sugar called glucose is $C_6H_{12}O_6$, which means that each glucose molecule consists of six atoms of carbon, twelve atoms of hydrogen, and six atoms of oxygen.

Usually, the atoms of each element form a specific number of bonds. Hydrogen atoms form single bonds, oxygen atoms form two bonds, nitrogen atoms form three bonds, and carbon atoms form four bonds. Symbols and lines can be used as follows:

$$-H \quad -O- \quad \diagdown\!\!\!N\!\!\!\diagup \quad -\underset{|}{\overset{|}{C}}-$$

These representations depict how atoms are joined and arranged in various molecules. Single lines represent single bonds, and double lines represent double bonds. Illustrations of this type are called **structural formulas** (struk´cher-ol for´mu-lahz) (fig. 2.7). Three-dimensional models of structural formulas use different colors for the different kinds of atoms (fig. 2.8).

Chemical Reactions

Chemical reactions form or break bonds between atoms, ions, or molecules, generating new chemical combinations. For example, when two or more atoms (reactants) bond to form a more complex structure (product), the reaction is called **synthesis** (sin´thĕ-sis). Such a reaction is symbolized in this way:

$$A + B \rightarrow AB$$

If the bonds within a reactant molecule break so that simpler molecules, atoms, or ions form, the reaction is called **decomposition** (de´´kom-po-zish´un). Decomposition is symbolized as follows:

$$AB \rightarrow A + B$$

Synthesis, which requires energy, is particularly important in the growth of body parts and the repair of worn or damaged tissues, which require buildup of larger molecules from smaller ones. In contrast, decomposition occurs when foods are digested and energy is released from them.

A third type of chemical reaction is an **exchange reaction.** In this reaction, parts of two different types

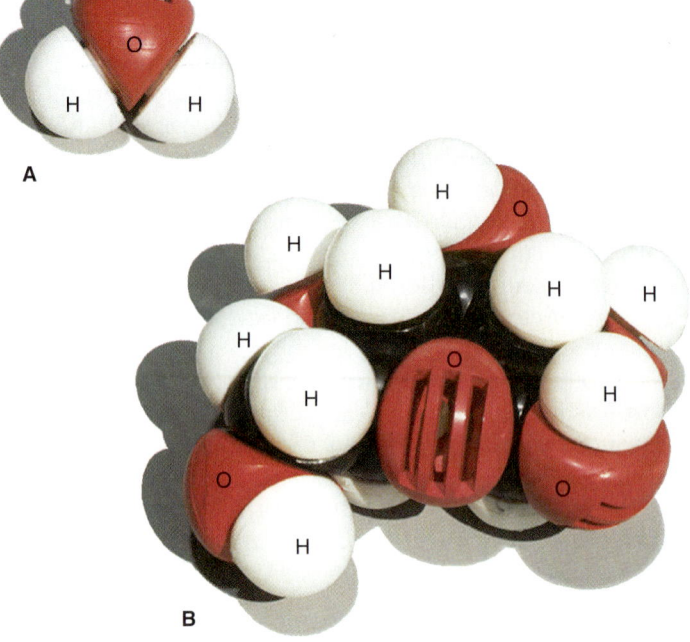

Figure 2.8
Three-dimensional molecular models. (*A*) A water molecule (H_2O), with the white parts depicting hydrogen atoms and the red parts representing oxygen. (*B*) A glucose molecule ($C_6H_{12}O_6$), in which the black parts represent carbon atoms.

of molecules trade positions. The reaction is symbolized as follows:

$$AB + CD \rightarrow AD + CB$$

An example of an exchange reaction is when an acid reacts with a base, producing water and a salt.

Many chemical reactions are reversible. This means that the product (or products) of the reaction can change back to the reactant (or reactants) that originally underwent the reaction. A **reversible reaction** is symbolized with a double arrow:

$$A + B \rightleftharpoons AB$$

Whether a reversible reaction proceeds in one direction or the other depends on such factors as the relative proportions of the reactant (or reactants) and product (or products), as well as the amount of energy available to the reaction. The speed of the reaction may be affected by the presence or absence of **catalysts**—particular atoms or molecules that can change the rate of a reaction without being consumed in the process.

Acids and Bases

Some compounds release ions when they dissolve in water or react with water molecules. Sodium chloride (NaCl), for example, releases sodium ions (Na^+) and chloride ions (Cl^-) when it dissolves:

$$NaCl \rightarrow Na^+ + Cl^-$$

Since the resulting solution contains electrically charged particles (ions), it will conduct an electric current. Substances that release ions in water are, therefore, called **electrolytes** (e-lek′tro-lītz). Electrolytes that release hydrogen ions (H⁺) in water are called **acids.** For example, in water, the compound hydrochloric acid (HCl) releases hydrogen ions (H⁺) and chloride ions (Cl⁻):

$$HCl \rightarrow H^+ + Cl^-$$

Electrolytes that release ions that combine with hydrogen ions are called **bases.** For example, the compound sodium hydroxide (NaOH) releases hydroxide ions (OH⁻) when placed in water:

$$NaOH \rightarrow Na^+ + OH^-$$

The hydroxide ions, in turn, can combine with hydrogen ions to form water; thus, sodium hydroxide is a base. (Note: Some ions, such as OH⁻, contain two or more atoms. However, such a group behaves like a single atom and usually remains unchanged during a chemical reaction.)

The concentrations of hydrogen ions (H⁺) and hydroxide ions (OH⁻) in body fluids greatly affect the chemical reactions that control certain physiological functions, such as blood pressure and breathing rate. Since their concentrations are inversely related (if one goes up, the other goes down), we need to keep track of only one of them. A value called **pH** measures hydrogen ion concentration.

The pH scale ranges from 0 to 14. A solution with a pH of 7.0, the midpoint of the scale, contains equal numbers of hydrogen and hydroxide ions and is said to be *neutral*. A solution that contains more hydrogen than hydroxide ions has a pH less than 7.0 and is *acidic*. A solution with fewer hydrogen than hydroxide ions has a pH greater than 7.0 and is *basic* (alkaline).

Figure 2.9 indicates the pH values of some common substances. Each whole number on the pH scale represents a tenfold difference in hydrogen ion concentration, and as the hydrogen ion concentration increases, the pH number gets smaller. Thus, a solution with a pH of 6 has ten times the hydrogen ion concentration of a solution with a pH of 7. This means that relatively small changes in pH can reflect large changes in hydrogen ion concentration. Chapter 18 (p. 489) discusses the regulation of the hydrogen ion concentration in body fluids.

Note in figure 2.9 that the pH of human blood is about 7.4 (the normal pH range is from 7.35 to 7.45). If this pH value drops below 7.35, the person is said to have *acidosis;* if it rises above 7.45, the condition is called *alkalosis*. Without medical intervention, a person usually cannot survive if blood pH drops to 6.9 or rises to 7.8 for more than a few hours.

CHECK YOUR RECALL

1. What is a molecular formula? A structural formula?
2. Describe three kinds of chemical reactions.
3. Compare the characteristics of an acid with those of a base.
4. What does pH measure?

2.3 Chemical Constituents of Cells

Chemicals that enter into metabolic reactions or are produced by them can be divided into two large groups. Those that contain both carbon and hydrogen atoms are called **organic** (or-gan′ik). The rest are **inorganic** (in′′or-gan′ik).

Generally, inorganic substances dissolve in water or react with water to release ions; thus, they are *electrolytes*. Some organic compounds also dissolve in

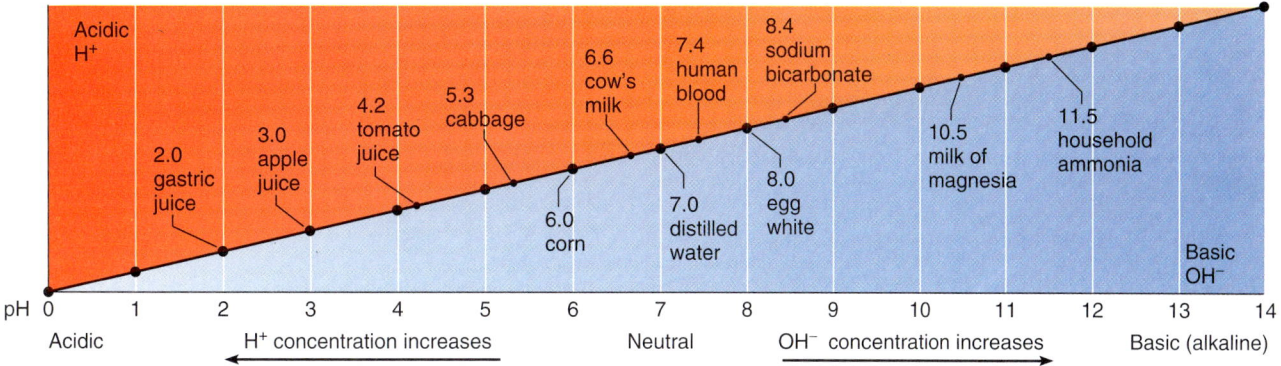

Figure 2.9
As the concentration of hydrogen ions (H⁺) increases, a solution becomes more acidic, and the pH value decreases. As the concentration of ions that combine with hydrogen ions (such as hydroxide ions) increases, a solution becomes more basic (alkaline), and the pH value increases. The pH values of some common substances are shown.

water, but they are more likely to dissolve in organic liquids, such as ether or alcohol. Organic substances that dissolve in water usually do not release ions and are therefore called *nonelectrolytes*.

Inorganic Substances

Among the inorganic substances common in cells are water, oxygen, carbon dioxide, and a group of compounds called salts.

Water

Water is the most abundant compound in living material and accounts for about two-thirds of the weight of an adult human. It is the major component of blood and other body fluids, including those within cells.

Water is an important *solvent* because many substances readily dissolve in it. When a substance (*solute*) dissolves in water, relatively large pieces of the solute break into smaller ones, and eventually, molecular-sized particles result. These particles, which may be ions, are much more likely to react with one another than were the original large pieces. Consequently, most metabolic reactions occur in water.

Water also plays an important role in moving chemicals within the body. The aqueous (watery) portion of blood, for example, carries many vital substances, such as oxygen, sugars, salts, and vitamins, from the organs of digestion and respiration to the body cells.

Water can absorb and transport heat. Blood carries heat released from muscle cells during exercise from deeper parts of the body to the surface, where it may be lost to the outside.

Oxygen

Molecules of oxygen (O_2) enter the body through the respiratory organs and are transported throughout the body by blood and red blood cells. Cellular organelles use oxygen to release energy from the sugar *glucose* and other nutrients. The released energy drives the cell's metabolic activities.

Carbon Dioxide

Carbon dioxide (CO_2) is a simple, carbon-containing compound of the inorganic group. It is produced as a waste product when certain metabolic processes release energy, and it is exhaled from the lungs.

Salts

Salts are abundant in tissues and fluids. They provide many necessary ions, including sodium (Na^+), chloride (Cl^-), potassium (K^+), calcium (Ca^{+2}), magnesium (Mg^{+2}), phosphate (PO_4^{-3}), carbonate (CO_3^{-2}), bicarbonate (HCO_3^-), and sulfate (SO_4^{-2}). These ions are important in metabolic processes, including transport of substances into and out of cells, muscle contraction, and nerve impulse conduction. Table 2.4 summarizes the functions of some of the inorganic substances in cells.

CHECK YOUR RECALL

1. How do inorganic and organic molecules differ?
2. How do electrolytes and nonelectrolytes differ?
3. Name the inorganic substances common in body fluids.

TABLE 2.4 — INORGANIC SUBSTANCES COMMON IN CELLS

SUBSTANCE	SYMBOL OR FORMULA	FUNCTIONS
I. Inorganic molecules		
Water	H_2O	Major component of body fluids (chapter 12, p. 314); medium in which most biochemical reactions occur; transports chemicals (chapter 12, p. 314); helps regulate body temperature (chapter 6, p. 120)
Oxygen	O_2	Used in energy release from glucose molecules (chapter 4, p. 78)
Carbon dioxide	CO_2	Waste product that results from metabolism (chapter 4, p. 78); reacts with water to form carbonic acid (chapter 16, p. 453)
II. Inorganic ions		
Bicarbonate ions	HCO_3^-	Helps maintain acid-base balance (chapter 18, p. 489)
Calcium ions	Ca^{+2}	Necessary for bone development (chapter 7, p. 130), muscle contraction (chapter 8, p. 176), and blood clotting (chapter 12, p. 318)
Carbonate ions	CO_3^{-2}	Component of bone tissue (chapter 7, p. 135)
Chloride ions	Cl^-	Helps maintain water balance (chapter 18, p. 484)
Magnesium ions	Mg^{+2}	Component of bone tissue (chapter 7, p. 135); required for certain metabolic processes (chapter 15, p. 426)
Phosphate ions	PO_4^{-3}	Required for synthesis of ATP, nucleic acids, and other vital substances (chapter 4, p. 79); component of bone tissue (chapter 7, p. 135); helps maintain polarization of cell membranes (chapter 9, p. 213)
Potassium ions	K^+	Required for polarization of cell membranes (chapter 9, p. 213)
Sodium ions	Na^+	Required for polarization of cell membranes (chapter 9, p. 213); helps maintain water balance (chapter 18, p. 484)
Sulfate ions	SO_4^{-2}	Helps maintain polarization of cell membranes (chapter 9, p. 213) and acid-base balance (chapter 18, p. 489)

Organic Substances

Important groups of organic substances in cells include carbohydrates, lipids, proteins, and nucleic acids.

Carbohydrates

Carbohydrates (kar″bo-hi´drātz) provide much of the energy that cells require. They supply materials to build certain cell structures and often are stored as reserve energy supplies.

Carbohydrate molecules contain atoms of carbon, hydrogen, and oxygen. These molecules usually have twice as many hydrogen as oxygen atoms—the same ratio of hydrogen to oxygen as in water molecules (H_2O). This ratio is easy to see in the molecular formulas of the carbohydrates glucose ($C_6H_{12}O_6$) and sucrose ($C_{12}H_{22}O_{11}$).

The carbon atoms of carbohydrate molecules join in chains whose lengths vary with the type of carbohydrate. For example, carbohydrates with shorter chains are called **sugars**.

Sugars with 6-carbon atoms (hexoses) are known as *simple sugars,* or **monosaccharides,** and they are the building blocks of more complex carbohydrate molecules. The simple sugars include glucose, fructose, and galactose. Figure 2.10 illustrates the structural formula of glucose.

In complex carbohydrates, a number of simple sugar molecules link to form molecules of varying sizes (fig. 2.11). Some complex carbohydrates, such as sucrose (table sugar) and lactose (milk sugar), are *double sugars,* or **disaccharides,** whose molecules each contain two simple-sugar building blocks. Others are made up of many simple-sugar units joined to form **polysaccharides,** such as plant starch. Animals, including humans, synthesize a polysaccharide similar to starch, called *glycogen.*

Lipids

Lipids (lip´idz) are organic substances that are insoluble in water but soluble in certain organic solvents, such as ether and chloroform. Lipids include a variety

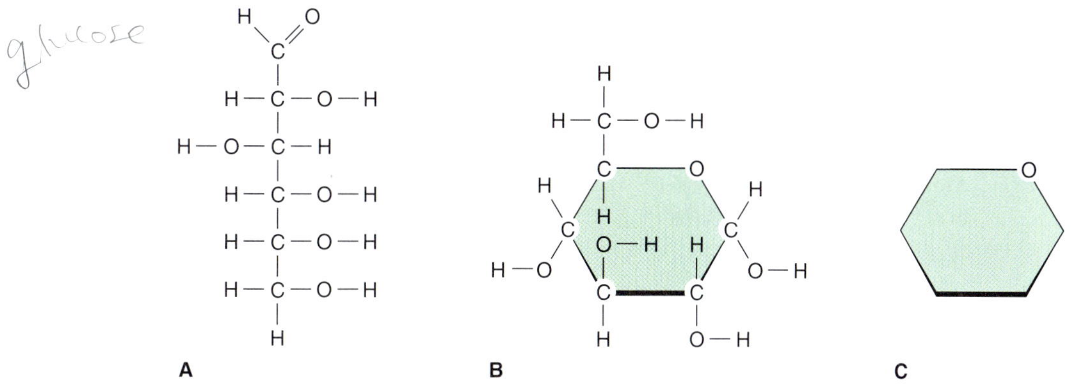

Figure 2.10
Structural formulas for glucose. (A) Some glucose molecules ($C_6H_{12}O_6$) have a straight chain of carbon atoms. (B) More commonly, glucose molecules form a ring structure. (C) This shape symbolizes the ring structure of a glucose molecule.

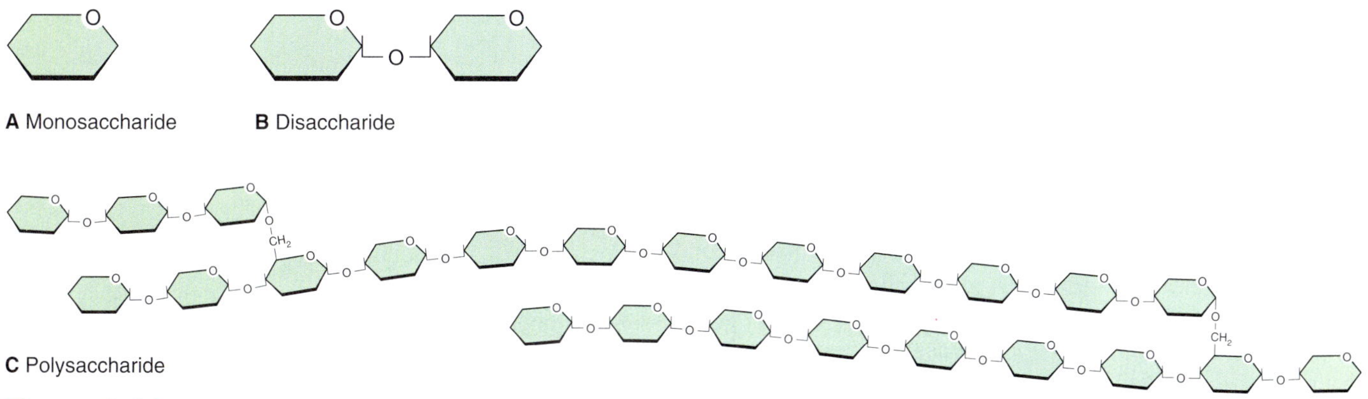

A Monosaccharide **B** Disaccharide

C Polysaccharide

Figure 2.11
Carbohydrates. (A) A monosaccharide molecule consists of one 6-carbon atom building block. (B) A disaccharide molecule consists of two of these building blocks. (C) A polysaccharide molecule consists of many building blocks.

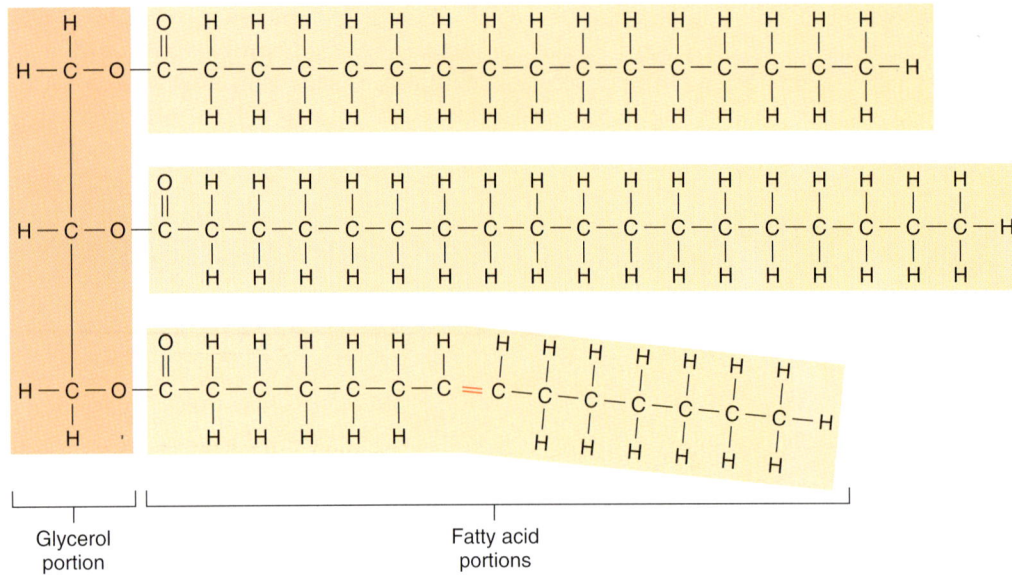

Figure 2.12
A triglyceride molecule (fat) consists of a glycerol portion and three fatty acid portions. This is an example of an unsaturated fat. The double bond between carbon atoms is shown in red.

of compounds—fats, phospholipids, and steroids—that have vital functions in cells. The most common lipids are fats.

Fats are used primarily to store energy for cellular activities. Fat molecules can supply more energy, gram for gram, than carbohydrate molecules.

Like carbohydrates, fat molecules are composed of carbon, hydrogen, and oxygen atoms. They contain, however, a much smaller proportion of oxygen than do carbohydrates. The formula for the fat tristearin, $C_{57}H_{110}O_6$, illustrates these characteristic proportions.

The building blocks of fat molecules are **fatty acids** and **glycerol.** Each glycerol molecule combines with three fatty acid molecules to produce a single fat, or *triglyceride,* molecule.

The glycerol portion of every fat molecule is the same, yet there are many kinds of fatty acids and, therefore, many kinds of fats. Fatty acid molecules differ in the lengths of their carbon atom chains, although such chains usually contain an even number of carbon atoms. The chains also may vary in the way the carbon atoms combine. In some cases, the carbon atoms all join by single carbon–carbon bonds. This type of fatty acid is said to be *saturated;* that is, each carbon atom is bound to as many hydrogen atoms as possible and is thus saturated with hydrogen atoms. Other fatty acid chains have not bound their maximum number of hydrogen atoms. Therefore, they have one or more double bonds between carbon atoms. Those with such double bonds are said to be *unsaturated,* and fatty acid molecules with many double-bonded carbon atoms are called *polyunsaturated.* Similarly, fat molecules that contain only saturated fatty acids are called *saturated fats,* and those that include unsaturated fatty acids are called *unsaturated fats* (fig. 2.12).

A **phospholipid** molecule is similar to a fat molecule in that it contains a glycerol portion and fatty acid chains. The phospholipid, however, has only two fatty acid chains; in place of the third is a portion containing a phosphate group. This phosphate portion is soluble in water (hydrophilic) and forms the "head" of the molecule, while the fatty acid portion is insoluble in water (hydrophobic) and forms a "tail" (fig. 2.13). Phospholipids are important in cellular structures.

Steroid molecules are complex structures that include four connected rings of carbon atoms (fig. 2.14). Among the more important steroids are cholesterol, which occurs in all body cells and is used to synthesize other steroids; sex hormones, such as estrogen, progesterone, and testosterone; and several hormones from the adrenal glands. Chapters 11 and 19 discuss

A diet high in saturated fats increases the chance of developing *atherosclerosis,* which obstructs arteries. For this reason, many nutritionists recommend that polyunsaturated fats replace some dietary saturated fats.

As a rule, saturated fats are more abundant in fatty foods that are solids at room temperature, such as butter, lard, and most other animal fats. Unsaturated fats, on the other hand, are likely to be plentiful in fatty foods that are liquids at room temperature, such as soft margarine and various seed oils, including corn, cottonseed, safflower, sesame, soybean, and sunflower oils. Exceptions include coconut and palm kernel oils, which are high in saturated fats.

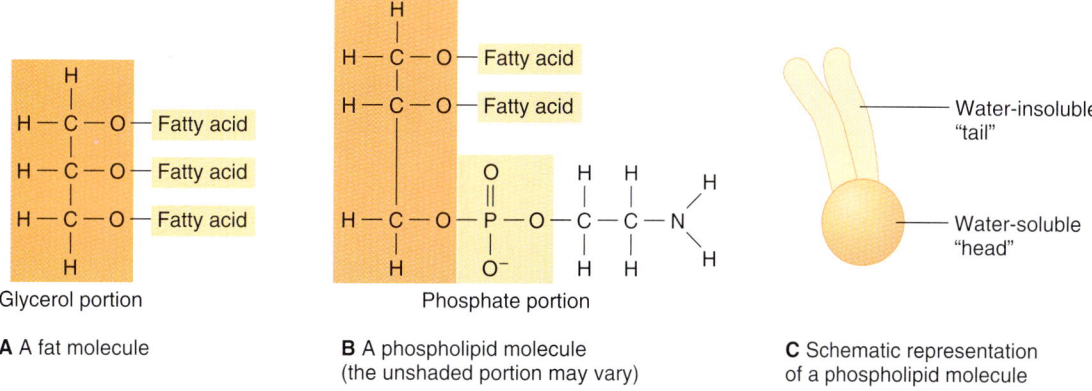

A A fat molecule

B A phospholipid molecule (the unshaded portion may vary)

C Schematic representation of a phospholipid molecule

Figure 2.13
Fats and phospholipids. (A) A fat molecule (triglyceride) contains a glycerol and three fatty acids. (B) In a phospholipid molecule, a phosphate-containing group replaces one fatty acid. (C) Schematic representation of a phospholipid.

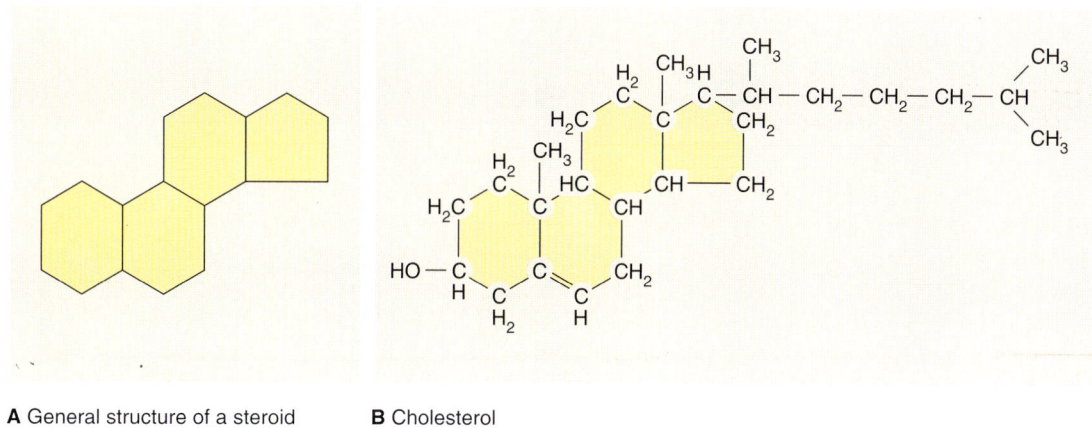

A General structure of a steroid

B Cholesterol

Figure 2.14
Steroid structure. (A) The general structure of a steroid. (B) The structural formula for cholesterol, a steroid widely distributed in the body.

these steroids. Table 2.5 lists three important groups of lipids and their characteristics.

Proteins

Some **proteins** (pro′tēnz, pro′te-inz) serve as structural materials, energy sources, and hormones. Others combine with carbohydrates (glycoproteins) and function on cell surfaces as **receptors** that are specialized to bond to particular kinds of molecules. Proteins called **antibodies** act against foreign substances that enter the body. Still others, known as **enzymes,** catalyze vital metabolic processes. That is, they speed specific chemical reactions without being consumed by these reactions. (Enzymes are discussed in more detail in chapter 4, pp. 75–76.)

TABLE 2.5 — IMPORTANT GROUPS OF LIPIDS

GROUP	BASIC MOLECULAR STRUCTURE	CHARACTERISTICS
Triglycerides	Three fatty acid molecules bound to a glycerol molecule	Most common lipids in body; stored in fat tissue as an energy supply; fat tissue also provides insulation beneath the skin
Phospholipids	Two fatty acid molecules and a phosphate group bound to a glycerol molecule	Used as structural components in cell membranes; abundant in liver and parts of nervous system
Steroids	Four connected rings of carbon atoms	Widely distributed in body and have a variety of functions; include cholesterol, hormones of adrenal cortex, sex hormones, bile acids, and vitamin D

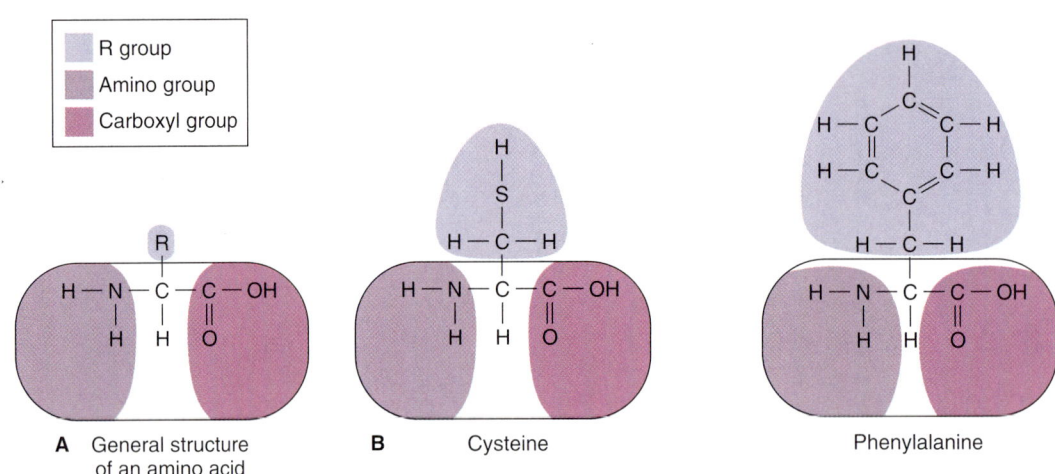

Figure 2.15
Amino acid structure. (*A*) The general structure of an amino acid. Note the amino group and carboxyl group that are common to all amino acid molecules. (*B*) Some representative amino acids and their structural formulas. Each amino acid molecule has a particular shape due to the different R groups.

Figure 2.16
The levels of protein structure.

A Each oblong shape in this chain represents an amino acid molecule. The whole chain represents a portion of a protein molecule.

B The amino acid chain of a protein molecule is often either pleated or twisted to form a coil. Dotted lines represent hydrogen bonds. R groups (see fig. 2.15) are indicated in bold.

Pleated structure

Coiled structure

C The pleated and coiled amino acid chain of a protein molecule folds into a unique three-dimensional structure.

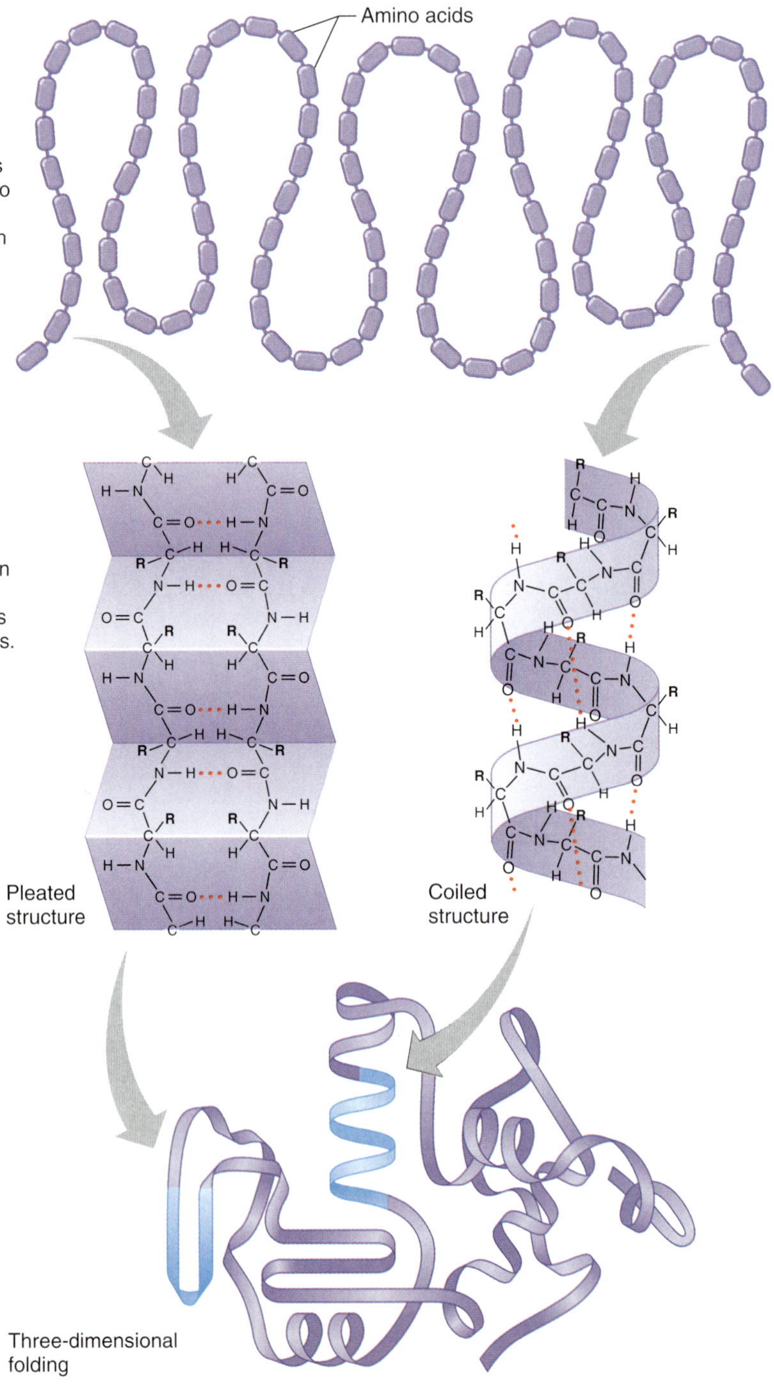

Three-dimensional folding

A human body has more than 100,000 different types of proteins.

Like carbohydrates and lipids, proteins are composed of atoms of carbon, hydrogen, and oxygen. In addition, proteins always contain nitrogen atoms and, sometimes, sulfur atoms. The building blocks of proteins are smaller molecules called **amino acids,** each of which has an *amino group* (—NH$_2$) at one end and a *carboxyl group* (—COOH) at the other (fig. 2.15). Amino acids also contain a *side chain,* or *R group* ("R" may be thought of as the "rest of the molecule"). The composition of the R group distinguishes one type of amino acid from another (fig. 2.15).

About twenty different kinds of amino acids occur commonly in the proteins of life. Within a protein molecule, the amino acids join in a chain (polypeptide chain) that varies in length from less than 100 to more than 5,000 amino acids.

Proteins have several levels of structure; primary, secondary, and tertiary levels are shown in figure 2.16. Hydrogen bonding and even covalent bonding between atoms in different parts of the polypeptide give the final protein a complicated three-dimensional shape, or **conformation** (fig. 2.17). The conformation of a protein determines its function. Some proteins are long and fibrous, such as the keratin protein that forms hair or the threads of fibrin protein that knit a blood clot. Many proteins are globular and function as enzymes, ion channels, carrier proteins, or receptors. Myoglobin and hemoglobin, which transport oxygen in muscle and blood, respectively, are globular.

When hydrogen bonds break as a result of exposure to excessive heat, radiation, electricity, pH changes, or various chemicals, this can change a protein's unique shape. Such proteins are said to be *denatured*. At the same time, the proteins lose their special properties. For example, heat denatures the protein in egg white (albumin), changing it from a liquid to a solid. This is an irreversible change—a hard-boiled egg cannot return to its uncooked, runny state. Similarly, cellular proteins that are denatured may be permanently changed and lose their functions.

Not all proteins are single polypeptide chains. Sometimes several polypeptide chains are connected in a fourth level, or *quaternary structure,* to form a very large protein. Hemoglobin is a quaternary protein made up of four separate polypeptide chains.

The conformation of a protein determines its function, and for most proteins, the conformation is always the same for a given primary structure (see Clinical Connection, p. 44). Thus, it is the amino acid sequence that ultimately determines the role of a protein in the body. Genes, made of nucleic acid, contain the information for the amino acid sequences of all the body's proteins in a form that the cell can decode.

Nucleic Acids

Nucleic acids (nu-kle´ik as´idz) form genes and take part in protein synthesis. These molecules are generally very large and complex. They contain atoms of carbon, hydrogen, oxygen, nitrogen, and phosphorus, which form building blocks called **nucleotides.** Each nucleotide consists of a 5-carbon *sugar* (ribose or deoxyribose), a *phosphate group,* and one of several *organic bases* (fig. 2.18). A nucleic acid molecule consists of a chain (polynucleotide chain) of many nucleotides.

Nucleic acids are of two major types. One type—**RNA** (ribonucleic acid)—is composed of molecules whose nucleotides contain ribose. RNA exists as a single

Figure 2.17
A model of a portion of the protein collagen.

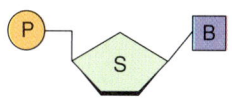

Figure 2.18
A nucleotide consists of a 5-carbon sugar (S = sugar), a phosphate group (P = phosphate), and an organic base (B = base).

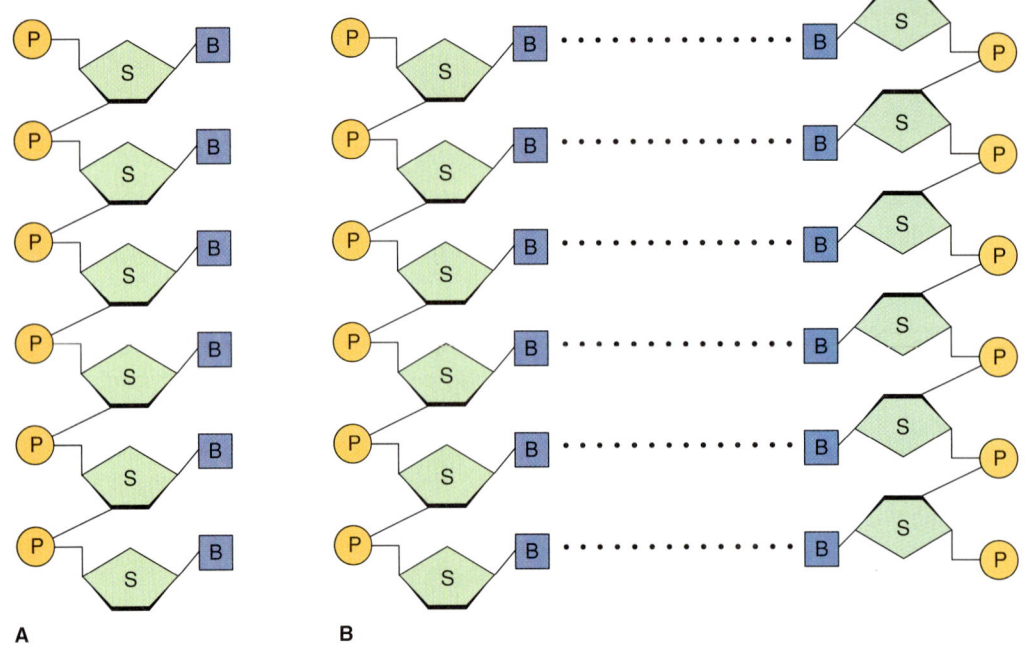

Figure 2.19
A schematic representation of nucleic acid structure. A nucleic acid molecule consists of (A) one (RNA) or (B) two (DNA) polynucleotide chains. DNA chains are held together by hydrogen bonds (dotted lines).

polynucleotide chain. The second type—**DNA** (deoxyribonucleic acid)—contains deoxyribose and forms a double polynucleotide chain. The two chains are held together by hydrogen bonds (fig. 2.19).

DNA molecules store information in a type of molecular code. Cells use this information to synthesize protein molecules. RNA molecules help in protein synthesis. (Nucleic acids are discussed in more detail in chapter 4, pp. 79–86.) Table 2.6 summarizes these four groups of organic compounds.

 CHECK YOUR RECALL

1. Compare the chemical composition of carbohydrates, lipids, proteins, and nucleic acids.
2. How does an enzyme affect a chemical reaction?
3. What is the chemical basis of the great diversity of proteins?
4. What are the functions of nucleic acids?

Clinical Connection

Discovery of prion protein overturned the long-held idea that a protein can assume only one three-dimensional shape, or conformation. A prion is a particular type of protein that can assume up to a dozen different conformations. Some are infectious, because when they bind to normal prions, they convert them into the infectious type. These abnormal prions cause diseases called transmissible spongiform encephalopathies. More than 80 species of mammals are subject to these disorders. Familiar ones are bovine spongiform encephalopathy, better known as "mad cow disease," and in humans, variant Creutzfeldt-Jakob disease. The prion disorders turn parts of the brain into spongy masses, with steady loss of mental functions. Other types of diseases may be caused by misfolded proteins that are not infectious, but form gummy plaques in the brain that disrupt functioning. Some forms of Alzheimer disease may be caused by protein misfolding.

TABLE 2.6 ORGANIC COMPOUNDS OF CELLS

COMPOUND	ELEMENTS PRESENT	BUILDING BLOCKS	FUNCTIONS	EXAMPLES
Carbohydrates	C,H,O	Simple sugars	Provide energy, cell structure	Glucose, starch
Lipids	C,H,O (often P)	Glycerol, fatty acids, phosphate groups	Provide energy, cell structure	Triglycerides, phospholipids, steroids
Proteins	C,H,O,N (often S)	Amino acids	Provide cell structure, enzymes, energy	Albumins, hemoglobin
Nucleic acids	C,H,O,N,P	Nucleotides	Store information for protein synthesis; control cell activities	RNA, DNA

SUMMARY OUTLINE

2.1 Introduction (p. 31)
Chemistry describes the composition of substances and how chemicals react with each other.

2.2 Structure of Matter (p. 31)
1. Elements and atoms
   a. Naturally occurring matter on earth is composed of 92 elements.
   b. Some elements occur in pure form, but many are found combined with other elements.
   c. Elements are composed of atoms, which are the smallest complete units of elements.
   d. Atoms of different elements vary in size, weight, and ways of interacting.
2. Atomic structure
   a. An atom consists of one or more electrons surrounding a nucleus, which contains one or more protons and usually one or more neutrons.
   b. Electrons are negatively charged, protons are positively charged, and neutrons are uncharged.
   c. A complete atom is electrically neutral.
   d. An element's atomic number is equal to the number of protons in each atom. The atomic weight is equal to the number of protons plus the number of neutrons in each atom.
3. Bonding of atoms
   a. When atoms combine, they gain, lose, or share electrons.
   b. Electrons occupy shells around a nucleus.
   c. Atoms with completely filled outer shells are inert, but atoms with incompletely filled outer shells tend to gain, lose, or share electrons and thus achieve stable structures.
   d. Atoms that lose electrons become positively charged ions. Atoms that gain electrons become negatively charged ions.
   e. Ions with opposite electrical charges attract and form ionic bonds. Atoms that share electrons form covalent bonds.
4. Molecules and compounds
   a. Two or more atoms of the same element may bond to form a molecule of that element. Atoms of different elements may bond to form a molecule of a compound.
   b. Molecules contain definite kinds and numbers of atoms.
5. Formulas
   a. A molecular formula represents the numbers and types of atoms in a molecule.
   b. A structural formula depicts the arrangement of atoms within a molecule.
6. Chemical reactions
   a. A chemical reaction breaks or forms bonds between atoms, ions, or molecules.
   b. Three types of chemical reactions are: synthesis, in which larger molecules form from smaller particles; decomposition, in which larger molecules are broken down into smaller particles; and exchange reactions, in which the parts of two different molecules trade positions.
   c. Many reactions are reversible. The direction of a reaction depends on the proportions of reactants and end products, the energy available, and the presence of catalysts.
7. Acids and bases
   a. Compounds that release ions when they dissolve in water are electrolytes.
   b. Electrolytes that release hydrogen ions are acids, and those that release hydroxyl or other ions that react with hydrogen ions are bases.
   c. A value called pH represents a solution's concentration of hydrogen ions (H^+) and hydroxide ions (OH^-).
   d. A solution with equal numbers of H^+ and OH^- is neutral and has a pH of 7.0. A solution with more H^+ than OH^- is acidic and has a pH less than 7.0. A solution with fewer H^+ than OH^- is basic and has a pH greater than 7.0.
   e. Each whole number on the pH scale represents a tenfold difference in the hydrogen ion concentration.

2.3 Chemical Constituents of Cells (p. 37)
Molecules containing carbon and hydrogen atoms are organic and are usually nonelectrolytes. Other molecules are inorganic and are usually electrolytes.

1. Inorganic substances
   a. Water is the most abundant compound in cells and is a solvent in which chemical reactions occur. Water transports chemicals and heat.
   b. Oxygen releases energy from glucose and other nutrients. This energy drives metabolism.
   c. Carbon dioxide is produced when metabolism releases energy.
   d. Salts provide a variety of ions that metabolic processes require.
2. Organic substances
   a. Carbohydrates provide much of the energy that cells require and also contribute to cell structure. Their basic building blocks are simple sugar molecules.
   b. Lipids, such as fats, phospholipids, and steroids, supply energy and build cell parts. The basic building blocks of fats—the most common lipid—are molecules of glycerol and fatty acids.
   c. Proteins serve as structural materials, energy sources, hormones, cell surface receptors, and enzymes.
      (1) Enzymes speed chemical reactions without being consumed.
      (2) The building blocks of proteins are amino acids.
      (3) Proteins vary in the numbers and types of amino acids they contain and in the sequence of these amino acids.
      (4) The amino acid chain of a protein molecule folds into a complex shape that is maintained largely by hydrogen bonds.
      (5) Excessive heat, radiation, electricity, altered pH, or various chemicals can denature proteins.
   d. Nucleic acids are the genetic material and control cellular activities.
      (1) Nucleic acid molecules are composed of nucleotides.
      (2) The two types of nucleic acids are RNA and DNA.
      (3) DNA molecules store information that cell parts use to construct specific protein molecules. RNA molecules help synthesize proteins.

REVIEW EXERCISES

1. Define *chemistry*. (p. 31)
2. Define *matter*. (p. 31)
3. Explain the relationship between elements and atoms. (p. 31)
4. List the four most abundant elements in the human body. (p. 31)
5. Describe the major parts of an atom. (p. 31)
6. Explain why a complete atom is electrically neutral. (p. 31)
7. Define *atomic number* and *atomic weight*. (p. 32)
8. Explain how electrons are arranged within an atom. (p. 32)
9. Distinguish between an ionic bond and a covalent bond. (p. 34)
10. Explain the relationship between molecules and compounds. (p. 35)
11. Distinguish between a molecular formula and a structural formula. (p. 35)

12. Explain what the formula $C_6H_{12}O_6$ means. (p. 35)
13. Describe three major types of chemical reactions. (p. 36)
14. Explain what a reversible reaction is. (p. 36)
15. Define *catalyst*. (p. 36)
16. Define *acid* and *base*. (p. 37)
17. Explain what pH measures, and describe the pH scale. (p. 37)
18. Distinguish between electrolytes and nonelectrolytes. (p. 37)
19. Distinguish between inorganic and organic substances. (p. 37)
20. Describe the roles water and oxygen play in the human body. (p. 38)
21. List several of the ions found in body fluids. (p. 38)
22. Describe the general characteristics of carbohydrates. (p. 39)
23. Distinguish between simple and complex carbohydrates. (p. 39)
24. Describe the general characteristics of lipids. (p. 39)
25. Define *triglyceride*. (p. 40)
26. Distinguish between saturated and unsaturated fats. (p. 40)
27. Describe the general characteristics of proteins. (p. 41)
28. Define *enzyme*. (p. 41)
29. Explain how protein molecules may denature. (p. 43)
30. Describe the structure of nucleic acids. (p. 43)
31. Explain the major functions of nucleic acids. (p. 44)

CRITICAL THINKING

1. If a shampoo is labeled "nonalkaline," would it more likely have a pH of 3, 7, or 12?
2. A topping for ice cream contains fructose, hydrogenated soybean oil, salt, and cellulose. What types of chemicals are in it?
3. An advertisement for a supposedly healthful cookie claims that it contains an "organic carbohydrate." Why is this statement silly?
4. At a restaurant, a waiter recommends a sparkling carbonated beverage, claiming that it contains no carbohydrates. The product label lists water and fructose as ingredients. Is the waiter correct?
5. A Horta is a fictional animal (from "Star Trek") whose biochemistry is based on the element silicon. Consult a periodic table. Is silicon a likely substitute for carbon in a life-form? Cite a reason for your answer.
6. A man on a very low-fat diet proclaims to his friend, "I'm going to get my cholesterol down to zero!" Why is this an impossible (and undesirable) goal?
7. What acidic and basic substances do you encounter in your everyday activities? What acidic foods do you eat regularly? What basic foods do you eat?
8. How would you explain the dietary importance of amino acids and proteins to a person who is following a diet composed primarily of carbohydrates?
9. What clinical laboratory tests have you encountered that require a knowledge of chemistry to understand the significance of what they measure?

WEB CONNECTIONS

Visit the website for additional study questions and more information about this chapter at:

http://www.mhhe.com/shieress8

chapter 3
Cells

CLONING AND STEM CELLS. On February 7, 1997, a sheep appeared on the cover of *Nature* magazine and changed the world. Dolly was a cloned sheep—one whose development began, not as a sperm meeting an egg, but as a packet of genes from a cell in the breast of an adult sheep placed into an egg cell whose own packet of genes had been removed. The transferred genes took over, and Dolly developed, much as any sheep would, in a surrogate sheep mother. The rest, as they say, is history.

Dolly's existence evoked visions of someday farming human clones for spare parts, and a flurry of legislation bloomed to prevent this fate. Meanwhile, receiving far less attention was another technology that may make it possible to grow human cells into tissues and organs in laboratory culture—made-to-order replacement parts.

Stem cell technology begins not with a fertilized egg or an engineered first cell, but with a normal cell that naturally retains the ability to specialize as any, or nearly any, cell type. Such cells from embryos of pigs, cows, rabbits, sheep, and monkeys have been nurtured into various tissues. In humans, stem cells taken from blood in the umbilical cords of newborns are used to generate new bone marrow in people suffering from various blood disorders, with greater success than bone marrow transplants. Tissues in adults harbor stem cells too, which one day may be used to fashion needed parts. Neural stem cells, for example, are being studied for their use as implants to treat degenerative disorders such as Parkinson disease and multiple sclerosis, and heal spinal cord injuries.

Photo:
Dolly the cloned sheep made headlines in the late 1990s, but stem cell technology is much more likely to lead to new medical treatments than cloning.

Chapter Objectives

After studying this chapter, you should be able to do the following:

3.1 Introduction
1. Explain how cells differ from one another. (p. 49)

3.2 Composite Cell
2. Describe the characteristics of a composite cell. (p. 49)
3. Explain how the structure of a cell membrane makes possible its functions. (p. 51)
4. Describe each type of cytoplasmic organelle, and explain its function. (p. 52)
5. Describe the parts of the cell nucleus and its function. (p. 56)

3.3 Movements Through Cell Membranes
6. Explain how substances move through cell membranes. (p. 58)

3.4 The Cell Cycle
7. Describe the cell cycle. (p. 64)
8. Explain how a cell divides. (p. 65)
9. Discuss what happens when a cell specializes. (p. 67)
10. Describe how cell death is a normal part of development. (p. 68)

Aids to Understanding Words

cyt- [cell] *cyt*oplasm: Fluid (cytosol) and organelles that occupy the space between the cell membrane and nuclear envelope.

endo- [within] *endo*plasmic reticulum: Complex of membranous structures within the cytoplasm.

hyper- [above] *hyper*tonic: Solution that has a greater osmotic pressure than body fluids.

hypo- [below] *hypo*tonic: Solution that has a lesser osmotic pressure than body fluids.

inter- [between] *inter*phase: Stage that occurs between mitotic divisions of a cell.

iso- [equal] *iso*tonic: Solution that has the same osmotic pressure as body fluids.

mit- [thread] *mit*osis: Process of cell division when threadlike chromosomes become visible within a cell.

phag- [to eat] *phag*ocytosis: Process by which a cell takes in solid particles.

pino- [to drink] *pino*cytosis: Process by which a cell takes in tiny droplets of liquid.

-som [body] ribo*som*e: Tiny, spherical structure that consists of protein and RNA.

Key Terms

active transport (ak´tiv trans´port)
apoptosis (ap´´o-to´-sis)
cell membrane (sel mem´-brayn)
centrosome (sent´tro-sōm)
chromosome (kro´mo-sōm)
cytoplasm (si´to-plazm)
cytoskeleton (si-to-skel´-e-tun)
differentiation (dif´´er-en´´she-a´shun)
diffusion (dĭ-fu´zhun)
endocytosis (en´´do-si-to´sis)
endoplasmic reticulum (en´do-plaz´mik rĕ-tik´u-lum)

equilibrium (e´´kwĭ-lib´re-um)
exocytosis (ex-o-si-to´sis)
facilitated diffusion (fah-sil´´ĭ-tāt´ed dĭ-fu´zhun)
filtration (fil-tra´shun)
Golgi apparatus (gol´je ap´´ah-ra´tus)
lysosome (li´so-sōm)
mitochondrion (mi´´to-kon´dre-un); plural: mitochondria (mi´´to-kon´dre-ah)
mitosis (mi-to´sis)

nucleolus (nu-kle´o-lus)
nucleus (nu´kle-us)
organelle (or-gan-el´)
osmosis (oz-mo´sis)
phagocytosis (fag´´o-si-to´sis)
pinocytosis (pi´´no-si-to´sis)
ribosome (ri´bo-sōm)
selectively permeable (se-lek´tiv-le per´me-ah-bl)
vesicle (ves´ĭ-k'l)

3.1 Introduction

Recipe for a human being: cells, their products, and fluids. A cell, as the unit of life, is a world unto itself. To build a human, trillions of cells connect and interact, forming dynamic tissues, organs, and organ systems.

The estimated 75 trillion cells that make up an adult human body have much in common. Yet, cells in different tissues vary considerably in size and shape, and typically, their shapes make possible their functions. For instance, nerve cells often have long, threadlike extensions that transmit electrical impulses from one part of the body to another. Epithelial cells that line the inside of the mouth are thin, flattened, and tightly packed, an arrangement that enables them to protect underlying cells. Muscle cells, which pull structures closer together, are slender and rodlike, with their ends attached to the structures they move (fig. 3.1).

3.2 Composite Cell

Because cells vary so greatly in size, shape, content, and function, describing a "typical" cell is impossible. The cell shown in figure 3.2 and described in this chapter is a composite cell that includes many known cell structures. In reality, cells have most, but not all, of these structures.

Under the light microscope, with a properly applied stain to make structures visible, the **nucleus** (nu´kle-us), **cytoplasm** (si´to-plazm), and **cell membrane** (sel mem´-brayn) are easily seen. The nucleus is often centrally located in the cell and is surrounded by a thin nuclear envelope. The cytoplasm surrounds the nucleus and is itself encircled by an even thinner cell membrane (also called the plasma membrane).

Within the cytoplasm are specialized structures called **organelles** (or-gan-elz´), which can be seen clearly only under the higher magnification of electron microscopes. Organelles, suspended in a liquid called *cytosol*, perform specific functions, in a sense dividing the labor of the cell. The nucleus, on the other hand, directs overall cell activities by housing the genetic material. The cell membrane determines which substances enter and leave the cell, and oversees how cells interact.

CHECK YOUR RECALL

1. Give two examples to illustrate how the shape of a cell makes possible its function.
2. Name the three major parts of a cell.
3. What are the general functions of the cytoplasm, nucleus, and cell membrane?
4. What are organelles?

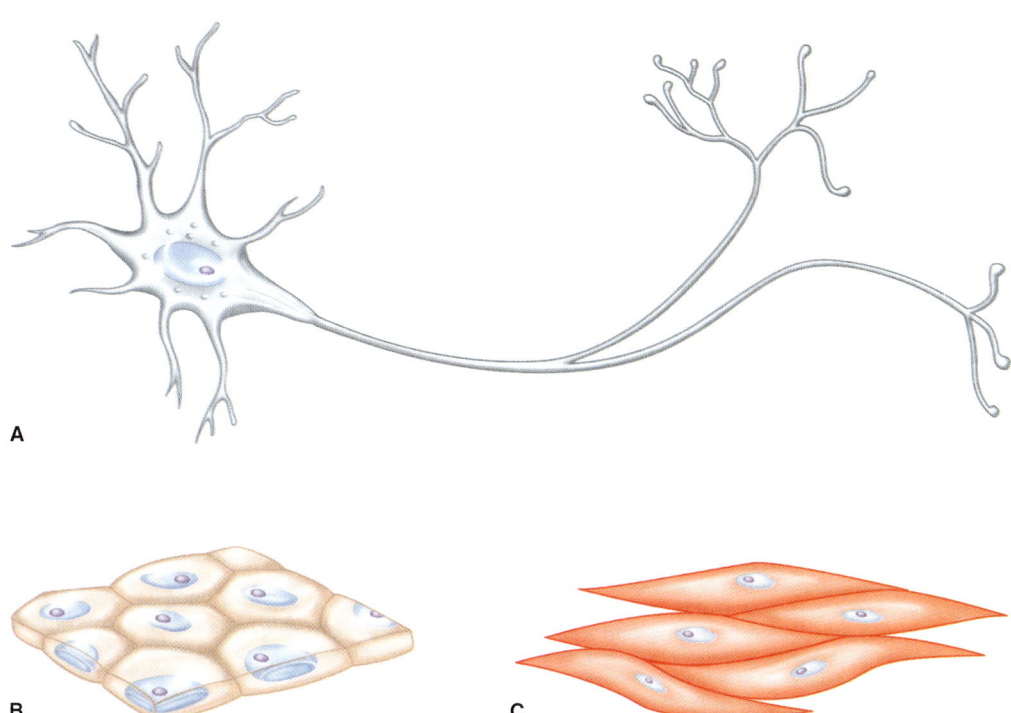

Figure 3.1
Cells vary in structure and function. (*A*) A nerve cell transmits impulses from one body part to another. (*B*) Epithelial cells protect underlying cells. (*C*) Muscle cells contract, pulling structures closer together.

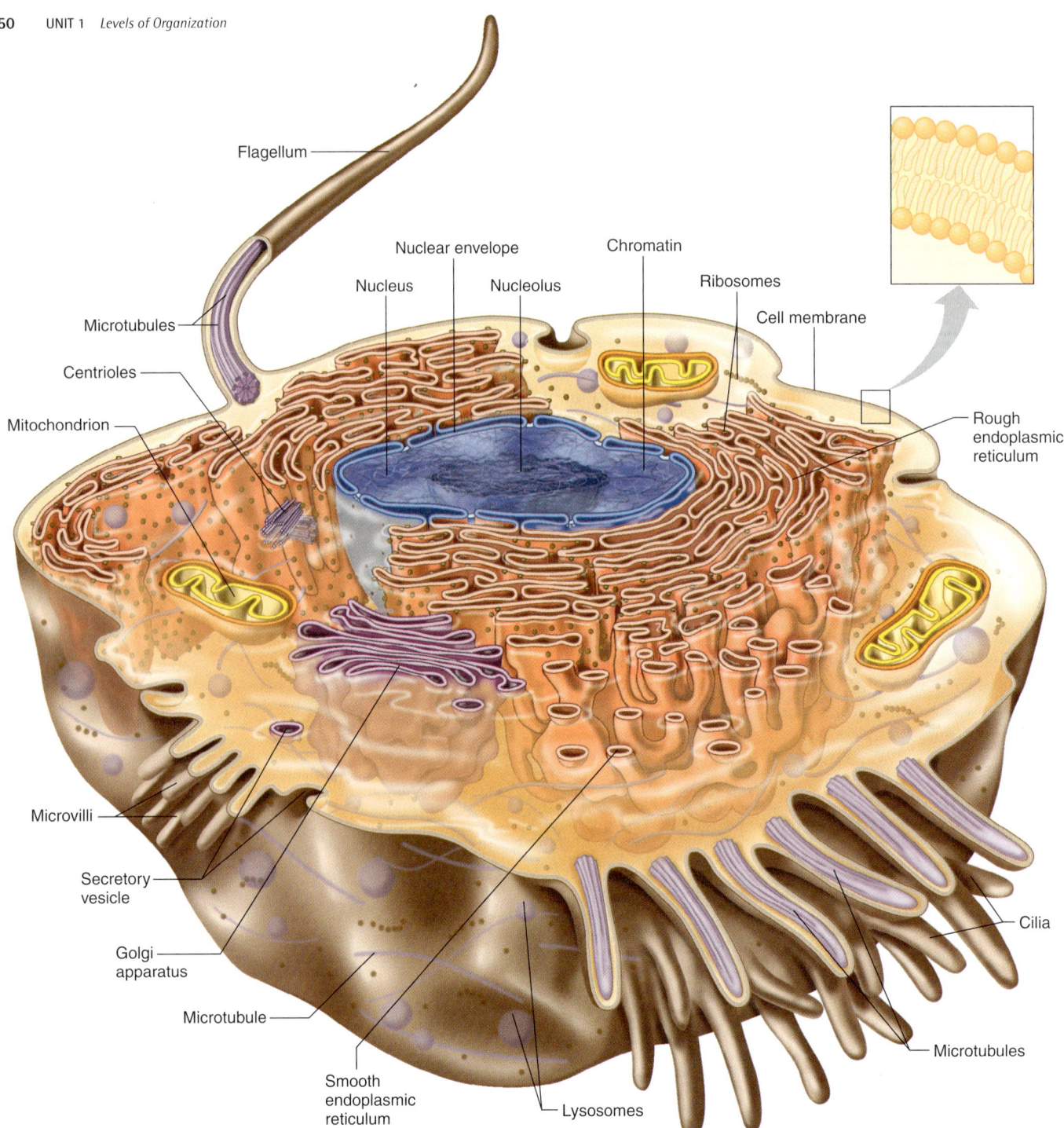

Figure 3.2
A composite cell. Organelles are not drawn to scale.

Cell Membrane

The cell membrane is more than a simple boundary surrounding the cellular contents. It is an actively functioning part of the living material. The cell membrane regulates movement of substances in and out of the cell and is the site of much biological activity. Molecules that are part of the cell membrane receive stimulation from outside the cell and transmit it into the cell, a process called *signal transduction*. The cell membrane also helps cells adhere to certain other cells, which is important in forming tissues.

General Characteristics

The cell membrane is extremely thin—visible only with the aid of an electron microscope—but it is flexible and somewhat elastic. It typically has complex surface features with many outpouchings and infoldings that increase surface area (fig. 3.2).

In addition to maintaining cell integrity, the cell membrane controls which substances exit and enter. A membrane that functions in this way is called **selectively permeable** (se-lek´tiv-le per´me-ah-bl) (also known as *semipermeable* or *differentially permeable*).

Cell Membrane Structure

A cell membrane is composed mainly of lipids, proteins, and a small quantity of carbohydrates. Its basic framework is a double layer, or *bilayer,* of phospholipid molecules. Each phospholipid molecule includes a phosphate group and two fatty acids bound to a glycerol molecule (see chapter 2, p. 40). The water-soluble phosphate "heads" form the surfaces of the membrane, and the water-insoluble fatty acid "tails" make up the interior of the membrane. The lipid molecules can move sideways within the plane of the membrane. Together they form a soft and flexible, but stable, fluid film.

Because the membrane's interior consists largely of the fatty acid portions of the phospholipid molecules (fig. 3.3), it is oily. Molecules such as oxygen and carbon dioxide, which are soluble in lipids, can pass through this bilayer easily. However, the bilayer is impermeable to water-soluble molecules, such as amino acids, sugars, proteins, nucleic acids, and various ions. Many cholesterol molecules embedded in the cell membrane's interior also help make the membrane less permeable to water-soluble substances. In addition, the relatively rigid structure of the cholesterol molecules helps stabilize the membrane.

A cell membrane includes a few types of lipid molecules, but many kinds of proteins, which provide special functions. Membrane proteins are classified according to their positions. For example, membrane-spanning (transmembrane) proteins extend through the lipid bilayer and may protrude from one or both faces. In contrast, peripheral membrane proteins are associated with one side of the bilayer. Membrane proteins also vary in shape—some may be globular, others rodlike.

Membrane proteins have a variety of functions. Some form receptors on the cell surface that bind incoming hormones or growth factors, starting signal transduction. The cell then responds. Other proteins transport ions or molecules across the cell membrane. Still other membrane proteins form selective channels that allow only particular ions to enter or leave. In nerve cells, for example, such selective channels control movement of sodium and potassium ions (see chapter 9, p. 212). The Genetics Connection (p. 52) discusses some inherited disorders that result from abnormal ion channels in cell membranes.

Proteins that extend from the inner face of the cell membrane anchor it to the rods and tubules that support the cell from within. Proteins that extend from the outer surface, where carbohydrate molecules may attach to them, mark the cell as part of a particular tissue or organ in a particular person. This identification is important for

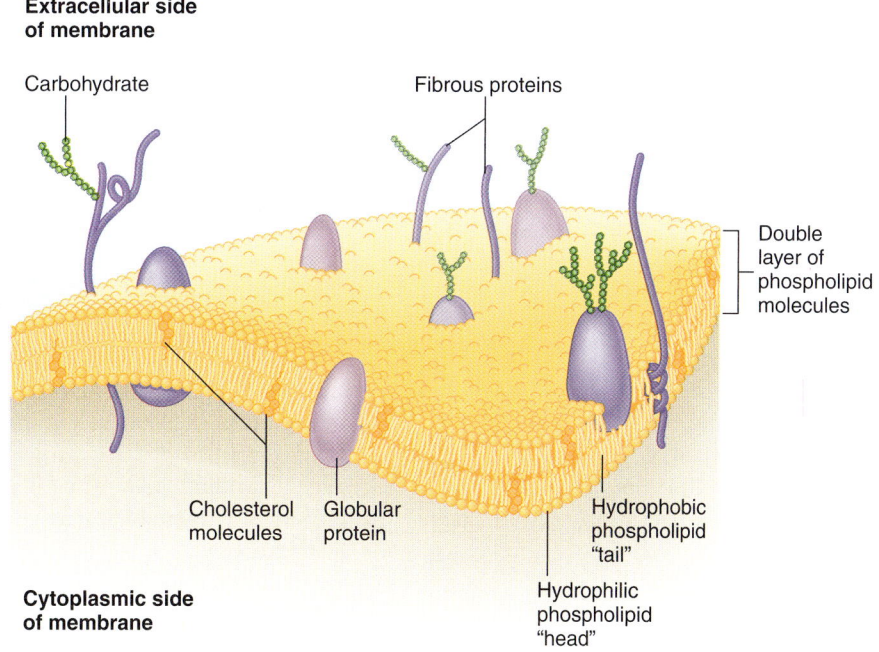

Figure 3.3
The cell membrane is composed primarily of phospholipids (and some cholesterol), with proteins scattered throughout the lipid bilayer and associated with its surfaces.

Genetics Connection

FAULTY ION CHANNELS CAUSE INHERITED DISEASE

What do collapsing horses, irregular heartbeats, and cystic fibrosis have in common? All result from abnormal ion channels in cell membranes.

Ion channels are protein-lined tunnels in the phospholipid bilayer of a biological membrane. These passageways permit electrical signals to pass in and out of membranes in the form of ions (charged particles). Many ion channels open or close like a gate in response to specific ions under specific conditions, such as a change in electrical forces across the membrane, binding of a molecule, or receiving biochemical messages from inside or outside the cell.

Ion channels are specific for calcium (Ca^{+2}), sodium (Na^+), potassium (K^+), or chloride (Cl^-). A cell membrane may have a few thousand ion channels specific for each of these ions. Ten million ions can pass through an ion channel in one second! Drugs may act by affecting ion channels, and abnormal ion channels cause certain disorders, including the following:

Hyperkalemic Periodic Paralysis and Sodium Channels

The quarter horse was originally bred in the 1600s to run the quarter mile, but one of the four very fast stallions used to establish much of today's population of 3 million animals inherited hyperkalemic periodic paralysis (HPP). The horse, otherwise a champion, sometimes collapsed from sudden attacks of weakness and paralysis.

HPP results from abnormal sodium channels in the cell membranes of muscle cells. But the trigger for the temporary paralysis is another ion: potassium. A rising blood potassium level, which may follow intense exercise, slightly alters the muscle cell membrane's electrical charge. Normally, this slight change would have no effect, but in horses with HPP, sodium channels open too widely, allowing too much sodium into the cell. For a short time, the muscle cell cannot respond to nervous stimulation. Although the effect is brief, it's long enough for the racehorse to fall.

People can inherit HPP too. In one family, several members collapsed suddenly after eating bananas! These fruits are very high in potassium, which caused the symptoms.

Long-QT Syndrome and Potassium Channels

Four children in a Norwegian family were born deaf, and three of them died at ages four, five, and nine. All of the children had inherited from unaffected "carrier" parents "long-QT syndrome associated with deafness." They had abnormal potassium channels in the heart muscle and in the inner ear. In the heart, the malfunctioning channels disrupted electrical activity, causing a fatal disturbance to the heart rhythm. In the inner ear, the abnormal channels caused an increase in the extracellular concentration of potassium ions, impairing hearing.

Cystic Fibrosis and Chloride Channels

A seventeenth-century English saying, "A child that is salty to taste will die shortly after birth," described the consequence of abnormal chloride channels in the inherited illness cystic fibrosis (CF). The disorder is inherited from carrier parents. The major symptoms—difficulty breathing, frequent severe respiratory infections, and a clogged pancreas that disrupts digestion—all result from buildup of extremely thick mucous secretions.

Abnormal chloride channels in cells lining the lung passageways and ducts of the pancreas cause the symptoms of CF. The primary defect in the chloride channels also causes sodium channels to malfunction. The result: Salt trapped inside cells draws moisture in and thickens surrounding mucus. Experimental gene therapy is attempting to supply patients' lung-lining cells with the instructions to produce normal chloride channels.

---

the functioning of the immune system (see chapter 14, p. 375). Another type of protein on a cell's surface is a cellular adhesion molecule (CAM), which determines a cell's interactions with other cells. For example, a series of CAMs helps a white blood cell move to the site of an injury, such as a splinter in the skin.

Cytoplasm

When viewed through a light microscope, cytoplasm usually appears as a clear jelly with specks scattered throughout. However, an electron microscope, which produces much greater magnification and the ability to distinguish fine detail (resolution), reveals that cytoplasm contains networks of membranes and organelles suspended in the clear liquid *cytosol*. Cytoplasm also includes abundant protein rods and tubules that form a framework, or **cytoskeleton** (si-to-skel´-e-tun), meaning "cell skeleton."

Cell activities occur mainly in the cytoplasm, where nutrients are received, processed, and used. The following organelles have specific functions:

1. **Endoplasmic reticulum** (en´do-plaz´mik rĕ-tik´u-lum) The endoplasmic reticulum (ER) is a complex organelle composed of membrane-bounded,

flattened sacs, elongated canals, and fluid-filled, bubblelike sacs called *vesicles*. These membranous parts are interconnected and communicate with the cell membrane, the nuclear envelope, and other organelles. The ER provides a vast tubular network that transports molecules from one cell part to another.

The endoplasmic reticulum participates in the synthesis of protein and lipid molecules. These molecules may leave the cell as secretions or be used within the cell for such functions as producing new ER or cell membrane as the cell grows.

In many places, the ER's outer membrane is studded with many tiny, spherical structures called *ribosomes*, which give the ER a textured appearance when viewed with an electron microscope (fig. 3.4A, B). These parts of the ER are called rough ER. The ribosomes are sites of protein synthesis and exist independently in the cytoplasm as well as associated with ER. Proteins being synthesized then move through ER tubules to another organelle, the Golgi apparatus, for further processing.

ER that lacks ribosomes is called *smooth ER* (fig. 3.4C). Smooth ER contains enzymes important in lipid synthesis, absorption of fats from the digestive tract, and the metabolism of drugs. Cells that break down drugs and alcohol, such as liver cells, have extensive networks of smooth ER.

2. **Ribosomes** (ri´bo-sōmz) Ribosomes are the sites of protein synthesis. Many ribosomes are attached to ER membranes; others are scattered throughout the cytoplasm. All ribosomes are composed of protein and RNA molecules. Ribosomes provide enzymes as well as a structural support for the RNA molecules that come together as the cell synthesizes proteins from amino acids. Chapter 4 (pp. 81–85) describes protein synthesis.

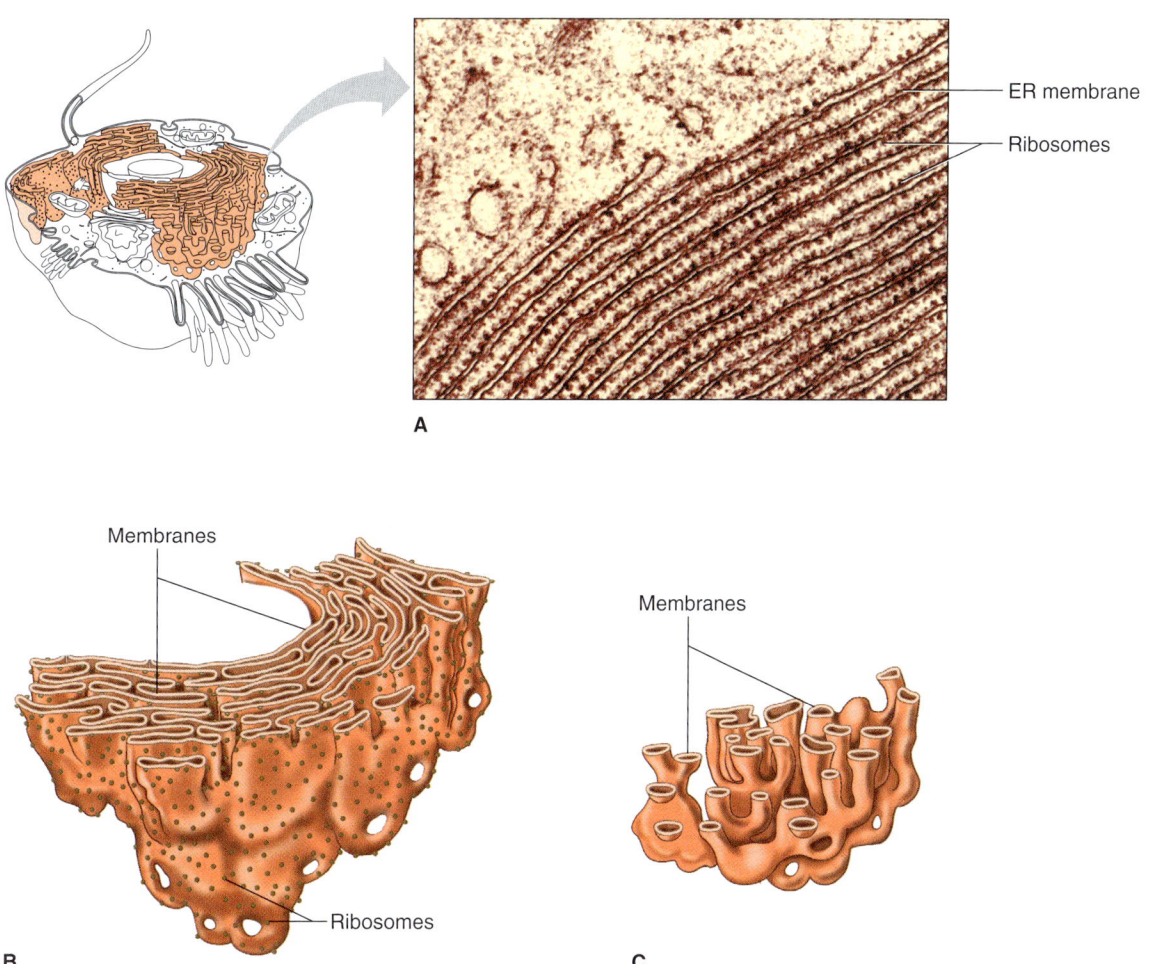

Figure 3.4
The endoplasmic reticulum. (A) A transmission electron micrograph of rough endoplasmic reticulum (ER) (28,500×). (B) Rough ER is dotted with ribosomes, whereas (C) smooth ER lacks ribosomes.

3. **Golgi apparatus** (gol´je ap´´ah-ra´tus) The Golgi apparatus is composed of a stack of about six flattened, membranous sacs. This organelle refines, packages, and delivers proteins synthesized on ribosomes associated with the ER. Proteins arrive at the Golgi apparatus enclosed in vesicles composed of the ER membrane. These vesicles fuse with the membrane at the innermost end of the Golgi apparatus, which is specialized to receive glycoproteins (sugars bound to proteins).

As the glycoproteins pass from layer to layer through the stacks of Golgi membrane, they are modified chemically. For example, sugar molecules may be added or removed. When the altered glycoproteins reach the outermost layer, they are packaged in bits of Golgi membrane, which bud off and form bubblelike structures called transport vesicles. Such a vesicle may then move to and fuse with the cell membrane, releasing its contents to the outside as a secretion (figs. 3.2 and 3.5). This process is called *exocytosis* (see page 62). Other vesicles, some of which bud off the cell membrane's inner face, may transport glycoproteins to organelles within the cell. The vesicles in a cell form a delivery service of sorts.

CHECK YOUR RECALL

1. What is a selectively permeable membrane?
2. Describe the chemical structure of a cell membrane.
3. What are the functions of the endoplasmic reticulum?
4. Describe the functions of the Golgi apparatus.

4. **Mitochondria** (mi´´to-kon´dre-ah; *sing*. mi´´to-kon´dre-un) Mitochondria are elongated, fluid-filled sacs that vary in size and shape. They can move slowly through the cytoplasm and reproduce by dividing. A mitochondrion has an outer and an inner layer (figs. 3.2 and 3.6). The inner layer folds extensively to form partitions called *cristae*. Connected to the cristae are enzymes that control some of the chemical reactions that release energy from certain nutrient molecules. Mitochondria are the major sites of chemical reactions that transform

Mitochondria may provide clues to the origin of life. According to the endosymbiont theory, mitochondria are the remnants of once free-living bacteria-like cells that were swallowed by more complex primitive cells. These bacterial passengers remain in our cells today, where they are crucial to extracting energy from nutrients. The bacterium that causes typhus, for example, is remarkably similar to a mitochondrion.

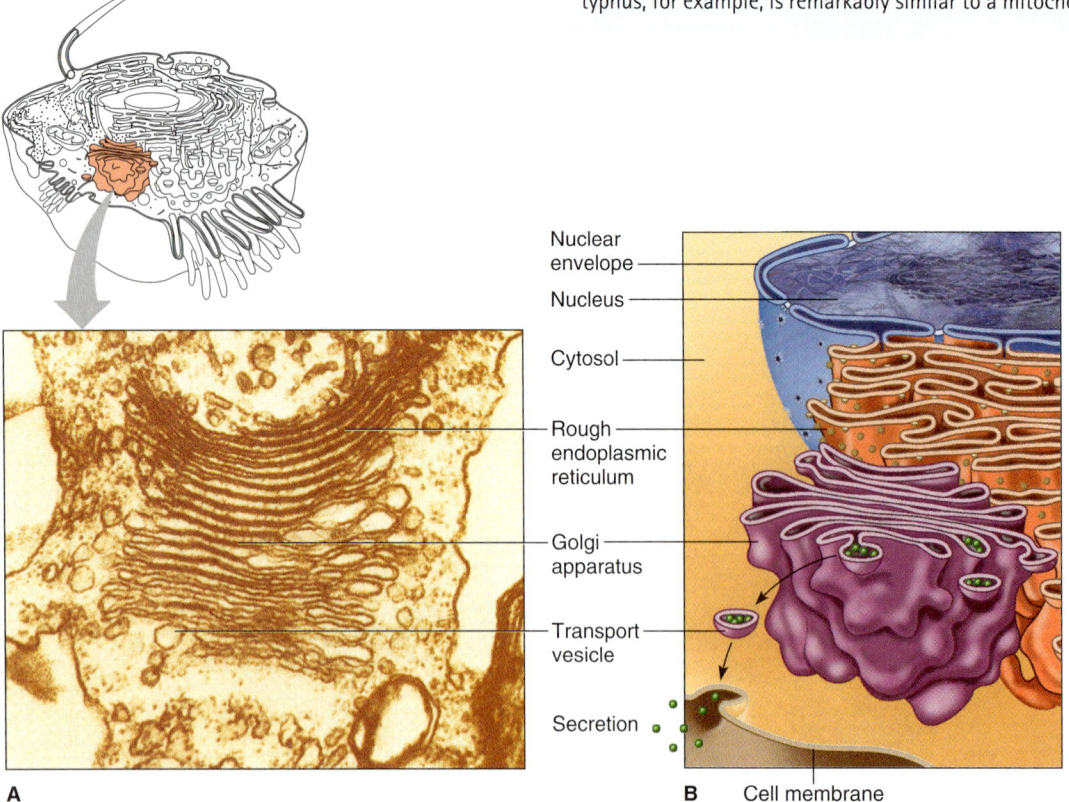

Figure 3.5
The Golgi apparatus. (*A*) A transmission electron micrograph of a Golgi apparatus (48,500×). (*B*) The Golgi apparatus consists of membranous sacs that continually receive vesicles from the endoplasmic reticulum and produce vesicles that enclose secretions.

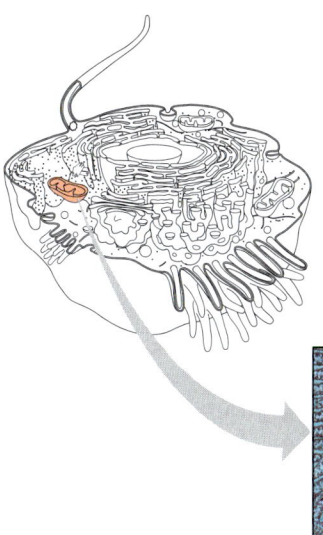

Figure 3.6
A mitochondrion. (*A*) A transmission electron micrograph of a mitochondrion (28,000×). (*B*) Cristae partition this saclike organelle.

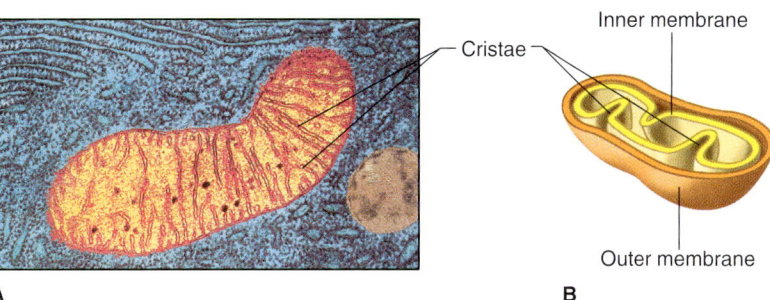

this energy into adenosine triphosphate (ATP), a chemical form the cell can use. Very active cells, such as muscle cells, contain many thousands of mitochondria. (Chapter 4 p. 78 describes this energy-releasing function in more detail.) Mitochondria resemble bacterial cells and even contain a small amount of their own DNA.

5. **Lysosomes** (li´so-sōmz) Lysosomes, the "garbage disposals of the cell," are tiny membranous sacs (see fig. 3.2). They contain powerful enzymes that break down nutrient molecules or foreign particles. Certain white blood cells, for example, can engulf bacteria, which are then digested by the lysosomal enzymes. This is one way that white blood cells fight bacterial infections. Lysosomes also destroy worn cellular parts.

6. **Peroxisomes** (pĕ-roks´ĭ-sōmz) These membranous sacs are abundant in liver and kidney cells. They house enzymes that catalyze (speed) a variety of biochemical reactions, including synthesis of bile acids (used to digest fats); detoxification of hydrogen peroxide, a by-product of metabolism; breakdown of certain lipids and rare biochemicals; and detoxification of alcohol.

7. **Microfilaments and microtubules** Microfilaments and microtubules are two types of thin, threadlike strands within the cytoplasm. They form the cytoskeleton and are also part of structures that have specialized activities.

    Microfilaments are tiny rods of actin protein that form meshworks or bundles. They provide cell motility (movement). In muscle cells, for example, microfilaments aggregate to form *myofibrils*, which help these cells contract (see chapter 8, p. 173).

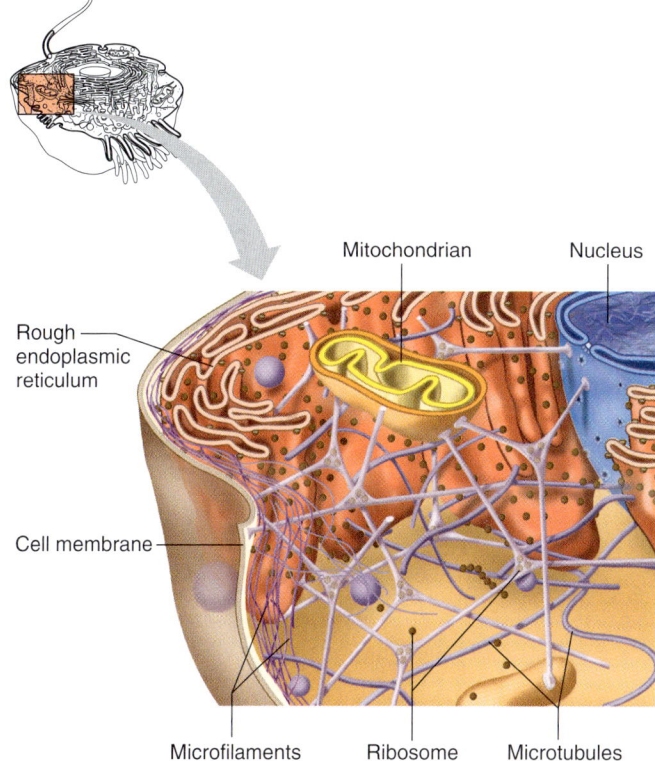

Figure 3.7
Microtubules built of tubulin and microfilaments built of actin help maintain the shape of a cell by forming a cytoskeleton within the cytoplasm.

Microtubules are long, slender tubes with diameters two or three times that of microfilaments. Microtubules are composed of globular tubulin proteins that are typically arrayed in a characteristic "9 + 2" pattern, in which nine outside tubules form a ring around two inner ones (fig. 3.7).

8. **Centrosome** (sent´tro-sōm) The centrosome is a structure near the Golgi apparatus and nucleus. It is nonmembranous and consists of two hollow cylinders, called *centrioles,* which are composed of microtubules (figs. 3.2 and 3.8). The centrioles lie at right angles to each other. During mitosis, they distribute chromosomes to newly forming cells.

9. **Cilia and flagella** Cilia and flagella, motile extensions from the surfaces of certain cells, are also composed of microtubules in a "9 + 2" array. They are similar structures that differ mainly in length and abundance.

   Cilia fringe the free surfaces of some epithelial (lining) cells. Each cilium is a tiny, hairlike structure that is attached beneath the cell membrane (see fig. 3.2). Cilia form in precise patterns, and they move in a "to-and-fro" manner, coordinated so that rows of cilia beat in succession, producing a wave of motion that sweeps over the ciliated surface. This wave moves fluids, such as mucus, over the surface of certain tissues, including those that form the inner linings of the respiratory tubes (fig. 3.9A).

   Flagella are considerably longer than cilia, and usually a cell has only a single flagellum. Flagella have an undulating wavelike motion, which begins at their base. The tail of a sperm cell is a flagellum that enables this motile cell to "swim" and is the only example of a flagellum in humans (fig. 3.9B).

10. **Vesicles** (ves´ĭ-k'lz) Vesicles (vacuoles) are membranous sacs formed by part of the cell membrane folding inward and pinching off. As a result, a tiny, bubblelike vesicle, containing some liquid or solid material formerly outside the cell, appears in the cytoplasm. The Golgi apparatus and ER also form vesicles that play a role in secretion (see fig. 3.2).

CHECK YOUR RECALL

1. Describe a mitochondrion.
2. What is the function of a lysosome?
3. How do microfilaments and microtubules differ?
4. What are some structures that consist of microtubules?

Cell Nucleus

The nucleus houses the genetic material (DNA), which directs all cell activities (figs. 3.2 and 3.10). It is a large, spherical structure enclosed in a double-layered **nuclear envelope,** which consists of inner and outer lipid bilayer membranes. The nuclear envelope has protein-lined channels called nuclear pores that allow certain molecules to exit the nucleus. A nuclear pore is not just a hole, but a complex opening formed from 100 or so types of proteins.

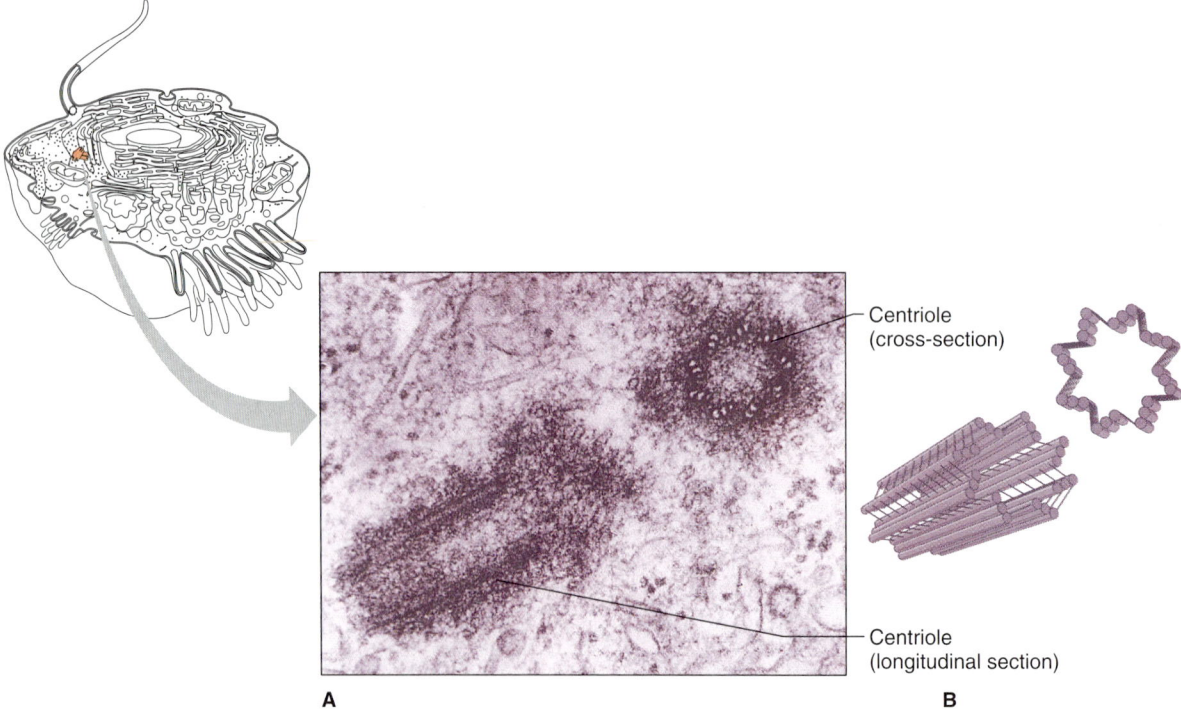

Figure 3.8
Centrioles. (*A*) Transmission electron micrograph of the two centrioles in a centrosome (120,000×). (*B*) The centrioles lie at right angles to one another.

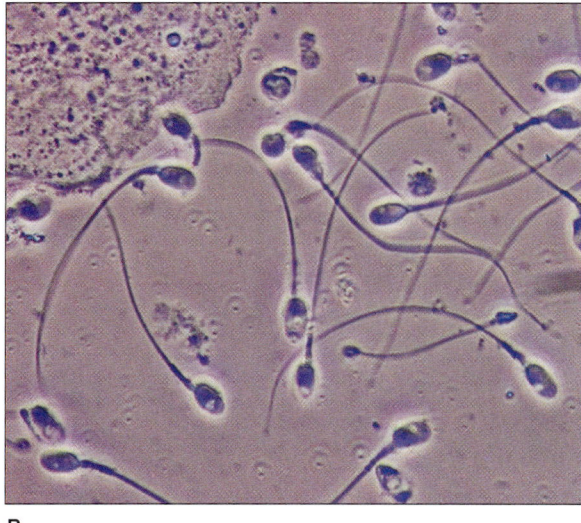

Figure 3.9
Cilia and flagella. (*A*) Cilia are common on the surfaces of certain cells, including those that form the inner lining of the respiratory tubes (5,800×). (*B*) Flagella form the tails of these human sperm cells (840×).

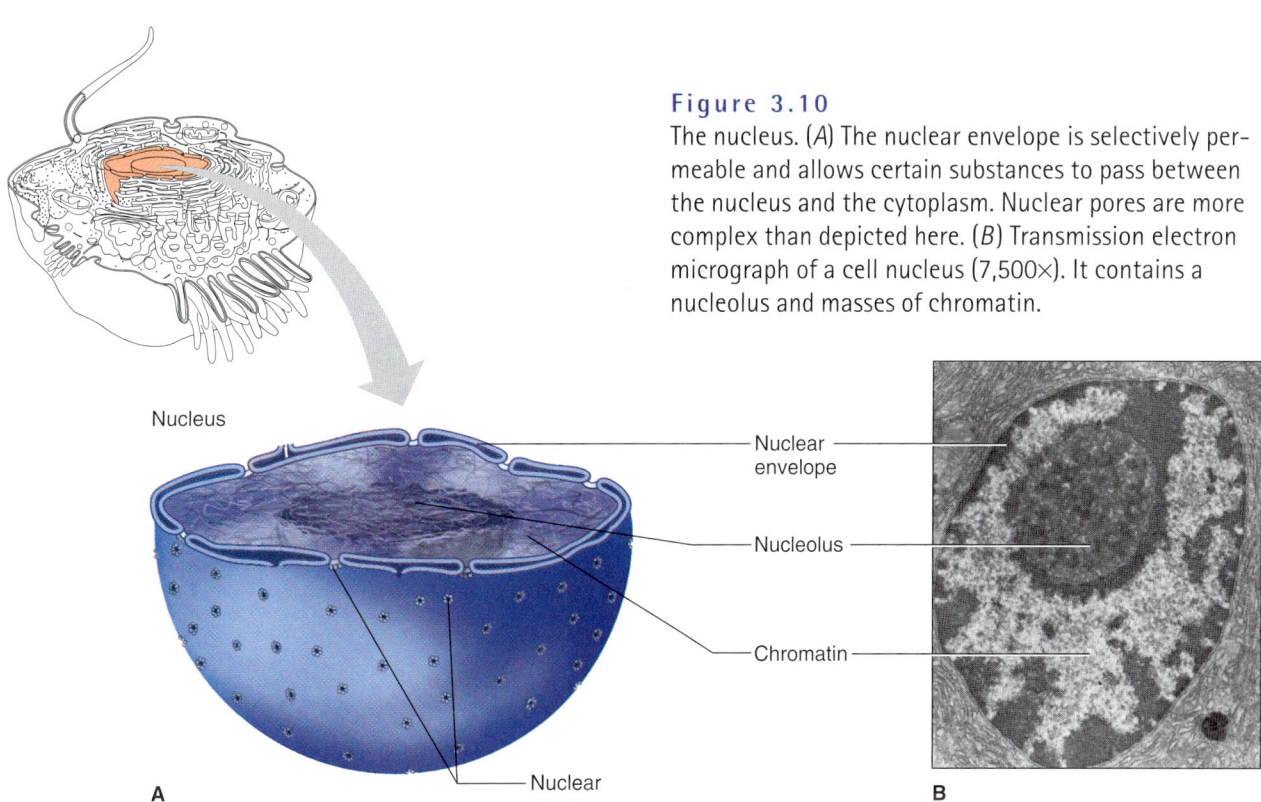

Figure 3.10
The nucleus. (*A*) The nuclear envelope is selectively permeable and allows certain substances to pass between the nucleus and the cytoplasm. Nuclear pores are more complex than depicted here. (*B*) Transmission electron micrograph of a cell nucleus (7,500×). It contains a nucleolus and masses of chromatin.

The nucleus contains a fluid in which the following structures are suspended:

1. **Nucleolus** (nu-kle´o-lus) A nucleolus ("little nucleus") is a small, dense body composed largely of RNA and protein. It has no surrounding membrane and forms in specialized regions of certain chromosomes. Ribosomes form in the nucleolus, then migrate through nuclear pores to the cytoplasm.

2. **Chromatin** Chromatin consists of loosely coiled fibers of DNA and protein called **chromosomes** (kro´mo-sōmz). The DNA contains the information for protein synthesis. When the cell begins to divide, chromatin fibers coil tightly and individual chromosomes become visible.

Table 3.1 summarizes the structures and functions of organelles.

CHECK YOUR RECALL

1. What structure separates the nuclear contents from the cytoplasm?
2. What is produced in the nucleolus?
3. What is chromatin?

3.3 Movements Through Cell Membranes

The cell membrane is a selective barrier that controls which substances enter and leave the cell. These movements include passive mechanisms that do not require cellular energy (diffusion, facilitated diffusion, osmosis, and filtration) and active mechanisms that use cellular energy (active transport, endocytosis, and exocytosis).

Passive Mechanisms

Diffusion

Diffusion (dĭ-fu´zhun) (also called *simple diffusion*) is the process by which molecules or ions scatter or spread spontaneously from regions where they are in higher concentrations toward regions where they are in lower concentrations. This difference in concentration is called a *concentration gradient*. Atoms, molecules, and ions are said to diffuse down their concentration gradients.

Under natural conditions, molecules and ions constantly move at high speeds. Each particle travels in a separate path along a straight line until it collides and bounces off some other particle, changing direction, colliding again, and changing direction once more. Such random motion mixes molecules. At body temperature, small molecules such as water move more than a thousand miles per hour. However, the internal environment is crowded, from a molecule's point of view. A single molecule may collide with other molecules a million times each second.

Consider sugar (a solute). When first put into a glass of water (a solvent), the sugar remains highly concentrated at the bottom of the glass (fig. 3.11). As a result of diffusion, the sugar molecules move away from the area of high concentration and spread into solution among the moving water molecules. Eventually the sugar molecules become uniformly distributed in the water, a state called **equilibrium** (e´´kwĭ-lib´re-um). Molecules continue to move after equilibrium occurs, but their concentrations no longer change.

Diffusion of a substance into or out of the cell can occur only if (1) the cell membrane is permeable to that substance, and (2) a concentration gradient exists such that the substance is at a higher concentration either out-

TABLE 3.1 STRUCTURES AND FUNCTIONS OF CYTOPLASMIC ORGANELLES

ORGANELLE(S)	STRUCTURE	FUNCTION
Cell membrane	Membrane composed of protein and lipid molecules	Maintains integrity of cell and controls passage of materials into and out of cell
Endoplasmic reticulum	Complex of interconnected membrane-bounded sacs and canals	Transports materials within cell, provides attachment for ribosomes, and synthesizes lipids
Ribosomes	Particles composed of protein and RNA molecules	Synthesize proteins
Golgi apparatus	Group of flattened, membranous sacs	Packages protein molecules for transport and secretion
Mitochondria	Membranous sacs with inner partitions	Release energy from nutrient molecules and transform energy into usable form
Lysosomes	Membranous sacs	Digest worn cellular parts or substances that enter cells
Peroxisomes	Membranous sacs	House enzymes that catalyze diverse reactions, including bile acid synthesis, lipid breakdown, and alcohol detoxification
Microfilaments and microtubules	Thin rods and tubules	Support the cytoplasm and help move substances and organelles within the cytoplasm
Centrosome	Nonmembranous structure composed of two rodlike centrioles	Helps distribute chromosomes to new cells during cell division
Cilia and flagella	Motile projections attached beneath the cell membrane	Cilia propel fluid over cellular surfaces, and a flagellum enables a sperm cell to move
Vesicles	Membranous sacs	Contain and transport various substances
Nuclear envelope	Double membrane that separates the nuclear contents from the cytoplasm	Maintains integrity of nucleus and controls passage of materials between nucleus and cytoplasm
Nucleolus	Dense, nonmembranous body composed of protein and RNA	Site of ribosome synthesis
Chromatin	Fibers composed of protein and DNA	Contains information for synthesizing proteins

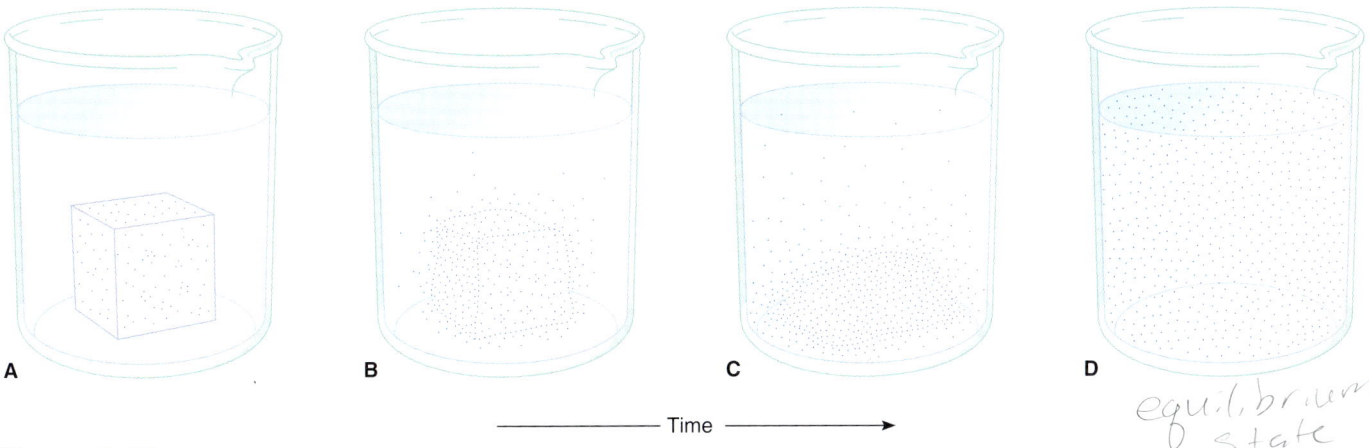

Figure 3.11
An example of diffusion. (*A, B,* and *C*) A sugar cube placed in water slowly disappears as the sugar molecules dissolve and then diffuse from regions where they are more concentrated toward regions where they are less concentrated. (*D*) Eventually, the sugar molecules are distributed evenly throughout the water.

side or inside the cell. Consider oxygen and carbon dioxide, two substances to which cell membranes are permeable. In the body, diffusion is the process whereby oxygen enters cells and carbon dioxide leaves cells, but equilibrium is never reached. Intracellular oxygen is always low because oxygen is constantly used up in metabolic reactions. Extracellular oxygen is maintained at a high level by homeostatic mechanisms in the respiratory and cardiovascular systems. Thus, a concentration gradient always allows oxygen to diffuse into the body's cells.

The level of carbon dioxide, produced as a waste product of metabolism, is always high inside cells. Homeostasis maintains a lower extracellular carbon dioxide level, so a concentration gradient always favors carbon dioxide diffusing out of cells (fig. 3.12).

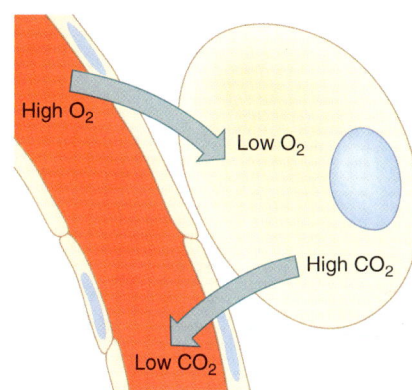

Figure 3.12
Diffusion is the process whereby oxygen enters the cells and carbon dioxide leaves the cells.

D*ialysis* is a chemical technique that uses diffusion to separate small molecules from larger ones in a liquid. The artificial kidney uses a variant of this process—*hemodialysis*—to treat patients suffering from kidney damage or failure. An artificial kidney (dialyzer) passes blood from a patient through a long, coiled tubing composed of porous cellophane. The size of the pores allows smaller molecules carried in the blood, such as the waste material urea, to exit through the tubing, while larger molecules, such as those of blood proteins, remain inside the tubing. The tubing is submerged in a tank of dialyzing fluid (wash solution), which contains varying concentrations of different chemicals. The fluid has low concentrations of substances that should leave the blood and higher concentrations of those that should remain in the blood.

Altering the concentrations of molecules in the dialyzing fluid can control which molecules diffuse out of blood and which remain in it. For example, to remove blood urea, the dialyzing fluid must have a lower urea concentration than the blood; to maintain blood glucose concentration, glucose concentration in the dialyzing fluid must be at least equal to that of the blood.

Facilitated Diffusion

Substances that are not able to pass through the lipid bilayer need the help of membrane proteins to get across, a process known as **facilitated diffusion** (fah-sil´´ĭ-tāt´ed dĭ-fu´zhun) (fig. 3.13). One form involves the ion channels and pores described earlier. Molecules like glucose and amino acids are not lipid soluble, yet are too large to pass through membrane channels. They enter cells by another form of facilitated diffusion. In this process, which occurs in most cells, the glucose molecule combines with a special protein carrier molecule at the surface of the cell membrane. This union of the glucose and carrier molecules changes the shape of the carrier, enabling it to move glucose to the other side of the membrane. The carrier releases the glucose and then returns to its original shape and picks up another glucose molecule. The hormone *insulin*, discussed in chapter 11 (p. 296), promotes facilitated diffusion of glucose through the membranes of certain cells.

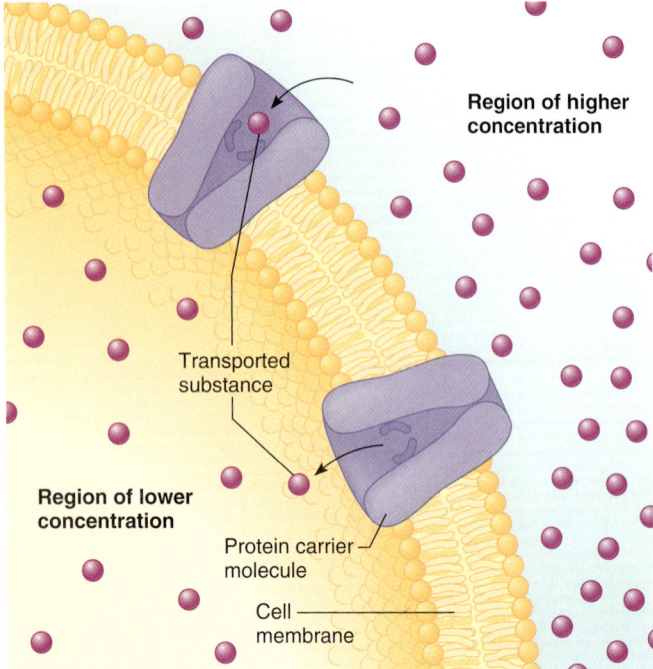

Figure 3.13
Some substances move into or out of cells by facilitated diffusion, transported by carrier molecules from a region of higher concentration to one of lower concentration.

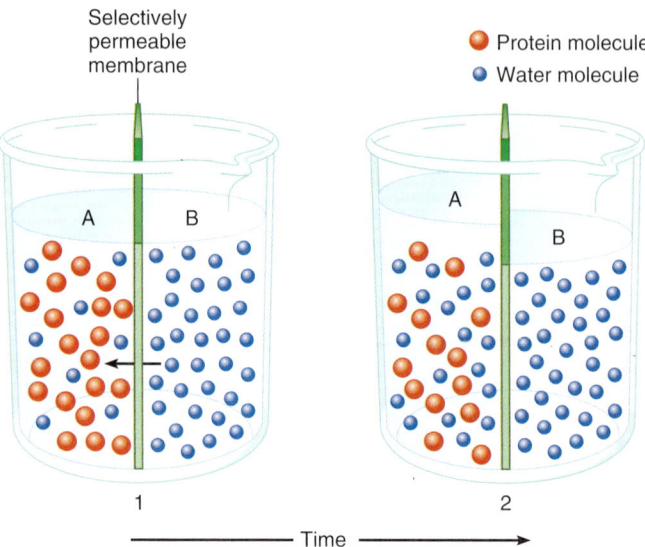

Figure 3.14
Osmosis. (*1*) A selectively permeable membrane separates the container into two compartments. At first, compartment *A* contains water and protein molecules, while compartment *B* contains only water. As a result of molecular motion, water diffuses by osmosis from compartment *B* into compartment *A*. Protein molecules remain in compartment *A* because they are too large to pass through the pores of the membrane. (*2*) Also, because more water is entering compartment *A* than is leaving it, water accumulates in this compartment. The level of liquid rises on this side.

Facilitated diffusion is similar to simple diffusion in that it only moves molecules from regions of higher concentration toward regions of lower concentration. The number of carrier molecules in the cell membrane limits the rate of facilitated diffusion.

Osmosis

Osmosis (oz-mo´sis) is a special case of diffusion. It occurs whenever water molecules diffuse from a region of higher water concentration to a region of lower water concentration across a selectively permeable membrane, such as a cell membrane. In the example that follows, assume that the selectively permeable membrane is permeable to water molecules (the solvent), but impermeable to protein molecules (the solute).

In solutions, a higher concentration of solute (protein in this case) means a lower concentration of water; a lower concentration of solute means a higher concentration of water. This is because solute molecules take up space that water molecules would otherwise occupy.

Just like molecules of other substances, molecules of water diffuse from areas of higher concentration to areas of lower concentration. In figure 3.14, the presence of protein in compartment *A* means that the water concentration there is less than the concentration of pure water in compartment *B*. Therefore, water diffuses from compartment *B* across the selectively permeable membrane and into compartment *A*. In other words, water moves from compartment *B* into compartment *A* by osmosis. Protein, on the other hand, cannot diffuse out of compartment *A* because the selectively permeable membrane is impermeable to it.

Note in figure 3.14 that, as osmosis occurs, the water level on side *A* rises. This ability of osmosis to generate enough pressure to lift a volume of water is called *osmotic pressure*. The greater the concentration of nonpermeable solute particles (protein in this case) in a solution, the *lower* the water concentration of that solution and the *greater* the osmotic pressure. Water always tends to diffuse toward solutions of greater osmotic pressure.

Since cell membranes are generally permeable to water, water equilibrates by osmosis throughout the body, and the concentration of water and solutes everywhere in the intracellular and extracellular fluids is essentially the same. Therefore, the osmotic pressure of the intracellular and extracellular fluids is the same. Any solution that has the same osmotic pressure as body fluids is called **isotonic.**

Solutions with a higher osmotic pressure than body fluids are called **hypertonic.** If cells are put into a hypertonic solution, there is a net movement of water by osmosis out of the cells into the surrounding solution, and the cells shrink. Conversely, cells put into a **hypotonic** solution, which has a lower osmotic pressure than body fluids, tend to gain water by osmosis and swell (fig. 3.15).

> The concentration of solute in solutions that are infused into body tissues or blood must be controlled. Otherwise, osmosis may cause cells to swell or shrink, impairing their function. For instance, if red blood cells are placed in distilled water (which is hypotonic to them), water diffuses into the cells, and they burst (hemolyze). On the other hand, red blood cells exposed to 0.9% NaCl solution (normal saline) do not change in shape because this solution is isotonic to human cells. A red blood cell placed in a hypertonic solution shrinks.

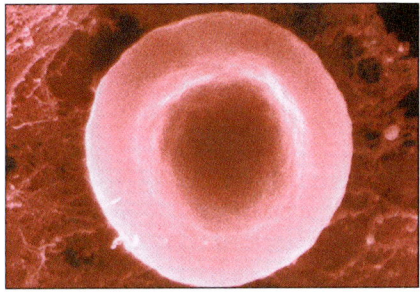
A

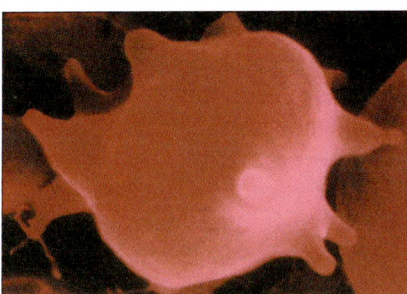
B

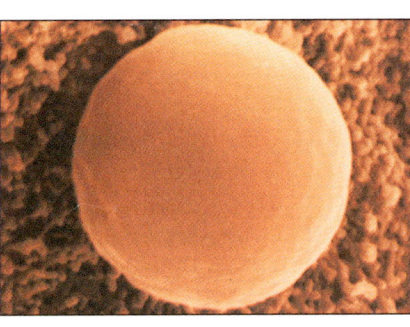
C

Figure 3.15
Red blood cells placed (A) in an isotonic solution, equal volumes of water enter and leave cells, size and shape remain unchanged. (B) In a hypertonic solution, more water leaves than enters, and cells shrink. (C) In a hypotonic solution, more water enters than leaves, cells swell and may burst (5,000×).

Filtration

Molecules pass through membranes by diffusion or osmosis because of random movements. In other instances, the process of **filtration** (fil-tra´shun) forces molecules through membranes.

Filtration is commonly used to separate solids from water. One method is to pour a mixture of solids and water onto filter paper in a funnel. The paper is a porous membrane through which the small water molecules can pass, leaving behind the larger solid particles. Hydrostatic pressure, which is created by the weight of water on the paper due to gravity, forces the water molecules through to the other side. A familiar example of filtration is making coffee by the drip method.

In the body, tissue fluid forms when water and small dissolved substances are forced out through the thin, porous walls of blood capillaries, but larger particles, such as blood protein molecules, are left inside (fig. 3.16). The force for this movement comes from blood pressure, generated mostly by heart action, which is greater within the vessel than outside it. However, the impermeable proteins tend to hold water in blood vessels by osmosis, thus preventing the formation of excess tissue fluid, a condition called **edema.**

 CHECK YOUR RECALL

1. What kinds of substances diffuse most readily through a cell membrane?
2. Explain the differences between diffusion and osmosis.
3. Distinguish among hypertonic, hypotonic, and isotonic solutions.
4. Explain how filtration occurs within the body.

Active Mechanisms

When molecules or ions pass through cell membranes by diffusion, facilitated diffusion, or osmosis, their net movements are from regions of higher concentration toward regions of lower concentration. Sometimes, how-

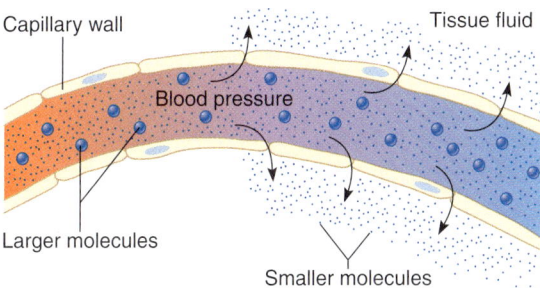

Figure 3.16
In this example of filtration, blood pressure forces smaller molecules through tiny openings in the capillary wall. The larger molecules remain inside.

ever, particles move from a region of lower concentration to one of higher concentration. This requires energy, which comes from cellular metabolism and, specifically, from a molecule called adenosine triphosphate (ATP).

Active Transport

Active transport (ak´tiv trans´port) is a process that moves particles through membranes from a region of lower concentration to a region of higher concentration. Sodium ions, for example, can diffuse passively into cells through protein channels in cell membranes, but their concentration typically remains much greater outside cells than inside. This occurs because active transport continually moves sodium ions through cell membranes from regions of lower concentration (inside) to regions of higher concentration (outside). Equilibrium is never reached.

Active transport is similar to facilitated diffusion in that it uses specific carrier molecules in cell membranes (fig. 3.17). It differs from facilitated diffusion in that particles move from areas of low concentration to areas of high concentration, and energy from ATP is required. Up to 40% of a cell's energy supply may be used to actively transport particles through cell membranes.

The carrier molecules in active transport are proteins with binding sites that combine with the particles being transported. Such a union triggers release of energy, and this alters the shape of the carrier protein. As a result, the "passenger" particles move through the membrane. Once on the other side, the transported particles are released, and the carriers can accept other passenger molecules at that binding site. Because these carrier proteins transport substances from regions of low concentration to regions of high concentration, they are sometimes called "pumps."

Particles that are actively transported across cell membranes include sugars and amino acids as well as sodium, potassium, calcium, and hydrogen ions. Active transport also absorbs nutrient molecules into cells that line intestinal walls.

Endocytosis and Exocytosis

Two processes use cellular energy to move substances into or out of a cell without actually crossing the cell membrane. In **endocytosis** (en´´do-si-to´sis), molecules or other particles that are too large to enter a cell by diffusion, facilitated diffusion, or active transport are conveyed within a vesicle formed from a section of the cell membrane. In **exocytosis** (ex-o-si-to´sis), the reverse process secretes a substance stored in a vesicle from the cell. Nerve cells use exocytosis to release the neurotransmitter chemicals that signal other nerve cells, muscle cells, or glands.

Endocytosis occurs in three forms: pinocytosis, phagocytosis, and receptor-mediated endocytosis. In **pinocytosis** (pi´´no-si-to´sis), meaning "cell drinking," cells take in tiny droplets of liquid from their surroundings, as a small portion of the cell membrane indents. The open end of the tubelike part that forms seals off and produces a small vesicle, which detaches from the surface and moves into the cytoplasm. Eventually, the vesicular membrane breaks down, and the liquid inside becomes part of the cytoplasm. In this way, a cell can take in water and the particles dissolved in it, such as proteins, that otherwise might be too large to enter.

Phagocytosis (fag´´o-si-to´sis), meaning "cell eating," is similar to pinocytosis, but the cell takes in solids rather than liquids. Certain kinds of white blood cells are called *phagocytes* because they can take in solid particles such as bacteria and cellular debris. When a phagocyte first encounters a particle, the particle attaches to the phagocyte's cell membrane. This stimulates a portion of the membrane to project outward, surround the particle, and slowly draw it inside the cell. The part of the membrane surrounding the particle detaches from the cell's surface, forming a vesicle that contains the particle (fig. 3.18).

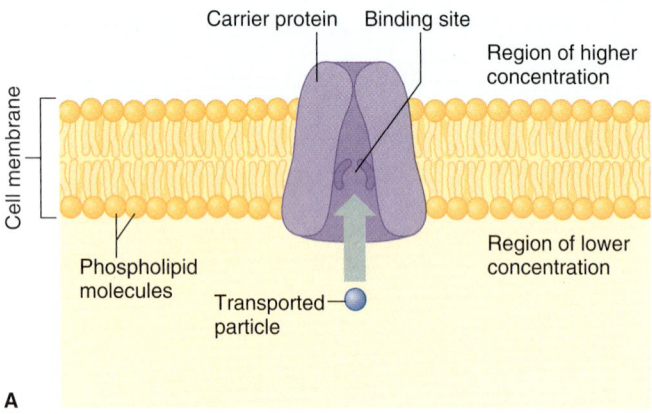

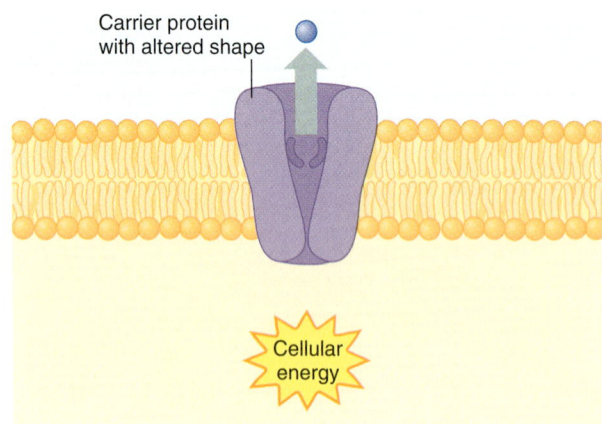

Figure 3.17
Active transport. (A) During active transport, a molecule or an ion combines with a carrier protein, whose shape changes as a result. (B) This process, which requires cellular energy, transports the particle across the cell membrane.

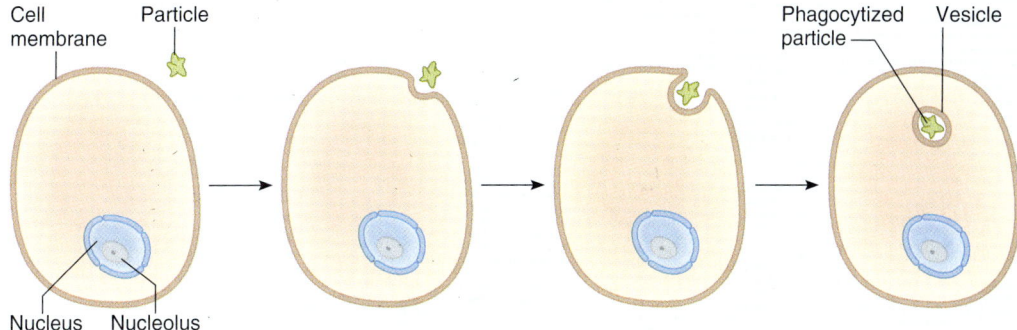

Figure 3.18
A cell may take in a solid particle from its surroundings by phagocytosis.

TABLE 3.2 — MOVEMENTS THROUGH CELL MEMBRANES

PROCESS	CHARACTERISTICS	SOURCE OF ENERGY	EXAMPLE
Passive mechanisms			
Diffusion	Molecules or ions move from regions of higher concentration toward regions of lower concentration.	Molecular motion	Exchange of oxygen and carbon dioxide in lungs
Facilitated diffusion	Carrier molecules move molecules through a membrane from a region of higher concentration to one of lower concentration.	Molecular motion	Movement of glucose through cell membrane
Osmosis	Water molecules move from regions of higher concentration toward regions of lower concentration through a selectively permeable membrane.	Molecular motion	Distilled water entering a cell
Filtration	Molecules are forced from regions of higher pressure to regions of lower pressure.	Hydrostatic pressure	Water molecules leaving blood capillaries
Active mechanisms			
Active transport	Carrier molecules move molecules or ions through membranes from regions of lower concentration toward regions of higher concentration.	Cellular energy (ATP)	Movement of various ions, sugars, and amino acids through membranes
Endocytosis			
Pinocytosis	Membrane engulfs droplets of liquid from surroundings.	Cellular energy	Membrane forming vesicles containing liquid and dissolved particles
Phagocytosis	Membrane engulfs particles from surroundings.	Cellular energy	White blood cell engulfing bacterial cell
Receptor-mediated endocytosis	Receptors bind specific ligands, and they are drawn into the cell.	Cellular energy	Cholesterol molecules entering cells
Exocytosis	Vesicle fuses with membrane to expel substances from cell.	Cellular energy	Secretion of certain hormones

Commonly, a lysosome then combines with such a newly formed vesicle, and the lysosomal digestive enzymes decompose the contents. The products of this decomposition may diffuse out of the lysosome and into the cytoplasm. Exocytosis usually expels remaining residue from the cell.

Pinocytosis and phagocytosis engulf any molecules in the vicinity of the cell membrane. In contrast, **receptor-mediated endocytosis** moves very specific kinds of particles into the cell (fig. 3.19) by binding them first. In this process, protein molecules extend through a portion of the cell membrane to the outer surface, where they serve as receptors to which only specific substances (ligands) from outside the cell can bind. Cholesterol molecules enter cells by this mechanism. Table 3.2 summarizes the types of movements into and out of cells.

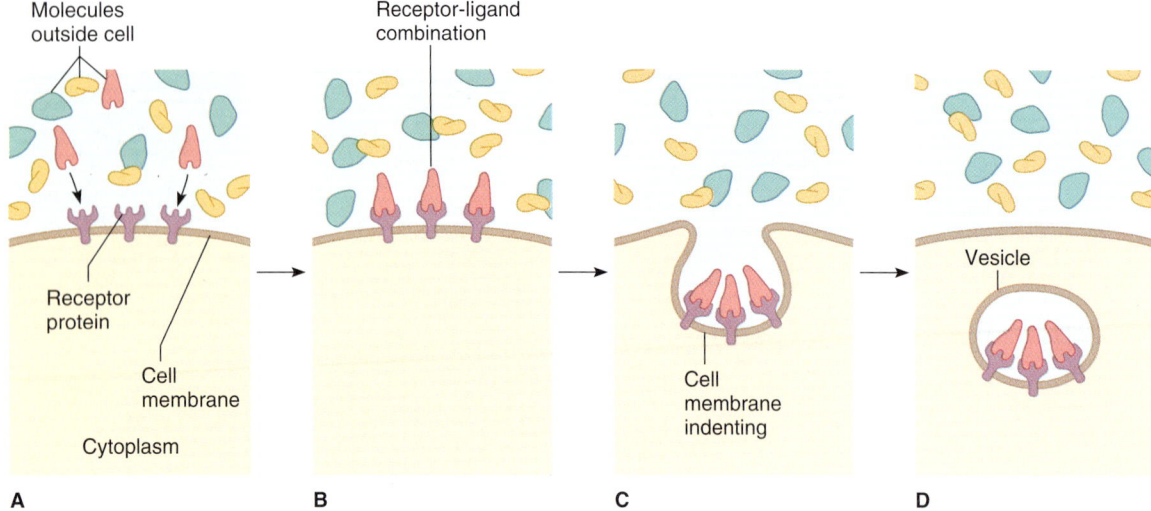

Figure 3.19
Receptor-mediated endocytosis. (*A*) A specific substance binds to a receptor site protein. (*B, C*) The combination of the substance with the receptor site protein stimulates the cell membrane to indent. (*D*) The resulting vesicle transports the substance into the cytoplasm.

 CHECK YOUR RECALL

1. What type of mechanism maintains unequal concentrations of ions on opposite sides of a cell membrane?
2. How are facilitated diffusion and active transport similar? How are they different?
3. What is the difference between endocytosis and exocytosis?
4. How is receptor-mediated endocytosis more specific than pinocytosis or phagocytosis?

3.4 The Cell Cycle

The series of changes that a cell undergoes from the time it forms until it divides is called the *cell cycle* (fig. 3.20). Superficially, this cycle seems rather simple: A newly formed cell grows for a time and then divides to form two new cells, which in turn may grow and divide. Yet, the stages of the cycle are quite complex and include interphase, mitosis, cytoplasmic division (cytokinesis), and differentiation. Several events called *checkpoints* control the cell cycle. Of particular importance is the restriction checkpoint that determines a cell's fate—whether it will continue in the cell cycle and divide, move into a nondividing stage as a specialized cell, or die.

The cell cycle is very highly regulated. Stimulation, such as from a hormone, may trigger cell division. This occurs, for example, when the breasts develop into milk-producing glands during pregnancy.

Cells do not normally divide continually. Most types of human cells, if grown in the laboratory, divide only

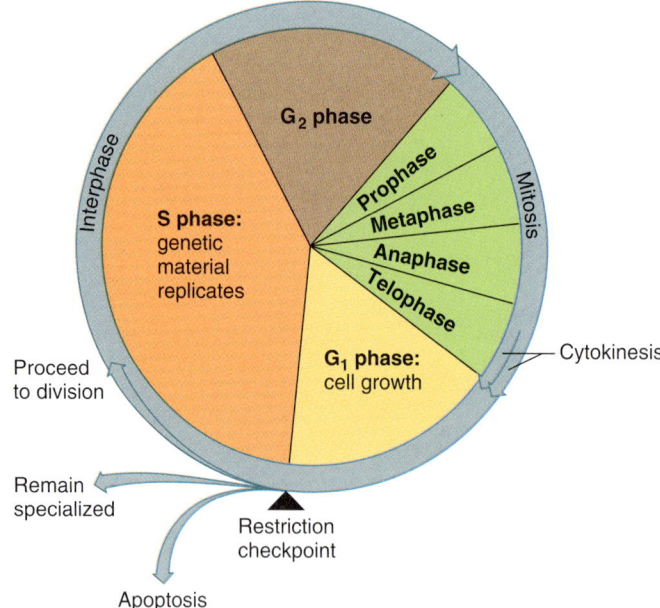

Figure 3.20
The cell cycle is divided into interphase, when cellular components duplicate, and cell division (mitosis and cytokinesis), when the cell splits in two, distributing its contents into two cells. Interphase is divided into two gap phases (G_1 and G_2), when specific molecules and structures duplicate, and a synthesis phase (S), when the genetic material replicates. Mitosis can be described as consisting of stages—prophase, metaphase, anaphase, and telophase.

forty to sixty times. Presumably, such limits operate in the body too. Some cells may divide the maximum number of times, such as cells that line the small intestine. Others normally do not divide, such as nerve cells. A cell "knows" when to stop dividing because of a built-in "clock" in the form of the chromosome tips. These structures, called *telomeres,* shorten with each mitosis. When they shorten to a certain length, the cell ceases dividing.

Interphase

Before a cell actively divides, it must grow and duplicate much of its contents, so that two cells can form from one. This period of preparedness is called **interphase.**

Once thought to be a time of rest, interphase is actually a time of great synthetic activity. During interphase, the cell obtains nutrients, utilizes them to manufacture new living material, and maintains routine "housekeeping" functions. The cell duplicates membranes, ribosomes, lysosomes, and mitochondria. Perhaps most importantly, the cell in interphase takes on the tremendous task of replicating its genetic material. This is important so that each of the two new cells will have a complete set of genetic instructions.

Interphase is considered in phases. DNA is replicated during the S, or synthesis phase, which is bracketed by two gap or growth periods, called G_1 and G_2, when other structures are duplicated.

Mitosis

A cell can divide in two ways. The first process, called *meiosis,* is part of *gametogenesis,* the formation of egg cells (in the female) and sperm cells (in the male). Since an egg fertilized by a sperm must have the normal complement of 46 chromosomes, both the egg and the sperm must first halve their normal chromosome number to 23 chromosomes. Meiosis, through a process called reduction division, accomplishes this. Chapter 19 (p. 500) describes meiosis in detail.

The second form of cell division increases cell number, which is necessary for growth and development and for wound healing. It consists of two separate processes: (1) division of the nucleus, called **mitosis** (mi-to´sis), and (2) division of the cytoplasm, called **cytokinesis** (si´´to-ki-ne´sis).

Division of the nucleus must be very precise because it contains the DNA. Each new cell resulting from mitosis must have a complete and accurate copy of this information to survive. DNA replicates during interphase, but is equally distributed into two new cells in mitosis.

Although mitosis is often described in terms of stages, the process is continuous, without marked changes between one step and the next (fig. 3.21). Stages, however, indicate the sequence of major events. They include:

1. **Prophase** One of the first indications that a cell is going to divide is that the chromosomes become visible in the nucleus. Because the cell has gone through S phase each prophase chromosome is composed of two identical portions (chromatids), which are temporarily attached at a region on each called the *centromere.*

    The centrioles of the centrosome replicate just before mitosis begins. During prophase, the two newly formed centriole pairs move to opposite ends of the cell. Soon, the nuclear envelope and the nucleolus break up, disperse, and are no longer visible. Microtubules are assembled from tubulin proteins in the cytoplasm and associate with the centrioles and chromosomes. A spindle-shaped array of microtubules (spindle fibers) forms between the centrioles as they move apart.

2. **Metaphase** The chromosomes line up about midway between the centrioles, as a result of microtubule activity. Spindle fibers attach to the centromeres of each chromosome so that a fiber from one pair of centrioles contacts one centromere, and a fiber from the other pair of centrioles attaches to the other centromere.

3. **Anaphase** Soon the centromeres are pulled apart. As the chromatids separate, they become individual chromosomes. The separated chromosomes now move in opposite directions, once again guided by microtubule activity. The spindle fibers shorten and pull their attached chromosomes toward the centrioles at opposite ends of the cell.

4. **Telophase** The final stage of mitosis begins when the chromosomes complete their migration toward the centrioles. It is much like prophase, but in reverse. As the chromosomes approach the centrioles, they begin to elongate and unwind from rodlike into the threadlike fibers of chromatin. A nuclear envelope forms around each chromosome set, and nucleoli appear within the newly formed nuclei. Finally, the microtubules disassemble into free tubulin molecules.

Cytoplasmic Division

Cytoplasmic division (cytokinesis) begins during anaphase, when the cell membrane starts to constrict down the middle of the cell. This constriction continues through telophase. Contraction of a ring of microfilaments, which assemble in the cytoplasm and attach to the inner surface of the cell membrane, divides the cytoplasm. The contractile ring lies at right angles to the microtubules that pulled the chromosomes to opposite sides of the cell. The ring pinches inward, separating the two newly formed nuclei and distributing about half

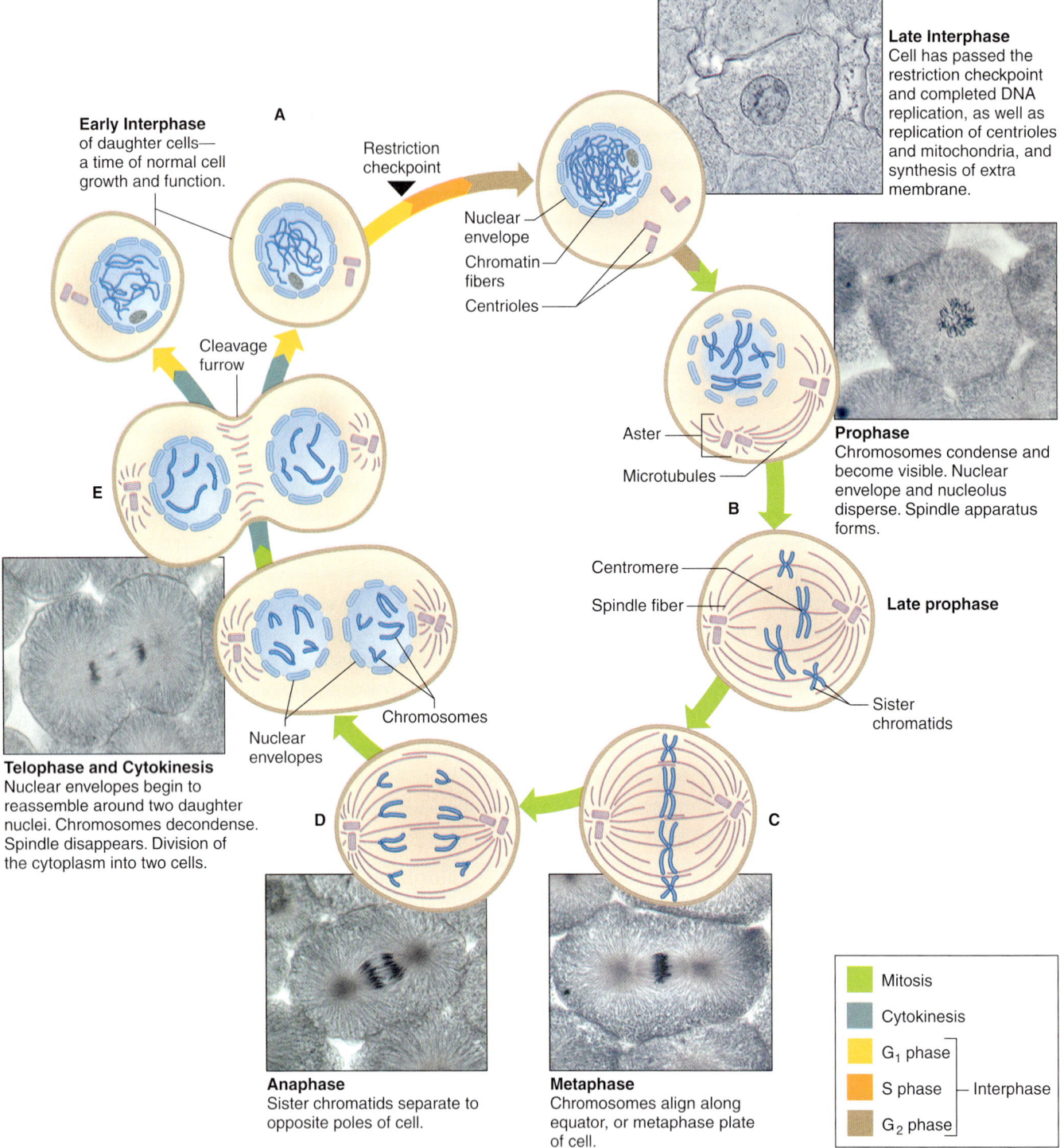

Figure 3.21
Mitosis and cytokinesis. (A) During interphase, before mitosis, chromosomes are visible only as chromatin fibers. A single pair of centrioles is present, but not visible at this magnification. (B) In prophase, as mitosis begins, chromosomes have condensed and are easily visible when stained. The centrioles have replicated, and each pair moves to an opposite end of the cell. The nuclear envelope and nucleolus disappear, and spindle fibers associate with the centrioles and the chromosomes. (C) In metaphase, the chromosomes line up midway between the centrioles. (D) In anaphase, the centromeres are pulled apart by the spindle fibers, and the chromatids, now individual chromosomes, move in opposite directions. (E) In telophase, chromosomes complete their migration and become chromatin, the nuclear envelope reforms, and microtubules disassemble. Cytokinesis, which actually began during anaphase, continues during telophase. Not all chromosomes are shown in these drawings. (Micrographs approximately 360×)

Topic of Interest

CANCER

Cancer is a group of closely related diseases that can occur in many different tissues. One in three of us will develop some form of cancer. These conditions result from changes in cells that alter the cell cycle. Cancers share the following characteristics:

1. **Hyperplasia** Hyperplasia is uncontrolled cell division. Normal cells divide a set number of times, signaled by the shortening of chromosome tips. Cancer cells activate an enzyme, called *telomerase*, that continually rebuilds chromosomes, so that cells are not signaled to stop dividing.
2. **Dedifferentiation** Cancer cells typically resemble undifferentiated cells, a state termed dedifferentiation. Cancer cells lose the specialized structures and functions of the normal type of cell from which they descend, and are therefore said to be dedifferentiated. Cancer cells also grow into disorganized masses, rather than forming orderly groups as normal cells do.
3. **Invasiveness** Cancer cells break through boundaries, called *basement membranes*, which separate cell types within some organs.
4. **Angiogenesis** Cancer cells induce the formation of blood vessels, which nourish them and remove wastes, enabling the cancer to persist, grow, and spread.
5. **Metastasis** Metastasis is a tendency to spread into other tissues. Normal cells usually aggregate in groups of similar kinds. Small numbers of cancer cells can detach from their original mass and move from their place of origin (fig. 3A), often into the bloodstream. They may establish new tumors elsewhere in the body.

Mutations in certain genes cause many cancers. A cancer-causing mutation may activate a cancer-causing oncogene (a gene that normally controls mitotic rate) or inactivate a protective gene called a tumor suppressor. Most often, a person inherits one abnormal cancer-causing gene. When the second copy of that gene changes (mutates), sometimes in response to an environmental trigger, cancer develops. Some cancers result from a series of genetic changes.

Figure 3A
The lack of cilia on these cancer cells is one sign of their dedifferentiation (2,250×).

of the organelles into each new cell. The new cells may differ slightly in size and number of organelles, but they contain identical genetic information.

Cell Differentiation

All body cells form by mitosis and contain the same DNA information, yet they do not look or function alike. A fertilized egg cell divides to form two new cells; they, in turn, divide into four cells, the four yield eight, and so forth. Then, sometime during development, the cells begin to *specialize* (fig. 3.22). That is, they develop special structures or begin to function in different ways. Some cells become muscle cells, others become bone cells, and still others become nerve cells.

The process by which cells develop different characteristics in structure and function is called **differentiation** (dif″er-en″she-a′shun). Cells differentiate by expressing some of the DNA information and repressing other DNA information. For example, the DNA information required for general cell activities is activated in both nerve and bone cells, but information important to specific bone cell functions is activated in bone cells, yet repressed in nerve cells. Similarly, the information necessary for specific nerve cell functions is repressed in bone cells. By birth, a human has more than 260 types of specialized cells. Throughout life, certain cells, called stem cells, remain relatively unspecialized. These cells retain the potential to produce new differentiated cells. Stem cells are important in growth and healing.

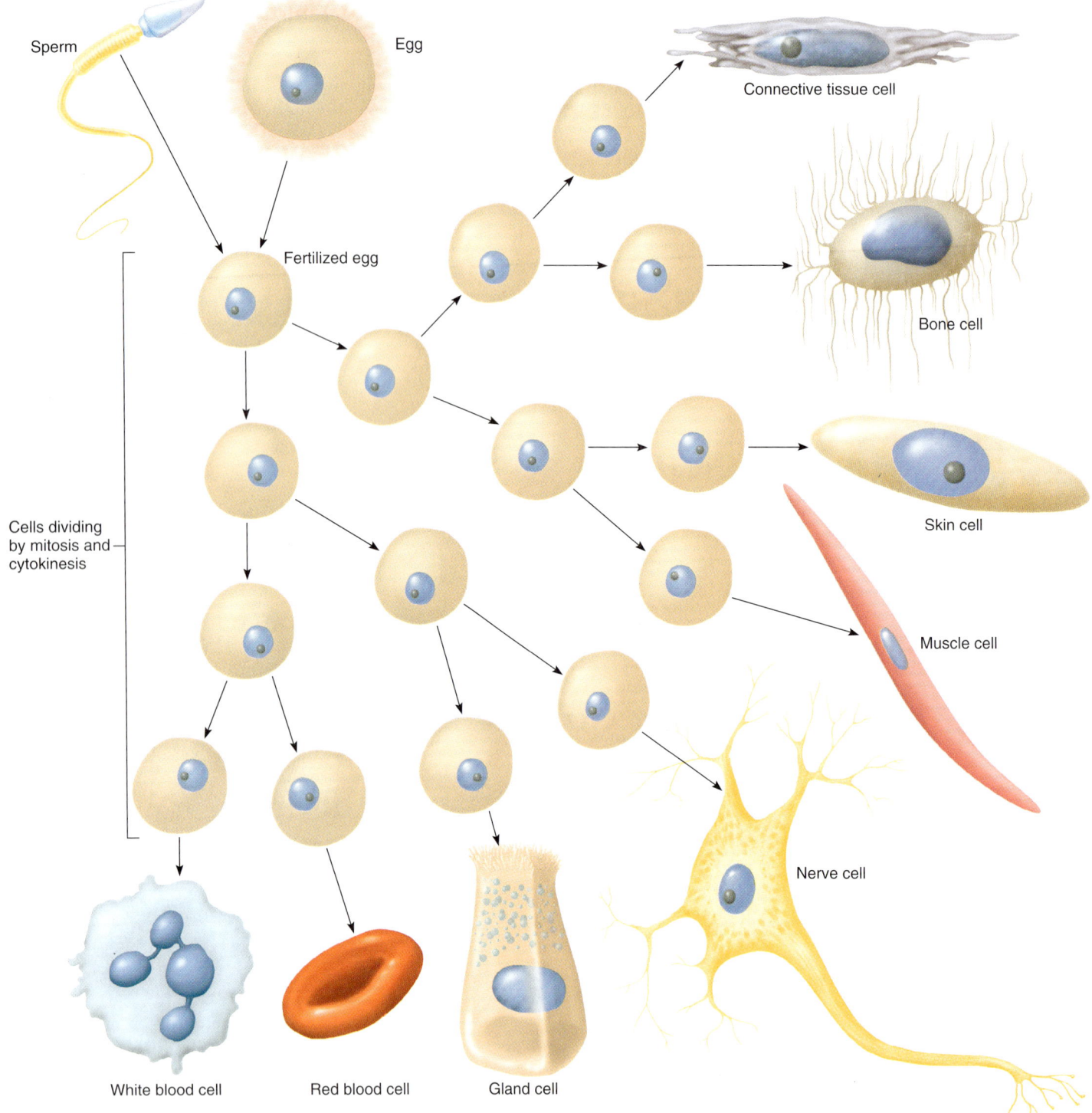

Figure 3.22
During development, many body cells are produced from a single fertilized egg cell by mitosis. As these cells differentiate, they become different kinds of cells with specialized functions. (Relative cell sizes are not to scale.)

Cell Death

A cell that does not divide or specialize has another option—it may die. **Apoptosis** (ap″o-to′-sis) is a form of cell death that is actually a normal part of development, sculpting organs from overgrown tissues. In the fetus, apoptosis carves away webbing between developing fingers and toes, removes extra brain cells, and preserves only those immune system cells that recognize the body's cell surfaces. After birth, apoptosis follows a sunburn—it literally peels away cells so damaged that they might otherwise turn cancerous.

A cell in the throes of apoptosis goes through characteristic steps. It rounds up and bulges, the nuclear membrane breaks down, chromatin condenses, and enzymes chop the chromosomes into many equal-sized pieces. Finally, the cell shatters into membrane-enclosed fragments, and a scavenger cell mops it up.

CHECK YOUR RECALL

1. Why is it important that the division of nuclear materials during mitosis be precise?
2. Describe the events that occur during mitosis.
3. Name the process by which cells specialize.
4. How is cell death a normal part of development?

Clinical Connection

In certain highly specialized cell types, endocytosis and exocytosis meet in a process called, appropriately, transcytosis. A particle enters such a cell by endocytosis, journeys through the cytoplasm, then exits the cell by exocytosis from the other end.

Transcytosis occurs in the lining of the small intestine, where rare cells called M cells sample bits of food, transporting the captured molecules through themselves, then ejecting them by exocytosis to be met by a gathering of immune system cells just beneath the cell's basal surface. From here, if the transported particles are recognized as posing a threat, they may stimulate an immune response, and other cells flood the small intestinal lining with specific antibodies against the potentially dangerous substance. This transcytosis through the M cell portals of the small intestinal lining thus ensures that what is eaten is safe. The infectious prion proteins, discussed in the Clinical Connection in Chapter 2 (p. 44), can enter the body by evading the M cell barrier in the small intestine. Variant Creutzfeldt-Jakob disease may be transmitted this way from infected beef. Before the 1970s, cannibals in Papua New Guinea contracted a prion disorder called kuru by consuming their war heroes.

Another example of transcytosis occurs in the female genital tract and in the rectum. HIV (the virus that causes AIDS) uses this route to cross the epithelium, and reach the bloodstream.

SUMMARY OUTLINE

3.1 Introduction (p. 49)
Cells vary considerably in size, shape, and function. The shapes of cells make possible their functions.

3.2 Composite Cell (p. 49)
A cell includes a nucleus, cytoplasm, and a cell membrane. Organelles perform specific functions; the nucleus controls overall cell activities because it contains DNA.

1. Cell membrane
   a. The cell membrane forms the outermost limit of the living material.
   b. It is a selectively permeable passageway that controls the entrance and exit of substances. Its molecules transmit signals.
   c. The cell membrane includes protein, lipid, and carbohydrate molecules.
   d. The cell membrane's framework is mainly a bilayer of phospholipid molecules.
   e. Molecules that are soluble in lipids pass through the cell membrane easily, but water-soluble molecules do not.
   f. Proteins function as receptors on membrane surfaces and form channels for the passage of ions and molecules. They can be classified by position.
   g. Carbohydrates associated with membrane proteins enable certain cells to recognize one another.
2. Cytoplasm
   a. Cytoplasm contains membranes, organelles, and the rods and tubules of the cytoskeleton, suspended in cytosol.
   b. The endoplasmic reticulum is a tubular communication system in the cytoplasm that transports lipids and proteins.
   c. Ribosomes function in protein synthesis.
   d. The Golgi apparatus packages glycoproteins for secretion.
   e. Mitochondria contain enzymes that catalyze reactions that release energy from nutrient molecules.
   f. Lysosomes contain digestive enzymes that decompose substances.
   g. Peroxisomes house enzymes that catalyze bile acid synthesis, hydrogen peroxide degradation, lipid breakdown, and detoxification of alcohol.
   h. Microfilaments (built of actin) and microtubules (built of tubulin) aid cellular movements and support and stabilize the cytoplasm and organelles. Together they form the cytoskeleton. Microtubules also form centrioles, cilia, and flagella.
   i. The centrosome contains centrioles that aid in distributing chromosomes during cell division.
   j. Cilia and flagella are motile extensions from cell surfaces.
   k. Vesicles contain substances that recently entered the cell or that are to be secreted from the cell.
3. Cell nucleus
   a. The nucleus is enclosed in a double-layered nuclear envelope.
   b. It contains a nucleolus, which is the site of ribosome production.
   c. It contains chromatin, which is composed of loosely coiled fibers of DNA and protein. As chromatin fibers condense, chromosomes become visible during cell division.

3.3 Movements Through Cell Membranes (p. 58)
The cell membrane is a barrier through which substances enter and leave a cell.

1. Passive mechanisms do not require cellular energy.
   a. Diffusion
      (1) Diffusion is the movement of molecules or ions from regions of higher concentration toward regions of lower concentration.
      (2) It exchanges oxygen and carbon dioxide.

b. Facilitated diffusion
   (1) In facilitated diffusion, special carrier molecules move substances through the cell membrane.
   (2) This process moves substances only from regions of higher concentration toward regions of lower concentration.
c. Osmosis
   (1) Osmosis is diffusion of water molecules from regions of higher water concentration toward regions of lower water concentration through a selectively permeable membrane.
   (2) Osmotic pressure increases as the number of impermeable particles dissolved in a solution increases.
   (3) A solution is isotonic to a cell when it has the same osmotic pressure as the cell.
   (4) Cells lose water when placed in hypertonic solutions and gain water when placed in hypotonic solutions.
d. Filtration
   (1) Filtration is the movement of molecules from regions of higher hydrostatic pressure toward regions of lower hydrostatic pressure.
   (2) Blood pressure causes filtration through porous capillary walls, forming tissue fluid.

2. Active mechanisms require cellular energy.
   a. Active transport
      (1) Active transport moves molecules or ions from regions of lower concentration toward regions of higher concentration.
      (2) It requires cellular energy from ATP and carrier molecules in the cell membrane.
   b. Endocytosis and exocytosis
      (1) Endocytosis may convey relatively large particles into a cell. Exocytosis is the reverse of endocytosis.
      (2) In pinocytosis, a cell membrane engulfs tiny droplets of liquid.
      (3) In phagocytosis, a cell membrane engulfs solid particles.
      (4) Receptor-mediated endocytosis moves specific types of particles into cells.

3.4 The Cell Cycle (p. 64)

The cell cycle includes interphase, mitosis, cytoplasmic division, and differentiation. It is highly regulated.

1. Interphase
   a. During interphase, a cell duplicates membranes, ribosomes, organelles, and DNA.
   b. Interphase terminates when mitosis begins.
2. Mitosis
   a. Meiosis is a form of cell division that forms sex cells.
   b. Mitosis is the division and distribution of genetic material, organelles, and other structures to new cells.
   c. The stages of mitosis are prophase, metaphase, anaphase, and telophase.
3. Cytoplasmic division distributes cytoplasm into two portions following mitosis.
4. Cell differentiation is the development of specialized structures and functions.
5. A cell that does not divide or differentiate may undergo apoptosis, a form of cell death that is a normal part of development.

REVIEW EXERCISES

1. Describe how the shapes of nerve and muscle cells are important for their functions. (p. 49)
2. Name the three major portions of a cell, and describe their relationships to one another. (p. 49)
3. Define *selectively permeable*. (p. 51)
4. Describe the chemical structure of a cell membrane. (p. 51)
5. Explain how the structure of a cell membrane determines which substances can pass through it. (p. 51)
6. Describe the structures and functions of each of the following (p. 52):
   a. Endoplasmic reticulum
   b. Ribosomes
   c. Golgi apparatus
   d. Mitochondria
   e. Lysosomes
   f. Peroxisomes
   g. Microfilaments
   h. Microtubules
   i. Centrosome
   j. Cilia and flagella
   k. Vesicles
7. Describe the structure and contents of the nucleus. (p. 56)
8. Define *diffusion*. (p. 58)
9. Explain how diffusion aids in the exchange of gases within the body. (p. 59)
10. Distinguish between diffusion and facilitated diffusion. (p. 59)
11. Define *osmosis*. (p. 60)
12. Define *osmotic pressure*. (p. 60)
13. Distinguish between solutions that are hypertonic, hypotonic, and isotonic. (p. 60)
14. Define *filtration*. (p. 61)
15. Explain how filtration moves substances through capillary walls. (p. 61)
16. Distinguish between facilitated diffusion and active transport. (p. 62)
17. Distinguish between pinocytosis and phagocytosis. (p. 62)
18. Identify the structures that make receptor-mediated endocytosis specific. (p. 63)
19. List the phases of the cell cycle. (p. 64)
20. Explain what happens during interphase. (p. 65)
21. Name the two types of cell division. (p. 65)
22. Describe the major events of mitosis. (p. 65)
23. Explain how the cytoplasm divides. (p. 65)
24. Define *differentiation*. (p. 67)
25. Describe apoptosis. (p. 68)

CRITICAL THINKING

1. Organelles compartmentalize a cell, much as a department store displays related items together. What advantage does such compartmentalization offer a large cell? Cite two examples of organelles and the activities they compartmentalize.
2. Liver cells are packed with glucose. What mechanism could be used to transport more glucose into a liver cell? Why would only this mode of transport work?
3. In an inherited condition called glycogen cardiomyopathy, teenagers develop muscle weakness, which affects the heart as well as other muscles. Samples of the affected muscle cells contain huge lysosomes, swollen with the carbohydrate glycogen. How might this condition arise?
4. Why does a muscle cell contain many mitochondria, and a white blood cell contain many lysosomes?

5. Many cancer drugs stop working because a large protein, called P-gp, in the cell membranes of some cancer cells pumps many types of drugs out of the cell before they can enter the cytoplasm. P-gp requires ATP to function, and it continually cycles from its site of synthesis in the cytoplasm, to the cell membrane, and then back inside the cell. What structures most likely transport P-gp?

6. Which process—diffusion, osmosis, or filtration—is utilized in the following situations?
   a. Injection of a drug that is hypertonic to the tissues stimulates pain.
   b. The urea concentration in the dialyzing fluid of an artificial kidney is decreased.

7. What characteristic of cell membranes may explain why fat-soluble substances like chloroform and ether rapidly affect cells?

8. Exposure to tobacco smoke immobilizes cilia, and they eventually disappear. How might this effect explain why smokers have an increased incidence of coughing and respiratory infections?

9. How would a drug that functions as an angiogenesis inhibitor be useful in treating cancer?

WEB CONNECTIONS

Visit the website for additional study questions and more information about this chapter at:

http://www.mhhe.com/shieress8

chapter 4

Cellular Metabolism

ARSENIC POISONING. In 1821, Napoleon Bonaparte died after six years in exile on the remote British island of St. Helena in the south Atlantic Ocean. The causes of death were officially listed as a perforated stomach ulcer and cancer. But strands of his hair that were saved suggest another scenario—arsenic poisoning. Arsenic kills by blocking a key step in the breakdown of nutrients so that they cannot release the energy held in their chemical bonds.

Arsenic has been a common way to poison people since the Middle Ages, because the arsenic oxide form that is left in the body is difficult to detect. So frequently was arsenic poisoning used to conveniently dispose of wealthy elders that it was once called "inheritance powder." Given in one large dose, arsenic poisoning causes chest pain, vomiting, diarrhea, shock, coma, and death. In contrast, giving many small doses causes tingly, dark skin lesions, numb hands and feet, and eventually skin cancers. Such gradual poisoning, called arsenicosis, may occur from contact with pesticides or environmental pollutants. Today, hundreds of thousands of people in India and Bangladesh are suffering from arsenicosis, poisoned from naturally occurring arsenic in well water. Ironically, the wells were dug to provide fresh drinking water, which was in short supply. The first cases of arsenic poisoning appeared in the early 1980s, and many more are expected.

Arsenic damages the body by binding to sulfur bonds that hold together certain proteins. It particularly targets an enzyme that helps the breakdown products of glucose enter the mitochondria, where energy is extracted. Deprived of this enzyme, cells run out of energy. Arsenic also has noticeable effects on hair, because the hair protein keratin has many sulfur atoms. A forensic analysis of hair can even reveal intermittent poisoning attempts by detecting arsenic in different parts of a long hair. In one famous case from the 1950s, hair from a dead woman showed arsenic corresponding to the times her husband had been home on leave from the military—he had slowly poisoned her!

Forensic analysis of Napoleon's hair in the 1960s revealed large amounts of arsenic. His will, which predicted that he would be "murdered by the British oligarchy," might have been accurate.

Photo:
Forensic evidence suggests that Napoleon Bonaparte died from arsenic poisoning. The arsenic blocked the transport of the breakdown products of glucose molecules into mitochondria. This effectively—and fatally—shut down his energy metabolism.

Chapter Objectives

After studying this chapter, you should be able to do the following:

4.1 Introduction
1. Briefly explain the role of genes in cellular metabolism. (p. 74)

4.2 Metabolic Reactions
2. Define anabolism and catabolism. (p. 74)

4.3 Control of Metabolic Reactions
3. Explain how enzymes control metabolic reactions. (p. 75)

4.4 Energy for Metabolic Reactions
4. Explain how cellular respiration releases chemical energy. (p. 77)
5. Describe how energy becomes available for cellular activities. (p. 78)

4.5 Metabolic Pathways
6. Describe the general metabolic pathways of carbohydrates, lipids, and proteins. (p. 79)

4.6 Nucleic Acids
7. Explain how nucleic acid molecules (DNA and RNA) store and carry genetic information. (p. 79)

4.7 Protein Synthesis
8. Explain how genetic information controls cellular processes. (p. 82)

4.8 DNA Replication
9. Describe how DNA molecules replicate. (p. 86)

Aids to Understanding Words

an- [without] *an*aerobic respiration: Respiratory process that proceeds without oxygen.

ana- [up] *ana*bolism: Cellular processes that use smaller molecules to build larger ones.

cata- [down] *cata*bolism: Cellular processes that break larger molecules into smaller ones.

mut- [change] *mut*ation: Change in a nucleic acid sequence.

-zym [causing to ferment] en*zym*e: Protein that initiates or speeds a chemical reaction without being consumed.

Key Terms

aerobic (a″er-o´bik)
anabolism (an″ah-bol´lizm)
anaerobic (an″a-er-o´bik)
anticodon (an´tĭ-ko´don)
catabolism (kat″ah-bol´lizm)
codon (ko´don)
dehydration synthesis (de″hi-dra´shun sin´thĕ-sis)
enzyme (en´zīm)
gene (jēn)
hydrolysis (hi-drol´ĭ-sis)
oxidation (ok″sĭ-da´shun)
replication (rĕ″plĭ-ka´shun)
substrate (sub´strāt)

4.1 Introduction

Cells require energy and information to build bodies. Cells house the many chemical reactions of metabolism. These reactions break down nutrients to release energy and also build molecules to store energy. Cells carry genetic information that encodes the amino acid sequences of proteins.

A cell is the site of many metabolic reactions that provide the energy to maintain life. A special type of protein called an **enzyme** (en´zīm) controls each of the interrelated reactions of metabolism. Genes carry the information that instructs a cell to manufacture particular proteins.

4.2 Metabolic Reactions

Metabolic reactions are of two major types. **Anabolism** (an´´ah-bol´lizm), the buildup of larger molecules from smaller ones, requires energy. **Catabolism** (kat´´ah-bol´lizm), the breakdown of larger molecules into smaller ones, releases energy.

Anabolism

Anabolism provides the biochemicals required for cell growth and repair. For example, cells join many simple sugar molecules (monosaccharides) into a chain to form larger molecules of glycogen, a carbohydrate, using an anabolic process called **dehydration synthesis** (de´´hi-dra´shun sin´thĕ-sis). As adjacent monosaccharide units join, an —OH (hydroxyl group) from one monosaccharide molecule and an —H (hydrogen atom) from an —OH group of another are removed. The —H and —OH react to produce a water molecule, and the monosaccharides are joined by a shared oxygen atom (fig. 4.1). As this process repeats, the molecular chain grows.

Dehydration synthesis also links glycerol and fatty acid molecules in fat (adipose) cells to form fat molecules (triglycerides). In this case, three hydrogen atoms are removed from a glycerol molecule, and an —OH group is removed from each of the three fatty acid molecules (fig. 4.2). The result is three water molecules and a single fat molecule, whose glycerol and fatty acid portions are bound by shared oxygen atoms.

Cells also use dehydration synthesis to join amino acid molecules, which eventually form protein molecules. When two amino acid molecules unite, an —OH from one and an —H from the —NH_2 group of another are removed. A water molecule forms, and the amino acid molecules are joined by a bond between a carbon atom and a nitrogen atom, called a *peptide bond* (fig. 4.3). Two bound amino acids form a *dipep-*

Figure 4.1
A disaccharide is formed from two monosaccharides in a dehydration synthesis reaction.

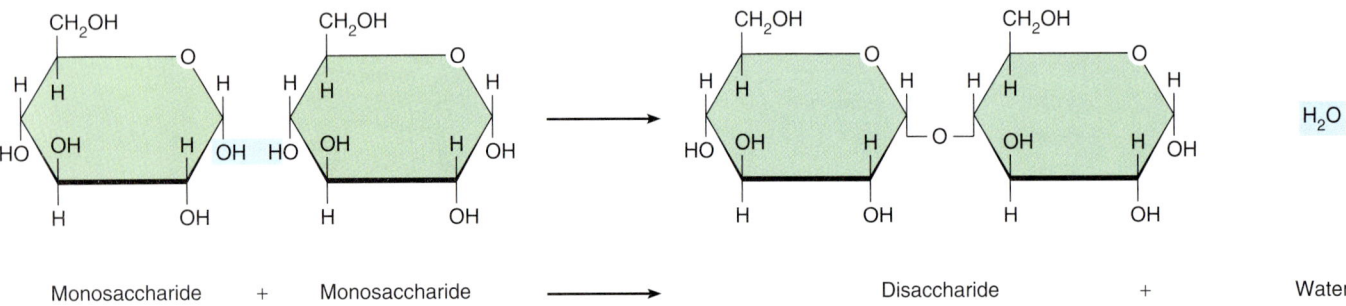

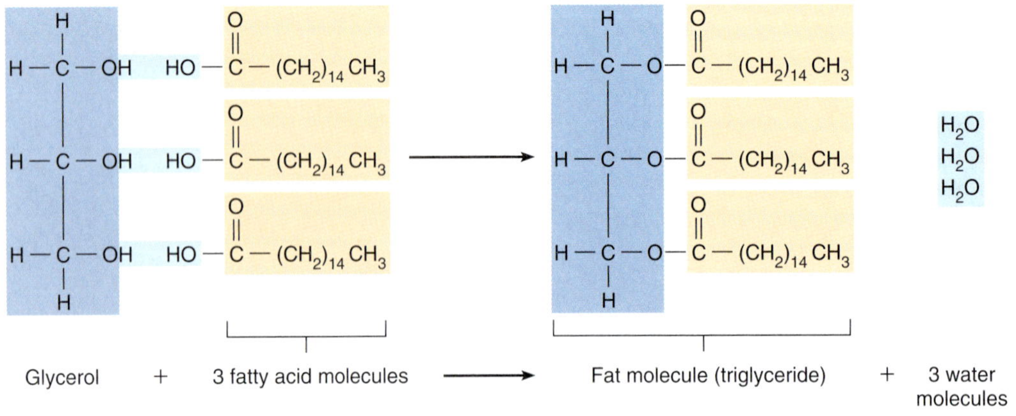

Figure 4.2
A glycerol molecule and three fatty acid molecules form a fat molecule (triglyceride) in a dehydration synthesis reaction.

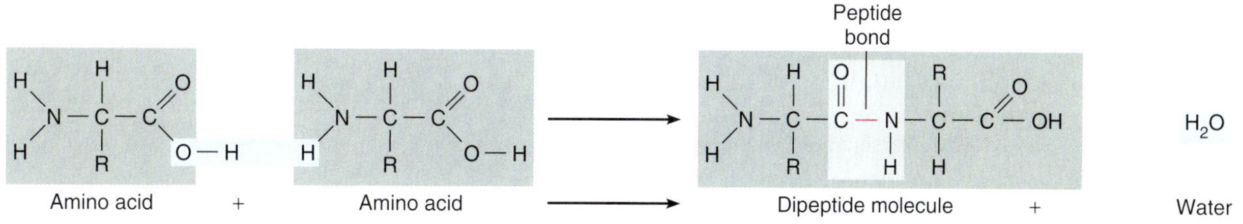

Peptide bond

Amino acid + Amino acid → Dipeptide molecule + Water

Figure 4.3
When dehydration synthesis unites two amino acid molecules, a peptide bond forms between a carbon atom and a nitrogen atom, resulting in a dipeptide.

tide, and many joined in a chain form a *polypeptide*. Generally, a polypeptide that has a specific function and consists of perhaps 100 or more amino acid molecules is called a *protein*, although the boundary distinguishing between polypeptides and proteins is not defined precisely.

Catabolism

Physiological processes that break larger molecules into smaller ones constitute catabolism. An example of catabolism is **hydrolysis** (hi-drol´ĭ-sis), which decomposes carbohydrates, lipids, and proteins, and splits a water molecule in the process. Hydrolysis of a disaccharide such as sucrose, for instance, yields two monosaccharides—glucose and fructose—as a molecule of water splits:

$$C_{12}H_{22}O_{11} + H_2O \rightarrow C_6H_{12}O_6 + C_6H_{12}O_6$$
(Sucrose) (Water) (Glucose) (Fructose)

In this case, the bond between the simple sugars within the sucrose molecule breaks, and the water molecule supplies a hydrogen atom to one sugar molecule and a hydroxyl group to the other. Thus, hydrolysis is the opposite of dehydration synthesis (see figs. 4.1, 4.2, and 4.3). Each of these reactions is reversible and can be summarized as follows:

Hydrolysis
Disaccharide + Water ⇌ Monosaccharide + Monosaccharide
Dehydration synthesis

Hydrolysis, which occurs during *digestion,* breaks down carbohydrates into monosaccharides, fats into glycerol and fatty acids, proteins into amino acids, and nucleic acids into nucleotides. (Chapter 15 discusses digestion in more detail.)

CHECK YOUR RECALL

1. What are the functions of anabolism? Of catabolism?
2. What is the product of anabolism of monosaccharides? Of glycerol and fatty acids? Of amino acids?
3. Distinguish between dehydration synthesis and hydrolysis.

*A*nabolic steroids are a group of lipids that stimulate anabolism and thus promote the growth of certain tissues. Doctors prescribe anabolic steroids to treat various diseases. Some individuals, however, use these powerful drugs illicitly to increase their muscle mass, often with the hope of enhancing athletic performance. Such nonmedical use of steroids can cause dangerous side effects, including adverse changes in liver functions, increased risk of heart and blood vessel diseases, upsets in normal hormonal balances, infertility, and severe psychological disorders.

4.3 Control of Metabolic Reactions

Specialized cells, such as nerve, muscle, or blood cells, have distinctive chemical reactions. However, all cells perform certain basic reactions, such as buildup and breakdown of carbohydrates, lipids, proteins, and nucleic acids. These reactions include hundreds of specific chemical changes that occur rapidly—yet in a coordinated fashion—thanks to enzymes.

Enzyme Action

Like other chemical reactions, metabolic reactions require energy to proceed. The temperature conditions in cells, however, are usually too mild to adequately promote the reactions that support life. Enzymes make these reactions possible.

Enzymes (en´zīmz) are complex molecules, almost always proteins, that promote chemical reactions within cells by lowering the amount of energy, called the activation energy, required to start these reactions. Thus, enzymes speed the rates of metabolic reactions. This acceleration is called catalysis, and an enzyme is a catalyst. Enzymes are used in very small quantities because, as they function, they are not consumed and can therefore be recycled.

The reaction a particular enzyme catalyzes is very specific. Each enzyme acts only on a particular chemical, which is called its **substrate** (sub´strāt). For example, the substrate of the enzyme *catalase* is hydrogen peroxide, a toxic by-product of certain metabolic reactions. This enzyme's only function is to speed up the breakdown of hydrogen peroxide into water and oxygen. Thus, catalase helps prevent an accumulation of hydrogen peroxide, which might damage cells.

Specific enzymes catalyze each of the hundreds of different chemical reactions comprising cellular metabolism. Thus, every cell contains hundreds of different enzymes, and each enzyme must "recognize" its specific substrate. This ability to identify its substrate depends on the shape of the enzyme molecule. That is, each enzyme's polypeptide chain twists and coils into a unique three-dimensional form that fits the specific shape of its substrate.

During an enzyme-catalyzed reaction, parts of the enzyme molecule called **active sites** temporarily combine with portions of the substrate, forming an enzyme-substrate complex (fig. 4.4). This interaction between the molecules distorts or strains chemical bonds within the substrate, which increases the likelihood that the reaction will occur. When the substrate changes, the reaction takes place, its products form, and the enzyme is released in its original shape.

An enzyme-catalyzed reaction can be summarized as follows:

$$\text{Substrate molecule} + \text{Enzyme molecule} \rightarrow \text{Enzyme-substrate complex} \rightarrow \text{Product (changed substrate)} + \text{Enzyme molecule}$$

The speed of an enzyme-controlled reaction depends partly on the number of enzyme and substrate molecules in the cell. Generally, the reaction occurs more rapidly if the concentrations of the enzyme or the substrate increase. Also, the efficiency of different kinds of enzymes varies greatly. Some enzymes can process only a few substrate molecules per second, while others can catalyze thousands or hundreds of thousands of chemical reactions per second. Many enzymatic reactions are reversible.

Factors That Alter Enzymes

Almost all enzymes are proteins, and like other proteins, they can be denatured by exposure to heat, radiation, electricity, certain chemicals, or fluids with extreme pH values. For example, many enzymes become inactive at 45°C, and nearly all of them are denatured at 55°C. Some poisons work by denaturing enzymes. Potassium cyanide, for instance, interferes with respiratory enzymes, impairing a cell's ability to release energy from nutrient molecules.

Cofactors and Coenzymes

Some enzymes are inactive until they combine with a nonprotein component. Such a substance, called a **cofactor,** may be an ion of an element, such as copper, iron, or zinc, or a small organic molecule, called a **coenzyme** (ko-en´zīm). Many coenzymes are vitamin molecules. An example of a coenzyme is coenzyme A, which is part of a participant in cellular respiration, discussed in the next section.

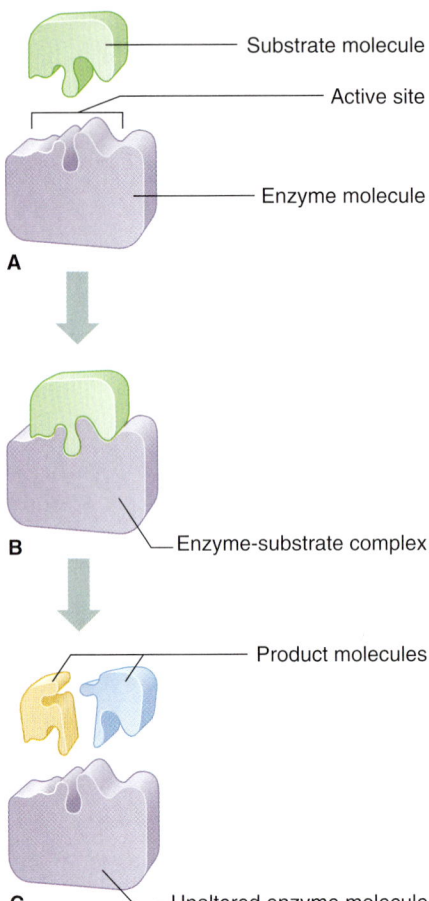

Figure 4.4
An enzyme catalyzed reaction. (*A*) The shape of a substrate molecule fits the shape of the enzyme's active site. (*B*) When the substrate molecule temporarily combines with the enzyme, a chemical reaction occurs. The result is (*C*) product molecules and an unaltered enzyme.

CHECK YOUR RECALL

1. What is an enzyme?
2. How does an enzyme recognize its substrate?
3. What factors affect the speed of an enzyme-controlled reaction?
4. What factors can denature enzymes?

4.4 Energy for Metabolic Reactions

Energy is the capacity to change or move matter; that is, energy is the ability to do work. Therefore, we recognize energy by what it has done—by whatever changes occur. Common forms of energy include heat, light, sound, electrical energy, mechanical energy, and chemical energy. Most metabolic processes use chemical energy.

Release of Chemical Energy

Chemical energy is held in the bonds between the atoms of molecules and is released when these bonds are broken, as in burning. Such a reaction usually starts by applying heat to activate the burning process. As the chemical burns, bonds break, and energy escapes as heat and light.

Cells "burn" glucose molecules in a process more correctly called **oxidation** (ok˝sĭ-da´shun). The energy released by the oxidation of glucose powers the reactions of cellular metabolism. However, the oxidation of biochemicals inside cells and the burning of substances outside cells differ in some ways.

Burning usually requires a relatively large amount of energy to begin, and most of the energy released escapes as heat or light. In cells, enzymes reduce the activation energy required for oxidation in *cellular respiration*. Also, by transferring energy to special energy-carrying molecules, cells can capture about half of the energy released from breaking chemical bonds. The rest escapes as heat, which helps maintain body temperature.

CHECK YOUR RECALL

1. What is energy?
2. How does cellular oxidation differ from burning?

Cellular Respiration

Cellular respiration occurs in three distinct, yet interconnected, series of reactions: **glycolysis,** the **citric acid cycle,** and the **electron transport chain,** shown schematically in figure 4.5. The products of these

Glycolysis

1. The 6-carbon sugar glucose is broken down into two 3-carbon pyruvic acid molecules with a net gain of 2 ATP and the release of high energy electrons.

Citric Acid Cycle

2. The 3-carbon pyruvic acids generated by glycolysis enter the mitochondria. Each loses a carbon (generating CO_2) and is combined with a coenzyme to form a 2-carbon acetyl Coenzyme A (acetyl CoA). More high energy electrons are released.

3. Each acetyl CoA combines with a 4-carbon oxaloacetic acid to form the 6-carbon citric acid, for which the cycle is named. For each citric acid a series of reactions removes 2 carbons (generating 2 CO_2's), synthesizes 1 ATP and releases more high energy electrons. The figure shows 2 ATP, resulting directly from 2 turns of the cycle per glucose molecule that enters glycolysis.

Electron Transport Chain

4. The high energy electrons still contain most of the chemical energy of the original glucose molecule. Special carrier molecules bring the high energy electrons to a series of enzymes that convert much of the remaining energy to more ATP molecules. The other products are heat and water. The requirement of oxygen in this last step is why the overall process is called aerobic respiration.

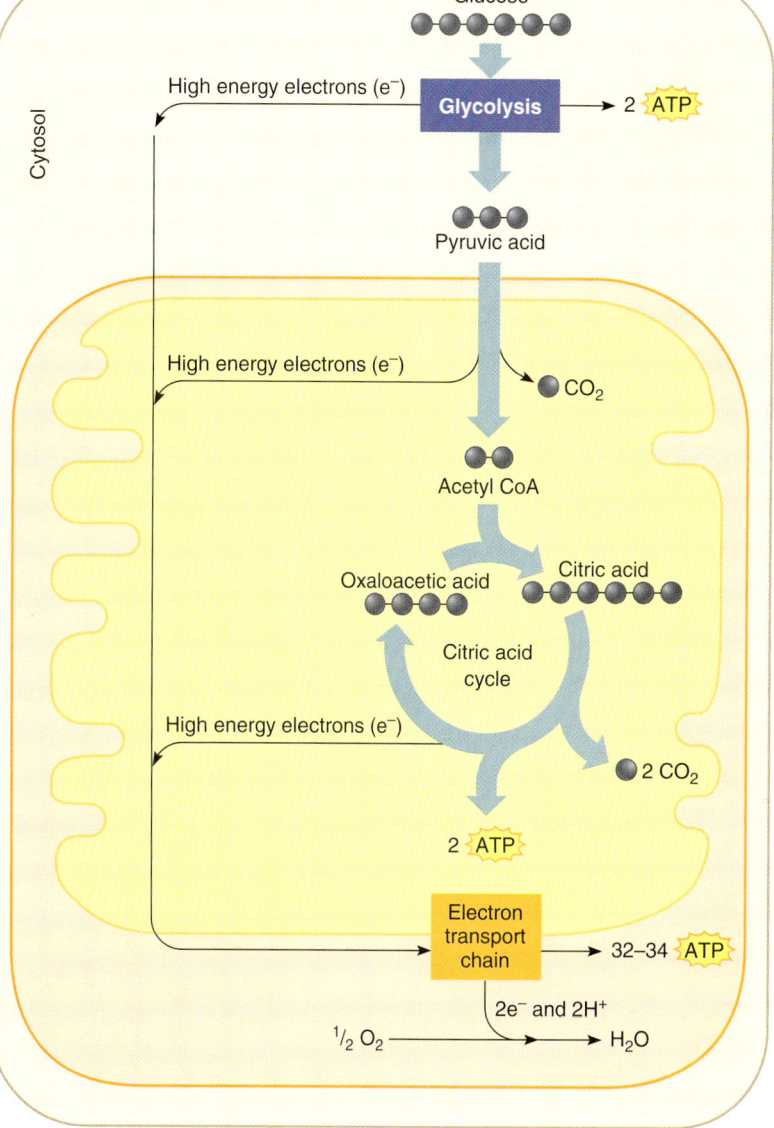

Figure 4.5

Glycolysis occurs in the cytosol and does not require oxygen. Aerobic respiration occurs in the mitochondria and only in the presence of oxygen. The products include ATP, heat, CO_2, and water.

reactions include CO_2, water, and energy. Although most of the energy is lost as heat, almost half is captured in the form of high energy electrons that the cell can use through the synthesis of **ATP (adenosine triphosphate).**

ATP

Each ATP molecule includes a chain of three chemical groups called phosphates (fig. 4.6). As energy is released during cellular respiration, some of it is captured in the bond of the end phosphate. When energy is required for a metabolic reaction, this terminal phosphate bond breaks, releasing the stored energy. The cell uses ATP for a variety of functions, including muscle contraction, active transport, and synthesis of various compounds.

An ATP molecule that has lost its terminal phosphate becomes an ADP (adenosine diphosphate) molecule. The ADP can be converted back into ATP by the addition of energy and a third phosphate. Thus, as figure 4.7 shows, ATP and ADP molecules shuttle back and forth between the energy-releasing reactions of cellular respiration and the energy-utilizing reactions of the cell.

Glycolysis

Cellular respiration begins with glycolysis, literally "the breaking of glucose." Glycolysis occurs in the cytosol (the liquid portion of the cytoplasm), and because it does not require oxygen, it is sometimes referred to as the **anaerobic** (an´´a-er-o´bik) phase of *cellular respiration* (see fig. 4.5).

Aerobic Respiration

What happens next to pyruvic acid depends on the availability of oxygen. If oxygen is available in sufficient quantity, the pyruvic acid generated by glycolysis can enter the pathways of **aerobic** (a´´er-o´bik) **respiration.** These reactions occur in the mitochondria and include the citric acid cycle and the electron transport chain. (Because the reactions of the electron transport chain add phosphates to form ATP, they are also known as oxidative phosphorylation.) The aerobic reactions yield up to 36 ATP molecules per glucose (see fig. 4.5).

For each glucose molecule that is decomposed completely, up to 38 molecules of ATP can be produced. Two of these ATP molecules are the result of glycolysis, and the rest form during the aerobic phase. About half the energy released goes to ATP synthesis, while the rest ends up as heat.

In addition to releasing energy, the complete oxidation of glucose produces carbon dioxide and water. The carbon dioxide is eventually exhaled by the lungs, and the water becomes part of the internal environment.

In humans, the amount of water produced by metabolism is far less than our daily water needs. Humans need to drink water to survive. However, the small desert rodent known as the kangaroo rat can survive entirely on the water produced by aerobic respiration.

CHECK YOUR RECALL

1. What happens during glycolysis?
2. What are the final products of cellular respiration?
3. What is the general function of ATP?

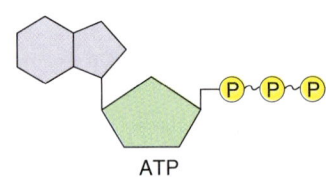

Figure 4.6
Phosphate bonds contain the energy stored in ATP.

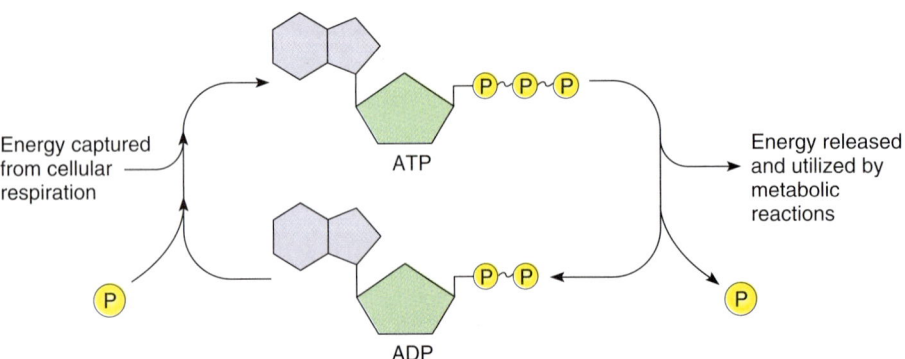

Figure 4.7
ATP provides energy for cellular reactions. Cellular respiration generates ATP.

4.5 Metabolic Pathways

Like cellular respiration, anabolic and catabolic reactions in general have a number of steps that must occur in a particular sequence. The enzymes that control the rates of these reactions must act in a specific sequence. Such coordination suggests that the enzymes are positioned in exactly the same sequence as that of the reactions they control. For example, the enzymes responsible for aerobic respiration are located in tiny, stalked particles on the membranes (cristae) within the mitochondria, in the sequence in which they function. A sequence of enzyme-controlled reactions is called a **metabolic pathway** (fig. 4.8).

Recall that the rate of an enzyme-controlled reaction usually increases if either the number of substrate molecules or the number of enzyme molecules increases. However, the rate of a metabolic pathway is often determined by a regulatory enzyme responsible for one of its steps. This regulatory enzyme is present in limited quantity. Consequently, it can become saturated with substrate molecules whenever the substrate concentration increases above a certain level. Once the enzyme is saturated, increasing the number of substrate molecules will no longer affect the reaction rate. In this way, a single enzyme in a pathway can control the whole pathway.

As a rule, a *rate-limiting enzyme* is the first enzyme in a series. This position is important because some intermediate chemical in the pathway might accumulate if an enzyme occupying another location in the sequence were rate limiting.

Although this section has dealt with the metabolism of glucose, lipids and proteins can also be broken down to release energy for ATP synthesis. In all three cases, the final process is aerobic respiration, and the most common entry point is into the citric acid cycle as acetyl coenzyme A (acetyl CoA) (fig. 4.9). These metabolic pathways and their regulation are described in detail in chapter 15 (pp. 421–422).

CHECK YOUR RECALL

1. What is a metabolic pathway?
2. What is a rate-limiting enzyme?

4.6 Nucleic Acids

Because enzymes control the metabolic processes that enable cells to survive, cells must have instructions for producing these specialized proteins, as well as many other proteins. **DNA (deoxyribonucleic acid)** molecules hold such information in the form of a *genetic code*. This code "instructs" cells how to synthesize enzymes and other specific protein molecules.

Genetic Information

Children resemble their parents because of inherited traits, but what actually passes from parents to child is *genetic information* in the form of DNA molecules from the parents' sex cells. The portions of DNA molecules that contain the genetic information for making particular proteins are called **genes** (jēnz). Thus, inherited traits are determined by the genes contained in the parents' sex cells, which fuse to form the first cell of an offspring's body. As an offspring develops, mitosis passes the information from cell to cell. Genetic information "tells" cells of the developing body how to construct specific protein molecules, which in turn function as structural materials, enzymes, or other vital biochemicals. In other words, genes instruct cells to synthesize the enzymes that control metabolic pathways.

All of the DNA in a cell constitutes the **genome.** Only a small proportion of the human genome encodes protein. Researchers do not yet understand the functions of much of the genome.

DNA Molecules

As described in chapter 2 (p. 43), the building blocks of nucleic acids are nucleotides joined so that the sugar and phosphate portions alternate. They form a long "backbone" to the polynucleotide chain (see fig. 2.19, p. 44).

In a DNA molecule, the organic bases project from this backbone and bind weakly to the bases of the second strand (fig. 4.10). The resulting structure is like a ladder in which the uprights represent the sugar and phosphate backbones of the two strands, and the rungs represent the organic bases. The organic base of a DNA nucleotide can be one of four types: *adenine* (A), *thymine* (T), *cytosine* (C), or *guanine* (G).

Both strands of a DNA molecule consist of nucleotides in a particular sequence. Because of their molecular shapes, the nucleotide bases pair in specific ways. An adenine will bond only to a thymine, and a cytosine will bond only to a guanine. As a consequence of this *complementary base pairing,* a DNA strand with the base sequence G, A, C, T joins a second strand with the complementary base sequence C, T, G, A (see the upper region of DNA in figure 4.10). The sequence of base pairs along a DNA molecule encodes the genetic information that specifies a particular protein's amino acid sequence.

The DNA molecule twists to form a double helix. A molecule of DNA is typically millions of base pairs long.

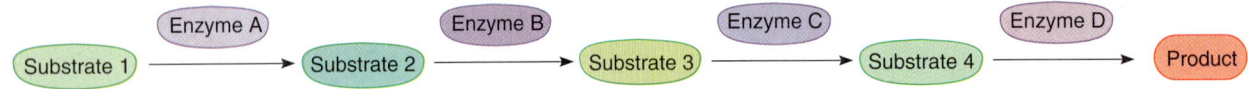

Figure 4.8
A metabolic pathway consists of a series of enzyme-controlled reactions leading to formation of a product.

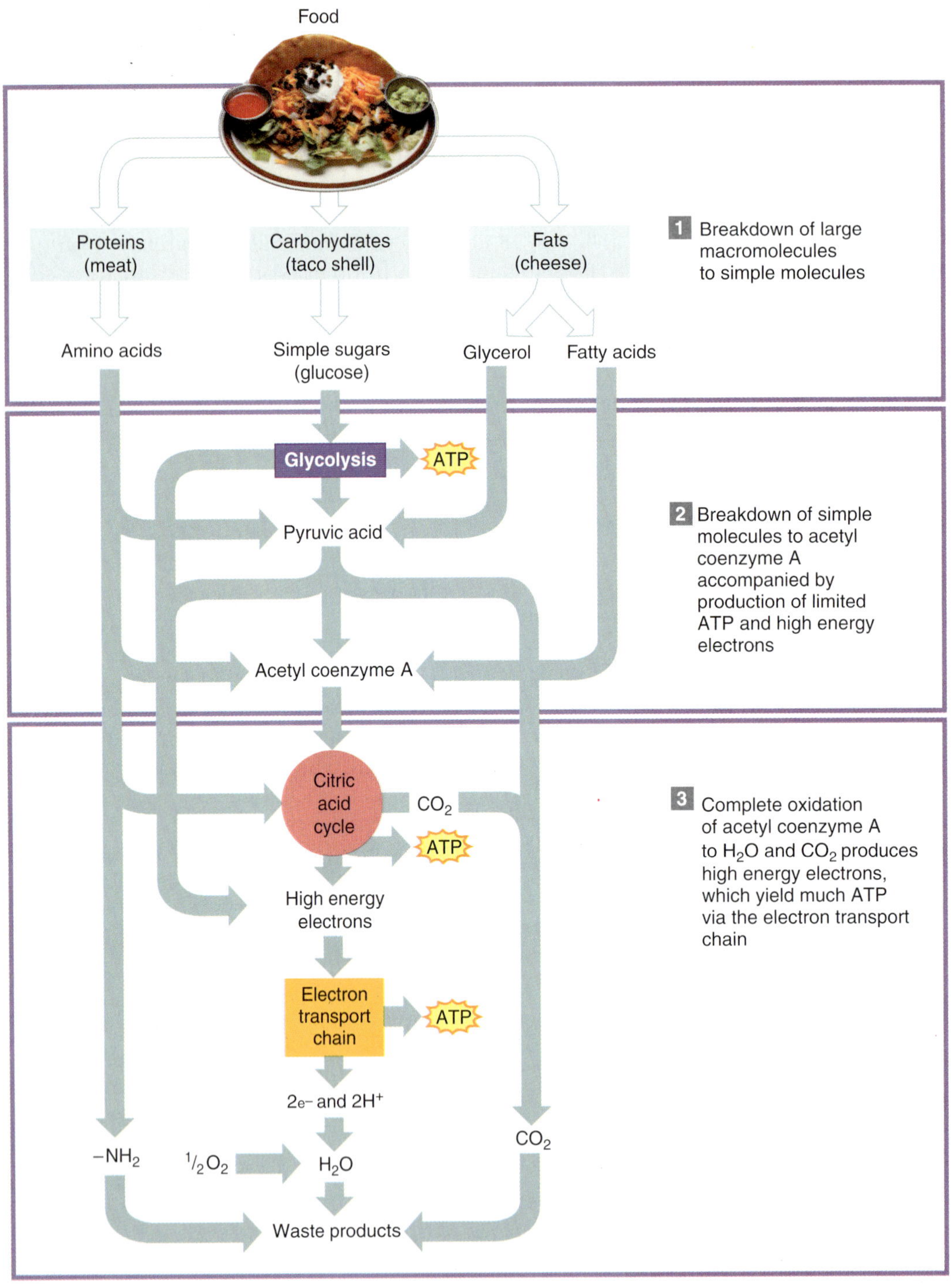

Figure 4.9
A summary of the breakdown (catabolism) of proteins, carbohydrates, and fats.

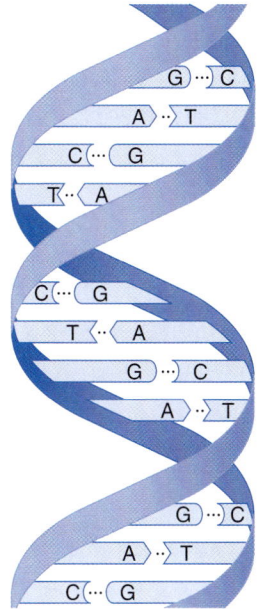

Figure 4.10
The molecular ladder of a double-stranded DNA molecule twists into a double helix, held together by "rungs" consisting of complementary base pairs—A with T (or T with A) and G with C (or C with G).

Because the sequence of DNA nucleotides in each person's cells is unique (with the exception of identical twins), a person can be matched to a sample of cells by analyzing the cells' DNA. This technique is called DNA fingerprinting. It was used extensively to identify victims of the World Trade Center collapse. DNA from human remains was matched to DNA from blood relatives and from skin and hair cells on victims' toothbrushes and hairbrushes.

4.7 Protein Synthesis

All four groups of organic molecules—proteins, carbohydrates, lipids, and nucleic acids—depend on the instructions written in the genes. This is true because proteins functioning as enzymes are necessary for chemical reactions within cells. Since proteins are involved directly in so many aspects of cell function, genes provide the instructions that are necessary for survival of the cell.

The Genetic Code—Instructions for Making Proteins

Genetic information contains the instructions for synthesizing proteins. Because proteins consist of twenty types of amino acids joined in specified sequences, the genetic information must tell how to position the amino acids correctly in a polypeptide chain.

Each of the amino acid types is represented in a DNA molecule by a particular sequence of three nucleotides. That is, the sequence G, G, T in a DNA strand represents one type of amino acid; the sequence G, C, A represents another type, and T, T, A still another type. This method of storing information for protein synthesis is the **genetic code.** Other nucleotide sequences represent the instructions for beginning or ending the synthesis of a protein molecule. Thus, the sequence of nucleotides in a DNA molecule denotes the order of amino acids of a protein molecule, as well as where to start or stop that protein's synthesis.

Transcription

Because DNA molecules are within a cell's nucleus and protein synthesis occurs in the cytoplasm, genetic information must be carried from the nucleus into the cytoplasm. Molecules of **messenger RNA (mRNA)** accomplish this information transfer in a process called **transcription.** RNA is a nucleic acid, and mRNA is the type that carries a gene's message.

RNA (ribonucleic acid) molecules differ from DNA molecules in several ways. RNA molecules are single-stranded, and their nucleotides contain ribose rather than deoxyribose sugar. Like DNA, each RNA nucleotide contains one of four organic bases, but whereas adenine, cytosine, and guanine nucleotides are in both DNA and RNA, thymine nucleotides are found only in DNA. In place of thymine nucleotides, RNA molecules contain *uracil* (U) nucleotides.

mRNA is synthesized by the enzyme RNA polymerase, according to the rules of complementary base pairing. For example, if the sequence of DNA bases is A, T, G, C, G, then the complementary bases in the developing mRNA molecule are U, A, C, G, C (fig. 4.11). RNA polymerase somehow "knows" which of the two DNA strands contains the information. It also knows where the gene begins, where it stops, and the correct direction to read the DNA. Just as a sentence has a beginning, an end, and a correct direction to be read, so does a gene. When the polymerase reaches the end of the gene, it releases the newly formed mRNA. Transcription is complete.

Translation

Each amino acid in a protein was originally represented by a series of three bases in DNA. Those amino acids, in the proper order, are represented by a series of three base sequences, called **codons** (ko´donz), in mRNA (Table 4.1). To complete the process of protein synthesis, mRNA must leave the nucleus and associate with a ribosome. There the series of codons on mRNA

DNA | RNA

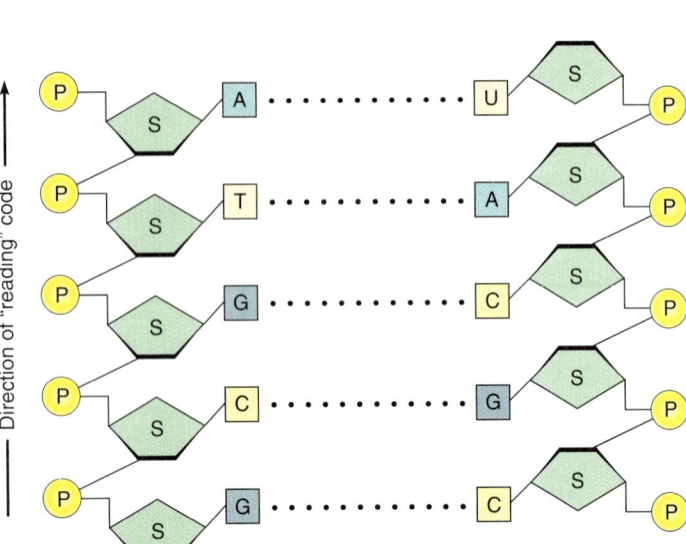

Figure 4.11
When an RNA molecule is synthesized beside a strand of DNA, complementary nucleotides bond as in a double-stranded DNA molecule, with one exception: RNA contains uracil nucleotides (U) in place of thymine nucleotides (T).

TABLE 4.1   SOME NUCLEOTIDE SEQUENCES OF THE GENETIC CODE

	DNA SEQUENCE	RNA SEQUENCE
Amino acids		
Alanine	CGT	GCA
Arginine	GCA	CGU
Asparagine	TTA	AAU
Aspartic acid	CTA	GAU
Cysteine	ACA	UGU
Glutamic acid	CTT	GAA
Glutamine	GTT	CAA
Glycine	CCG	GGC
Histidine	GTA	CAU
Isoleucine	TAG	AUC
Leucine	GAA	CUU
Lysine	TTT	AAA
Methionine	TAC	AUG
Phenylalanine	AAA	UUU
Proline	GGA	CCU
Serine	AGG	UCC
Threonine	TGC	ACG
Tryptophan	ACC	UGG
Tyrosine	ATA	UAU
Valine	CAA	GUU
Instructions		
Start protein synthesis	TAC	AUG
Stop protein synthesis	ATT	UAA

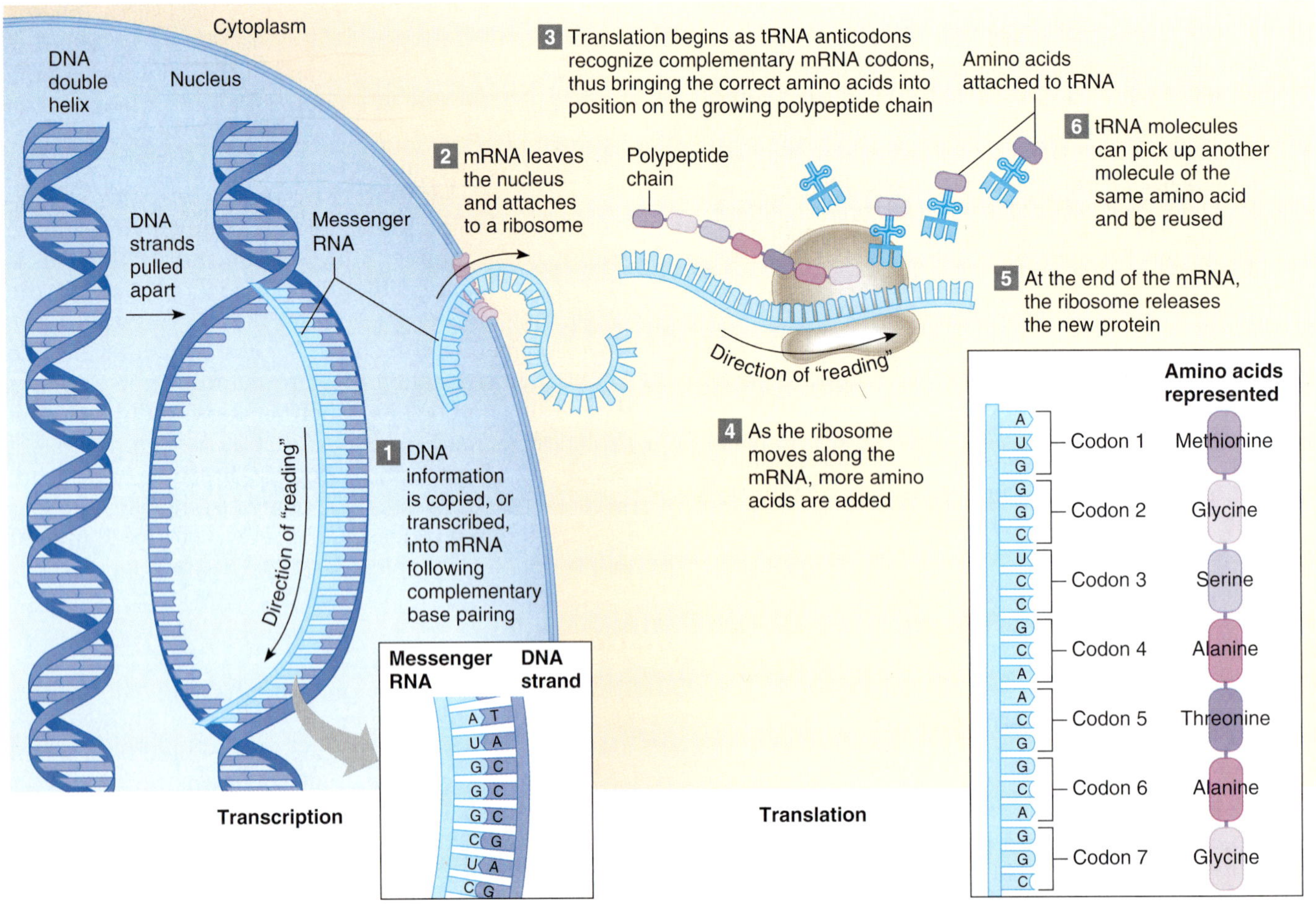

Figure 4.12
DNA information is transcribed into mRNA, which in turn is translated into a sequence of amino acids.

will be converted from the "language" of nucleic acids to the "language" of amino acids. This is the process of **translation.**

Building a protein molecule requires that the correct amino acids be present in the cytoplasm and positioned in the proper locations along a strand of mRNA. A second kind of RNA molecule, synthesized in the nucleus and called **transfer RNA (tRNA),** correctly aligns amino acids to form proteins.

Because twenty different types of amino acids form biological proteins, at least twenty different types of tRNA molecules must be available to serve as guides, one for each type of amino acid. Each type of tRNA has a region at one end that consists of three nucleotides in a particular sequence. These nucleotides bond a specific complementary set of three nucleotides of an mRNA molecule—a codon. Thus, the set of three nucleotides in the tRNA is called an **anticodon** (an´´tĭ-ko´don). In this way, tRNA carries its amino acid to a correct position on an mRNA strand. This action occurs in close association with a ribosome (fig. 4.12).

Shortly after protein synthesis begins, a ribosome is bound to an mRNA molecule. A tRNA molecule with the complementary anticodon holds its amino acid and the growing polypeptide chain. A second tRNA complementarily binds to the next codon, bringing its amino acid to an adjacent site on the ribosome. Then a peptide bond forms, adding the new amino acid to the chain. The first tRNA molecule is released from its amino acid and returns to the cytoplasm (fig. 4.13).

This process repeats again and again as the ribosome moves along the mRNA molecule. The amino acids released by the tRNA molecules are added one at a time to the developing protein. Enzymes associated with the ribosome control this addition of amino acids.

As the protein molecule forms, it folds into its unique shape and is then released to become a separate functional molecule. The tRNA molecules can pick up other amino acids from the cytoplasm, and like the mRNA molecules, can function repeatedly. Table 4.2 compares RNA and DNA, and table 4.3 summarizes protein synthesis.

A gene that is transcribed and translated into a protein is said to be *expressed*. The proteins that result determine the function a cell performs in the body. Gene expression is the basis for cell differentiation, described in chapter 3 (p. 67). The Topic of Interest (pp. 84–85) considers the roles that analyzing gene expression will play in health care.

CHECK YOUR RECALL

1. What is the function of DNA?
2. How is information carried from the nucleus to the cytoplasm?
3. How are protein molecules synthesized?

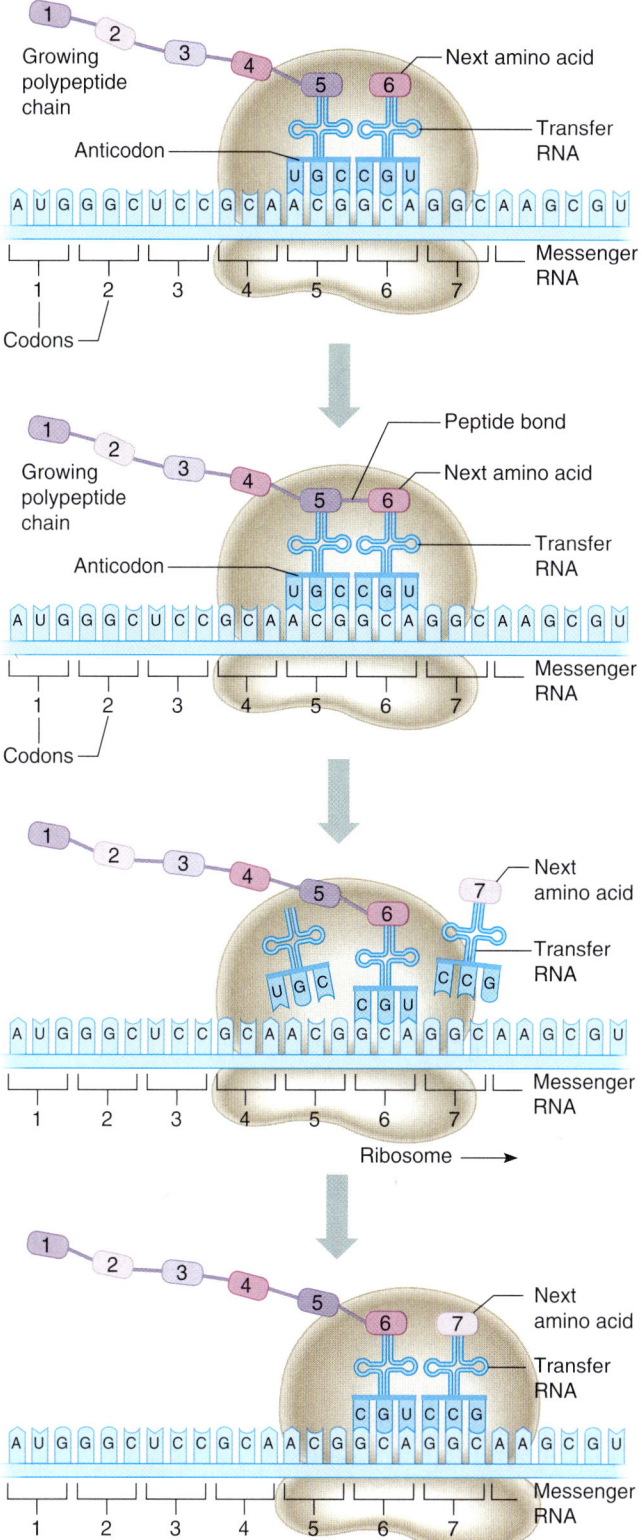

Figure 4.13
Molecules of transfer RNA (tRNA) attach to and carry specific amino acids, aligning them in the sequence determined by the codons of mRNA. These amino acids, connected by peptide bonds, form the polypeptide chain of a protein molecule.

Topic of Interest

OF GENOMES, CHIPS AND SNPs

Researchers unveiled a "first draft" of the 3 billion DNA base sequence that comprises the human genome in June, 2000. The achievement marked a crucial milestone in the decade-long project, which was conceived in 1985 and carried out by hundreds of researchers. The project is actually the continuation of investigations, most of them centered on single genes, undertaken throughout the past century.

Genome sequencing grew out of methods invented in the late 1970s to determine the order of bases in DNA pieces. To do this, multiple copies of the same short sequence are cut from one end, using enzymes, to generate a set of pieces that differ by the one base that caps one end. All four DNA base types are labeled with a different fluorescent dye. When the pieces are aligned, a sequencing device reads off the end base, of each piece, generating a sequence. Powerful computer algorithms made it possible to sequence DNA on a chromosomal scale. To obtain the first draft genome sequence, one group of researchers assigned overlapped DNA pieces to known sites on the chromosomes. The other group shattered, or "shotgunned," the entire genome, and aligned the overlaps in sequence, an approach that they first proved could work on the genomes of several microorganisms and the fruit fly. The shotgun group finished slightly ahead of the chromosome-by-chromosome group, but both announced and published results at the same time to share credit.

Now that the human genome sequence first draft is known, research is veering in two major directions: profiling gene expression, and identifying the differences among us.

Proteomics is a new name for the study of gene expression—that is, which cell types synthesize which proteins. Researchers use small squares of glass or nylon called DNA microarrays or "chips" to immobilize many genes of interest. A DNA chip for diabetes mellitus might bear gene variants that reflect how the body handles glucose transport and uptake into cells. A chip for cardiovascular disease includes thousands of genes whose protein products control blood pressure, blood clotting, and the synthesis, transport and metabolism of cholesterol and other lipids. Then, the cell type affected in the condition is sampled, and its messenger RNA molecules collected and copied, using an enzyme, into DNA molecules. The mRNAs from a specialized cell reflect the types of proteins manufactured. The DNA copies are labeled and then added to the DNA chip. The resulting pattern of fluorescent spots reveals, at a glance (to a computer), which genes are expressed.

DNA chips allow researchers to make compelling comparisons. A muscle cell from a person with diabetes mellitus expresses different genes than the same cell type from a person who does not have this metabolic disorder. DNA chips to diagnose and monitor cancer are particularly valuable. They can reveal not only the subtype of cancer that a person has, refining diagnosis, but predict which drugs will be effective. In one study, DNA chip tests for breast cancer were performed on tumor cells of 20 women who had advanced breast cancer, before and after a three-month regimen of chemotherapy. The gene expression pattern remained consistent with cancer in 17 of the women after treatment, but had become normal in three of the women. Those were the only three women in whom the chemotherapy was successful, demonstrating that gene expression can correlate to clinical signs and symptoms.

The first draft human genome sequence indicates that we humans are 99.9 percent similar to each other. Researchers are focusing on that tiny percentage of difference. Specifically, they are studying "single nucleotide polymorphisms," or SNPs. A SNP is a single base site in the DNA sequence that differs in more than one percent of a population. ("Polymorphism" means "many forms".) The human genome is riddled with up to 20 million SNPs. Researchers have identified more than 3 million SNPs, spread apart enough to enable them to track any gene of interest.

A SNP by itself can be helpful, harmful, or, in most instances, have no effect at all. But the patterns generated by many SNPs are valuable for their predictive power, because different SNP patterns correlate very strongly to different disorders. DNA chips are used to detect SNP patterns. A technique called an association study examines DNA variants in populations and detects particular combinations of SNPs that are found almost exclusively among people with a particular disorder, but not in others. The correlations between certain SNP patterns and elevated disease risks can guide medical decisions, including the ability to make more meaningful prognoses, and to predict an individual's likely response to a particular drug. SNP association studies are well underway in many biotech and pharmaceutical companies as well as academic laboratories. It is likely that in the years to come, SNP tests will become a standard part of medical practice, beginning with predicting disease susceptibilities at birth or even before.

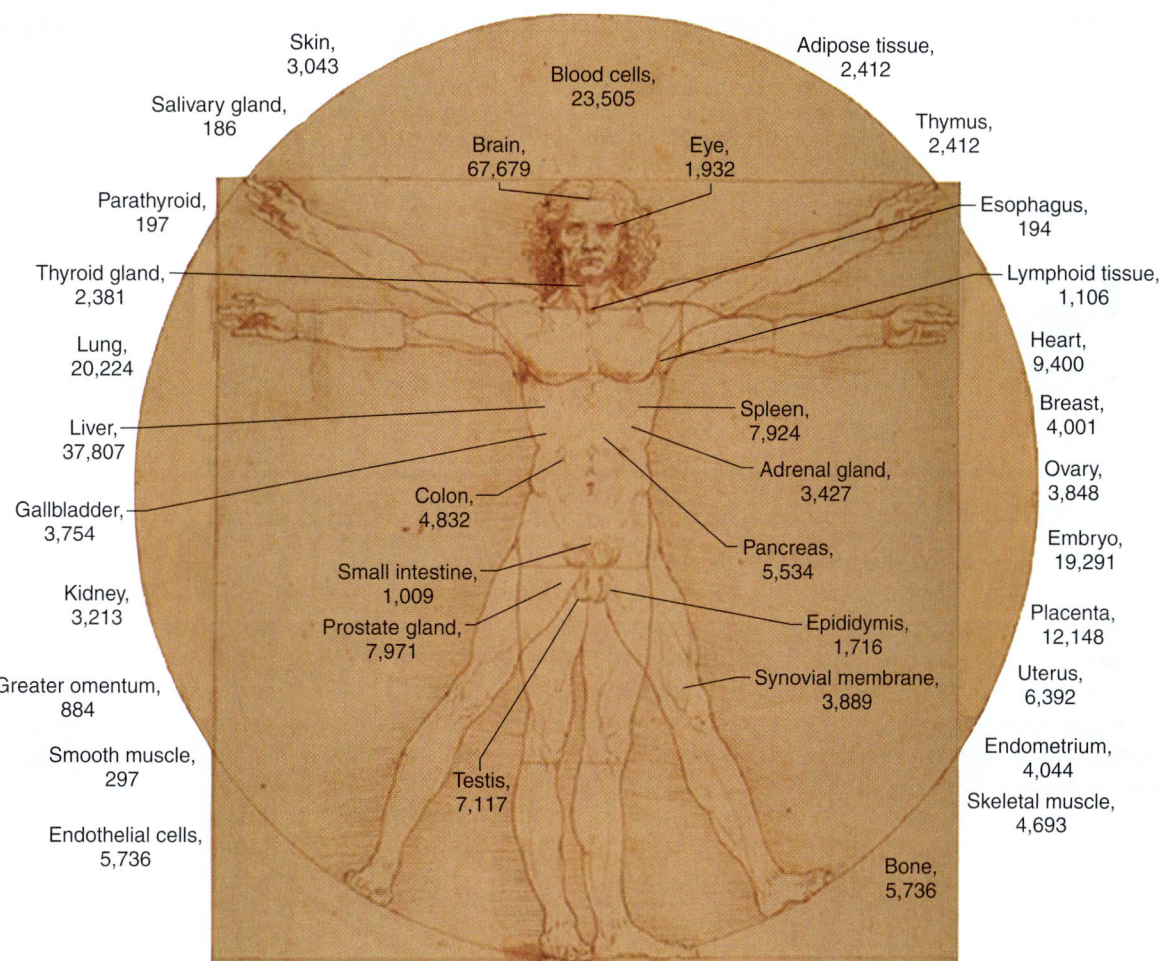

Figure 4A
One way to study the human genome is to determine which genes are expressed in different tissues.

TABLE 4.2 — A COMPARISON OF DNA AND RNA MOLECULES

	DNA	RNA
Main location	Part of chromosomes in nucleus	In the cytoplasm
5-carbon sugar	Deoxyribose	Ribose
Basic molecular structure	Double-stranded	Single-stranded
Organic bases included	Adenine, thymine, cytosine, guanine	Adenine, uracil, cytosine, guanine
Major functions	Contains genetic code for protein synthesis; replicates prior to cell division	mRNA carries transcribed DNA information to cytoplasm and acts as template for synthesis of protein molecules; tRNA carries amino acids to mRNA

TABLE 4.3 — PROTEIN SYNTHESIS

TRANSCRIPTION (OCCURS IN THE NUCLEUS)

1. RNA polymerase associates with the base sequence of one strand of a gene.
2. Other enzymes unwind the DNA molecule, exposing a portion of the gene.
3. RNA polymerase moves along the exposed gene and polymerizes an mRNA molecule, whose nucleotides are complementary to those of the gene strand.
4. When the RNA polymerase reaches the end of the gene, the newly formed mRNA molecule is released.
5. The mRNA molecule passes through a pore in the nuclear envelope and enters the cytoplasm.

TRANSLATION (OCCURS IN THE CYTOPLASM)

1. A ribosome binds to the mRNA molecule near the codon at the beginning of the messenger strand.
2. A tRNA molecule with the complementary anticodon associates with the ribosome, and the amino acid it carries becomes part of the chain.
3. This process repeats for each codon in the mRNA sequence as the ribosome moves along the mRNA's length.
4. Enzymes associated with the ribosome catalyze peptide bonds, forming a chain of amino acids.
5. As the chain of amino acids grows, it folds into the unique shape of a functional protein molecule.
6. The completed protein molecule is released. The mRNA molecule, ribosome, and tRNA molecules can function repeatedly to synthesize other protein molecules.

4.8 DNA Replication

When a cell divides, each newly formed cell requires a copy of the parent cell's genetic information so that the new cell can synthesize the proteins necessary to maintain life functions, build cell parts, and metabolize. To accomplish this, DNA molecules are duplicated, or **replicated** (rĕ′′plĭ-ka′tĭd), during interphase of the cell cycle.

As replication begins, hydrogen bonds between complementary base pairs of the double strands in each DNA molecule break (fig. 4.14). Then the double-stranded structure pulls apart and unwinds, exposing the organic bases of its nucleotides. New DNA nucleotides form complementary pairs with the exposed bases, and enzymes knit together the sugar-phosphate backbone. In this way, a new strand of complementary nucleotides forms along each of the old strands, producing two complete DNA molecules, each with one old strand of the original molecule and one new strand. These two DNA molecules become incorporated into replicate copies of a chromosome and separate during mitosis so that one passes to each of the newly forming cells. The Topic of Interest (p. 87) describes errors in DNA replication.

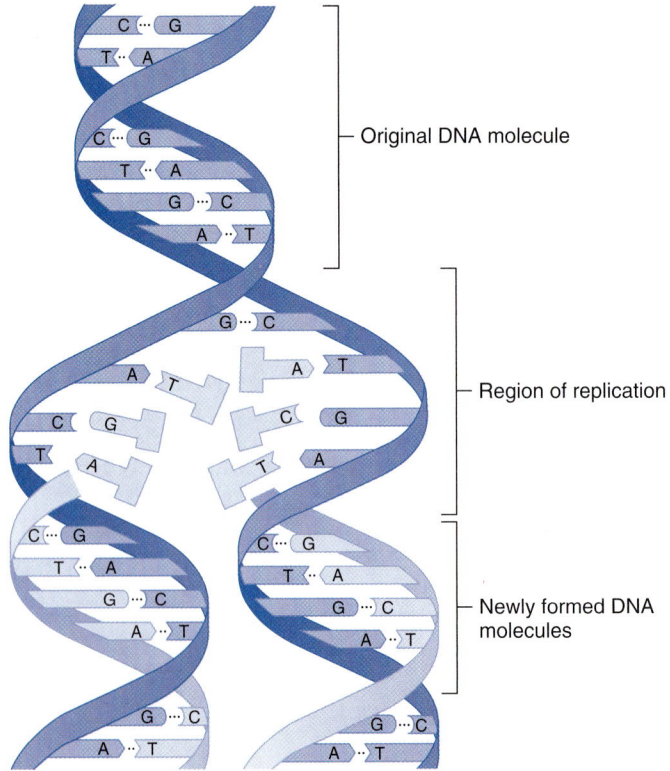

Figure 4.14
When a DNA molecule replicates, its original strands separate locally. A new strand of complementary nucleotides forms along each original strand.

CHECK YOUR RECALL

1. Why must DNA molecules be replicated?
2. How is replication accomplished?

Topic of Interest

MUTATIONS

It is easy to make an error when typing a sentence consisting of several hundred letters. DNA replication, which is similar to copying such a sentence, is also error-prone. A newly replicated gene may have too many or too few bases, or an "A" where the parent DNA sequence indicates there should be a "C." Fortunately, cells have several mechanisms to scan newly replicated DNA, detect such changes (*mutations*) in genetic information, and correct them. When this DNA repair fails, health may suffer. Mutations may occur spontaneously or may be induced by agents called mutagens, such as certain toxic chemicals and ionizing radiation.

Hundreds of inherited illnesses result from mutations. Certain genetic tests detect a particular mutation before symptoms of the associated illness begin. This is possible because the mutated gene is present in an individual from the time of conception. For example, a genetic test may identify the neurological disorder Huntington disease in an eighteen-year-old, even though the symptoms—personality changes and uncontrollable, constant movements—probably will not appear for another two decades. Predictive tests are also available for inherited forms of breast cancer. Such testing is controversial, particularly when the illness is not treatable.

Not all mutations are harmful. About one percent of the individuals in caucasian populations have a mutation that makes HIV infection impossible. The gene that is mutant normally encodes a protein to which the virus must bind on immune system cells to enter them. Without this protein, the virus cannot bind to and enter human cells. Asian and African populations do not seem to have this mutation.

Clinical Connection

Sperm cells that cannot swim cannot fertilize an egg cell. One cause of such nonmotile sperm is the presence of white blood cells in semen. The blood cells produce toxic compounds called reactive oxygen species, which bind to sperm cell membranes. Within the membranes are some of the enzymes that function in glycolysis and cellular respiration. When the reactive oxygen species destroy these enzymes, the sperm cell cannot manufacture enough ATP to effectively move. If many sperm cells are affected in this way, fertility declines. Clinical trials are underway to test various drug candidates that exert anti-oxidant effects to treat this form of infertility in the male.

SUMMARY OUTLINE

4.1 Introduction (p. 74)
A cell continuously carries on metabolic reactions.

4.2 Metabolic Reactions (p. 74)
1. Anabolism
   a. Anabolism builds large molecules from smaller molecules.
   b. In dehydration synthesis, water forms, and smaller molecules attach by sharing atoms.
   c. Carbohydrates are synthesized from monosaccharides, fats are synthesized from glycerol and fatty acids, and proteins are synthesized from amino acids.
2. Catabolism
   a. Catabolism breaks down larger molecules into smaller ones.
   b. In hydrolysis, a water molecule is split as an enzyme breaks the bond between two portions of a molecule.
   c. Hydrolysis breaks down carbohydrates into monosaccharides, fats into glycerol and fatty acids, proteins into amino acids, and nucleic acids into nucleotides.

4.3 Control of Metabolic Reactions (p. 75)
Enzymes control metabolic reactions, which consist of many specific chemical changes.
1. Enzyme action
   a. Enzymes are molecules that promote metabolic reactions without being consumed.
   b. An enzyme acts upon a specific substrate.
   c. The shape of an enzyme molecule fits the shape of its substrate molecule.
   d. When an enzyme combines with its substrate, the substrate changes, lowering the energy necessary for a reaction to proceed. A product forms, and the enzyme is released in its original form.
   e. The speed of an enzyme-controlled reaction depends partly upon the number of enzyme and substrate molecules present and the enzyme's efficiency.
2. Factors that alter enzymes
   a. Almost all enzymes are proteins. Harsh conditions cause them to lose their shape, or denature.
   b. Factors that may denature enzymes include heat, radiation, electricity, certain chemicals, and extreme pH values.

4.4 Energy for Metabolic Reactions (p. 76)
Energy is the capacity to do work. Common forms of energy include heat, light, sound, electrical energy, mechanical energy, and chemical energy.
1. Release of chemical energy
   a. Most metabolic processes use chemical energy released when molecular bonds break.
   b. The energy released from glucose breakdown during cellular respiration drives the reactions of cellular metabolism.

2. Cellular respiration
   a. ATP molecules
      (1) Energy is captured in the bond of the terminal phosphate of each ATP molecule.
      (2) When a cell requires energy, the terminal phosphate bond of an ATP molecule breaks, releasing stored energy.
      (3) An ATP molecule that loses its terminal phosphate becomes an ADP molecule.
      (4) An ADP molecule that captures energy and a phosphate becomes ATP.
   b. Glycolysis
      (1) The first phase of glucose decomposition does not require oxygen.
      (2) Some of the energy released is transferred to ATP molecules.
   c. Aerobic respiration
      (1) The second phase of glucose decomposition requires oxygen.
      (2) Considerably more ATP molecules form during this phase than during the anaerobic phase.
      (3) The final products of glucose breakdown are carbon dioxide, water, and energy.

4.5 Metabolic Pathways (p. 79)

Metabolic processes consist of chemical reactions that occur in a certain sequence. A sequence of enzyme-controlled reactions constitutes a metabolic pathway.

1. Regulation of metabolic pathways
   a. Rate-limiting enzymes present in limited quantities determine the rates of metabolic pathways.
   b. Rate-limiting enzymes become saturated when substrate concentrations increase above a certain level.

4.6 Nucleic Acids (p. 79)

DNA molecules contain information that instructs a cell how to synthesize enzymes and other proteins.

1. Genetic information
   a. Inherited traits result from DNA information passed from parents to offspring.
   b. A gene is a portion of a DNA molecule that contains the genetic information for making a particular protein.
2. DNA molecules
   a. A DNA molecule consists of two strands of nucleotides twisted into a double helix.
   b. The nucleotides of a DNA strand are in a particular sequence.
   c. The nucleotides of each strand pair with those of the other strand in a complementary fashion.

4.7 Protein Synthesis (p. 81)

Genes provide instructions for making proteins, which are involved directly in many aspects of cell function.

1. The genetic code—Instructions for making proteins
   a. The sequence of nucleotides in a DNA molecule encodes the sequence of amino acids in a protein molecule.
   b. RNA molecules transfer genetic information from the nucleus to the cytoplasm.
   c. Transcription
      (1) RNA molecules are usually single-stranded; they contain ribose instead of deoxyribose and uracil nucleotides in place of thymine nucleotides.
      (2) Messenger RNA (mRNA) molecules consist of nucleotide sequences that are complementary to those of exposed strands of DNA.
      (3) Messenger RNA molecules associate with ribosomes and provide patterns for the synthesis of protein molecules.
   d. Translation
      (1) A ribosome binds to a mRNA molecule and allows a transfer RNA (tRNA) molecule to recognize its correct position on the mRNA.
      (2) Molecules of tRNA position amino acids along a strand of mRNA.
      (3) Amino acids released from the tRNA molecules join and form a protein molecule that folds into a unique shape.

4.8 DNA Replication (p. 86)

When a cell divides, each new cell requires a copy of the parent cell's genetic information. DNA molecules replicate during interphase of the cell cycle. Each new DNA molecule contains one old strand and one new strand.

REVIEW EXERCISES

1. Distinguish between anabolism and catabolism. (p. 74)
2. Distinguish between dehydration synthesis and hydrolysis. (p. 74)
3. Define *enzyme*. (p. 75)
4. Describe how an enzyme interacts with its substrate. (p. 75)
5. Explain how an enzyme can be denatured. (p. 76)
6. Explain how oxidation of molecules inside cells differs from the burning of substances outside cells. (p. 77)
7. Explain the importance of ATP to cell processes. (p. 78)
8. Describe the relationship between ATP and ADP molecules. (p. 78)
9. Distinguish between glycolysis and aerobic respiration. (p. 78)
10. Explain what a *metabolic pathway* is. (p. 79)
11. Explain how one enzyme can control the rate of a metabolic pathway. (p. 79)
12. Explain how DNA encodes genetic information. (p. 79)
13. Describe the relationship between a DNA molecule and a gene. (p. 82)
14. Distinguish between messenger RNA and transfer RNA. (p. 83)
15. Define *transcription*. (p. 83)
16. Define *translation*. (p. 84)
17. Distinguish between a codon and an anticodon. (p. 84)
18. Calculate the number of amino acids that a DNA sequence of twenty-seven nucleotides encodes. (p. 84)
19. Explain the function of a ribosome in protein synthesis. (p. 85)
20. Explain why a new cell requires a copy of the parent cell's genetic information. (p. 86)
21. Describe how DNA molecules replicate. (p. 86)

CRITICAL THINKING

1. How can the same biochemical be both a reactant (starting material) and a product in aerobic respiration?
2. After finishing a grueling marathon, a runner exclaims, "Whew, I think I've used up all my ATP!" Could this be possible?
3. Mutations in proteins that participate in ATP formation in mitochondria are never seen. Why might this be?
4. A mutation that deletes one or two DNA nucleotides changes gene function more drastically than a substitution of one nucleotide for another type, or removal or addition of three DNA nucleotides. Why?
5. A portion of a protein molecule has the following amino acid sequence: serine-lysine-glycine-proline-tyrosine-alanine-

glutamine-valine. Write DNA and RNA sequences that can specify this chain of amino acids.
6. What effect might changes in the pH of body fluids that accompany illness have on enzymes?
7. Some weight-reduction diets drastically limit intake of carbohydrates but allow foods high in fat and protein. What changes would such a diet cause in the dieter's cellular metabolism? What changes might be noted in this person's urine?

WEB CONNECTIONS

Visit the website for additional study questions and more information about this chapter at:

http://www.mhhe.com/shieress8

chapter 5

Tissues

BUILDING A BLOOD VESSEL. Cells aggregate to form tissues, and tissues interact to form organs. Dissecting a complex organ to observe how tissues comprise it is a commonly performed exercise; attempting to build an organ from its component cells and tissues is much more challenging.

The field of tissue engineering uses cells, synthetic materials, or combinations of them to fashion human body parts. Consider the task facing graduate student Nicolas L'Heureux, who recreated a small-diameter human blood vessel as his thesis project. Such a vessel has three layers: an innermost layer of tilelike endothelial cells that secrete anti-clotting agents, a middle layer of smooth muscle and elastic connective tissue, and an outer layer of fibroblasts and the collagen protein they secrete.

Previous attempts at producing a small blood vessel combined natural and synthetic ingredients in various ways, with mixed results. The goal is to keep the inner lining smooth enough to prevent blood clots from forming, but construct outer layers that are strong enough to keep the vessel open under the pressure of circulating blood. The trick, L'Heureux found, was to let the cells do the work—with a little help.

L'Heureux and his co-workers grew fibroblasts and smooth muscle cells in sheets. They then rolled the sheets around tubes through which nutrients circulated in, and cellular wastes circulated out. Then the researchers seeded endothelial cells onto the inner surface, where the cells knit a smooth inner lining. By allowing the fibroblasts to secrete the collagen, rather than supplying the protein directly, the vessels formed in a more natural way and persisted. Blood vessels engineered in this way may eventually be used to treat the thousands of people who need vascular grafts in their legs or new coronary arteries.

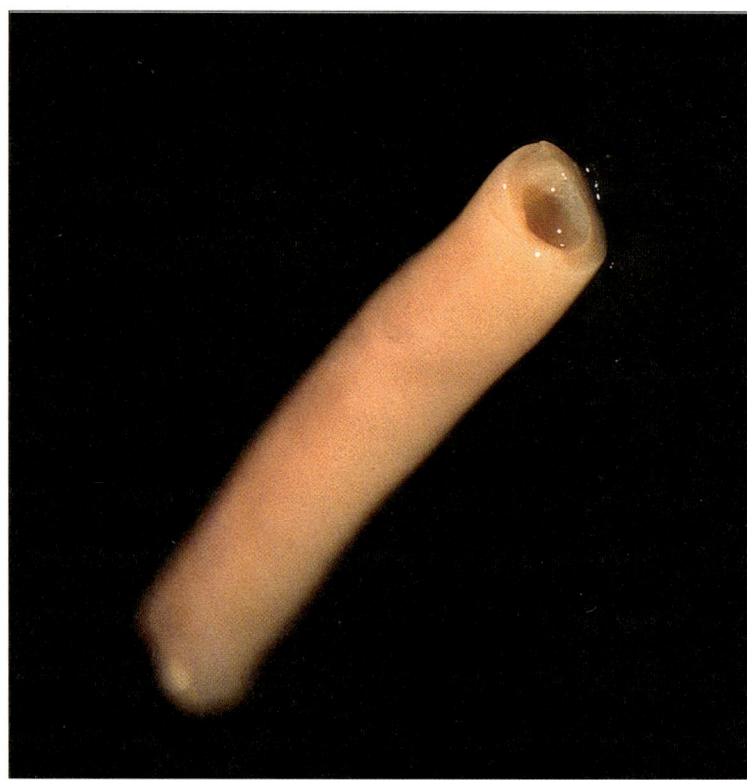

Photo:
Recipe for a lab-built small-diameter blood vessel: Seed lining cells onto the inner surfaces of tubes of collagen-secreting cells and smooth muscle. These engineered blood vessels may someday replace damaged vessels in people's legs and hearts.

Chapter Objectives

After studying this chapter, you should be able to do the following:

5.1 Introduction
1. List the four major tissue types, and provide examples of where each occurs in the body. (p. 92)

5.2 Epithelial Tissues
2. Describe the general characteristics and functions of epithelial tissues. (p. 92)
3. Name the types of epithelium, and identify an organ in which each is found. (p. 92)
4. Explain how glands are classified. (p. 97)

5.3 Connective Tissues
5. List the types of connective tissues within the body. (p. 98)
6. Describe the general cellular components, structures, fibers, and matrix (where applicable) of each type of connective tissue. (p. 100)
7. Describe the major functions of each type of connective tissue. (p. 100)

5.4 Muscle Tissues
8. Distinguish among the three types of muscle tissues. (p. 105)

5.5 Nervous Tissues
9. Describe the general characteristics and functions of nervous tissues. (p. 107)

Aids to Understanding Words

adip- [fat] *adip*ose tissue: Tissue that stores fat.
chondr- [cartilage] *chondr*ocyte: Cartilage cell.
-cyt [cell] osteo*cyt*e: Bone cell.
epi- [upon] *epi*thelial tissue: Tissue that covers all free body surfaces.
-glia [glue] neuro*glia*: Cells that bind nervous tissue together.
inter- [between] *inter*calated disc: Band between cardiac muscle cells.
macro- [large] *macro*phage: Large phagocytic cell.
oss- [bone] *oss*eous tissue: Bone tissue.
pseudo- [false] *pseudo*stratified epithelium: Tissue whose cells appear to be in layers, but are not.
squam- [scale] *squam*ous epithelium: Tissue whose cells appear flattened or scalelike.
strat- [layer] *strat*ified epithelium: Tissue whose cells occur in layers.

Key Terms

adipose tissue (ad´ĭ-pōs tish´u)
cartilage (kar´tĭ-lij)
chondrocyte (kon´dro-sīt)
connective tissue (kŏ-nek´tiv tish´u)
epithelial tissue (ep´´ĭ-the´le-al tish´u)
fibroblast (fi´bro-blast)
macrophage (mak´ro-fāj)
muscle tissue (mus´el tish´u)
nervous tissue (ner´vus tish´u)
neuron (nu´ron)
osteocyte (os´te-o-sīt´´)
osteon (os´te-on)

5.1 Introduction

Cells, the basic units of structure and function within the human organism, are organized into groups and layers called **tissues** (tish´uz). Each type of tissue is composed of similar cells specialized to carry on a particular function. The tissues of the human body include four major types: epithelial, connective, muscle, and nervous. Epithelial tissues form protective coverings and function in secretion and absorption. Connective tissues support softer body parts and bind structures together. Muscle tissues produce body movements, and nervous tissues conduct impulses that help control and coordinate body activities. In addition to cells, all tissues contain a nonliving portion called the *matrix*, or intercellular substance. This material varies in composition from tissue to tissue and supports the cells within. Table 5.1 compares the four major tissue types. Throughout this chapter, simplified line drawings (for example, fig. 5.1*A*) are included with each micrograph (for example, fig. 5.1*B*) to emphasize the distinguishing characteristics of the specific tissue.

CHECK YOUR RECALL

1. What is a tissue?
2. List the four major types of tissues.

5.2 Epithelial Tissues

General Characteristics

Epithelial (ep´´ĭ-the´le-al) tissues are widespread throughout the body, covering organs, forming the inner linings of body cavities, and lining hollow organs. This tissue, because it forms linings, always has a free surface—one that is exposed to the outside or to an open internal space. The underside of this tissue is anchored to connective tissue by a thin, nonliving layer, called the **basement membrane.**

As a rule, epithelial tissues lack blood vessels. However, nutrients diffuse to epithelium from underlying connective tissues, which have abundant blood vessels.

Epithelial cells readily divide. As a result, injuries heal rapidly as new cells replace lost or damaged ones. Skin cells and cells that line the stomach and intestines are continually damaged and replaced.

Epithelial cells are tightly packed, with little intercellular material between them. Consequently, these cells form effective protective barriers in such structures as the outer layer of the skin and the lining of the mouth. Other epithelial functions include secretion, absorption, excretion, and sensory reception.

Epithelial tissues are classified according to shape and number of layers of cells. Epithelial tissues that are composed of single layers of cells are *simple;* those with two or more layers of cells are *stratified;* those with thin, flattened cells are *squamous;* those with cube-shaped cells are *cuboidal;* and those with elongated cells are *columnar.* In the following descriptions, note that the free surfaces of epithelial cells are modified to reflect their specialized functions.

CHECK YOUR RECALL

1. List the general characteristics of epithelial tissues.
2. Describe the structure of epithelium as it relates to the shape and number of layers of cells.

Simple Squamous Epithelium

 (skwa´mus) **epithelium** consists of a single layer of thin, flattened cells. These cells fit tightly together, somewhat like floor tiles, and their nuclei are usually broad and thin (fig. 5.1).

Substances pass rather easily through simple squamous epithelium, which is common at sites of diffusion and filtration. For instance, simple squamous epithelium lines the air sacs (alveoli) of the lungs where oxygen and carbon dioxide are exchanged. It also forms the walls of capillaries, lines the insides of blood and lymph

TABLE 5.1			TISSUES
TYPE	FUNCTION	LOCATION	DISTINGUISHING CHARACTERISTICS
Epithelial	Protection, secretion, absorption, excretion	Cover body surfaces, cover and line internal organs, compose glands	Lack blood vessels, readily divide; cells are tightly packed
Connective	Bind, support, protect, fill spaces, store fat, produce blood cells	Widely distributed throughout body	Mostly have good blood supply; cells are farther apart than cells of epithelia, with matrix in between
Muscle	Movement	Attached to bones, in the walls of hollow internal organs, heart	Contractile
Nervous	Transmit impulses for coordination, regulation, integration, and sensory reception	Brain, spinal cord, nerves	Cells connect to each other and other body parts

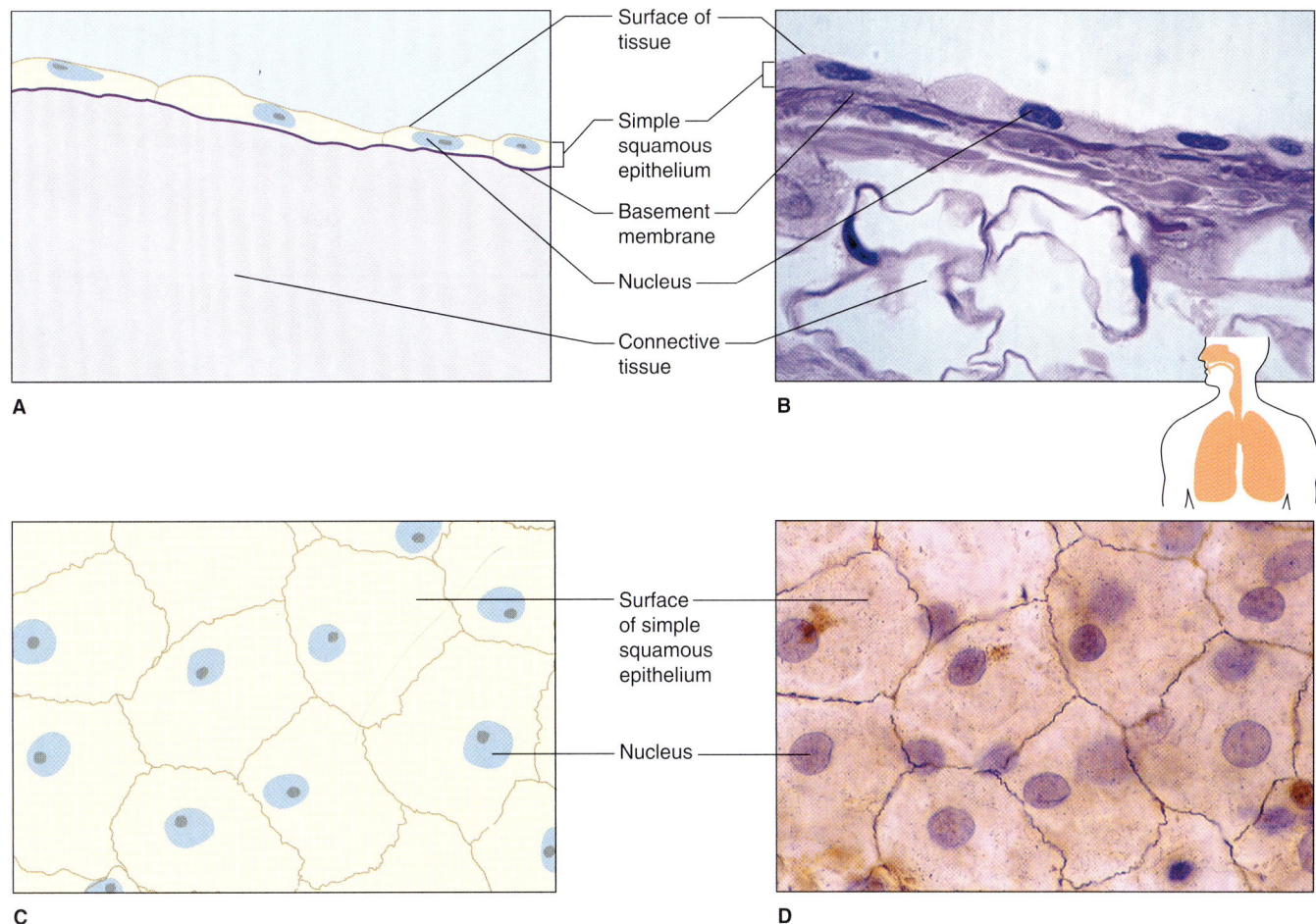

Figure 5.1
Simple squamous epithelium consists of a single layer of tightly packed, flattened cells (670×). (A) and (B) side view, (C) and (D) surface view.

vessels, and covers membranes that line body cavities. However, because it is so thin and delicate, simple squamous epithelium is easily damaged.

Simple Cuboidal Epithelium

Simple cuboidal epithelium consists of a single layer of cube-shaped cells. These cells usually have centrally located, spherical nuclei (fig. 5.2).

Simple cuboidal epithelium covers the ovaries and lines most of the kidney tubules and the ducts of certain glands, such as the salivary glands, thyroid gland, pancreas, and liver. In the kidneys, this tissue functions in secretion and absorption; in glands, it secretes glandular products.

Simple Columnar Epithelium

The cells of **simple columnar epithelium** are elongated; that is, they are longer than they are wide. This tissue is composed of a single layer of cells whose nuclei are usually located at about the same level, near the basement membrane (fig. 5.3). The cells of this tissue can be ciliated or nonciliated. *Cilia* extend from the free surfaces of cells and move constantly (see chapter 3, p. 56). In the female reproductive tubes, cilia aid in moving egg cells to the uterus.

Nonciliated simple columnar epithelium lines the uterus and most organs of the digestive tract, including the stomach and the small and large intestines. Because its cells are elongated, this tissue is thick, which enables it to protect underlying tissues. Simple columnar epithelium also secretes digestive fluids and absorbs nutrients from digested food.

Simple columnar cells, specialized for absorption, often have many minute, cylindrical processes extending from their surfaces. These processes, called *microvilli*, increase the surface area of the cell membrane where it is exposed to substances being absorbed.

Typically, specialized, flask-shaped glandular cells are scattered among the columnar cells of simple columnar epithelium. These cells, called *goblet cells*, secrete a protective fluid (*mucus*) onto the free surface of the tissue (fig. 5.3).

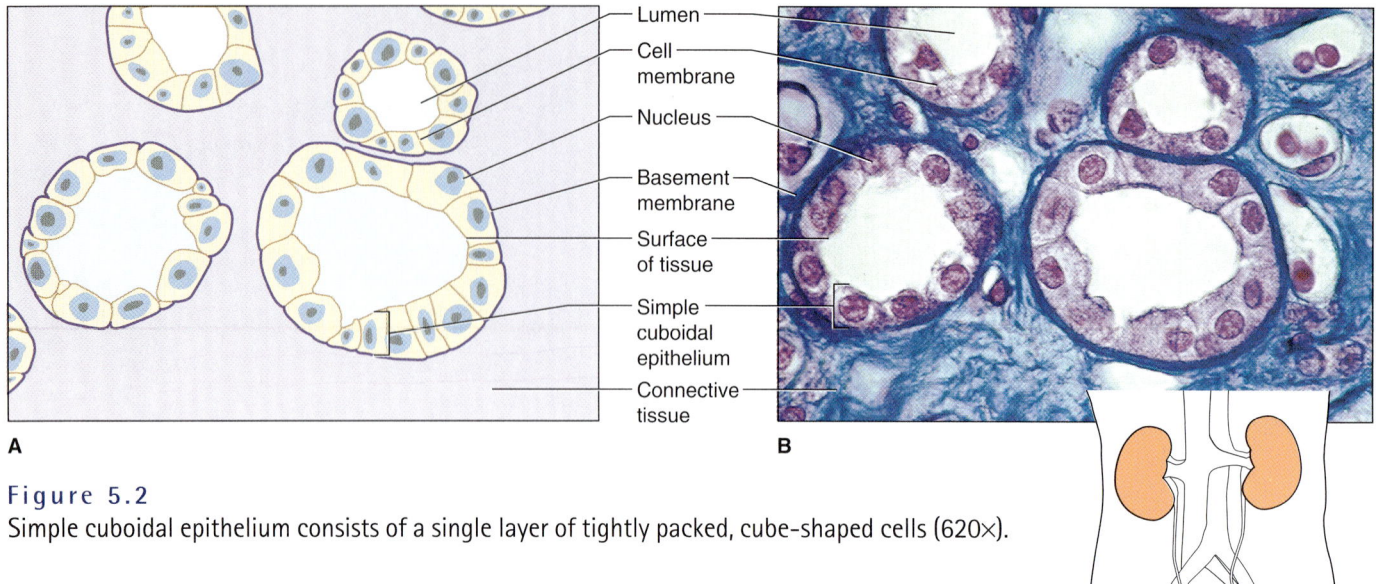

Figure 5.2
Simple cuboidal epithelium consists of a single layer of tightly packed, cube-shaped cells (620×).

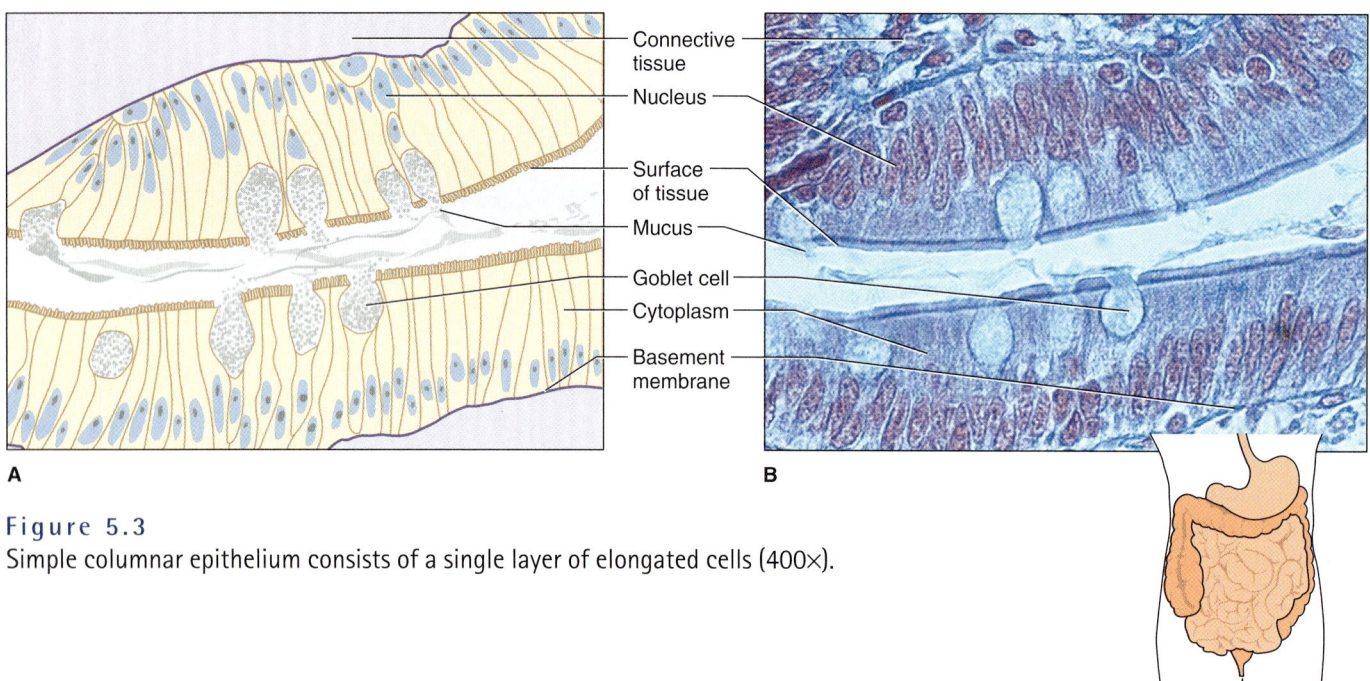

Figure 5.3
Simple columnar epithelium consists of a single layer of elongated cells (400×).

Pseudostratified Columnar Epithelium

The cells of **pseudostratified** (soo˝do-strat´ĭ-fid) **columnar epithelium** appear stratified or layered, but they are not. A layered effect occurs because the cell nuclei are at two or more levels in the row of aligned cells (fig. 5.4).

Pseudostratified columnar epithelial cells commonly have cilia, which extend from the free surfaces of the cells. Goblet cells scattered throughout this tissue secrete mucus, which the cilia sweep away.

Pseudostratified columnar epithelium lines the passages of the respiratory system. Here, the mucus-covered linings are sticky and trap dust and microorganisms that enter with the air. The cilia move the mucus and its captured particles upward and out of the airways.

Stratified Squamous Epithelium

The many cell layers of **stratified squamous epithelium** make this tissue relatively thick. Cells divide in the deeper layers, and newer cells push older ones farther outward, where they become flattened (fig. 5.5).

Stratified squamous epithelium forms the outer layer of the skin (*epidermis*). As skin cells age, they accumulate a protein called *keratin* and then harden and die. This "keratinization" produces a covering of dry, tough, protective material that prevents water and other

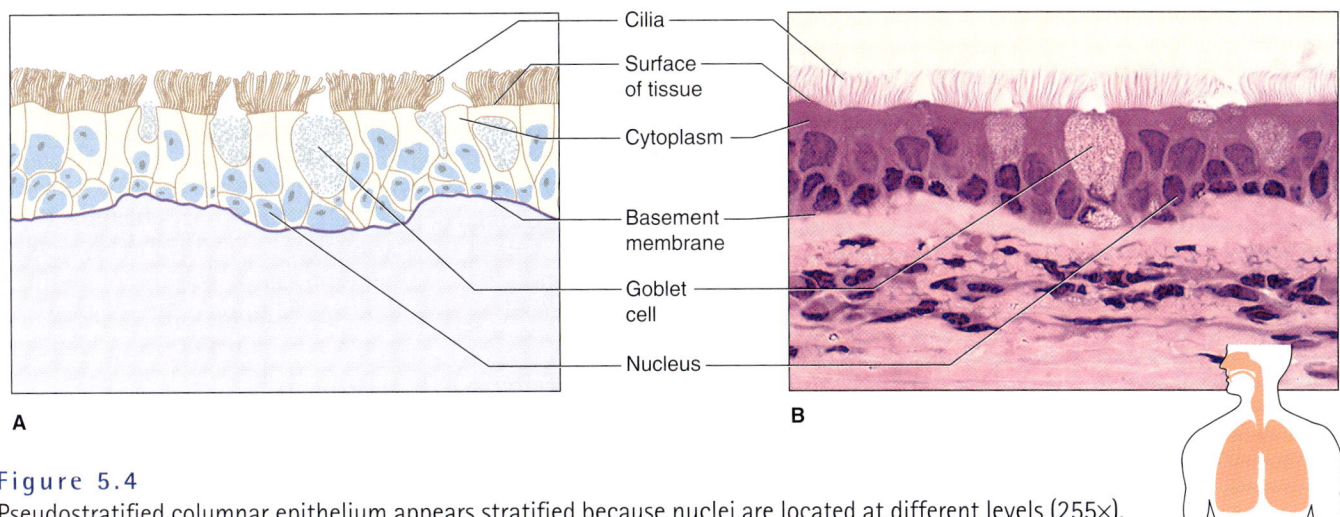

Figure 5.4
Pseudostratified columnar epithelium appears stratified because nuclei are located at different levels (255×).

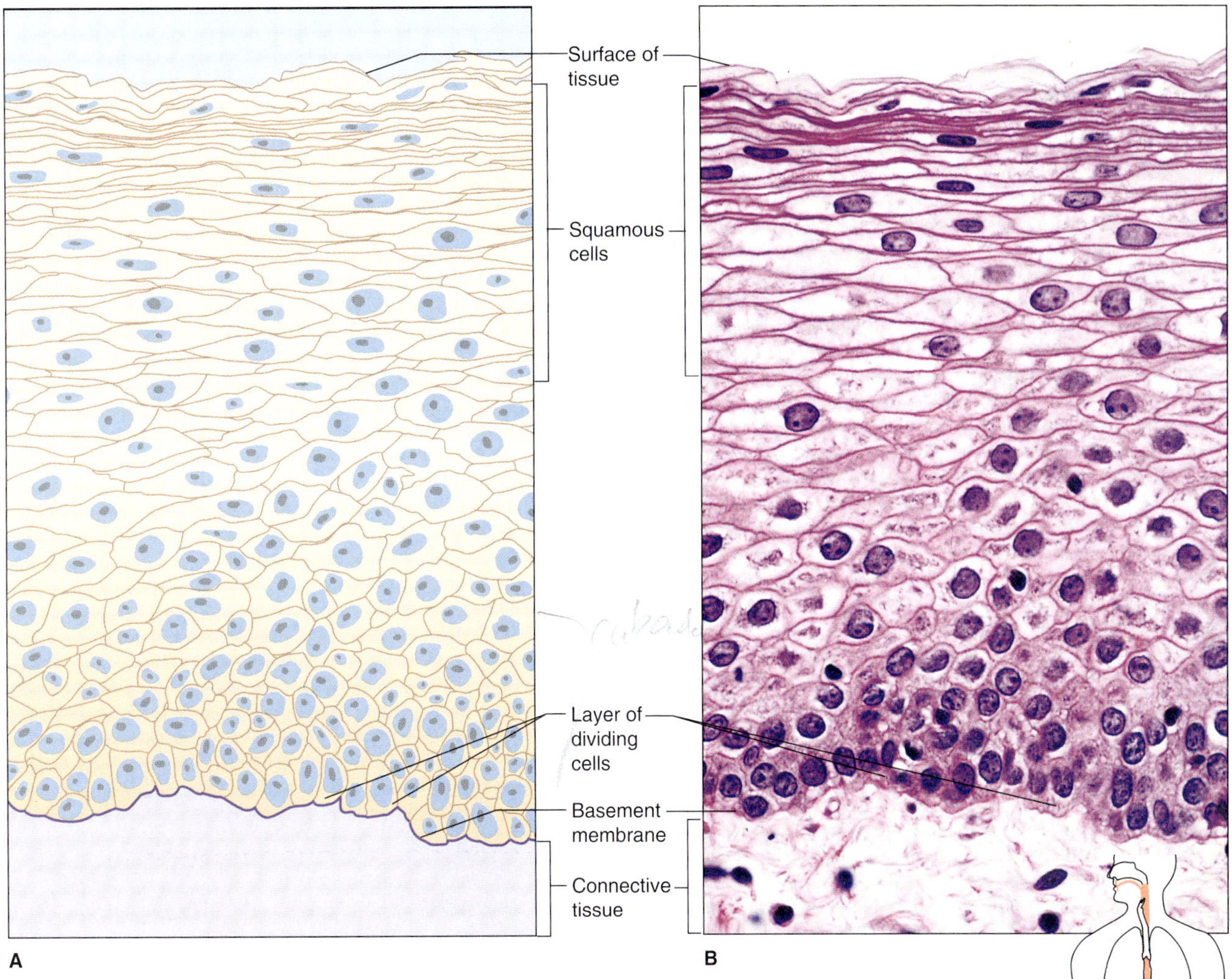

Figure 5.5
Stratified squamous epithelium consists of many layers of cells (385×).

substances from escaping underlying tissues and blocks various chemicals and microorganisms from entering.

Stratified squamous epithelium also lines the mouth, throat, vagina, and anal canal. In these parts, the tissue is not keratinized; it stays soft and moist, and the cells on its free surfaces remain alive.

Stratified Cuboidal Epithelium

Stratified cuboidal epithelium consists of two or three layers of cuboidal cells that form the lining of a lumen. The layering of the cells provides more protection than the single layer affords (fig. 5.6).

Stratified cuboidal epithelium lines the larger ducts of the mammary glands, sweat glands, salivary glands, and pancreas. It also forms the lining of developing ovarian follicles and seminiferous tubules, which are parts of the female and male reproductive systems, respectively.

Stratified Columnar Epithelium

Stratified columnar epithelium consists of several layers of cells (fig. 5.7). The superficial cells are elongated, whereas the basal layers consist of cube-shaped cells. Stratified columnar epithelium is found in the male urethra and vas deferens and in parts of the pharynx.

Transitional Epithelium

Transitional epithelium is specialized to change in response to increased tension. It forms the inner lining of the urinary bladder and lines the ureters and part of

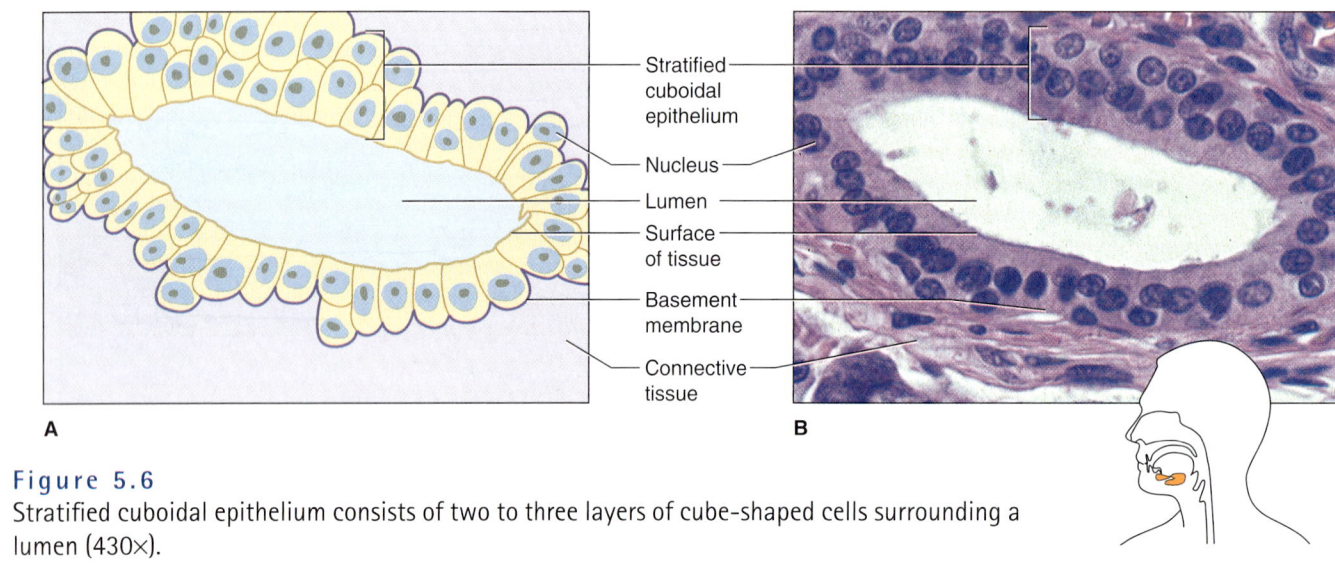

Figure 5.6
Stratified cuboidal epithelium consists of two to three layers of cube-shaped cells surrounding a lumen (430×).

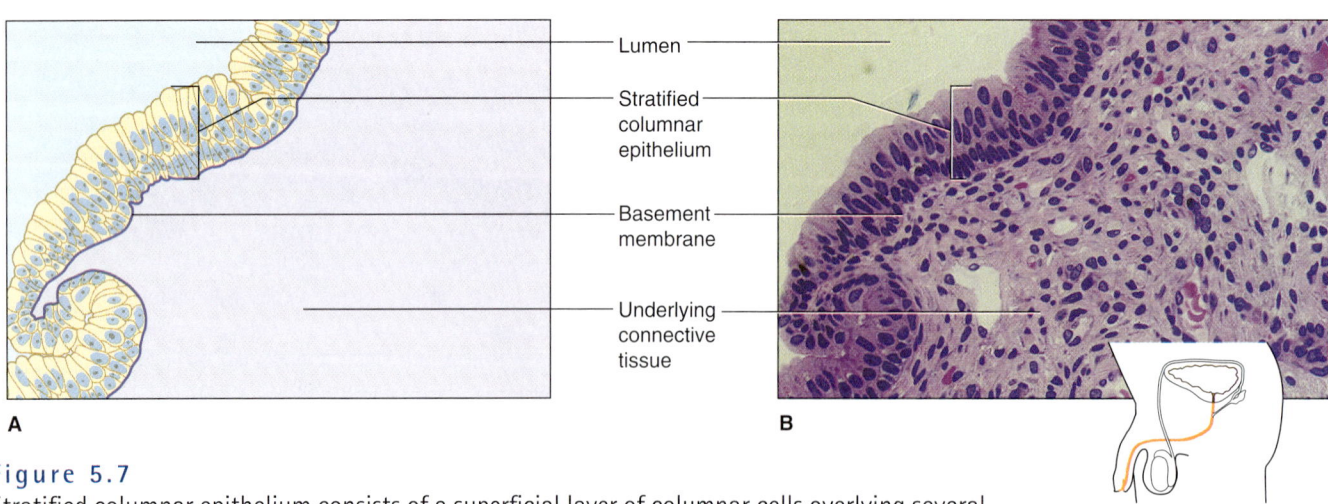

Figure 5.7
Stratified columnar epithelium consists of a superficial layer of columnar cells overlying several layers of cuboidal cells (220×).

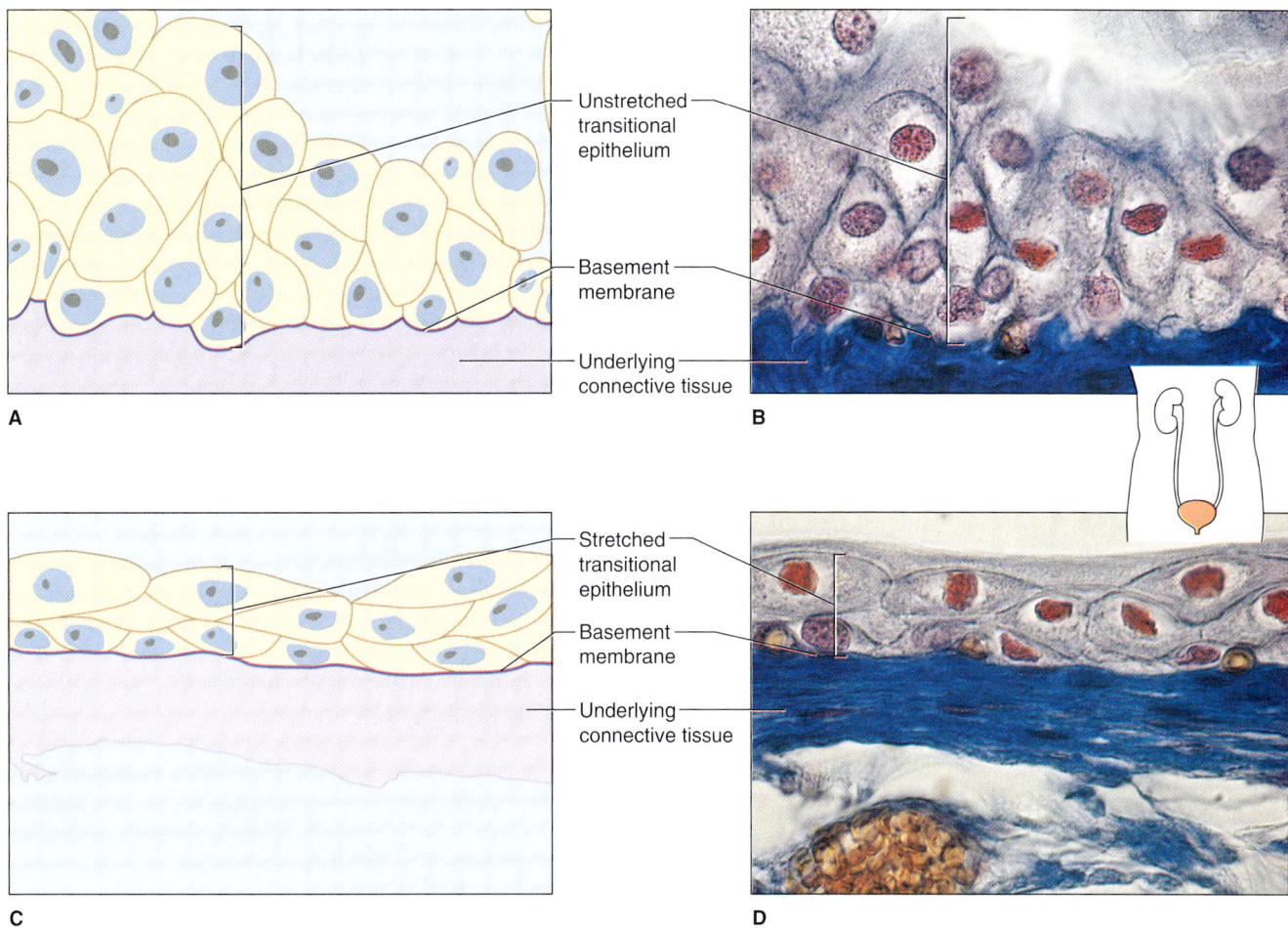

Figure 5.8
Transitional epithelium. (*A* and *B*) When the organ wall contracts, transitional epithelium is unstretched and consists of many layers (675×). (*C* and *D*) When the organ is distended, the tissue stretches and appears thinner (675×).

the urethra. When the walls of one of these organs contract, the tissue consists of several layers of cuboidal cells; however, when the organ is distended, the tissue stretches, and the physical relationships among the cells change (fig. 5.8). In addition to providing an expandable lining, transitional epithelium forms a barrier that helps prevent the contents of the urinary tract from diffusing back into the internal environment.

U p to 90% of all human cancers are *carcinomas,* which are growths that originate in epithelium. Most carcinomas begin on surfaces that contact the external environment, such as skin, linings of the airways in the respiratory tract, or linings of the stomach or intestine in the digestive tract. This observation suggests that the more common cancer-causing agents may not penetrate tissues very deeply.

Glandular Epithelium

Glandular epithelium is composed of cells that are specialized to produce and secrete substances into ducts or into body fluids. Such cells are usually found within columnar and cuboidal epithelia, and one or more of these cells constitute a *gland*. Glands that secrete their products into ducts that open onto some internal or external surface are called **exocrine glands.** Glands that secrete their products into tissue fluid or blood are called **endocrine glands.** (Endocrine glands are discussed in chapter 11.)

Exocrine glands are classified according to the ways these glands secrete their products (fig. 5.9). Glands that release watery, protein-rich fluids by exocytosis are called **merocrine** (mer´o-krin) **glands.** Glands that lose small portions of their glandular cell bodies during secretion are called **apocrine** (ap´o-krin) **glands.** **Holocrine** (ho´lo-krin) **glands** are those in which the

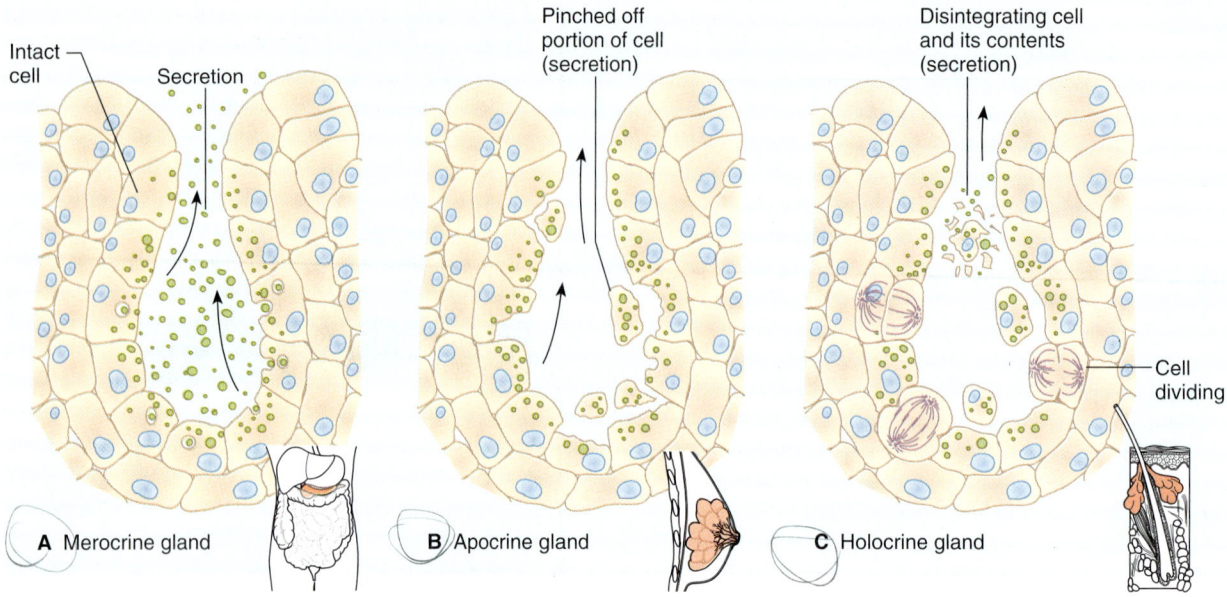

Figure 5.9
Glandular secretions. (A) Merocrine glands release secretions without losing cytoplasm. (B) Apocrine glands lose small portions of their cell bodies during secretion. (C) Holocrine glands release entire cells filled with secretory products.

entire cell lyses during secretion. Table 5.2 summarizes these glands and their secretions.

Most exocrine secretory cells are merocrine, and they can be further subdivided as either *serous cells* or *mucous cells*. The secretion of serous cells is typically watery, has a high concentration of enzymes, and is called *serous fluid*. Serous cells are common in the linings of the body cavities. Mucous cells secrete a thicker fluid called *mucus*. This substance is rich in the glycoprotein *mucin* and is abundantly secreted from the inner linings of the digestive and respiratory systems. Table 5.3 summarizes the characteristics of the different types of epithelial tissues.

 CHECK YOUR RECALL

1. Describe the special functions of each type of epithelium.
2. Distinguish between exocrine glands and endocrine glands.
3. Explain how exocrine glands are classified.
4. Distinguish between a serous cell and a mucous cell.

5.3 Connective Tissues

General Characteristics

Connective (kŏ-nek´tiv) **tissues** bind structures, provide support and protection, serve as frameworks, fill spaces, store fat, produce blood cells, protect against infections, and help repair tissue damage. Connective tissue cells are farther apart than epithelial cells, and they have an abundance of intercellular material, or **matrix** (ma´triks), between them. This matrix consists of *fibers* and a *ground substance* whose consistency varies from fluid to semisolid to solid.

Connective tissue cells can usually divide. These tissues have varying degrees of vascularity, but in most cases, they have good blood supplies and are well nourished. Some connective tissues, such as bone and cartilage, are quite rigid. Loose connective tissue (areolar), adipose tissue, and dense connective tissue are more flexible.

TABLE 5.2   TYPES OF GLANDULAR SECRETIONS

TYPE OF GLAND	DESCRIPTION OF SECRETION	EXAMPLE
Merocrine glands	A fluid product released through the cell membrane by exocytosis	Salivary glands, pancreatic glands, sweat glands of the skin
Apocrine glands	Cellular product and portions of the free ends of glandular cells pinched off during secretion	Mammary glands, ceruminous glands lining the external ear canal
Holocrine glands	Entire cells filled with secretory products disintegrate	Sebaceous glands of the skin

TABLE 5.3 — EPITHELIAL TISSUES

TYPE	FUNCTION	LOCATION
Simple squamous epithelium	Filtration, diffusion, osmosis; covers surface	Air sacs of the lungs, walls of capillaries, linings of blood and lymph vessels
Simple cuboidal epithelium	Secretion, absorption	Surface of ovaries, linings of kidney tubules, and linings of ducts of certain glands
Simple columnar epithelium	Absorption, secretion, protection	Linings of uterus, stomach, and intestine
Pseudostratified columnar epithelium	Protection, secretion, movement of mucus	Linings of respiratory passages
Stratified squamous epithelium	Protection	Outer layer of skin, linings of oral cavity, throat, vagina, and anal canal
Stratified cuboidal epithelium	Protection	Linings of larger ducts of mammary glands, sweat glands, salivary glands, and pancreas
Stratified columnar epithelium	Protection, secretion	Vas deferens, part of the male urethra, parts of the pharynx
Transitional epithelium	Distensibility, protection	Inner lining of urinary bladder and linings of ureters and part of urethra
Glandular epithelium	Secretion	Salivary glands, sweat glands, endocrine glands

Major Cell Types

Connective tissues contain a variety of cell types. Some of them are called *fixed cells* because they are usually present in stable numbers. These include fibroblasts and mast cells. Other cells, such as macrophages, are *wandering cells*. They appear in tissues temporarily, usually in response to an injury or infection.

Fibroblasts (fi´bro-blastz) are the most common type of fixed cell in connective tissue. These large, star-shaped cells produce fibers by secreting proteins into the matrix of connective tissues (fig. 5.10).

Macrophages (mak´ro-fājez), or histiocytes, originate as white blood cells (see chapter 14, p. 374) and are almost as numerous as fibroblasts in some connective tissues. They are specialized to carry on phagocytosis. Macrophages can move about and function as scavenger and defensive cells that clear foreign particles from tissues (fig. 5.11).

Mast cells are large and widely distributed in connective tissues. They are usually located near blood vessels (fig. 5.12). Mast cells release heparin, which prevents blood clotting, and histamine, which promotes some of the reactions associated with inflammation and allergies (see chapter 14, pp. 374 and 382).

Connective Tissue Fibers

Fibroblasts produce three types of connective tissue fibers: collagenous fibers, elastic fibers, and reticular fibers. Of these, collagenous and elastic fibers are the most abundant (fig. 5.13).

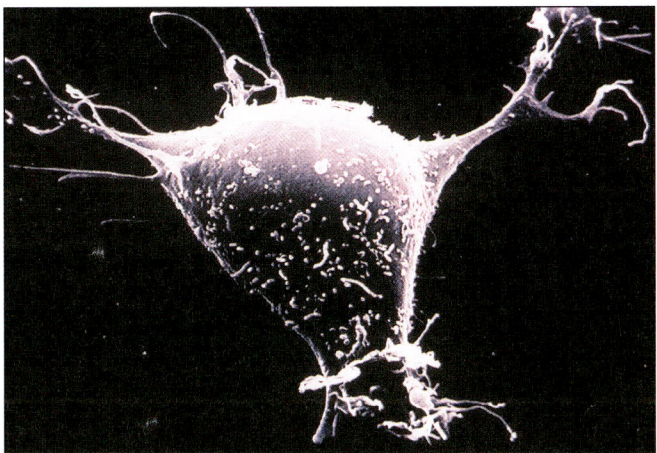

Figure 5.10
Scanning electron micrograph of a fibroblast (3,800×).

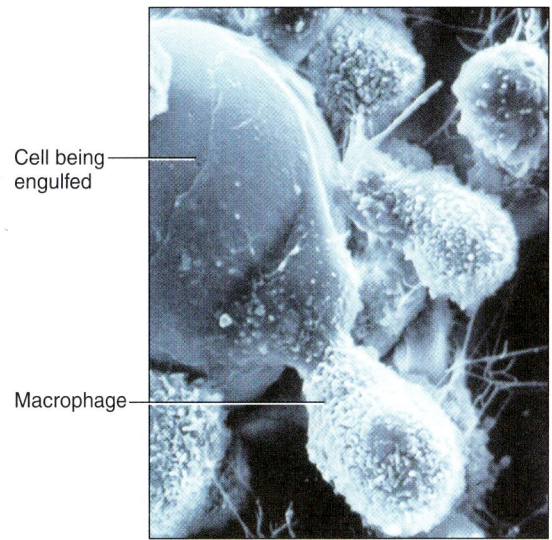

Figure 5.11
Macrophages are scavenger cells common in connective tissues. This scanning electron micrograph shows a number of macrophages engulfing a larger cell (3,300×).

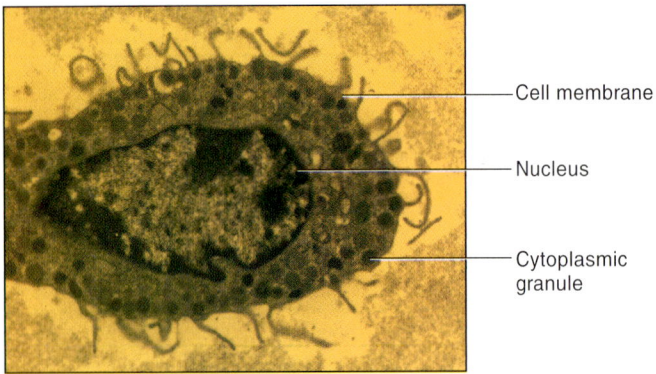

Figure 5.12
Transmission electron micrograph of a mast cell (5,000×).

When skin is exposed to prolonged and intense sunlight, connective tissue fibers lose elasticity, and the skin stiffens and becomes leathery. In time, the skin may sag and wrinkle. Collagen injections may temporarily smooth out wrinkles. However, collagen applied as a cream to the skin does not combat wrinkles because collagen molecules are far too large to actually penetrate the skin.

Collagenous (kol-laj´ĕ-nus) **fibers** are thick threads of the protein *collagen*. Collagenous fibers are grouped in long, parallel bundles, and they are flexible but only slightly elastic. More importantly, they have great tensile strength—that is, they resist considerable pulling force. Thus, collagenous fibers are important components of body parts that hold structures together, such as **ligaments** (which connect bones to bones) and **tendons** (which connect muscles to bones).

Tissue containing abundant collagenous fibers is called *dense connective tissue*. Such tissue appears white, and for this reason, collagenous fibers are sometimes called *white fibers*.

Elastic fibers are composed of a protein called *elastin*. These thin fibers branch, forming complex networks. Elastic fibers are weaker than collagenous fibers, but they stretch easily and can resume their original lengths and shapes. Elastic fibers are common in body parts that are frequently stretched, such as the vocal cords. They are sometimes called *yellow fibers* because tissues well supplied with them appear yellowish.

Reticular fibers are very thin collagenous fibers. They are highly branched and form delicate supporting networks in a variety of tissues.

Categories of Connective Tissue

Connective tissue is broken down into two categories. *Connective tissue proper* includes loose connective tissue, adipose tissue, and dense connective tissue. The *specialized connective tissues* include cartilage, bone, and blood.

Loose Connective Tissue

Loose connective tissue, or **areolar** (ah-re´o-lar) **tissue,** forms delicate, thin membranes throughout the body. The cells of this tissue, mainly fibroblasts, are located some distance apart and are separated by a gel-like matrix containing many collagenous and elastic fibers that fibroblasts secrete (fig. 5.13).

Loose connective tissue binds the skin to the underlying organs and fills spaces between muscles. It lies beneath most layers of epithelium, where its many blood vessels nourish nearby epithelial cells.

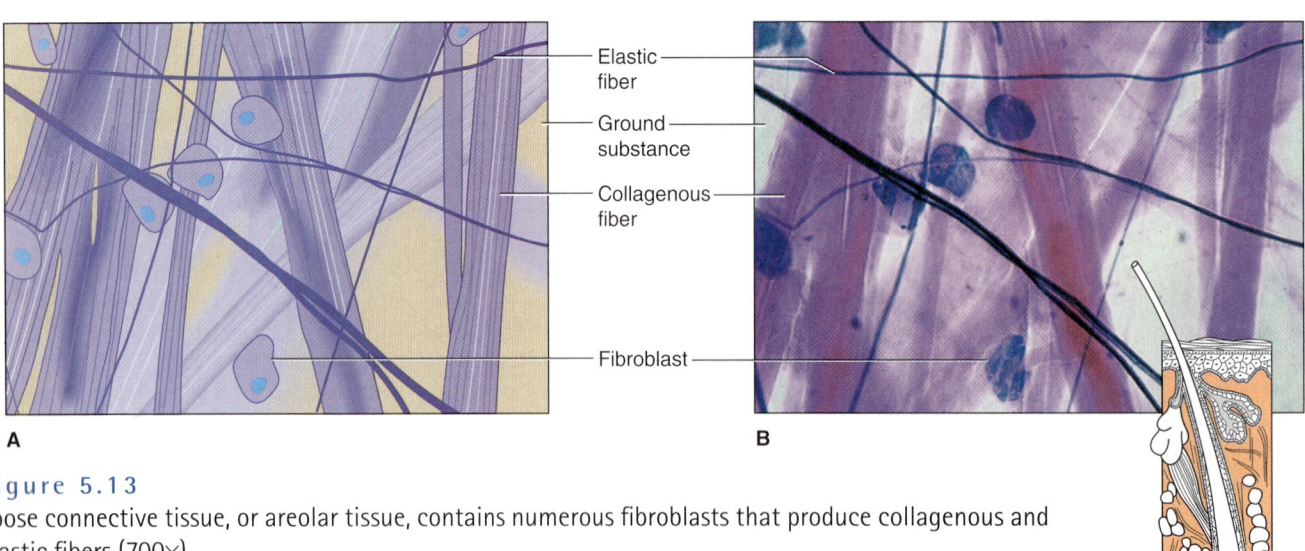

Figure 5.13
Loose connective tissue, or areolar tissue, contains numerous fibroblasts that produce collagenous and elastic fibers (700×).

Adipose Tissue

Adipose (ad´ĭ-pōs) **tissue,** or fat, is a specialized form of loose connective tissue that develops when certain cells (adipocytes) store fat in droplets within their cytoplasm and enlarge (fig. 5.14). When such cells are so numerous that they crowd other cell types, they form adipose tissue. Adipose tissue lies beneath the skin, in spaces between muscles, around the kidneys, behind the eyeballs, in certain abdominal membranes, on the surface of the heart, and around certain joints.

Adipose tissue cushions joints and some organs, such as the kidneys. It also insulates beneath the skin, and it stores energy in fat molecules.

 The average adult has between 40 and 50 billion fat cells.

Overeating and lack of exercise can increase the size of adipose cells, leading to becoming overweight or obese. During periods of fasting, however, fat supplies energy, and adipocytes lose fat, shrink, and become more like fibroblasts.

Dense Connective Tissue

Dense connective tissue consists of many closely packed, thick, collagenous fibers and a fine network of elastic fibers. It has relatively few cells, most of which are fibroblasts (fig. 5.15).

Collagenous fibers of dense connective tissue are very strong, enabling the tissue to withstand pulling forces. Dense connective tissue often binds body parts together as parts of tendons and ligaments. This type of tissue is also in the protective white layer of the eyeball

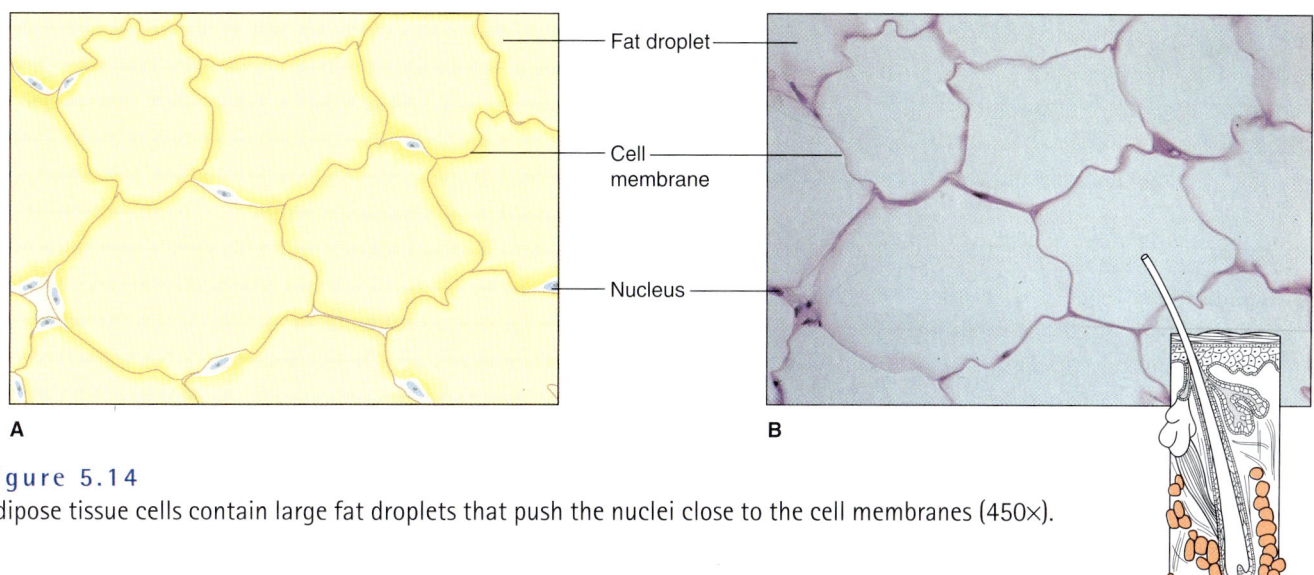

Figure 5.14
Adipose tissue cells contain large fat droplets that push the nuclei close to the cell membranes (450×).

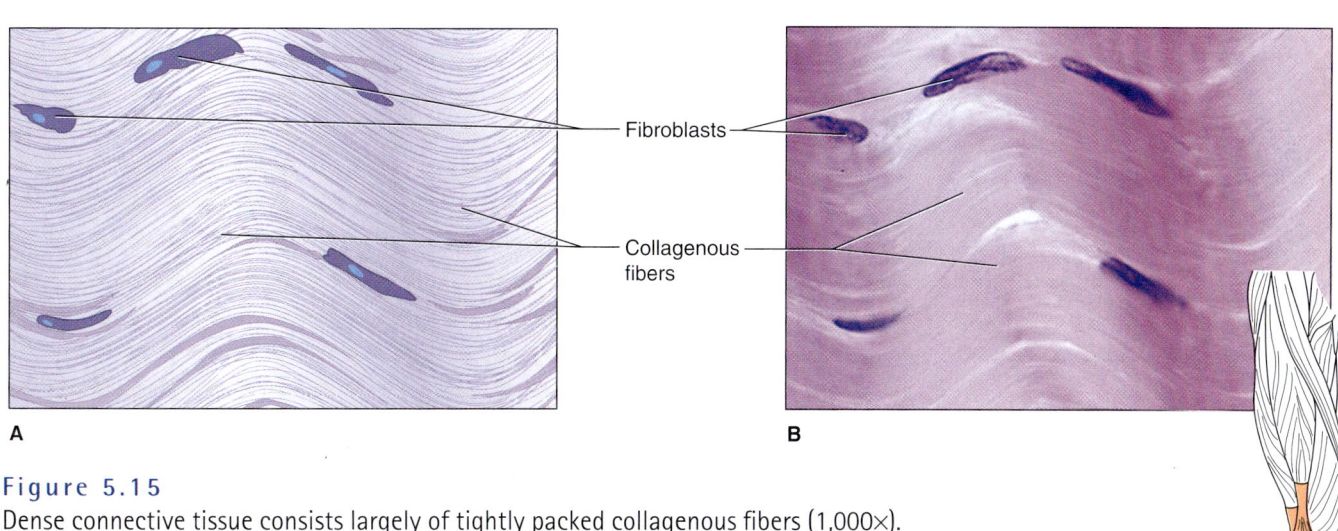

Figure 5.15
Dense connective tissue consists largely of tightly packed collagenous fibers (1,000×).

and in the deeper skin layers. The blood supply to dense connective tissue is poor, slowing tissue repair.

 CHECK YOUR RECALL

1. What are the general characteristics of connective tissues?
2. What are the characteristics of collagen and elastin?
3. What feature distinguishes adipose tissue from other connective tissues?
4. Explain the difference between loose connective tissue and dense connective tissue.

Cartilage

Cartilage (kar´ti-lij) is a rigid connective tissue. It provides support, frameworks, and attachments; protects underlying tissues; and forms structural models for many developing bones.

Cartilage matrix is abundant and is largely composed of collagenous fibers embedded in a gel-like ground substance. Cartilage cells, or **chondrocytes** (kon´dro-sītz), occupy small chambers called *lacunae* and thus lie completely within the matrix (fig. 5.16).

A cartilaginous structure is enclosed in a covering of connective tissue called the *perichondrium*. The perichondrium contains blood vessels that provide cartilage cells with nutrients by diffusion. The lack of a direct blood supply to cartilage tissue is why torn cartilage heals slowly and why chondrocytes do not divide frequently.

Different types of intercellular material (matrix) distinguish three types of cartilage. **Hyaline cartilage,** the most common type, has very fine collagenous fibers in its matrix and looks somewhat like white glass (fig. 5.16). It is found on the ends of bones in many joints, in the soft part of the nose, and in the supporting rings of the respiratory passages. Hyaline cartilage is also important in the growth of most bones (see chapter 7, p. 130).

Elastic cartilage contains a dense network of elastic fibers and thus is more flexible than hyaline cartilage (fig. 5.17). It provides the framework for the external ears and for parts of the larynx.

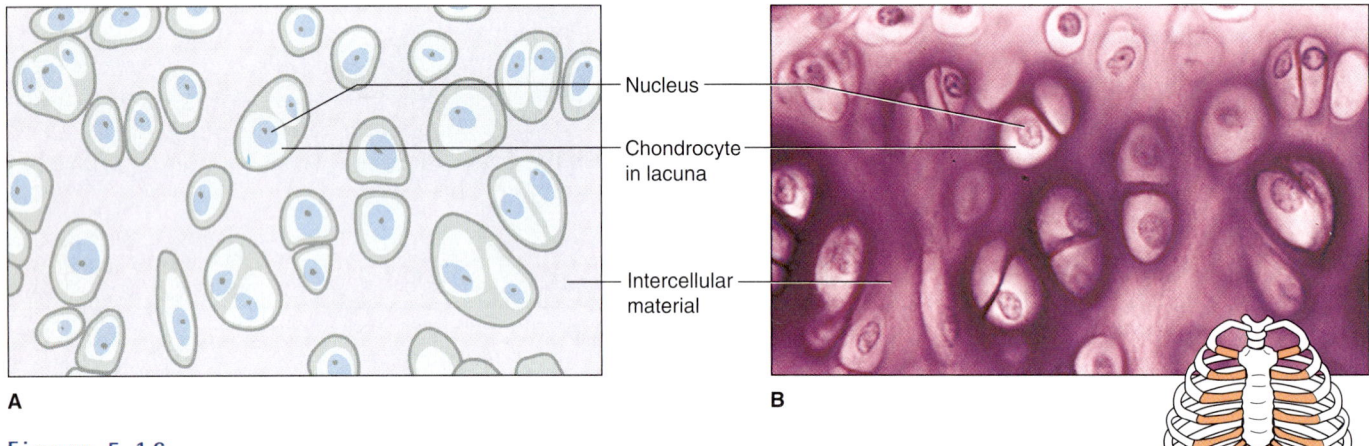

Figure 5.16
Hyaline cartilage cells (chondrocytes) are located in lacunae, which are in turn surrounded by intercellular material containing very fine collagenous fibers (610×).

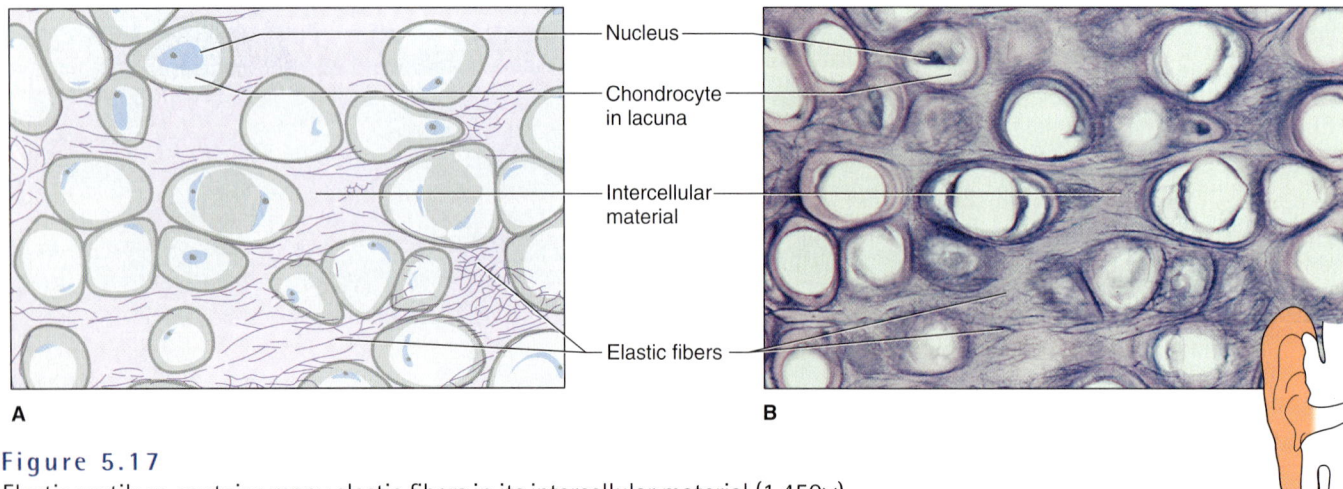

Figure 5.17
Elastic cartilage contains many elastic fibers in its intercellular material (1,450×).

Between ages 30 and 70, a nose may lengthen and widen by as much as half an inch, and the ears may lengthen by a quarter inch due to the fact that cartilage is one of the few tissues that continues to grow as we age.

Fibrocartilage, a very tough tissue, contains many collagenous fibers (fig. 5.18). It is a shock absorber for structures that are subjected to pressure. For example, fibrocartilage forms pads (intervertebral discs) between the individual bones (vertebrae) of the spinal column. It also cushions bones in the knees and in the pelvic girdle.

Bone

Bone is the most rigid connective tissue. Its hardness is largely due to mineral salts, such as calcium phosphate and calcium carbonate, between cells. This matrix also contains abundant collagen fibers, which are flexible and reinforce the mineral components of bone.

Bone internally supports body structures. It protects vital parts in the cranial and thoracic cavities, and is an attachment for muscles. Bone also contains red marrow, which forms blood cells, and it stores and releases inorganic chemicals such as calcium and phosphorus.

Bone matrix is deposited in thin layers called *lamellae,* which form concentric patterns around tiny longitudinal tubes called *central canals,* or Haversian canals (fig. 5.19). Bone cells, or **osteocytes** (os´te-o-sītz), are located in lacunae, which are rather evenly spaced between the lamellae. Consequently, osteocytes too form concentric circles.

In a bone, the osteocytes and layers of intercellular material, which are concentrically clustered around a central canal, form a cylinder-shaped unit called an **osteon** (os´te-on), or Haversian system. Many osteons cemented together form the substance of bone.

Each central canal contains a blood vessel, which places every bone cell near a nutrient supply. In addition, bone cells have many cytoplasmic processes that extend outward and pass through very small tubes in the matrix called *canaliculi.* These cellular processes connect with the membranes of nearby cells. As a result, materials can move rapidly between blood vessels and bone cells. Thus, in spite of its inert appearance, bone is a very active tissue that heals much more rapidly than does injured cartilage. (The microscopic structure of bone is described in more detail in chapter 7, p. 129.)

Blood

Blood transports a variety of materials between interior body cells and those that exchange substances with the external environment. In this way, blood helps maintain stable internal environmental conditions. Blood is composed of *formed elements* suspended in a fluid matrix called *blood plasma.* The formed elements include *red blood cells, white blood cells,* and cell fragments called *platelets* (fig. 5.20). Most blood cells form in red marrow within the hollow parts of certain long bones. Chapter 12 describes blood in detail. Table 5.4 lists the characteristics of the connective tissues.

CHECK YOUR RECALL

1. Describe the general characteristics of cartilage.
2. Explain why injured bone heals more rapidly than injured cartilage.
3. What are the major components of blood?

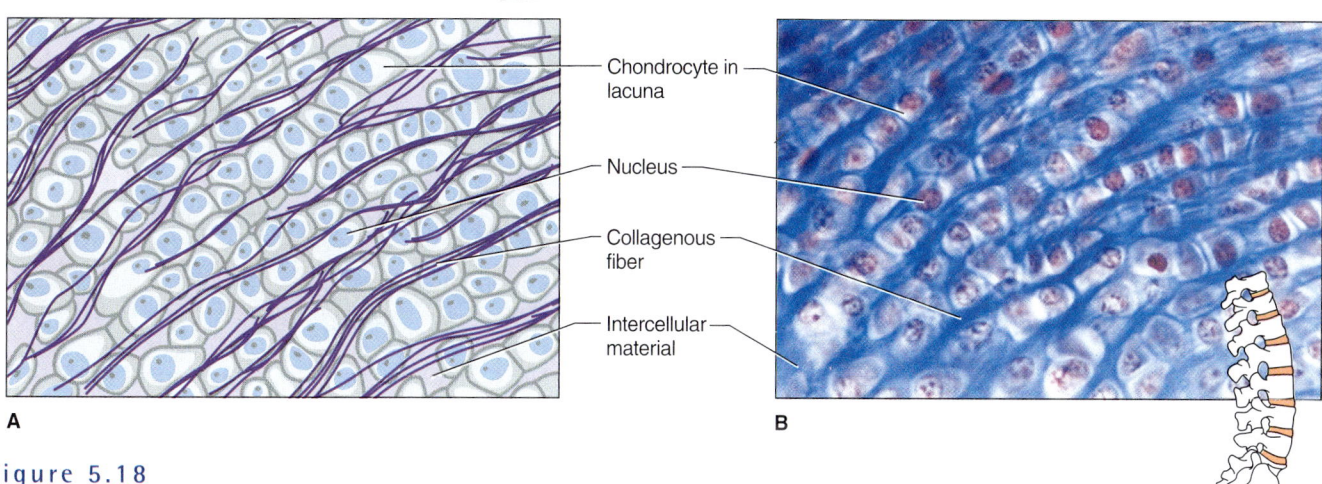

Figure 5.18
Fibrocartilage contains many large collagenous fibers in its intercellular material (1,800×).

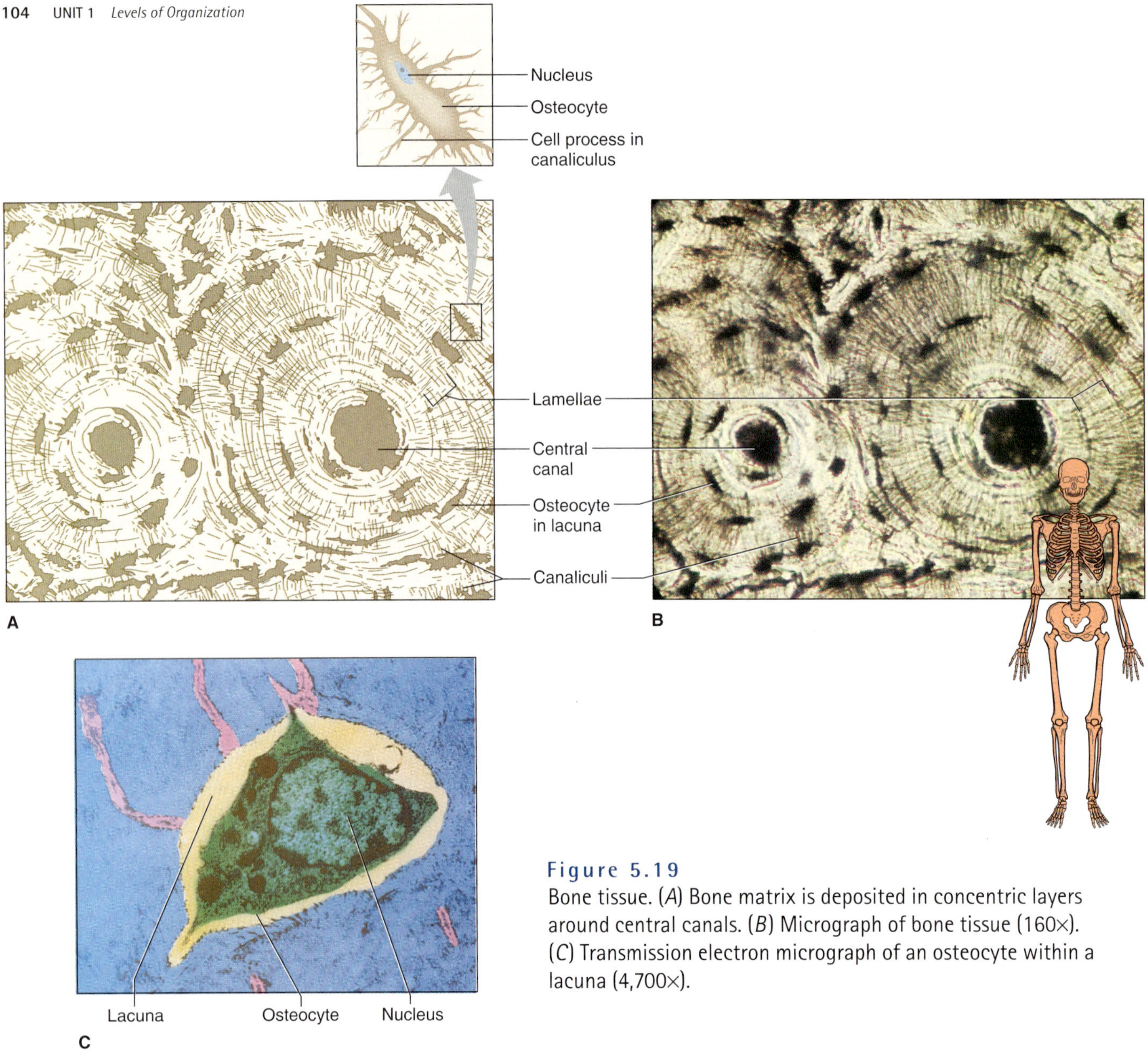

104 UNIT 1 *Levels of Organization*

Figure 5.19
Bone tissue. (*A*) Bone matrix is deposited in concentric layers around central canals. (*B*) Micrograph of bone tissue (160×). (*C*) Transmission electron micrograph of an osteocyte within a lacuna (4,700×).

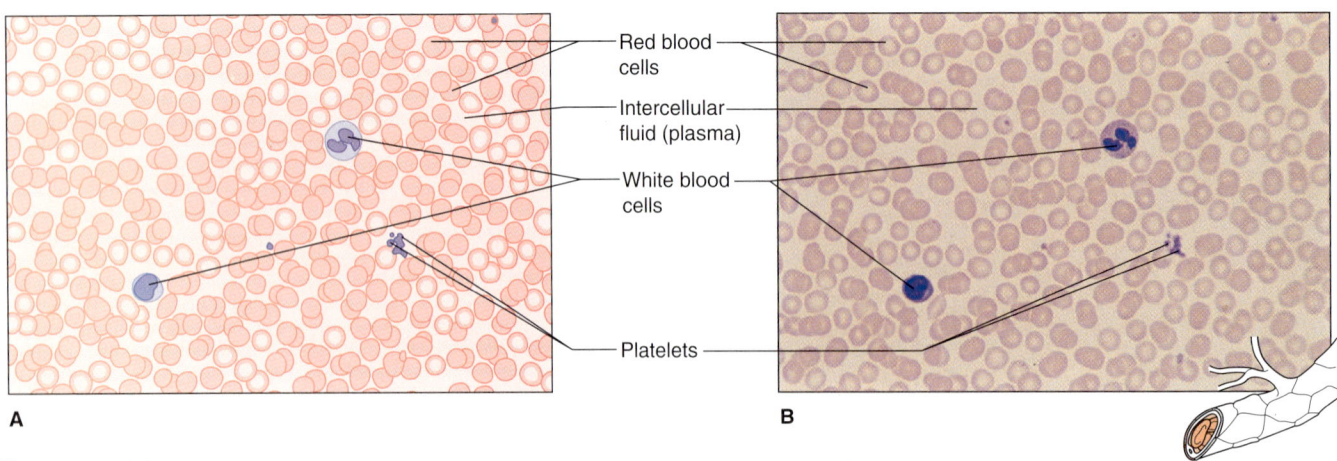

Figure 5.20
Blood tissue consists of red blood cells, white blood cells, and platelets suspended in an intercellular fluid (425×).

TABLE 5.4 CONNECTIVE TISSUES

TYPE	FUNCTION	LOCATION
Loose connective tissue	Binds organs together, holds tissue fluids	Beneath skin, between muscles, beneath epithelial tissues
Adipose tissue	Protects, insulates, stores fat	Beneath skin, around kidneys, behind eyeballs, on surface of heart
Dense connective tissue	Binds organs together	Tendons, ligaments, deeper layers of skin
Hyaline cartilage	Supports, protects, provides framework	Nose, ends of bones, rings in the walls of respiratory passages
Elastic cartilage	Supports, protects, provides flexible framework	Framework of external ear and parts of larynx
Fibrocartilage	Supports, protects, absorbs shock	Between bony parts of spinal column, parts of pelvic girdle and knee
Bone	Supports, protects, provides framework	Bones of skeleton
Blood	Transports substances, helps maintain stable internal environment	Throughout body within a closed system of blood vessels and heart chambers

5.4 Muscle Tissues

General Characteristics

Muscle (mus´el) **tissues** are contractile; that is, their elongated cells, or *muscle fibers,* can shorten. As they contract, muscle fibers pull at their attached ends, and this action moves body parts. The three types of muscle tissue—skeletal, smooth, and cardiac—are discussed in chapter 8.

> Approximately 40% of the body is skeletal muscle, and almost another 10% is smooth and cardiac muscle.

Skeletal Muscle Tissue

Skeletal muscle tissue is found in muscles that attach to bones and are controlled by conscious effort. For this reason, it is often called *voluntary* muscle tissue. The long, threadlike cells of skeletal muscle have alternating light and dark cross-markings called *striations.* Each cell has many nuclei just beneath its cell membrane (fig. 5.21). A nerve impulse can stimulate the muscle fiber to contract and then relax.

The muscles containing skeletal muscle tissue move the head, trunk, and limbs. They enable us to make facial expressions, write, talk, sing, chew, swallow, and breathe.

Smooth Muscle Tissue

Smooth muscle tissue is called smooth because its cells do not have striations. Smooth muscle cells are shorter than those of skeletal muscle and are spindle-shaped, each with a single, centrally located nucleus (fig. 5.22). This tissue comprises the walls of hollow internal organs, such as the stomach, intestine, urinary bladder, uterus, and blood vessels. Unlike skeletal muscle, smooth muscle usually cannot be stimulated to

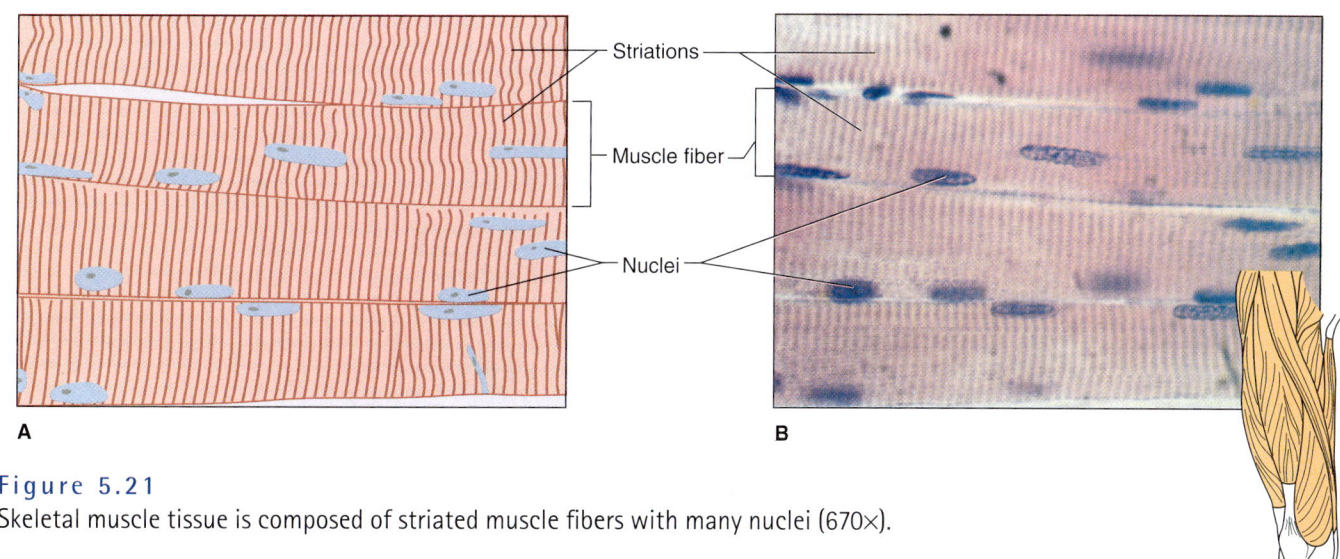

Figure 5.21
Skeletal muscle tissue is composed of striated muscle fibers with many nuclei (670×).

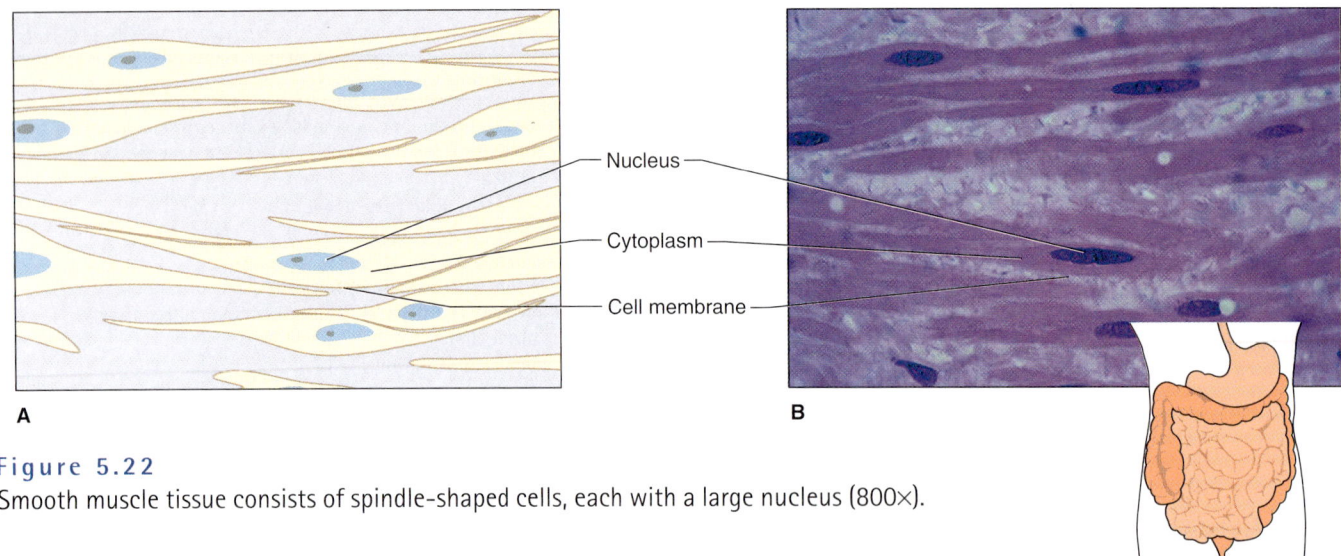

Figure 5.22
Smooth muscle tissue consists of spindle-shaped cells, each with a large nucleus (800×).

contract by conscious efforts. Thus, its actions are *involuntary*. For example, smooth muscle tissue moves food through the digestive tract, constricts blood vessels, and empties the urinary bladder.

Cardiac Muscle Tissue

Cardiac muscle tissue is only in the heart. Its cells, which are striated, are joined end-to-end. The resulting muscle fibers are branched and connected in complex networks. Each cell within a cardiac muscle fiber has a single nucleus (fig. 5.23). Where it touches another cell is a specialized intercellular junction called an *intercalated disc*, discussed further in chapter 8, p. 184.

Cardiac muscle, like smooth muscle, is controlled involuntarily. This tissue makes up the bulk of the heart and pumps blood through the heart chambers and into blood vessels.

CHECK YOUR RECALL

1. List the general characteristics of muscle tissues.
2. Distinguish among skeletal, smooth, and cardiac muscle tissues.

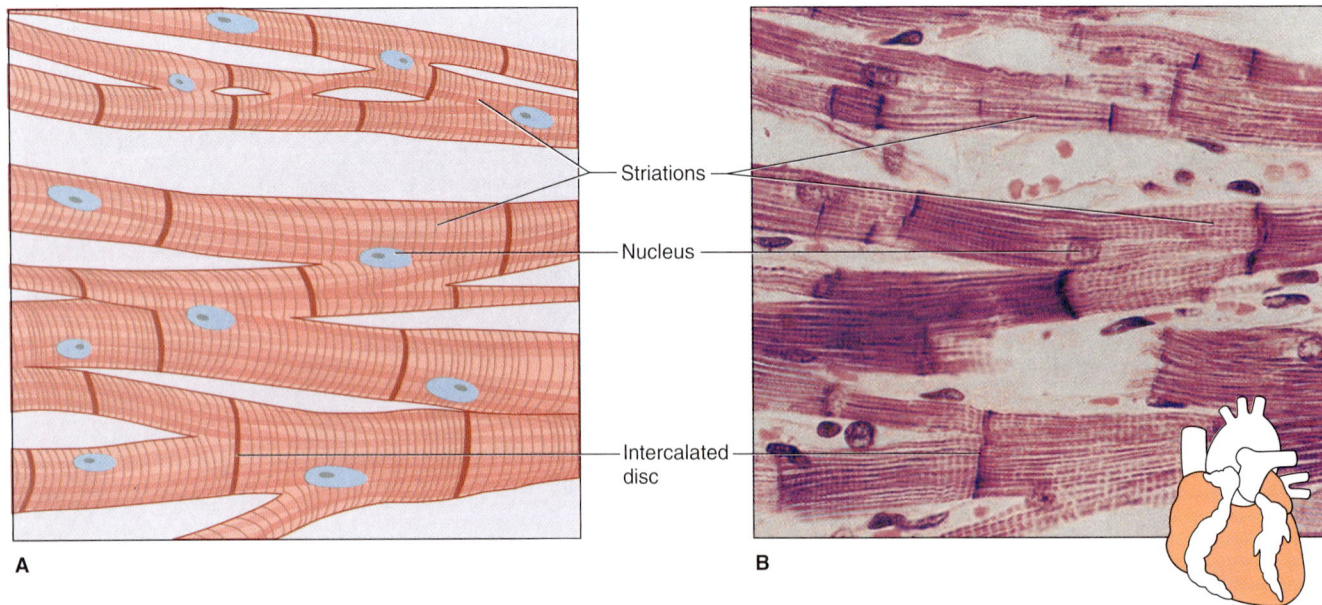

Figure 5.23
Cardiac muscle fibers are branched and interconnected, with a single nucleus each (360×).

The cells of different tissues vary greatly in their abilities to divide. Cells that divide continuously include the epithelial cells of the skin and inner lining of the digestive tract and the connective tissue cells that form blood cells in red bone marrow. However, skeletal and cardiac muscle cells and nerve cells do not usually divide at all after differentiating.

Fibroblasts respond rapidly to injuries by increasing in number and fiber production. They are often the principal agents of repair in tissues that have limited abilities to regenerate. For instance, cardiac muscle tissue typically degenerates in the regions damaged by a heart attack. Fibroblasts then, over time, knit connective tissue that replaces the damaged cardiac muscle. A scar is formed.

TABLE 5.5		MUSCLE AND NERVOUS TISSUES
TYPE	FUNCTION	LOCATION
Skeletal muscle tissue (striated)	Voluntary movements of skeletal parts	Muscles usually attached to bones
Smooth muscle tissue (lacks striations)	Involuntary movements of internal organs	Walls of hollow internal organs
Cardiac muscle tissue (striated)	Heart movements	Heart muscle
Nervous tissue	Sensory reception and conduction of nerve impulses	Brain, spinal cord, and peripheral nerves

5.5 Nervous Tissues

Nervous (ner´vus) **tissues** are found in the brain, spinal cord, and peripheral nerves. The basic cells are called **neurons** (nu´ronz), or nerve cells (fig. 5.24). Neurons sense certain types of changes in their surroundings. They respond by transmitting nerve impulses along cytoplasmic extensions (cellular processes) to other neurons or to muscles or glands. Because neurons communicate with each other and with muscle and gland cells, they can coordinate, regulate, and integrate many body functions.

In addition to neurons, nervous tissue includes **neuroglial cells** (nu-rog´le-ahl selz) or supporting cells, shown in figure 5.24. These cells support and bind the components of nervous tissue, carry on phagocytosis, and help supply nutrients to neurons by connecting them to blood vessels. Nervous tissue is discussed in chapter 9. Table 5.5 summarizes the general characteristics of muscle and nervous tissues.

CHECK YOUR RECALL

1. Describe the general characteristics of nervous tissues.
2. Distinguish between neurons and neuroglial cells.

Clinical Connection

Canavan disease is an inherited illness that illustrates what can happen when communication between the cells of a tissue fails. A child with Canavan disease lags in acquiring developmental skills, such as sitting and standing. Vision worsens, and the child does not make eye contact or react much to the surroundings. He or she may have seizures, require tube feeding, and muscle control may be so poor that the head cannot even be held erect. The child usually dies before adolescence.

The cause of Canavan disease is disruption of the interaction between certain neurons in the brain and

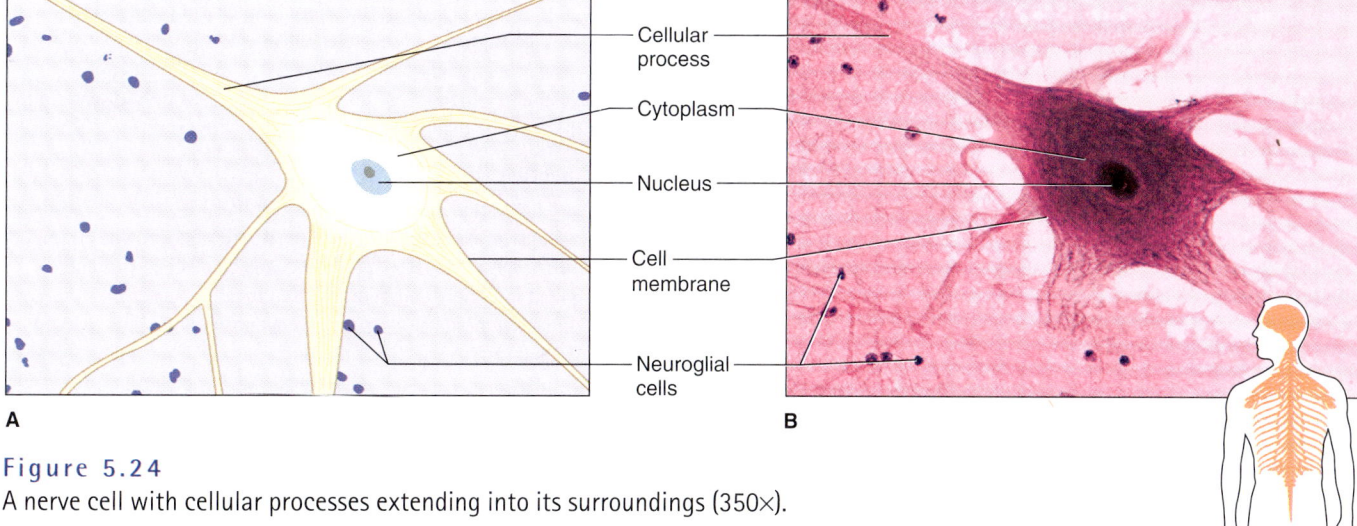

Figure 5.24
A nerve cell with cellular processes extending into its surroundings (350×).

the neighboring neuroglial cells, that produce the fatty myelin that coats the neurons and makes nerve impulse transmission fast enough for the brain to function. Specifically, the neurons release N-acetylaspartate (NAA), which is broken down into harmless compounds by an enzyme, aspartoacylase, that the neuroglia produce. Canavan disease is aspartoacylase deficiency. NAA builds up, which eventually destroys the neuroglia. Without sufficient myelin, the neurons cease to function, and symptoms begin.

Canavan disease has no conventional treatment. However, experimental gene therapy delivers the healthy gene for aspartoacylase into neurons, enabling them to secrete their own enzyme, which in turn stimulates the neuroglial cells to make myelin. Although the gene therapy is invasive—delivered through holes bored into the skull—and must be repeated every few years, it appears to work. Three months after treatment, the first recipient, 18-month-old Lindsay Karlin, looked around, moved, and vocalized, when previously she could barely open her eyes and did not interact with anyone. "It was as though she had awakened," wrote her mother. A magnetic resonance image of Lindsay's brain showed that myelination of neurons had begun in regions where it had vanished.

SUMMARY OUTLINE

5.1 Introduction (p. 92)
Tissues are groups of cells with specialized structural and functional roles. Intercellular materials, varying from solid to liquid, separate cells. The four major types of human tissue are epithelial tissues, connective tissues, muscle tissues, and nervous tissues.

5.2 Epithelial Tissues (p. 92)
1. General characteristics
   a. Epithelial tissue covers all free body surfaces and is the major tissue of glands.
   b. Epithelium is anchored to connective tissue by a basement membrane, lacks blood vessels, contains little intercellular material (matrix), and is replaced continuously.
   c. It functions in protection, secretion, absorption, and excretion.
   d. Epithelial tissues are classified according to shape and number of layers of cells.
2. Simple squamous epithelium
   a. This tissue consists of a single layer of thin, flattened cells.
   b. It functions in gas exchange in the lungs and lines the blood and lymph vessels and various body cavities.
3. Simple cuboidal epithelium
   a. This tissue consists of a single layer of cube-shaped cells.
   b. It carries on secretion and absorption in the kidneys and various glands.
4. Simple columnar epithelium
   a. This tissue is composed of elongated cells whose nuclei are located near the basement membrane.
   b. It lines the uterus and digestive tract.
   c. Absorbing cells often possess microvilli.
   d. This tissue contains goblet cells that secrete mucus.
5. Pseudostratified columnar epithelium
   a. This tissue appears stratified because the nuclei are located at two or more levels.
   b. Its cells may have cilia that move mucus over the surface of the tissue.
   c. It lines passageways of the respiratory system.
6. Stratified squamous epithelium
   a. This tissue is composed of many layers of cells.
   b. It protects underlying cells.
   c. It covers the skin and lines the mouth, throat, vagina, and anal canal.
7. Stratified cuboidal epithelium
   a. This tissue is composed of two or three layers of cube-shaped cells.
   b. It lines the larger ducts of the mammary glands, sweat glands, salivary glands, and pancreas.
   c. It functions in protection.
8. Stratified columnar epithelium
   a. The top layer of cells in this tissue contains elongated columns. Cube-shaped cells make up the bottom layers.
   b. It is in the vas deferens, parts of the male urethra, and parts of the pharynx.
   c. This tissue protects and secretes.
9. Transitional epithelium
   a. This tissue is specialized to become distended.
   b. It is in the walls of various organs of the urinary tract.
10. Glandular epithelium
    a. Glandular epithelium is composed of cells that are specialized to secrete substances.
    b. A gland consists of one or more cells.
       (1) Exocrine glands secrete into ducts.
       (2) Endocrine glands secrete into tissue fluid or blood.
    c. Exocrine glands are classified according to composition of their secretions.
       (1) Merocrine glands secrete fluid without loss of cytoplasm.
          (a) Serous cells secrete watery fluid with a high enzyme content.
          (b) Mucous cells secrete mucus.
       (2) Apocrine glands lose portions of their cells during secretion.
       (3) Holocrine glands release cells filled with secretory products.

5.3 Connective Tissues (p. 98)
1. General characteristics
   a. Connective tissue connects, supports, protects, provides frameworks, fills spaces, stores fat, produces blood cells, protects against infection, and helps repair damaged tissues.
   b. Connective tissue cells usually have considerable intercellular material (matrix) between them.
   c. This intercellular matrix consists of fibers and a ground substance.
   d. Major cell types
      (1) Fibroblasts produce collagenous and elastic fibers.
      (2) Macrophages are phagocytes.
      (3) Mast cells may release heparin and histamine, and usually are near blood vessels.

e. Connective tissue fibers
   (1) Collagenous fibers are composed of collagen and have great tensile strength.
   (2) Elastic fibers are composed of microfibrils embedded in elastin and are very elastic.
   (3) Reticular fibers are very fine, collagenous fibers.
2. Categories of connective tissue
   a. Connective tissue proper includes loose connective tissue, adipose tissue, and dense connective tissue.
   b. Specialized connective tissue includes cartilage, bone, and blood.
3. Loose connective tissue
   a. This tissue forms thin membranes between organs and binds them.
   b. It is beneath the skin and between muscles.
4. Adipose tissue
   a. This tissue is a specialized form of connective tissue that stores fat.
   b. It is found beneath the skin, in certain abdominal membranes, and around the kidneys, heart, and various joints.
5. Dense connective tissue
   a. This tissue is largely composed of strong, collagenous fibers.
   b. It is found in the tendons, ligaments, white portions of the eyes, and the deep layer of the skin.
6. Cartilage
   a. Cartilage provides a supportive framework for various structures.
   b. Its intercellular material is composed of fibers and a gel-like ground substance.
   c. Cartilaginous structures are enclosed in a perichondrium.
   d. Cartilage lacks a direct blood supply and is slow to heal.
   e. Major types are hyaline cartilage, elastic cartilage, and fibrocartilage.
7. Bone
   a. The intercellular matrix of bone contains mineral salts and collagen.
   b. Its cells are usually organized in concentric circles around central canals. Canaliculi connect them.
   c. Bone is an active tissue that heals rapidly.
8. Blood
   a. Blood transports substances and helps maintain a stable internal environment.
   b. Blood is composed of red cells, white cells, and platelets suspended in plasma.
   c. Blood cells develop in red marrow in the hollow parts of long bones.

5.4 Muscle Tissues (p. 105)

1. General characteristics
   a. Muscle tissues contract, moving structures that are attached to them.
   b. Three types are skeletal, smooth, and cardiac muscle tissues.
2. Skeletal muscle tissue
   a. Muscles containing this tissue usually are attached to bones and controlled by conscious effort.
   b. Cells, or muscle fibers, are long and threadlike.
   c. Muscle fibers contract when stimulated by nerve impulses, then immediately relax.
3. Smooth muscle tissue
   a. This tissue is in the walls of hollow internal organs.
   b. Usually it is involuntarily controlled.
4. Cardiac muscle tissue
   a. This tissue is found only in the heart.
   b. Cells are joined by intercalated discs and form branched networks.

5.5 Nervous Tissues (p. 107)

1. Nervous tissue is in the brain, spinal cord, and peripheral nerves.
2. Neurons (nerve cells)
   a. Neurons sense changes and respond by transmitting nerve impulses to other neurons or to muscles or glands.
   b. They coordinate, regulate, and integrate body activities.
3. Supporting cells
   a. Some of these cells bind and support nervous tissue.
   b. Others carry on phagocytosis.
   c. Still others connect neurons to blood vessels.

REVIEW EXERCISES

1. Define *tissue*. (p. 92)
2. Name the four major types of tissues in the human body. (p. 92)
3. Describe the general characteristics of epithelial tissues. (p. 92)
4. Explain how the structure of simple squamous epithelium provides its function. (p. 92)
5. Name an organ in which each of the following tissues is found, and give the function of each tissue. (p. 92)
   a. Simple squamous epithelium
   b. Simple cuboidal epithelium
   c. Simple columnar epithelium
   d. Pseudostratified columnar epithelium
   e. Stratified squamous epithelium
   f. Stratified cuboidal epithelium
   g. Stratified columnar epithelium
   h. Transitional epithelium
   i. Glandular epithelium
6. Define *gland*. (p. 97)
7. Distinguish between an exocrine gland and an endocrine gland. (p. 97)
8. Explain how glandular secretions differ. (p. 97)
9. Define *mucus*. (p. 97)
10. Describe the general characteristics of connective tissues. (p. 98)
11. Define *matrix*. (p. 98)
12. Describe three major types of connective tissue cells. (p. 99)
13. Distinguish between collagen and elastin. (p. 100)
14. Explain the relationship between loose connective tissue and adipose tissue. (p. 101)
15. Define *dense connective tissue*. (p. 101)
16. Explain why injured dense connective tissue and cartilage are usually slow to heal. (p. 102)
17. Name the types of cartilage, and describe their differences and similarities. (p. 102)
18. Describe how bone cells are organized in bone tissue. (p. 103)
19. Describe the composition of blood. (p. 103)
20. Describe the general characteristics of muscle tissues. (p. 105)
21. Distinguish among skeletal, smooth, and cardiac muscle tissues. (p. 105)
22. Describe the general characteristics of nervous tissues. (p. 107)
23. Distinguish between neurons and neuroglial cells. (p. 107)

CRITICAL THINKING

1. Tissue engineering combines living cells with synthetic materials to create functional substitutes for human tissues. What components would you use to engineer replacement (*a*) skin, (*b*) blood, (*c*) bone, and (*d*) muscle?

2. Collagen and elastin are added to many beauty products. What type of tissue are they normally part of?

3. Joints such as the elbow, shoulder, and knee contain considerable amounts of cartilage and dense connective tissue. How does this explain the fact that joint injuries are often very slow to heal?

4. Cancer-causing agents (carcinogens) usually act on cells that are dividing. Which of the four tissues would carcinogens most influence? Least influence?

WEB CONNECTIONS

Visit the website for additional study questions and more information about this chapter at:

http://www.mhhe.com/shieress8

Unit 2
Support and Movement

chapter 6

Skin and the Integumentary System

TOO MUCH SWEAT. Sweating is a highly effective mechanism for cooling the body. Becoming drenched with sweat following heavy exertion or an intense workout can feel good. But for people with hyperhidrosis, sweating is profuse, uncontrollable, unpredictable, and acutely embarrassing.

Sweat consists of water released from about 5 million eccrine glands in the skin, in response to stimulation by the nervous system. About 2 million of these glands are in the hands, which explains why our palms become sweaty when we are nervous. For the 1% of the population with hyperhidrosis, the body, often for no apparent reason, breaks out in torrents of sweat. An affected person cannot grasp a pen, clothes become drenched, and social interactions become very difficult. Some people may inherit the condition, but usually the cause isn't known.

Jeffrey Schweitzer, a surgeon at Northwestern University Medical School in Chicago, has developed a treatment for hyperhidrosis. He inserts an endoscope (a small lit tube) through an opening in the patient's chest wall and removes the nerves that signal sweat glands in the palms. The success rate is greater than 80% in alleviating the sweaty palms.

Photo:
Sweating in response to exertion is a normal way for the body to cool itself.

Chapter Objectives

After studying this chapter, you should be able to do the following:

6.1 Introduction
1. Describe what constitutes an organ, particularly as it relates to membranes and systems. (p. 113)

6.2 Types of Membranes
2. Describe the four major types of membranes. (p. 113)

6.3 Skin and Its Tissues
3. Describe the structure of the layers of the skin. (p. 113)
4. List the general functions of each layer of skin. (p. 114)
5. Summarize the factors that determine skin color. (p. 116)

6.4 Accessory Organs of the Skin
6. Describe the accessory organs associated with the skin. (p. 117)

6.5 Regulation of Body Temperature
7. Explain how the skin helps regulate body temperature. (p. 120)

6.6 Healing of Wounds
8. Describe the events that are part of wound healing. (p. 120)

Aids to Understanding Words

cut- [skin] sub*cut*aneous: Beneath the skin.
derm- [skin] *derm*is: Inner layer of the skin.
epi- [upon] *epi*dermis: Outer layer of the skin.
follic- [small bag] hair *follic*le: Tubelike depression in which a hair develops.
kerat- [horn] *kerat*in: Protein produced as epidermal cells die and harden.
melan- [black] *melan*in: Dark pigment produced by certain cells.
seb- [grease] *seb*aceous gland: Gland that secretes an oily substance.

Key Terms

cutaneous membrane (ku-ta′ne-us mem′brān)
dermis (der′mis)
epidermis (ep′′ĭ-der′mis)
hair follicle (hār fol′ĭ-kl)
integumentary (in-teg-u-men′tar-e)
keratinization (ker′′ah-tin′′ĭ-za′shun)
melanin (mel′ah-nin)
mucous membrane (mu′kus mem′brān)
sebaceous gland (se-ba′shus gland)
serous membrane (se′rus mem′brān)
subcutaneous layer (sub′′ku-ta′ne-us la′er)
sweat gland (swet gland)
synovial membrane (sĭ-no′ve-al mem′brān)

6.1 Introduction

Chemicals, cells, tissues, organs, and finally, organ systems build a human body. Two or more kinds of tissues grouped together and performing specialized functions constitute an organ. For example, the thin, sheetlike membranes composed of epithelium and connective tissue that cover body surfaces and line body cavities are organs. The cutaneous membrane (commonly called *skin*) together with certain accessory organs make up the **integumentary** (in-teg-u-men´tar-e) **system.**

6.2 Types of Membranes

The four major types of membranes are serous, mucous, synovial, and cutaneous. **Serous membranes** (se´rus mem´branz) line body cavities that lack openings to the outside. They form the inner linings of the thorax (parietal pleura) and abdomen (parietal peritoneum), and they cover the organs within these cavities (visceral pleura and visceral peritoneum, respectively). A serous membrane consists of a layer of simple squamous epithelium and a thin layer of loose connective tissue. Cells of a serous membrane secrete watery *serous fluid,* which lubricates membrane surfaces.

Mucous (mu´kus) **membranes** line cavities and tubes that open to the outside of the body. These include the oral and nasal cavities and the tubes of the digestive, respiratory, urinary, and reproductive systems. A mucous membrane consists of epithelium overlying a layer of loose connective tissue. Goblet cells within a mucous membrane secrete *mucus.*

Synovial (sĭ-no´ve-al) **membranes** form the inner linings of the joint cavities between the ends of bones at freely movable joints (synovial joints). These membranes usually include dense connective tissue overlying loose connective tissue and adipose tissue. Cells of a synovial membrane secrete a thick, colorless *synovial fluid* into the joint cavity, which lubricates the ends of the bones within the joint. The **cutaneous** (ku-ta´ne-us) **membrane,** or skin, is described in detail in this chapter.

6.3 Skin and Its Tissues

The skin is one of the larger and more versatile organs of the body, and it is vital in maintaining homeostasis. The skin is a protective covering, helps regulate body temperature, retards water loss from deeper tissues, houses sensory receptors, synthesizes various biochemicals, and excretes small quantities of wastes.

The skin includes two distinct tissue layers (fig. 6.1). The outer layer, called the **epidermis** (ep´´ĭ-der´mis), is composed of stratified squamous epithelium. The inner layer, or **dermis** (der´mis), is thicker than the epidermis, and it contains connective tissue consisting of collagenous and elastic fibers, epithelial tissue, smooth muscle tissue, nervous tissue, and blood. A *basement membrane* that is anchored to the dermis separates the two skin layers.

Beneath the dermis are masses of loose connective and adipose tissues that bind the skin to the underlying organs. These tissues form the **subcutaneous** (sub´´ku-ta´ne-us) **layer** (hypodermis).

If the skin of a 150-pound person were spread out flat, it would cover approximately 20 square feet.

CHECK YOUR RECALL

1. Name the four types of membranes, and explain how they differ.
2. List the general functions of the skin.
3. Name the tissue in the outer layer of the skin.
4. Name the tissues in the inner layer of the skin.

Epidermis

Since the epidermis is composed of stratified squamous epithelium, it lacks blood vessels. However, the deepest layer of epidermal cells, called the *stratum basale* (stratum germinativum), is close to the dermis and is nourished by dermal blood vessels (fig. 6.1A). As the cells of this layer divide and grow, the older epidermal cells are pushed away from the dermis toward the skin surface. The farther the cells move, the poorer their nutrient supply becomes, and in time, they die.

The older cells (keratinocytes) harden in a process called **keratinization** (ker´´ah-tin´´ĭ-za´shun). The cytoplasm fills with strands of a tough, fibrous, waterproof *keratin* protein. As a result, many layers of tough, tightly

Subcutaneous injections are administered through a hollow needle into the subcutaneous layer beneath the skin. *Intradermal injections* are injected within the skin. Subcutaneous injections and *intramuscular injections,* administered into muscles, are sometimes called hypodermic injections.

Some substances are introduced through the skin by means of an adhesive transdermal patch that includes a small reservoir containing a drug. The drug passes from the reservoir through a permeable membrane at a known rate. It then diffuses into the epidermis and enters the blood vessels of the dermis. Transdermal patches are used to protect against motion sickness, chest pain associated with heart disease, and elevated blood pressure. A transdermal patch that delivers nicotine is used to help people stop smoking.

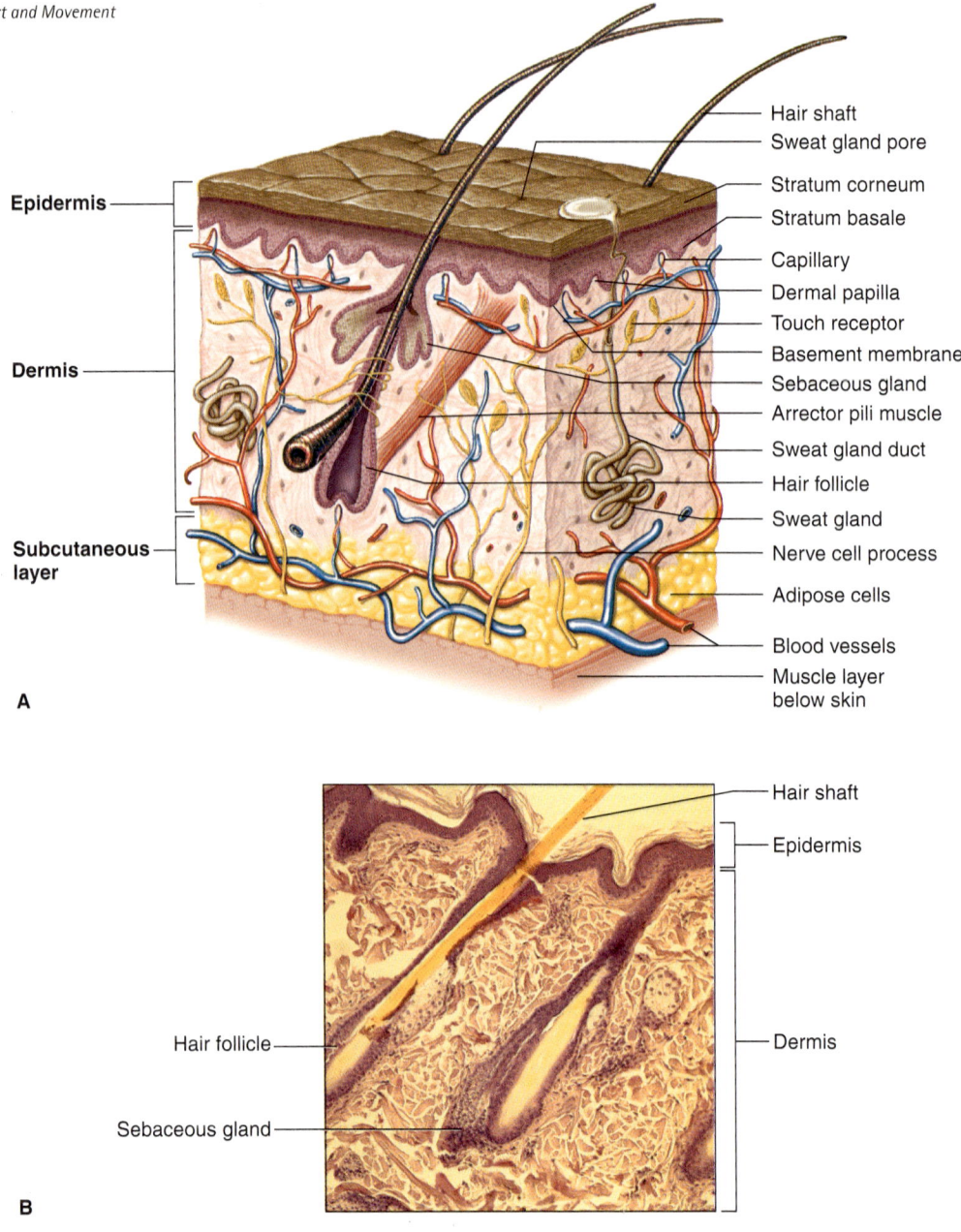

Figure 6.1
Skin. (A) A section of skin. (B) A light micrograph depicting the layered structure of the skin (75×).

packed dead cells accumulate in the outer epidermis, forming an outermost layer called the *stratum corneum*. Dead cells that compose it are eventually shed.

The thickness of the epidermis varies from region to region. In most areas, only four layers can be distinguished. They are the *stratum basale, stratum spinosum, stratum granulosum,* and *stratum corneum*. An additional layer, the *stratum lucidum,* is in the thickened skin of the palms and soles. The stratum granulosum may be missing where the epidermis is thin (fig. 6.2).

In healthy skin, production of epidermal cells is closely balanced with loss of dead cells from the stratum corneum, so that the skin does not wear away completely. In fact, the rate of cell division increases where the skin is rubbed or pressed regularly, causing growth of thickened areas called *calluses* on the palms and soles, and keratinized conical masses on the toes called *corns*.

The epidermis has important protective functions. It shields the moist underlying tissues against excessive water loss, mechanical injury, and the effects of harmful chemicals. When unbroken, the epidermis also keeps out disease-causing microorganisms.

Specialized cells in the epidermis called *melanocytes* produce **melanin** (mel´ah-nin), a dark pigment that provides skin color (fig. 6.3A). Melanin absorbs ultraviolet radiation in sunlight, preventing mutations in the DNA of skin cells and other damaging effects. Melanocytes lie in the deepest portion of the epidermis. Although they are the only cells that can produce

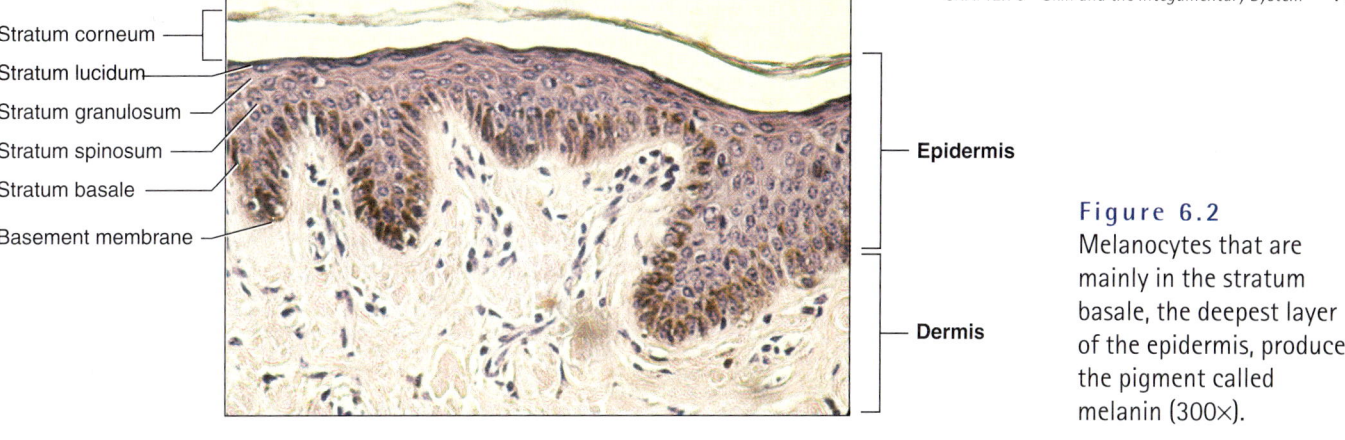

Figure 6.2
Melanocytes that are mainly in the stratum basale, the deepest layer of the epidermis, produce the pigment called melanin (300×).

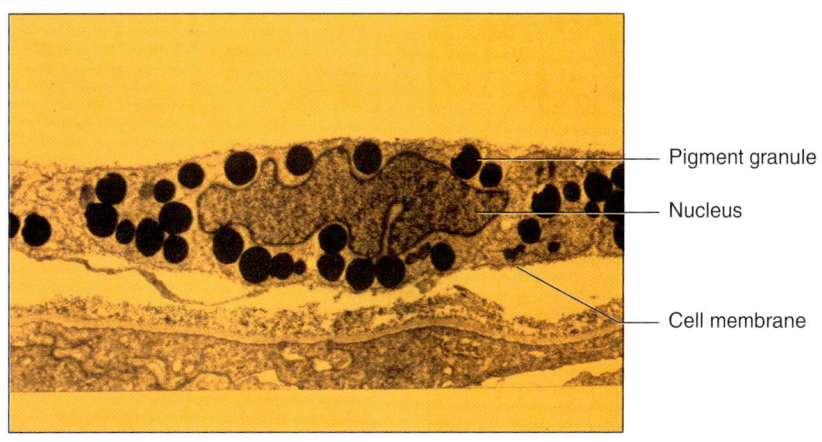

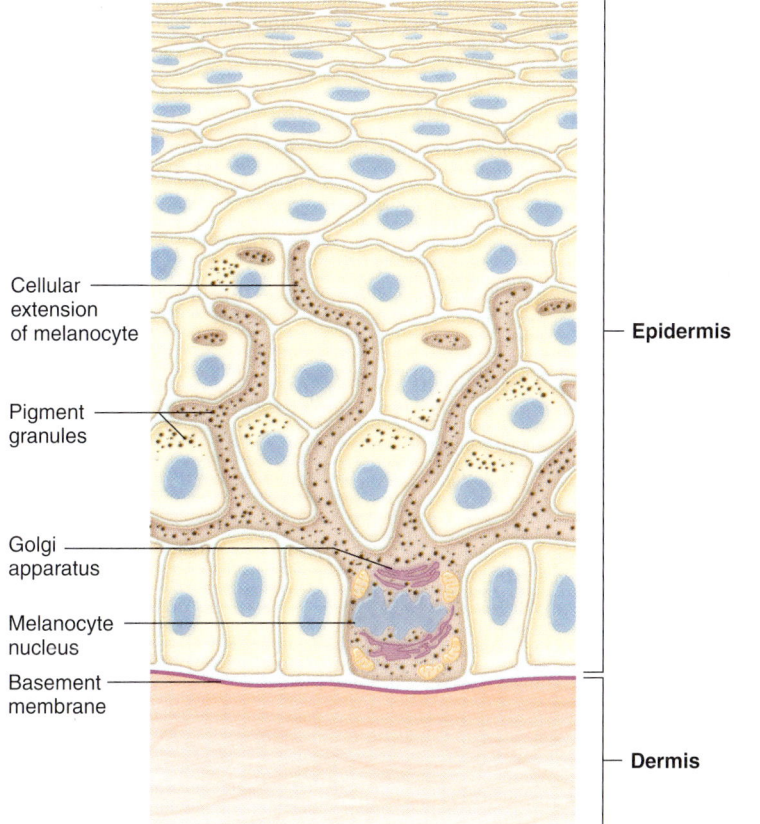

Figure 6.3
Melanocyte. (A) Transmission electron micrograph of a melanocyte with pigment-containing granules (10,600×). (B) A melanocyte may have pigment-containing extensions that pass between epidermal cells and transfer pigment into them.

melanin, the pigment also may be present in other epidermal cells nearby. This happens because melanocytes have long, pigment-containing cellular extensions that pass upward between epidermal cells. The extensions transfer melanin granules into these other cells by a process called *cytocrine secretion.* Nearby epidermal cells may contain more melanin than the melanocytes (fig. 6.3B).

Skin Color

Skin color is due largely to melanin. All people have about the same number of melanocytes in their skin. Differences in skin color result from differences in the amount of melanin that melanocytes produce and in the distribution and size of the pigment granules. Skin color is mostly genetically determined. If genes instruct melanocytes to produce abundant melanin, the skin is dark.

Environmental and physiological factors also influence skin color. Sunlight, ultraviolet light from sunlamps, or X rays stimulate production of additional pigment. Blood in the dermal vessels may affect skin color as physiological changes occur. When blood is well oxygenated, the blood pigment (hemoglobin) is bright red, making the skin of light-complexioned people appear pinkish. On the other hand, when blood oxygen concentration is low, hemoglobin is dark red, and the skin appears bluish—a condition called *cyanosis.* Other physiological factors affecting skin color include diet (carotene in yellow vegetables) and chemicals (bilirubin accumulation in skin of newborns).

CHECK YOUR RECALL

1. Explain how the epidermis is formed.
2. Distinguish between the stratum basale and the stratum corneum.
3. What is the function of melanin?
4. What factors influence skin color?

Dermis

Epidermal ridges projecting inward and conical projections of dermis called dermal papillae passing into the spaces between the ridges cause the boundary between the epidermis and dermis to be uneven (see fig. 6.1A). Fingerprints form from these undulations of the skin at the distal end of the palmar surface of a finger. Genes determine fingerprint patterns but they can change slightly as a fetus moves and presses the forming ridges against the uterine wall. For this reason, the fingerprints of identical twins are usually not exactly alike.

The dermis binds the epidermis to underlying tissues (see fig. 6.1A). It is largely composed of dense connective tissue that includes tough collagenous fibers and elastic fibers within a gel-like ground substance. Networks of these fibers give the skin toughness and elasticity.

Because dermal blood vessels supply nutrients to the epidermis, interference with blood flow may kill epidermal cells. For example, when a person lies in one position for a prolonged period, the weight of the body pressing against the bed blocks the skin's blood supply. If cells die, the tissues begin to break down (necrosis), and a *pressure ulcer* (also called a decubitus ulcer or bedsore) may appear.

Pressure ulcers usually occur in the skin overlying bony projections, such as on the hip, heel, elbow, or shoulder. Frequently changing body position or massaging the skin to stimulate blood flow in regions associated with bony prominences can prevent ulcers.

Dermal blood vessels supply nutrients to all skin cells. These vessels also help regulate body temperature, as explained later in this chapter.

Nerve cell processes are scattered throughout the dermis. Motor processes carry impulses out from the brain or spinal cord to dermal muscles and glands. Sensory processes carry impulses away from specialized sensory receptors, such as touch receptors located within the dermis, and into the brain or spinal cord. The dermis also contains hair follicles, sebaceous (oil-producing) glands, and sweat glands, which are discussed later in the chapter (see fig. 6.1A).

Skin cells help produce vitamin D, which is necessary for normal bone and tooth development. This vitamin can form from a substance (dehydrocholesterol) that is synthesized by cells in the digestive system or obtained in the diet. When dehydrocholesterol reaches the skin by means of the blood and is exposed to ultraviolet light from the sun, it is converted to another chemical, which becomes vitamin D.

Certain skin cells (keratinocytes) assist the immune system by producing hormonelike substances that stimulate development of certain white blood cells (T lymphocytes) that defend against infection by disease-causing bacteria and viruses (see chapter 14, p. 377).

Subcutaneous Layer

The subcutaneous layer (hypodermis) beneath the dermis consists of loose connective and adipose tissues (see fig. 6.1A). The collagenous and elastic fibers of this layer are continuous with those of the dermis. Most of these fibers run parallel to the surface of the skin, extending in all directions. As a result, no sharp boundary separates the dermis and the subcutaneous layer.

The adipose tissue of the subcutaneous layer insulates, helping to conserve body heat and impeding the entrance of heat from the outside. The subcutaneous layer also contains the major blood vessels that supply the skin and underlying adipose tissue.

CHECK YOUR RECALL

1. What kinds of tissues make up the dermis?
2. What are the functions of these tissues?
3. What are the functions of the subcutaneous layer?

6.4 Accessory Organs of the Skin

Hair Follicles

Hair is present on all skin surfaces except the palms, soles, lips, nipples, and parts of the external reproductive organs. Each hair develops from a group of epidermal cells at the base of a tubelike depression called a **hair follicle** (hār fol′i-kl) (figs. 6.1 and 6.4). This follicle extends from the surface into the dermis and contains the hair *root*. The epidermal cells at its base are nourished from dermal blood vessels in a projection of connective tissue at the deep end of the follicle. As these epidermal cells divide and grow, older cells are pushed toward the surface. The cells that move upward and away from their nutrient supply become keratinized and die. Their remains constitute the structure of a developing hair, whose *shaft* extends away from the skin surface (fig. 6.5). In other words, a hair is composed of dead epidermal cells.

Genes determine hair color by directing the type and amount of pigment that epidermal melanocytes produce. If these cells, which lie at the deep end of a follicle, produce an abundance of melanin, the hair is dark; if an intermediate quantity of pigment is produced, the hair is blond; if no pigment appears, the hair is white. Another pigment, trichosiderin, is found only in red hair. A mixture of pigmented and unpigmented hair usually appears gray.

A bundle of smooth muscle cells, forming the *arrector pili muscle,* attaches to each hair follicle (see figs. 6.1A and 6.4A). This muscle is positioned so that a short hair within the follicle stands on end when the muscle contracts. If a person is emotionally upset or very cold, nerve impulses may stimulate the arrector pili muscles to contract, causing gooseflesh or goose bumps.

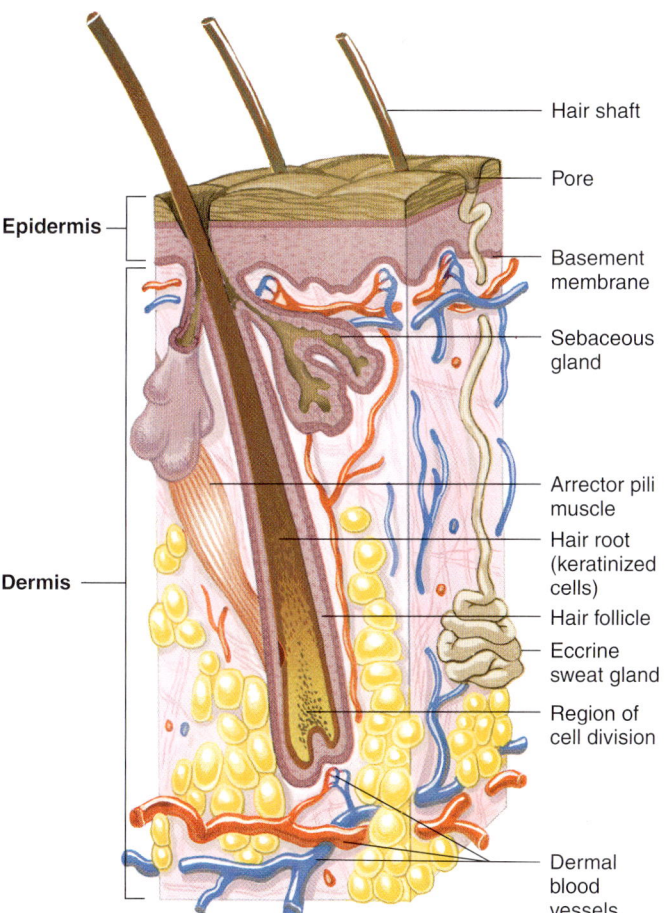

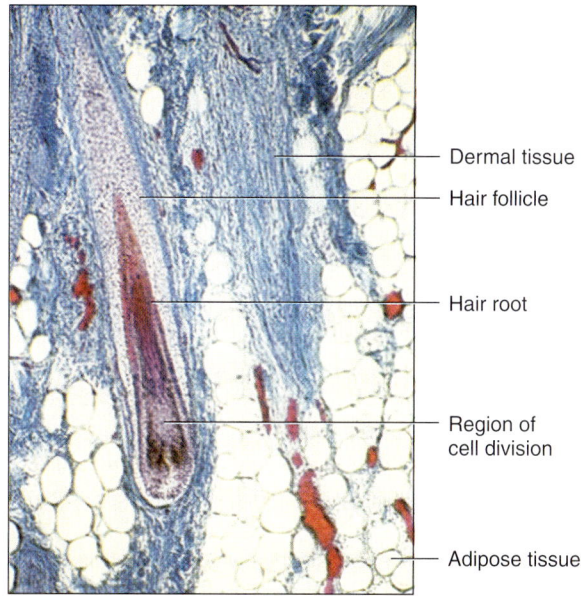

Figure 6.4
Hair follicle. (*A*) A hair grows from the base of a hair follicle when epidermal cells divide and older cells move outward and become keratinized. (*B*) Light micrograph of a hair follicle (160×).

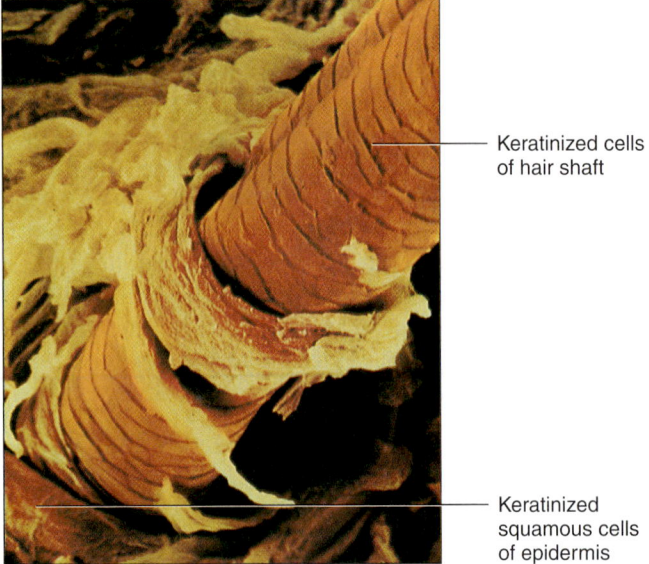

Figure 6.5
Scanning electron micrograph of a hair emerging from the epidermis (875×).

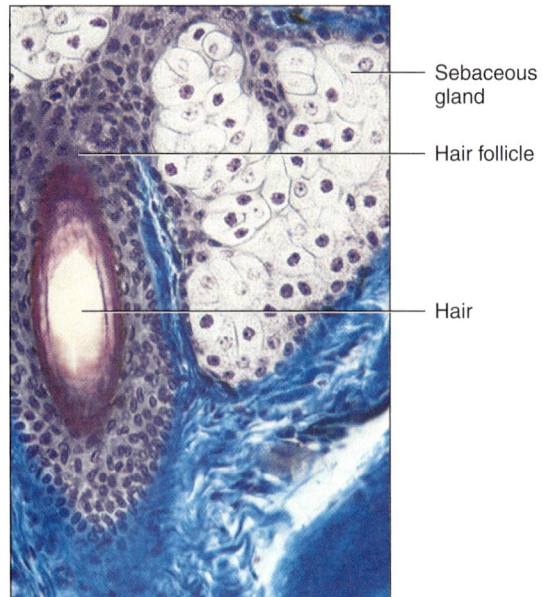

Figure 6.6
A sebaceous gland secretes sebum into a hair follicle, shown here in oblique section (300×).

> Just above the "bulge" region at the base of a hair follicle are stem cells that can give rise to hair as well as epidermal cells. The first clue to the existence of these "young transient amplifying cells" was that new skin in burn patients arises from hair follicles. Then, experiments in mice that mark stem cells and their descendants showed that the young transient amplifying cells give rise to both hair and skin.

Sebaceous Glands

Sebaceous glands (se-ba´shus glandz) contain groups of specialized epithelial cells and are usually associated with hair follicles (figs. 6.4A and 6.6). They are holocrine glands (see chapter 5, p. 97) that secrete an oily mixture of fatty material and cellular debris called *sebum* through small ducts into the hair follicles. Sebum helps keep the hair and skin soft, pliable, and waterproof.

Nails

Nails are protective coverings on the ends of the fingers and toes. Each nail consists of a *nail plate* that overlies a surface of skin called the *nail bed*. Specialized epithelial cells that are continuous with the epithelium of the skin produce the nail bed. The whitish, thickened, half-moon–shaped region (lunula) at the base of a nail plate is the most active growing region. The epithelial cells here divide, and the newly formed cells are keratinized. This gives rise to tiny, keratinized scales that become part of the nail plate, pushing it forward over the nail bed. In time, the plate extends beyond the end of the nail bed and with normal use gradually wears away (fig. 6.7).

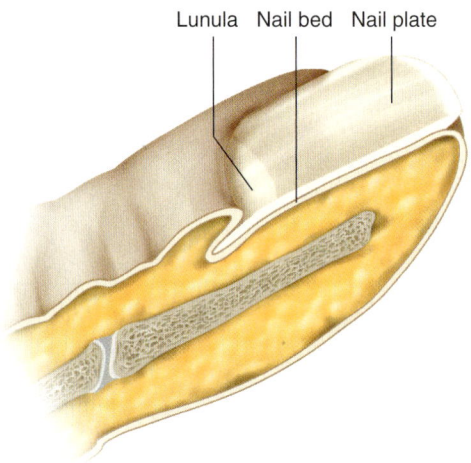

Figure 6.7
Nails grow from epithelial cells that divide and become keratinized as the rest of the nail.

> Many teens are all too familiar with a disorder of the sebaceous glands called *acne* (acne vulgaris). Overactive and inflamed glands in some body regions become plugged and surrounded by small red elevations containing blackheads (comedones) or pimples (pustules).

 The thumbnail grows the slowest; the middle nail grows the fastest.

Topic of Interest

SKIN CANCER

Skin cancer usually arises from nonpigmented epithelial cells within the deep layer of the epidermis or from melanocytes. Skin cancers originating from epithelial cells are called *cutaneous carcinomas* (basal cell carcinoma or squamous cell carcinoma); those arising from melanocytes are *cutaneous melanomas* (melanocarcinomas or malignant melanomas) (fig. 6A).

Cutaneous carcinomas are the most common type of skin cancer, occurring most frequently in light-skinned people over forty years of age. These cancers usually appear in individuals who are regularly exposed to sunlight, such as farmers, sailors, athletes, and sunbathers, and may be the result of failure of normally protective apoptosis—peeling of sun-damaged cells.

Cutaneous carcinomas often develop from hard, dry, scaly growths (lesions) that have reddish bases. Such lesions may be either flat or raised, and they firmly adhere to the skin. Fortunately, cutaneous carcinomas are typically slow growing and can usually be cured completely by surgical removal or radiation treatment.

Because melanomas develop from melanocytes, they are pigmented with melanin, often with a variety of colored areas, such as variegated brown, black, gray, or blue. They usually have irregular rather than smooth outlines, and may feel bumpy.

Cutaneous melanomas may appear in people of any age, and seem to be caused by short, intermittent exposure to high-intensity sunlight. Risk is highest in people who stay indoors but occasionally sustain blistering sunburns. Melanoma is not associated with sustained sun exposure, as are the other types of skin cancers.

A cutaneous melanoma may arise from normal-appearing skin or from a mole (nevus). The lesion spreads horizontally through the skin, but eventually may thicken and grow downward into the skin, invading deeper tissues. If the melanoma is surgically removed while it is in its horizontal growth phase, it may be arrested. Once it thickens and spreads into deeper tissues, unfortunately, it becomes difficult to treat, and the survival rate is very low. A type of gene therapy called a "cancer vaccine" attempts to stimulate a person's immune system to locate and destroy melanoma cells that have spread.

To reduce the chances of developing skin cancer, avoid exposing the skin to high-intensity sunlight, use sunscreens and sunblocks, and examine the skin regularly. Report any unusual lesions—particularly those that change in color, shape, or surface texture—to a physician.

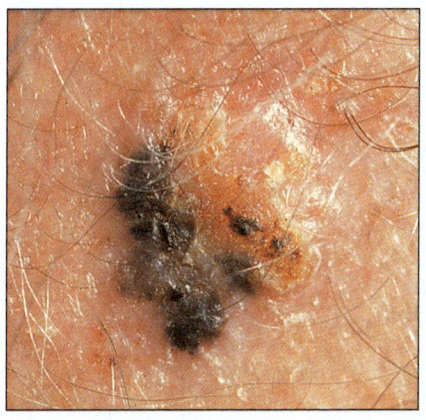 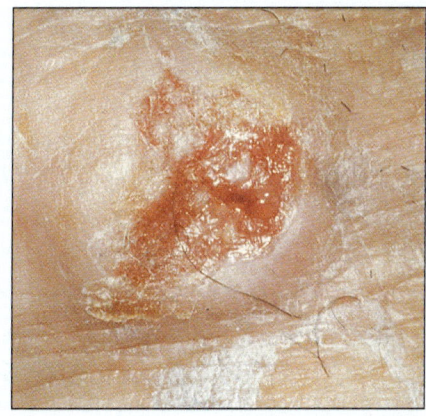 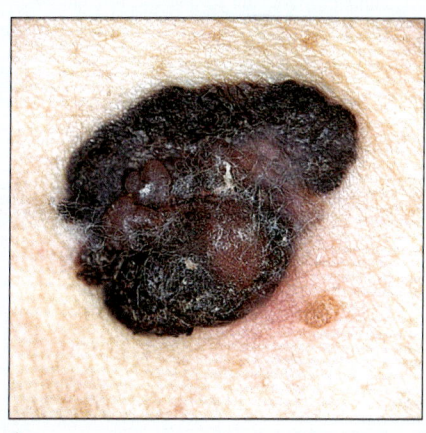

A B C

Figure 6A
Skin cancer. (A) Squamous cell carcinoma. (B) Basal cell carcinoma. (C) Malignant melanoma.

Sweat Glands

Sweat glands (swet glandz), or sudoriferous glands, are exocrine glands that are widespread in the skin. Each gland consists of a tiny tube that originates as a ball-shaped coil in the deeper dermis or superficial subcutaneous layer. The coiled portion of the gland is closed at its deep end and is lined with sweat-secreting epithelial cells.

The most numerous sweat glands, the *eccrine glands*, respond throughout life to body temperature elevated by environmental heat or physical exercise (see fig. 6.4A). These glands are common on the forehead, neck, and back, where they produce profuse sweat on hot days or during intense physical activity.

The fluid (sweat) that eccrine glands secrete is carried away in a duct that opens at the surface as a *pore*. Sweat is mostly water, but it also contains small quantities of salt and wastes, such as urea and uric acid. Thus, sweating is also an excretory function.

Other sweat glands, known as *apocrine glands*, become active when a person is emotionally upset, frightened, or in pain. Although they are currently called apocrine, these glands secrete by the same mechanism as eccrine glands. They are most numerous in the axillary regions and groin, and usually connect to hair follicles.

Other sweat glands are structurally and functionally modified to secrete specific fluids, such as the ceruminous glands of the external ear canal that secrete earwax. The female mammary glands that secrete milk are another example of modified sweat glands.

 The average square inch of skin holds 650 sweat glands, 20 blood vessels, 60,000 melanocytes, and more than a thousand nerve endings.

CHECK YOUR RECALL
1. Explain how a hair forms.
2. What is the function of the sebaceous glands?
3. Distinguish between the eccrine and apocrine sweat glands.

6.5 Regulation of Body Temperature

Regulation of body temperature is vitally important because even slight shifts can disrupt rates of metabolic reactions. Normally, the temperature of deeper body parts remains close to a set point of 37°C (98.6°F). Maintenance of a stable temperature requires that the amount of heat the body loses be balanced by the amount it produces. The skin plays a key role in the homeostatic mechanism that regulates body temperature.

Heat is a product of cellular metabolism; thus, the more active cells of the body are the major heat producers. These cells include skeletal and cardiac muscle cells and the cells of certain glands, such as the liver.

In intense heat, nerve impulses stimulate structures in the skin and other organs to release heat. For example, during physical exercise, active muscles release heat, which the blood carries away. The warmed blood reaches the part of the brain (the hypothalamus) that controls the body's temperature set point, which signals muscles in the walls of specialized dermal blood vessels to relax. As these vessels dilate (vasodilation), more blood enters them, and some of the heat the blood carries escapes to the outside.

At the same time the skin loses heat, the nervous system stimulates the eccrine sweat glands to become active and to release sweat onto the skin surface. As this fluid evaporates (changes from a liquid to a gas), it carries heat away from the surface, cooling the skin further.

If too much heat is lost, as may occur in a very cold environment, the brain triggers different responses in the skin structures. Muscles in the walls of dermal blood vessels are stimulated to contract; this decreases the flow of heat-carrying blood through the skin and helps reduce heat loss. Also, the sweat glands remain inactive, decreasing heat loss by evaporation. If body temperature continues to drop, the nervous system may stimulate muscle fibers in the skeletal muscles throughout the body to contract slightly. This action requires an increase in the rate of cellular respiration and produces heat as a by-product. If this response does not raise body temperature to normal, small groups of muscles may contract rhythmically with still greater force, and the person begins to shiver, generating more heat. Figure 6.8 summarizes the body's temperature-regulating mechanism.

 Most of the body's heat (80%) escapes through the head.

CHECK YOUR RECALL
1. Why is regulation of body temperature so important?
2. How does the body lose excess heat?
3. Which actions help the body conserve heat?

6.6 Healing of Wounds

A wound and the area surrounding it usually become red and painfully swollen. This is the result of **inflammation,** which is a normal response to injury or stress. Blood vessels in affected tissues dilate and become more permeable, forcing fluids to leave the blood vessels and enter the damaged tissues. Inflamed skin may become reddened, warm, swollen, and painful to touch (table 6.1). However, the dilated blood vessels provide the tissues with more nutrients and oxygen, which aids healing.

The specific events in healing depend on the nature and extent of the injury. If a break in the skin is shallow, epithelial cells along its margin are stimulated to divide more rapidly than usual, and the newly formed cells fill the gap.

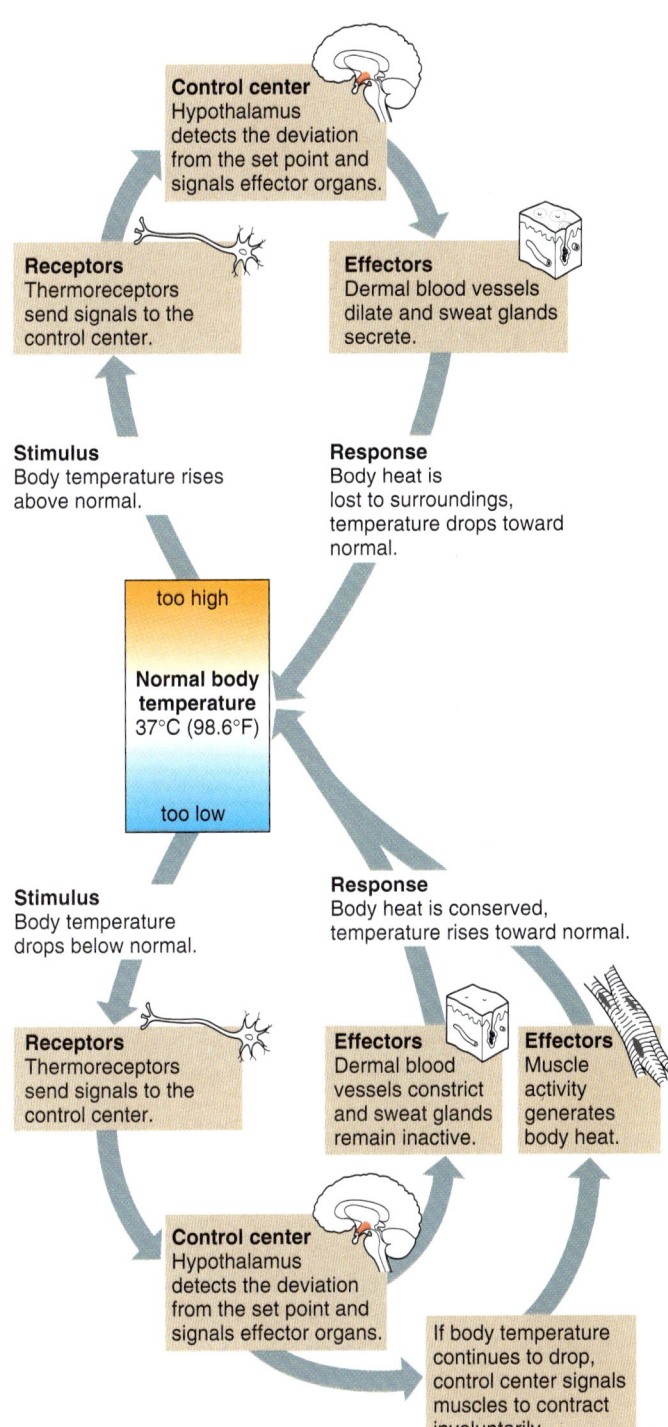

Figure 6.8
Body temperature regulation is an example of homeostasis.

TABLE 6.1	INFLAMMATION
SYMPTOM	CAUSE
Redness	Increased vasodilation, more blood in area
Heat	By-product of increased metabolic activity in tissue as white blood cells attempt to destroy invaders
Swelling	Increased interstitial fluid in area due to change in osmotic pressure of tissues caused by increased numbers of white blood cells
Pain	Swelling puts pressure on nerve endings in area

fibers that bind the edges of the wound together. Suturing or otherwise closing a large break in the skin speeds this process.

As healing continues, blood vessels extend into the area beneath the scab. Phagocytic cells remove dead cells and other debris. Eventually, the damaged tissues are replaced, and the scab sloughs off. If the wound is extensive, the newly formed connective tissue may appear on the surface as a *scar*.

In large, open wounds, healing may be accompanied by formation of small, rounded masses called *granulations* in the exposed tissues. A granulation consists of a new branch of a blood vessel and a cluster of collagen-secreting fibroblasts that the vessel nourishes. In time, some of the blood vessels are resorbed, and the fibroblasts migrate away, leaving a scar.

CHECK YOUR RECALL

1. Describe how inflammation helps a wound heal.
2. Distinguish between the activities necessary to heal a wound in the epidermis and those necessary to heal a wound in the dermis.
3. Explain the role of phagocytic cells in wound healing.
4. Define granulation.

Common Skin Disorders

acne (ak′ne) Disease of the sebaceous glands that produces blackheads and pimples.
alopecia (al′′o-pe′she-ah) Hair loss, usually sudden.
athlete's foot (ath′-lētz foot) Fungus infection (*Tinea pedis*) usually in the skin of the toes and soles.
birthmark (berth′ mark) Congenital blemish or spot on the skin, visible at birth or soon after.
boil (boil) Bacterial infection (furuncle) of the skin, produced when bacteria enter a hair follicle.
carbuncle (kar′bung-kl) Bacterial infection, similar to a boil, that spreads into the subcutaneous tissues.
cyst (sist) Liquid-filled sac or capsule.
dermatitis (der′′mah-ti′tis) Inflammation of the skin.
eczema (ek′zĕ-mah) Noncontagious skin rash that produces itching, blistering, and scaling.

If the injury extends into the dermis or subcutaneous layer, blood vessels break, and the escaping blood forms a clot in the wound. The blood clot and dried tissue fluids form a *scab* that covers and protects underlying tissues. Before long, fibroblasts migrate into the injured region and begin forming new collagenous

Topic of Interest

ELEVATED BODY TEMPERATURE

It was a warm June morning when the harried and hurried father strapped his five-month-old son Bryan into the backseat of his car and headed for work. Tragically, the father forgot to drop his son off at the babysitter's. When his wife called him at work late that afternoon to ask why the child was not at the sitter's, the shocked father realized his mistake and hurried down to his parked car. But it was too late—Bryan had died. Left for 10 hours in the car in the sun, all windows shut, the baby's temperature had quickly soared. Two hours after he was discovered, the child's temperature still exceeded 41°C (106°F).

Sarah's elevated body temperature was more typical. She awoke with a fever of 40°C (104°F) and a terribly painful sore throat. Peering down the five-year-old's throat with a flashlight, her mother spotted the whitish lesions that can indicate a *Streptococcus* infection. Sarah indeed had strep throat, and the fever was her body's attempt to fight the infection.

The true cases of young Bryan and Sarah illustrate two reasons why body temperature may rise: (1) inability of the temperature homeostatic mechanism to handle an extreme environment and (2) an immune system response to infection.

In Bryan's case, sustained exposure to very high heat overwhelmed the temperature-regulating mechanism, resulting in hyperthermia. Body heat built up faster than it could dissipate, and body temperature rose, even though the set point of the thermostat was normal. His blood vessels dilated so greatly in an attempt to dissipate the excess heat that after a few hours, his circulatory system collapsed.

Fever is a special case of hyperthermia, in which molecules on the surfaces of the infectious agents (usually bacteria or viruses) stimulate phagocytes to release a substance called interleukin-1 (IL-1, also called endogenous pyrogen, meaning "fire maker from within"). The bloodstream carries IL-1 to the hypothalamus, where it raises the set point controlling temperature. In response, the brain signals skeletal muscles to increase heat production, blood flow to the skin to decrease, and sweat glands to decrease secretion. As a result, body temperature rises to the new set point, and a fever develops. The increased body temperature helps the immune system kill the pathogens.

Rising body temperature requires different treatments, depending on the degree of elevation. Hyperthermia in response to exposure to intense, sustained heat should be rapidly treated by administering liquids to replace lost body fluids and electrolytes, sponging the skin with water to increase cooling by evaporation, and covering the person with a refrigerated blanket. Fever can be lowered with ibuprofen or acetaminophen, or with aspirin in adults. Some health professionals believe that a slight fever should not be reduced (with medication or cold baths) because it may be part of a normal immune response. A high or prolonged fever, however, requires medical attention.

---

erythema (er″ĭ-the′mah) Reddening of the skin due to dilation of dermal blood vessels in response to injury or inflammation.

herpes (her′pēz) Infectious disease of the skin, usually caused by the herpes simplex virus and characterized by recurring formations of small clusters of vesicles.

impetigo (im″pĕ-ti′go) Contagious disease of bacterial origin, characterized by pustules that rupture and become covered with loosely held crusts.

keloid (ke′loid) Elevated, enlarging fibrous scar usually initiated by an injury.

mole (mōl) Fleshy skin tumor (nevus) that is usually pigmented; colors range from brown to black.

pediculosis (pĕ-dik″u-lo′sis) Disease produced by an infestation of lice.

pruritus (proo-ri′tus) Itching of the skin.

psoriasis (so-ri′ah-sis) Chronic skin disease characterized by red patches covered with silvery scales.

pustule (pus′tūl) Elevated, pus-filled area on the skin.

scabies (ska′bēz) Disease resulting from an infestation of mites.

seborrhea (seb″o-re′ah) Hyperactivity of the sebaceous glands, causing greasy skin and dandruff.

ulcer (ul′ser) Open sore.

urticaria (ur″tĭ-ka′re-ah) Allergic reaction of the skin that produces reddish, elevated patches (hives).

wart (wort) Flesh-colored, raised area caused by a viral infection.

Organization

INTEGUMENTARY SYSTEM

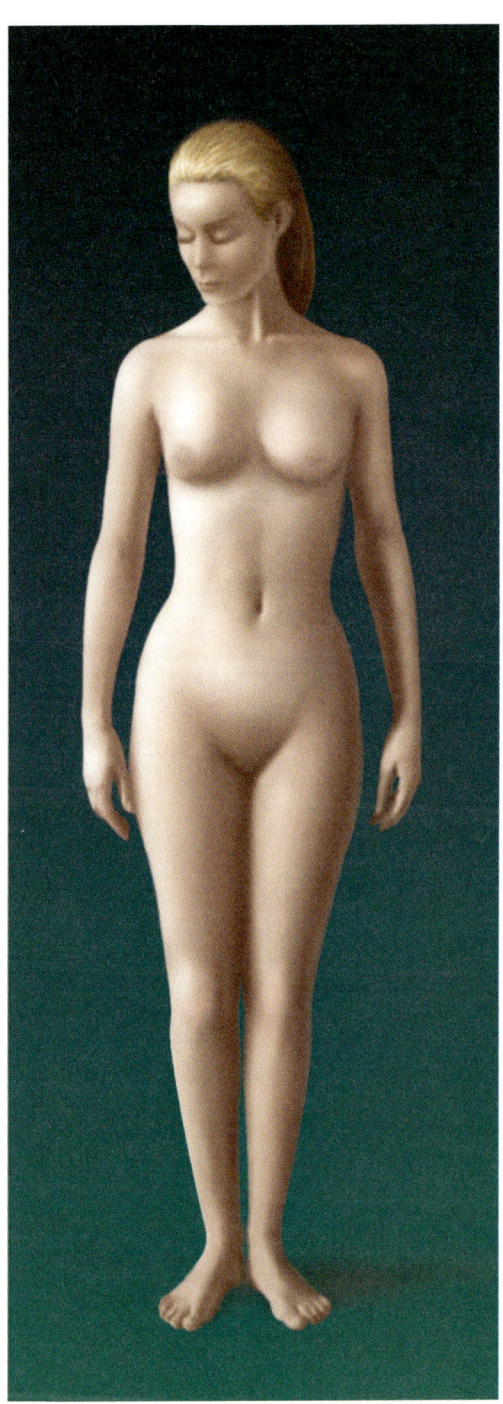

The skin provides protection, contains sensory organs, and helps control body temperature.

Skeletal System

Vitamin D activated by the skin helps provide calcium for bone matrix.

Muscular System

Involuntary muscle contractions (shivering) work with the skin to control body temperature. Muscles act on facial skin to create expressions.

Nervous System
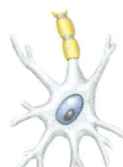
Sensory receptors provide information about the outside world to the nervous system. Nerves control the activity of sweat glands.

Endocrine System

Hormones help to increase skin blood flow during exercise. Other hormones stimulate either the synthesis or the decomposition of subcutaneous fat.

Cardiovascular System
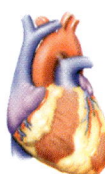
Skin blood vessels play a role in regulating body temperature.

Lymphatic System

The skin provides an important first line of defense for the immune system.

Digestive System
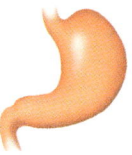
Excess calories may be stored as subcutaneous fat. Vitamin D activated by the skin stimulates dietary calcium absorption.

Respiratory System
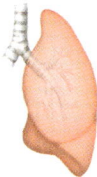
Stimulation of skin receptors may alter respiratory rate.

Urinary System

The kidneys help compensate for water and electrolytes lost in sweat.

Reproductive System
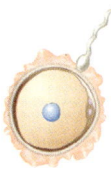
Sensory receptors play an important role in sexual activity and in the suckling reflex.

Clinical Connection

When skin must heal from a severe ulcer, differentiated cells can *de*differentiate, reverting to stem cells that can help to fill in destroyed tissue. Researchers treated eight patients with leg ulcers with epidermal growth factor, and compared stem cells in their skin to those of seven patients with leg ulcers who had not received the treatment. They detected the stem cells by staining the cells for varieties of integrin and keratin proteins unique to skin stem cells. In the patients who were not treated with the growth factor, scattered stem cells appeared in one layer at the bottom of the basement membrane. However, in the patients who had been treated, stem cells were considerably more abundant, grouped into "stem cell islands" that traverse more than one layer, particularly where the epidermis dips down into the region of the dermis. In areas of healthy skin in all the patients, stem cells were quite scarce, appearing in hair follicles and near the basement membrane. The researchers think that the new stem cells arise from differentiated cells losing their specialization, rather than from stem cells proliferating. Discovering how skin stem cells function can lead to new ways to treat burns.

SUMMARY OUTLINE

6.1 Introduction (p. 113)
Organs, such as membranes, are composed of two or more kinds of tissues. The skin is an organ. Together with its accessory organs, it constitutes the integumentary system.

6.2 Types of Membranes (p. 113)
1. Serous membranes
   a. Serous membranes line body cavities that lack openings to the outside.
   b. Cells of serous membranes secrete watery serous fluid that lubricates membrane surfaces.
2. Mucous membranes
   a. Mucous membranes line cavities and tubes that open to the outside of the body.
   b. Cells of mucous membranes secrete mucus.
3. Synovial membranes
   a. Synovial membranes line joint cavities.
   b. They secrete synovial fluid that lubricates the ends of the bones at joints.
4. The cutaneous membrane is the external body covering commonly called the skin.

6.3 Skin and Its Tissues (p. 113)
Skin is a protective covering, helps regulate body temperature, retards water loss, houses sensory receptors, synthesizes various biochemicals, and excretes wastes. It is composed of an epidermis and a dermis separated by a basement membrane.
1. Epidermis
   a. The deepest layer of the epidermis, called the stratum basale, contains cells that divide.
   b. Epidermal cells undergo keratinization as they mature and are pushed toward the surface.
   c. The outermost layer, called the stratum corneum, is composed of dead epidermal cells.
   d. The epidermis protects underlying tissues against water loss, mechanical injury, and the effects of harmful chemicals.
   e. Melanin protects underlying cells from the effects of ultraviolet light.
   f. Melanocytes transfer melanin to nearby epidermal cells.
2. Skin color
   a. All people have about the same concentration of melanocytes.
   b. Skin color is due largely to the amount of melanin and the distribution and size of the pigment granules in the epidermis.
   c. Environmental and physiological factors, as well as genes, influence skin color.
3. Dermis
   a. The dermis binds the epidermis to underlying tissues.
   b. Dermal blood vessels supply nutrients to all skin cells and help regulate body temperature.
   c. Nerve fibers are scattered throughout the dermis.
      (1) Some dermal nerve fibers carry impulses to muscles and glands of the skin.
      (2) Other dermal nerve fibers are associated with sensory receptors in the skin, and carry impulses to the brain and spinal cord.
   d. The dermis also contains hair follicles, sebaceous glands, and sweat glands.
4. Subcutaneous layer
   a. The subcutaneous layer beneath the dermis consists of loose connective and adipose tissues.
   b. Adipose tissue helps conserve body heat.
   c. The subcutaneous layer contains blood vessels that supply the skin and underlying adipose tissue.

6.4 Accessory Organs of the Skin (p. 117)
1. Hair follicles
   a. Each hair develops from epidermal cells at the base of a tubelike hair follicle.
   b. As newly formed cells develop and grow, older cells are pushed toward the surface and undergo keratinization.
   c. A bundle of smooth muscle cells is attached to each hair follicle.
   d. Hair color is determined by genes that direct the amount of melanin that melanocytes associated with hair follicles produce.
2. Sebaceous glands
   a. Sebaceous glands are usually associated with hair follicles.
   b. Sebaceous glands secrete sebum, which helps keep the skin and hair soft and waterproof.
3. Nails
   a. Nails are protective covers on the ends of fingers and toes.
   b. Specialized epidermal cells that are keratinized make up nails.
   c. The keratin of nails is harder than that produced by the skin's epidermal cells.
4. Sweat glands
   a. Each sweat gland is a coiled tube.
   b. Sweat is primarily water but also contains salts and waste products.
   c. Eccrine sweat glands respond to elevated body temperature, whereas apocrine glands respond to emotional stress.

6.5 Regulation of Body Temperature (p. 120)

Regulation of body temperature is vital because heat affects the rates of metabolic reactions. The normal temperature of deeper body parts is close to a set point of 37°C (98.6°F).

1. When body temperature rises above the normal set point, dermal blood vessels dilate, and sweat glands secrete sweat.
2. If body temperature drops below the normal set point, dermal blood vessels constrict, and sweat glands become inactive.
3. Excessive heat loss stimulates skeletal muscles to contract involuntarily.
4. Fever results from an elevated temperature set point.

6.6 Healing of Wounds (p. 120)

Skin injuries trigger inflammation. The affected area becomes red, warm, swollen, and tender.

1. Dividing epithelial cells fill in shallow cuts in the epidermis.
2. Clots close deeper cuts, sometimes leaving a scar where connective tissue replaces skin.
3. Granulations form in large, open wounds as part of the healing process.

REVIEW EXERCISES

1. Explain why a membrane is an organ. (p. 113)
2. Define *integumentary system*. (p. 113)
3. Distinguish between serous and mucous membranes. (p. 113)
4. Explain the functions of synovial membranes. (p. 113)
5. List six functions of skin. (p. 113)
6. Distinguish between the epidermis and the dermis. (p. 113)
7. Explain what happens to epidermal cells as they undergo keratinization. (p. 113)
8. Describe the function of melanocytes. (p. 114)
9. List the factors that affect skin color. (p. 116)
10. Review the functions of dermal nervous tissue. (p. 116)
11. Describe the subcutaneous layer and its functions. (p. 116)
12. Explain how blood is supplied to various skin layers. (p. 116)
13. Distinguish between a hair and a hair follicle. (p. 117)
14. Explain the function of sebaceous glands. (p. 118)
15. Describe how nails are formed. (p. 118)
16. Distinguish between eccrine and apocrine sweat glands. (p. 119)
17. Explain how body heat is produced. (p. 120)
18. Explain how sweat glands help regulate body temperature. (p. 120)
19. Describe the body's responses to decreasing body temperature. (p. 120)
20. Distinguish between the healing of shallow and deeper breaks in the skin. (p. 120)

CRITICAL THINKING

1. Everyone's skin contains about the same number of melanocytes even though people are of many different colors. How is this possible?
2. Why would collagen and elastin added to skin creams be unlikely to penetrate the skin—as some advertisements imply they do?
3. A severe form of the inherited illness epidermolysis bullosa causes extreme blistering of the skin. The person lacks a protein called laminin, which normally attaches the dermis to the epidermis. Explain how lack of this protein disrupts the skin's structure.
4. How is skin peeling after a severe sunburn protective? How might a fever be protective?
5. What special problems would result from the loss of 50% of a person's functional skin surface? How might this person's environment be modified to compensate partially for such a loss?
6. A premature infant typically lacks subcutaneous adipose tissue. Also, the surface area of an infant's small body is relatively large compared to its volume. How do you think these factors affect the ability of an infant to regulate its body temperature?
7. Which of the following would result in the more rapid absorption of a drug: a subcutaneous injection or an intradermal injection? Why?
8. How would you explain to an athlete the importance of keeping the body hydrated when exercising in warm weather?

WEB CONNECTIONS

Visit the website for additional study questions and more information about this chapter at:

http://www.mhhe.com/shieress8

chapter 7

Skeletal System

CLUES FROM SKELETONS PAST. As the hardest and therefore most enduring of human tissues, bone has persisted over time to provide clues to early humans and their forebears. Some glimpses into the past, courtesy of skeletal remains or fossils, include:

7300–6220 B.C. Skulls with circular holes are the earliest evidence of trepanation, a technique used to relieve pressure following a skull fracture or as a spiritual treatment for headache, tumors, or mental illness. A few of the people treated with trepanation were lucky—they survived, as evidenced by new bony growth over the holes made in their skulls. However, most trepanated skulls have gaping, drilled holes, indicating that the "treatment" was lethal.

2.8–2.6 million years ago "Mr. Ples" is the name anthropologists have given to the face and left side of a skull from Sterkfontein, South Africa, which once belonged to a member of *Australopithecus africanus* (see photo), a type of primate that preceded *Homo sapiens*. Using computer modeling to fashion a "virtual endocast" of the entire skull contents, researchers have estimated the cranial capacity of *A. africanus* at 515 cubic centimeters (cc). By comparison, a chimp's cranial capacity averages 370 cc, and a modern human's, 1,350. Expanded cranial capacity correlates to increase in intelligence.

3.5 million years ago Not all evidence of a skeletal system is in the form of preserved bone. On the Serengeti Plain are clues to our ancestors who first began to walk upright, a stance that freed their hands, perhaps making possible the development of tools. This evidence consists of shallow footprints where an animal called *Australopithecus afarensis* once lived. The prints reveal that it had long big toes and arched feet.

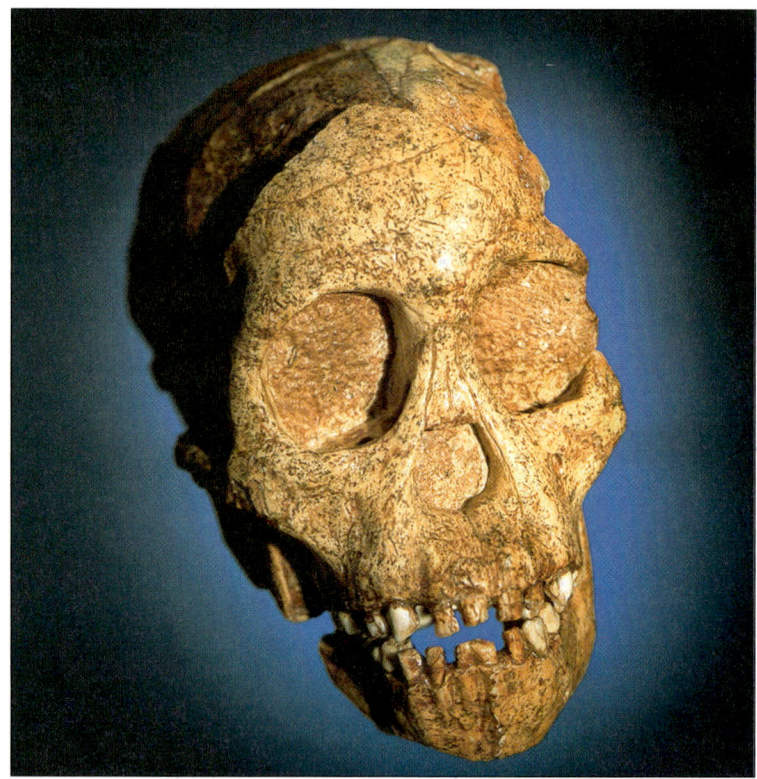

Photo:
Australopithecus africanus lived from 2.8 to 2.6 million years ago. Our knowledge of this primate comes from skeletal evidence.

Chapter Objectives

After studying this chapter, you should be able to do the following:

7.1 Introduction
1. List the active tissues in a bone. (p. 128)

7.2 Bone Structure
2. Describe the general structure of a bone, and list the functions of its parts. (p. 128)

7.3 Bone Development and Growth
3. Distinguish between intramembranous and endochondral bones, and explain how such bones develop and grow. (p. 130)

7.4 Bone Function
4. Discuss the major functions of bones. (p. 131)

7.5 Skeletal Organization
5. Distinguish between the axial and appendicular skeletons, and name the major parts of each. (p. 135)

7.6–7.12 Skull—Lower Limb
6. Locate and identify the bones and the major features of the bones that comprise the skull, vertebral column, thoracic cage, pectoral girdle, upper limb, pelvic girdle, and lower limb. (p. 136)

7.13 Joints
7. List three classes of joints, describe their characteristics, and name an example of each. (p. 156)
8. List six types of synovial joints, and describe the actions of each. (p. 157)
9. Explain how skeletal muscles produce movements at joints, and identify several types of joint movements. (p. 159)

Aids to Understanding Words

acetabul- [vinegar cup] *acetabul*um: Depression of the coxa that articulates with the head of the femur.
ax- [axis] *ax*ial skeleton: Upright portion of the skeleton that supports the head, neck, and trunk.
-blast [budding] osteo*blast*: Cell that will form bone tissue.
carp- [wrist] *carp*als: Wrist bones.
-clast [break] osteo*clast*: Cell that breaks down bone tissue.
condyl- [knob] *condyl*e: Rounded, bony process.
corac- [a crow's beak] *corac*oid process: Beaklike process of the scapula.
cribr- [sieve] *cribr*iform plate: Portion of the ethmoid bone with many small openings.
crist- [crest] *crist*a galli: Bony ridge that projects upward into the cranial cavity.
fov- [pit] *fov*ea capitis: Pit in the head of a femur.
glen- [joint socket] *glen*oid cavity: Depression in the scapula that articulates with the head of the humerus.
hema- [blood] *hema*toma: Blood clot.
inter- [among, between] *inter*vertebral disc: Structure located between adjacent vertebrae.
intra- [inside] *intra*membranous bone: Bone that forms within sheetlike masses of connective tissue.
meat- [passage] auditory *meat*us: Canal of the temporal bone that leads inward to parts of the ear.
odont- [tooth] *odont*oid process: Toothlike process of the second cervical vertebra.
poie- [make, produce] hemato*poie*sis: Process by which blood cells are formed.

Key Terms

articular cartilage (ar-tik´u-lar kar´tĭ-lij)
bursa (ber´sah)
cartilaginous joint (kar´´tĭ-lah´jin-us joint)
compact bone (kom´pakt bōn)
diaphysis (di-af´ĭ-sis)
endochondral bone (en´´do-kon´dral bōn)
epiphyseal plate (ep´´ĭ-fiz´e-al plāt)
epiphysis (e-pif´ĭ-sis)
fibrous joint (fi´brus joint)
hemopoiesis (he´´mo-poi-e´sis)
intramembranous bone (in´´trah-mem´brah-nus bōn)
lever (lev´er)
marrow (mar´o)
medullary cavity (med´u-lār´´e kav´ĭ-te)
meniscus (mĕ-nis´kus)
osteoblast (os´te-o-blast)
osteoclast (os´te-o-klast)
osteocyte (os´te-o-sīt)
periosteum (per´´e-os´te-um)
spongy bone (spun´je bōn)
synovial joint (sĭ-no´ve-al joint)

7.1 Introduction

Halloween skeletons and the skull-and-crossbones symbol of poison and pirates may make bones seem like lifeless objects, but in actuality, bones are not only very much alive but also multifunctional. Bones, the organs of the skeletal system, provide points of attachment for muscles, protect and support softer tissues, house blood-producing cells, store inorganic salts, and contain passageways for blood vessels and nerves. Bone contains a variety of very active tissues: bone tissue, cartilage, dense connective tissue, blood, and nervous tissue.

7.2 Bone Structure

The bones of the skeletal system differ greatly in size and shape, yet they are similar in structure, development, and functions.

Parts of a Long Bone

The femur, a long bone in the thigh, illustrates the structure of bone (fig. 7.1). At each end of such a bone is an expanded portion called an **epiphysis** (e-pif´ĭ-sis) (plural, *epiphyses*), which articulates (forms a joint) with another bone. On its outer surface, the articulating portion of the epiphysis is coated with a layer of hyaline cartilage called **articular cartilage** (ar-tik´u-lar kar´tĭ-lij). The shaft of the bone, which is located between the epiphyses, is called the **diaphysis** (di-af´ĭ-sis).

A tough, vascular covering of fibrous tissue called the **periosteum** (per´´e-os´te-um) completely encloses the bone, except for the articular cartilage on the bone's ends. The periosteum is firmly attached to the bone, and periosteal fibers are continuous with ligaments and tendons that connect to the membrane. The periosteum also helps form and repair bone tissue.

A bone's shape makes possible its functions. For example, bony projections called *processes* provide sites for ligaments and tendons to attach; grooves and openings form passageways for blood vessels and nerves; and a depression of one bone may articulate with a process of another.

The wall of the diaphysis is mainly composed of tightly packed tissue called **compact bone** (kom´pakt bōn), or cortical bone. This type of bone has a continuous matrix with no gaps. The epiphyses, in contrast, are composed largely of **spongy bone** (spun´je bōn), or cancellous bone, with thin layers of compact bone on their surfaces. Spongy bone consists of numerous branching bony plates. Irregular connecting spaces between these plates help reduce the bone's weight. The bony plates are most highly developed in the regions of the epiphyses that are subjected to compressive forces. Both compact and spongy bone are strong and resist bending (fig. 7.2).

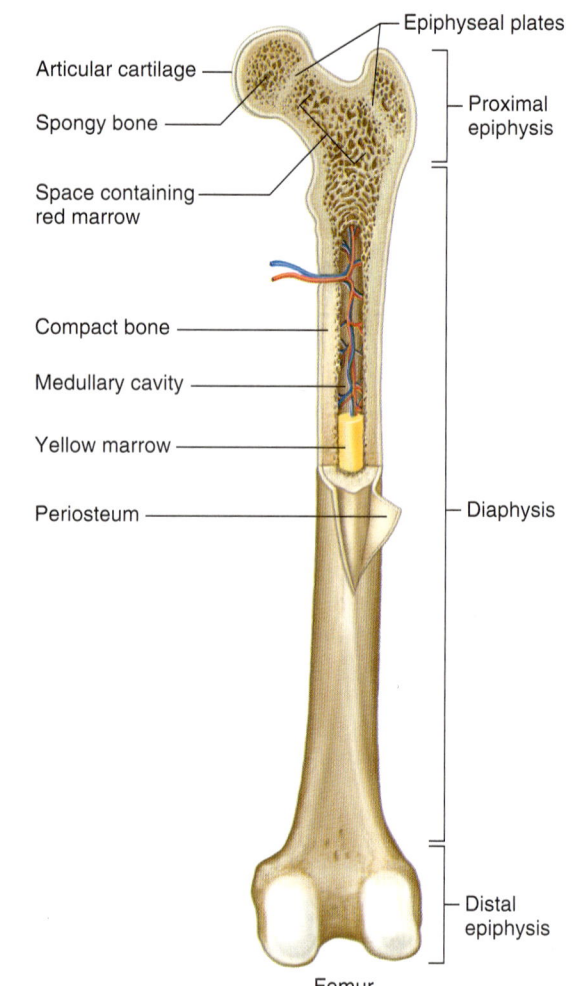

Figure 7.1
Major parts of a long bone.

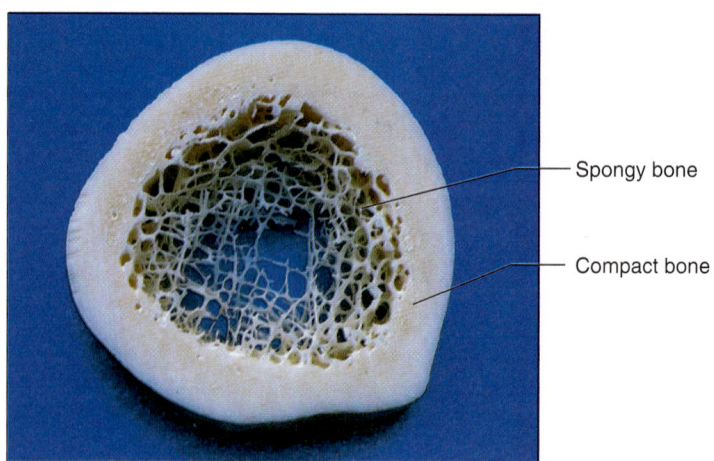

Figure 7.2
This cross section of a long bone contains a layer of spongy bone beneath a layer of compact bone.

Compact bone in the diaphysis of a long bone forms a semirigid tube with a hollow chamber called the **medullary cavity** (med´u-lar˝e kav´ĭ-te) that is continuous with the spaces of the spongy bone. A thin layer of cells called **endosteum** (en-dos´te-um) lines these areas, and a specialized type of soft connective tissue called **marrow** (mar´o) fills them.

Microscopic Structure

Recall from chapter 5 (p. 103) that bone cells called **osteocytes** (os´te-o-sītz) are located in very small, bony chambers called *lacunae,* which form concentric circles around *central canals* (Haversian canals) (fig. 7.3; see fig. 5.19). Osteocytes communicate with nearby cells by means of cellular processes passing through canaliculi. The intercellular material of bone tissue is largely collagen and inorganic salts. Collagen gives bone its strength and resilience, and inorganic salts make it hard and resistant to crushing.

In compact bone, the osteocytes and layers of intercellular material concentrically clustered around a central canal form a cylinder-shaped unit called an *osteon* (Haversian system). Many of these units cemented together form the substance of compact bone.

Each central canal contains blood vessels (usually capillaries) and nerve fibers surrounded by loose connective tissue. Blood in these vessels nourishes bone cells associated with the central canal.

Central canals extend longitudinally through bone tissue, and transverse *perforating canals* (Volkmann's canals) connect them. Perforating canals contain larger blood vessels and nerves by which the smaller blood vessels and nerve fibers in central canals communicate with the surface of the bone and the medullary cavity (fig. 7.3).

Spongy bone is also composed of osteocytes and intercellular material, but the bone cells do not aggregate around central canals. Instead, substances diffusing into canaliculi that lead to the surface of these thin, bony plates nourish the cells.

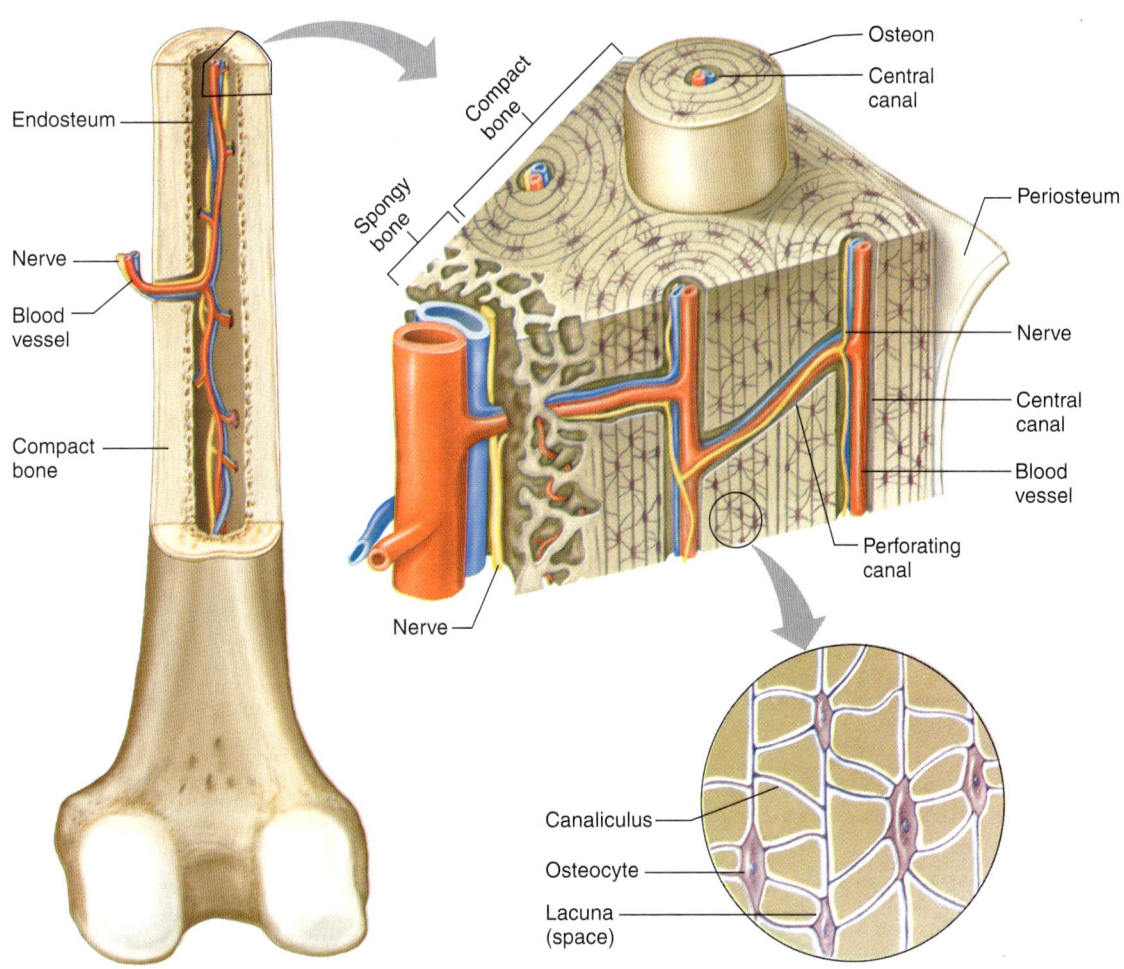

Figure 7.3
Compact bone is composed of osteons cemented together by bone matrix.

CHECK YOUR RECALL

1. List five major parts of a long bone.
2. How do compact and spongy bone differ in structure?
3. Describe the microscopic structure of compact bone.

7.3 Bone Development and Growth

Parts of the skeletal system begin to form during the first few weeks of prenatal development, and bony structures continue to develop and grow into adulthood. Bones form by replacing existing connective tissues in either of two ways: (1) Intramembranous bones originate between sheetlike layers of connective tissues. (2) Endochondral bones begin as masses of cartilage that bone tissue later replaces.

Intramembranous Bones

The broad, flat bones of the skull are **intramembranous bones** (in´´trah-mem´brah-nus bōnz) (fig. 7.4). During their development, membranelike layers of connective tissues appear at the sites of the future bones. Then, some of the primitive connective tissue cells enlarge and differentiate into bone-forming cells called **osteoblasts** (os´te-o-blastz). The osteoblasts become active within the membranes and deposit bony matrix around themselves. As a result, spongy bone tissue forms in all directions within the layers of primitive connective tissues. Eventually, cells of the membranous tissues that persist outside the developing bone give rise to the periosteum. Osteoblasts on the inside of the periosteum form a layer of compact bone over the surface of the newly formed spongy bone. When matrix completely surrounds osteoblasts, they are called osteocytes.

Endochondral Bones

Most of the bones of the skeleton are **endochondral bones** (en´´do-kon´dral bōnz). They develop from masses of hyaline cartilage shaped like future bony structures (fig. 7.4). These cartilaginous models grow rapidly for a time and then begin to extensively change. In a long bone, for example, the changes begin in the center of the diaphysis, where the cartilage slowly breaks down and disappears (fig. 7.5). At about the same time, a periosteum forms from connective tissue that encircles the developing diaphysis. Blood vessels and osteoblasts from the periosteum invade the disintegrating cartilage, and spongy bone forms in its place. This region of bone formation is called the *primary ossification center,* and bone tissue develops from it toward the ends of the cartilaginous structure.

Meanwhile, osteoblasts from the periosteum deposit a thin layer of compact bone around the primary ossification center. The epiphyses of the developing bone remain cartilaginous and continue to grow. Later, *secondary ossification centers* appear in the epiphyses, and spongy bone forms in all directions from them. As spongy bone is deposited in the diaphysis and in the epiphysis, a band of cartilage called the **epiphyseal plate** (ep´´ĭ-fiz´e-al plāt), or metaphysis, remains between these two ossification centers.

The cartilaginous cells of the epiphyseal plate include layers of young cells that are undergoing mitosis and producing new cells. As these cells enlarge and matrix forms around them, the cartilaginous plate thickens, lengthening the bone. At the same time, calcium salts accumulate in the matrix adjacent to the oldest cartilaginous cells, and as the matrix calcifies, the cells begin to die.

In time, large, multinucleated cells called **osteoclasts** (os´te-o-klastz) break down the calcified matrix. These large cells originate in bone marrow when certain single-nucleated white blood cells (monocytes) fuse (see chapter 12, p. 312).

Osteoclasts secrete an acid that dissolves the inorganic component of the calcified matrix, and their lysosomal enzymes digest the organic components. After osteoclasts remove the matrix, bone-building osteoblasts invade the region and deposit new bone tissue in place of the calcified cartilage.

A long bone continues to lengthen while the cartilaginous cells of the epiphyseal plates are active. However, once the ossification centers of the diaphysis

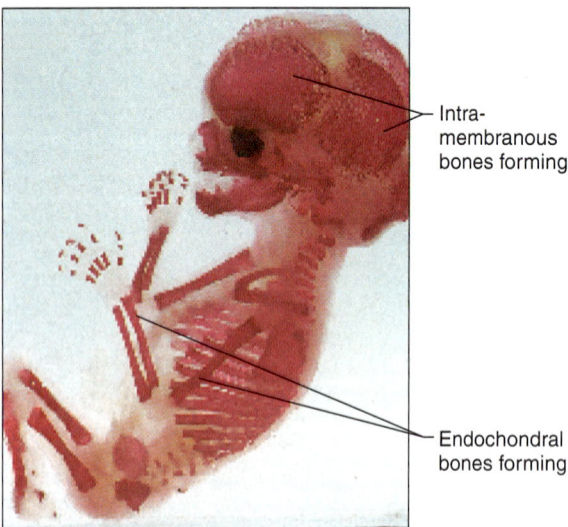

Figure 7.4
Note the stained, developing bones of this fourteen-week fetus.

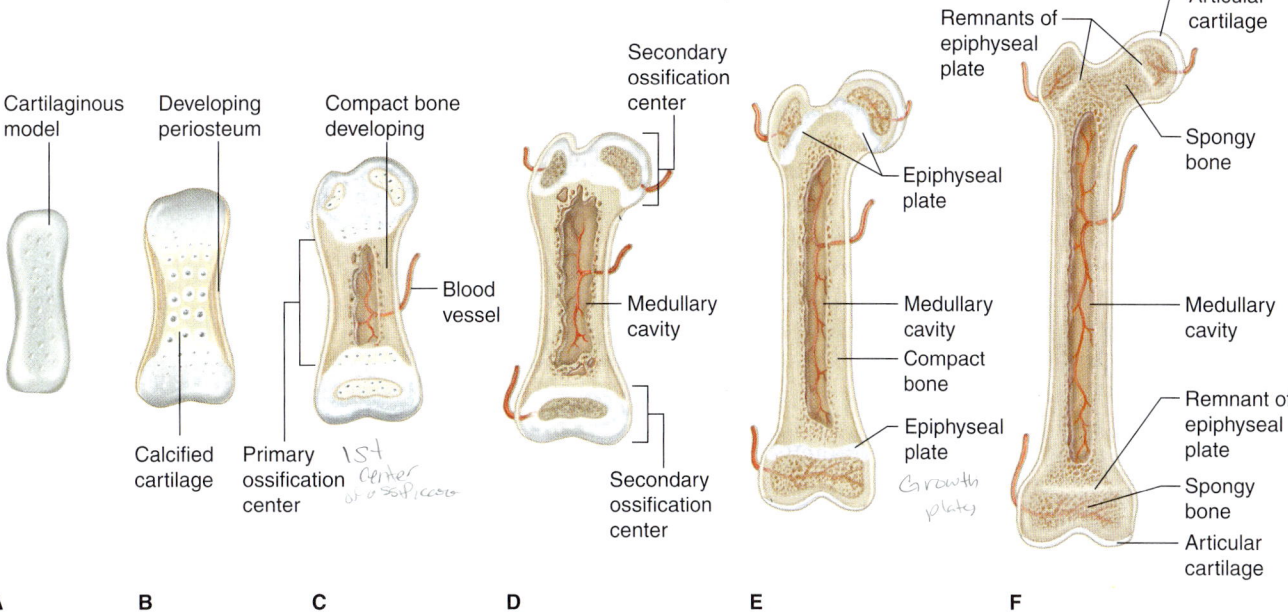

Figure 7.5
Major stages (*A–F*) in the development of an endochondral bone. (Relative bone sizes are not to scale.)

and epiphyses meet and the epiphyseal plates ossify, lengthening is no longer possible in that end of the bone.

A developing long bone thickens as compact bone is deposited on the outside, just beneath the periosteum. As this compact bone forms on the surface, osteoclasts erode other bone tissue on the inside. The resulting space becomes the medullary cavity of the diaphysis, which later fills with marrow. The bone in the central regions of the epiphyses and diaphysis remains spongy, and hyaline cartilage on the ends of the epiphyses persists throughout life as articular cartilage.

> If an epiphyseal plate is damaged before it ossifies, elongation of the long bone may cease prematurely, or growth may be uneven. For this reason, injuries to the epiphyses of a young person's bones are of special concern. Surgeons can alter an epiphysis to equalize the growth rate of bones developing at very different rates.

Homeostasis of Bone Tissue

After the intramembranous and endochondral bones form, the actions of osteoclasts and osteoblasts continually remodel them. Throughout life, osteoclasts resorb bone matrix, and osteoblasts replace it. Hormones regulate these opposing processes of *resorption* and *deposition* of calcium (see chapter 11, p. 291). As a result, the total mass of bone tissue of an adult skeleton normally remains nearly constant, even though 3–5% of bone calcium is exchanged each year.

> In bone cancers, abnormally active osteoclasts destroy bone tissue. Interestingly, cancer of the prostate gland can have the opposite effect if the cancer cells reach the bone marrow (as they do in most cases of advanced prostatic cancer). These cells stimulate osteoblast activity, which promotes formation of new bone on the surfaces of the bony plates.

CHECK YOUR RECALL

1. Describe the development of an intramembranous bone.
2. Explain how an endochondral bone develops.
3. Explain how osteoclasts and osteoblasts remodel bone.

7.4 Bone Function

Bones shape, support, and protect body structures. They also aid body movements, house tissues that produce blood cells, and store various inorganic salts.

Support and Protection

Bones give shape to such structures as the head, face, thorax, and limbs and also provide support and protection. For example, the bones of the lower limbs, pelvis, and backbone support the body's weight. The bones of the skull protect the eyes, ears, and brain. Those of the rib cage and shoulder girdle protect the heart and lungs, whereas bones of the pelvic girdle protect the lower abdominal and internal reproductive organs.

Topic of Interest

REPAIR OF A BONE FRACTURE

A *fracture* is a break in a bone. Whenever a bone breaks, blood vessels within it and its periosteum rupture, and the periosteum is likely to tear. Blood escaping from the broken vessels spreads through the damaged area and soon forms a blood clot, or *hematoma*. Vessels in surrounding tissues dilate, swelling and inflaming the tissues.

Within days or weeks, developing blood vessels and large numbers of osteoblasts from the periosteum invade the hematoma. The osteoblasts rapidly divide in the regions close to the new blood vessels, building spongy bone nearby. Granulation tissue develops, and in regions farther from a blood supply, fibroblasts produce masses of fibrocartilage. Meanwhile, phagocytic cells begin to remove the blood clot, as well as any dead or damaged cells in the affected area. Osteoclasts also appear and resorb bone fragments, aiding in "cleaning up" debris.

In time, fibrocartilage fills the gap between the ends of the broken bone. This mass, termed a *cartilaginous callus*, is later replaced by bone tissue in much the same way that the hyaline cartilage of a developing endochondral bone is replaced. That is, the cartilaginous callus breaks down, blood vessels and osteoblasts invade the area, and a *bony callus* fills the space.

Typically, more bone is produced at the site of a healing fracture than is required to replace the damaged tissues. Osteoclasts remove the excess, and the final result is a bone shaped very much like the original (fig. 7A).

Physicians can help the bone-healing process. The first casts to immobilize fractured bones were introduced in Philadelphia in 1876, and soon after, doctors began using screws and plates internally to align healing bone parts. Today, orthopedic surgeons also use rods, wires, and nails. These devices have become lighter and smaller; many are built of titanium. A new approach, called a hybrid fixator, treats a broken leg using metal pins internally to align bone pieces. The pins are anchored to a metal ring device worn outside the leg.

Figure 7A

Major steps (A–D) in repair of a fracture.

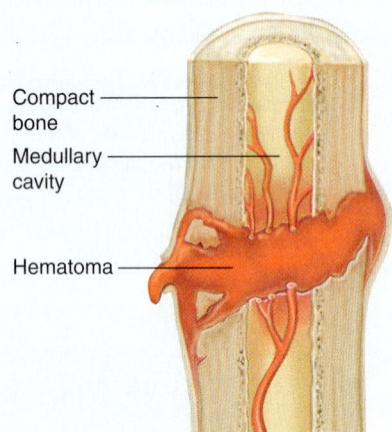

A Blood escapes from ruptured blood vessels and forms a hematoma.

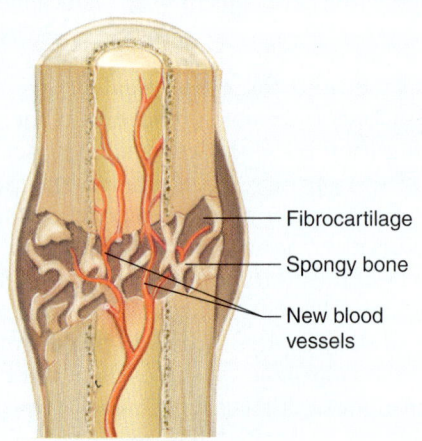

B Spongy bone forms in regions close to developing blood vessels, and fibrocartilage forms in more distant regions.

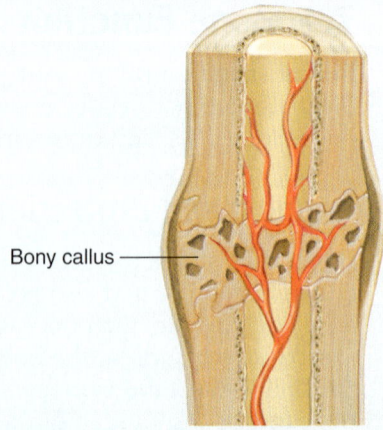

C A bony callus replaces fibrocartilage.

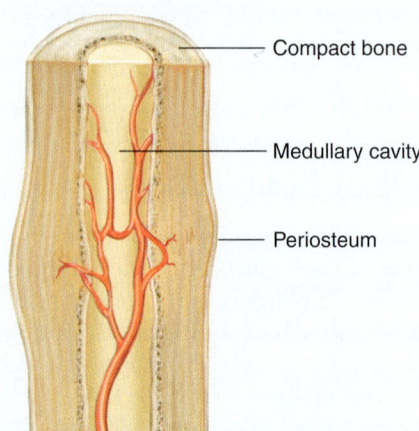

D Osteoclasts remove excess bony tissue, restoring new bone structure much like the original.

Body Movement

Whenever limbs or other body parts move, bones and muscles interact as simple mechanical devices called **levers** (lev´erz). A lever has four basic components: (1) a rigid rod or bar, (2) a fulcrum or pivot on which the bar turns, (3) an object that is moved against resistance, and (4) a force that supplies energy for the movement of the bar.

The actions of bending and straightening the upper limb at the elbow illustrate bones and muscles functioning as levers (fig. 7.6). When the upper limb bends, the forearm bones represent the rigid rod, the elbow joint is the fulcrum, the hand is moved against the resistance provided by its weight, and the force is supplied by muscles on the anterior side of the arm. One of these muscles, the *biceps brachii,* is attached by a tendon to a projection on a bone (radius) in the forearm, a short distance below the elbow.

When the upper limb straightens at the elbow, the forearm bones again serve as the rigid rod, and the elbow joint serves as the fulcrum. However, this time, the *triceps brachii,* a muscle located on the posterior side of the arm, supplies the force. A tendon of this muscle attaches to a projection on a bone (ulna) at the point of the elbow.

Blood Cell Formation

Very early in life, the process of blood cell formation, called **hemopoiesis** (he´´mo-poi-e´sis), begins in the *yolk sac,* which lies outside the human embryo (see chapter 20). Later in development, blood cells are manufactured in the liver and spleen, and still later, they form in bone marrow.

Marrow is a soft, netlike mass of connective tissue within the medullary cavities of long bones, in the irregular spaces of spongy bone, and in the larger central canals of compact bone tissue. The two kinds of marrow are red marrow and yellow marrow. *Red marrow* functions in the formation of red blood cells (erythrocytes), white blood cells (leukocytes), and blood platelets. It is red because of the red, oxygen-carrying pigment **hemoglobin** in the red blood cells.

Red marrow occupies the cavities of most bones in an infant. With increasing age, however, yellow marrow replaces much of it. *Yellow marrow* stores fat and is inactive in blood cell production. In an adult, red marrow is primarily found in the spongy bone of the skull, ribs, sternum, clavicles, vertebrae, and pelvis. Chapter 12 (pp. 308 and 311) describes blood cell formation in more detail.

Storage of Inorganic Salts

The intercellular matrix of bone tissue is rich in calcium salts, mostly in the form of calcium phosphate. Vital metabolic processes require calcium. When the blood is low in calcium, parathyroid hormone stimulates osteoclasts to break down bone tissue, which releases calcium salts from the intercellular matrix into the blood. A high blood calcium level inhibits osteoclast activity, and calcitonin from the thyroid gland stimulates osteoblasts to form bone tissue, storing excess calcium in the

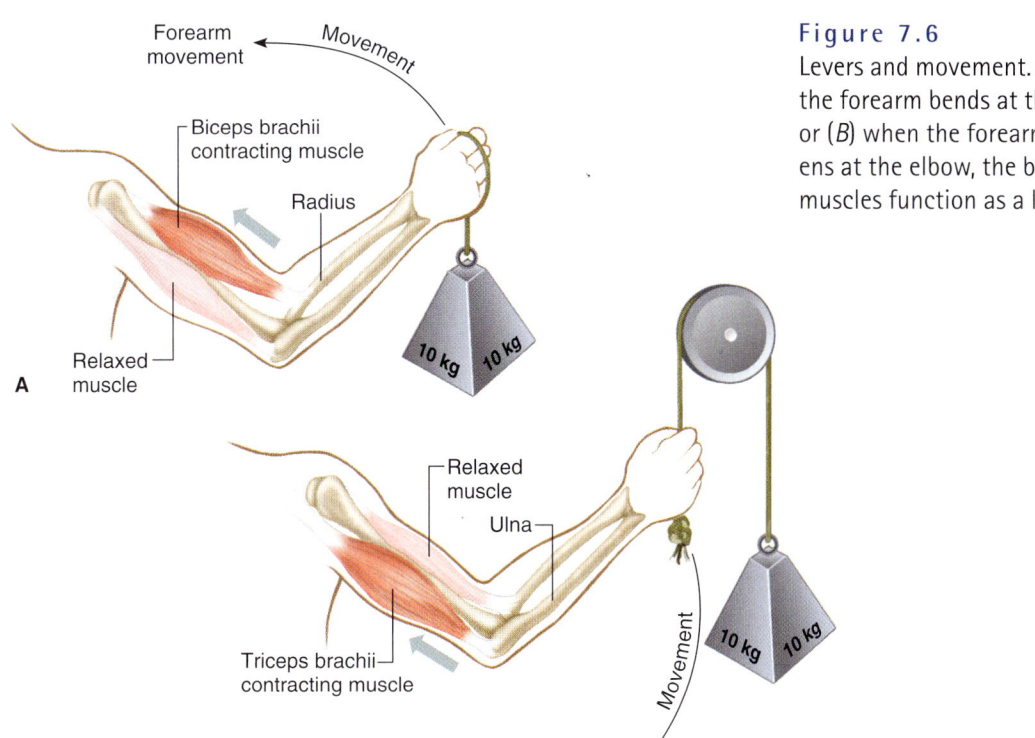

Figure 7.6
Levers and movement. (*A*) When the forearm bends at the elbow or (*B*) when the forearm straightens at the elbow, the bones and muscles function as a lever.

matrix. Chapter 11 (p. 290) describes the details of this homeostatic mechanism (fig. 7.7). Maintaining sufficient blood calcium levels is important in muscle contraction, nervous impulse conduction, blood clotting, and other physiological processes. The Topic of Interest on page 135 discusses bone mass loss due to less calcium in the bones.

Bone tissue contains lesser amounts of magnesium, sodium, potassium, and carbonate ions than the other constituents. Bones also accumulate certain harmful metallic elements, such as lead, radium, or strontium. These are not normally present in the body, but are sometimes ingested accidentally.

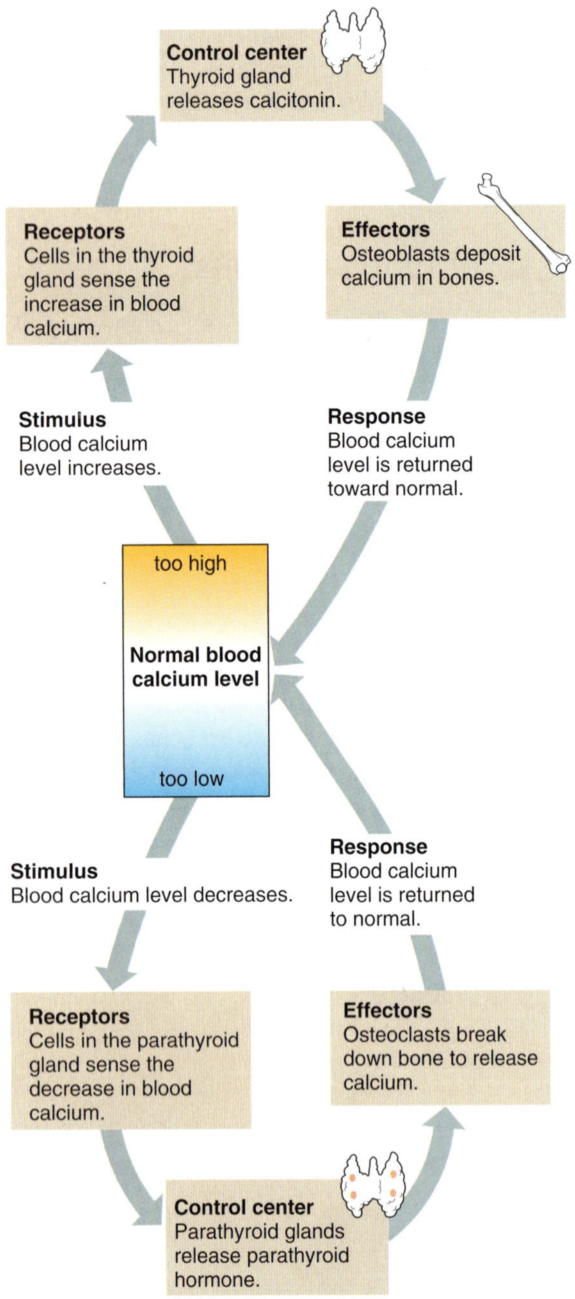

CHECK YOUR RECALL

1. Name three major functions of bones.
2. Distinguish between the functions of red marrow and yellow marrow.
3. List the substances normally stored in bone tissue.

A bone marrow transplant (BMT) is a life-saving, but risky, procedure. A hollow needle and syringe are used to remove normal red marrow cells from the spongy bone of a donor, or stem cells (which can give rise to specialized blood cells) are separated out from the donor's bloodstream. Stem cells from the umbilical cord of a newborn can also be used in place of bone marrow.

The donor is selected because the pattern of molecules on his or her cell surfaces closely matches that of the recipient. In 30% of BMTs, the donor is a blood relative. The cells are injected into the bloodstream of the recipient, whose own marrow has been intentionally destroyed with radiation or chemotherapy. If all goes well, the donor cells travel to the spaces within bones that red marrow normally occupies and replenish the blood supply—with healthy cells. About 15% of the time, the patient dies from infection because the immune system rejects the transplant, or because the transplanted tissue attacks the recipient, which is a condition called graft-versus-host disease.

BMT is used to treat more than sixty types of illnesses, mostly blood disorders such as sickle cell disease and leukemias. In cancer treatment, BMTs enable a patient to withstand high doses of radiation or chemotherapy, which usually destroys bone marrow along with cancer cells. BMT is used when other cancer treatments have failed. In the future, bone marrow may become a major part of "regenerative medicine," because it contains a variety of stem cells that can replenish many types of tissues.

7.5 Skeletal Organization

For purposes of study, it is convenient to divide the skeleton into two major portions—an axial skeleton and an appendicular skeleton (fig. 7.8). The **axial skeleton** consists of the bony and cartilaginous parts that support and protect the organs of the head, neck, and trunk. These parts include:

1. **Skull** The skull is composed of the **cranium** (kra´ne-um), or brain case, and the *facial bones*.
2. **Hyoid bone** The hyoid (hi´oid) bone is located in the neck between the lower jaw and the larynx. It supports the tongue and is an attachment for certain muscles that help move the tongue during swallowing.
3. **Vertebral column** The vertebral column (backbone) consists of many vertebrae separated by cartilaginous *intervertebral discs*. Near its distal

Figure 7.7
Hormonal regulation of bone calcium resorption and deposition.

Topic of Interest

OSTEOPOROSIS

In *osteoporosis*, the skeletal system loses bone volume and mineral content. The affected bones develop spaces and canals that enlarge and fill with fibrous and fatty tissues. Such bones easily fracture and may spontaneously break because they are no longer able to support body weight. For example, a person with osteoporosis may suffer a spontaneous fracture of the thighbone (femur) at the hip or a collapse of sections of the backbone (vertebrae).

Osteoporosis is associated with aging and causes many fractures in persons over age forty-five. It is most common in light-complexioned females after menopause.

Factors that increase the risk of osteoporosis include low intake of dietary calcium, lack of physical exercise (particularly during the early growing years), and in females, decrease in blood estrogen concentration. (Estrogen is a hormone the ovaries produce until menopause.) Drinking alcohol, smoking cigarettes, and inheriting certain genes may also increase a person's risk of developing osteoporosis.

Bone mass usually peaks at about age thirty-five. Thereafter, bone loss may exceed bone formation in both males and females. To reduce such loss, people in their mid-twenties and older should take in 1,000–1,500 milligrams of calcium daily. In addition, people should regularly engage in exercise, such as walking or jogging, that requires the bones to support the body weight. Postmenopausal women may also require estrogen replacement therapy to prevent osteoporosis, or drugs to slow the disease process.

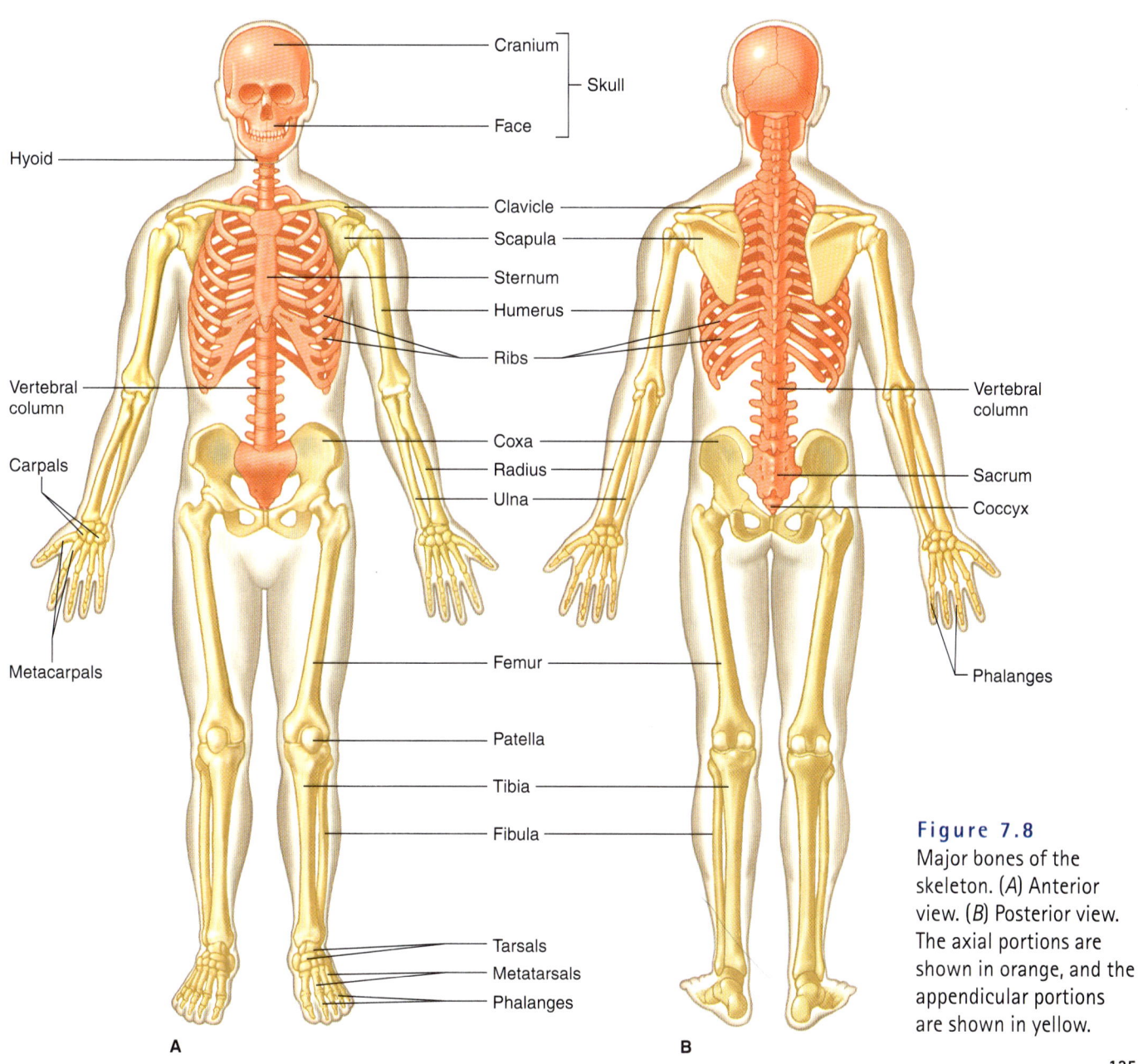

Figure 7.8
Major bones of the skeleton. (*A*) Anterior view. (*B*) Posterior view. The axial portions are shown in orange, and the appendicular portions are shown in yellow.

end, several vertebrae fuse to form the **sacrum** (sa´krum), which is part of the pelvis. A small, rudimentary tailbone called the **coccyx** (kok´siks) is attached to the end of the sacrum.

4. **Thoracic cage** The thoracic cage protects the organs of the thoracic cavity and the upper abdominal cavity. It is composed of twelve pairs of **ribs,** which articulate posteriorly with thoracic vertebrae. The thoracic cage also includes the **sternum** (ster´num), to which most of the ribs attach anteriorly.

The **appendicular skeleton** consists of the bones of the upper and lower limbs and the bones that anchor the limbs to the axial skeleton. It includes:

1. **Pectoral girdle** A **scapula** (scap´u-lah) and a **clavicle** (klav´ĭ-k'l) bone form the pectoral girdle on both sides of the body. The pectoral girdle connects the bones of the upper limbs to the axial skeleton and aids in upper limb movements.
2. **Upper limbs** Each upper limb consists of a **humerus** (hu´mer-us), or arm bone, two forearm bones—a **radius** (ra´de-us) and an **ulna** (ul´nah)—and a wrist and hand. The humerus, radius, and ulna articulate with each other at the elbow joint. At the distal end of the radius and ulna is the wrist. There are eight **carpals** (kar´pals), or wrist bones. The five bones of the palm are called **metacarpals** (met´´ah-kar´pals), and the fourteen finger bones are called **phalanges** (fah-lan´jēz).
3. **Pelvic girdle** Two coxae (kok´se), or hipbones, form the pelvic girdle and are attached to each other anteriorly and to the sacrum posteriorly. They connect the bones of the lower limbs to the axial skeleton and, with the sacrum and coccyx, form the **pelvis.**
4. **Lower limbs** Each lower limb consists of a **femur** (fe´mur), or thighbone, two leg bones—a large **tibia** (tib´e-ah) and a slender **fibula** (fib´u-lah)—an ankle and a foot. The femur and tibia articulate with each other at the knee joint, where the **patella** (pah-tel´ah) covers the anterior surface. At the distal ends of the tibia and fibula is the ankle. There are seven **tarsals** (tahr´sals), or anklebones. The five bones of the instep are called **metatarsals** (met´´ah-tahr´sals), and the fourteen bones of the toes (like the fingers) are called **phalanges.**

Table 7.1 lists the bones of the adult skeleton, and table 7.2 lists terms that describe skeletal structures.

 The skeleton of an average 160-pound body weighs about 29 pounds.

 CHECK YOUR RECALL

1. Distinguish between the axial and appendicular skeletons.
2. List the bones of the axial skeleton and the appendicular skeleton.

7.6 Skull

A human skull usually consists of twenty-two bones that, except for the lower jaw, are firmly interlocked along lines called *sutures* (soo´cherz) (fig. 7.9). Eight of these interlocked bones make up the cranium, and thirteen form the facial skeleton. The **mandible** (man´dĭ-b'l), or lower jawbone, is a movable bone held to the cranium by ligaments. (Three other bones found in each middle ear are discussed in chapter 10, p. 261.) Reference plates 8–11 on pages 167–169 are a set of photographs of the human skull and its parts.

Cranium

The **cranium** encloses and protects the brain, and its surface provides attachments for muscles that make chewing and head movements possible. Some of the cranial bones contain air-filled cavities called *sinuses*, which are lined with mucous membranes and connected by passageways to the nasal cavity (fig. 7.10). Sinuses reduce the skull's weight and increase the intensity of the voice by serving as resonant sound chambers.

The eight bones of the cranium are (figs. 7.9 and 7.11):

1. **Frontal bone** The frontal (frun´tal) bone forms the anterior portion of the skull above the eyes. On the upper margin of each orbit (the bony socket of the eye), the frontal bone is marked by a *supraorbital foramen* (or *supraorbital notch* in some skulls), through which blood vessels and nerves pass to the tissues of the forehead. Within the frontal bone are two *frontal sinuses,* one above each eye near the midline (fig. 7.10).
2. **Parietal bones** One parietal (pah-ri´ĕ-tal) bone is located on each side of the skull just behind the frontal bone (fig. 7.11). Together, the parietal bones form the bulging sides and roof of the cranium. They are fused at the midline along the *sagittal suture,* and they meet the frontal bone along the *coronal suture.*
3. **Occipital bone** The occipital (ok-sip´ĭ-tal) bone joins the parietal bones along the *lambdoidal* (lam´doid-al) *suture* (figs. 7.11 and 7.12). It forms the back of the skull and the base of the cranium. Through a large opening on its lower surface called the *foramen magnum* pass nerve fibers from the brain, which enter the vertebral canal to become part of the spinal cord. Rounded processes called

TABLE 7.1 — BONES OF THE ADULT SKELETON

1. **Axial Skeleton**
   a. Skull
      8 cranial bones
      - frontal 1
      - parietal 2
      - occipital 1
      - temporal 2
      - sphenoid 1
      - ethmoid 1

      14 facial bones
      - maxilla 2
      - zygomatic 2
      - palatine 2
      - inferior nasal concha 2
      - mandible 1
      - lacrimal 2
      - nasal 2
      - vomer 1

      22 bones

   b. Middle ear bones
      - malleus 2
      - incus 2
      - stapes 2

      6 bones

   c. Hyoid
      - hyoid bone 1

      1 bone

   d. Vertebral column
      - cervical vertebrae 7
      - thoracic vertebrae 12
      - lumbar vertebrae 5
      - sacrum 1
      - coccyx 1

      26 bones

   e. Thoracic cage
      - rib 24
      - sternum 1

      25 bones

2. **Appendicular Skeleton**
   a. Pectoral girdle
      - scapula 2
      - clavicle 2

      4 bones

   b. Upper limbs
      - humerus 2
      - radius 2
      - ulna 2
      - carpal 16
      - metacarpal 10
      - phalanx 28

      60 bones

   c. Pelvic girdle
      - coxal bone 2

      2 bones

   d. Lower limbs
      - femur 2
      - tibia 2
      - fibula 2
      - patella 2
      - tarsal 14
      - metatarsal 10
      - phalanx 28

      60 bones

   Total **206 bones**

TABLE 7.2 — TERMS USED TO DESCRIBE SKELETAL STRUCTURES

TERM	DEFINITION	EXAMPLES
Condyle (kon´dīl)	A rounded process that usually articulates with another bone	Occipital condyle of occipital bone (fig. 7.12)
Crest (krest)	A narrow, ridgelike projection	Iliac crest of ilium (fig. 7.27)
Epicondyle (ep´´ĭ-kon´dīl)	A projection situated above a condyle	Medial epicondyle of humerus (fig. 7.23)
Facet (fas´et)	A small, nearly flat surface	Rib facet of thoracic vertebra (fig. 7.16)
Fontanel (fon´´tah-nel´)	A soft spot in the skull where membranes cover the space between bones	Anterior fontanel between frontal and parietal bones (fig. 7.15)
Foramen (fo-ra´men)	An opening through a bone that usually is a passageway for blood vessels, nerves, or ligaments	Foramen magnum of occipital bone (fig. 7.12)
Fossa (fos´ah)	A relatively deep pit or depression	Olecranon fossa of humerus (fig. 7.23)
Fovea (fo´ve-ah)	A tiny pit or depression	Fovea capitis of femur (fig. 7.29)
Head (hed)	An enlargement on the end of a bone	Head of humerus (fig. 7.23)
Meatus (me-a´tus)	A tubelike passageway within a bone	External auditory meatus of ear (fig. 7.11)
Process (pros´es)	A prominent projection on a bone	Mastoid process of temporal bone (fig. 7.11)
Sinus (si´nus)	A cavity within a bone	Frontal sinus of frontal bone (fig. 7.14)
Spine (spīn)	A thornlike projection	Spine of scapula (fig. 7.22)
Suture (soo´cher)	An interlocking line of union between bones	Lambdoidal suture between occipital and parietal bones (fig. 7.11)
Trochanter (tro-kan´ter)	A relatively large process	Greater trochanter of femur (fig. 7.29)
Tubercle (tu´ber-kl)	A small, knoblike process	Greater tubercle of humerus (fig. 7.23)
Tuberosity (tu´´bĕ-ros´ĭ-te)	A knoblike process usually larger than a tubercle	Radial tuberosity of radius (fig. 7.24)

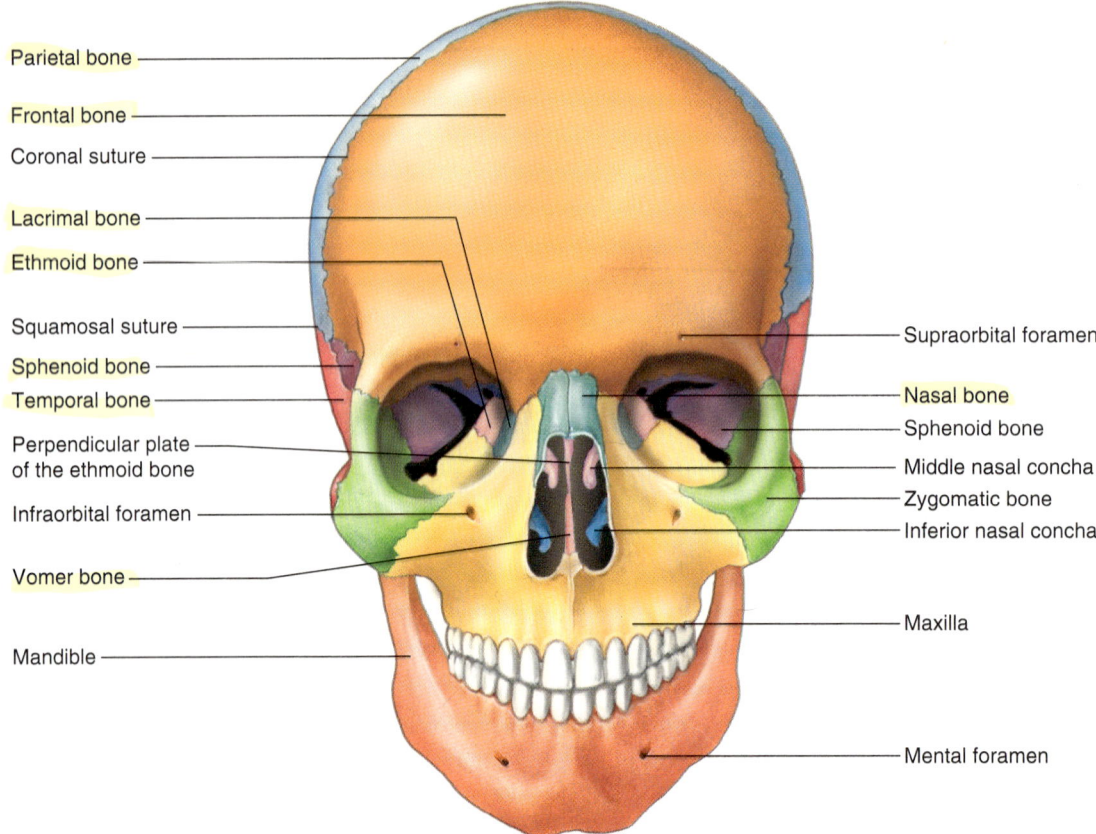

Figure 7.9
Anterior view of the skull.

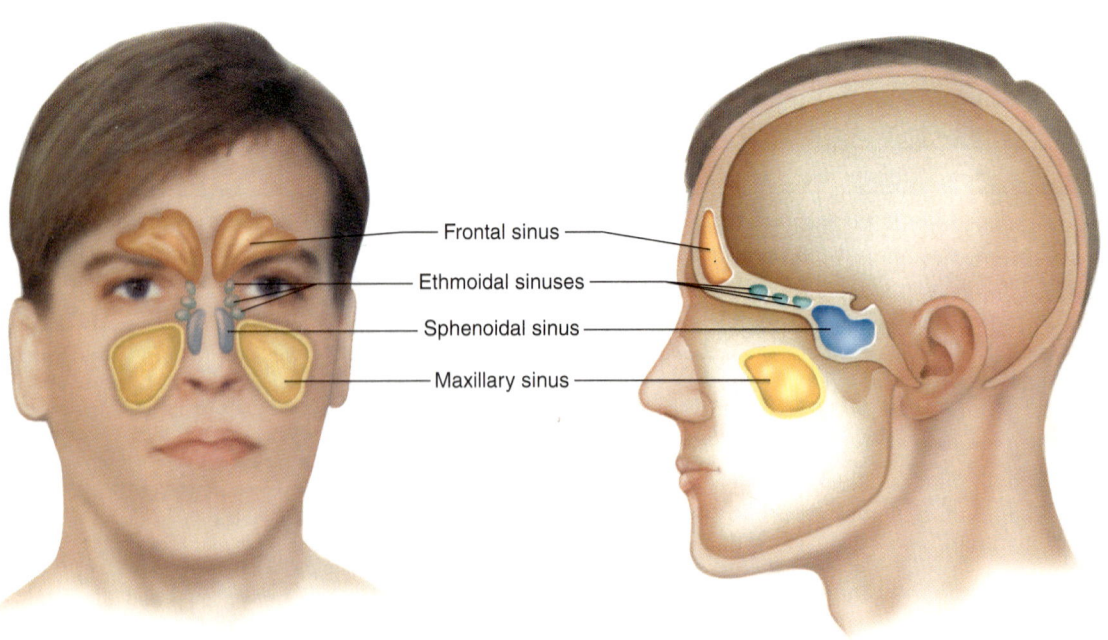

Figure 7.10
Locations of the sinuses.

CHAPTER 7 *Skeletal System* 139

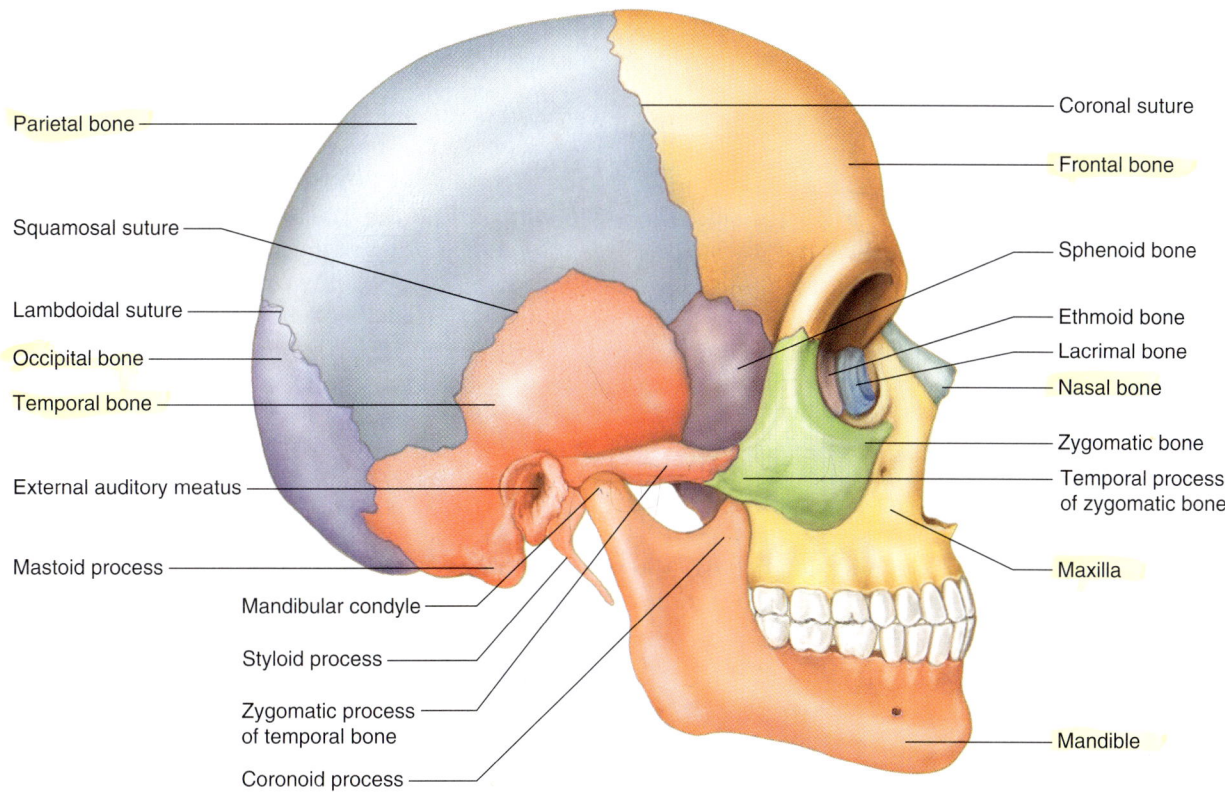

Figure 7.11
Lateral view of the skull.

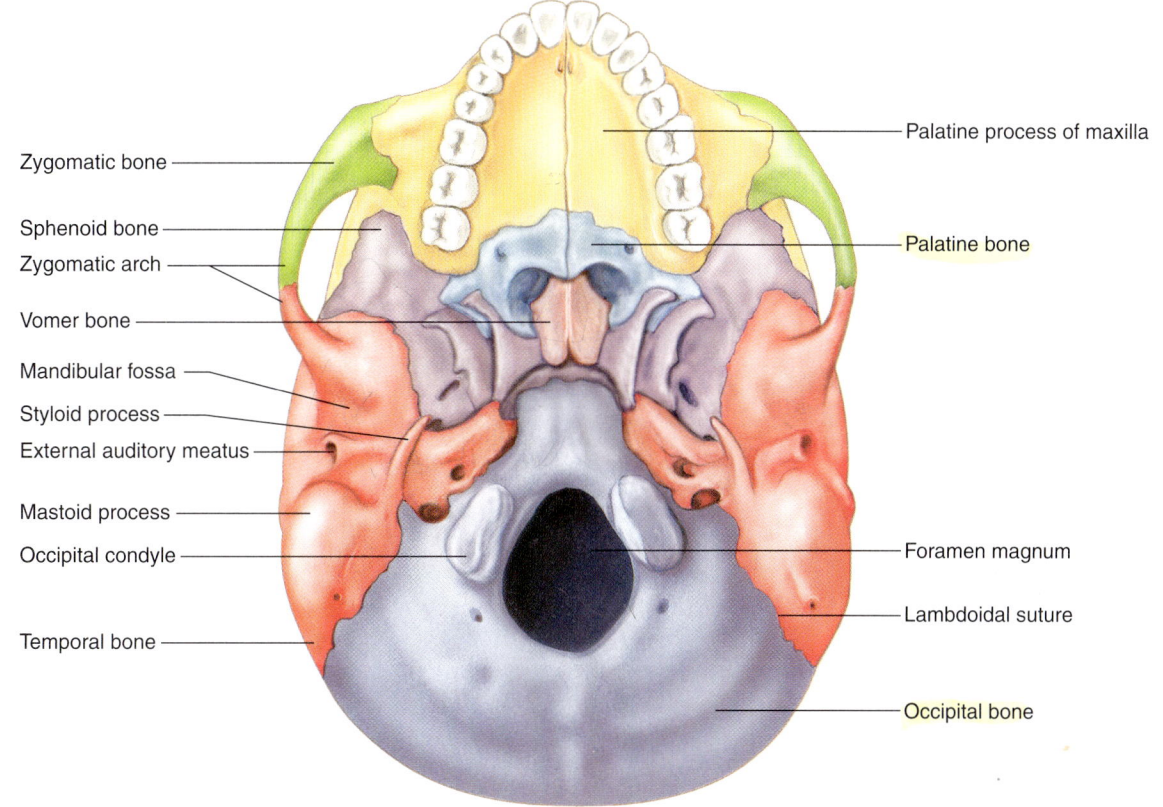

Figure 7.12
Inferior view of the skull.

occipital condyles, located on each side of the foramen magnum, articulate with the first vertebra (atlas) of the vertebral column.

4. **Temporal bones** A temporal (tem´po-ral) bone on each side of the skull joins the parietal bone along a *squamosal* (skwa-mo´sal) *suture* (see figs. 7.9 and 7.11). The temporal bones form parts of the sides and the base of the cranium. Located near the inferior margin is an opening, the *external auditory meatus,* which leads inward to parts of the ear. The temporal bones have depressions called the *mandibular fossae* that articulate with condyles of the mandible. Below each external auditory meatus are two projections—a rounded *mastoid process* and a long, pointed *styloid process.* The mastoid process provides an attachment for certain muscles of the neck, whereas the styloid process anchors muscles associated with the tongue and pharynx. A *zygomatic process* projects anteriorly from the temporal bone, joins the *zygomatic bone,* and helps form the prominence of the cheek.

5. **Sphenoid bone** The sphenoid (sfe´noid) bone is wedged between several other bones in the anterior portion of the cranium (figs. 7.11 and 7.12). It consists of a central part and two winglike structures that extend laterally toward each side of the skull. This bone helps form the base of the cranium, the sides of the skull, and the floors and sides of the orbits. Along the midline within the cranial cavity, a portion of the sphenoid bone indents to form the saddle-shaped *sella turcica* (sel´ah tur´si-ka). The pituitary gland occupies this depression. The sphenoid bone also contains two *sphenoidal sinuses* (see fig. 7.10).

6. **Ethmoid bone** The ethmoid (eth´moid) bone is located in front of the sphenoid bone (figs. 7.11 and 7.13). It consists of two masses, one on each side of the nasal cavity, which are joined horizontally by thin *cribriform* (krib´rĭ-form) *plates.* These plates form part of the roof of the nasal cavity (fig. 7.13).

   Projecting upward into the cranial cavity between the cribriform plates is a triangular process of the ethmoid bone called the *crista galli* (kris´tă gal´li) (cock's comb). Membranes that enclose the brain attach to this process (figs. 7.13 and 7.14). Portions of the ethmoid bone also form sections of the cranial floor, the orbital walls, and the nasal cavity walls. A *perpendicular plate* projects downward in the midline from the cribriform plates and forms most of the nasal septum (fig. 7.14).

   Delicate scroll-shaped plates called the *superior nasal concha* (kong´kah) and the *middle nasal concha* project inward from the lateral portions of the ethmoid bone toward the perpendicular plate (see fig. 7.9). The lateral portions of the ethmoid bone contain many small air spaces, the *ethmoidal sinuses* (see fig. 7.10).

Figure 7.13
Floor of the cranial cavity, viewed from above.

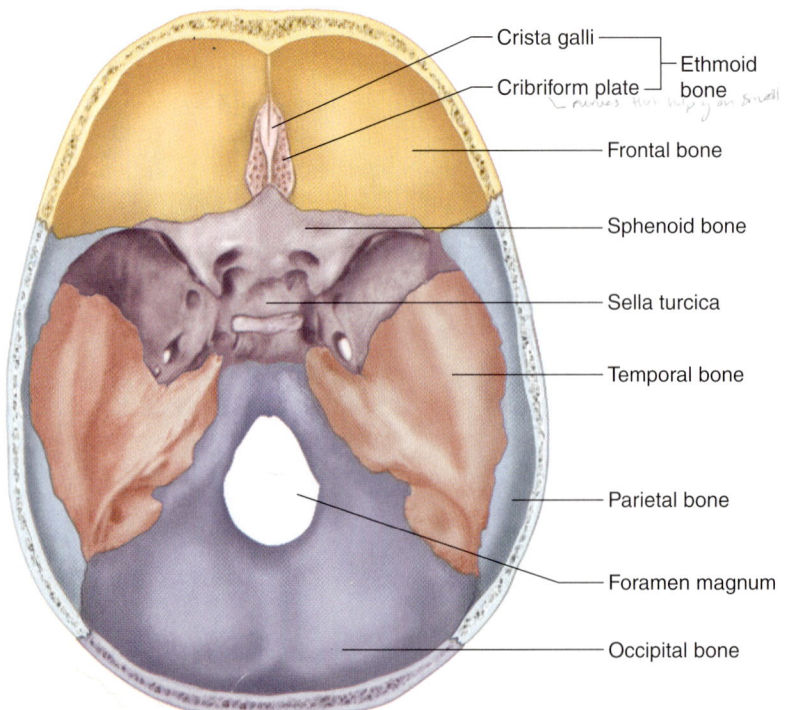

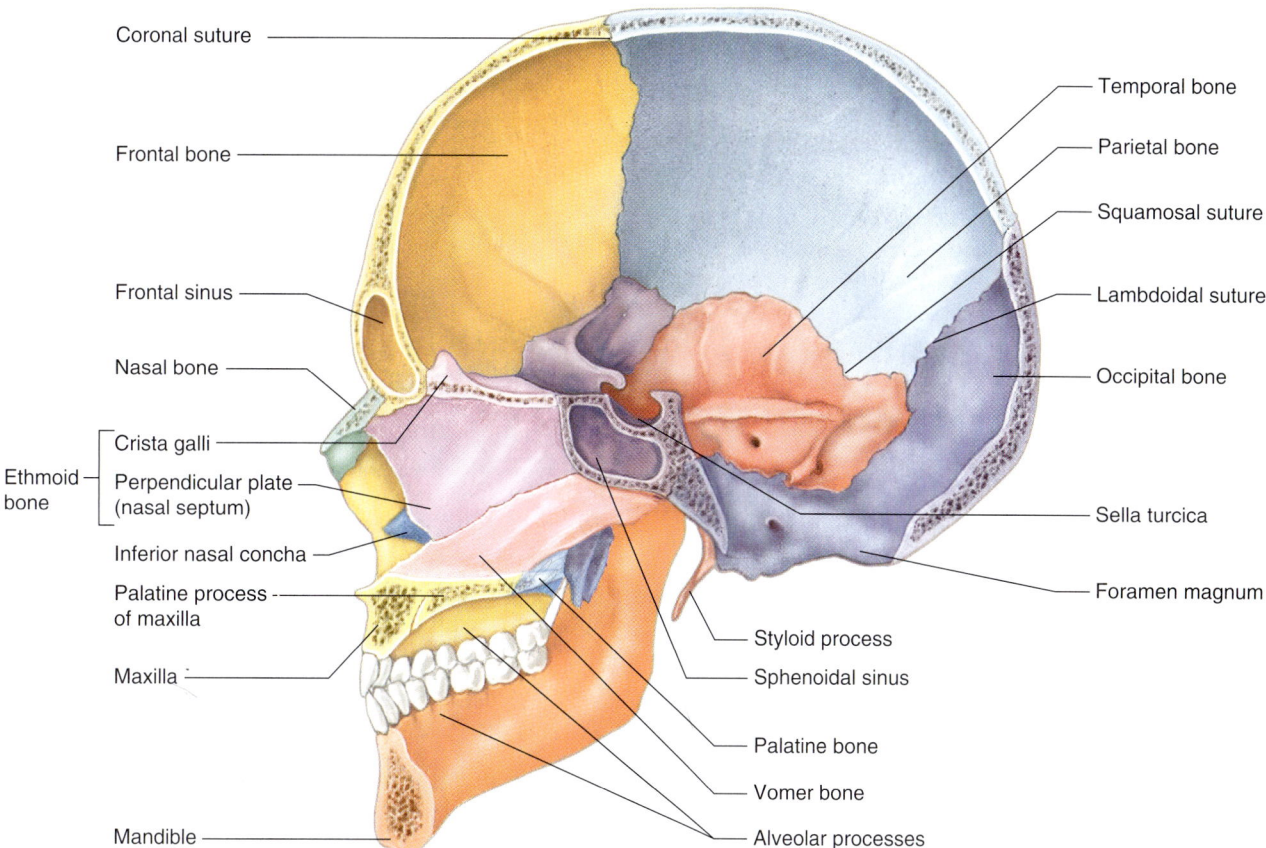

Figure 7.14
Sagittal section of the skull.

Facial Skeleton

The **facial skeleton** consists of thirteen immovable bones and a movable lower jawbone. These bones form the basic shape of the face and provide attachments for muscles that move the jaw and control facial expressions.

The bones of the facial skeleton are:

1. **Maxillae** The maxillae (mak-sil´e; singular, *maxilla,* mak-sil´ah) form the upper jaw (see figs. 7.11 and 7.12). Portions of these bones comprise the anterior roof of the mouth (*hard palate*), the floors of the orbits, and the sides and floor of the nasal cavity. They also contain the sockets of the upper teeth. Inside the maxillae, lateral to the nasal cavity, are *maxillary sinuses,* the largest of the sinuses (see fig. 7.10).

    During development, portions of the maxillae called *palatine processes* grow together and fuse along the midline to form the anterior section of the hard palate. The inferior border of each maxillary bone projects downward, forming an *alveolar* (al-ve´o-lar) *process.* Together, these processes form a horseshoe-shaped *alveolar arch* (dental arch) (fig. 7.14). Teeth occupy cavities in this arch (dental alveoli). Dense connective tissue binds teeth to the bony sockets.

    Sometimes, fusion of the palatine processes of the maxillae is incomplete at birth; the result is a *cleft palate.* Infants with a cleft palate may have trouble suckling because of the opening between the oral and nasal cavities. A temporary prosthetic device (artificial palate) may be inserted within the mouth, or a special type of nipple can be placed on bottles until surgery can be performed to correct the condition.

2. **Palatine bones** The L-shaped palatine (pal´ah-tīn) bones are located behind the maxillae (see figs. 7.12 and 7.14). The horizontal portions form the posterior section of the hard palate and the floor of the nasal cavity. The perpendicular portions help form the lateral walls of the nasal cavity.

3. **Zygomatic bones** The zygomatic (zi´´go-mat´ik) bones form the prominences of the cheeks below

and to the sides of the eyes (see figs. 7.11 and 7.12). These bones also help form the lateral walls and the floors of the orbits. Each bone has a *temporal process*, which extends posteriorly to join the zygomatic process of a temporal bone. Together, these processes form a *zygomatic arch*.

4. **Lacrimal bones** A lacrimal (lak´rĭ-mal) bone is a thin, scalelike structure located in the medial wall of each orbit between the ethmoid bone and the maxilla (see figs. 7.9 and 7.11).

5. **Nasal bones** The nasal (na´zal) bones are long, thin, and nearly rectangular (see figs. 7.9 and 7.11). They lie side by side and are fused at the midline, where they form the bridge of the nose.

6. **Vomer bone** The thin, flat vomer (vo´mer) bone is located along the midline within the nasal cavity (see figs. 7.9 and 7.14). Posteriorly, it joins the perpendicular plate of the ethmoid bone, and together they form the nasal septum.

7. **Inferior nasal conchae** The inferior nasal conchae (kong´ke) are fragile, scroll-shaped bones attached to the lateral walls of the nasal cavity (see figs. 7.9 and 7.14). Like the ethmoidal conchae, the inferior conchae support mucous membranes within the nasal cavity.

8. **Mandible** The mandible is a horizontal, horseshoe-shaped body with a flat portion projecting upward at each end (see figs. 7.9 and 7.11). This projection is divided into two processes—a posterior *mandibular condyle* and an anterior *coronoid process*. The mandibular condyles articulate with the mandibular fossae of the temporal bones (see fig. 7.12), whereas the coronoid processes provide attachments for muscles used in chewing. Other large chewing muscles insert on the lateral surface of the mandible. A curved bar of bone on the superior border of the mandible, the *alveolar arch*, contains the hollow sockets (dental alveoli) that bear the lower teeth (fig. 7.14).

Infantile Skull

At birth, the skull is incompletely developed, with fibrous membranes connecting the cranial bones. These membranous areas are called **fontanels** (fon´´tah-nels) or, more commonly, soft spots (fig. 7.15). They permit some movement between the bones, so that the developing skull is partially compressible and can slightly change shape. This enables an infant's skull to more easily pass through the birth canal. Eventually, the fontanels close as the cranial bones grow together.

Other characteristics of an infantile skull include a relatively small face with a prominent forehead and large orbits. The jaw and nasal cavity are small, the sinuses are incompletely formed, and the frontal bone is in two parts. The skull bones are thin, but they are also somewhat flexible and thus are less easily fractured than adult bones.

CHECK YOUR RECALL

1. Locate and name each of the bones of the cranium.
2. Locate and name each of the facial bones.
3. Explain how an adult skull differs from that of an infant.

7.7 Vertebral Column

The **vertebral column** extends from the skull to the pelvis and forms the vertical axis of the skeleton. It is composed of many bony parts called **vertebrae** (ver´tĕ-brā) that are separated by masses of fibrocartilage called *intervertebral discs* and are connected to one another by ligaments (fig. 7.16). The vertebral column supports the head and trunk of the body. It also protects the spinal cord, which passes through a *vertebral canal* formed by openings in the vertebrae.

A Typical Vertebra

Vertebrae in different regions of the vertebral column have special characteristics, but they also have features in common. A typical vertebra has a drum-shaped *body*, which forms the thick, anterior portion of the bone (fig. 7.17). A longitudinal row of these vertebral bodies supports the weight of the head and trunk. The intervertebral discs, which separate adjacent vertebral bodies, cushion and soften the forces from movements such as walking and jumping.

Projecting posteriorly from each vertebral body are two short stalks called *pedicles* (ped´ĭ-k'lz). Two plates called *laminae* (lam'i-ne) arise from the pedicles and fuse in the back to become a *spinous process*. The pedicles, laminae, and spinous process together complete a bony *vertebral arch* around the *vertebral foramen*, through which the spinal cord passes.

> If the laminae of the vertebrae fail to unite during development, the vertebral arch remains incomplete, resulting in a condition called *spina bifida*. The contents of the vertebral canal protrude outward. This problem occurs most frequently in the lumbosacral region.

Between the pedicles and laminae of a typical vertebra is a *transverse process*, which projects laterally and posteriorly. Ligaments and muscles are attached to the dorsal spinous process and the transverse processes. Projecting upward and downward from each vertebral arch are *superior* and *inferior articulating processes*. These processes bear cartilage-covered facets that join each vertebra to the ones above and below it.

On the lower surfaces of the vertebral pedicles are notches that align to form openings called *interverte-

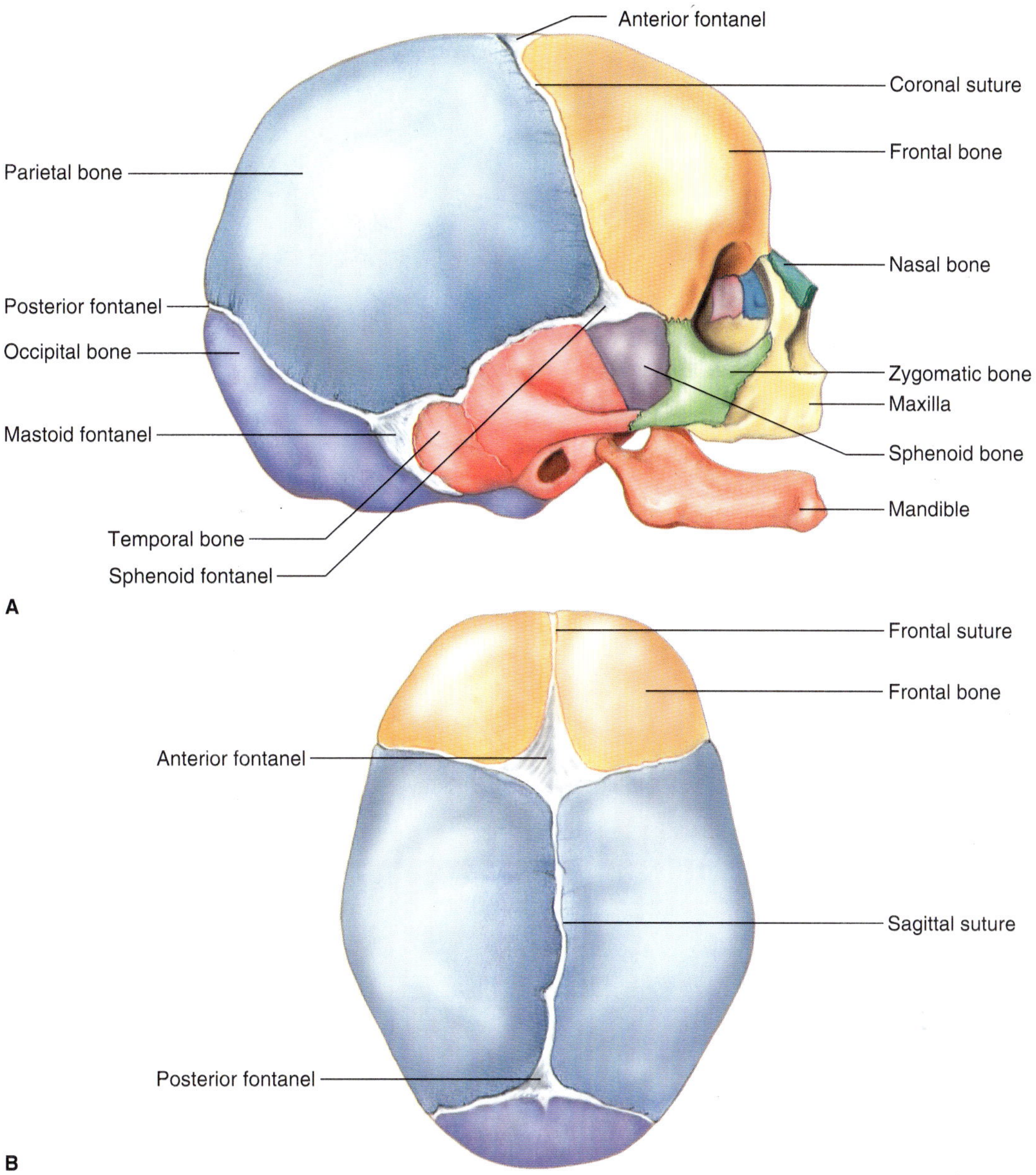

Figure 7.15
Fontanels. (A) Lateral view and (B) superior view of the infantile skull.

bral foramina (in″ter-ver′tĕ-bral fo-ram′ĭ-nah). These openings provide passageways for spinal nerves that proceed between vertebrae and connect to the spinal cord (see fig. 7.16).

Cervical Vertebrae

Seven **cervical vertebrae** comprise the bony axis of the neck (see fig. 7.16). The transverse processes of these vertebrae are distinctive because they have *transverse foramina,* which are passageways for arteries leading to the brain (see fig. 7.17). Also, the spinous processes of the second through the fifth cervical vertebrae are uniquely forked (bifid). These processes provide attachments for muscles.

Two cervical vertebrae are of special interest (fig. 7.18). The first vertebra, or **atlas** (at′las), supports the head. On its superior surface are two kidney-shaped *facets* that articulate with the occipital condyles.

The second cervical vertebra, or **axis** (ak′sis), bears a toothlike *dens* (odontoid process) on its body. This process projects upward and lies in the ring of the atlas. As the head is turned from side to side, the atlas pivots around the dens.

 Giraffes and humans have the same number of vertebrae in their necks . . . 7.

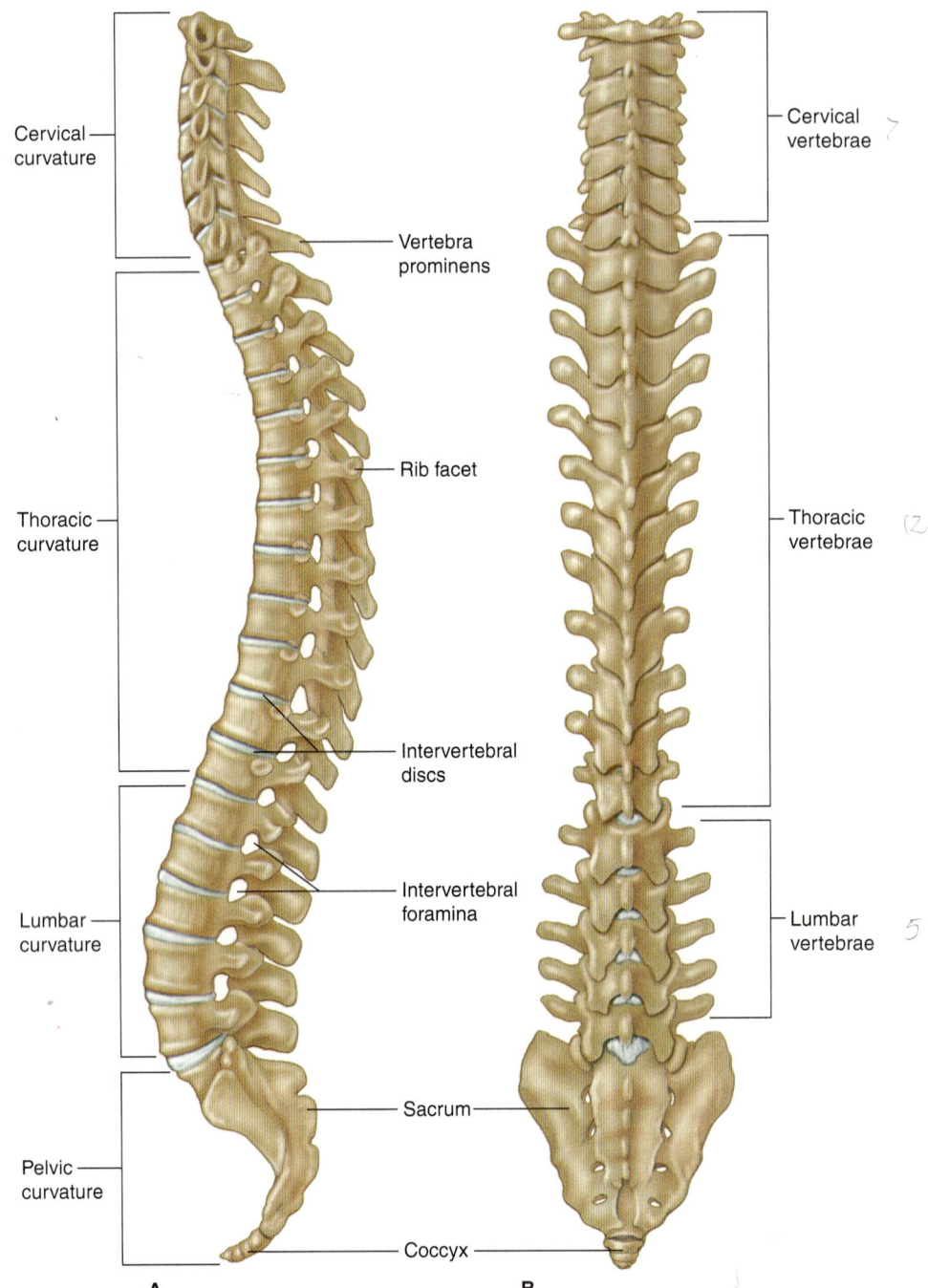

Figure 7.16
The curved vertebral column consists of many vertebrae separated by intervertebral discs. (A) Left lateral view. (B) Posterior view.

Thoracic Vertebrae

The twelve **thoracic vertebrae** are larger than the cervical vertebrae (see fig. 7.16). Each vertebra has a long, pointed spinous process, which slopes downward, and facets on the sides of its body, which articulate with a rib.

Beginning with the third thoracic vertebra and moving inferiorly, the bodies of these bones increase in size. Thus, they are adapted to bear increasing loads of body weight.

Lumbar Vertebrae

Five **lumbar vertebrae** are in the small of the back (loin) (see fig. 7.16). These vertebrae are adapted with larger and stronger bodies to support more weight than the vertebrae above them.

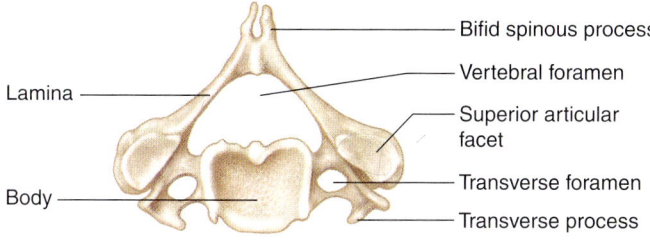

A Cervical vertebra

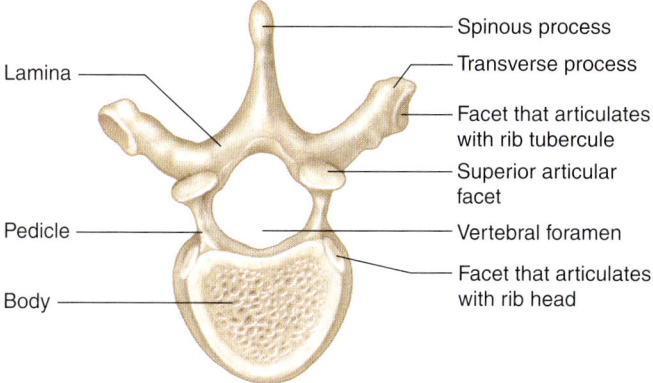

B Thoracic vertebra

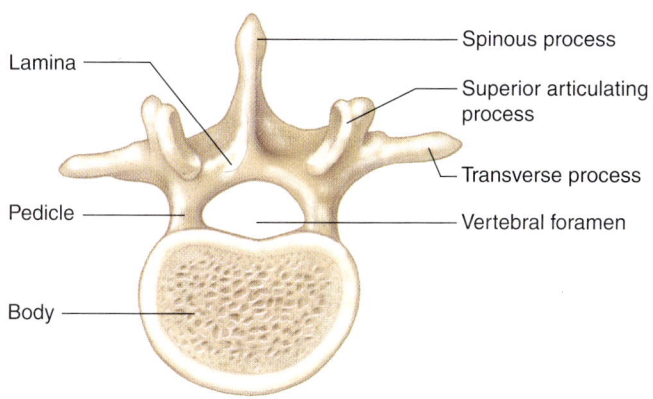

C Lumbar vertebra

Figure 7.17
Superior view of (A) a cervical vertebra, (B) a thoracic vertebra, and (C) a lumbar vertebra.

Sacrum

The **sacrum** (sa´krum) is a triangular structure, composed of five fused vertebrae, that forms the base of the vertebral column (fig. 7.19). The spinous processes of these fused bones form a ridge of *tubercles.* To the sides of the tubercles are rows of openings, the *dorsal sacral foramina,* through which nerves and blood vessels pass.

The vertebral foramina of the sacral vertebrae form the *sacral canal,* which continues through the sacrum to an opening of variable size at the tip, called the *sacral hiatus* (sa´kral hi-a´tus). On the ventral surface of the sacrum, four pairs of *pelvic sacral foramina* provide passageways for nerves and blood vessels.

Coccyx

The **coccyx** (kok´siks), the lowest part of the vertebral column, is usually composed of four fused vertebrae (fig. 7.19). Ligaments attach it to the margins of the sacral hiatus. Table 7.3 summarizes the various features of the different vertebrae.

Changes in the intervertebral discs can cause back problems. Each disc is composed of a tough outer layer of fibrocartilage and an elastic central mass. With age, these discs degenerate—the central masses lose firmness, and the outer layers thin and weaken, developing cracks. Extra pressure, as when a person falls or lifts a heavy object, can break the outer layer of a disc, squeezing out the central mass. Such a rupture may press on the spinal cord or on a spinal nerve that branches from it. This condition—a ruptured or herniated disc—may cause back pain and numbness or the loss of muscular function in the parts innervated by the affected spinal nerve.

 CHECK YOUR RECALL

1. Describe the structure of the vertebral column.
2. Describe a typical vertebra.
3. How do the structures of cervical, thoracic, and lumbar vertebrae differ?

TABLE 7.3 FEATURES OF VERTEBRAE

TYPE OF VERTEBRAE	FEATURES THAT ARE PRESENT						
	BODY	DENS	SPINOUS PROCESS	TRANSVERSE PROCESS	TRANSVERSE FORAMEN	FACETS FOR RIBS	PEDICLES AND LAMINAE
Cervical (7)			Some are bifid				
a. C-1					✓		
b. C-2	✓	✓	✓	✓	✓		✓
c. C-3 through C-7	✓		✓	✓	✓		✓
Thoracic (12)	✓		✓	✓		✓	✓
Lumbar (5)	✓		✓	✓			✓
Sacrum (5 fused)	✓			✓ Fused			
Coccyx (4 fused)							

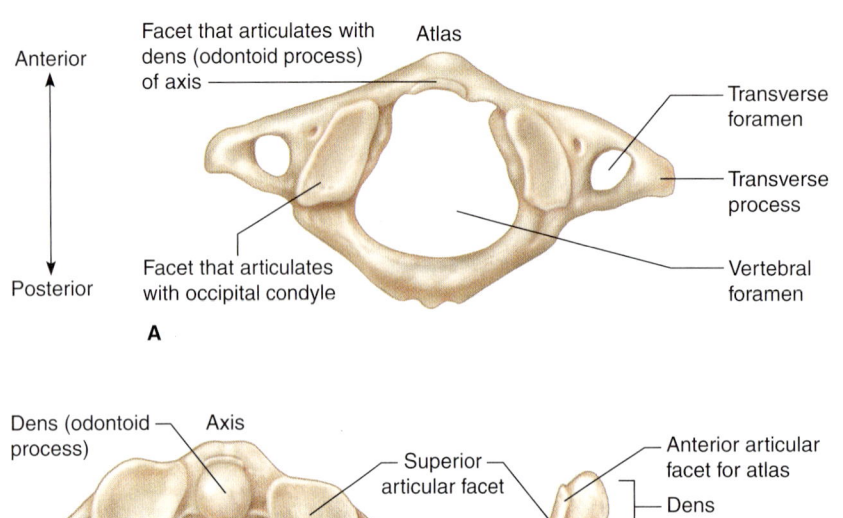
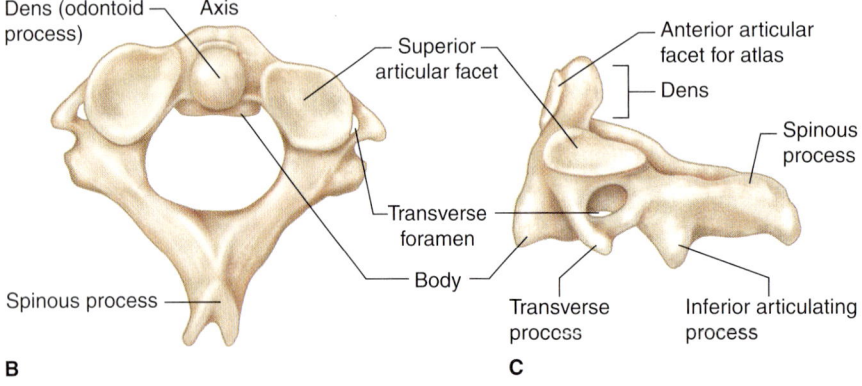

Figure 7.18
Superior view of the (A) atlas and (B) axis. (C) Lateral view of the axis.

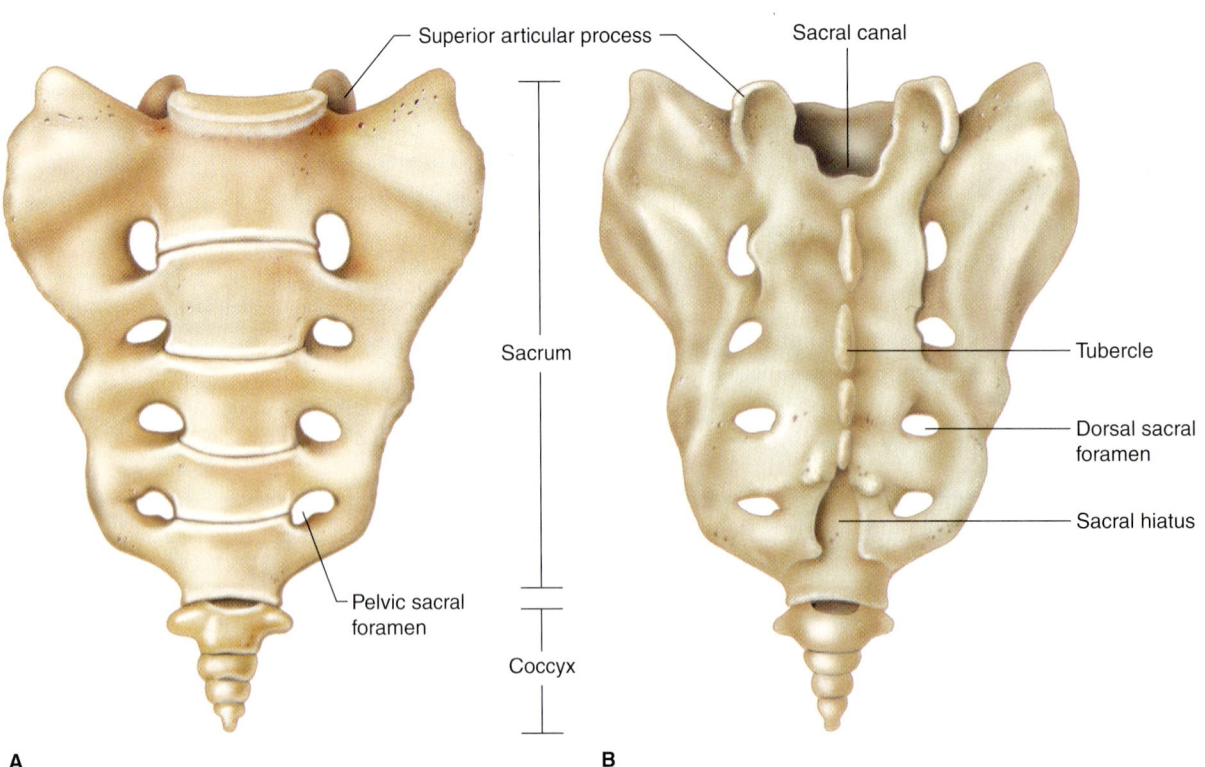

Figure 7.19
Sacrum and coccyx. (A) Anterior view and (B) posterior view.

7.8 Thoracic Cage

The **thoracic cage** includes the ribs, thoracic vertebrae, sternum, and costal cartilages that attach the ribs to the sternum (fig. 7.20). These bones support the shoulder girdle and upper limbs, protect the viscera in the thoracic and upper abdominal cavities, and play a role in breathing.

Ribs

The usual number of **ribs** is twelve—one pair attached to each of the twelve thoracic vertebrae. The first seven rib pairs, *true ribs* (vertebrosternal ribs), join the sternum directly by their costal cartilages. The remaining five pairs are called *false ribs*, because their cartilages do not reach the sternum directly. Instead, the cartilages of the upper three false ribs (vertebrochondral ribs) join the cartilages of the seventh rib. The last two (or sometimes three) rib pairs are called *floating ribs* (vertebral ribs) because they have no cartilaginous attachments to the sternum.

A typical rib has a long, slender shaft, which curves around the chest and slopes downward. On the posterior end is an enlarged *head* by which the rib articulates with a *facet* on the body of its own vertebra and with the body of the next higher vertebra. A *tubercle,* close to the head of the rib, articulates with the transverse process of the vertebra.

Sternum

The **sternum,** or breastbone, is located along the midline in the anterior portion of the thoracic cage (fig. 7.20). This flat, elongated bone develops in three parts—an upper *manubrium* (mah-nu´bre-um), a middle *body,* and a lower *xiphoid* (zīf´oid) *process* that projects downward. The manubrium articulates with the clavicles by facets on its superior border.

CHECK YOUR RECALL

1. List the components of the thoracic cage.
2. Distinguish between true ribs and false ribs.
3. State the three parts of the sternum.

7.9 Pectoral Girdle

The **pectoral** (pek´to-ral) **girdle,** or shoulder girdle, is composed of four parts—two clavicles and two scapulae (fig. 7.21). Although the word *girdle* suggests a ring-shaped structure, the pectoral girdle is an incomplete ring. It is open in the back between the scapulae, and the sternum separates its bones in front. The pectoral girdle supports the upper limbs and is an attachment for several muscles that move them.

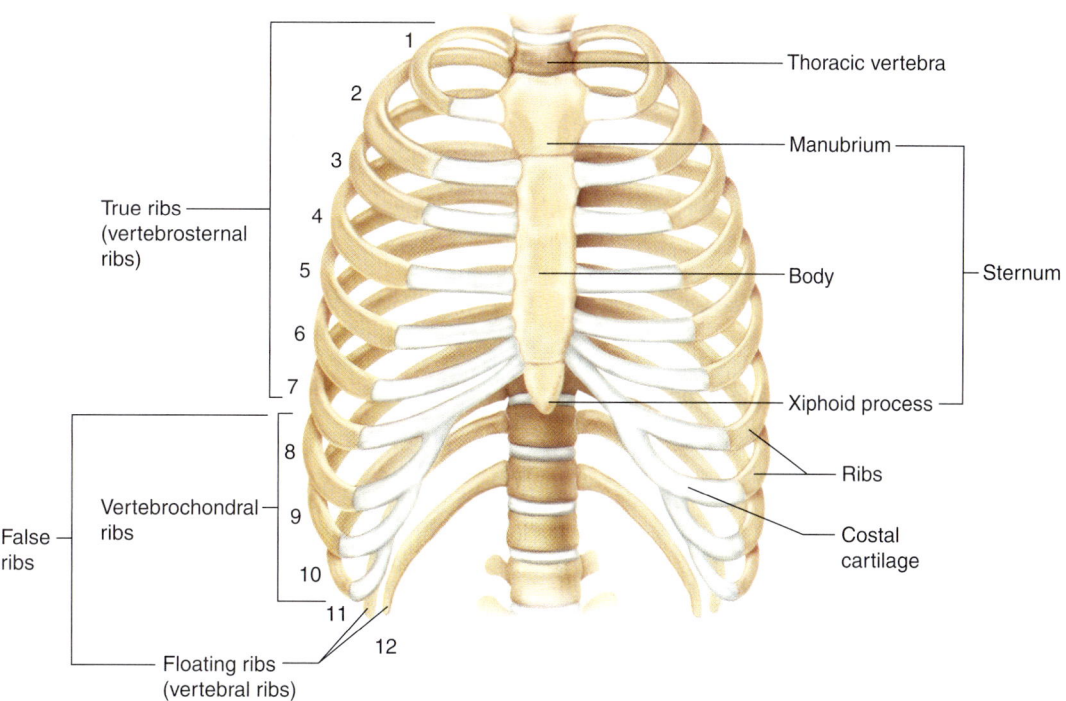

Figure 7.20
The thoracic cage includes the thoracic vertebrae, the sternum, the ribs, and the costal cartilages that attach the ribs to the sternum.

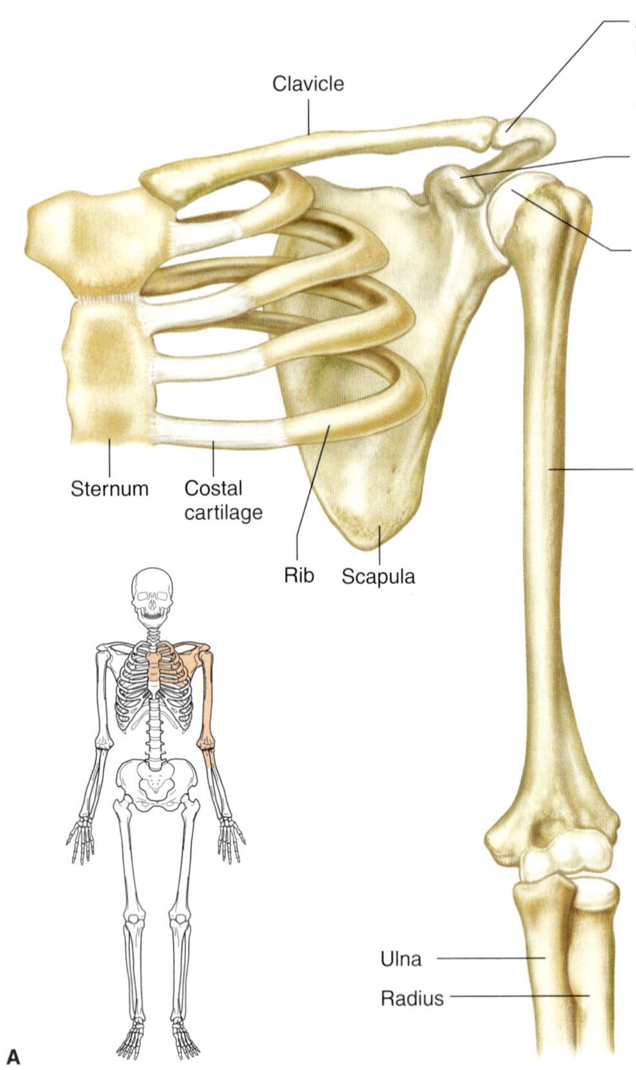

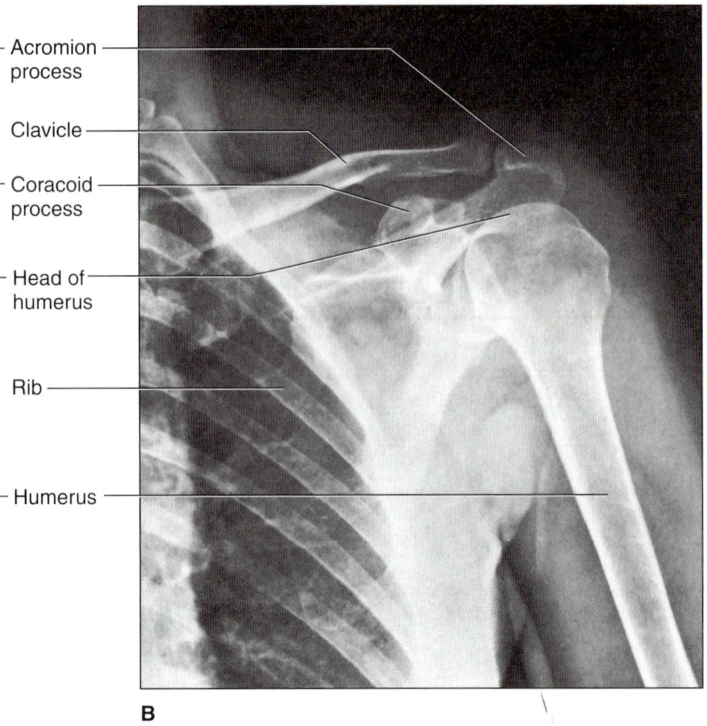

Figure 7.21
The pectoral girdle (A), to which the upper limbs are attached, consists of a clavicle and a scapula on each side. (B) Radiograph of the left shoulder region, viewed from the front.

Clavicles

The **clavicles,** or collarbones, are slender, rodlike bones with elongated S shapes (fig. 7.21). Located at the base of the neck, they run horizontally between the manubrium and scapulae.

The clavicles brace the freely movable scapulae, helping to hold the shoulders in place. They also provide attachments for muscles of the upper limbs, chest, and back.

Scapulae

The **scapulae** (skap´u-le), or shoulder blades, are broad, somewhat triangular bones located on either side of the upper back (figs. 7.21 and 7.22). A *spine* divides the posterior surface of each scapula into unequal portions. This spine leads to two processes—an *acromion* (ah-kro´me-on) *process* that forms the tip of the shoulder and a *coracoid* (kor´ah-koid) *process* that curves anteriorly and inferiorly to the clavicle. The acromion process articulates with the clavicle and provides attachments for muscles of the upper limb and chest. The coracoid process also provides attachments for upper limb and chest muscles. Between the processes is a depression called the *glenoid cavity* (glenoid fossa of the scapula) that articulates with the head of the arm bone (humerus).

CHECK YOUR RECALL

1. Which bones form the pectoral girdle?
2. What is the function of the pectoral girdle?

7.10 Upper Limb

The bones of the upper limb form the framework of the arm, forearm, wrist, and hand. They also provide attachments for muscles, and they function in levers that move limb parts. These bones include a humerus, a radius, an ulna, carpals, metacarpals, and phalanges (see fig. 7.8).

Humerus

The **humerus** is a heavy bone that extends from the scapula to the elbow (fig. 7.23). At its upper end is a smooth, rounded *head* that fits into the glenoid cavity

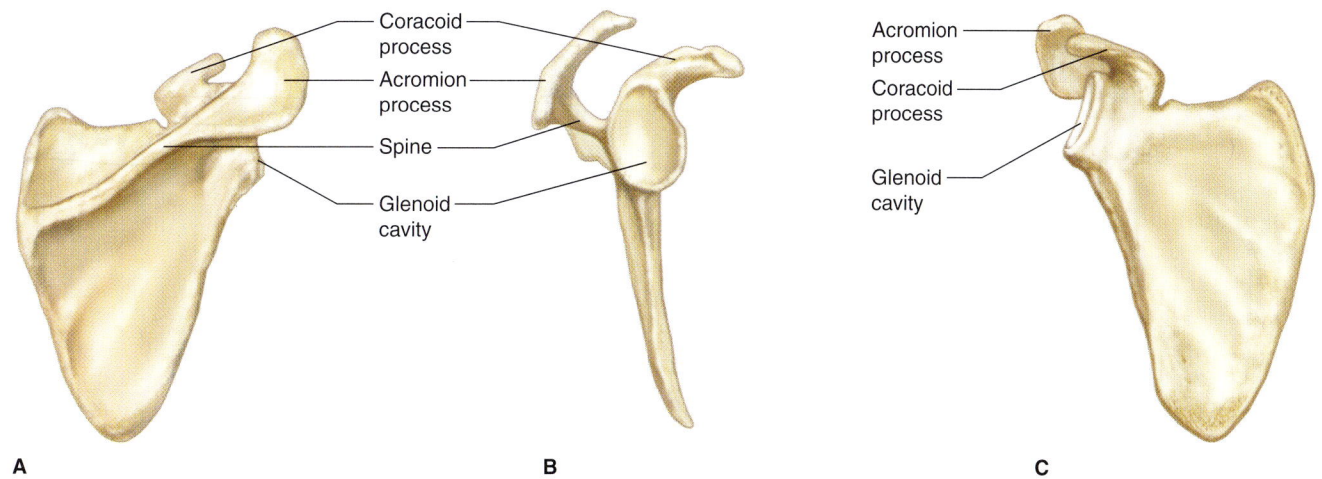

Figure 7.22
Scapula. (*A*) Posterior surface of the right scapula. (*B*) Lateral view showing the glenoid cavity that articulates with the head of the humerus. (*C*) Anterior surface.

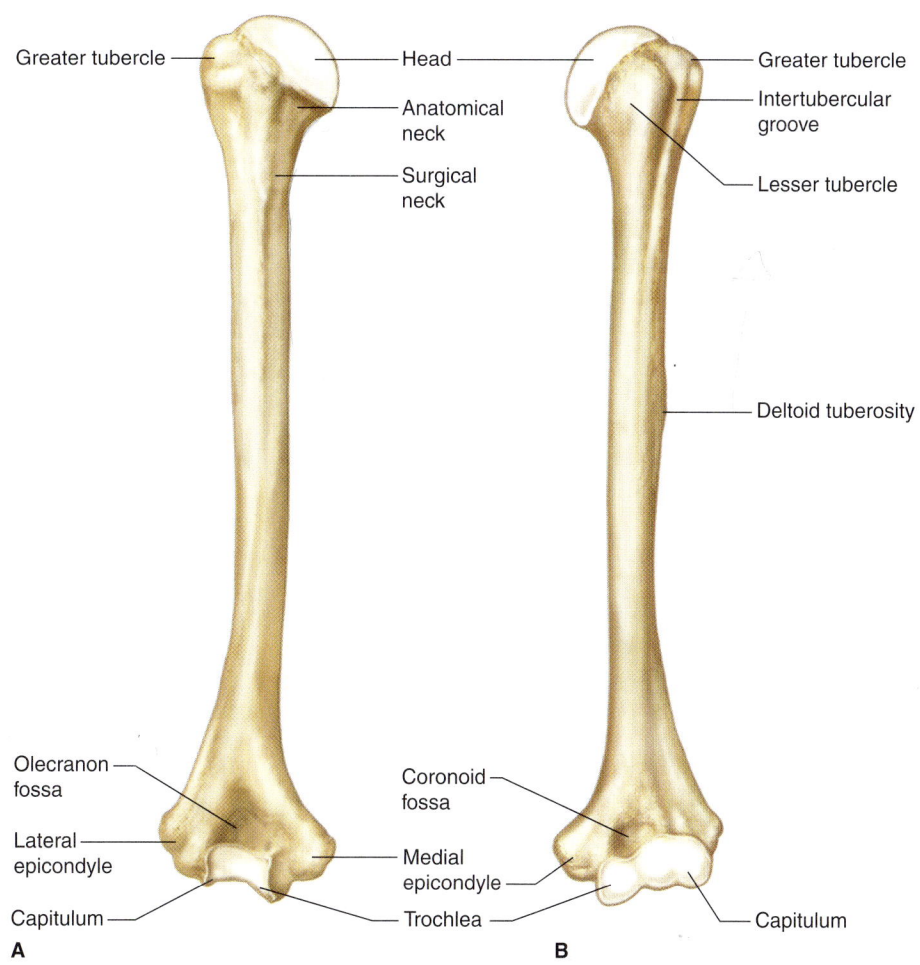

Figure 7.23
Humerus. (*A*) Posterior surface and (*B*) anterior surface of the left humerus.

of the scapula. Just below the head are two processes—a *greater tubercle* on the lateral side and a *lesser tubercle* on the anterior side. These tubercles provide attachments for muscles that move the upper limb at the shoulder. Between them is a narrow furrow, the *intertubercular groove*.

The narrow depression along the lower margin of the humerus head separates it from the tubercles and is called the *anatomical neck*. Just below the head and tubercles is a tapering region called the *surgical neck*, so named because fractures commonly occur there. Near the middle of the bony shaft on the lateral side is a rough V-shaped area called the *deltoid tuberosity*. It provides an attachment for the muscle (deltoid) that raises the upper limb horizontally to the side.

At the lower end of the humerus are two smooth *condyles* (a lateral *capitulum* and a medial *trochlea*) that articulate with the radius on the lateral side and the ulna on the medial side. Above the condyles on either side are *epicondyles*, which provide attachments for muscles and ligaments of the elbow. Between the epicondyles anteriorly is a depression, the *coronoid* (kor´o-noid) *fossa*, that receives a process of the ulna (coronoid process) when the elbow bends. Another depression on the posterior surface, the *olecranon* (o´´lek´ra-non) *fossa*, receives an ulnar process (olecranon process) when the upper limb straightens at the elbow.

Radius

The **radius**, located on the thumb side of the forearm, extends from the elbow to the wrist and crosses over the ulna when the hand is turned so that the palm faces backward (fig. 7.24). A thick, disclike *head* at the upper end of the radius articulates with the humerus and a notch of the ulna (radial notch). This arrangement allows the radius to rotate freely.

On the radial shaft just below the head is a process called the *radial tuberosity*. It is an attachment for a muscle (biceps brachii) that bends the upper limb at the elbow. At the distal end of the radius, a lateral *styloid* (sti´loid) *process* provides attachments for ligaments of the wrist.

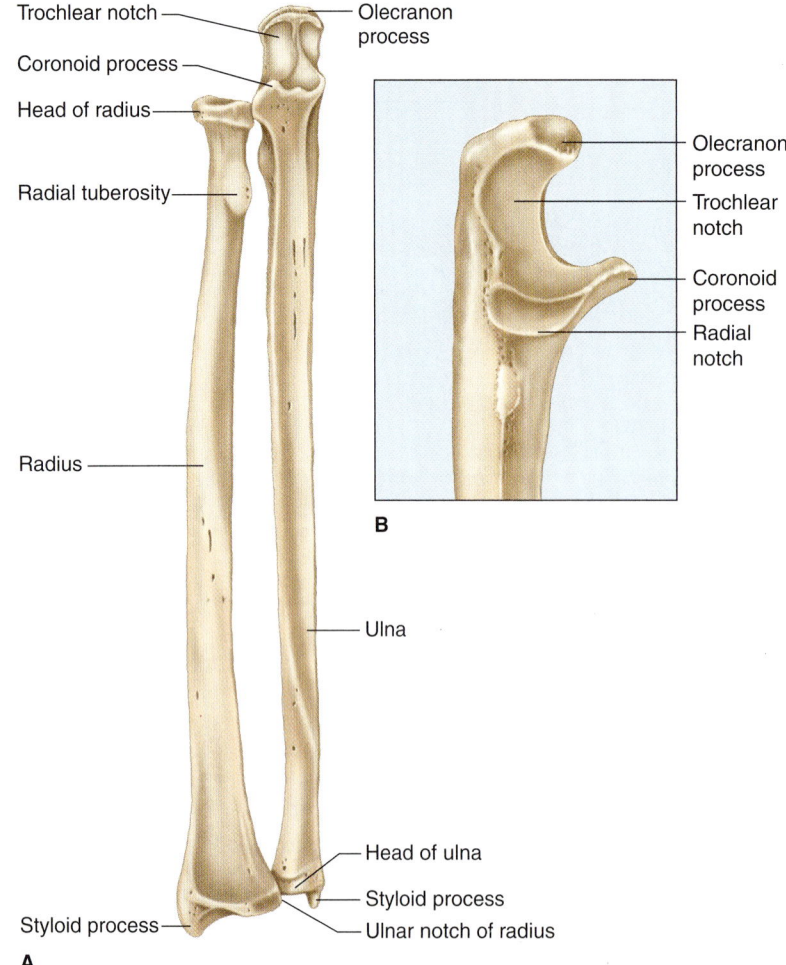

Figure 7.24
Radius and ulna. (*A*) The head of the right radius articulates with the radial notch of the ulna, and the head of the ulna articulates with the ulnar notch of the radius. (*B*) Lateral view of the proximal end of the ulna.

Ulna

The **ulna** is longer than the radius and overlaps the end of the humerus posteriorly (fig. 7.24). At its proximal end, the ulna has a wrenchlike opening, the *trochlear* (trok´le-ar) *notch,* that articulates with the humerus. Two processes on either side of this notch, the *olecranon process* and the *coronoid process,* provide attachments for muscles.

At the distal end of the ulna, its knoblike *head* articulates laterally with a notch of the radius (ulnar notch) and with a disc of fibrocartilage inferiorly. This disc, in turn, joins a wrist bone (triquetrum). A medial *styloid process* at the distal end of the ulna provides attachments for wrist ligaments.

Wrist and Hand

The skeleton of the wrist consists of eight small **carpal bones** that are firmly bound in two rows of four bones each. The resulting compact mass is called a *carpus* (kar´pus). The carpus articulates with the radius and with the fibrocartilaginous disc on the ulnar side. Its distal surface articulates with the metacarpal bones. Figure 7.25 names the individual bones of the carpus.

Five **metacarpal bones,** one in line with each finger, form the framework of the palm of the hand. These bones are cylindrical, with rounded distal ends that form the knuckles of a clenched fist. They are numbered 1–5, beginning with the metacarpal of the thumb (fig. 7.25). The metacarpals articulate proximally with the carpals and distally with the phalanges.

The **phalanges** are the finger bones. Each finger has three phalanges—a proximal, a middle, and a distal phalanx—except the thumb, which has two (it lacks a middle phalanx).

CHECK YOUR RECALL

1. Locate and name each of the bones of the upper limb.
2. Explain how the bones of the upper limb articulate with one another.

7.11 Pelvic Girdle

The **pelvic girdle** consists of two **coxae** (hipbones or innominate bones) which articulate with each other anteriorly and with the sacrum posteriorly. The sacrum,

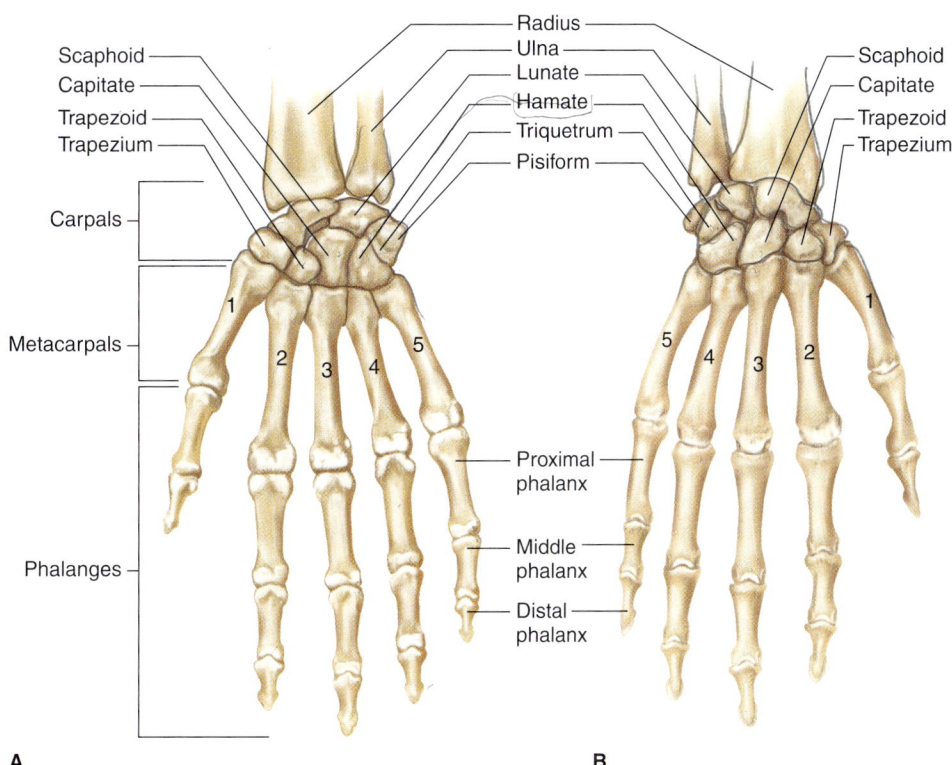

Figure 7.25
Wrist and hand. (*A*) Anterior view and (*B*) posterior view of the right wrist and hand.

coccyx, and pelvic girdle together form the bowl-shaped **pelvis** (fig. 7.26). The pelvic girdle supports the trunk of the body, provides attachments for the lower limbs, and protects the urinary bladder, the distal end of the large intestine, and the internal reproductive organs.

Each coxa develops from three parts—an ilium, an ischium, and a pubis (fig. 7.27). These parts fuse in the region of a cup-shaped cavity called the *acetabulum* (as″ĕ-tab´u-lum). This depression, on the lateral surface of the hipbone, receives the rounded head of the femur (thighbone).

The **ilium** (il´e-um), which is the largest and uppermost portion of the coxa, flares outward, forming the prominence of the hip. The margin of this prominence is called the *iliac crest*.

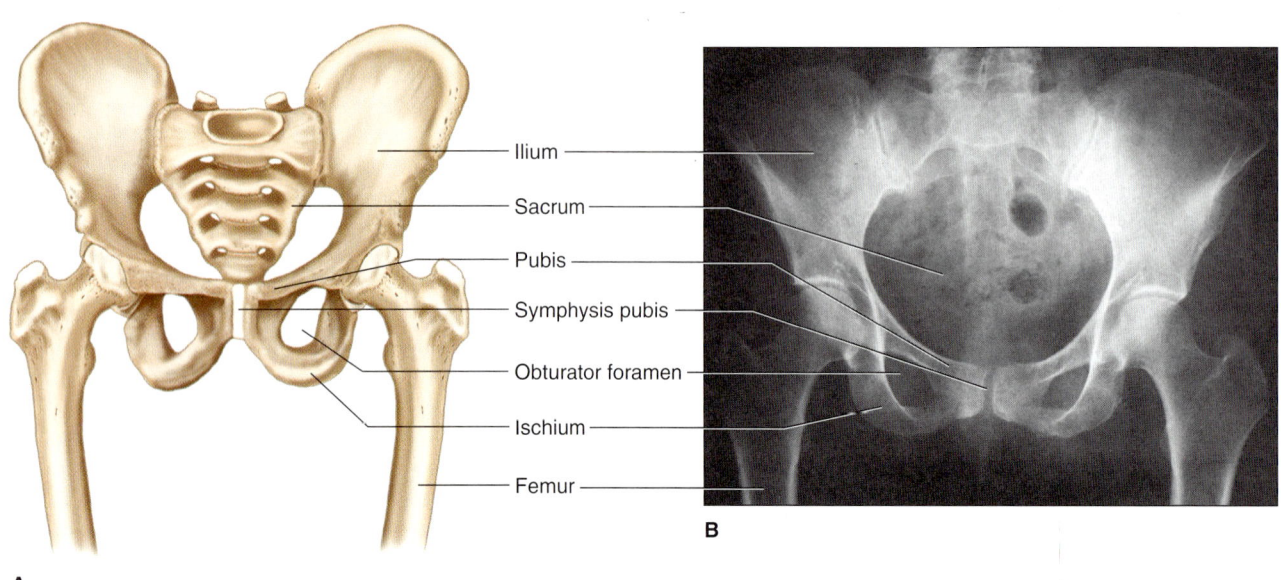

Figure 7.26
Pelvic girdle. (*A*) Formed by two coxae, the sacrum, and coccyx. (*B*) Radiograph of the pelvic girdle.

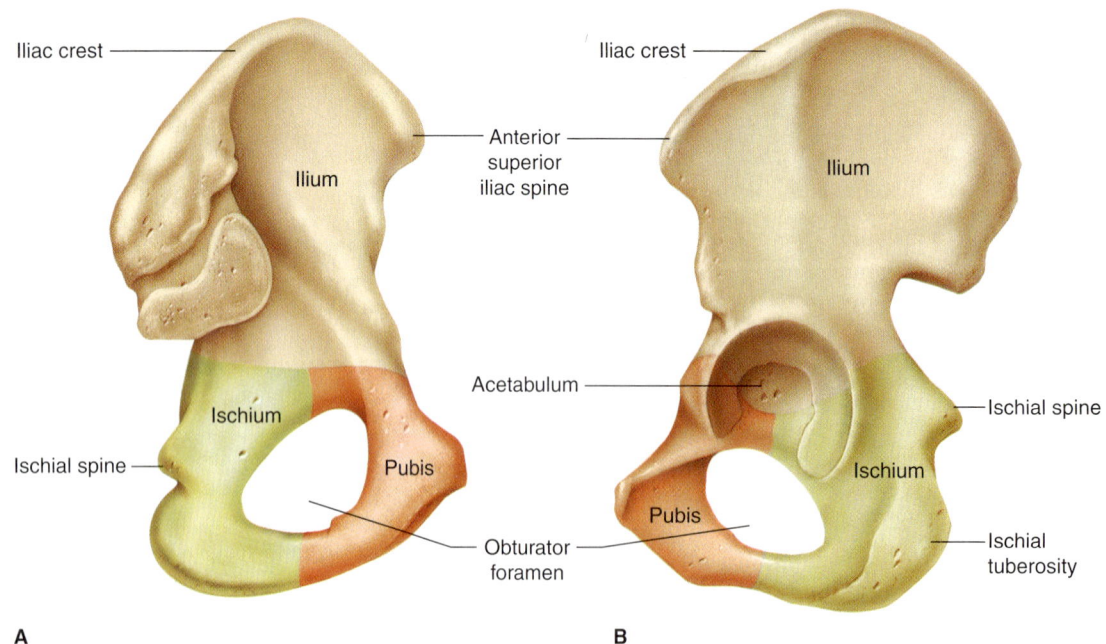

Figure 7.27
Coxa. (*A*) Medial surface of the left coxa. (*B*) Lateral view.

Posteriorly, the ilium joins the sacrum at the *sacroiliac* (sa˝kro-il´e-ak) *joint*. A projection of the ilium, the *anterior superior iliac spine,* can be felt lateral to the groin and provides attachments for ligaments and muscles.

The **ischium** (is´ke-um), which forms the lowest portion of the coxa, is L-shaped, with its angle, the *ischial tuberosity,* pointing posteriorly and downward. This tuberosity has a rough surface that provides attachments for ligaments and lower limb muscles. It also supports the weight of the body during sitting. Above the ischial tuberosity, near the junction of the ilium and ischium, is a sharp projection called the *ischial spine.*

The **pubis** (pu´bis) constitutes the anterior portion of the coxa. The two pubic bones join at the midline, forming a joint called the *symphysis pubis* (sim´fi-sis pu´bis). The angle these bones form below the symphysis is the *pubic arch* (fig. 7.28).

A portion of each pubis passes posteriorly and downward to join an ischium. Between the bodies of these bones on either side is a large opening, the *obturator foramen,* which is the largest foramen in the skeleton (see figs. 7.26 and 7.27).

If a line were drawn along each side of the pelvis from the sacral promontory downward and anteriorly to the upper margin of the symphysis pubis, it would mark the *pelvic brim* (linea terminalis) (fig. 7.28). This margin separates the lower, or lesser (true), pelvis from the upper, or greater (false), pelvis. Table 7.4 summarizes some differences in the female and male pelves and other skeletal structures.

CHECK YOUR RECALL

1. Locate and name each bone that forms the pelvis.
2. Name the bones that fuse to form a coxa.

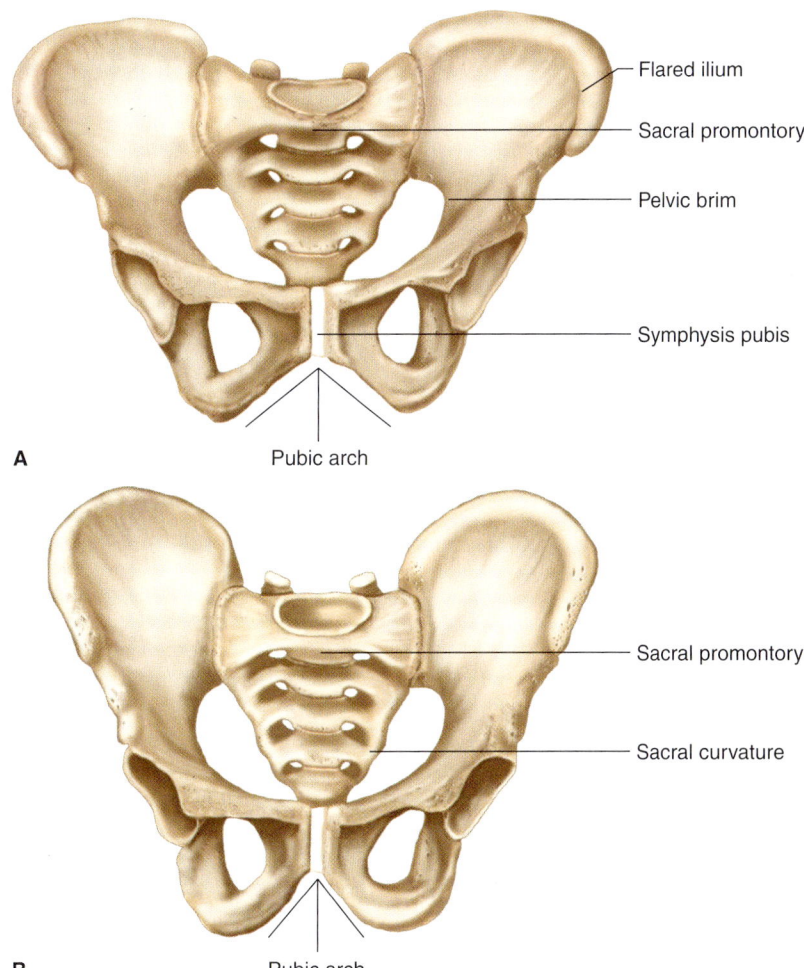

Figure 7.28
The female pelvis is usually wider in all diameters and roomier than that of the male. (*A*) Female pelvis. (*B*) Male pelvis.

TABLE 7.4 DIFFERENCES BETWEEN THE FEMALE AND MALE SKELETONS

PART	DIFFERENCES
Skull	Female skull is smaller and lighter, with less conspicuous muscular attachments. Female facial area is rounder, jaw is smaller, and mastoid process is less prominent than those of a male.
Pelvic girdle	Female coxae are lighter, thinner, and have less obvious muscular attachments. The obturator foramina and acetabula are smaller and farther apart than those of a male.
Pelvic cavity	Female pelvic cavity is wider in all diameters and is shorter, roomier, and less funnel-shaped. The distances between the ischial spines and ischial tuberosities are greater than in a male.
Sacrum	Female sacrum is wider, the first sacral vertebra projects forward to a lesser degree, and sacral curvature is bent more sharply posteriorly than in a male.
Coccyx	Female coccyx is more movable than that of a male.

7.12 Lower Limb

Bones of the lower limb form the frameworks of the thigh, leg, ankle, and foot. They include a femur, a tibia, a fibula, tarsals, metatarsals, and phalanges (see fig. 7.8).

Femur

The **femur,** or thigh bone, is the longest bone in the body and extends from the hip to the knee (fig. 7.29). A large, rounded *head* at its proximal end projects medially into the acetabulum of the coxa. On the head, a pit called the *fovea capitis* marks the attachment of a ligament (ligamentum capitis). Just below the head are a constriction, or *neck,* and two large processes—a superior, lateral *greater trochanter* and an inferior, medial *lesser trochanter.* These processes provide attachments for muscles of the lower limbs and buttocks.

The strongest bone in the body, the femur, is hollow. Ounce for ounce, it has greater pressure tolerance and bearing strength than a rod of equivalent size in cast steel.

At the distal end of the femur, two rounded processes, the *lateral* and *medial condyles,* articulate with the tibia of the leg. A **patella,** or kneecap, also articulates with the femur on its distal anterior surface (see fig. 7.8). It is located in a tendon that passes anteriorly over the knee.

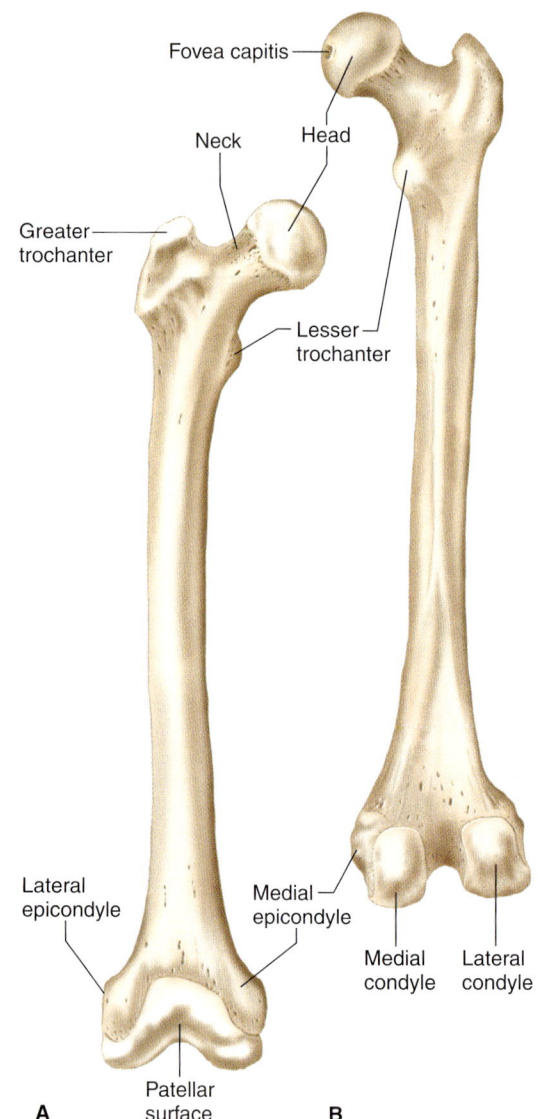

Figure 7.29
Femur. (*A*) Anterior surface and (*B*) posterior surface of the right femur.

Hip fracture is one of the more serious causes of hospitalization among elderly persons. The site of hip fracture is most commonly the neck of a femur or the region between the trochanters of a femur. Falls often cause hip fracture, especially in people who have osteoporosis.

Tibia

The **tibia,** or shin bone, is the larger of the two leg bones and is located on the medial side (fig. 7.30). Its proximal end is expanded into *medial* and *lateral condyles,* which have concave surfaces and articulate with the condyles of the femur. Below the condyles, on the anterior surface, is a process called the *tibial*

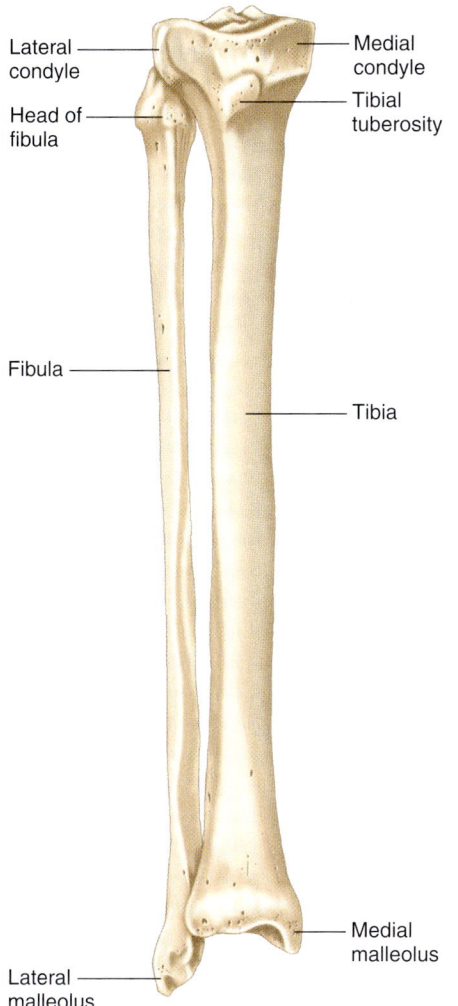

Figure 7.30
Bones of the right leg, viewed from the front.

tuberosity, which provides an attachment for the *patellar ligament*—a continuation of the patella-bearing tendon.

At its distal end, the tibia expands to form a prominence on the inner ankle called the *medial malleolus* (mah-le´o-lus), which is an attachment for ligaments. On its lateral side is a depression that articulates with the fibula. The inferior surface of the tibia's distal end articulates with a large bone (the talus) in the ankle.

Fibula

The **fibula** is a long, slender bone located on the lateral side of the tibia (fig. 7.30). Its ends are slightly enlarged into a proximal *head* and a distal *lateral malleolus*. The head articulates with the tibia just below the lateral condyle; however, it does not enter into the knee joint and does not bear any body weight. The lateral malleolus articulates with the ankle and protrudes on the lateral side.

Ankle and Foot

The ankle and foot consist of a *tarsus* (tahr´sus), a *metatarsus* (met´´ah-tar´sus), and five toes. The tarsus is composed of seven **tarsal bones** (figs. 7.31 and 7.32). These bones are arranged so that one of them, the **talus** (ta´lus), can move freely where it joins the tibia and fibula. The remaining tarsal bones are bound firmly together, forming a mass supporting the talus. Figure 7.32 names the individual bones of the tarsus.

The largest of the tarsals, the **calcaneus** (kal-ka´ne-us), or heel bone, is located below the talus, where it projects backward to form the base of the heel. The calcaneus helps support body weight and provides an attachment for muscles that move the foot.

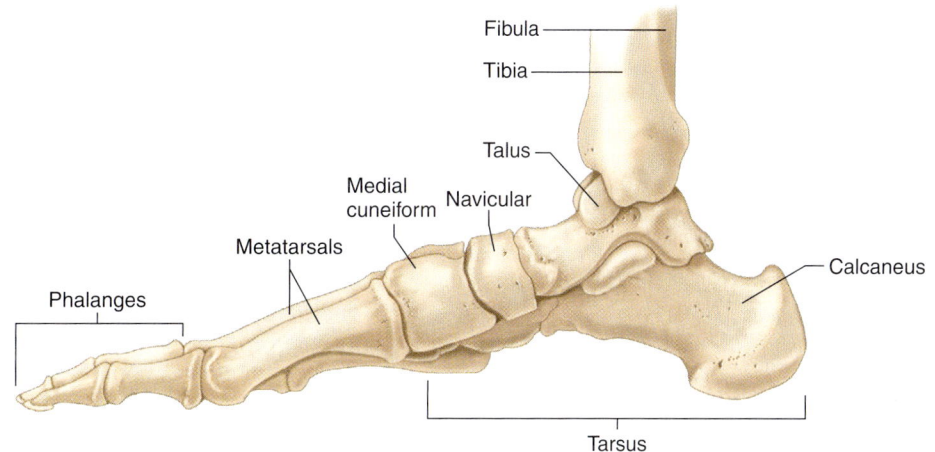

Figure 7.31
The talus moves freely where it articulates with the tibia and fibula.

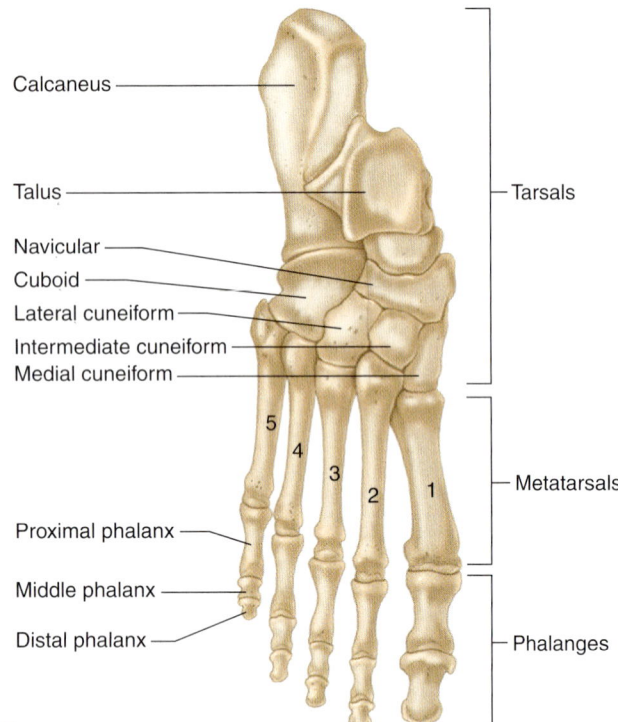

Figure 7.32
The right foot viewed superiorly.

The metatarsus consists of five elongated **metatarsal bones,** that articulate with the tarsus. They are numbered 1–5, beginning on the medial side (fig. 7.32). The heads at the distal ends of these bones form the ball of the foot. The tarsals and metatarsals are arranged and bound by ligaments to form the arches of the foot. A longitudinal arch extends from the heel to the toe, and a transverse arch stretches across the foot. These arches provide a stable, springy base for the body. Sometimes, however, the tissues that bind the metatarsals weaken, producing fallen arches, or flat feet.

The **phalanges** of the toes are similar to those of the fingers and align and articulate with the metatarsals. Each toe has three phalanges—a proximal, a middle, and a distal phalanx—except the great toe, which lacks a middle phalanx.

 CHECK YOUR RECALL

1. Locate and name each of the bones of the lower limb.
2. Explain how the bones of the lower limb articulate with one another.
3. Describe how the foot is adapted to support the body.

7.13 Joints

Joints (articulations) are functional junctions between bones. They bind parts of the skeletal system, make possible bone growth, permit parts of the skeleton to change shape during childbirth, and enable the body to move in response to skeletal muscle contractions. Joints vary considerably in structure and function. If classified according to the degree of movement they make possible, joints can be immovable, slightly movable, or freely movable. Joints also can be grouped by the type of tissue (fibrous, cartilaginous, or synovial) that binds the bones together at each junction. Currently, structural classification by tissue type is more commonly used.

 There are 230 joints in the body.

Fibrous Joints

Fibrous (fi´brus) **joints** lie between bones that closely contact one another. A thin layer of dense connective tissue joins the bones at such joints, as in the case of a *suture* between a pair of flat bones of the skull (fig. 7.33). No appreciable movement takes place at a fibrous joint. Some fibrous joints, such as the joint in the leg between the distal ends of the tibia and fibula, have limited movement.

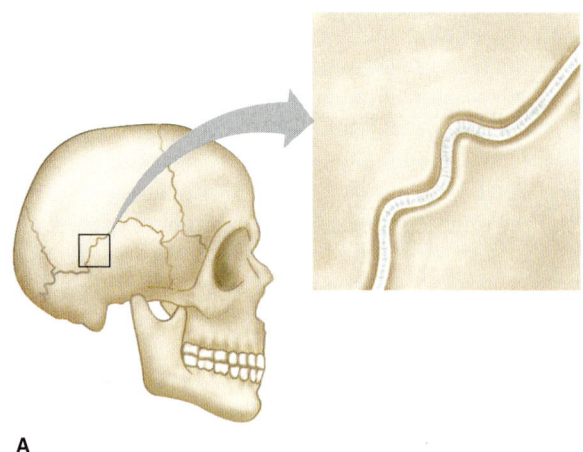

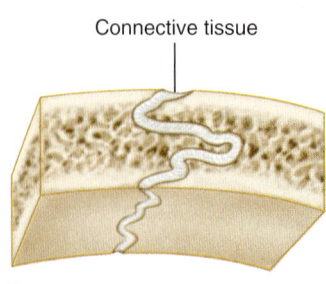

Figure 7.33
Fibrous joints. (*A*) The fibrous joints between the bones of the skull are immovable and are called sutures. (*B*) A thin layer of connective tissue connects the bones at the suture.

Cartilaginous Joints

Hyaline cartilage, or fibrocartilage, connects the bones of **cartilaginous** (kar´´tĭ-lah´jin-us) **joints.** For example, joints of this type separate the vertebrae of the vertebral column. Each intervertebral disc is composed of a band of fibrocartilage (annulus fibrosus) surrounding a pulpy or gelatinous core (nucleus pulposus). The disc absorbs shocks and helps equalize pressure between adjacent vertebrae when the body moves (see fig. 7.16).

Due to the slight flexibility of the discs, cartilaginous joints allow limited movement, as when the back is bent forward or to the side or is twisted. Other examples of cartilaginous joints include the symphysis pubis and the first rib with the sternum.

Synovial Joints

Most joints within the skeletal system are **synovial** (sĭ-no´ve-al) **joints** and allow free movement. They are more complex structurally than fibrous or cartilaginous joints.

The articular ends of the bones in a synovial joint are covered with hyaline cartilage (articular cartilage), and a surrounding, tubular capsule of dense connective tissue holds them together (fig. 7.34). This *joint capsule* is composed of an outer layer of ligaments and an inner lining of *synovial membrane,* which secretes synovial fluid. With a consistency similar to uncooked egg white, synovial fluid lubricates joints.

Some synovial joints have flattened, shock-absorbing pads of fibrocartilage called **menisci** (mĕ-nis´ke) (singular, *meniscus*) between the articulating surfaces of the bones (fig. 7.35). Such joints may also have fluid-filled sacs called **bursae** (ber´se) associated with them. Each bursa is lined with synovial membrane, which may be continuous with the synovial membrane of a nearby joint cavity.

Bursae are commonly located between the skin and underlying bony prominences, as in the case of the patella of the knee or the olecranon process of the elbow. They aid movement of tendons that glide over these bony parts or over other tendons. Figures 7.35 and 7.36 show and name some of the bursae associated with the knee and shoulder.

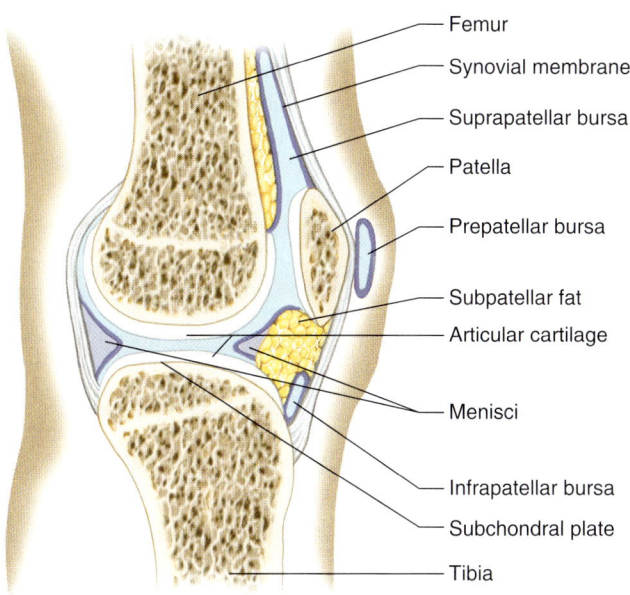

Figure 7.35
Menisci separate the articulating surfaces of the femur and tibia. Several bursae are associated with the knee joint.

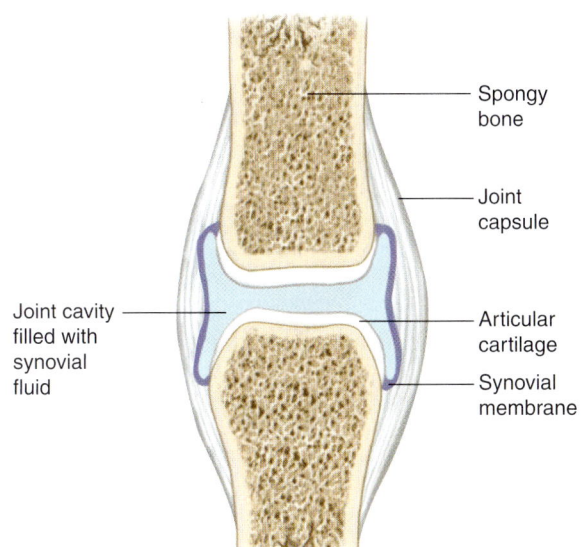

Figure 7.34
The generalized structure of a synovial joint.

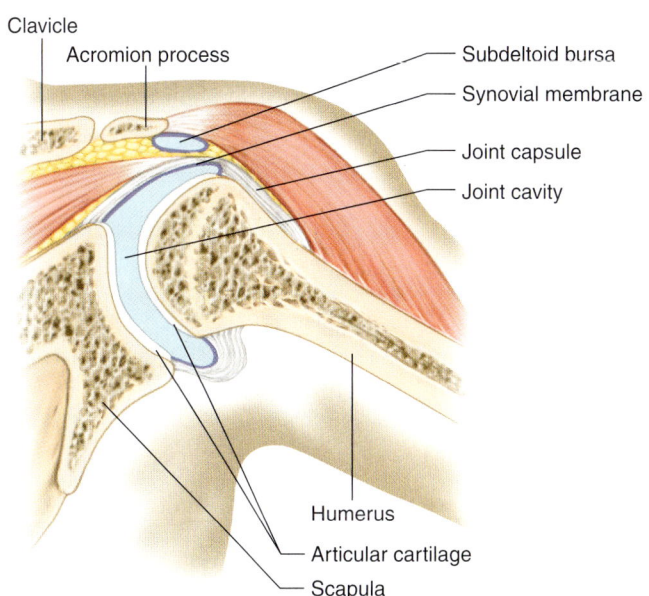

Figure 7.36
The shoulder joint allows movements in all directions. Several bursae are associated with the shoulder joint (not all are shown).

Based on the shapes of their parts and the movements they permit, synovial joints are classified as follows:

1. A **ball-and-socket joint** consists of a bone with a ball-shaped head that articulates with the cup-shaped cavity of another bone. Such a joint allows a wider range of motion than does any other kind, permitting movements in all planes, as well as rotational movement around a central axis. The shoulder and hip contain joints of this type (figs. 7.36 and 7.37).

2. In a **condyloid joint,** an oval-shaped condyle of one bone fits into an elliptical cavity of another bone, as in the joints between the metacarpals and phalanges (see fig. 7.25). This type of joint permits a variety of movements in different planes; rotational movement, however, is not possible.

3. The articulating surfaces of **gliding joints** are nearly flat or slightly curved. Most of the joints within the wrist (see fig. 7.25) and ankle, as well as those between the articular processes of adjacent vertebrae, belong to this group. They allow sliding and twisting movements. The sacroiliac joints and the joints formed by ribs 2–7 connecting with the sternum are also gliding joints.

4. In a **hinge joint,** the convex surface of one bone fits into the concave surface of another, as in the elbow (fig. 7.38) and the joints of the phalanges. Such a joint resembles the hinge of a door in that it permits movement in one plane only.

5. In a **pivot joint,** the cylindrical surface of one bone rotates within a ring formed of bone and ligament. Movement is limited to the rotation around a central axis. The joint between the proximal ends of the radius and the ulna is of this type (see fig. 7.24).

6. A **saddle joint** forms between bones whose articulating surfaces have both concave and convex regions. The surface of one bone fits the complementary surface of the other. This physical relationship permits a variety of movements, as in the joint between the carpal (trapezium) and metacarpal of the thumb (see fig. 7.25).

Table 7.5 summarizes the types of joints.

Arthritis is a disease that causes inflamed, swollen, and painful joints. More than a hundred different types of arthritis affect 50 million people in the United States. The most common forms are *rheumatoid arthritis* and *osteoarthritis*.

In rheumatoid arthritis, which is the most painful and debilitating of the arthritic diseases, the synovial membrane of a freely movable joint becomes inflamed and thickened. Then the articular cartilage is damaged, and fibrous tissue infiltrates, interfering with joint movement. In time, the joint may ossify, fusing the articulating bones. Rheumatoid arthritis is an autoimmune disorder in which the immune system attacks the body's healthy tissues.

Osteoarthritis is a degenerative disorder that occurs as a result of aging, but an inherited form may appear as early as one's thirties. In osteoarthritis, articular cartilage softens and disintegrates gradually, roughening the articular surfaces. Joints become painful, and movement is restricted. Osteoarthritis most often affects joints that are used the most over a lifetime, such as those of the fingers, hips, knees, and lower parts of the vertebral column.

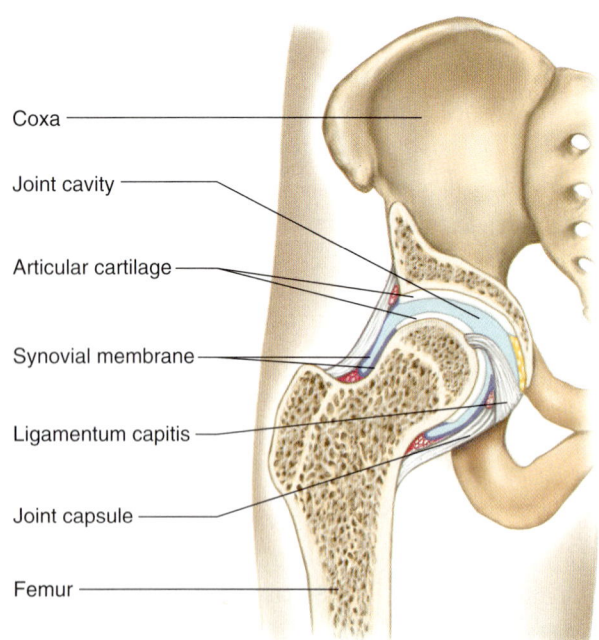

Figure 7.37
The hip is a ball-and-socket joint.

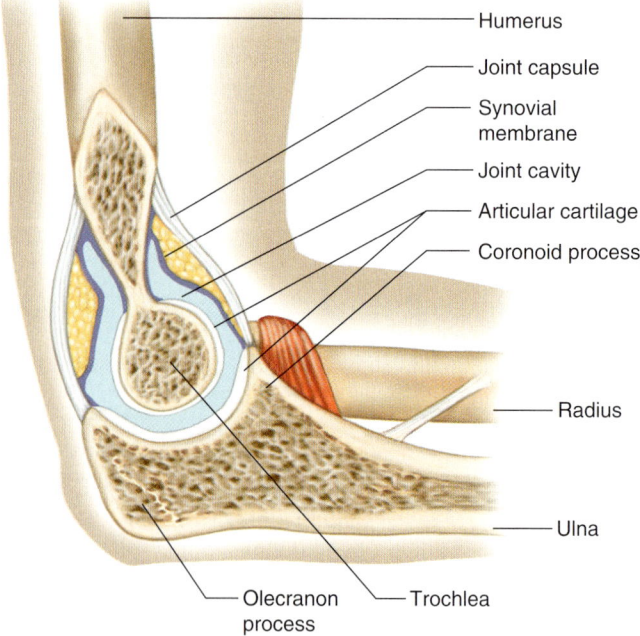

Figure 7.38
The elbow is a hinge joint.

TABLE 7.5 TYPES OF JOINTS

TYPE OF JOINT	DESCRIPTION	POSSIBLE MOVEMENTS	EXAMPLE
Fibrous	Articulating bones are fastened together by a thin layer of dense connective tissue.	None	Suture between bones of skull, joint between the distal ends of tibia and fibula
Cartilaginous	Articulating bones are connected by hyaline cartilage or fibrocartilage.	Limited movement, as when back is bent or twisted	Joints between the bodies of vertebrae, symphysis pubis
Synovial	Articulating bones are surrounded by a joint capsule of ligaments and synovial membranes; ends of articulating bones are covered by hyaline cartilage and separated by synovial fluid.	Allow free movement (see the following list)	
1. Ball-and-socket	Ball-shaped head of one bone articulates with cup-shaped cavity of another.	Movements in all planes and rotation	Shoulder, hip
2. Condyloid	Oval-shaped condyle of one bone articulates with elliptical cavity of another.	Variety of movements in different planes, but no rotation	Joints between the metacarpals and phalanges
3. Gliding	Articulating surfaces are nearly flat or slightly curved.	Sliding or twisting	Joints between various bones of wrist and ankle, sacroiliac joints, joints between ribs 2–7 and sternum
4. Hinge	Convex surface of one bone articulates with concave surface of another.	Flexion and extension	Elbow, joints of phalanges
5. Pivot	Cylindrical surface of one bone articulates with ring of bone and ligament.	Rotation around a central axis	Joint between the proximal ends of radius and ulna
6. Saddle	Articulating surfaces have both concave and convex regions; the surface of one bone fits the complementary surface of another.	Variety of movements	Joint between the carpal and metacarpal of thumb

Types of Joint Movements

Skeletal muscle action produces movements at synovial joints. Typically, one end of a muscle is attached to a relatively immovable or fixed part on one side of a joint, and the other end of the muscle is fastened to a movable part on the other side. When the muscle contracts, its fibers pull its movable end **(insertion)** toward its fixed end **(origin),** and a movement occurs at the joint.

The following terms describe movements at joints (figs. 7.39, 7.40, and 7.41):

flexion (flek´shun) Bending parts at a joint so that the angle between them decreases and the parts come closer together (bending the lower limb at the knee).

extension (ek-sten´shun) Straightening parts at a joint so that the angle between them increases and the parts move farther apart (straightening the lower limb at the knee).

dorsiflexion (dor´´sĭ-flek´shun) Bending the foot at the ankle toward the shin (bending the foot upward).

plantar flexion (plan´tar flek´shun) Bending the foot at the ankle toward the sole (bending the foot downward).

hyperextension (hi´´per-ek-sten´shun) Excess extension of the parts at a joint, beyond the anatomical position (bending the head back beyond the upright position).

abduction (ab-duk´shun) Moving a part away from the midline (lifting the upper limb horizontally to form a right angle with the side of the body).

adduction (ah-duk´shun) Moving a part toward the midline (returning the upper limb from the horizontal position to the side of the body).

rotation (ro-ta´shun) Moving a part around an axis (twisting the head from side to side).

circumduction (ser´´kum-duk´shun) Moving a part so that its end follows a circular path (moving the finger in a circular motion without moving the hand).

pronation (pro-na´shun) Turning the hand so that the palm is downward or facing posteriorly (in anatomical position).

supination (soo´´pĭ-na´shun) Turning the hand so that the palm is upward or facing anteriorly (in anatomical position).

eversion (e-ver´zhun) Turning the foot so that the sole faces laterally.

inversion (in-ver´zhun) Turning the foot so that the sole faces medially.

retraction (re-trak´shun) Moving a part backward (pulling the chin backward).

protraction (pro-trak´shun) Moving a part forward (thrusting the chin forward).

elevation (el´´ĕ-va´shun) Raising a part (shrugging the shoulders).

depression (de-presh´un) Lowering a part (drooping the shoulders).

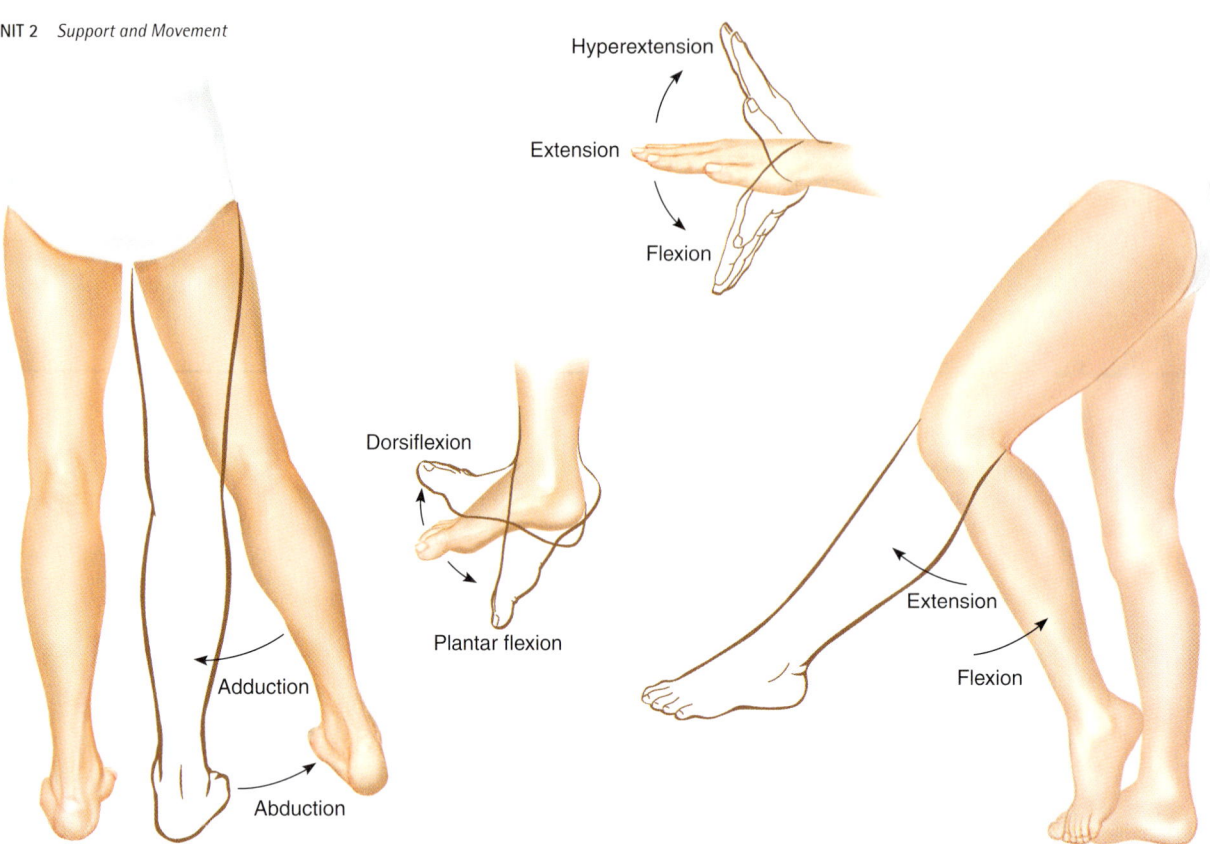

Figure 7.39
Joint movements illustrating adduction, abduction, dorsiflexion, plantar flexion, hyperextension, extension, and flexion.

Figure 7.40
Joint movements illustrating rotation, circumduction, pronation, and supination.

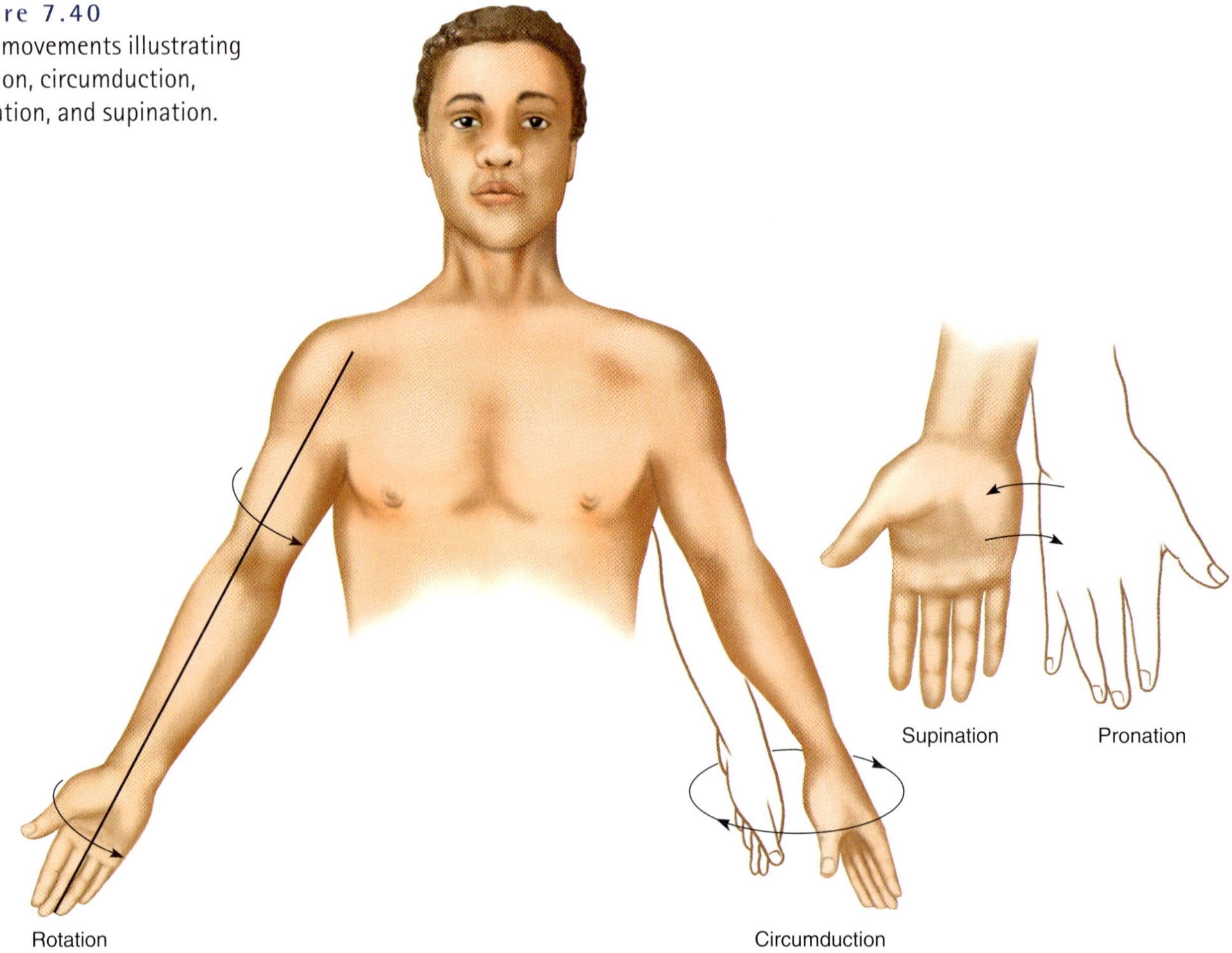

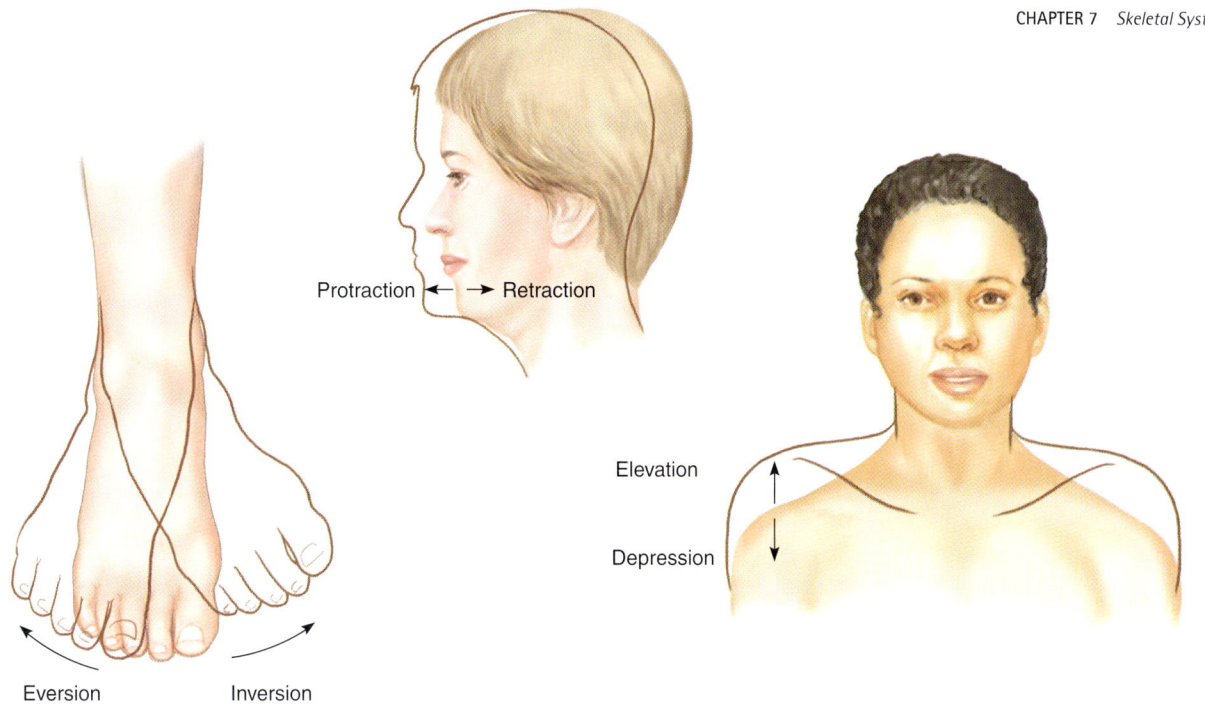

Figure 7.41
Joint movements illustrating eversion, inversion, retraction, protraction, elevation, and depression.

 CHECK YOUR RECALL

1. Describe the characteristics of the three major types of joints.
2. Name six types of synovial joints.
3. What terms describe movements possible at synovial joints?

Injuries to the elbow, shoulder, and knee are commonly diagnosed and treated using a procedure called *arthroscopy*. Arthroscopy enables a surgeon to visualize a joint's interior and even perform diagnostic or therapeutic procedures, guided by the image on a video screen. An arthroscope is a thin, tubular instrument about 25 centimeters long containing optical fibers that transmit an image. The surgeon inserts the device through a small incision in the joint capsule. Arthroscopy is far less invasive than conventional surgery. Many runners have undergone uncomplicated arthroscopy and raced just weeks later.

Clinical Terms Related to the Skeletal System

acromegaly (ak´´ro-meg´ah-le) Abnormal enlargement of facial features, hands, and feet in adults as a result of overproduction of growth hormone.

ankylosis (ang´´kĭ-lo´sis) Abnormal stiffness of a joint or fusion of bones at a joint, often due to damage to the joint membranes from chronic rheumatoid arthritis.

arthralgia (ar-thral´je-ah) Pain in a joint.

arthrocentesis (ar´´thro-sen-te´sis) Puncture of and removal of fluid from a joint cavity.

arthrodesis (ar´´thro-de´sis) Surgery to fuse the bones at a joint.

arthroplasty (ar´thro-plas´´te) Surgery to make a joint more movable.

Colles fracture (kol´ēz frak´ture) Fracture at the distal end of the radius that displaces the smaller fragment posteriorly.

epiphysiolysis (ep´´ĭ-fiz´´e-ol´ĭ-sis) Separation or loosening of the epiphysis from the diaphysis of a bone.

hemarthrosis (hem´´ar-thro´sis) Blood in a joint cavity.

laminectomy (lam´´ĭ-nek´to-me) Surgical removal of the posterior arch of a vertebra, usually to relieve symptoms of a ruptured intervertebral disc.

lumbago (lum-ba´go) Dull ache in the lumbar region of the back.

orthopedics (or´´tho-pe´diks) Medical specialty that prevents, diagnoses, and treats diseases and abnormalities of the skeletal and muscular systems.

ostealgia (os´´te-al´je-ah) Pain in a bone.

ostectomy (os-tek´to-me) Surgical removal of a bone.

osteitis (os´´te-i´tis) Inflammation of bone tissue.

osteochondritis (os´´te-o-kon-dri´tis) Inflammation of bone and cartilage tissues.

osteogenesis (os´´te-o-jen´ĕ-sis) Bone development.

osteogenesis imperfecta (os´´te-o-jen´ĕ-sis im-per-fek´ta) Inherited condition of deformed and abnormally brittle bones.

osteoma (os´´te-o´mah) Tumor composed of bone tissue.

osteomalacia (os´´te-o-mah-la´she-ah) Softening of adult bone due to a disorder in calcium and phosphorus metabolism, usually caused by vitamin D deficiency.

Organization

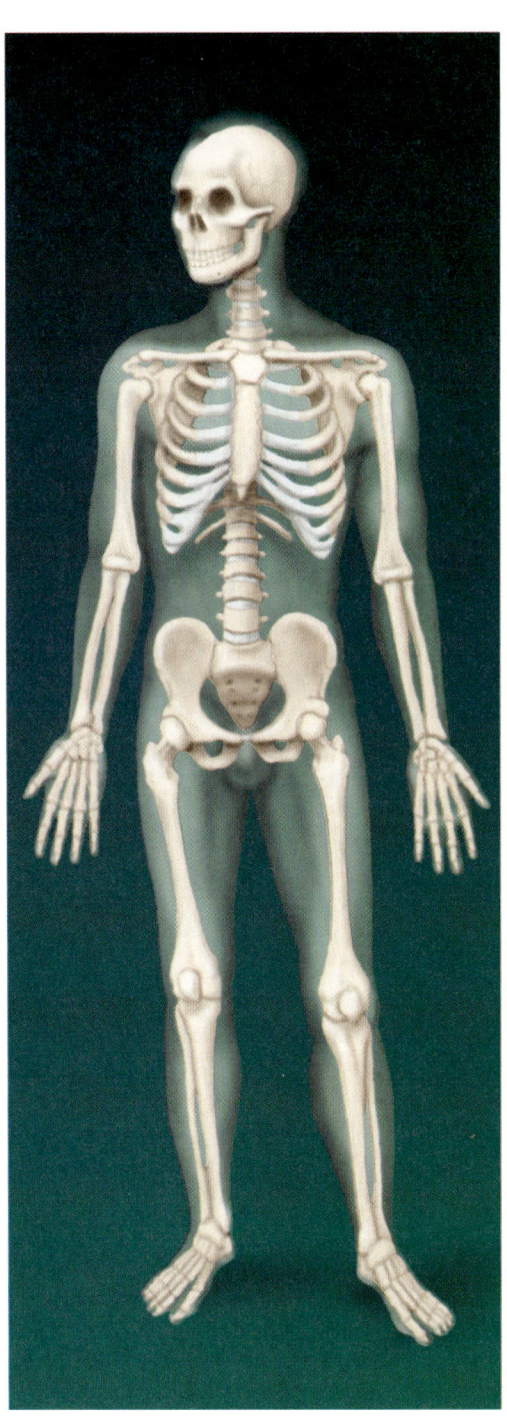

Skeletal System
Bones provide support, protection, and movement and also play a role in calcium balance.

Integumentary System
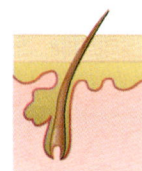
Vitamin D, activated in the skin, plays a role in calcium availability for bone matrix.

Lymphatic System

Cells of the immune system originate in the bone marrow.

Muscular System

Muscles pull on bones to cause movement.

Digestive System

Absorption of dietary calcium provides material for bone matrix.

Nervous System
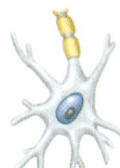
Proprioceptors sense the position of body parts. Pain receptors warn of trauma to bone. Bones protect the brain and spinal cord.

Respiratory System
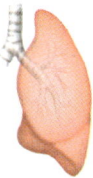
Ribs and muscles work together in breathing.

Endocrine System

Some hormones act on bone to help regulate blood calcium levels.

Urinary System

The kidneys and bones work together to help regulate blood calcium levels.

Cardiovascular System
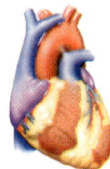
Blood transports nutrients to bone cells. Bone helps regulate plasma calcium levels, important to heart function.

Reproductive System
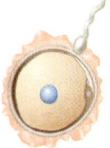
The pelvis helps support the uterus during pregnancy. Bone may provide a source of calcium during lactation.

osteomyelitis (os´´te-o-mi´´ĕ-li´tis) Bone inflammation caused by the body's reaction to bacterial or fungal infection.

osteonecrosis (os´´te-o-ne-kro´sis) Death of bone tissue. This condition occurs most commonly in the femur head in elderly persons and may be due to obstructed arteries supplying the bone.

osteopathology (os´´te-o-pah-thol´o-je) Study of bone diseases.

osteotomy (os´´te-ot´o-me) Cutting a bone.

synovectomy (sin´´o-vek´to-me) Surgical removal of the synovial membrane of a joint.

Clinical Connection

When the 20-year-old professional soccer player jammed his left toe at high speed against the ball and howled in pain, he thought it would just get better in a few days, as such injuries usually do. This time, the injured toe started to turn bluish red immediately, as a hematoma formed beneath the nail. The pain continued, for weeks. Pus swelled from beneath the darkened nail. Finally, barely able to walk let alone continuing playing his sport, the athlete consulted a physician, who, assuming the wound was infected, prescribed antibiotics and an anti-inflammatory cream. But the unrelenting pain was not due to infection. The young man finally went to an emergency room, where a sample of the pus revealed no bacteria. X-rays instead clearly indicated an osteochondroma, a spike of bone emerging 4 millimeters from the dorsal terminal phalanx of the left great toe, capped with cartilage. Usually an osteochondroma is a benign bone tumor that arises during fetal development. The physician in charge, however, suspected that the soccer player's spike was a response to trauma—and then failure to rest afterwards. Surgery removed the spike, and a month later, the athlete was back on the field.

SUMMARY OUTLINE

7.1 Introduction (p. 128)
Individual bones are the organs of the skeletal system. A bone contains very active tissues.

7.2 Bone Structure (p. 128)
Bone structure reflects its function.
1. Parts of a long bone
   a. Epiphyses at each end are covered with articular cartilage and articulate with other bones.
   b. The shaft of a bone is called the diaphysis.
   c. Except for the articular cartilage, a bone is covered by a periosteum.
   d. Compact bone has a continuous matrix with no gaps.
   e. Spongy bone has irregular interconnecting spaces between bony plates that reduce the weight of bone.
   f. Both compact and spongy bone are strong and resist bending.
   g. The diaphysis contains a medullary cavity filled with marrow.
2. Microscopic structure
   a. Compact bone contains osteons cemented together.
   b. Central canals contain blood vessels that nourish the cells of osteons.
   c. Diffusion from the surface of the thin, bony plates nourishes the cells of spongy bone.

7.3 Bone Development and Growth (p. 130)
1. Intramembranous bones
   a. Intramembranous bones develop from layers of connective tissues.
   b. Osteoblasts within the membranous layers form bone tissue.
   c. Mature bone cells are called osteocytes.
2. Endochondral bones
   a. Endochondral bones develop as hyaline cartilage that is later replaced by bone tissue.
   b. The primary ossification center appears in the diaphysis, whereas secondary ossification centers appear in the epiphyses.
   c. An epiphyseal plate remains between the primary and secondary ossification centers.
   d. The epiphyseal plates are responsible for lengthening.
   e. Long bones continue to lengthen until the epiphyseal plates are ossified.
   f. Growth in thickness is due to intramembranous ossification beneath the periosteum.
3. Homeostasis of bone tissue
   a. Osteoclasts and osteoblasts continually remodel bone.
   b. The total mass of bone remains nearly constant.

7.4 Bone Function (p. 131)
1. Support and protection
   a. Bones shape and form body structures.
   b. Bones support and protect softer, underlying tissues.
2. Body movement
   a. Bones and muscles function together as levers.
   b. A lever consists of a rod, a pivot (fulcrum), a resistance, and a force that supplies energy.
3. Blood cell formation
   a. At different ages, hemopoiesis occurs in the yolk sac, liver and spleen, and red bone marrow.
   b. Red marrow houses developing red blood cells, white blood cells, and blood platelets. Yellow marrow stores fat.
4. Storage of inorganic salts
   a. The intercellular material of bone tissue contains large quantities of calcium phosphate.
   b. When blood calcium is low, osteoclasts break down bone, releasing calcium salts. When blood calcium is high, osteoblasts form bone tissue and store calcium salts.
   c. Bone stores small amounts of magnesium, sodium, potassium, and carbonate ions.

7.5 Skeletal Organization (p. 135)
1. The skeleton can be divided into axial and appendicular portions.
2. The axial skeleton consists of the skull, hyoid bone, vertebral column, and thoracic cage.

3. The appendicular skeleton consists of the pectoral girdle, upper limbs, pelvic girdle, and lower limbs.

7.6 Skull (p. 136)

The skull consists of twenty-two bones: eight cranial bones, fourteen facial bones.

1. Cranium
   a. The cranium encloses and protects the brain.
   b. Some cranial bones contain air-filled sinuses.
   c. Cranial bones include the frontal bone, parietal bones, occipital bone, temporal bones, sphenoid bone, and ethmoid bone.
2. Facial skeleton
   a. Facial bones form the basic shape of the face and provide attachments for muscles.
   b. Facial bones include the maxillae, palatine bones, zygomatic bones, lacrimal bones, nasal bones, vomer bone, inferior nasal conchae, and mandible.
3. Infantile skull
   a. Fontanels connect incompletely developed bones.
   b. The proportions of the infantile skull are different from those of an adult skull.

7.7 Vertebral Column (p. 142)

The vertebral column extends from the skull to the pelvis and protects the spinal cord. It is composed of vertebrae, separated by intervertebral discs.

1. A typical vertebra
   a. A typical vertebra consists of a body and a bony vertebral arch, which surrounds the spinal cord.
   b. Notches on the upper and lower surfaces provide intervertebral foramina through which spinal nerves pass.
2. Cervical vertebrae
   a. Transverse processes bear transverse foramina.
   b. The atlas (first vertebra) supports and balances the head.
   c. The dens of the axis (second vertebra) provides a pivot for the atlas when the head is turned from side to side.
3. Thoracic vertebrae
   a. Thoracic vertebrae are larger than cervical vertebrae.
   b. Facets on the sides articulate with the ribs.
4. Lumbar vertebrae
   a. The vertebral bodies are large and strong.
   b. They support more body weight than other vertebrae.
5. Sacrum
   a. The sacrum is a triangular structure formed of five fused vertebrae.
   b. Vertebral foramina form the sacral canal.
6. Coccyx
   a. The coccyx, composed of four fused vertebrae, forms the lowest part of the vertebral column.
   b. It acts as a shock absorber when a person sits.

7.8 Thoracic Cage (p. 147)

The thoracic cage includes the ribs, thoracic vertebrae, sternum, and costal cartilages. It supports the pectoral girdle and upper limbs, protects viscera, and functions in breathing.

1. Ribs
   a. Twelve pairs of ribs attach to the twelve thoracic vertebrae.
   b. Costal cartilages of the true ribs join the sternum directly. Those of the false ribs join it indirectly or not at all.
   c. A typical rib has a shaft, a head, and tubercles that articulate with the vertebrae.
2. Sternum
   a. The sternum consists of a manubrium, body, and xiphoid process.
   b. It articulates with the clavicles.

7.9 Pectoral Girdle (p. 147)

The pectoral girdle is composed of two clavicles and two scapulae. It forms an incomplete ring that supports the upper limbs and provides attachments for muscles.

1. Clavicles
   a. Clavicles are rodlike bones located between the manubrium and scapulae.
   b. They hold the shoulders in place and provide attachments for muscles.
2. Scapulae
   a. The scapulae are broad, triangular bones.
   b. They articulate with the humerus of each upper limb and provide attachments for muscles.

7.10 Upper Limb (p. 148)

Bones of the upper limb provide the frameworks and attachments of muscles, and function in levers that move the limb and its parts.

1. Humerus
   a. The humerus extends from the scapula to the elbow.
   b. It articulates with the radius and ulna at the elbow.
2. Radius
   a. The radius is located on the thumb side of the forearm between the elbow and the wrist.
   b. It articulates with the humerus, ulna, and wrist.
3. Ulna
   a. The ulna is longer than the radius and overlaps the humerus posteriorly.
   b. It articulates with the radius laterally and with a disc of fibrocartilage inferiorly.
4. Wrist and hand
   a. The wrist is composed of eight carpal bones that form a carpus.
   b. The hand includes five metacarpal bones and fourteen phalanges.

7.11 Pelvic Girdle (p. 152)

The pelvic girdle consists of two coxae that articulate with each other anteriorly and with the sacrum posteriorly.

1. The sacrum, coccyx, and pelvic girdle form the bowl-shaped pelvis.
2. Each coxa consists of an ilium, ischium, and pubis, which are fused in the region of the acetabulum.
   a. The ilium
      (1) The ilium is the largest portion of the coxa.
      (2) It joins the sacrum at the sacroiliac joint.
   b. The ischium
      (1) The ischium is the lowest portion of the coxa.
      (2) It supports body weight when sitting.
   c. The pubis
      (1) The pubis is the anterior portion of the coxa.
      (2) The pubic bones are fused anteriorly at the symphysis pubis.

7.12 Lower Limb (p. 154)

Bones of the lower limb provide frameworks for the thigh, leg, ankle, and foot.

1. Femur
   a. The femur extends from the hip to the knee.
   b. The patella articulates with the femur's anterior surface.
2. Tibia
   a. The tibia is located on the medial side of the leg.
   b. It articulates with the talus of the ankle.
3. Fibula
   a. The fibula is located on the lateral side of the tibia.
   b. It articulates with the ankle but does not bear body weight.

    4. Ankle and foot
        a. The ankle and foot consist of the tarsus, metatarsus, and five toes.
        b. Included are the talus that helps form the ankle, six other tarsals, five metatarsals, and fourteen phalanges.

7.13 Joints (p. 156)
Joints can be classified according to the type of tissue that binds the bones together.
    1. Fibrous joints
        a. Bones at fibrous joints are tightly joined by a layer of dense connective tissue.
        b. Little or no movement occurs at a fibrous joint.
    2. Cartilaginous joints
        a. A layer of cartilage joins bones of cartilaginous joints.
        b. Such joints allow limited movement.
    3. Synovial joints
        a. The bones of a synovial joint are covered with hyaline cartilage and held together by a fibrous joint capsule.
        b. The joint capsule consists of an outer layer of ligaments and an inner lining of synovial membrane.
        c. Bursae are located between the skin and underlying bony prominences.
        d. Types of synovial joints include: ball-and-socket, condyloid, gliding, hinge, pivot, and saddle.
    4. Types of joint movements
        a. Muscles fastened on either side of a joint produce movements of synovial joints.
        b. Joint movements include flexion, extension, dorsiflexion, plantar flexion, hyperextension, abduction, adduction, rotation, circumduction, pronation, supination, eversion, inversion, retraction, protraction, elevation, and depression.

REVIEW EXERCISES

Part A
1. Sketch a typical long bone, and label its epiphyses, diaphysis, medullary cavity, periosteum, and articular cartilages. (p. 128)
2. Distinguish between spongy and compact bone. (p. 128)
3. Explain how central canals and perforating canals are related. (p. 129)
4. Explain how the development of intramembranous bone differs from that of endochondral bone. (p. 130)
5. Distinguish between osteoblasts and osteocytes. (p. 130)
6. Explain the function of an epiphyseal plate. (p. 130)
7. Explain how a bone thickens. (p. 131)
8. Provide several examples to illustrate how bones support and protect body parts. (p. 131)
9. Describe a lever. (p. 133)
10. Explain how upper limb movements function as levers. (p. 133)
11. Describe the functions of red and yellow bone marrow. (p. 133)
12. Explain the mechanism that regulates the concentration of blood calcium ions. (p. 134)
13. Distinguish between the axial and appendicular skeletons. (p. 135)
14. Name the bones of the cranium and the facial skeleton. (p. 136)
15. Explain the importance of fontanels. (p. 142)
16. Describe a typical vertebra. (p. 142)
17. Explain the differences among the cervical, thoracic, and lumbar vertebrae. (p. 143)
18. Name the bones that comprise the thoracic cage. (p. 147)
19. List the bones that form the pectoral and pelvic girdles. (p. 147)
20. Name the bones of the upper limb. (p. 148)
21. Name the bones that comprise a coxa. (p. 152)
22. List the bones of the lower limb. (p. 154)
23. Define *joint*. (p. 156)
24. Describe a fibrous joint, a cartilaginous joint, and a synovial joint. (p. 156)
25. Define *bursa*. (p. 157)
26. List six types of synovial joints, and name an example of each type. (p. 158)

Part B
Match the parts listed in column I with the bones listed in column II.

I	II
1. Coronoid process	a. Ethmoid bone
2. Cribriform plate	b. Frontal bone
3. Foramen magnum	c. Mandible
4. Mastoid process	d. Maxilla
5. Palatine process	e. Occipital bone
6. Sella turcica	f. Temporal bone
7. Supraorbital foramen	g. Sphenoid bone
8. Temporal process	h. Zygomatic bone
9. Acromion process	i. Femur
10. Deltoid tuberosity	j. Fibula
11. Greater trochanter	k. Humerus
12. Lateral malleolus	l. Radius
13. Medial malleolus	m. Scapula
14. Olecranon process	n. Sternum
15. Radial tuberosity	o. Tibia
16. Xiphoid process	p. Ulna

Part C
Match the movements in column I with the descriptions in column II.

I	II
1. Rotation	a. Turning palm upward
2. Supination	b. Decreasing angle between parts
3. Extension	c. Moving part forward
4. Eversion	d. Moving part around an axis
5. Protraction	e. Turning sole of foot to face laterally
6. Flexion	f. Increasing angle between parts
7. Pronation	g. Lowering a part
8. Abduction	h. Turning palm downward
9. Depression	i. Moving part away from midline

CRITICAL THINKING

1. How does the structure of a bone make it strong yet lightweight?
2. Archaeologists discover skeletal remains of humanlike animals in Ethiopia. Examination of the bones suggests that the remains represent four types of individuals. Two of the skeletons have bone densities that are 30% less than those of the other two skeletons. The skeletons with the lower bone mass also have broader front

pelvic bones. Within the two groups defined by bone mass, smaller skeletons have bones with evidence of epiphyseal plates, but larger bones have only a thin line where the epiphyseal plates should be. Give the age group and gender of the individuals in this find.

3. When a child's bone is fractured, growth may be stimulated at the epiphyseal plate of that bone. What problems might this extra growth cause in an upper or lower limb before the growth of the other limb compensates for the difference in length?

4. Compared to the shoulder and hip joints, in what way is the knee joint poorly protected and thus especially vulnerable to injuries?

WEB CONNECTIONS

Visit the website for additional study questions and more information about this chapter at:

http://www.mhhe.com/shieress8

Reference Plates

HUMAN SKULL

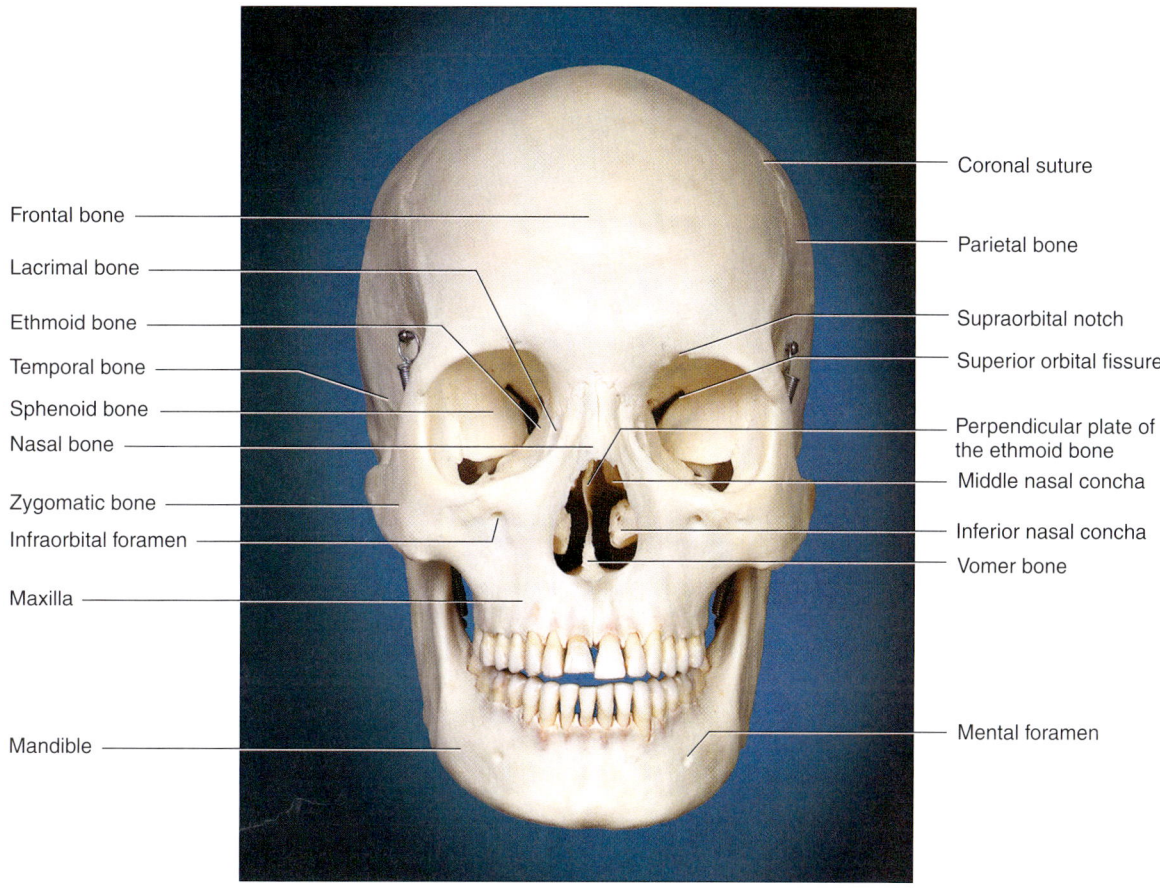

PLATE 8
The skull, anterior view.

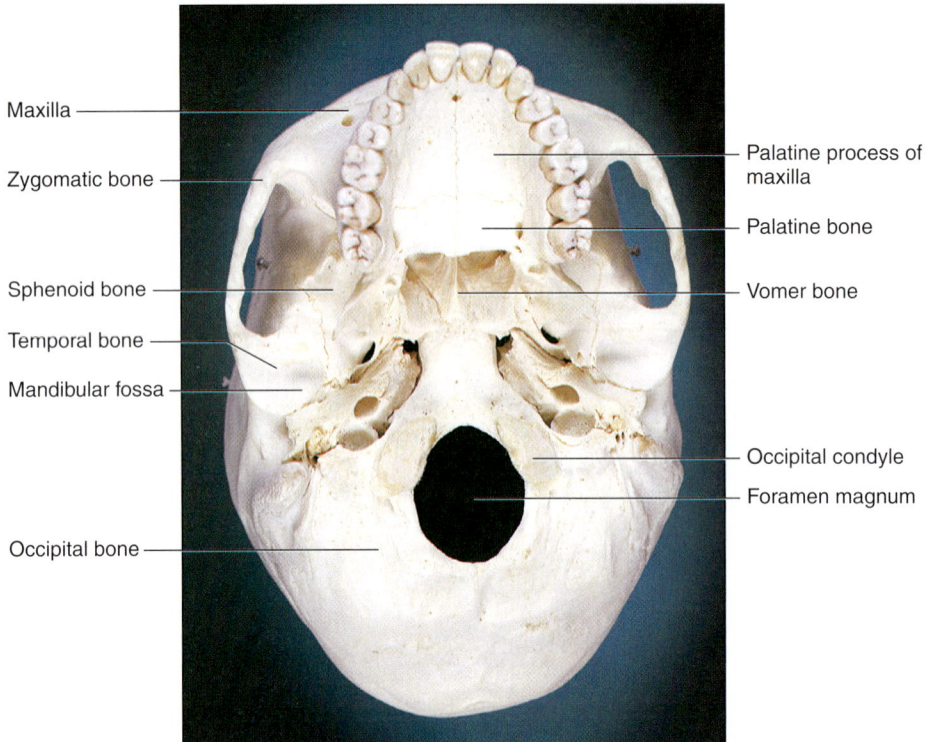

PLATE 9
The skull, inferior view.

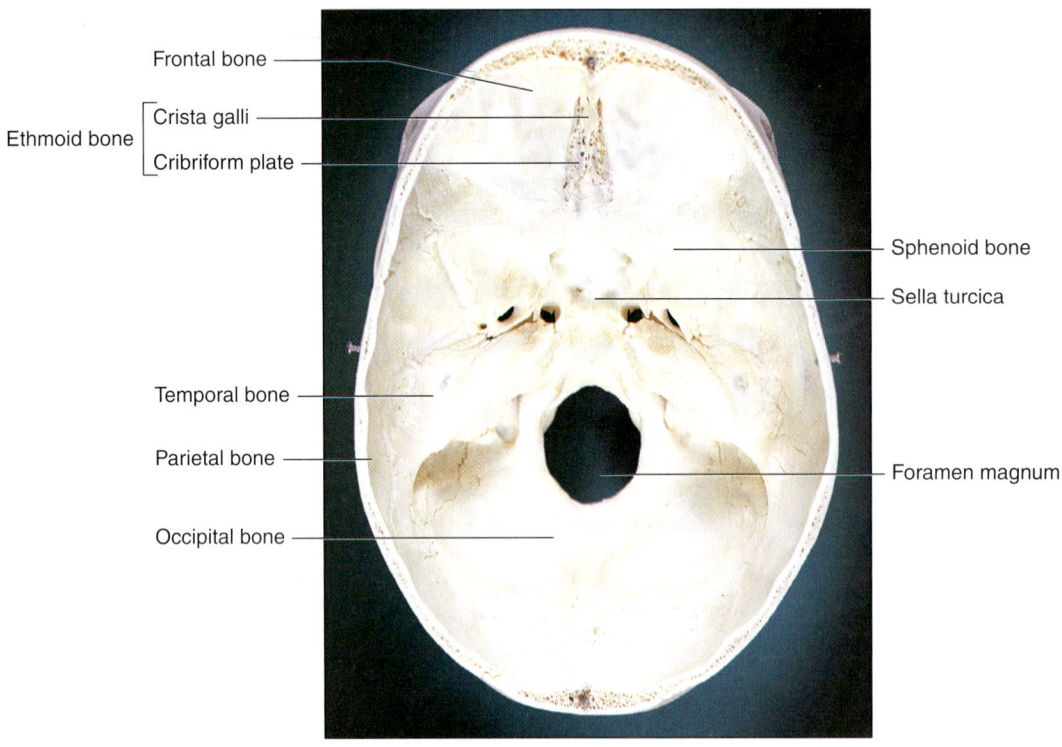

PLATE 10
The skull, floor of the cranial cavity.

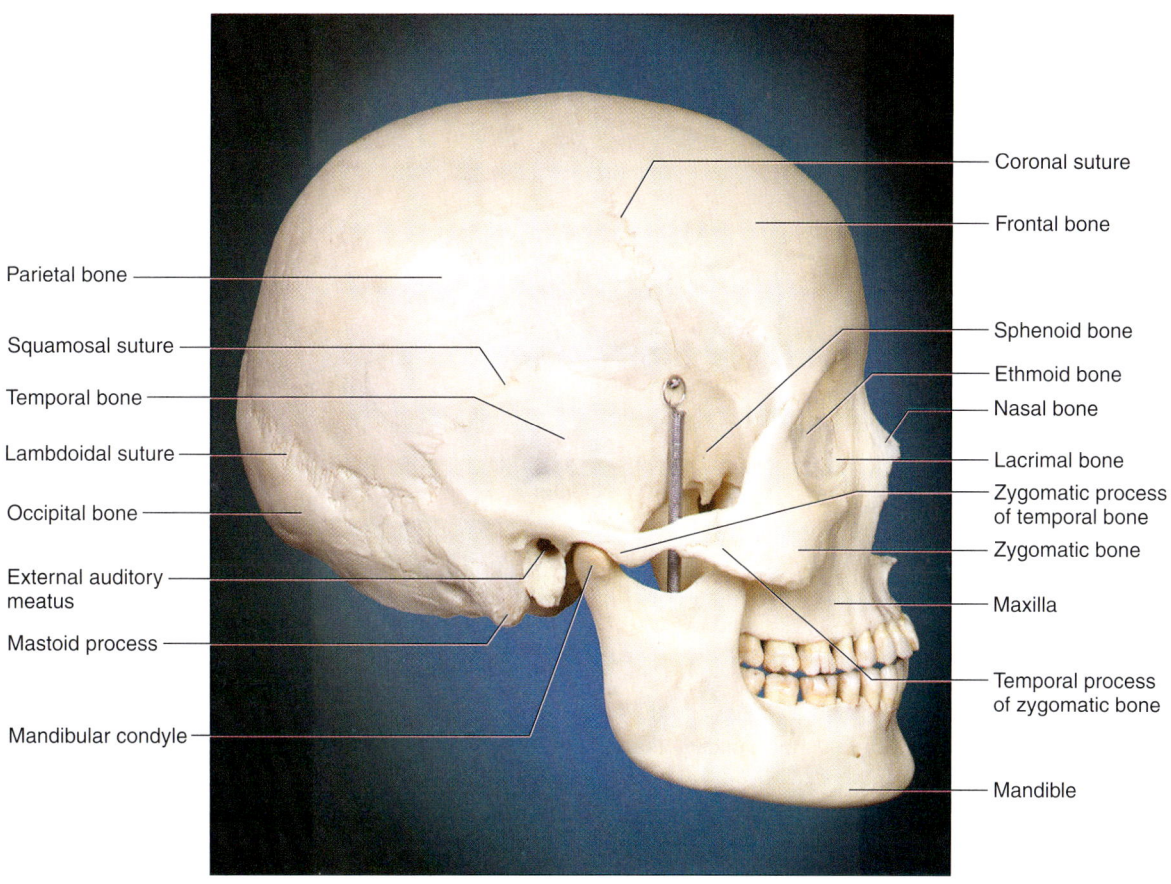

PLATE 11
The skull, lateral view.

chapter 8

Muscular System

SLEEPING LIKE AN APE. Modern humans are plagued with bad backs and aching joints, especially in the morning. Perhaps we could avoid such problems by taking cues from other primates, suggests British physiotherapist Michael Tetley. He has discovered that various monkeys and apes sleep in positions that accomplish several feats: keeping joints flexible, protecting private parts, conserving heat, maintaining readiness to fight or flee, and last but not least, keeping bugs out of the mouth.

Gorillas, chimps, and gibbons sleep on their sides, without pillows. In this position, the animal can become quickly alert, because the bottom shoulder keeps the ears unobstructed. The head is down, with the mouth shut. The stretching movements that accompany breathing repeatedly realign the joints between the vertebrae, which is impossible if a pillow is used. Pillow use also keeps the feet dorsiflexed, which rotates the knees in a way that alters the angle between the patellar tendon and the quadriceps muscle, paving the way for eventual pain. Without a pillow, if a male brings his knee up to the elbow, a quite comfortable position, the penis extends along the flexed leg, protecting it in the curve of the body. Members of several native African tribes taught Tetley to sleep in this position.

For more than fifty years, physicians and anthropologists have reported increases in the incidence of osteoarthritis as people became more "civilized." Tetley advises us to learn from the sleeping positions of indigenous peoples and nonhuman primates while we can still study them, as a "natural" approach to preventing or alleviating back pain. "All we have to do is be good primates and use these preventive techniques," he concludes.

Photo:
We humans can avoid backaches and other troubles, such as bugs in the mouth, by taking cues from our slumbering primate cousins.

Chapter Objectives

After studying this chapter, you should be able to do the following:

8.1 Introduction
1. List various outcomes of muscular actions. (p. 172)

8.2 Structure of a Skeletal Muscle
2. Describe how connective tissue is part of a skeletal muscle. (p. 172)
3. Name the major parts of a skeletal muscle fiber, and describe the function of each. (p. 172)

8.3 Skeletal Muscle Contraction
4. Explain the major events of skeletal muscle fiber contraction. (p. 176)
5. Explain how the muscle fiber contraction mechanism obtains energy. (p. 179)
6. Describe how oxygen debt develops and how a muscle may become fatigued. (p. 179)

8.4 Muscular Responses
7. Distinguish between a twitch and a sustained contraction. (p. 182)
8. Explain how the types of muscular contractions produce body movements and help maintain posture. (p. 183)

8.5 Smooth Muscle
9. Distinguish between the structures and functions of a multiunit smooth muscle and a visceral smooth muscle. (p. 184)

10. Compare the contraction mechanisms of skeletal and smooth muscle fibers. (p. 184)

8.6 Cardiac Muscle
11. Compare the contraction mechanism of skeletal and cardiac muscle fibers. (p. 186)

8.7 Skeletal Muscle Actions
12. Explain how the locations and interactions of skeletal muscles make possible certain movements. (p. 187)

8.8 Major Skeletal Muscles
13. Describe the locations and actions of the major skeletal muscles of each body region. (p. 187)

Aids to Understanding Words

calat- [something inserted] inter*calat*ed disc: Membranous band that connects cardiac muscle cells.

erg- [work] syn*erg*ist: Muscle that works with a prime mover to produce a movement.

hyper- [over, more] muscular *hyper*trophy: Enlargement of muscle fibers.

inter- [between] *inter*calated disc: Membranous band that connects cardiac muscle cells.

laten- [hidden] *laten*t period: Time between application of a stimulus and the beginning of a muscle contraction.

myo- [muscle] *myo*fibril: Contractile structure within a muscle cell.

sarco- [flesh] *sarco*plasm: Material (cytoplasm) within a muscle fiber.

syn- [together] *syn*ergist: Muscle that works with a prime mover to produce a movement.

tetan- [stiff] *tetan*ic contraction: Sustained muscular contraction.

-troph [well fed] muscular hyper*troph*y: Enlargement of muscle fibers.

Key Terms

actin (ak´tin)
antagonist (an-tag´o-nist)
aponeurosis (ap´´o-nu-ro´sis)
fascia (fash´e-ah)
insertion (in-ser´shun)
motor neuron (mo´tor nu´ron)
motor unit (mo´tor u´nit)

muscle impulse (mus´el im´puls)
myofibril (mi´´o-fi´bril)
myosin (mi´o-sin)
neurotransmitter (nu´´ro-trans´mit-er)
origin (or´ĭ-jin)
oxygen debt (ok´sĭ-jen det)

prime mover (prīm moōv´er)
recruitment (re-kroōt´ment)
sarcomere (sar´ko-mēr)
synergist (sin´er-jist)
threshold stimulus (thresh´old stim´u-lus)

8.1 Introduction

Talking and walking, breathing and sneezing—in fact, all movements—require muscles. Muscles are organs composed of specialized cells that use the chemical energy stored in nutrients to contract. Muscular actions also provide muscle tone, propel body fluids and food, generate the heartbeat, and distribute heat.

Muscles are of three types—skeletal muscle, smooth muscle, and cardiac muscle. This chapter focuses on skeletal muscle, which attaches to bones and is under conscious control.

8.2 Structure of a Skeletal Muscle

A skeletal muscle is an organ of the muscular system. It is composed of skeletal muscle tissue, nervous tissue, blood, and connective tissues.

Connective Tissue Coverings

Layers of fibrous connective tissue called **fascia** (fash´e-ah) separate an individual skeletal muscle from adjacent muscles and hold it in position (fig. 8.1). This connective tissue surrounds each muscle and may project beyond its end to form a cordlike tendon. Fibers in a tendon may intertwine with those in a bone's periosteum, attaching the muscle to the bone. In other cases, the connective tissue forms broad fibrous sheets called **aponeuroses** (ap´´o-nu-ro´sez), which may attach to the coverings of adjacent muscles (see figs. 8.17 and 8.19).

The layer of connective tissue that closely surrounds a skeletal muscle is called *epimysium* (fig. 8.1). Other layers of connective tissue, called *perimysium*, extend inward from the epimysium and separate the muscle tissue into small compartments. These compartments contain bundles of skeletal muscle fibers called *fascicles* (fasciculi). Each muscle fiber within a fascicle

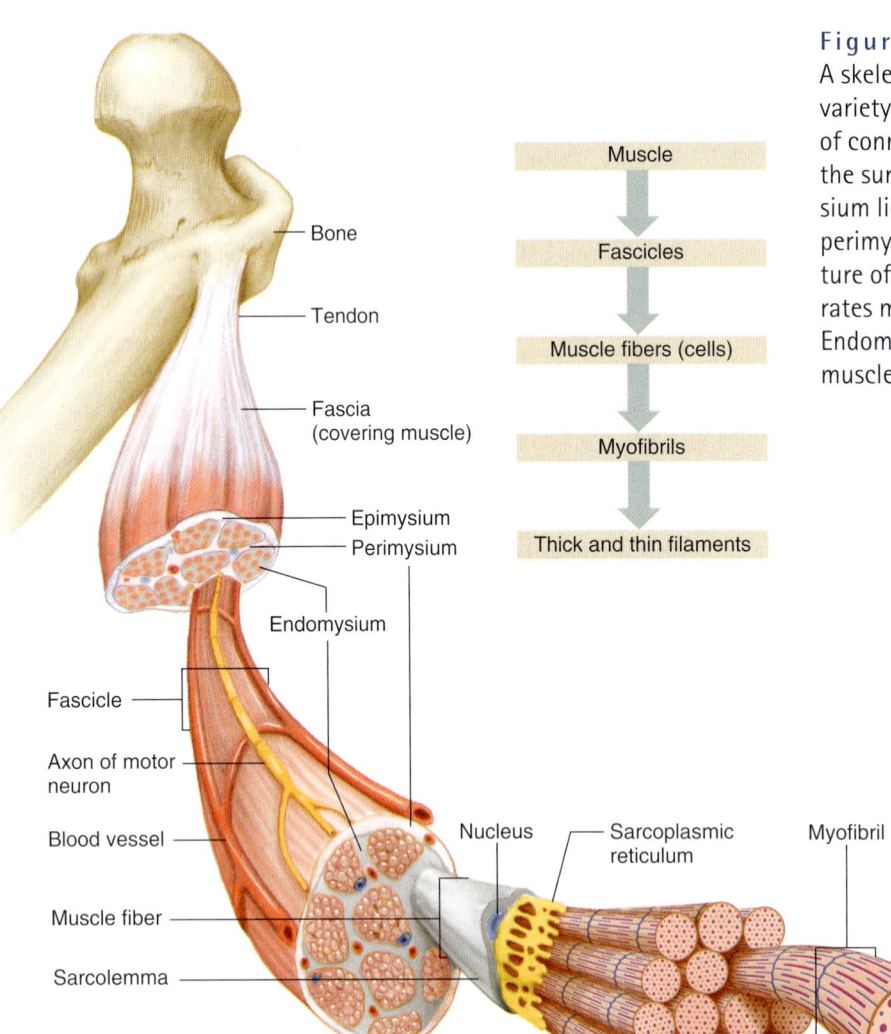

Figure 8.1
A skeletal muscle is composed of a variety of tissues, including layers of connective tissue. Fascia covers the surface of the muscle, epimysium lies beneath the fascia, and perimysium extends into the structure of the muscle, where it separates muscle cells into fascicles. Endomysium separates individual muscle fibers.

(fasciculus) lies within a layer of connective tissue in the form of a thin covering called *endomysium*. Thus, all parts of a skeletal muscle are enclosed in layers of connective tissue, which form a network extending throughout the muscular system.

A tendon, the attachment of a muscle to a bone, or the connective tissue sheath of a tendon (called the tenosynovium), may become painfully inflamed and swollen following injury or the repeated stress of athletic activity. These conditions are called *tendinitis* and *tenosynovitis,* respectively. The tendons most commonly affected are those associated with the joint capsules of the shoulder, elbow, and hip and those that move the wrist, hand, thigh, and foot.

Skeletal Muscle Fibers

A skeletal muscle fiber is a single cell that contracts in response to stimulation and then relaxes when the stimulation ends. Each skeletal muscle fiber is a thin, elongated cylinder with rounded ends, and it may extend the full length of the muscle. Just beneath its cell membrane (or *sarcolemma*), the cytoplasm (or *sarcoplasm*) of the fiber contains many small, oval nuclei and mitochondria (fig. 8.1). The sarcoplasm also contains many threadlike **myofibrils** (mi´´o-fi´brilz) that lie parallel to one another.

Myofibrils play a fundamental role in muscle contraction. They contain two kinds of protein filaments—thick ones composed of the protein **myosin** (mi´o-sin) and thin ones mainly composed of the protein **actin** (ak´tin)(figs. 8.2 and 8.3). (Two other thin filament proteins, troponin and tropomyosin, will be discussed later.) The organization of these filaments produces the characteristic alternating light and dark *striations*, or bands, of a skeletal muscle fiber.

 Muscle cells are stuffed with actin and myosin, but several other types of protein are important in contractibility too. Discovery of the gene that causes the two most common forms of muscular dystrophy took many years because the protein that is absent or incomplete, called dystrophin, comprises only 0.002 percent of the protein in skeletal muscle.

The striation pattern of skeletal muscle fibers has two main parts. The first, the *I bands* (the light bands),

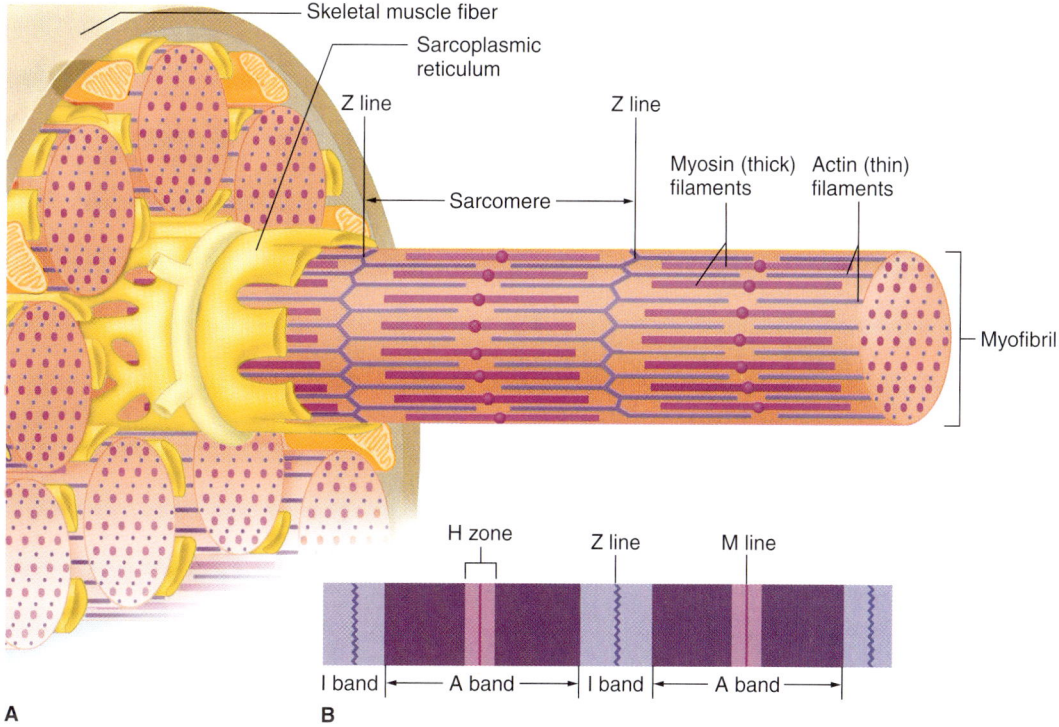

Figure 8.2
Skeletal muscle fiber. (*A*) A skeletal muscle fiber contains numerous myofibrils, each consisting of (*B*) repeating units called sarcomeres. The characteristic striations of a sarcomere are due to the arrangement of actin and myosin filaments.

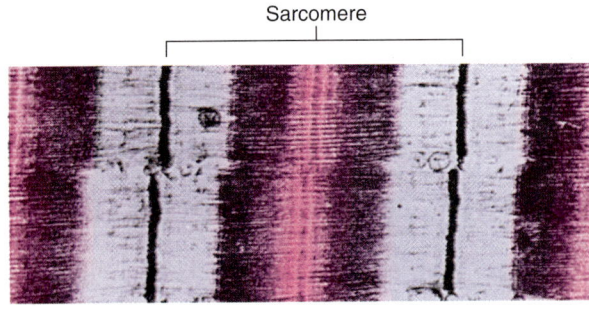

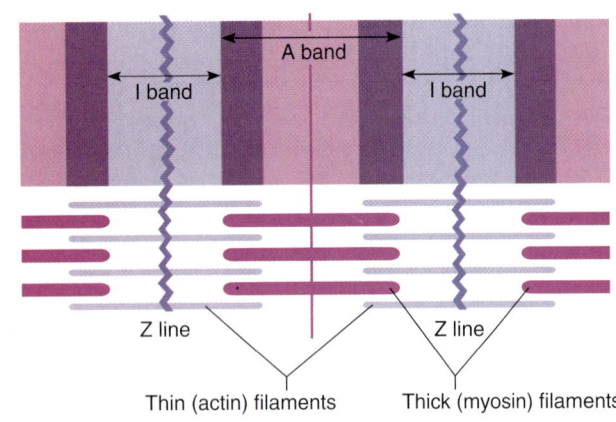

Figure 8.3
A sarcomere (16,000×).

are composed of thin actin filaments directly attached to structures called *Z lines*.

The second part of the striation pattern consists of the *A bands* (the dark bands), which are composed of thick myosin filaments overlapping thin actin filaments. The A band consists not only of a region where the thick and thin filaments overlap, but also a central region (*H zone*) consisting only of thick filaments, plus a thickening known as the *M line* (fig. 8.2). The segment of a myofibril that extends from one Z line to the next Z line is called a **sarcomere** (sar´ko-mēr) (figs. 8.2 and 8.3).

Within the sarcoplasm of a muscle fiber is a network of membranous channels that surrounds each myofibril and runs parallel to it (fig. 8.4). These membranes form the **sarcoplasmic reticulum,** which corresponds to the endoplasmic reticulum of other cells. Another set of membranous channels, called **transverse tubules** (T tubules), extends inward as invaginations from the fiber's membrane and passes all the way through the fiber. Thus, each tubule opens to the outside of the muscle fiber and contains extracellular fluid. Furthermore, each transverse tubule lies between two enlarged portions of the sarcoplasmic reticulum called *cisternae,* near the region where the actin and myosin filaments overlap. The sarcoplasmic reticulum and transverse tubules activate the muscle contraction mechanism when the fiber is stimulated.

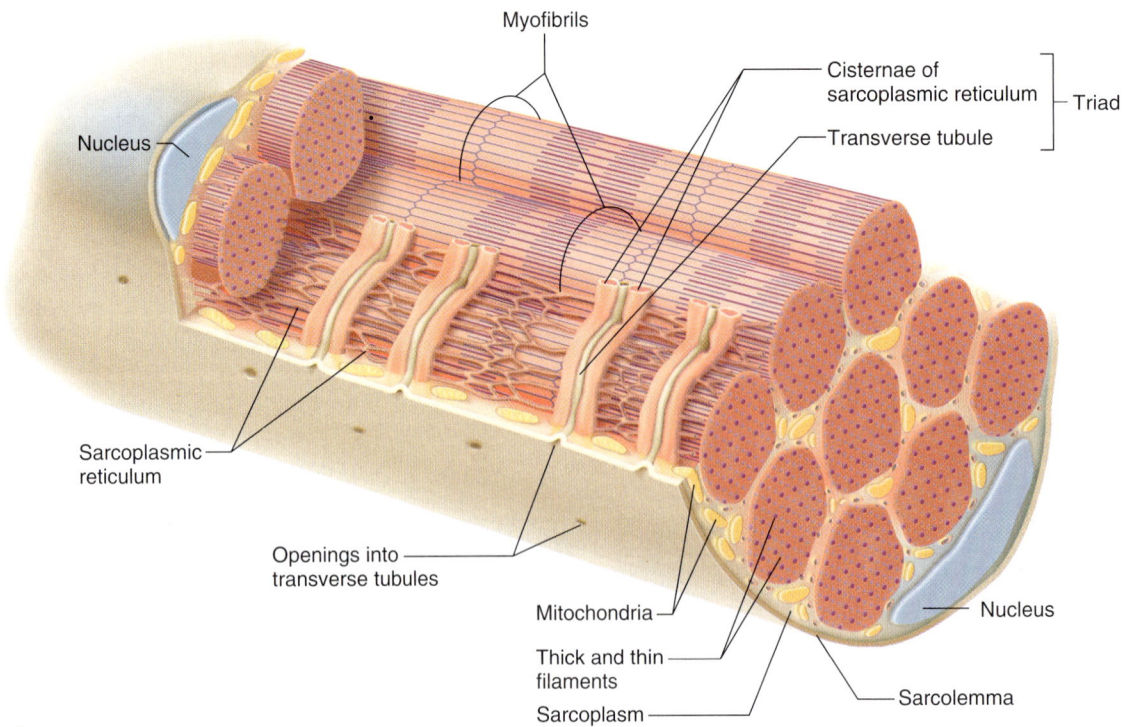

Figure 8.4
Within the sarcoplasm of a skeletal muscle fiber are a network of sarcoplasmic reticulum and a system of transverse tubules.

Muscle fibers and their associated connective tissues are flexible but can tear if overstretched. This type of injury, common in athletes, is called *muscle strain*. The seriousness of the injury depends on the degree of damage the tissues sustain. A mild strain injures only a few muscle fibers, the fascia remains intact, and loss of function is minimal. In a severe strain, however, many muscle fibers as well as the fascia tear, and muscle function may be completely lost. Such a severe strain is painful and produces discoloration and swelling.

CHECK YOUR RECALL

1. Describe how connective tissue is part of a skeletal muscle.
2. Describe the general structure of a skeletal muscle fiber.
3. Explain why skeletal muscle fibers appear striated.
4. Explain the relationship between the sarcoplasmic reticulum and the transverse tubules.

Neuromuscular Junction

Each skeletal muscle fiber connects to an axon from a nerve cell, called a **motor neuron** (mo´tor nu´ron). This axon extends outward from the brain or spinal cord, and the muscle fiber contracts only when the motor neuron stimulates it.

The connection between the motor neuron and muscle fiber is called a **neuromuscular junction.** Here, the muscle fiber membrane is specialized to form a **motor end plate.** In this region of the muscle fiber, nuclei and mitochondria are abundant, and the cell membrane (sarcolemma) is extensively folded (fig. 8.5).

The end of the motor neuron branches and projects into recesses of the muscle fiber membrane. The cytoplasm at the distal ends of these motor neuron axons is rich in mitochondria and contains many tiny vesicles (synaptic vesicles) that store chemicals called **neurotransmitters** (nu´´ro-trans´mit-erz).

When a nerve impulse traveling from the brain or spinal cord reaches the end of a motor neuron axon, some of the vesicles release a neurotransmitter into the gap (synaptic cleft) between the neuron and the motor end plate of the muscle fiber. This action stimulates the muscle fiber to contract.

Motor Units

A muscle fiber usually has a single motor end plate. The axons of motor neurons, however, are densely branched. By means of these branches, one motor neuron may connect to many muscle fibers. When a motor neuron transmits an impulse, all of the muscle fibers it links to are stimulated to contract simultaneously. Together, a motor neuron and the muscle fibers that it controls constitute a **motor unit** (mo´tor u´nit) (fig. 8.6).

CHECK YOUR RECALL

1. Which two structures approach each other at a neuromuscular junction?
2. Describe a motor end plate.
3. What is the function of a neurotransmitter?
4. What is a motor unit?

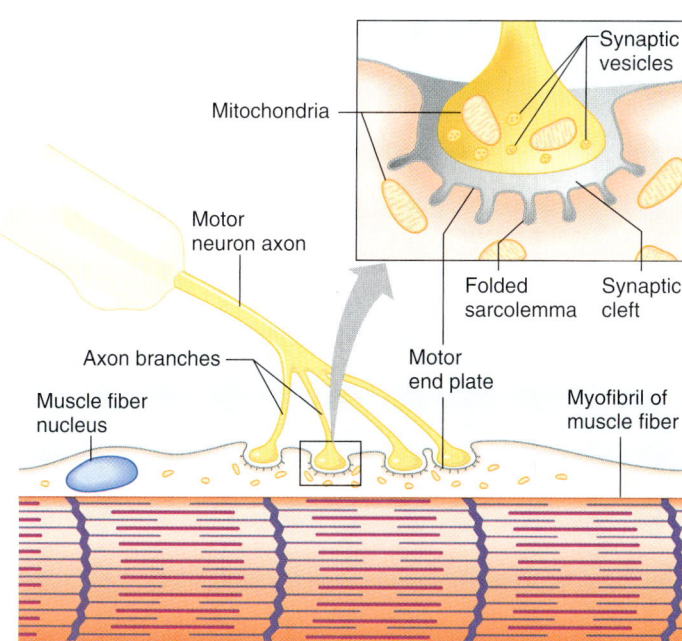

Figure 8.5
A neuromuscular junction includes the end of a motor neuron and the motor end plate of a muscle fiber.

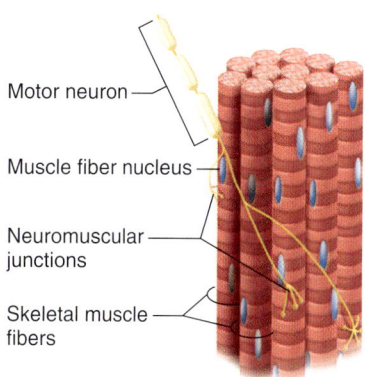

Figure 8.6
Muscle fibers within a motor unit may be distributed throughout the muscle.

8.3 Skeletal Muscle Contraction

A muscle fiber contraction is a complex interaction of organelles and molecules. The result is a movement within the myofibrils in which the filaments of actin and myosin slide past one another. This action shortens the muscle fiber so that it pulls on its attachments.

Role of Myosin and Actin

A myosin molecule is composed of two twisted protein strands with globular parts called cross-bridges projecting outward along their lengths. Many of these molecules comprise a myosin filament. An actin molecule is a globular structure with a binding site to which the myosin cross-bridges can attach. Many actin molecules twist into a double strand (helix), forming an actin filament. The proteins **troponin** and **tropomyosin** are also part of the actin filament (fig. 8.7).

According to the **sliding filament model**, sarcomeres shorten because cross-bridges pull on the thin filaments. A myosin cross-bridge can attach to an actin binding site and bend slightly, pulling on the actin filament. Then the head can release, straighten, combine with another binding site further down the actin filament, and pull again.

The globular portions of the myosin filaments contain an enzyme, **ATPase**, which catalyzes the breakdown of ATP to ADP and phosphate, releasing energy (see chapter 4, p. 78) that puts the myosin cross-bridge in a "cocked" position. When a cocked cross-bridge binds to actin, it pulls on the thin filament. After the cross-bridge pulls, another ATP binding to the cross-bridge causes it to be released from actin even before the ATP splits.

Presumably, this cycle repeats, as long as ATP is available as an energy source and as long as the muscle fiber is stimulated to contract. As the cross-bridges pull, the actin filament moves toward the center of the sarcomere, and the sarcomere shortens (figs. 8.8 and 8.9).

Stimulus for Contraction

A skeletal muscle fiber normally does not contract until a neurotransmitter stimulates it. In skeletal muscle, the neurotransmitter is **acetylcholine.** This neurotransmitter is synthesized in the cytoplasm of the motor neuron and stored in vesicles at the distal end of the motor neuron axons. When a nerve impulse (described in chapter 9, p. 213) reaches the end of a motor neuron axon, some of the vesicles release their acetylcholine into the space (synaptic cleft) between the motor neuron axon and the motor end plate (see fig. 8.5).

Acetylcholine diffuses rapidly across the synaptic cleft and combines with certain protein molecules (receptors) in the muscle fiber membrane, stimulating a **muscle impulse** (mus´el im´puls), which is very much like a nerve impulse. The impulse passes in all directions over the surface of the muscle fiber membrane and travels through the transverse tubules, deep into the fiber, and reaches the sarcoplasmic reticulum (see fig. 8.4).

The sarcoplasmic reticulum contains a high concentration of calcium ions. In response to a muscle impulse, the membranes of the cisternae become more permeable to these ions, and the calcium ions diffuse into the sarcoplasm of the muscle fiber.

When a high concentration of calcium ions is present in the sarcoplasm, troponin and tropomyosin interact in a way that exposes binding sites on actin. As a result, linkages form between the actin and myosin filaments, and the muscle fiber contracts (see figs. 8.8 and 8.9). The contraction, which also requires ATP, continues as long as nerve impulses release acetylcholine. When the nerve impulses cease, two events lead to muscle relaxation.

First, the acetylcholine that stimulated the muscle fiber is rapidly decomposed by the enzyme **acetylcholinesterase.** This enzyme is present at the neuromuscular junction on the membranes of the motor end

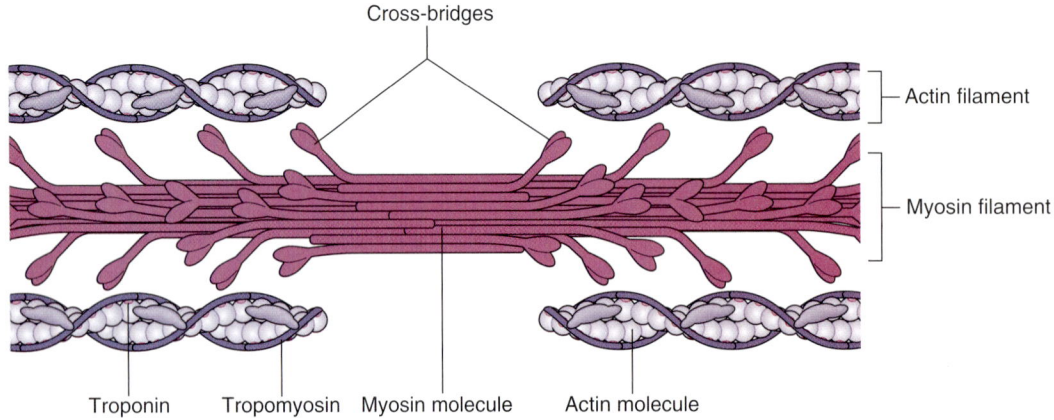

Figure 8.7
Thick filaments are composed of the protein myosin, and thin filaments are composed of actin. Myosin molecules have cross-bridges that extend toward nearby actin filaments.

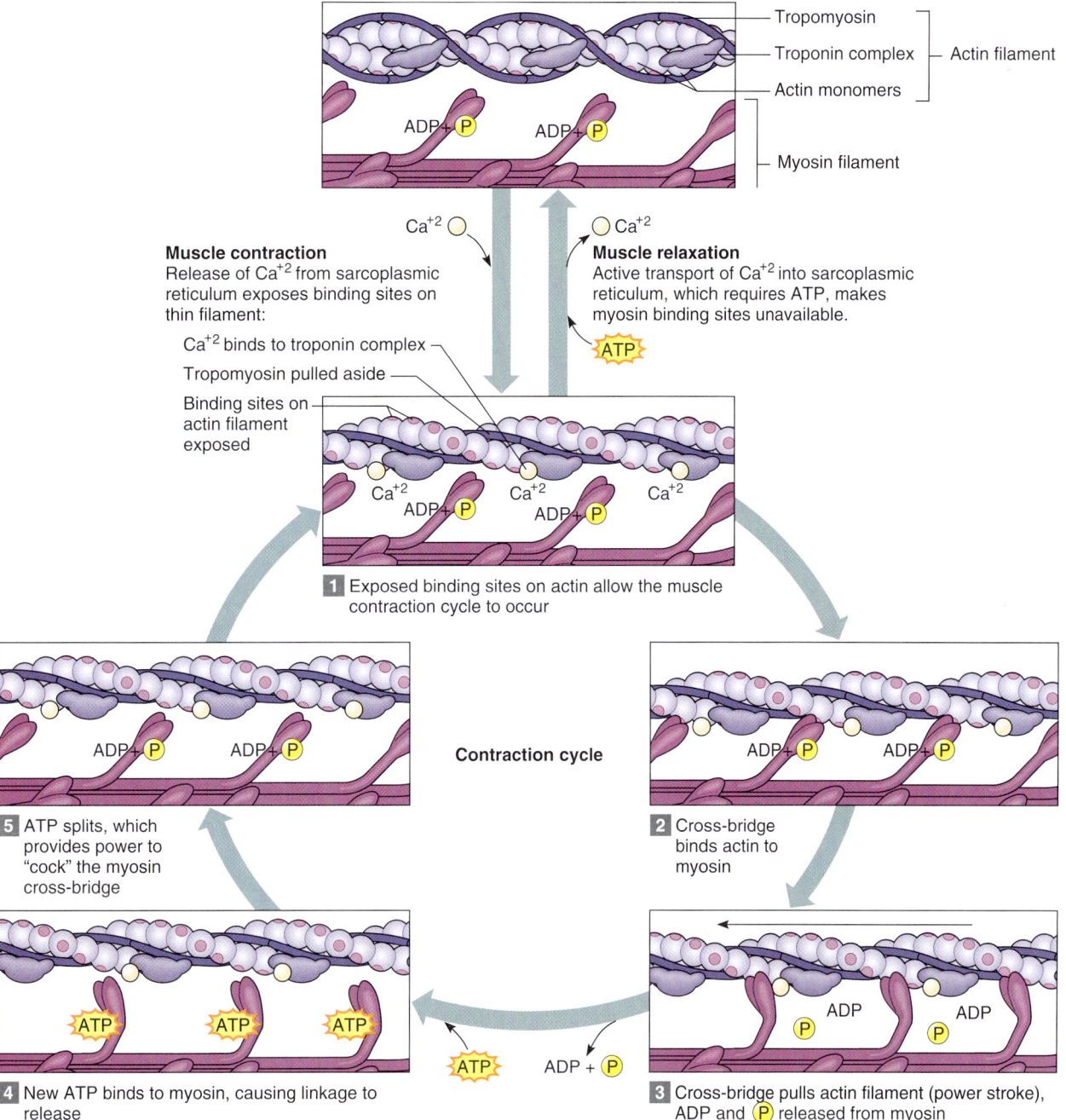

Figure 8.8
According to the sliding filament theory, (*1–2*) When calcium ion concentration rises, binding sites on actin filaments open, and cross-bridges attach. (*3*) Upon binding to actin, cross-bridges spring from the cocked position and pull on actin filaments. (*4*) ATP binds to the cross-bridge (but is not yet broken down), causing it to release from the actin filament. (*5*) ATP breakdown provides energy to "cock" the unattached myosin cross-bridge. As long as ATP and calcium ions are present, the cycle continues. When calcium ion concentration is low, the muscle remains relaxed.

plate. Acetylcholinesterase prevents a single nerve impulse from continuously stimulating the muscle fiber.

When acetylcholine is broken down, the stimulus to the muscle fiber ceases, allowing the second event in relaxation to occur. Calcium ions are actively transported back into the sarcoplasmic reticulum, which decreases the calcium ion concentration of the sarcoplasm. The linkages between actin and myosin filaments break, and consequently, the muscle fiber relaxes. Table 8.1 summarizes the major events leading to muscle contraction and relaxation.

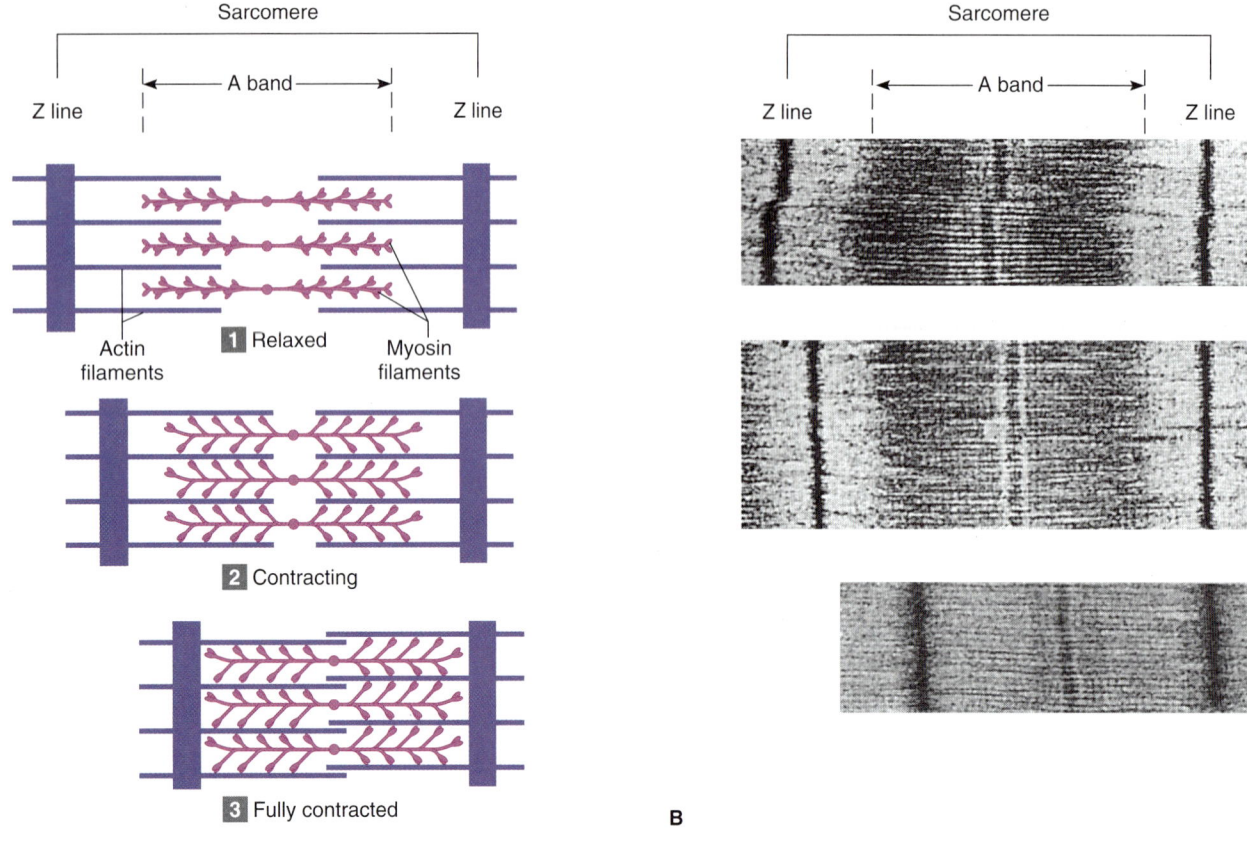

Figure 8.9
When a skeletal muscle contracts, (A) individual sarcomeres shorten as thick and thin filaments slide past one another. (B) Transmission electron micrograph showing a sarcomere shortening during muscle contraction (23,000×).

The bacterium *Clostridium botulinum* produces a poison, called botulinum toxin, that can prevent the release of acetylcholine from motor neuron axons at neuromuscular junctions, causing *botulism*, a very serious form of food poisoning. This condition is most likely to result from eating home-processed food that has not been heated enough to kill the bacteria in it or to inactivate the toxin.

Botulinum toxin blocks stimulation of muscle fibers, paralyzing muscles, including those responsible for breathing. Without prompt medical treatment, the fatality rate for botulism is high.

CHECK YOUR RECALL

1. Describe a neuromuscular junction.
2. Define *motor unit*.
3. Explain how the filaments of a myofibril interact during muscle contraction.
4. Explain how a motor nerve impulse can trigger a muscle contraction.

TABLE 8.1 — MAJOR EVENTS OF MUSCLE CONTRACTION AND RELAXATION

MUSCLE FIBER CONTRACTION	MUSCLE FIBER RELAXATION
1. The distal end of a motor neuron releases acetylcholine.	1. Acetylcholinesterase decomposes acetylcholine, and the muscle fiber membrane is no longer stimulated.
2. Acetylcholine diffuses across the gap at the neuromuscular junction.	2. Calcium ions are actively transported into the sarcoplasmic reticulum.
3. The muscle fiber membrane is stimulated, and a muscle impulse travels deep into the fiber through the transverse tubules and reaches the sarcoplasmic reticulum.	3. ATP causes linkages between actin and myosin filaments to break without being broken down itself.
4. Calcium ions diffuse from the sarcoplasmic reticulum into the sarcoplasm and bind to troponin molecules.	4. Troponin and tropomyosin interact, blocking binding sites on actin.
5. Troponin and tropomyosin interact to expose binding sites on actin.	5. The muscle fiber relaxes.
6. Actin and myosin filaments form linkages.	6. ATP breakdown "cocks" myosin cross-bridges. The muscle fiber remains ready for further stimulation.
7. Myosin cross-bridges pull actin filaments inward.	
8. The muscle fiber shortens as a contraction occurs.	

Energy Sources for Contraction

ATP molecules supply the energy for muscle fiber contraction. However, a muscle fiber has only enough ATP to contract for a very short time, so when a fiber is active, ATP must be regenerated.

The initial source of energy available to a contracting muscle comes from existing ATP molecules in the cell. Almost immediately, however, cells must regenerate ATP from ADP and phosphate. The molecule that makes this possible is **creatine phosphate.** Like ATP, creatine phosphate contains high-energy phosphate bonds, and it is four to six times more abundant in muscle fibers than ATP. Creatine phosphate, however, cannot directly supply energy to a cell's energy-utilizing reactions. Instead, it stores excess energy released from the mitochondria. When ATP supply is sufficient, an enzyme in the mitochondria (creatine phosphokinase) catalyzes the synthesis of creatine phosphate, which stores excess energy in its phosphate bonds (fig. 8.10).

As ATP decomposes, the energy from creatine phosphate can be transferred to ADP molecules, converting them back into ATP. Active muscle, however, rapidly exhausts the supply of creatine phosphate. When this happens, the muscle fibers use cellular respiration of glucose as an energy source for synthesizing ATP.

Oxygen Supply and Cellular Respiration

As chapter 4 describes (p. 77), glycolysis can take place in the absence of oxygen. The more complete breakdown of glucose, however, occurs in the mitochondria and requires oxygen. The blood carries the oxygen required to support this aerobic respiration from the lungs to body cells. Red blood cells carry the oxygen, loosely bound to molecules of **hemoglobin,** the pigment responsible for the red color of blood.

Another pigment, **myoglobin,** is synthesized in muscle cells and imparts the reddish-brown color of skeletal muscle tissue. Like hemoglobin, myoglobin can combine loosely with oxygen. This ability to temporarily store oxygen reduces a muscle's requirement for a continuous blood supply during muscular contraction (fig. 8.11).

Oxygen Debt

When a person is resting or is moderately active, the respiratory and circulatory systems can usually supply sufficient oxygen to skeletal muscles to support aerobic respiration. However, this is not the case when skeletal muscles are used strenuously for even a minute or two. In this situation, muscle fibers must increasingly use anaerobic respiration to obtain energy.

In one form of anaerobic respiration, glucose molecules are broken down by glycolysis to yield *pyruvic acid* (see chapter 4, p. 77). Because the oxygen supply is low, however, the pyruvic acid reacts to produce *lactic acid* that may accumulate in the muscles (fig. 8.11). Lactic acid diffuses into the bloodstream and eventually reaches the liver. In liver cells, reactions requiring ATP synthesize glucose from lactic acid.

During strenuous exercise, available oxygen is used primarily to synthesize the ATP the muscle fiber requires to contract, rather than to make ATP for synthesizing glucose from lactic acid. Consequently, as lactic acid accumulates, a person develops an **oxygen debt** (ok´sĭ-jen det) that must be repaid. Oxygen debt equals the amount of oxygen liver cells require to convert the accumulated lactic acid into glucose, plus the amount muscle cells require to restore ATP and creatine phosphate to their original concentrations.

The conversion of lactic acid back into glucose is slow. Repaying an oxygen debt following vigorous exercise may take several hours.

The metabolic capacity of a muscle may change with training. With high-intensity exercise that depends more on glycolysis for ATP, a muscle synthesizes more glycolytic enzymes, and its capacity for glycolysis increases. With aerobic exercise, more capillaries and mitochondria form, and the muscle's capacity for aerobic respiration is greater.

Muscle Fatigue

A muscle exercised strenuously for a prolonged period may lose its ability to contract, a condition called *fatigue.* Interruption in the muscle's blood supply or, rarely, lack of acetylcholine in motor neuron axons may cause fatigue. However, fatigue is most likely to arise from accumulation of lactic acid in the muscle as a

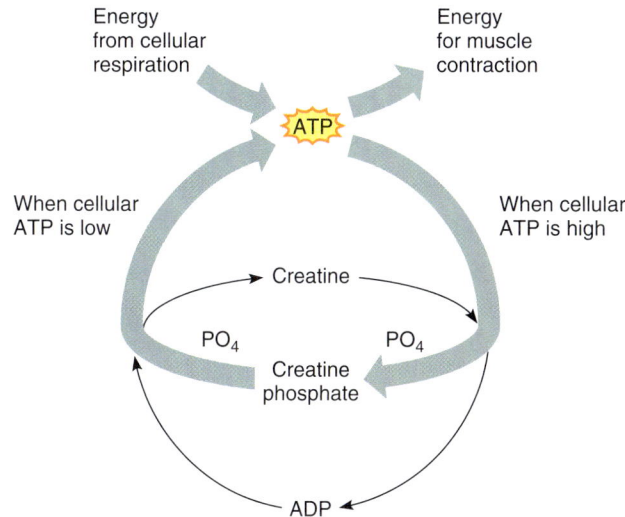

Figure 8.10
A muscle cell uses energy released in cellular respiration to synthesize ATP. ATP is then used to power muscle contraction or to synthesize creatine phosphate. Later, creatine phosphate may be used to synthesize ATP.

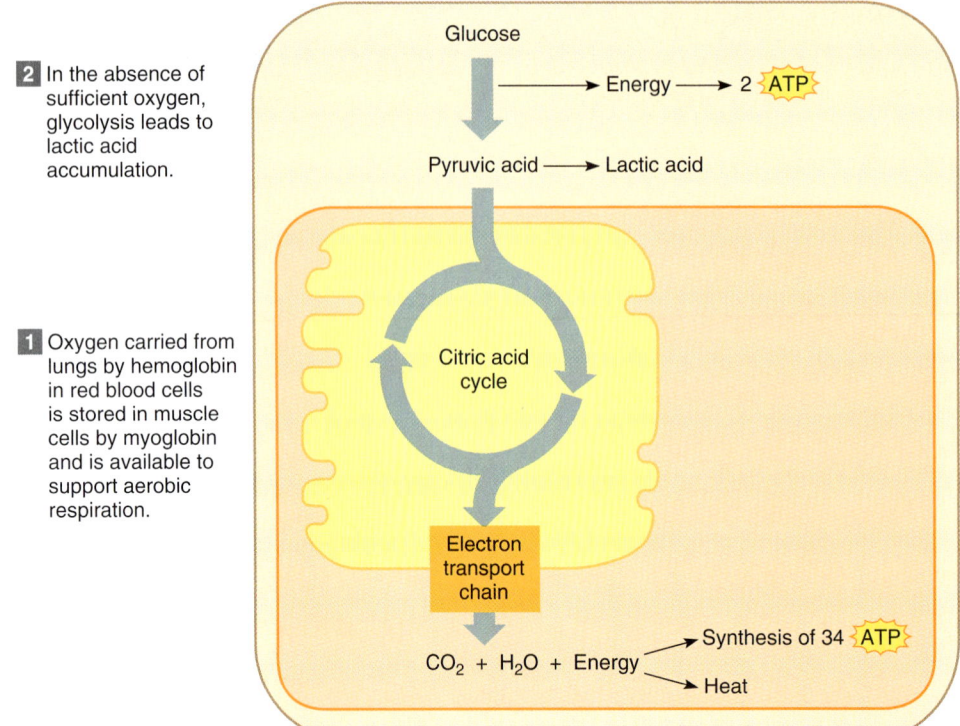

Figure 8.11
The oxygen required to support aerobic respiration is carried in the blood and stored in myoglobin. In the absence of sufficient oxygen, pyruvic acid is converted to lactic acid by anaerobic respiration. The maximum number of ATPs generated per glucose molecule varies with cell type; in skeletal muscle, it is 36 (2 + 34).

result of anaerobic respiration. The lactic acid buildup lowers pH, and as a result, muscle fibers no longer respond to stimulation.

Occasionally, a muscle becomes fatigued and cramps at the same time. A cramp is a painful condition in which a muscle undergoes a sustained involuntary contraction. Cramps are thought to occur when changes in the extracellular fluid surrounding the muscle fibers and their motor neurons somehow trigger uncontrolled stimulation of the muscle.

> Several hours after death, the skeletal muscles undergo a partial contraction that fixes the joints. This condition, *rigor mortis*, may continue for 72 hours or more. It results from an increase in membrane permeability to calcium ions and a decrease in ATP in muscle fibers, which prevents relaxation. Thus, the actin and myosin filaments of the muscle fibers remain linked until the muscles begin to decompose.

Heat Production

Less than half of the energy released in cellular respiration is available for use in metabolic processes; the rest becomes heat. Although all active cells generate heat, muscle tissue is a major heat source because muscle is such a large proportion of the total body mass. Blood transports heat generated in muscle to other tissues, which helps maintain body temperature.

 CHECK YOUR RECALL

1. Which biochemicals provide the energy to regenerate ATP?
2. What are the sources of oxygen for aerobic respiration?
3. How are lactic acid, oxygen debt, and muscle fatigue related?
4. What is the relationship between cellular respiration and heat production?

8.4 Muscular Responses

One way to observe muscle contraction is to remove a single muscle fiber from a skeletal muscle and connect it to a device that records changes in the fiber's length. Such experiments usually require the use of an electrical device that can produce stimuli of varying strengths and frequencies.

Threshold Stimulus

When an isolated muscle fiber is exposed to a series of stimuli of increasing strength, the fiber remains unresponsive until a certain strength of stimulation is applied.

Topic of Interest

STEROIDS AND ATHLETES—
AN UNHEALTHY COMBINATION

Canadian track star Ben Johnson flew past his competitors in the 100-meter run at the 1988 Summer Olympics in Seoul, Korea. But 72 hours later, officials rescinded the gold medal awarded for his record-breaking time of 9.79 seconds after a urine test revealed traces of the anabolic steroid drug stanozolol (fig. 8A).

Stanozolol is one of several synthetic versions of the steroid hormone testosterone. Like testosterone, these drugs promote signs of masculinity (their androgenic effect) and increased synthesis of muscle proteins (their anabolic effect).

Athletes call anabolic steroids 'roids, juice, pump, or hype. Abusers may take one large dose to obtain instant strength, or gradually increase the dose, a strategy called pyramiding. In a different approach called stacking, an athlete combines steroid types.

Steroid abusers may improve their performances and physiques in the short term, but in the long run, they may suffer. Steroids hasten maturation, stunting height and causing early baldness. In males, steroid abuse leads to breast development. In females steroids cause a deepening of the voice, mood swings, acne, tendons and ligaments that tear as muscles overgrow, and replacement of fat-padded curves with a more masculine physique. In both sexes, steroid abuse may damage the kidneys, liver, and heart, and atherosclerosis may develop because steroids raise LDL and lower HDL—the opposite of a healthy cholesterol profile. In males, the body mistakes the synthetic steroids for the natural hormone and lowers its own testosterone production. The price of athletic prowess today may be infertility later. Steroid abuse also causes psychiatric symptoms, including depression, delusions, and violent tendencies, sometimes called 'roid rage.

Anabolic steroids were created for medical purposes. They were first used clinically in the 1930s to treat underdevelopment of the testes and the resulting deficit in testosterone. In the 1950s, physicians used anabolic steroids to treat anemia, muscle-wasting disorders, and to bulk up patients whose muscles had atrophied from extended bed rest. In the 1960s, anabolic steroids were used to treat some forms of short stature and dwarfism, which was discontinued when pure preparations of human growth hormone became available using recombinant DNA technology. Today, anabolic steroids are being studied for their use in treating the wasting associated with AIDS.

Steroid abuse began in Nazi Germany, where Hitler used the drugs to fashion his "super race." Ironically, steroids were used shortly after to build up the emaciated bodies of concentration camp survivors. In the 1950s, Soviet athletes began using steroids in the Olympics, and a decade later, U.S. athletes did the same. In 1976, the International Olympic Committee banned steroid use and required urine tests to detect the drugs.

Such a test caught Ben Johnson in his tracks. A urine test can detect part-per-billion traces of synthetic steroids even weeks after they are taken. Johnson at first claimed the stanozolol in his urine was the result of a spiked drink of an approved anti-inflammatory drug used on his ankle, but a test showed his natural testosterone level to be only 15% of normal—a sure sign that this athlete had been taking steroids for a long time. But he was not the last Olympic athlete to abuse steroids and get caught. In the 1992 summer games in Barcelona, Spain, several athletes were dismissed for using drugs that they thought would have steroid-like effects. And in the 2000 summer games, a urine test on U.S. shot-putter C.J. Hunter revealed 1000 times the allowable limit of nandrolone, a breakdown product of testosterone. Today, about 30 percent of college and professional athletes use anabolic steroids, as do 10 to 20 percent of high school athletes. Use is highest among bodybuilders, shot putters, discus throwers, wrestlers, and swimmers.

Figure 8A

Canadian track star Ben Johnson ran away with the gold medal in the 100-meter race at the 1988 Summer Olympics—but had to return it after a urine test revealed traces of a steroid drug.

This minimal strength required to cause a contraction is called the **threshold stimulus** (thresh´old stim´u-lus). An impulse in a motor neuron normally releases enough acetylcholine to bring the muscle fibers in its motor unit to threshold.

All-or-None Response

A skeletal muscle fiber exposed to a stimulus of threshold strength (or above) responds to its fullest extent. Increasing the strength of the stimulus does not affect the fiber's degree of contraction. In other words, a skeletal muscle fiber normally does not contract partially; if it contracts at all, it contracts fully. (The extent of shortening depends, of course, on the resistance.) This phenomenon is called the **all-or-none response.**

Recording a Muscle Contraction

A skeletal muscle removed from a frog or other small animal can show how a whole muscle responds to stimulation. The muscle is stimulated electrically, and when it contracts, its movement is recorded. The resulting pattern is called a **myogram.**

If a muscle is exposed to a single stimulus of sufficient strength to activate some of its motor units, the muscle will contract and then relax. This action—a single contraction that lasts only a fraction of a second—is called a **twitch.** A twitch produces a myogram like that in figure 8.12. The delay between the time the stimulus was applied and the time the muscle responded is the **latent period.** In a frog muscle, the latent period lasts about 0.01 second; in a human muscle, it is even shorter. The latent period is followed by a *period of contraction* when the muscle pulls at its attachments and a *period of relaxation* when it returns to its former length. The Topic of Interest box on page 183 describes two types of twitches—the fatigue-resistent slow twitch and the fatiguable fast twitch. Muscle fibers are either slow twitch or fast twitch.

 The skeletal muscles of an average person contain about half fast twitch and half slow twitch muscle fibers. In contrast, the muscles of an Olympic sprinter typically have more than 80 percent fast twitch muscle fibers, and those of an Olympic marathoner, more than 90 percent slow twitch muscle fibers.

Summation

The force that a muscle fiber can generate is not limited to the maximum force of a single twitch. A muscle fiber exposed to a series of stimuli of increasing frequency reaches a point when it is unable to completely relax before the next stimulus in the series arrives. When this happens, the force of individual twitches combines by the process of **summation.** When the resulting forceful, sustained contraction lacks even partial relaxation, it is called a **tetanic** (tĕ-tan´ik) **contraction,** or tetanus (fig. 8.13).

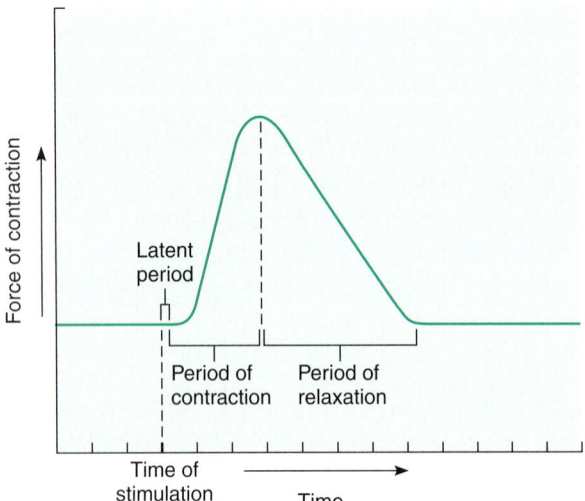

Figure 8.12
A myogram of a single muscle twitch.

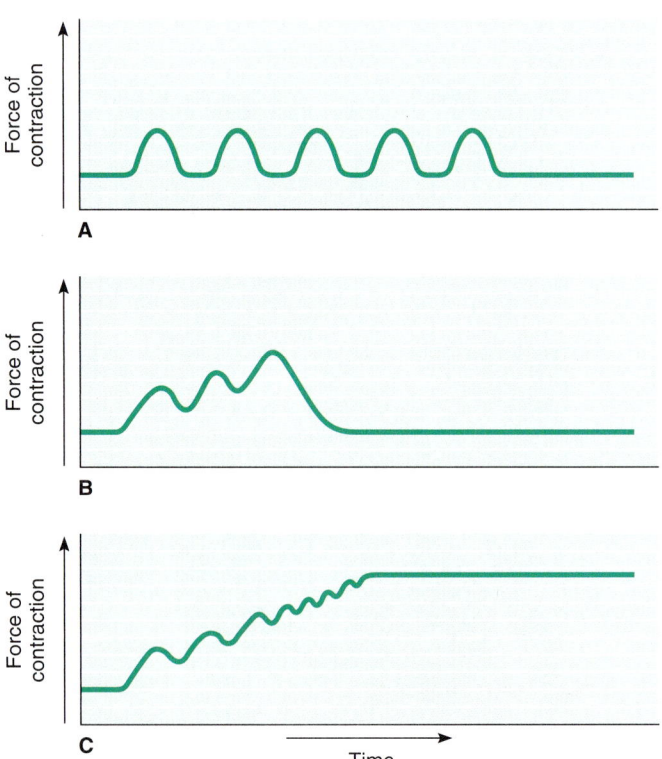

Figure 8.13
Myograms of (A) a series of twitches, (B) summation, and (C) a tetanic contraction. Note that stimulation frequency increases from one myogram to the next.

Topic of Interest

USE AND DISUSE OF SKELETAL MUSCLES

Skeletal muscles are very responsive to use and disuse. Forcefully exercised muscles enlarge, a phenomenon called *muscular hypertrophy*. Conversely, an unused muscle undergoes *atrophy*, decreasing in size and strength.

The way a muscle responds to use also depends on the type of exercise. A muscle contracting weakly, such as during swimming and running, activates a specialized group of muscle fibers called *slow fibers*, which are fatigue-resistant. With use, these specialized muscle fibers develop more mitochondria, and more extensive capillary networks envelop them. Such changes increase slow fibers' ability to resist fatigue during prolonged exercise, although their sizes and strengths may remain unchanged.

Forceful exercise, such as weight lifting, in which a muscle exerts more than 75% of its maximum tension, utilizes another group of specialized muscle fibers called *fast fibers*, which are fatigable. In response to strenuous exercise, these fibers produce new filaments of actin and myosin, the diameters of the muscle fibers increase, and the entire muscle enlarges. However, the muscular hypertrophy does not produce new muscle fibers.

The strength of a muscular contraction is directly proportional to the diameter of the activated muscle fibers. Consequently, an enlarged muscle can produce stronger contractions than before. Such a change, however, does not increase the muscle's ability to resist fatigue during activities like swimming or running.

If regular exercise stops, the capillary networks shrink, and the number of mitochondria within the muscle fibers drops. The number of actin and myosin filaments decreases, and the entire muscle atrophies. Such atrophy commonly occurs when accidents or diseases interfere with motor nerve impulses and prevent them from reaching muscle fibers. An unused muscle may decrease to less than half its usual size within a few months.

The fibers of muscles whose motor neurons are severed not only shrink, but also may fragment and, in time, be replaced by fat or fibrous connective tissue. However, reinnervation within the first few months following an injury may restore muscle function.

Recruitment of Motor Units

Since the muscle fibers within a muscle are organized into motor units and a single motor neuron controls each motor unit, all the muscle fibers in a motor unit are stimulated at the same time. Therefore, a motor unit also responds in an all-or-none manner. A whole muscle, however, does not respond like this because it is composed of many motor units controlled by different motor neurons, which respond to different thresholds of stimulation. If only the motor neurons with low thresholds are stimulated, few motor units contract. At higher intensities of stimulation, other motor neurons respond, and more motor units are activated. Such an increase in the number of motor units being activated is called **recruitment** (re-krōōt´ment). As the intensity of stimulation increases, recruitment of motor units continues until, finally, all possible motor units in that muscle are activated and the muscle contracts with maximal tension.

Sustained Contractions

At the same time twitches combine, the strength of the contractions may increase due to the recruitment of motor units. The smaller motor units, which have finer fibers, are most easily stimulated and tend to respond earlier in the series of stimuli. The larger motor units, which contain thicker fibers, respond later and more forcefully. Summation and recruitment together can produce a *sustained contraction* of increasing strength.

Although twitches may occur occasionally in human skeletal muscles, as when an eyelid twitches, such contractions are of limited use. More commonly, muscular contractions are sustained. For example, lifting a weight or walking maintains sustained contractions in the upper limb or lower limb muscles for varying lengths of time. These contractions are responses to a rapid series of stimuli transmitted from the brain and spinal cord on motor neuron axons.

Even when a muscle appears to be at rest, a certain amount of sustained contraction is occurring in its fibers. This is called **muscle tone** (tonus). Muscle tone is a response to nerve impulses that originate repeatedly from the spinal cord and stimulate a few muscle fibers.

Muscle tone is particularly important in maintaining posture. If muscle tone is suddenly lost, as happens when a person loses consciousness, the body collapses.

When skeletal muscles contract very forcefully, they may generate up to 50 pounds of pull for each square inch of muscle cross section. Consequently, large muscles, such as those in the thigh, can pull with several hundred pounds of force. Occasionally, this force is so great that the tendons of muscles tear away from their attachments to the bones (*muscle pull*).

CHECK YOUR RECALL

1. Define *threshold stimulus*.
2. What is an all-or-none response?
3. Distinguish between a twitch and a sustained contraction.
4. How is muscle tone maintained?

8.5 Smooth Muscle

The contractile mechanism of smooth muscles is essentially the same as for skeletal muscles. The cells of smooth muscle, however, have some important structural and functional differences from the other types of muscle.

Smooth Muscle Fibers

Recall from chapter 5 (p. 105) that smooth muscle cells are elongated, with tapering ends. They contain filaments of actin and myosin in myofibrils that extend the lengths of the cells. However, these filaments are organized differently and more randomly than those in skeletal muscle. Consequently, smooth muscle cells lack striations. The sarcoplasmic reticulum in these cells is not well developed.

The two major types of smooth muscles are multiunit and visceral. In **multiunit smooth muscle,** the muscle fibers are separate rather than organized into sheets. Smooth muscle of this type is found in the irises of the eyes and in the walls of blood vessels. Typically, multiunit smooth muscle tissue contracts only in response to stimulation by motor nerve impulses or certain hormones.

Visceral smooth muscle is composed of sheets of spindle-shaped cells in close contact with one another (see fig. 5.22, p. 106). This more common type of smooth muscle is found in the walls of hollow organs, such as the stomach, intestines, urinary bladder, and uterus.

Fibers of visceral smooth muscles can stimulate each other. When one fiber is stimulated, the impulse moving over its surface may excite adjacent fibers, which in turn stimulate still others. Visceral smooth muscles also display *rhythmicity,* a pattern of repeated contractions. Rhythmicity is due to self-exciting fibers that deliver spontaneous impulses periodically into surrounding muscle tissue. These two features—transmission of impulses from cell to cell and rhythmicity—are largely responsible for the wavelike motion, called **peristalsis,** that occurs in certain tubular organs, such as the intestines, and helps force the contents of these organs along their lengths.

Smooth Muscle Contraction

Smooth muscle contraction resembles skeletal muscle contraction in a number of ways. Both mechanisms include reactions of actin and myosin, both are triggered by membrane impulses and an increase in intracellular calcium ions, and both use energy from ATP. These two types of muscle tissue, however, also have significant differences.

Recall that acetylcholine is the neurotransmitter in skeletal muscle. Two neurotransmitters affect smooth muscle—acetylcholine and norepinephrine. Each of these neurotransmitters stimulates contractions in some smooth muscles and inhibits contractions in others (see chapter 9, p. 217). Also, a number of hormones affect smooth muscle, stimulating contractions in some cases and altering the degree of response to neurotransmitters in others.

Smooth muscle is slower to contract and to relax than skeletal muscle. On the other hand, smooth muscle can maintain a forceful contraction longer with a given amount of ATP. Also, unlike skeletal muscle, smooth muscle fibers can change length without changing tautness; therefore, smooth muscles in the stomach and intestinal walls can stretch as these organs fill, maintaining the pressure inside these organs.

CHECK YOUR RECALL

1. Describe two major types of smooth muscle.
2. What special characteristics of visceral smooth muscle make peristalsis possible?
3. How does smooth muscle contraction differ from that of skeletal muscle?

8.6 Cardiac Muscle

Cardiac muscle is found only in the heart. Its mechanism of contraction is essentially the same as that of skeletal and smooth muscle, but with some important differences. Cardiac muscle is composed of striated cells joined end to end, forming fibers (see fig. 5.23, p. 106). These fibers interconnect in branching, three-dimensional networks. Each cell contains many filaments of actin and myosin, similar to those in skeletal muscle. A cardiac muscle cell also has a sarcoplasmic reticulum, many mitochondria, and a system of transverse tubules. The cisternae of cardiac muscle fibers, however, are less well developed and store less calcium than those of skeletal muscle. On the other hand, the transverse tubules of cardiac muscle are larger, and they release large numbers of calcium ions into the sarcoplasm in response to muscle impulses. This extra calcium from the transverse tubules comes from fluid outside the muscle fibers and causes cardiac muscle twitches to be longer than skeletal muscle twitches.

The opposing ends of cardiac muscle cells are connected by cross-bands called *intercalated discs.* These bands form from elaborate junctions between cell membranes. The discs help to join cells and to transmit the force of contraction from cell to cell. Intercalated discs also allow muscle impulses to pass freely so that they travel rapidly from cell to cell.

When one portion of the cardiac muscle network is stimulated, the resulting impulse passes to the other fibers of the network, and the whole structure contracts as a unit; that is, the network responds to stimulation in an all-or-none manner. Cardiac muscle is also self-exciting and rhythmic. Consequently, a pattern of contraction and relaxation repeats again and again and causes the rhythmic contractions of the heart.

Table 8.2 summarizes the characteristics of the three types of muscle tissue. The Genetics Connection on page 186 considers several inherited diseases that affect the muscular system.

CHECK YOUR RECALL

1. How is cardiac muscle similar to smooth muscle?
2. How is cardiac muscle similar to skeletal muscle?
3. What is the function of intercalated discs?
4. What characteristic of cardiac muscle causes contraction of the heart as a unit?

8.7 Skeletal Muscle Actions

Skeletal muscles provide a variety of body movements, as described in chapter 7 (p. 159). Each muscle's movement depends largely on the kind of joint it is associated with and the way the muscle attaches on either side of that joint.

Origin and Insertion

Recall that bones forming movable joints function as levers (see chapter 7, p. 133). One end of a skeletal muscle usually fastens to a relatively immovable or fixed part at a movable joint, and the other end connects to a movable part on the other side of that joint. The immovable end of the muscle is called its **origin** (or´ĭ-jin), and the movable one is its **insertion** (in-ser´shun). When a muscle contracts, its insertion is pulled toward its origin.

Some muscles have more than one origin or insertion. The *biceps brachii* in the arm, for example, has two origins. This is reflected in the name *biceps*, which means *two heads*. (Note: The head of a muscle is the part nearest its origin.) One head of the muscle attaches to the coracoid process of the scapula, and the other head arises from a tubercle above the glenoid cavity of the scapula. The muscle extends along the front surface of the humerus and is inserted by means of a tendon on the radial tuberosity of the radius. When the biceps brachii contracts, its insertion is pulled toward its origin, and the forearm flexes at the elbow (fig. 8.14).

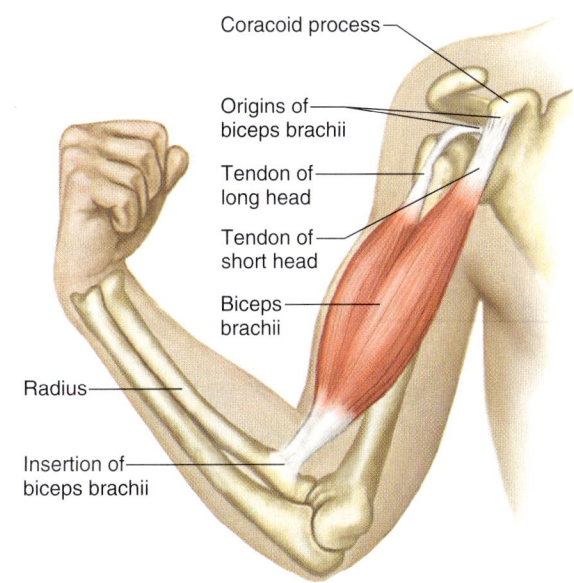

Figure 8.14
The biceps brachii has two heads that originate on the scapula. A tendon inserts this muscle on the radius.

TABLE 8.2   TYPES OF MUSCLE TISSUE

	SKELETAL	SMOOTH	CARDIAC
Major location	Skeletal muscles	Walls of hollow viscera, blood vessels	Wall of the heart
Major function	Movement of bones at joints, maintenance of posture	Movement of viscera, peristalsis, vasoconstriction	Pumping action of the heart
Cellular characteristics			
Striations	Present	Absent	Present
Nucleus	Many nuclei	Single nucleus	Single nucleus
Special features	Well-developed transverse tubule system	Lacks transverse tubules	Well-developed transverse tubule system; intercalated discs separating adjacent cells
Mode of control	Voluntary	Involuntary	Involuntary
Contraction characteristics	Contracts and relaxes rapidly	Contracts and relaxes slowly; self-exciting; rhythmic	Network of fibers contracts as a unit; self-exciting; rhythmic

Genetics Connection

INHERITED DISEASES OF MUSCLE

A variety of inherited conditions affect muscle tissue. These disorders differ in the nature of the genetic defect, the type of protein that is abnormal in form or function, and the particular muscles in the body that are impaired.

The Muscular Dystrophies—Missing Proteins

A muscle cell is packed with filaments of actin and myosin. Less abundant, but no less important, is a protein called *dystrophin*. It literally holds skeletal muscle cells together by linking actin in the cell to glycoproteins (called *dystrophin-associated glycoproteins*, or DAGs) that are part of the cell membrane. This helps attach the cell to the surrounding matrix. Missing or abnormal dystrophin or DAGs cause muscular dystrophies. These illnesses vary in severity and age of onset, but in all cases, muscles weaken and degenerate. Eventually, fat and connective tissue replace muscle.

Duchenne muscular dystrophy (DMD) is the most severe type of the illness (fig. 8B). Symptoms begin by age five and affect only boys. By age thirteen, the person cannot walk, and by early adulthood he usually dies from failure of the respiratory muscles. In DMD, dystrophin is often missing. In Becker muscular dystrophy, symptoms begin in early adulthood, are less severe, and result from underproduction of dystrophin. Limb-girdle muscular dystrophy causes weakness in the upper limbs, usually noticeable in a person's thirties. This form of muscular dystrophy is often the result of a missing or abnormal DAG, which causes the other DAGs to be deficient too.

Charcot-Marie-Tooth Disease—A Duplicate Gene

Charcot-Marie-Tooth disease causes a slowly progressing weakness in the muscles of the hands and feet and a decrease in tendon reflexes in these parts. In this illness, an extra gene impairs the insulating sheath around affected nerve cells, so that nerve cells cannot adequately stimulate the involved muscles. Symptoms resemble those of diverse other conditions, including AIDS, alcoholism, vitamin B_{12} deficiency, diabetes mellitus, and heavy metal poisoning. Therefore, physicians perform two tests—electromyography and nerve conduction velocities—to diagnose Charcot-Marie-Tooth disease. It is also possible to test for the gene mutation to establish a diagnosis.

Myotonic Dystrophy—An Expanding Gene

Myotonic dystrophy causes delayed muscle relaxation following contraction (myotonia), which causes facial and limb weakness, cataracts, and an irregular heartbeat. It is caused by inheriting either of two "expanding genes" that actually grow with each generation. As the gene enlarges, symptoms increase in severity or begin at an earlier age. For example, a grandfather might experience only mild weakness in his forearms, but his daughter might have more noticeable arm and leg weakness. By the third generation, affected children might suffer severe muscle impairment. For many years, physicians attributed the worsening of symptoms over generations to psychological causes. We know now there is a physical basis for the phenomenon. The expanded genes are actually transcribed into messenger RNA molecules that are too large to leave the nucleus, so that the proteins that they encode are not synthesized.

Hereditary Idiopathic Dilated Cardiomyopathy— A Tiny Glitch

This very rare inherited form of heart failure begins usually in a person's forties and is lethal in 50% of cases within five years of diagnosis, unless a heart transplant can be performed. The condition is caused by a tiny genetic error in a form of actin found only in cardiac muscle, where it is the predominant component of the thin filaments. Change of a single DNA building block (nucleotide base) apparently disturbs actin's ability to anchor to the Z bands in heart muscle cells. The mutation prevents actin from effectively transmitting the force of contraction, which gradually causes the heart chambers to enlarge and eventually to fail to function.

Figure 8B
This young man has Duchenne muscular dystrophy. The condition has not yet severely limited his activities, but he shows the hypertrophied (overdeveloped) calf muscles that result from his inability to rise from a sitting position the usual way—an early sign of the illness.

The movements termed "flexion" and "extension" describe changes in the angle between bones that meet at a joint. For example, flexion at the elbow refers to a movement of the forearm that bends the elbow, or decreases the angle. Alternatively, one could say that flexion at the elbow results from the action of the biceps brachii on the radius of the forearm.

Since students often find it helpful to think of movements in terms of the specific actions of the muscles involved, we may also describe flexion and extension in these terms. Thus, the action of the biceps brachii may be described as "flexion of the forearm at the elbow," and the action of the quadriceps group as "extension of the leg at the knee." We believe this occasional departure from strict anatomical terminology eases understanding and learning.

Interaction of Skeletal Muscles

Skeletal muscles almost always function in groups. Consequently, for a particular body movement to occur, a person must do more than contract a single muscle; instead, after learning to make a particular movement, the person wills the movement to occur, and the nervous system stimulates the appropriate group of muscles.

Careful observation of body movements indicates the special roles of muscles. For instance, when the upper limb is lifted horizontally away from the side, a contracting *deltoid* muscle provides most of the movement and is said to be the **prime mover** (prīm mōōv´er), also referred to as an **agonist** (ag´o-nist). However, while a prime mover is acting, certain nearby muscles are also contracting. In the case of the contracting deltoid muscle, nearby muscles help hold the shoulder steady and in this way make the prime mover's action more effective. Muscles that contract and assist the prime mover are called **synergists** (sin´er-jistz).

Still other muscles act as **antagonists** (an-tag´o-nistz) to prime movers. These muscles can resist a prime mover's action and cause movement in the opposite direction. For example, the antagonist of the prime mover that raises the upper limb can lower the upper limb, or the antagonist of the prime mover that bends the upper limb can straighten it (see fig. 7.6, p. 133). If both a prime mover and its antagonist contract simultaneously, the part they act upon remains rigid. Consequently, smooth body movements depend on antagonists relaxing and, thus, giving way to the prime movers whenever the prime movers contract. Once again, the nervous system controls these complex actions, as chapter 9 describes.

CHECK YOUR RECALL

1. Distinguish between the origin and insertion of a muscle.
2. Define *prime mover*.
3. What is the function of a synergist? An antagonist?

8.8 Major Skeletal Muscles

The section that follows discusses the locations, actions, and attachments of some of the major skeletal muscles. (Figures 8.15 and 8.16 and reference plates 1 and 2, pp. 22–23, show the locations of the superficial skeletal muscles—those near the surface.)

Note that the names of these muscles often describe them. A name may indicate a muscle's relative size, shape, location, action, number of attachments, or the direction of its fibers, as in the following examples:

pectoralis major Of large size (major) located in the pectoral region (chest).
deltoid Shaped like a delta or triangle.
extensor digitorum Extends the digits (fingers or toes).
biceps brachii Having two heads (biceps) or points of origin and located in the brachium (arm).
sternocleidomastoid Attached to the sternum, clavicle, and mastoid process.
external oblique Located near the outside with fibers that run obliquely (in a slanting direction).

Muscles of Facial Expression

A number of small muscles that lie beneath the skin of the face and scalp enable us to communicate feelings through facial expression (fig. 8.17). Many of these muscles, located around the eyes and mouth, are responsible for such expressions as surprise, sadness, anger, fear, disgust, and pain. As a group, the muscles of facial expression join the bones of the skull to connective tissue in various regions of the overlying skin. They include:

epicranius (ep´´ĭ-kra´ne-us) Composed of two parts, the *frontalis* (frun-ta´lis) and the *occipitalis* (ok-sip´´ĭ-ta´lis).
orbicularis oculi (or-bik´u-la-rus ok´u-li)
orbicularis oris (or-bik´u-la-rus o´ris)
buccinator (buk´sĭ-na´´tor)
zygomaticus (zi´´go-mat´ik-us)
platysma (plah-tiz´mah)

Table 8.3 lists the origins, insertions, and actions of the muscles of facial expression. (The muscles that move the eyes are listed in chapter 10, p. 269.)

 The human body contains more than 600 distinct skeletal muscles. Just the face includes 60 muscles, more than 40 of which are used to frown, and 20 to smile. Thinner than a thread and barely visible, the stapedius in the middle ear is the body's smallest muscle. In contrast is the gluteus maximus, the largest muscle, located in the buttock. The sartorius, which pulls on the thigh, is the longest muscle in the body.

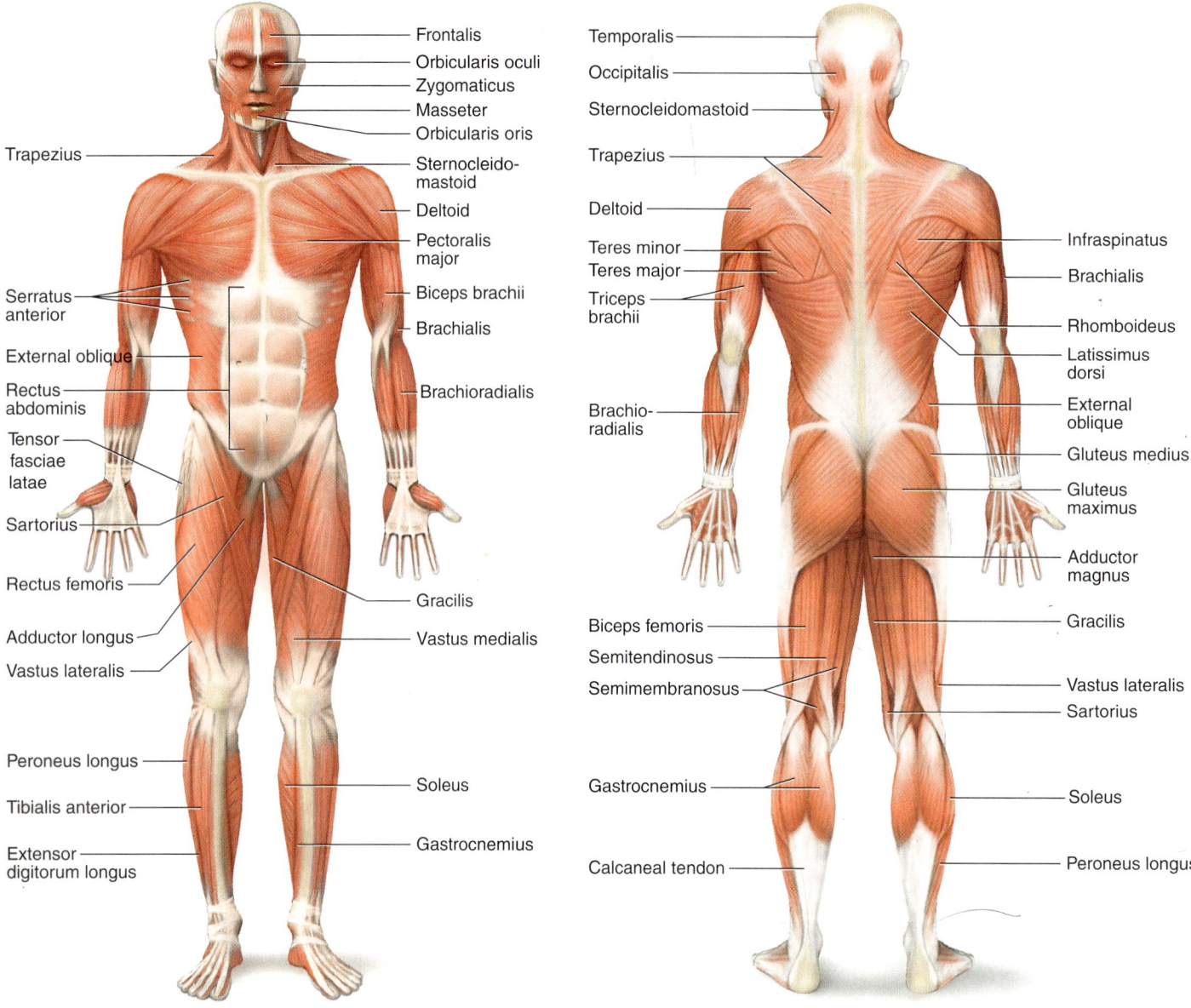

Figure 8.15
Anterior view of superficial skeletal muscles.

Figure 8.16
Posterior view of superficial skeletal muscles.

TABLE 8.3			MUSCLES OF FACIAL EXPRESSION
MUSCLE	ORIGIN	INSERTION	ACTION
Epicranius	Occipital bone	Skin and muscles around eye	Raises eyebrow
Orbicularis oculi	Maxillary and frontal bones	Skin around eye	Closes eye
Orbicularis oris	Muscles near the mouth	Skin of lips	Closes and protrudes lips
Buccinator	Outer surfaces of maxilla and mandible	Orbicularis oris	Compresses cheeks inward
Zygomaticus	Zygomatic bone	Orbicularis oris	Raises corner of mouth
Platysma	Fascia in upper chest	Lower border of mandible	Draws angle of mouth downward

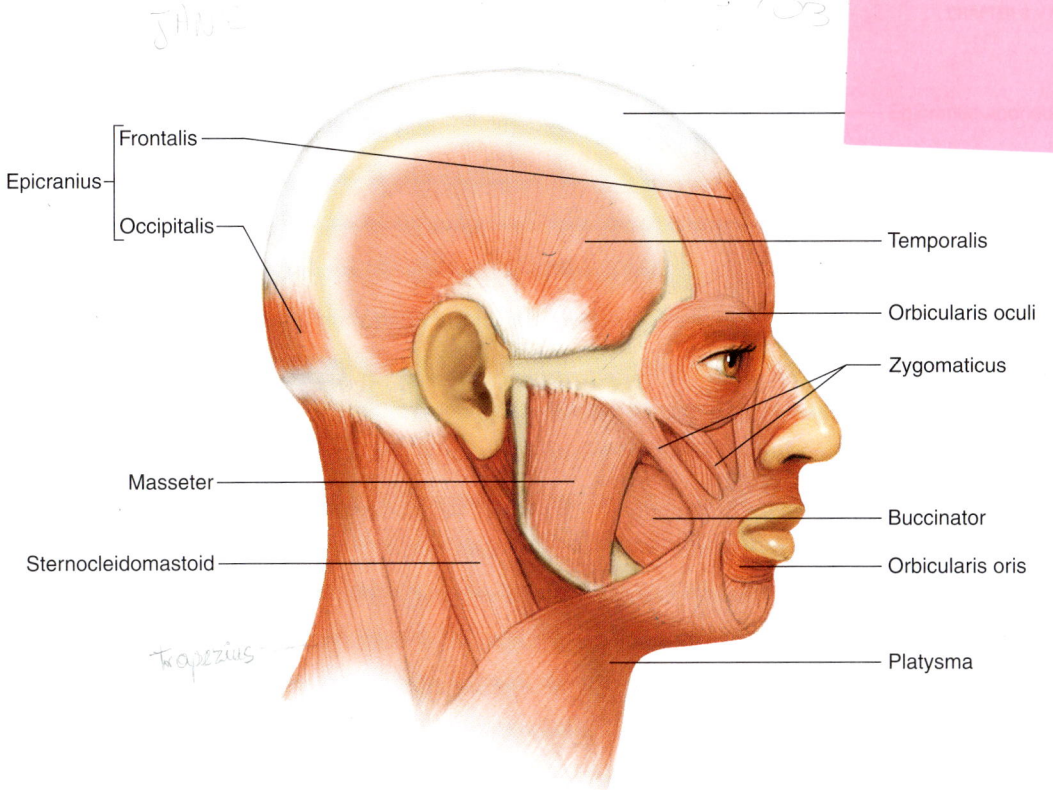
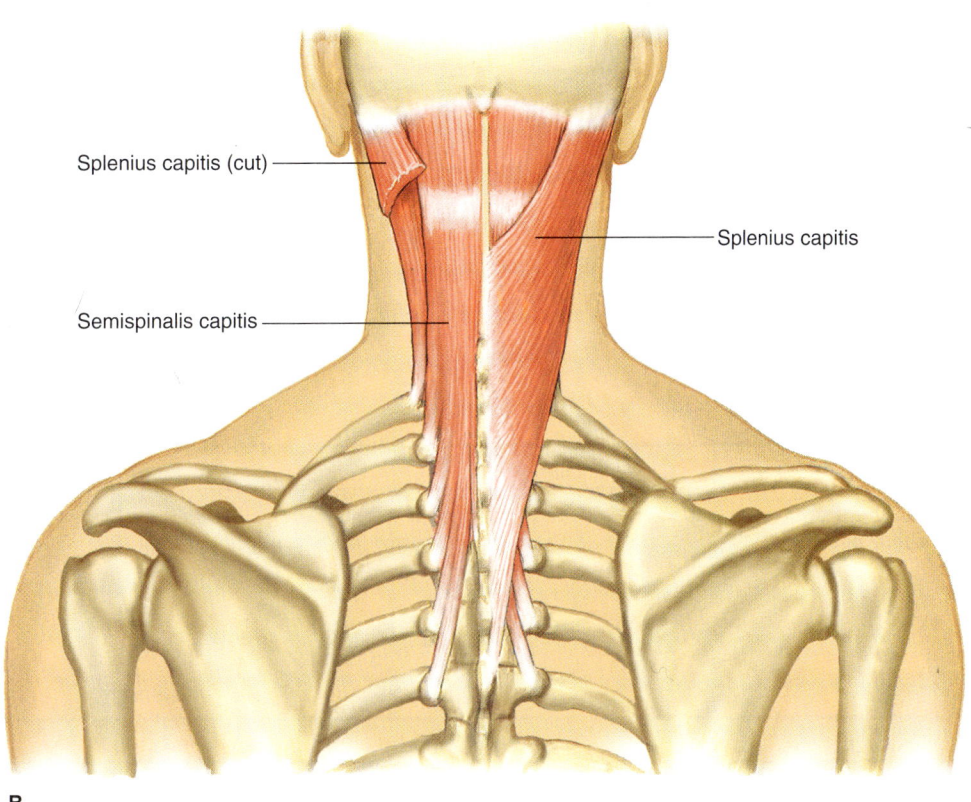

Figure 8.17
Muscles of face and neck. (A) Muscles of facial expression and mastication. (B) Posterior view of muscles that move the head.

Topic of Interest

A NEW MUSCLE DISCOVERED?

An unusual view of a cadaver provided a new perspective that may have revealed a previously undiscovered muscle. In 1995, two dentists were examining a cadaver's skull whose eyes had been dissected out when they discovered what they believe is a new muscle in the head. The muscle, named the *sphenomandibularis,* extends about an inch and a half from behind the eyes to the inside of the jawbone and may help produce the movements of chewing. The muscle has a unique combination of the five characteristics of muscles: origin, insertion, innervation, blood vessel supply, and specific function.

In traditional dissection from the side, the new muscle's origin and insertion are not visible, so it may have appeared to be part of the larger and overlying temporalis muscle. Although the sphenomandibularis inserts on the inner side of the jawbone, as does the temporalis, it originates differently, on the sphenoid bone.

Following their discovery of the sphenomandibularis in the cadaver head, the dentists quickly identified it in twenty-five other cadavers. Other researchers soon found it in live patients undergoing magnetic resonance imaging scans. If yet other researchers confirm that the muscle is newly identified, it will certainly change the commonly held view that anatomy is a "dead" science.

Muscles of Mastication

Muscles attached to the mandible produce chewing movements. Two pairs of these muscles close the lower jaw, a motion used in biting. These muscles are the *masseter* (mas-se´ter) and the *temporalis* (tem-po-ra´lis) (fig. 8.17). Table 8.4 lists the origins, insertions, and actions of muscles of mastication. The Topic of Interest above describes a newly identified muscle thought to be associated with chewing.

> Grinding the teeth, a common response to stress, may strain the temporomandibular joint—the articulation between the mandibular condyle of the mandible and the mandibular fossa of the temporal bone. This condition, called temporomandibular joint syndrome (TMJ syndrome), may produce headache, earache, and pain in the jaw, neck, or shoulder.

Muscles That Move the Head

Head movements result from the actions of paired muscles in the neck and upper back. These muscles flex, extend, and rotate the head. They include (fig. 8.17):

sternocleidomastoid (ster´´no-kli´´do-mas´toid)
splenius capitis (sple´ne-us kap´ĭ-tis)
semispinalis capitis (sem´´e-spi-na´lis kap´ĭ-tis)

Table 8.5 lists the origins, insertions, and actions of muscles that move the head.

Muscles That Move the Pectoral Girdle

Muscles that move the pectoral girdle are closely associated with those that move the arm. A number of these chest and shoulder muscles connect the scapula to nearby bones and move the scapula upward, down-

TABLE 8.4 — MUSCLES OF MASTICATION

MUSCLE	ORIGIN	INSERTION	ACTION
Masseter	Lower border of zygomatic arch	Lateral surface of mandible	Closes jaw
Temporalis	Temporal bone	Coronoid process and lateral surface of mandible	Closes jaw

TABLE 8.5 — MUSCLES THAT MOVE THE HEAD

MUSCLE	ORIGIN	INSERTION	ACTION
Sternocleidomastoid	Anterior surface of sternum and upper surface of clavicle	Mastoid process of temporal bone	Pulls head to one side, pulls head toward chest, or raises sternum
Splenius capitis	Spinous processes of lower cervical and upper thoracic vertebrae	Mastoid process of temporal bone	Rotates head, bends head to one side, or brings head into an upright position
Semispinalis capitis	Processes of lower cervical and upper thoracic vertebrae	Occipital bone	Extends head, bends head to one side, or rotates head

ward, forward, and backward. They include (figs. 8.18 and 8.19):

trapezius (trah-pe´ze-us)
rhomboideus major (rom-boid´e-us)
levator scapulae (le-va´tor scap´u-lē)
serratus anterior (ser-ra´tus an-te´re-or)
pectoralis minor (pek´´to-ra´lis)

Table 8.6 lists the origins, insertions, and actions of muscles that move the pectoral girdle.

Muscles That Move the Arm

The arm is one of the more freely movable parts of the body. Muscles that connect the humerus to various regions of the pectoral girdle, ribs, and vertebral column make these movements possible (figs. 8.18, 8.19, 8.20, and 8.21). These muscles can be grouped according to their primary actions—flexion, extension, abduction, and rotation—as follows:

Flexors
coracobrachialis (kor´´ah-ko-bra´ke-al-is)
pectoralis major (pek´´to-ra´lis)
Extensors
teres major (te´rēz)
latissimus dorsi (lah-tis´ĭ-mus dor´si)
Abductors
supraspinatus (su´´prah-spi´na-tus)
deltoid (del´toid)
Rotators
subscapularis (sub-scap´u-lar-is)
infraspinatus (in´´frah-spi´na-tus)
teres minor (te´rēz)

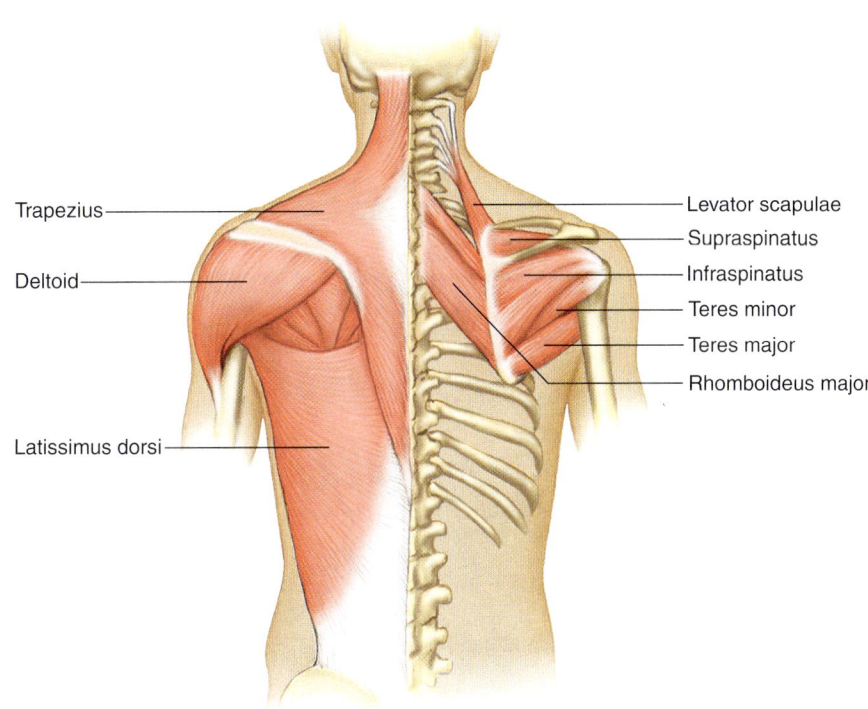

Figure 8.18
Muscles of the posterior shoulder. The right trapezius is removed to show underlying muscles.

TABLE 8.6   MUSCLES THAT MOVE THE PECTORAL GIRDLE

MUSCLE	ORIGIN	INSERTION	ACTION
Trapezius	Occipital bone and spines of cervical and thoracic vertebrae	Clavicle; spine and acromion process of scapula	Rotates scapula and raises arm; raises scapula; pulls scapula medially; or pulls scapula and shoulder downward
Rhomboideus major	Spines of upper thoracic vertebrae	Medial border of scapula	Raises and adducts scapula
Levator scapulae	Transverse processes of cervical vertebrae	Medial margin of scapula	Elevates scapula
Serratus anterior	Outer surfaces of upper ribs	Ventral surface of scapula	Pulls scapula anteriorly and downward
Pectoralis minor	Sternal ends of upper ribs	Coracoid process of scapula	Pulls scapula anteriorly and downward or raises ribs

192 UNIT 2 *Support and Movement*

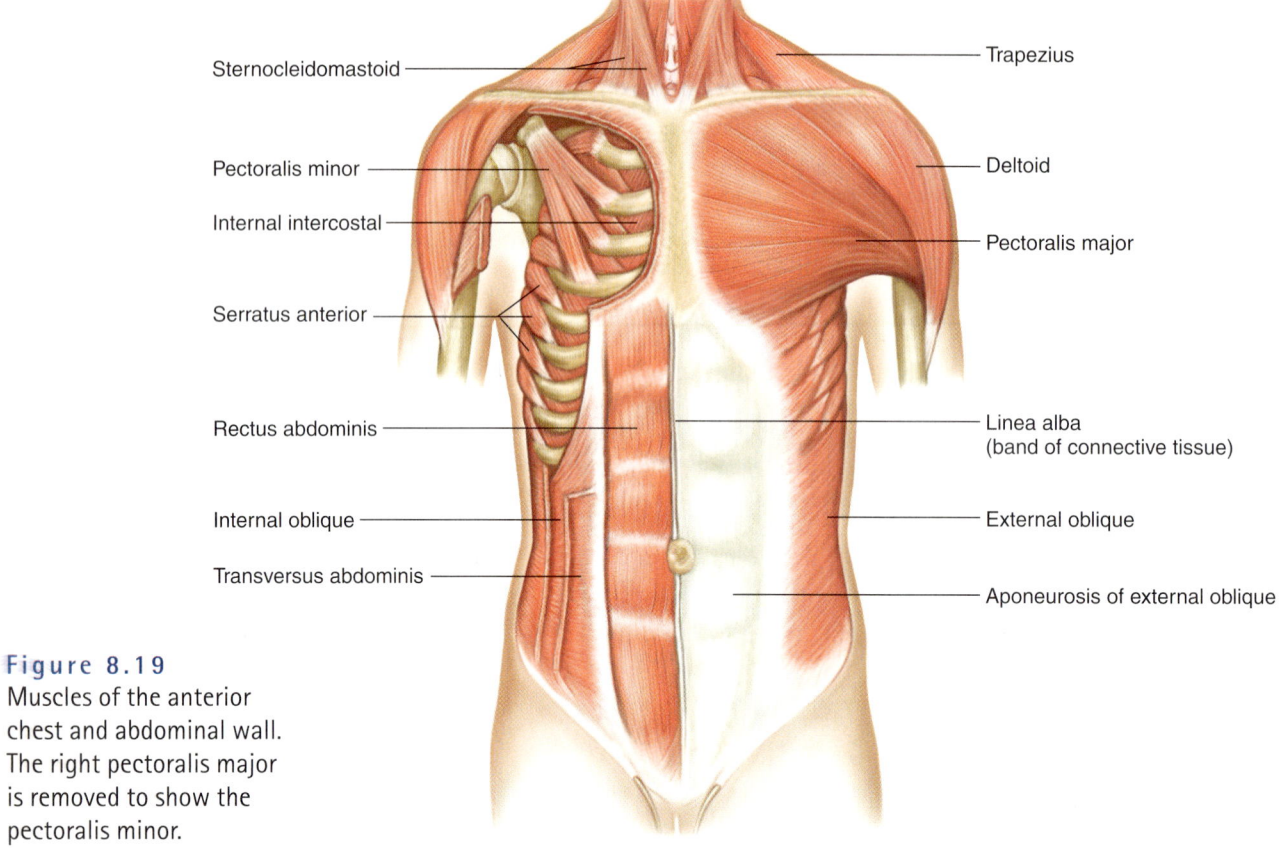

Figure 8.19
Muscles of the anterior chest and abdominal wall. The right pectoralis major is removed to show the pectoralis minor.

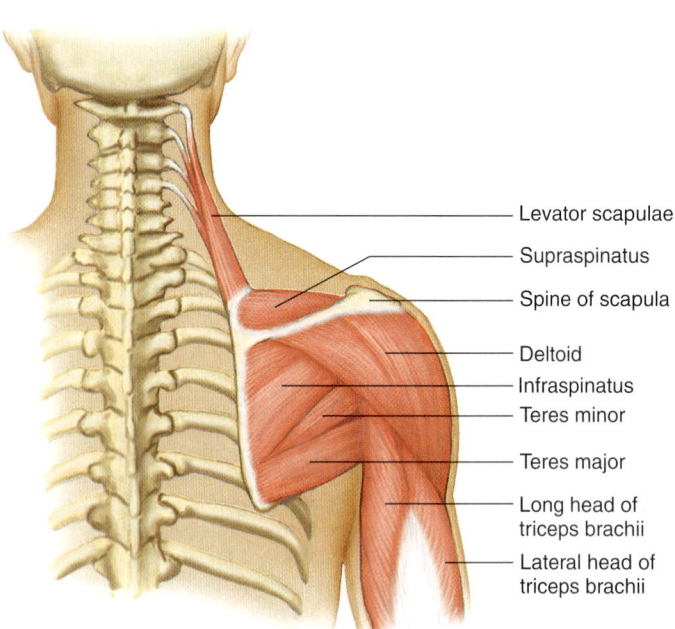

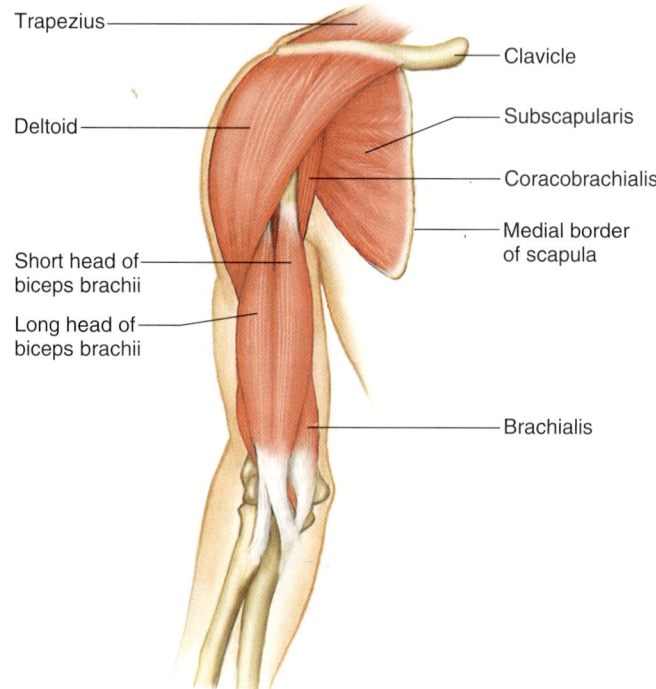

Figure 8.20
Muscles of the posterior surface of the scapula and arm.

Figure 8.21
Muscles of the anterior shoulder and arm, with the rib cage removed.

Frontalis
Orbicularis oculi
Zygomaticus
Masseter
Orbicularis oris
Sternocleidomastoid
Platysma
Trapezius
Deltoid
Pectoralis major
Pectoralis minor
Serratus anterior
Rectus abdominis
External oblique
Internal oblique
Transversus abdominis
Subscapularis
Coracobrachialis
Long head of biceps brachii
Short head of biceps brachii
Levator scapulae
Supraspinatus
Infraspinatus
Teres minor
Teres major
Long head of Triceps brachii
Lateral head -ll-
Biceps brachii
Brachialis
Supinator
Pronator teres
Brachioradialis
Extensor carpi radialis longus
Flexor carpi radialis
Palmaris longus
Flexor carpi ulnaris
Pronator quadratus

Psoas major
Iliacus
Tensor fasciae latae
Rectus femoris
Sartorius
Gracilis
Adductor longus
Adductor magnus
Vastus lateralis
Vastus medialis

back
Triceps brachii
Brachioradialis
Extensor carpi radialis longus
Extensor carpi radialis brevis
Extensor digitorum
Flexor carpi ulnaris
Extensor carpi ulnaris

Gluteus medius
Gluteus maximus
Sartorius
Gracilis
Biceps femoris

- Frontalis
- orbicularis oculi
- orbicularis oris
- sternocleidomastoid
- trapezius
- Deltoid
- Pectoralis major
- Pectoralis minor

TABLE 8.7 — MUSCLES THAT MOVE THE ARM

MUSCLE	ORIGIN	INSERTION	ACTION
Coracobrachialis	Coracoid process of scapula	Shaft of humerus	Flexes and adducts arm
Pectoralis major	Clavicle, sternum, and costal cartilages of upper ribs	Intertubercular groove of humerus	Pulls arm anteriorly and across chest, rotates humerus, or adducts arm
Teres major	Lateral border of scapula	Intertubercular groove of humerus	Extends humerus or adducts and rotates arm medially
Latissimus dorsi	Spines of sacral, lumbar, and lower thoracic vertebrae, iliac crest, and lower ribs	Intertubercular groove of humerus	Extends and adducts arm and rotates humerus inwardly, or pulls shoulder downward and posteriorly
Supraspinatus	Posterior surface of scapula	Greater tubercle of humerus	Abducts arm
Deltoid	Acromion process, spine of scapula, and clavicle	Deltoid tuberosity of humerus	Abducts arm, extends or flexes humerus
Subscapularis	Anterior surface of scapula	Lesser tubercle of humerus	Rotates arm medially
Infraspinatus	Posterior surface of scapula	Greater tubercle of humerus	Rotates arm laterally
Teres minor	Lateral border of scapula	Greater tubercle of humerus	Rotates arm laterally

Table 8.7 lists the origins, insertions, and actions of muscles that move the arm.

Muscles That Move the Forearm

Muscles that connect the radius or ulna to the humerus or pectoral girdle produce most forearm movements. A group of muscles located along the anterior surface of the humerus flexes the elbow, and a single posterior muscle extends this joint. Other muscles move the radioulnar joint and rotate the forearm.

Muscles that move the forearm include (figs. 8.20, 8.21, and 8.22):

Flexors
biceps brachii (bi´seps bra´ke-i)
brachialis (bra´ke-al-is)
brachioradialis (bra´´ke-o-ra´´de-a´lis)

Extensor
triceps brachii (tri´seps bra´ke-i)

Rotators
supinator (su´pĭ-na-tor)
pronator teres (pro-na´tor te´rēz)
pronator quadratus (pro-na´tor kwod-ra´tus)

Table 8.8 lists the origins, insertions, and actions of muscles that move the forearm.

Muscles That Move the Wrist, Hand, and Fingers

Many muscles move the wrist, hand, and fingers. They originate from the distal end of the humerus and from the radius and ulna. The two major groups of these muscles are flexors on the anterior side of the forearm and extensors on the posterior side. These muscles include (figs. 8.22 and 8.23):

Flexors
flexor carpi radialis (flex´sor kar-pi´ ra´´de-a´lis)
flexor carpi ulnaris (flex´sor kar-pi´ ul-na´ris)

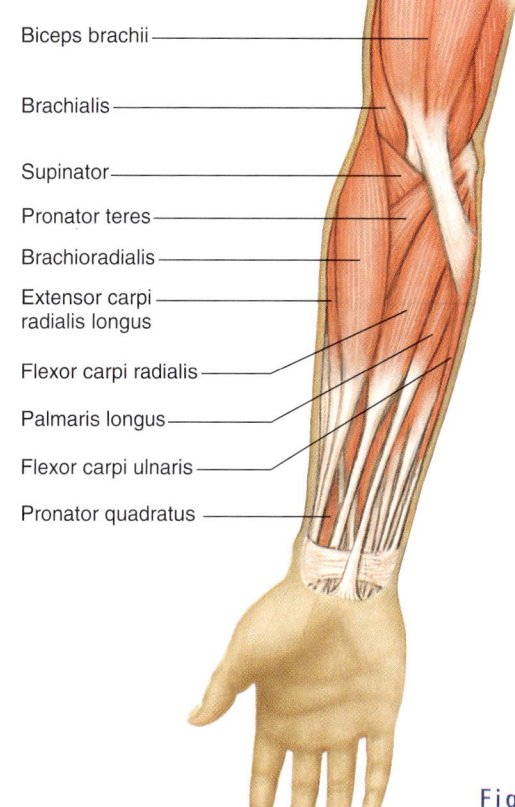

Figure 8.22
Muscles of the anterior forearm.

palmaris longus (pal-ma´ris long´gus)
flexor digitorum profundus (flex´sor dij´´ĭ-to´rum pro-fun´dus)

Extensors
extensor carpi radialis longus (eks-ten´sor kar-pi´ ra´´de-a´lis long´gus)
extensor carpi radialis brevis (eks-ten´sor kar-pi´ ra´´de-a´lis brev´ĭs)

TABLE 8.8			MUSCLES THAT MOVE THE FOREARM
MUSCLE	ORIGIN	INSERTION	ACTION
Biceps brachii	Coracoid process and tubercle above glenoid cavity of scapula	Radial tuberosity of radius	Flexes forearm at elbow and rotates hand laterally
Brachialis	Anterior shaft of humerus	Coronoid process of ulna	Flexes forearm at elbow
Brachioradialis	Distal lateral end of humerus	Lateral surface of radius above styloid process	Flexes forearm at elbow
Triceps brachii	Tubercle below glenoid cavity and lateral and medial surfaces of humerus	Olecranon process of ulna	Extends forearm at elbow
Supinator	Lateral epicondyle of humerus and crest of ulna	Lateral surface of radius	Rotates forearm laterally
Pronator teres	Medial epicondyle of humerus and coronoid process of ulna	Lateral surface of radius	Rotates forearm medially
Pronator quadratus	Anterior distal end of ulna	Anterior distal end of radius	Rotates forearm medially

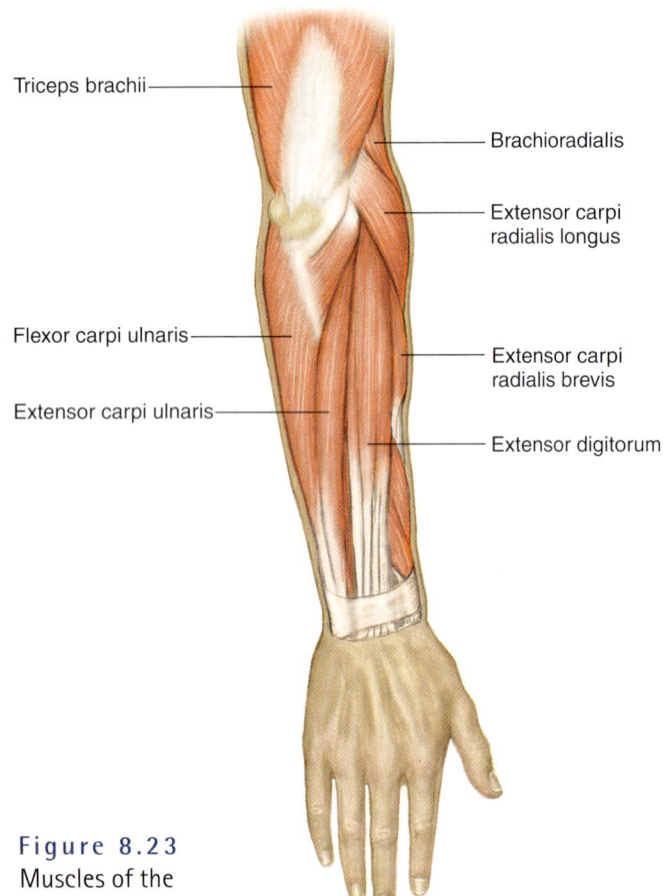

Figure 8.23
Muscles of the posterior forearm.

extensor carpi ulnaris (eks-ten´sor kar-pi´ ul-na´ris)
extensor digitorum (eks-ten´sor dij´´ĭ-to´rum)

Table 8.9 lists the origins, insertions, and actions of muscles that move the wrist, hand, and fingers.

Muscles of the Abdominal Wall

Bone supports the walls of the chest and pelvic regions, but not those of the abdomen. Instead, the anterior and lateral walls of the abdomen are composed of layers of broad, flattened muscles. These muscles connect the rib cage and vertebral column to the pelvic girdle. A band of tough connective tissue called the **linea alba** extends from the xiphoid process of the sternum to the symphysis pubis (see fig. 8.19). It is an attachment for some of the abdominal wall muscles.

Contraction of these muscles decreases the size of the abdominal cavity and increases the pressure inside. These actions help press air out of the lungs during forceful exhalation and aid in the movements of defecation, urination, vomiting, and childbirth.

The abdominal wall muscles include (see fig. 8.19):

external oblique (eks-ter´nal o-blēk´)
internal oblique (in-ter´nal o-blēk´)
transversus abdominis (trans-ver´sus ab-dom´ĭ-nis)
rectus abdominis (rek´tus ab-dom´ĭ-nis)

Table 8.10 lists the origins, insertions, and actions of muscles of the abdominal wall.

Muscles of the Pelvic Outlet

Two muscular sheets—a deeper **pelvic diaphragm** and a more superficial **urogenital diaphragm**—span the outlet of the pelvis. The pelvic diaphragm forms the floor of the pelvic cavity, and the urogenital diaphragm fills the space within the pubic arch (see fig. 7.26, p. 152). The muscles of the male and female pelvic outlets include (fig. 8.24):

Pelvic diaphragm
 levator ani (le-va´tor ah-ni´)
Urogenital diaphragm
 superficial transversus perinei (su´´per-fish´al trans-ver´sus per´´ĭ-ne´i)
 bulbospongiosus (bul´´bo-spon´´je-o´sus)
 ischiocavernosus (is´´ke-o-kav´´er-no´sus)

Table 8.11 lists the origins, insertions, and actions of pelvic outlet muscles.

TABLE 8.9 — MUSCLES THAT MOVE THE WRIST, HAND AND FINGERS

MUSCLE	ORIGIN	INSERTION	ACTION
Flexor carpi radialis	Medial epicondyle of humerus	Base of second and third metacarpals	Flexes and abducts wrist
Flexor carpi ulnaris	Medial epicondyle of humerus and olecranon process	Carpal and metacarpal bones	Flexes and adducts wrist
Palmaris longus	Medial epicondyle of humerus	Fascia of palm	Flexes wrist
Flexor digitorum profundus	Anterior surface of ulna	Bases of distal phalanges in fingers 2-5	Flexes distal joints of fingers
Extensor carpi radialis longus	Distal end of humerus	Base of second metacarpal	Extends wrist and abducts hand
Extensor carpi radialis brevis	Lateral epicondyle of humerus	Base of second and third metacarpals	Extends wrist and abducts hand
Extensor carpi ulnaris	Lateral epicondyle of humerus	Base of fifth metacarpal	Extends and adducts wrist
Extensor digitorum	Lateral epicondyle of humerus	Posterior surface of phalanges in fingers 2-5	Extends fingers

TABLE 8.10 — MUSCLES OF THE ABDOMINAL WALL

MUSCLE	ORIGIN	INSERTION	ACTION
External oblique	Outer surfaces of lower ribs	Outer lip of iliac crest and linea alba	Tenses abdominal wall and compresses abdominal contents
Internal oblique	Crest of ilium and inguinal ligament	Cartilages of lower ribs, linea alba, and crest of pubis	Tenses abdominal wall and compresses abdominal contents
Transversus abdominis	Costal cartilages of lower ribs, processes of lumbar vertebrae, lip of iliac crest, and inguinal ligament	Linea alba and crest of pubis	Tenses abdominal wall and compresses abdominal contents
Rectus abdominis	Crest of pubis and symphysis pubis	Xiphoid process of sternum and costal cartilages	Tenses abdominal wall and compresses abdominal contents; also flexes vertebral column

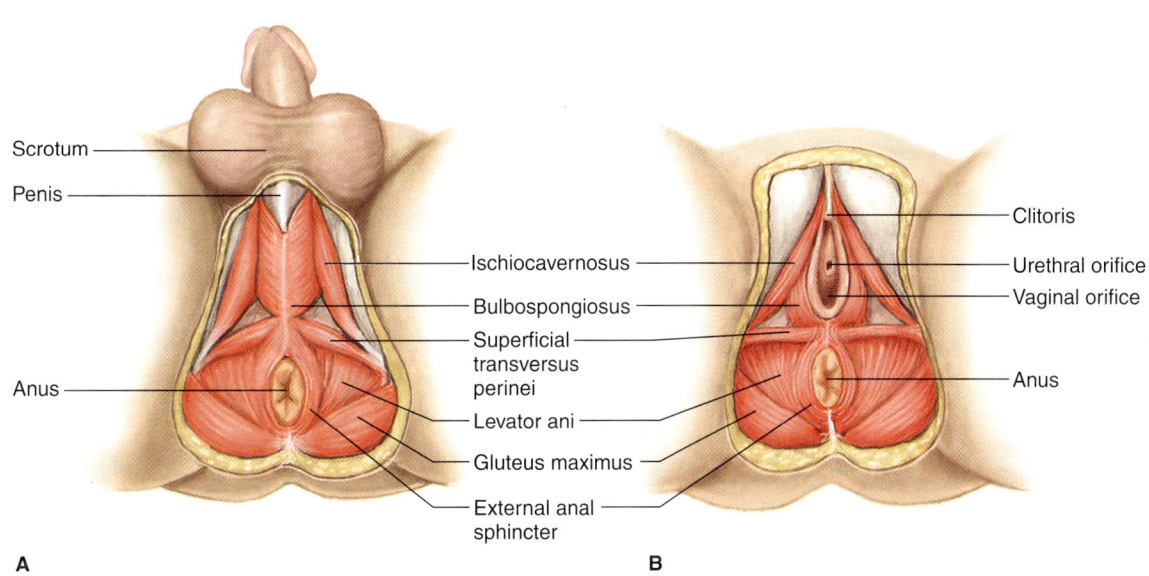

Figure 8.24
External view of muscles of (A) the male pelvic outlet and (B) the female pelvic outlet.

TABLE 8.11 MUSCLES OF THE PELVIC OUTLET

MUSCLE	ORIGIN	INSERTION	ACTION
Levator ani	Pubic bone and ischial spine	Coccyx	Supports pelvic viscera and provides sphincterlike action in anal canal and vagina
Superficial transversus perinei	Ischial tuberosity	Central tendon	Supports pelvic viscera
Bulbospongiosus	Central tendon	Males: Urogenital diaphragm and fascia of the penis Females: Pubic arch and root of clitoris	Males: Assists emptying of urethra Females: Constricts vagina
Ischiocavernosus	Ischial tuberosity	Pubic arch	Assists function of bulbospongiosus

Muscles That Move the Thigh

Muscles that move the thigh are attached to the femur and to some part of the pelvic girdle. These muscles occur in anterior and posterior groups. Muscles of the anterior group primarily flex the thigh; those of the posterior group extend, abduct, or rotate the thigh. The muscles in these groups include (figs. 8.25, 8.26, and 8.27):

Anterior group
psoas major (so´as)
iliacus (il´e-ak-us)

Posterior group
Gluteus maximus (gloo´te-us mak´si-mus)
gluteus medius (gloo´te-us me´de-us)
gluteus minimus (gloo´te-us min´ĭ-mus)
tensor fasciae latae (ten´sor fash´e-e lah-tē)

Still another group of muscles attached to the femur and pelvic girdle adduct the thigh. They include (figs. 8.25 and 8.27):

adductor longus (ah-duk´tor long´gus)
adductor magnus (ah-duk´tor mag´nus)
gracilis (gras´il-is)

Table 8.12 lists the origins, insertions, and actions of muscles that move the thigh.

Muscles That Move the Leg

Muscles that move the leg connect the tibia or fibula to the femur or to the pelvic girdle. They can be separated into two major groups—those that flex the knee and those that extend the knee. Muscles of these groups include the hamstring group and the quadriceps femoris group (figs. 8.25, 8.26, and 8.27):

Flexors
biceps femoris (bi´seps fem´or-is)
semitendinosus (sem´´e-ten´dĭ-no-sus)
semimembranosus (sem´´e-mem´brah-no-sus)
sartorius (sar-to´re-us)

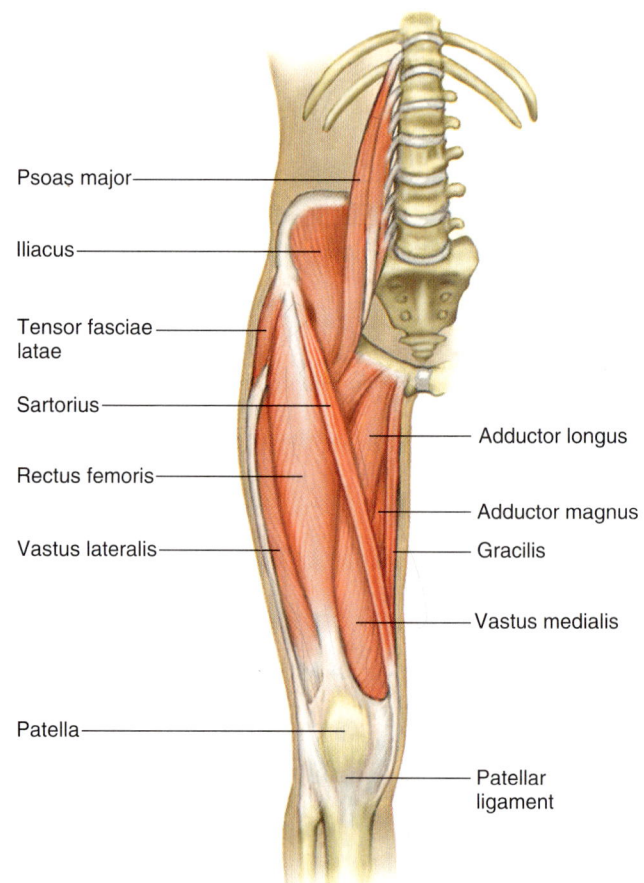

Figure 8.25
Muscles of the anterior right thigh. (Note that the vastus intermedius is a deep muscle not visible in this view.)

Extensor
quadriceps femoris group (kwod´rĭ-seps fem´or-is)
Composed of four parts—the rectus femoris, vastus lateralis, vastus medialis, and vastus intermedius.

Table 8.13 lists the origins, insertions, and actions of muscles that move the leg.

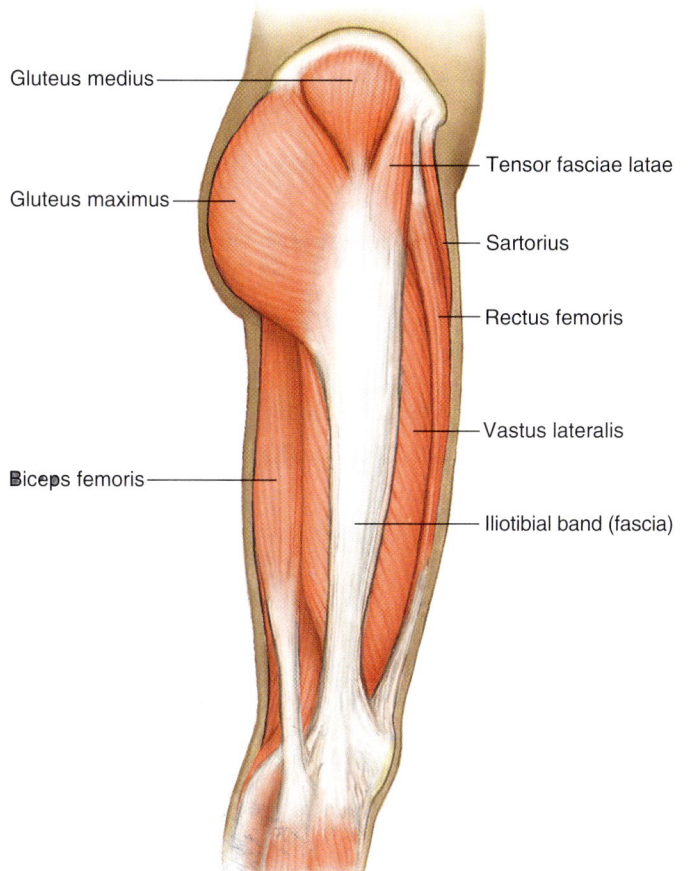

Figure 8.26
Muscles of the lateral right thigh.

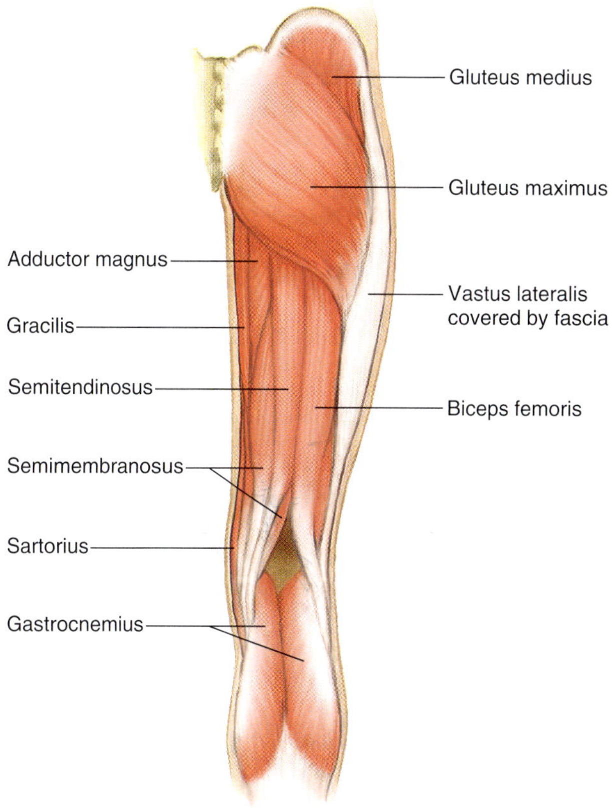

Figure 8.27
Muscles of the posterior right thigh.

TABLE 8.12			MUSCLES THAT MOVE THE THIGH
MUSCLE	ORIGIN	INSERTION	ACTION
Psoas major	Lumbar intervertebral discs, bodies and transverse processes of lumbar vertebrae	Lesser trochanter of femur	Flexes thigh
Iliacus	Iliac fossa of ilium	Lesser trochanter of femur	Flexes thigh
Gluteus maximus	Sacrum, coccyx, and posterior surface of ilium	Posterior surface of femur and fascia of thigh	Extends thigh
Gluteus medius	Lateral surface of ilium	Greater trochanter of femur	Abducts and rotates thigh medially
Gluteus minimus	Lateral surface of ilium	Greater trochanter of femur	Abducts and rotates thigh medially
Tensor fasciae latae	Anterior iliac crest	Fascia of thigh	Abducts, flexes, and rotates thigh medially
Adductor longus	Pubic bone near symphysis pubis	Posterior surface of femur	Adducts, flexes, and rotates thigh laterally
Adductor magnus	Ischial tuberosity	Posterior surface of femur	Adducts, extends, and rotates thigh laterally
Gracilis	Lower edge of symphysis pubis	Medial surface of tibia	Adducts thigh, flexes and rotates lower limb medially

TABLE 8.13 — MUSCLES THAT MOVE THE LEG

MUSCLE	ORIGIN	INSERTION	ACTION
Sartorius	Anterior superior iliac spine	Medial surface of tibia	Flexes leg and thigh, abducts thigh, rotates thigh laterally, and rotates leg medially
Hamstring group			
Biceps femoris	Ischial tuberosity and posterior surface of femur	Head of fibula and lateral condyle of tibia	Flexes leg, extends thigh
Semitendinosus	Ischial tuberosity	Medial surface of tibia	Flexes leg, extends thigh
Semimembranosus	Ischial tuberosity	Medial condyle of tibia	Flexes leg, extends thigh
Quadriceps femoris group			
Rectus femoris	Spine of ilium and margin of acetabulum	Patella by the tendon, which continues as patellar ligament to tibial tuberosity	Extends leg at knee
Vastus lateralis	Greater trochanter and posterior surface of femur	Patella by the tendon, which continues as patellar ligament to tibial tuberosity	Extends leg at knee
Vastus medialis	Medial surface of femur	Patella by the tendon, which continues as patellar ligament to tibial tuberosity	Extends leg at knee
Vastus intermedius	Anterior and lateral surfaces of femur	Patella by the tendon, which continues as patellar ligament to tibial tuberosity	Extends leg at knee

Muscles That Move the Ankle, Foot, and Toes

A number of muscles that move the ankle, foot, and toes are located in the leg. They attach the femur, tibia, and fibula to bones of the foot, move the foot upward (dorsiflexion) or downward (plantar flexion), and turn the sole of the foot medial (inversion) or lateral (eversion). These muscles include (figs. 8.28, 8.29, and 8.30):

Dorsal flexors
- *tibialis anterior* (tib´´e-a´lis an-te´re-or)
- *peroneus (fibularis) tertius* (per´´o-ne´us ter´shus)
- *extensor digitorum longus* (eks-ten´sor dij´´ĭ-to´rum long´gus)

Plantar flexors
- *gastrocnemius* (gas´´trok-ne´me-us)
- *soleus* (so´le-us)
- *flexor digitorum longus* (flek´sor dij´´ĭ-to´rum long´gus)

Invertor
- *tibialis posterior* (tib´´e-a´lis pos-tēr´e-or)

Evertor
- *peroneus (fibularis) longus* (per´´o-ne´us long´gus)

Table 8.14 lists the origins, insertions, and actions of muscles that move the ankle, foot, and toes.

CHECK YOUR RECALL

1. What information is imparted in a muscle's name?
2. Which muscles provide facial expressions, ability to chew, and head movements?
3. Which muscles move the pectoral girdle, abdominal wall, pelvic outlet, the limbs, wrist and hand, ankle and foot, and digits?

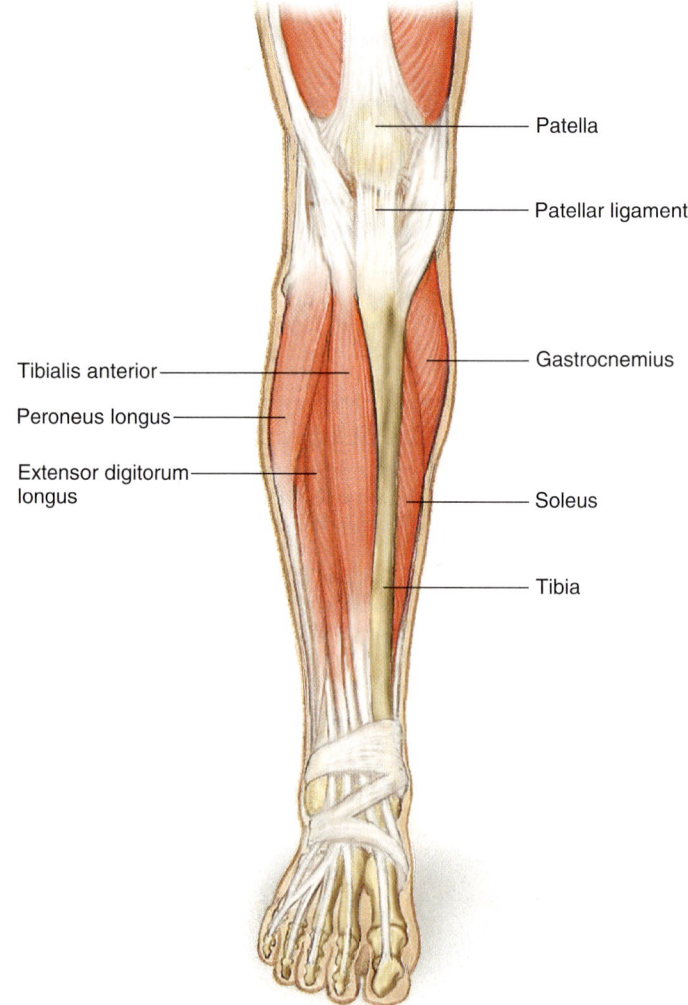

Figure 8.28
Muscles of the anterior right leg.

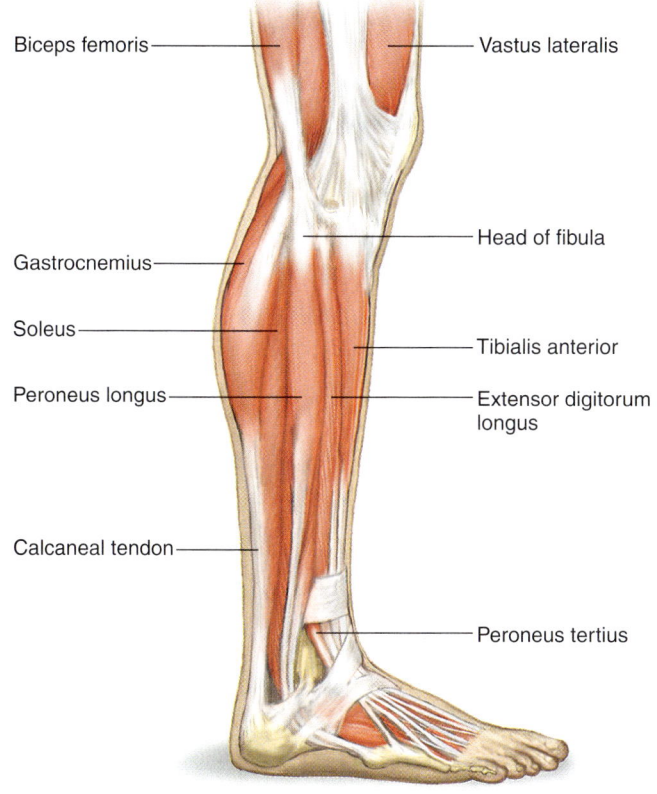

Figure 8.29
Muscles of the lateral right leg. (Note that the tibialis posterior is a deep muscle not visible in this view.)

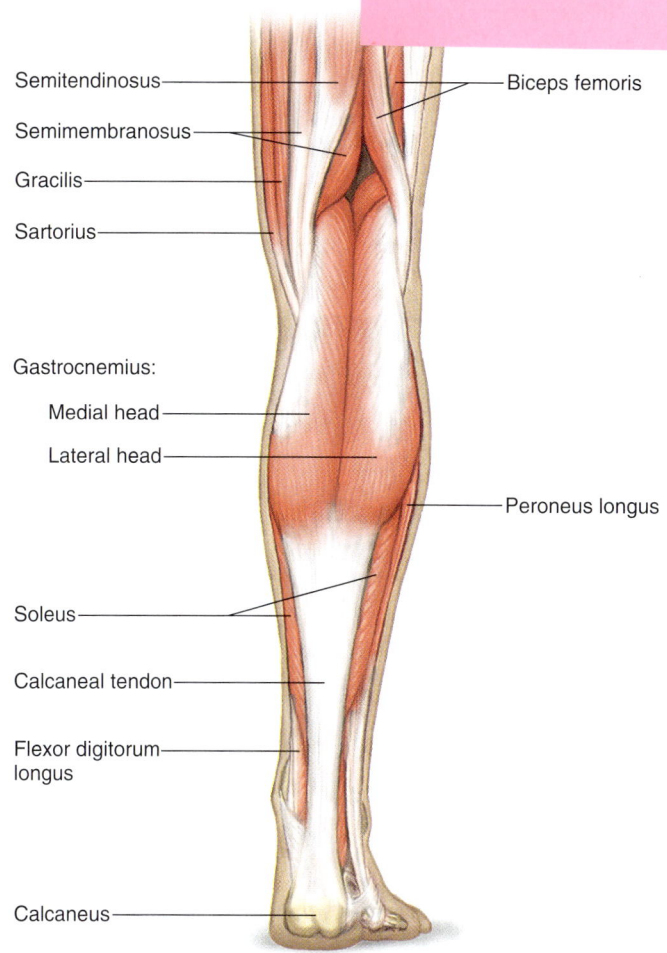

Figure 8.30
Muscles of the posterior right leg.

TABLE 8.14   MUSCLES THAT MOVE THE ANKLE, FOOT, AND TOES

MUSCLE	ORIGIN	INSERTION	ACTION
Tibialis anterior	Lateral condyle and lateral surface of tibia	Tarsal bone (cuneiform) and first metatarsal	Dorsiflexion and inversion of foot
Peroneus tertius	Anterior surface of fibula	Dorsal surface of fifth metatarsal	Dorsiflexion and eversion of foot
Extensor digitorum longus	Lateral condyle of tibia and anterior surface of fibula	Dorsal surfaces of second and third phalanges of the four lateral toes	Dorsiflexion and eversion of foot and extension of toes
Gastrocnemius	Lateral and medial condyles of femur	Posterior surface of calcaneus	Plantar flexion of foot and flexion of leg at knee
Soleus	Head and shaft of fibula and posterior surface of tibia	Posterior surface of calcaneus	Plantar flexion of foot
Flexor digitorum longus	Posterior surface of tibia	Distal phalanges of the four lateral toes	Plantar flexion and inversion of foot, and flexion of the four lateral toes
Tibialis posterior	Lateral condyle and posterior surface of tibia, and posterior surface of fibula	Tarsal and metatarsal bones	Plantar flexion and inversion of foot
Peroneus longus	Lateral condyle of tibia and head and shaft of fibula	Tarsal and metatarsal bones	Plantar flexion and eversion of foot; also supports arch

Organization

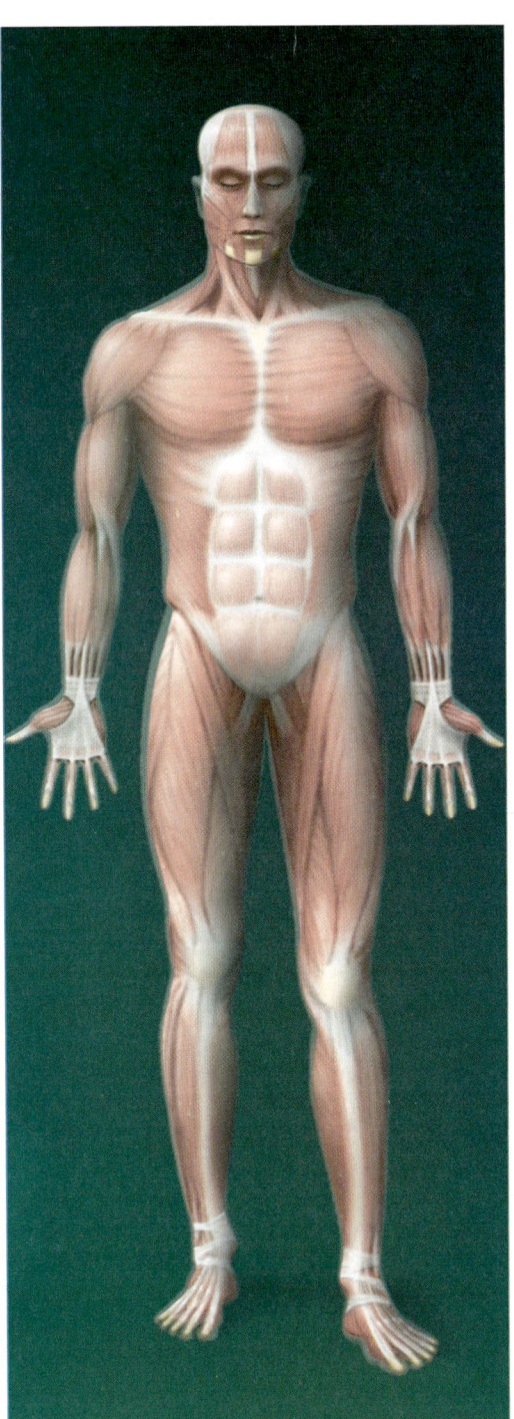

Muscular System
Muscles provide the force for moving body parts.

Integumentary System
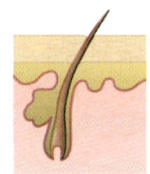
The skin increases heat loss during skeletal muscle activity. Sensory receptors function in the reflex control of skeletal muscles.

Skeletal System
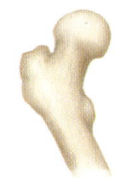
Bones provide attachments that allow skeletal muscles to cause movement.

Nervous System
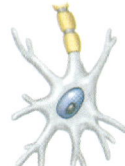
Neurons control muscle contractions.

Endocrine System

Hormones help increase blood flow to exercising skeletal muscles.

Cardiovascular System
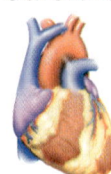
Blood flow delivers oxygen and nutrients and removes wastes.

Lymphatic System

Muscle action pumps lymph through lymphatic vessels.

Digestive System
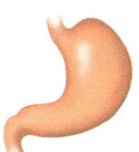
Skeletal muscles are important in swallowing. The digestive system absorbs needed nutrients.

Respiratory System
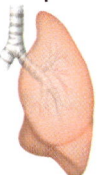
Breathing depends on skeletal muscles. The lungs provide oxygen for body cells and excrete carbon dioxide.

Urinary System
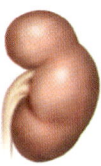
Skeletal muscles help control expulsion of urine from the urinary bladder.

Reproductive System

Skeletal muscles are important in sexual activity.

Clinical Terms Related to the Muscular System

contracture (kon-trak´tur) Condition of great resistance to the stretch of a muscle.
convulsion (kun-vul´shun) Series of involuntary contractions of various voluntary muscles.
electromyography (e-lek´´tro-mi-og´rah-fe) Technique for recording electrical changes in muscle tissues.
fibrillation (fi´´brĭ-la´shun) Spontaneous contractions of individual muscle fibers, producing rapid and uncoordinated activity within a muscle.
fibrosis (fi-bro´sis) Degenerative disease in which fibrous connective tissue replaces skeletal muscle tissue.
fibrositis (fi´´bro-si´tis) Inflammation of fibrous connective tissues, especially in the muscle fascia. This disease is also called *muscular rheumatism.*
muscular dystrophies (mus´ku-lar dis´tro-fez) Group of inherited disorders in which deficiency of cytoskeletal protein (or glycoprotein) collapses muscle cells, leading to progressive loss of function.
myalgia (mi-al´je-ah) Pain from any muscular disease or disorder.
myasthenia gravis (mi´´as-the´ne-ah gra´vis) Chronic disease in which muscles are weak and easily fatigued because of malfunctioning neuromuscular junctions.
myokymia (mi´´o-ki´me-ah) Persistent quivering of a muscle.
myology (mi-ol´o-je) Study of muscles.
myoma (mi-o´mah) Tumor composed of muscle tissue.
myopathy (mi-op´ah-the) Any muscular disease.
myositis (mi´´o-si´tis) Inflammation of skeletal muscle tissue.
myotomy (mi-ot´o-me) Cutting of muscle tissue.
myotonia (mi´´o-to´ne-ah) Prolonged muscular spasm.
paralysis (pah-ral´ĭ-sis) Loss of ability to move a body part.
paresis (pah-re´sis) Partial or slight paralysis of muscles.
shin splints (shin splints) Soreness on the front of the leg due to straining the flexor digitorum longus, often as a result of walking up and down hills.
torticollis (tor´´tĭ-kol´is) Condition in which the neck muscles, such as the sternocleidomastoids, contract involuntarily. It is more commonly called *wryneck.*

Clinical Connection

During summer and fall in the 1940s and early 1950s, thousands of children in the U.S. developed a viral infection called *acute paralytic poliomyelitis.* Usually, the virus remained in the throat or small intestine lining, or traveled to the tonsils and lymph nodes, but when it entered the spinal cord and concentrated in cells that control muscle contraction, paralysis could develop in just days. When fever first struck a child, there was no way to predict the consequences.

Polio survivors vividly recall their treatment. Because of the infectious nature of polio, patients were quarantined. Many had their limbs splinted or entire bodies immobilized in casts, or wore braces or had surgery to restore muscle function. An early type of respirator called an iron lung enabled patients to breathe when their respiratory muscles could not work. The survivors learned to live with permanent disabilities by training other muscles to take over the function of damaged ones. But a few decades later, symptoms of muscle weakness, great fatigue, muscle and joint pain, difficulty sleeping and breathing, and headache began to plague people who'd had polio as children. They have *postpolio syndrome.* The precise cause of this new collection of symptoms isn't known.

Despite decades of vaccination against polio in many nations, the disease still exists, in places where vaccine was not available, and possibly from vaccine strains that have mutated into pathogenic strains. The goal of the World Health Organization was to achieve eradication of polio by the year 2000. In September of that year, the organization extended the goal to 2005.

SUMMARY OUTLINE

8.1 Introduction (p. 172)
The three types of muscle tissue are skeletal, smooth, and cardiac.

8.2 Structure of a Skeletal Muscle (p. 172)
Individual muscles are the organs of the muscular system. They contain skeletal muscle tissue, nervous tissue, blood, and connective tissues.
1. Connective tissue coverings
   a. Fascia covers skeletal muscles.
   b. Other connective tissues attach muscles to bones or to other muscles.
   c. A network of connective tissue extends throughout the muscular system.
2. Skeletal muscle fibers
   a. Each skeletal muscle fiber is a single muscle cell, which is the unit of contraction.
   b. The cytoplasm contains mitochondria, sarcoplasmic reticulum, and myofibrils of actin and myosin.
   c. The organization of actin and myosin filaments produces striations.
   d. Transverse tubules extend inward from the cell membrane and associate with the sarcoplasmic reticulum.

3. Neuromuscular junction
   a. Motor neurons stimulate muscle fibers to contract.
   b. In response to a nerve impulse, the end of a motor neuron axon secretes a neurotransmitter, which stimulates the muscle fiber to contract.
4. Motor units
   a. One motor neuron and the muscle fibers associated with it constitute a motor unit.
   b. All the muscle fibers of a motor unit contract together.

8.3 Skeletal Muscle Contraction (p. 176)

Muscle fiber contraction results from a sliding movement of actin and myosin filaments.

1. Role of myosin and actin
   a. Cross-bridges of myosin filaments form linkages with actin filaments.
   b. The reaction between actin and myosin filaments generates the force of contraction.
2. Stimulus for contraction
   a. Acetylcholine released from the distal end of a motor neuron axon stimulates a skeletal muscle fiber.
   b. Acetylcholine causes the muscle fiber to conduct an impulse over the surface of the fiber that reaches deep within the fiber through the transverse tubules.
   c. A muscle impulse signals the sarcoplasmic reticulum to release calcium ions.
   d. Linkages form between actin and myosin, and the myosin cross-bridges pull on actin filaments, shortening the fiber.
   e. The muscle fiber relaxes when cross-bridges release from actin (ATP is needed, but is not broken down) and when calcium ions are actively transported (requiring ATP breakdown) back into the sarcoplasmic reticulum.
   f. Acetylcholinesterase breaks down acetylcholine.
3. Energy sources for contraction
   a. ATP supplies the energy for muscle fiber contraction.
   b. Creatine phosphate stores energy that can be used to synthesize ATP.
   c. ATP is needed for muscle relaxation.
4. Oxygen supply and cellular respiration
   a. Aerobic respiration requires oxygen.
   b. Red blood cells carry oxygen to body cells.
   c. Myoglobin in muscle cells temporarily stores oxygen.
5. Oxygen debt
   a. During rest or moderate exercise, muscles receive enough oxygen to respire aerobically.
   b. During strenuous exercise, oxygen deficiency may cause lactic acid to accumulate.
   c. Oxygen debt is the amount of oxygen required to convert accumulated lactic acid to glucose and to restore supplies of ATP and creatine phosphate.
6. Muscle fatigue
   a. A fatigued muscle loses its ability to contract.
   b. Muscle fatigue is usually due to accumulation of lactic acid.
7. Heat production
   a. More than half of the energy released in cellular respiration is lost as heat.
   b. Muscle action is an important source of body heat.

8.4 Muscular Responses (p. 180)

1. Threshold stimulus is the minimal stimulus required to elicit a muscular contraction.
2. All-or-none response
   a. If a skeletal muscle fiber contracts at all, it will contract completely.
   b. Motor units respond in an all-or-none manner.
3. Recording a muscle contraction
   a. A myogram is a recording of an electrically stimulated isolated muscle.
   b. A twitch is a single, short contraction reflecting stimulation of some motor units in a muscle.
   c. The latent period, the time between stimulus and responding muscle contraction, is followed by a period of contraction and a period of relaxation.
4. Summation
   a. A rapid series of stimuli may produce summation of twitches.
   b. Forceful, sustained contraction without relaxation is a tetanic contraction.
5. Recruitment of motor units
   a. At a low intensity of stimulation, small numbers of motor units contract.
   b. At increasing intensities of stimulation, other motor units are recruited until the muscle contracts with maximal tension.
6. Sustained contractions
   a. Summation and recruitment together can produce a sustained contraction of increasing strength.
   b. Even when a muscle is at rest, its fibers usually remain partially contracted.

8.5 Smooth Muscle (p. 184)

The contractile mechanism of smooth muscle is similar to that of skeletal muscle.

1. Smooth muscle fibers
   a. Smooth muscle cells contain filaments of actin and myosin.
   b. Types include multiunit smooth muscle and visceral smooth muscle.
   c. Visceral smooth muscle displays rhythmicity and is self-exciting.
2. Smooth muscle contraction
   a. Two neurotransmitters—acetylcholine and norepinephrine—and hormones affect smooth muscle function.
   b. Smooth muscle can maintain a contraction longer with a given amount of energy than can skeletal muscle.
   c. Smooth muscles can change length without changing tension.

8.6 Cardiac Muscle (p. 184)

1. Cardiac muscle twitches last longer than skeletal muscle twitches.
2. Intercalated discs connect cardiac muscle cells.
3. A network of fibers contracts as a unit and responds to stimulation in an all-or-none manner.
4. Cardiac muscle is self-exciting and rhythmic.

8.7 Skeletal Muscle Actions (p. 185)

The type of movement a skeletal muscle produces depends on the way the muscle attaches on either side of a joint.

1. Origin and insertion
   a. The movable end of a skeletal muscle is its insertion, and the immovable end is its origin.
   b. Some muscles have more than one origin.
2. Interaction of skeletal muscles
   a. Skeletal muscles function in groups.
   b. A prime mover is responsible for most of a movement. Synergists aid prime movers. Antagonists can resist the action of a prime mover.
   c. Smooth movements depend on antagonists giving way to the actions of prime movers.

8.8 Major Skeletal Muscles (p. 187)

1. Muscles of facial expression
   a. These muscles lie beneath the skin of the face and scalp and are used to communicate feelings through facial expression.
   b. They include the epicranius, orbicularis oculi, orbicularis oris, buccinator, zygomaticus, and platysma.
2. Muscles of mastication
   a. These muscles attach to the mandible and are used in chewing.
   b. They include the masseter and temporalis.
3. Muscles that move the head
   a. Muscles in the neck and upper back move the head.
   b. They include the sternocleidomastoid, splenius capitis, and semispinalis capitis.
4. Muscles that move the pectoral girdle
   a. Most of these muscles connect the scapula to nearby bones and closely associate with muscles that move the arm.
   b. They include the trapezius, rhomboideus major, levator scapulae, serratus anterior, and pectoralis minor.
5. Muscles that move the arm
   a. These muscles connect the humerus to various regions of the pectoral girdle, ribs, and vertebral column.
   b. They include the coracobrachialis, pectoralis major, teres major, latissimus dorsi, supraspinatus, deltoid, subscapularis, infraspinatus, and teres minor.
6. Muscles that move the forearm
   a. These muscles connect the radius and ulna to the humerus or pectoral girdle.
   b. They include the biceps brachii, brachialis, brachioradialis, triceps brachii, supinator, pronator teres, and pronator quadratus.
7. Muscles that move the wrist, hand, and fingers
   a. These muscles arise from the distal end of the humerus and from the radius and ulna.
   b. They include the flexor carpi radialis, flexor carpi ulnaris, palmaris longus, flexor digitorum profundus, extensor carpi radialis longus, extensor carpi radialis brevis, extensor carpi ulnaris, and extensor digitorum.
8. Muscles of the abdominal wall
   a. These muscles connect the rib cage and vertebral column to the pelvic girdle.
   b. They include the external oblique, internal oblique, transversus abdominis, and rectus abdominis.
9. Muscles of the pelvic outlet
   a. These muscles form the floor of the pelvic cavity and fill the space within the pubic arch.
   b. They include the levator ani, superficial transversus perinei, bulbospongiosus, and ischiocavernosus.
10. Muscles that move the thigh
    a. These muscles attach to the femur and to some part of the pelvic girdle.
    b. They include the psoas major, iliacus, gluteus maximus, gluteus medius, gluteus minimus, tensor fasciae latae, adductor longus, adductor magnus, and gracilis.
11. Muscles that move the leg
    a. These muscles connect the tibia or fibula to the femur or pelvic girdle.
    b. They include the biceps femoris, semitendinosus, semimembranosus, sartorius, and the quadriceps femoris group.
12. Muscles that move the ankle, foot, and toes
    a. These muscles attach the femur, tibia, and fibula to bones of the foot.
    b. They include the tibialis anterior, peroneus tertius, extensor digitorum longus, gastrocnemius, soleus, flexor digitorum longus, tibialis posterior, and peroneus longus.

REVIEW EXERCISES

Part A

1. List the three types of muscle tissue. (p. 172)
2. Distinguish between a tendon and an aponeurosis. (p. 172)
3. Describe how connective tissue associates with skeletal muscle. (p. 172)
4. List the major parts of a skeletal muscle fiber, and describe the function of each part. (p. 173)
5. Describe a neuromuscular junction. (p. 175)
6. Explain the function of a neurotransmitter. (p. 175)
7. Define *motor unit*. (p. 175)
8. Describe the major events of muscle fiber contraction. (p. 176)
9. Explain how ATP and creatine phosphate interact. (p. 179)
10. Describe how muscles obtain oxygen. (p. 179)
11. Describe how an oxygen debt may develop. (p. 179)
12. Explain how muscles may become fatigued. (p. 179)
13. Explain how skeletal muscle function affects the maintenance of body temperature. (p. 180)
14. Define *threshold stimulus*. (p. 180)
15. Explain an *all-or-none response*. (p. 182)
16. Sketch a myogram of a single muscular twitch, and identify the latent period, period of contraction, and period of relaxation. (p. 182)
17. Explain *motor unit recruitment*. (p. 183)
18. Explain how skeletal muscle stimulation produces a sustained contraction. (p. 183)
19. Distinguish between tetanic contraction and muscle tone. (p. 183)
20. Distinguish between multiunit and visceral smooth muscle fibers. (p. 184)
21. Compare smooth and skeletal muscle contractions. (p. 184)
22. Compare the structure of cardiac and skeletal muscle fibers. (p. 184)
23. Distinguish between a muscle's origin and its insertion. (p. 185)
24. Define *prime mover, synergist,* and *antagonist*. (p. 187)

Part B

Match the muscles in column I with the descriptions and functions in column II.

I	II
1. Buccinator	a. Inserted on coronoid process of mandible
2. Epicranius	b. Draws corner of mouth upward
3. Orbicularis oris	c. Can raise and adduct scapula
4. Platysma	d. Can pull head into an upright position
5. Rhomboideus major	e. Raises eyebrow
6. Splenius capitis	f. Compresses cheeks
7. Temporalis	g. Extends over neck from chest to face
8. Zygomaticus	h. Closes lips
9. Biceps brachii	i. Extends forearm at elbow
10. Brachialis	j. Pulls shoulder back and downward
11. Deltoid	k. Abducts arm
12. Latissimus dorsi	
13. Pectoralis major	
14. Pronator teres	
15. Teres minor	

16. Triceps brachii
17. Biceps femoris
18. External oblique
19. Gastrocnemius
20. Gluteus maximus
21. Gluteus medius
22. Gracilis
23. Rectus femoris

l. Inserted on radial tuberosity
m. Pulls arm forward and across chest
n. Rotates forearm medially
o. Inserted on coronoid process of ulna
p. Rotates arm laterally
q. Inverts foot

24. Tibialis anterior

r. Member of quadriceps femoris group
s. Plantar flexor of foot
t. Compresses contents of abdominal cavity
u. Extends thigh
v. Hamstring muscle
w. Adducts thigh
x. Abducts thigh

Part C
Which muscles can you identify in the bodies of these models?

CRITICAL THINKING

1. A person with severe, lifelong constipation finally receives an accurate diagnosis: He is missing some nerves in his lower digestive tract, resulting in sluggish and intermittent peristalsis (rhythmic contractions of the digestive system wall). How can a problem with muscles really be a problem with nerves?
2. A man exercises extensively, building up his muscles. He believes he will pass this hypertrophy on to his future children. Why is he mistaken?
3. A woman takes her daughter to a sports medicine specialist and requests that she determine the percent of fast and slow fibers in the girl's leg muscles. The parent wants to know if the healthy girl should try out for soccer or cross-country running. Do you think this is a valid reason to test muscle tissue? Why or why not?
4. Why do you think athletes generally perform better if they warm up by exercising before a competitive event?
5. What steps might minimize atrophy of the skeletal muscles in patients confined to bed for prolonged times?
6. Lactic acid and other biochemicals accumulating in an active muscle stimulate pain receptors, and the muscle may feel sore. How might the application of heat or substances that dilate blood vessels relieve such soreness?
7. A nerve injury may paralyze the muscle it supplies. How would you explain to a patient the importance of moving the disabled muscles passively or contracting them using electrical stimulation?

WEB CONNECTIONS

Visit the website for additional study questions and more information about this chapter at:

http://www.mhhe.com/shieress8

Unit 3
Integration and Coordination

chapter 9

Nervous System

DEADLY MERCURY. Cases of mercury poisoning are fortunately rare, for it is a terrible way to die. The Mad Hatter, a character in Lewis Carroll's *Alice in Wonderland,* made light of this serious type of poisoning that was once an occupational hazard for hat makers. It is also a risk chemists face when they work with mercury-containing compounds.

On August 14, 1996, Dartmouth College chemist Karen Wetterhahn was working with an organic compound called dimethylmercury when a few drops spilled onto her latex glove. She cleaned up the spill and then took off the glove. Although she recorded the accident in her notebook, she didn't think too much about it—until January, when symptoms of mercury poisoning began. The chemical had penetrated her glove and skin, but taken a long time to affect her brain.

At first, Dr. Wetterhahn felt tingles in her fingers and toes, and developed poor balance and slurred speech. Within days, her visual field narrowed, she saw flashes of light, and her hearing diminished. By the end of January, following chemical analysis of her blood, urine, and hair, the diagnosis was in: mercury poisoning. The level of the deadly compound in her blood exceeded that required for toxicity eighty-fold. By the end of February, she no longer responded to light, sound, or light touch, although her eyes opened, she responded to painful stimuli, and she had bouts of crying. She never regained consciousness and died on June 8, 1997.

Dimethylmercury is a "supertoxin" that clings to brain neurons, destroying them. The neurons, which communicate with each other, gradually die as the more supportive cells, called neuroglia, overgrow. On autopsy, Dr. Wetterhahn's brain was a mere vestige of normal. Her cerebral cortex, which controls thinking, sensation, and perception, was reduced to a thin lining, and her cerebellum, which controls balance, appeared eaten away.

Some good, however, came out of Dr. Wetterhahn's tragic death. Between the time she was diagnosed and the time she lost consciousness, she did everything possible to prevent someone else from suffering her fate. Said her doctor, "She was concerned that people immediately increase efforts to use better and safer handling techniques, and that physicians learn to pick up the signs earlier. She was so sick, yet concerned above all about her colleagues."

Photo:
Mercury poisoning is uncommon today, because we are aware of compounds and situations that may pose a threat. In the nineteenth century, people who made hats frequently displayed the neurological disturbances that resulted from exposure to mercury, and this was the basis for the "Mad Hatter" character in Lewis Carroll's *Alice in Wonderland*. Mercuric nitrate was sprayed onto the felt used to make hats shiny, and the workers would inhale the toxic fumes. After many years of exposure, they experienced symptoms such as difficulty thinking and twitching hands.

Chapter Objectives

After studying this chapter, you should be able to do the following:

9.1 Introduction
1. Distinguish between the two types of cells that comprise nervous tissue. (p. 207)
2. Name the two major groups of nervous system organs. (p. 207)

9.2 General Functions of the Nervous System
3. Explain the general functions of the nervous system. (p. 207)

9.3 Neuroglial Cells
4. State the functions of neuroglial cells in the central nervous system. (p. 208)
5. Distinguish among the types of neuroglial cells in the central nervous system. (p. 208)
6. Describe the Schwann cells of the peripheral nervous system. (p. 209)

9.4 Neurons
7. Describe the general structure of a neuron. (p. 209)
8. Explain how differences in structure and function are used to classify neurons. (p. 211)

9.5 Cell Membrane Potential
9. Explain how a membrane becomes polarized. (p. 212)

9.6 Nerve Impulse
10. Describe the events that lead to the conduction of a nerve impulse. (p. 215)

9.7 The Synapse
11. Explain how information passes from one neuron to another. (p. 216)

9.8 Impulse Processing
12. Describe the general ways in which the nervous system processes information (p. 219)

9.9 Types of Nerves
13. Describe how nerve fibers in peripheral nerves are classified. (p. 220)

9.10 Nerve Pathways
14. Name the parts of a reflex arc, and describe the function of each part. (p. 220)

9.11 Meninges
15. Describe the coverings of the brain and spinal cord. (p. 222)

9.12 Spinal Cord
16. Describe the structure of the spinal cord and its major functions. (p. 224)

9.13 Brain
17. Name the major parts and functions of the brain. (p. 225)
18. Distinguish among motor, sensory, and association areas of the cerebral cortex. (p. 228)
19. Describe the formation and function of cerebrospinal fluid. (p. 230)

9.14 Peripheral Nervous System
20. List the major parts of the peripheral nervous system. (p. 235)
21. Name the cranial nerves, and list their major functions. (p. 236)
22. Describe the structure of a spinal nerve. (p. 238)

9.15 Autonomic Nervous System
23. Describe the functions of the autonomic nervous system. (p. 240)
24. Distinguish between the sympathetic and parasympathetic divisions of the autonomic nervous system. (p. 240)
25. Describe a sympathetic and a parasympathetic nerve pathway. (p. 241)

Aids to Understanding Words

ax- [axle] *ax*on: Cylindrical nerve fiber that carries impulses away from a neuron cell body.

dendr- [tree] *dendr*ite: Branched nerve cell process that serves as a receptor surface of a neuron.

funi- [small cord or fiber] *funi*culus: Major nerve tract or bundle of myelinated nerve cell axons within the spinal cord.

gangli- [a swelling] *gangli*on: Mass of neuron cell bodies.

-lemm [rind or peel] neuri*lemm*a: Sheath that surrounds the myelin of a nerve cell axon.

mening- [membrane] *mening*es: Membranous coverings of the brain and spinal cord.

moto- [moving] *mot*or neuron: Neuron that stimulates a muscle to contract or a gland to secrete.

peri- [all around] *peri*pheral nervous system: Portion of the nervous system that consists of nerves branching from the brain and spinal cord.

plex- [interweaving] choroid *plex*us: Mass of specialized capillaries associated with spaces in the brain.

sens- [feeling] *sens*ory neuron: Sensory receptor that stimulates a neuron to conduct impulses into the brain or spinal cord.

syn- [together] *syn*apse: Junction between two neurons.

ventr- [belly or stomach] *ventr*icle: Fluid-filled space within the brain.

Key Terms

action potential (ak′shun po-ten′shal)
autonomic nervous system (aw″to-nom′ik ner′vus sis′tem)
axon (ak′son)
central nervous system (CNS) (sen′tral ner′vus sis′tem)
chromatophilic substance (kro-mat-o-fill-ik substance)
convergence (kon-ver′jens)
dendrite (den′drīt)
divergence (di-ver′jens)
effector (e-fek′tor)
facilitation (fah-sil″ĭ-ta′shun)
ganglion (gang′gle-on)
meninges (me-nin′jēz)
myelin (mi′ĕ-lin)
neurilemma (nu″rĭ-lem′mah)
neurotransmitter (nu″ro-trans-mit′er)
parasympathetic nervous system (par″ah-sim″pah-thet′ik ner′vus sis′tem)
peripheral nervous system (PNS) (pĕ-rif′er-al ner′vus sis′tem)
plexus (plek′sus)
reflex (re′fleks)
sensory receptor (sen′sor-e re-sep′tor)
somatic nervous system (so-mat′ik ner′vus sis′tem)
sympathetic nervous system (sim″pah-thet′ik ner′vus sis′tem)
synapse (sin′aps)

9.1 Introduction

Feeling, thinking, remembering, moving, and being aware of the world require activity from the nervous system. This vast collection of cells also helps coordinate all other body functions to maintain homeostasis and to enable the body to respond to changing conditions. Sensory receptors bring information from within and outside the body to the brain and spinal cord, which then stimulate responses from muscles and glands.

Recall from chapter 5 (p. 107) that nervous tissue consists of masses of nerve cells, or **neurons.** These cells are the structural and functional units of the nervous system and are specialized to react to physical and chemical changes in their surroundings (fig. 9.1). Neurons transmit information in the form of electrochemical changes, called **nerve impulses,** to other neurons and to cells outside the nervous system.

Neurons typically have a rounded area called the **cell body,** and two types of extensions: dendrites and axons. **Dendrites,** which may be numerous, receive electrochemical messages. **Axons** are extensions that send information. Usually a neuron has only one axon. Figure 9.1 depicts these major parts of a neuron.

Nerves are bundles of axons. Nervous tissue also includes supporting cells called **neuroglial cells** that provide neurons with certain physiological requirements and function similarly to the connective tissue cells in other organ systems.

The organs of the nervous system can be divided into two groups. One group, consisting of the brain and spinal cord, forms the **central nervous system (CNS),** and the other, composed of the nerves (peripheral nerves) that connect the central nervous system to other body parts, is called the **peripheral nervous system (PNS).** Together, these systems provide three general functions: sensory, integrative, and motor.

CHECK YOUR RECALL

1. What are the cells that communicate in the nervous system?
2. What are the two major subdivisions of the nervous system?

9.2 General Functions of the Nervous System

The **sensory function** of the nervous system derives from **sensory receptors** (sen´sor-e re-sep´torz) at the ends of peripheral neurons (see chapter 10, p. 253). These receptors gather information by detecting changes inside and outside the body. Sensory receptors monitor external environmental factors, such as light and sound intensities, and conditions of the body's

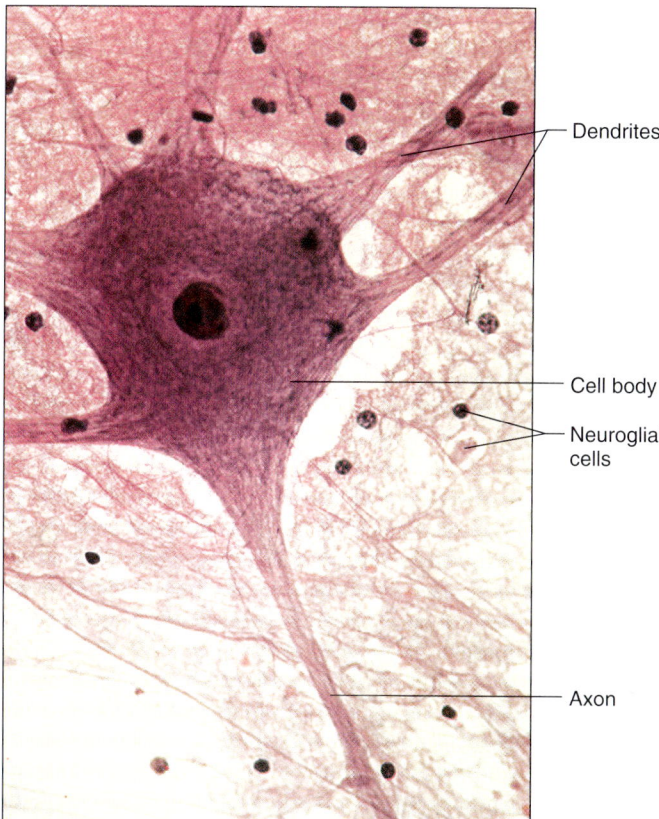

Figure 9.1
Neurons are the structural and functional units of the nervous system (600×). The dark spots in the area surrounding the neuron are neuroglial cells. Note the dendrites and single axon.

internal environment, such as temperature and oxygen concentration.

Sensory receptors convert environmental information into nerve impulses, which are then transmitted over peripheral nerves to the central nervous system. There, the signals are integrated; that is, they are brought together, creating sensations, adding to memory, or helping produce thoughts that translate sensations into perceptions. As a result of this **integrative function,** we make conscious or subconscious decisions, and then we use **motor functions** to act on them.

The motor functions of the nervous system employ peripheral neurons, which carry impulses from the central nervous system to responsive structures called **effectors** (e-fek´torz). Effectors are outside the nervous system and include muscles that contract and glands that secrete when stimulated by nerve impulses.

Motor functions of the nervous system can be divided into two categories. Those that are consciously controlled comprise the **somatic nervous system** of the peripheral nervous system, which controls skeletal muscle. Effectors that are involuntary, such as the heart, smooth muscle in blood vessels, and various glands, are controlled by the **autonomic nervous system.**

The nervous system can detect changes outside and within the body, make decisions based on the information received, and stimulate muscles or glands to respond. Typically, these responses counteract the effects of the changes detected, and in this way, the nervous system helps maintain homeostasis.

CHECK YOUR RECALL

1. How do sensory receptors collect information?
2. How does the central nervous system integrate incoming information?
3. What are the two types of motor functions of the nervous system?

9.3 Neuroglial Cells

A number of cells other than neurons are important in nervous tissue. These **neuroglial cells** (neuroglia) fill spaces, provide structural frameworks, produce myelin, and carry on phagocytosis.

In the central nervous system, neuroglial cells greatly outnumber neurons and are of the following types (fig. 9.2):

1. **Microglial cells** are scattered throughout the central nervous system. They support neurons and phagocytize bacterial cells and cellular debris.
2. **Oligodendrocytes** occur in rows along nerve fibers and they provide insulating layers of myelin

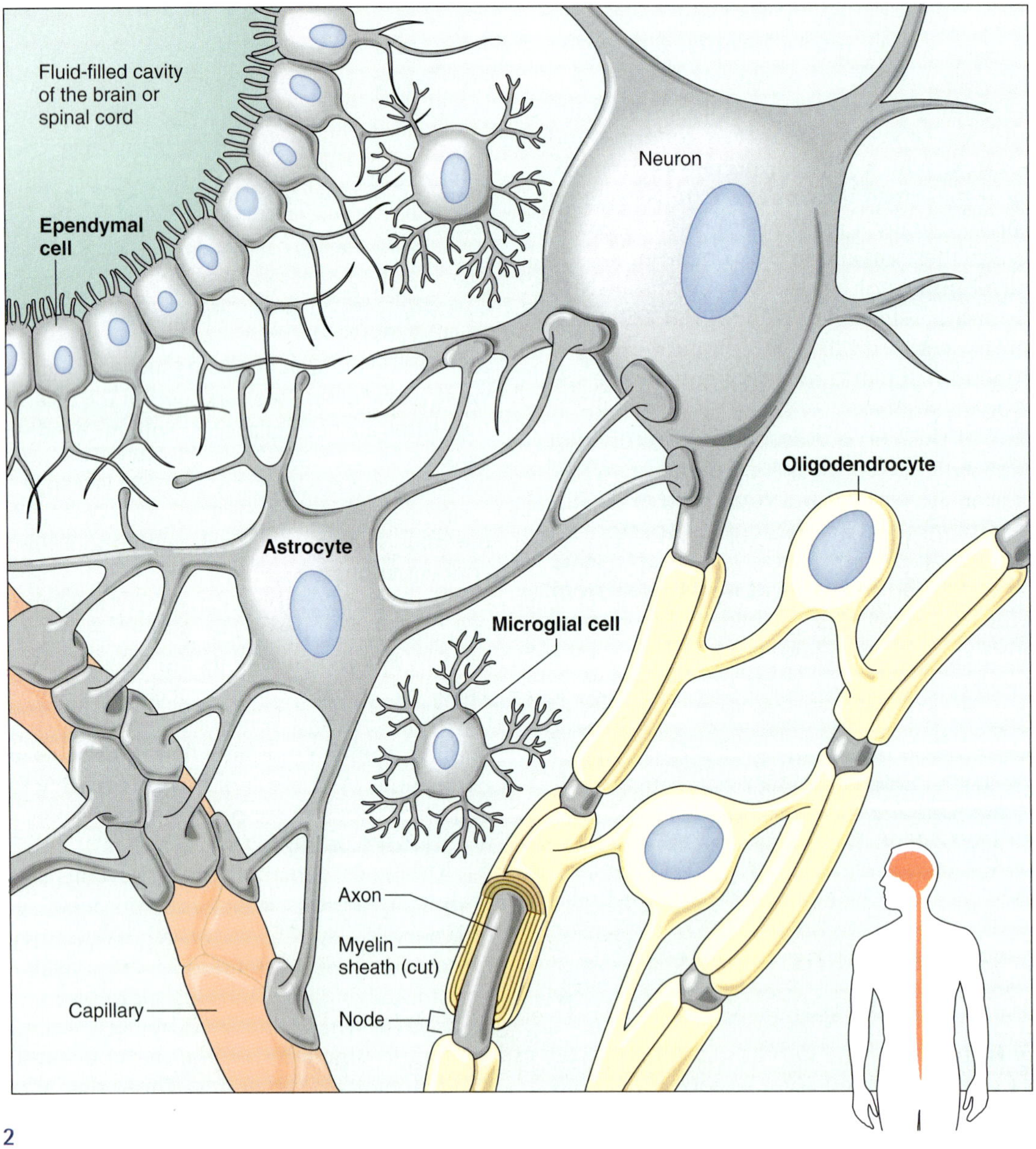

Figure 9.2
Types of neuroglial cells in the central nervous system include the microglial cell, oligodendrocyte, astrocyte, and ependymal cell. Cilia are found on most all ependymal cells during development and early childhood, but in the adult are found mostly on ependymal cells in the ventricles of the brain.

(*myelin sheaths*) around axons within the brain and spinal cord.

3. **Astrocytes,** commonly found between neurons and blood vessels, provide structural support, join parts by numerous cellular processes, and help regulate the concentrations of nutrients and ions within the tissue. Astrocytes also form scar tissue that fills spaces following injury to the CNS.
4. **Ependymal cells** form an epithelial-like membrane that covers specialized brain parts (choroid plexuses) and forms the inner linings that enclose spaces within the brain (ventricles) and spinal cord (central canal).

The peripheral nervous system contains neuroglial cells called **Schwann cells.** Schwann cells form a covering called a **myelin sheath** around axons.

CHECK YOUR RECALL

1. List the functions of cells that support neurons.
2. Distinguish among the types of neuroglia in the central nervous system.
3. What is the function of Schwann cells in the peripheral nervous system?

9.4 Neurons

Neuron Structure

Neurons vary considerably in size and shape, but they all have common features. These include a cell body and the tubular, cytoplasm-filled dendrites which conduct nerve impulses to the neuron cell body and axons which conduct impulses away (fig. 9.3).

The neuron cell body consists of granular cytoplasm, a cell membrane, and organelles such as mitochondria, lysosomes, a Golgi apparatus, and a network of fine threads called **neurofibrils,** which extend into the axons. Scattered throughout the cytoplasm are many membranous sacs called **chromatophilic substance** (Nissl bodies), which are similar to rough endoplasmic reticulum in other cells. Ribosomes attached to chromatophilic substance function in protein synthesis, as they do elsewhere. Near the center of the cell body is a large, spherical nucleus with a conspicuous nucleolus. Mature neurons do not divide. However, some parts of the nervous system include neural stem cells that can divide to give rise to neurons or neuroglia.

Dendrites are usually short and highly branched. These processes, together with the membrane of the cell body, are the neuron's main receptive surfaces with which axons from other neurons communicate.

The axon usually arises from a slight elevation of the cell body called the *axonal hillock*. It conducts nerve impulses away from the cell body. Many mitochondria, microtubules, and neurofibrils are in the axon cytoplasm. An axon originates as a single structure but may give off side branches. Near its end, it may have many fine extensions that contact the receptive surfaces of other cells.

Larger axons of peripheral neurons are enclosed in sheaths composed of many Schwann cells (fig. 9.4). These cells wind tightly around axons, somewhat like a bandage wrapped around a finger, coating them with many layers of cell membrane that have little or no cytoplasm between them. These membrane layers are composed largely of a lipid-protein (lipoprotein) called **myelin** (mi´ĕ-lin). The layers, which have a higher proportion of lipid than other cell membranes, form a *myelin sheath* around an axon. In addition, the portions of the Schwann cells that contain most of the cytoplasm and the nuclei remain outside the myelin sheath and comprise a **neurilemma** (nu´´rĭ-lem´mah), neurilemmal sheath, which surrounds the myelin sheath. Narrow gaps in the myelin sheath between Schwann cells are called **nodes of Ranvier.**

> Myelin begins to form on nerve fibers during the fourteenth week of prenatal development. Yet, many of the nerve fibers in newborns are not completely myelinated. Consequently, the infants' nervous systems are unable to function as effectively as those of older children or adults. Their responses to stimuli are coarse and undifferentiated, and may involve the whole body. All myelinated fibers have begun to develop sheaths by the time a child starts to walk, and myelination continues into adolescence. Interference with the supply of essential nutrients during the developmental years may limit myelin formation, which may impair nervous system function later in life.

Axons with myelin sheaths are called *myelinated*, and those that lack sheaths are *unmyelinated*. Myelin is also found in the CNS. There, groups of myelinated axons appear white, and masses of such axons form the *white matter* in the central nervous system. Unmyelinated axons and neuron cell bodies form *gray matter* within the central nervous system.

When peripheral nerves are damaged, their axons often regenerate. The outer, living layer of the Schwann cell, the neurilemma, plays an important role in this process. In contrast, central nervous system axons are myelinated by oligodendrocytes, which do not provide a neurilemma. When central nervous system neurons are damaged, they do not usually regenerate.

 To picture the relative sizes of a typical neuron's parts, imagine that the cell body is the size of a tennis ball. The axon would then be a mile long and half an inch thick. The dendrites would fill a large bedroom.

Figure 9.3
A common neuron.

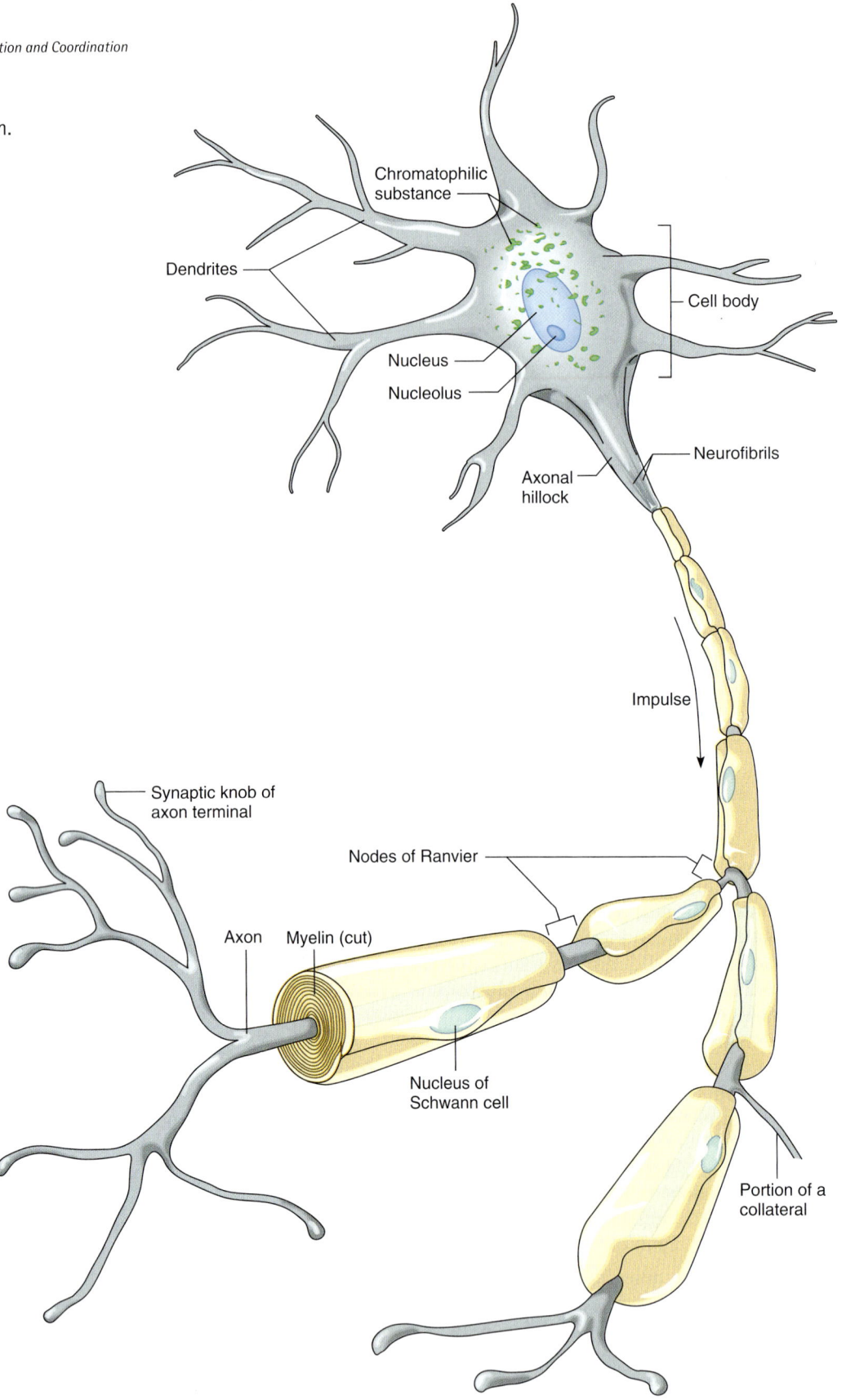

Neurons in the CNS are generally thought to be incapable of dividing. The ability of neurons to form new synapses has been viewed as sufficient to repair damaged areas. The brain, however, does harbor small collections of neural stem cells, which are capable of dividing to give rise to new neurons or glia, depending upon their chemical surroundings. Neural stem cells are found in the hippocampus, in the brain's ventricles, and possibly elsewhere, including the peripheral nervous system. Their possible role in repairing damage is an exciting prospect, but it remains to be established.

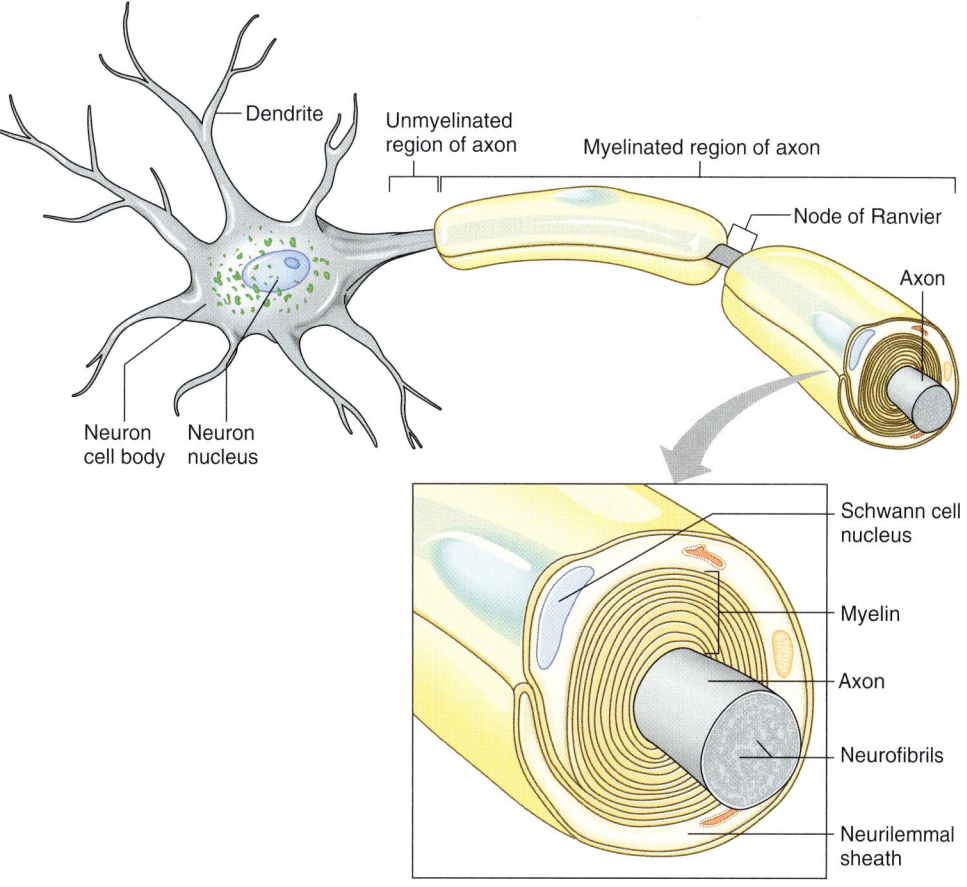

Figure 9.4
The portion of a Schwann cell that winds tightly around an axon forms a myelin sheath, and the cytoplasm and nucleus of the Schwann cell, remaining outside, form a neurilemmal sheath.

Classification of Neurons

Neurons differ in the structure, size, and shape of their cell bodies. They also vary in the length and size of their axons and dendrites and in the number of connections they make with other neurons.

Neurons vary in function. Some neurons carry impulses into the brain or spinal cord, others transmit impulses out of the brain or spinal cord, and still others conduct impulses from neuron to neuron within the brain or spinal cord.

On the basis of structural differences, neurons are classified into the following three major groups (fig. 9.5):

1. **Bipolar neurons** The cell body of a bipolar neuron has only two processes, one arising from each end. Although these processes are structurally similar, one is an axon and the other is a dendrite. Neurons within specialized parts of the eyes, nose, and ears are bipolar.

2. **Unipolar neurons** Each unipolar neuron has a single process extending from its cell body. A short distance from the cell body, this process divides into two branches, which really function as a single axon. One branch (the peripheral process) is associated with dendrites near a peripheral body part. The other branch (the central process) enters the brain or spinal cord. The cell bodies of some unipolar neurons aggregate in specialized masses of nervous tissue called **ganglia** (gang´gle-ah) (singular, *ganglion*), which are located outside the brain and spinal cord.

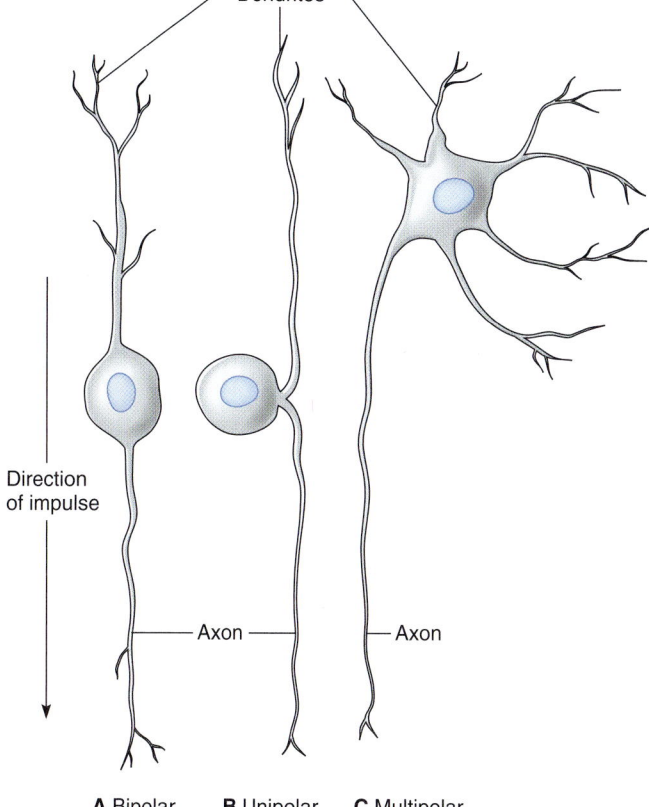

Figure 9.5
Structural types of neurons include (*A*) the bipolar neuron, (*B*) the unipolar neuron, and (*C*) the multipolar neuron.

3. **Multipolar neurons** Multipolar neurons have many processes arising from their cell bodies. Only one process of each neuron is an axon; the rest are dendrites. Most neurons whose cell bodies lie within the brain or spinal cord are multipolar.

On the basis of functional differences, neurons are grouped as follows (fig. 9.6):

1. **Sensory neurons** (afferent neurons) These carry nerve impulses from peripheral body parts into the brain or spinal cord. Sensory neurons either have specialized *receptor ends* at the tips of their dendrites, or they have dendrites that are closely associated with *receptor cells* in the skin or in sensory organs.

    Changes that occur inside or outside the body stimulate receptor ends or receptor cells, triggering sensory nerve impulses. The impulses travel along the sensory neuron axons, which lead to the brain or spinal cord, where other neurons process the impulses. Most sensory neurons are unipolar, although some are bipolar.

2. **Interneurons** (also called *association* or *internuncial neurons*) These neurons lie entirely within the brain or spinal cord. They are multipolar and link other neurons. Interneurons transmit impulses from one part of the brain or spinal cord to another. That is, they may direct incoming sensory impulses to appropriate parts for processing and interpreting. Other incoming impulses are transferred to motor neurons.

3. **Motor neurons** (efferent neurons) Motor neurons are multipolar and carry nerve impulses out of the brain or spinal cord to effectors. Motor impulses stimulate muscles to contract and glands to release secretions.

Neurons deprived of oxygen change shape as their nuclei shrink, and they eventually disintegrate. Oxygen deficiency can result from lack of blood flow through nerve tissue (ischemia), an abnormally low blood oxygen concentration (hypoxemia), or toxins that block aerobic respiration.

CHECK YOUR RECALL

1. Describe the components of a neuron.
2. Distinguish between a dendrite and an axon.
3. Describe how a myelin sheath forms.
4. Why can axons of peripheral nerves regenerate, but axons of central nervous system nerves cannot regenerate?

9.5 Cell Membrane Potential

The surface of a cell membrane (including a nonstimulated or *resting* neuron) is usually electrically charged, or *polarized,* with respect to the inside. This polarization arises from an unequal distribution of positive and negative ions between sides of the membrane, and it is particularly important in the conduction of muscle and nerve impulses.

Distribution of Ions

The distribution of ions inside and outside cell membranes is determined in part by pores or channels in those membranes (see chapter 3, p. 51). Some channels are always open, and others can be opened or closed. Furthermore, channels can be selective; that is, a chan-

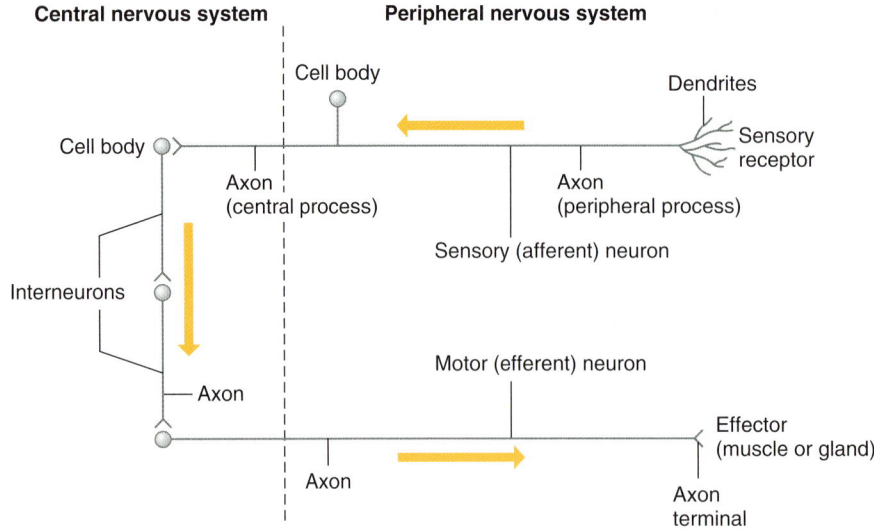

Figure 9.6
Sensory (afferent) neurons carry information into the central nervous system (CNS), interneurons are completely within the CNS, and motor (efferent) neurons carry instructions to the peripheral nervous system (PNS).

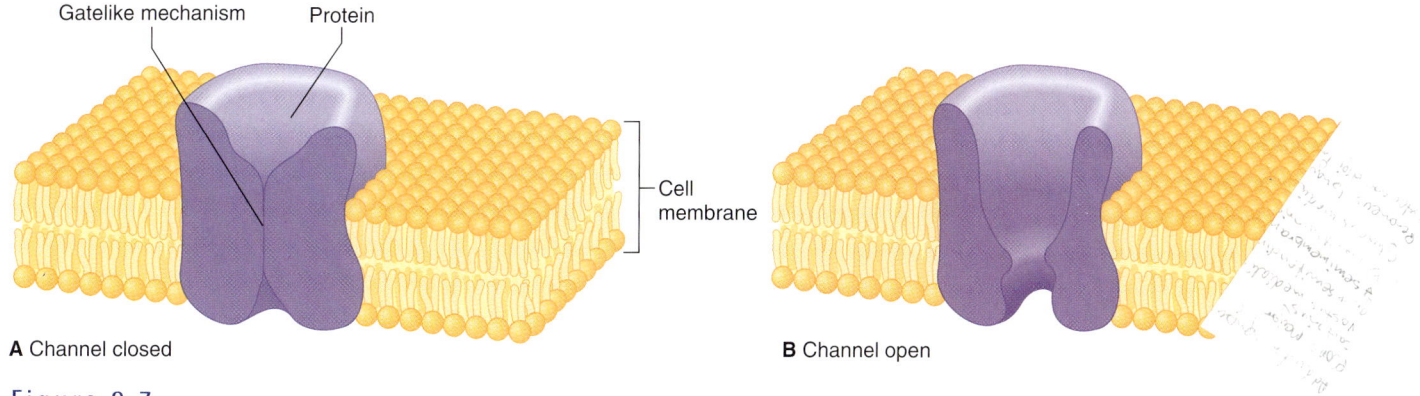

Figure 9.7
A gatelike mechanism can (A) close or (B) open some of the channels in cell membranes through which ions pass.

nel may allow one kind of ion to pass through and exclude other kinds (fig. 9.7).

Potassium ions pass through cell membranes much more easily than sodium ions. This makes potassium ions a major contributor to membrane polarization. Calcium ions are less able to cross the resting cell membrane than either sodium ions or potassium ions, and have a special role in nerve function, described later.

Resting Potential

Because of the active transport of sodium and potassium ions, cells throughout the body have a greater concentration of sodium ions (Na^+) outside and a greater concentration of potassium ions (K^+) inside. The cytoplasm of these cells has many large, negatively charged particles, including phosphate ions (PO_4^{-3}), sulfate ions (SO_4^{-2}), and proteins, that cannot diffuse across the cell membranes (fig. 9.8A).

Sodium and potassium ions follow the laws of diffusion discussed in chapter 2 (p. 58) and show a net movement from high concentration to low concentration as permeabilities permit. Because a resting cell membrane is more permeable to potassium ions than to sodium ions, potassium ions diffuse out of the cell more rapidly than sodium ions can diffuse in (fig. 9.8A). Every millisecond, more positive charges leave the cell by diffusion than enter it. As a result, the outside of the cell membrane gains a slight surplus of positive charges, and the inside is left with a slight surplus of impermeable negative charges, (fig. 9.8B). At the same time, the cell continues to expend energy to drive the Na^+/K^+ "pumps" that actively transport sodium and potassium ions in the opposite directions. The pump maintains the concentration gradients responsible for diffusion of these ions in the first place.

The difference in electrical charge between two regions is called a *potential difference*. In a resting nerve cell, the potential difference between the region inside the membrane and the region outside the membrane is called a **resting potential.** As long as a nerve cell membrane is undisturbed, the membrane remains in this polarized state (fig. 9.9A).

Potential Changes

Nerve cells are excitable; that is, they can respond to changes in their surroundings. Some nerve cells, for example, are specialized to detect changes in temperature, light, or pressure occurring outside the body. Many neurons respond to signals from other neurons. Such changes (or stimuli) usually affect the resting potential in a particular region of a nerve cell membrane. If the membrane's resting potential decreases (as the inside of the membrane becomes less negative when compared to the outside), the membrane is said to be *depolarizing*.

Changes in the resting potential of a membrane are graded. This means that the change in potential is directly proportional to the intensity of stimulation. If additional stimulation arrives before the effect of previous stimulation subsides, the change in potential is still greater. This additive phenomenon is called *summation*, and as a result of summated potentials, a level called **threshold potential** may be reached. Many subthreshold potential changes must combine to reach threshold, and once threshold is achieved, an event called an **action potential** (ak´shun po-ten´shal) occurs.

Action Potential

At the threshold potential, permeability suddenly changes in the region of the cell membrane being stimulated. Channels highly selective for sodium ions open and allow sodium to diffuse freely inward (fig. 9.9B). This movement is aided by the negative electrical condition on the inside of the membrane, which attracts the positively charged sodium ions.

As sodium ions diffuse inward, the membrane loses its negative electrical charge and becomes depolarized. At almost the same time, however, membrane channels

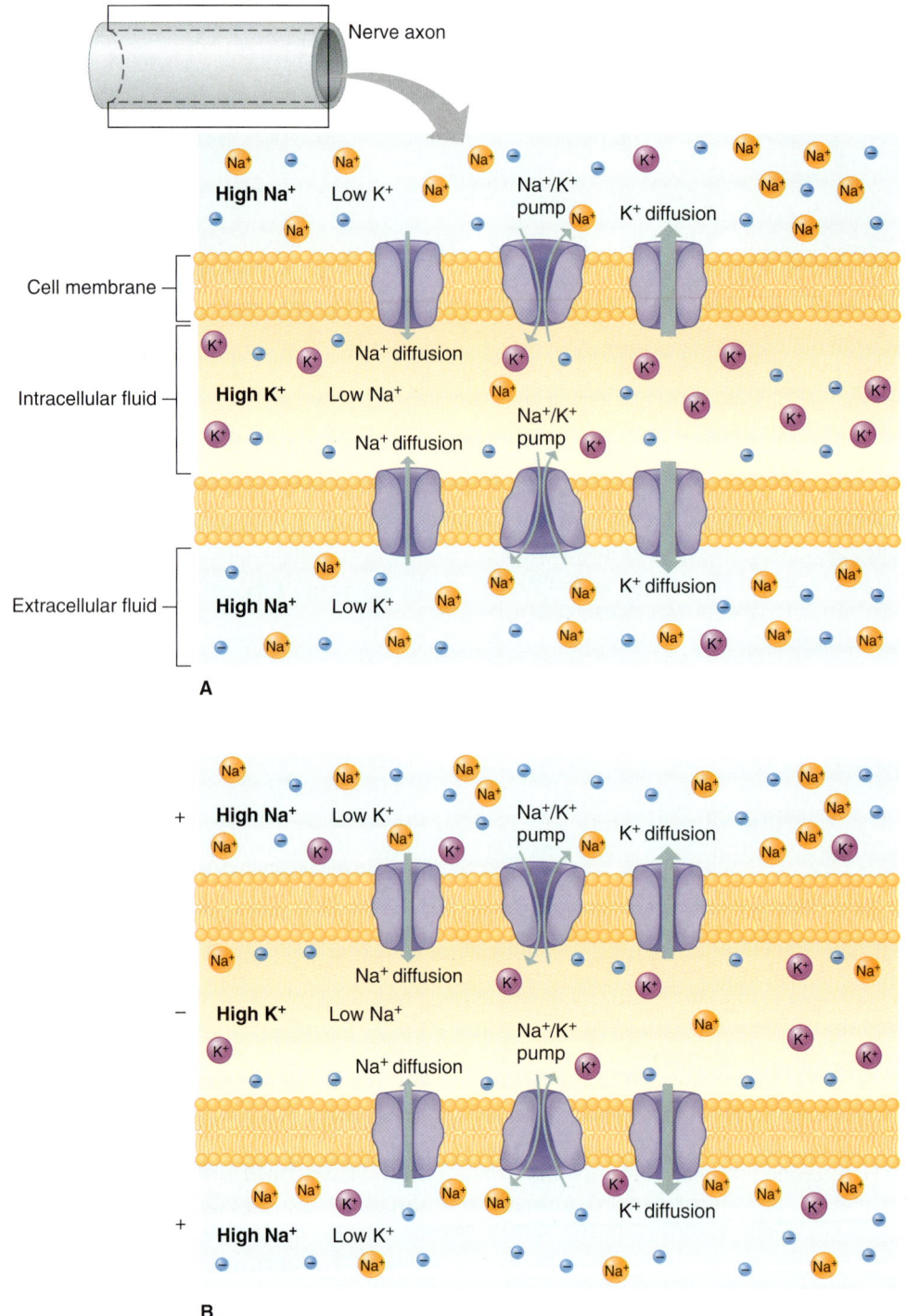

Figure 9.8
Development of the resting membrane potential. (A) Active transport creates a concentration gradient across the cell membrane for sodium ions (Na⁺) and potassium ions (K⁺). K⁺ diffuses out of the cell rather slowly, but nonetheless faster than Na⁺ can diffuse in. (B) This unequal diffusion results in a net loss of positive charge and a resultant excess of negative charge inside the membrane. (Count the ions. Note the loss of positive charges from inside the membrane.)

open that allow potassium ions to pass through, and as these positive ions diffuse outward, the inside of the membrane becomes negatively charged once more (fig. 9.9C). Thus, the membrane becomes *repolarized,* and it remains in this state until stimulated again.

This rapid sequence of depolarization and repolarization takes about one-thousandth of a second and is an action potential. Because only a small proportion of the sodium and potassium ions move through a membrane during an action potential, many action potentials

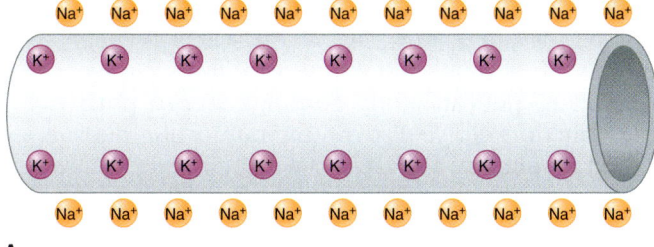

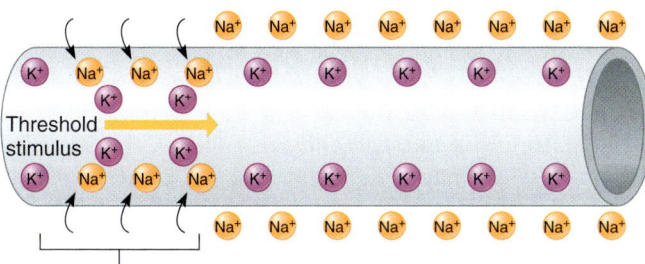

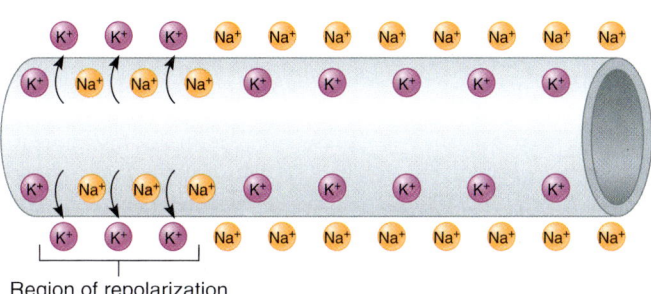

Figure 9.9
Action potential. (*A*) At rest, the membrane potential is negative. (For simplicity, negative ions are not shown.) (*B*) When the membrane reaches threshold, sodium channels open, some sodium (Na⁺) diffuses in, and the membrane is depolarized. (*C*) Soon afterward, potassium channels open, potassium (K⁺) diffuses out, and the membrane is repolarized. (Negative ions are not shown.)

9.6 Nerve Impulse

When an action potential occurs in one region of a nerve cell membrane, it causes a bioelectric current to flow to adjacent portions of the membrane. This *local current* stimulates the adjacent membrane to its threshold level and triggers another action potential. This, in turn, stimulates the next adjacent region. A wave of action potentials moves down the axon to the end. This propagation of action potentials along a nerve axon constitutes a **nerve impulse** (fig. 9.10). Table 9.1 summarizes the events leading to the conduction of a nerve impulse.

> Certain local anesthetic drugs, such as those used by dentists, decrease membrane permeability to sodium ions. Such a drug in the fluids surrounding an axon prevents impulses from passing through the affected region. This keeps impulses from reaching the brain, preventing sensations of touch and pain.

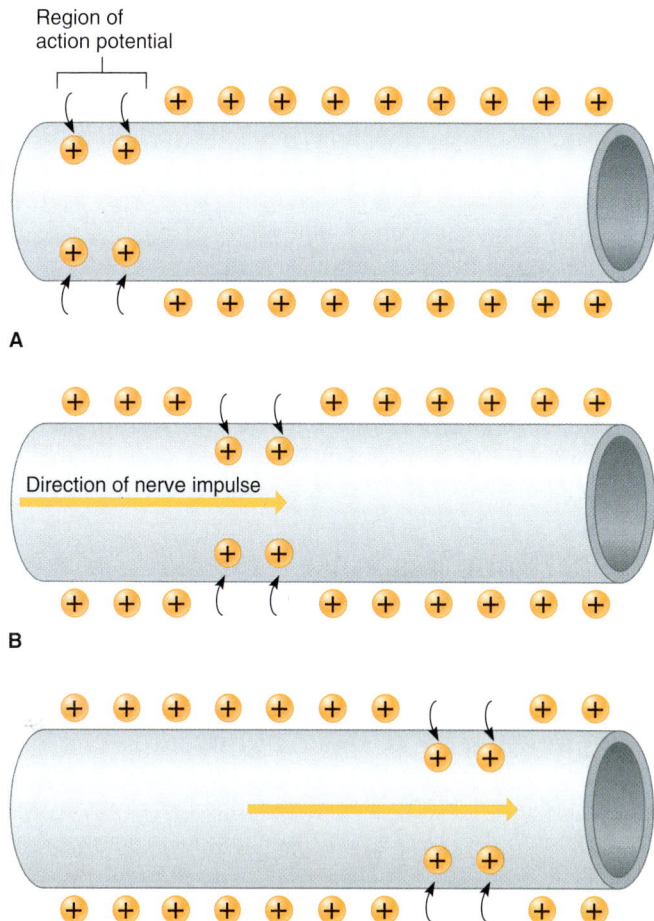

Figure 9.10
Nerve impulse. (*A*) An action potential in one region stimulates the adjacent region, and (*B, C*) a wave of action potentials (a nerve impulse) moves along the axon. (Negative ions are not shown.)

can occur before the original concentrations of these ions change significantly. Also, the active transport mechanism within the membrane works to maintain the original concentrations of sodium and potassium ions on either side; thus, the resting potential quickly returns.

✓ CHECK YOUR RECALL

1. Summarize how a nerve fiber becomes polarized.
2. List the major events of an action potential.

TABLE 9.1 EVENTS LEADING TO THE CONDUCTION OF A NERVE IMPULSE

1. Neuron membrane maintains resting potential.
2. Threshold stimulus is received.
3. Sodium channels in a local region of the membrane open.
4. Sodium ions diffuse inward, depolarizing the membrane.
5. Potassium channels in the membrane open.
6. Potassium ions diffuse outward, repolarizing the membrane.
7. The resulting action potential causes a local bioelectric current that stimulates adjacent portions of the membrane.
8. Wave of action potentials travels the length of the axon as a nerve impulse.

Impulse Conduction

An unmyelinated axon conducts an impulse over its entire surface. A myelinated axon functions differently because myelin is an insulator and prevents almost all ion flow through the membrane it encloses. The myelin sheath would prevent the conduction of a nerve impulse altogether if the sheath were continuous. However, nodes of Ranvier between adjacent Schwann cells interrupt the sheath (see fig. 9.3). Action potentials occur at these nodes, where the exposed axon membrane contains sodium and potassium channels. Thus, a nerve impulse traveling along a myelinated axon appears to jump from node to node. This type of impulse conduction (saltatory) is many times faster than conduction on an unmyelinated axon.

The speed of nerve impulse conduction is proportional to the diameter of the axon—the greater the diameter, the faster the impulse. For example, an impulse on a relatively thick, myelinated axon, such as that of a motor neuron associated with a skeletal muscle, might travel 120 meters per second. An impulse on a thin, unmyelinated axon, such as that of a sensory neuron associated with the skin, might move only 0.5 meter per second.

All-or-None Response

Like muscle fiber contraction, nerve impulse conduction is an *all-or-none response*. That is, if a neuron responds at all, it responds completely. Thus, a nerve impulse is conducted whenever a stimulus of threshold intensity or above is applied to an axon, and all impulses carried on that axon are of the same strength. A greater intensity of stimulation does not produce a stronger impulse, but rather, more impulses per second.

CHECK YOUR RECALL

1. What is the relationship between action potentials and nerve impulses?
2. Explain how impulse conduction differs in myelinated and unmyelinated nerve fibers.
3. Define *all-or-none response*.

9.7 The Synapse

Within the nervous system, nerve impulses travel from neuron to neuron along complex **nerve pathways.** The junction between two communicating neurons is called a **synapse** (sin´aps). Two such neurons are not in direct contact at a synapse. A gap called a *synaptic cleft* separates them. An impulse continuing along a nerve pathway must cross this gap (fig. 9.11).

Synaptic Transmission

Within a neuron (*presynaptic neuron*), an impulse travels from a dendrite to the cell body and then moves along the axon to the end. There, the impulse encounters a synapse separating it from a dendrite or cell body of another neuron (*postsynaptic neuron*). The process of crossing the synapse is called *synaptic transmission*.

Transmission from an axon of one neuron to a dendrite or cell body of another neuron is one-way because axons usually have several rounded *synaptic knobs* at

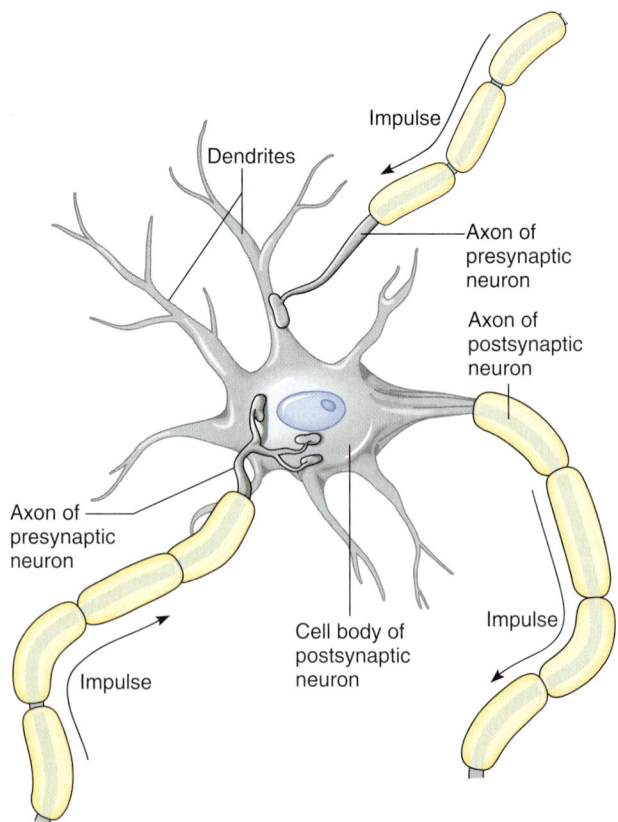

Figure 9.11
For an impulse to continue from one neuron to another, it must cross the synaptic cleft at a synapse. A synapse usually occurs between an axon and a dendrite or between an axon and a cell body.

their distal ends, which dendrites lack. These knobs contain many membranous sacs, called *synaptic vesicles*. When a nerve impulse reaches a synaptic knob, some of the synaptic vesicles release a biochemical called a **neurotransmitter** (nu´ro-trans-mit´er) (figs. 9.12 and 9.13). The neurotransmitter diffuses across the synaptic cleft and reacts with specific receptors on the postsynaptic neuron membrane.

Excitatory and Inhibitory Actions

Neurotransmitters that increase postsynaptic membrane permeability to sodium ions may trigger nerve impulses and thus are said to be **excitatory.** Other neurotransmitters decrease membrane permeability to sodium ions, thus making it less likely that threshold will be reached. This action is called **inhibitory** because it lessens the chance that a nerve impulse will occur.

The synaptic knobs of a thousand or more neurons may communicate with the dendrites and cell body of a single postsynaptic neuron. Neurotransmitters released by some of these knobs have an excitatory action, but those from other knobs have an inhibitory action. The effect on the postsynaptic neuron depends on which presynaptic knobs are activated from moment to moment. In other words, if more excitatory than inhibitory neurotransmitters are released, the postsynaptic neuron's threshold may be reached, and a nerve impulse will be triggered. Conversely, if most of the neurotransmitters released are inhibitory, no impulse will be initiated.

Neurotransmitters

About fifty types of neurotransmitters have been identified in the nervous system. Some neurons release only one type, while others produce two or three kinds. The neurotransmitters include *acetylcholine,* which stimulates skeletal muscle contractions (see chapter 8, p. 176); a group of compounds called *monoamines* (such as epinephrine, norepinephrine, dopamine, and serotonin), which form from modified amino acids; several *amino acids* (such as glycine, glutamic acid, aspartic acid, and gamma-aminobutyric acid—GABA); and a large group of *neuropeptides,* which are short chains of amino acids. Acetylcholine and norepinephrine are excitatory. Dopamine, GABA, and glycine are inhibitory. Neurotransmitters are usually synthesized in the cytoplasm of the synaptic knobs and stored in the synaptic vesicles. Some neurotransmitters and their actions are listed in Table 9.2.

When an action potential reaches the membrane of a synaptic knob, it increases the membrane's permeability to calcium ions by opening the membrane's calcium ion channels. Consequently, calcium ions diffuse inward, and in response, some synaptic vesicles fuse with the membrane and release their contents into the

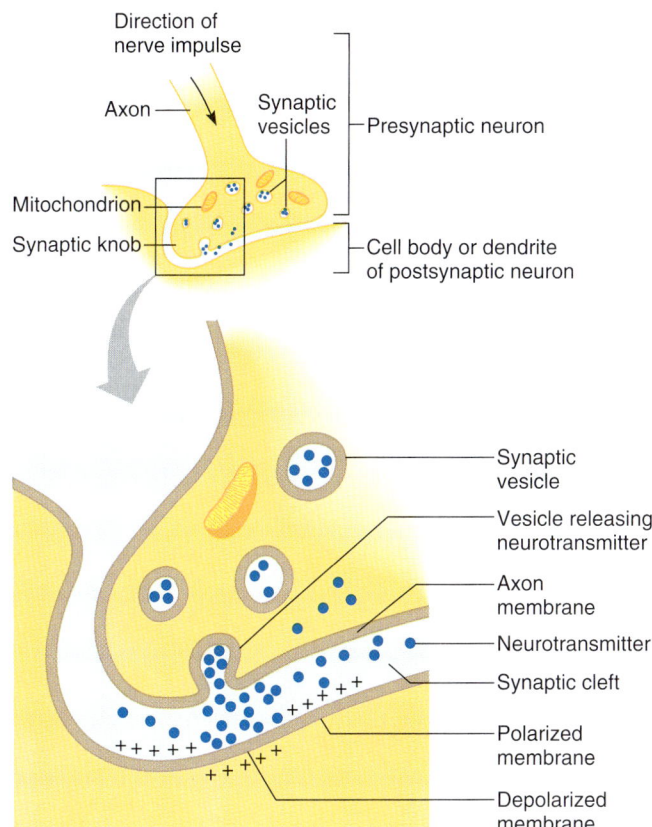

Figure 9.12
When a nerve impulse reaches the synaptic knob at the end of an axon, synaptic vesicles release a neurotransmitter that diffuses across the synaptic cleft.

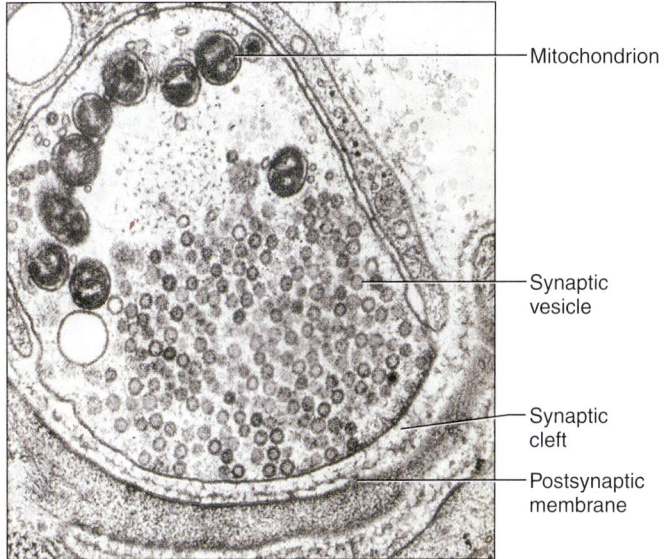

Figure 9.13
Transmission electron micrograph of a synaptic knob filled with synaptic vesicles (37,500×).

Topic of Interest

FACTORS AFFECTING SYNAPTIC TRANSMISSION

Nerve impulses reaching synaptic knobs too rapidly can exhaust neurotransmitter supplies, and impulse conduction ceases until more neurotransmitters are synthesized. This happens during an epileptic seizure. Abnormal and too rapid impulses originate from certain brain cells and reach skeletal muscle fibers, stimulating violent contractions. In time, the synaptic knobs run out of neurotransmitters, and the seizure subsides.

A drug called Dilantin (diphenylhydantoin) treats seizure disorders by increasing the effectiveness of the sodium active transport mechanism. More sodium ions transported from inside the neurons stabilize membrane thresholds against too rapid stimulation.

Many other drugs affect synaptic transmission. For example, caffeine in coffee, tea, and cola drinks stimulates nervous system activity by lowering the thresholds at synapses so that neurons are more easily excited. Antidepressants called "selective serotonin reuptake inhibitors" keep the neurotransmitter serotonin in synapses longer, compensating for a still-little-understood deficit that presumably causes depression. Prozac, Paxil, and Zoloft are three such drugs.

TABLE 9.2 SOME NEUROTRANSMITTERS AND REPRESENTATIVE ACTIONS

NEUROTRANSMITTER	LOCATION	MAJOR ACTIONS
Acetylcholine	CNS	Controls skeletal muscle actions
	PNS	Stimulates skeletal muscle contraction at neuromuscular junctions. May excite or inhibit at autonomic nervous system synapses
Monoamines		
Norepinephrine	CNS	Creates a sense of feeling good; low levels may lead to depression
	PNS	May excite or inhibit autonomic nervous system actions, depending on receptors
Dopamine	CNS	Creates a sense of feeling good; deficiency in some brain areas is associated with Parkinson disease
	PNS	Limited actions in autonomic nervous system; may excite or inhibit, depending on receptors
Serotonin	CNS	Primarily inhibitory; leads to sleepiness; action is blocked by LSD, enhanced by selective serotonin reuptake inhibitor drugs
Histamine	CNS	Release in hypothalamus promotes alertness
Amino acids		
GABA	CNS	Generally inhibitory
Glutamate	CNS	Generally excitatory
Neuropeptides		
Substance P	PNS	Excitatory; pain perception
Endorphins, enkephalins	CNS	Generally inhibitory; reduce pain by inhibiting substance P release
Gases		
Nitric oxide	PNS	Vasodilation
	CNS	May play a role in memory

synaptic cleft. A vesicle that has released its neurotransmitter eventually breaks away from the membrane and reenters the cytoplasm, where it can pick up more neurotransmitter.

After being released, some neurotransmitters are decomposed by enzymes in the synaptic cleft. Other neurotransmitters are transported back into the synaptic knob that released them (reuptake) or into nearby neurons or neuroglial cells. The enzyme *acetylcholinesterase,* for example, decomposes acetylcholine and is present at synapses that separate neurons using this neurotransmitter. Similarly, the enzyme *monoamine oxidase* inactivates monoamines. Decomposition or removal of neurotransmitters prevents continuous stimulation of postsynaptic neurons. Table 9.3 summarizes the events leading to the release of a neurotransmitter.

CHECK YOUR RECALL

1. Describe the events that occur at a synapse.
2. Distinguish between the actions of excitatory and inhibitory neurotransmitters.
3. What types of chemicals function as neurotransmitters?
4. What are possible fates of neurotransmitters?

TABLE 9.3	EVENTS LEADING TO THE RELEASE OF A NEUROTRANSMITTER

1. Action potential passes along an axon and over the surface of its synaptic knob.
2. Synaptic knob membrane becomes more permeable to calcium ions, and they diffuse inward.
3. In the presence of calcium ions, synaptic vesicles fuse to synaptic knob membrane.
4. Synaptic vesicles release their neurotransmitter into synaptic cleft.
5. Synaptic vesicles reenter cytoplasm of axon and pick up more neurotransmitter.

9.8 Impulse Processing

The way the nervous system processes and responds to nerve impulses reflects, in part, the organization of neurons and their axons within the brain and spinal cord.

Neuronal Pools

Neurons within the central nervous system are organized into **neuronal pools.** These are groups of neurons that make hundreds of synaptic connections with each other and work together to perform a common function. Each pool receives input from neurons (which may be part of other pools), and each pool generates output. Neuronal pools may have excitatory or inhibitory effects on other pools or on peripheral effectors.

Facilitation

As a result of incoming impulses and neurotransmitter release, a particular neuron of a neuronal pool may receive excitatory and inhibitory input. If the net effect of the input is excitatory, threshold may be reached, and an outgoing impulse triggered. If the net effect is excitatory but subthreshold, an impulse is not triggered, but the neuron is more excitable to incoming stimulation than before, a state called **facilitation** (fah-sil´´ĭ-ta´shun).

Convergence

Any single neuron in a neuronal pool may receive impulses from two or more incoming axons. Axons originating from different parts of the nervous system and leading to the same neuron exhibit **convergence** (kon-ver´jens) (fig. 9.14A).

Convergence makes it possible for impulses arriving from different sources to have an additive effect on a neuron. For example, if a neuron is facilitated by receiving subthreshold stimulation from one input neuron, it may reach threshold if it receives additional stimulation from a second input neuron. As a result, a nerve impulse may travel to a particular effector and evoke a response.

Incoming impulses often bring information from several sensory receptors that detect changes. Convergence allows the nervous system to collect a variety of kinds of information, process it, and respond to it in a special way.

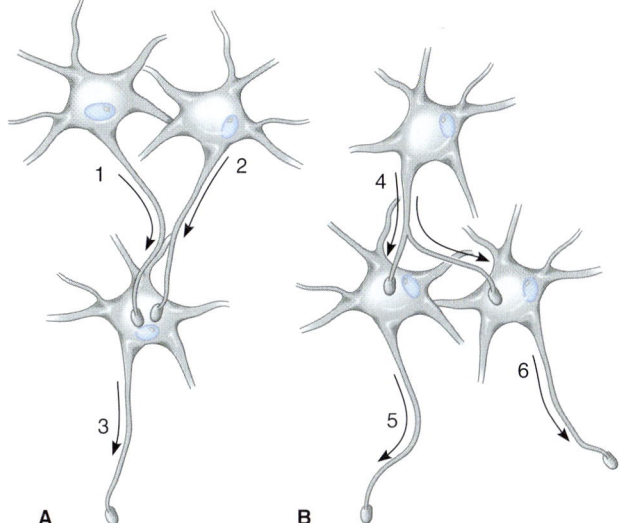

Figure 9.14
Impulse processing in neuronal pools. (A) Axons of neurons 1 and 2 converge to the cell body of neuron 3. (B) The axon of neuron 4 diverges to the cell bodies of neurons 5 and 6.

Divergence

Impulses leaving a neuron of a neuronal pool often exhibit **divergence** (di-ver´jens) by passing into several other output neurons (fig. 9.14B). For example, an impulse from one neuron may stimulate two others; each of these, in turn, may stimulate several others, and so forth. Divergence can amplify an impulse—that is, spread it to more neurons within the pool. As a result of divergence, an impulse originating from a single neuron in the central nervous system may be amplified so that impulses reach enough motor units within a skeletal muscle to cause forceful contraction (see chapter 8, p. 183). Similarly, an impulse originating from a sensory receptor may diverge and reach several different regions of the central nervous system, where the resulting impulses are processed and acted upon.

 **CHECK YOUR RECALL**

1. Define *neuronal pool*.
2. Distinguish between convergence and divergence.

9.9 Types of Nerves

An axon is often referred to as a nerve fiber. Because of this, and for simplicity, we will refer to the neuron processes that bring sensory information into the CNS as **sensory fibers,** or **afferent fibers.** A nerve is a cordlike bundle (or group of bundles) of nerve fibers held together by layers of connective tissue (fig. 9.15).

 The term "muscle fiber" refers to a muscle cell, whereas the term "nerve fiber" refers to a cellular process, or part of a cell. The terminology for the connective tissue holding them together, however, is quite similar. In both cases, for example, fibers are bundled into fascicles, whereas epineurium in nerves corresponds to epimysium in muscles, and so forth (fig. 9.15).

Like neurons, nerves that conduct impulses into the brain or spinal cord are called **sensory nerves,** and those that carry impulses to muscles or glands are termed **motor nerves.** Most nerves include both sensory and motor fibers and are called **mixed nerves.**

 CHECK YOUR RECALL

1. What is a nerve?
2. How does a mixed nerve differ from a sensory nerve? From a motor nerve?

9.10 Nerve Pathways

Recall from section 9.7 that the routes nerve impulses follow as they travel through the nervous system are called *nerve pathways*. The simplest of these pathways includes only a few neurons and is called a **reflex** (re´fleks) **arc.** It constitutes the structural and functional basis for involuntary actions called reflexes.

Reflex Arcs

A reflex arc begins with a receptor at the end of a sensory fiber. This fiber usually leads to several interneurons within the CNS, which serve as a processing center, or *reflex center*. These interneurons may connect with interneurons in other parts of the nervous system. They also communicate with motor neurons, whose fibers pass outward from the CNS to effectors.

Reflex Behavior

Reflexes are automatic subconscious responses to changes (stimuli) within or outside the body. They help maintain homeostasis by controlling many involuntary processes, such as heart rate, breathing rate, blood pressure, and digestion. Reflexes also carry out the automatic actions of swallowing, sneezing, coughing, and vomiting.

The *knee-jerk reflex* (patellar tendon reflex) is an example of a simple reflex that employs only two neurons—a sensory neuron communicating directly with a motor neuron. Striking the patellar ligament just below

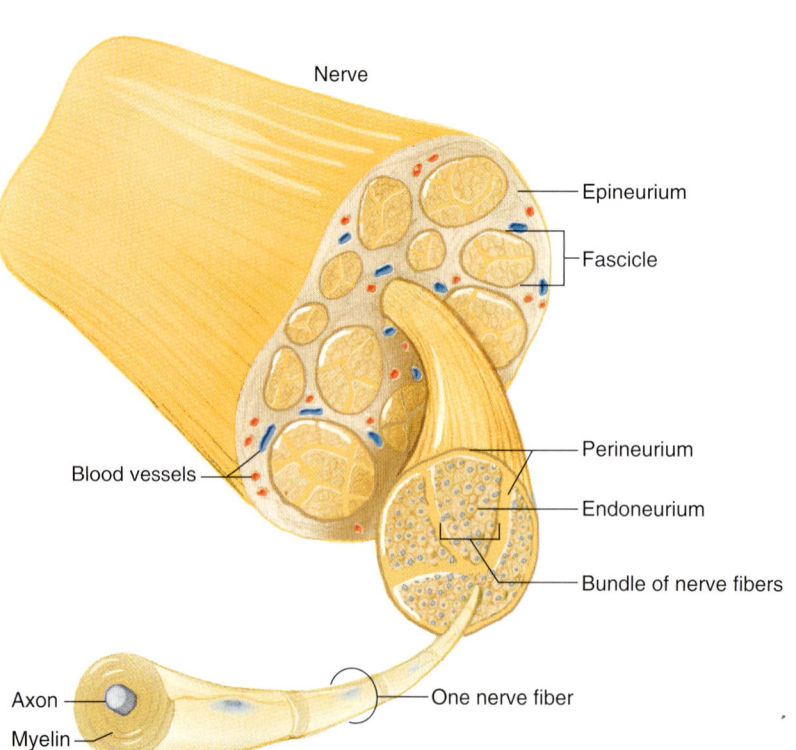

Figure 9.15
Connective tissue holds together a bundle of nerve fibers, forming a fascicle. Many fascicles form a nerve.

the patella initiates this reflex. As a result, the quadriceps femoris muscle group, which is attached to the patella by a tendon, is pulled slightly, stimulating stretch receptors within the muscles. These receptors, in turn, trigger impulses that pass along the fibers of a sensory neuron into the spinal cord. Within the spinal cord, the sensory axon forms a synapse with a motor neuron. The impulse then continues along the axon of the motor neuron and travels back to the quadriceps femoris. The muscle group contracts in response, and the reflex is completed as the leg extends (fig. 9.16).

The knee-jerk reflex helps maintain upright posture. If the knee begins to bend from the force of gravity when a person is standing still, the quadriceps femoris is stretched, the reflex is triggered, and the leg straightens again.

Another type of reflex, called a *withdrawal reflex*, occurs when a person unexpectedly touches a body part to something painful. This activates skin receptors and sends sensory impulses to the spinal cord. There, the impulses pass to the interneurons of a reflex center and are directed to motor neurons. The motor neurons transmit signals to flexor muscles in the arm, and the muscles contract in response. At the same time, the antagonistic extensor muscles are inhibited, and the hand is rapidly and unconsciously withdrawn from the painful stimulation. Concurrent with the withdrawal reflex, other interneurons carry sensory impulses to the brain, and the person becomes aware of the experience and may feel pain (fig. 9.17). A withdrawal reflex is protective because it may limit tissue damage caused by touching something potentially harmful. Table 9.4 summarizes the parts of a reflex arc.

Because normal reflexes depend on normal neuron functions, reflexes provide information about the condition of the nervous system. For instance, an anesthesiologist may try to initiate a reflex in a patient being anesthetized to determine how well the anesthetic drug is affecting nerve functions. Also, in the case of injury to some part of the nervous system, reflexes may be tested to discover the location and extent of damage.

CHECK YOUR RECALL

1. What is a nerve pathway?
2. List the parts of a reflex arc.
3. Define *reflex*.
4. Review the actions that occur during a withdrawal reflex.

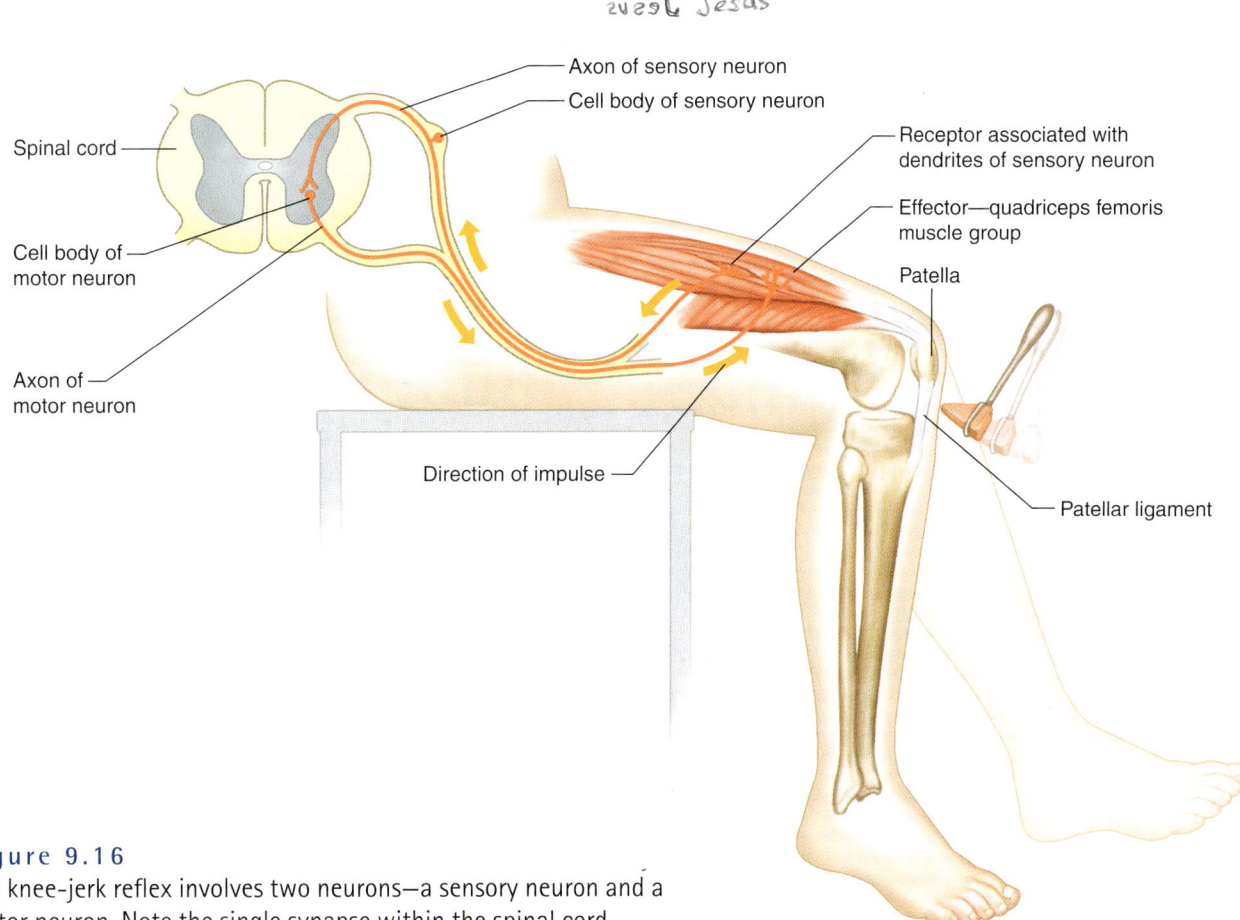

Figure 9.16
The knee-jerk reflex involves two neurons—a sensory neuron and a motor neuron. Note the single synapse within the spinal cord.

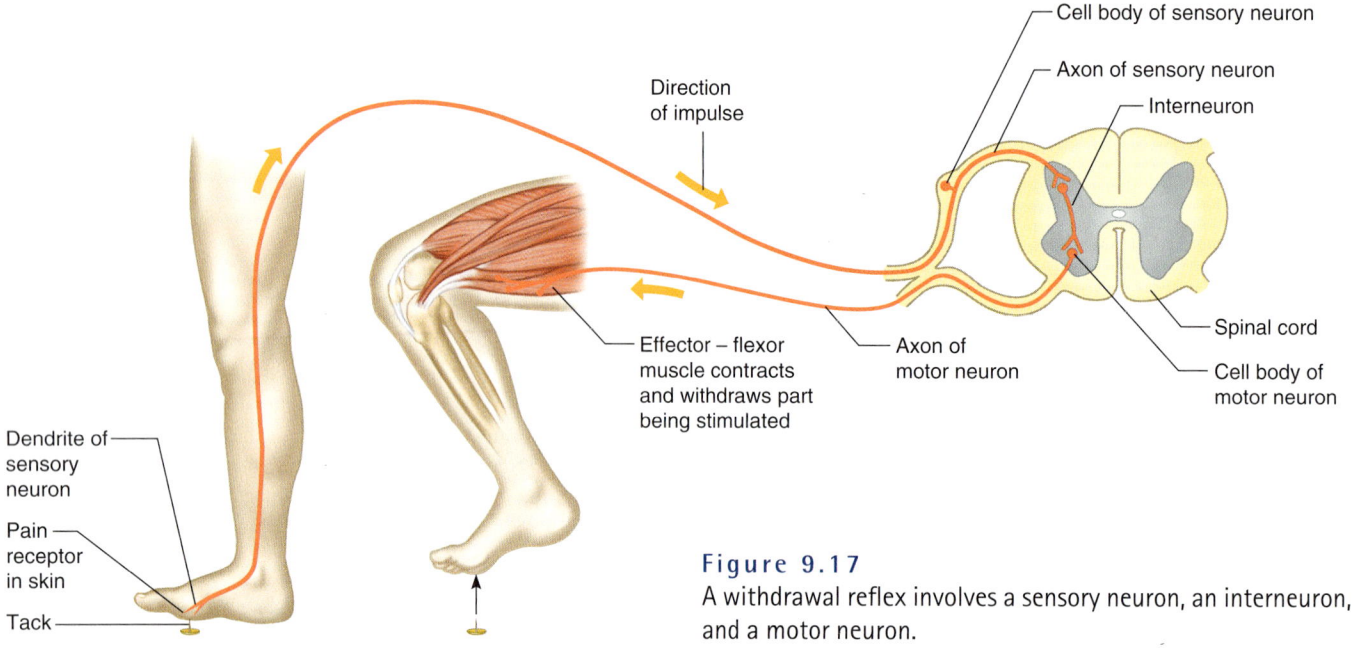

Figure 9.17
A withdrawal reflex involves a sensory neuron, an interneuron, and a motor neuron.

TABLE 9.4 — PARTS OF A REFLEX ARC

PART	DESCRIPTION	FUNCTION
Receptor	Receptor end of a dendrite or a specialized receptor cell in a sensory organ	Senses specific type of internal or external change
Sensory neuron	Dendrite, cell body, and axon of a sensory neuron	Transmits nerve impulse from receptor into brain or spinal cord
Interneuron	Dendrite, cell body, and axon of a neuron within the brain or spinal cord	Conducts nerve impulse from sensory neuron to motor neuron
Motor neuron	Dendrite, cell body, and axon of a motor neuron	Transmits nerve impulse from brain or spinal cord out to effector
Effector	Muscle or gland	Responds to stimulation by motor neuron and produces reflex or behavioral action

9.11 Meninges

Bones, membranes, and fluid surround the organs of the CNS. The brain lies within the cranial cavity of the skull, and the spinal cord occupies the vertebral canal within the vertebral column. Beneath these bony coverings, membranes called **meninges** (mě-nin′jēz) (singular, *meninx*), located between the bone and soft tissues of the nervous system, protect the brain and spinal cord (fig. 9.18A).

The meninges have three layers—dura mater, arachnoid mater, and pia mater (fig. 9.18B). The **dura mater** is the outermost layer of the meninges. It is composed primarily of tough, white, fibrous connective tissue and contains many blood vessels and nerves. It attaches to the inside of the cranial cavity and forms the internal periosteum of the surrounding skull bones. In some regions, the dura mater extends inward between lobes of the brain and forms partitions that support and protect these parts.

The dura mater continues into the vertebral canal as a strong, tubular sheath that surrounds the spinal cord. It terminates as a blind sac below the end of the cord. The membrane around the spinal cord is not attached directly to the vertebrae but is separated by an *epidural space,* which lies between the dural sheath and the bony walls (fig. 9.19). This space contains loose connective and adipose tissues, which provide a protective pad around the spinal cord.

The **arachnoid mater** is a thin, weblike membrane that lacks blood vessels and is located between the dura and pia maters. It spreads over the brain and spinal cord but generally does not dip into the grooves and depressions on their surfaces.

Between the arachnoid and pia maters is a *subarachnoid space* that contains the clear, watery **cerebrospinal fluid (CSF).** The **pia mater** is very thin and contains many nerves and blood vessels that nourish underlying cells of the brain and spinal cord. This layer hugs the surfaces of these organs and follows their irregular contours, passing over high areas and dipping into depressions.

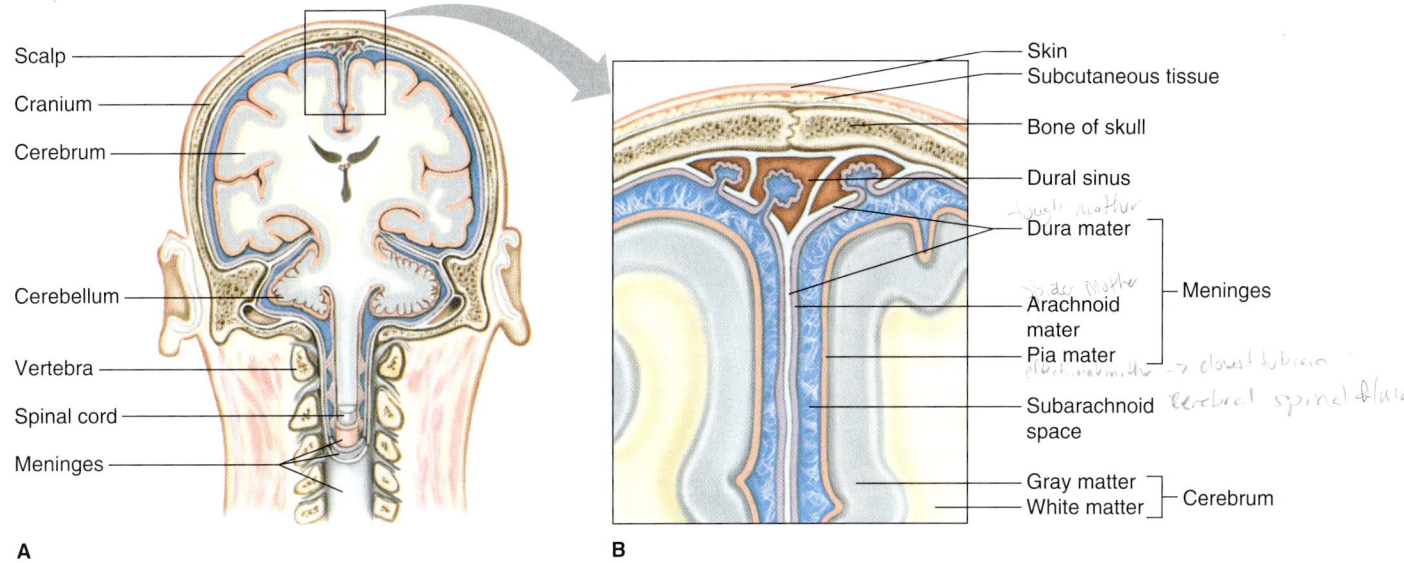

Figure 9.18
Meninges. (*A*) Membranes called meninges enclose the brain and spinal cord. (*B*) The meninges include three layers: dura mater, arachnoid mater, and pia mater.

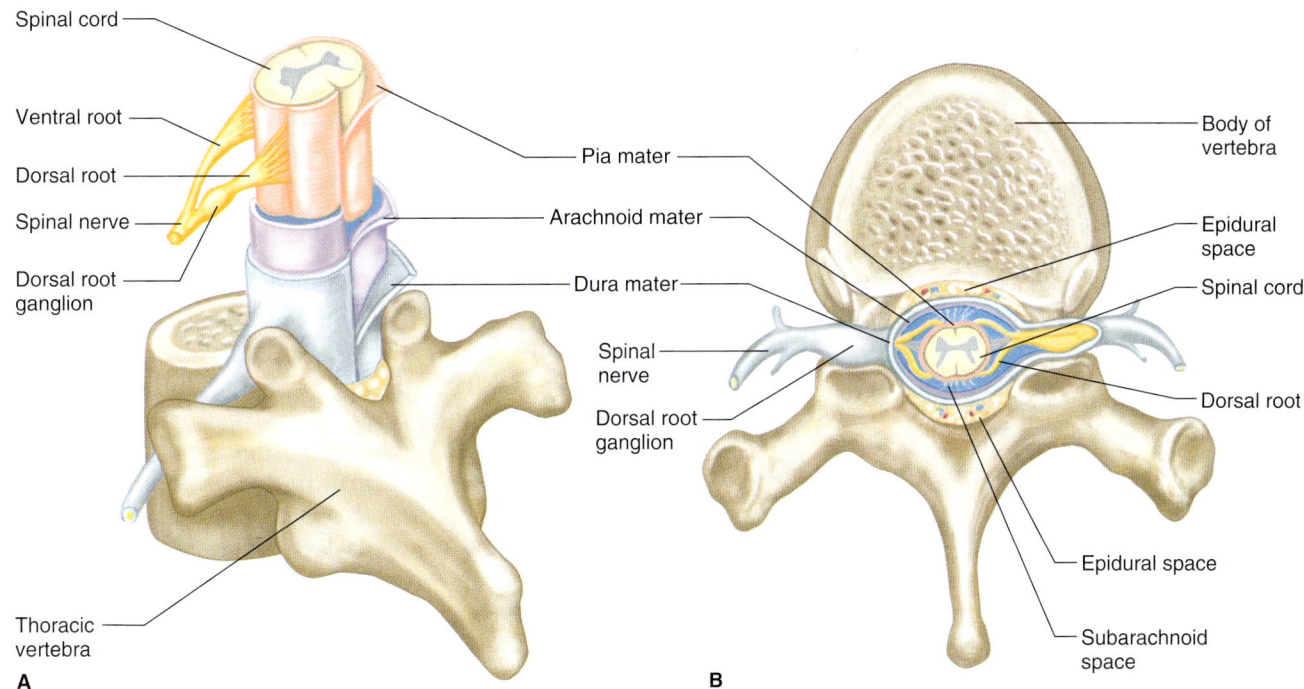

Figure 9.19
Meninges of the spinal cord. (*A*) The dura mater ensheaths the spinal cord. (*B*) Tissues forming a protective pad around the cord fill the epidural space between the dural sheath and the bone of the vertebra.

A blow to the head may break some blood vessels associated with the brain, and escaping blood may collect in the space beneath the dura mater. Such a *subdural hematoma* increases pressure between the rigid bones of the skull and the soft tissues of the brain. Unless the accumulating blood is evacuated, compression of the brain may lead to functional losses or even death.

CHECK YOUR RECALL

1. Describe the meninges.
2. Name the layers of the meninges.
3. State the location of cerebrospinal fluid.

9.12 Spinal Cord

The **spinal cord** is a slender nerve column that passes downward from the brain into the vertebral canal. Although continuous with the brain, the spinal cord is said to begin where nervous tissue leaves the cranial cavity at the level of the foramen magnum. The cord tapers to a point and terminates near the intervertebral disc that separates the first and second lumbar vertebrae (fig. 9.20).

Structure of the Spinal Cord

The spinal cord consists of thirty-one segments, each of which gives rise to a pair of **spinal nerves.** These nerves branch to various body parts and connect them with the central nervous system.

In the neck region, a thickening in the spinal cord, called the *cervical enlargement,* supplies nerves to the upper limbs. A similar thickening in the lower back, the *lumbar enlargement,* gives off nerves to the lower limbs (fig. 9.20).

Two grooves, a deep *anterior median fissure* and a shallow *posterior median sulcus,* extend the length of the spinal cord, dividing it into right and left halves (fig. 9.21). A cross section of the cord reveals that it consists of a core of gray matter within white matter. The pattern the gray matter produces roughly resembles a butterfly with its wings spread. The upper and lower wings of gray matter are called the *posterior horns* and *anterior horns,* respectively. Between them on either side in the thoracic and upper lumbar segments is a protrusion of gray matter called the *lateral horn.*

Neurons with large cell bodies located in the anterior horns give rise to motor fibers that pass out through spinal nerves to skeletal muscles. The majority of neurons in the gray matter of the spinal cord, however, are interneurons.

Gray matter divides the white matter of the spinal cord into three regions on each side—the *anterior, lateral,* and *posterior funiculi* (fig. 9.21A). Each funiculus consists of longitudinal bundles of myelinated nerve fibers that comprise major nerve pathways called **nerve tracts.**

A horizontal bar of gray matter in the middle of the spinal cord, the *gray commissure,* connects the wings of the gray matter on the right and left sides. This bar surrounds the **central canal,** which contains cerebrospinal fluid.

Functions of the Spinal Cord

The spinal cord has two major functions—conducting nerve impulses and serving as a center for spinal reflexes. The nerve tracts of the spinal cord consist of axons that provide a two-way communication system between the brain and body parts outside the nervous system. The tracts that carry sensory information to the brain are called **ascending tracts** (fig. 9.22); those that conduct motor impulses from the brain to muscles and glands are called **descending tracts** (fig. 9.23).

Typically, all the axons within a given tract originate from neuron cell bodies located in the same part of the nervous system and terminate together in some other part. The names that identify nerve tracts often reflect these common origins and terminations. For example, a *spinothalamic tract* begins in the spinal cord and carries sensory impulses associated with the sensations of pain, touch, and temperature to the thalamus of the brain. A *corticospinal tract* originates in the cortex of the brain and carries motor impulses downward through the spinal cord and spinal nerves. These impulses control skeletal muscle movements.

Corticospinal tracts are also called *pyramidal tracts* after the pyramid-shaped areas in the medulla oblongata of the brain through which they pass. Other descending tracts, called *extrapyramidal tracts,* control motor activities associated with maintaining balance and posture.

In addition to providing a pathway for nerve tracts, the spinal cord functions in many reflexes, including

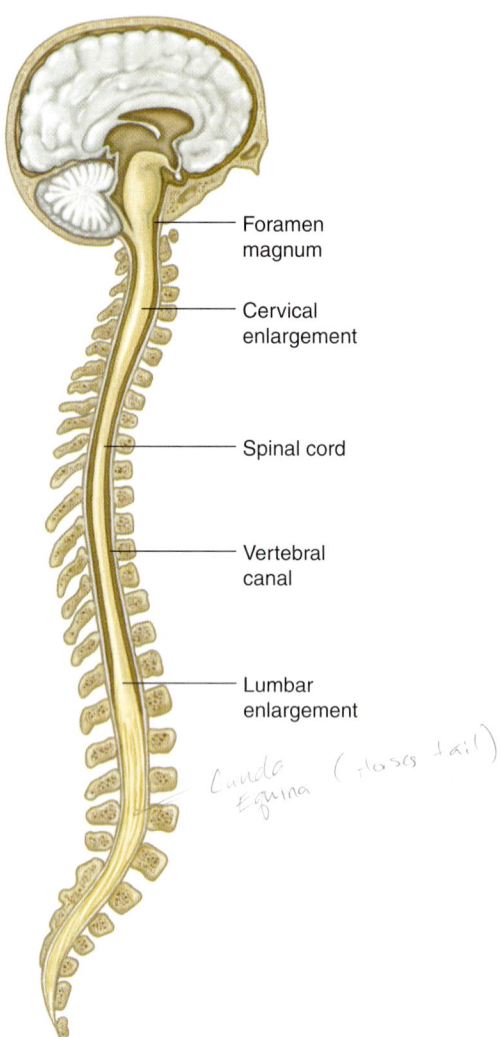

Figure 9.20
The spinal cord begins at the level of the foramen magnum.

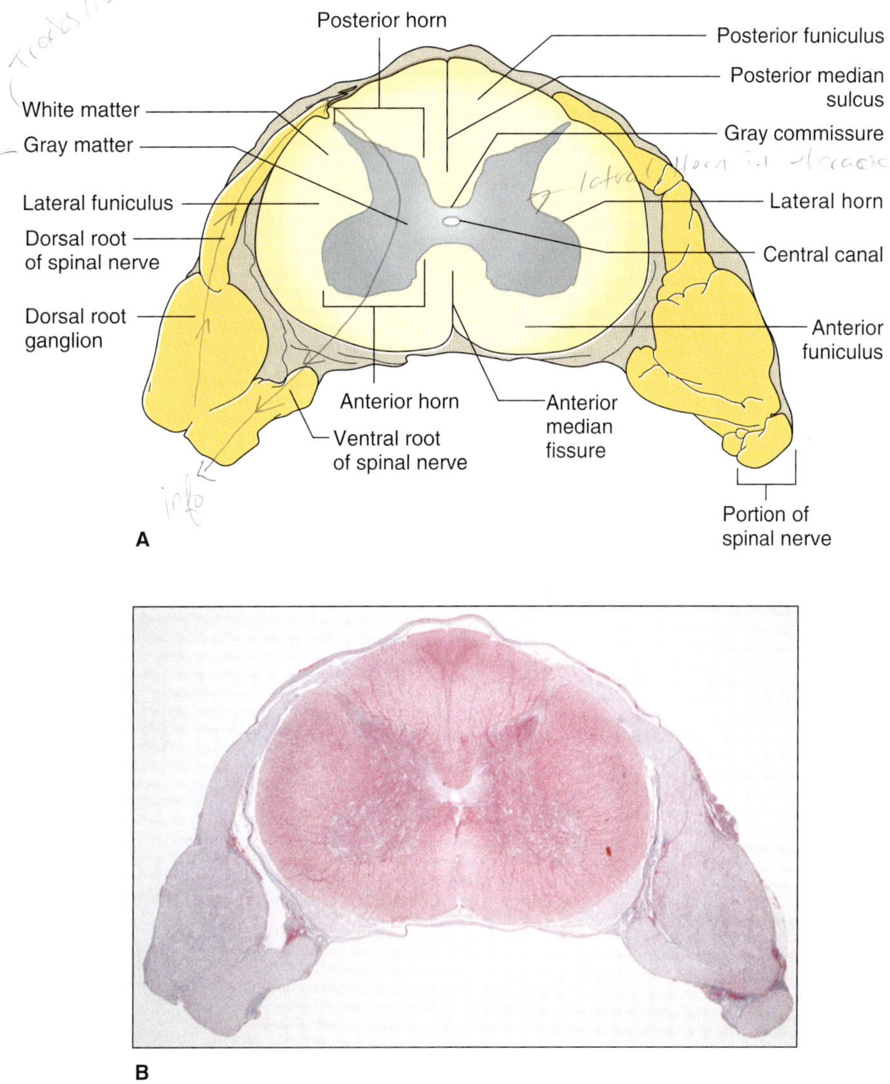

Figure 9.21
Spinal cord. (*A*) Cross section of the spinal cord. (*B*) Artificially stained micrograph of the cord (7.5×).

the knee-jerk and withdrawal reflexes described previously. These are called **spinal reflexes** because their reflex arcs pass through the spinal cord.

 Axons extend from the base of the spinal cord to the toes. If you stub your toe, a sensory message reaches the spinal cord in less than one hundredth of a second.

CHECK YOUR RECALL

1. Describe the structure of the spinal cord.
2. Describe the general functions of the spinal cord.
3. Distinguish between an ascending and a descending tract.

9.13 Brain

The **brain** is composed of about 100 billion (10^{11}) multipolar neurons and even more nerve fibers by which these neurons communicate with one another and with neurons in other parts of the nervous system. As figure 9.24 shows, the brain can be divided into three major portions—the cerebrum, the cerebellum, and the brain stem. The **cerebrum,** the largest part, contains nerve centers associated with sensory and motor functions and provides higher mental functions, including memory and reasoning. The **diencephalon** also processes sensory information. The **cerebellum** includes centers that coordinate voluntary muscular movements. Nerve pathways in the **brain stem** connect various parts of the nervous system and regulate certain visceral activities.

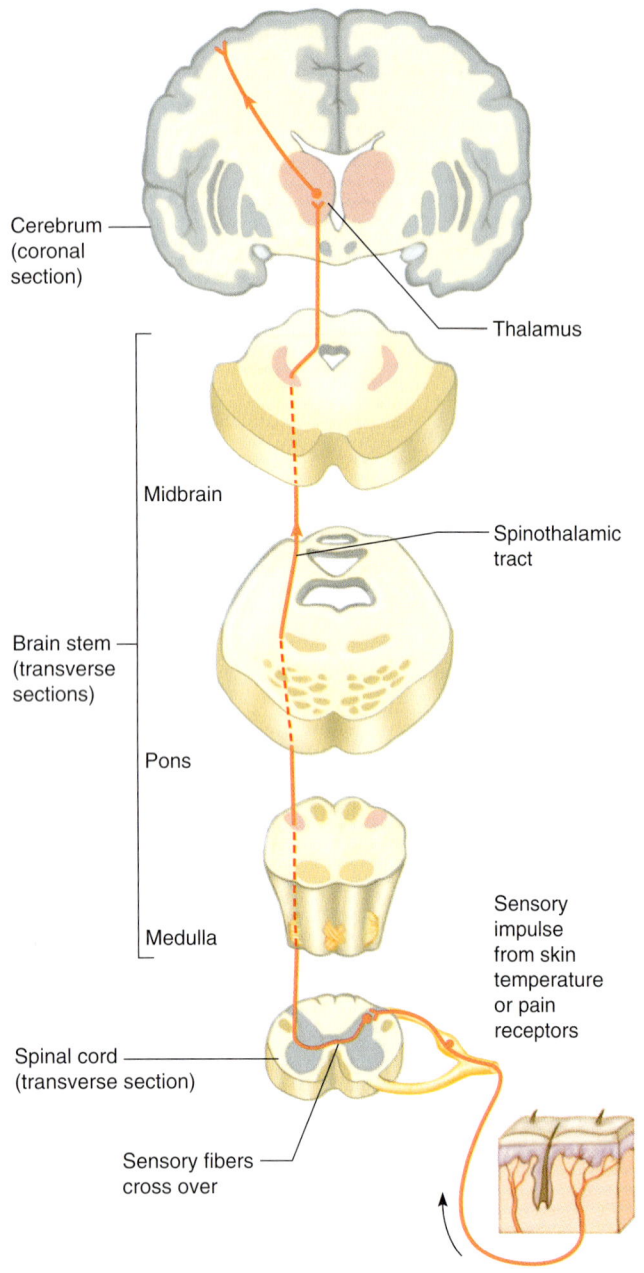

Figure 9.22
Sensory impulses originating in skin receptors cross over in the spinal cord and ascend to the brain. Other sensory tracts cross over in the medulla oblongata.

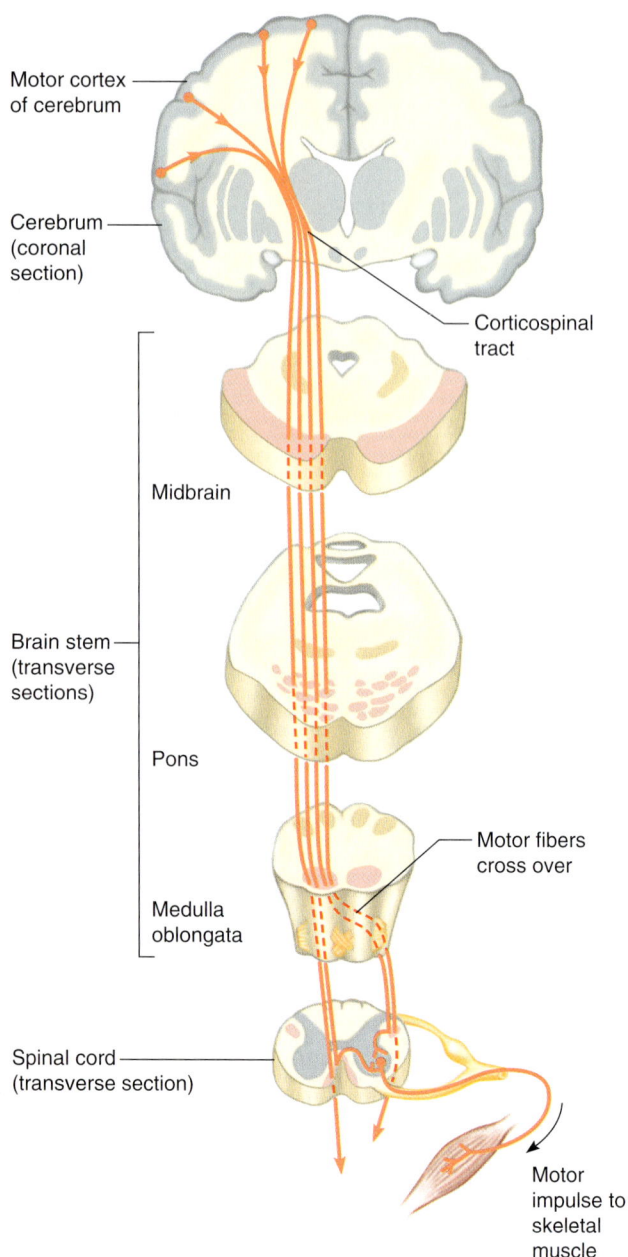

Figure 9.23
Motor fibers of the corticospinal tract begin in the cerebral cortex, cross over in the medulla oblongata, and descend in the spinal cord. There, they synapse with neurons whose fibers lead to the spinal nerves that supply skeletal muscles.

Structure of the Cerebrum

The cerebrum consists of two large masses called the left and right **cerebral hemispheres,** which are essentially mirror images of each other. A deep bridge of nerve fibers called the **corpus callosum** connects the cerebral hemispheres. A layer of dura mater (falx cerebri) separates them.

The surface of the cerebrum has many ridges, or **convolutions** (gyri), separated by grooves. A shallow groove is called a **sulcus,** and a deep groove is called a **fissure.** Although the arrangement of these elevations and depressions is complex, they form distinct patterns in all normal brains. For example, a *longitudinal fissure* separates the right and left cerebral hemispheres, a

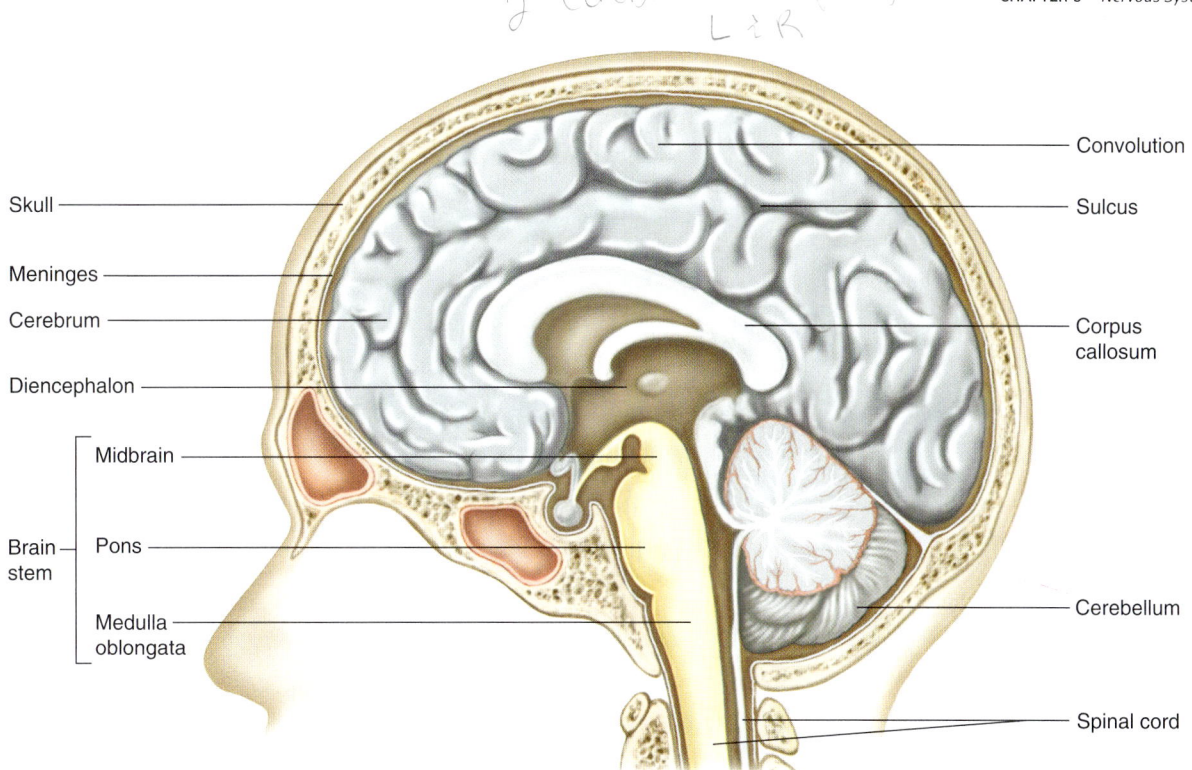

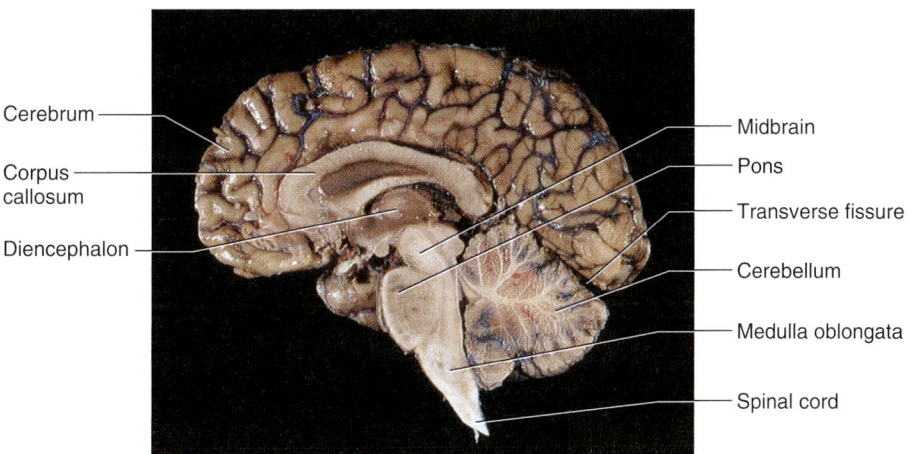

Figure 9.24
The major portions of the brain include the cerebrum, the diencephalon, the cerebellum, and the brain stem.

transverse fissure separates the cerebrum from the cerebellum, and several sulci divide each hemisphere into lobes.

The lobes of the cerebral hemispheres are named after the skull bones they underlie (fig. 9.25). They include:

1. **Frontal lobe** The frontal lobe forms the anterior portion of each cerebral hemisphere. It is bordered posteriorly by a *central sulcus,* which extends from the longitudinal fissure at a right angle, and inferiorly by a *lateral sulcus,* which extends from the undersurface of the brain along its sides.

2. **Parietal lobe** The parietal lobe is posterior to the frontal lobe and separated from it by the central sulcus.

3. **Temporal lobe** The temporal lobe lies below the frontal and parietal lobes and is separated from them by the lateral sulcus.

4. **Occipital lobe** The occipital lobe forms the posterior portion of each cerebral hemisphere and is separated from the cerebellum by a shelflike extension of dura mater (tentorium cerebelli). The boundary between the occipital lobe and the parietal and temporal lobes is not distinct.

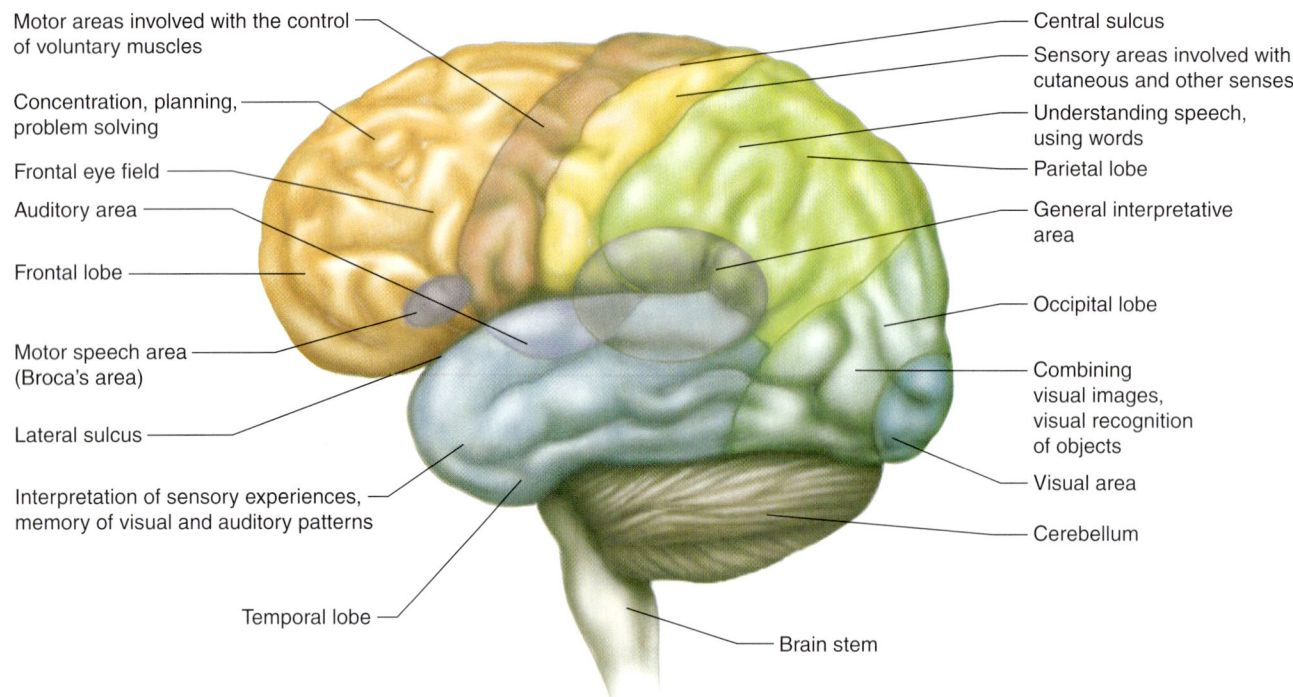

Figure 9.25
Some motor, sensory, and association areas of the left cerebral cortex.

5. **Insula** The insula is located deep within the lateral sulcus and is covered by parts of the frontal, parietal, and temporal lobes. A *circular sulcus* separates the insula from the lobes.

   A thin layer of gray matter called the **cerebral cortex** is the outermost portion of the cerebrum. This layer covers the convolutions and dips into the sulci and fissures. It contains nearly 75% of all the neuron cell bodies in the nervous system.

   Just beneath the cerebral cortex is a mass of white matter that makes up the bulk of the cerebrum. This mass contains bundles of myelinated axons that connect neuron cell bodies of the cortex with other parts of the nervous system. Some of these fibers pass from one cerebral hemisphere to the other by way of the corpus callosum, and others carry sensory or motor impulses from portions of the cortex to nerve centers in the brain or spinal cord.

Functions of the Cerebrum

The cerebrum provides higher brain functions. It contains centers for interpreting sensory impulses arriving from sense organs and centers for initiating voluntary muscular movements. The cerebrum stores the information of memory and utilizes it to reason. Intelligence and personality also stem from cerebral activity.

Functional Regions of the Cerebral Cortex

Specific regions of the cerebral cortex perform specific functions. Although functions overlap among regions, the cortex can be divided into motor, sensory, and association areas.

The primary **motor areas** of the cerebral cortex lie in the frontal lobes, just in front of the central sulcus (fig. 9.25). The nervous tissue in these regions contains numerous large *pyramidal cells,* named for their pyramid-shaped cell bodies.

Impulses from the pyramidal cells travel downward through the brain stem and into the spinal cord on the corticospinal tracts (see fig. 9.23). Most of the axons in these tracts cross over from one side of the brain to the other within the brain stem. As a result, the motor area of the right cerebral hemisphere generally controls skeletal muscles on the left side of the body, and vice versa.

In addition to the primary motor areas, certain other regions of the frontal lobe affect motor functions. For example, a region called the *motor speech area (Broca's area)* is just anterior to the primary motor cortex and superior to the lateral sulcus (fig. 9.25). It coordinates the complex muscular actions of the mouth, tongue, and larynx that make speech possible. Above Broca's area is a region called the *frontal eye field.* The motor cortex in this area controls voluntary movements of the eyes and eyelids. Another region just in front of the primary motor area controls the muscular movements of the hands and fingers that make skills such as writing possible.

Sensory areas located in several lobes of the cerebrum interpret impulses that arrive from sensory receptors, producing feelings or sensations. For example, sensations from all parts of the skin (cutaneous senses) arise in the anterior portions of the parietal lobes along the central sulcus (fig. 9.25). The posterior parts of the occipital lobes affect vision (visual area), and the temporal lobes contain the centers for hearing (auditory area). The sensory areas for taste are located near the bases of the central sulci along the lateral sulci, and the sense of smell arises from centers deep within the cerebrum.

Like motor fibers, sensory fibers cross over either in the spinal cord or in the brain stem (see fig. 9.22). Thus, the centers in the right cerebral hemisphere interpret impulses originating from the left side of the body, and vice versa.

Association areas are neither primarily sensory nor primarily motor. They connect with one another and with other brain structures. These areas analyze and interpret sensory experiences and oversee memory, reasoning, verbalizing, judgment, and emotion. Association areas occupy the anterior portions of the frontal lobes and are widespread in the lateral portions of the parietal, temporal, and occipital lobes (fig. 9.25).

Association areas of the frontal lobes control a number of higher intellectual processes. These include concentrating, planning, complex problem solving, and judging the possible consequences of behavior. Association areas of the parietal lobes help in understanding speech and choosing words to express thoughts and feelings. Association areas of the temporal lobes and regions at the posterior ends of the lateral fissures interpret complex sensory experiences, such as those needed to understand speech and to read. These regions also provide memory of visual scenes, music, and other complex sensory patterns. Association areas of the occipital lobes that are adjacent to the visual centers are important in analyzing visual patterns and combining visual images with other sensory experiences, as when one recognizes another person or an object.

The parietal, temporal, and occipital association areas meet near the posterior end of the lateral sulcus. This important region is called the *general interpretative area,* and it plays the primary role in complex thought processing (fig. 9.25).

CHECK YOUR RECALL

1. List the major divisions of the brain.
2. Describe the cerebral cortex.
3. What are the major functions of the cerebrum?
4. Locate the major functional regions of the cerebral cortex.

The effects of injuries to the cerebral cortex depend on the location and extent of the damage. The abilities that become impaired can indicate the site of damage. For example, injury to the motor areas of one frontal lobe causes partial or complete paralysis on the opposite side of the body.

A person with damage to the association areas of the frontal lobe may have difficulty concentrating on complex mental tasks and may appear disorganized and easily distracted. A person who suffers damage to association areas of the temporal lobes may have difficulty recognizing printed words or arranging words into meaningful thoughts.

Hemisphere Dominance

Both cerebral hemispheres participate in basic functions, such as receiving and analyzing sensory impulses, controlling skeletal muscles, and storing memory. However, in most persons, one side of the cerebrum is the **dominant hemisphere,** controlling the ability to use and understand language.

In more than 90% of the population, the left hemisphere is dominant for the language-related activities of speech, writing, and reading, and for complex intellectual functions requiring verbal, analytical, and computational skills. In other persons, the right hemisphere is dominant for the language-related abilities, or the hemispheres are equally dominant. Broca's area in the dominant hemisphere controls the muscles that function in speaking.

In addition to carrying on basic functions, the nondominant hemisphere specializes in nonverbal functions, such as motor tasks that require orientation of the body in space, understanding and interpreting musical patterns, and nonverbal visual experiences. The nondominant hemisphere also controls emotional and intuitive thinking.

Nerve fibers of the **corpus callosum,** which connect the cerebral hemispheres, allow the dominant hemisphere to control the motor cortex of the nondominant hemisphere (see fig. 9.24). These fibers also transfer sensory information reaching the nondominant hemisphere to the dominant one, where the information can be used in decision making.

Deep within each cerebral hemisphere are several masses of gray matter called **basal nuclei** (basal ganglia) (fig. 9.26). They are the *caudate nucleus,* the *putamen,* and the *globus pallidus.* Their neuron cell bodies serve as relay stations for motor impulses originating in the cerebral cortex and passing into the brain stem and spinal cord. In the process, the basal nuclei modify the pattern of these motor impulses and thereby help control various skeletal muscle activities. Neurons of the basal nuclei respond to the inhibitory neurotransmitter dopamine released from nearby cells.

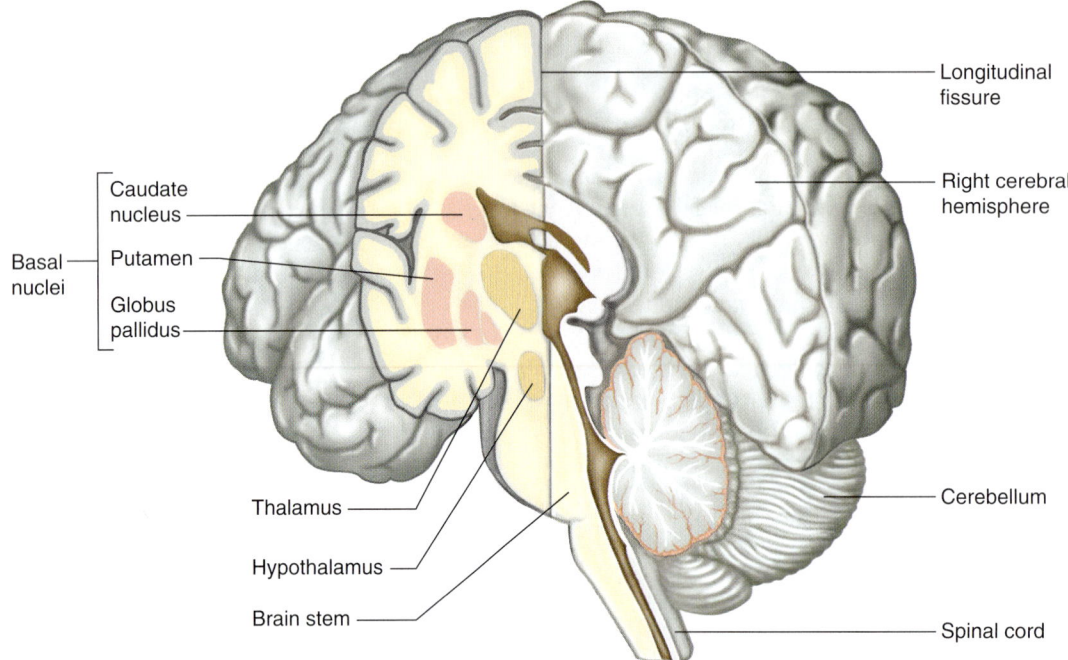

Figure 9.26
A coronal section of the left cerebral hemisphere reveals some of the basal nuclei.

The signs of Parkinson disease and Huntington disease result from altered activity of basal nuclei neurons. In the case of Parkinson disease, nearby neurons release less dopamine, and the basal nuclei become overactive, inhibiting movement. In Huntington disease, basal nuclei neurons gradually deteriorate, resulting in unrestrained movement. The cause is build-up of extra-long proteins in the cell nuclei.

Ventricles and Cerebrospinal Fluid

Within the cerebral hemispheres and brain stem is a series of interconnected cavities called **ventricles** (fig. 9.27). These spaces are continuous with the central canal of the spinal cord, and like it, they contain cerebrospinal fluid.

The largest ventricles are the *lateral ventricles* (first and second ventricles), which extend into the cerebral hemispheres and occupy portions of the frontal, temporal, and occipital lobes. A narrow space that constitutes the *third ventricle* is in the midline of the brain, beneath the corpus callosum. This ventricle communicates with the lateral ventricles through openings (interventricular foramina) in its anterior end. The *fourth ventricle* is in the brain stem just anterior to the cerebellum. A narrow canal, the *cerebral aqueduct,* connects it to the third ventricle and passes lengthwise through the brain stem. This ventricle is continuous with the central canal of the spinal cord and has openings in its roof that lead into the subarachnoid space of the meninges.

Tiny, reddish, cauliflower-like masses of specialized capillaries from the pia mater, called **choroid plexuses** (plek´sus-ez), secrete cerebrospinal fluid (fig. 9.28). These structures project into the ventricles. Most of the cerebrospinal fluid arises in the lateral ventricles. From there, it circulates slowly into the third and fourth ventricles and into the central canal of the spinal cord. Cerebrospinal fluid also enters the subarachnoid space of the meninges through the wall of the fourth ventricle near the cerebellum and completes its circuit by being reabsorbed into the blood.

Cerebrospinal fluid completely surrounds the brain and spinal cord because it occupies the subarachnoid space of the meninges. In effect, these organs float in the fluid, which supports and protects them by absorbing forces that might otherwise jar and damage them. Cerebrospinal fluid also maintains a stable ionic concentration in the central nervous system and provides a pathway to the blood for wastes.

Because cerebrospinal fluid is secreted and reabsorbed continuously, the fluid pressure in the ventricles normally remains relatively constant. An infection, a tumor, or a blood clot can interfere with fluid circulation, increasing pressure within the ventricles and thus in the cranial cavity (intracranial pressure). This can injure the brain by forcing it against the rigid skull.

A *lumbar puncture* (spinal tap) is used to measure the pressure of cerebrospinal fluid. In this procedure, a fine, hollow needle is inserted into the subarachnoid space between the third and fourth or between the fourth and fifth lumbar vertebrae. An instrument called a *manometer* measures the pressure.

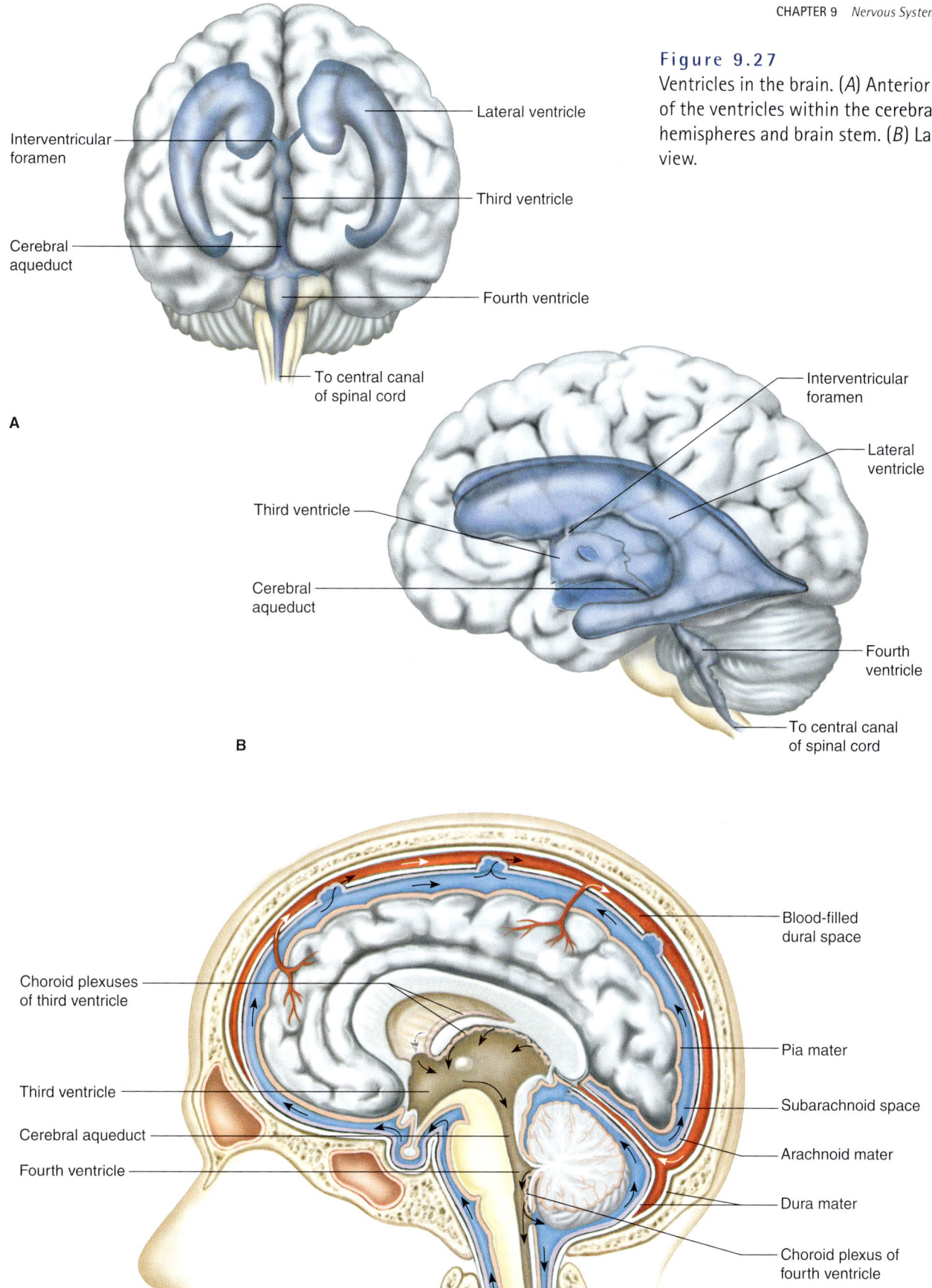

Figure 9.27
Ventricles in the brain. (*A*) Anterior view of the ventricles within the cerebral hemispheres and brain stem. (*B*) Lateral view.

Figure 9.28
The choroid plexuses in the walls of the ventricles secrete cerebrospinal fluid. The fluid circulates through the ventricles and central canal, enters the subarachnoid space, and is reabsorbed into the blood.

CHECK YOUR RECALL

1. What is hemisphere dominance?
2. What are the major functions of the dominant hemisphere? The nondominant one?
3. Where are the ventricles of the brain?
4. Describe the circulation of cerebrospinal fluid.

Diencephalon

The **diencephalon** is located between the cerebral hemispheres and above the midbrain. It surrounds the third ventricle and is composed largely of gray matter. Within the diencephalon, a dense mass called the **thalamus** bulges into the third ventricle from each side (figs. 9.26 and 9.29). Another region of the diencephalon that includes many nuclei is the **hypothalamus.** It lies below the thalamus and forms the lower walls and floor of the third ventricle.

Other parts of the diencephalon include: (1) the **optic tracts** and the **optic chiasma** that is formed by optic nerve fibers crossing over each other; (2) the **infundibulum,** a conical process behind the optic chiasma to which the pituitary gland attaches; (3) the **posterior pituitary gland,** which hangs from the floor of the hypothalamus; (4) the **mammillary bodies,** which appear as two rounded structures behind the infundibulum; and (5) the **pineal gland,** a cone-shaped structure attached to the upper portion of the diencephalon (see chapter 11, p. 297).

The thalamus is a central relay station for sensory impulses ascending from other parts of the nervous system to the cerebral cortex. It receives all sensory impulses (except those associated with the sense of smell) and channels them to the appropriate regions of the cortex for interpretation. In addition, all regions of the cerebral cortex can communicate with the thalamus by means of descending fibers.

The cerebral cortex pinpoints the origin of sensory stimulation. The thalamus produces a general awareness of certain sensations, such as pain, touch, and temperature.

Nerve fibers connect the hypothalamus to the cerebral cortex, thalamus, and other parts of the brain stem. The hypothalamus maintains homeostasis by regulating a variety of visceral activities and by linking the nervous and endocrine systems.

The hypothalamus regulates:

1. Heart rate and arterial blood pressure
2. Body temperature
3. Water and electrolyte balance
4. Control of hunger and body weight

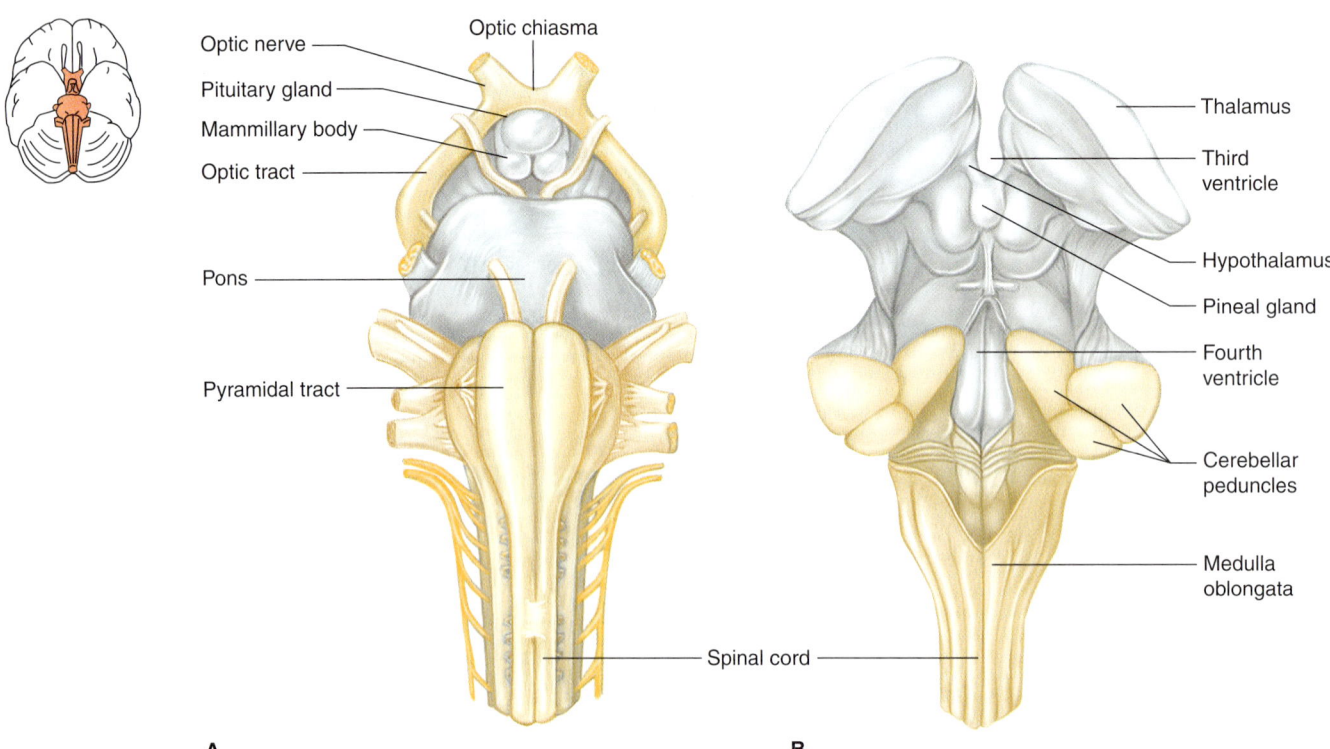

Figure 9.29

Brain stem. (*A*) Ventral view of the brain stem. (*B*) Dorsal view of the brain stem with the cerebellum removed, exposing the fourth ventricle.

5. Control of movements and glandular secretions of the stomach and intestines
6. Production of neurosecretory substances that stimulate the pituitary gland to secrete hormones
7. Sleep and wakefulness

Structures in the general region of the diencephalon also control emotional responses. For example, portions of the cerebral cortex in the medial parts of the frontal and temporal lobes interconnect with a number of deep masses of gray matter called *nuclei,* including the hypothalamus, thalamus, and basal nuclei. Together, these structures comprise a complex called the **limbic system.**

The limbic system controls emotional experience and expression. It can modify the way a person acts by producing such feelings as fear, anger, pleasure, and sorrow. The limbic system recognizes upsets in a person's physical or psychological condition that might threaten life. By causing pleasant or unpleasant feelings about experiences, the limbic system guides a person into behavior that is likely to increase the chance of survival.

Brain Stem

The **brain stem** is a bundle of nervous tissue that connects the cerebrum to the spinal cord. It consists of numerous tracts of nerve fibers and several nuclei. The parts of the brain stem include the midbrain, pons, and medulla oblongata (figs. 9.24 and 9.29).

Midbrain

The **midbrain** is a short section of the brain stem between the diencephalon and the pons (see fig. 9.24). It contains bundles of myelinated axons that join lower parts of the brain stem and spinal cord with higher parts of the brain. Two prominent bundles of axons on the underside of the midbrain include the corticospinal tracts and are the main motor pathways between the cerebrum and lower parts of the nervous system.

The midbrain includes several masses of gray matter that serve as reflex centers. For example, the midbrain contains the centers for certain visual reflexes, such as those responsible for moving the eyes to view something as the head turns. It also contains the auditory reflex centers that enable a person to move the head to hear sounds more distinctly.

Pons

The **pons** is a rounded bulge on the underside of the brain stem, where it separates the midbrain from the medulla oblongata (see fig. 9.24). The dorsal portion of the pons consists largely of longitudinal nerve fibers, which relay impulses to and from the medulla oblongata and the cerebrum. The ventral portion of the pons contains large bundles of transverse nerve fibers, which transmit impulses from the cerebrum to centers within the cerebellum.

Several nuclei of the pons relay sensory impulses from peripheral nerves to higher brain centers. Other nuclei function with centers of the medulla oblongata to regulate the rate and depth of breathing (see chapter 16, p. 447).

Medulla Oblongata

The **medulla oblongata** extends from the pons to the foramen magnum of the skull (see fig. 9.24). Its dorsal surface flattens to form the floor of the fourth ventricle, and its ventral surface is marked by the corticospinal tracts, most of whose fibers cross over at this level (see fig. 9.23).

All the ascending and descending nerve fibers connecting the brain and spinal cord must pass through the medulla oblongata because of its location. As in the spinal cord, the white matter of the medulla oblongata surrounds a central mass of gray matter. Here, however, the gray matter breaks into nuclei separated by nerve fibers. Some of these nuclei relay ascending impulses to the other side of the brain stem and then on to higher brain centers.

Other nuclei within the medulla oblongata control vital visceral activities. These centers include:

1. **Cardiac center** Impulses originating in the cardiac center are transmitted to the heart on peripheral nerves, altering heart rate.
2. **Vasomotor center** Certain cells of the vasomotor center initiate impulses that travel to smooth muscles in the walls of certain blood vessels and stimulate them to contract. This constricts the blood vessels (vasoconstriction), elevating blood pressure. Other cells of the vasomotor center produce the opposite effect—dilating blood vessels (vasodilation) and consequently dropping blood pressure.
3. **Respiratory center** The respiratory center acts with centers in the pons to regulate the rate, rhythm, and depth of breathing.

Still other nuclei within the medulla oblongata are centers for the reflexes associated with coughing, sneezing, swallowing, and vomiting.

Reticular Formation

Scattered throughout the medulla oblongata, pons, and midbrain is a complex network of nerve fibers associated with tiny islands of gray matter. This network, the **reticular formation** (reticular activating system), extends from the upper portion of the spinal cord into the diencephalon. Its nerve fibers join centers of the hypothalamus, basal nuclei, cerebellum, and cerebrum with fibers in all the major ascending and descending tracts.

When sensory impulses reach the reticular formation, it responds by activating the cerebral cortex into a state of wakefulness. Without this arousal, the cortex remains unaware of stimulation and cannot interpret sensory information or carry on thought processes. Thus, decreased activity in the reticular formation results in sleep. If the reticular formation is injured so that it cannot function, the person remains unconscious and cannot be aroused, even with strong stimulation. This is called a comatose state.

CHECK YOUR RECALL

1. What are the major functions of the thalamus? The hypothalamus?
2. How may the limbic system influence behavior?
3. List the structures of the brain stem.
4. What vital reflex centers are located in the brain stem?
5. What is the function of the reticular formation?

Cerebellum

The **cerebellum** is a large mass of tissue located below the occipital lobes of the cerebrum and posterior to the pons and medulla oblongata (see fig. 9.24). It consists of two lateral hemispheres partially separated by a layer of dura mater (falx cerebelli) and connected in the midline by a structure called the *vermis*. Like the cerebrum, the cerebellum is composed primarily of white matter, with a thin layer of gray matter, the **cerebellar cortex,** on its surface.

The cerebellum communicates with other parts of the central nervous system by means of three pairs of nerve tracts called *cerebellar peduncles* (figs. 9.29 and 9.30). One pair (the inferior peduncles) brings sensory information concerning the position of the limbs, joints, and other body parts to the cerebellum. Another pair (the middle peduncles) transmits signals from the cerebral cortex to the cerebellum concerning the desired positions of these parts. After integrating and analyzing this information, the cerebellum sends correcting impulses via a third pair (the superior peduncles) to the midbrain. These corrections are incorporated into motor impulses that travel downward through the pons, medulla oblongata, and spinal cord in the appropriate patterns to move the body in the desired way.

The cerebellum is a reflex center for integrating sensory information concerning the position of body parts and for coordinating complex skeletal muscle movements. It also helps maintain posture. Damage to the cerebellum is likely to result in tremors, inaccurate movements of voluntary muscles, loss of muscle tone, a reeling walk, and loss of equilibrium.

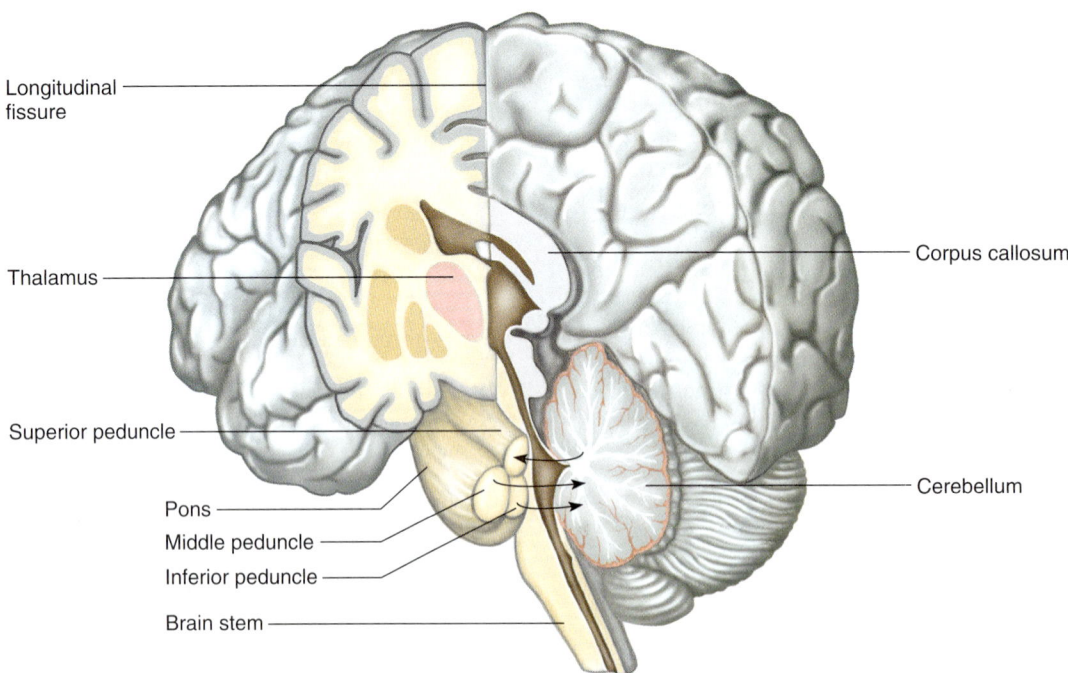

Figure 9.30

The cerebellum, which is located below the occipital lobes of the cerebrum, communicates with other parts of the nervous system by means of the cerebellar peduncles.

Topic of Interest

DRUG ABUSE

Drug abuse is the chronic self-administration of a drug in doses high enough to cause *addiction*—a physical or psychological dependence in which the user is preoccupied with locating and taking the drug. Stopping drug use causes intense, unpleasant withdrawal symptoms. Prolonged and repeated abuse of a drug may also result in *drug tolerance,* in which the physiological response to a particular dose of the drug becomes less intense over time. Drug tolerance results as the drug increases synthesis of certain liver enzymes, which metabolize the drug more rapidly, so that the addict needs the next dose sooner. Drug tolerance also arises from physiological changes that lessen the drug's effect on its target cells.

The most commonly abused drugs are central nervous system depressants ("downers"), CNS stimulants ("uppers"), hallucinogens, and anabolic steroids.

CNS depressants include barbiturates, benzodiazepines, opiates, and cannabinoids. *Barbiturates,* such as amytal, nembutal, and seconal, act uniformly throughout the brain; however, the reticular formation is particularly sensitive to their effects. CNS depression occurs due to inhibited secretion of certain excitatory and inhibitory neurotransmitters. Effects range from mild calming of the nervous system (sedation) to sleep, loss of sensory sensations (anesthesia), respiratory distress, cardiovascular collapse, and death.

The *benzodiazepines,* such as diazepam (Valium), depress activity in the limbic system and the reticular formation. Low doses are commonly prescribed to relieve anxiety. Higher doses cause sedation, sleep, or anesthesia. These drugs increase either the activity or release of the inhibitory neurotransmitter GABA. When benzodiazepines are metabolized, they may form other biochemicals that have depressing effects.

The *opiates* include heroin (which has no legal use in the United States), codeine, hydromorphone (Dilaudid), meperidine (Demerol), and methadone. These drugs stimulate certain receptors (opioid receptors) in the CNS, and when taken in prescribed dosages, they sedate and relieve pain (analgesia). Opiates cause both physical and psychological dependence. Effects of overdose include a feeling of well-being (euphoria), respiratory distress, convulsions, coma, and possible death. On the other hand, these drugs are very important in treating chronic, severe pain. For example, cancer patients find pain relief with OxyContin (oxycodone), which is taken twice daily in a timed-release pill. Many people abuse this drug, and by breaking the pills, release high doses rapidly, which can be deadly.

The *cannabinoids* include marijuana and hashish, both derived from the hemp plant. Hashish is several times more potent than marijuana. These drugs depress higher brain centers and release lower brain centers from the normal inhibitory influence of the higher centers. This induces an anxiety-free state, characterized by euphoria and a distorted perception of time and space. *Hallucinations* (sensory perceptions that have no external stimuli), respiratory distress, and vasomotor depression may occur with higher doses.

CNS stimulants include amphetamines and cocaine (including "crack"). These drugs have great abuse potential and may quickly produce psychological dependence. Cocaine, especially when smoked or inhaled, produces euphoria but may also change personality, cause seizures, and constrict certain blood vessels, leading to sudden death from stroke or cardiac arrhythmia. Cocaine's very rapid effect, and perhaps its addictiveness, reflect its rapid entry and metabolism in the brain. Cocaine arrives at the basal ganglia in 4 to 6 minutes and is cleared mostly within 30 minutes. The drug inhibits transporter molecules that remove dopamine from synapses after it is released. "Ecstasy" is a form of methamphetamine.

Hallucinogens alter perceptions. They cause *illusions,* which are distortions of vision, hearing, taste, touch, and smell; *synesthesia,* such as "hearing" colors or "feeling" sounds; and hallucinations. The most commonly abused and most potent hallucinogen is lysergic acid diethylamide (LSD). LSD may act as an excitatory neurotransmitter. A person under the influence of LSD may greatly overestimate physical capabilities, such as believing he or she can fly off the top of a high building. Phencyclidine (PCP) is another commonly abused hallucinogen. Its use can lead to prolonged psychosis that may provoke assault, murder, and suicide.

CHECK YOUR RECALL

1. Where is the cerebellum located?
2. What are the major functions of the cerebellum?

9.14 Peripheral Nervous System

The peripheral nervous system (PNS) consists of nerves that branch out from the CNS and connect it to other body parts. The PNS includes the cranial nerves, which arise from the brain, and the spinal nerves, which arise from the spinal cord.

The PNS can also be subdivided into the somatic and autonomic nervous systems. Generally, the **somatic** (so-mat´ik) **nervous system** consists of the cranial and spinal nerve fibers that connect the CNS to the skin and skeletal muscles; it oversees conscious activities. The **autonomic** (aw´´to-nom´ik) **nervous system** includes fibers that connect the CNS to viscera,

TABLE 9.5 — SUBDIVISIONS OF THE NERVOUS SYSTEM

1. Central nervous system (CNS)
   a. Brain
   b. Spinal cord
2. Peripheral nervous system (PNS)
   a. Cranial nerves arising from the brain
      (1) Somatic fibers connecting to skin and skeletal muscles
      (2) Autonomic fibers connecting to viscera
   b. Spinal nerves arising from the spinal cord
      (1) Somatic fibers connecting to skin and skeletal muscles
      (2) Autonomic fibers connecting to viscera

such as the heart, stomach, intestines, and glands; it controls unconscious activities. Table 9.5 outlines the subdivisions of the nervous system.

Cranial Nerves

Twelve pairs of **cranial nerves** arise from the underside of the brain (fig. 9.31). Except for the first pair, which begins within the cerebrum, these nerves originate from the brain stem. They pass from their sites of origin through foramina of the skull and lead to parts of the head, neck, and trunk.

Most of the cranial nerves are mixed nerves, but some of those associated with special senses, such as smell and vision, contain only sensory fibers. Other cranial nerves that affect muscles and glands are composed primarily of motor fibers.

Sensory fibers present in the cranial nerves have neuron cell bodies that are outside the brain, usually in groups called *ganglia*. On the other hand, motor neuron cell bodies are typically located within the gray matter of the brain.

Numbers or names designate the cranial nerves. The numbers indicate the order in which the nerves arise from the front to the back of the brain, and the names describe their primary functions or the general distribution of their fibers (fig. 9.31).

The first pair of cranial nerves, the **olfactory nerves (I)**, are associated with the sense of smell and contain only axons of sensory neurons. These bipolar neurons are located in the lining of the upper nasal cavity where they serve as *olfactory receptor cells*. Axons from these receptors pass upward through the cribriform plates of the ethmoid bone. Their impulses reach the olfactory neurons in the *olfactory bulbs,* which are extensions of the cerebral cortex located just beneath the frontal lobes (see fig. 10.4, p. 257). Sensory impulses travel from the olfactory bulbs along *olfactory tracts* to cerebral centers where they are interpreted. The result of this interpretation is the sensation of smell.

The second pair of cranial nerves, the **optic nerves (II)**, lead from the eyes to the brain and are associated with vision. The sensory nerve cell bodies of these nerve fibers are in ganglion cell layers within the eyes, and their axons pass through the *optic foramina* of the orbits and continue into the visual nerve pathways of the brain (see chapter 10, p. 275). Sensory impulses

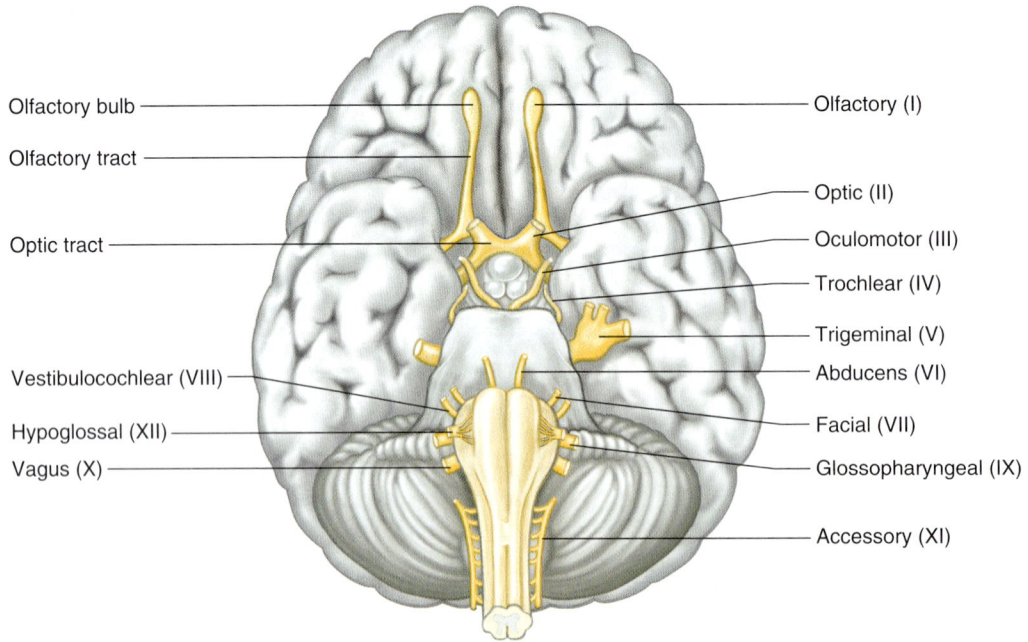

Figure 9.31
The cranial nerves, except for the first pair, arise from the brain stem. They are identified either by numbers indicating their order, their function, or the general distribution of their fibers.

transmitted on the optic nerves are interpreted in the visual cortices of the occipital lobes.

The third pair of cranial nerves, the **oculomotor nerves (III),** arise from the midbrain and pass into the orbits of the eyes. One component of each nerve connects to the voluntary muscles that raise the eyelid and to four of the six muscles that move the eye. A second component of each oculomotor nerve is part of the autonomic nervous system and supplies involuntary muscles within the eyes. These muscles adjust the amount of light entering the eyes and focus the lenses.

The fourth pair, the **trochlear nerves (IV),** arise from the midbrain and are the smallest cranial nerves. Each nerve carries motor impulses to a fifth voluntary muscle that moves the eye and is not innervated by the oculomotor nerve.

The fifth pair, the **trigeminal nerves (V),** are the largest cranial nerves and arise from the pons. They are mixed nerves, with the sensory portions more extensive than the motor portions. Each sensory component includes three large branches, called the ophthalmic, maxillary, and mandibular divisions.

The *ophthalmic division* of the trigeminal nerves consists of sensory fibers that bring impulses to the brain from the surface of the eyes, the tear glands, and the skin of the anterior scalp, forehead, and upper eyelids. The fibers of the *maxillary division* carry sensory impulses from the upper teeth, upper gum, and upper lip, as well as from the mucous lining of the palate and the skin of the face. The *mandibular division* includes both motor and sensory fibers. The sensory branches transmit impulses from the scalp behind the ears, the skin of the jaw, the lower teeth, the lower gum, and the lower lip. The motor branches supply the muscles of mastication and certain muscles in the floor of the mouth.

The sixth pair of cranial nerves, the **abducens nerves (VI),** are quite small and originate from the pons near the medulla oblongata. Each nerve enters the orbit of the eye and supplies motor impulses to the remaining muscle that moves the eye.

The seventh pair of cranial nerves, the **facial nerves (VII),** arise from the lower part of the pons and emerge on the sides of the face. Their sensory branches are associated with taste receptors on the anterior two-thirds of the tongue, and some of their motor fibers transmit impulses to muscles of facial expression. Still other motor fibers of these nerves function in the autonomic nervous system and stimulate secretions from tear glands and salivary glands.

The eighth pair of cranial nerves, the **vestibulocochlear nerves (VIII),** are sensory nerves that arise from the medulla oblongata. Each of these nerves has two distinct parts—a vestibular branch and a cochlear branch.

The neuron cell bodies of the *vestibular branch* fibers are located in ganglia associated with parts of the inner ear. These parts contain the receptors involved with reflexes that help maintain equilibrium. The neuron cell bodies of the *cochlear branch* fibers are located in the parts of the inner ear that house the hearing receptors. Impulses from these branches pass through the pons and medulla oblongata on their way to the temporal lobes, where they are interpreted.

The ninth pair of cranial nerves, the **glossopharyngeal nerves (IX),** are associated with the tongue and pharynx. These mixed nerves arise from the medulla oblongata, with predominantly sensory fibers. These sensory fibers carry impulses from the linings of the pharynx, tonsils, and posterior third of the tongue to the brain. Fibers in the motor component innervate muscles of the pharynx that function in swallowing.

The tenth pair of cranial nerves, the **vagus nerves (X),** originate in the medulla oblongata and extend downward through the neck into the chest and abdomen. These nerves are mixed, containing both somatic and autonomic branches, with autonomic fibers predominant. Certain somatic motor fibers carry impulses to muscles of the larynx that are associated with speech and swallowing. Autonomic motor fibers of the vagus nerves supply the heart and many smooth muscles and glands in the thorax and abdomen.

The eleventh pair of cranial nerves, the **accessory nerves (XI),** originate in the medulla oblongata and the spinal cord; thus, they have both cranial and spinal branches. Each *cranial branch* joins a vagus nerve and carries impulses to muscles of the soft palate, pharynx, and larynx. The *spinal branch* descends into the neck and supplies motor fibers to the trapezius and sternocleidomastoid muscles.

The twelfth pair of cranial nerves, the **hypoglossal nerves (XII),** arise from the medulla oblongata and pass into the tongue. They include motor fibers that carry impulses to muscles that move the tongue in speaking, chewing, and swallowing. Table 9.6 summarizes the functions of the cranial nerves.

CHECK YOUR RECALL

1. Define *peripheral nervous system.*
2. Distinguish between somatic and autonomic nerve fibers.
3. Name the cranial nerves, and list the major functions of each.

The consequences of a cranial nerve injury depend on the injury's location and extent. For example, damage to one member of a nerve pair limits loss of function to the affected side, but injury to both nerves affects both sides. Also, if a nerve is severed completely, the functional loss is total; if the cut is incomplete, the loss may be partial.

TABLE 9.6 — FUNCTIONS OF CRANIAL NERVES

NERVE	TYPE	FUNCTION
I Olfactory	Sensory	Sensory fibers transmit impulses associated with the sense of smell.
II Optic	Sensory	Sensory fibers transmit impulses associated with the sense of vision.
III Oculomotor	Primarily motor	Motor fibers transmit impulses to muscles that raise eyelids, move eyes, adjust the amount of light entering the eyes, and focus lenses. Some sensory fibers transmit impulses associated with the condition of muscles.
IV Trochlear	Primarily motor	Motor fibers transmit impulses to muscles that move the eyes. Some sensory fibers transmit impulses associated with the condition of muscles.
V Trigeminal	Mixed	
Ophthalmic division		Sensory fibers transmit impulses from the surface of the eyes, tear glands, scalp, forehead, and upper eyelids.
Maxillary division		Sensory fibers transmit impulses from the upper teeth, upper gum, upper lip, lining of the palate, and skin of the face.
Mandibular division		Sensory fibers transmit impulses from the skin of the jaw, lower teeth, lower gum, and lower lip. Motor fibers transmit impulses to muscles of mastication and to muscles in the floor of the mouth.
VI Abducens	Primarily motor	Motor fibers transmit impulses to muscles that move the eyes. Some sensory fibers transmit impulses associated with the condition of muscles.
VII Facial	Mixed	Sensory fibers transmit impulses associated with taste receptors of the anterior tongue. Motor fibers transmit impulses to muscles of facial expression, tear glands, and salivary glands.
VIII Vestibulocochlear	Sensory	
Vestibular branch		Sensory fibers transmit impulses associated with the sense of equilibrium.
Cochlear branch		Sensory fibers transmit impulses associated with the sense of hearing.
IX Glossopharyngeal	Mixed	Sensory fibers transmit impulses from the pharynx, tonsils, posterior tongue, and carotid arteries. Motor fibers transmit impulses to muscles of the pharynx used in swallowing and to salivary glands.
X Vagus	Mixed	Somatic motor fibers transmit impulses to muscles associated with speech and swallowing; autonomic motor fibers transmit impulses to the heart, smooth muscles, and glands in the thorax and abdomen. Sensory fibers transmit impulses from the pharynx, larynx, esophagus, and viscera of the thorax and abdomen.
XI Accessory	Primarily motor	
Cranial branch		Motor fibers transmit impulses to muscles of the soft palate, pharynx, and larynx.
Spinal branch		Motor fibers transmit impulses to muscles of the neck and back.
XII Hypoglossal	Primarily motor	Motor fibers transmit impulses to muscles that move the tongue.

Spinal Nerves

Thirty-one pairs of **spinal nerves** originate from the spinal cord (fig. 9.32). They are mixed nerves that provide two-way communication between the spinal cord and parts of the upper and lower limbs, neck, and trunk.

Spinal nerves are not named individually, but are grouped according to the level from which they arise. Each nerve is numbered in sequence. Thus, there are eight pairs of *cervical nerves* (numbered C1 to C8), twelve pairs of *thoracic nerves* (numbered T1 to T12), five pairs of *lumbar nerves* (numbered L1 to L5), five pairs of *sacral nerves* (numbered S1 to S5), and one pair of *coccygeal nerves* (Co).

The adult spinal cord ends at the level between the first and second lumbar vertebrae. The lumbar, sacral, and coccygeal nerves descend beyond the end of the cord, forming a structure called the *cauda equina* (horse's tail).

Each spinal nerve emerges from the cord by two short branches, or *roots,* which lie within the vertebral column. The **dorsal root** (posterior or sensory root) can be identified by an enlargement called the *dorsal root ganglion* (see fig. 9.21). This ganglion contains the cell bodies of the sensory neurons whose dendrites conduct impulses inward from the peripheral body parts. The axons of these neurons extend through the dorsal root and into the spinal cord, where they form synapses with dendrites of other neurons. The **ventral**

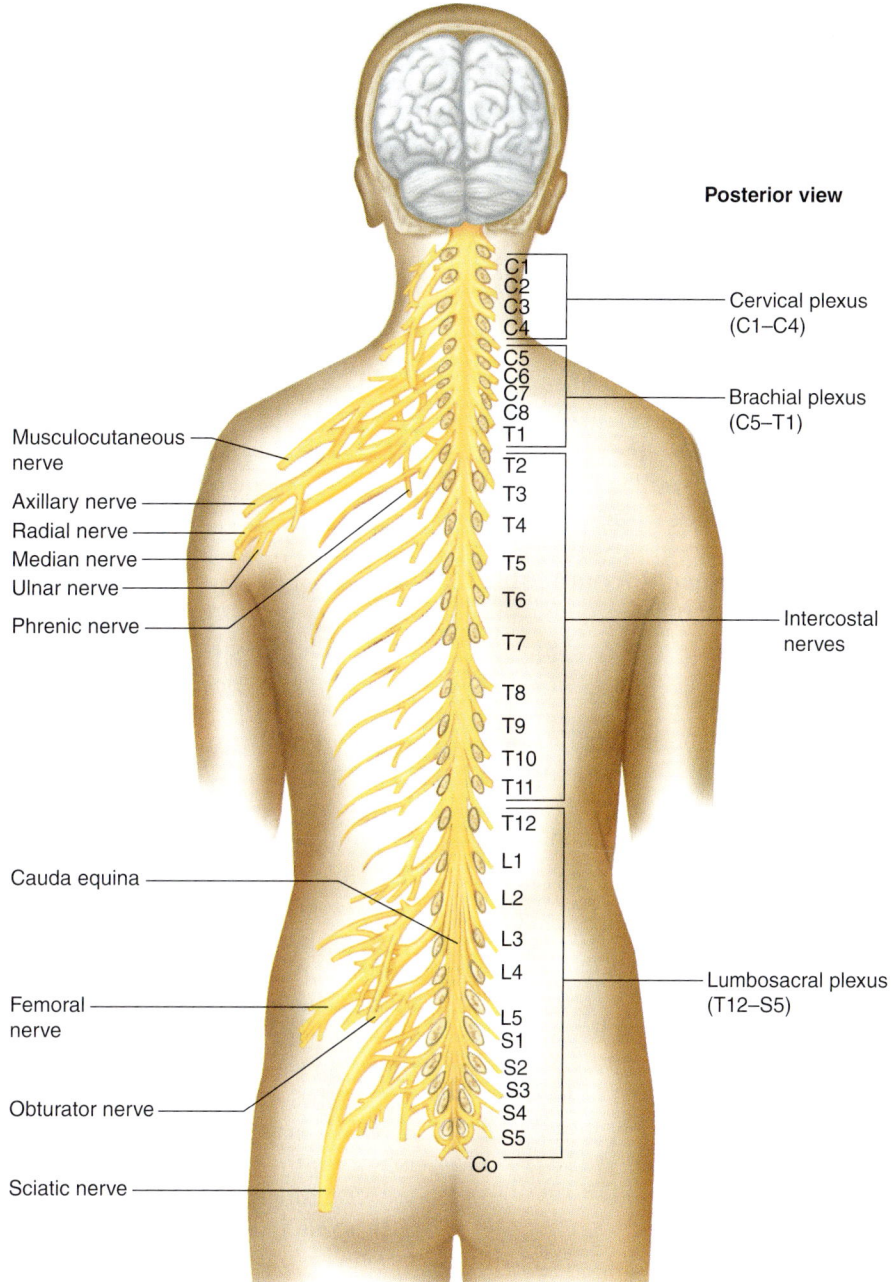

Figure 9.32
The anterior branches of the spinal nerves in the thoracic region give rise to intercostal nerves. Those in other regions combine to form complex networks called plexuses.

root (anterior or motor root) of each spinal nerve consists of axons from the motor neurons whose cell bodies are located within the gray matter of the cord.

A ventral root and a dorsal root unite to form a spinal nerve, which extends outward from the vertebral canal through an *intervertebral foramen* (see fig. 7.16, p. 144). Just beyond its foramen, each spinal nerve divides into several parts.

Except in the thoracic region, the main portions of the spinal nerves combine to form complex networks called **plexuses** instead of continuing directly to peripheral body parts (fig. 9.32). In a plexus, spinal nerve fibers are sorted and recombined so that fibers that innervate a particular peripheral body part reach it in the same nerve, even though the fibers originate from different spinal nerves.

Cervical Plexuses

The **cervical plexuses** lie deep in the neck on either side and form from the branches of the first four cervical nerves. Fibers from these plexuses supply the muscles and skin of the neck. In addition, fibers from the third, fourth, and fifth cervical nerves pass into the right and left **phrenic nerves,** which conduct motor impulses to the muscle fibers of the diaphragm.

Brachial Plexuses

Branches of the lower four cervical nerves and the first thoracic nerve give rise to the **brachial plexuses.** These networks of nerve fibers are located deep within the shoulders between the neck and axillae (armpits). The major branches emerging from the brachial plexuses supply the muscles and skin of the arm, forearm, and hand, and include the **musculocutaneous, ulnar, median, radial,** and **axillary nerves.**

Lumbosacral Plexuses

The **lumbosacral plexuses** are formed on either side by the last thoracic nerve and the lumbar, sacral, and coccygeal nerves. These networks of nerve fibers extend from the lumbar region of the back into the pelvic cavity, giving rise to a number of motor and sensory fibers associated with the muscles and skin of the lower abdominal wall, external genitalia, buttocks, thighs, legs, and feet. The major branches of these plexuses include the **obturator, femoral,** and **sciatic nerves.**

The anterior branches of the thoracic spinal nerves do not enter a plexus. Instead, they enter spaces between the ribs and become **intercostal nerves.** These nerves supply motor impulses to the intercostal muscles and the upper abdominal wall muscles. They also receive sensory impulses from the skin of the thorax and abdomen.

CHECK YOUR RECALL

1. How are spinal nerves grouped?
2. Describe how a spinal nerve joins the spinal cord.
3. Name and locate the major nerve plexuses.

Spinal nerves may be injured in a variety of ways, including stabs, gunshot wounds, birth injuries, dislocations and fractures of the vertebrae, and pressure from tumors in surrounding tissues. The nerves of the cervical plexuses, for example, are sometimes compressed by a sudden bending of the neck called *whiplash,* which may occur during rear-end automobile collisions. Whiplash may cause continuing headaches and pain in the neck and skin, which are supplied by the cervical nerves.

9.15 Autonomic Nervous System

The **autonomic nervous system** is the portion of the PNS that functions independently (autonomously) and continuously without conscious effort. This system controls visceral functions by regulating the actions of smooth muscles, cardiac muscles, and glands. It regulates heart rate, blood pressure, breathing rate, body temperature, and other visceral activities that maintain homeostasis. Portions of the autonomic nervous system respond to emotional stress and prepare the body to meet the demands of strenuous physical activity.

General Characteristics

Reflexes in which sensory signals originate from receptors within the viscera and the skin regulate autonomic activities. Nerve fibers transmit these signals to nerve centers within the brain or spinal cord. In response, motor impulses travel out from these centers on peripheral nerve fibers within cranial and spinal nerves.

Typically, peripheral nerve fibers lead to ganglia outside the CNS. The impulses they carry are integrated within these ganglia and relayed to viscera (muscles and glands) that respond by contracting, releasing secretions, or being inhibited. The integrative function of the ganglia provides the autonomic system with a degree of independence from the brain and spinal cord.

The autonomic nervous system includes two divisions—the **sympathetic** (sim´´pah-thet´ik) and **parasympathetic** (par´´ah-sim´´pah-thet´ik) **divisions.** Some viscera have nerve fibers from each division. In such cases, impulses on one set of fibers may activate an organ, while impulses on the other set inhibit it. Thus, the divisions may act antagonistically, alternately activating or inhibiting the actions of some viscera.

The functions of the autonomic divisions are mixed; that is, each activates some organs and inhibits others. However, the divisions have important functional differences. The sympathetic division prepares the body for energy-expending, stressful, or emergency situations. Conversely, the parasympathetic division is most active under ordinary, restful conditions. It also counterbalances the effects of the sympathetic division and restores the body to a resting state following a stressful experience. For example, during an emergency, the sympathetic division increases heart and breathing rates; following the emergency, the parasympathetic division decreases these activities.

Autonomic Nerve Fibers

The nerve fibers of the autonomic nervous system are motor fibers. Unlike the motor pathways of the somatic nervous system, however, which usually include a single neuron between the brain or spinal cord and a

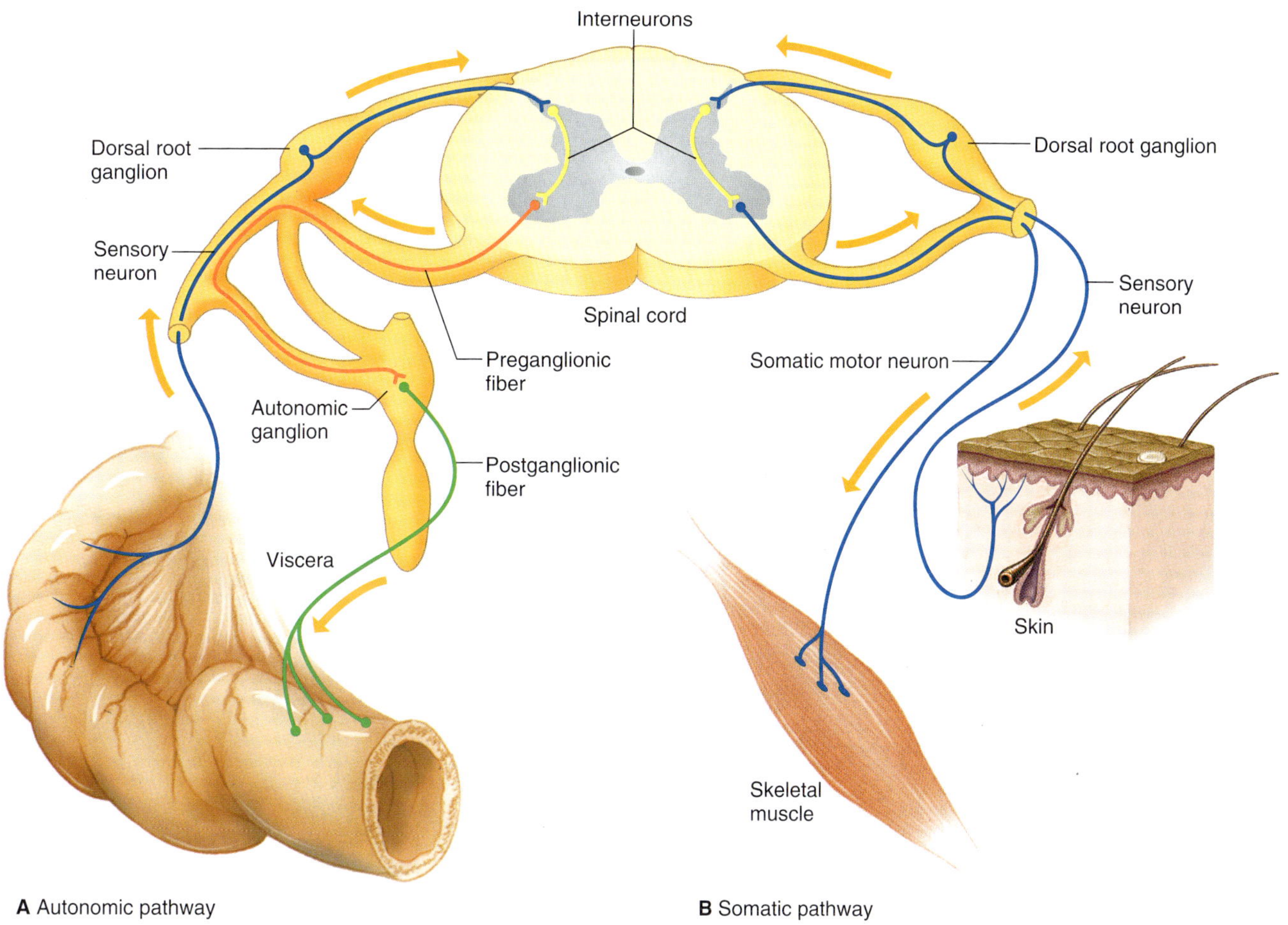

Figure 9.33
Motor pathways. (A) Autonomic pathways include two neurons between the central nervous system and an effector. (B) Somatic pathways usually have a single neuron between the central nervous system and an effector.

skeletal muscle, those of the autonomic system include two neurons (fig. 9.33). The cell body of one neuron is located in the brain or spinal cord. Its axon, the **preganglionic fiber,** leaves the CNS and synapses with one or more neurons whose cell bodies are housed within an autonomic ganglion. The axon of such a second neuron is called a **postganglionic fiber,** and it extends to a visceral effector.

Sympathetic Division

Within the sympathetic division, the preganglionic fibers originate from neurons in the gray matter of the spinal cord (fig. 9.34). Their axons leave the cord through the ventral roots of spinal nerves in the first thoracic through the second lumbar segments. After traveling a short distance, these fibers leave the spinal nerves, and each enters a member of a chain of *paravertebral ganglia*. One of these chains extends longitudinally along each side of the vertebral column.

Within a paravertebral ganglion, a preganglionic fiber forms a synapse with a second neuron. The axon of this neuron, the postganglionic fiber, typically returns to a spinal nerve and extends with it to a visceral effector.

Parasympathetic Division

The preganglionic fibers of the parasympathetic division arise from the brain stem and sacral region of the spinal cord (fig. 9.35). From there, they lead outward in cranial or sacral nerves to ganglia located near or within various viscera. The relatively short postganglionic fibers continue from the ganglia to specific muscles or glands within these viscera.

CHECK YOUR RECALL

1. What parts of the nervous system are included in the autonomic nervous system?
2. How are the divisions of the autonomic nervous system distinguished?
3. Describe a sympathetic nerve pathway and a parasympathetic nerve pathway.

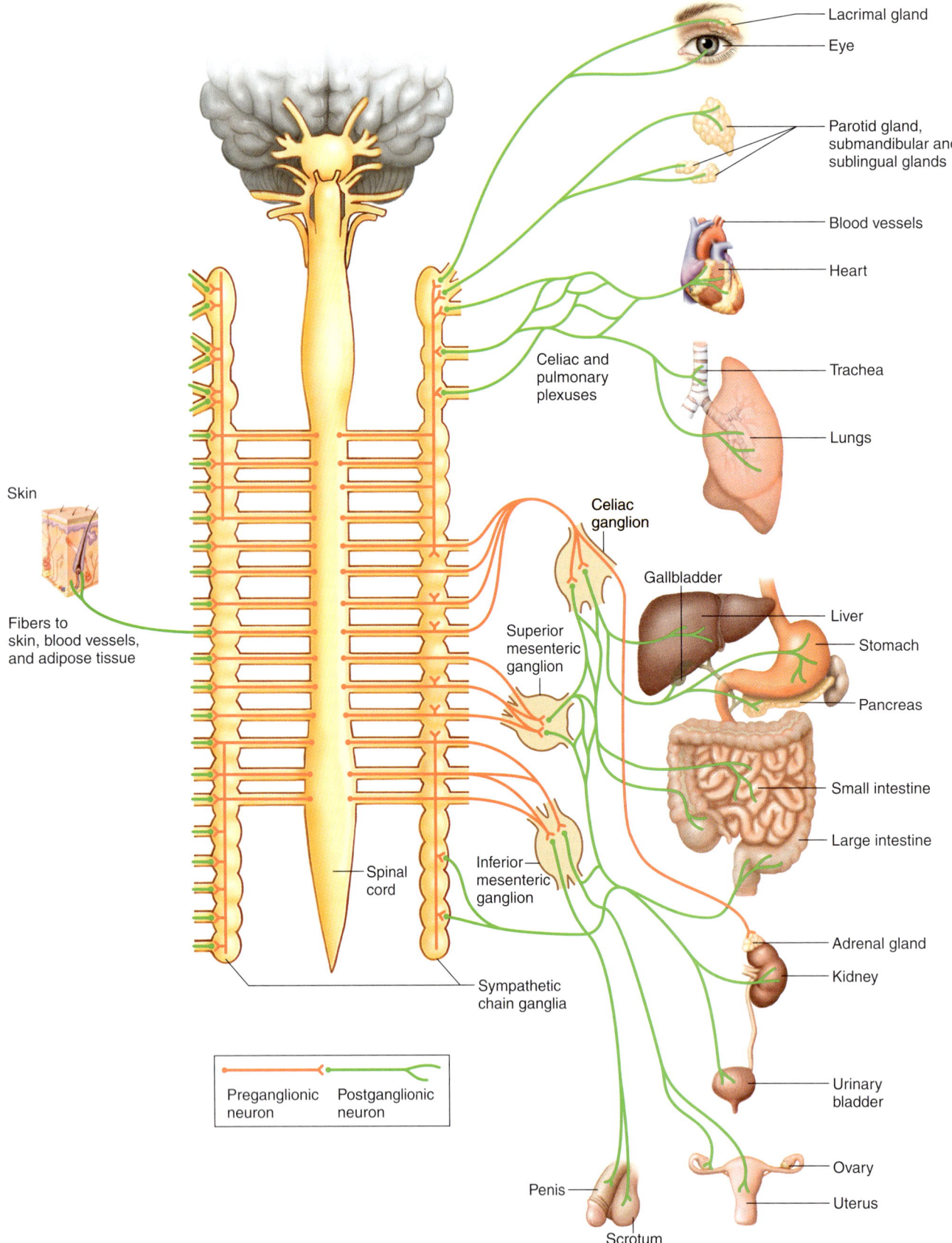

Figure 9.34
The preganglionic fibers of the sympathetic division of the autonomic nervous system arise from the thoracic and lumbar regions of the spinal cord. Note that the adrenal medulla is innervated directly by a preganglionic fiber.

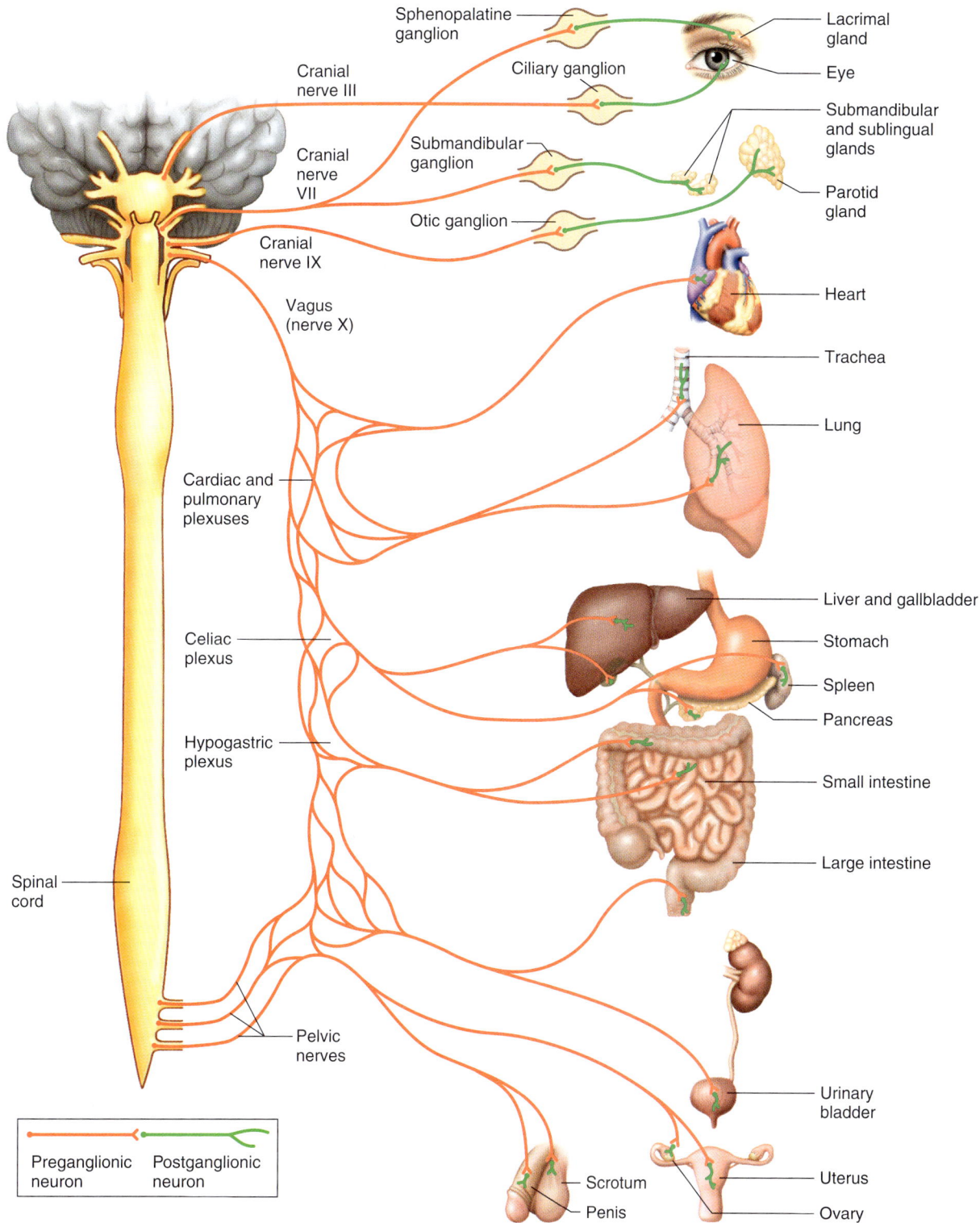

Figure 9.35
The preganglionic fibers of the parasympathetic division of the autonomic nervous system arise from the brain and sacral region of the spinal cord.

Autonomic Neurotransmitters

The preganglionic fibers of the sympathetic and parasympathetic divisions secrete *acetylcholine* and are therefore called **cholinergic fibers.** The parasympathetic postganglionic fibers are also cholinergic fibers (fig. 9.36). Most sympathetic postganglionic fibers, however, secrete *norepinephrine* (noradrenalin) and are called **adrenergic fibers.** The different postganglionic neurotransmitters cause the different effects that the sympathetic and parasympathetic divisions have on their effector organs.

Although each division of the autonomic nervous system can activate some effectors and inhibit others, most viscera are controlled primarily by one division or the other. That is, the divisions are not always actively antagonistic. For example, the sympathetic division regulates the diameter of most blood vessels, which lack parasympathetic innervation. Smooth muscles in the walls of these vessels are continuously stimulated and thus are in a state of partial contraction (sympathetic tone). Decreasing sympathetic stimulation increases (dilates) the diameter of the vessels, which relaxes their muscular walls. Conversely, increasing sympathetic stimulation constricts the vessels.

Similarly, the parasympathetic division dominates in controlling movements in the digestive system. Parasympathetic impulses stimulate stomach and intestinal motility. When the impulses decrease, motility lessens. Table 9.7 summarizes the effects of adrenergic and cholinergic fibers on some visceral effectors.

Control of Autonomic Activity

The brain and spinal cord largely control the autonomic nervous system, despite the system's independence resulting from the integrative function of its ganglia. Consider control centers in the medulla oblongata for cardiac, vasomotor, and respiratory activities. These reflex centers receive sensory impulses from viscera on vagus nerve fibers, and they employ autonomic nerve pathways to stimulate motor responses in muscles and glands. Similarly, the hypothalamus helps regulate body temperature, hunger, thirst, and water and electrolyte balance by influencing autonomic pathways.

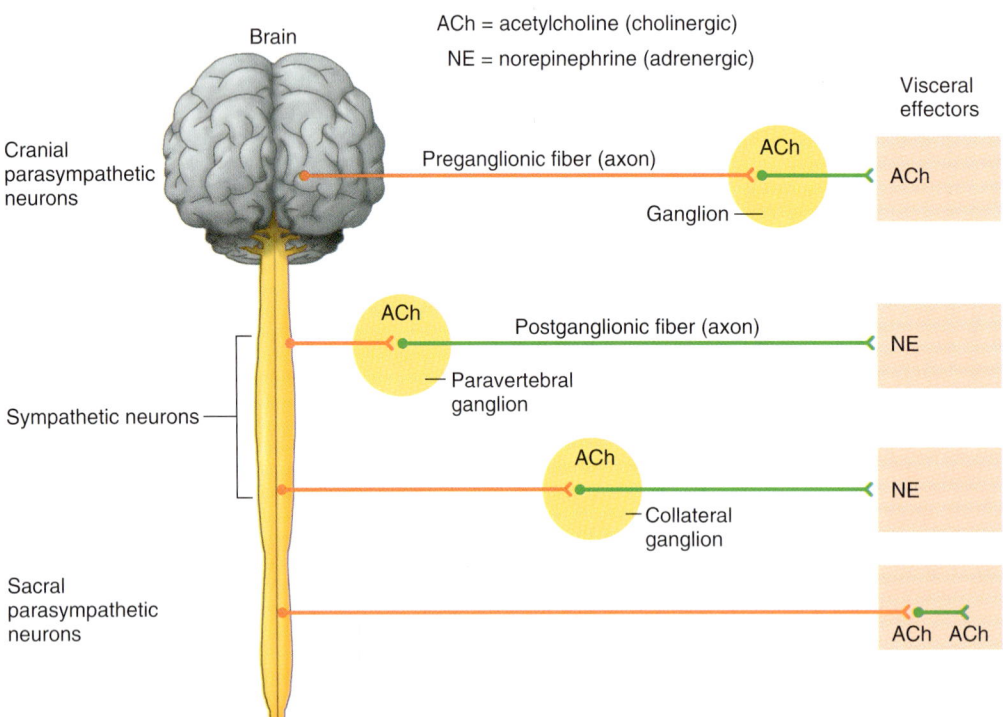

Figure 9.36
Most sympathetic fibers are adrenergic and secrete norepinephrine at the ends of the postganglionic fiber; parasympathetic fibers are cholinergic and secrete acetylcholine at the ends of the postganglionic fibers. Two arrangements of parasympathetic postganglionic fibers are seen in both cranial and sacral portions. Similarly, sympathetic paravertebral and collateral ganglia are seen in both the thoracic and lumbar portions of the nervous system. (Note, in this diagrammatic representation, dendrites are not shown.)

TABLE 9.7 — EFFECTS OF NEUROTRANSMITTER SUBSTANCES ON VISCERAL EFFECTORS OR ACTIONS

VISCERAL EFFECTOR OR ACTION	RESPONSE TO ADRENERGIC STIMULATION (SYMPATHETIC)	RESPONSE TO CHOLINERGIC STIMULATION (PARASYMPATHETIC)
Pupil of the eye	Dilation	Constriction
Heart rate	Increases	Decreases
Bronchioles of lungs	Dilation	Constriction
Muscles of intestinal wall	Slows peristaltic action	Speeds peristaltic action
Intestinal glands	Secretion decreases	Secretion increases
Blood distribution	More blood to skeletal muscles; less blood to digestive organs	More blood to digestive organs; less blood to skeletal muscles
Blood glucose concentration	Increases	Decreases
Salivary glands	Secretion decreases	Secretion increases
Tear glands	No action	Secretion
Muscles of gallbladder	Relaxation	Contraction
Muscles of urinary bladder	Relaxation	Contraction

More complex centers within the brain, including the limbic system and the cerebral cortex, control the autonomic nervous system during emotional stress. These structures utilize autonomic pathways to regulate emotional expression and behavior.

CHECK YOUR RECALL

1. Which neurotransmitters operate in the autonomic nervous system?
2. How do the divisions of the autonomic nervous system regulate visceral activities?
3. How are autonomic activities controlled?

Clinical Connection

In the 1970s, researchers discovered that the body produces its own compounds that bind to the same receptors on brain neurons as do opiate drugs such as morphine. The opiates presumably exert their effects by interfering with this system. In 1992, researchers discovered a compound in the body that binds the same receptors on brain neurons as the active ingredient in marijuana. The existence of this compound, called anandamide, explains why marijuana exerts psychoactive and other effects. Anandamide normally is released from postsynaptic neurons stimulated by calcium ion influx, then binds to the presynaptic cell, temporarily shutting down neurotransmission. When a person takes in the plant version of anandamide through smoke, these neural connections are overwhelmed, and, somehow, this interference produces marijuana's effects on mood and thinking.

Like the opiates, which come from the poppy plant, marijuana has medicinal applications, and is available by prescription in several states to treat pain and to stimulate appetite. In 1985, the Food and Drug Administration approved Marinol, a drug based on the most active chemical in marijuana, to treat nausea and vomiting resulting from cancer chemotherapy. In 1992, the agency approved Marinol to treat AIDS-related anorexia. A drug is being developed from a different chemical found in marijuana smoke called ajulemic acid. It relieves pain, but without psychoactive effects.

Organization

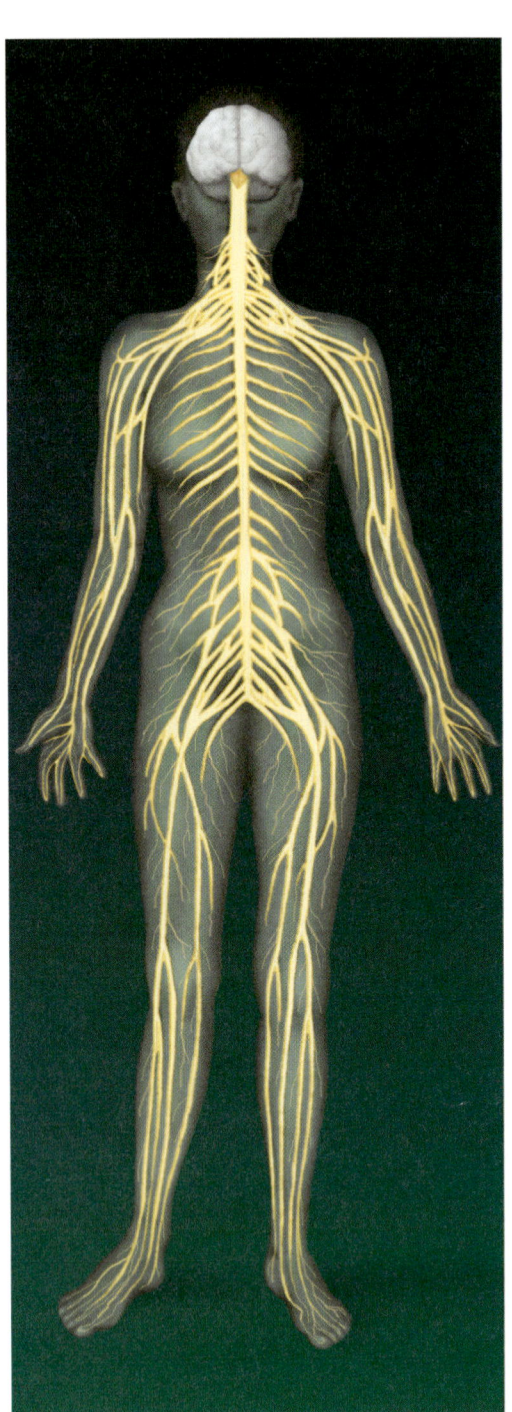

Nervous System
Neurons carry impulses that allow body systems to communicate.

Integumentary System
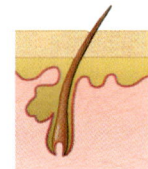
Sensory receptors provide the nervous system with information about the outside world.

Skeletal System

Bones protect the brain and spinal cord and help maintain plasma calcium, which is important to neuron function.

Muscular System

Nerve impulses control movement and carry information about the position of body parts.

Endocrine System

The hypothalamus controls secretion of many hormones.

Cardiovascular System
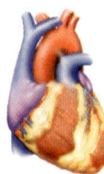
Nerve impulses help control blood flow and blood pressure.

Lymphatic System

Stress may impair the immune response.

Digestive System
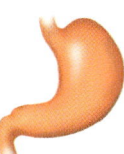
The nervous system can influence digestive function.

Respiratory System
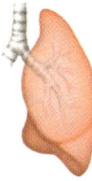
The nervous system alters respiratory activity to control oxygen levels and blood pH.

Urinary System

Nerve impulses affect urine production and elimination.

Reproductive System

The nervous system plays a role in egg and sperm formation, sexual pleasure, childbirth, and nursing.

Clinical Terms Related to the Nervous System

analgesia (an″al-je´ze-ah) Loss or reduction in the ability to sense pain, but without loss of consciousness.
analgesic (an″al-je´sik) Pain-relieving drug.
anesthesia (an″es-the´ze-ah) Loss of feeling.
aphasia (ah-fa´ze-ah) Disturbance or loss of the ability to use words or to understand them, usually due to damage to cerebral association areas.
apraxia (ah-prak´se-ah) Impairment in the ability to use objects.
ataxia (ah-tak´se-ah) Partial or complete inability to coordinate voluntary movements.
cerebral palsy (ser´ĕ-bral pawl´ze) Partial paralysis and lack of muscular coordination caused by damage to the cerebrum.
coma (ko´mah) Unconscious condition in which a person does not respond to stimulation.
cordotomy (kor-dot´o-me) Surgical procedure that severs a nerve tract within the spinal cord to relieve intractable pain.
craniotomy (kra″ne-ot´o-me) Surgical procedure that opens part of the skull.
electroencephalogram (EEG) (e-lek″tro-en-sef´ah-lo-gram″) Recording of the brain's electrical activity.
encephalitis (en″sef-ah-li´tis) Inflammation of the brain and meninges, producing drowsiness and apathy.
epilepsy (ep´ĭ-lep´´se) Disorder of the central nervous system that temporarily disturbs brain impulses, producing convulsive seizures and loss of consciousness.
hemiplegia (hem″ĭ-ple´je-ah) Paralysis of one side of the body and the limbs on that side.
Huntington disease (hunt´ing-tun diz-ēz´) Inherited disorder of the brain causing involuntary, dancelike movements and personality changes.
laminectomy (lam″ĭ-nek´to-me) Surgical removal of the posterior arch of a vertebra, usually to relieve symptoms of a ruptured intervertebral disc pressing on a spinal nerve.
monoplegia (mon″o-ple´je-ah) Paralysis of a single limb.
multiple sclerosis (mul´tĭ-pl skle-ro´sis) Loss of myelin and the appearance of scarlike patches throughout the brain or spinal cord or both.
neuralgia (nu-ral´je-ah) Sharp, recurring pain associated with a nerve; usually caused by inflammation or injury.
neuritis (nu-ri´tis) Inflammation of a nerve.
paraplegia (par″ah-ple´je-ah) Paralysis of both lower limbs.
quadriplegia (kwod″rĭ-ple´je-ah) Paralysis of all four limbs.
vagotomy (va-got´o-me) Surgical severing of a vagus nerve—for example, to reduce acid secretion in a patient with ulcers nonresponsive to other treatment.

SUMMARY OUTLINE

9.1 Introduction (p. 207)
Organs of the nervous system are divided into the central and peripheral nervous systems. These structures provide sensory, integrative, and motor functions. Nervous tissue includes neurons, which are the structural and functional units of the nervous system, and neuroglial cells. A neuron has a cell body, usually one axon, and several dendrites.

9.2 General Functions of the Nervous System (p. 207)
1. Sensory functions employ receptors that detect internal and external changes.
2. Integrative functions collect sensory information and make decisions that motor functions carry out.
3. Motor functions stimulate effectors to respond.

9.3 Neuroglial Cells (p. 208)
1. Central nervous system neuroglial cells include microglial cells, oligodendrocytes, astrocytes and ependymal cells.
2. Peripheral nervous system contains Schwann cells that form myelin sheaths.

9.4 Neurons (p. 209)
1. A neuron includes a cell body, dendrites and axons.
2. Dendrites and the cell body provide receptive surfaces.
3. A single axon arises from the cell body and may be enclosed in a myelin sheath and a neurilemma.
4. Classification of neurons
    a. On the basis of structure, neurons can be classified as multipolar, bipolar, or unipolar.
    b. On the basis of function, neurons can be classified as sensory neurons, interneurons, or motor neurons.

9.5 Cell Membrane Potential (p. 212)
A cell membrane is usually polarized as a result of unequal ion distribution.
1. Distribution of ions
    a. Ion distribution is due to pores and channels in the membranes that allow passage of some ions but not others.
    b. Potassium ions pass more easily through cell membranes than do sodium ions.
2. Resting potential
    a. A high concentration of sodium ions is on the outside of a membrane, and a high concentration of potassium ions is on the inside.
    b. Many negatively charged ions are inside a cell.
    c. In a resting cell, more positive ions leave than enter, so the outside of the cell membrane develops a positive charge, while the inside develops a negative charge.

3. Potential changes
   a. Stimulation of a membrane affects the membrane's resting potential.
   b. When its resting potential becomes more positive, a membrane becomes depolarized.
   c. Potential changes are subject to summation.
   d. Achieving threshold potential triggers an action potential.
4. Action potential
   a. At threshold, sodium channels open, and sodium ions diffuse inward, depolarizing the membrane.
   b. About the same time, potassium channels open, and potassium ions diffuse outward, repolarizing the membrane.
   c. This rapid change in potential is an action potential.
   d. Many action potentials can occur before active transport reestablishes the resting potential.

9.6 Nerve Impulse (p. 215)
A wave of action potentials is a nerve impulse.
1. Impulse conduction
   a. Unmyelinated axons conduct impulses over their entire surfaces.
   b. Myelinated axons conduct impulses more rapidly.
   c. Axons with larger diameters conduct impulses faster than those with smaller diameters.
2. All-or-none response
   a. A nerve impulse is conducted in an all-or-none manner whenever a stimulus of threshold intensity is applied to an axon.
   b. All the impulses conducted on an axon are of the same strength.

9.7 The Synapse (p. 216)
A synapse is a junction between two neurons.
1. Synaptic transmission
   a. Impulses usually travel from a dendrite to a cell body, then along the axon to a synapse.
   b. Axons have synaptic knobs at their distal ends, which secrete neurotransmitters.
   c. A neurotransmitter is released when a nerve impulse reaches the end of an axon.
   d. A neurotransmitter reaching the postsynaptic neuron membrane triggers a nerve impulse.
2. Excitatory and inhibitory actions
   a. Neurotransmitters that trigger nerve impulses are excitatory. Those that inhibit impulses are inhibitory.
   b. The net effect of synaptic knobs communicating with a neuron depends on which knobs are activated from moment to moment.
3. Neurotransmitters
   a. The nervous system produces many different neurotransmitters.
   b. Neurotransmitters include acetylcholine, monoamines, amino acids, and peptides.
   c. A synaptic knob releases neurotransmitters when an action potential increases membrane permeability to calcium ions.
   d. After being released, neurotransmitters are decomposed or removed from synaptic clefts.

9.8 Impulse Processing (p. 219)
How the nervous system processes and responds to nerve impulses reflects the organization of neurons in the brain and spinal cord.
1. Neuronal pools
   a. Neurons form pools within the central nervous system.
   b. Each pool receives impulses, processes them, and conducts impulses away.
2. Facilitation
   a. Each neuron in a pool may receive excitatory and inhibitory stimuli.
   b. A neuron is facilitated when it receives subthreshold stimuli and becomes more excitable.
3. Convergence
   a. Impulses from two or more incoming axons may converge on a single neuron.
   b. Convergence enables impulses from different sources to have an additive effect on a neuron.
4. Divergence
   a. Impulses leaving a pool may diverge by passing into several output neurons.
   b. Divergence amplifies impulses.

9.9 Types of Nerves (p. 220)
1. Nerves are cordlike bundles of nerve fibers.
2. Nerves can be classified as sensory nerves, motor nerves, or mixed nerves, depending on which type of fibers they contain.

9.10 Nerve Pathways (p. 220)
A nerve pathway is the route an impulse follows through the nervous system.
1. A reflex arc usually includes a sensory neuron, a reflex center composed of interneurons, and a motor neuron.
2. Reflex behavior
   a. Reflexes are automatic, subconscious responses to changes.
   b. They help maintain homeostasis.
   c. The knee-jerk reflex employs only two neurons.
   d. Withdrawal reflexes are protective.

9.11 Meninges (p. 222)
1. Bone and meninges surround the brain and spinal cord.
2. The meninges consist of the dura mater, arachnoid mater, and pia mater.
3. Cerebrospinal fluid occupies the space between the arachnoid and pia maters.

9.12 Spinal Cord (p. 224)
The spinal cord is a nerve column that extends from the brain into the vertebral canal.
1. Structure of the spinal cord
   a. The spinal cord is composed of thirty-one segments, each of which gives rise to a pair of spinal nerves.
   b. The spinal cord has a cervical enlargement and a lumbar enlargement.
   c. It has a central core of gray matter within white matter.
   d. The white matter is composed of bundles of myelinated axons.
2. Functions of the spinal cord
   a. The spinal cord provides a two-way communication system between the brain and other body parts.
   b. Ascending tracts carry sensory impulses to the brain. Descending tracts carry motor impulses to muscles and glands.

9.13 Brain (p. 225)
The brain is subdivided into the cerebrum, cerebellum, and brain stem.
1. Structure of the cerebrum
   a. The cerebrum consists of two cerebral hemispheres connected by the corpus callosum.
   b. The cerebral cortex is a thin layer of gray matter near the surface.
   c. White matter consists of myelinated axons that connect neurons within the nervous system and communicate with other body parts.

2. Functions of the cerebrum
   a. The cerebrum provides higher brain functions.
   b. The cerebral cortex consists of sensory, motor, and association areas.
   c. One cerebral hemisphere usually dominates for certain intellectual functions.
3. Ventricles and cerebrospinal fluid
   a. Ventricles are interconnected cavities within the cerebral hemispheres and brain stem.
   b. Cerebrospinal fluid fills the ventricles.
   c. The choroid plexuses in the walls of the ventricles secrete cerebrospinal fluid.
4. Diencephalon
   a. The diencephalon contains the thalamus, which is a central relay station for incoming sensory impulses, and the hypothalamus, which maintains homeostasis.
   b. The limbic system produces emotions and modifies behavior.
5. Brain stem
   a. The brain stem consists of the midbrain, pons, and medulla oblongata.
   b. The midbrain contains reflex centers associated with eye and head movements.
   c. The pons transmits impulses between the cerebrum and other parts of the nervous system and contains centers that help regulate the rate and depth of breathing.
   d. The medulla oblongata transmits all ascending and descending impulses and contains several vital and nonvital reflex centers.
   e. The reticular formation filters incoming sensory impulses, arousing the cerebral cortex into wakefulness when significant impulses arrive.
6. Cerebellum
   a. The cerebellum consists of two hemispheres.
   b. It functions primarily as a reflex center for integrating sensory information required in the coordination of skeletal muscle movements and maintenance of equilibrium.

9.14 Peripheral Nervous System (p. 235)

The peripheral nervous system consists of cranial and spinal nerves that branch from the brain and spinal cord to all body parts. It is subdivided into the somatic and autonomic systems.

1. Cranial nerves
   a. Twelve pairs of cranial nerves connect the brain to parts in the head, neck, and trunk.
   b. Most cranial nerves are mixed, but some are purely sensory, and others are primarily motor.
   c. The names of the cranial nerves indicate their primary functions or the general distributions of their fibers.
   d. Some cranial nerves are somatic, and others are autonomic.
2. Spinal nerves
   a. Thirty-one pairs of spinal nerves originate from the spinal cord.
   b. These mixed nerves provide a two-way communication system between the spinal cord and parts of the upper and lower limbs, neck, and trunk.
   c. Spinal nerves are grouped according to the levels from which they arise, and they are numbered in sequence.
   d. Each spinal nerve emerges by a dorsal and a ventral root.
   e. Each spinal nerve divides into several branches just beyond its foramen.
   f. Most spinal nerves combine to form plexuses in which nerve fibers are sorted and recombined so that those fibers associated with a particular part reach it together.

9.15 Autonomic Nervous System (p. 240)

The autonomic nervous system functions without conscious effort. It regulates the visceral activities that maintain homeostasis.

1. General characteristics
   a. Autonomic functions are reflexes controlled from nerve centers in the brain and spinal cord.
   b. The autonomic nervous system consists of two divisions—sympathetic and parasympathetic.
   c. The sympathetic division responds to stressful and emergency conditions.
   d. The parasympathetic division is most active under ordinary conditions.
2. Autonomic nerve fibers
   a. Autonomic nerve fibers are motor fibers.
   b. Sympathetic fibers leave the spinal cord and synapse in paravertebral ganglia.
   c. Parasympathetic fibers begin in the brain stem and sacral region of the spinal cord and synapse in ganglia near viscera.
3. Autonomic neurotransmitters
   a. Sympathetic and parasympathetic preganglionic fibers secrete acetylcholine.
   b. Parasympathetic postganglionic fibers secrete acetylcholine. Sympathetic postganglionic fibers secrete norepinephrine.
   c. The different effects of the autonomic divisions are due to the different neurotransmitters the postganglionic fibers release.
   d. One division predominantly controls most viscera.
4. Control of autonomic activity
   a. The autonomic nervous system is somewhat independent.
   b. Control centers in the medulla oblongata and hypothalamus utilize autonomic nerve pathways.
   c. The limbic system and cerebral cortex control the autonomic system during emotional stress.

REVIEW EXERCISES

1. List three general functions of the nervous system. (p. 207)
2. Explain the relationship between the central nervous system and the peripheral nervous system. (p. 207)
3. Describe the generalized structure of a neuron, and explain the functions of its parts. (p. 207)
4. Distinguish between neurons and neuroglial cells. (p. 207)
5. Discuss the functions of each type of neuroglial cell. (p. 208)
6. Distinguish between myelinated and unmyelinated axons. (p. 209)
7. Explain how neurons can be classified on the basis of their structure. (p. 211)
8. Explain how a membrane becomes polarized. (p. 212)
9. Describe how ions associated with nerve cell membranes are distributed. (p. 212)
10. Define *resting potential*. (p. 213)
11. Explain how threshold potential may be achieved. (p. 213)
12. List the events during an action potential. (p. 213)
13. Explain how nerve impulses are related to action potentials. (p. 215)
14. Explain how impulses are conducted on unmyelinated and myelinated axons. (p. 216)
15. Define *synapse*. (p. 216)
16. Explain how information passes from one neuron to another. (p. 216)

17. Distinguish between excitatory and inhibitory actions of neurotransmitters. (p. 217)
18. List four types of neurotransmitters. (p. 217)
19. Explain what happens to neurotransmitters after they are released. (p. 218)
20. Describe a neuronal pool. (p. 219)
21. Distinguish between convergence and divergence. (p. 219)
22. Distinguish among sensory, motor, and mixed nerves. (p. 220)
23. Define *reflex*. (p. 220)
24. Describe a reflex arc that consists of two neurons. (p. 220)
25. Name the layers of the meninges, and explain their functions. (p. 222)
26. Describe the structure of the spinal cord. (p. 224)
27. Distinguish between the ascending and descending tracts of the spinal cord. (p. 224)
28. Name the three major portions of the brain, and describe the general functions of each. (p. 225)
29. Describe the general structure of the cerebrum. (p. 226)
30. Describe the location of the motor, sensory, and association areas of the cerebral cortex, and describe the general functions of each. (p. 228)
31. Define *hemisphere dominance*. (p. 229)
32. Explain the function of the corpus callosum. (p. 229)
33. Distinguish between the cerebral cortex and basal nuclei. (p. 229)
34. Describe the location of the ventricles of the brain. (p. 230)
35. Explain how cerebrospinal fluid is produced and how it functions. (p. 230)
36. Define *limbic system*, and explain its functions. (p. 233)
37. Name the parts of the brain stem, and describe the general functions of each part. (p. 233)
38. Name the parts of the midbrain, and describe the general functions of each part. (p. 233)
39. Describe the pons and its functions. (p. 233)
40. Describe the medulla oblongata and its functions. (p. 233)
41. Describe the functions of the cerebellum. (p. 234)
42. Name, locate, and describe the major functions of each pair of cranial nerves. (p. 236)
43. Explain how the spinal nerves are grouped and numbered. (p. 238)
44. Describe the structure of a spinal nerve. (p. 238)
45. Define *plexus*, and locate the major plexuses of the spinal nerves. (p. 239)
46. Describe the general functions of the autonomic nervous system. (p. 240)
47. Distinguish between the sympathetic and parasympathetic divisions of the autonomic nervous system. (p. 240)
48. Distinguish between preganglionic and postganglionic nerve fibers. (p. 241)
49. Explain why the effects of the sympathetic and parasympathetic autonomic divisions differ. (p. 241)
50. Describe how portions of the central nervous system control autonomic activities. (p. 244)

CRITICAL THINKING

1. List four skills encountered in everyday life that depend on nervous system function. Then list the part of the nervous system responsible for each.
2. What is the role of the cerebellum in athletics? The cerebrum?
3. A fetus or newborn with anencephaly lacks higher brain structures, possessing only a brain stem. What functions would such an individual have? What functions would he or she lack?
4. Narcolepsy is a condition in which a person suddenly falls fast asleep, even in the midst of an activity or conversation. What part of the brain is probably responsible for narcolepsy?
5. What nervous system functions contribute to thinking?
6. What functional losses would you expect in a patient who has suffered injury to the right occipital lobe of the cerebral cortex? The right temporal lobe?
7. A reflex called the *biceps-jerk reflex* employs motor neurons that exit the spinal cord in the fifth spinal nerve (C5)—that is, fifth from the top of the cord. Another reflex, called the *triceps-jerk reflex*, utilizes motor neurons in the seventh spinal nerve (C7). How might these reflexes help locate the site of damage in a patient with a neck injury?
8. In multiple sclerosis, nerve fibers in the central nervous system lose their myelin. Why would this loss affect skeletal muscle function?
9. Why are rapidly growing cancers that originate in nervous tissue most likely composed of neuroglial cells rather than neurons?
10. Intravenous drug abusers sometimes dissolve and then inject tablets that contain fillers, such as talc or cornstarch, in addition to the drug. These fillers may obstruct tiny blood vessels in the cerebrum. What problems might such obstructions create?

WEB CONNECTIONS

Visit the website for additional study questions and more information about this chapter at:

http://www.mhhe.com/shieress8

chapter 10

Somatic and Special Senses

GETTING A COCHLEAR IMPLANT. Yolanda Santana, of Rochester, New York, probably lost her hearing when she was only eight weeks old and suffered a high fever. But it wasn't until she was nine months old, when Yolanda didn't babble like her age-mates, that her parents first suspected she might be deaf. She was fitted with hearing aids and did well at a preschool for the deaf. Then Yolanda's parents read about cochlear implants on the Internet and decided to pursue this option for their daughter. They learned that a cochlear implant does not magically restore hearing, but enables a person to hear certain sounds. Teamed with speech therapy and use of sign language, though, the cochlear implant enables a person to make enough sense of sounds to speak.

At the time Carlos and Beth Santana read about the implants, Yolanda was already approaching three years old. Before age three is the best time to receive a cochlear implant because this is when the brain is rapidly processing speech and hearing as a person masters language. Of the 22,000 people in the United States who have received cochlear implants since they became available in 1984, about half have had them since early childhood.

The implant consists of a part placed under the skin above the ear that leads to two dozen electrodes placed near the auditory nerve in the cochlea, the snail-shaped part of the inner ear. Yolanda wears a headset that includes a microphone lodged at the back of her ear to pick up incoming sounds and a fanny pack containing a speech processor that digitizes the sounds into coded signals. A transmitter on the headset sends the coded signals, as FM radio waves, to the implant, which changes them to electrical signals and delivers them to the cochlea. Here, the auditory nerve is stimulated and sends neural messages to the brain's cerebral cortex, which interprets the input as sound.

Yolanda's audiologist turned on the speech processor a month after the surgery. At first, the youngster heard low sounds and sometimes responded with a low hum. She would grab at the processor, which meant that she realized it was the source of the sound. Still, sounds had no meaning, for she had never heard them before. But, gradually, the little girl learned from context. One day when Carlos signed "father" and said "poppy," Yolanda signed back and tried to say the word! Able to connect mouth movements to sounds to concepts, Yolanda was well on her way to hearing.

Photo:
Yolanda Santana received a cochlear implant when she was three years old. The device enables her to detect enough sounds to communicate effectively.

Chapter Objectives

After studying this chapter, you should be able to do the following:

10.1 Introduction
1. Distinguish between somatic senses and special senses. (p. 253)

10.2 Receptors and Sensations
2. Name five kinds of receptors, and explain their functions. (p. 253)
3. Explain how a sensation arises. (p. 253)

10.3 Somatic Senses
4. Describe the receptors associated with the senses of touch, pressure, temperature, and pain. (p. 253)
5. Describe how the sense of pain is produced. (p. 254)

10.4 Special Senses
6. Identify the location of the receptors associated with the special senses. (p. 257)

10.5 Sense of Smell
7. Explain the relationship between the senses of smell and taste. (p. 257)
8. Explain one hypothesized mechanism for smell. (p. 258)

10.6 Sense of Taste
9. Explain the mechanism for taste. (p. 259)

10.7 Sense of Hearing
10. Name the parts of the ear, and explain the function of each part. (p. 260)

10.8 Sense of Equilibrium
11. Distinguish between static and dynamic equilibrium. (p. 265)

10.9 Sense of Sight
12. Name the parts of the eye, and explain the function of each part. (p. 267)
13. Explain how the eye refracts light. (p. 273)
14. Describe the visual nerve pathway. (p. 273)

Aids to Understanding Word

choroid [skinlike] *choroid* coat: Middle, vascular layer of the eye.
cochlea [snail] *cochlea*: Coiled tube in the inner ear.
iris [rainbow] *iris*: Colored, muscular part of the eye.
labyrinth [maze] *labyrinth*: Complex system of connecting chambers and tubes of the inner ear.
lacri- [tears] *lacri*mal gland: Tear gland.
macula [spot] *macula* lutea: Yellowish spot on the retina.
olfact- [to smell] *olfact*ory: Pertaining to the sense of smell.
scler- [hard] *scler*a: Tough, outer protective layer of the eye.
tympan- [drum] *tympan*ic membrane: Eardrum.
vitre- [glass] *vitre*ous humor: Clear, jelly-like substance within the eye.

Key Terms

accommodation (ah-kom″o-da´shun)
ampulla (am-pul´lah)
chemoreceptor (ke´mo-re-sep´tor)
cochlea (kok´le-ah)
cornea (kor´ne-ah)
dynamic equilibrium (di-nam´ik e″kwĭ-lib´re-um)
labyrinth (lab´i-rinth)
macula (mak´u-lah)
mechanoreceptor (mek″ah-no-re-sep´tor)
pain receptor (pān re-sep´tor)
photoreceptor (fo″to-re-sep´tor)
projection (pro-jek´shun)
referred pain (re-furd´ pān)
refraction (re-frak´shun)
retina (ret´ĭ-nah)
rhodopsin (ro-dop´sin)
sclera (skle´rah)
sensory adaptation (sen´so-re ad″ap-ta´shun)
static equilibrium (stat´ik e″kwĭ-lib´re-um)
thermoreceptor (therm´o-re-sep´tor)

10.1 Introduction

How dull life would be without sight and sound, smell and taste, touch and balance. Our senses are not only necessary for us to enjoy life, but to survive. *Sensory receptors* detect environmental changes and trigger nerve impulses that travel on sensory pathways into the central nervous system for processing and interpretation. The body reacts with a particular feeling or sensation.

Sensory receptors vary greatly but fall into two major categories. Receptors associated with the *somatic senses* of touch, pressure, temperature, and pain form one group. These receptors are widely distributed throughout the skin and deeper tissues, and are structurally simple. Receptors of the second type are parts of complex, specialized sensory organs that provide the *special senses* of smell, taste, hearing, equilibrium, and vision.

10.2 Receptors and Sensations

The many kinds of sensory receptors share common features. However, each type of receptor is particularly sensitive to a distinct type of environmental change and is much less sensitive to other forms of stimulation. This selective response distinguishes the senses.

Types of Receptors

Sensory receptors are categorized into five types according to their sensitivities: **Chemoreceptors** (ke˝´mo-re-sep´torz) are stimulated by changes in the chemical concentration of substances; **pain receptors** (pān re-sep´torz) by tissue damage; **thermoreceptors** (therm´o-re-sep˝´torz) by changes in temperature; **mechanoreceptors** (mek˝´ah-no-re-sep´torz) by changes in pressure or movement; and **photoreceptors** (fo˝´to-re-sep´torz) by light energy.

Sensations

A **sensation** (perception) is a feeling that occurs when the brain interprets sensory impulses. Because all the nerve impulses that travel away from sensory receptors into the central nervous system are alike, the resulting sensation depends on which region of the brain receives the impulse. For example, impulses reaching one region are always interpreted as sounds, and those reaching another are always sensed as touch.

At the same time that a sensation forms, the cerebral cortex causes the feeling to seem to come from the stimulated receptors. This process is called **projection** (pro-jek´shun) because the brain projects the sensation back to its apparent source. Projection allows a person to pinpoint the region of stimulation; thus, the eyes seem to see, and the ears seem to hear.

Sensory Adaptation

When sensory receptors are continuously stimulated, many of them undergo an adjustment called **sensory adaptation** (sen´so-re ad˝´ap-ta´shun). As receptors adapt, impulses leave them at decreasing rates, until finally these receptors may stop sending signals. Once receptors have adapted, impulses can be triggered only if the stimulus strength changes. A person entering a room that has a strong odor experiences sensory adaptation. At first, the scent seems intense, but it becomes less and less noticeable as the smell (olfactory) receptors adapt.

CHECK YOUR RECALL

1. List five general types of sensory receptors.
2. Explain how a sensation occurs.
3. What is sensory adaptation?

10.3 Somatic Senses

Somatic senses are associated with receptors in the skin, muscles, joints, and viscera. They include the senses of touch and pressure, temperature, and pain.

Touch and Pressure Senses

The senses of touch and pressure derive from three kinds of receptors (fig. 10.1). These receptors sense mechanical forces that deform or displace tissues. Touch and pressure receptors include:

1. **Sensory nerve fibers** These receptors are common in epithelial tissues, where their free ends extend between epithelial cells. They are associated with the sensations of touch and pressure.
2. **Meissner's corpuscles** These are small, oval masses of flattened connective tissue cells within connective tissue sheaths. Two or more sensory nerve fibers branch into each corpuscle and end within it as tiny knobs.

    Meissner's corpuscles are abundant in the hairless portions of the skin, such as the lips, fingertips, palms, soles, nipples, and external genital organs. They respond to the motion of objects that barely contact the skin, interpreting impulses from them as the sensation of light touch.
3. **Pacinian corpuscles** These sensory bodies are relatively large structures composed of connective tissue fibers and cells. They are common in the deeper subcutaneous tissues and in muscle tendons and joint ligaments. Pacinian corpuscles respond to heavy pressure and are associated with the sensation of deep pressure.

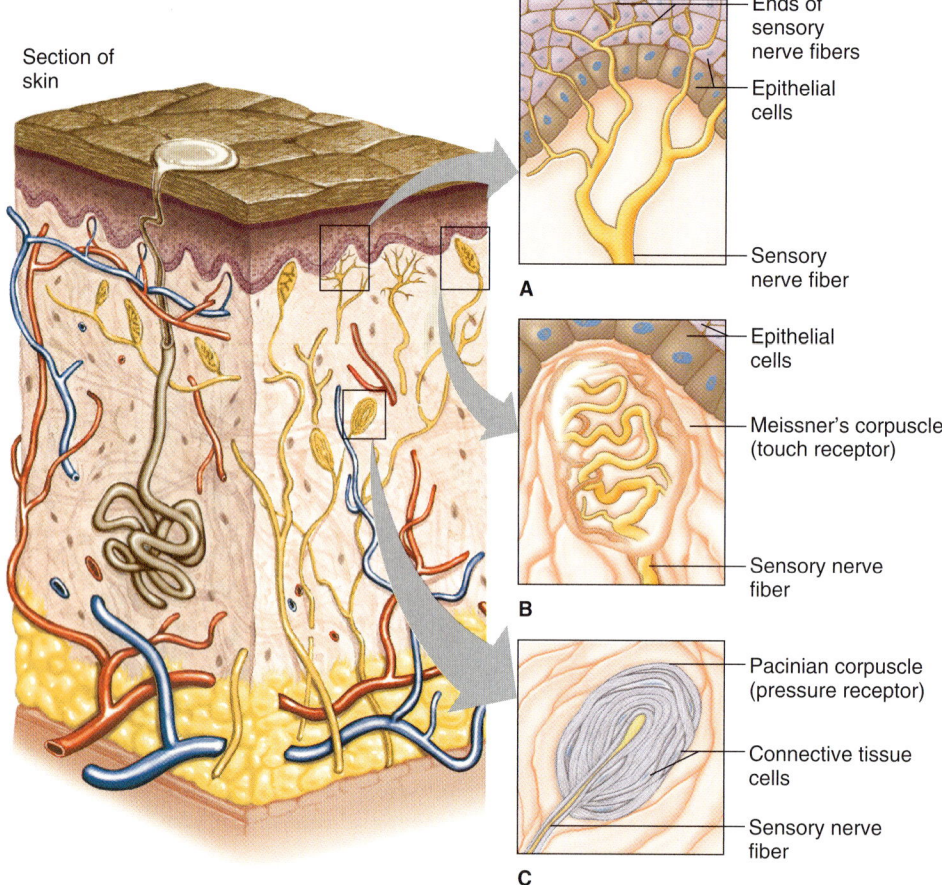

Figure 10.1
Touch and pressure receptors include (A) free ends of sensory nerve fibers, (B) Meissner's corpuscles, and (C) Pacinian corpuscles.

Temperature Senses

Temperature sensation depends on two types of free nerve endings in the skin. Those that respond to warmer temperatures are called *warm receptors,* and those that respond to colder temperatures are called *cold receptors.*

Warm receptors are most sensitive to temperatures above 25°C (77°F) and become unresponsive at temperatures above 45°C (113°F). Temperatures near and above 45°C stimulate pain receptors, producing a burning sensation.

Cold receptors are most sensitive to temperatures between 10°C (50°F) and 20°C (68°F). Temperatures below 10°C stimulate pain receptors, producing a freezing sensation.

Both warm and cold receptors rapidly adapt. Within about a minute of continuous stimulation, the sensation of warmth or cold begins to fade.

Sense of Pain

Other receptors that consist of free nerve endings sense pain. These receptors are widely distributed throughout the skin and internal tissues, except in the nervous tissue of the brain, which lacks pain receptors.

Pain receptors protect the body because tissue damage stimulates them. Pain sensation is usually perceived as unpleasant, and it signals a person to take action to remove the stimulation. Pain receptors adapt poorly, if at all. Thus, once a pain receptor is activated, even by a single stimulus, it may send impulses into the central nervous system for some time. Thus, pain may persist.

The way in which tissue damage stimulates pain receptors is poorly understood. Injuries are believed to promote the release of certain chemicals that build up and stimulate pain receptors. Deficiency of oxygen-rich blood (ischemia) in a tissue or stimulation of certain mechanoreceptors also triggers pain sensations. The pain elicited during a muscle cramp, for example, stems from sustained contraction that squeezes capillaries and interrupts blood flow. Stimulation of mechanical-sensitive pain receptors also contributes to the sensation.

> Injuries to bones, tendons, or ligaments stimulate pain receptors that may also contract nearby skeletal muscles. The contracting muscles may become ischemic, which may trigger still other pain receptors within the muscle tissue, further increasing muscular contraction in a "vicious circle."

Visceral Pain

As a rule, pain receptors are the only receptors in viscera whose stimulation produces sensations. Pain receptors in these organs respond differently to stimulation than those associated with surface tissues. For example, localized damage to intestinal tissue during surgical procedures may not elicit pain sensations, even in a conscious person. However, when visceral tissues are subjected to more widespread stimulation, as when intestinal tissues are stretched or smooth muscles in intestinal walls undergo spasms, a strong pain sensation may follow. Once again, the resulting pain seems to stem from stimulation of mechanoreceptors and from decreased blood flow accompanied by lower tissue oxygen concentration and accumulation of pain-stimulating chemicals.

Visceral pain may feel as if it is coming from some part of the body other than the part being stimulated, a phenomenon called **referred pain.** For example, pain originating in the heart may be referred to the left shoulder or left upper limb (fig. 10.2). Referred pain may arise from common nerve pathways that carry sensory impulses from skin areas as well as viscera. In other words, pain impulses from the heart travel over the same nerve pathways as those from the skin of the left shoulder and left upper limb (fig. 10.3). Consequently, during a heart attack, the cerebral cortex may incorrectly interpret the source of pain impulses as the left shoulder or upper limb, rather than the heart.

CHECK YOUR RECALL

1. Describe the three types of touch and pressure receptors.
2. Describe the receptors that sense temperature.
3. What types of stimuli excite pain receptors?
4. What is referred pain?

Pain Nerve Fibers

Nerve fibers that conduct impulses away from pain receptors are of two main types: acute pain fibers and chronic pain fibers. *Acute pain fibers* are relatively thin, myelinated nerve fibers. They conduct nerve impulses rapidly and are associated with the sensation of sharp pain, which typically originates from a restricted area of the skin and seldom continues after the pain-producing stimulus stops. *Chronic pain fibers* are thin, unmyelinated

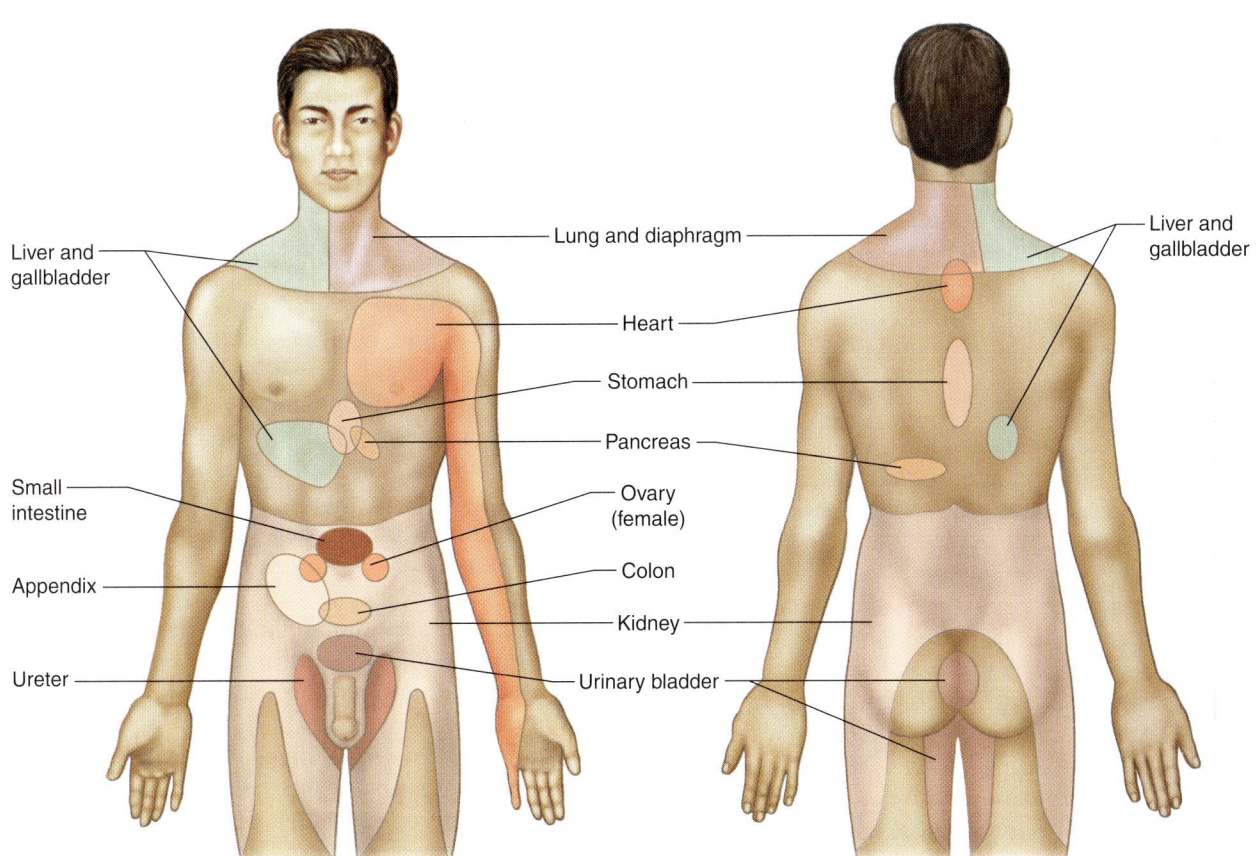

Figure 10.2
Surface regions to which visceral pain may be referred.

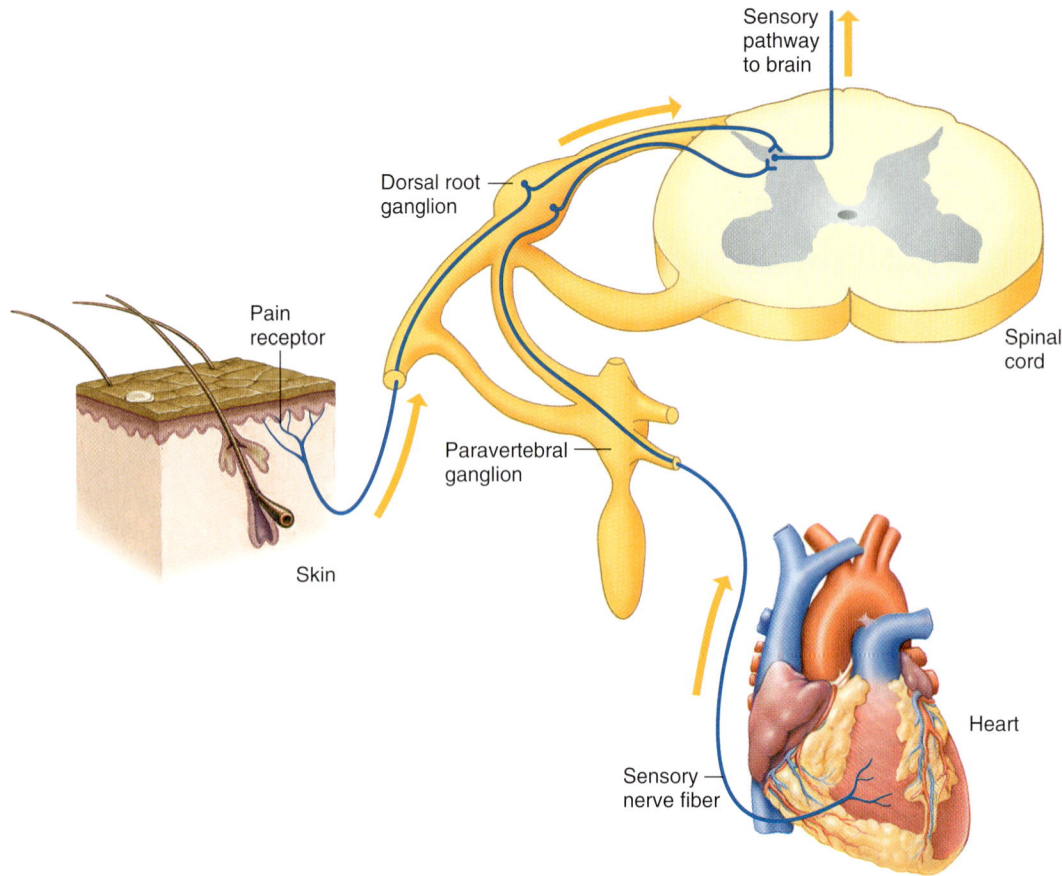

Figure 10.3
Pain originating in the heart may feel as if it is coming from the skin because sensory impulses from those two regions follow common nerve pathways to the brain.

nerve fibers. They conduct impulses more slowly and produce a dull, aching pain sensation that may be diffuse and difficult to pinpoint. Such pain may continue for some time after the original stimulus ceases. Acute pain is usually sensed as coming only from the skin; chronic pain is felt in deeper tissues as well.

An event that stimulates pain receptors usually triggers impulses on both acute and chronic pain fibers. This causes a dual sensation—a sharp, pricking pain, followed shortly by a dull, aching one. The aching pain is usually more intense and may worsen with time. Chronic pain can cause prolonged suffering.

Pain impulses that originate from the head reach the brain on sensory fibers of cranial nerves. All other pain impulses travel on the sensory fibers of spinal nerves, and they pass into the spinal cord by way of the dorsal roots of these spinal nerves. Within the spinal cord, neurons process pain impulses in the gray matter of the dorsal horn, and the impulses are transmitted to the brain. Here, most pain fibers terminate in the reticular formation (see chapter 9, p. 233). From there, other neurons conduct impulses to the thalamus, hypothalamus, and cerebral cortex.

Regulation of Pain Impulses

Awareness of pain arises when pain impulses reach the thalamus—that is, even before they reach the cerebral cortex. The cerebral cortex, however, determines pain intensity, locates the pain source, and mediates emotional and motor responses to the pain.

Areas of gray matter in the midbrain, pons, and medulla oblongata regulate movement of pain impulses from the spinal cord. Impulses from special neurons in these brain areas descend in the lateral funiculus (see chapter 9, p. 233) to various levels of the spinal cord. These impulses stimulate the ends of certain nerve fibers to release biochemicals that can block pain signals by inhibiting presynaptic nerve fibers in the posterior horn of the spinal cord.

The inhibiting substances released in the posterior horn include neuropeptides called *enkephalins* and the monoamine *serotonin* (see chapter 9, p. 218). Enkephalins can suppress acute and chronic pain impulses and thus can relieve severe pain, much as morphine and other opiate drugs do. In fact, enkephalins bind to the same receptor sites on neuron membranes as does morphine. Serotonin stimulates other neurons to release enkephalins.

Endorphins are another group of neuropeptides with pain-suppressing, morphinelike actions. Endorphins are found in the pituitary gland and the hypothalamus. Enkephalins and endorphins are released in response to extreme pain and provide natural pain control.

CHECK YOUR RECALL

1. Describe two types of pain fibers.
2. How do acute pain and chronic pain differ?
3. What parts of the brain interpret pain impulses?
4. How do neuropeptides help control pain?

10.4 Special Senses

Special senses are those whose sensory receptors are within large, complex sensory organs in the head. These senses and their respective organs include the following:

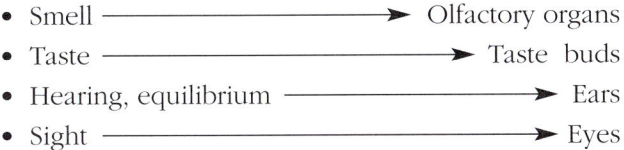

- Smell ⟶ Olfactory organs
- Taste ⟶ Taste buds
- Hearing, equilibrium ⟶ Ears
- Sight ⟶ Eyes

10.5 Sense of Smell

The sense of smell is associated with complex sensory structures in the upper region of the nasal cavity.

Olfactory Receptors

Smell (olfactory) receptors and taste receptors are chemoreceptors, which means that chemicals dissolved in liquids stimulate them. Smell and taste function closely together and aid in food selection because we usually smell food at the same time we taste it.

Olfactory Organs

The **olfactory organs,** which contain the olfactory receptors, are yellowish-brown masses that cover the upper parts of the nasal cavity, the superior nasal conchae, and a portion of the nasal septum. **Olfactory receptor cells** are bipolar neurons surrounded by columnar epithelial cells (fig. 10.4). Hairlike cilia cover tiny knobs at the distal ends of these neurons' dendrites. The cilia project into the nasal cavity and harbor the 500 types of olfactory receptor proteins. Chemicals that stimulate olfactory receptors, called odorant molecules, bind to the receptors in different patterns. Odorant molecules enter the nasal cavity as gases, but they must dissolve at least partially in the watery fluids that surround the cilia before receptors can detect them.

Olfactory Nerve Pathways

Stimulated olfactory receptor cells send nerve impulses along their axons. These fibers (which form the first cranial nerves) synapse with neurons located in enlargements called **olfactory bulbs.** These structures lie on either side of the crista galli of the ethmoid bone (see fig. 7.12, p. 140). Within the olfactory bulbs, the impulses are analyzed, and as a result, additional impulses travel along the **olfactory tracts** to the limbic

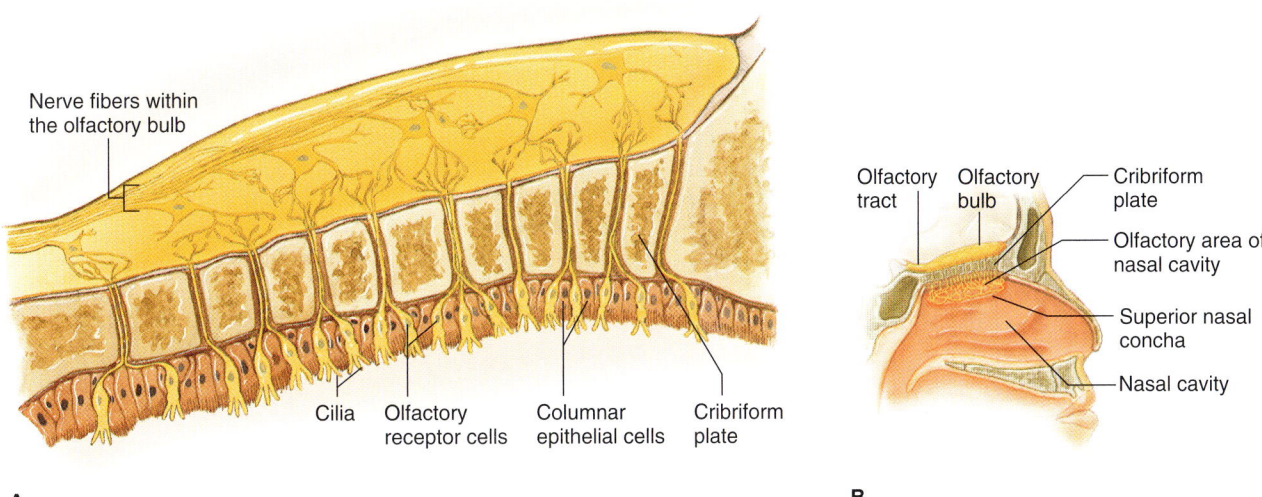

Figure 10.4
Olfactory receptors. (*A*) Columnar epithelial cells support olfactory receptor cells, which have cilia at their distal ends. (*B*) The olfactory area is associated with the superior nasal concha.

Topic of Interest

HEADACHE

Headaches are a common type of pain. Although the nervous tissue of the brain lacks pain receptors, nearly all the other tissues of the head, including the meninges and blood vessels, are richly innervated.

Many headaches are associated with stressful life situations that cause fatigue, emotional tension, anxiety, or frustration. These conditions can trigger various physiological changes, such as prolonged contraction of the skeletal muscles in the forehead, sides of the head, or back of the neck. Such contractions may stimulate pain receptors and produce a *tension headache*. More severe *vascular headaches* accompany constriction or dilation of the cranial blood vessels. For example, the throbbing headache of a "hangover" from drinking too much alcohol may be due to blood pulsating through dilated cranial vessels.

Migraine is another form of vascular headache. In this disorder, certain cranial blood vessels constrict, producing a localized cerebral blood deficiency. This causes a variety of symptoms, such as seeing patterns of bright light that obstruct vision or feeling numbness in the limbs or face. Typically, vasoconstriction subsequently leads to vasodilation of the affected vessels, causing a severe headache, which usually affects one side of the head and may last for several hours. Effective drug treatment is now available for migraines.

Other causes of headaches include sensitivity to food additives, high blood pressure, increased intracranial pressure due to a tumor or to blood escaping from a ruptured vessel, decreased cerebrospinal fluid pressure following a lumbar puncture, or sensitivity to or withdrawal from certain drugs.

---

system (see chapter 9, p. 233). The major interpreting areas (olfactory cortex) for these impulses are located within the temporal lobes and at the bases of the frontal lobes, anterior to the hypothalamus.

 Humans smell the world using about 12 million olfactory receptor cells. Bloodhounds have 4 billion such cells—and hence a much better sense of smell. Specially trained dogs were used to find victims of the World Trade Center collapse in the days after September 11, 2001.

Olfactory Stimulation

Biologists are uncertain how odorant molecules stimulate olfactory receptor cells. One hypothesis suggests that the shapes of gaseous molecules fit complementary shapes of membrane receptor sites on these receptor cells. According to this idea, a molecule binding to its particular receptor site triggers a nerve impulse. (Recall that membrane receptors are molecules such as glycoproteins on cell membranes. Sensory receptors, on the other hand, may be as small as cells or as large as organs, such as the eye.)

A few receptors can detect many odors if each odor stimulates a distinct set of receptor subtypes. The brain then interprets different receptor combinations as an *olfactory code*. For example, if there are ten odor receptors, parsley might stimulate receptors 3, 4, and 8, while chocolate might stimulate receptors 1, 5, and 10.

Because the olfactory organs are located high in the nasal cavity above the usual pathway of inhaled air, a person may have to sniff and force air up to the receptor areas to smell a faint odor. Also, olfactory receptors undergo sensory adaptation rather rapidly, but even if they have adapted to one scent, their sensitivity to other odors persists. For example, a person visiting a fish market might at first be acutely aware of the fishy smell, but then that odor seems to fade. But if a second person enters the fish market wearing a strong perfume, the scent will be quite noticeable to those already present.

Partial or complete loss of smell is called *anosmia*. It may result from a variety of factors, including inflammation of the nasal cavity lining due to a respiratory infection, tobacco smoking, or using certain drugs, such as cocaine.

CHECK YOUR RECALL

1. Where are olfactory receptors located?
2. Trace the pathway of an olfactory impulse from a receptor to the cerebrum.

10.6 Sense of Taste

Taste buds are the special organs of taste (fig. 10.5). They are located primarily on the surface of the tongue and are associated with tiny elevations called *papillae*. Taste buds are also scattered in the roof of the mouth and walls of the throat.

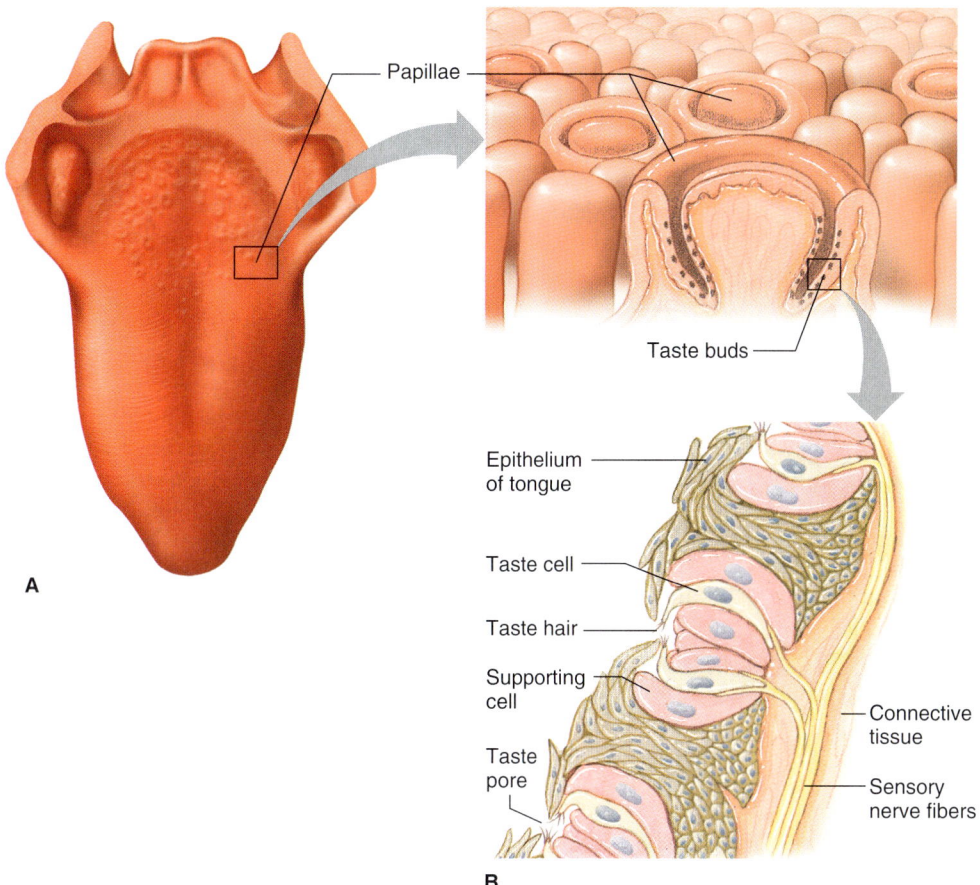

Figure 10.5
Taste receptors. (*A*) Taste buds on the surface of the tongue are associated with nipplelike elevations called papillae. (*B*) A taste bud contains taste cells and has an opening, the taste pore, at its free surface.

Taste Receptors

Each taste bud includes a group of modified epithelial cells, the **taste cells** (gustatory cells), which function as receptors. Each taste bud has 50 to 150 receptor cells, each replaced every three days. The taste bud also includes epithelial supporting cells. The entire structure is somewhat spherical with an opening, the **taste pore,** on its free surface. Tiny projections called **taste hairs** protrude from the outer ends of the taste cells and extend from the taste pore. These taste hairs are believed to be the sensitive parts of the receptor cells.

Interwoven among the taste cells and wrapped around them is a network of nerve fibers. Stimulation of a receptor cell triggers an impulse on a nearby nerve fiber, and the impulse then travels into the brain.

 About 1,000 of your 10,000 taste buds are found not on your tongue, but scattered throughout the roof of your mouth and the back of your throat.

Before a particular chemical can be tasted, it must dissolve in the watery fluid surrounding the taste buds. The salivary glands provide this fluid. As is the case for smell, researchers hypothesize that food molecules combine with specific receptor sites on taste hair surfaces, stimulating the sense of taste. Such a combination could then generate sensory impulses on nearby nerve fibers.

The taste cells in all taste buds appear alike microscopically but are of at least four types. Each type is most sensitive to a particular kind of chemical stimulus, producing at least four primary taste (gustatory) sensations.

Taste Sensations

The four primary taste sensations are:

1. *Sweet,* such as table sugar
2. *Sour,* such as a lemon
3. *Salty,* such as table salt
4. *Bitter,* such as caffeine or quinine

Some investigators recognize other taste sensations—*alkaline, metallic,* and most recently, *umami,* which detects monosodium glutamate (MSG), used as a flavor enhancer in many prepared foods.

A flavor results from one of the primary sensations or from a combination of two or more of them. Experiencing flavors involves taste (concentrations of stimulating chemicals), as well as the sensations of odor, texture (touch), and temperature. Furthermore, the chemicals in some foods—chili peppers and ginger, for instance—may stimulate pain receptors, which cause a burning sensation.

Current evidence suggests that all taste cells are responsive to at least two taste sensations, although for a given taste cell, one taste sensation is likely to predominate. Due to the distribution of taste cells, responsiveness to particular sensations may vary from one region of the tongue to another. In a given individual, responsiveness to a sweet stimulus may be greatest at the tip of the tongue, responsiveness to sour greatest at the margins of the tongue, responsiveness to bitter at the back of the tongue, and responsiveness to salt quite widely distributed. However, due to the fact that all taste receptor cells are sensitive to all of these stimuli to some degree, there is a wide range of individual variation in these responses. More importantly, it may be the pattern of these responses from differentially sensitive receptor cells that provide the brain with the information necessary to create what we call taste.

Taste receptors, like olfactory receptors, undergo sensory adaptation rapidly. Moving bits of food over the surface of the tongue to stimulate different receptors at different moments avoids the loss of taste resulting from sensory adaptation.

Taste Nerve Pathways

Sensory impulses from taste receptors in the tongue travel on fibers of the facial, glossopharyngeal, and vagus nerves into the medulla oblongata. From there, the impulses ascend to the thalamus and are directed to the gustatory cortex, which is located in the parietal lobe of the cerebrum, along a deep portion of the lateral sulcus (see fig. 9.25, p. 228).

CHECK YOUR RECALL

1. Why is saliva necessary for the sense of taste?
2. Name the four primary taste sensations.
3. Trace a sensory impulse from a taste receptor to the cerebral cortex.

10.7 Sense of Hearing

The organ of hearing, the ear, has external, middle, and inner parts. The ear also functions in the sense of equilibrium.

External Ear

The external ear consists of two parts. The first is an outer, funnel-like structure called the **auricle** (pinna). The second is an S-shaped tube called the **external auditory meatus** (me-a′tus), or external auditory canal, that leads inward through the temporal bone for about 2.5 centimeters (fig. 10.6).

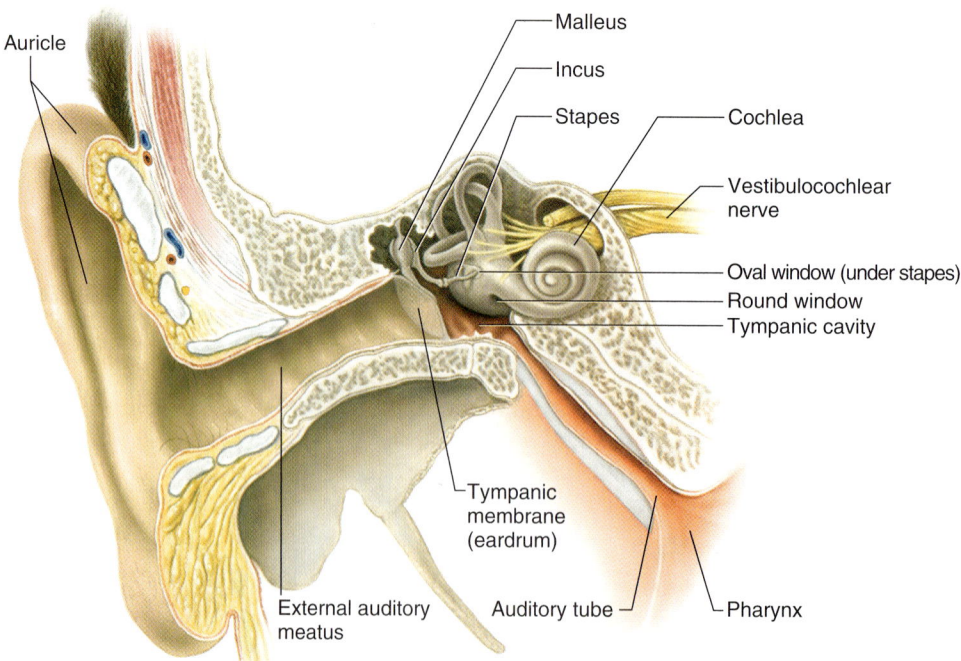

Figure 10.6
Major parts of the ear.

Vibrating objects produce sounds, and the vibrations are transmitted through matter in the form of sound waves. For example, vibrating strings or reeds produce the sounds of some musical instruments, and vibrating vocal folds in the larynx produce the voice. The auricle of the ear helps collect sound waves traveling through air and directs them into the external auditory meatus.

Middle Ear

The middle ear includes an air-filled space in the temporal bone called the *tympanic cavity,* an eardrum (tympanic membrane), and three small bones called auditory ossicles. The **eardrum** is a semitransparent membrane covered by a thin layer of skin on its outer surface and by mucous membrane on the inside. It has an oval margin and is cone-shaped, with the apex of the cone directed inward. The attachment of one of the auditory ossicles (the malleus) maintains the eardrum's cone shape. Sound waves that enter the external auditory meatus change the pressure on the eardrum, which moves back and forth in response and thus reproduces the vibrations of the sound wave source.

The three **auditory ossicles** are the *malleus, incus,* and *stapes* (fig. 10.7). Tiny ligaments attach them to the wall of the tympanic cavity, and they are covered by mucous membrane. These bones bridge the eardrum and the inner ear, transmitting vibrations between these parts. Specifically, the malleus attaches to the eardrum, and when the eardrum vibrates, the malleus vibrates in unison. The malleus causes the incus to vibrate, and the incus passes the movement onto the stapes. Ligaments hold the stapes to an opening in the wall of the tympanic cavity called the **oval window,** which leads into the inner ear. Vibration of the stapes at the oval window moves a fluid within the inner ear, which stimulates the hearing receptors.

In addition to transmitting vibrations, the auditory ossicles help increase (amplify) the force of vibrations as they pass from the eardrum to the oval window. Because the ossicles transmit vibrations from the relatively large surface of the eardrum to a much smaller area at the oval window, the vibrational force concentrates as it moves from the external to the inner ear. As a result, the pressure (per square millimeter) that the stapes applies on the oval window is many times greater than the pressure that sound waves exert on the eardrum.

Auditory Tube

An **auditory tube** (eustachian tube) connects each middle ear to the throat. This tube conducts air between the tympanic cavity and the outside of the body by way of the throat (nasopharynx) and mouth. The auditory tube helps maintain equal air pressure on both sides of the eardrum, which is necessary for normal hearing.

The function of the auditory tube is noticeable during rapid changes in altitude. As a person moves from a high altitude to a lower one, air pressure on the outside of the eardrum increases. This may push the eardrum inward, impairing hearing. When the air pressure difference is great enough, air movement through the auditory tube equalizes the pressure on both sides of the eardrum, and the membrane moves back into its regular position. This produces a popping sound, which restores normal hearing.

Because auditory tube mucous membranes connect directly with middle ear linings, mucous membrane infections of the throat may spread through these tubes and cause middle ear infection. For this reason, pinching a nostril when blowing the nose is poor practice because the pressure in the nasal cavity may force material from the throat up the auditory tube and into the middle ear.

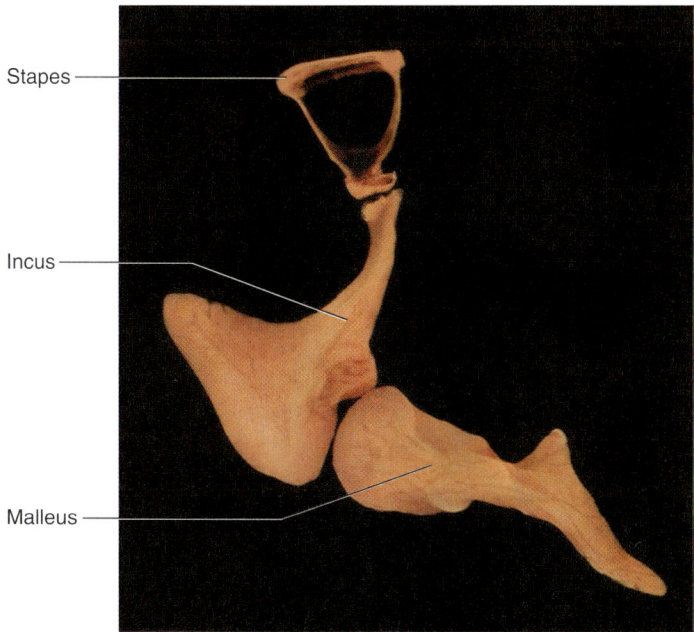

Figure 10.7
Auditory ossicles—malleus, incus, and stapes. These bones bridge the tympanic membrane and the inner ear (5×) (see fig. 10.6).

Inner Ear

The inner ear is a complex system of communicating chambers and tubes called a **labyrinth** (lab´i-rinth). Each ear has two such structures—the *osseous labyrinth* and the *membranous labyrinth* (fig. 10.8). The osseous labyrinth is a bony canal in the temporal bone. The membranous labyrinth is a tube that lies within the osseous labyrinth and has a similar shape. Between the osseous and membranous labyrinths is a fluid called *perilymph,* which is secreted by cells in the wall of the bony canal. The membranous labyrinth contains another fluid called *endolymph*.

The parts of the labyrinths include three **semicircular canals,** which provide a sense of equilibrium (discussed in section 10.8), and a **cochlea** (kok´le-ah), which functions in hearing. The cochlea contains a bony core and a thin, bony shelf that winds around the core like the threads of a screw. The shelf divides the osseous labyrinth of the cochlea into upper and lower compartments. The upper compartment, called the *scala vestibuli,* leads from the oval window to the apex of the spiral. The lower compartment, the *scala tympani,* extends from the apex of the cochlea to a membrane-covered opening in the wall of the inner ear called the **round window** (fig. 10.8).

The portion of the membranous labyrinth within the cochlea is called the *cochlear duct*. It lies between the two bony compartments and ends as a closed sac at the apex of the cochlea. The cochlear duct is separated from the scala vestibuli by a *vestibular membrane* (Reissner's membrane) and from the scala tympani by a *basilar membrane* (fig. 10.9).

The basilar membrane contains many thousands of stiff, elastic fibers, which lengthen from the base of the cochlea to its apex. Sound vibrations entering the perilymph at the oval window travel along the scala vestibuli and pass through the vestibular membrane and into the endolymph of the cochlear duct, where they move the basilar membrane.

After passing through the basilar membrane, the vibrations enter the perilymph of the scala tympani. Movements of the membrane covering the round window dissipate the vibrations into the tympanic cavity.

The **organ of Corti,** which contains the hearing receptors, is located on the upper surface of the basilar membrane and stretches from the apex to the base of

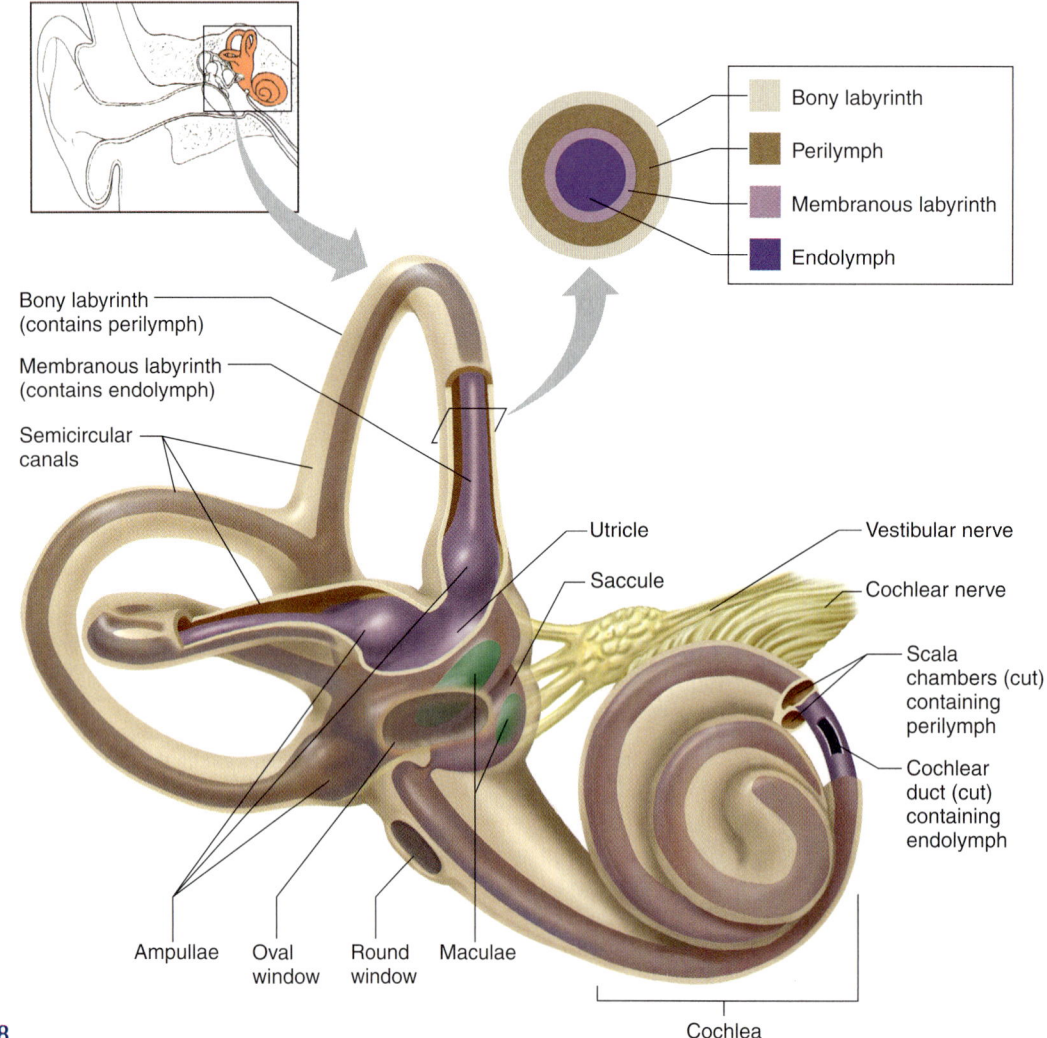

Figure 10.8
Perilymph separates the osseous labyrinth of the inner ear from the membranous labyrinth, which contains endolymph.

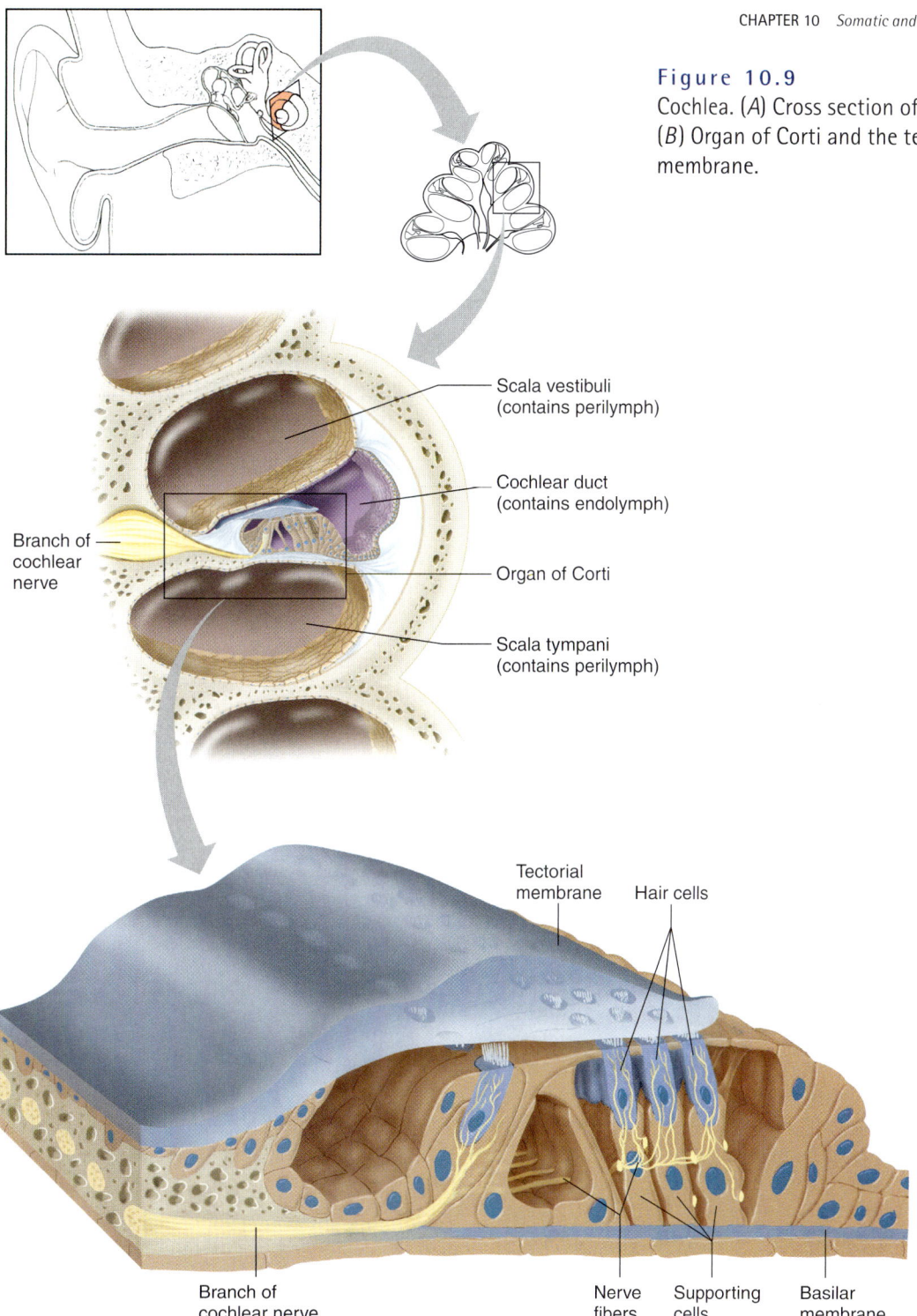

Figure 10.9
Cochlea. (*A*) Cross section of the cochlea. (*B*) Organ of Corti and the tectorial membrane.

the cochlea (fig. 10.9). Its receptor cells (*hair cells*) are organized in rows and have many hairlike processes that project into the endolymph of the cochlear duct. Above these hair cells is a *tectorial membrane* attached to the bony shelf of the cochlea, passing over the receptor cells and contacting the tips of their hairs.

As sound vibrations pass through the inner ear, the hairs shear back and forth against the tectorial membrane, and the resulting mechanical deformation of the hairs stimulates the receptor cells (figs. 10.9 and 10.10). Various receptor cells, however, have slightly different sensitivities to deformation. Thus, a particular sound frequency excites certain receptor cells, and a sound of another frequency stimulates a different set of hair cells.

Hearing receptor cells are epithelial, but function somewhat like neurons. For example, when a hearing receptor cell is at rest, its membrane is polarized. When it is stimulated, selective ion channels open, depolarizing the membrane and making it more permeable to calcium ions. The receptor cell has no axon or dendrites; however, it has neurotransmitter-containing vesicles near its base. As calcium ions diffuse into the cell, some of these vesicles fuse with the cell membrane and release neurotransmitter. The neurotransmitter stimulates the ends of nearby sensory nerve fibers, and in response, they transmit impulses along the cochlear branch of the vestibulocochlear nerve to the auditory cortex of the temporal lobe of the brain.

The ear of a young person with normal hearing can detect sound waves with frequencies ranging from 20 to 20,000 or more vibrations per second. The range of greatest sensitivity is 2,000–3,000 vibrations per second. Table 10.1 summarizes the steps of hearing.

Units called *decibels* (dB) measure sound intensity. The decibel scale begins at 0 dB, which is the intensity of the sound that is least perceptible by a normal human ear. The decibel scale is logarithmic. Thus, a sound of 10 dB is 10 times as intense as the least perceptible sound; a sound of 20 dB is 100 times as intense; and a sound of 30 dB is 1,000 times as intense.

On this scale, a whisper has an intensity of about 40 dB, normal conversation measures 60–70 dB, and heavy traffic produces about 80 dB. A sound of 120 dB, common at a rock concert, produces discomfort, and a sound of 140 dB, such as that emitted by a jet plane at takeoff, causes pain. Frequent or prolonged exposure to sounds with intensities above 90 dB can damage hearing receptors and cause permanent hearing loss. Many rock musicians suffer hearing loss from years of exposure to loud sounds.

Auditory Nerve Pathways

The nerve fibers associated with hearing enter the auditory nerve pathways, which pass into the auditory cortices of the temporal lobes of the cerebrum, where they are interpreted. On the way, some of these fibers cross over, so that impulses arising from each ear are interpreted on both sides of the brain. Consequently, damage to a temporal lobe on one side of the brain does not necessarily cause complete hearing loss in the ear on that side.

A variety of factors can cause partial or complete hearing loss, including interference with the transmission of vibrations to the inner ear (*conductive deafness*) or damage to the cochlea, auditory nerve, or auditory nerve pathways (*sensorineural deafness*). Conductive deafness may be due to plugging of the external auditory meatus or to changes in the eardrum or auditory ossicles. For example, the eardrum may harden as a result of disease and thus be less responsive to sound waves, or disease or injury may tear or perforate the eardrum. Sensorineural deafness can be caused by loud sounds, tumors in the central nervous system, brain damage as a result of vascular accidents, or use of certain drugs.

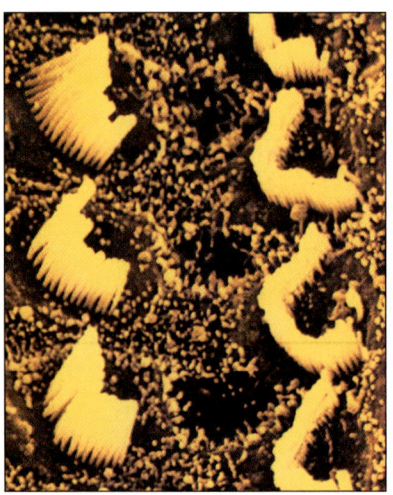

Figure 10.10
Scanning electron micrograph of hair cells in the organ of Corti (3,800×).

TABLE 10.1   STEPS IN THE GENERATION OF SENSORY IMPULSES FROM THE EAR

1. Sound waves enter external auditory meatus.
2. Waves of changing pressures cause eardrum to reproduce vibrations coming from sound wave source.
3. Auditory ossicles amplify and transmit vibrations to end of stapes.
4. Movement of stapes at oval window transmits vibrations to perilymph in scala vestibuli.
5. Vibrations pass through vestibular membrane and enter endolymph of cochlear duct.
6. Different frequencies of vibration in endolymph stimulate different sets of receptor cells.
7. As a receptor cell depolarizes, its membrane becomes more permeable to calcium ions.
8. Inward diffusion of calcium ions causes vesicles at the base of the receptor cell to release neurotransmitter.
9. Neurotransmitter stimulates ends of nearby sensory neurons.
10. Sensory impulses are triggered on fibers of cochlear branch of vestibulocochlear nerve.
11. Auditory cortex of temporal lobe interprets sensory impulses.

CHECK YOUR RECALL

1. How are sound waves transmitted through the external, middle, and inner ears?
2. Distinguish between the osseous and membranous labyrinths.
3. Describe the organ of Corti.

10.8 Sense of Equilibrium

The sense of equilibrium is really two senses—static equilibrium and dynamic equilibrium—that come from different sensory organs. The organs of **static** (stat´ik) **equilibrium** sense the position of the head, maintaining stability and posture when the head and body are still. When the head and body suddenly move or rotate, the organs of **dynamic** (di-nam´ik) **equilibrium** detect such motion and aid in maintaining balance.

Static Equilibrium

The organs of static equilibrium are located within the **vestibule,** a bony chamber between the semicircular canals and the cochlea. The membranous labyrinth inside the vestibule consists of two expanded chambers—a **utricle** and a **saccule** (see fig. 10.8).

Each of these chambers has a tiny structure called a **macula** (mak´u-lah). Maculae contain numerous hair cells, which serve as sensory receptors. When the head is upright, the hairs of the hair cells project upward into a mass of gelatinous material, which has grains of calcium carbonate (otoliths) embedded in it. These particles add weight to the gelatinous structure.

The head bending forward, backward, or to one side stimulates hair cells. Such movements tilt the gelatinous masses of the maculae, and as they sag in response to gravity, the hairs projecting into them bend. This action stimulates the hair cells, and they signal the nerve fibers associated with them in a manner similar to that of hearing receptors. The resulting nerve impulses travel into the central nervous system on the vestibular branch of the vestibulocochlear nerve. These impulses inform the brain of the head's new position. The brain responds by sending motor impulses to skeletal muscles, which contract or relax to maintain balance (fig. 10.11).

Dynamic Equilibrium

The organs of dynamic equilibrium are the three **semicircular canals** located in the labyrinth. They detect motion of the head and aid in balancing the head and body during sudden movement. These canals lie at right angles to each other, and each corresponds to a different anatomical plane (see fig. 10.8).

Suspended in the perilymph of the bony portion of each semicircular canal is a membranous canal that

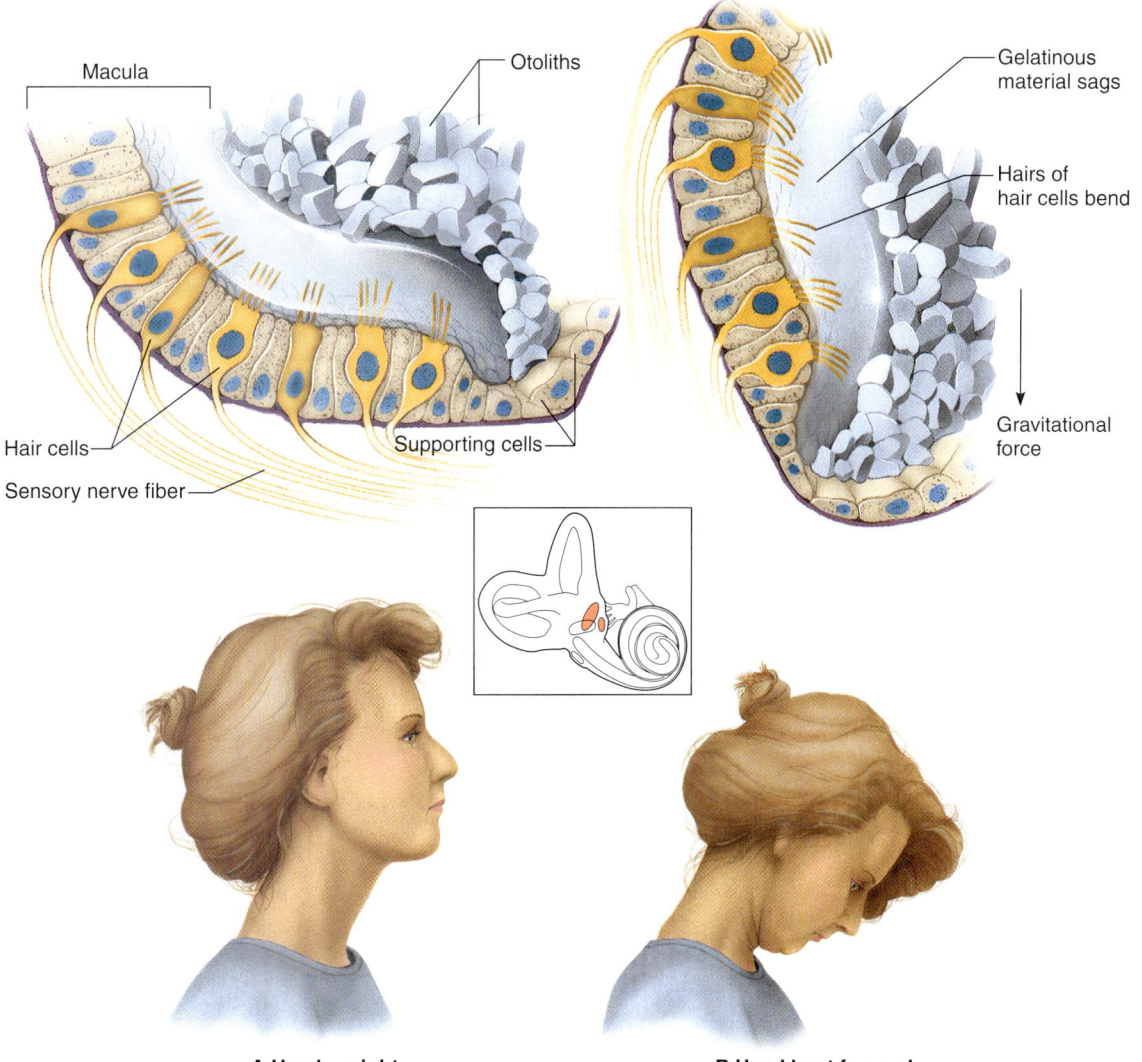

A Head upright   **B Head bent forward**

Figure 10.11
The macula responds to changes in head position. (*A*) Macula with the head in an upright position. (*B*) Macula with the head bent forward.

266 UNIT 3 Integration and Coordination

ends in a swelling called an **ampulla** (am-pul´lah), which contains the sensory organs of the semicircular canals. Each of these organs, called a **crista ampullaris,** contains a number of sensory hair cells and supporting cells (fig. 10.12). As in the hairs of the maculae, the hair cells extend upward into a dome-shaped gelatinous mass called the *cupula*.

Rapid turns of the head or body stimulate the hair cells of the crista ampullaris (fig. 10.13). At such times, the semicircular canals move with the head or body, but the fluid inside the membranous canals remains stationary. This bends the cupula in one or more of the canals in a direction opposite that of the head or body movement, and the hairs embedded in it also bend. The stimulated hair cells signal their associated nerve fibers, sending impulses to the brain.

Parts of the cerebellum are particularly important in interpreting impulses from the semicircular canals. Analysis of such information allows the brain to predict the consequences of rapid body movements, and by modifying signals to appropriate skeletal muscles, the cerebellum can maintain balance.

A Head in still position

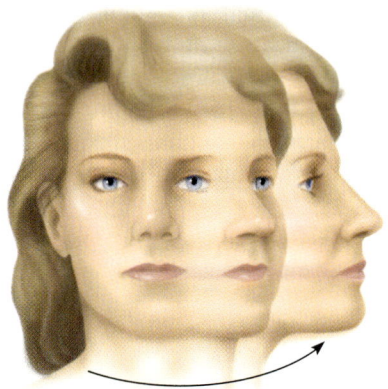

B Head rotating

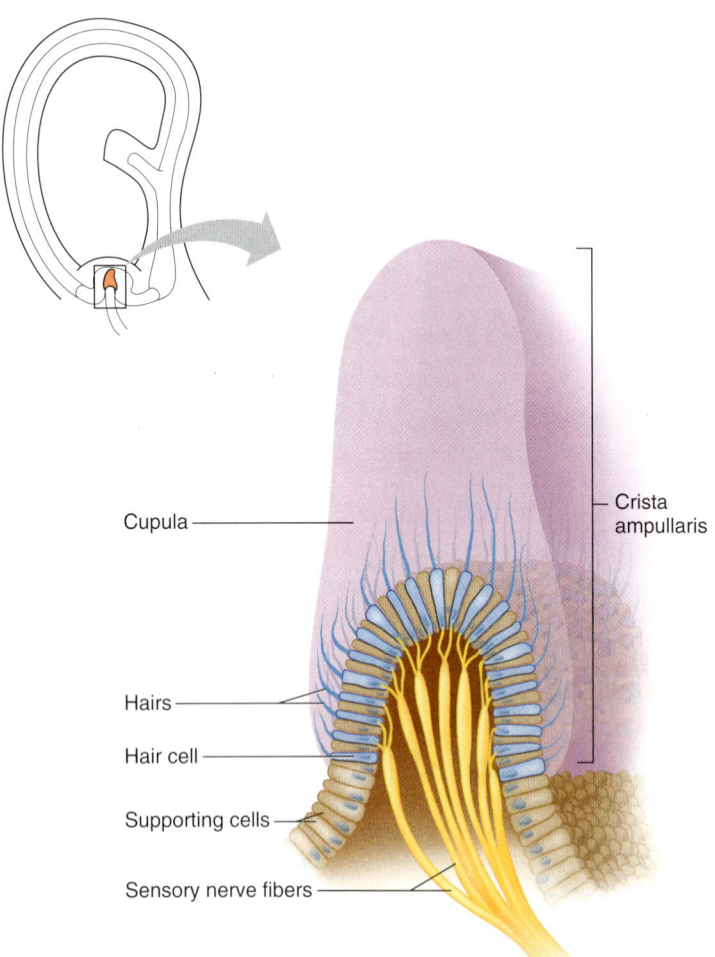

Figure 10.12
A crista ampullaris is located within the ampulla of each semicircular canal.

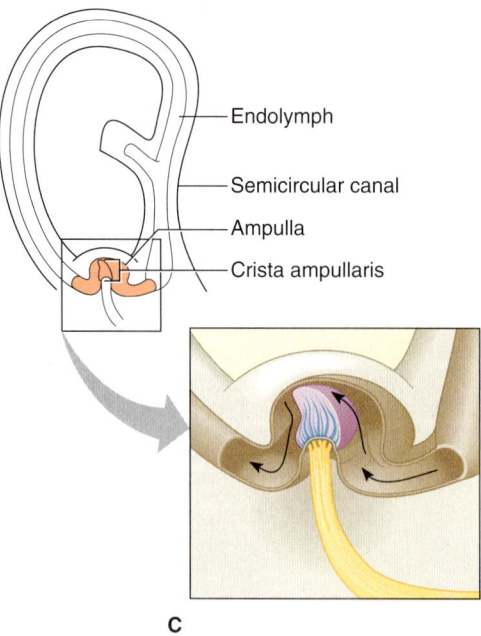

Figure 10.13
Equilibrium. (*A*) When the head is stationary, the cupula of the crista ampullaris remains upright. (*B*) When the head is moving rapidly, (*C*) the cupula bends opposite the motion of the head, stimulating sensory receptors.

Other sensory structures aid in maintaining equilibrium. For example, certain mechanoreceptors (proprioceptors), particularly those associated with the joints of the neck, inform the brain about the position of body parts. In addition, the eyes detect changes in posture that result from body movements. Such visual information is so important that even if the organs of equilibrium are damaged, a person may be able to maintain normal balance by keeping the eyes open and moving slowly.

Some people experience *motion sickness* in a moving boat, airplane, or automobile. The cause is abnormal and irregular body motions that disturb the organs of equilibrium. Symptoms of motion sickness include nausea, vomiting, dizziness, headache, and prostration.

CHECK YOUR RECALL

1. Distinguish between static and dynamic equilibrium.
2. Which structures provide the sense of static equilibrium? Of dynamic equilibrium?
3. How does sensory information from other receptors help maintain equilibrium?

10.9 Sense of Sight

The eye, the organ containing visual receptors, provides vision, with the assistance of *accessory organs*. These accessory organs include the eyelids and lacrimal apparatus, which protect the eye, and a set of extrinsic muscles, which move the eye.

Visual Accessory Organs

The eye, lacrimal gland, and associated extrinsic muscles are housed within the pear-shaped orbital cavity of the skull. This orbit, which is lined with the periosteum of various bones, also contains fat, blood vessels, nerves, and connective tissues.

Each **eyelid** has four layers—skin, muscle, connective tissue, and conjunctiva. The skin of the eyelid, which is the thinnest skin of the body, covers the lid's outer surface and fuses with its inner lining near the margin of the lid. The eyelids are moved by the *orbicularis oculi* muscle (see fig. 8.17, p. 189), which acts as a sphincter and closes the lids when it contracts, and by the *levator palpebrae superioris* muscle, which raises the upper lid and thus helps open the eye (fig. 10.14). The **conjunctiva** is a mucous membrane that lines the inner surfaces of the eyelids and folds back to cover the anterior surface of the eyeball, except for its central portion (cornea).

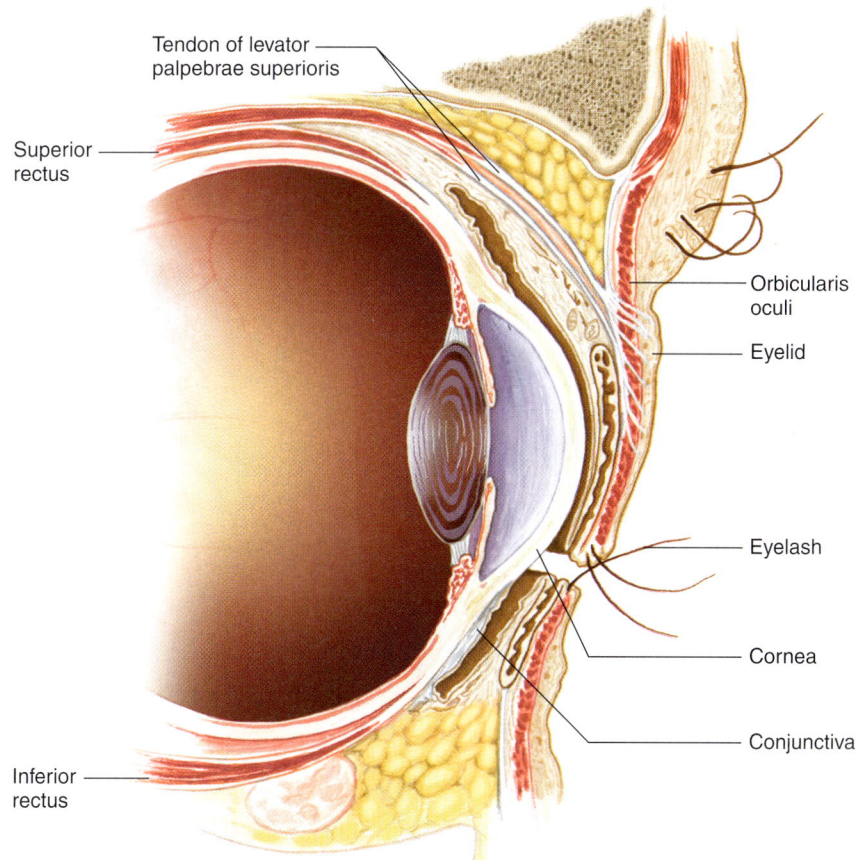

Figure 10.14
Sagittal section of the closed eyelids and anterior portion of the eye.

The *lacrimal apparatus* consists of the **lacrimal gland,** which secretes tears, and a series of ducts that carry tears into the nasal cavity (fig. 10.15). The gland is located in the orbit and secretes tears continuously. The tears exit through tiny tubules and flow downward and medially across the eye.

Two small ducts (the superior and inferior canaliculi) collect tears, which flow into the *lacrimal sac,* located in a deep groove of the lacrimal bone, and then into the *nasolacrimal duct,* which empties into the nasal cavity. Secretion of the lacrimal gland moistens and lubricates the surface of the eye and the lining of the lids. Tears also contain an enzyme (*lysozyme*) that is an antibacterial agent, reducing the risk of eye infections.

The **extrinsic muscles** arise from the bones of the orbit and insert by broad tendons on the eye's tough outer surface. Six extrinsic muscles move the eye in various directions. Any given eye movement may utilize more than one extrinsic muscle, but each muscle is associated with one primary action. Figure 10.16 illustrates the locations of these extrinsic muscles, and table 10.2 lists their functions, as well as the functions of the eyelid muscles.

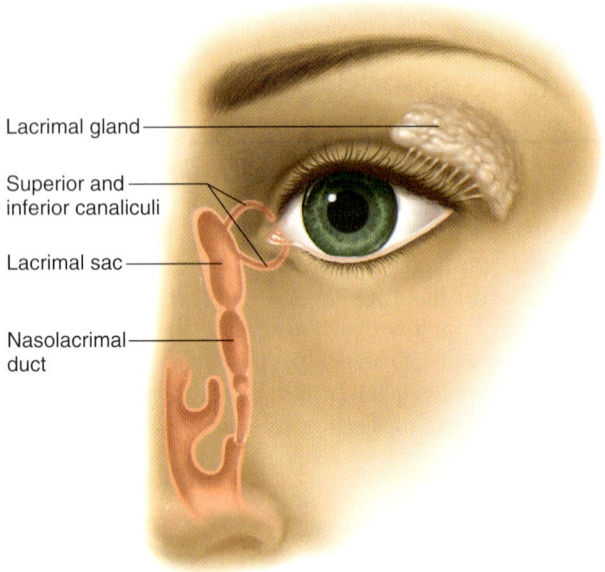

Figure 10.15
The lacrimal apparatus consists of a tear-secreting gland and a series of ducts.

One eye deviating from the line of vision may result in double vision (diplopia). If this condition persists, the brain may suppress the image from the deviated eye. As a result, the turning eye may become blind (suppression amblyopia). Treating eye deviation early in life with exercises, eyeglasses, and surgery can prevent such monocular (one eye) blindness.

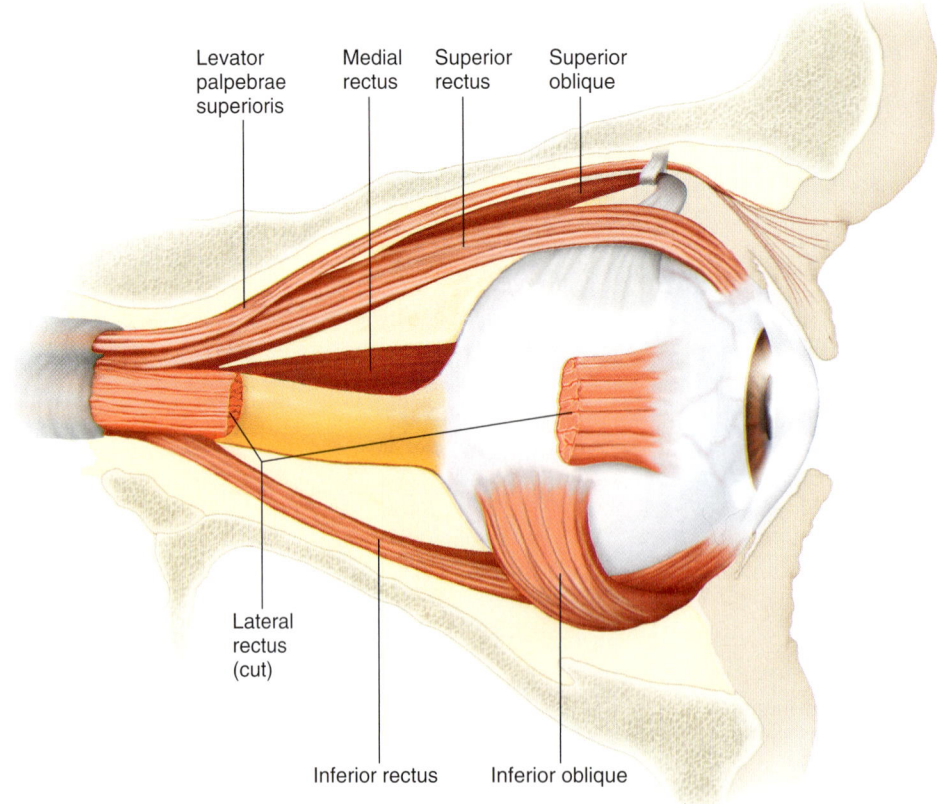

Figure 10.16
Extrinsic muscles of the right eye (lateral view).

TABLE 10.2   MUSCLES ASSOCIATED WITH THE EYELIDS AND EYES

NAME	INNERVATION	FUNCTION
Muscles of the eyelids		
Orbicularis oculi	Facial nerve (VII)	Closes eye
Levator palpebrae superioris	Oculomotor nerve (III)	Opens eye
Extrinsic muscles of the eyes		
Superior rectus	Oculomotor nerve (III)	Rotates eye upward and toward midline
Inferior rectus	Oculomotor nerve (III)	Rotates eye downward and toward midline
Medial rectus	Oculomotor nerve (III)	Rotates eye toward midline
Lateral rectus	Abducens nerve (VI)	Rotates eye away from midline
Superior oblique	Trochlear nerve (IV)	Rotates eye downward and away from midline
Inferior oblique	Oculomotor nerve (III)	Rotates eye upward and away from midline

CHECK YOUR RECALL

1. Explain how the eyelid moves.
2. Describe the conjunctiva.
3. What is the function of the lacrimal apparatus?

Structure of the Eye

The eye is a hollow, spherical structure about 2.5 centimeters in diameter. Its wall has three distinct layers—a fibrous *outer tunic,* a vascular *middle tunic,* and a nervous *inner tunic.* The spaces within the eye are filled with fluids that support its wall and internal parts and help maintain its shape. Figure 10.17 shows the major parts of the eye.

Outer Tunic

The anterior sixth of the outer tunic (fibrous tunic) bulges forward as the transparent **cornea** (kor´ne-ah), which is the window of the eye and helps focus entering light rays. The cornea is composed largely of connective tissue with a thin layer of epithelium on its surface. It is transparent because it contains few cells and no blood vessels, and its cells and collagenous fibers form unusually regular patterns.

Along its circumference, the cornea is continuous with the **sclera** (skle´rah), the white portion of the eye. The sclera makes up the posterior five-sixths of the outer tunic and is opaque due to many large, disorganized, collagenous and elastic fibers. The sclera protects the eye and is an attachment for the extrinsic muscles. In the back of the eye, the **optic nerve** and certain blood vessels pierce the sclera.

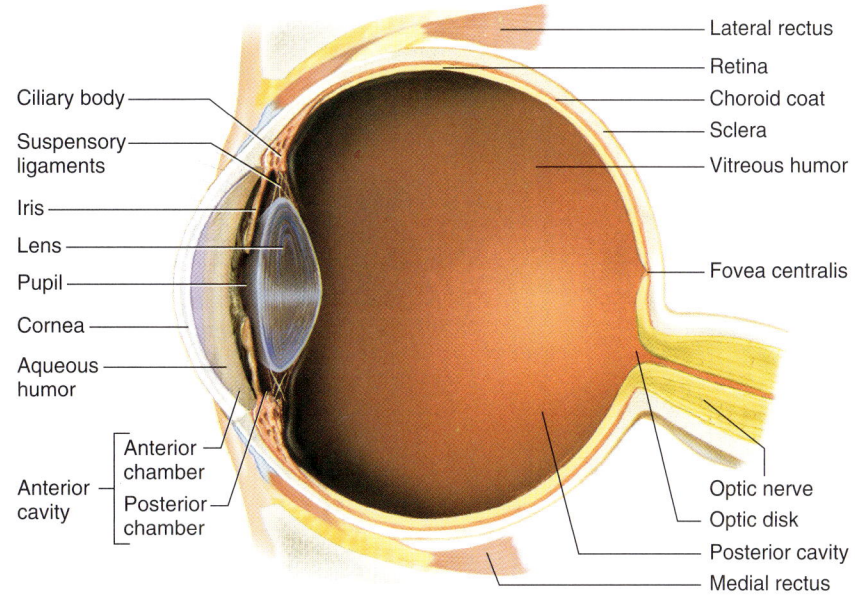

Figure 10.17
Transverse section of the right eye (superior view).

Worldwide, the most common cause of blindness is loss of transparency of the cornea. A corneal transplant (penetrating keratoplasty) can treat this condition by replacing the central two-thirds of the defective cornea with a similar-sized portion of cornea from a donor eye. Because corneal tissues lack blood vessels, transplanted tissue is usually not rejected. The success rate of the procedure is very high.

Middle Tunic

The middle tunic (vascular tunic) includes the choroid coat, ciliary body, and iris (fig. 10.17). The **choroid coat,** in the posterior five-sixths of the globe of the eye, is loosely joined to the sclera and is honeycombed with blood vessels, which nourish surrounding tissues. The choroid coat also contains many pigment-producing melanocytes. The melanin that these cells produce absorbs excess light and thus helps keep the inside of the eye dark.

The **ciliary body,** which is the thickest part of the middle tunic, extends forward from the choroid coat and forms an internal ring around the front of the eye. Within the ciliary body are many radiating folds called *ciliary processes* and groups of muscle fibers that constitute the *ciliary muscles.*

Many strong but delicate fibers, called *suspensory ligaments,* extend inward from the ciliary processes and hold the transparent **lens** in position (fig. 10.18). The distal ends of these fibers attach along the margin of a thin capsule that surrounds the lens. The body of the lens lies directly behind the iris and pupil and is composed of differentiated epithelial cells called *lens fibers.* The cytoplasm of these cells is the transparent substance of the lens.

The lens capsule is a clear, membranelike structure composed largely of intercellular material. Its elastic nature keeps it under constant tension. As a result, the lens can assume a globular shape. The suspensory ligaments attached to the margin of the capsule are also under tension. They pull outward, flattening the capsule and the lens inside (fig. 10.19).

A common eye disorder, particularly in older people, is *cataract*. The lens or its capsule slowly becomes cloudy and opaque. Without treatment, cataracts eventually cause blindness. In the past, cataracts were treated surgically, with a two-week recovery period. Today, cataracts are treated on an outpatient basis with a laser.

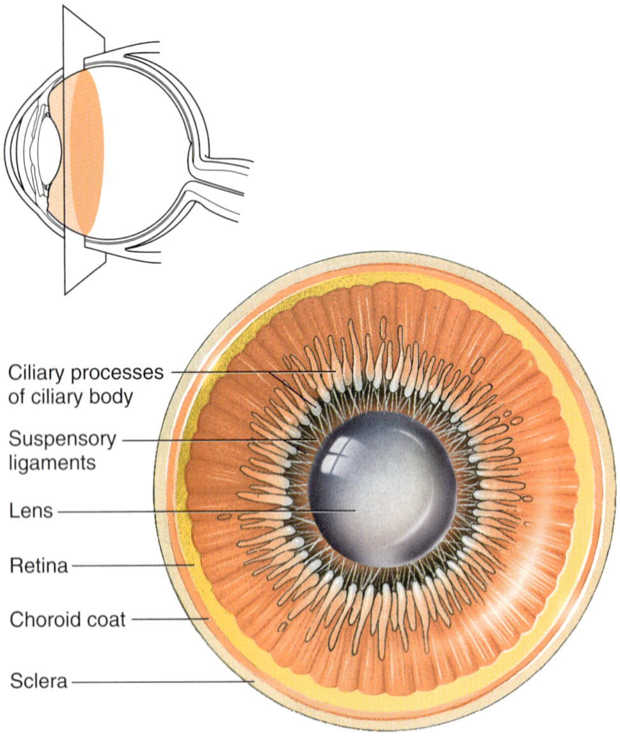

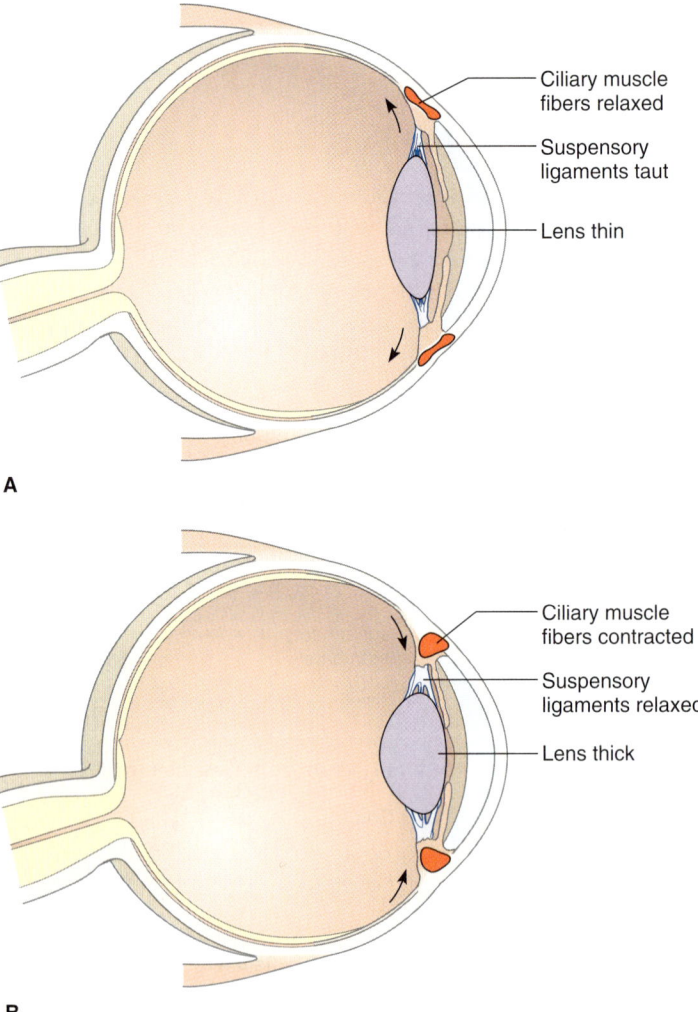

Figure 10.18
Lens and ciliary body viewed from behind.

Figure 10.19
In accommodation, (*A*) the lens thins as ciliary muscle fibers relax. (*B*) The lens thickens as ciliary muscle fibers contract.

If the tension on the suspensory ligaments relaxes, the elastic capsule rebounds, and the lens surface becomes more convex. The ciliary muscles accomplish this. For example, one set of these muscle fibers extends back from fixed points in the sclera to the choroid coat. When the fibers contract, the choroid coat is pulled forward, and the ciliary body shortens. This relaxes the suspensory ligaments; the lens thickens in response and is now focused for viewing closer objects than before. To focus on more distant objects, the ciliary muscles relax, tension on the suspensory ligaments increases, and the lens becomes thinner and less convex again. This ability of the lens to adjust shape to facilitate focusing is called **accommodation** (ah-kom″o-da′shun) (fig. 10.19).

An eye disorder called *glaucoma* develops when the rate of aqueous humor formation exceeds the rate of its removal. As fluid accumulates in the anterior chamber of the eye, fluid pressure rises and is transmitted to all parts of the eye. In time, the building pressure squeezes shut blood vessels that supply the receptor cells of the retina. Cells that are robbed of nutrients and oxygen in this way may die, and permanent blindness can result.

When diagnosed early, glaucoma can usually be treated successfully with drugs, laser therapy, or surgery, all of which promote the outflow of aqueous humor. Since glaucoma in its early stages typically produces no symptoms, discovery of the condition usually depends on measuring intraocular pressure, using an instrument called a *tonometer*.

CHECK YOUR RECALL

1. Describe the outer and middle tunics of the eye.
2. What factors contribute to the transparency of the cornea?
3. How does the shape of the lens change during accommodation?

The **iris** is a thin diaphragm composed mostly of connective tissue and smooth muscle fibers. From the outside, the iris is the colored portion of the eye. The iris extends forward from the periphery of the ciliary body and lies between the cornea and lens (see fig. 10.17). The iris divides the space (anterior cavity) separating these parts into an *anterior chamber* (between the cornea and iris) and a *posterior chamber* (between the iris and vitreous body and containing the lens).

The epithelium on the inner surface of the ciliary body secretes a watery fluid called **aqueous humor** into the posterior chamber. The fluid circulates from this chamber through the **pupil,** a circular opening in the center of the iris, and into the anterior chamber. Aqueous humor fills the space between the cornea and lens, helps nourish these parts, and aids in maintaining the shape of the front of the eye. It subsequently leaves the anterior chamber through veins and a special drainage canal, the scleral venous sinus (canal of Schlemm) located in its wall at the junction of the cornea and the sclera.

The smooth muscle fibers of the iris are organized into two groups, a *circular set* and a *radial set.* These muscles control the size of the pupil, through which light passes as it enters the eye. The circular set of muscle fibers acts as a sphincter. When it contracts, the pupil gets smaller, and less light enters. When the radial muscle fibers contract, the pupil's diameter increases, and more light enters (fig. 10.20).

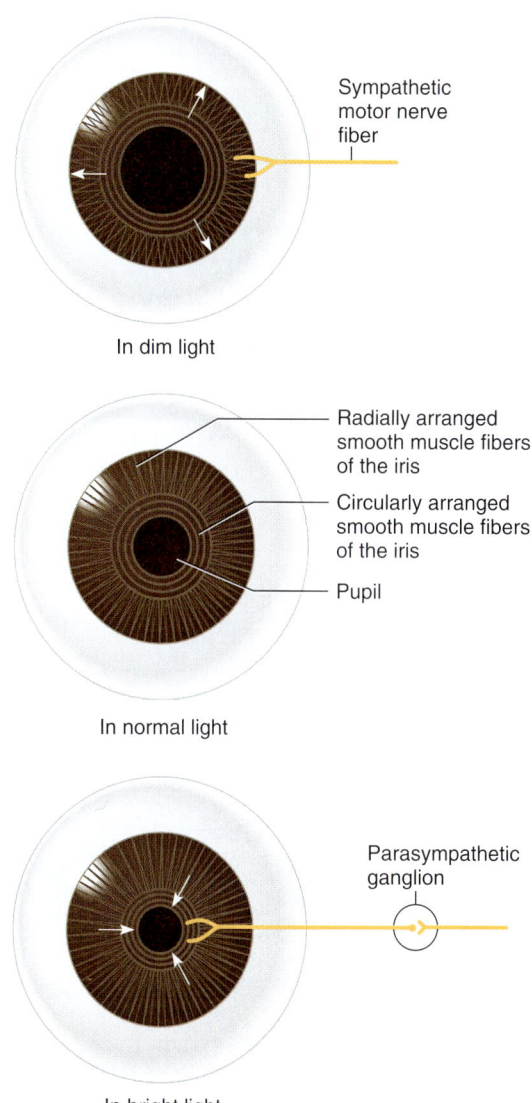

Figure 10.20
Dim light stimulates the radial muscles of the iris to contract, and the pupil dilates. Bright light stimulates the circular muscles of the iris to contract, and the pupil constricts.

Inner Tunic

The inner tunic consists of the **retina** (ret´ĭ-nah), which contains the visual receptor cells (photoreceptors). This nearly transparent sheet of tissue is continuous with the optic nerve in the back of the eye and extends forward as the inner lining of the eyeball. It ends just behind the margin of the ciliary body.

The retina is thin and delicate, but its structure is quite complex. It has a number of distinct layers, as figures 10.21 and 10.22 illustrate.

In the central region of the retina is a yellowish spot called the *macula lutea*. A depression in its center, called the **fovea centralis**, is in the region of the retina that produces the sharpest vision (see fig. 10.17).

 The fovea centralis of the human eye has 150,000 cones per square millimeter. In contrast, a bird of prey's eye has about a million cones per square millimeter.

Just medial to the fovea centralis is an area called the **optic disc** (fig. 10.23). Here, nerve fibers from the retina leave the eye and join the optic nerve. A central artery and vein also pass through the optic disc. These vessels are continuous with the capillary networks of the retina, and with vessels in the underlying choroid coat, they supply blood to the cells of the inner tunic. Because the optic disc region lacks receptor cells, it is commonly known as the *blind spot* of the eye.

The space bounded by the lens, ciliary body, and retina is the largest compartment of the eye and is called the *posterior cavity* (see fig. 10.17). The posterior cavity is filled with a transparent, jellylike fluid called **vitreous humor**, which along with collagenous fibers comprises the *vitreous body*. The vitreous body supports the internal parts of the eye and helps maintain its shape.

As a person ages, tiny, dense clumps of gel or deposits of crystal-like substances form in the vitreous humor. When these clumps cast shadows on the retina, the person sees small, moving specks in the field of vision. Such specks, known as *floaters*, are most apparent when looking at a plain background, such as the sky or a wall.

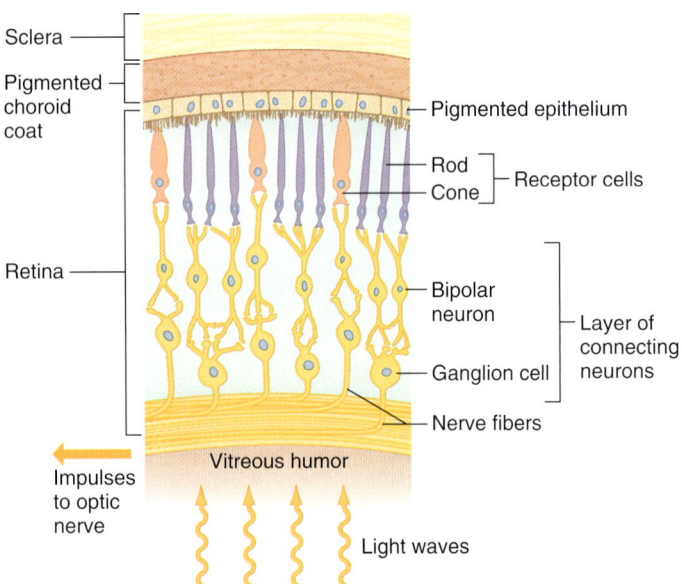

Figure 10.21
The retina consists of several cell layers.

 CHECK YOUR RECALL

1. Explain the origin of aqueous humor, and trace its path through the eye.
2. How is the size of the pupil regulated?
3. Describe the structure of the retina.

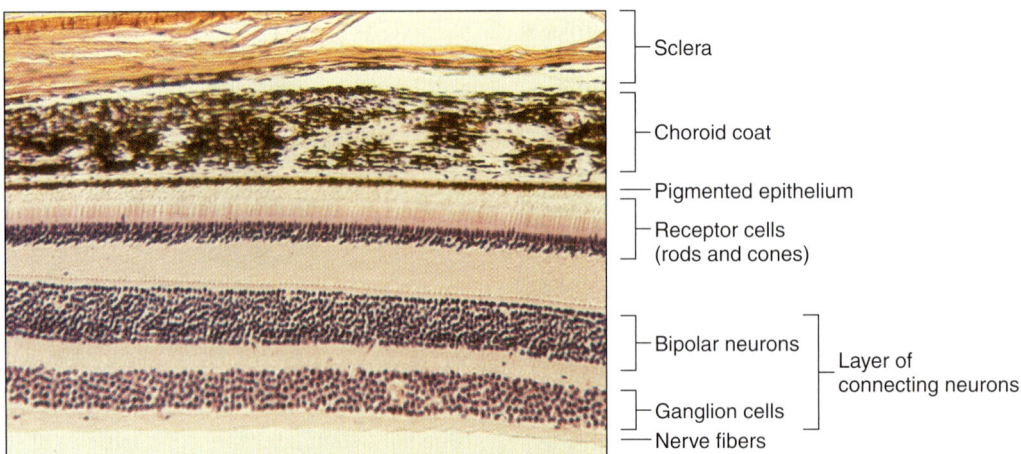

Figure 10.22
Note the layers of cells and nerve fibers in this light micrograph of the retina (75×).

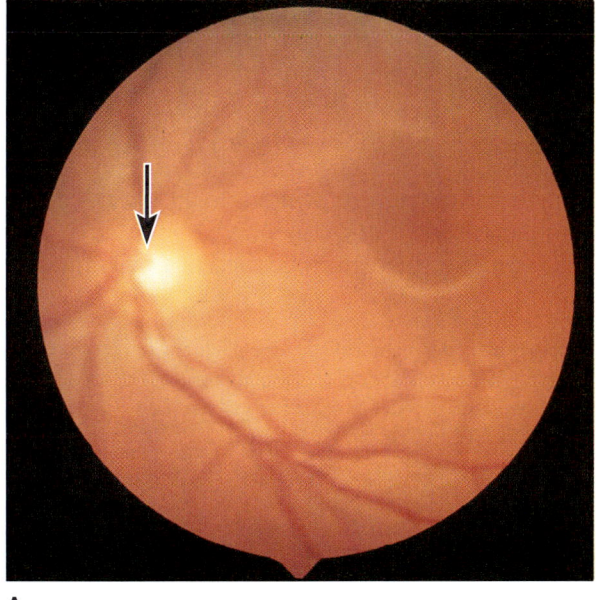
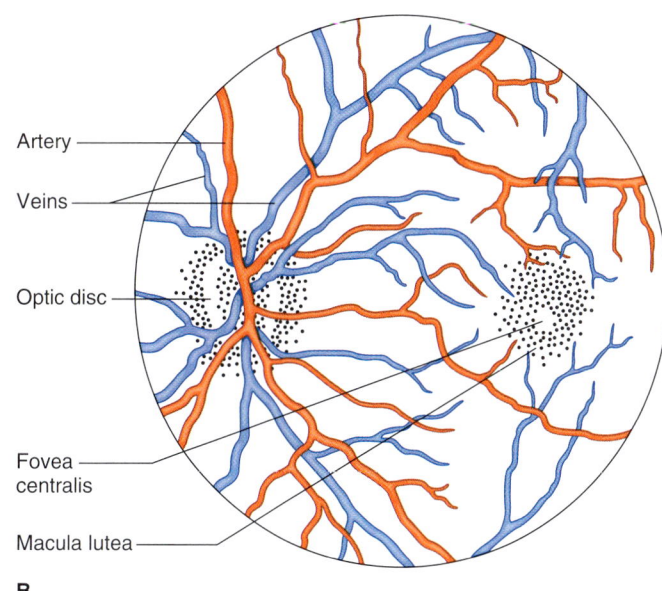

Figure 10.23
Retina. (*A*) Nerve fibers leave the eye in the area of the optic disc (arrow) to form the optic nerve (53×). (*B*) Major features of the retina.

Light Refraction

When a person sees something, either the object is giving off light, or light waves are reflected from it. These light waves enter the eye, and an image of the object is focused on the retina. Focusing bends the light waves, a phenomenon called **refraction** (re-frak´shun).

Refraction occurs when light waves pass at an oblique angle from a medium of one optical density into a medium of a different optical density. This occurs at the curved surface between the air and the cornea and at the curved surface of the lens itself. A lens with a *convex* surface (such as in the eye) causes light waves to converge (fig. 10.24).

The convex surface of the cornea refracts light waves from outside objects. The convex surface of the lens and, to a lesser extent, the surfaces of the fluids within the chambers of the eye then refract the light again.

If eye shape is normal, light waves focus sharply on the retina, much as a motion picture image is focused on a screen for viewing. Unlike the motion picture image, however, an image that forms on the retina is upside down and reversed from left to right. The visual cortex interprets the image in its proper position.

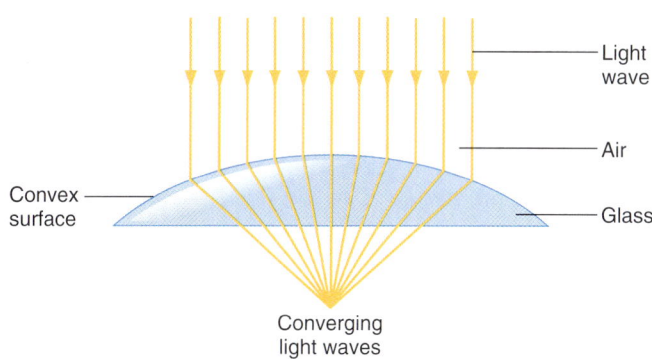

Figure 10.24
A lens with a convex surface causes light waves to converge.

Visual Receptors

Visual receptor cells are modified neurons of two distinct kinds, as figure 10.21 illustrates. One group, called *rods,* have long, thin projections at their ends, while the other group, *cones,* have short, blunt projections.

 A human eye has 125 million rods and 7 million cones.

Rods and cones are in a deep portion of the retina, closely associated with a layer of pigmented epithelium (see fig. 10.22). The epithelial pigment absorbs light

CHECK YOUR RECALL

1. What is refraction?
2. What parts of the eye provide refracting surfaces?

waves not absorbed by the receptor cells, and together with the pigment of the choroid coat, keeps light from reflecting off surfaces inside the eye. Projections from receptors, which are loaded with light-sensitive visual pigments, extend into this pigmented layer.

Visual receptors are stimulated only when light reaches them. A light image focused on an area of the retina stimulates some receptors, and impulses travel from them to the brain. The impulse leaving each activated receptor, however, provides only a fragment of the information required for the brain to interpret a total scene.

Rods and cones function differently. Rods are hundreds of times more sensitive to light than cones and therefore can provide vision in dim light. Rods produce colorless vision, whereas cones detect color.

Rods and cones also differ in visual acuity—the sharpness of the perceived images. Cones provide sharp images, and rods provide more general outlines of objects. Rods give less precise images because nerve fibers from many rods converge, their impulses transmitted to the brain on the same nerve fiber (fig. 10.25A). Thus, if a point of light stimulates a rod, the brain cannot tell which one of many receptors has been stimulated. Convergence of impulses is less common among cones. When a cone is stimulated, the brain can pinpoint the stimulation more accurately (fig. 10.25B).

The fovea centralis, the area of sharpest vision, lacks rods but contains densely packed cones with few or no converging fibers (see fig. 10.17). Also in the fovea centralis, the overlying layers of the retina and the retinal blood vessels are displaced to the sides, more fully exposing receptors to incoming light. Consequently, to view something in detail, a person moves the eyes so that the important part of an image falls on the fovea centralis.

Visual Pigments

Both rods and cones contain light-sensitive pigments that decompose when they absorb light energy. The light-sensitive biochemical in rods is called **rhodopsin** (ro-dop´sin) or *visual purple*. In the presence of light, rhodopsin molecules are broken down into a colorless protein called *opsin* and a yellowish substance called *retinal* (retinene) that is synthesized from vitamin A.

> Poor vision in dim light, called night blindness, results from vitamin A deficiency. Lack of the vitamin reduces the supply of retinal, rhodopsin production falls, and rod sensitivity is low. Vitamin A is used to treat the condition.

Decomposition of rhodopsin molecules activates an enzyme that initiates a series of reactions altering the permeability of the rod cell membrane. As a result, a complex pattern of nerve impulses originates in the retina. The impulses travel away from the retina along the optic nerve into the brain, where they are interpreted as vision.

In bright light, nearly all of the rhodopsin in the rods of the retina decomposes, greatly reducing rod sensitivity. In dim light, however, regeneration of rhodopsin from opsin and retinal is faster than rhodopsin breakdown. This regeneration requires cellular energy, provided by ATP (see chapter 4, p. 78).

The light-sensitive pigments of the cones are similar to rhodopsin in that they are composed of retinal combined with a protein; the protein, however, differs from the protein in the rods. Three different sets of cones each contain an abundance of one of three different visual pigments.

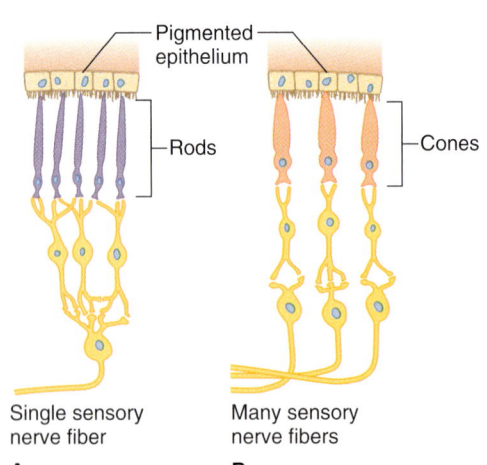 

Figure 10.25
Rods and cones. (*A*) A single sensory nerve fiber transmits impulses from several rods to the brain. (*B*) Separate sensory nerve fibers transmit impulses from cones to the brain. (*C*) Scanning electron micrograph of rods and cones (1,350×).

The wavelength of light determines the color that the brain perceives from it. For example, the shortest wavelengths of visible light are perceived as violet, and the longest are perceived as red. One type of cone pigment (erythrolabe) is most sensitive to red light waves, another (chlorolabe) to green light waves, and a third (cyanolabe) to blue light waves. The color a person perceives depends on which set of cones or combination of sets the light in a given image stimulates. If all three sets of cones are stimulated, the person senses the light as white, and if none are stimulated, the person senses black. Different forms of color blindness result from lack of different types of cone pigments.

Visual Nerve Pathways

The axons of the retinal neurons leave the eyes to form the *optic nerves* (fig. 10.26). Just anterior to the pituitary gland, these nerves give rise to the X-shaped *optic chiasma,* and within the chiasma, some of the fibers cross over. More specifically, the fibers from the nasal (medial) half of each retina cross over, but those from the temporal (lateral) sides do not. Thus, fibers from the nasal half of the left eye and the temporal half of the right eye form the *right optic tract,* and fibers from the nasal half of the right eye and the temporal half of the left eye form the *left optic tract.*

Just before the nerve fibers reach the thalamus, a few of them enter nuclei that function in various visual reflexes. Most of the fibers, however, enter the thalamus and synapse in its posterior portion (lateral geniculate body). From this region, the visual impulses enter nerve pathways called *optic radiations,* which lead to the visual cortex of the occipital lobes.

CHECK YOUR RECALL

1. Distinguish between the rods and cones of the retina.
2. Explain the roles of visual pigments.
3. Trace a nerve impulse from the retina to the visual cortex.

Clinical Terms Related to the Senses

amblyopia (am´´ble-o´pe-ah) Dim vision due to a cause other than a refractive disorder or lesion.
anopia (an-o´pe-ah) Absence of an eye.
audiometry (aw´´de-om´ĕ-tre) Measurement of auditory acuity for various frequencies of sound waves.
blepharitis (blef´´ah-ri´tis) Inflammation of the eyelid margins.
causalgia (kaw-zal´je-ah) Persistent, burning pain usually associated with injury to a limb.
conjunctivitis (kon-junk´´tĭ-vi´tis) Inflammation of the conjunctiva.
diplopia (dĭ-plo´pe-ah) Double vision.
emmetropia (em´´ĕ-tro´pe-ah) Normal condition of the eyes; eyes with no refractive defects.
enucleation (e-nu´´kle-a´shun) Removal of the eyeball.
exophthalmos (ek´´sof-thal´mos) Abnormal protrusion of the eyes.
hemianopsia (hem´´e-an-op´se-ah) Defective vision affecting half of the visual field.
hyperalgesia (hi´´per-al-je´ze-ah) Heightened sensitivity to pain.
iridectomy (ir´´ĭ-dek´to-me) Surgical removal of part of the iris.
iritis (i-ri´tis) Inflammation of the iris.
keratitis (ker´´ah-ti´tis) Inflammation of the cornea.
labyrinthectomy (lab´´ĭ-rin-thek´to-me) Surgical removal of the labyrinth.
labyrinthitis (lab´´ĭ-rin-thi´tis) Inflammation of the labyrinth.
Ménière's disease (men´´e-ārz´ dĭ-zēz´) Inner ear disorder that causes ringing in the ears, increased sensitivity to sounds, dizziness, and hearing loss.
neuralgia (nu-ral´je-ah) Pain resulting from inflammation of a nerve or a group of nerves.

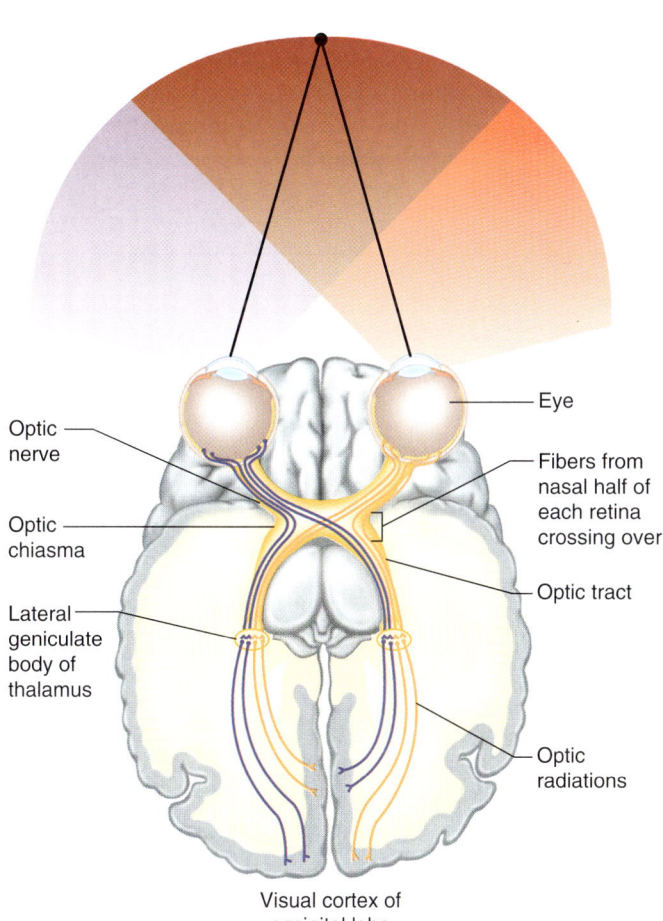

Figure 10.26
The visual pathway includes the optic nerve, optic chiasma, optic tract, and optic radiations.

neuritis (nu-ri´tis) Inflammation of a nerve.
nystagmus (nis-tag´mus) Involuntary oscillation of the eyes.
otitis media (o-ti´tis me´de-ah) Inflammation of the middle ear.
otosclerosis (o´´to-skle-ro´sis) Formation of spongy bone in the inner ear, which often causes deafness by fixing the stapes to the oval window.
pterygium (tĕ-rij´e-um) Abnormally thickened patch of conjunctiva that extends over part of the cornea.
retinitis pigmentosa (ret´´ĭ-ni´tis pig´´men-to´sa) Inherited, progressive retinal sclerosis characterized by pigment deposits in the retina and by retinal atrophy.
retinoblastoma (ret´´ĭ-no-blas-to´mah) Inherited, highly malignant tumor arising from immature retinal cells.
tinnitus (tĭ-ni´tus) Ringing or buzzing noise in the ears.
trachoma (trah-ko´mah) Bacterial disease of the eye that causes conjunctivitis, which may lead to blindness.
tympanoplasty (tim´´pah-no-plas´te) Surgical reconstruction of the middle ear bones and the establishment of continuity from the eardrum to the oval window.
uveitis (u´´ve-i´tis) Inflammation of the uvea, the region of the eye that includes the iris, ciliary body, and choroid coat.
vertigo (ver´tĭ-go) Sensation of dizziness.

Clinical Connection

"The song was full of glittering orange diamonds."
"The paint smelled blue."
"The sunset was salty."
"The pickle tasted like a rectangle."

To 1 in 500,000 peope with a condition called synesthesia, sensation and perception mix, so that the brain perceives a stimulus to one sense as coming from another. Most commonly, letters, numbers, or periods of time evoke specific colors. These associations are involuntary, are very specific, and persist over a lifetime. For example, a person might report that 3 is always mustard yellow, or Thursday brown.

Synesthesia seems to be inherited and is more common in women. One of the authors of this book (R. L.) has it—to her, days of the week are specific colors. People have reported the condition to psychologists and physicians for at least 200 years.

PET (positron emission tomography) scanning reveals a physical basis to synesthesia. Brain scans of six nonsynesthetes were compared with those of six synesthetes who associate words with colors. The researchers monitored blood flow in the cerebral cortex while a list of words was read aloud to both groups. While blood flow was increased in word-processing areas for both groups, the scans revealed that areas important in vision and color processing were also lit up in those with synesthesia.

SUMMARY OUTLINE

10.1 Introduction (p. 253)
Sensory receptors sense changes in their surroundings.

10.2 Receptors and Sensations (p. 253)
1. Types of receptors
   a. Each type of receptor is most sensitive to a distinct type of stimulus.
   b. The major types of receptors are chemoreceptors, pain receptors, thermoreceptors, mechanoreceptors, and photoreceptors.
2. Sensations
   a. Sensations are feelings resulting from sensory stimulation.
   b. A particular part of the sensory cortex interprets every impulse reaching it in the same way.
   c. The cerebral cortex projects a sensation back to the region of stimulation.
3. Sensory adaptation
   Sensory adaptations are adjustments of sensory receptors to continuous stimulation. Impulses are triggered at decreasing rates.

10.3 Somatic Senses (p. 253)
Somatic senses are associated with receptors in the skin, muscles, joints, and viscera.
1. Touch and pressure senses
   a. Free ends of sensory nerve fibers are receptors for the sensations of touch and pressure.
   b. Meissner's corpuscles are receptors for the sensation of light touch.
   c. Pacinian corpuscles are receptors for the sensation of heavy pressure.
2. Temperature senses
   Temperature receptors include two sets of free nerve endings that are warm and cold receptors.
3. Sense of pain
   a. Pain receptors are free nerve endings that tissue damage stimulates.
   b. Visceral pain
      (1) Pain receptors are the only receptors in viscera that provide sensations.
      (2) Sensations produced from visceral receptors feel as if they are coming from some other body part.
      (3) Visceral pain may be referred because sensory impulses from the skin and viscera travel on common nerve pathways.
   c. Pain nerve fibers
      (1) The two main types of pain fibers are acute pain fibers and chronic pain fibers.
      (2) Acute pain fibers conduct nerve impulses rapidly. Chronic pain fibers conduct impulses more slowly.
      (3) Pain impulses are processed in the gray matter of the spinal cord and ascend to the brain.
      (4) Within the brain, pain impulses pass through the reticular formation before being conducted to the cerebral cortex.

d. Regulation of pain impulses
   (1) Awareness of pain occurs when pain impulses reach the thalamus.
   (2) The cerebral cortex determines pain intensity and locates its source.
   (3) Impulses descending from the brain stimulate neurons to release pain-relieving neuropeptides, such as enkephalins.

10.4 Special Senses (p. 257)
Special senses are those whose receptors are within relatively large, complex sensory organs of the head.

10.5 Sense of Smell (p. 257)
1. Olfactory receptors
   a. Olfactory receptors are chemoreceptors that are stimulated by chemicals dissolved in liquid.
   b. Olfactory receptors function with taste receptors and aid in food selection.
2. Olfactory organs
   a. Olfactory organs consist of receptors and supporting cells in the nasal cavity.
   b. Olfactory receptor cells are bipolar neurons with cilia.
3. Olfactory nerve pathways
   Nerve impulses travel from the olfactory receptor cells through the olfactory nerves, olfactory bulbs, and olfactory tracts to interpreting centers in the temporal and frontal lobes of the cerebrum.
4. Olfactory stimulation
   a. Olfactory impulses may result when gaseous molecules combine with specific sites on cilia of receptor cells.
   b. Olfactory receptor cells adapt rapidly.

10.6 Sense of Taste (p. 258)
1. Taste receptors
   a. Taste buds consist of taste (receptor) cells and supporting cells.
   b. Taste cells have taste hairs.
   c. Taste hair surfaces have receptor sites to which chemicals combine.
2. Taste sensations
   a. The four primary taste sensations are sweet, sour, salty, and bitter.
   b. Various taste sensations result from the stimulation of at least two sets of taste receptors.
3. Taste nerve pathways
   a. Sensory impulses from taste receptors travel on fibers of the facial, glossopharyngeal, and vagus nerves.
   b. These impulses are carried to the medulla oblongata and then ascend to the thalamus, from which they travel to the gustatory cortex in the parietal lobes.

10.7 Sense of Hearing (p. 260)
1. External ear
   The external ear collects sound waves of vibrating objects.
2. Middle ear
   Auditory ossicles of the middle ear conduct sound waves from the eardrum to the oval window of the inner ear.
3. Auditory tube
   Auditory tubes connect the middle ears to the throat and help maintain equal air pressure on both sides of the eardrums.
4. Inner ear
   a. The inner ear is a complex system of connected tubes and chambers—the osseous and membranous labyrinths.
   b. The organ of Corti contains hearing receptors that are stimulated by vibrations in the fluids of the inner ear.
   c. Different frequencies of vibrations stimulate different sets of receptor cells.
5. Auditory nerve pathways
   a. Auditory nerves carry impulses to the auditory cortices of the temporal lobes.
   b. Some auditory nerve fibers cross over, so that impulses arising from each ear are interpreted on both sides of the brain.

10.8 Sense of Equilibrium (p. 265)
1. Static equilibrium
   Static equilibrium maintains the stability of the head and body when they are motionless.
2. Dynamic equilibrium
   a. Dynamic equilibrium balances the head and body when they are moved or rotated suddenly.
   b. Other structures that help maintain equilibrium include the eyes and mechanoreceptors associated with certain joints.

10.9 Sense of Sight (p. 267)
1. Visual accessory organs
   Visual accessory organs include the eyelids, lacrimal apparatus, and extrinsic muscles of the eyes.
2. Structure of the eye
   a. The wall of the eye has an outer, a middle, and an inner layer (tunic).
      (1) The outer tunic (sclera) is protective, and its transparent anterior portion (cornea) refracts light entering the eye.
      (2) The middle tunic (choroid coat) is vascular and contains pigments that keep the inside of the eye dark.
      (3) The inner tunic (retina) contains the visual receptor cells.
   b. The lens is a transparent, elastic structure. Ciliary muscles control its shape.
   c. The lens must thicken to focus on close objects.
   d. The iris is a muscular diaphragm that controls the amount of light entering the eye.
   e. Spaces within the eye are filled with fluids that help maintain its shape.
3. Light refraction
   The cornea and lens refract light waves to focus an image on the retina.
4. Visual receptors
   a. Visual receptors are rods and cones.
   b. Rods are responsible for colorless vision in dim light, and cones provide color vision.
5. Visual pigments
   a. A light-sensitive pigment in rods decomposes in the presence of light and triggers a complex series of reactions that initiate nerve impulses.
   b. Color vision comes from three sets of cones containing different light-sensitive pigments.
6. Visual nerve pathways
   a. Nerve fibers from the retina form the optic nerves.
   b. Some fibers cross over in the optic chiasma.
   c. Most of the fibers enter the thalamus and synapse with others that continue to the visual cortex in the occipital lobes.

REVIEW EXERCISES

1. List five groups of sensory receptors, and name the kind of change to which each is sensitive. (p. 253)
2. Define *sensation*. (p. 253)
3. Explain projection of a sensation. (p. 253)

4. Define *sensory adaptation*, and provide an example. (p. 253)
5. Describe the functions of sensory nerve fibers, Meissner's corpuscles, and Pacinian corpuscles. (p. 253)
6. Define *referred pain*, and provide an example. (p. 255)
7. Explain why pain may be referred. (p. 255)
8. Describe the olfactory organs and their functions. (p. 257)
9. Trace a nerve impulse from an olfactory receptor to the interpreting center of the cerebrum. (p. 257)
10. Explain how salivary glands aid the function of taste receptors. (p. 259)
11. Name the four primary taste sensations. (p. 260)
12. Trace the pathway of a taste impulse from a taste receptor to the cerebral cortex. (p. 260)
13. Distinguish among the external, middle, and inner ears. (p. 260)
14. Trace the path of a sound wave from the eardrum to the hearing receptors. (p. 261)
15. Describe the functions of the auditory ossicles. (p. 261)
16. Explain the function of the auditory tube. (p. 261)
17. Distinguish between the osseous and membranous labyrinths. (p. 261)
18. Describe the cochlea and its function. (p. 262)
19. Describe a hearing receptor. (p. 263)
20. Explain how a hearing receptor stimulates a sensory neuron. (p. 264)
21. Trace a nerve impulse from the organ of Corti to the interpreting centers of the cerebrum. (p. 264)
22. Describe the organs of static and dynamic equilibrium and their functions. (p. 265)
23. List the visual accessory organs, and describe the functions of each organ. (p. 267)
24. Name the three layers of the eye wall, and describe the functions of each layer. (p. 269)
25. Describe how accommodation is accomplished. (p. 271)
26. Explain how the iris functions. (p. 271)
27. Distinguish between the aqueous humor and the vitreous humor. (p. 271)
28. Distinguish between the fovea centralis and the optic disc. (p. 272)
29. Explain how light waves are focused on the retina. (p. 273)
30. Distinguish between rods and cones. (p. 273)
31. Explain why cone vision is generally more acute than rod vision. (p. 274)
32. Describe the function of rhodopsin. (p. 274)
33. Describe the relationship between light wavelengths and color vision. (p. 275)
34. Trace a nerve impulse from the retina to the visual cortex. (p. 275)

CRITICAL THINKING

1. Loss of the sense of smell often precedes the major symptoms of Alzheimer disease and Parkinson disease. What additional information is needed to use this association to prevent or treat these diseases?
2. Why is dietary vitamin A good for eyesight?
3. PET (positron emission tomography) scans of the brains of people blind since birth reveal high neural activity in the visual centers of the cerebral cortex when these people read Braille. However, when sighted individuals run their fingers over the raised letters of Braille, the visual centers do not show increased activity. Explain these experimental results.
4. People who are deaf due to cochlea damage do not suffer from motion sickness. Why not?
5. We have relatively few sensory systems. How, then, do we experience such a huge and diverse number of sensory perceptions?
6. We humans love sucrose (table sugar), but armadillos, hedgehogs, lions, and seagulls do not respond to it. Opossums love lactose (milk sugar), but rats avoid it. Chickens hate the sugar xylose, while cattle love it, and humans are indifferent. In what way might these diverse tastes in the animal kingdom help an organism survive?
7. Why are astronauts unable to taste their food while eating in zero-gravity conditions?
8. Why does a fish market at first seem to have a strong odor that in time becomes less offensive?
9. Why are some serious injuries, such as a bullet entering the abdomen, relatively painless, but others, such as a burn, considerably more painful?
10. Labyrinthitis is an inflammation of the inner ear. What symptoms would you expect in a patient with this disorder?
11. A patient with heart disease experiences pain at the base of the neck and in the left shoulder and upper limb during exercise. How would you explain the probable origin of this pain to the patient?

WEB CONNECTIONS

Visit the website for additional study questions and more information about this chapter at:

http://www.mhhe.com/shieress8

chapter 11

Endocrine System

SMELLY TEE SHIRTS. The endocrine system produces hormones, which are biochemicals that spread messages within an individual. Less well studied are pheromones, which are chemical signals sent between individuals. In insects and rodents, pheromones often stimulate mating behavior. Preliminary experiments suggest that this may be the case with humans, too.

Mice and rats choose mates that are genetically dissimilar to themselves—specifically, in a group of genes that encode proteins involved with providing immunity. Their sense of smell helps them discern appropriate mates. Biologists think this phenomenon may protect offspring in two ways—it prevents close relatives from mating, and it may bring together immune systems with different types of strengths.

To test whether heterosexual humans use the sense of smell to respond to pheromones in mate selection, researchers in Switzerland recruited forty-nine young women and forty-four young men. Each participant donated DNA, which was typed for the same set of genes that affect mating in rodents. The women used nasal spray for two weeks to keep their nasal passages clear. The men wore the same tee shirt on two consecutive days, using no deodorant or soap and avoiding contact with anything smelly that could linger. Each woman was then given three tee shirts from men genetically similar to her and three tee shirts from men genetically dissimilar to her. She did not know which shirts came from which men.

The women rated the shirts on intensity, pleasantness, and sexiness. And, like the mice and rats, women preferred the sweaty tees from the men least like them, at least by this genetic criterion. Further experiments are necessary to show the existence of pheromones in humans.

Photo:
The endocrine system produces hormones, which act within an individual. Humans may also produce pheromones, which affect other individuals and may play a role in mate selection, as they do in rodents and insects.

Chapter Objectives

After studying this chapter, you should be able to do the following:

11.1 Introduction
1. Define hormone. (p. 281)
2. Distinguish between paracrine and autocrine secretions. (p. 281)
3. Distinguish between endocrine and exocrine glands. (p. 281)

11.2 General Characteristics of the Endocrine System
4. Explain how the nervous and endocrine systems are alike and how they are different. (p. 281)
5. Describe the source of specificity of the endocrine system. (p. 281)
6. Name some functions of hormones. (p. 281)

11.3 Hormone Action
7. Explain how steroid and nonsteroid hormones affect target cells. (p. 282)

11.4 Control of Hormonal Secretions
8. Discuss how negative feedback mechanisms regulate hormonal secretions. (p. 285)
9. Explain how the nervous system controls secretion. (p. 285)

11.5–11.10 Pituitary Gland—Other Endocrine Glands
10. Name and describe the location of the major endocrine glands, and list the hormones they secrete. (p. 286)
11. Describe the general functions of the hormones that endocrine glands secrete. (p. 287)
12. Explain how the secretion of each hormone is regulated. (p. 287)

11.11 Stress and Health
13. Define stress, and describe how the body responds to it. (p. 299)

Aids to Understanding Words

-crin [to secrete] endo*crin*e: Pertaining to internal secretions.
diuret- [to pass urine] *diuret*ic: Substance that promotes urine production.
endo- [within] *endo*crine gland: Gland that releases its secretion internally into a body fluid.
exo- [outside] *exo*crine gland: Gland that releases its secretion to the outside through a duct.

hyper- [above] *hyper*thyroidism: Condition resulting from an above-normal secretion of thyroid hormone.
hypo- [below] *hypo*thyroidism: Condition resulting from a below-normal secretion of thyroid hormone.

para- [beside] *para*thyroid glands: Set of glands located on the surface of the thyroid gland.
toc- [birth] oxy*toc*in: Hormone that stimulates the uterine muscles to contract during childbirth.

Key Terms

adrenal cortex (ah-dre´nal kor´teks)
adrenal medulla (ah-dre´nal me-dul´ah)
anterior pituitary (an-ter´e-or pĭ-tu´ĭ-tar´´e)
cyclic AMP or **cAMP** (si´ klik ay em pee)
hormone (hor´mōn)

negative feedback system (neg´ah-tiv fēd´bak sis´tem)
pancreas (pan´kre-as)
parathyroid gland (par´´ah-thi´roid gland)
pineal gland (pin´e-al gland)

posterior pituitary (pos-ter´e-or pĭ-tu´ĭ-tar´´e)
prostaglandin (pros´´tah-glan´din)
target cell (tar´get sel)
thymus gland (thi´mus gland)
thyroid gland (thi´roid gland)

11.1 Introduction

Regulating the functions of the human body to maintain homeostasis is an enormous job. Two organ systems function coordinately to enable body parts to communicate with each other and to adjust constantly to changing incoming signals. The nervous system is one biological communication system; it utilizes nerve impulses carried on nerve fibers. The other is the endocrine system.

The **endocrine system** includes cells, tissues, and organs, collectively called endocrine glands, that secrete substances, called **hormones** (hor´mōnz), into the internal environment. The hormones diffuse from the interstitial fluid into the bloodstream, and eventually act on cells called **target cells** (tar´get selz) some distance away.

Some glands secrete substances into the interstitial fluid, but because these secretions are rapidly broken down, they do not reach the bloodstream and are not hormones by the traditional definition. However, they do function similarly as messenger molecules and are sometimes referred to as "local hormones." These include **paracrine** secretions, which affect only neighboring cells, and **autocrine** secretions, which affect only the secreting cell itself.

A different group of glands, called exocrine glands, secrete outside the body through tubes or ducts that lead to the surface. Sweat, secreted by sweat glands and reaching the surface of the skin, is one example of an exocrine secretion (see chapter 5, p. 97).

CHECK YOUR RECALL
1. What are the components of the endocrine system?
2. How do paracrine and autocrine secretions function differently from traditionally defined hormones?
3. Distinguish between endocrine and exocrine glands.

11.2 General Characteristics of the Endocrine System

Both the endocrine system and the nervous system are involved in communication between cells using chemical signals that bind to receptor molecules. Some similarities and differences between the two systems are summarized in table 11.1. In contrast to the nervous system, which releases neurotransmitter molecules into synapses, the endocrine system releases hormones into the bloodstream, which carries these messenger molecules everywhere. However, the endocrine system is no less precise, because only target cells can respond to a hormone. A hormone's target cells have specific receptors that other cells lack. These receptors are proteins or glycoproteins with binding sites for a specific hormone.

Endocrine glands and their hormones help regulate metabolic processes. They control the rates of certain chemical reactions, aid in transport of substances across membranes, and help regulate water and electrolyte balances. They also play vital roles in reproduction, development, and growth.

Specialized small groups of cells produce some hormones. However, the larger endocrine glands—the pituitary gland, thyroid gland, parathyroid glands, adrenal glands, and pancreas—are the subject of this chapter (fig. 11.1). Subsequent chapters discuss several other hormone-secreting glands and tissues, such as those that participate in digestion and reproduction.

CHECK YOUR RECALL
1. Explain how the nervous and endocrine systems are alike and how they are different.
2. State some functions of hormones.

11.3 Hormone Action

Chemically, most hormones are either steroids (or steroidlike substances) synthesized from cholesterol, or they are amines, peptides, proteins, or glycoproteins synthesized from amino acids. Hormones can stimulate changes in target cells even if present in extremely low concentrations (table 11.2).

TABLE 11.1		A COMPARISON BETWEEN THE NERVOUS SYSTEM AND THE ENDOCRINE SYSTEM
	NERVOUS SYSTEM	ENDOCRINE SYSTEM
Cells	Neurons	Glandular epithelium
Chemical signal	Neurotransmitter	Hormone
What determines which cells respond?	Receptors on postsynaptic cell	Receptors on target cell
Speed of onset	Seconds	Seconds to hours
Duration of action	Very brief unless neuronal activity continues	May be brief or may last for days even if secretion ceases

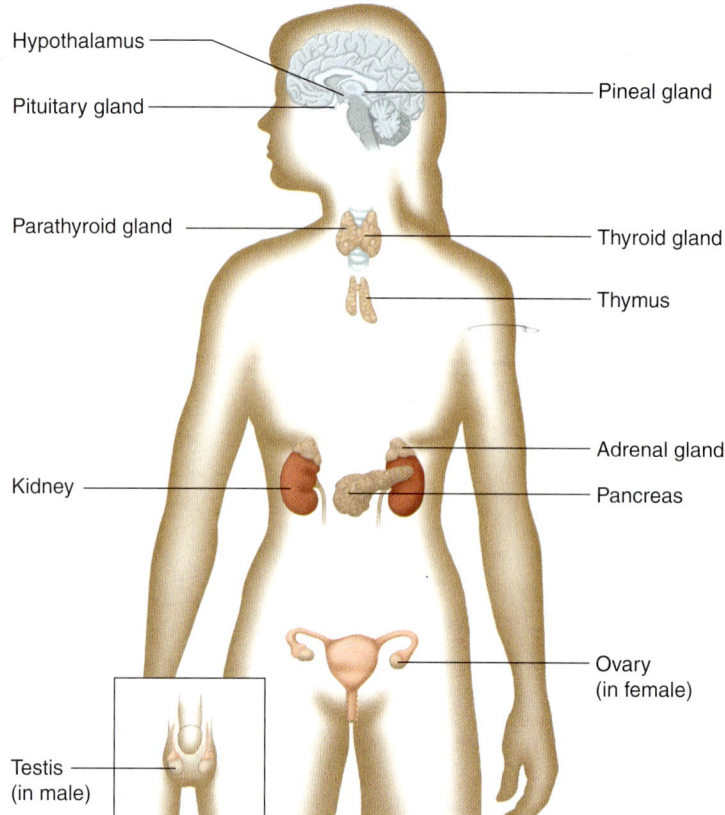

Figure 11.1
Locations of major endocrine glands.

TABLE 11.2		TYPES OF HORMONES
TYPE OF COMPOUND	FORMED FROM	EXAMPLES
Amines	Amino acids	Norepinephrine, epinephrine
Peptides	Amino acids	Antidiuretic hormone, oxytocin, thyrotropin-releasing hormone
Proteins	Amino acids	Parathyroid hormone, growth hormone, prolactin
Glycoproteins	Protein and carbohydrate	Follicle-stimulating hormone, luteinizing hormone, thyroid-stimulating hormone
Steroids	Cholesterol	Estrogen, testosterone, aldosterone, cortisol

Steroid Hormones

Steroids are compounds whose molecules contain complex rings of carbon and hydrogen atoms. Steroids differ according to the kinds and numbers of atoms attached to these rings and the ways they are joined.

Steroid hormones are insoluble in water. They are carried in the bloodstream weakly bound to plasma proteins in such a way that allows them to be released in sufficient quantity in the vicinity of their target cells. However, unlike amine, peptide, and protein hormones, steroid hormones are lipid-soluble.

Because lipids make up the bulk of cell membranes, steroid molecules can diffuse into cells relatively easily and may enter any cell in the body. When a steroid hormone molecule enters a target cell, the following events may occur (fig. 11.2):

1. The lipid soluble steroid hormone diffuses through the cell membrane.
2. The steroid hormone may combine with a specific protein molecule—the receptor for that hormone.
3. The resulting hormone-receptor complex binds within the nucleus to particular regions of the target cell's DNA and activates transcription of specific genes into messenger RNA (mRNA) molecules.
4. The mRNA molecules enter the cytoplasm.
5. mRNA molecules associate with ribosomes to direct the synthesis of specific proteins.

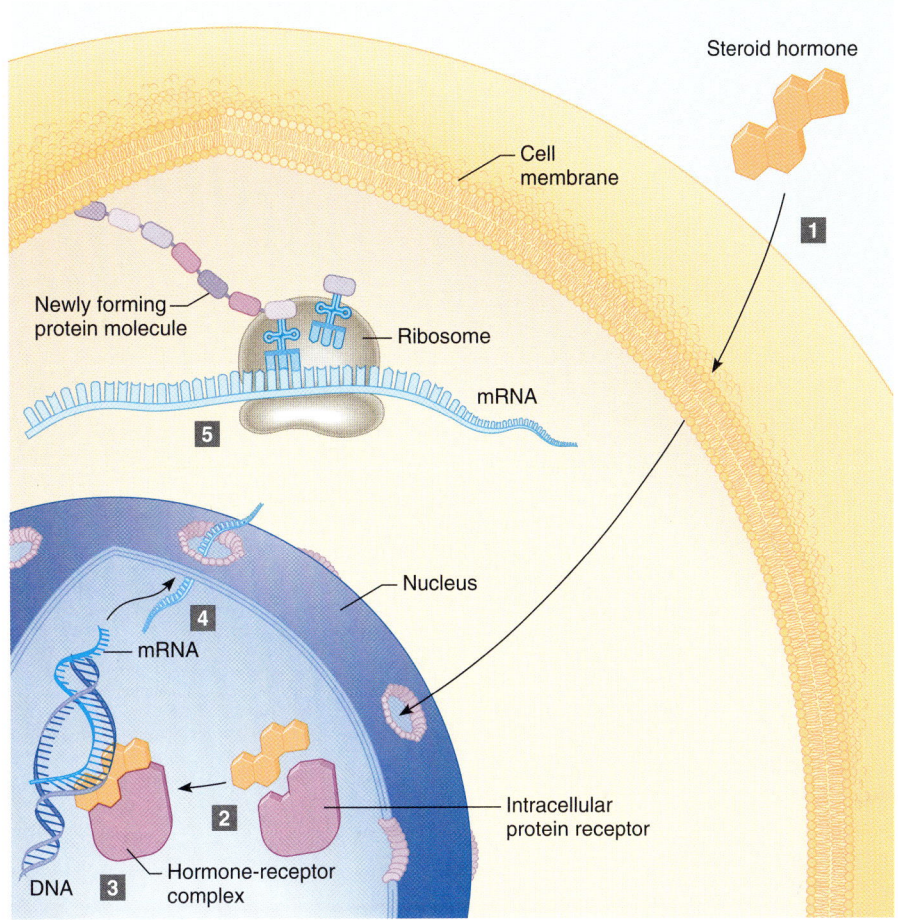

Figure 11.2
Steroid hormones. (*1*) A steroid hormone crosses a cell membrane and (*2*) combines with a protein receptor, usually in the nucleus. (*3*) The hormone-receptor complex activates synthesis of specific messenger RNA (mRNA) molecules. (*4*) The mRNA molecules leave the nucleus and enter the cytoplasm (*5*) where they guide synthesis of the encoded proteins.

The newly synthesized proteins, which may be enzymes, transport proteins, or even hormone receptors, exert the characteristic effects of that particular steroid hormone.

Nonsteroid Hormones

Nonsteroid hormones, such as amines, peptides, and proteins, usually combine with receptors in target cell membranes. Each of these receptor molecules is a protein with a *binding site* and an *activity site*. A hormone molecule delivers its message to its target cell by uniting with the binding site of its receptor. This combination stimulates the receptor's activity site to interact with other membrane proteins. Receptor binding may alter the function of enzymes or membrane transport mechanisms, changing the concentrations of still other cellular components. The hormone that triggers this cascade of biochemical activity is called a *first messenger*. The biochemicals in the cell that induce changes in response to the hormone's binding are called *second messengers*.

The second messenger associated with one group of hormones is *cyclic adenosine monophosphate,* or **cyclic AMP (cAMP)** (si′klik ay em pee). This mechanism works as follows:

1. A hormone binds to its receptor.
2. The resulting hormone-receptor complex activates a protein called a *G protein*.
3. The G protein activates an enzyme called *adenylate cyclase,* which is bound to the inside of the cell membrane.
4. Activated adenylate cyclase causes ATP molecules within the cytoplasm to take on a circular form, making cAMP.
5. The cAMP activates another set of enzymes, called *protein kinases,* which transfer phosphate groups from ATP to their substrate molecules, specific proteins within the cell. This action, called phosphorylation, alters the shapes of these substrate molecules, thereby activating them.

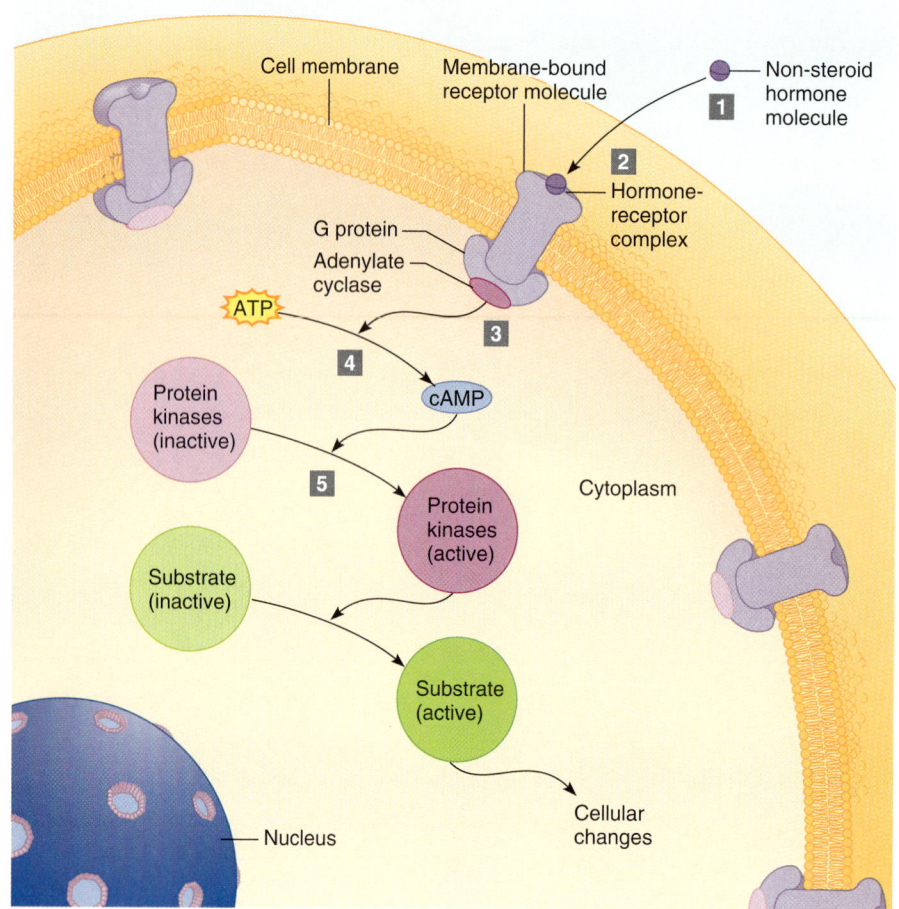

Figure 11.3
One mechanism of nonsteroid hormones. (1) Body fluids carry nonsteroid hormone molecules to the target cell, where (2) they combine with receptor molecules on the cell membrane. (3) This activates molecules of adenylate cyclase, which (4) catalyze conversion of ATP into cyclic adenosine monophosphate (cAMP). (5) The cAMP promotes a series of reactions leading to the cellular changes associated with the hormone's action.

The activated proteins then alter various cellular processes to bring about the effect of that particular hormone (fig. 11.3).

The type of membrane receptors present and also the kinds of protein substrate molecules that a cell contains determine the cell's response to a hormone. Cellular responses to second messenger activation include altering membrane permeabilities, activating enzymes, promoting synthesis of certain proteins, stimulating or inhibiting specific metabolic pathways, moving the cell, and initiating secretion of hormones or other substances.

Another enzyme (phosphodiesterase) quickly inactivates cAMP, so that its action is short-lived. For this reason, a continuing response within a target cell depends on a continuing signal produced by hormone molecules combining with the target cell's membrane receptors.

A number of other second messengers work in much the same way. These include diacylglycerol (DAG) and inositol triphosphate (IP_3).

 Abnormal or missing G proteins cause a variety of disorders. They include color blindness, precocious puberty, retinitis pigmentosa, and several thyroid problems.

Prostaglandins

A group of biochemicals called **prostaglandins** (pros´´tah-glan´dinz) also regulates cells. Prostaglandins are lipids synthesized from a fatty acid (arachidonic acid) in cell membranes. A great variety of cells produce prostaglandins, including those of the liver, kidneys, heart, lungs, thymus gland, pancreas, brain, and reproductive organs. Prostaglandins usually act more locally than hormones, often affecting only the organ where they are produced.

Prostaglandins are potent and are present in very small quantities. They are not stored in cells but rather synthesized just before release. They are rapidly inactivated.

Prostaglandins produce diverse and even opposite effects. Some, for example, relax smooth muscles in the airways of the lungs and in blood vessels, while others contract smooth muscles in the walls of the uterus and intestines. Prostaglandins stimulate hormone secretion from the adrenal cortex and inhibit secretion of hydrochloric acid from the stomach wall. They also influence the movements of sodium ions and water molecules in the kidneys, help regulate blood pressure, and have powerful effects on male and female reproductive physiology.

CHECK YOUR RECALL

1. How does a steroid hormone promote cellular changes? A nonsteroid hormone?
2. What is a second messenger?
3. What are prostaglandins?
4. What kinds of effects do prostaglandins produce?

11.4 Control of Hormonal Secretions

Hormones are continually excreted in the urine and broken down by various enzymes, primarily in the liver. Therefore, increasing or decreasing blood levels of a hormone requires increased or decreased secretion. Hormone secretion is precisely regulated.

Hormone secretion is controlled in three ways, all of which employ **negative feedback** (see chapter 1, p. 5). In each case, an endocrine gland or the system controlling it is sensitive to the concentration of the hormone the gland secretes, a process the hormone controls, or an action the hormone has on the internal environment (fig. 11.4).

1. The hypothalamus controls the anterior pituitary gland's release of hormones that stimulate other endocrine glands to release hormones. Its location near the thalamus and the third ventricle allows the hypothalamus to constantly receive information about the internal environment from neural connections and cerebrospinal fluid (fig. 11.4A).
2. The nervous system stimulates some glands directly. The adrenal medulla, for example, secretes its hormones in response to sympathetic nerve impulses (fig. 11.4B).
3. Another group of glands responds directly to changes in the composition of the internal environment. For example, when the blood glucose level rises, the pancreas secretes insulin, and when the blood glucose level falls, it secretes glucagon, as discussed later in the chapter (fig. 11.4C).

In each of the above cases, as hormone levels rise in the blood and the hormone exerts its effects, the system is inhibited by negative feedback and hormone secretion decreases. Then, as hormone levels in the blood decrease and the hormone's effects are no longer taking place, inhibition of the system ceases, and secretion of that hormone increases again. As a result of negative feedback, hormone levels in the bloodstream remain relatively stable, tending to fluctuate slightly above and below an average value (fig. 11.5).

CHECK YOUR RECALL

1. Explain three mechanisms that help control hormonal secretions.
2. Describe a negative feedback system.

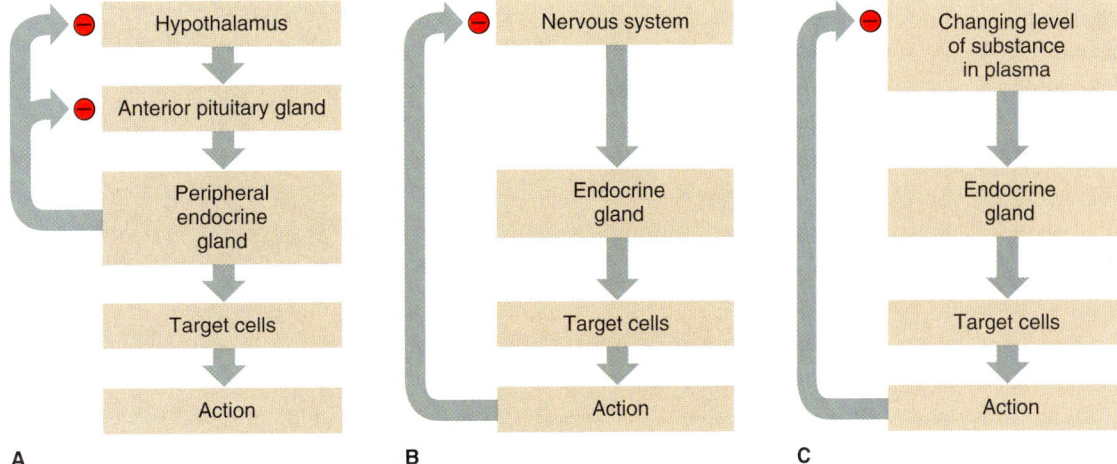

Figure 11.4
Control of the endocrine system occurs in three ways: (A) the hypothalamus and anterior pituitary, (B) the nervous system directly, and (C) glands that respond directly to changes in the internal environment. Negative feedback inhibition is indicated by ⊖.

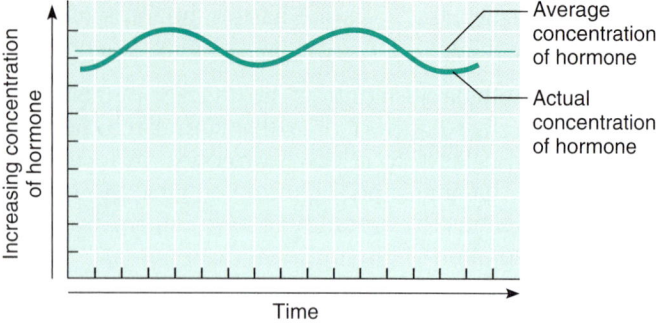

Figure 11.5
As a result of negative feedback, hormone concentrations remain relatively stable, although they may fluctuate slightly above and below average concentrations.

11.5 Pituitary Gland

The **pituitary gland** (hypophysis) is located at the base of the brain, where a pituitary stalk (infundibulum) attaches it to the hypothalamus. The gland is about 1 centimeter in diameter and consists of an **anterior pituitary** (an-ter´e-or pĭ-tu´ĭ-tar´´e), or anterior lobe, and a **posterior pituitary** (pos-ter´e-or pĭ-tu´ĭ-tar´´e), or posterior lobe (fig. 11.6).

During fetal development, a narrow region develops between the anterior and posterior lobes of the pituitary gland. Called the *intermediate lobe* (pars intermedia), it produces melanocyte-stimulating hormone (MSH), which regulates the synthesis of melanin—the pigment in skin and in portions of the eyes and brain. The intermediate lobe atrophies during prenatal development and appears only as a vestige in adults.

The brain controls most of the pituitary gland's activities. For example, the posterior pituitary releases hormones when nerve impulses from the hypothalamus signal the axon ends of neurosecretory cells in the posterior pituitary (fig. 11.7). On the other hand, **releasing hormones** from the hypothalamus control secretion from the anterior pituitary (fig. 11.7). These releasing hormones travel in a capillary network associated with the hypothalamus. The capillaries merge to form the **hypophyseal portal veins,** which pass downward along the pituitary stalk and give rise to a capillary net in the anterior pituitary. Thus, the hypothalamus releases substances that the blood carries directly to the anterior pituitary.

Upon reaching the anterior pituitary, each of the hypothalamic releasing hormones acts on a specific population of cells there. Some of the resulting actions are inhibitory, but most stimulate the anterior pituitary to release hormones that stimulate secretions from peripheral endocrine glands. In many of these cases, important negative feedback relationships regulate hormone levels in the bloodstream.

CHECK YOUR RECALL

1. Where is the pituitary gland located?
2. Explain how the hypothalamus controls the actions of the posterior and anterior lobes of the pituitary gland.

Acromegaly is the overproduction of growth hormone in adulthood. The many symptoms attest to the wide effects of this hormone. They include enlarged heart, bones, thyroid, facial features, hands, feet, and head. Early symptoms include headache, joint pain, fatigue and depression.

Figure 11.6
The pituitary gland is attached to the hypothalamus and lies in the sella turcica of the sphenoid bone.

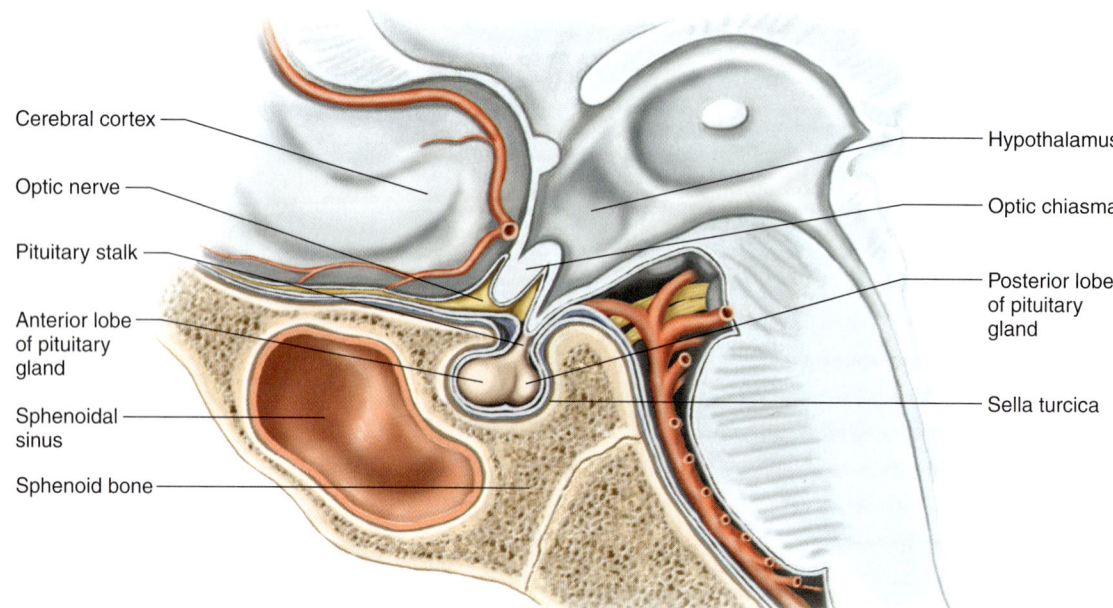

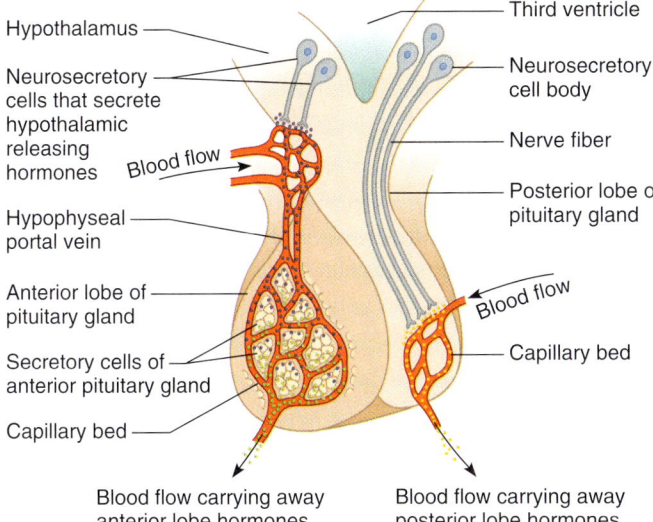

Figure 11.7
The two parts of the pituitary gland secrete hormones by different mechanisms. Hypothalamic releasing hormones stimulate cells of the anterior lobe to secrete hormones. Nerve impulses originating in the hypothalamus cause nerve endings in the posterior lobe of the pituitary gland to release hormones.

Anterior Pituitary Hormones

The anterior pituitary is enclosed in a capsule of dense, collagenous connective tissue and consists largely of epithelial tissue organized in blocks around many thin-walled blood vessels. So far, researchers have identified five types of secretory cells within the epithelium. Four of these each secrete a different hormone—growth hormone (GH), prolactin (PRL), thyroid-stimulating hormone (TSH), and adrenocorticotropic hormone (ACTH). The fifth type of cell secretes follicle-stimulating hormone (FSH) and luteinizing hormone (LH). (Note: In males, luteinizing hormone is known as interstitial cell stimulating hormone, or ICSH.)

Growth hormone (GH) stimulates cells to increase in size and divide more frequently. It also enhances the movement of amino acids across cell membranes and speeds the rate at which cells utilize carbohydrates and fats. The hormone's effect on amino acids is important in stimulating growth.

Two hormones from the hypothalamus control GH secretion: GH releasing hormone and GH release-inhibiting hormone. Nutritional state also influences control of GH. For example, more GH is released during periods of protein deficiency and abnormally low blood glucose concentration. Conversely, when blood protein and glucose concentrations increase, GH secretion decreases.

Prolactin (PRL) stimulates and sustains a woman's milk production following the birth of an infant (see chapter 20, p. 544). The effect of PRL in males is less well understood, although excess PRL secretion may cause a deficiency of male sex hormones.

Thyroid-stimulating hormone (TSH) controls thyroid gland secretions, described later in this chapter. The hypothalamus partially regulates TSH secretion by producing *thyrotropin-releasing hormone (TRH)* (fig. 11.8). Circulating thyroid hormones inhibit release of TRH and TSH. As the blood concentration of thyroid hormones increases, secretions of TRH and TSH decrease.

CHECK YOUR RECALL

1. How does growth hormone affect protein synthesis?
2. What is the function of prolactin?
3. How is secretion of thyroid-stimulating hormone regulated?

Insufficient secretion of growth hormone (GH) during childhood limits growth, causing a type of *dwarfism* called hypopituitary dwarfism. In this condition, body parts are usually correctly proportioned, and mental development is normal. However, abnormally low GH secretion usually accompanies deficient secretion of other anterior pituitary hormones, leading to additional hormone deficiency symptoms. Hormone therapy can help.

Oversecretion of GH during childhood causes gigantism, in which the height may exceed 8 feet. This rare condition is usually a result of a pituitary gland tumor, which may also cause oversecretion of other pituitary hormones. As a result, a person with gigantism often has several metabolic disturbances.

Acromegaly is the overproduction of growth hormone in adulthood. The many symptoms attest to the wide effects of this hormone. They include enlarged heart, bones, thyroid, facial features, hands, feet, and head. Early symptoms include headache, joint pain, fatigue, and depression.

Adrenocorticotropic hormone (ACTH) controls the manufacture and secretion of certain hormones from the outer layer, or *cortex,* of the adrenal gland. These hormones are discussed later in the chapter.

ACTH secretion is regulated in part by *corticotropin-releasing hormone (CRH),* which the hypothalamus releases in response to decreased concentrations of adrenal cortical hormones. Also, stress may increase ACTH secretion by stimulating the release of CRH.

Follicle-stimulating hormone (FSH) and luteinizing hormone (LH) are *gonadotropins,* which means they exert their actions on the gonads or reproductive organs. Gonads include the **testes** in the male and the **ovaries** in the female. Chapter 19, pages 498 and 507, discusses the functions of these gonadotropins and the ways they interact.

CHECK YOUR RECALL

1. What is the function of adrenocorticotropic hormone?
2. What is a gonadotropin?

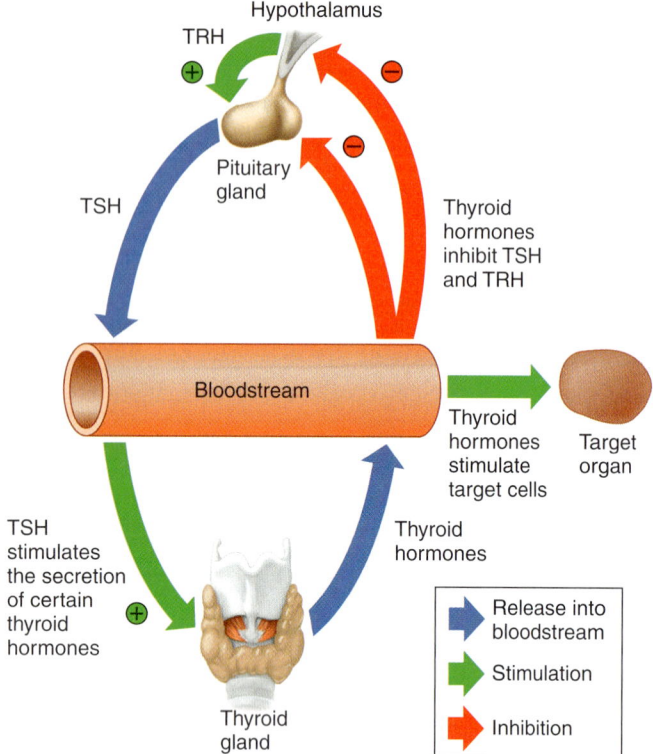

Figure 11.8
Thyrotropin-releasing hormone (TRH) from the hypothalamus stimulates the anterior pituitary gland to release thyroid-stimulating hormone (TSH), which stimulates the thyroid gland to release hormones. These thyroid hormones reduce the secretion of TSH and TRH. (Note: ⊕ = stimulation; ⊖ = inhibition.)

Posterior Pituitary Hormones

Unlike the anterior pituitary, which is composed primarily of glandular epithelial cells, the posterior pituitary consists largely of nerve fibers and neuroglial cells. The neuroglial cells support the nerve fibers, which originate in the hypothalamus.

Specialized neurons in the hypothalamus produce the two hormones associated with the posterior pituitary—**antidiuretic hormone (ADH)** and **oxytocin (OT)** (see fig. 11.7). These hormones travel down axons through the pituitary stalk to the posterior lobe, and vesicles (secretory granules) near the ends of the axons store them. Nerve impulses from the hypothalamus release the hormones into the blood.

A *diuretic* is a chemical that increases urine production, whereas an *antidiuretic* decreases urine formation. ADH produces an antidiuretic effect by reducing the volume of water the kidneys excrete. In this way, ADH regulates the water concentration of body fluids.

The hypothalamus regulates ADH secretion. Certain neurons in this part of the brain, called *osmoreceptors,* sense changes in the osmotic pressure of body fluids. Dehydration due to lack of water intake increasingly concentrates blood solutes. Osmoreceptors, sensing the resulting increase in osmotic pressure, signal the posterior pituitary to release ADH, which travels in the blood to the kidneys. As a result, the kidneys produce less urine, conserving water. On the other hand, drinking too much water dilutes body fluids, inhibiting ADH release. The kidneys excrete more dilute urine until the water concentration of body fluids returns to normal.

> If an injury or tumor damages any parts of the ADH-regulating mechanism, too little ADH may be synthesized or released, producing *diabetes insipidus*. An affected individual may produce as much as 25–30 liters of very dilute urine per day, and solute concentrations in body fluids rise.

OT contracts smooth muscles in the uterine wall and stimulates uterine contractions in the later stages of childbirth. Stretching of uterine and vaginal tissues late in pregnancy triggers OT release during childbirth. In the breast, OT contracts certain cells associated with the milk-producing glands and their ducts. In lactating breasts, this action forces liquid from the milk glands into the milk ducts and ejects the milk from the breasts for breast-feeding. In addition, OT is an antidiuretic, but it is much weaker than ADH. Table 11.3 reviews the hormones of the pituitary gland.

> If the uterus is not contracting sufficiently to expel a fully developed fetus, commercial preparations of oxytocin are sometimes used to stimulate uterine contractions, inducing labor. Such preparations are also often administered to the mother following childbirth to ensure that uterine muscles contract enough to squeeze broken blood vessels closed, minimizing the risk of hemorrhage.

CHECK YOUR RECALL

1. What is the function of antidiuretic hormone?
2. How is secretion of antidiuretic hormone controlled?
3. What effects does oxytocin produce in females?

11.6 Thyroid Gland

The **thyroid gland** (thi´roid gland) is a very vascular structure that consists of two large lobes connected by a broad isthmus (fig. 11.9 and reference plate 4, p. 25). It is just below the larynx on either side and in front of the trachea.

Structure of the Gland

A capsule of connective tissue covers the thyroid gland, which is made up of many secretory parts called *folli-*

TABLE 11.3 HORMONES OF THE PITUITARY GLAND

HORMONE	ACTION	SOURCE OF CONTROL
Anterior lobe		
Growth hormone (GH)	Stimulates an increase in the size and division rate of body cells; enhances movement of amino acids across membranes	Growth hormone-releasing hormone and growth hormone release-inhibiting hormone from hypothalamus
Prolactin (PRL)	Sustains milk production after birth	Secretion restrained by prolactin release-inhibiting hormone and stimulated by prolactin-releasing factor from hypothalamus
Thyroid-stimulating hormone (TSH)	Controls secretion of hormones from thyroid gland	Thyrotropin-releasing hormone (TRH) from hypothalamus
Adrenocorticotropic hormone (ACTH)	Controls secretion of certain hormones from adrenal cortex	Corticotropin-releasing hormone (CRH) from hypothalamus
Follicle-stimulating hormone (FSH)	In females, responsible for the development of egg-containing follicles in ovaries and stimulates follicular cells to secrete estrogen; in males, stimulates production of sperm cells	Gonadotropin-releasing hormone from hypothalamus
Luteinizing hormone (LH)	Promotes secretion of sex hormones; plays a role in releasing an egg cell in females	Gonadotropin-releasing hormone from hypothalamus
Posterior lobe		
Antidiuretic hormone (ADH)	Causes kidneys to conserve water; in high concentration, increases blood pressure	Hypothalamus in response to changes in water concentration in body fluids
Oxytocin (OT)	Contracts muscles in the uterine wall; contracts muscles associated with milk-secreting glands	Hypothalamus in response to stretching of uterine and vaginal walls and stimulation of breasts

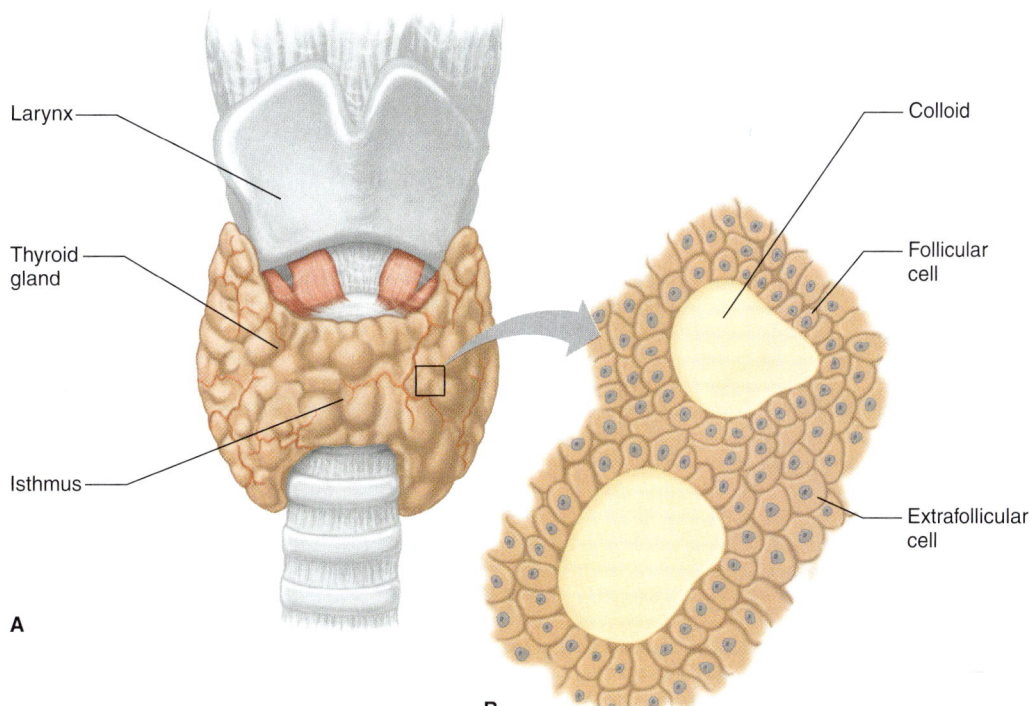

Figure 11.9
Thyroid gland. (A) The thyroid gland consists of two lobes connected anteriorly by an isthmus. (B) Follicular cells secrete thyroid hormones.

cles. The cavities within these follicles are lined with a single layer of cuboidal epithelial cells and filled with a clear, viscous substance called *colloid*. The follicular cells produce and secrete hormones that may either be stored in the colloid or released into the blood in nearby capillaries.

Thyroid Hormones

The follicular cells of the thyroid gland synthesize two hormones—**thyroxine** (tetraiodothyronine), also known as T_4 because it contains four atoms of iodine, and **triiodothyronine**, known as T_3 because it includes three atoms of iodine. Thyroxine and triiodothyronine have similar actions, although triiodothyronine is five times more potent. These hormones help regulate the metabolism of carbohydrates, lipids, and proteins. They increase the rate at which cells release energy from carbohydrates; they increase the rate of protein synthesis; and they stimulate breakdown and mobilization of lipids. They are the major factors determining how many calories the body must consume at rest in order to maintain life, the *basal metabolic rate (BMR)*. Thyroid hormones are required for normal growth and development, and are essential to nervous system maturation.

 Up to 80 percent of the iodine in the body is in the thyroid gland. Here, the concentration of iodine is 25 times that in the bloodstream.

Follicular cells require iodine salts (iodides) to produce thyroxine and triiodothyronine. Foods normally provide iodides, and after the iodides have been absorbed from the intestine, blood transports them to the thyroid gland. An efficient active transport mechanism moves the iodides into the follicular cells, where they are used to synthesize the hormones. The hypothalamus and pituitary gland control release of thyroid hormones. Once in the blood, thyroxine and triiodothyronine combine with proteins in the blood (plasma proteins) and are transported to body cells.

A third hormone, **calcitonin,** is often not considered a thyroid hormone because the gland's extrafollicular (other than follicle) cells produce it. Along with parathyroid hormone (PTH) from the parathyroid glands, calcitonin regulates the concentrations of blood calcium and phosphate ions.

Blood concentration of calcium ions regulates calcitonin release. As this concentration increases, so does calcitonin secretion. Calcitonin inhibits the bone-resorbing activity of osteoclasts (see chapter 7, p. 130) and increases the kidneys' excretion of calcium and phosphate ions—actions that lower the blood calcium and phosphate ion concentrations. Table 11.4 reviews the actions and controls of the thyroid hormones.

CHECK YOUR RECALL

1. Where is the thyroid gland located?
2. Which hormones of the thyroid gland affect carbohydrate metabolism and protein synthesis?
3. How does the thyroid gland influence the concentrations of blood calcium and phosphate ions?

Many thyroid disorders produce overactivity (*hyperthyroidism*) or underactivity (*hypothyroidism*) of the glandular cells. One form of hypothyroidism appears in infants whose thyroid glands do not function normally. An affected child may appear normal at birth because the mother provided an adequate supply of thyroid hormones for the child *in utero*. But when the infant's own thyroid gland does not produce sufficient quantities of these hormones, a condition called *cretinism* develops. Symptoms include stunted growth, abnormal bone formation, retarded mental development, low body temperature, and sluggishness. Without treatment within a month or so following birth, the child may suffer permanent mental retardation.

Hyperthyroidism produces an elevated metabolic rate, restlessness, and overeating. The eyes protrude (exophthalmos) because of swelling in the tissues behind them, and the thyroid gland enlarges, producing a bulge in the neck called a *goiter*.

11.7 Parathyroid Glands

The **parathyroid glands** (par˝ah-thi´roid glandz) are on the posterior surface of the thyroid gland, as figure 11.10 shows. Usually, there are four parathyroid glands—a superior and an inferior gland associated with each of the thyroid's lateral lobes.

Structure of the Glands

A thin capsule of connective tissue covers each small, yellowish-brown parathyroid gland. The body of the

TABLE 11.4 — HORMONES OF THE THYROID GLAND

HORMONE	ACTION	SOURCE OF CONTROL
Thyroxine (T_4)	Increases rate of energy release from carbohydrates; increases rate of protein synthesis; accelerates growth; stimulates activity in nervous system	Thyroid-stimulating hormone from the anterior pituitary gland
Triiodothyronine (T_3)	Same as above, but five times more potent than thyroxine	Thyroid-stimulating hormone from the anterior pituitary gland
Calcitonin	Lowers blood calcium and phosphate ion concentrations by inhibiting release of calcium and phosphate ions from bones and by increasing excretion of these ions by kidneys	Blood calcium concentration

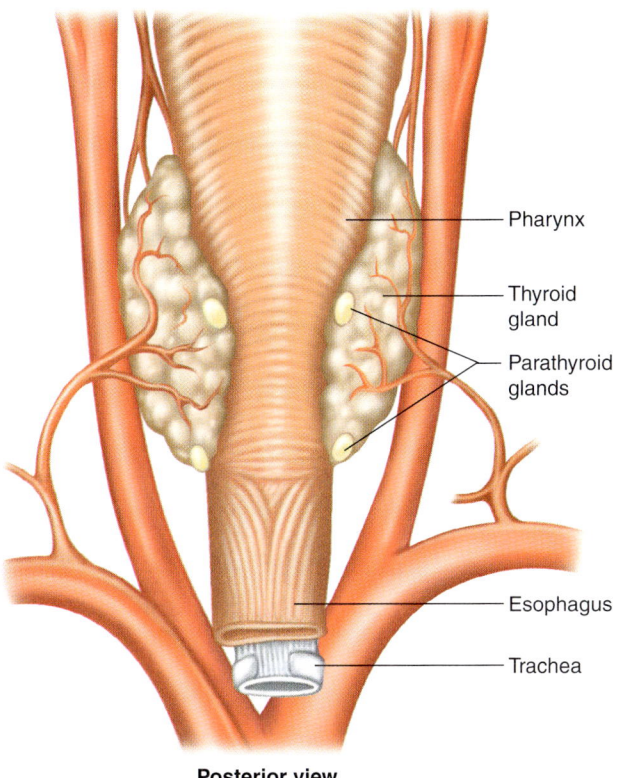

Figure 11.10
The parathyroid glands are embedded in the posterior surface of the thyroid gland.

Parathyroid Hormone

The parathyroid glands secrete **parathyroid hormone (PTH),** which increases blood calcium concentration and decreases blood phosphate ion concentration. PTH affects the bones, kidneys, and intestine.

The intercellular matrix of bone tissue is rich in mineral salts, including calcium phosphate (see chapter 7, p. 134). PTH inhibits the activity of osteoblasts and stimulates osteocytes and osteoclasts to resorb bone and release calcium and phosphate ions into the blood. At the same time, PTH causes the kidneys to conserve blood calcium and to excrete more phosphate ions in the urine. It also stimulates calcium absorption from food in the intestine, further increasing blood calcium concentration.

Negative feedback between the parathyroid glands and the blood calcium concentration regulates PTH secretion. As blood calcium concentration drops, more PTH is secreted; as blood calcium concentration rises, less PTH is released (fig. 11.11).

To summarize, calcitonin and PTH activities maintain stable blood calcium concentration. Calcitonin decreases an above-normal blood calcium concentration, while PTH increases a below-normal blood calcium concentration.

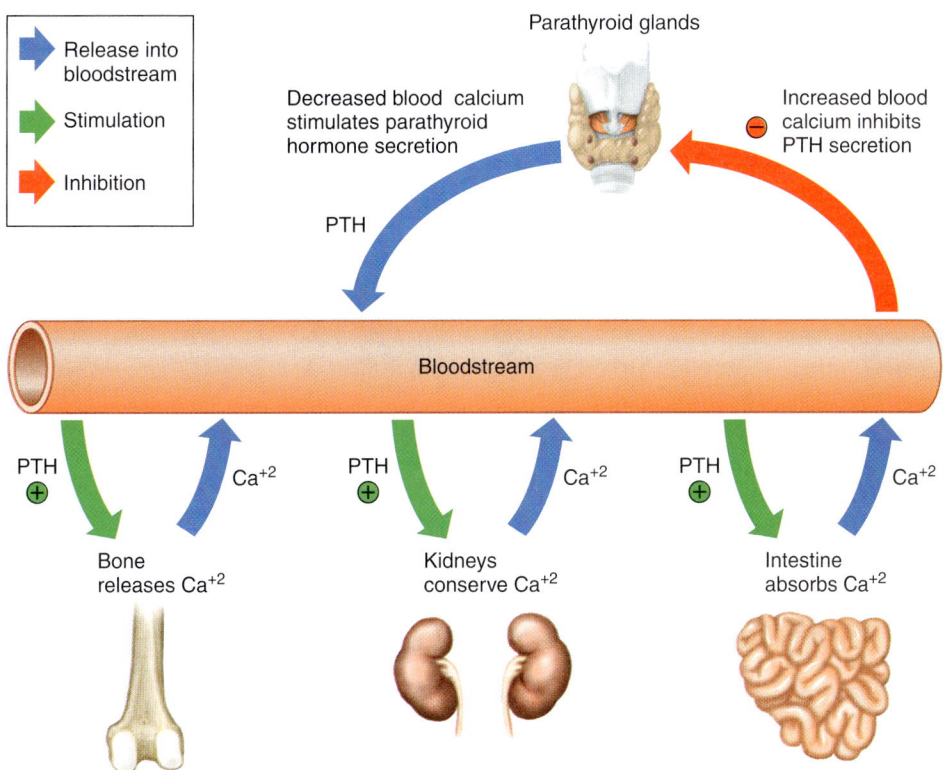

Figure 11.11
Parathyroid hormone (PTH) stimulates bone to release calcium (Ca^{+2}) and the kidneys to conserve calcium. It indirectly stimulates the intestine to absorb calcium. The resulting increase in blood calcium concentration inhibits secretion of PTH. (Note: ⊕ = stimulation; ⊖ = inhibition.)

CHECK YOUR RECALL

1. Where are the parathyroid glands?
2. How does parathyroid hormone help regulate concentrations of blood calcium and phosphate ions?

A tumor in a parathyroid gland may cause *hyperparathyroidism,* which increases PTH secretion. This stimulates osteoclast activity, and as bone tissue is resorbed, the bones soften, deform, and more easily fracture spontaneously. In addition, excess calcium and phosphate released into body fluids may be deposited in abnormal places, causing new problems, such as kidney stones.

Injury to the parathyroids or their surgical removal can cause *hypoparathyroidism,* in which decreased PTH secretion reduces osteoclast activity. Although the bones remain strong, the blood calcium concentration decreases. The nervous system may become abnormally excitable, triggering spontaneous impulses. As a result, muscles may undergo tetanic contractions, possibly leading to respiratory failure and death.

11.8 Adrenal Glands

The **adrenal glands** are closely associated with the kidneys (fig. 11.12 and reference plate 6, p. 27). A gland sits atop each kidney like a cap and is embedded in the mass of adipose tissue that encloses the kidney.

Structure of the Glands

Each adrenal gland is very vascular and consists of two parts: The central portion is the **adrenal medulla** (ah-dre′nal me-dul′ah), and the outer part is the **adrenal cortex** (ah-dre′nal kor′teks). These regions are not sharply divided, but they are functionally distinct glands that secrete different hormones.

The adrenal medulla consists of irregularly shaped cells organized in groups around blood vessels. These cells are intimately connected with the sympathetic division of the autonomic nervous system. Adrenal medullary cells are actually modified postganglionic neurons. Preganglionic autonomic nerve fibers lead to them from the central nervous system (see chapter 9, p. 242).

The adrenal cortex, which makes up the bulk of the adrenal gland, is composed of closely packed masses of epithelial cells, organized in layers. These layers form an outer, middle, and inner zone of the cortex (fig. 11.12*B*). As in the adrenal medulla, the cells of the adrenal cortex are well supplied with blood vessels.

CHECK YOUR RECALL

1. Where are the adrenal glands?
2. Describe the two portions of an adrenal gland.

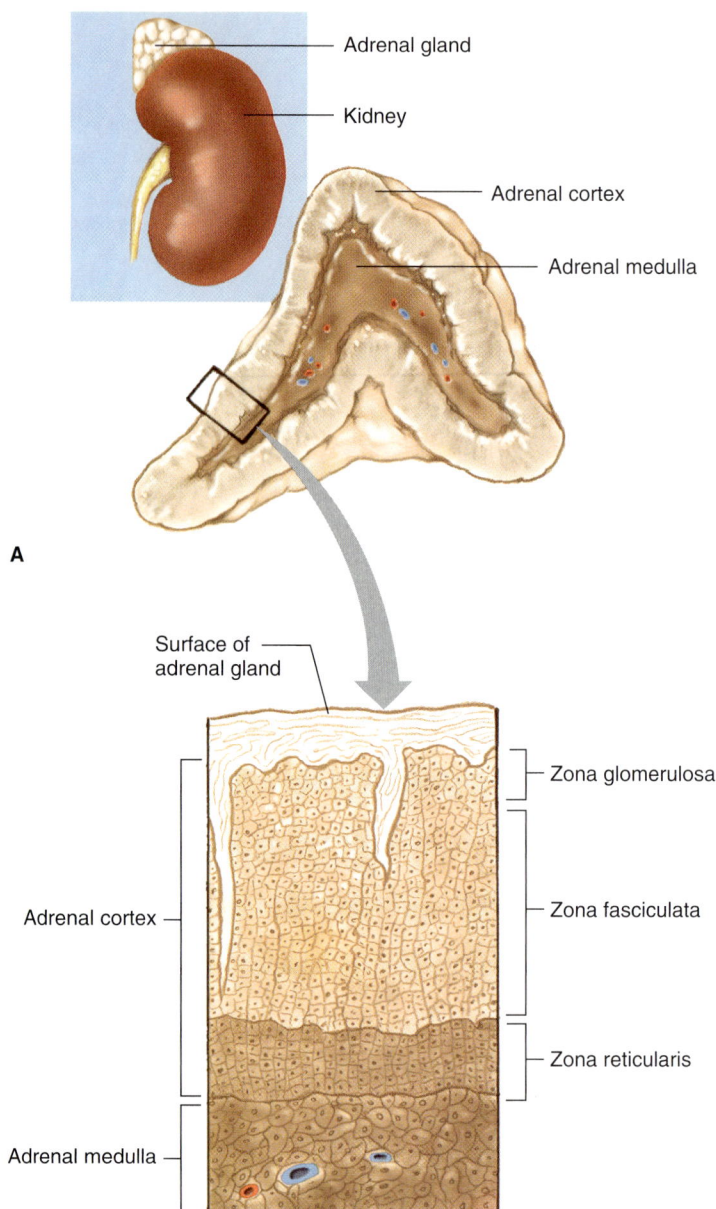

Figure 11.12
Adrenal glands. (*A*) An adrenal gland consists of an outer cortex and an inner medulla. (*B*) The cortex consists of the three layers, or zones, of cells shown.

Hormones of the Adrenal Medulla

The cells of the adrenal medulla secrete two closely related hormones—**epinephrine** (adrenalin) and **norepinephrine** (noradrenalin). These hormones have similar molecular structures and physiological functions. In fact, epinephrine, which makes up 80% of the adrenal medullary secretion, is synthesized from norepinephrine.

The effects of the adrenal medullary hormones resemble those of sympathetic neurons stimulating their effectors. Hormonal effects, however, last up to ten times longer because hormones are broken down more slowly than neurotransmitters. The effects of epinephrine and norepinephrine include increased heart rate, increased force of cardiac muscle contraction, increased breathing rate, elevated blood pressure, increased blood glucose, and decreased digestive activity.

Impulses arriving on sympathetic nerve fibers stimulate the adrenal medulla to release its hormones at the same time that sympathetic impulses are stimulating other effectors. These sympathetic impulses originate in the hypothalamus in response to stress. Thus, adrenal medullary secretions function with the sympathetic division of the autonomic nervous system in preparing the body for energy-expending action, sometimes called "fight or flight responses." Table 11.5 compares some of the effects of the adrenal medullary hormones.

Tumors in the adrenal medulla can increase hormonal secretion. Release of norepinephrine usually predominates, prolonging sympathetic responses—high blood pressure, increased heart rate, elevated blood sugar, and so forth. Surgical removal of the tumor corrects the condition.

CHECK YOUR RECALL

1. Name the hormones the adrenal medulla secretes.
2. What effects do hormones from the adrenal medulla produce?
3. What stimulates release of hormones from the adrenal medulla?

Hormones of the Adrenal Cortex

The cells of the adrenal cortex produce more than thirty different steroids, including several hormones. Unlike the adrenal medullary hormones, without which a person can still survive, some adrenal cortical hormones are vital. Without adrenal cortical secretions, a person usually dies within a week unless extensive electrolyte therapy is provided. The most important adrenal cortical hormones are aldosterone, cortisol, and certain sex hormones.

Aldosterone

Cells in the outer zone of the adrenal cortex synthesize **aldosterone.** Aldosterone is a *mineralocorticoid* because it helps regulate the concentration of mineral electrolytes. More specifically, aldosterone causes the kidney to conserve sodium ions and excrete potassium ions. By conserving sodium ions, aldosterone stimulates water retention indirectly by osmosis. This helps maintain blood volume and blood pressure.

A decrease in the blood concentration of sodium ions or an increase in the blood concentration of potassium ions stimulates the cells that secrete aldosterone. The kidneys also indirectly stimulate aldosterone secretion if blood pressure falls (see chapter 17, p. 471).

Cortisol

Cortisol (hydrocortisone) is a *glucocorticoid,* which means it affects glucose metabolism. It is produced in the middle zone of the adrenal cortex and, like aldosterone, is a steroid. Cortisol also influences protein and fat metabolism.

The more important actions of cortisol include:

1. Inhibition of protein synthesis in tissues, increasing the blood concentration of amino acids.
2. Promotion of fatty acid release from adipose tissue, increasing the utilization of fatty acids as an energy source and decreasing the use of glucose.
3. Stimulation of liver cells to synthesize glucose from noncarbohydrates, such as circulating amino acids and glycerol, increasing the blood glucose concentration.

These actions of cortisol help keep blood glucose concentration within the normal range between meals, because a few hours without food can exhaust the supply of liver glycogen, a major source of glucose.

Negative feedback controls cortisol release. This is much like control of thyroid hormones, involving the hypothalamus, anterior pituitary gland, and adrenal cortex. The hypothalamus secretes corticotropin-releasing hormone (CRH) into the hypophyseal portal veins,

TABLE 11.5		COMPARATIVE EFFECTS OF EPINEPHRINE AND NOREPINEPHRINE
PART OR FUNCTION AFFECTED	EPINEPHRINE	NOREPINEPHRINE
Heart	Rate increases; force of contraction increases	Rate increases; force of contraction increases
Blood vessels	Vessels in skeletal muscle dilate, decreasing resistance to blood flow	Blood flow to skeletal muscles increases, resulting from constriction of blood vessels in skin and viscera
Systemic blood pressure	Some increase due to increased cardiac output	Great increase due to vasoconstriction
Airways	Dilation	Some dilation
Reticular formation of brain	Activated	Little effect
Liver	Promotes breakdown of glycogen to glucose, increasing blood sugar concentration	Little effect on blood sugar concentration
Metabolic rate	Increases	Increases

which carry CRH to the anterior pituitary, stimulating it to secrete ACTH. In turn, ACTH stimulates the adrenal cortex to release cortisol. Cortisol inhibits the release of CRH and ACTH, and as concentrations of these fall, cortisol production drops (fig. 11.13).

The set point of the feedback mechanism controlling cortisol secretion changes from time to time, altering hormone output to meet the demands of changing conditions. For example, under stress—injury, disease, extreme temperature, or emotional upset—nerve impulses send the brain information concerning the stressful condition. In response, brain centers signal the hypothalamus to release more CRH, leading to a higher cortisol concentration until the stress subsides (fig. 11.13).

Adrenal Sex Hormones

Cells in the inner zone of the adrenal cortex produce sex hormones. These hormones are male types (adrenal androgens), but some are converted to female hormones (estrogens) in the skin, liver, and adipose tissue. Adrenal sex hormones may supplement the supply of sex hormones from the gonads and stimulate early development of reproductive organs. Table 11.6 summarizes the characteristics of the adrenal cortical hormones.

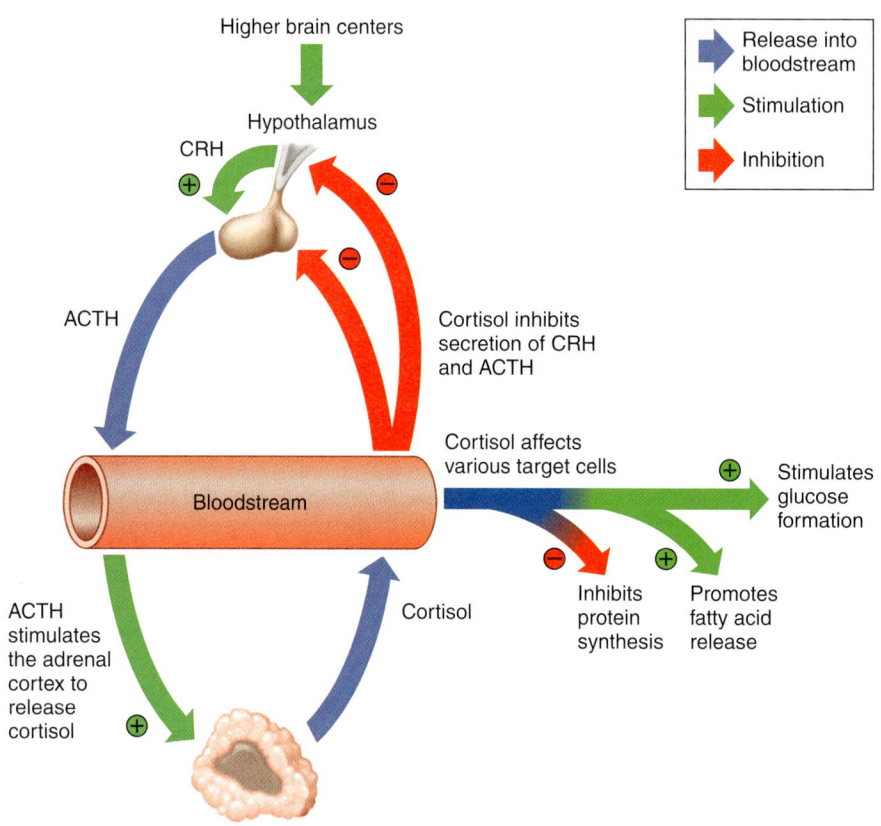

Figure 11.13
Negative feedback regulates cortisol secretion, similar to the regulation of thyroid hormone secretion (see fig. 11.8). (Note: ⊕ = stimulation; ⊖ = inhibition.)

TABLE 11.6		HORMONES OF THE ADRENAL CORTEX
HORMONE	ACTION	FACTOR REGULATING SECRETION
Aldosterone	Helps regulate concentration of extracellular electrolytes by conserving sodium ions and excreting potassium ions	Electrolyte concentrations in body fluids
Cortisol	Decreases protein synthesis, increases fatty acid release, and stimulates glucose synthesis from noncarbohydrates	Corticotropin-releasing hormone from hypothalamus and adrenocorticotropic hormone from anterior pituitary
Adrenal androgens	Supplement sex hormones from the gonads; may be converted to estrogens in females	

Hyposecretion of adrenal cortical hormones leads to *Addison disease,* a condition characterized by decreased blood sodium, increased blood potassium, low blood glucose concentration (hypoglycemia), dehydration, low blood pressure, and increased skin pigmentation. Without treatment with mineralocorticoids and glucocorticoids, Addison disease can be lethal in days because of severe disturbances in electrolyte balance.

Hypersecretion of adrenal cortical hormones, which may be associated with an adrenal tumor or with the anterior pituitary oversecreting ACTH, causes *Cushing syndrome.* This condition alters carbohydrate and protein metabolism and electrolyte balance. For example, when mineralocorticoids and glucocorticoids are overproduced, blood glucose concentration remains high, depleting tissue protein. Also, too much sodium is retained, increasing tissue fluids, and the skin becomes puffy. At the same time, increase in adrenal sex hormone production may cause masculinizing effects in a female, such as beard growth and deepening of the voice.

CHECK YOUR RECALL

1. Name the most important hormones of the adrenal cortex.
2. What is the function of aldosterone?
3. What actions does cortisol produce?
4. How are the blood concentrations of aldosterone and cortisol regulated?

11.9 Pancreas

The **pancreas** (pan´kre-as) consists of two major types of secretory tissues. This organization reflects its dual function as an exocrine gland that secretes digestive juice and as an endocrine gland that releases hormones (fig. 11.14 and reference plate 6, p. 27).

Structure of the Gland

The pancreas is an elongated, somewhat flattened organ posterior to the stomach and behind the parietal peritoneum. A duct joins the pancreas to the duodenum (the first section of the small intestine) and transports pancreatic digestive juice to the intestine.

The endocrine portion of the pancreas consists of groups of cells that are closely associated with blood vessels. These groups form islands of cells (*islets of Langerhans*) that include two distinct types of cells—alpha cells, which secrete the hormone glucagon, and beta cells, which secrete the hormone insulin (fig. 11.15). Chapter 15 (p. 404) discusses the digestive functions of the pancreas.

Hormones of the Islets of Langerhans

Glucagon stimulates the liver to break down glycogen and certain noncarbohydrates, such as amino acids, into glucose, raising blood sugar concentration. Glucagon

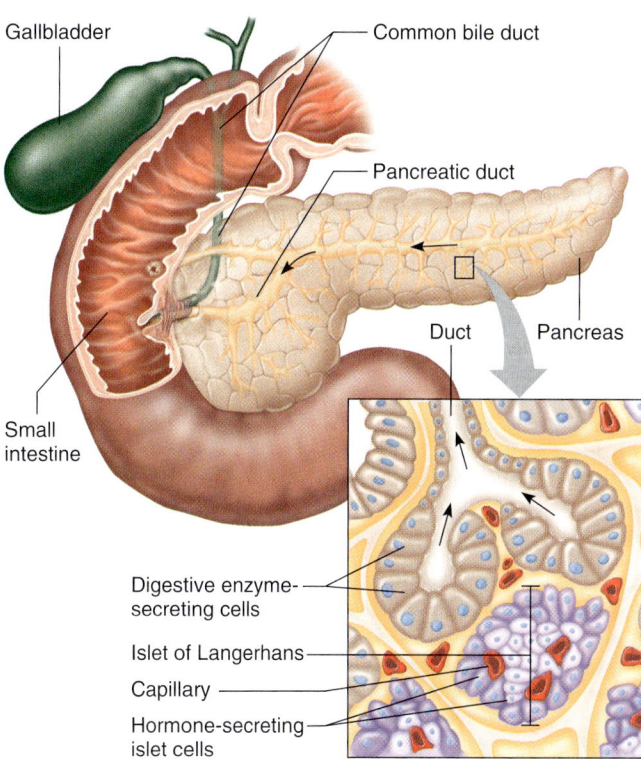

Figure 11.14
The hormone-secreting cells of the pancreas are grouped in clusters, or islets, that are closely associated with blood vessels. Other pancreatic cells secrete digestive enzymes into ducts.

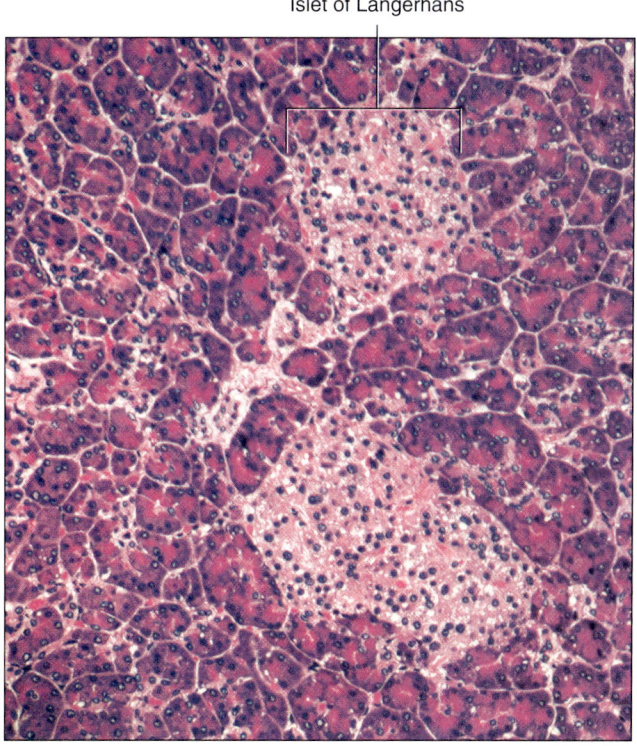

Figure 11.15
Light micrograph of an islet of Langerhans within the pancreas (200×).

much more effectively elevates blood glucose than does epinephrine.

A negative feedback system regulates glucagon secretion. Low blood glucose concentration stimulates alpha cells to release glucagon. When blood glucose concentration rises, glucagon secretion falls. This control prevents hypoglycemia when glucose concentration is relatively low, such as between meals, or when glucose is used rapidly, such as during periods of exercise.

The main effect of **insulin** is exactly opposite that of glucagon. Insulin stimulates the liver to form glycogen from glucose and inhibits conversion of noncarbohydrates into glucose. Insulin also has the special effect of promoting facilitated diffusion (see chapter 3, p. 59) of glucose across cell membranes that have insulin receptors. Cells that admit glucose include those of cardiac muscle, adipose tissue, and resting skeletal muscle. (Glucose uptake by exercising skeletal muscle does not depend on insulin.) These actions of insulin decrease blood glucose concentration. In addition, insulin secretion promotes transport of amino acids into cells, increases protein synthesis, and stimulates adipose cells to synthesize and store fat.

A negative feedback system sensitive to blood glucose concentration regulates insulin secretion. When blood glucose concentration is high, as after a meal, beta cells release insulin. Insulin helps prevent too high a blood glucose concentration by promoting glycogen formation in the liver and entrance of glucose into adipose and muscle cells. When glucose concentration falls, such as between meals or during the night, insulin secretion decreases.

As insulin output decreases, less and less glucose enters adipose and muscle cells. Cells that lack insulin receptors, such as nerve cells, can then use the glucose that remains in the blood. At the same time that insulin is decreasing, glucagon secretion is increasing. Therefore, insulin and glucagon function coordinately to maintain a relatively stable blood glucose concentration, despite great variation in the amount of carbohydrates a person eats (fig. 11.16).

Nerve cells, including those of the brain, obtain glucose by a facilitated diffusion mechanism that is not dependent on insulin but rather only on the blood glucose concentration. For this reason, nerve cells are particularly sensitive to changes in blood glucose concentration, and conditions that cause such changes—oversecretion of insulin leading to decreased blood glucose, for example—are likely to alter brain functions.

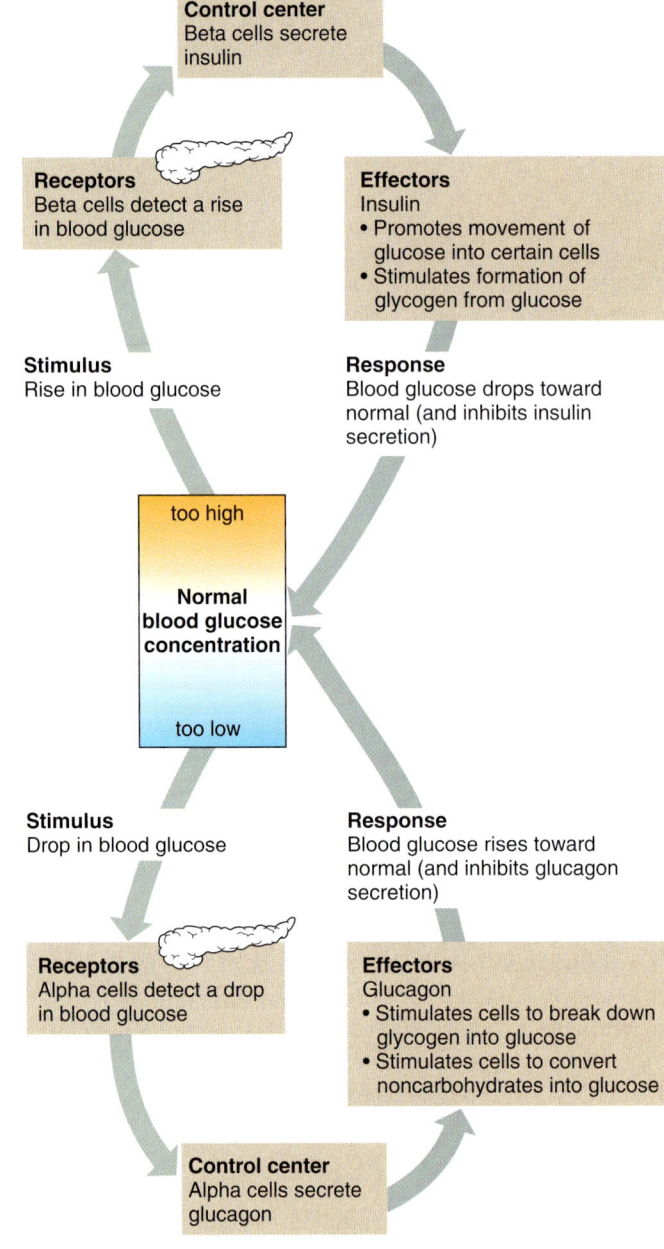

Figure 11.16

Insulin and glucagon function together to help maintain a relatively stable blood glucose concentration. Negative feedback responding to blood glucose concentration controls the levels of both hormones.

Cancer cells that develop from nonendocrine tissues sometimes inappropriately synthesize and secrete great amounts of peptide hormones or peptide hormonelike substances. For example, a cancer patient may develop an endocrine disorder that seems unrelated to the cancer (endocrine paraneoplastic syndrome), usually overproducing ADH, ACTH, a PTH-like substance, or an insulin-like substance.

CHECK YOUR RECALL

1. What is the endocrine portion of the pancreas called?
2. What is the function of glucagon?
3. What is the function of insulin?
4. How are glucagon and insulin secretions controlled?
5. Why are nerve cells particularly sensitive to changes in blood glucose concentration?

Topic of Interest

DIABETES MELLITUS

Diabetes mellitus is a condition resulting from insulin deficiency. It disturbs carbohydrate, protein, and fat metabolism. More specifically, since insulin helps glucose cross some cell membranes, movement of glucose into adipose and resting skeletal muscle cells decreases in diabetes. At the same time, glycogen formation decreases. As a result, blood sugar concentration rises (hyperglycemia). At a certain high concentration, the kidneys begin to excrete the excess. Glucose appearing in the urine (glycosuria) raises the urine's osmotic pressure, and more water and electrolytes than usual are excreted. Excess urine output causes the affected person to become dehydrated and extremely thirsty (polydipsia).

Diabetes mellitus also hampers protein and fat synthesis. Glucose-starved cells increase their use of proteins as an energy source. Tissues waste away, and the person loses weight, is very hungry, fatigues easily, and has a decreasing ability to grow and repair tissues. Changes in fat metabolism cause fatty acids and ketone bodies to accumulate in the blood, which lowers pH (acidosis). The dehydration and acidosis may harm brain cells, and the person may become disoriented, slip into a coma, and die.

The two common forms of diabetes mellitus are type 1 (also called *insulin-dependent diabetes mellitus* or juvenile-onset diabetes mellitus) and type 2 (also called *non-insulin-dependent diabetes mellitus* or maturity-onset diabetes mellitus). Type 1 diabetes mellitus usually appears before age twenty and is an autoimmune disease. This means that the immune system destroys the beta cells of the pancreas (see chapter 14, p. 386). Treatment involves injections of insulin, usually several times daily, or implanting an insulin pump to deliver the hormone. Figure 11A shows one of the first recipients of insulin treatment—a three-year-old boy who weighed only 15 pounds. Just two months of treatment doubled his weight. Until 1978, people with diabetes were treated with insulin extracted from pig pancreases. Since then, the human version of the hormone has been obtained from bacteria modified to have the human insulin gene. People are less likely to be allergic to the human hormone. In the near future, implants of insulin-producing cells may treat people with diabetes.

Type 2 diabetes mellitus is the type seen in 70–80% of people with diabetes. It usually develops gradually after

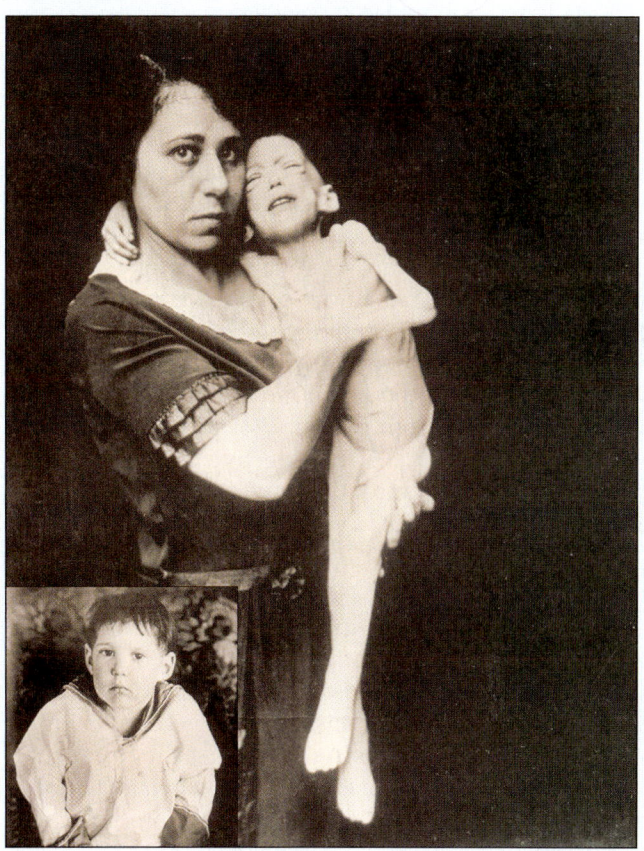

Figure 11A
Before and after insulin treatment: The boy in his mother's arms is three years old but weighs only 15 pounds because of diabetes mellitus. The inset shows the same child after just two months of receiving insulin.

age forty and produces milder symptoms than type 1 diabetes. Most affected individuals are overweight when they first experience symptoms. In type 2 diabetes, the beta cells of the pancreas function, but body cells lose sensitivity to insulin. Treatment includes controlling the diet, exercising, and maintaining a desirable body weight. Drugs are available too.

People with diabetes must monitor their blood glucose levels several times daily. Failure to control blood glucose concentration increases the chance of developing complications of diabetes mellitus, including coronary artery disease and retinal and nerve damage.

11.10 Other Endocrine Glands

Other glands that produce hormones and thus are parts of the endocrine system include the pineal gland, thymus gland, reproductive glands, and certain glands of the digestive tract, heart, and kidneys.

Pineal Gland

The **pineal gland** (pin´e-al gland) is a small structure located deep between the cerebral hemispheres, where it attaches to the upper portion of the thalamus near the roof of the third ventricle (see fig. 11.1). The pineal gland secretes the hormone **melatonin** in response to light conditions outside the body. Nerve impulses originating in the retinas of the eyes send this information to the pineal gland. In the dark, nerve impulses from the eyes decrease, and melatonin secretion increases.

Melatonin acts on certain brain regions that function as a "biological clock", and may thereby help to regulate **circadian rhythms,** which are patterns of repeated activity associated with the environmental cycles of day and night. The changing levels of melatonin throughout the 24-hour day may enable the body to distinguish day from night. Examples of circadian rhythms include the sleep-wake rhythm and seasonal cycles of fertility in many mammals. The study of circadian and other rhythms is called chronobiology.

Although the mechanism of melatonin action is poorly understood, the hormone inhibits the secretion of gonadotropins from the anterior pituitary and may help regulate the female reproductive cycle (menstrual cycle). It may also control the onset of puberty.

Thymus Gland

The **thymus gland** (thi´mus gland), which lies in the mediastinum posterior to the sternum and between the lungs, is relatively large in young children but shrinks with age (see fig. 11.1). This gland secretes a group of hormones called **thymosins** that affect the production and differentiation of certain white blood cells (lymphocytes). In this way, the thymus plays an important role in immunity, discussed in chapter 14 (p. 371).

Reproductive Glands

The reproductive organs that secrete important hormones include the ovaries, which produce estrogens and progesterone; the **placenta,** which produces estrogens, progesterone, and gonadotropin; and the testes, which produce testosterone. These glands and their secretions are discussed in chapter 19 (pp. 499 and 507) and chapter 20 (p. 532).

Digestive Glands

The digestive glands that secrete hormones are associated with the linings of the stomach and small intestine. Chapter 15 (pp. 403 and 405) describes these structures and their secretions.

Other Hormone-Producing Organs

Other organs outside of the endocrine system produce hormones. The heart, for example, secretes *atrial natriuretic peptide,* a hormone that stimulates urinary sodium excretion (see chapter 17, p. 468). The kidneys secrete a red blood cell growth hormone called *erythropoietin* (see chapter 12, p. 308).

CHECK YOUR RECALL
1. Where is the pineal gland located?
2. What is the function of the pineal gland?
3. Where is the thymus gland located?
4. Which reproductive organs secrete hormones?
5. Which other organs secrete hormones?

11.11 Stress and Health

Survival depends on the maintenance of homeostasis. Therefore, factors that change the body's internal environment can threaten life. When the body senses danger, nerve impulses to the hypothalamus trigger physiological responses that preserve homeostasis. These responses include increased activity in the sympathetic division of the autonomic nervous system and increased secretion of adrenal and other hormones. A factor that can stimulate such a response is called a *stressor,* and the condition it produces in the body is called *stress.*

Types of Stress

Stressors include physical factors, such as exposure to extreme heat or cold, decreased oxygen concentration, infections, injuries, prolonged heavy exercise, and loud sounds. Stressors also include psychological factors, such as thoughts about real or imagined dangers, personal losses, and unpleasant social interactions. Feelings of anger, fear, grief, anxiety, depression, and guilt can also produce psychological stress. Sometimes, even pleasant stimuli, such as friendly social contact, feelings of joy and happiness, or sexual arousal, may be stressful.

Responses to Stress

Physiological responses to stress consist of reactions called the *general stress syndrome* (general adaptation syndrome), which is under hypothalamic control. Typically, the hypothalamus activates mechanisms that prepare the body for "fight or flight." These responses include raising blood concentrations of glucose, glycerol, and fatty acids; increasing heart rate, blood pressure, and breathing rate; dilating air passages; shunting blood from the skin and digestive organs to the skeletal muscles; and increasing epinephrine secretion from the adrenal medulla (fig. 11.17).

The hypothalamus also releases CRH, which in turn stimulates the anterior pituitary to secrete ACTH. ACTH causes the adrenal cortex to increase cortisol secretion. Cortisol increases blood amino acid concentration, fatty acid release, and glucose formation from noncarbohydrates. Thus, while the body prepares for physical activity, cortisol supplies cells with biochemicals required during stress (fig. 11.17).

Topic of Interest

BIOLOGICAL RHYTHMS

Biological rhythms are changes that systematically recur in organisms. In complex animals, they include the daily ebb and flow of biochemical levels in blood, reproductive cycles, and migration schedules. The period of any rhythm is the duration of one complete cycle. The frequency of a rhythm is the number of cycles per time unit.

Three common types of rhythms in humans are ultradian, infradian, and circadian rhythms. *Ultradian rhythms* have periods shorter than 24 hours and include the cardiac cycle and the breathing cycle. Periods of *infradian rhythms,* such as the menstrual cycle, are longer than 24 hours. Periods of *circadian rhythms,* such as the sleep-wake cycle, are approximately 24 hours.

Both external (exogenous) and internal (endogenous) factors regulate human biological rhythms. Exogenous factors are environmental components, such as daily temperature changes and the light-dark cycle. Endogenous factors are genetically programmed internal "clocks" found in all types of organisms. Researchers have identified such "clock" genes in a range of species, including fruit flies, mice, nematode worms, bread mold, and plants.

The sleep-wake cycle is the most obvious circadian rhythm in humans. Most individuals sleep 6–8 hours each night, but culture largely determines this pattern. As we age, many people require less sleep. However, environmental cues are not the only determinants of the sleep-wake cycle. Under laboratory conditions of constant light or dark, the human body eventually follows an approximately 25 hour a day cycle.

Other circadian rhythms in humans affect body temperature, cardiovascular functioning, and hormone secretion. Body temperature is, for the most part, endogenously regulated, but light exposure and physical activity also help keep this rhythm on a 24- rather than 25-hour cycle. Body temperature is usually lowest between 4 and 6 A.M., then increases and peaks between 5 and 11 P.M. It drops during the late evening hours and into the night.

Cardiovascular functioning is least efficient between 6 and 9 A.M. Platelet cohesion, blood pressure, and pulse rate are typically highest 2 hours after awakening, which may explain why heart attacks and strokes are more likely between 8 and 10 A.M. than at other times.

Hormones may have ultradian and infradian as well as circadian rhythms. Plasma cortisol, for example, surges and peaks at about 6 A.M., then gradually declines to its minimum level in late evening before increasing again in the early morning. Growth hormone secretion peaks during the night. Antidiuretic hormone is greater at night, decreasing urine formation then.

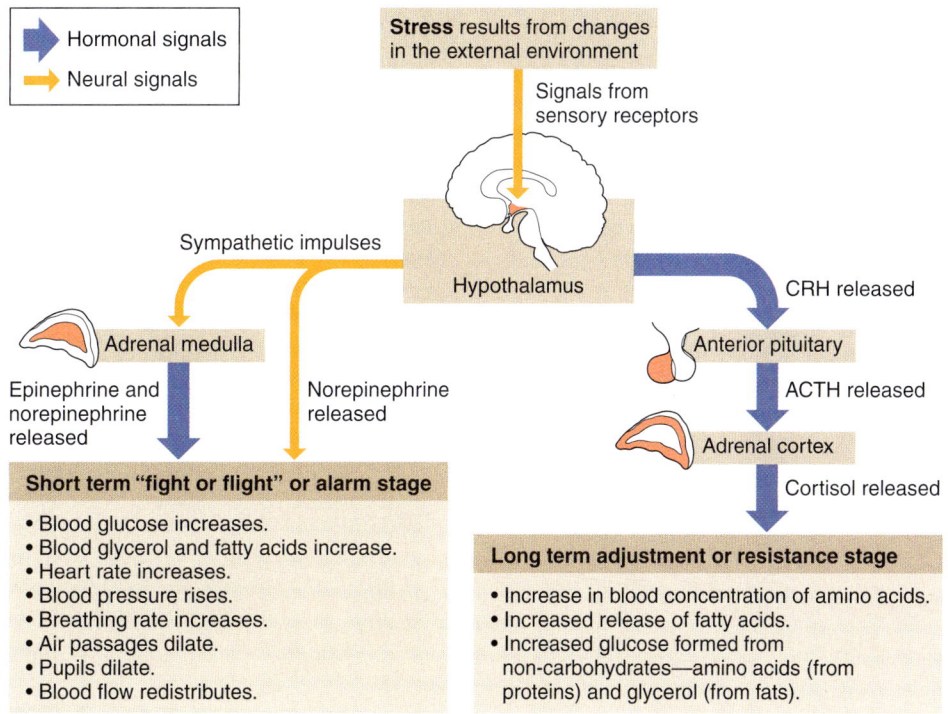

Figure 11.17
During stress, the hypothalamus helps prepare the body for "fight or flight" by triggering sympathetic impulses to various organs. It also stimulates epinephrine release, intensifying the sympathetic responses.

Other hormones whose secretions increase with stress include glucagon, GH, and ADH. Glucagon and GH mobilize energy sources, such as glucose, glycerol, fatty acids, and amino acids. ADH stimulates the kidneys to retain water, which increases blood volume—particularly important if a person is bleeding or sweating heavily.

Increased cortisol secretion may be accompanied by a decrease in the number of certain white blood cells (lymphocytes), which lowers resistance to infectious diseases and some cancers. Also, excess cortisol production may raise the risk of developing high blood pressure, atherosclerosis, and gastrointestinal ulcers.

CHECK YOUR RECALL

1. What is stress?
2. Distinguish between physical stress and psychological stress.
3. Describe the general stress syndrome.

Clinical Terms Related to the Endocrine System

adrenalectomy (ah-dre´´nah-lek´to-me) Surgical removal of the adrenal glands.
adrenogenital syndrome (ah-dre´´no-jen´ĭ-tal sin´drōm) A group of symptoms associated with changes in sexual characteristics as a result of increased secretion of adrenal androgens.
diabetes insipidus (di´´ah-be´tēz in-sip´ĭdus) Metabolic disorder, not involving blood sugar, characterized by a large output of dilute urine and caused by the posterior pituitary's decreased secretion of antidiuretic hormone.
diabetes mellitus (di´´ah-be´tēz mel´ĭ-tus) Condition due to insulin deficiency or the inability to respond to insulin. This condition disturbs carbohydrate, protein, and lipid metabolism.
exophthalmos (ek´´sof-thal´mos) Abnormal protrusion of the eyes.
goiter (goi´ter) Bulge in the neck resulting from an enlarged thyroid gland.
hirsutism (her´sūt-izm) Excess hair growth, especially in women.
hypercalcemia (hi´´per-kal-se´me-ah) Excess blood calcium.
hyperglycemia (hi´´per-gli-se´me-ah) Excess blood glucose.
hypocalcemia (hi´´po-kal-se´me-ah) Deficiency of blood calcium.
hypoglycemia (hi´´po-gli-se´me-ah) Deficiency of blood glucose.
hypophysectomy (hi-pof´´ĭ-sek´to-me) Surgical removal of the pituitary gland.
parathyroidectomy (par´´ah-thi´´roi-dek´to-me) Surgical removal of the parathyroid glands.
pheochromocytoma (fe-o-kro´´mo-si-to´mah) Type of tumor in the adrenal medulla usually associated with high blood pressure.
polyphagia (pol´´e-fa´je-ah) Excessive eating.
thymectomy (thi-mek´to-me) Surgical removal of the thymus gland.
thyroidectomy (thi´´roi-dek´to-me) Surgical removal of the thyroid gland.
thyroiditis (thi´´roi-di´tis) Inflammation of the thyroid gland.
virilism (vir´ĭ-lizm) Masculinization of a female.

Clinical Connection

Ads for melatonin sold as a "food supplement" claim that the hormone treats a staggering variety of ills, including AIDS, autism, Alzheimer disease, cancer, depression, heart disease, influenza, Parkinson disease, seizures and schizophrenia. In fact, evidence supports only one use—insomnia. Although taking melatonin pills is widely reported to overcome the fatigue of jet lag, the hormone only resets the body's clock by an hour. Just because melatonin can easily be obtained in a supermarket does not mean it has been shown to be safe. In species that are seasonal breeders, such as hamsters and deer, additional melatonin disrupts reproduction. In the U.S. melatonin supplements are not regulated or marketed as if they are drugs, even though hormones such as estrogen are considered to be prescription drugs. Therefore, much about melatonin simply isn't known, including its effects on reproduction, its long-term effects, dangers of overdose, whether people react differently to different doses or if some people should avoid the supplements, or whether melatonin interacts with compounds that are considered to be drugs.

Organization

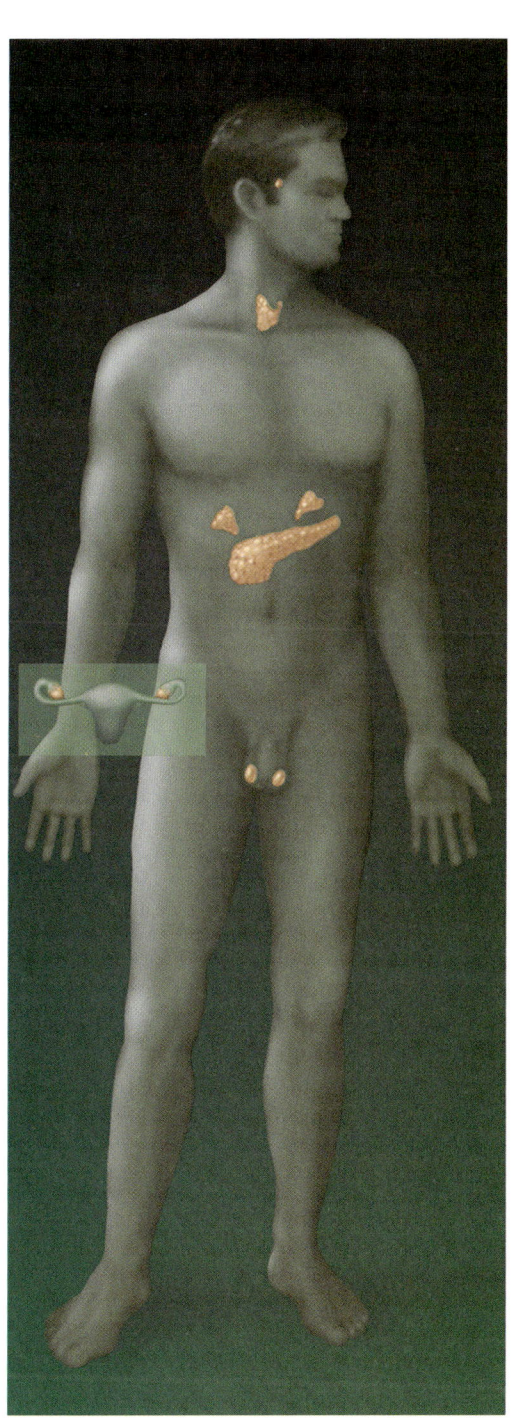

Endocrine System
Glands secrete hormones that have a variety of effects on cells, tissues, organs, and organ systems.

Integumentary System

Melanocytes produce skin pigment in response to hormonal stimulation.

Lymphatic System

Hormones stimulate lymphocyte production.

Skeletal System

Hormones act on bones to control calcium balance.

Digestive System

Hormones help control digestive system activity.

Muscular System

Hormones help increase blood flow to exercising muscles.

Respiratory System

Decreased oxygen causes hormonal stimulation of red blood cell production; red blood cells transport oxygen and carbon dioxide.

Nervous System
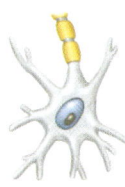
Neurons control the secretions of the anterior and posterior pituitary glands and the adrenal medulla.

Urinary System

Hormones act on the kidneys to help control water and electrolyte balance.

Cardiovascular System
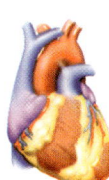
Hormones are carried in the bloodstream; some have direct actions on the heart and blood vessels.

Reproductive System

Sex hormones play a major role in development of secondary sex characteristics, egg, and sperm.

SUMMARY OUTLINE

11.1 Introduction (p. 281)
The endocrine and nervous systems maintain homeostasis.
1. The endocrine system is a network of glands that secrete hormones, which travel in the bloodstream and affect the functioning of target cells.
2. Paracrine secretions act locally and autocrine secretions act on the cells that produce them.
3. Exocrine glands secrete through tubes or ducts.

11.2 General Characteristics of the Endocrine System (p. 281)
1. The nervous and endocrine systems both exert precise effects.
2. Hormones are secreted from glands or from specialized groups of cells.

11.3 Hormone Action (p. 282)
Endocrine glands secrete hormones that affect target cells with specific receptors. Hormones are very potent.
1. Chemically, hormones are steroids, amines, peptides, proteins, or glycoproteins.
2. Steroid hormones
   a. Steroid hormones enter a target cell and combine with receptors to form complexes within the nucleus.
   b. These complexes activate specific genes, which cause protein synthesis.
3. Nonsteroid hormones
   a. Nonsteroid hormones combine with receptors in the target-cell membrane.
   b. The hormone-receptor complex signals a G protein to stimulate a membrane protein, such as adenylate cyclase, to induce formation of second messenger molecules.
   c. A second messenger, such as cyclic adenosine monophosphate (cAMP), diacylglycerol (DAG), or inositol triphosphate (IP_3), activates protein kinases.
   d. Protein kinases activate protein substrate molecules, which in turn change a cellular process.
4. Prostaglandins
   a. Prostaglandins act on the cells of the organs that produce them.
   b. Prostaglandins are present in small quantities and have powerful hormonelike effects.

11.4 Control of Hormonal Secretions (p. 285)
The concentration of each hormone in body fluids is regulated.
1. Some endocrine glands secrete hormones in response to releasing hormones the hypothalamus secretes.
2. Other glands secrete their hormones in response to nerve impulses.
3. Some glands respond to levels of a substance in the bloodstream.
4. All three of the above control mechanisms employ negative feedback.
   a. In a negative feedback system, a gland is sensitive to the concentration of a substance it regulates.
   b. When the concentration of the regulated substance reaches a certain point, it inhibits the gland.
   c. As the gland secretes less hormone, the controlled substance also decreases.
   d. Negative feedback systems maintain relatively stable hormone concentrations.

11.5 Pituitary Gland (p. 286)
The pituitary gland has an anterior lobe and a posterior lobe. The hypothalamus controls most pituitary secretions.

1. Anterior pituitary hormones
   a. The anterior pituitary secretes growth hormone (GH), prolactin (PRL), thyroid-stimulating hormone (TSH), adrenocorticotropic hormone (ACTH), follicle-stimulating hormone (FSH), and luteinizing hormone (LH).
   b. Growth hormone
      (1) GH stimulates cells to increase in size and divide more frequently.
      (2) GH releasing hormone and GH release-inhibiting hormone from the hypothalamus control GH secretion.
   c. PRL stimulates and sustains a woman's milk production.
   d. Thyroid-stimulating hormone
      (1) TSH controls secretion of hormones from the thyroid gland.
      (2) The hypothalamus secretes thyrotropin-releasing hormone (TRH), which regulates TSH secretion.
   e. Adrenocorticotropic hormone
      (1) ACTH controls secretion of hormones from the adrenal cortex.
      (2) The hypothalamus secretes corticotropin-releasing hormone (CRH), which regulates ACTH secretion.
   f. FSH and LH are gonadotropins.
2. Posterior pituitary hormones
   a. The posterior lobe of the pituitary gland consists largely of neuroglial cells and nerve fibers.
   b. The hypothalamus produces the hormones of the posterior pituitary.
   c. Antidiuretic hormone (ADH)
      (1) ADH reduces the amount of water the kidneys excrete.
      (2) The hypothalamus regulates ADH secretion.
   d. Oxytocin (OT)
      (1) OT contracts muscles in the uterine wall.
      (2) OT also contracts cells associated with producing and ejecting milk.

11.6 Thyroid Gland (p. 289)
The thyroid gland is located in the neck and consists of two lobes.
1. Structure of the gland
   a. The thyroid gland consists of many follicles.
   b. The follicles are fluid-filled and store hormones.
2. Thyroid hormones
   a. Thyroxine and triiodothyronine increase the metabolic rate of cells, enhance protein synthesis, and stimulate lipid utilization.
   b. Calcitonin helps regulate concentrations of blood calcium and phosphate ions.

11.7 Parathyroid Glands (p. 291)
The parathyroid glands are on the posterior surface of the thyroid gland.
1. Structure of the glands
   Each parathyroid gland consists of secretory cells that are well supplied with capillaries.
2. Parathyroid hormone (PTH)
   a. PTH increases blood calcium level and decreases blood phosphate ion concentration.
   b. A negative feedback mechanism operates between the parathyroid glands and the blood.

11.8 Adrenal Glands (p. 292)
The adrenal glands are located atop the kidneys.
1. Structure of the glands
   a. Each gland consists of an adrenal medulla and an adrenal cortex.
   b. The adrenal medulla and adrenal cortex are functionally distinct glands that secrete different hormones.

2. Hormones of the adrenal medulla
   a. The adrenal medulla secretes epinephrine and norepinephrine, which have similar effects.
   b. Sympathetic impulses stimulate secretion of these hormones.
3. Hormones of the adrenal cortex
   a. The adrenal cortex produces several steroid hormones.
   b. Aldosterone is a mineralocorticoid that causes the kidneys to conserve sodium ions and water and to excrete potassium ions.
   c. Cortisol is a glucocorticoid that affects carbohydrate, protein, and fat metabolism.
   d. Adrenal sex hormones
      (1) These hormones are of the male type but may be converted to female hormones.
      (2) They may supplement the sex hormones the gonads produce.

11.9 Pancreas (p. 295)

The pancreas secretes digestive juices as well as hormones.

1. Structure of the gland
   a. The pancreas is attached to the small intestine.
   b. The islets of Langerhans secrete glucagon and insulin.
2. Hormones of the islets of Langerhans
   a. Glucagon stimulates the liver to produce glucose from glycogen and noncarbohydrates.
   b. Insulin moves glucose across some cell membranes, stimulates glucose and fat storage, and promotes protein synthesis.
   c. Nerve cells are not dependent on insulin for a glucose supply.

11.10 Other Endocrine Glands (p. 297)

1. Pineal gland
   a. The pineal gland attaches to the thalamus.
   b. It secretes melatonin in response to varying light conditions.
   c. Melatonin may help regulate the female reproductive cycle by inhibiting gonadotropin secretion from the anterior pituitary.
2. Thymus gland
   a. The thymus gland lies behind the sternum and between the lungs.
   b. It secretes thymosins, which affect the production of certain lymphocytes that function in immunity.
3. Reproductive glands
   a. The ovaries secrete estrogens and progesterone.
   b. The placenta secretes estrogens, progesterone, and gonadotropin.
   c. The testes secrete testosterone.
4. Digestive glands
   Certain glands of the stomach and small intestine secrete hormones.
5. Other hormone-producing organs
   Other organs, such as the heart and the kidneys, also produce hormones.

11.11 Stress and Health (p. 299)

Stress occurs when the body responds to stressors that threaten the maintenance of homeostasis. Stress responses include increased activity of the sympathetic nervous system and increased secretion of adrenal hormones.

1. Types of stress
   a. Physical stress results from environmental factors that are harmful or potentially harmful to tissues.
   b. Psychological stress results from thoughts about real or imagined dangers.
2. Responses to stress
   a. Responses to stress maintain homeostasis.
   b. The hypothalamus controls a general stress syndrome.

REVIEW EXERCISES

1. Define *endocrine gland*. (p. 281)
2. Define *hormone* and *target cell*. (p. 281)
3. Explain how steroid hormones produce their effects. (p. 282)
4. Explain how nonsteroid hormones employ second messenger molecules. (p. 283)
5. Describe how prostaglandins are similar to hormones. (p. 284)
6. Describe a negative feedback system. (p. 285)
7. Describe the location and structure of the pituitary gland. (p. 286)
8. Explain how the brain controls pituitary gland activity. (p. 286)
9. Define *releasing hormone*, and give an example of one. (p. 286)
10. List the hormones that the anterior pituitary secretes. (p. 287)
11. Explain how growth hormone produces its effects. (p. 287)
12. List the major factors that affect secretion of growth hormone. (p. 287)
13. Summarize the function of prolactin. (p. 287)
14. Describe the mechanism that regulates the concentrations of circulating thyroid hormones. (p. 287)
15. Explain the control of adrenocorticotropic hormone secretion. (p. 288)
16. Compare the cellular structures of the anterior and posterior lobes of the pituitary gland. (p. 288)
17. Describe the functions of the posterior pituitary hormones. (p. 288)
18. Explain the regulation of antidiuretic hormone release. (p. 288)
19. Describe the location and structure of the thyroid gland. (p. 289)
20. Name the hormones the thyroid gland secretes, and list the general functions of each. (p. 289)
21. Describe the location and structure of the parathyroid glands. (p. 291)
22. Explain the general functions of parathyroid hormone. (p. 291)
23. Describe the mechanism that regulates parathyroid hormone secretion. (p. 291)
24. Distinguish between the adrenal medulla and the adrenal cortex. (p. 292)
25. List the hormones the adrenal medulla produces, and describe their general functions. (p. 292)
26. Name the most important hormones of the adrenal cortex, and describe the general functions of each. (p. 293)
27. Describe how the pituitary gland controls the secretion of cortisol. (p. 294)
28. Describe the location and structure of the pancreas. (p. 295)
29. List the hormones the islets of Langerhans secrete, and describe their functions. (p. 296)
30. Summarize the regulation of hormone secretion from the pancreas. (p. 296)
31. Describe the location and general function of the pineal gland. (p. 297)
32. Describe the location and general function of the thymus gland. (p. 297)
33. Name five additional hormone-secreting organs. (p. 297)

CRITICAL THINKING

1. When reactor 4 at the Chernobyl Nuclear Power Station in Ukraine exploded at 1:23 P.M. on April 26, 1986, a great plume of radioactive isotopes erupted into the air and spread for thousands of miles.

Most of the isotopes emitted immediately following the blast were of the element iodine. Which of the glands of the endocrine system would be most seriously—and immediately—affected by the blast, and how do you think this would become evident in the nearby population?

2. Human growth hormone and human insulin today are manufactured using recombinant DNA technology, which means that bacteria use human genes to produce the human proteins. Why is this safer than the former method of extracting medically useful hormones from pigs, cows, or human cadavers?

3. A young woman feels shaky, distracted, and generally ill. She lives with her mother, who is dying. A friend tells the young woman, "It's just stress, it's all in your head." Is it?

4. How does a pheromone differ from a hormone?

5. Growth hormone is administered to people who have pituitary dwarfism. Parents wanting their normal children to be taller have requested the treatment for them. Do you think this is a wise request? Why or why not?

6. A sixty-two-year-old woman recently gave birth after receiving hormone treatments that mimicked the hormonal environment of a woman of childbearing age. Do you think such treatments should become part of medical practice, enabling women past menopause to have children?

7. What hormone supplements would an adult whose anterior pituitary has been removed require?

8. How might the environment of a patient with hyperthyroidism be modified to minimize the drain on body energy resources?

9. The adrenal cortex of a patient who has lost a large volume of blood will increase secretion of aldosterone. What effect will this increased secretion have on the patient's blood concentrations of sodium and potassium ions?

10. Why might oversecretion of insulin actually reduce glucose uptake by nerve cells?

WEB CONNECTIONS

Visit the website for additional study questions and more information about this chapter at:

http://www.mhhe.com/shieress8

Unit 4
Transport

chapter 12

Blood

BLOOD SUBSTITUTES. Finding a substitute for human blood is a tall order, especially trying to mimic the major functions of this connective tissue. Blood carries oxygen (red blood cells), provides protection against infection (white blood cells), promotes clotting (platelets), and carries other vital substances. Efforts to replace blood in the past have sought to fill in the missing fluid volume or to reproduce blood's oxygen-carrying role. The search for blood substitutes intensified after each of the two world wars, because injured soldiers had desperately needed transfusions, and again when the AIDS pandemic made transfusions dangerous unless blood is properly screened.

A red blood cell substitute must meet several requirements: It must carry oxygen and give it up to tissues, be nontoxic, be storeable, function until the body can take over, and not provoke an immune response. The nine red blood cell substitutes currently in clinical trials are of two basic types. Perfluorocarbons are synthetic chemicals that carry dissolved oxygen. These were actually developed in the 1960s, and a famous photo shows a mouse apparently drowning in a beaker of the chemical—but still breathing even though submerged.

The second type of red blood cell substitute dismantles red blood cells and isolates the oxygen-carrying hemoglobin molecules, which are then linked in various ways. The starting material is usually cow's blood or old stored human blood. Red blood cell substitutes used in times past include wine, ale, milk, plant resins, urine, and opium!

Photo:
At this biotechnology company, hemoglobin is purified in efforts to develop a red blood cell substitute.

Chapter Objectives

After studying this chapter, you should be able to do the following:

12.1 Introduction
1. Describe the general characteristics of blood, and discuss its major functions. (p. 307)

12.2 Blood and Blood Cells
2. Distinguish among the formed elements of the blood. (p. 307)
3. Explain the control of red blood cell production. (p. 308)
4. Distinguish among the five types of white blood cells, and give the function(s) of each type. (p. 311)

12.3 Blood Plasma
5. List the major components of blood plasma, and describe the functions of each. (p. 314)

12.4 Hemostasis
6. Define *hemostasis*, and explain the mechanisms that help achieve it. (p. 317)

7. Review the major steps in blood coagulation. (p. 318)

12.5 Blood Groups and Transfusions
8. Explain blood typing and how it is used to avoid adverse reactions following blood transfusions. (p. 320)
9. Describe how blood reactions may occur between fetal and maternal tissues. (p. 323)

Aids to Understanding Words

agglutin- [to glue together] *agglutin*ation: Clumping together of red blood cells.
bil- [bile] *bil*irubin: Pigment excreted in the bile.
embol- [stopper] *embol*ism: Obstruction of a blood vessel.
erythr- [red] *erythr*ocyte: Red blood cell.
hem- [blood] *hem*oglobin: Red pigment responsible for the color of blood.

leuko- [white] *leuko*cyte: White blood cell.
-osis [abnormal condition] leuko*cytosis*: Condition in which white blood cells are overproduced.
-poie [make, produce] erythro*poie*tin: Hormone that stimulates the production of red blood cells.

-sta [halt] hemo*sta*sis: Arrest of bleeding from damaged blood vessels.
thromb- [clot] *thromb*ocyte: Blood platelet involved in the formation of a blood clot.

Key Terms

albumin (al-bu´min)
antibody (an´ti-bod´´e)
antigen (an´ti-jen)
basophil (ba´so-fil)
coagulation (ko-ag´´u-la´shun)
eosinophil (e´´o-sin´o-fil)

erythrocyte (ĕ-rith´ro-sīt)
erythropoietin (ĕ-rith´´ro-poi´ĕ-tin)
fibrinogen (fi-brin´o-jen)
globulin (glob´u-lin)
hemostasis (he´´mo-sta´sis)
leukocyte (lu´ko-sīt)

lymphocyte (lim´fo-sīt)
monocyte (mon´o-sīt)
neutrophil (nu´tro-fil)
plasma (plaz´mah)
platelet (plāt´let)

12.1 Introduction

Blood signifies life, and for good reason—it has many vital functions. This complex mixture of cells, cell fragments, and dissolved biochemicals transports nutrients, oxygen, wastes, and hormones; helps maintain the stability of the interstitial fluid; and distributes heat. The blood, heart, and blood vessels form the cardiovascular system and link the body's internal and external environments.

Blood is a type of connective tissue whose cells are suspended in a liquid material. Blood is vital in transporting substances between body cells and the external environment, thereby promoting homeostasis.

12.2 Blood and Blood Cells

Whole blood is slightly heavier and three to four times more viscous than water. Its cells, which form mostly in red bone marrow, include red blood cells and white blood cells. Blood also contains cellular fragments called blood platelets (fig. 12.1). The cells and platelets are termed "formed elements" of the blood, in contrast to the liquid portion.

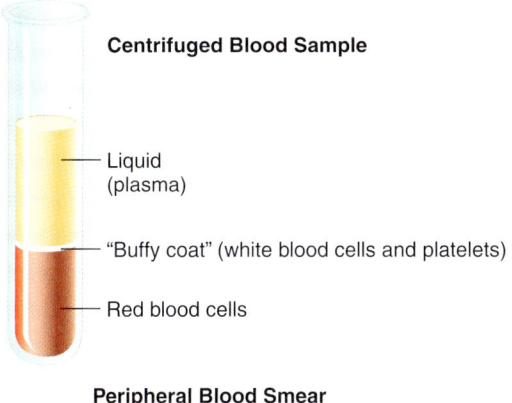

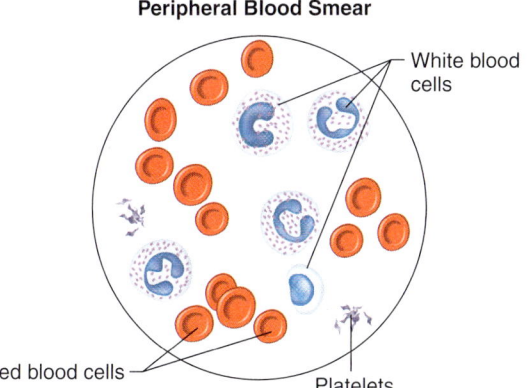

Figure 12.1
Blood consists of a liquid portion called plasma and a solid portion that includes red blood cells, white blood cells, and platelets. (Note: When blood components are separated, the white blood cells and platelets form a thin layer, called the "buffy coat," between the plasma and the red blood cells.)

Blood Volume and Composition

Blood volume varies with body size, changes in fluid and electrolyte concentrations, and the amount of adipose tissue. An average-sized adult has a blood volume of about 5.3 quarts (5 liters).

 Men have more blood than women. Men have 1.500 gallons, compared to 0.875 gallons for women.

A blood sample is usually about 45% cells by volume. This percentage is called the **hematocrit (HCT).** Most blood cells are red cells, with much smaller numbers of white cells and blood platelets. The remaining 55% of a blood sample is a clear, straw-colored liquid called **plasma** (plaz´mah) (fig. 12.1). Plasma is a complex mixture of water, amino acids, proteins, carbohydrates, lipids, vitamins, hormones, electrolytes, and cellular wastes.

CHECK YOUR RECALL

1. What factors affect blood volume?
2. What are the major components of blood?

Red Blood Cells

Red blood cells, or **erythrocytes** (ĕ-rith´ro-sītz), are biconcave discs. This shape is an adaptation for transporting gases; it increases the surface area through which gases can diffuse (fig. 12.2). The red blood cell's shape also places the cell membrane closer to oxygen-carrying *hemoglobin* within the cell.

Each red blood cell is about one-third hemoglobin by volume. This protein is responsible for the color of the blood. When hemoglobin combines with oxygen, the resulting *oxyhemoglobin* is bright red, and when oxygen is released, the resulting *deoxyhemoglobin* is darker.

Red blood cells have nuclei during their early stages of development but extrude them as the cells mature, providing more space for hemoglobin. Since they lack nuclei, red blood cells cannot synthesize proteins or divide.

A person experiencing prolonged oxygen deficiency (hypoxia) may become *cyanotic.* The skin and mucous membranes appear bluish due to an abnormally high blood concentration of deoxyhemoglobin. Exposure to low temperature may also result in cyanosis. Such exposure constricts superficial blood vessels, which slows blood flow and removes more oxygen than usual from blood flowing through the vessels.

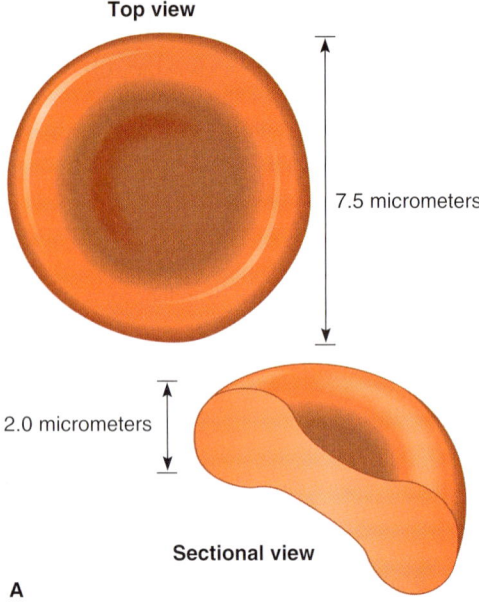

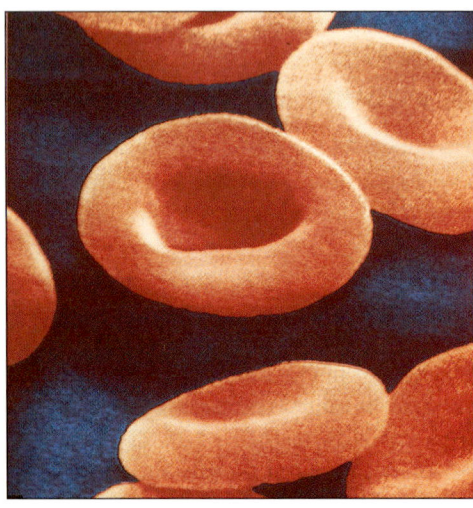

Figure 12.2
Red blood cells. (*A*) The biconcave shape of a red blood cell makes possible its function. (*B*) Scanning electron micrograph of human red blood cells (falsely colored) (5,000×).

Red Blood Cell Counts

The number of red blood cells in a cubic millimeter (mm^3) of blood is called the *red blood cell count* (*RBCC* or *RCC*). The typical range for adult males is 4,600,000–6,200,000 cells per mm^3, and that for adult females is 4,200,000–5,400,000 cells per mm^3.

Since increasing the number of circulating red blood cells increases the blood's *oxygen-carrying capacity*, changes in this number may affect health. For this reason, red blood cell counts are routinely consulted to help diagnose and evaluate the courses of various diseases.

CHECK YOUR RECALL

1. Describe a red blood cell.
2. What is the function of hemoglobin?
3. How does a red blood cell change as it matures?
4. What is the typical red blood cell count for an adult male? For an adult female?

Red Blood Cell Production and Its Control

Recall from chapter 7 (p. 133) that red blood cell formation (hemopoiesis) initially occurs in the yolk sac, liver, and spleen. After an infant is born, these cells are produced almost exclusively in tissue lining the spaces in bones, the red bone marrow.

The average life span of a red blood cell is 120 days. Many of these cells are removed from the circulation each day, yet the number of cells in the circulating blood remains relatively stable. This observation suggests a *homeostatic* control of the rate of red blood cell production.

 The combined surface area of all the red blood cells in the human body is roughly 2,000 times as great as the body's exterior surface.

A *negative feedback mechanism* utilizing the hormone **erythropoietin** (ĕ-rith´´ro-poi´ĕ-tin) controls the rate of red blood cell formation. The kidneys, and to a lesser extent the liver, release erythropoietin in response to prolonged oxygen deficiency (fig. 12.3). At high altitudes, for example, where the percentage of oxygen in the air is reduced, the amount of oxygen delivered to the tissues initially decreases. This drop in oxygen triggers the release of erythropoietin, which travels via the blood to the red bone marrow and stimulates red blood cell production.

After a few days, many newly formed red blood cells appear in the circulating blood. The increased rate of production continues until the number of erythrocytes in the circulation is sufficient to supply these tissues with their oxygen requirements. When the oxygen level in the air returns to normal, erythropoietin release decreases, and the rate of red blood cell production returns to normal as well. Figure 12.4 illustrates the stages in the development and differentiation of red blood cells from *hemocytoblasts* (stem cells).

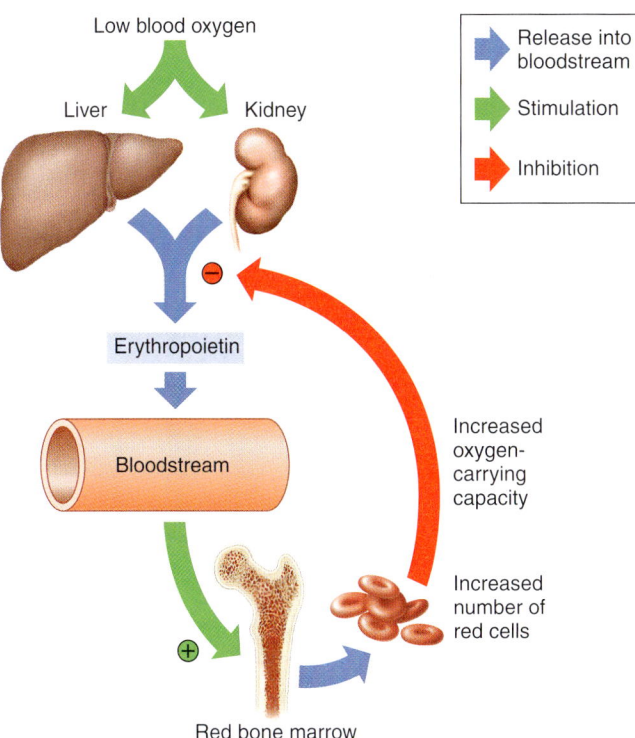

Figure 12.3
Low blood oxygen causes the kidneys and liver to release erythropoietin, which stimulates the production of red blood cells that carry oxygen to tissues.

 **CHECK YOUR RECALL**

1. Where are red blood cells produced?
2. How is red blood cell production controlled?
3. Which vitamins are necessary for red blood cell production?
4. Why is iron required for the development of red blood cells?

In *sickle cell disease*, a single DNA base change causes an incorrect amino acid to be incorporated into globin, causing hemoglobin to crystallize in a low-oxygen environment. This bends the red blood cells containing the hemoglobin into a sickle shape, which blocks circulation in small blood vessels, causing excruciating joint pain and damaging many organs. As the spleen works harder to recycle the abnormally short-lived red blood cells, infection becomes likely.

Most children with sickle cell disease are diagnosed at birth and receive antibiotics daily for years to prevent infection. Hospitalization for blood transfusions may be necessary if the person experiences painful sickling "crises" of blocked circulation.

A bone marrow transplant can completely cure sickle cell disease but has a 15% risk of causing death. A new treatment is an old drug, used to treat cancer, called hydroxyurea. It activates a slightly different form of hemoglobin in the fetus. Because of the presence of the functional fetal hemoglobin, the sickle hemoglobin cannot crystallize as quickly as it otherwise would. Sickling becomes delayed, which enables red blood cells carrying sickled hemoglobin to more quickly reach the lungs—where fresh oxygen restores the cells' normal shapes.

Dietary Factors Affecting Red Blood Cell Production

B-complex vitamins—*vitamin B₁₂* and *folic acid*—significantly influence red blood cell production. These vitamins are necessary for DNA synthesis, so all cells with nuclei require them to grow and divide. Since cell division occurs frequently in the blood-cell-forming (hemopoietic) tissue, this tissue is especially vulnerable to deficiency of either of these vitamins.

Hemoglobin synthesis and normal red blood cell production require iron. The small intestine absorbs iron slowly from food. The body reuses much of the iron released during decomposition of hemoglobin from damaged red blood cells. Therefore, the diet need only supply small quantities of iron.

Too few red blood cells or too little hemoglobin causes *anemia*. This reduces the oxygen-carrying capacity of the blood, and the affected person may appear pale and lack energy. A pregnant woman may become anemic if she doesn't eat iron-rich foods because her blood volume increases due to fluid retention to accommodate the requirements of the fetus. This increased blood volume decreases the hematocrit.

Destruction of Red Blood Cells

Red blood cells are quite elastic and flexible, and they readily bend as they pass through small blood vessels. With age, however, these cells become more fragile, and they are frequently damaged simply by passing through capillaries, particularly those in active muscles. **Macrophages** phagocytize and destroy damaged red blood cells, primarily in the liver and spleen (see chapter 5, p. 99).

Hemoglobin molecules liberated from the red blood cells are broken down into subunits of *heme*, an iron-containing portion, and *globin*, a protein. The heme further decomposes into iron and a greenish pigment called **biliverdin.** The blood may transport the iron, combined with a protein, to the hemopoietic tissue in red bone marrow to be reused in synthesizing new hemoglobin. Otherwise, the liver stores iron in the form of an iron-protein complex. Biliverdin eventually is converted to an orange pigment called **bilirubin.** Biliverdin and bilirubin are excreted in the bile as bile pigments (see chapter 15, p. 409). Figure 12.5 summarizes the life cycle of a red blood cell.

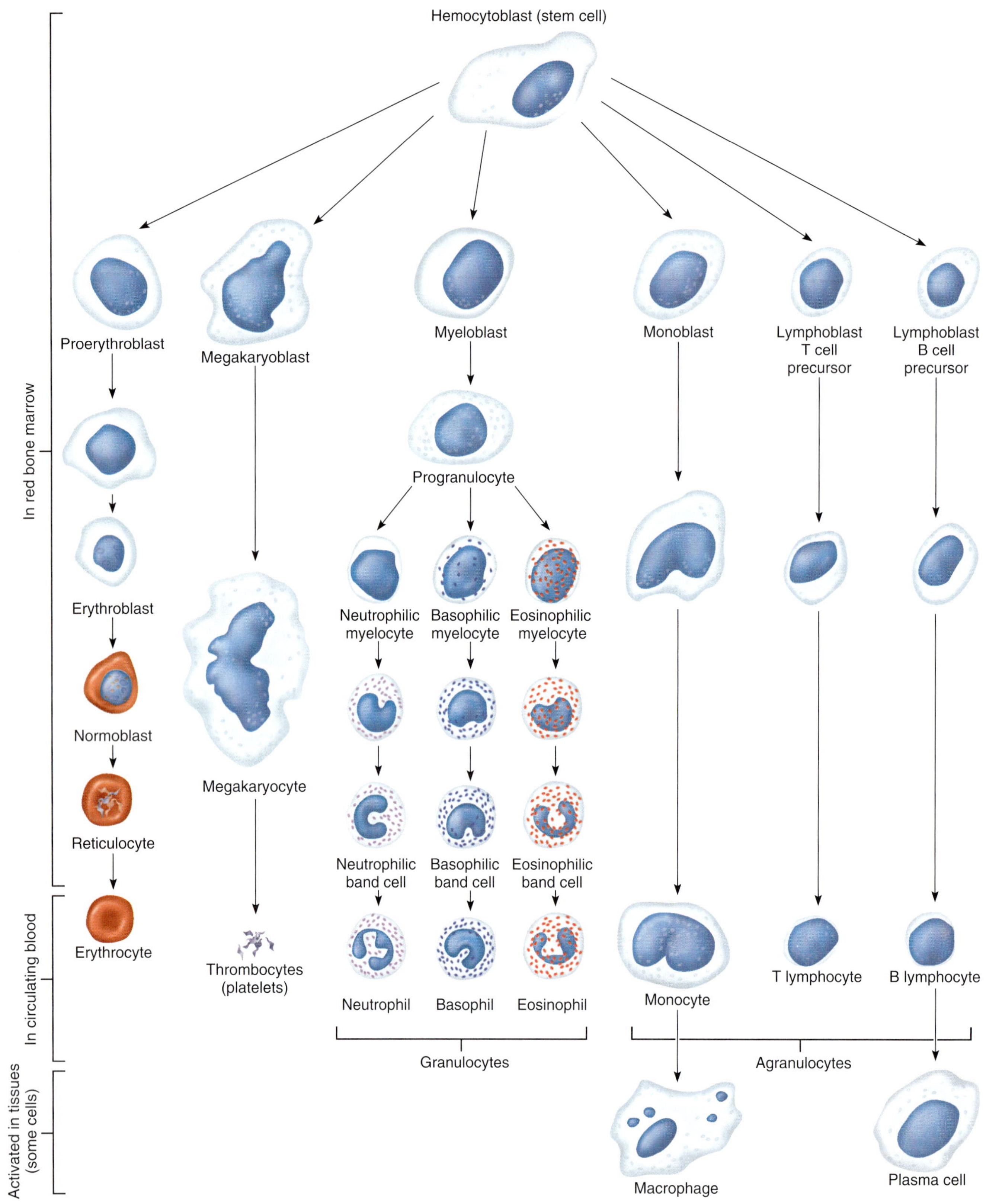

Figure 12.4
Origin and development of blood cells from hemocytoblasts (stem cells) in bone marrow.

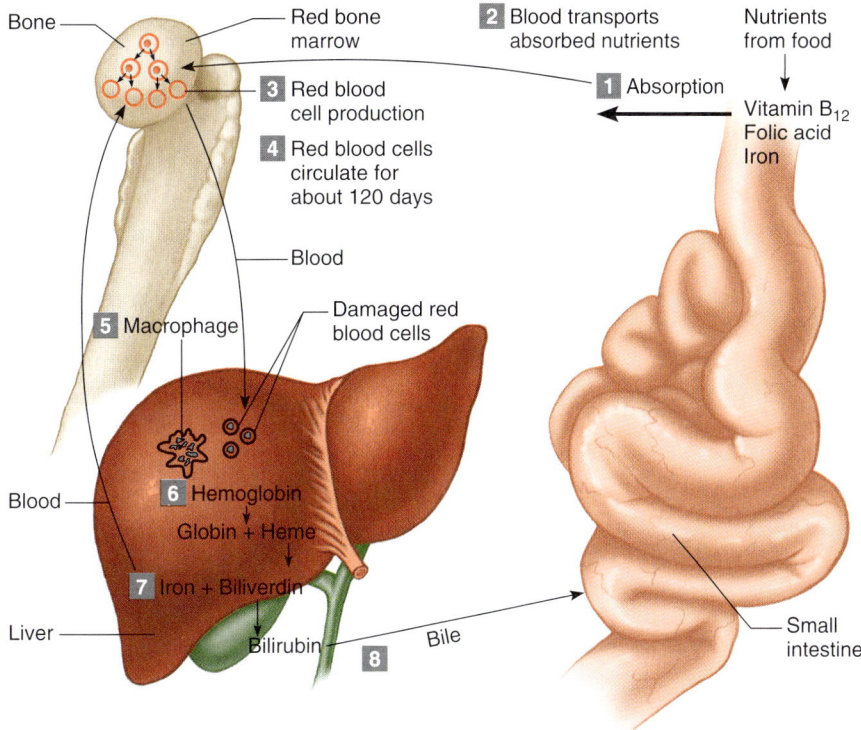

Figure 12.5
Life cycle of a red blood cell. (*1*) The small intestine absorbs essential nutrients. (*2*) Blood transports nutrients to red bone marrow. (*3*) In the marrow, red blood cells arise from the division of less specialized cells. (*4*) Mature red blood cells are released into the bloodstream, where they circulate for about 120 days. (*5*) Macrophages destroy damaged red blood cells in the spleen and liver. (*6*) Hemoglobin liberated from red blood cells is broken down into heme and globin. (*7*) Iron from heme returns to red bone marrow and is reused. (*8*) Biliverdin and bilirubin are excreted in bile.

Newborns can develop *physiologic jaundice* a few days after birth. In this condition and other forms of jaundice (icterus), accumulation of bilirubin turns the skin and eyes yellowish.

Physiologic jaundice may be the result of immature liver cells that ineffectively excrete bilirubin into the bile. Treatment includes exposure to fluorescent light, which breaks down bilirubin in the tissues, and feedings that promote bowel movements. In hospital nurseries, babies being treated for physiologic jaundice lie under "bili lights," clad only in diapers and protective goggles. The healing effect of fluorescent light was discovered in the 1950s, when an astute nurse noted that jaundiced babies improved after sun exposure, except in the areas their diapers covered.

CHECK YOUR RECALL

1. What happens to damaged red blood cells?
2. What are the products of hemoglobin breakdown?

White Blood Cells

White blood cells, or **leukocytes** (lu´ko-sītz), protect against disease. Leukocytes develop from *hemocyto-* *blasts* (see fig. 12.4) in response to hormones also. These hormones fall into two groups—**interleukins** and **colony-stimulating factors (CSFs).** Interleukins are numbered, while most colony-stimulating factors are named for the cell population they stimulate. Blood transports white blood cells to sites of infection. White blood cells may then leave the bloodstream as described later in this chapter.

Normally, five types of white blood cells are in circulating blood. They differ in size, the nature of their cytoplasm, the shape of the nucleus, and their staining characteristics. For example, leukocytes with granular cytoplasm are called **granulocytes,** whereas those without cytoplasmic granules are called **agranulocytes** (see fig. 12.4).

A typical granulocyte is about twice the size of a red blood cell. Members of this group include neutrophils, eosinophils, and basophils. Granulocytes develop in red bone marrow as do red blood cells, but have short life spans, averaging about 12 hours.

Neutrophils (nu´tro-filz) have fine cytoplasmic granules that stain light purple in neutral stain. The nucleus of a neutrophil is lobed and consists of two to five sections connected by thin strands of chromatin (fig. 12.6). Neutrophils account for 54–62% of the leukocytes in a typical blood sample from an adult.

Eosinophils (e˝o-sin´o-filz) contain coarse, uniformly sized cytoplasmic granules that stain deep red in acid stain (fig. 12.7). The nucleus usually has only two lobes (termed bilobed). These cells make up 1–3% of the total number of circulating leukocytes.

Basophils (ba´so-filz) are similar to eosinophils in size and in the shape of their nuclei, but they have fewer, more irregularly shaped cytoplasmic granules that stain deep blue in basic stain (fig. 12.8). Basophils usually account for less than 1% of the circulating leukocytes.

The leukocytes of the agranulocyte group include monocytes and lymphocytes. Monocytes generally arise from red bone marrow. Lymphocytes are formed in the organs of the lymphatic system, as well as in the red bone marrow (see chapter 14, p. 371).

Monocytes (mon´o-sītz), the largest blood cells, are two to three times greater in diameter than red blood cells (fig. 12.9). Their nuclei vary in shape and are round, kidney-shaped, oval, or lobed. They usually make up 3–9% of the leukocytes in a blood sample and live for several weeks or even months.

Lymphocytes (lim´fo-sītz) are usually only slightly larger than red blood cells. A typical lymphocyte contains a large, round nucleus surrounded by a thin rim of cytoplasm (fig. 12.10). These cells account for 25–33% of circulating leukocytes. They may live for years.

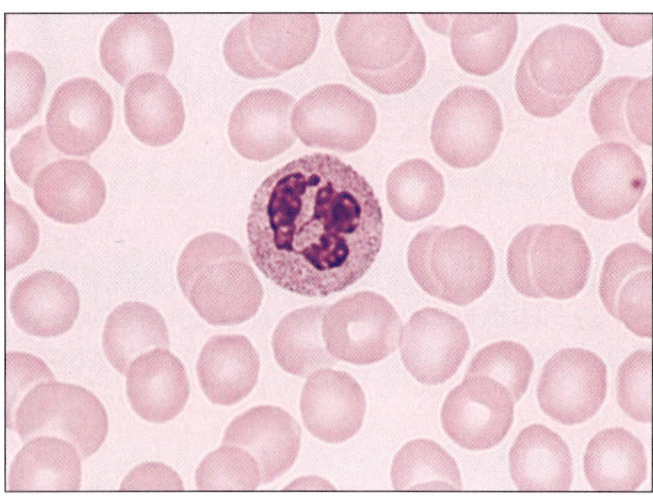

Figure 12.6
The neutrophil has a lobed nucleus with two to five components (1,060×).

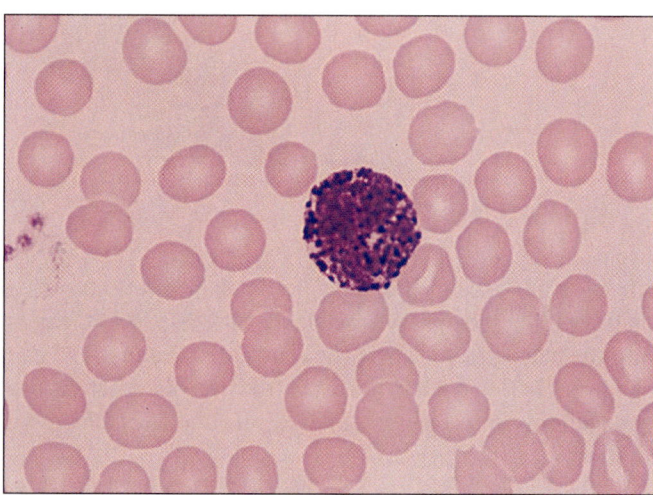

Figure 12.8
The basophil has cytoplasmic granules that stain deep blue (1,060×).

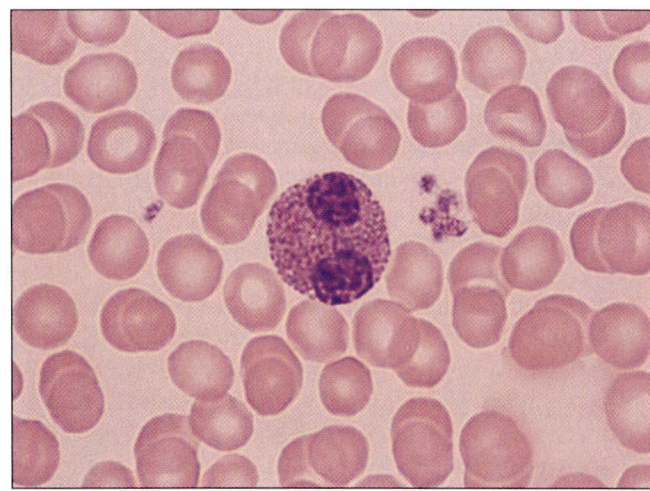

Figure 12.7
The eosinophil has red-staining cytoplasmic granules (1,060×).

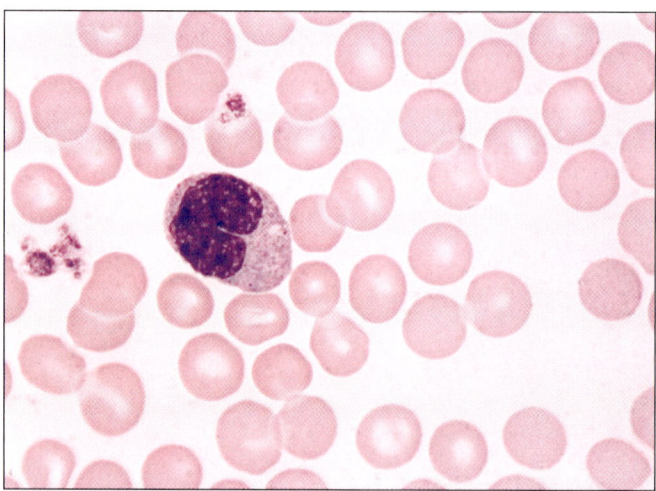

Figure 12.9
A monocyte may leave the bloodstream and become a macrophage (1,060×).

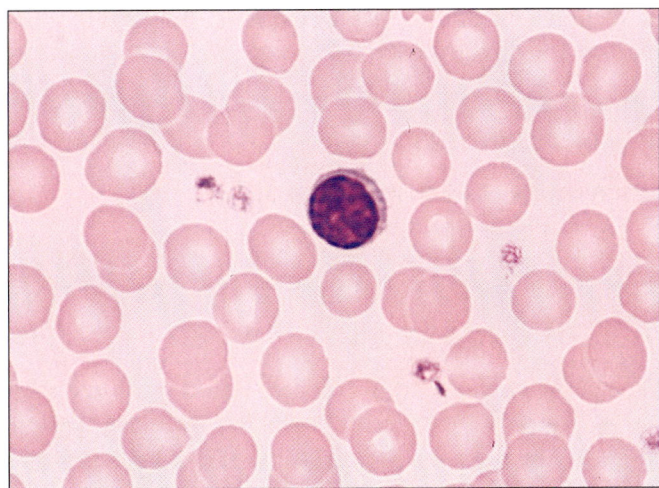

Figure 12.10
The lymphocyte contains a large, round nucleus (1,060×).

 CHECK YOUR RECALL

1. Which hormones are necessary for development of white blood cells from hemocytoblasts in the red bone marrow?
2. Distinguish between granulocytes and agranulocytes.
3. List the five types of white blood cells, and explain how they differ from one another.

Functions of White Blood Cells

White blood cells protect against infection in various ways. Some leukocytes phagocytize bacterial cells in the body, and others produce proteins (*antibodies*) that destroy or disable foreign particles.

Leukocytes can squeeze between the cells that form blood vessel walls. This movement, called *diapedesis*, allows the white blood cells to leave the circulation (fig. 12.11). Once outside the blood, they move through interstitial spaces using a form of self-propulsion called *amoeboid motion*.

The most mobile and active phagocytic leukocytes are neutrophils and monocytes. Neutrophils cannot ingest particles much larger than bacterial cells, but monocytes can engulf large objects. Both of these phagocytes contain many *lysosomes*, which are organelles filled with digestive enzymes that break down organic molecules in captured bacteria. Neutrophils and monocytes often become so engorged with digestive products and bacterial toxins that they also die.

Eosinophils are only weakly phagocytic, but they are attracted to and can kill certain parasites. Eosinophils also help control inflammation and allergic reactions by removing biochemicals associated with these reactions.

Some of the cytoplasmic granules of basophils contain a blood-clot-inhibiting substance called *heparin*, and other granules contain *histamine*. Basophils release

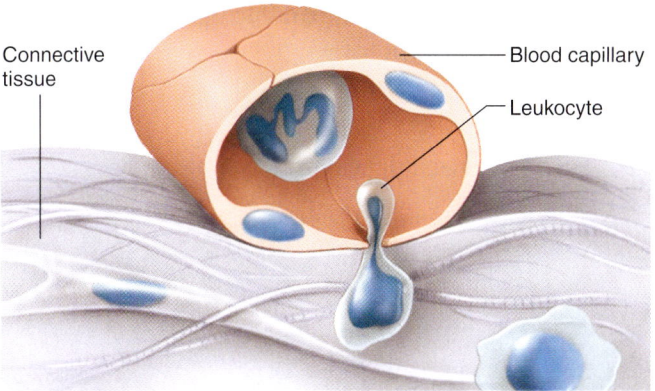

Figure 12.11
In a type of movement called diapedesis, leukocytes squeeze between the cells of a capillary wall and enter the tissue space outside the blood vessel.

heparin which helps prevent intravascular blood clot formation, and they release histamines which increase blood flow to injured tissues. Basophils also play major roles in certain allergic reactions.

Lymphocytes are important in *immunity*. Some, for example, produce antibodies that attack specific foreign substances that enter the body. Chapter 14 discusses immunity.

 CHECK YOUR RECALL

1. What are the primary functions of white blood cells?
2. Which white blood cells are the most active phagocytes?
3. What are the functions of eosinophils and basophils?

White Blood Cell Counts

The number of white blood cells in a cubic millimeter of human blood, called the *white blood cell count (WBCC)*, normally includes 5,000–10,000 cells. Because this number may change in response to abnormal conditions, white blood cell counts are of clinical interest. For example, a rise in the number of circulating white blood cells may indicate infection. A total number of white blood cells exceeding 10,000 per mm^3 of blood constitutes **leukocytosis,** indicating acute infection, such as appendicitis.

A total white blood cell count below 5,000 per mm^3 of blood is called **leukopenia.** Such a deficiency may accompany typhoid fever, influenza, measles, mumps, chicken pox, AIDS, or poliomyelitis.

A *differential white blood cell count (DIFF)* lists percentages of the types of leukocytes in a blood sample. This test is useful because the relative proportions of white blood cells may change in particular diseases. The number of neutrophils, for instance, usually increases during bacterial infections, and the number of eosinophils may increase during certain parasitic infections and allergic reactions. In AIDS, the number of a certain type of lymphocyte drops sharply.

CHECK YOUR RECALL

1. What is the normal human white blood cell count?
2. Distinguish between leukocytosis and leukopenia.
3. What is a differential white blood cell count?

Blood Platelets

Platelets (plāt´letz), or **thrombocytes,** are not complete cells. They arise from very large cells in red bone marrow, called **megakaryocytes,** that fragment like a shattered plate, releasing small sections of cytoplasm—the platelets—into the circulation. The larger fragments of the megakaryocytes shrink and become platelets as they pass through blood vessels in the lungs. Megakaryocytes and therefore platelets develop from *hemocytoblasts* (see fig. 12.4) in response to the hormone **thrombopoietin.**

Each platelet lacks a nucleus and is less than half the size of a red blood cell. It is capable of amoeboid movement and may circulate for about ten days. In normal blood, the *platelet count* varies from 130,000 to 360,000 per mm^3. Platelets help close breaks in damaged blood vessels and initiate formation of blood clots, as section 12.4 of this chapter explains. Table 12.1 summarizes the characteristics of blood cells and platelets.

CHECK YOUR RECALL

1. What is the normal blood platelet count?
2. What is the function of blood platelets?

12.3 Blood Plasma

Plasma is the clear, straw-colored, liquid portion of the blood in which the cells and platelets are suspended. It is approximately 92% water and contains a complex mixture of organic and inorganic biochemicals. Functions of plasma constituents include transporting nutrients, gases, and vitamins; helping regulate fluid and electrolyte balance; and maintaining a favorable pH.

Plasma Proteins

Plasma proteins are the most abundant of the dissolved substances (solutes) in plasma. These proteins remain in the blood and interstitial fluids, and ordinarily are not used as energy sources. The three main plasma protein groups—albumins, globulins, and fibrinogen—differ in chemical composition and physiological function.

Albumins (al-bu´minz) are the smallest of the plasma proteins, yet account for about 60% of these pro-

TABLE 12.1 — CELLULAR COMPONENTS OF BLOOD

COMPONENT	DESCRIPTION	NUMBER PRESENT	FUNCTION
Red blood cell (erythrocyte)	Biconcave disc without a nucleus; about one-third hemoglobin	4,200,000–6,200,000 per mm^3	Transports oxygen and carbon dioxide
White blood cell (leukocyte)		5,000–10,000 per mm^3	Destroys pathogenic microorganisms and parasites and removes worn cells
Granulocytes	About twice the size of red blood cells; cytoplasmic granules are present		
1. Neutrophil	Nucleus with two to five lobes; cytoplasmic granules stain light purple in neutral stain	54–62% of white blood cells present	Phagocytizes small particles
2. Eosinophil	Bilobed nucleus, cytoplasmic granules stain red in acid stain	1–3% of white blood cells present	Kills parasites and helps control inflammation and allergic reactions
3. Basophil	Bilobed nucleus, cytoplasmic granules stain blue in basic stain	Less than 1% of white blood cells present	Releases heparin and histamine
Agranulocytes	Cytoplasmic granules are absent		
1. Monocyte	Two to three times larger than a red blood cell; nuclear shape varies from spherical to lobed	3–9% of white blood cells present	Phagocytizes large particles
2. Lymphocyte	Only slightly larger than a red blood cell; its nucleus nearly fills cell	25–33% of white blood cells present	Provides immunity
Platelet (thrombocyte)	Cytoplasmic fragment	130,000–360,000 per mm^3	Helps control blood loss from broken vessels

Topic of Interest

LEUKEMIA

The young woman at first noticed fatigue and headaches, which she attributed to studying for final exams. She had frequent colds and bouts of fever, chills, and sweats that she thought were just minor infections. When she developed several bruises and bone pain and noticed that her blood did not clot very quickly after cuts and scrapes, she consulted her physician, who examined her and took a blood sample. One glance at a blood smear under a microscope alarmed the doctor—there were far too few red blood cells and platelets and too many white blood cells. She sent the sample to a laboratory to diagnose the type of *leukemia,* or cancer of the white blood cells, that was causing her patient's symptoms.

The young woman had *myeloid leukemia*. Her red bone marrow was producing too many granulocytes, but they were immature cells, unable to fight infection (fig. 12A). This explained the frequent illnesses. The leukemic cells were crowding out red blood cells and their precursors in the red marrow, causing anemia and resulting fatigue. Platelet deficiency (thrombocytopenia) led to an increased tendency to bleed. Finally, spread of the cancer cells outside the marrow painfully weakened the surrounding bone. Eventually, if the patient was not treated, the cancer cells would spread outside the cardiovascular system, causing other tissues that would normally not produce white blood cells to do so.

A second type of leukemia, distinguished by the source of the cancer cells, is *lymphoid leukemia*. These cancer cells are lymphocytes, produced in lymph nodes. Many of the symptoms are similar to those of myeloid leukemia. Sometimes a person has no leukemia symptoms at all, and a routine blood test detects the condition.

Leukemia is also classified as acute or chronic. An acute condition appears suddenly, symptoms progress rapidly, and without treatment, death occurs in a few months. Chronic forms begin more slowly and may remain undetected for months or even years or, in rare cases, decades. Without treatment, life expectancy after symptoms develop is about three years. With treatment, 50–80% of patients enter remission, a period of stability that may become a cure. Chemotherapy may be necessary for a year or two to increase the chances of long remission.

Leukemia treatment includes correcting the symptoms with blood transfusions, treating infections, and using drugs that kill cancer cells. Several drugs in use for many years have led to spectacular increases in cure rates, particularly for acute lymphoid leukemia in children. A new drug called Gleevec has had spectacular success in treating a type of chronic leukemia. If other treatments fail, a bone marrow transplant can cure leukemia, but the procedure is very risky. Stem cell transplants, using cells from donated umbilical cord blood, can also cure leukemia. Therefore, people with leukemia often have many treatment options.

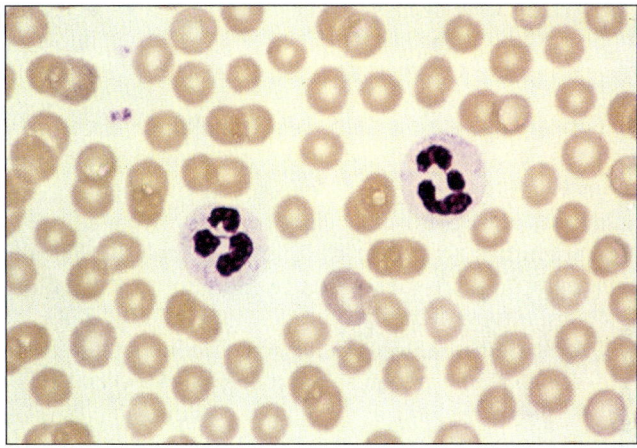

A

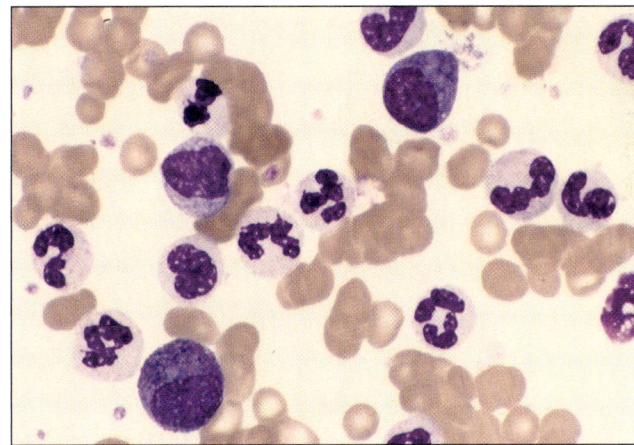

B

Figure 12A

Leukemia and blood cells. (A) Normal blood cells (700×). (B) Blood cells from a person with granulocytic leukemia, a type of myeloid leukemia (700×). Note the increased number of leukocytes.

teins by weight. They are synthesized in the liver, and because they are so plentiful, albumins are an important determinant of the *osmotic pressure* of the plasma.

Recall from chapter 3 (p. 60) that the presence of solute that cannot cross a selectively permeable membrane creates an osmotic pressure and that water always diffuses toward a greater osmotic pressure.

Because plasma proteins are too large to pass through the capillary walls, they create an osmotic pressure that tends to hold water in the capillaries, despite the fact that blood pressure tends to force water out of capillaries by filtration (see chapter 3, p. 61). The term *colloid osmotic pressure* is often used to describe this osmotic effect due to the plasma proteins.

By maintaining the colloid osmotic pressure of plasma, albumins and other plasma proteins help regulate water movement between the blood and the tissues. In doing so, they help control blood volume, which in turn is directly related to blood pressure (see chapter 13, p. 348).

If the concentration of plasma proteins falls, tissues swell—a condition called *edema*. This may result from starvation or a protein-deficient diet, either of which requires the body to use protein for energy, or from an impaired liver that cannot synthesize plasma proteins. As the concentration of plasma proteins drops, so does colloid osmotic pressure, sending fluids into intercellular spaces.

Globulins (glob´u-linz), which make up about 36% of the plasma proteins, can be further subdivided into *alpha, beta,* and *gamma globulins.* The liver synthesizes alpha and beta globulins. They have a variety of functions, including transport of lipids and fat-soluble vitamins. Lymphatic tissues produce the gamma globulins, which are a type of antibody (see chapter 14, p. 378).

Fibrinogen (fi-brin´o-jen), which constitutes about 4% of the plasma proteins, functions in blood coagulation, as discussed later in the chapter. Synthesized in the liver, fibrinogen is the largest of the plasma proteins. Table 12.2 summarizes the characteristics of the plasma proteins.

CHECK YOUR RECALL

1. List three types of plasma proteins.
2. How do albumins help maintain water balance between blood and tissues?
3. What are the functions of the globulins?
4. What is the role of fibrinogen?

TABLE 12.2 PLASMA PROTEINS

PROTEIN	PERCENTAGE OF TOTAL	ORIGIN	FUNCTION
Albumin	60%	Liver	Helps maintain colloid osmotic pressure
Globulin	36%		
Alpha globulins		Liver	Transport lipids and fat-soluble vitamins
Beta globulins		Liver	Transport lipids and fat-soluble vitamins
Gamma globulins		Lymphatic tissues	Constitute a type of antibody
Fibrinogen	4%	Liver	Blood coagulation

Gases and Nutrients

The most important *blood gases* are oxygen and carbon dioxide. Plasma also contains a considerable amount of dissolved nitrogen, which ordinarily has no physiological function. Chapter 16 (p. 451) discusses the blood gases and their transport.

The *plasma nutrients* include amino acids, simple sugars, nucleotides, and lipids absorbed from the digestive tract. For example, plasma transports glucose from the small intestine to the liver, where glucose can be stored as glycogen or converted to fat. If blood glucose concentration drops below the normal range, glycogen may be broken down into glucose, as chapter 11 (p. 296) describes. Plasma also carries recently absorbed amino acids to the liver, where they can be used to manufacture proteins, or deaminated and used as an energy source (see chapter 15, p. 422).

Plasma lipids include fats (triglycerides), phospholipids, and cholesterol. Because lipids are not water soluble and plasma is almost 92% water, these lipids combine with proteins in **lipoprotein** complexes. Lipoprotein molecules are relatively large and consist of a surface layer of phospholipid, cholesterol, and protein surrounding a triglyceride core. The protein constituents of lipoproteins in the outer layer, called *apoproteins,* can combine with receptors on the membranes of specific target cells. Lipoprotein molecules vary in the proportions of lipids they contain.

Because lipids are less dense than proteins, as the proportion of lipids in a lipoprotein increases, the density of the particle decreases. Conversely, as the proportion of lipids decreases, the density increases. Lipoproteins are classified on the basis of their densities, which reflect their composition. *Chylomicrons* mainly consist of triglycerides absorbed from the small intestine (see chapter 15, p. 414). *Very low-density lipoproteins (VLDL)* have a relatively high concentration of triglycerides. *Low-density lipoproteins (LDL)* have a relatively high concentration of cholesterol and are the major cholesterol-carrying lipoproteins. *High-density lipoproteins (HDL)* have a relatively high concentration of protein and a lower concentration of lipids. Chylomicrons transport dietary fats to muscle and adipose cells. Table 12.3 summarizes the characteristics and functions of lipoproteins.

Nonprotein Nitrogenous Substances

Molecules that contain nitrogen atoms but are not proteins comprise a group called **nonprotein nitrogenous substances.** In plasma, this group includes amino acids, urea, and uric acid. Amino acids come from protein digestion and amino acid absorption. Urea and uric acid are products of protein and nucleic acid catabolism, respectively, and are excreted in the urine.

TABLE 12.3 PLASMA LIPOPROTEINS

LIPOPROTEIN	CHARACTERISTICS	FUNCTIONS
Chylomicron	High concentration of triglycerides	Transports dietary fats to muscle and adipose cells
Very low-density lipoprotein (VLDL)	Relatively high concentration of triglycerides; produced in the liver	Transports triglycerides from the liver to adipose cells
Low-density lipoprotein (LDL)	Relatively high concentration of cholesterol; formed from remnants of VLDL molecules that have given up their triglycerides	Delivers cholesterol to various cells, including liver cells
High-density lipoprotein (HDL)	Relatively high concentration of protein and low concentration of lipids	Transports to the liver remnants of chylomicrons that have given up their triglycerides

Plasma Electrolytes

Blood plasma contains a variety of *electrolytes* that are absorbed from the intestine or released as by-products of cellular metabolism. They include sodium, potassium, calcium, magnesium, chloride, bicarbonate, phosphate, and sulfate ions. Of these, sodium and chloride ions are the most abundant. Bicarbonate ions are important in maintaining the osmotic pressure and pH of plasma, and like other plasma constituents, they are regulated so that their blood concentrations remain relatively stable. Chapter 18 (p. 482) discusses these electrolytes in connection with water and electrolyte balance.

CHECK YOUR RECALL

1. Which gases are in plasma?
2. Which nutrients are in plasma?
3. What is a nonprotein nitrogenous substance?
4. What are the sources of plasma electrolytes?

12.4 Hemostasis

Hemostasis (he″mo-sta′sis) is the stoppage of bleeding, which is vitally important when blood vessels are damaged. Following an injury to the blood vessels, several actions may help limit or prevent blood loss, including blood vessel spasm, platelet plug formation, and blood coagulation.

Blood Vessel Spasm

Cutting or breaking a smaller blood vessel stimulates the smooth muscles in its walls to contract, and blood loss lessens almost immediately. In fact, the ends of a severed vessel may close completely by such a **vasospasm.**

Vasospasm may last only a few minutes, but the effect of the direct stimulation usually continues for about 30 minutes. By then, a *platelet plug* has formed, and blood is coagulating. Also, platelets release **serotonin,** which contracts smooth muscles in the blood vessel walls. This vasoconstriction further helps reduce blood loss.

Platelet Plug Formation

Platelets adhere to any rough surface and to the collagen in connective tissue. Consequently, when a blood vessel breaks, platelets adhere to the collagen underlying the endothelial lining of blood vessels. Platelets also adhere to each other, forming a platelet plug in the vascular break. A plug may control blood loss from a small break, but a larger break may require a blood clot to halt bleeding. Figure 12.12 shows the steps in platelet plug formation.

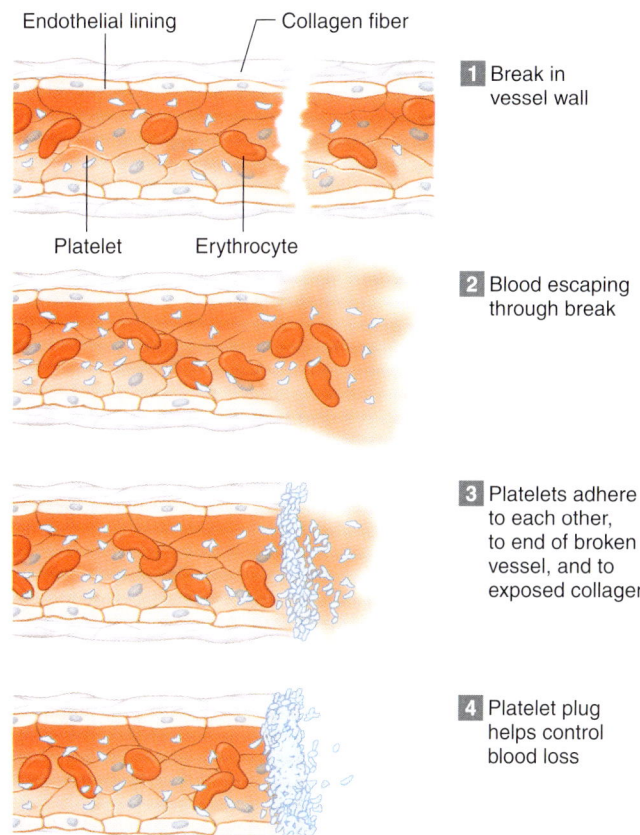

Figure 12.12
Steps in platelet plug formation.

CHECK YOUR RECALL

1. What is hemostasis?
2. How does a blood vessel spasm help control bleeding?
3. Describe the formation of a platelet plug.

Blood Coagulation

Coagulation (ko-ag´´u-la´shun), the most effective hemostatic mechanism, causes formation of a *blood clot*. Blood coagulation is complex and utilizes many biochemicals called *clotting factors*. Some of these factors promote coagulation, and others inhibit it. Whether or not blood coagulates depends on the balance between these two groups of factors. Normally, anticoagulants prevail, and the blood does not clot. However, as a result of injury (trauma), biochemicals that favor coagulation may increase in concentration, and the blood may coagulate.

The major event in blood clot formation is the conversion of the soluble plasma protein fibrinogen into insoluble threads of the protein **fibrin.** Damaged tissues release *tissue thromboplastin,* initiating a series of reactions that results in the production of *prothrombin activator.* This series of changes depends on the presence of calcium ions as well as certain proteins and phospholipids for completion.

Prothrombin is an alpha globulin that the liver continually produces and is thus a normal constituent of plasma. In the presence of calcium ions, prothrombin activator converts prothrombin into **thrombin.** Thrombin, in turn, catalyzes a reaction that fragments fibrinogen. The fibrinogen pieces join, forming long threads of fibrin. Certain other proteins also enhance fibrin formation.

Once fibrin threads form, they stick to the exposed surfaces of damaged blood vessels, creating a meshwork that entraps blood cells and platelets (fig. 12.13). The resulting mass is a blood clot, which may block a vascular break and prevent further blood loss. The clear, yellow liquid that remains after the clot forms is called *serum*. Serum is the same as plasma, minus clotting factors.

The amount of prothrombin activator that appears in the blood is directly proportional to the degree of tissue damage. Once a blood clot begins to form, it promotes more clotting because thrombin also acts directly on blood clotting factors other than fibrinogen, causing prothrombin to form more thrombin. This is an example of a **positive feedback system,** in which the original action stimulates more of the same type of action. Such a positive feedback mechanism produces unstable conditions and can operate for only a short time in a living system because life depends on the maintenance of a stable internal environment (see chapter 1, p. 5).

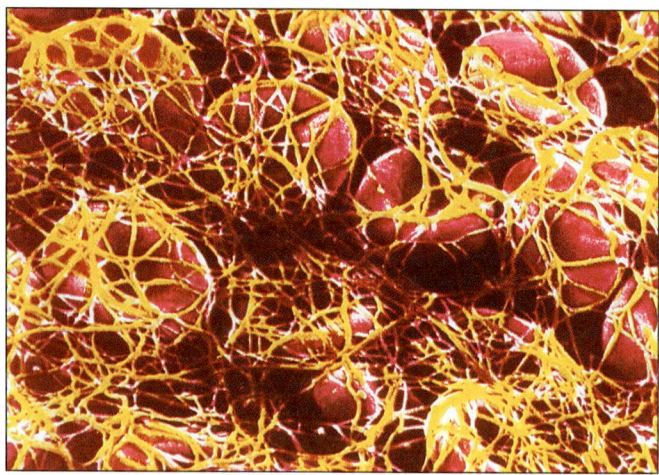

Figure 12.13
A scanning electron micrograph of fibrin threads (2,700×).

Laboratory tests commonly used to evaluate the blood coagulation mechanisms include *prothrombin time (PT)* and *partial thromboplastin time (PTT)*. Both of these tests measure the time it takes for fibrin threads to form in a sample of plasma.

Normally, blood flow throughout the body prevents the formation of a massive clot within the cardiovascular system by rapidly carrying excess thrombin away and keeping its concentration too low to enhance further clotting. Consequently, blood coagulation is usually limited to blood that is standing still (or moving relatively slowly), and clotting ceases where a clot contacts circulating blood.

Fibroblasts (see chapter 5, p. 99) invade blood clots that form in ruptured vessels, producing connective tissue with numerous fibers throughout the clots, which helps strengthen and seal vascular breaks. Many clots, including those that form in tissues as a result of blood leakage (hematomas), disappear in time. This dissolution depends on activation of a plasma protein that can digest fibrin threads and other proteins associated with clots. Clots that fill large blood vessels, however, are seldom removed naturally.

A blood clot abnormally forming in a vessel is a **thrombus** (throm´bus). If the clot dislodges or if a fragment of it breaks loose and is carried away by the blood flow, it is called an **embolus** (em´bo-lus). Generally, emboli continue to move until they reach narrow places in vessels, where they may lodge and block blood flow.

Abnormal clot formations are often associated with conditions that change the endothelial linings of vessels. For example, in *atherosclerosis,* accumulations of fatty deposits change the arterial linings, sometimes initiating inappropriate clotting (fig. 12.14). Figure 12.15 summarizes the three primary hemostatic mechanisms.

Genetics Connection

COAGULATION DISORDERS

Hemophilia

In 1962, five-year-old Bob Massie developed uncontrollable bleeding in his left knee, a symptom of his *hemophilia A*, an inherited clotting disorder. It took thirty transfusions of plasma over the next three months to stop the bleeding. Because the knee joint had swelled and locked into place during that time, Bob was unable to walk for the next seven years. Today, Bob still suffers from painful joint bleeds, but he injects himself with factor VIII, the coagulation protein his body cannot make. Factor VIII controls his bleeding.

Hemophilia has left its mark on history. One of the earliest descriptions is in the Talmud, a second century B.C. Jewish document, which reads, "If she circumcised her first child and he died, and a second one also died, she must not circumcise her third child." England's Queen Victoria (1819–1901) passed the hemophilia gene to several of her children, eventually spreading the condition to the royal families of Russia, Germany, and Spain. Hemophilia achieved notoriety when factor VIII pooled from blood donations was discovered to transmit HIV in 1985. Ninety percent of people with severe hemophilia who used such pooled factor VIII in the few years prior to that time have developed AIDS.

Abnormalities of different clotting factors cause different forms of hemophilia, but hemophilia A is the most common. Symptoms of the hemophilias include severe hemorrhage following minor injuries, frequent nosebleeds, large intramuscular hematomas, and blood in the urine.

von Willebrand Disease

The tendency to bleed and bruise easily is a sign of *von Willebrand disease*, another inherited clotting disorder that is usually far less severe than hemophilia. Affected persons lack a plasma protein, von Willebrand factor, that is secreted by endothelial cells lining blood vessels and enables platelets to adhere to damaged blood vessel walls, a key step preceding actual clotting. Sometimes, the condition can cause spontaneous bleeding from the mucous membranes of the gastrointestinal and urinary tracts.

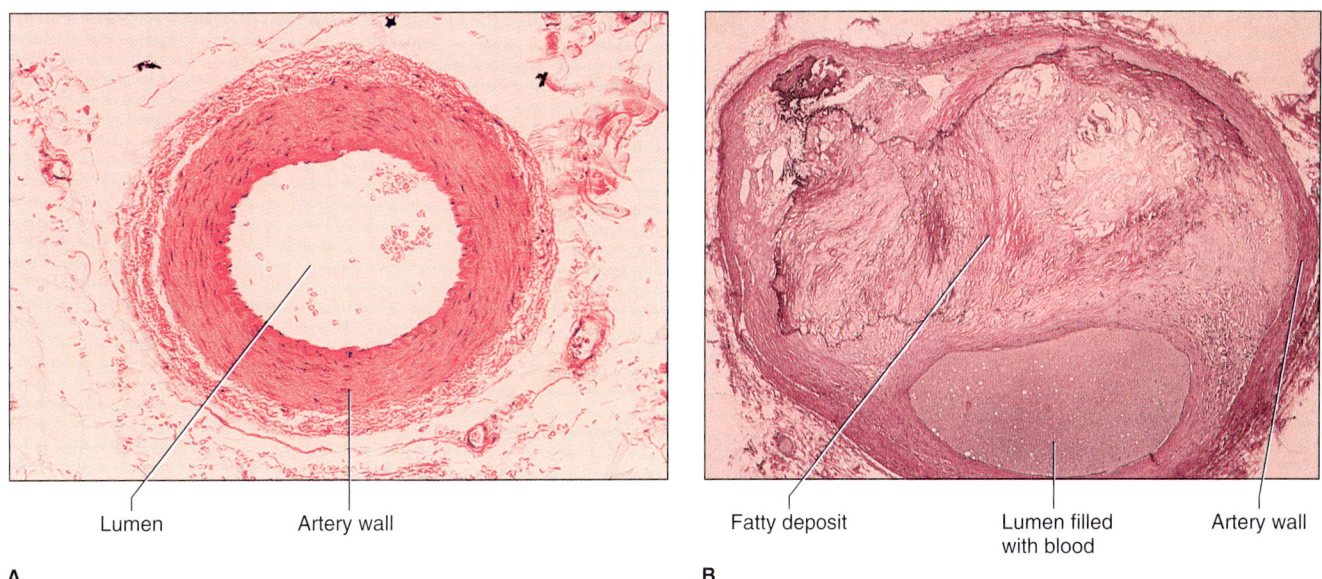

Figure 12.14
Artery cross sections. (*A*) Light micrograph of a normal artery (90×). (*B*) The inner wall of this artery changed as a result of atherosclerosis (55×).

A blood clot forming in a vessel that supplies a vital organ, such as the heart (coronary thrombosis) or the brain (cerebral thrombosis), kills tissues the vessel serves (*infarction*) and may be fatal. A blood clot that travels and then blocks a vessel that supplies a vital organ, such as the lungs (pulmonary embolism), affects the portion of the organ the blocked blood vessel supplies. Tissue plasminogen activators (tPA) break up abnormal blood clots and are used to treat heart attacks and strokes.

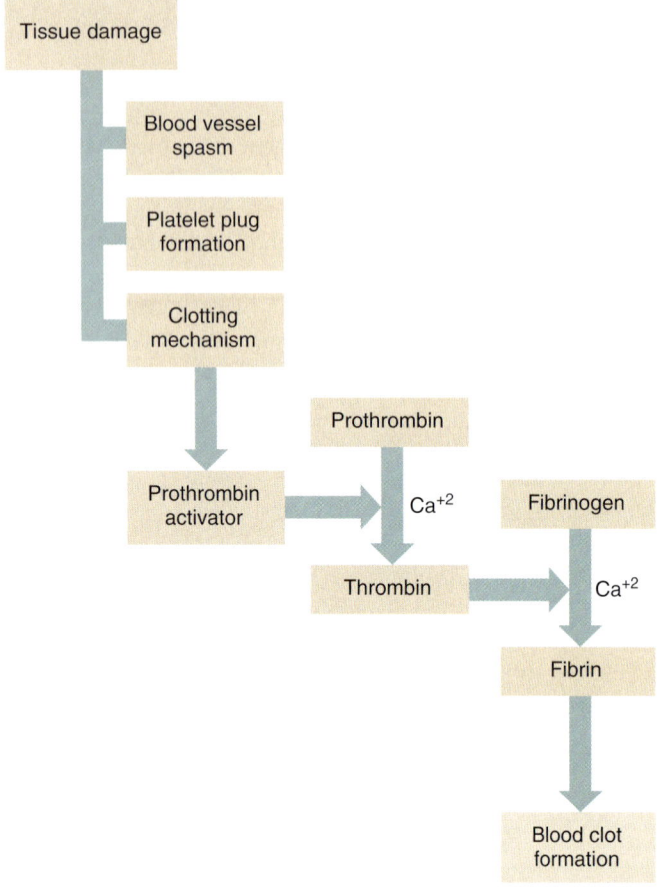

Figure 12.15
Blood vessel spasm, platelet plug formation, and blood coagulation provide hemostasis following tissue damage.

CHECK YOUR RECALL

1. Review the major steps in blood clot formation.
2. What prevents the formation of massive clots throughout the cardiovascular system?
3. Distinguish between a thrombus and an embolus.

12.5 Blood Groups and Transfusions

Early attempts to transfer blood from one person to another produced varied results. Sometimes, the recipient improved. Other times, the recipient suffered a blood reaction in which the red blood cells clumped, obstructing vessels and producing great pain and organ damage.

Eventually, scientists determined that blood is of differing types and that only certain combinations of blood types are compatible. These discoveries led to the development of procedures for typing blood. Today, safe transfusions of whole blood depend on properly matching the blood types of donors and recipients.

Antigens and Antibodies

Agglutination is the clumping of red blood cells following a transfusion reaction. This phenomenon is due to a reaction between red blood cell surface molecules called **antigens** (an´ti-jenz), formerly called *agglutinogens,* and protein **antibodies** (an´ti-bod´´ēz), formerly called *agglutinins,* carried in plasma.

Only a few of the many antigens on red blood cell membranes can produce serious transfusion reactions. These include the antigens of the ABO group and those of the Rh group. Avoiding the mixture of certain kinds of antigens and antibodies prevents adverse transfusion reactions.

A mismatched blood tranfusion quickly produces telltale signs of agglutination—anxiety, breathing difficulty, facial flushing, headache, and severe pain in the neck, chest, and lumbar area. Red blood cells burst, releasing free hemoglobin. Macrophages phagocytize the hemoglobin, converting it to bilirubin, which may sufficiently accumulate to cause the yellow skin of jaundice. Free hemoglobin in the kidneys may ultimately cause them to fail.

ABO Blood Group

The *ABO blood group* is based on the presence (or absence) of two major antigens on red blood cell membranes—antigen A and antigen B. A person's erythrocytes have on their surfaces one of four antigen combinations as a result of inheritance: only A, only B, both A and B, or neither A nor B.

A person with only antigen A has *type A blood.* A person with only antigen B has *type B blood.* An individual with both antigen A and B has *type AB blood.* A person with neither antigen A nor B has *type O blood.* Thus, all humans have one of four possible ABO blood types—A, B, AB, or O.

 In the U.S., the most common ABO blood types are O (47%) and A (41%). Rarer are type B (9%) and type AB (3%).

Certain antibodies are synthesized in the plasma about two to eight months following birth. Specifically, whenever antigen A is absent in red blood cells, an antibody called *anti-A* is produced, and whenever antigen B is absent, an antibody called *anti-B* is produced. Therefore, persons with type A blood have antibody anti-B in their plasma; those with type B blood have antibody anti-A; those with type AB blood have neither antibody; and those with type O blood have both antibody anti-A and antibody anti-B (fig. 12.16 and table

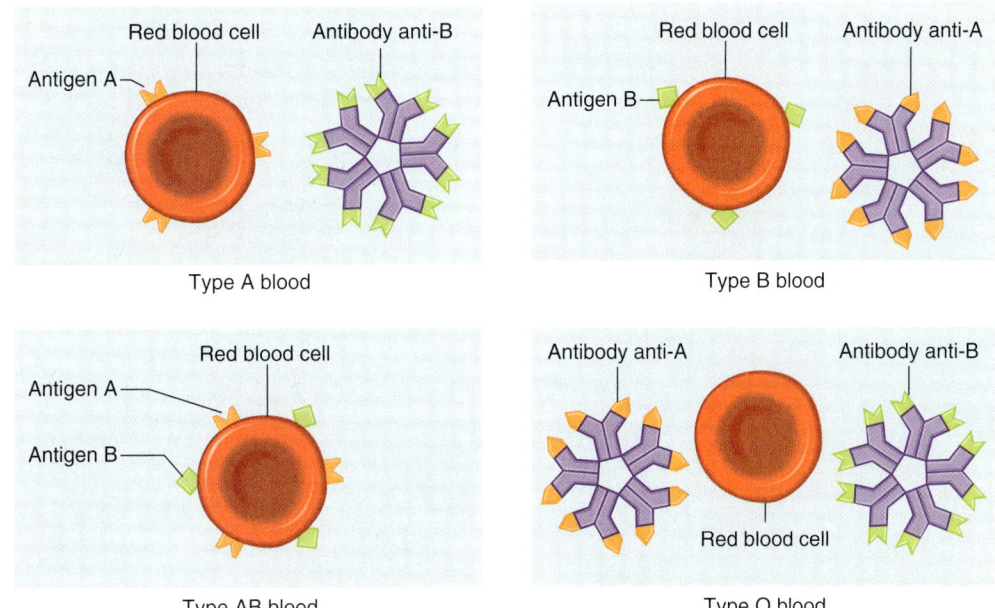

Figure 12.16
Different combinations of antigens and antibodies distinguish blood types. Cells and antibodies not drawn to scale.

TABLE 12.4		ANTIGENS AND ANTIBODIES OF THE ABO BLOOD GROUP
BLOOD TYPE	ANTIGEN	ANTIBODY
A	A	Anti-B
B	B	Anti-A
AB	A and B	Neither anti-A nor anti-B
O	Neither A nor B	Both anti-A and anti-B

12.4). The antibodies anti-A and anti-B are large and do not cross the placenta. Thus, a pregnant woman and her fetus may be of different ABO blood types, and agglutination in the fetus does not occur.

An antibody of one type will react with an antigen of the same type and clump red blood cells; therefore, such combinations must be avoided. The major concern in blood transfusion procedures is that the cells in the donated blood not clump due to antibodies in the recipient's plasma. For this reason, a person with type A (anti-B) blood must not receive blood of type B or AB, either of which would clump in the presence of anti-B in the recipient's type A blood. Likewise, a person with type B (anti-A) blood must not receive type A or AB blood, and a person with type O (anti-A and anti-B) blood must not receive type A, B, or AB blood (fig. 12.17).

Because type AB blood lacks both anti-A and anti-B antibodies, an AB person can receive a transfusion of blood of any type. For this reason, type AB persons are sometimes called *universal recipients*. However, type A (anti-B) blood, type B (anti-A) blood, and type O (anti-A and anti-B) blood still contain antibodies (either anti-A and/or anti-B) that could agglutinate type AB cells if transfused rapidly. Consequently, even for AB individuals, using donor blood of the same type as the recipient is best (table 12.5).

Because type O blood lacks antigens A and B, this type could theoretically be transfused into persons with blood of any other type. Therefore, persons with type O blood are sometimes called *universal donors*. Type O blood, however, does contain both anti-A and anti-B antibodies. If type O blood is given to a person with blood type A, B, or AB, it should be transfused slowly so that the recipient's larger blood volume will dilute it, minimizing the chance of an adverse reaction.

TABLE 12.5		PREFERRED AND PERMISSIBLE BLOOD TYPES FOR TRANSFUSIONS
BLOOD TYPE OF RECIPIENT	PREFERRED BLOOD TYPE OF DONOR	PERMISSIBLE BLOOD TYPE OF DONOR (IN EXTREME EMERGENCY)
A	A	A, O
B	B	B, O
AB	AB	AB, A, B, O
O	O	O

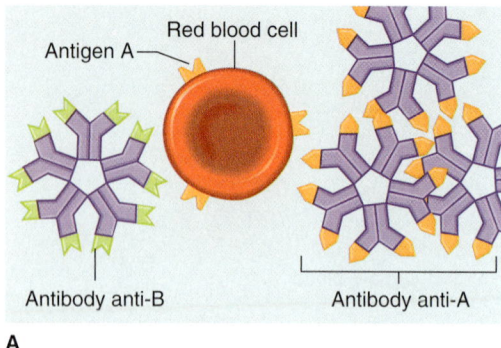
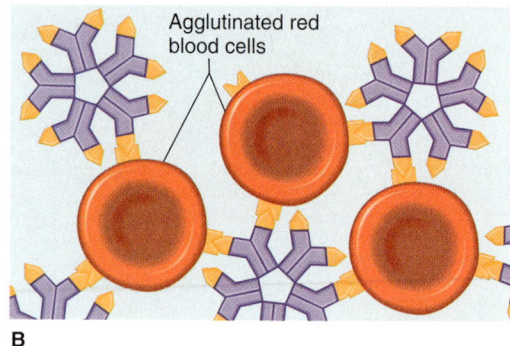
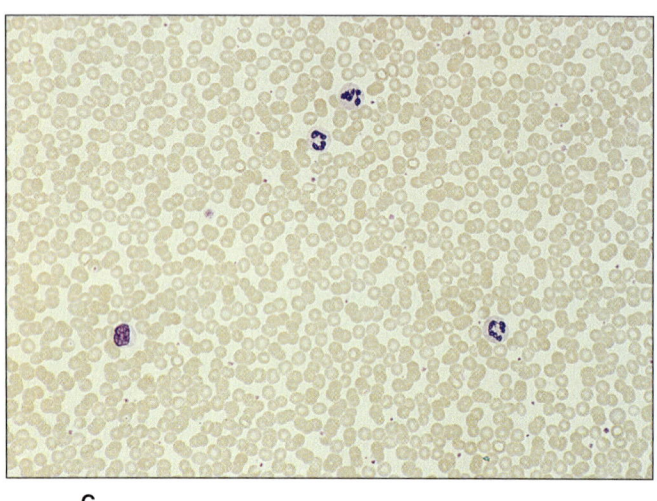
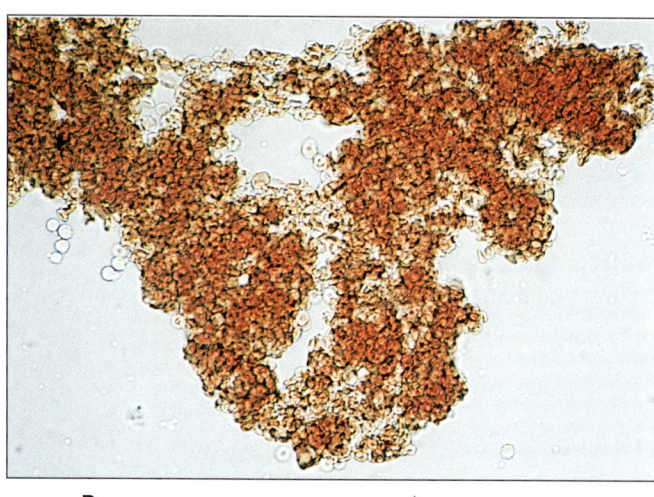

Figure 12.17
Agglutination. (*A*) If red blood cells with antigen A are added to blood containing antibody anti-A, (*B*) the antibodies react with the antigens, causing clumping (agglutination). (*C*) Nonagglutinated blood (210×). (*D*) Agglutinated blood (220×).

CHECK YOUR RECALL

1. Distinguish between antigens and antibodies.
2. What is the main concern when blood is transfused from one individual to another?
3. Why is a type AB person called a universal recipient?
4. Why is a type O person called a universal donor?

Rh Blood Group

The *Rh blood group* was named after the rhesus monkey in which it was first studied. In humans, this group includes several Rh antigens (factors). The most important of these is *antigen D;* however, if any of the antigen D and other Rh antigens are present on the red blood cell membranes, the blood is said to be *Rh-positive.* Conversely, if the red blood cells lack Rh antigens, the blood is called *Rh-negative.*

 Only 15% of the U.S. population is Rh-negative. Therefore, AB⁻ blood is the rarest type, and O⁺ the most common.

As in the case of antigens A and B, the presence (or absence) of Rh antigens is an inherited trait. Unlike anti-A and anti-B, antibodies for Rh (*anti-Rh*) do not appear spontaneously. Instead, they form only in Rh-negative persons in response to special stimulation.

If an Rh-negative person receives a transfusion of Rh-positive blood, the Rh antigens stimulate the recipient's antibody-producing cells to begin producing anti-Rh antibodies. Generally, this initial transfusion has no serious consequences, but if the Rh-negative person—who is now sensitized to Rh-positive blood—receives another transfusion of Rh-positive blood some months later, the donated red cells are likely to agglutinate.

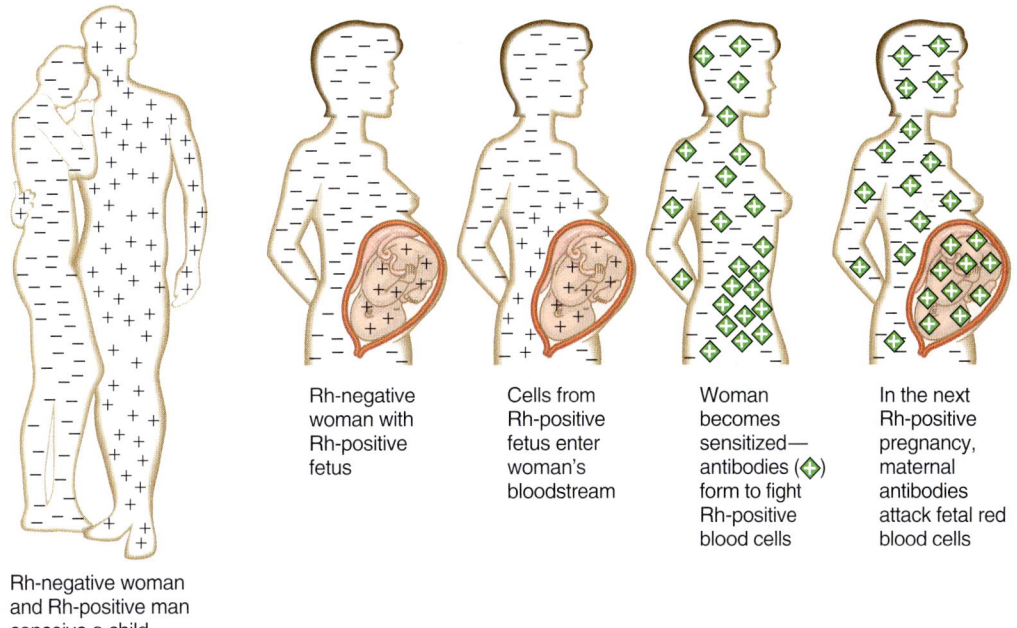

Figure 12.18
If a man who is Rh positive and a woman who is Rh negative conceive a child who is Rh positive, the woman's body may manufacture antibodies that attack future Rh-positive offspring.

A related condition may occur when an Rh-negative woman is pregnant with an Rh-positive fetus for the first time. Such a pregnancy may be uneventful; however, at birth (or if a miscarriage occurs), the placental membranes that separated the maternal blood from the fetal blood during the pregnancy tear, and some of the infant's Rh-positive blood cells may enter the maternal circulation. These Rh-positive cells may then stimulate the maternal tissues to begin producing anti-Rh antibodies.

If a woman who has already developed anti-Rh antibodies becomes pregnant with a second Rh-positive fetus, these anti-Rh antibodies, called hemolysins, cross the placental membrane and destroy the fetal red blood cells (fig. 12.18). The fetus then develops a condition called **erythroblastosis fetalis** (hemolytic disease of the newborn).

Erythroblastosis fetalis is extremely rare today because physicians carefully track Rh status. An Rh-negative woman who might carry an Rh-positive fetus is given an injection of a drug called RhoGAM. This injection is actually composed of anti-Rh antibodies, which bind to and shield any Rh-positive fetal cells that might contact the woman's cells, sensitizing her immune system. RhoGAM must be given within 72 hours of possible contact with Rh-positive cells—including giving birth, terminating a pregnancy, miscarrying, or undergoing amniocentesis (a prenatal test in which a needle is inserted into the uterus).

CHECK YOUR RECALL

1. What is the Rh blood group?
2. What are two ways that Rh incompatibility can arise?

Clinical Terms Related to the Blood

anisocytosis (an-i´´so-si-to´sis) Abnormal variation in the size of erythrocytes.

antihemophilic plasma (an´´ti-he´´mo-fil´ik plaz´mah) Normal blood plasma that has been processed to preserve an antihemophilic factor.

citrated whole blood (sit´rāt-ed hōl blud) Normal blood to which a solution of acid citrate has been added to prevent coagulation.

dried plasma (drīd plaz´mah) Normal blood plasma that has been vacuum-dried to prevent the growth of microorganisms.

hemorrhagic telangiectasia (hem´´o-raj´ik tel-an´´je-ek-ta´ze-ah) Inherited tendency to bleed from localized lesions of the capillaries.

heparinized whole blood (hep´er-ĭ-nīzd´´ hōl blud) Normal blood to which a solution of heparin has been added to prevent coagulation.

macrocytosis (mak´´ro-si-to´sis) Abnormally large erythrocytes.

microcytosis (mi´´kro-si-to´sis) Abnormally small erythrocytes.

neutrophilia (nu″tro-fil′e-ah) Increase in the number of circulating neutrophils.

packed red cells Concentrated suspension of red blood cells from which the plasma has been removed.

pancytopenia (pan″si-to-pe′ne-ah) Abnormal depression of all the cellular components of blood.

poikilocytosis (poi″kĭ-lo-si-to′sis) Irregularly shaped erythrocytes.

purpura (per′pu-rah) Spontaneous bleeding into the tissues and through the mucous membranes.

septicemia (sep″ti-se′me-ah) Presence of disease-causing microorganisms or their toxins in the blood.

spherocytosis (sfēr″o-si-to′sis) Hemolytic anemia caused by defective proteins supporting the cell membranes of red blood cells. The cells are abnormally spherical.

thalassemia (thal″ah-se′me-ah) Group of hereditary hemolytic anemias resulting from very thin, fragile erythrocytes.

Clinical Connection

Thrombotic thrombocytopenic purpura (TTP) was first described in 1924, but the underlying cause was not discovered until 2001. In this rare disorder, platelets adhere to abnormally large clumps of the plasma protein von Willebrand factor. When the clumps lodge in narrow blood vessels in major organs, symptoms begin, usually in young adulthood. Neurological symptoms include headache, confusion, changes in speech and altered consciousness. The kidneys may fail. The deficiency of platelets causes bleeding beneath the skin, which leads to characteristic red bruises. Anemia results from the shattering of red blood cells, which overtaxes the spleen. Treatment, which is 80 to 90 percent effective, includes removal of the spleen, and cleansing the plasma using a technique called plasmaphoresis.

A hereditary from of TTP results from absence or malfunction of an enzyme that normally cuts von Willebrand factor protein aggregates down to size. It is likely that noninherited forms of the illness also affect the size of these plasma protein aggregates. The missing enzyme is very similar, in structure and apparently in function, to a component of snake venom that causes bleeding in a bite victim.

SUMMARY OUTLINE

12.1 Introduction (p. 307)
Blood is a type of connective tissue whose cells are suspended in a liquid intercellular material. It transports substances between body cells and the external environment, and helps maintain a stable internal environment.

12.2 Blood and Blood Cells (p. 307)
Blood contains red blood cells, white blood cells, and platelets.
1. Blood volume and composition
    a. Blood volume varies with body size, fluid and electrolyte balance, and adipose tissue content.
    b. Blood can be separated into formed elements and liquid portions.
        (1) The formed elements portion is mostly red blood cells.
        (2) The liquid plasma includes water, gases, nutrients, hormones, electrolytes, and cellular wastes.
2. Characteristics of red blood cells
    a. Red blood cells are biconcave discs with shapes that increase surface area.
    b. Red blood cells contain hemoglobin, which combines with oxygen.
3. Red blood cell counts
    a. The red blood cell count equals the number of cells per cubic millimeter (mm^3) of blood.
    b. The average count ranges from approximately 4 to 6 million cells per mm^3 of blood.
    c. Red blood cell count is related to the oxygen-carrying capacity of the blood, which is used to diagnose and evaluate the courses of diseases.
4. Red blood cell production and its control
    a. Red bone marrow produces red blood cells.
    b. The number of red blood cells remains relatively stable.
    c. A negative feedback mechanism utilizing erythropoietin controls the rate of red blood cell production.
5. Dietary factors affecting red blood cell production
    a. Availability of vitamin B_{12} and folic acid influences red blood cell production.
    b. Hemoglobin synthesis requires iron.
6. Destruction of red blood cells
    a. Macrophages in the liver and spleen phagocytize damaged red blood cells.
    b. Hemoglobin molecules decompose, and the iron they contain is recycled.
    c. Hemoglobin releases biliverdin and bilirubin pigments.
7. Types of white blood cells
    a. White blood cells develop from hemocytoblasts, in response to interleukins and colony-stimulating factors, to protect against disease.
    b. Granulocytes include neutrophils, eosinophils, and basophils.
    c. Agranulocytes include monocytes and lymphocytes.
8. Functions of white blood cells
    a. Neutrophils and monocytes phagocytize foreign particles.
    b. Eosinophils kill parasites and help control inflammation and allergic reactions.
    c. Basophils release heparin, which inhibits blood clotting, and histamine to increase blood flow to injured tissues.
    d. Lymphocytes produce antibodies that attack specific foreign substances.

9. White blood cell counts
   a. Normal total white blood cell counts vary from 5,000 to 10,000 cells per mm^3 of blood.
   b. The number of white blood cells may change in response to abnormal conditions, such as infections, emotional disturbances, or excessive loss of body fluids.
   c. A differential white blood cell count indicates the percentages of various types of leukocytes present.
10. Blood platelets
    a. Blood platelets, which develop in the red bone marrow in response to thrombopoietin, are fragments of giant cells.
    b. The normal platelet count varies from 130,000 to 360,000 platelets per mm^3 of blood.
    c. Platelets help close breaks in blood vessels.

12.3 Plasma (p. 314)

Plasma transports gases and nutrients, helps regulate fluid and electrolyte balance, and helps maintain stable pH.

1. Plasma proteins
   a. Plasma proteins remain in blood and interstitial fluids, and are not normally used as energy sources.
   b. Three major groups exist.
      (1) Albumins help maintain the colloid osmotic pressure.
      (2) Globulins include antibodies. They provide immunity and transport lipids and fat-soluble vitamins.
      (3) Fibrinogen functions in blood clotting.
2. Gases and nutrients
   a. Gases in plasma include oxygen, carbon dioxide, and nitrogen.
   b. Plasma nutrients include simple sugars, amino acids, and lipids.
      (1) The liver stores glucose as glycogen and releases glucose whenever blood glucose concentration falls.
      (2) Amino acids are used to synthesize proteins and are deaminated for use as energy sources.
      (3) Lipoproteins function in the transport of lipids.
3. Nonprotein nitrogenous substances
   a. Nonprotein nitrogenous substances are composed of molecules that contain nitrogen atoms but are not proteins.
   b. They include amino acids, urea, and uric acid.
4. Plasma electrolytes
   a. Plasma electrolytes include ions of sodium, potassium, calcium, magnesium, chloride, bicarbonate, phosphate, and sulfate.
   b. Bicarbonate ions are important in maintaining osmotic pressure and pH of plasma.

12.4 Hemostasis (p. 317)

Hemostasis is the stoppage of bleeding.

1. Blood vessel spasm
   a. Smooth muscles in blood vessel walls reflexly contract following injury.
   b. Platelets release serotonin, which stimulates vasoconstriction and helps maintain vessel spasm.
2. Platelet plug formation
   a. Platelets adhere to rough surfaces and exposed collagen.
   b. Platelets adhere to each other at injury sites and form platelet plugs in broken vessels.
3. Blood coagulation
   a. Blood clotting is the most effective means of hemostasis.
   b. Clot formation depends on the balance between factors that promote clotting and those that inhibit clotting.
   c. The basic event of coagulation is the conversion of soluble fibrinogen into insoluble fibrin.
   d. Biochemicals that promote clotting include prothrombin activator, prothrombin, and calcium ions.
   e. A thrombus is an abnormal blood clot in a vessel. An embolus is a clot or fragment of a clot that moves in a vessel.

12.5 Blood Groups and Transfusions (p. 320)

Blood can be typed on the basis of cell surface antigens.

1. Antigens and antibodies
   a. Agglutination is the clumping of red blood cells following a transfusion reaction.
   b. Red blood cell membranes may contain specific antigens, and blood plasma may contain antibodies against certain of these antigens.
2. ABO blood group
   a. Blood is grouped according to the presence or absence of antigens A and B.
   b. Mixing red blood cells that contain an antigen with plasma that contains the corresponding antibody results in an adverse transfusion reaction.
3. Rh blood group
   a. Rh antigens are present on the red blood cell membranes of Rh-positive blood. They are absent in Rh-negative blood.
   b. Mixing Rh-positive red blood cells with plasma that contains anti-Rh antibodies agglutinates the positive cells.
   c. Anti-Rh antibodies in maternal blood may cross the placental tissues and react with the red blood cells of an Rh-positive fetus.

REVIEW EXERCISES

1. List the major components of blood. (p. 307)
2. Describe a red blood cell. (p. 307)
3. Distinguish between oxyhemoglobin and deoxyhemoglobin. (p. 307)
4. Describe the life cycle of a red blood cell. (p. 308)
5. Define *erythropoietin,* and explain its function. (p. 308)
6. Explain how vitamin B_{12} and folic acid deficiencies affect red blood cell production. (p. 309)
7. Distinguish between biliverdin and bilirubin. (p. 309)
8. Distinguish between granulocytes and agranulocytes. (p. 311)
9. Name five types of leukocytes, and list the major functions of each type. (p. 311)
10. Explain the significance of white blood cell counts as aids to diagnosing diseases. (p. 313)
11. Describe a blood platelet, and explain its functions. (p. 314)
12. Name three types of plasma proteins, and list the major functions of each type. (p. 314)
13. Define *lipoprotein.* (p. 316)
14. Describe the relative densities of lipids and proteins. (p. 316)
15. Distinguish between low-density lipoprotein and high-density lipoprotein. (p. 316)
16. Describe how lipoproteins are removed from plasma. (p. 316)
17. Define *nonprotein nitrogenous substances,* and name those commonly present in plasma. (p. 316)
18. Name several plasma electrolytes. (p. 317)
19. Define *hemostasis.* (p. 317)
20. Explain how blood vessel spasms are stimulated following an injury. (p. 317)
21. Explain how a platelet plug forms. (p. 317)

22. List the major steps leading to the formation of a blood clot. (p. 318)
23. Distinguish between fibrinogen and fibrin. (p. 318)
24. Explain how positive feedback operates during blood clotting. (p. 318)
25. Distinguish between a thrombus and an embolus. (p. 318)
26. Distinguish between an antigen and an antibody. (p. 320)
27. Explain the basis of ABO blood types. (p. 320)
28. Explain why an exact match between donor and recipient blood is best. (p. 321)
29. Distinguish between Rh-positive and Rh-negative blood. (p. 322)
30. Describe how a person may become sensitized to Rh-positive blood. (p. 322)
31. Define *erythroblastosis fetalis,* and explain how this condition may develop. (p. 323)

CRITICAL THINKING

1. Erythropoietin is available as a drug. Why would athletes abuse it?
2. How might a technique to remove A and B antigens from red blood cells be used to increase the supply of donated blood?
3. Why can a person receive platelets donated by anyone, but must receive a particular type of whole blood?
4. Researchers are developing several types of chemicals to be used as temporary red blood cell substitutes. What characteristics should a red blood cell substitute have?
5. If a patient with an inoperable cancer is treated using a drug that reduces the rate of cell division, how might the patient's white blood cell count change? How might the patient's environment be modified to compensate for the effects of these changes?
6. Hypochromic (iron-deficiency) anemia is common among aging persons who are admitted to hospitals for other conditions. What environmental and sociological factors might promote this form of anemia?
7. Why do patients with liver diseases commonly develop blood-clotting disorders?
8. How would you explain to a patient with leukemia, who has a greatly elevated white blood count, the importance of avoiding bacterial infections?

WEB CONNECTIONS

Visit the website for additional study questions and more information about this chapter at:

http://www.mhhe.com/shieress8

chapter 13

Cardiovascular System

A BURST BLOOD VESSEL. David Cone was one of the all-time great baseball pitchers when, in the spring of his eleventh season, he began to feel a nagging numbness in his right hand, the one he uses to pitch. He and his doctors were relieved when dye injected into his cardiovascular system showed no blood clot, but his symptoms persisted, even though he was taking blood-thinning drugs. Further investigation revealed an aneurysm, the sudden ballooning of a weakened area, in the subclavian artery in his right shoulder. Years of pitching had built up the muscle in the area, which began to press on the artery, eventually injuring it. Repeated trauma caused this aneurysm. Cone was lucky—the aneurysm did not burst and was surgically repaired.

The rock band R.E.M. was about halfway through a show on their European tour when drummer Bill Berry suddenly developed an excruciating headache in the middle of a song. Holding his head, he stopped playing. Berry's bandmates quickly ended the show and rushed him to a hospital. Berry had an aneurysm in his head. The weakened area had probably been present since birth. Thanks to quick surgery, Berry recovered, and the band was touring the United States two months later.

Photo:
A famous pitcher developed an injury in his pitching arm from repetitive motions, which pressed muscle against an artery, eventually causing an aneurysm.

Chapter Objectives

After studying this chapter, you should be able to do the following:

13.1 Introduction
1. Name the organs of the cardiovascular system, and discuss their functions. (p. 329)

13.2 Structure of the Heart
2. Name and describe the locations and functions of the major parts of the heart. (p. 331)
3. Trace the pathway of the blood through the heart and the vessels of the coronary circulation. (p. 333)

13.3 Heart Actions
4. Discuss the cardiac cycle, and explain how it is controlled. (p. 336)

5. Identify the parts of a normal ECG pattern, and discuss the significance of this pattern. (p. 339)

13.4 Blood Vessels
6. Compare the structures and functions of the major types of blood vessels. (p. 342)
7. Describe how substances are exchanged between blood in capillaries and the tissue fluid surrounding body cells. (p. 345)
8. Describe the mechanisms that aid in returning venous blood to the heart. (p. 346)

13.5 Blood Pressure
9. Explain how blood pressure is produced and controlled. (p. 347)

13.6 Paths of Circulation
10. Compare the pulmonary and systemic circuits of the cardiovascular system. (p. 351)

13.7–13.8 Arterial System—Venous System
11. Identify and locate the major arteries and veins of the pulmonary and systemic circuits. (p. 351)

Aids to Understanding Words

brady- [slow] *brady*cardia: Abnormally slow heartbeat.
diastol- [dilation] *diastol*ic pressure: Blood pressure when the ventricle of the heart is relaxed.
-gram [something written] electrocardio*gram*: Recording of the electrical changes in the heart muscle during a cardiac cycle.
papill- [nipple] *papill*ary muscle: Small mound of muscle within a ventricle of the heart.
syn- [together] *syn*cytium: Mass of merging cells that function together.

systol- [contraction] *systol*ic pressure: Blood pressure resulting from a ventricular contraction.
tachy- [rapid] *tachy*cardia: Abnormally fast heartbeat.

Key Terms

arteriole (ar-te´re-ōl)
atrium (a´tre-um)
capillary (kap´i-lar˝e)
cardiac conduction system (kar´de-ak kon-duk´shun sis´tem)
cardiac cycle (kar´de-ak si´kl)
cardiac output (kar´de-ak owt´poot)
diastole (di-as´to-le)
electrocardiogram (ECG) (e-lek˝tro-kar´de-o-gram˝)
endocardium (en˝do-kar´de-um)
epicardium (ep˝i-kar´de-um)
functional syncytium (funk´-shun-al sin-sish´e-um)
myocardium (mi˝o-kar´de-um)
pericardium (per˝i-kar´de-um´)
peripheral resistance (pĕ-rif´er-al re-zis´tans)
pulmonary circuit (pul´mo-ner˝e sur´kit)

systemic circuit (sis-tem´ik sur´kit)
systole (sis´to-le)
vasoconstriction (vas˝o-kon-strik´shun)
vasodilation (vas˝o-di-la´shun)
ventricle (ven´tri-kl)
venule (ven´ūl)
viscosity (vis-kos´ĭ-te)

13.1 Introduction

The heart pumps 7,000 liters of blood through the body each day, contracting some 2.5 billion times in an average lifetime. This muscular pump forces blood through arteries, which connect to smaller-diameter vessels. The tiniest tubes, the capillaries, are the sites of nutrient, electrolyte, gas, and waste exchange. Capillaries converge into venules, which in turn converge into veins that return blood to the heart, completing the closed system of blood circulation. These structures—the pump and its vessels—form the cardiovascular system.

The cardiovascular system brings oxygen and nutrients to all body cells and removes wastes. A functional cardiovascular system is vital for survival because, without circulation, tissues lack a supply of oxygen and nutrients, and wastes accumulate. Under such conditions, the cells soon begin irreversible change, which quickly leads to death. Figure 13.1 shows the general pattern of blood transport in the cardiovascular system.

13.2 Structure of the Heart

The heart is a hollow, cone-shaped, muscular pump located within the thoracic cavity and resting on the diaphragm (fig. 13.2).

Size and Location of the Heart

Heart size varies with body size. An average adult's heart is about 14 centimeters long and 9 centimeters wide.

The heart is within the mediastinum, bordered laterally by the lungs, posteriorly by the vertebral column, and anteriorly by the sternum. The *base* of the heart, which attaches to several large blood vessels, lies beneath the second rib. The heart's distal end extends downward and to the left, terminating as a bluntly pointed *apex* at the level of the fifth intercostal space.

Coverings of the Heart

The **pericardium** (per´´ĭ-kar´de-um) encloses the heart and the proximal ends of the large blood vessels to which it attaches. The pericardium consists of an outer bag, the fibrous pericardium, that surrounds a more delicate, double-layered sac. The innermost layer of this sac, the *visceral pericardium* (epicardium), covers the heart. At the base of the heart, the visceral pericardium turns back on itself to become the *parietal pericardium*, which forms the inner lining of the fibrous pericardium (figs. 13.2 and 13.3; reference plate 3, p. 24).

The fibrous pericardium is dense connective tissue. It is attached to the central portion of the diaphragm, the posterior of the sternum, the vertebral column, and the large blood vessels emerging from the heart.

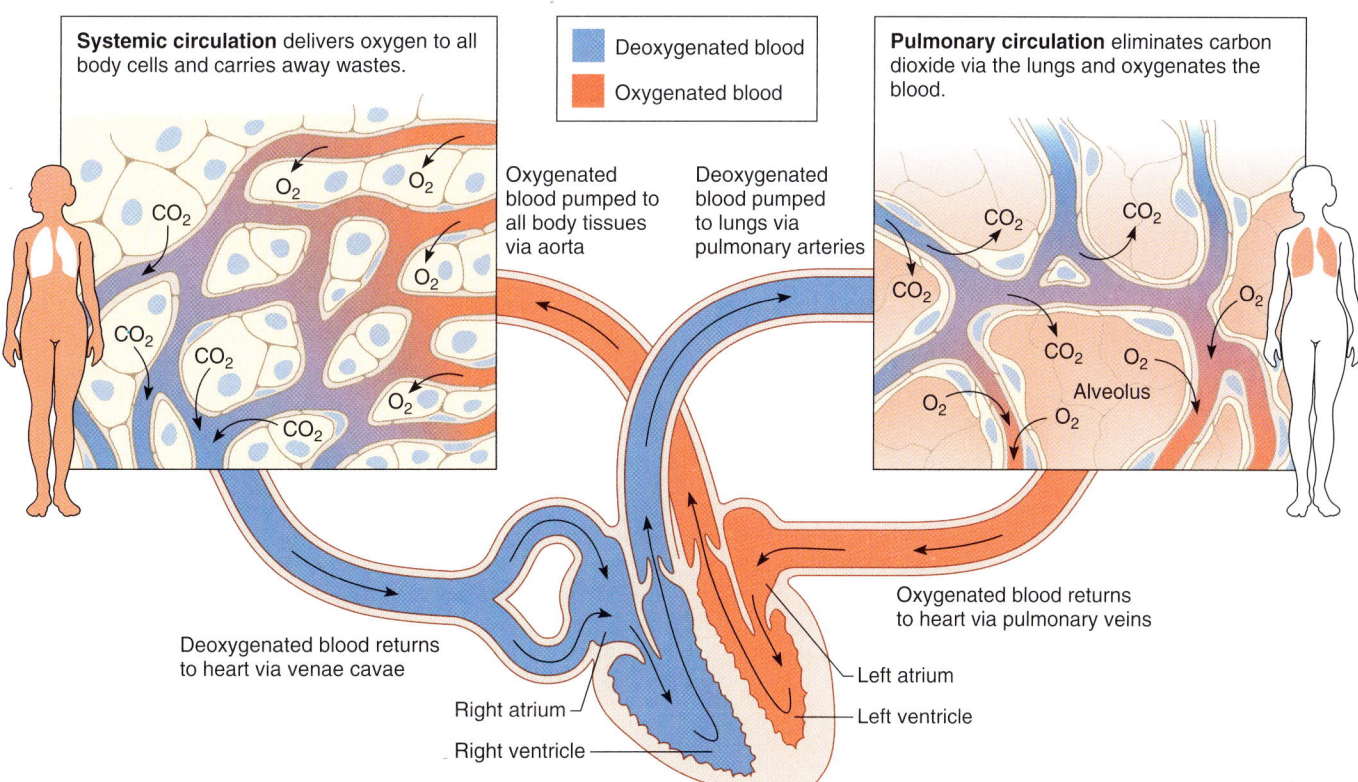

Figure 13.1
The cardiovascular system transports blood between the body cells and organs such as the lungs, intestines, and kidneys that communicate with the external environment.

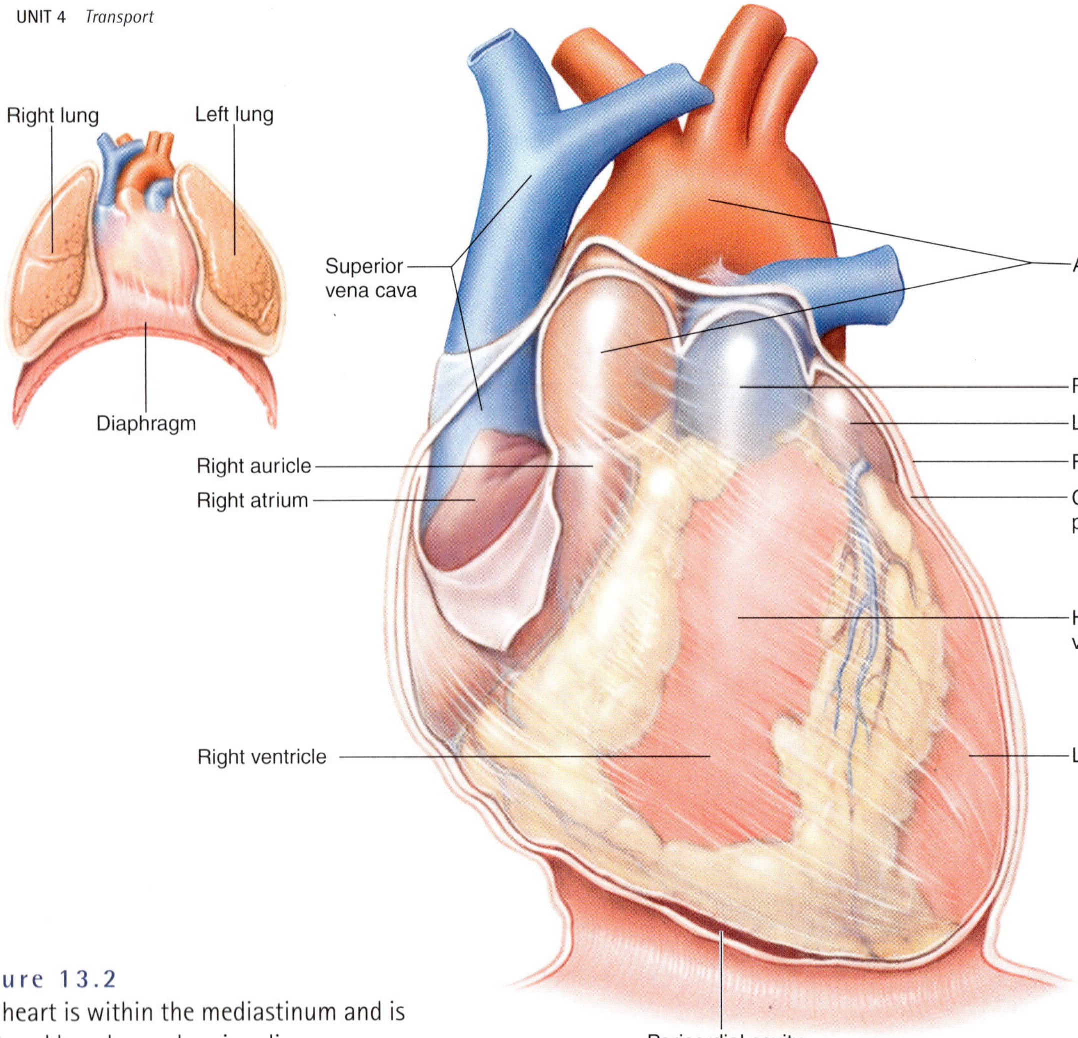

330 UNIT 4 Transport

Figure 13.2
The heart is within the mediastinum and is enclosed by a layered pericardium.

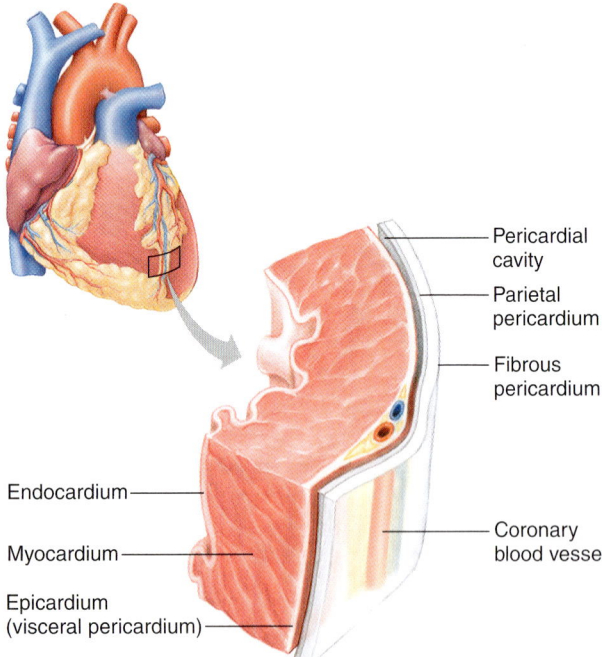

Figure 13.3
The heart wall has three layers: an endocardium, a myocardium, and an epicardium.

Between the parietal and visceral layers of the pericardium is a space, the *pericardial cavity,* that contains a small volume of serous fluid (fig. 13.3). This fluid reduces friction between the pericardial membranes as the heart moves within them.

In *pericarditis,* inflammation of the pericardium due to viral or bacterial infection produces adhesions that attach the layers of the pericardium to each other. This condition is very painful and interferes with heart movements.

CHECK YOUR RECALL

1. Where is the heart located?
2. Distinguish between the visceral pericardium and the parietal pericardium.

Wall of the Heart

The wall of the heart is composed of three distinct layers—an outer epicardium, a middle myocardium, and an inner endocardium (fig. 13.3). The **epicardium** (ep″ĭ-kar′de-um), which corresponds to the visceral pericardium, protects the heart by reducing friction. It is a serous membrane that consists of connective tissue beneath epithelium. Its deeper portion often contains adipose tissue, particularly along the paths of coronary arteries and cardiac veins that carry blood through the myocardium.

The thick middle layer, or **myocardium** (mi″o-kar′de-um), consists mostly of cardiac muscle tissue that pumps blood out of the heart chambers. The muscle fibers are organized in planes, separated by connective tissue richly supplied with blood capillaries, lymph capillaries, and nerve fibers.

The inner layer, or **endocardium** (en″do-kar′de-um), consists of epithelium and connective tissue that contains many elastic and collagenous fibers. The endocardium also contains blood vessels and some specialized cardiac muscle fibers, called *Purkinje fibers,* described later in this chapter. The endocardium is continuous with the inner linings of blood vessels attached to the heart.

Heart Chambers and Valves

Internally, the heart is divided into four hollow chambers—two on the left and two on the right (fig. 13.4). The upper chambers, called **atria** (a′tre-ah; singular, *atrium*), have thin walls and receive blood returning to the heart. Small, earlike projections called *auricles* extend anteriorly from the atria. The lower chambers, the **ventricles** (ven′tri-klz), receive blood from the atria and contract to force blood out of the heart into arteries.

A solid, wall-like **septum** separates the atrium and ventricle on the right side from their counterparts on the left. As a result, blood from one side of the heart never mixes with blood from the other side (except in the fetus, see chapter 20, p. 540). An *atrioventricular valve* (A-V valve), the tricuspid on the right and the bicuspid on the left, ensures one-way blood flow between the atria and ventricles.

The right atrium receives blood from two large veins—the *superior vena cava* and the *inferior vena cava.* A smaller vein, the *coronary sinus,* also drains blood into the right atrium from the myocardium of the heart itself.

A large **tricuspid valve,** which has three *cusps* as its name implies, lies between the right atrium and the right ventricle (fig. 13.4). The valve permits blood to move from the right atrium into the right ventricle and prevents backflow.

Strong, fibrous strings called *chordae tendineae* attach to the cusps of the tricuspid valve on the ventricular side. These strings originate from small mounds of cardiac muscle tissue, the **papillary muscles,** that project inward from the walls of the ventricle. The papillary

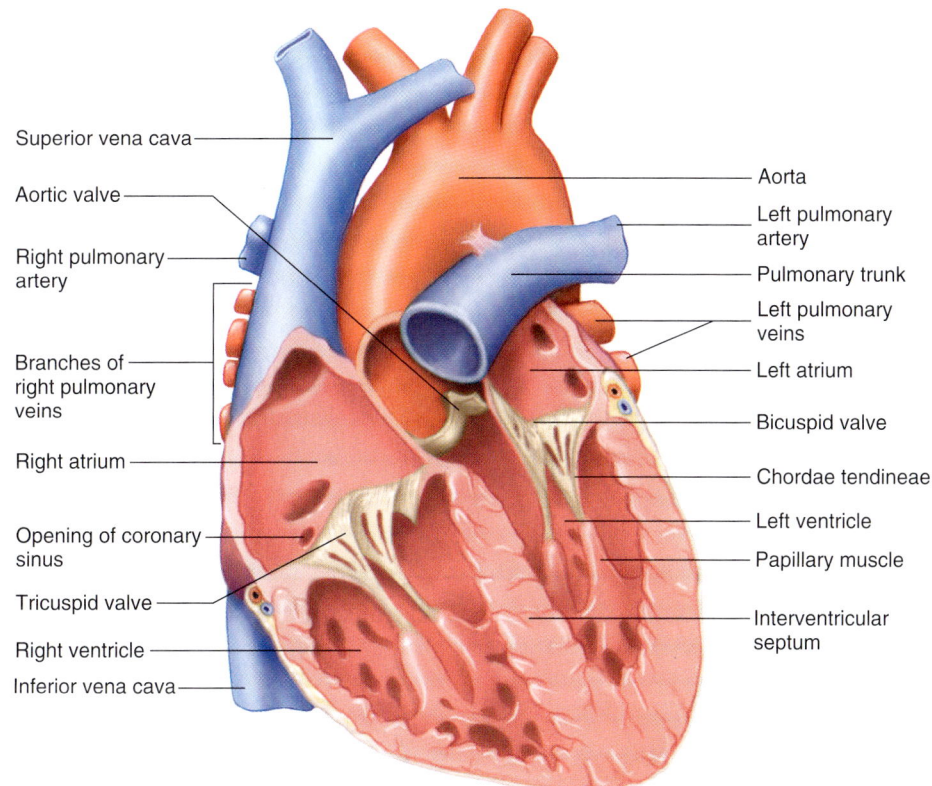

Figure 13.4
Coronal section of the heart showing the connection between the left ventricle and the aorta as well as the four hollow chambers.

muscles contract when the ventricle contracts. As the tricuspid valve closes, these muscles pull on the chordae tendineae and prevent the cusps from swinging back into the atrium.

The right ventricle has a thinner muscular wall than the left ventricle (fig. 13.4). This right chamber pumps blood a short distance to the lungs against a relatively low resistance to blood flow. The left ventricle, on the other hand, must force blood to all the other parts of the body against a much greater resistance to flow.

When the muscular wall of the right ventricle contracts, the blood inside its chamber is put under increasing pressure, and the tricuspid valve closes passively. As a result, the only exit for the blood is through the *pulmonary trunk,* which divides to form the left and right *pulmonary arteries* that lead to the lungs. At the base of this trunk is a **pulmonary valve** with three cusps. This valve allows blood to leave the right ventricle and prevents backflow into the ventricular chamber (fig. 13.5).

The left atrium receives blood from the lungs through four *pulmonary veins*—two from the right lung and two from the left lung. Blood passes from the left atrium into the left ventricle through the **bicuspid (mitral) valve,** which prevents blood from flowing back into the left atrium from the ventricle. As with the tricuspid valve, the papillary muscles and the chordae tendineae prevent the cusps of the bicuspid valve from swinging back into the left atrium during ventricular contraction.

When the left ventricle contracts, the bicuspid valve closes passively, and the only exit is through a large artery, the **aorta.** At the base of the aorta is the **aortic valve,** which has three cusps. The aortic valve opens and allows blood to leave the left ventricle as it contracts. When the ventricular muscles relax, this valve closes and prevents blood from backing up into the ventricle (fig. 13.5).

The bicuspid and tricuspid valves are called atrioventricular valves because they are between atria and ventricles. The pulmonary and aortic valves are called "semilunar" because of the half-moon shapes of their cusps. Table 13.1 summarizes the locations and functions of the heart valves.

Mitral valve prolapse (MVP) affects up to 6% of the U.S. population. In this condition, one (or both) of the cusps of the mitral valve stretches and bulges into the left atrium during ventricular contraction. The valve usually continues to function adequately, but sometimes blood regurgitates into the left atrium. Through a stethoscope, a regurgitating MVP sounds like a click at the end of ventricular contraction, then a murmur as blood goes back through the valve into the left atrium. Symptoms of MVP include chest pain, palpitations, fatigue, and anxiety.

The mitral valve can be damaged by certain species of *Streptococcus* bacteria. Endocarditis, an inflammation of the endocardium due to an infection, appears as a plantlike growth on the valve. People with MVP are particularly susceptible to endocarditis. Individuals with MVP must take antibiotics before undergoing dental work to prevent infection by *Streptococcus* bacteria in the mouth.

CHECK YOUR RECALL

1. Describe the layers of the heart wall.
2. Name and locate the four chambers and valves of the heart.
3. Describe the function of each heart valve.

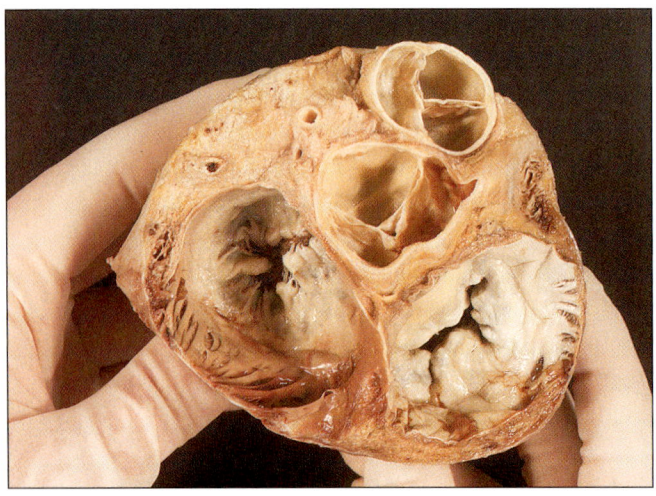

Figure 13.5
Photograph of a transverse section through the heart showing the four valves (superior view, see also fig. 13.6).

TABLE 13.1   HEART VALVES

VALVE	LOCATION	FUNCTION
Tricuspid valve	Opening between right atrium and right ventricle	Prevents blood from moving from right ventricle into right atrium during ventricular contraction
Pulmonary valve	Entrance to pulmonary trunk	Prevents blood from moving from pulmonary trunk into right ventricle during ventricular relaxation
Bicuspid (mitral) valve	Opening between left atrium and left ventricle	Prevents blood from moving from left ventricle into left atrium during ventricular contraction
Aortic valve	Entrance to aorta	Prevents blood from moving from aorta into left ventricle during ventricular relaxation

Skeleton of the Heart

Rings of dense connective tissue surround the pulmonary trunk and aorta at their proximal ends (fig. 13.6). These rings provide firm attachments for the heart valves and for muscle fibers; they also prevent the outlets of the atria and ventricles from dilating during contraction. The fibrous rings, together with other masses of dense connective tissue in the portion of the septum between the ventricles (interventricular septum), constitute the *skeleton of the heart*.

Path of Blood Through the Heart

Blood that is low in oxygen and high in carbon dioxide enters the right atrium through the venae cavae and coronary sinus. As the right atrial wall contracts, the blood passes through the tricuspid valve and enters the chamber of the right ventricle (fig. 13.7). When the right ventricular wall contracts, the tricuspid valve closes, and blood moves through the pulmonary valve and into the pulmonary trunk and its branches (pulmonary arteries).

From the pulmonary arteries, blood enters the capillaries associated with the alveoli of the lungs. Gas exchanges occur between blood in the capillaries and air in the alveoli. The freshly oxygenated blood, which is now relatively low in carbon dioxide, returns to the heart through the pulmonary veins that lead to the left atrium.

The left atrial wall contracts, and blood moves through the bicuspid valve and into the chamber of the left ventricle. When the left ventricular wall contracts, the bicuspid valve closes, and blood moves through the aortic valve and into the aorta and its branches.

Blood Supply to the Heart

The first two branches of the aorta, called the right and left **coronary arteries,** supply blood to the tissues of the heart. Their openings lie just beyond the aortic valve (fig. 13.8).

A thrombus or embolus that blocks or narrows a coronary artery branch deprives myocardial cells of oxygen, producing ischemia and a painful condition called *angina pectoris*. The pain usually occurs during physical activity, when oxygen requirements exceed oxygen supply. Pain lessens with rest. Emotional disturbance may also trigger angina pectoris.

Angina pectoris may cause a sensation of heavy pressure, tightening, or squeezing in the chest. The pain is usually felt in the region behind the sternum or in the anterior portion of the upper thoracic cavity, but may radiate to the neck, jaw, throat, upper limb, shoulder, elbow, back, or upper abdomen. Other symptoms include profuse perspiration (diaphoresis), difficulty breathing (dyspnea), nausea, or vomiting.

A blood clot completely obstructing a coronary artery or one of its branches (coronary thrombosis) kills part of the heart. This is a *myocardial infarction (MI)*, more commonly known as a heart attack.

The heart must beat continually to supply blood to body tissues. Therefore, myocardial cells require a constant supply of freshly oxygenated blood. Branches of

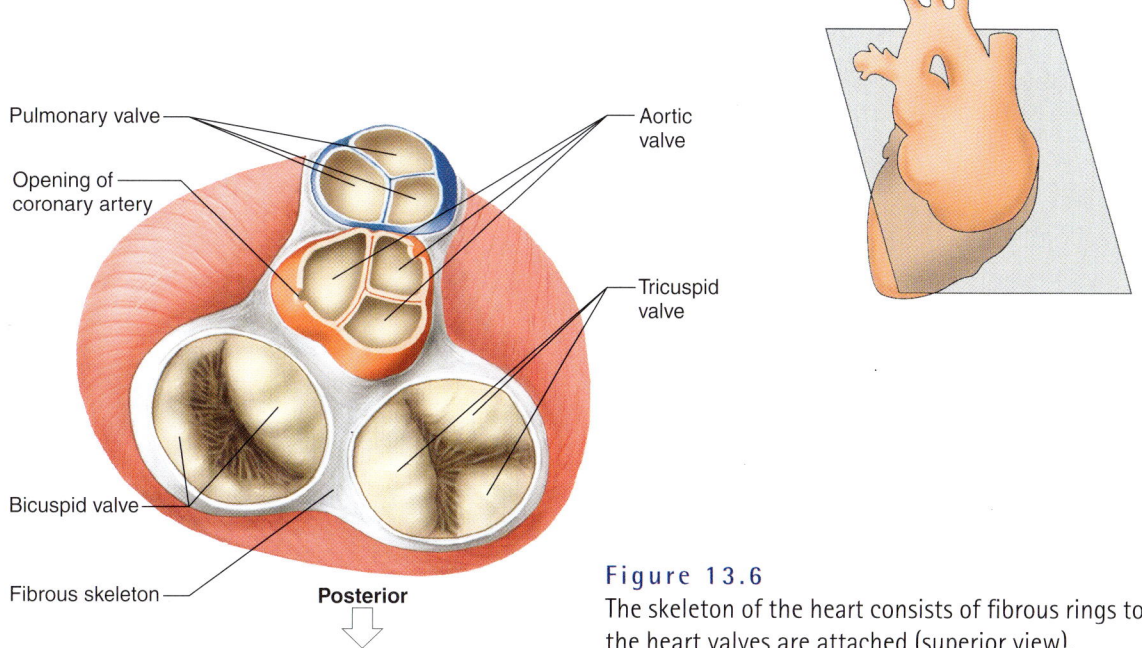

Figure 13.6
The skeleton of the heart consists of fibrous rings to which the heart valves are attached (superior view).

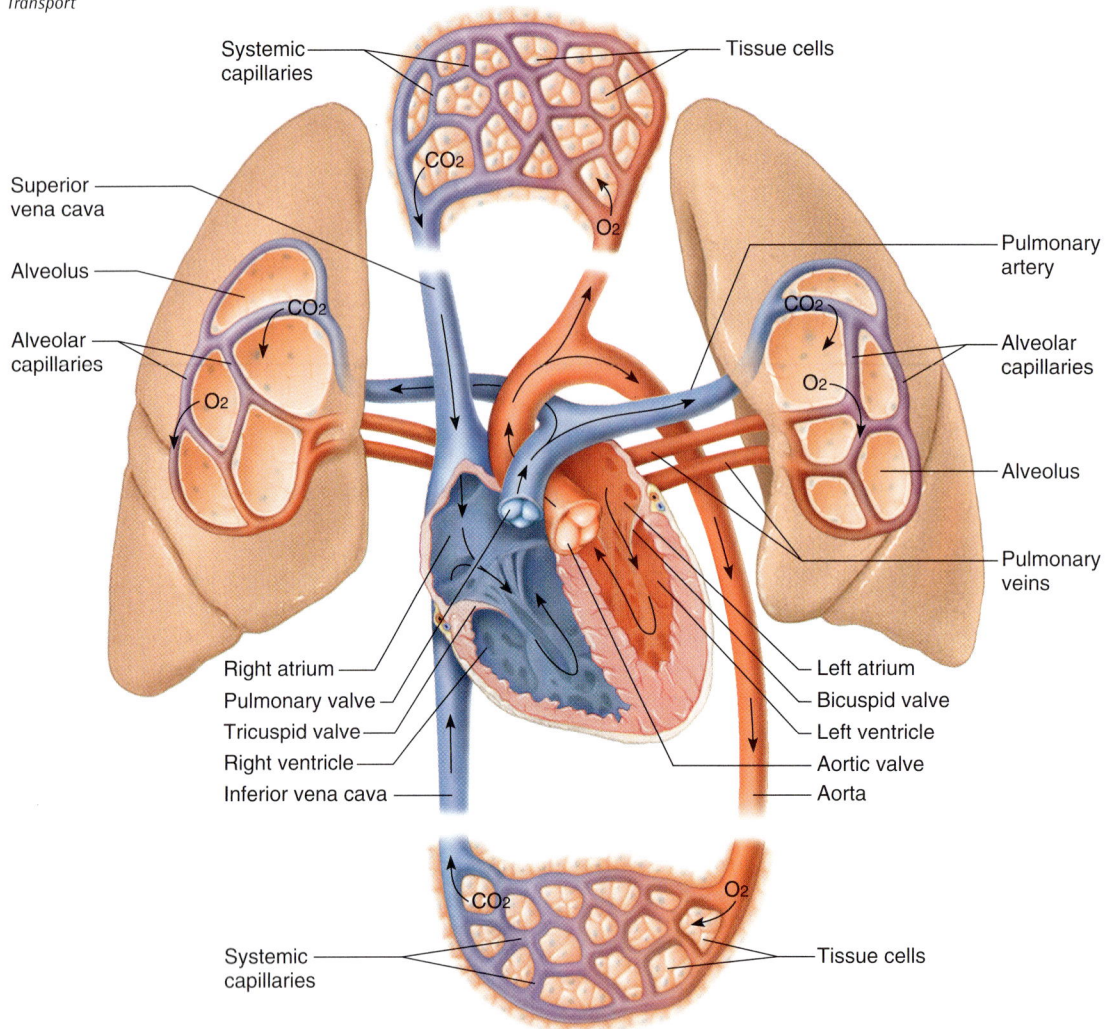

Figure 13.7
The right ventricle forces blood to the lungs, whereas the left ventricle forces blood to all other body parts. (Structures are not drawn to scale.)

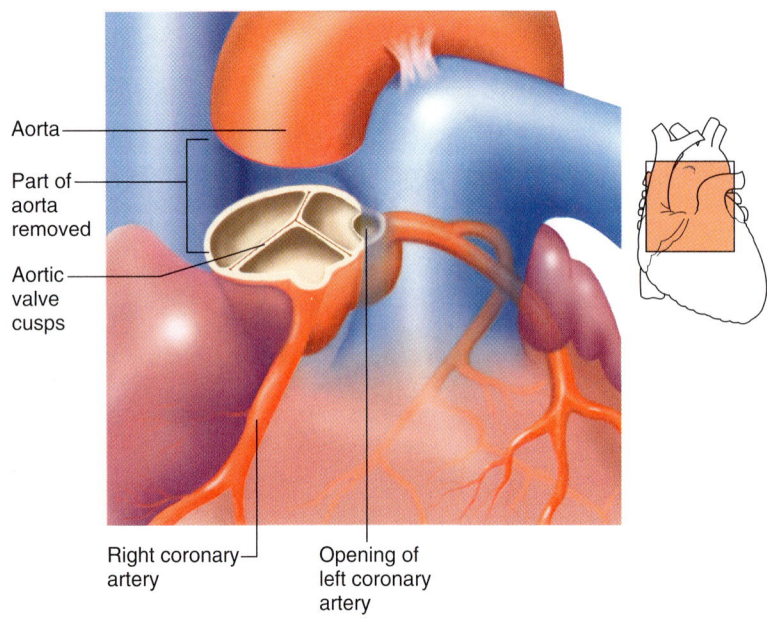

Figure 13.8
The openings of the coronary arteries lie just beyond the aortic valve.

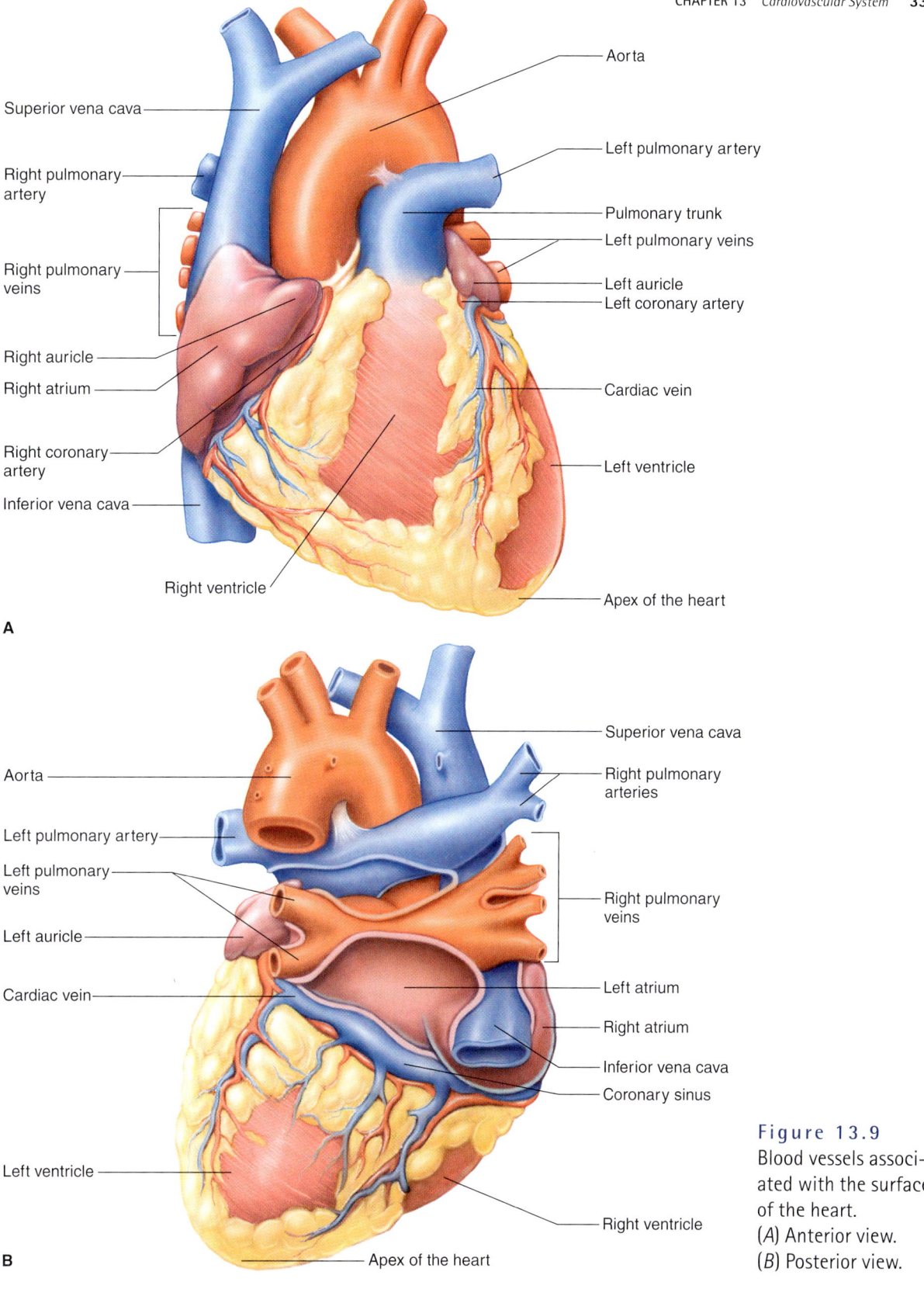

Figure 13.9
Blood vessels associated with the surface of the heart. (*A*) Anterior view. (*B*) Posterior view.

the coronary arteries feed the many capillaries of the myocardium (fig. 13.9). The smaller branches of these arteries usually have connections (anastomoses) between vessels that provide alternate pathways for blood, called collateral circulation. These detours in circulation may supply sufficient oxygen and nutrients to the myocardium when a coronary artery is blocked.

Branches of the **cardiac veins,** whose paths roughly parallel those of the coronary arteries, drain blood that has passed through myocardial capillaries.

As figure 13.9B shows, these veins join an enlarged vein on the heart's posterior surface—the **coronary sinus**—which empties into the right atrium (see fig. 13.4).

In *heart transplantation,* the recipient's failing heart is removed, except for the posterior walls of the right and left atria and their connections to the venae cavae and pulmonary veins. The donor heart is prepared similarly and is attached to the atrial cuffs remaining in the recipient's thoracic cavity. Finally, the recipient's aorta and pulmonary arteries are connected to those of the donor heart. The Clinical Connection at the end of this chapter describes a promising alternative to a heart transplant.

CHECK YOUR RECALL

1. Which structures make up the skeleton of the heart?
2. Review the path of blood through the heart.
3. Which vessels supply blood to the myocardium?
4. How does blood return from the cardiac tissues to the right atrium?

13.3 Heart Actions

The heart chambers function in coordinated fashion. Their actions are regulated so that atria contract, called atrial **systole** (sis´to-le), while ventricles relax, called ventricular **diastole** (di-as´to-le); then ventricles contract (ventricular systole) while atria relax (atrial diastole). Then the atria and ventricles both relax for a brief interval. This series of events constitutes a complete heartbeat, or **cardiac cycle** (kar´de-ak si´kl).

Cardiac Cycle

During a cardiac cycle, pressure within the heart chambers rises and falls. Pressure in the ventricles is low early in diastole, and the pressure difference between the atria and ventricles causes the A-V valves to open and the ventricles to fill. About 70% of the returning blood enters the ventricles prior to contraction, and ventricular pressure gradually increases. When the atria contract, the remaining 30% of returning blood is pushed into the ventricles, and ventricular pressure increases a bit more. Then, as the ventricles contract, ventricular pressure rises sharply, and as soon as the ventricular pressure exceeds the atrial pressure, the A-V valves close. At the same time, the papillary muscles contract, and by pulling on the chordae tendineae, they prevent the cusps of the A-V valves from bulging too far into the atria.

During ventricular contraction, the A-V valves remain closed. The atria are now relaxed, and pressure in the atria is quite low, even lower than venous pressure. As a result, blood flows into the atria from the large, attached veins. That is, as the ventricles are contracting, the atria are filling, already preparing for the next cardiac cycle (fig. 13.10).

As ventricular systole progresses, ventricular pressure continues to increase until it exceeds the pressure in the pulmonary trunk (right side) and aorta (left side). At this point, the pressure difference across the semilunar valves causes the pulmonary and aortic valves to open, and blood is ejected from each valve's respective ventricle into these arteries.

As blood flows out of the ventricles, ventricular pressure begins to drop, and it drops even further as the ventricles begin to relax. When ventricular pressure is lower than blood pressure in the aorta and pulmonary trunk, the pressure difference is reversed, and the semilunar valves close. The ventricles continue to relax, and as soon as ventricular pressure is less than atrial pressure, the A-V valves open, and the ventricles begin to fill once more. Atria and ventricles are both relaxed for a brief interval. The graph in figure 13.11 summarizes some of the changes that occur during a cardiac cycle.

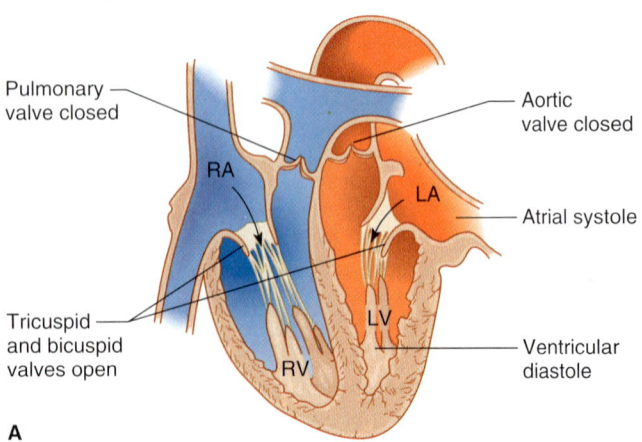
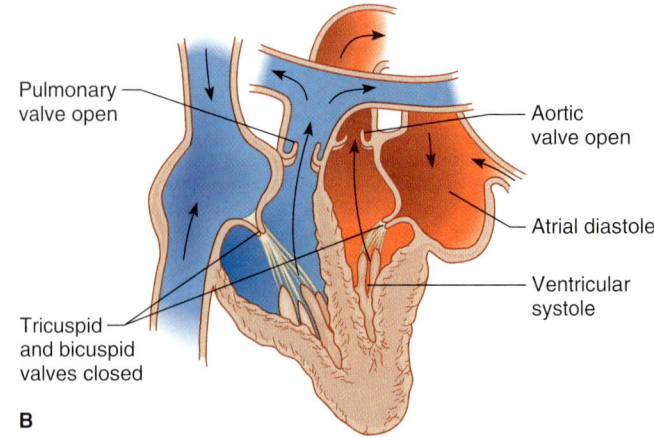

Figure 13.10
The atria (*A*) empty during atrial systole and (*B*) fill with blood during atrial diastole.

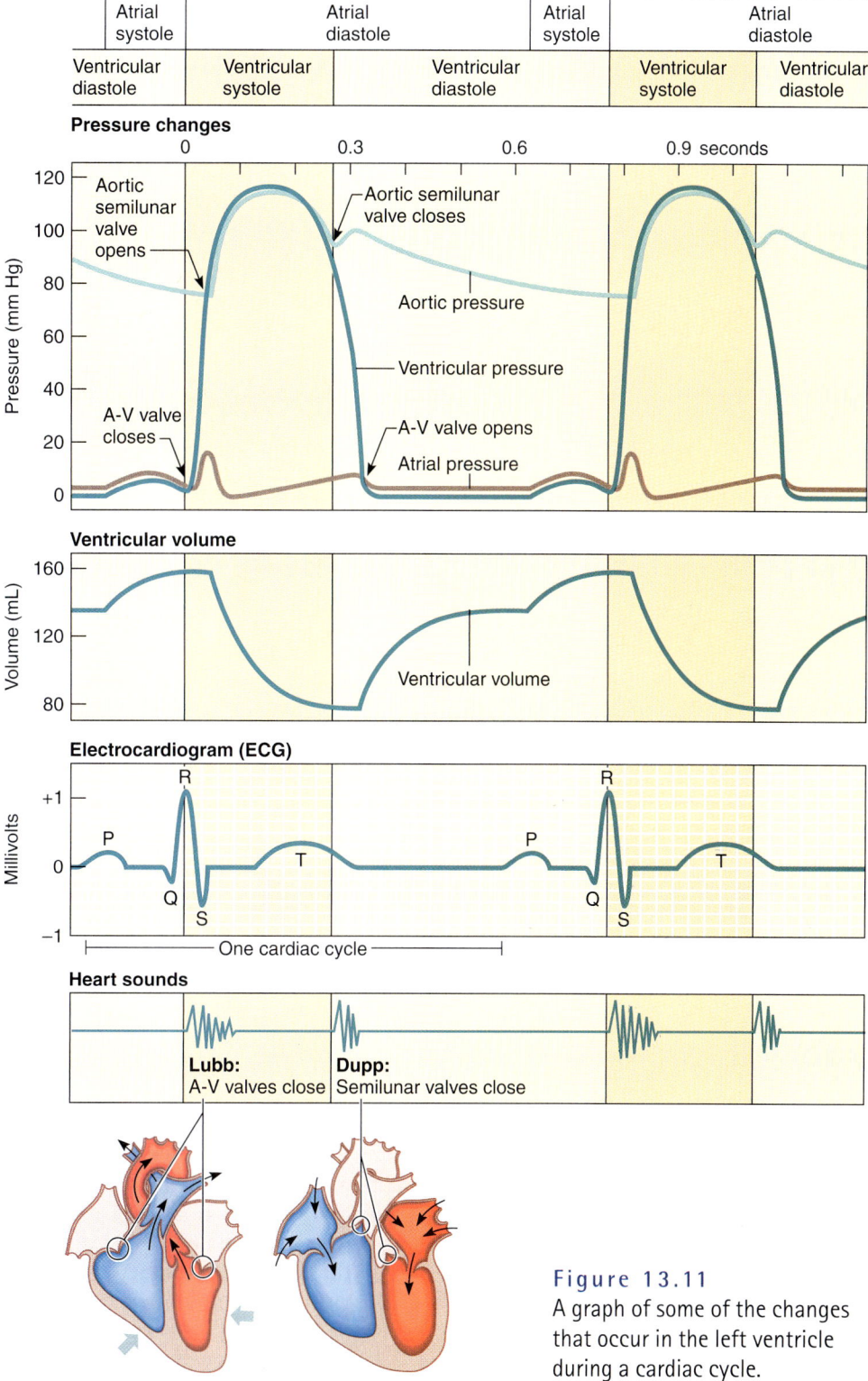

Figure 13.11
A graph of some of the changes that occur in the left ventricle during a cardiac cycle.

Heart Sounds

A heartbeat heard through a stethoscope sounds like *lubb-dupp*. These sounds are due to vibrations in the heart tissues associated with the closing of the valves.

The first part of a heart sound (*lubb*) occurs during ventricular contraction, when the A-V valves are closing. The second part (*dupp*) occurs during ventricular relaxation, when the pulmonary and aortic valves are closing (fig. 13.11).

Heart sounds provide information concerning the condition of the heart valves. For example, inflammation of the endocardium (endocarditis) may erode the edges of the valvular cusps. As a result, the cusps may not close completely, and some blood may leak back through the valve, producing an abnormal sound called a *murmur*. The seriousness of a murmur depends on the amount of valvular damage. Open heart surgery may repair or replace seriously damaged valves.

Cardiac Muscle Fibers

Cardiac muscle fibers function much like those of skeletal muscles, but the fibers connect in branching networks. Stimulation to any part of the network sends impulses throughout the heart, which contracts as a unit.

A mass of merging cells that function as a unit is called a **functional syncytium** (funk´shun-al sin-sish´e-um). Two such structures are in the heart—in the atrial walls and in the ventricular walls. Portions of the heart's fibrous skeleton separate these masses of cardiac muscle fibers from each other, except for a small area in the right atrial floor. In this region, the *atrial syncytium* and the *ventricular syncytium* are connected by fibers of the cardiac conduction system.

CHECK YOUR RECALL

1. Describe the pressure changes in the atria and ventricles during a cardiac cycle.
2. What causes heart sounds?
3. What is a functional syncytium?

Cardiac Conduction System

Throughout the heart are clumps and strands of specialized cardiac muscle tissue whose fibers contain only a few myofibrils. Instead of contracting, these areas initiate and distribute impulses throughout the myocardium. They comprise the **cardiac conduction system** (kar´de-ak kon-duk´shun sis´tem), which coordinates events of the cardiac cycle (fig. 13.12).

A key portion of this conduction system is the **sinoatrial node** (S-A node), a small, elongated mass of specialized cardiac muscle tissue just beneath the epicardium. It is located in the right atrium near the opening of the superior vena cava, and its fibers are continuous with those of the atrial syncytium.

The cells of the S-A node can reach threshold on their own, and their membranes contact one another. Without stimulation from nerve fibers or any other outside agents, the nodal cells initiate impulses that spread into the surrounding myocardium and stimulate cardiac muscle fibers to contract.

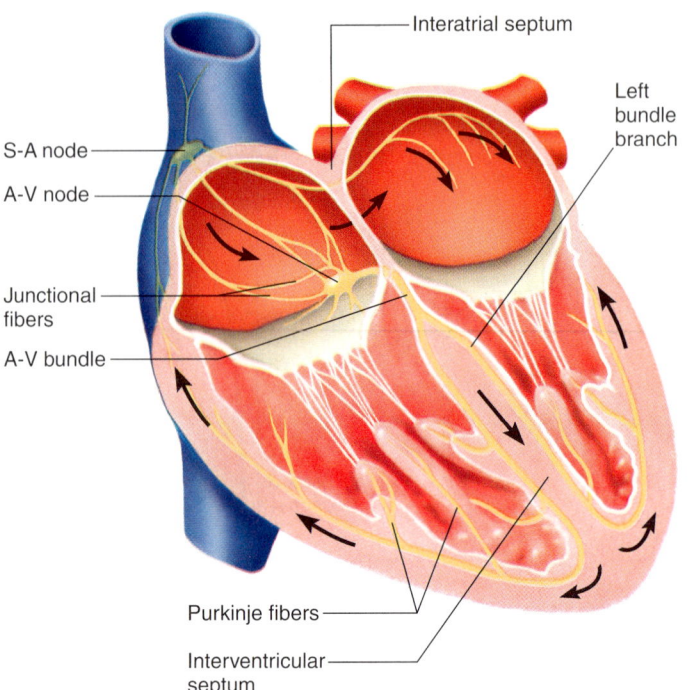

Figure 13.12
The cardiac conduction system.

S-A node activity is rhythmic. The S-A node initiates one impulse after another, seventy to eighty times a minute in an adult. Because it generates the heart's rhythmic contractions, it is often called the **pacemaker.**

As a cardiac impulse travels from the S-A node into the atrial syncytium, the right and left atria begin to contract almost simultaneously. The cardiac impulse does not pass directly into the ventricular syncytium, which is separated from the atrial syncytium by the fibrous skeleton of the heart. Instead, the impulse passes along fibers of the conduction system that lead to a mass of specialized cardiac muscle tissue called the **atrioventricular node** (A-V node). This node is located in the inferior portion of the septum that separates the atria (interatrial septum) and just beneath the endocardium. It provides the only normal conduction pathway between the atrial and ventricular syncytia.

The fibers that conduct the cardiac impulse into the A-V node (junctional fibers) have very small diameters, and because small fibers conduct impulses slowly, they delay impulse transmission. The impulse is delayed further as it moves through the A-V node, allowing more time for the atria to contract completely so they empty all their blood into the ventricles prior to ventricular contraction.

Once the cardiac impulse reaches the distal side of the A-V node, it passes into a group of large fibers that make up the **A-V bundle** (bundle of His). The A-V bundle enters the upper part of the interventricular septum and divides into right and left bundle branches that lie just beneath the endocardium. About halfway down the septum, the branches give rise to enlarged **Purkinje fibers** (pur-kin´je fi´berz).

Purkinje fibers spread from the interventricular septum into the papillary muscles, which project inward from ventricular walls and then continue downward to the apex of the heart. There they curve around the tips of the ventricles and pass upward over the lateral walls of these chambers. Along the way, the Purkinje fibers give off many small branches, which become continuous with cardiac muscle fibers (fig. 13.13).

The muscle fibers in ventricular walls form irregular whorls. When impulses on the Purkinje fibers stimulate these muscle fibers, the ventricular walls contract with a twisting motion (fig. 13.14). This action squeezes blood out of ventricular chambers and forces it into the aorta and pulmonary trunk.

CHECK YOUR RECALL

1. What kinds of tissues make up the cardiac conduction system?
2. How is a cardiac impulse initiated?
3. How is a cardiac impulse transmitted from the right atrium to the other heart chambers?

Electrocardiogram

An **electrocardiogram** (e-lek´´tor-kar´de-o-gram´´), or **ECG,** is a recording of the electrical changes that occur in the myocardium during a cardiac cycle. (This pattern occurs as action potentials stimulate cardiac muscle fibers to contract, but it is not the same as individual action potentials.) Because body fluids can conduct electrical currents, such changes can be detected on the surface of the body.

To record an ECG, electrodes are placed on the skin and connected by wires to an instrument that responds to very weak electrical changes by moving a pen or stylus on a moving strip of paper. Up-and-down movements of the pen correspond to electrical changes in the myocardium. Because the paper moves past the pen at a known rate, the distance between pen deflections indicates the time between phases of the cardiac cycle.

As figure 13.15A illustrates, a normal ECG pattern includes several deflections, or *waves,* during each cardiac cycle. Between cycles, the muscle fibers remain polarized, with no detectable electrical changes, and the pen does not move but simply marks along the baseline. When the S-A node triggers a cardiac impulse, atrial fibers depolarize, producing an electrical change. The pen moves, and at the end of the electrical change, returns to the base position. This first pen movement produces a *P wave,* corresponding to depolarization of the atrial fibers that will lead to contraction of the atria (fig. 13.15B).

When the cardiac impulse reaches ventricular fibers, they rapidly depolarize. Because ventricular walls are thicker than those of the atria, the electrical change is greater, and the pen deflects more. When the electrical change ends, the pen returns to the baseline, leaving a mark called the *QRS complex.* This mark consists of a *Q wave,* an *R wave,* and an *S wave,* and corresponds to depolarization of ventricular fibers just prior to the contraction of the ventricular walls.

The electrical changes occurring as the ventricular muscle fibers repolarize slowly produce a *T wave* as the

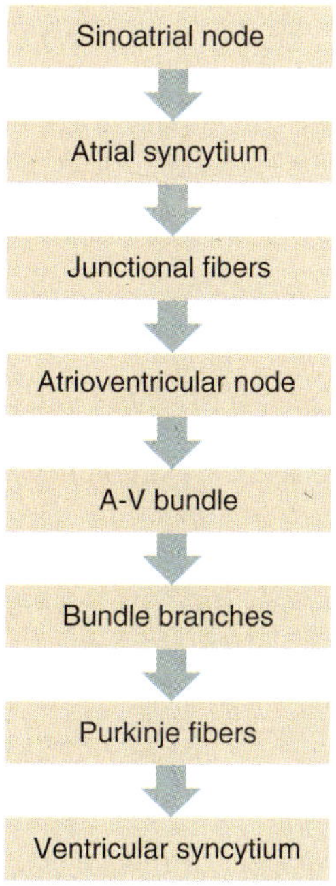

Figure 13.13
Components of the cardiac conduction system.

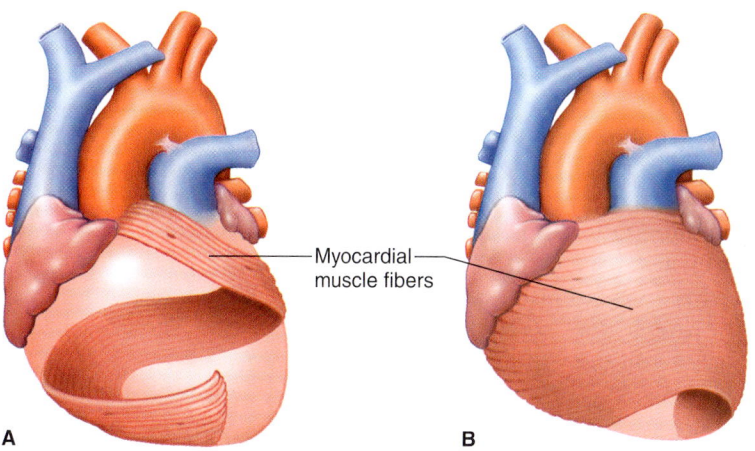

Figure 13.14
The muscle fibers within the ventricular walls are arranged in patterns of whorls. The fibers of groups (A) and (B) surround both ventricles in these anterior views of the heart.

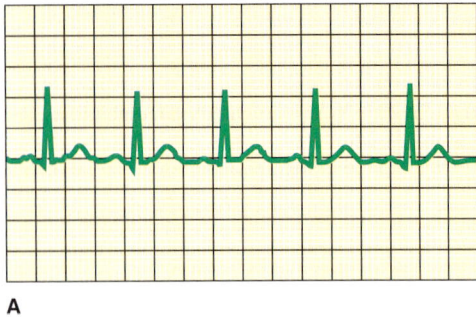

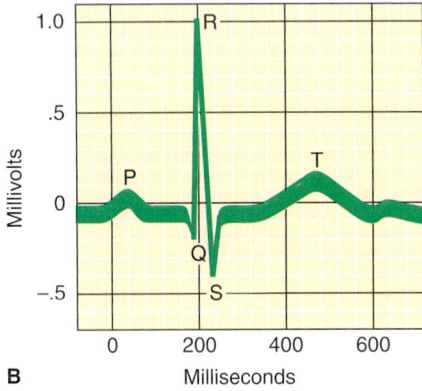

Figure 13.15
Electrocardiogram. (*A*) A normal ECG. (*B*) In an ECG pattern, the P wave results from a depolarization of the atria, the QRS complex results from a depolarization of the ventricles, and the T wave results from a repolarization of the ventricles.

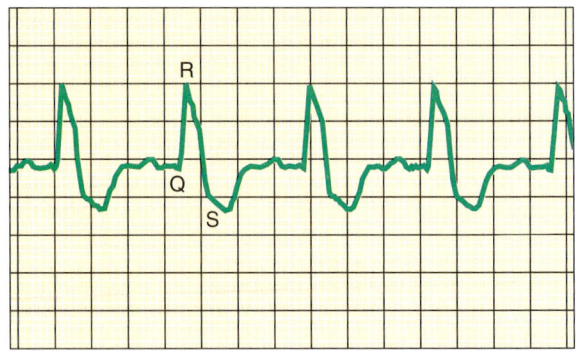

Figure 13.16
A prolonged QRS complex may result from damage to the A-V bundle fibers.

pen deflects again, ending the ECG pattern. The record of atrial repolarization seems to be missing from the pattern because atrial fibers repolarize at the same time that ventricular fibers depolarize. Thus, the QRS complex obscures the recording of atrial repolarization.

Physicians use ECG patterns to assess the heart's ability to conduct impulses. For example, the time period between the beginning of a P wave and the beginning of a QRS complex, *P-Q interval* (or if the initial portion of the QRS complex is upright, P-R interval), indicates the time for the cardiac impulse to travel from the S-A node through the A-V node. Ischemia or other problems affecting the fibers of the A-V conduction pathways can increase this P-Q interval. Similarly, injury to the A-V bundle can extend the QRS complex, because it may take longer for an impulse to spread throughout the ventricular walls (fig. 13.16).

✓ CHECK YOUR RECALL

1. What is an electrocardiogram?
2. Which cardiac events do the P wave, QRS complex, and T wave represent?

Regulation of the Cardiac Cycle

The volume of blood pumped changes to accommodate cellular requirements. For example, during strenuous exercise, skeletal muscles require more blood, and heart rate increases in response. Since the S-A node normally controls heart rate, changes in this rate are often a response to factors that affect the S-A node, such as the motor impulses carried on the parasympathetic and sympathetic nerve fibers (see chapter 9, p. 245).

The parasympathetic fibers that innervate the heart arise from neurons in the medulla oblongata (fig. 13.17). Most of these fibers branch to the S-A and A-V nodes. When the nerve impulses reach nerve fiber endings, they secrete acetylcholine, which decreases S-A and A-V nodal activity. As a result, heart rate decreases.

Parasympathetic fibers carry impulses continually to the S-A and A-V nodes, "braking" heart action. Consequently, parasympathetic activity can change heart rate in either direction. An increase in the impulses slows the heart rate, and a decrease in the impulses releases the parasympathetic "brake" and increases heart rate.

Sympathetic fibers reach the heart and join the S-A and A-V nodes as well as other areas of the atrial and ventricular myocardium. The endings of these fibers secrete norepinephrine in response to nerve impulses, which increases the rate and force of myocardial contractions.

The *cardiac control center* of the medulla oblongata maintains balance between the inhibitory effects of parasympathetic fibers and the excitatory effects of sympathetic fibers. This center receives sensory impulses from throughout the cardiovascular system and relays motor impulses to the heart in response. For example, receptors sensitive to stretch are located in certain regions of the aorta (aortic arch) and in the carotid arteries (carotid sinuses) (fig. 13.17). These receptors, called *baroreceptors* (pressoreceptors), can detect changes in blood pressure. Rising pressure stretches the receptors, and they signal the cardioin-

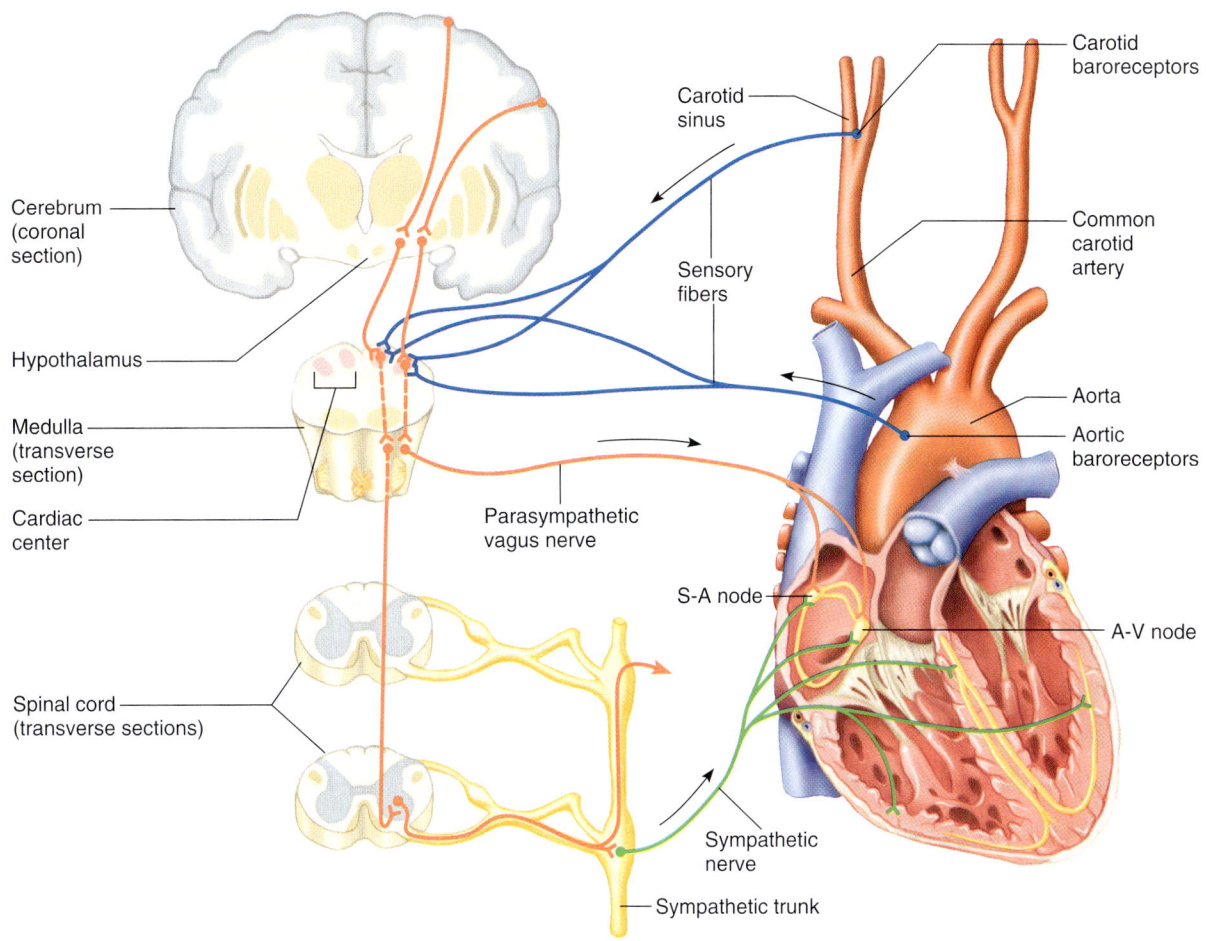

Figure 13.17
Autonomic nerve impulses alter the activities of the S-A and A-V nodes.

hibitor center in the medulla oblongata. In response, the medulla oblongata sends parasympathetic impulses to the heart via the vagus nerve, decreasing heart rate. This action helps lower blood pressure toward normal.

Impulses from the cerebrum or hypothalamus also influence the cardiac control center. Such impulses may decrease heart rate, as occurs when a person faints following an emotional upset, or they may increase heart rate during a period of anxiety.

Two other factors that influence heart rate are temperature change and certain ions. Rising body temperature increases heart action, which is why heart rate usually increases during fever. On the other hand, abnormally low body temperature decreases heart action.

Of the ions that influence heart action, the most important are potassium (K^+) and calcium (Ca^{+2}) ions. Excess potassium ions (hyperkalemia) decrease the rate and force of contractions. If potassium concentration drops below normal (hypokalemia), the heart may develop a potentially life-threatening abnormal rhythm (arrhythmia).

Excess extracellular calcium ions (hypercalcemia) increase heart actions, posing the danger that the heart will contract for a prolonged time. Conversely, low calcium concentration (hypocalcemia) depresses heart action.

CHECK YOUR RECALL

1. How do parasympathetic and sympathetic impulses help control heart rate?
2. How do changes in body temperature affect heart rate?
3. Describe the effects on the heart of abnormal concentrations of potassium and calcium ions.

13.4 Blood Vessels

The blood vessels form a closed circuit of tubes that carries blood from the heart to cells and back again. These vessels include arteries, arterioles, capillaries, venules, and veins.

There are about 62,000 miles of blood vessels in the body—2 1/2 times around the world.

Arteries and Arterioles

Arteries are strong, elastic vessels that are adapted for carrying blood away from the heart under high pressure. These vessels subdivide into progressively thinner tubes and eventually give rise to finer, branched **arterioles** (ar-te´re-olz).

The wall of an artery consists of three distinct layers (fig. 13.18A). The innermost layer (*tunica interna*) is composed of a layer of simple squamous epithelium, called *endothelium*, which rests on a connective tissue membrane that is rich in elastic and collagenous fibers. Endothelium helps prevent blood clotting by providing a smooth surface that allows blood cells and platelets to flow through without being damaged and by secreting biochemicals that inhibit platelet aggregation. Endothelium also may help regulate local blood flow by secreting substances that dilate or constrict blood vessels. For example, endothelium releases the gas nitric oxide, which causes the smooth muscle of the vessel to relax.

The middle layer (*tunica media*) makes up the bulk of the arterial wall. It includes smooth muscle fibers, which encircle the tube, and a thick layer of elastic connective tissue.

The outer layer (*tunica externa*) is relatively thin and chiefly consists of connective tissue with irregularly organized elastic and collagenous fibers. This layer attaches the artery to the surrounding tissues.

The sympathetic branches of the autonomic nervous system innervate smooth muscle in artery and arteriole walls. Impulses on these *vasomotor fibers* stimulate the smooth muscles to contract, reducing the diameter of the vessel. This is called **vasoconstriction** (vas´´o-kon-strik´shun). If vasomotor impulses are inhibited, the muscle fibers relax, and the diameter of the vessel increases. This is called **vasodilation** (vas´´o-di-la´shun). Changes in the diameters of arteries and arterioles greatly influence blood flow and blood pressure.

The walls of the larger arterioles have three layers similar to those of arteries. These walls thin as arterioles approach capillaries. The wall of a very small arteriole consists only of an endothelial lining and some smooth muscle fibers, surrounded by a small amount of connective tissue (fig. 13.19).

CHECK YOUR RECALL

1. Describe the wall of an artery.
2. What is the function of smooth muscle in the arterial wall?
3. How is the structure of an arteriole different from that of an artery?

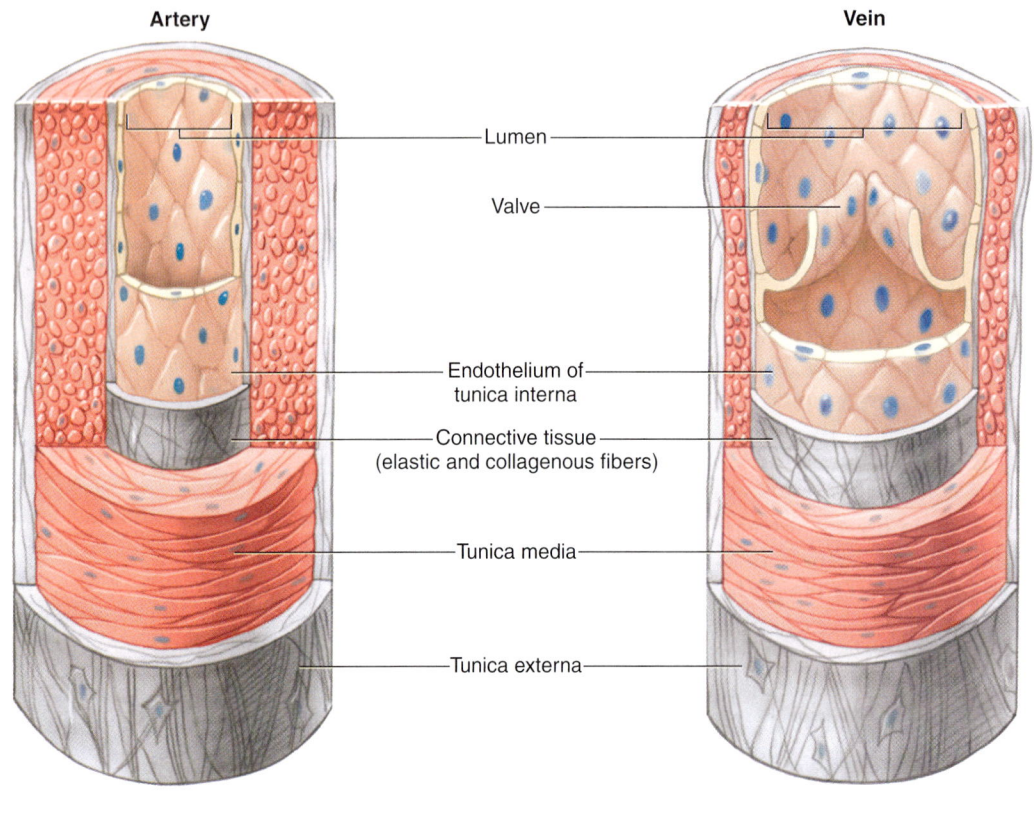

Figure 13.18

Blood vessels. (*A*) The wall of an artery. (*B*) The wall of a vein.

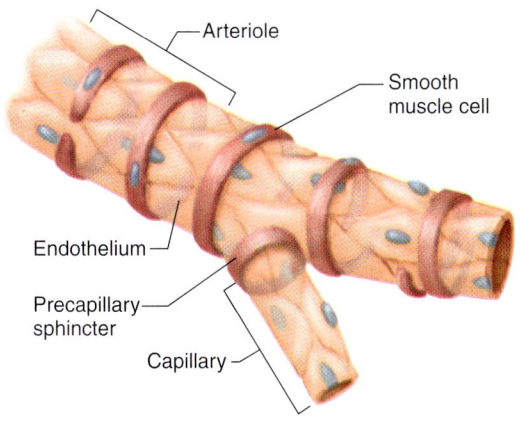

Figure 13.19
The smallest arterioles have only a few smooth muscle fibers in their walls. Capillaries lack these fibers.

Capillaries

Capillaries (kap´i-lar´´ez), the smallest-diameter blood vessels, connect the smallest arterioles and the smallest venules. Capillaries are extensions of the inner linings of arterioles in that their walls are endothelium (fig. 13.19). These thin walls form the semipermeable layer through which substances in the blood are exchanged for substances in the tissue fluid surrounding body cells.

The openings in capillary walls are thin slits where endothelial cells overlap (fig. 13.20). The sizes of these openings and, consequently, the permeability of the capillary wall vary from tissue to tissue. For example, the openings are relatively small in capillaries of smooth, skeletal, and cardiac muscle, whereas those in capillaries associated with endocrine glands, the kidneys, and the lining of the small intestine are larger.

Capillary density within tissues varies directly with tissues' rates of metabolism. Muscle and nerve tissues, which use large quantities of oxygen and nutrients, are richly supplied with capillaries. Tissues with slow metabolic rates, such as cartilaginous tissues, the epidermis, and the cornea, lack capillaries.

Patterns of capillary arrangement also differ in various body parts. For example, some capillaries pass directly from arterioles to venules, but others lead to highly branched networks (fig. 13.21).

Smooth muscles that encircle capillary entrances regulate blood distribution in capillary pathways. These muscles form *precapillary sphincters,* which may close a capillary by contracting or open it by relaxing (see fig. 13.19). A precapillary sphincter responds to the demands of the cells supplied by the capillary. When these cells have low concentrations of oxygen and nutrients, the sphincter relaxes; when cellular requirements have been met, the sphincter may contract again. Thus, blood can follow different pathways through a tissue to meet the changing requirements of cells.

Routing of blood flow to different parts of the body is due to vasoconstriction and vasodilation of arterioles and precapillary sphincters. During exercise, for example, blood enters the capillary networks of the skeletal muscles, where the cells have increased oxygen and nutrient requirements. At the same time, blood can bypass some of the capillary networks in the digestive tract tissues, where demand for blood is less immediate.

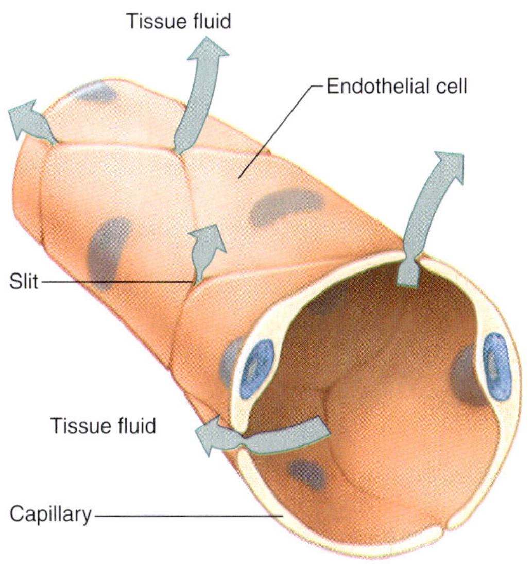

Figure 13.20
In capillaries, substances are exchanged between the blood and tissue fluid through openings (slits) separating endothelial cells.

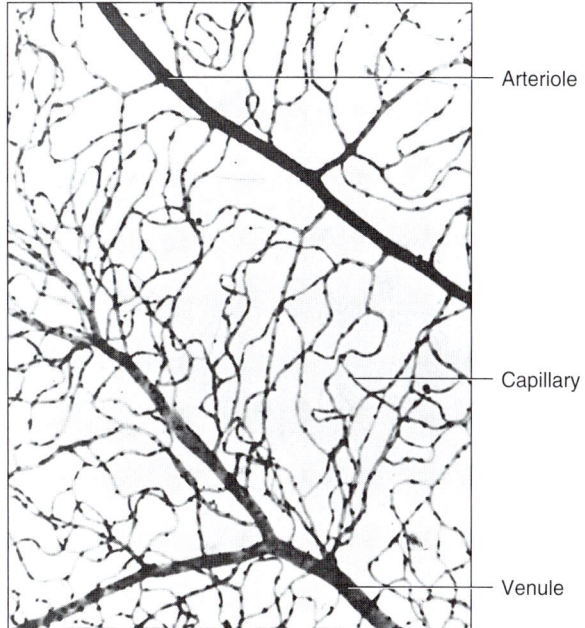

Figure 13.21
Light micrograph of a capillary network (100×).

Topic of Interest

ATHEROSCLEROSIS

In the arterial disease *atherosclerosis,* deposits of fatty materials, particularly cholesterol, are formed within and on the inner lining of the arterial walls. Such deposits, called *plaque,* protrude into the lumens of vessels and interfere with blood flow (fig. 13A). Furthermore, plaque often forms a surface texture that can initiate formation of a blood clot, increasing the risk of developing thrombi or emboli that cause blood deficiency (ischemia) or tissue death (necrosis) downstream from the obstruction.

The walls of affected arteries may degenerate, losing their elasticity and becoming hardened, or *sclerotic.* In this stage of the disease, called *arteriosclerosis,* a sclerotic vessel may rupture under the force of blood pressure.

Risk factors for developing atherosclerosis include a fatty diet, elevated blood pressure, tobacco smoking, obesity, and lack of physical exercise. Emotional and genetic factors may also increase susceptibility to atherosclerosis.

Several treatments attempt to clear clogged arteries. In *percutaneous transluminal angioplasty,* a thin, plastic catheter is passed through a tiny incision in the skin and into the lumen of the affected blood vessel. The catheter, with a tiny deflated balloon at its tip, is pushed along the vessel and into the blocked region. Once in position, the balloon is inflated for several minutes, with a pressure high enough to compress the atherosclerotic plaque against the arterial wall, widening the arterial lumen and restoring blood flow. However, blockage can recur if the underlying cause is not addressed.

Laser energy is also used to destroy atherosclerotic plaque and to channel through arterial obstructions to increase blood flow. In *laser angioplasty,* the light energy of a laser is transmitted through a bundle of optical fibers passed through a small incision in the skin and into the lumen of an obstructed artery.

Another procedure for treating arterial obstruction is *bypass graft surgery.* A surgeon uses a portion of a vein from the patient's lower limb or elsewhere to connect a healthy artery to the affected artery at a point beyond the obstruction. This allows blood from the healthy artery to bypass the narrowed region of the affected artery and supply the tissues downstream. The vein is connected backward, so that its valves do not impede blood flow.

A new treatment for atherosclerosis is *fibroblast growth factor,* a body chemical given as a drug. It stimulates new blood vessels to grow in the heart, a process called angiogenesis.

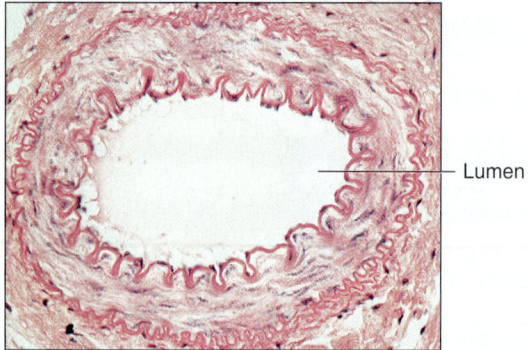

A

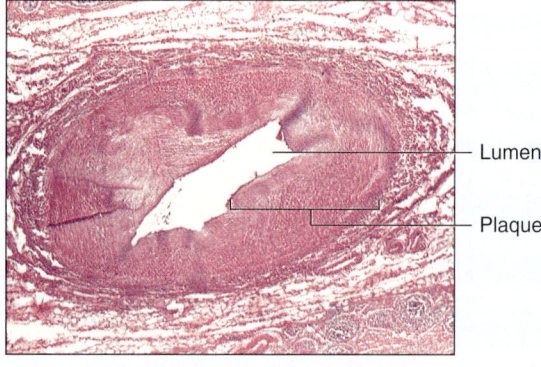

B

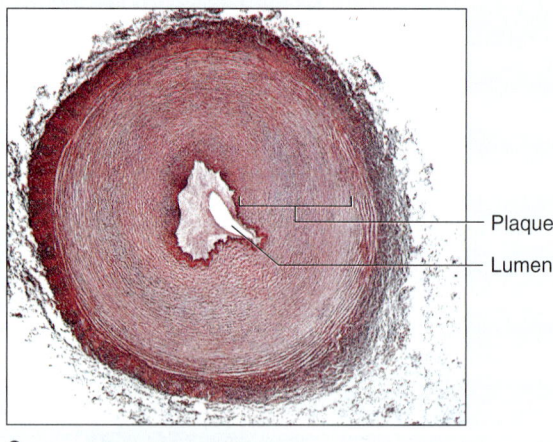

C

Figure 13A
Development of atherosclerosis. (A) Normal arteriole. (B, C) Accumulation of plaque on the inner wall of an arteriole.

CHECK YOUR RECALL

1. Describe a capillary wall.
2. What is the function of a capillary?
3. What controls blood flow into capillaries?

Exchanges in Capillaries

Gases, nutrients, and metabolic by-products are exchanged between the blood in capillaries and the tissue fluid surrounding body cells. The substances exchanged move through capillary walls by diffusion, filtration, and osmosis (see chapter 3, pp. 58–62).

Because blood entering systemic capillaries carries high concentrations of oxygen and nutrients, these substances diffuse through the capillary walls and enter the tissue fluid. Conversely, the concentrations of carbon dioxide and other wastes are generally greater in the tissues, and such wastes tend to diffuse into the capillary blood.

In the brain, the endothelial cells of capillary walls are more tightly fused than in other body regions. This arrangement forms a blood-brain barrier that protects the brain by keeping toxins out and preventing great biochemical fluctuations. Neuroglial cells also help form the blood-brain barrier. Unfortunately, the barrier keeps out many useful drugs as well. Researchers are developing ways to attach certain drugs to molecules that can cross the blood-brain barrier.

Plasma proteins generally remain in the blood because they are too large to diffuse through the membrane pores or slitlike openings between the endothelial cells of most capillaries. Also, these bulky proteins are not soluble in the lipid portions of capillary cell membranes.

Whereas diffusion depends on concentration gradients, filtration forces molecules through a membrane with hydrostatic pressure. In capillaries, the blood pressure generated when ventricle walls contract provides the force for filtration.

Blood pressure also moves blood through the arteries and arterioles. This pressure decreases as the distance from the heart increases, because of friction (peripheral resistance) between the blood and the vessel walls. For this reason, blood pressure is greater in the arteries than in the arterioles, and greater in the arterioles than in the capillaries. Blood pressure is similarly greater at the arteriolar end of a capillary than at the venular end. Therefore, the filtration effect occurs primarily at the arteriolar ends of capillaries.

The presence of a solute on one side of a cell membrane that cannot cross it creates an osmotic pressure. Because plasma proteins are trapped within the capillaries, they create an osmotic pressure that draws water into the capillaries. The term *colloid osmotic pressure* is often used to describe this osmotic effect due solely to the plasma proteins.

The effect of capillary blood pressure, favoring filtration, and the plasma colloid osmotic pressure, favoring reabsorption, have opposite actions. At the arteriolar end of capillaries, the blood pressure is higher than the colloid osmotic pressure, so at the arteriolar end of the capillary, filtration predominates. At the venular end, the colloid osmotic pressure is essentially unchanged, but the blood pressure has decreased due to resistance through the capillary. Thus, at the venular end, reabsorption predominates (fig. 13.22).

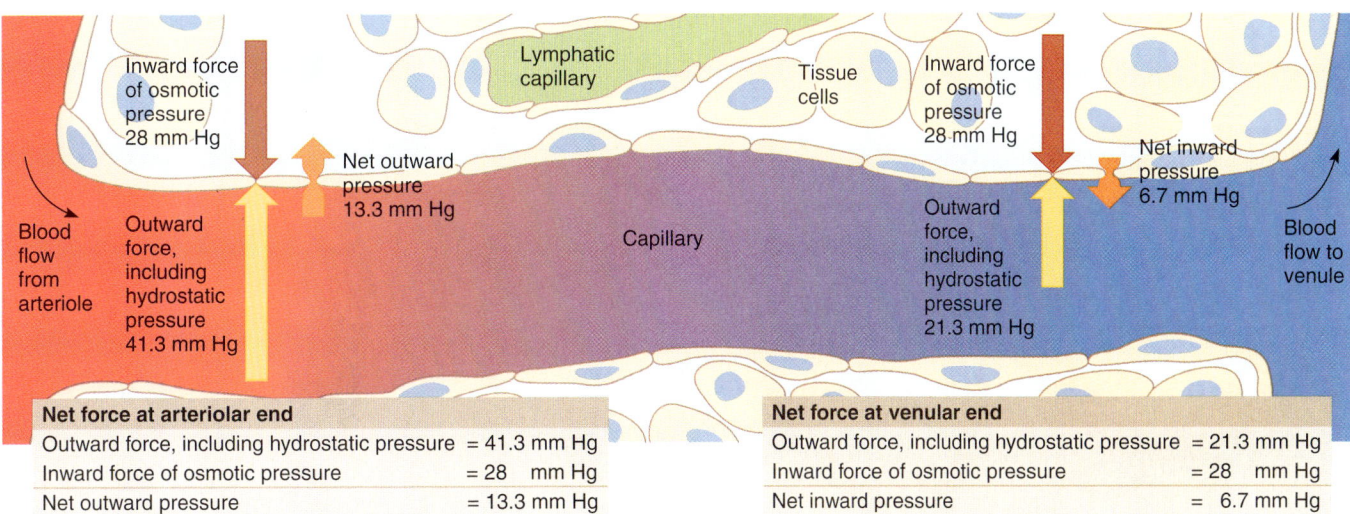

Figure 13.22
Water and other substances leave capillaries because of a net outward filtration pressure at the capillaries' arteriolar ends. Water enters at the capillaries' venular ends because of a net inward force of osmotic pressure. Substances move in and out along the length of the capillaries according to their respective concentration gradients.

Normally, more fluid leaves the capillaries than returns to them. Lymphatic vessels collect the excess fluid and return it to the venous circulation. Chapter 14 (p. 369) discusses this mechanism.

Sometimes unusual events may increase blood flow to capillaries, and excess fluid enters spaces between tissue cells. This may occur in response to certain chemicals, such as histamine, that vasodilate the arterioles near capillaries and increase capillary permeability. Enough fluid may leak out of the capillaries to overwhelm lymphatic drainage. Affected tissues become swollen (edematous) and painful.

CHECK YOUR RECALL

1. What forces are responsible for the exchange of substances between blood and tissue fluid?
2. Why is the fluid movement out of a capillary greater at its arteriolar end than at its venular end?

Venules and Veins

Venules (ven'ūlz) are the microscopic vessels that continue from the capillaries and merge to form **veins.** The veins, which carry blood back to the atria, follow pathways that roughly parallel those of the arteries.

The walls of veins are similar to those of arteries in that they are composed of three distinct layers (see fig. 13.18*B*). However, the middle layer of the venous wall is poorly developed. Consequently, veins have thinner walls that contain less smooth muscle and less elastic tissue than those of comparable arteries, but their lumens have greater diameters (fig. 13.23).

Many veins, particularly those in the upper and lower limbs, contain flaplike *valves,* which project inward from their linings. Valves are usually composed of two leaflets that close if blood begins to back up in a vein (fig. 13.24). These valves aid in returning blood to the heart because they open if blood flow is toward the heart, but close if it is in the opposite direction.

Veins also function as blood reservoirs. For example, in hemorrhage accompanied by a drop in arterial blood pressure, sympathetic nerve impulses reflexly stimulate the muscular walls of the veins. The resulting venous constrictions help maintain blood pressure by returning more blood to the heart. This mechanism ensures a nearly normal blood flow even when as much as 25% of blood volume is lost. Table 13.2 summarizes the characteristics of blood vessels.

CHECK YOUR RECALL

1. How does the structure of a vein differ from that of an artery?
2. How does venous circulation help maintain blood pressure when hemorrhaging causes blood loss?

13.5 Blood Pressure

Blood pressure is the force blood exerts against the inner walls of blood vessels. Although this force occurs throughout the vascular system, the term *blood pressure* most commonly refers to pressure in arteries supplied by branches of the aorta.

 The human heart creates enough pressure to squirt blood 30 feet.

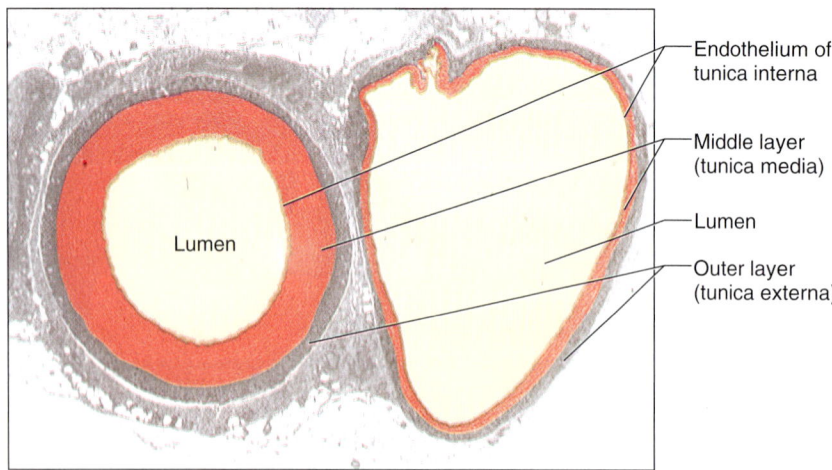

Figure 13.23
Note the structural differences in these cross sections of an artery and a vein (90×).

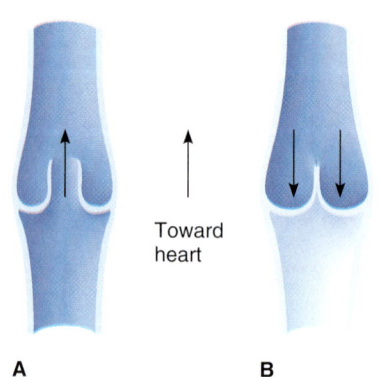

Figure 13.24
Venous valves (*A*) allow blood to move toward the heart, but (*B*) prevent blood from moving backward away from the heart.

TABLE 13.2		CHARACTERISTICS OF BLOOD VESSELS
VESSEL	TYPE OF WALL	FUNCTION
Artery	Thick, strong wall with three layers—an endothelial lining, a middle layer of smooth muscle and elastic tissue, and an outer layer of connective tissue	Carries blood under relatively high pressure from heart to arterioles
Arteriole	Thinner wall than an artery but with three layers; smaller arterioles have an endothelial lining, some smooth muscle tissue, and a small amount of connective tissue	Connects an artery to a capillary; helps control blood flow into a capillary by vasoconstricting or vasodilating
Capillary	Single layer of squamous epithelium	Provides a membrane through which nutrients, gases, and wastes are exchanged between the blood and tissue fluid; connects an arteriole to a venule
Venule	Thinner wall, less smooth muscle and elastic tissue than in an arteriole	Connects a capillary to a vein
Vein	Thinner wall than an artery but with similar layers; the middle layer is more poorly developed; some have flaplike valves	Carries blood under relatively low pressure from a venule to the heart; serves as blood reservoir; valves prevent a backflow of blood

Arterial Blood Pressure

Arterial blood pressure rises and falls in a pattern corresponding to the phases of the cardiac cycle. That is, contracting ventricles (ventricular systole) squeeze blood out and into the pulmonary trunk and aorta, which sharply increases the pressures in these arteries. The maximum pressure during ventricular contraction is called the **systolic pressure.** When the ventricles relax (ventricular diastole), the arterial pressure drops, and the lowest pressure that remains in the arteries before the next ventricular contraction is termed the **diastolic pressure** (fig. 13.25).

The surge of blood entering the arterial system during a ventricular contraction distends the elastic arterial walls, but the pressure drops almost immediately as the contraction ends, and the arterial walls recoil. This alternate expanding and recoiling of the arterial wall can be felt as a *pulse* in an artery that runs close to the surface.

CHECK YOUR RECALL

1. What is *blood pressure*?
2. Distinguish between systolic and diastolic blood pressure.
3. What causes a pulse in an artery?

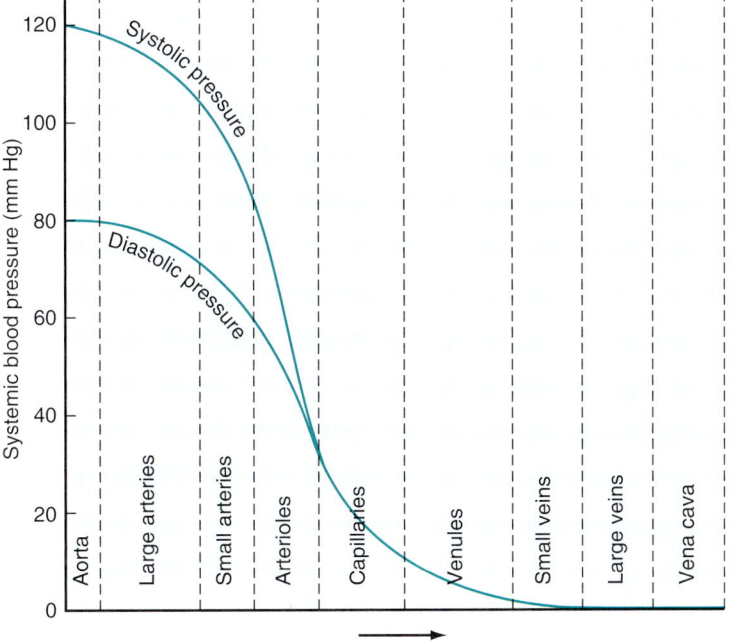

Figure 13.25
Blood pressure decreases as the distance from the left ventricle increases. Systolic pressure occurs during maximal ventricular contraction. Diastolic pressure occurs when the ventricles relax.

Factors That Influence Arterial Blood Pressure

Arterial blood pressure depends on a variety of factors. These include heart action, blood volume, peripheral resistance, and blood viscosity (fig. 13.26).

Heart Action

In addition to producing blood pressure by forcing blood into the arteries, heart action determines how much blood enters the arterial system with each ventricular contraction. The volume of blood discharged from the left ventricle with each contraction is called the **stroke volume** and equals about 70 milliliters in an average-weight male at rest. The volume discharged from the left ventricle per minute is called the **cardiac output,** calculated by multiplying the stroke volume by the heart rate in beats per minute (cardiac output = stroke volume × heart rate). Thus, if the stroke volume is 70 milliliters and the heart rate is 72 beats per minute, the cardiac output is 5,040 milliliters per minute.

Blood pressure varies with cardiac output. If either stroke volume or heart rate increases, so does cardiac output, and as a result, blood pressure initially rises. Conversely, if stroke volume or heart rate decreases, cardiac output decreases, and blood pressure also initially decreases.

Blood Volume

Blood volume equals the sum of the formed elements and plasma volumes in the vascular system. Although the blood volume varies somewhat with age, body size, and sex, it is usually about 5 liters for adults, or 8% of body weight in kilograms.

Blood pressure is normally directly proportional to blood volume within the cardiovascular system. Thus, any changes in blood volume can initially alter blood pressure. For example, if a hemorrhage reduces blood volume, blood pressure initially drops. If a transfusion restores normal blood volume, normal blood pressure may be reestablished. Blood volume can also fall if the fluid balance is upset, as happens in dehydration. Fluid replacement can reestablish normal blood volume and pressure.

 Licorice can raise your blood pressure.

Peripheral Resistance

Friction between the blood and the walls of blood vessels produces a force called **peripheral resistance** (per-rif´er-al re-zis´tans), which hinders blood flow. Blood pressure must overcome this force if the blood is to continue flowing. Therefore, factors that alter the peripheral resistance change blood pressure. For example, contracting smooth muscles in arteriolar walls increase the peripheral resistance by constricting these vessels. Blood tends to back up into the arteries supplying the arterioles, and the arterial pressure rises. Dilation of arterioles has the opposite effect—peripheral resistance lessens, and arterial blood pressure drops in response.

Blood Viscosity

Viscosity (vis-kos´ĭ-te) is the ease with which a fluid's molecules flow past one another. The greater the viscosity, the greater the resistance to flowing.

Blood cells and plasma proteins increase blood viscosity. Since the greater the blood's resistance to flowing, the greater the force needed to move it through the vascular system, it is not surprising that blood pressure rises as blood viscosity increases and drops as viscosity decreases.

 CHECK YOUR RECALL

1. What is the relationship between cardiac output and blood pressure?
2. How does blood volume affect blood pressure?
3. What is the relationship between peripheral resistance and blood pressure? Between blood viscosity and blood pressure?

Control of Blood Pressure

Two important mechanisms for maintaining normal arterial pressure are regulation of cardiac output and regulation of peripheral resistance. As mentioned previously, cardiac output depends on the volume of blood discharged from the left ventricle with each contraction (stroke volume) and heart rate. For example, the blood volume entering the ventricle affects the stroke volume. Entering blood mechanically stretches myocardial fibers in the ventricular wall. Within limits, the longer these

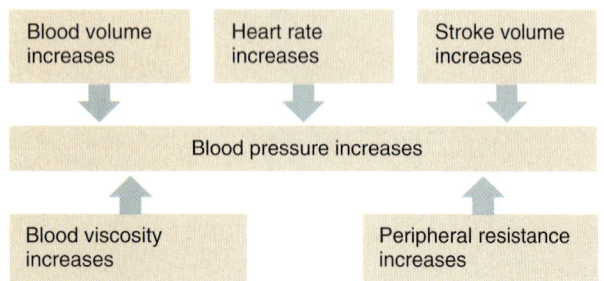

Figure 13.26
Some of the factors that influence arterial blood pressure.

fibers, the greater the force with which they contract. This relationship between fiber length (due to stretching of the cardiac muscle cell just before contraction) and force of contraction is called *Starling's law of the heart*. Because of it, the heart can respond to the immediate demands placed on it by the varying volumes of blood that return from the venous system. In other words, the more blood that enters the heart from the veins, the more the ventricular distends, the stronger it contracts, the greater the stroke volume, and the greater the cardiac output. The less blood that returns from the veins, the less the ventricle distends, and the lesser the stroke volume, and cardiac output, and the weaker the ventricular contraction. This mechanism ensures that the volume of blood discharged from the heart is equal to the volume entering its chambers.

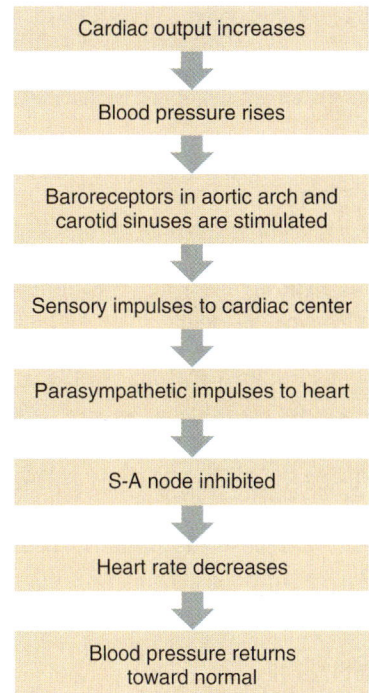

Figure 13.27
If blood pressure rises, baroreceptors initiate the cardio-inhibitor reflex to lower the blood pressure.

H*ypertension*, or high blood pressure, is persistently elevated arterial pressure. It is one of the more common diseases of the cardiovascular system.

High blood pressure with unknown cause is called *essential* (also primary or idiopathic) *hypertension*. Elevated pressure can be caused by another problem, such as kidney disease, high sodium intake, obesity, psychological stress, and arteriosclerosis. In arteriosclerosis, decreased elasticity of arterial walls and narrowed arterial lumens increase blood pressure.

The consequences of prolonged, uncontrolled hypertension can be very serious. As the left ventricle works overtime to pump sufficient blood, the myocardium thickens, enlarging the heart. If coronary blood vessels cannot support this overgrowth, parts of the heart muscle die and are replaced with fibrous tissue. Eventually, the enlarged and weakened heart dies.

Hypertension also contributes to the development of atherosclerosis. Plaque accumulation in arteries may cause a *coronary thrombosis* or *coronary embolism*. Similar changes in brain arteries increase the chances of a *cerebral vascular accident (CVA)*, or *stroke*, due to a cerebral thrombosis, embolism, or hemorrhage.

Treatment of hypertension varies among patients and may include exercising regularly, controlling body weight, reducing stress, and limiting sodium in the diet. Drug treatment includes diuretics and/or inhibitors of sympathetic nerve activity.

Baroreceptors in the walls of the aorta and carotid arteries sense changes in blood pressure. If arterial pressure increases, nerve impulses travel from the baroreceptors to the cardiac center of the medulla oblongata. This center relays parasympathetic impulses to the S-A node in the heart, and the heart rate decreases in response. As a result of this *cardioinhibitor reflex*, cardiac output falls, and blood pressure decreases toward the normal level (fig. 13.27). Conversely, decreasing arterial blood pressure initiates the *cardioaccelerator reflex*, which sends sympathetic impulses to the S-A node. As a result, the heart beats faster, increasing cardiac output and arterial pressure. Other factors that increase heart rate and blood pressure include physical exercise, a rise in body temperature, and emotional responses, such as fear and anger.

Peripheral resistance also controls blood pressure. Changes in arteriole diameters regulate peripheral resistance. Because blood vessels with smaller diameters offer a greater resistance to blood flow, factors that cause arteriole vasoconstriction increase peripheral resistance, and factors causing vasodilation decrease resistance.

The *vasomotor center* of the medulla oblongata continually sends sympathetic impulses to smooth muscles in the arteriole walls, keeping them in a state of tonic contraction, which helps maintain the peripheral resistance associated with normal blood pressure. Because the vasomotor center responds to changes in blood pressure, it can increase peripheral resistance by increasing its outflow of sympathetic impulses, or it can decrease such resistance by decreasing its sympathetic outflow. In the latter case, the vessels vasodilate as sympathetic stimulation falls.

Whenever arterial blood pressure suddenly increases, baroreceptors in the aorta and carotid arteries signal the vasomotor center, and the sympathetic outflow to the arterioles falls. The resulting vasodilation decreases peripheral resistance, and blood pressure decreases toward the normal level.

Certain chemicals, including carbon dioxide, oxygen, and hydrogen ions, also influence peripheral resistance

Topic of Interest

EXERCISE AND THE CARDIOVASCULAR SYSTEM

The cardiovascular system adapts to exercise. The conditioned athlete experiences increases in heart pumping efficiency, blood volume, blood hemoglobin concentration, and the number of mitochondria in muscle fibers. All of these adaptations improve oxygen delivery to, and utilization by, muscle tissue.

An athlete's heart typically changes in response to these increased demands, and may enlarge 40% or more. Myocardial mass increases, ventricular cavities expand, and the ventricle walls thicken. Stroke volume increases, and heart rate decreases, as does blood pressure. The lowest heart rate recorded in an athlete was 25 beats per minute! To a physician unfamiliar with a conditioned cardiovascular system, a trained athlete may appear abnormal!

The cardiovascular system responds beautifully to a slow, steady buildup in exercise frequency and intensity. It does not react well to sudden demands—such as a person who never exercises suddenly shoveling snow or running 3 miles.

A recent study confirmed age-old anecdotal reports of unaccustomed exercise causing heart failure. Researchers in the United States and Germany asked about 1,000 patients hospitalized for heart attacks about their exercise habits and what they were doing in the hour before the attack. They also questioned the same number of people who had not had heart attacks about their activities during the same hours as the ill people. The people with heart attacks were much more likely to have been engaging in unaccustomed strenuous exercise. But the study also turned up good news for those who exercise regularly: Although sedentary people have a two- to sixfold increased risk of cardiac arrest while exercising than when not, people in shape have little or no excess risk while exercising.

For exercise to benefit the cardiovascular system, the heart rate must be elevated to 70–85% of its "theoretical maximum" for at least half an hour three times a week. You can calculate your theoretical maximum by subtracting your age from 220. If you are eighteen years old, your theoretical maximum is 202 beats per minute. Then, 70–85% of this value is 141–172 beats per minute. Some good activities for raising the heart rate are tennis, skating, skiing, handball, vigorous dancing, hockey, basketball, biking, and fast walking.

It is wise to consult a physician before starting an exercise program. People over age thirty are advised to have a stress test, which is an electrocardiogram taken during exercise. (The standard electrocardiogram is taken at rest.) An arrhythmia that appears only during exercise may indicate heart disease that has not yet produced symptoms.

---

by affecting precapillary sphincters and smooth muscle in arteriole walls. For example, increasing blood carbon dioxide, decreasing blood oxygen, and lowering blood pH relaxes smooth muscle in the systemic circulation. This increases local blood flow to tissues with high metabolic rates, such as exercising skeletal muscles. In addition, epinephrine and norepinephrine vasoconstrict many systemic vessels, increasing peripheral resistance.

CHECK YOUR RECALL

1. What factors affect cardiac output?
2. What is the function of baroreceptors in the walls of the aorta and carotid arteries?
3. How does the vasomotor center control peripheral resistance?

Venous Blood Flow

Blood pressure decreases as blood moves through the arterial system and into the capillary networks, so that little pressure remains at the venular ends of capillaries (see fig. 13.25). Instead, blood flow through the venous system is only partly the direct result of heart action and depends on other factors, such as skeletal muscle contraction, breathing movements, and vasoconstriction of veins (*venoconstriction*).

Contracting skeletal muscles press on nearby vessels, squeezing the blood inside. As skeletal muscles press on veins with valves, some blood moves from one valve section to another (see fig. 13.24). This massaging action of contracting skeletal muscles helps push blood through the venous system toward the heart.

Respiratory movements also move venous blood. During inspiration, the pressure within the thoracic cavity is reduced as the diaphragm contracts and the rib cage moves upward and outward. At the same time, the pressure within the abdominal cavity is increased as the diaphragm presses downward on the abdominal viscera. Consequently, blood is squeezed out of abdominal veins and forced into thoracic veins. During exercise, these respiratory movements act with skeletal muscle contractions to increase return of venous blood to the heart.

Venoconstriction also returns venous blood to the heart. When venous pressure is low, sympathetic reflexes stimulate smooth muscles in the walls of veins to contract. The veins also provide a blood reservoir that can adapt its capacity to changes in blood volume. If some blood is lost and blood pressure falls, venoconstriction can force blood out of this reservoir. In both of these examples, venoconstriction helps maintain blood pressure by forcing more blood toward the heart.

CHECK YOUR RECALL

1. What is the function of venous valves?
2. How do skeletal muscles and respiratory movements affect venous blood flow?
3. What factors stimulate venoconstriction?

13.6 Paths of Circulation

The blood vessels can be divided into two major pathways. The **pulmonary circuit** (pul´mo-ner´´e sur´kit) consists of vessels that carry blood from the heart to the lungs and back to the heart. The **systemic circuit** (sis-tem´ik sur´kit) carries blood from the heart to all other parts of the body and back again (see fig. 13.1). The systemic circuit includes the coronary circulation, which supplies the heart itself and has already been described.

The circulatory pathways described in the following sections are those of an adult. Chapter 20 (p. 538) describes the fetal pathways, which are somewhat different.

Pulmonary Circuit

Blood enters the pulmonary circuit as it leaves the right ventricle through the pulmonary trunk. The pulmonary trunk extends upward and posteriorly from the heart. About 5 centimeters above its origin, it divides into the right and left pulmonary arteries (see fig. 13.4), which penetrate the right and left lungs, respectively. After repeated divisions, the pulmonary arteries give rise to arterioles that continue into the capillary networks associated with the walls of the alveoli (air sacs), where gas is exchanged between the blood and the air (see chapter 16, p. 450).

From the pulmonary capillaries, blood enters the venules, which merge to form small veins, and these veins in turn converge to form still larger veins. Four pulmonary veins, two from each lung, return blood to the left atrium. This completes the vascular loop of the pulmonary circuit.

Systemic Circuit

Freshly oxygenated blood moves from the left atrium into the left ventricle. Contraction of the left ventricle forces this blood into the systemic circuit. This circuit includes the aorta and its branches that lead to all the body tissues, as well as the companion system of veins that returns blood to the right atrium.

CHECK YOUR RECALL

1. Distinguish between the pulmonary and systemic circuits of the cardiovascular system.
2. Trace the path of blood through the pulmonary circuit from the right ventricle.

13.7 Arterial System

The **aorta** is the largest-diameter artery in the body. It extends upward from the left ventricle, arches over the heart to the left, and descends just anterior and to the left of the vertebral column. Figure 13.28 shows the aorta and its main branches.

Principal Branches of the Aorta

The first portion of the aorta is called the *ascending aorta*. Located at its base are the three cusps of the aortic valve, and opposite each cusp is a swelling in the aortic wall called an **aortic sinus.** The right and left coronary arteries arise from two of these sinuses (see fig. 13.8).

Three major arteries originate from the *arch of the aorta* (aortic arch): the **brachiocephalic** (brāk´´e-o-sĕ-fal´ik) **artery,** the left **common carotid** (kah-rot´id) **artery,** and the left **subclavian** (sub-kla´ve-an) **artery.**

The upper part of the *descending aorta* is positioned to the left of the midline. It gradually moves medially and finally lies directly in front of the vertebral column at the level of the twelfth thoracic vertebra. The portion of the descending aorta above the diaphragm is the **thoracic aorta.** It gives off many small branches to the thoracic wall and thoracic visceral organs.

Below the diaphragm, the descending aorta becomes the **abdominal aorta,** and it gives off branches to the abdominal wall and various abdominal organs. Branches to abdominal organs include: the **celiac** (se´le-ak) **artery,** which gives rise to the *gastric, splenic,* and *hepatic arteries;* the **superior** (supplies small intestine and superior portion of large intestine) and **inferior** (supplies inferior portion of large intestine) **mesenteric** (mes´´en-ter´ik) **arteries;** and the **suprarenal** (soo´´prah-re´nal) **arteries, renal** (re´nal) **arteries,** and **gonadal** (go´nad-al) **arteries,** which supply blood to the adrenal glands, kidneys, and ovaries or testes, respectively. The abdominal aorta ends near the brim of the pelvis, where it divides into right and left **common iliac** (il´e-ak) **arteries.** These vessels supply blood to lower regions of the abdominal wall, the pelvic organs, and the lower extremities. Table 13.3 summarizes the main branches of the aorta.

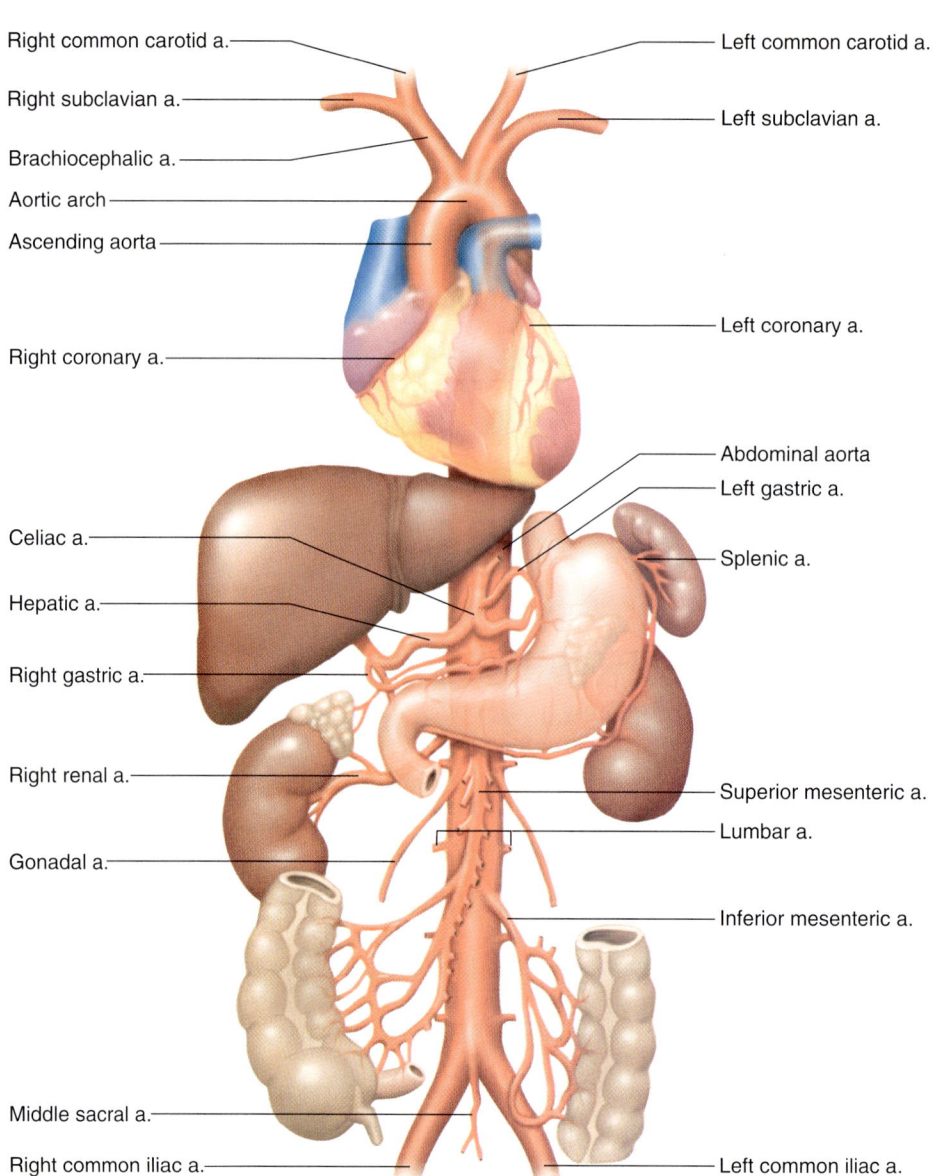

Figure 13.28
Principal branches of the aorta. (*a.* stands for *artery.*)

TABLE 13.3   PRINCIPAL BRANCHES OF THE AORTA

PORTION OF AORTA	MAJOR BRANCH	GENERAL REGIONS OR ORGANS SUPPLIED
Ascending aorta	Right and left coronary arteries	Heart
Arch of the aorta	Brachiocephalic artery	Right upper limb, right side of head
	Left common carotid artery	Left side of head
	Left subclavian artery	Left upper limb
Descending aorta	Bronchial artery	Bronchi
Thoracic aorta	Pericardial artery	Pericardium
	Esophageal artery	Esophagus
	Mediastinal artery	Mediastinum
	Posterior intercostal artery	Thoracic wall
Abdominal aorta	Celiac artery	Organs of upper digestive tract
	Phrenic artery	Diaphragm
	Superior mesenteric artery	Portions of small and large intestines
	Suprarenal artery	Adrenal gland
	Renal artery	Kidney
	Gonadal artery	Ovary or testis
	Inferior mesenteric artery	Lower portions of large intestine
	Lumbar artery	Posterior abdominal wall
	Middle sacral artery	Sacrum and coccyx
	Common iliac artery	Lower abdominal wall, pelvic organs, and lower limb

Arteries to the Neck, Head, and Brain

Branches of the subclavian and common carotid arteries supply blood to structures within the neck, head, and brain (fig. 13.29). The main divisions of the subclavian artery to these regions include the vertebral and thyrocervical arteries. The common carotid artery communicates with these regions by means of the internal and external carotid arteries.

The **vertebral arteries** pass upward through the foramina of the transverse processes of the cervical vertebrae and enter the skull through the foramen magnum. These vessels supply blood to the vertebrae and to their associated ligaments and muscles.

Within the cranial cavity, the vertebral arteries unite to form a single *basilar artery*. This vessel passes along the ventral brain stem and gives rise to branches leading to the pons, midbrain, and cerebellum. The basilar artery ends by dividing into two *posterior cerebral arteries* that supply portions of the occipital and temporal lobes of the cerebrum. The posterior cerebral arteries also help form the **cerebral arterial circle** (circle of Willis) at the base of the brain, which connects the vertebral artery and internal carotid artery systems (fig. 13.30). The union of these systems provides alternate pathways through which blood can reach brain tissues in the event of an arterial occlusion. It also equalizes blood pressure in the brain's blood supply.

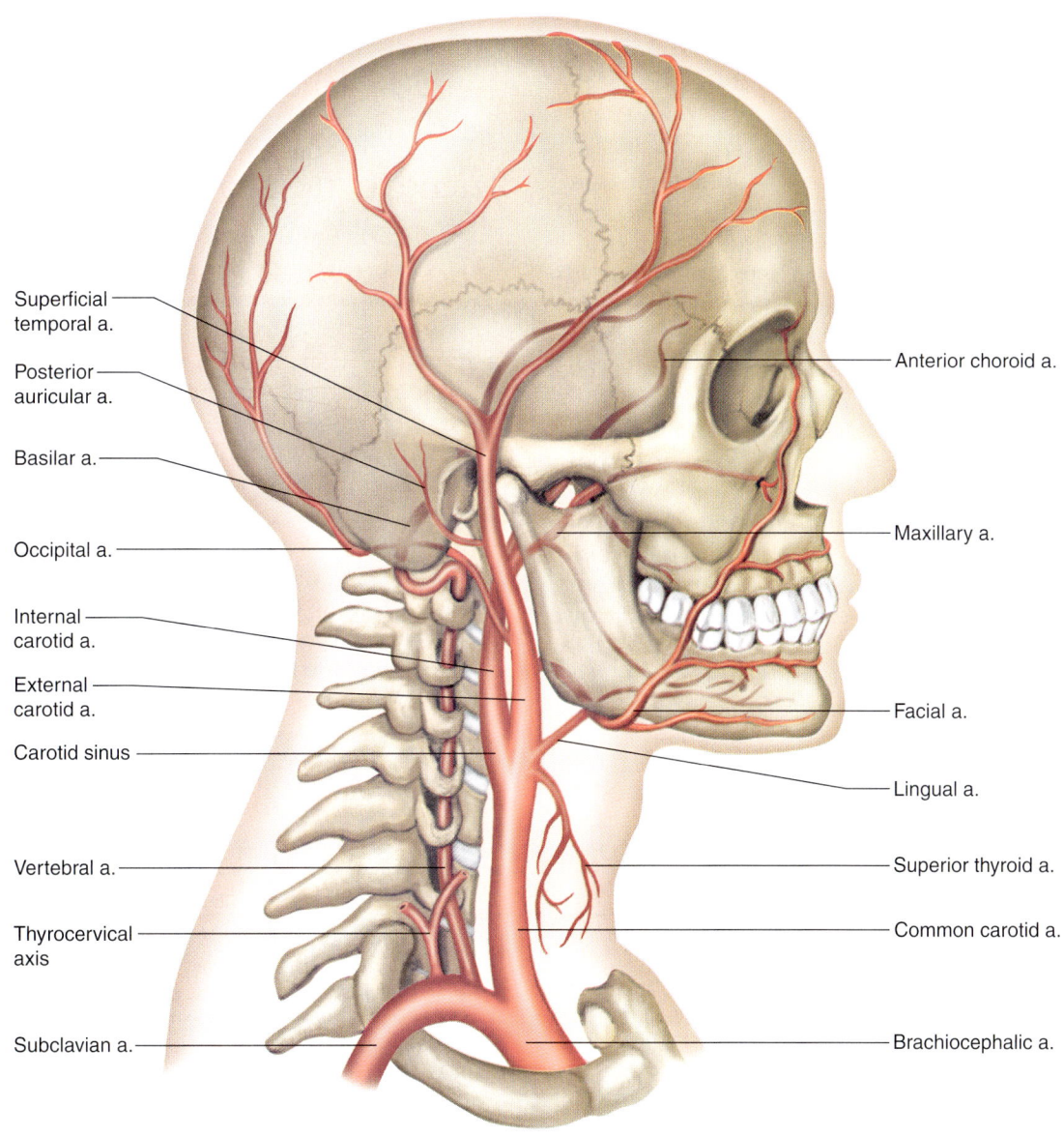

Figure 13.29
The main arteries of the head and neck. Note that the clavicle has been removed. (*a.* stands for *artery.*)

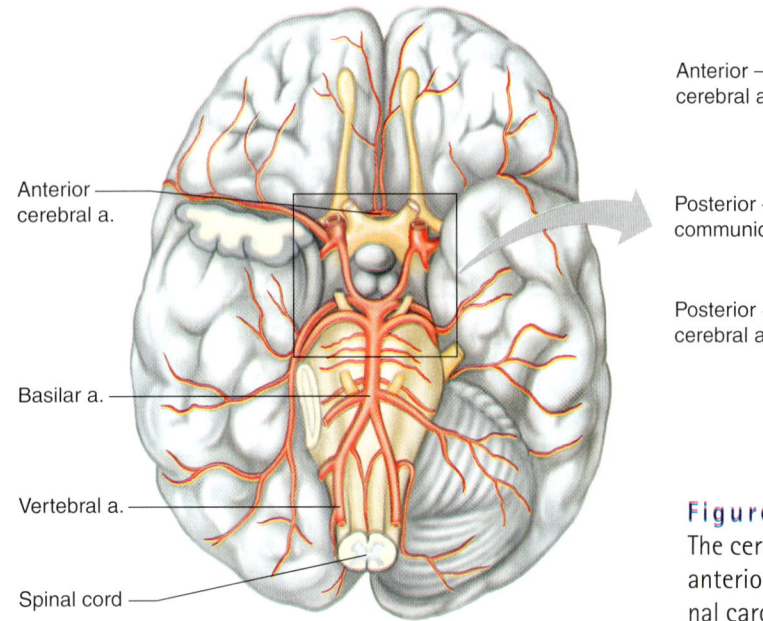
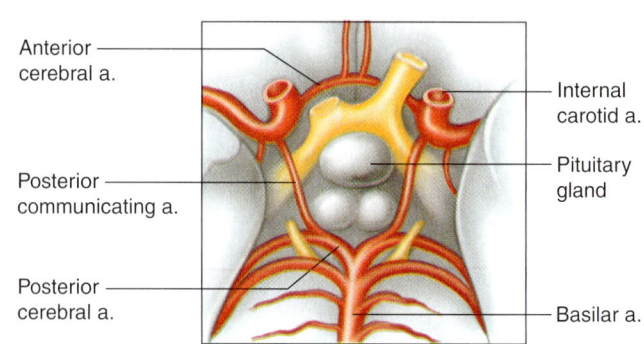

Figure 13.30
The cerebral arterial circle (circle of Willis) is formed by the anterior and posterior cerebral arteries, which join the internal carotid arteries. (*a.* stands for *artery*.)

The **thyrocervical** (thi″ro-ser′vĭ-kal) **arteries** are short vessels. At the thyrocervical axis, these vessels give off branches to the thyroid gland, parathyroid glands, larynx, trachea, esophagus, and pharynx, as well as to muscles in the neck, shoulder, and back.

The left and right *common carotid arteries* diverge into the internal and external carotid arteries. The **external carotid artery** courses upward on the side of the head, giving off branches to structures in the neck, face, jaw, scalp, and base of the skull. The **internal carotid artery** follows a deep course upward along the pharynx to the base of the skull. Entering the cranial cavity, it provides the major blood supply to the brain. Near the base of the internal carotid arteries are enlargements called **carotid sinuses** that, like aortic sinuses, contain baroreceptors controlling blood pressure. Table 13.4 summarizes the major branches of the external and internal carotid arteries.

Arteries to the Shoulder and Upper Limb

The subclavian artery, after giving off branches to the neck, continues into the arm (fig. 13.31). It passes between the clavicle and the first rib, and becomes the axillary artery. The **axillary artery** supplies branches to structures in the axilla and chest wall and becomes the **brachial artery,** which follows the humerus to the elbow. It gives rise to a *deep brachial artery* that curves posteriorly around the humerus and supplies the triceps brachii. Within the elbow, the brachial artery divides into an ulnar artery and a radial artery.

The **ulnar artery** leads downward on the ulnar side of the forearm to the wrist. Some of its branches supply the elbow joint, whereas others supply blood to muscles in the forearm.

The **radial artery** travels along the radial side of the forearm to the wrist, supplying the lateral muscles

TABLE 13.4 — MAJOR BRANCHES OF THE EXTERNAL AND INTERNAL CAROTID ARTERIES

ARTERY	MAJOR BRANCH	GENERAL REGION OR ORGANS SUPPLIED
External carotid artery	Superior thyroid artery	Larynx and thyroid gland
	Lingual artery	Tongue and salivary glands
	Facial artery	Pharynx, palate, chin, lips, and nose
	Occipital artery	Posterior scalp, meninges, and neck muscles
	Posterior auricular artery	Ear and lateral scalp
	Maxillary artery	Teeth, jaw, cheek, and eyelids
	Superficial temporal artery	Parotid salivary gland and surface of the face and scalp
Internal carotid artery	Ophthalmic artery	Eye and eye muscles
	Anterior choroid artery	Choroid plexus and brain
	Anterior cerebral artery	Frontal and parietal lobes of the brain

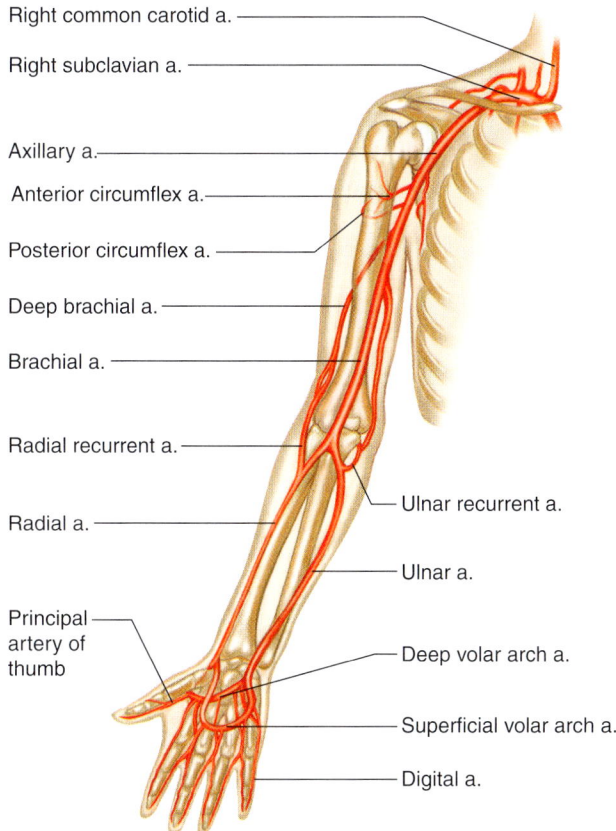

Figure 13.31
The main arteries to the shoulder and upper limb. (*a.* stands for *artery*.)

of the forearm. As the radial artery nears the wrist, it approaches the surface and provides a convenient vessel for taking the pulse (radial pulse).

At the wrist, the branches of the ulnar and radial arteries join to form a network of vessels. Arteries arising from this network supply blood to the wrist, hand, and fingers.

Arteries to the Thoracic and Abdominal Walls

Blood reaches the thoracic wall through several vessels. The **internal thoracic artery,** a branch of the subclavian artery, gives off two *anterior intercostal* (in″ter-kos′tal) *arteries* that supply the intercostal muscles and mammary glands. The *posterior intercostal arteries* arise from the thoracic aorta and enter the intercostal spaces. They supply the intercostal muscles, the vertebrae, the spinal cord, and the deep muscles of the back.

Branches of the *internal thoracic* and *external iliac arteries* provide blood to the anterior abdominal wall. Paired vessels originating from the abdominal aorta, including the *phrenic* and *lumbar arteries,* supply blood to structures in the posterior and lateral abdominal wall.

Arteries to the Pelvis and Lower Limb

The abdominal aorta divides to form the **common iliac** (il′e-ak) **arteries** at the level of the pelvic brim, and these vessels provide blood to the pelvic organs, gluteal region, and lower limbs (fig. 13.32). Each common iliac artery divides into an internal and an external branch. The **internal iliac artery** gives off many branches to pelvic muscles and visceral structures, as well as to the gluteal muscles and the external reproductive organs. The **external iliac artery** provides the main blood supply to the lower limbs. It passes downward along the brim of the pelvis and branches to supply the muscles and skin in the lower abdominal wall. Midway between the symphysis pubis and the anterior superior iliac spine of the ilium, the external iliac artery becomes the femoral artery.

The **femoral** (fem′or-al) **artery,** which approaches the anterior surface of the upper thigh, branches to muscles and superficial tissues of the thigh. These branches also supply the skin of the groin and the lower abdominal wall.

As the femoral artery reaches the proximal border of the space behind the knee, it becomes the **popliteal** (pop-lit′e-al) **artery.** Branches of this artery supply blood to the knee joint and to certain muscles in the thigh and calf. The popliteal artery diverges into the anterior and posterior tibial arteries.

The **anterior tibial artery** passes downward between the tibia and fibula, giving off branches to the skin and muscles in anterior and lateral regions of the leg. This vessel continues into the foot as the *dorsalis pedis artery* (dorsal pedis artery), which supplies blood to the foot and toes. The **posterior tibial artery,** the larger of the two popliteal branches, descends beneath the calf muscles, and branches to the skin, muscles, and other tissues of the leg along the way.

CHECK YOUR RECALL

1. Name the portions of the aorta.
2. Name the vessels that arise from the aortic arch.
3. Name the branches of the thoracic and abdominal aorta.
4. Which vessels supply blood to the head? To the upper limb? To the abdominal wall? To the lower limb?

13.8 Venous System

Venous circulation returns blood to the heart after blood and body cells exchange gases, nutrients, and wastes.

Characteristics of Venous Pathways

Venous vessels begin as capillaries merge into venules, venules merge into small veins, and small veins meet to form larger ones. Unlike the arterial pathways, however,

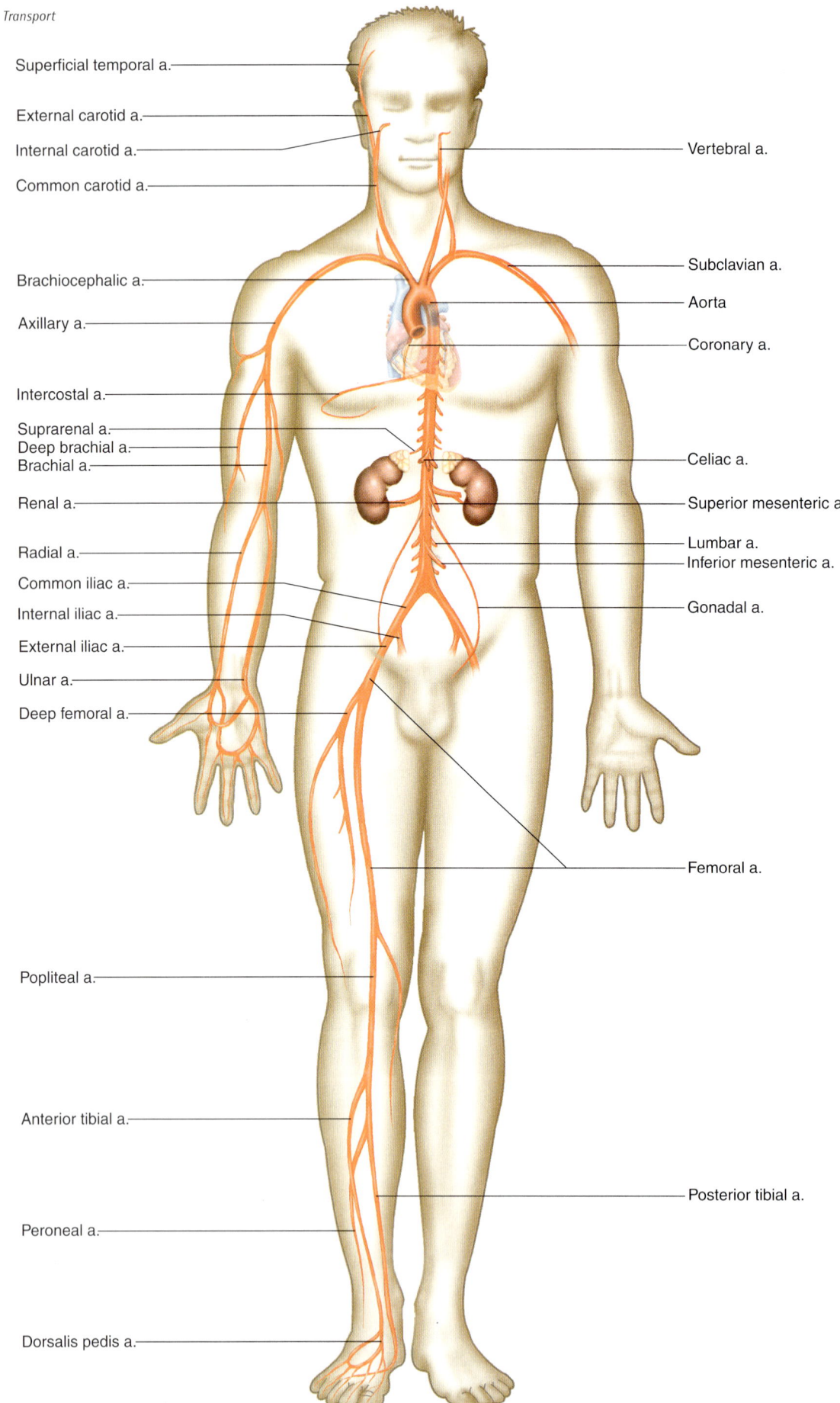

Figure 13.32
Major vessels of the arterial system. (*a.* stands for *artery.*)

the vessels of the venous system are difficult to follow. This is because they connect in irregular networks, so many unnamed tributaries may join to form a large vein.

On the other hand, the larger veins typically parallel the courses of named arteries, and these veins often have the same names as their arterial counterparts. For example, the renal vein parallels the renal artery, and the common iliac vein accompanies the common iliac artery.

The veins that carry blood from the lungs and myocardium back to the heart have already been described. The veins from all the other parts of the body converge into two major pathways, the superior and inferior venae cavae, which lead to the right atrium.

Veins from the Brain, Head, and Neck

The **external jugular** (jug´u-lar) **veins** drain blood from the face, scalp, and superficial regions of the neck. These vessels descend on either side of the neck and empty into the *right* and *left subclavian veins* (fig. 13.33).

The **internal jugular veins,** which are somewhat larger than the external jugular veins, arise from numerous veins and venous sinuses of the brain and from deep veins in parts of the face and neck. They descend through the neck and join the subclavian veins. These unions of the internal jugular and subclavian veins form large **brachiocephalic veins** on each side. The vessels then merge and give rise to the superior vena cava, which enters the right atrium.

Veins from the Upper Limb and Shoulder

A set of deep veins and a set of superficial ones drain the upper limb. The deep veins generally parallel the arteries in each region and have similar names, such as the *radial vein, ulnar vein, brachial vein,* and *axillary vein.* The superficial veins connect in complex networks just beneath the skin. They also communicate with the deep vessels of the upper limb, providing many alternate pathways through which blood can leave the tissues (fig. 13.34). The main vessels of the superficial network are the basilic and cephalic veins.

The **basilic** (bah-sil´ik) **vein** ascends from the forearm to the middle of the arm, where it penetrates

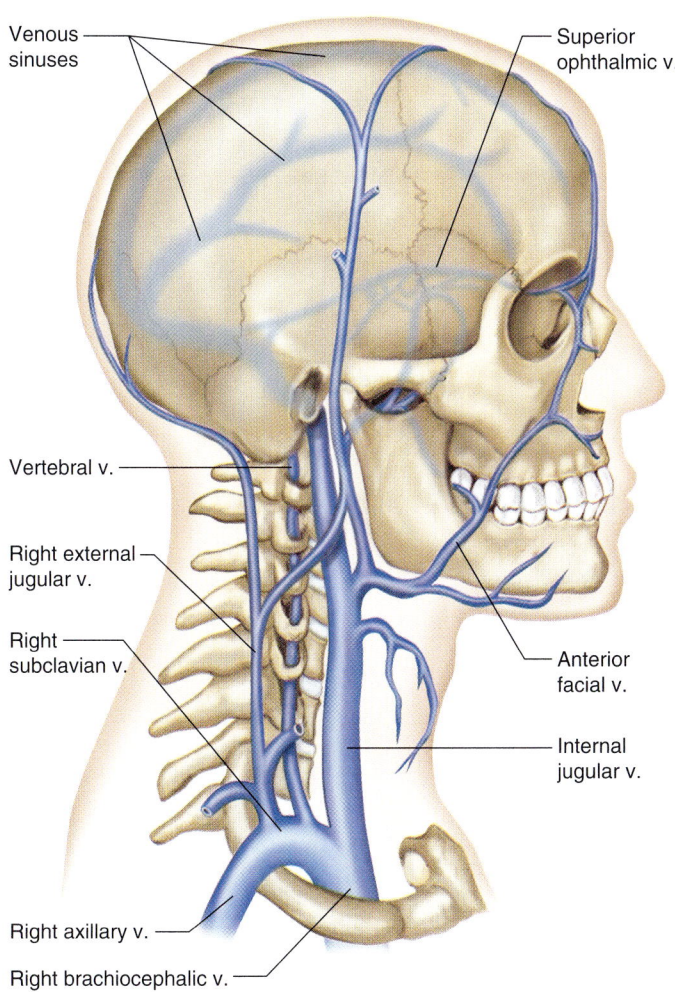

Figure 13.33
The major veins of the brain, head, and neck. Note that the clavicle has been removed. (*v.* stands for *vein.*)

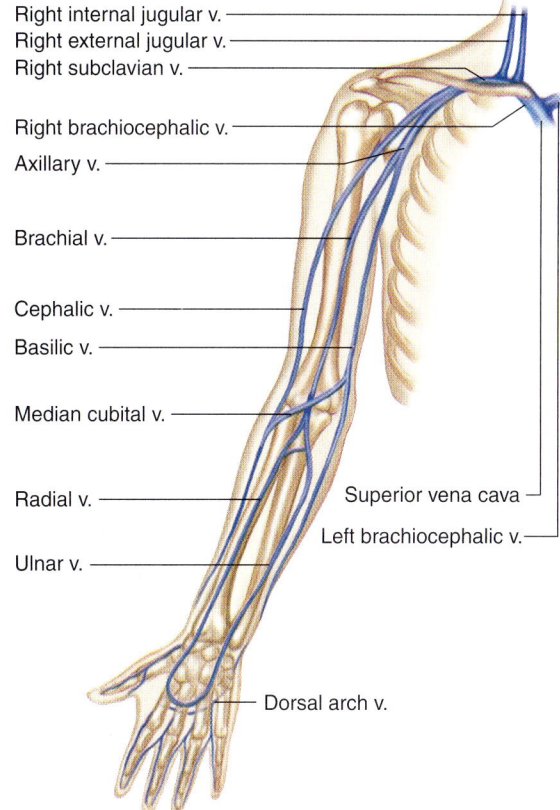

Figure 13.34
The main veins of the upper limb and shoulder. (*v.* stands for *vein.*)

deeply and joins the *brachial vein*. The basilic and brachial veins merge, forming the *axillary vein*.

The **cephalic** (sĕ-fal´ik) **vein** courses upward from the hand to the shoulder. In the shoulder, it pierces the tissues and empties into the axillary vein. Beyond the axilla, the axillary vein becomes the subclavian vein.

> In the bend of the elbow, a *median cubital vein* ascends from the cephalic vein on the lateral side of the forearm to the basilic vein on the medial side. This large vein is usually visible. It is often used as a site for *venipuncture*, when it is necessary to remove a blood sample for examination or to add fluids to blood.

Veins from the Abdominal and Thoracic Walls

Tributaries of the brachiocephalic and azygos veins drain the abdominal and thoracic walls. For example, the *brachiocephalic vein* receives blood from the *internal thoracic vein*, which generally drains the tissues the internal thoracic artery supplies. Some *intercostal veins* also empty into the brachiocephalic vein.

The **azygos** (az´ĭ-gos) **vein** originates in the dorsal abdominal wall and ascends through the mediastinum on the right side of the vertebral column to join the superior vena cava. It drains most of the muscular tissue in the abdominal and thoracic walls.

Tributaries of the azygos vein include the *posterior intercostal veins* on the right side, which drain the intercostal spaces, and the *superior* and *inferior hemiazygos veins*, which receive blood from the posterior intercostal veins on the left. The right and left *ascending lumbar veins*, with tributaries that include vessels from the lumbar and sacral regions, also connect to the azygos system.

Veins from the Abdominal Viscera

Veins usually carry the blood directly to the atria of the heart. However, those that drain the abdominal viscera are exceptions (fig. 13.35). They originate in the capillary networks of the stomach, intestines, pancreas, and spleen and carry blood from these organs through a

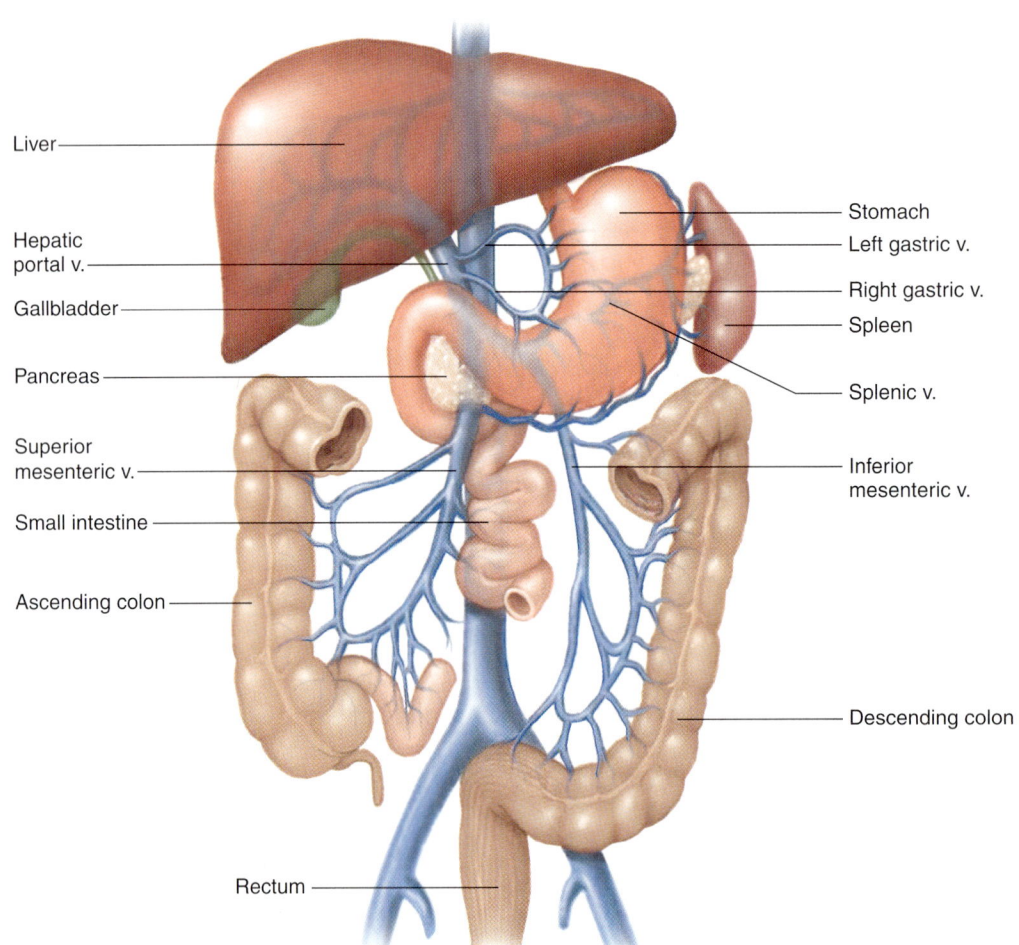

Figure 13.35
Veins that drain the abdominal viscera. (*v*. stands for *vein*.)

hepatic portal (por´tal) **vein** to the liver. This unique venous pathway is called the **hepatic portal system.**

Tributaries of the hepatic portal vein include:

1. Right and left *gastric veins* from the stomach.
2. *Superior mesenteric vein* from the small intestine, ascending colon, and transverse colon.
3. *Splenic vein* from a convergence of several veins draining the spleen, the pancreas, and a portion of the stomach. Its largest tributary, the *inferior mesenteric vein,* brings blood upward from the descending colon, sigmoid colon, and rectum.

About 80% of the blood flowing to the liver in the hepatic portal system comes from capillaries in the stomach and intestines, and is oxygen-poor but rich in nutrients. As discussed in chapter 15 (p. 408), the liver handles these nutrients in a variety of ways. It regulates blood glucose concentration by polymerizing excess glucose into glycogen for storage or by breaking down glycogen into glucose when blood glucose concentration drops below normal. The liver helps regulate blood concentrations of recently absorbed amino acids and lipids by modifying their molecules into forms cells can use, by oxidizing them, or by changing them into storage forms. The liver also stores certain vitamins and detoxifies harmful substances. Blood in the hepatic portal vein nearly always contains bacteria that have entered through intestinal capillaries. Large *Kupffer cells* lining small vessels in the liver called hepatic sinusoids phagocytize these microorganisms, removing them from portal blood before it leaves the liver.

After passing through the hepatic sinusoids of the liver, blood in the hepatic portal system travels through a series of merging vessels into **hepatic veins.** These veins empty into the inferior vena cava, returning the blood to the general circulation.

Veins from the Lower Limb and Pelvis

As in the upper limb, veins that drain blood from the lower limb are divided into deep and superficial groups (fig. 13.36). The deep veins of the leg, such as the *anterior* and *posterior tibial veins,* are named for the arteries they accompany. At the level of the knee, these vessels form a single trunk, the **popliteal vein.** This vein continues upward through the thigh as the **femoral vein,** which in turn becomes the **external iliac vein.**

The superficial veins of the foot, leg, and thigh connect to form a complex network beneath the skin. These vessels drain into two major trunks—the small and great saphenous veins. The **small saphenous** (sah-fe´nus) **vein** ascends along the back of the calf, enters the popliteal fossa, and joins the popliteal vein. The **great saphenous vein,** which is the longest vein in the body, ascends in front of the medial malleolus and extends upward along the medial side of the leg and thigh. In the thigh, it penetrates deeply and joins the femoral vein. Near its termination, the great saphenous vein receives tributaries from a number of vessels that drain the upper thigh, groin, and lower abdominal wall.

In addition to communicating freely with each other, the saphenous veins communicate extensively with the deep veins of the leg and thigh. Blood can thus return to the heart from the lower extremities by several routes.

In the pelvic region, vessels leading to the **internal iliac vein** carry blood away from the organs of the reproductive, urinary, and digestive systems. The internal iliac veins unite with the right and left external iliac veins to form the **common iliac veins.** These vessels, in turn, merge to produce the inferior vena cava.

Varicose veins have abnormal dilations. They result from increased blood pressure in the saphenous veins due to gravity, as occurs when a person stands for a prolonged period.

CHECK YOUR RECALL

1. Name the veins that return blood to the right atrium.
2. Which major veins drain blood from the head? From the upper limb? From the abdominal viscera? From the lower limb?

Clinical Terms Related to the Cardiovascular System

anastomosis (ah-nas´´to-mo´sis) Connection between two blood vessels, sometimes produced surgically.

angiospasm (an´je-o-spazm´´) Muscular spasm in the wall of a blood vessel.

arteriography (ar´´te-re-og´rah-fe) Injection of radiopaque solution into the vascular system for X-ray examination of arteries.

asystole (a-sis´to-le) Condition in which the myocardium fails to contract.

cardiac tamponade (kar´de-ak tam´´po-nād´) Compression of the heart by fluid accumulating within the pericardial cavity.

congestive heart failure (kon-jes´tiv hart fāl´yer) Inability of the left ventricle to pump adequate blood to cells.

cor pulmonale (kor pul-mo-na´le) Heart-lung disorder characterized by pulmonary hypertension and hypertrophy of the right ventricle.

embolectomy (em´´bo-lek´to-me) Removal of an embolus through an incision in a blood vessel.

endarterectomy (en´´dar-ter-ek´to-me) Removal of the inner wall of an artery to reduce an arterial occlusion.

palpitation (pal´´pĭ-ta´shun) Awareness of a heartbeat that is unusually rapid, strong, or irregular.

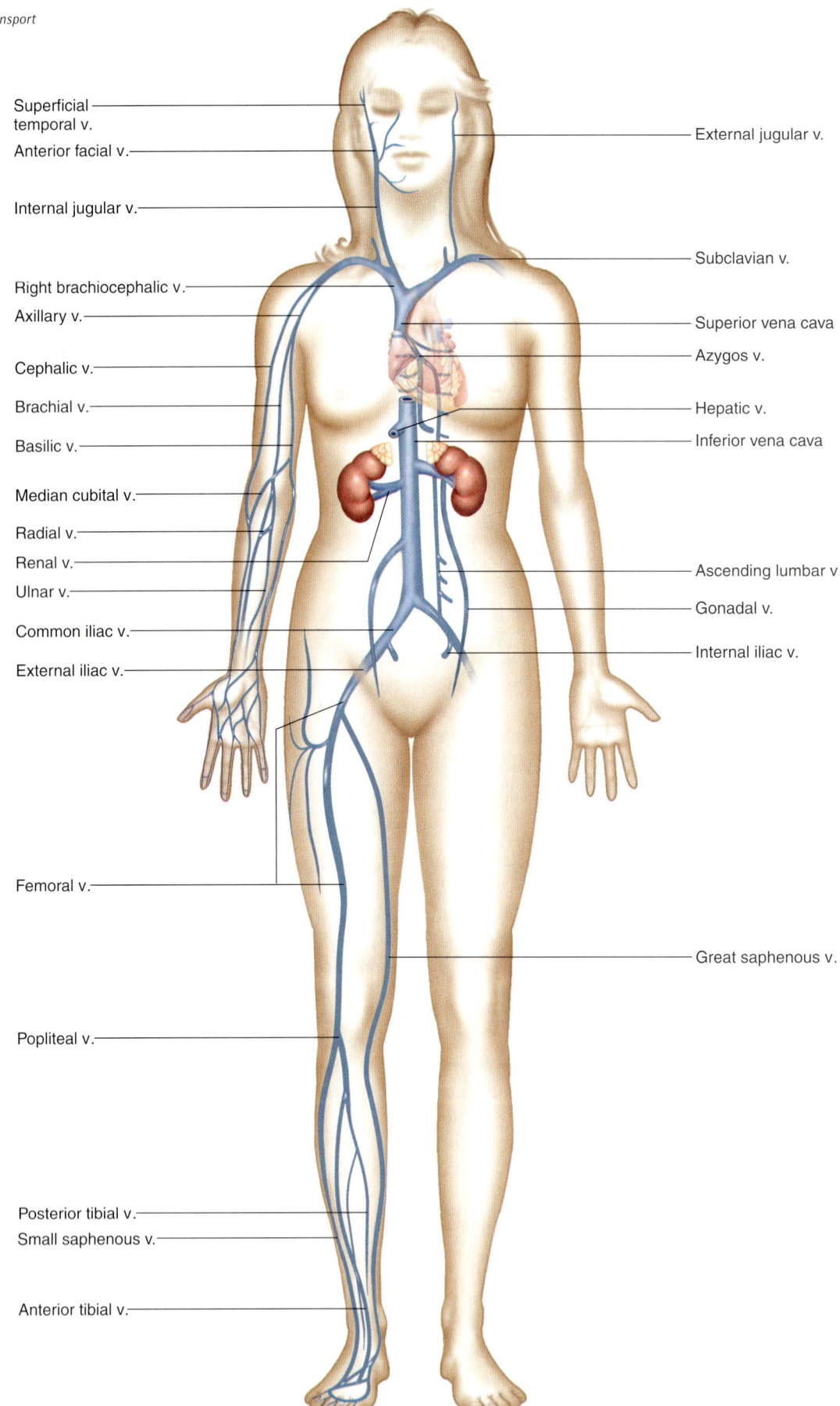

Figure 13.36
Major vessels of the venous system. (*v.* stands for *vein.*)

Organization

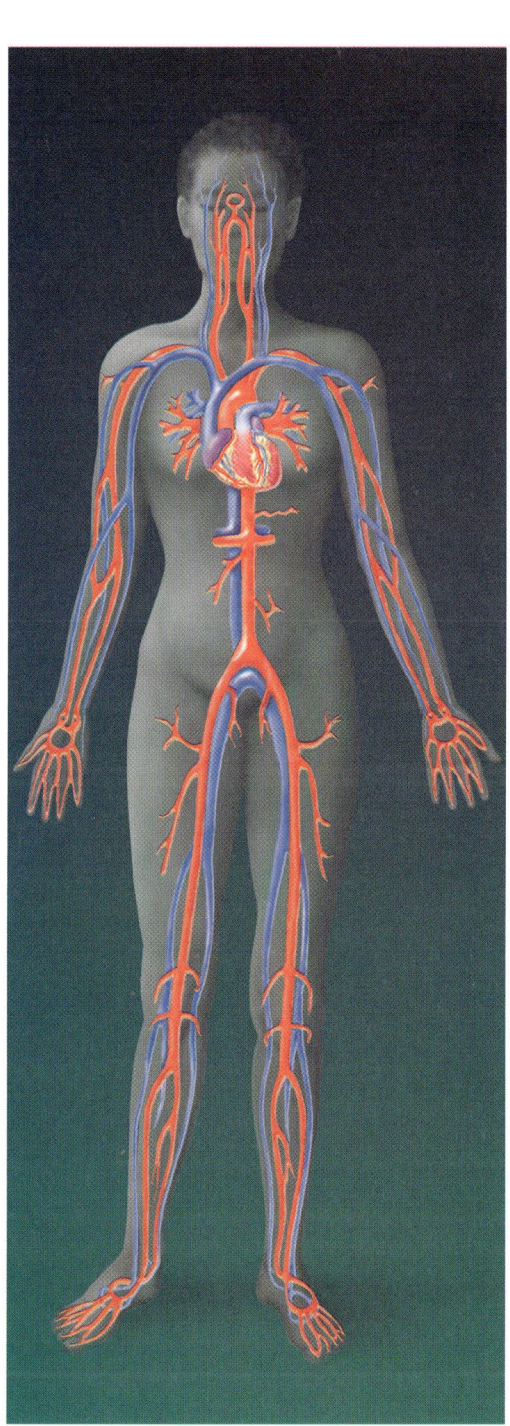

Cardiovascular System
The heart pumps blood through as many as 60,000 miles of blood vessels delivering nutrients to, and removing wastes from, all body cells.

Integumentary System

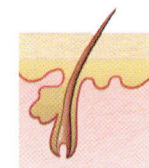

Changes in skin blood flow are important in temperature control.

Skeletal System

Bones help control plasma calcium levels.

Muscular System

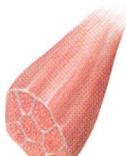

Blood flow increases to exercising skeletal muscle, delivering oxygen and nutrients and removing wastes. Muscle actions help the blood circulate.

Nervous System

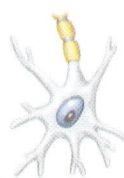

The brain depends on blood flow for survival. The nervous system helps control blood flow and blood pressure.

Endocrine System

Hormones are carried in the bloodstream. Some hormones directly affect the heart and blood vessels.

Lymphatic System

The lymphatic system returns tissue fluids to the bloodstream.

Digestive System

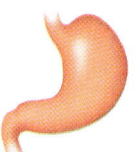

The digestive system breaks down nutrients into forms readily absorbed by the bloodstream.

Respiratory System

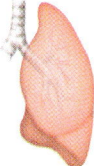

The respiratory system oxygenates the blood and removes carbon dioxide. Respiratory movements help the blood circulate.

Urinary System

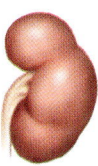

The kidneys clear the blood of wastes and substances present in the body. The kidneys help control blood pressure and blood volume.

Reproductive System

Blood pressure is important in normal function of the sex organs.

pericardiectomy (per˝ĭ-kar˝de-ek´to-me) Excision of the pericardium.
phlebitis (flĕ-bi´tis) Inflammation of a vein, usually in the lower limbs.
phlebotomy (flĕ-bot´o-me) Incision or puncture of a vein to withdraw blood.
sinus rhythm (si´nus rithm) The normal cardiac rhythm regulated by the S-A node.
thrombophlebitis (throm˝bo-flĕ-bi´tis) Formation of a blood clot in a vein in response to inflammation of the venous wall.
valvotomy (val-vot´o-me) Incision of a valve.
venography (ve-nog´rah-fe) Injection of radiopaque solution into the vascular system for X-ray examination of veins.

Clinical Connection

On July 2, 2001, 50 year old Bob Tools became the first person to receive an implantable artificial heart, developed at the University of Louisville. Weeks from death due to congestive heart failure at the time of the seven hour surgery, Tools enjoyed five more months of life, thanks to the device that replaced the functioning of his ventricles. The two-pound, titanium and plastic cardiac stand-in consists of an internal motor-driven hydraulic pump, battery and electronic package, and an external battery pack. The electronics component manages the rate and force of the pump's actions, tailoring them to the patient's condition. Several other patients have survived an average of two months with the implantable artificial heart.

Farther in the future, treatment for heart failure may consist of implants of stem cells that divide to produce new cardiac tissue. Evidence that this happens naturally comes from heart transplants from women to men. In one study, at varying times after the transplant, and after the recipients had died (of a variety of causes), researchers detected cells in the donor hearts that had the telltale Y chromosome of males. This meant that the recipient's cells had migrated to and mingled with donor heart cells. Some of the cells had the markings of stem cells, and some had already specialized into connective tissue, cardiac muscle tissue, and epithelium—precisely what was required to accept the new part. These experiments showed that a recipient's cells infiltrate a transplanted organ and eventually provide new specialized cells. The stem cells may come from the bit of recipient tissue to which the new organ is stitched, or migrate from the bone marrow and then differentiate into exactly what is needed to heal.

SUMMARY OUTLINE

13.1 Introduction (p. 329)
The cardiovascular system provides oxygen and nutrients to tissues and removes wastes.

13.2 Structure of the Heart (p. 329)
1. Size and location of the heart
   a. The heart is about 14 centimeters long and 9 centimeters wide.
   b. It is located within the mediastinum and rests on the diaphragm.
2. Coverings of the heart
   a. A layered pericardium encloses the heart.
   b. The pericardial cavity is a space between the parietal and visceral layers of the pericardium.
3. Wall of the heart
   The wall of the heart has three layers—an epicardium, a myocardium, and an endocardium.
4. Heart chambers and valves
   a. The heart is divided into two atria and two ventricles.
   b. Right chambers and valves
      (1) The right atrium receives blood from the venae cavae and coronary sinus.
      (2) The tricuspid valve separates the right atrium from the right ventricle.
      (3) A pulmonary valve guards the base of the pulmonary trunk.
   c. Left chambers and valves
      (1) The left atrium receives blood from the pulmonary veins.
      (2) The bicuspid valve separates the left atrium from the left ventricle.
      (3) An aortic valve guards the base of the aorta.
5. Skeleton of the heart
   The skeleton of the heart consists of fibrous rings that enclose the bases of the pulmonary artery and aorta.
6. Path of blood through the heart
   a. Blood low in oxygen and high in carbon dioxide enters the right side of the heart and is pumped into the pulmonary circulation.
   b. After blood is oxygenated in the lungs and some carbon dioxide is removed, it returns to the left side of the heart.
7. Blood supply to the heart
   a. The coronary arteries supply blood to the myocardium.
   b. Blood returns to the right atrium through the cardiac veins and coronary sinus.

13.3 Heart Actions (p. 336)
1. Cardiac cycle
   a. The atria contract while the ventricles relax. The ventricles contract while the atria relax.
   b. Pressure within the chambers rises and falls in repeated cycles.
2. Heart sounds
   Heart sounds are due to the vibrations the valve movements produce.
3. Cardiac muscle fibers
   a. Cardiac muscle fibers connect to form a functional syncytium.
   b. If any part of the syncytium is stimulated, the whole structure contracts as a unit.
4. Cardiac conduction system
   a. This system initiates and conducts impulses throughout the myocardium.
   b. Impulses from the S-A node pass slowly to the A-V node. Impulses travel rapidly along the A-V bundle and Purkinje fibers.

5. Electrocardiogram (ECG)
   a. An ECG records electrical changes in the myocardium during a cardiac cycle.
   b. The pattern contains several waves.
      (1) The P wave represents atrial depolarization.
      (2) The QRS complex represents ventricular depolarization.
      (3) The T wave represents ventricular repolarization.
6. Regulation of the cardiac cycle
   a. Physical exercise, body temperature, and the concentration of various ions affect heartbeat.
   b. Branches of sympathetic and parasympathetic nerve fibers innervate the S-A and A-V nodes.
   c. The cardiac center in the medulla oblongata regulates autonomic impulses to the heart.

13.4 Blood Vessels (p. 341)
Blood vessels form a closed circuit of tubes that carry blood from the heart to body cells and back again.
1. Arteries and arterioles
   a. Arteries are adapted to carry blood under high pressure away from the heart.
   b. The walls of arteries and arterioles consist of layers of endothelium, smooth muscle, and connective tissue.
   c. Autonomic fibers that can stimulate vasoconstriction or vasodilation innervate smooth muscle in vessel walls.
2. Capillaries
   a. Capillaries connect arterioles and venules.
   b. The capillary wall is a single layer of cells that forms a semipermeable membrane.
   c. Openings in capillary walls, where endothelial cells overlap, vary in size from tissue to tissue.
   d. Precapillary sphincters regulate capillary blood flow.
3. Exchanges in capillaries
   a. Capillary blood and tissue fluid exchange gases, nutrients, and metabolic by-products.
   b. Diffusion provides the most important means of transport.
   c. Filtration, which is due to the hydrostatic pressure of blood, causes a net outward movement of fluid at the arteriolar end of a capillary.
   d. Osmosis due to colloid osmotic pressure causes a net inward movement of fluid at the venular end of a capillary.
4. Venules and veins
   a. Venules continue from capillaries and merge to form veins.
   b. Veins carry blood to the heart.
   c. Venous walls are similar to arterial walls, but are thinner and contain less smooth muscle and elastic tissue.

13.5 Blood Pressure (p. 346)
Blood pressure is the force blood exerts against the insides of blood vessels.
1. Arterial blood pressure
   a. Arterial blood pressure rises and falls with the phases of the cardiac cycle.
   b. Systolic pressure is produced when the ventricle contracts. Diastolic pressure is the pressure in the arteries when the ventricle relaxes.
2. Factors that influence arterial blood pressure
   Arterial blood pressure increases as cardiac output, blood volume, peripheral resistance, or blood viscosity increases.
3. Control of blood pressure
   a. Blood pressure is controlled in part by the mechanisms that regulate cardiac output and peripheral resistance.
   b. The more blood that enters the heart, the stronger the ventricular contraction, the greater the stroke volume, and the greater the cardiac output.
   c. The cardiac center of the medulla oblongata regulates heart rate.
4. Venous blood flow
   a. Venous blood flow depends on skeletal muscle contraction, breathing movements, and venoconstriction.
   b. Many veins contain flaplike valves that prevent blood from backing up.
   c. Venoconstriction can increase venous pressure and blood flow.

13.6 Paths of Circulation (p. 351)
1. Pulmonary circuit
   The pulmonary circuit consists of vessels that carry blood from the right ventricle to the lungs and back to the left atrium.
2. Systemic circuit
   a. The systemic circuit consists of vessels that lead from the heart to the body cells (including those of the heart itself) and back to the heart.
   b. It includes the aorta and its branches.

13.7 Arterial System (p. 351)
1. Principal branches of the aorta
   a. The aorta is the largest artery with respect to diameter.
   b. Its major branches include the coronary, brachiocephalic, left common carotid, and left subclavian arteries.
   c. The branches of the descending aorta include the thoracic and abdominal groups.
   d. The abdominal aorta diverges into the right and left common iliac arteries.
2. Arteries to the neck, head, and brain
   These include branches of the subclavian and common carotid arteries.
3. Arteries to the shoulder and upper limb
   a. The subclavian artery passes into the upper limb, and in various regions is called the axillary and brachial artery.
   b. Branches of the brachial artery include the ulnar and radial arteries.
4. Arteries to the thoracic and abdominal walls
   a. Branches of the subclavian artery and thoracic aorta supply the thoracic wall.
   b. Branches of the abdominal aorta and other arteries supply the abdominal wall.
5. Arteries to the pelvis and lower limb
   The common iliac arteries supply the pelvic organs, gluteal region, and lower limbs.

13.8 Venous System (p. 355)
1. Characteristics of venous pathways
   a. Veins return blood to the heart.
   b. Larger veins usually parallel the paths of major arteries.
2. Veins from the brain, head, and neck
   a. Jugular veins drain these regions.
   b. Jugular veins unite with subclavian veins to form the brachiocephalic veins.
3. Veins from the upper limb and shoulder
   a. Sets of superficial and deep veins drain these regions.
   b. Deep veins parallel arteries with similar names.
4. Veins from the abdominal and thoracic walls
   Tributaries of the brachiocephalic and azygos veins drain these walls.
5. Veins from the abdominal viscera
   a. Blood from the abdominal viscera enters the hepatic portal system and is carried to the liver.
   b. From the liver, hepatic veins carry blood to the inferior vena cava.
6. Veins from the lower limb and pelvis
   a. Sets of deep and superficial veins drain these regions.
   b. The deep veins include the tibial veins, and the superficial veins include the saphenous veins.

REVIEW EXERCISES

1. Describe the general structure, function, and location of the heart. (p. 329)
2. Describe the pericardium. (p. 329)
3. Compare the layers of the cardiac wall. (p. 330)
4. Identify and describe the locations of the chambers and the valves of the heart. (p. 331)
5. Describe the skeleton of the heart, and explain its function. (p. 333)
6. Trace the path of blood through the heart. (p. 333)
7. Trace the path of blood through the coronary circulation. (p. 333)
8. Describe a cardiac cycle. (p. 336)
9. Describe the pressure changes in the atria and ventricles during a cardiac cycle. (p. 336)
10. Explain the origin of heart sounds. (p. 337)
11. Distinguish between the roles of the S-A node and the A-V node. (p. 338)
12. Explain how the cardiac conduction system controls the cardiac cycle. (p. 338)
13. Describe and explain the normal ECG pattern. (p. 339)
14. Discuss how the nervous system regulates the cardiac cycle. (p. 340)
15. Distinguish between an artery and an arteriole. (p. 342)
16. Explain control of vasoconstriction and vasodilation. (p. 342)
17. Describe the structure and function of a capillary. (p. 343)
18. Explain control of blood flow through a capillary. (p. 343)
19. Explain how diffusion functions in the exchange of substances between blood plasma and tissue fluid. (p. 345)
20. Explain why water and dissolved substances leave the arteriolar end of a capillary and enter the venular end. (p. 345)
21. Distinguish between a venule and a vein. (p. 346)
22. Explain how veins function as blood reservoirs. (p. 346)
23. Distinguish between systolic and diastolic blood pressures. (p. 347)
24. Name several factors that influence blood pressure, and explain how each produces its effect. (p. 348)
25. Describe the control of blood pressure. (p. 348)
26. List the major factors that promote the flow of venous blood. (p. 350)
27. Distinguish between the pulmonary and systemic circuits of the cardiovascular system. (p. 351)
28. Trace the path of blood through the pulmonary circuit. (p. 351)
29. Describe the aorta, and name its principal branches. (p. 351)
30. Describe the relationship between the major venous pathways and the major arterial pathways. (p. 355)

CRITICAL THINKING

1. How might the results of a cardiovascular exam differ for an athlete in top condition and a sedentary, overweight individual?
2. If you were asked to invent a blood vessel substitute, what materials might you use to build it? Include synthetic as well as natural materials.
3. What structures and properties should an artificial heart have?
4. Cigarette smoke contains thousands of chemicals, including nicotine and carbon monoxide. Nicotine constricts blood vessels. Carbon monoxide prevents oxygen from binding to hemoglobin. How do these two components of smoke affect the cardiovascular system?
5. Given the way capillary blood flow is regulated, do you think it is wiser to rest or to exercise following a heavy meal? Explain.
6. If a patient develops a blood clot in the femoral vein of the left lower limb and a portion of the clot breaks loose, where is the blood flow likely to carry the embolus? What symptoms are likely?
7. Cirrhosis of the liver, a disease commonly associated with alcoholism, obstructs blood flow through hepatic blood vessels. As a result, blood backs up, and capillary pressure greatly increases in organs the hepatic portal system drains. What effects might this increasing capillary pressure produce, and which organs would it affect?
8. If a cardiologist inserts a catheter into a patient's right femoral artery, which arteries would the tube have to pass through in order to reach the entrance to the left coronary artery?

WEB CONNECTIONS

Visit the website for additional study questions and more information about this chapter at:

http://www.mhhe.com/shieress8

chapter 14

Lymphatic System and Immunity

THE IMMUNE SYSTEM ACCEPTS A TRANSPLANT. Organ transplants succeed only when the recipient's immune system accepts the healing foreign tissue. Sometimes people in dire need of organ transplants get them from unexpected sources. Here are two transplant tales.

Bobbie diSabatino, fifty-six years old, had been awaiting a heart transplant at the University of Maryland Medical Center for four months, after suffering a heart attack. She awoke Valentine's Day, 1998, with a new heart—thanks to a friendship her daughter had struck up with another hospital visitor, Bob Bradshaw. Bob's wife, thirty-eight-year-old Cheryl, had a cluster of abnormal blood vessels in her brain that caused her death, following a one-month hospital stay. During that time, Bob and Cheryl made the difficult decision to bequeath Cheryl's heart to Bobbie. The organ was a perfect match, and the transplant saved the older woman's life.

Peter F. was sixteen years old in 1996 when an automobile accident left him paralyzed from the waist down. Much of his small intestine had to be removed. As a result, Peter had to be fed intravenously, but developed liver failure and recurrent infections at the site where the catheter was inserted. A vicious cycle set in. He lost so much weight because of his deficient small intestine that doctors could no longer find veins to deliver nutrients. The next step was to transplant a small intestine from a cadaver, but no match was found. His doctors at the University of Minnesota then looked for a living donor—Peter's father, who was a close enough match immunologically. The father donated 200 centimeters of his small intestine to his son, and both are now healthy. Peter eats normally again.

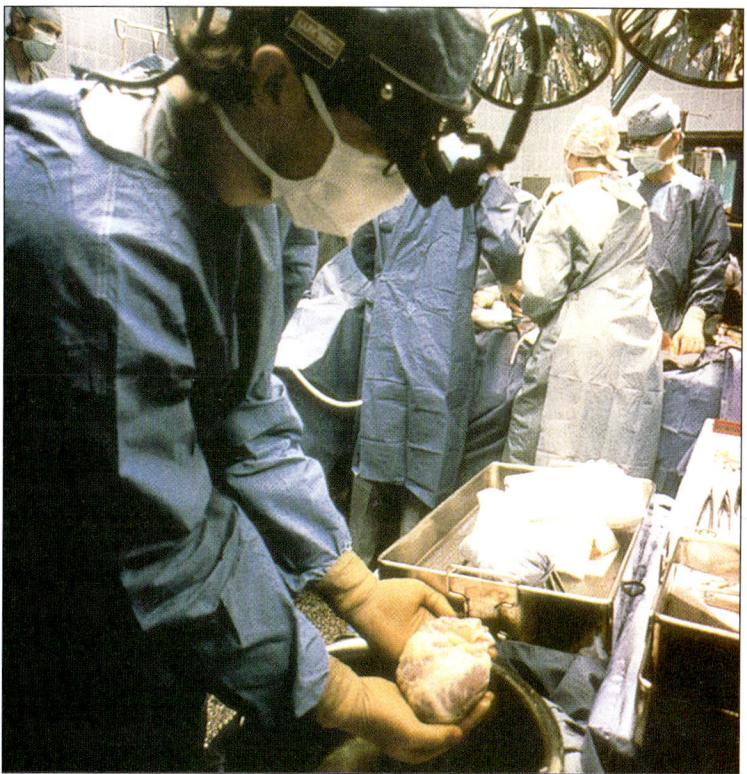

Photo:
Through heart transplantation, a heart that might have died with its donor years ago can provide life for a recipient, thanks to our understanding of the immune system—and a well-trained medical team!

Chapter Objectives

After studying this chapter, you should be able to do the following:

14.1 Introduction
1. Describe the general functions of the lymphatic system. (p. 367)

14.2 Lymphatic Pathways
2. Identify the locations of the major lymphatic pathways. (p. 367)

14.3 Tissue Fluid and Lymph
3. Describe how tissue fluid and lymph form, and explain the function of lymph. (p. 369)

14.4 Lymph Movement
4. Explain how lymphatic circulation is maintained. (p. 369)

14.5 Lymph Nodes
5. Describe a lymph node and its major functions. (p. 370)

14.6 Thymus and Spleen
6. Discuss the functions of the thymus and spleen. (p. 371)

14.7 Body Defenses Against Infection
7. Distinguish between specific and nonspecific defenses, and provide examples of each. (p. 373)

14.8 Nonspecific Defenses
8. List six nonspecific body defense mechanisms, and describe the action of each mechanism. (p. 373)

14.9 Specific Defenses (Immunity)
9. Explain how two major types of lymphocytes are formed and activated, and how they function in immune mechanisms. (p. 375)
10. Name the major types of immunoglobulins, and discuss their origins and actions. (p. 378)
11. Distinguish between primary and secondary immune responses. (p. 380)
12. Distinguish between active and passive immunity. (p. 381)
13. Explain how allergic reactions, tissue rejection reactions, and autoimmunity arise from immune mechanisms. (p. 381)

Aids to Understanding Words

gen- [be produced] aller*gen*: Substance that stimulates an allergic response.

humor- [fluid] *humor*al immunity: Immunity resulting from antibodies in body fluids.

immun- [free] *immun*ity: Resistance to (freedom from) a specific disease.

inflamm- [set on fire] *inflamm*ation: Localized redness, heat, swelling, and pain in tissues.

nod- [knot] *nod*ule: Small mass of lymphocytes surrounded by connective tissue.

path- [disease] *path*ogen: Disease-causing agent.

Key Terms

allergen (al´-er-jen)
antibody (an´tĭ-bod´´e)
antigen (an´tĭ-jen)
clone (klōn)
complement (kom´plĕ-ment)
hapten (hap´ten)
immunity (ĭmu´nĭ-te)
immunoglobulin (im´´u-no-glob´-u-lin)
lymph (limf)
lymphatic pathway (lim-fat´ik path´wa)
lymph node (limf nōd)
lymphocyte (lim´fo-sīt)
macrophage (mak´ro-fāj)
pathogen (path´o-jen)
reticuloendothelial tissue (rĕ-tik´´u-lo-en´´do-the´le-al tish´u)
spleen (splēn)
thymus (thī´mus)

14.1 Introduction

Like the cardiovascular system, the lymphatic system includes a network of vessels that transports fluids. The lymphatic system is a vast collection of cells and biochemicals that travel in lymphatic vessels, and the organs and glands that produce them. Lymphatic vessels carry away excess fluid from interstitial spaces in most tissues and return it to the bloodstream (fig. 14.1). Without the lymphatic system, this fluid would accumulate in tissue spaces. Special lymphatic capillaries, called lacteals, are located in the lining of the small intestine, where they absorb digested fats and transport them to the venous circulation.

The lymphatic system has a major second function—it enables us to live in a world filled with different types of organisms, some of which take up residence in the human body and cause infectious diseases. Cells and biochemicals of the lymphatic system launch both generalized and targeted attacks against "foreign" particles, enabling the body to destroy infectious microorganisms and viruses. This immunity function of the lymphatic system also attacks toxins and cancer cells, and when abnormal, can cause cancer, autoimmune disorders in which the body attacks itself, and allergies.

14.2 Lymphatic Pathways

The **lymphatic pathways** (lim-fat´ik path´wāz) begin as lymphatic capillaries. These tiny tubes merge to form larger lymphatic vessels. These, in turn, lead to larger vessels that unite with the veins in the thorax.

Lymphatic Capillaries

Lymphatic capillaries are microscopic, closed-ended tubes (fig. 14.2). They extend into interstitial spaces,

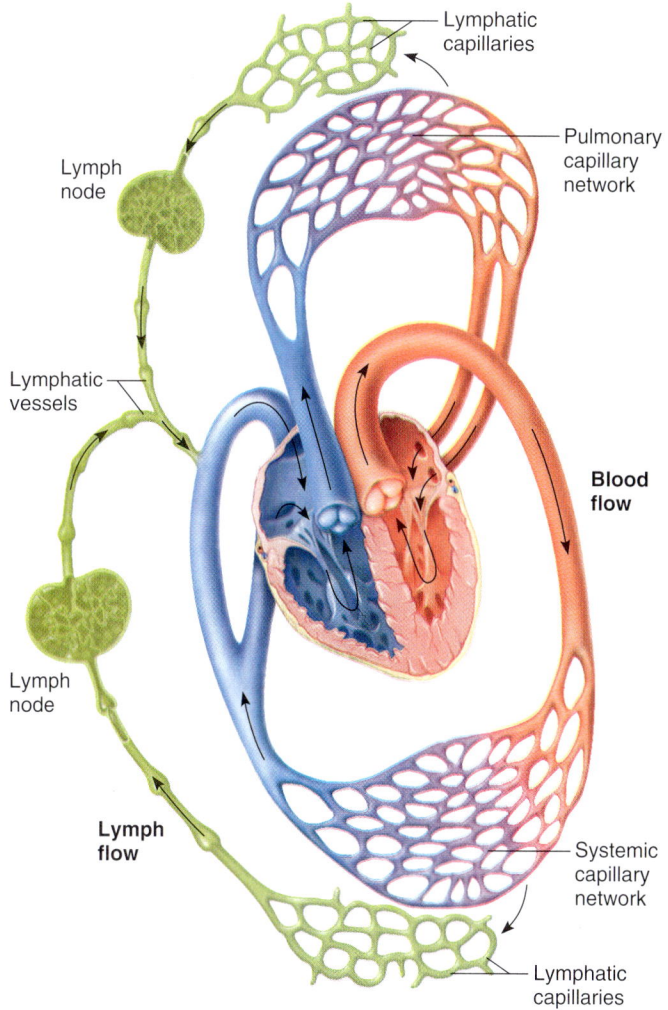

Figure 14.1
Schematic representation of lymphatic vessels transporting fluid from interstitial spaces to the bloodstream.

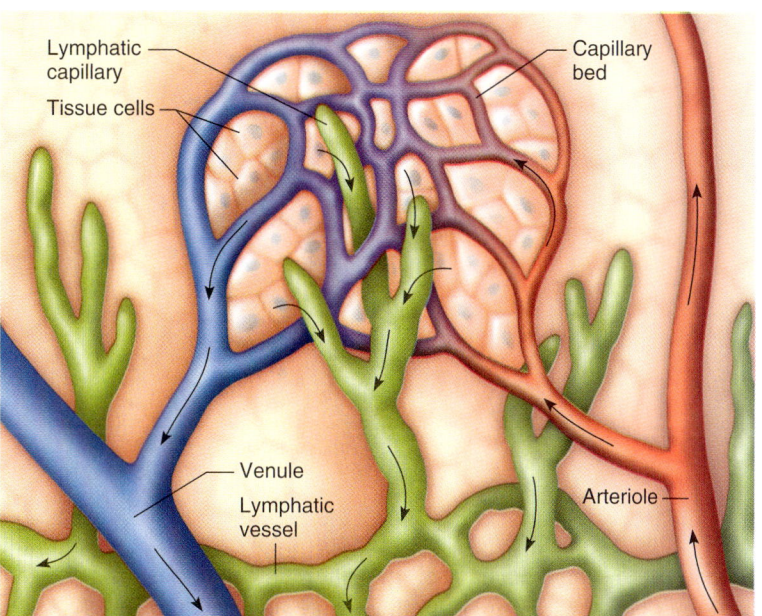

Figure 14.2
Lymphatic capillaries are microscopic, closed-ended tubes that begin in the interstitial spaces of most tissues.

forming complex networks that parallel those of blood capillaries. The walls of lymphatic capillaries, like those of blood capillaries, are formed from a single layer of squamous epithelial cells. These thin walls make it possible for tissue fluid to enter lymphatic capillaries. Fluid inside lymphatic capillaries is called **lymph** (limf).

Lymphatic Vessels

The walls of **lymphatic vessels** are similar to those of veins, but thinner. Also like veins, lymphatic vessels have flaplike valves that help prevent backflow of lymph (fig. 14.3).

The larger lymphatic vessels lead to specialized organs called **lymph nodes** (limf nōdz). After leaving the nodes, the vessels merge to form still larger **lymphatic trunks.**

Lymphatic Trunks and Collecting Ducts

Lymphatic trunks, which drain lymph, are named for the regions they serve. They join one of two **collecting ducts**—the thoracic duct or the right lymphatic duct (fig. 14.4A).

The **thoracic duct** is the larger and longer collecting duct. It receives lymph from the lower limbs and abdominal regions, left upper limb, and left side of the thorax, head, and neck, and empties into the left sub-

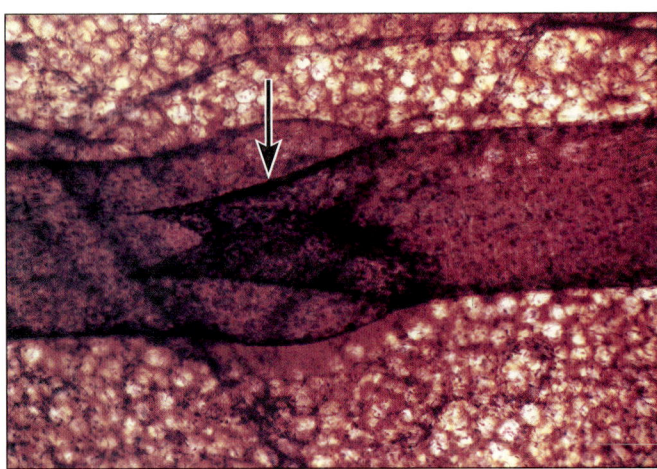

Figure 14.3
Light micrograph of the flaplike valve (arrow) within a lymphatic vessel (25×).

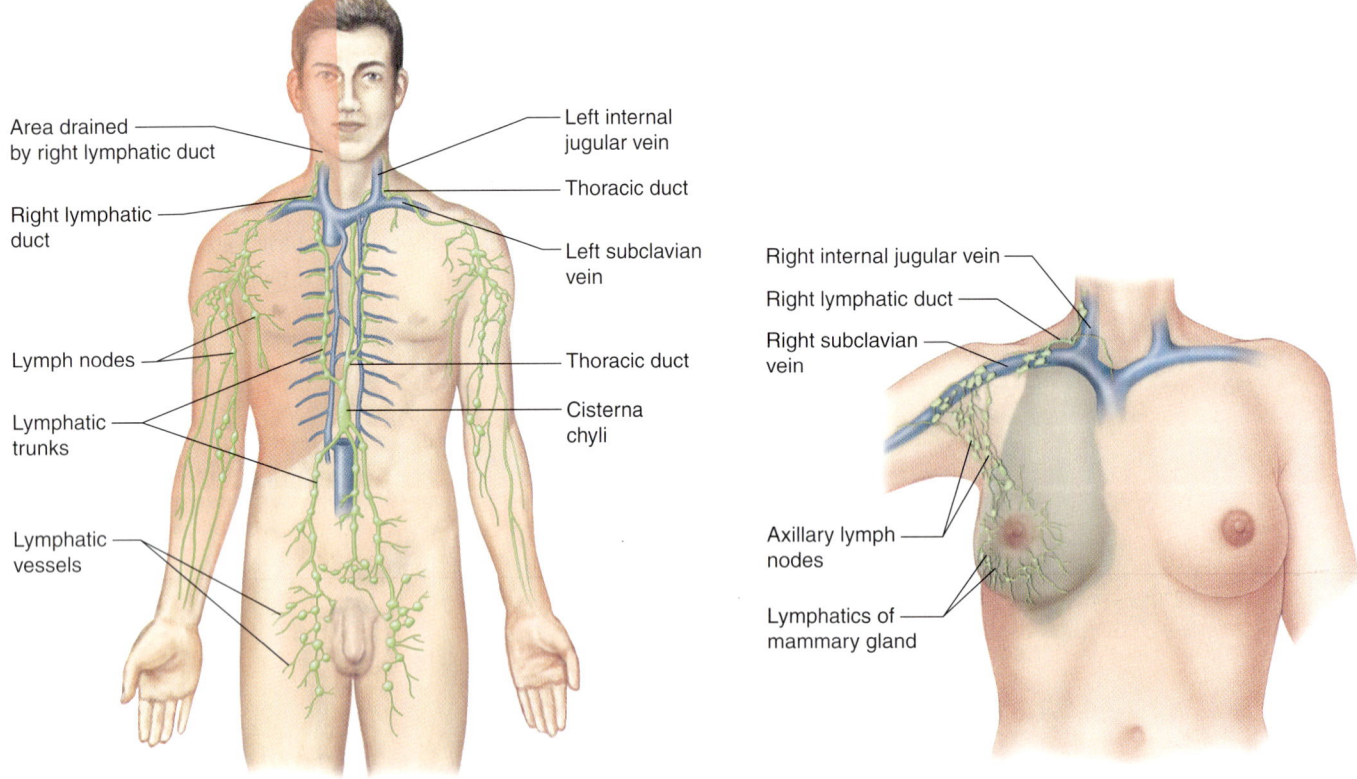

Figure 14.4
Lymphatic pathways. (A) The right lymphatic duct drains lymph from the upper right side of the body, whereas the thoracic duct drains lymph from the rest of the body. (B) Lymph drainage of the right breast illustrates a localized function of the lymphatic system. Surgery to remove a cancerous breast can disrupt this drainage, causing painful swelling.

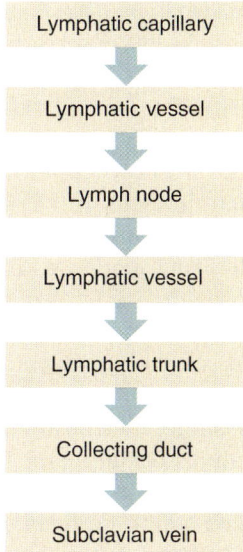

Figure 14.5
The lymphatic pathway.

clavian vein near the junction of the left jugular vein. The **right lymphatic duct** receives lymph from the right side of the head and neck, right upper limb, and right thorax, and empties into the right subclavian vein near the junction of the right jugular vein.

After leaving the collecting ducts, lymph enters the venous system and becomes part of the plasma just before blood returns to the right atrium. Figure 14.5 summarizes the typical lymphatic pathway.

> The skin has many lymphatic capillaries. Consequently, if the skin is broken or if something is injected into it (such as venom from a stinging insect), foreign substances rapidly enter the lymphatic system.

 **CHECK YOUR RECALL**

1. What are the general functions of the lymphatic system?
2. Distinguish between the thoracic duct and the right lymphatic duct.

14.3 Tissue Fluid and Lymph

Lymph is essentially tissue fluid that has entered a lymphatic capillary. Thus, lymph formation depends upon tissue fluid formation.

Tissue Fluid Formation

Recall from chapter 13 (p. 345) that tissue fluid originates from blood plasma and is composed of water and dissolved substances that leave blood capillaries. Capillary blood pressure causes filtration of water and small molecules from the plasma. The resulting fluid has much the same composition as the blood plasma (including nutrients, gases, and hormones), with the important exception of the plasma proteins, which are generally too large to pass through the capillary walls. The osmotic effect of these (called the *plasma colloid osmotic pressure*) helps draw fluid back into the capillaries by osmosis.

Lymph Formation and Function

Filtration from the plasma normally exceeds reabsorption, leading to the net formation of tissue fluid (interstitial fluid). This increases the interstitial fluid hydrostatic pressure somewhat, favoring movement of tissue fluid into lymphatic capillaries, forming lymph (see fig. 14.2). Lymph returns to the bloodstream most of the small proteins that leak out of blood capillaries. At the same time, lymph transports foreign particles, such as bacteria or viruses, to lymph nodes.

 CHECK YOUR RECALL

1. What is the relationship between tissue fluid and lymph?
2. How do plasma proteins in tissue fluid affect lymph formation?
3. What are the major functions of lymph?

14.4 Lymph Movement

The hydrostatic pressure of tissue fluid drives the entry of lymph into lymphatic capillaries. However, muscular activity largely influences the movement of lymph through the lymphatic vessels.

Lymph, like venous blood, is under low hydrostatic pressure and may not flow readily through lymphatic vessels without outside help. These forces include contraction of skeletal muscles, contraction of the smooth muscle in the walls of the larger lymphatic trunks, and pressure changes associated with breathing.

Contracting skeletal muscles compress lymphatic vessels and move the lymph inside lymphatic vessels. These vessels contain valves that prevent backflow, so lymph can only move toward a collecting duct. Additionally, the smooth muscle in the walls of larger lymphatic trunks can contract and compress the lymph inside, forcing the fluid onward.

Breathing aids lymph circulation by creating a relatively low pressure in the thoracic cavity during inhalation. At the same time, the contracting diaphragm increases the pressure in the abdominal cavity. Together, these actions squeeze lymph out of the abdominal vessels and force it into the thoracic vessels. Once again, valves within lymphatic vessels prevent lymph backflow.

The continuous movement of fluid from interstitial spaces into blood and lymphatic capillaries stabilizes the volume of fluid in these spaces. Conditions that interfere with lymph movement cause tissue fluids to accumulate within the interstitial spaces, producing *edema,* or swelling. This may happen when surgery removes lymphatic tissue, obstructing certain lymphatic vessels. For example, a surgeon removing a cancerous breast tumor also usually removes nearby axillary lymph nodes to prevent associated lymphatic vessels from transporting cancer cells to other sites. Removing the lymphatic tissue can obstruct drainage from the upper limb, causing edema (fig. 14.4*B*).

CHECK YOUR RECALL

1. What factors promote lymph flow?
2. What is the consequence of lymphatic obstruction?

14.5 Lymph Nodes

Lymph nodes (lymph glands) are located along the lymphatic pathways. They contain large numbers of **lymphocytes** (lim′fo-sītz) and **macrophages** (mak′ro-fājez) that fight invading microorganisms.

Structure of a Lymph Node

Lymph nodes vary in size and shape, but are usually less than 2.5 centimeters long and somewhat bean-shaped (figs. 14.6 and 14.7). Blood vessels and nerves join a lymph node through the indented region of the node, called the **hilum.** The lymphatic vessels leading to a node (afferent vessels) enter separately at various points on its convex surface, but the lymphatic vessels leaving the node (efferent vessels) exit from the hilum.

A *capsule* of connective tissue encloses each lymph node and subdivides it into compartments that contain dense masses of lymphocytes and macrophages. These masses, called **lymph nodules,** are the structural units of the lymph node. The spaces within a node, called **lymph sinuses,** provide a complex network of chambers and channels through which lymph circulates. Macrophages are most highly concentrated in the lymph sinuses.

Nodules occur singly or in groups associated with the mucous membranes of the respiratory and digestive tracts. The *tonsils,* described in chapter 15 (p. 396), are partially encapsulated lymph nodules. Also, aggregations of nodules called *Peyer's patches* are scattered throughout the mucosal lining of the ileum of the small intestine.

Locations of Lymph Nodes

Lymph nodes generally occur in groups or chains along the paths of the larger lymphatic vessels throughout the body, but are absent in the central nervous system. Figure 14.8 shows the locations of the major lymph nodes.

Functions of Lymph Nodes

Lymph nodes have two primary functions: (1) filtering potentially harmful particles from lymph before returning it to the bloodstream, and (2) immune surveillance, provided by lymphocytes and macrophages. Along with

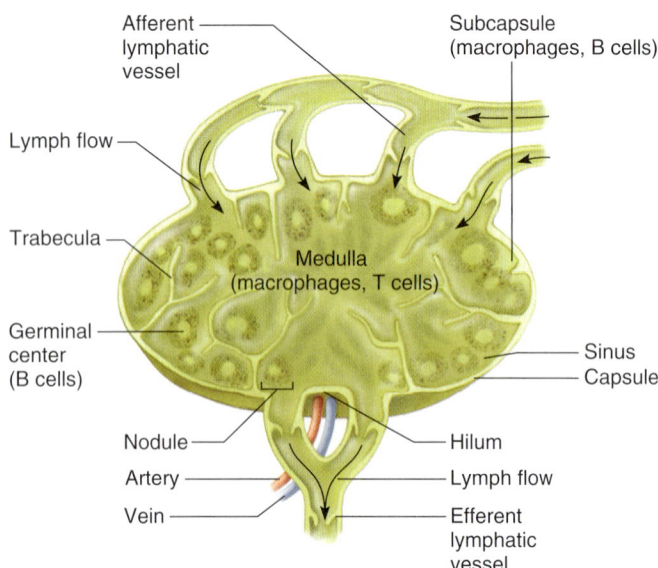

Figure 14.6
A section of a lymph node.

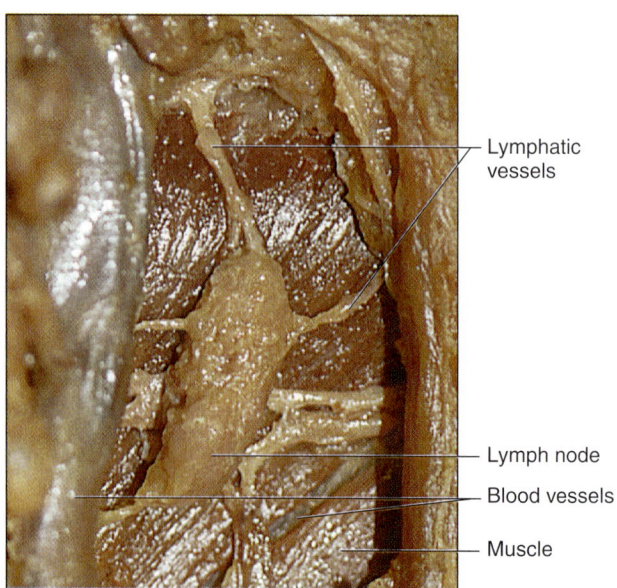

Figure 14.7
Lymph enters and leaves a lymph node through lymphatic vessels.

14.6 Thymus and Spleen

Two other lymphatic organs whose functions are similar to those of the lymph nodes are the thymus and the spleen.

Thymus

The **thymus** (thī′mus) gland is a soft, bilobed structure enclosed in a connective tissue capsule and located anterior to the aorta and posterior to the upper part of the sternum (fig. 14.9A). The thymus is relatively large during infancy and early childhood, but shrinks after puberty and may be quite small in an adult. In elderly persons, adipose and connective tissues replace lymphatic tissue in the thymus.

Connective tissues extend inward from the thymus surface, subdividing the thymus into *lobules* (fig. 14.9B). The lobules contain abundant lymphocytes. Most of these cells (thymocytes) are inactive; however, some mature into **T cells** (T lymphocytes), which leave the thymus and provide immunity. Epithelial cells in the thymus secrete hormones called *thymosins,* which stimulate maturation of T cells after they leave the thymus and migrate to other lymphatic tissues.

> By age 70 years, the thymus is one-tenth the size it was at the age of 10, and the immune system is only 25% as powerful.

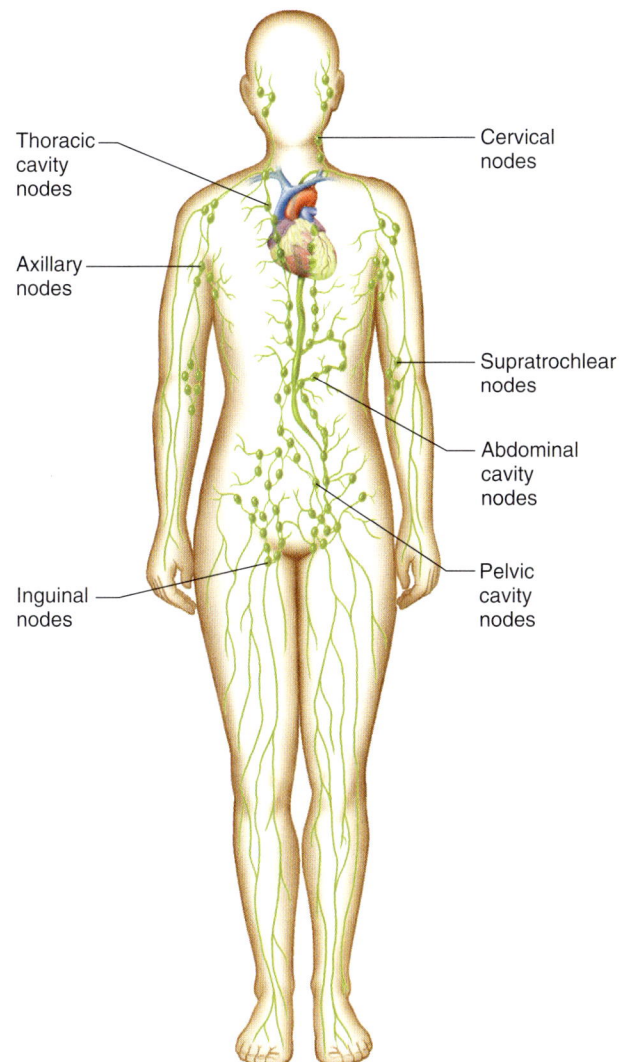

Figure 14.8
Major locations of lymph nodes.

Spleen

The **spleen** (splēn), the largest lymphatic organ, is in the upper left portion of the abdominal cavity, just inferior to the diaphragm and posterior and lateral to the stomach (fig. 14.9A). The spleen resembles a large lymph node and is subdivided into lobules. However, unlike the sinuses of a lymph node, the spaces (venous sinuses) of the spleen contain blood instead of lymph (fig. 14.10).

The tissues within splenic lobules are of two types. The *white pulp* is distributed throughout the spleen in tiny islands. This tissue is composed of splenic nodules, which are similar to those in lymph nodes and contain many lymphocytes. The *red pulp,* which fills the remaining spaces of the lobules, surrounds the venous sinuses. This pulp contains numerous red blood cells, which impart its color, plus many lymphocytes and macrophages.

Blood capillaries within the red pulp are quite permeable. Red blood cells can squeeze through the pores in these capillary walls and enter the venous sinuses. The older, more fragile red blood cells may rupture as they make this passage, and the resulting cellular debris

red bone marrow, the lymph nodes are centers for lymphocyte production. Lymphocytes attack invading viruses, bacteria, and other parasitic cells that lymphatic vessels bring to the nodes. Macrophages in the nodes engulf and destroy foreign substances, damaged cells, and cellular debris.

Superficial lymphatic vessels inflamed by bacterial infection appear as red streaks beneath the skin, a condition called *lymphangitis.* Inflammation of the lymph nodes, called *lymphadenitis,* often follows. Affected nodes enlarge and may be quite painful.

CHECK YOUR RECALL

1. Distinguish between a lymph node and a lymph nodule.
2. What are the major functions of the lymph nodes?

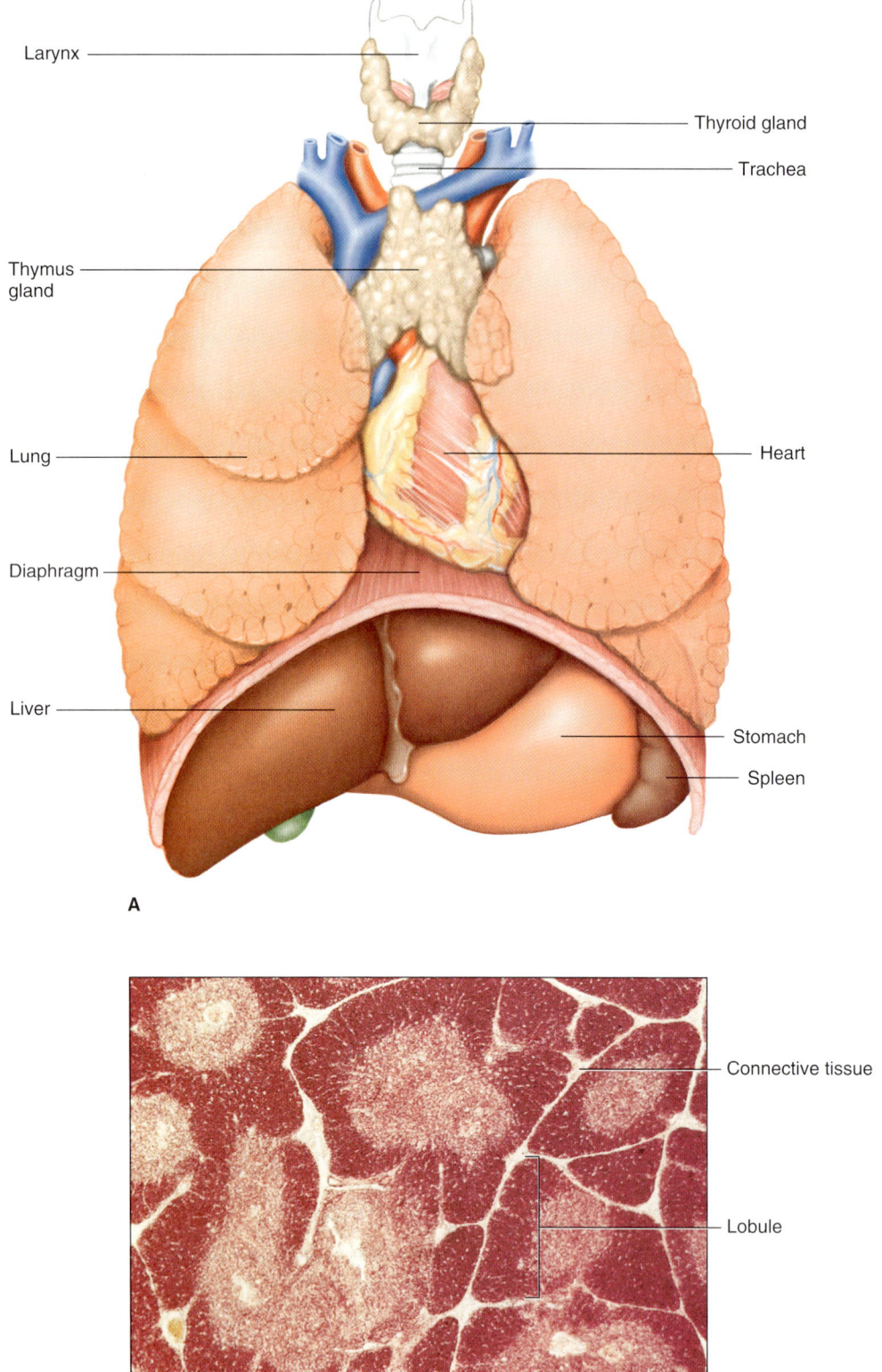

Figure 14.9
Thymus and spleen. (*A*) The thymus gland is bilobed and located between the lungs and superior to the heart. The spleen is located inferior to the diaphragm and posterior and lateral to the stomach. (*B*) A cross section of the thymus (20×). Note how the gland is subdivided into lobules.

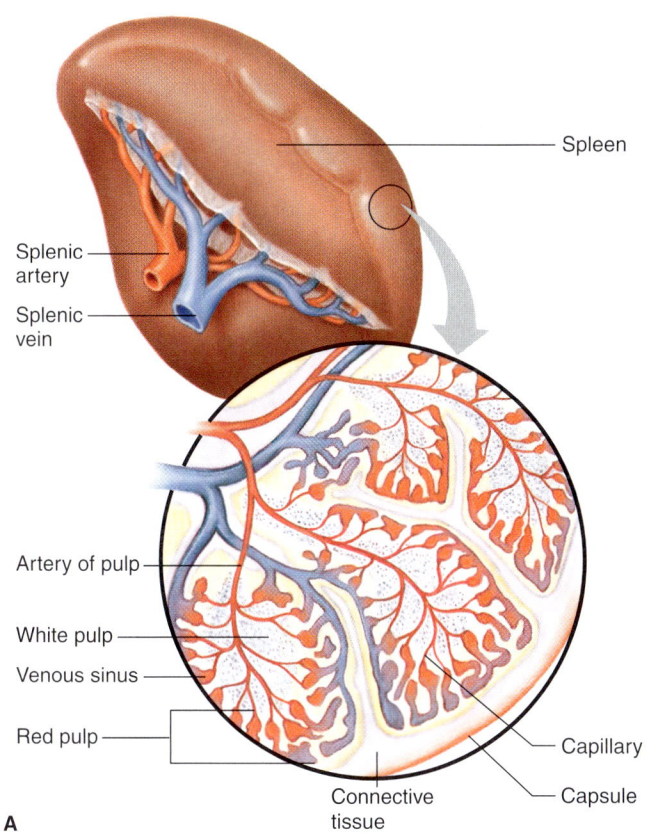
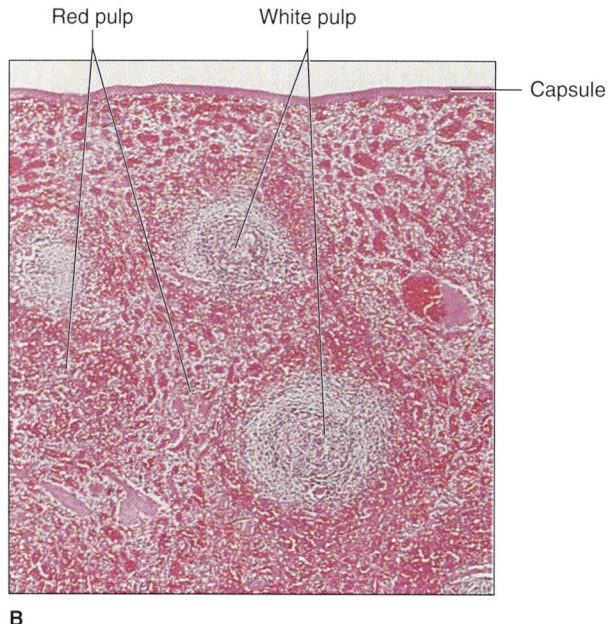

Figure 14.10
Spleen. (*A*) The spleen resembles a large lymph node. (*B*) Light micrograph of the spleen (15×).

is removed by phagocytic macrophages within the splenic sinuses. These macrophages also engulf and destroy foreign particles, such as bacteria, that may be carried in the blood as it flows through the sinuses. Thus, the spleen filters blood much as the lymph nodes filter lymph.

CHECK YOUR RECALL

1. Why are the thymus and spleen considered organs of the lymphatic system?
2. What are the major functions of the thymus and the spleen?

14.7 Body Defenses Against Infection

The presence and multiplication of a disease-causing agent, or **pathogen** (path´o-jen), causes an **infection.** Pathogens include viruses, bacteria, fungi, and protozoans.

The human body can prevent entry of pathogens or destroy them if they enter. Some mechanisms are quite general and protect against many types of pathogens, providing **nonspecific defense.** These mechanisms include species resistance, mechanical barriers, chemical barriers (enzyme action and interferon), fever, inflammation, and phagocytosis. Other defense mechanisms are very precise, targeting certain pathogens and providing **specific defense,** or **immunity** (ĭmu´nĭ-te). Specialized lymphocytes that recognize foreign molecules (nonself antigens) in the body and respond to them execute specific defense mechanisms. Nonspecific and specific defense mechanisms work together to protect the body against infection. While the nonspecific defenses, which respond quite rapidly, are being activated, the slower-to-respond specific defenses are being activated as well.

A word on terminology: nonspecific and specific immunity are also termed innate and acquired immunity. "Innate" refers to inborn protection mechanisms, whereas "acquired" aspects of immunity are stimulated by environmental factors. Acquired immunity is increasingly called "adaptive immunity."

14.8 Nonspecific Defenses

Species Resistance

Species resistance refers to the fact that a given kind of organism, or *species* (such as the human species, *Homo sapiens*), develops a set of diseases that is unique to it. At the same time, a species may be resistant to diseases that affect other species because its tissues somehow fail to provide the temperature or chemical environment that a particular pathogen requires. For example, the infectious agents that cause measles, mumps, gonorrhea, and syphilis infect humans, but not other animal species.

Mechanical Barriers

The skin and mucous membranes lining the passageways of the respiratory, digestive, urinary, and reproductive systems create **mechanical barriers** that prevent entry of some infectious agents. Along with the hair that traps infectious agents associated with the skin and mucous membranes is the fluid (sweat and mucus) that rinses away microorganisms. These barriers provide a *first line of defense*. As long as the skin and mucous membranes remain intact, they can keep out many pathogens. The rest of the nonspecific defenses discussed in this section are part of the *second line of defense*.

Chemical Barriers

Enzymes in body fluids provide a **chemical barrier** to pathogens. Gastric juice, for example, contains the protein-splitting enzyme pepsin and has a low pH due to the presence of hydrochloric acid (HCl) (see chapter 15, p. 403). The combined effect of pepsin and HCl is lethal to many pathogens that enter the stomach. Similarly, tears contain the enzyme lysozyme, which has an antibacterial action against certain pathogens that may get onto eye surfaces. Finally, the accumulation of salt from perspiration kills certain bacteria on the skin.

Certain cells, including lymphocytes and fibroblasts, produce hormonelike peptides called **interferons** in response to viruses or tumor cells. Once released from the virus-infected cell, interferon binds to receptors on uninfected cells, stimulating them to synthesize proteins that block replication of a variety of viruses. Thus, the effect of interferon is nonspecific. Interferons also stimulate phagocytosis and enhance the activity of other cells that help the body resist infections and the growth of tumors.

Fever

Elevated body temperature due to **fever** offers powerful protection. Higher body temperature causes the liver and spleen to sequester iron, which reduces the level of iron in the blood. Since bacteria and fungi require more iron as temperature rises, their growth and reproduction in a fever-ridden body slow and may cease. Also, phagocytic cells attack more vigorously when the temperature rises. For these reasons, low-grade fever of short duration may be a desired response, not something to be treated aggressively with medications.

Inflammation

Inflammation is a tissue response to injury or infection, producing localized redness, swelling, heat, and pain. The redness is a result of blood vessel dilation and the consequent increase in blood volume within the affected tissues. This effect, coupled with an increase in the permeability of nearby capillaries, swells tissues (edema). The heat comes from blood from deeper body parts, which is generally warmer than that near the surface. Pain results from stimulation of nearby pain receptors.

Inflammation reactions result in walling off the site so infection cannot spread and bringing more blood with circulating phagocytes to remove the microorganisms from the site. Local heat speeds up phagocytic activity.

Infected cells release chemicals that attract white blood cells to inflammation sites, where they phagocytize pathogens. In bacterial infections, the resulting mass of white blood cells, bacterial cells, and damaged tissue may form a thick fluid called **pus.**

Body fluids also collect in inflamed tissues. These fluids contain fibrinogen and other blood-clotting factors. Clotting forms a network of fibrin threads within the affected region. Later, fibroblasts may arrive and secrete fibers until the area is enclosed in a sac of connective tissue containing many fibers. This action inhibits the spread of pathogens and toxic substances to adjacent tissues.

Phagocytosis

Recall from chapter 12 (p. 313) that blood's most active phagocytic cells are *neutrophils* and *monocytes*. These cells can leave the bloodstream by squeezing between the cells of blood vessel walls (diapedesis). Chemicals released from injured tissues attract these cells (chemotaxis). Neutrophils engulf and digest smaller particles; monocytes phagocytize larger ones.

Monocytes give rise to *macrophages* (histiocytes), which become fixed in various tissues and attach to the inner walls of blood and lymphatic vessels. These relatively nonmotile phagocytic cells, which can divide and produce new macrophages, are found in the lymph nodes, spleen, liver, and lungs. This diffuse group of phagocytic cells constitutes the **mononuclear phagocytic system** (reticuloendothelial system).

Phagocytosis removes foreign particles from the lymph as it moves from the interstitial spaces to the bloodstream. Phagocytes in the blood vessels and in the tissues of the spleen, liver, or bone marrow remove particles that reach the blood.

✓ CHECK YOUR RECALL

1. What is an infection?
2. Explain six nonspecific defense mechanisms.

14.9 Specific Defenses (Immunity)

The *third line of defense*, immunity, is resistance to particular pathogens or to their toxins or metabolic byproducts. Lymphocytes and macrophages that recognize and remember specific foreign molecules carry out immune responses.

Antigens

Antigens (an´tĭ-jenz) may be proteins, polysaccharides, glycoproteins, or glycolipids, usually located on a cell's surface. Before birth, body cells inventory the proteins and other large molecules in the body, learning to recognize them as "self." The lymphatic system responds to nonself, or foreign antigens, but not normally to self antigens. Receptors on lymphocyte surfaces enable these cells to recognize foreign antigens.

The antigens that are most effective in eliciting an immune response are large and complex, with few repeating parts. Sometimes, a smaller molecule that cannot by itself stimulate an immune response combines with a larger one, which makes it able to do so. Such a small molecule is called a **hapten** (hap´ten). Stimulated lymphocytes react either to the hapten or to the larger molecule of the combination. Haptens are found in certain drugs such as penicillin, in household and industrial chemicals, in dust particles, and in products of animal skins (dander).

Lymphocyte Origins

During fetal development (before birth), red bone marrow releases undifferentiated lymphocytes into the circulation. About half of these cells reach the thymus, where they specialize into T cells. Later, some of these T cells comprise 70–80% of the circulating lymphocytes in blood. Other T cells reside in lymphatic organs and are particularly abundant in the lymph nodes, thoracic duct, and spleen.

Other lymphocytes are thought to remain in the red bone marrow until they differentiate into **B cells** (B lymphocytes). The blood distributes B cells, which constitute 20–30% of circulating lymphocytes. B cells settle in lymphatic organs along with T cells and are abundant in the lymph nodes, spleen, bone marrow, and intestinal lining (figs. 14.11 and 14.12).

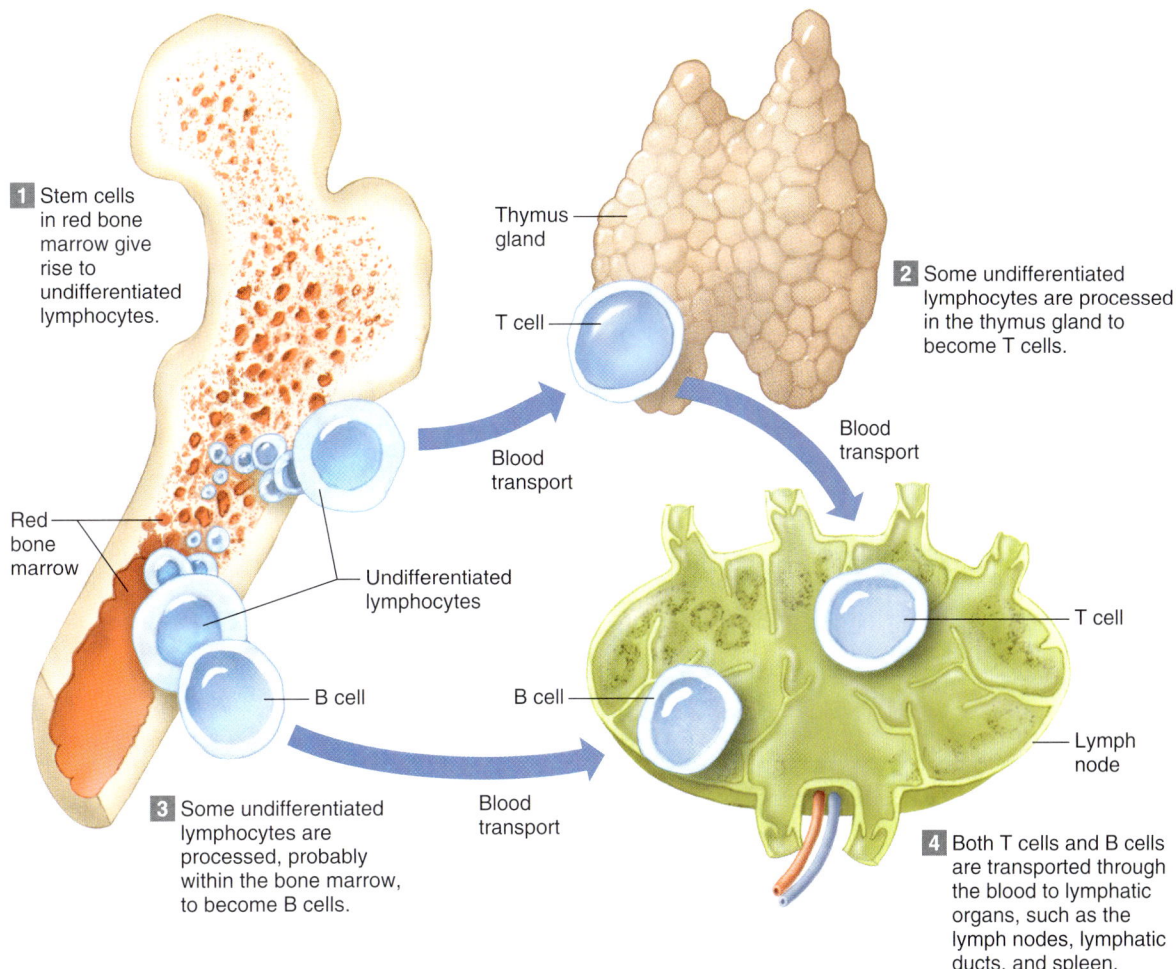

Figure 14.11
Bone marrow releases undifferentiated lymphocytes, which after processing become T cells (T lymphocytes) or B cells (B lymphocytes). Note that in the fetus the medullary cavity contains red marrow.

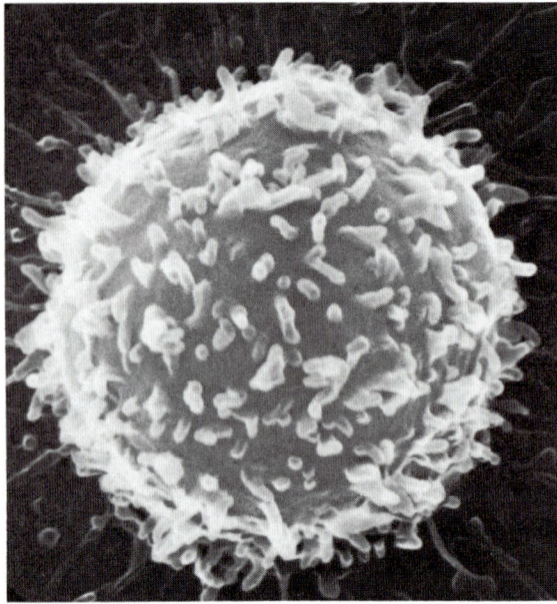

Figure 14.12
Scanning electron micrograph of a human circulating lymphocyte (7,000×).

CHECK YOUR RECALL

1. What is immunity?
2. What is the difference between an antigen and a hapten?
3. How do T cells and B cells originate?

Lymphocyte Functions

T cells and B cells respond to antigens they recognize in different ways. T cells attach to foreign, antigen-bearing cells, such as bacterial cells, and interact directly—that is, by cell-to-cell contact. This is called the **cellular immune response** or cell-mediated immunity.

T cells (and some macrophages) also synthesize and secrete polypeptides called *cytokines* (or, more specifically, lymphokines) that enhance certain cellular responses to antigens. For example, *interleukin-1* and *interleukin-2* stimulate synthesis of several cytokines from other T cells. In addition, interleukin-1 helps activate T cells, whereas interleukin-2 causes T cells to proliferate and activates another type of T cell (cytotoxic T cells). Other cytokines called *colony stimulating factors (CSFs)* stimulate leukocyte production in red bone marrow, cause B cells to grow and mature, and activate macrophages. T cells may also secrete toxins that kill their antigen-bearing target cells: growth-inhibiting factors that prevent target-cell growth or interferon that inhibits the proliferation of viruses and tumor cells.

B cells attack foreign antigens in a different way. They differentiate into **plasma cells,** which produce and secrete large globular proteins called **antibodies** (an′tĭ-bod′′ēz), or **immunoglobulins** (im′′u-no-glob′u-linz). Body fluids carry antibodies, which then react in various ways to destroy specific antigens or antigen-bearing particles. This antibody-mediated immune response is called the **humoral immune response** ("humoral" refers to fluid).

Each person has millions of varieties of T and B cells. Because the members of each variety originate from a single early cell, they are all alike, forming a **clone** (klōn) of cells (identical cells originating from division of a single cell). The members of each variety have a particular type of antigen receptor on their cell membranes that can respond only to a specific antigen. Table 14.1 compares the characteristics of T cells and B cells.

T Cells and the Cellular Immune Response

A lymphocyte must be activated before it can respond to an antigen. T cell activation requires the presence of processed fragments of antigen attached to the surface of another kind of cell, called an **antigen-presenting cell** (accessory cell). Macrophages, B cells, and several other cell types can be antigen-presenting cells.

T cell activation begins when a macrophage phagocytizes a bacterium, digesting it in its lysosomes. Some bacterial antigens exit the lysosomes and move to the macrophage's surface. Here, they are displayed on the cell membrane near certain protein molecules that are part of a group of proteins called the *major histocompatibility complex (MHC)*. A specialized type of T cell,

TABLE 14.1		A COMPARISON OF T CELLS AND B CELLS
CHARACTERISTIC	T CELLS	B CELLS
Origin of undifferentiated cell	Red bone marrow	Red bone marrow
Site of differentiation	Thymus	Probably the red bone marrow
Primary locations	Lymphatic tissues, 70–80% of the circulating lymphocytes	Lymphatic tissues, 20–30% of the circulating lymphocytes
Primary functions	Provides cellular immune response in which T cells interact directly with the antigens or antigen-bearing agents to destroy them	Provides humoral immune response in which B cells interact indirectly, producing antibodies that destroy the antigens or antigen-bearing agents

called a *helper T cell*, contacts a displayed foreign antigen. If the displayed antigen fits and combines with the helper T cell's antigen receptors, the helper cell becomes activated. Once activated, the helper T cell stimulates the B cell to produce antibodies that are specific for the displayed antigen.

A second type of T cell is a *cytotoxic T cell*, which recognizes nonself antigens that cancerous cells or virally infected cells display on their surfaces near certain MHC proteins. A cytotoxic T cell becomes activated when it combines with an antigen that fits its receptors. Next, the T cell proliferates, enlarging its clone of cells. Cytotoxic T cells then bind to the surfaces of antigen-bearing cells, where they release a protein that cuts porelike openings, destroying these cells. In this way, cytotoxic T cells continually monitor body cells, recognizing and eliminating tumor cells and cells infected with viruses.

Some of the T cells do not respond to the antigen on first exposure. Rather, they remain as *memory cells* that immediately differentiate into cytotoxic T cells upon subsequent exposure to the same antigen.

CHECK YOUR RECALL

1. What are the functions of T cells and B cells?
2. How do T cells become activated?
3. What is the function of cytokines?
4. How do cytotoxic T cells destroy antigen-bearing cells?

B Cells and the Humoral Immune Response

A B cell may become activated when it encounters an antigen whose molecular shape fits the shape of the B cell's antigen receptors. In response to the receptor-antigen combination, the B cell divides repeatedly, expanding its clone. However, most antigens require T cell "help" to activate B cells.

When an activated helper T cell encounters a B cell that has already combined with an identical foreign antigen, the helper cell releases certain cytokines. These cytokines stimulate the B cell to proliferate, thus enlarging its clone of antibody-producing cells (figs. 14.13 and 14.14). The cytokines also attract macrophages and leukocytes into inflamed tissues and help keep them there.

Some members of the activated B cell's clone differentiate further into *memory cells*. Like memory T cells, these memory B cells respond rapidly to subsequent exposure to a specific antigen.

Other members of the activated B cell's clone differentiate further into *plasma cells*, which secrete antibodies. These antibodies are similar in structure to the antigen receptor molecules on the original B cell's surface. Thus, antibodies can combine with the antigen-bearing agent that has invaded the body, and react against it. Table 14.2 summarizes the steps leading to antibody production as a result of B cell and T cell actions.

 A plasma cell, during its brief lifespan, secretes up to 2,000 identical antibodies per second.

An individual's B cells can produce an estimated 10 million to 1 billion different varieties of antibodies, each reacting against a specific antigen. The enormity and diversity of the antibody response defends against many pathogens.

TABLE 14.2 — STEPS IN ANTIBODY PRODUCTION

B CELL ACTIVITIES

1. Antigen-bearing agents enter tissues.
2. B cell becomes activated when it encounters an antigen that fits its antigen receptors, either alone or more often in conjunction with helper T cells.
3. Activated B cell proliferates, enlarging its clone.
4. Some of the newly formed B cells differentiate further to become plasma cells.
5. Plasma cells synthesize and secrete antibodies whose molecular structure is similar to the activated B cell's antigen receptors.
6. Antibodies combine with antigen-bearing agents, helping to destroy them.

T CELL ACTIVITIES

1. Antigen-bearing agents enter tissues.
2. Accessory cell, such as a macrophage, phagocytizes antigen-bearing agent, and the macrophage's lysosomes digest the agent.
3. Antigens from the digested antigen-bearing agents are displayed on the surface membrane of the accessory cell.
4. Helper T cell becomes activated when it encounters a displayed antigen that fits its antigen receptors.
5. Activated helper T cell releases cytokines when it encounters a B cell that has previously combined with an identical antigen-bearing agent.
6. Cytokines stimulate the B cell to proliferate.
7. Some of the newly formed B cells differentiate into antibody-secreting plasma cells.
8. Antibodies combine with antigen-bearing agents, helping to destroy them.

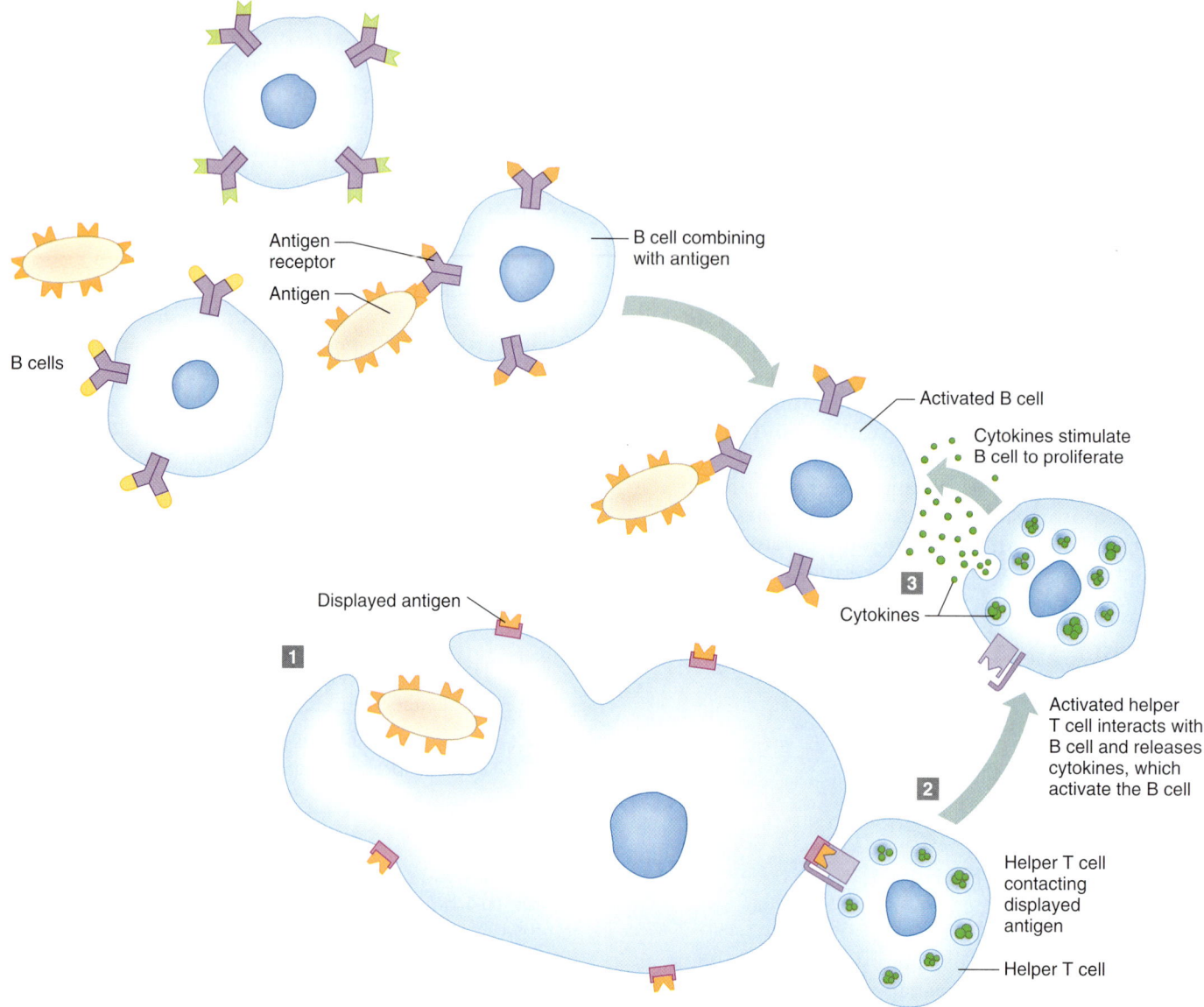

Figure 14.13
T cell and B cell activation. (*1*) After digesting antigen-bearing agents, a macrophage displays antigens on its surface. (*2*) Helper T cells become activated when they contact displayed antigens that fit their antigen receptors. (*3*) An activated helper T cell interacts with a B cell that has combined with an identical antigen and causes the B cell to proliferate.

Types of Antibodies

Antibodies (immunoglobulins) are soluble, globular proteins that constitute the *gamma globulin* fraction of plasma proteins (see chapter 12, p. 316). Of the five major types of immunoglobulins, the most abundant are immunoglobulin G, immunoglobulin A, and immunoglobulin M.

Immunoglobulin G (IgG) is in plasma and tissue fluids and is particularly effective against bacteria, viruses, and toxins. It also activates a group of immune system enzymes called complement, which is described later in this section.

Immunoglobulin A (IgA) is commonly found in exocrine gland secretions. It is in breast milk, tears, nasal fluid, gastric juice, intestinal juice, and bile. It is also in urine.

Immunoglobulin M (IgM) is a type of antibody that develops in the blood plasma in response to contact with certain antigens in foods or bacteria. The antibodies anti-A and anti-B, described in chapter 12 (p. 320), are examples of IgM. IgM also activates complement.

Immunoglobulin D (IgD) is found on the surfaces of most B cells, especially those of infants. IgD is important in activating B cells.

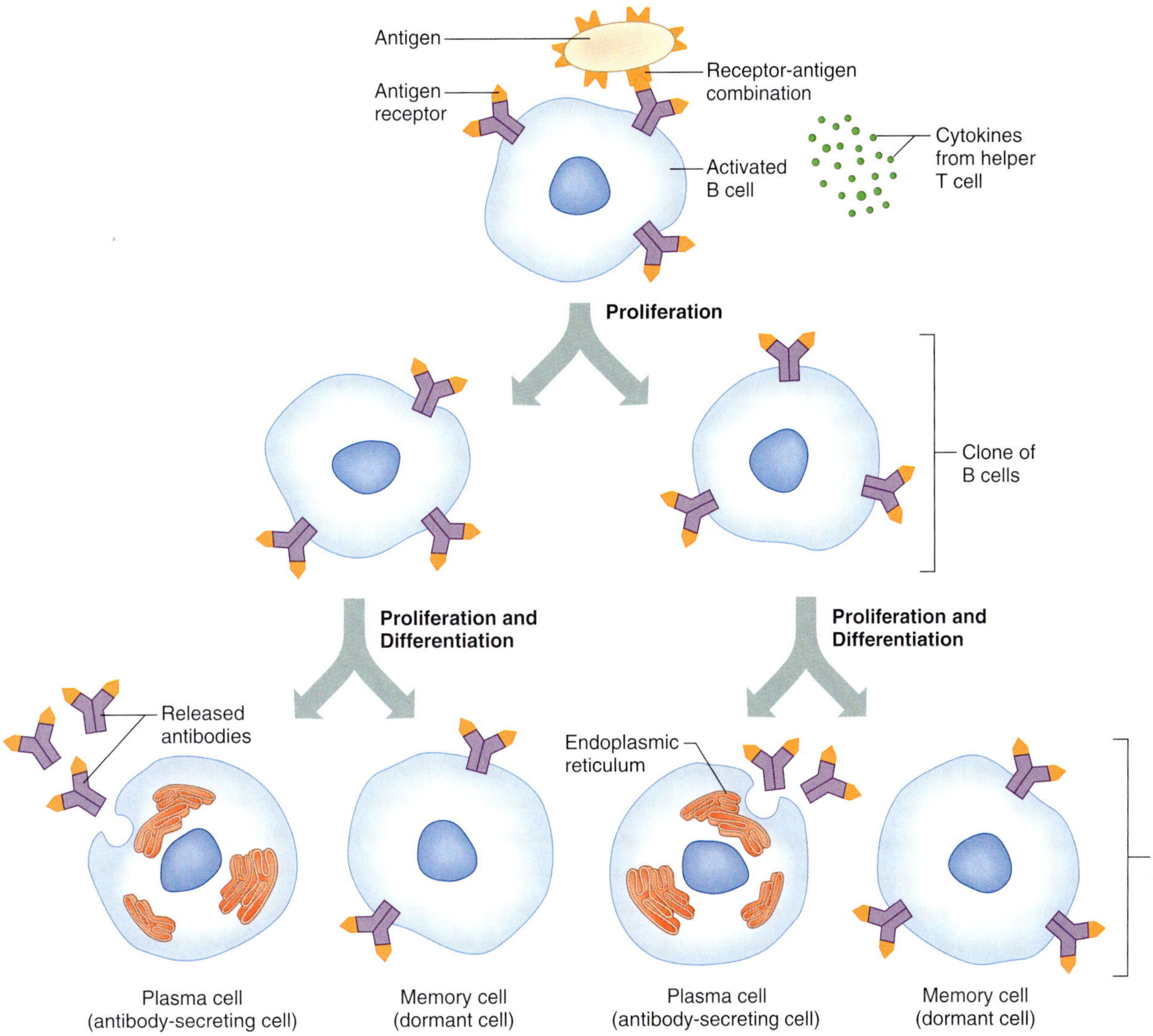

Figure 14.14
An activated B cell proliferates after stimulation by cytokines released by helper T cells. The B cell's clone enlarges. Some cells of the clone give rise to antibody-secreting plasma cells and others to dormant memory cells.

Immunoglobulin E (IgE) appears in exocrine secretions along with IgA. It is associated with allergic reactions, which are described later in this chapter.

A newborn does not yet have its own antibodies, but does retain IgG that passed through the placenta from the mother. These maternal antibodies protect the infant against some illnesses to which the mother is immune. Just as the maternal antibody supply falls, the infant begins to manufacture its own. The newborn receives IgA from colostrum, a substance secreted from the mother's breasts for the first few days after birth. Antibodies in colostrum protect against certain digestive and respiratory infections.

CHECK YOUR RECALL

1. How are B cells activated?
2. How does the antibody response protect against diverse infections?
3. Which immunoglobulins are most abundant, and how do they differ from each other?

Antibody Actions

In general, antibodies directly attack antigens, activate complement to attack the antigens, or stimulate changes in local areas that help prevent the spread of the antigens.

In a direct attack, antibodies combine with antigens and cause them to clump together (agglutinate) or to form insoluble substances (precipitate). Such actions make it easier for phagocytic cells to engulf the antigen-bearing agents and eliminate them. In other instances, antibodies cover the toxic portions of antigen molecules and neutralize their effects. However, under normal conditions, complement activation is more important in protecting against infection than is direct antibody attack.

Complement (kom´plĕ-ment) is a group of proteins in plasma and other body fluids. When certain IgG or IgM antibodies combine with antigens, they expose reactive sites on antibody molecules. This triggers a series of reactions, leading to activation of the complement proteins, which in turn produce a variety of effects. These include: coating the antigen-antibody complexes (opsonization), making them more susceptible to phagocytosis; attracting macrophages and neutrophils into the region (chemotaxis); clumping antigen-bearing agents; rupturing membranes of foreign cells (lysis); and altering the molecular structure of viruses, rendering them harmless. Other proteins promote inflammation, which helps prevent the spread of infectious agents (fig. 14.15).

CHECK YOUR RECALL

1. In what general ways do antibodies function?
2. What is the function of complement?
3. How is complement activated?

Immune Responses

When B cells or T cells become activated after first encountering the antigens for which they are specialized to react, their actions constitute a **primary immune response.** During such a response, plasma cells release antibodies (IgM, followed by IgG) into lymph. The antibodies are transported to the blood and then throughout the body, where they help destroy antigen-bearing agents. Production and release of antibodies continues for several weeks.

Following a primary immune response, some of the B cells produced during proliferation of the clone remain dormant as memory cells (fig. 14.14). If the identical antigen is encountered in the future, the clones of these memory cells enlarge, and they can respond rapidly with IgG to the antigen to which they were previously sensitized. These memory B cells, along with the memory T cells, produce a **secondary immune response.**

As a result of a primary immune response, detectable concentrations of antibodies usually appear in the body fluids within five to ten days following an exposure to antigens. If the identical antigen is encountered some time later, a secondary immune response may produce additional antibodies within a day or two (fig. 14.16). Although newly formed antibodies may persist in the body for only a few months or years, memory cells live much longer. Consequently, the ability to produce a secondary immune response may be long-lasting.

Superantigens are foreign antigens that elicit unusually vigorous lymphocyte responses. The bacterium *Staphylococcus aureus* produces two such superantigens. One type causes food poisoning until digestive enzymes destroy it. The second type causes toxic shock syndrome, a potentially fatal condition producing high fever, diarrhea, vomiting, confusion, and plummeting blood pressure.

CHECK YOUR RECALL

1. How do primary and secondary immune responses differ?

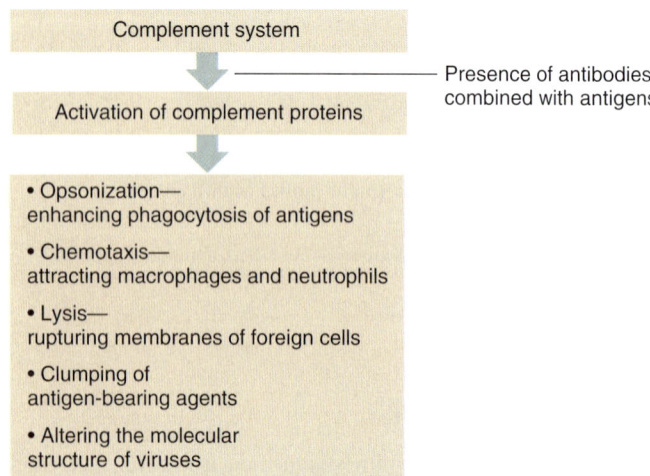

Figure 14.15
Actions of the complement system.

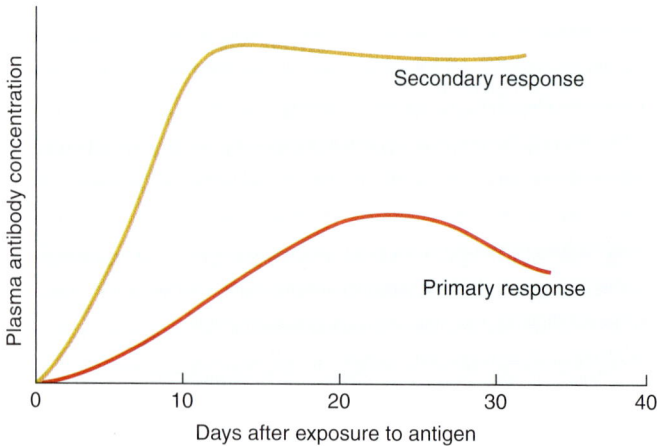

Figure 14.16
A primary immune response produces a lesser concentration of antibodies than does a secondary immune response.

Practical Classification of Immunity

Acquired immunity can arise in response to natural events or be induced artificially by injecting or orally administering a suspension of killed or weakened pathogens. Both naturally and artificially acquired immunities can be either active or passive. Active immunity results when the person produces an immune response (including memory cells) to the antigen; it is long-lasting. Passive immunity occurs when a person receives antibodies produced by another individual. Since the person does not produce an immune response, passive immunity is short-term, and the individual will be susceptible upon exposure to the antigen at some later date.

Naturally acquired active immunity occurs when a person exposed to a pathogen develops a disease. Resistance to that pathogen is the result of a primary immune response.

A **vaccine** produces another type of active immunity. A vaccine might consist of bacteria or viruses that have been killed or weakened so that they cannot cause a serious infection, or a toxoid, which is a toxin from an infectious organism that has been chemically altered to destroy its toxic effects. Whatever its composition, a vaccine includes the antigens that stimulate a primary immune response but does not produce the severe symptoms of disease. A vaccine causes a person to develop *artificially acquired active immunity.*

Vaccines stimulate active immunity against a variety of diseases, including typhoid fever, cholera, whooping cough, diphtheria, tetanus, polio, chicken pox, measles (rubeola), German measles (rubella), mumps, influenza, hepatitis B, and bacterial pneumonia. Vaccines have virtually eliminated natural smallpox from the world, but vaccination may resume in light of the possibility of smallpox being used as a bioweapon. Unfortunately, vaccine distribution is not equitable worldwide. Many thousands of people in underdeveloped countries die of infectious diseases for which vaccines are available in other nations.

 Some people were afraid of the first vaccinations, which were derived from cows. They were afraid that their vaccinated children might behave like cows.

Sometimes a person who has been exposed to infection needs protection against a disease-causing microorganism but lacks the time to develop active immunity. An injection of antiserum (ready-made antibodies) may help. These antibodies may be obtained from gamma globulin (see chapter 12, p. 316) separated from the plasma of persons who have already developed immunity against the particular disease. A gamma globulin injection provides *artificially acquired passive immunity.*

During pregnancy, certain antibodies (IgG) pass from the maternal blood into the fetal bloodstream. As a result, the fetus acquires limited immunity against pathogens that the pregnant woman has developed active immunities against. The fetus thus has *naturally acquired passive immunity,* which may last for six months to a year after birth. Table 14.3 summarizes the types of acquired immunity.

 **CHECK YOUR RECALL**

1. Distinguish between active and passive immunity.

Allergic Reactions

An allergic response is an immune attack against a nonharmful substance, such as chocolate. Allergic reactions are similar to immune responses because they sensitize lymphocytes, and antibodies may bind antigens. However, unlike normal immune responses, allergic reactions can damage tissues. Antigens that trigger allergic responses are called **allergens** (al´er-jenz).

A *delayed-reaction allergy* may affect anyone. It results from repeated exposure of the skin to certain chemicals—commonly, household or industrial chemicals

TABLE 14.3 — PRACTICAL CLASSIFICATION OF IMMUNITY

TYPE	MECHANISM	RESULT
Naturally acquired active immunity	Exposure to live pathogens	Symptoms of a disease and stimulation of an immune response
Artificially acquired active immunity	Exposure to a vaccine containing weakened or dead pathogens or their components	Stimulation of an immune response without the severe symptoms of a disease
Artificially acquired passive immunity	Injection of gamma globulin containing antibodies	Short-term immunity without stimulating an immune response
Naturally acquired passive immunity	Antibodies passed to fetus from pregnant woman with active immunity	Short-term immunity for infant without stimulating an immune response

Topic of Interest

IMMUNITY BREAKDOWN: AIDS

Infection by the *human immunodeficiency virus (HIV)* causes *acquired immune deficiency syndrome (AIDS)*, a progressive breakdown of the immune system. The virus attacks lymphocytes by attaching to receptors on helper T cells and sending in its RNA. Within the cell's nucleus, a viral enzyme, *reverse transcriptase,* catalyzes construction of a DNA strand complementary to the viral RNA. The initial viral DNA strand replicates to form a DNA double helix.

Using the invaded cell's protein-synthesizing machinery, HIV replicates, filling the cell with its RNA and proteins. Not only can the dying T cell no longer release cytokines or stimulate B cells to manufacture antibodies, but it bursts, unleashing new HIV particles. HIV replicates rapidly and soon overwhelms the immune system. Specific symptoms in different individuals reflect the infectious agents to which they are exposed. Common AIDS-related conditions include:

Persistent lymphadenopathy (swollen lymph glands)
Constant low-grade fever
Nausea and vomiting
Fatigue
Night sweats
Headaches
Wasting syndrome (persistent diarrhea, severe weight loss, weakness, fever)
Dementia (confusion, apathy, inability to concentrate, memory loss, insomnia, disorientation, sudden strong emotions)
Cancers (Kaposi sarcoma, cervical cancer, lymphoma, others)
Opportunistic infections (pneumonia, brain infection, diarrhea, spinal meningitis, tuberculosis, fungal infections, many others)

HIV infection has three stages: initial symptoms, a latency period, and AIDS. The initial, acute stage may include weakness, recurrent fever, night sweats, swollen neck glands, and weight loss. This stage varies in duration and severity. Often it lasts only a few days, and the person may think it is the flu. Then comes a latency period, typically lasting five to ten years, during which the person feels well. This well-being is deceptive because the immune system is struggling to contain the growing HIV population, first in the lymph nodes and then in the bloodstream. The third stage, AIDS, brings opportunistic infections, so called because they appear when the immune system is compromised.

Modes of Transmission

AIDS transmission requires contact with a body fluid containing abundant HIV, such as blood or semen. Although the virus has been detected in sweat, tears, and saliva, levels are so low that transmission is highly unlikely. Whether or not a person becomes infected appears to depend on the amount of infected fluid contacted, the site of exposure in the body, and the individual's health. Table 14A lists some of the ways that HIV infection can and cannot spread.

---

or some cosmetics. After repeated contacts, the presence of the foreign substance activates T cells, many of which collect in the skin. The T cells and the macrophages they attract release chemical factors, which in turn cause eruptions and inflammation of the skin (dermatitis). This reaction is called *delayed* because it usually takes about 48 hours to occur.

An *immediate-reaction allergy* occurs within minutes after contact with an allergen. Persons with this type of allergy have an inherited tendency to overproduce IgE antibodies in response to certain antigens. IgE normally comprises a minute fraction of plasma proteins.

An immediate-reaction allergy activates B cells, which become sensitized when the allergen is first encountered. Subsequent exposures to the allergen trigger allergic reactions. In the initial exposure, IgE attaches to the membranes of widely distributed mast cells and basophils. When a subsequent allergen-antibody reaction occurs, these cells release allergy mediators such as *histamine, prostaglandin D_2,* and *leukotrienes.* These substances cause a variety of physiological effects, including dilation of blood vessels, increased vascular permeability that swells tissues, contraction of bronchial and intestinal smooth muscles, and increased mucus production. The result is a severe inflammation reaction that is responsible for the symptoms of the allergy, such as hives, hay fever, asthma, eczema, or gastric disturbances.

Anaphylactic shock is a severe form of immediate-reaction allergy in which mast cells release allergy mediators throughout the body. The person may at first feel an inexplicable apprehension, and then suddenly the entire body itches and breaks out in red hives. Vomiting and diarrhea may follow. The face, tongue, and larynx begin to swell, and breathing becomes difficult. Unless the person receives an injection of epinephrine (adrenalin) and sometimes a tracheotomy (an incision into the windpipe to restore breathing), he or she will lose consciousness and may die within 5 minutes. Anaphylactic shock most often results from an allergy to penicillin or insect stings. Fortunately, thanks to prompt medical attention and avoidance of allergens by people who know they have allergies, fewer than 100 people a year actually die from anaphylactic shock.

Progress

Three groups of people are providing the clues that may lead to conquering HIV infection:

1. Infected individuals who never develop symptoms ("long-term nonprogressors"). These people have a weakened strain of HIV that lacks part of a gene HIV normally uses to replicate. Enough of the virus remains to alert the immune system to protect against other strains. A vaccine might be based on this weakened strain.
2. People exposed repeatedly who never become infected. About 1% of the population has a gene variant that protects them from becoming infected with HIV. The cells of these individuals lack either of two receptor molecules that HIV requires to enter cells.
3. Infected individuals who apparently become uninfected. Several infants infected at birth have apparently lost the virus as their immune systems matured.

Researchers are identifying how these people survive, evade, or vanquish HIV infection, and will use this knowledge to develop prevention and treatment strategies.

Several classes of drugs target HIV infection at various stages. The first drugs, such as AZT, ddI, ddC, and 3TC, block viral replication. Drugs called protease inhibitors prevent HIV from processing its proteins to a functional size, crippling the virus. A third class of drugs, called fusion inhibitors, block the binding of HIV to T cell surfaces. Combining drugs can keep viral load low and delay symptom onset and progression, although viral variants emerge that resist the drugs. The goal is to enable infected people to live normal life spans in relatively good health. More than 200 drugs are also available to treat AIDS-associated opportunistic infections and cancers.

Better understanding of the biology of HIV infection, plus new drug weapons and clues from survivors, are providing what has long been lacking in the global fight against AIDS—hope. Slowly, HIV infection is becoming a chronic illness, rather than a death sentence.

TABLE 14A	HIV TRANSMISSION
HOW HIV IS TRANSMITTED	
Sexual contact, particularly anal intercourse, but including vaginal intercourse and oral sex	
Contaminated needles (intravenous drug use, injection of anabolic steroids, accidental needle stick in medical setting)	
During birth from infected mother	
Receiving infected blood or other tissue (precautions usually prevent this)	
HOW HIV IS NOT TRANSMITTED	
Casual contact (social kissing, hugging, handshakes)	
Objects (toilet seats, deodorant sticks, doorknobs)	
Mosquitoes	
Sneezing and coughing	
Sharing food	
Swimming in the same water	
Donating blood	

Transplantation and Tissue Rejection

Transplantation of tissues or an organ, such as the skin, kidney, heart, or liver, from one person to another can replace a nonfunctional, damaged, or lost body part. The danger the immune system poses to transplanted tissue is that the recipient's cells may recognize the donor's tissues as foreign and attempt to destroy the transplanted tissue. Such a response is called a **tissue rejection reaction.**

Tissue rejection resembles the cellular immune response against a nonself antigen. The greater the antigenic difference between the cell surface molecules of the recipient's tissues and the donor's tissues, the more rapid and severe the rejection reaction. Matching donor and recipient tissues can minimize the rejection reaction. This means locating a donor whose tissues are antigenically similar to those of the person needing a transplant—a procedure much like matching the blood of a donor with that of a recipient before giving a blood transfusion.

Immunosuppressive drugs are used to reduce rejection of transplanted tissues. These drugs interfere with the recipient's immune response by suppressing formation of antibodies or production of T cells, thereby dampening the humoral and cellular immune responses. Unfortunately, the use of immunosuppressive drugs can leave a recipient unprotected against infections. It is not uncommon for a patient to survive a transplant, but die of infection because of a weakened immune system.

 Donated organs need to be transplanted quickly. How long can donated organs last outside the body?

- A heart lasts 3 to 5 hours.
- A liver lasts 10 hours.
- A kidney lasts 24 to 48 hours.

Autoimmunity

Sometimes the immune system fails to distinguish self from nonself, producing **autoantibodies** and cytotoxic T cells that attack and damage tissues and organs. This attack against self is called **autoimmunity.** The specific

Genetics Connection

CONQUERING INHERITED IMMUNE DEFICIENCY— CHILDREN WHO MADE MEDICAL HISTORY

T cells are the linchpins of the immune system. If they cannot function, they cannot activate B cells, and both the cellular and humoral immune responses shut down. We know well the importance of T cells in establishing and maintaining immunity from AIDS, which is acquired. Immune deficiency can also be inherited, resulting from mutations (changes in genes) that adversely affect receptors on T cells or cytokine production. We know of more than twenty forms of this severe combined immune deficiency, or SCID. A look at five children born with SCID also provides a compelling view of how quickly technology is making their lives easier, although it came too late for one young man, David Vetter.

David's Story

David Vetter was born in Texas in 1971 without a thymus gland. He therefore could not make mature T cells or activate his B cells. He spent his short life in a vinyl bubble, awaiting a treatment that never came (fig. 14A). David appeared on news programs, and a TV movie starring John Travolta depicted his life. In the era before AIDS, living without immunity was very unusual.

As David reached adolescence, he desperately wanted to leave the bubble. After receiving a bone marrow transplant, he did so. But the transplant hadn't worked, and within days, David began vomiting and developed diarrhea, both signs of infection. He soon died.

Laura's Story

For the first few years of her life, Laura Cay Boren didn't know what it felt like to be well (fig. 14B). From her birth in July 1982, she fought infection after infection. Colds would land her in the hospital with pneumonia, and routine vaccines caused severe skin abscesses. Laura had inherited a form of SCID in which the body lacks an enzyme, adenosine deaminase (ADA). Lack of ADA blocks a biochemical pathway that breaks down a metabolic toxin, which instead builds up and destroys T cells. The T cells in turn can no longer activate B cells. Immunity fails.

Laura spent her first and second birthdays at the Duke University Medical Center and then underwent two bone marrow transplants, which temporarily restored her immune defenses. Transfusions also helped. But by the end of 1985, Laura was near death. Then she was chosen to receive an experimental treatment, injections of ADA altered in a way that causes it to remain in the bloodstream long enough to help T cells survive. It worked! Within hours of the first treatment, Laura's ADA level increased twenty-fold. After three months, her blood was free of toxins, although her immunity was still suppressed. After six months, though, her immune function neared normal for the first time ever—and stayed that way, with weekly ADA shots. By the summer of 1988, she could play with other children, and by the following year, began school. She is healthy today.

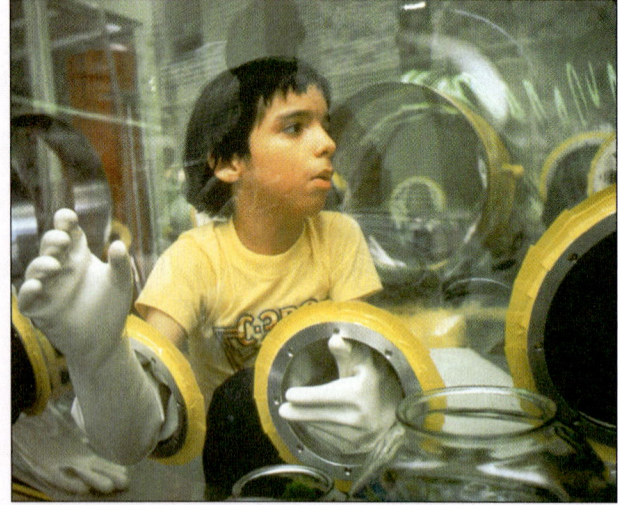

Figure 14A

David Vetter was born without a thymus gland. Because his T cells could not mature, he was virtually defenseless against infection.

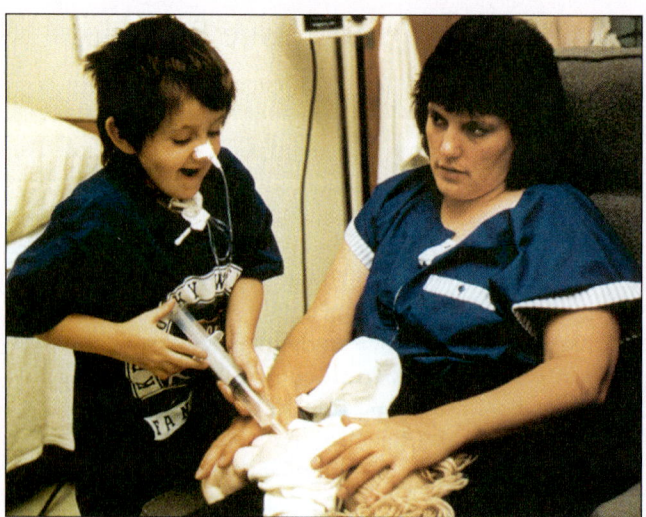

Figure 14B

Laura Cay Boren spent much of her life in hospitals until she received the enzyme her body lacks, adenosine deaminase (ADA). Here, she pretends to inject her doll as her mother looks on.

Ashanti's and Cynthia's Stories

On September 14, 1990, at 12:52 P.M., four-year-old Ashanti DaSilva sat up in bed at the National Institutes of Health in Bethesda, Maryland, and began receiving her own white blood cells intravenously. Earlier, doctors had removed the cells and patched them with normal ADA genes, which she lacked. Soon after that, an eight-year-old, Cynthia Cutshall, received the same treatment. Both girls also received ADA injections to keep them healthy, and doctors monitored their immune responses frequently. Because the gene therapy replaced mature T cells, the procedure had to be repeated as these cells lived out their natural life spans. This gene therapy, although successful (fig. 14C), was still not a permanent cure.

Andrew's Story

Crystal and Leonard Gobea had already lost a five-month-old baby to ADA deficiency when they learned that the fetus Crystal was carrying had also inherited the condition. They and two other couples were asked to allow their newborns to participate in a new type of gene therapy that would replace defective T cells with T cells taken from their umbilical cord blood and bolstered with the missing ADA gene (fig. 14D). The Gobea's baby, Andrew, was featured on the May 31, 1993, cover of *Time* magazine, where his parents told their story.

The three children were also given ADA injections to maintain health, as doctors monitored their immune functions. Since the defective cells were replaced early, the hope was that they would take over the immune system. They did, but slowly. After a few months, each child had about 1 in 10,000 T cells of the healthy type. After a year, that number rose to 1 in 100. Doctors lowered the ADA doses, and the children remained well. By the summer of 1995, the three toddlers had 3 in 100 T cells carrying the ADA gene, and at four years of age, 1–10% of the T cells were the repaired type. However, when the researchers discontinued giving ADA to one child, immune function declined. So although the gene therapy is working on the cellular level, on the organ system level it is not yet a complete cure. Researchers now need to determine what percentage of the T cell population must be "fixed" to completely restore and maintain immunity.

Figure 14C
Three years after receiving her own white blood cells, genetically altered to contain the ADA gene they lack, Ashanti DaSilva rides her bike—something she thought would never be possible.

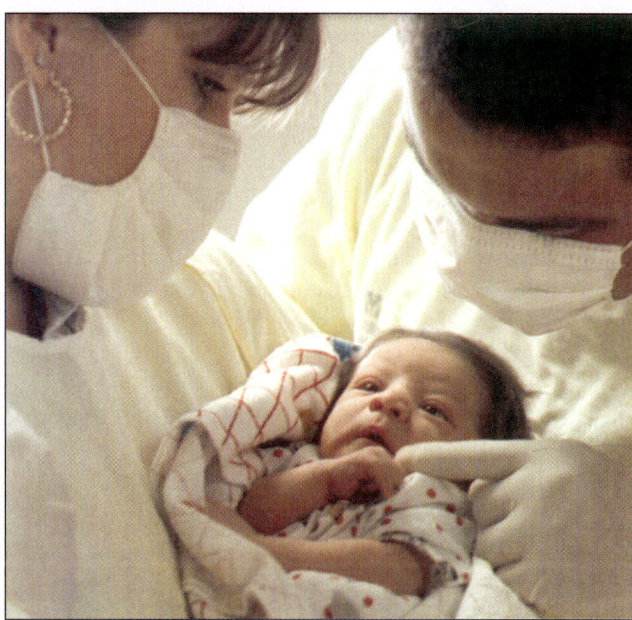

Figure 14D
Newborn Andrew Gobea received the ADA gene in stem cells taken from his umbilical cord. The percentage of T cells in his blood that carry the needed gene is steadily increasing. He is now healthy.

nature of an autoimmune disorder depends on the cell types that are the target of the immune attack. Juvenile diabetes, rheumatoid arthritis, and systemic lupus erythematosus are some autoimmune disorders.

Why might the immune system attack body tissues? Perhaps a virus, while replicating within a human cell, "borrows" proteins from the host cell's surface and incorporates them onto its own surface. When the immune system "learns" the surface of the virus in order to destroy it, it also learns to attack the human cells that normally bear the particular protein. Another explanation of autoimmunity is that somehow T cells never learn to distinguish self from nonself. A third possible route of autoimmunity is when a nonself antigen coincidentally resembles a self antigen. This is what happens when an infection by *Streptococcus* bacteria triggers inflammation of heart valves, as mentioned in chapter 13 (p. 332).

CHECK YOUR RECALL

1. How are allergic reactions and immune reactions similar yet different?
2. How does a tissue rejection reaction involve an immune response?
3. How is autoimmunity an abnormal functioning of the immune response?

Clinical Terms Related to the Lymphatic System and Immunity

allograft (al´o-graft) Transplantation of tissue from an individual of one species to another individual of that species.
asplenia (ah-sple´ne-ah) Absence of a spleen.
autograft (aw´to-graft) Transplantation of tissue from one part of the body to another part of the same body.
immunocompetence (im´´u-no-kom´pe-tens) Ability to produce an immune response to the presence of antigens.
immunodeficiency (im´´u-no-de-fish´en-se) Inability to produce an immune response.
lymphadenectomy (lim-fad´´ĕ-nek´to-me) Surgical removal of lymph nodes.
lymphadenopathy (lim-fad´´ĕ-nop´ah-the) Enlargement of lymph nodes.
lymphadenotomy (lim-fad´´ĕ-not´o-me) Incision of a lymph node.
lymphocytopenia (lim´´fo-si´´to-pe´ne-ah) Too few lymphocytes in blood.
lymphocytosis (lim´´fo-si´´to´sis) Too many lymphocytes in blood.
lymphoma (lim-fo´mah) Tumor composed of lymphatic tissue.
lymphosarcoma (lim´´fo-sar-ko´mah) Cancer within the lymphatic tissue.
splenectomy (sple-nek´to-me) Surgical removal of the spleen.
splenitis (sple-ni´tis) Inflammation of the spleen.
splenomegaly (splĕ´´no-meg´ah-le) Abnormal enlargement of the spleen.
splenotomy (sple-not´o-me) Incision of the spleen.
thymectomy (thi-mek´to-me) Surgical removal of the thymus.
thymitis (thi-mi´tis) Inflammation of the thymus.

Clinical Connection

Some disorders thought to be autoimmune may have a more bizarre cause—fetal cells persisting in a woman's circulation, for decades. In response to an as yet unknown trigger, the fetal cells, perhaps "hiding" in a tissue such as skin, emerge, and stimulate antibody production. If we didn't know the fetal cells were there, the resulting antibodies and symptoms would appear to be an autoimmune disorder. This mechanism, called microchimerism ("small mosaic"), may explain the higher prevalence of autoimmune disorders among women. It was discovered in a disorder called scleroderma, which means "hard skin".

Patients describe scleroderma, which typically begins between ages 45 and 55, as "the body turning to stone." Symptoms include fatigue, swollen joints, stiff fingers, and a masklike face. The hardening may affect blood vessels, the lungs, and the esophagus, too. Clues that scleroderma is a delayed response to persisting fetal cells include the following observations:

- It is much more common among women.
- Symptoms resemble those of graft-versus-host disease (GVHD), in which transplanted tissue produces chemicals that destroy the body. Antigens on cells in scleroderma lesions match those involved in GVHD.
- Mothers who have scleroderma and their sons have cell surfaces that are more similar than those of unaffected mothers and their sons. Perhaps the similarity of cell surfaces enabled the fetal cells to escape destruction by the woman's immune system. (Female fetal cells can theoretically cause scleroderma too, but they are harder to detect because male cells can be distinguished by the Y chromosome.)

Perhaps other disorders considered to be autoimmune may actually reflect an immune system response to lingering fetal cells.

Organization

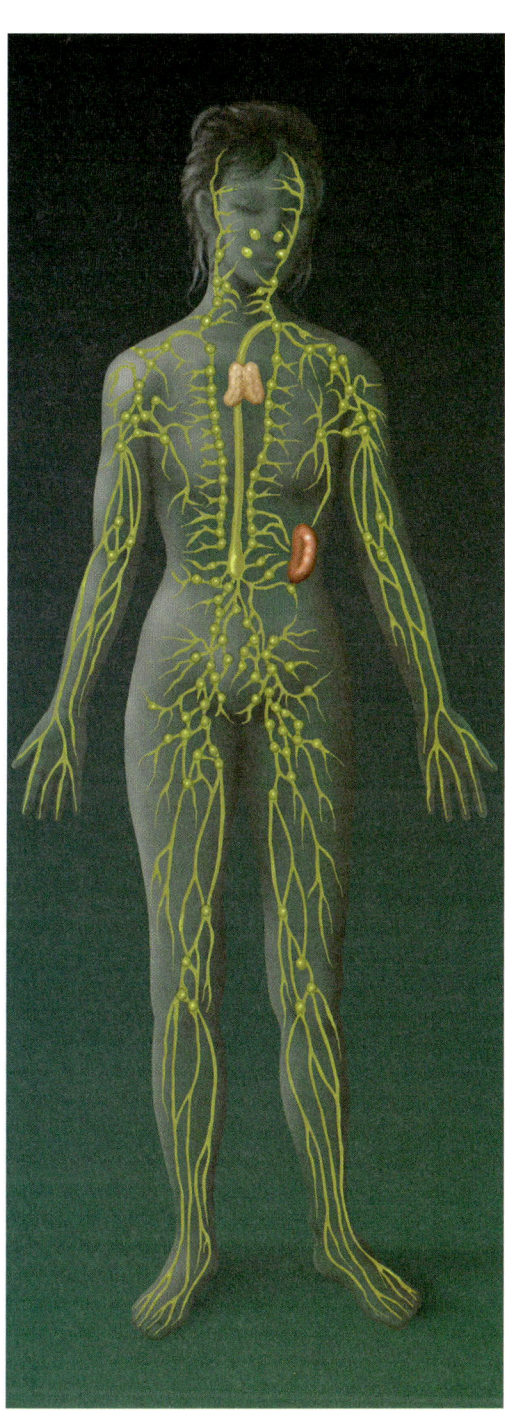

Lymphatic System
The lymphatic system is an important link between the interstitial fluid and the plasma; it also plays a major role in the response to infection.

Integumentary System
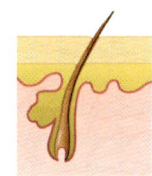
The skin is a first line of defense against infection.

Cardiovascular System
The lymphatic system returns tissue fluid to the bloodstream. Lymph originates as interstitial fluid, formed by the action of blood pressure.

Skeletal System

Cells of the immune system originate in the bone marrow.

Digestive System

Lymph plays a major role in the absorption of fats.

Muscular System
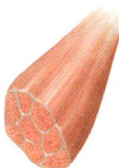
Muscle action helps pump lymph through the lymphatic vessels.

Respiratory System
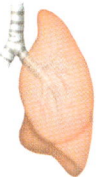
Cells of the immune system patrol the respiratory system to defend against infection.

Nervous System
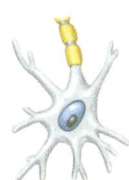
Stress may impair the immune response.

Urinary System

The kidneys control the volume of extracellular fluid, including lymph.

Endocrine System

Hormones stimulate lymphocyte production.

Reproductive System
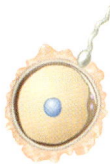
Special mechanisms inhibit the female immune system in its attack of sperm as foreign invaders.

SUMMARY OUTLINE

14.1 Introduction (p. 367)
The lymphatic system is closely associated with the cardiovascular system. It transports excess fluid to the bloodstream, absorbs fats, and helps defend the body against disease-causing agents.

14.2 Lymphatic Pathways (p. 367)
1. Lymphatic capillaries
   a. Lymphatic capillaries are microscopic, closed-ended tubes that extend into interstitial spaces.
   b. They receive lymph through their thin walls.
2. Lymphatic vessels
   a. Lymphatic vessels have walls similar to those of veins, only thinner, and possess valves that prevent backflow of lymph.
   b. Larger lymphatic vessels lead to lymph nodes and then merge into lymphatic trunks.
3. Lymphatic trunks and collecting ducts
   a. Lymphatic trunks lead to two collecting ducts—the thoracic duct and the right lymphatic duct.
   b. Collecting ducts join the subclavian veins.

14.3 Tissue Fluid and Lymph (p. 369)
1. Tissue fluid formation
   a. Tissue fluid originates from blood plasma.
   b. It generally lacks large proteins, but some smaller proteins leak into interstitial spaces.
   c. As the protein concentration of tissue fluid increases, colloid osmotic pressure increases.
2. Lymph formation and function
   a. Increasing pressure within interstitial spaces forces some tissue fluid into lymphatic capillaries, and this fluid becomes lymph.
   b. Lymph returns protein molecules to the bloodstream and transports foreign particles to lymph nodes.

14.4 Lymph Movement (p. 369)
1. Lymph is under low pressure and may not flow readily without external aid.
2. Lymph is moved by the contraction of skeletal muscles and low pressure in the thorax created by breathing movements.

14.5 Lymph Nodes (p. 370)
1. Structure of a lymph node
   a. Lymph nodes are subdivided into nodules.
   b. Nodules contain masses of lymphocytes and macrophages.
2. Locations of lymph nodes
   Lymph nodes aggregate in groups or chains along the paths of larger lymphatic vessels.
3. Functions of lymph nodes
   a. Lymph nodes filter potentially harmful foreign particles from lymph.
   b. Lymph nodes are centers for the production of lymphocytes, and they also contain phagocytic cells.

14.6 Thymus and Spleen (p. 371)
1. Thymus
   a. The thymus is composed of lymphatic tissue subdivided into lobules.
   b. It slowly shrinks after puberty.
   c. Some lymphocytes leave the thymus and provide immunity.
2. Spleen
   a. The spleen resembles a large lymph node subdivided into lobules.
   b. Spaces within splenic lobules are filled with blood.
   c. The spleen contains many macrophages, which filter foreign particles and damaged red blood cells from blood.

14.7 Body Defenses Against Infection (p. 373)
The body has nonspecific and specific defenses against infection.

14.8 Nonspecific Defenses (p. 373)
1. Species resistance
   Each species is resistant to certain diseases that may affect other species.
2. Mechanical barriers
   Mechanical barriers include the skin and mucous membranes, which block entrance of some pathogens.
3. Chemical barriers
   a. Enzymes in gastric juice and tears kill some pathogens.
   b. Interferons stimulate uninfected cells to synthesize antiviral proteins that stimulate phagocytosis, block proliferation of viruses, and enhance activity of cells that help resist infections and stifle tumor growth.
4. Fever
   Higher body temperature and the resulting decrease in blood iron level and increase in phagocytic activity hamper infection.
5. Inflammation
   a. Inflammation is a tissue response to injury or infection, and includes localized redness, swelling, heat, and pain.
   b. Chemicals released by damaged tissues attract white blood cells to the site.
   c. Connective tissue containing many fibers may form a sac around injured tissue and thus block the spread of pathogens.
6. Phagocytosis
   a. The most active phagocytes in blood are neutrophils and monocytes. Monocytes give rise to macrophages, which remain fixed in tissues.
   b. Phagocytic cells are associated with the linings of blood vessels in the bone marrow, liver, spleen, lungs, and lymph nodes.

14.9 Specific Defenses (Immunity) (p. 375)
1. Antigens
   a. Before birth, body cells inventory "self" proteins and other large molecules.
   b. After inventory, lymphocytes develop receptors that allow them to differentiate between nonself (foreign) and self antigens.
   c. Nonself antigens combine with T cell and B cell surface receptors and stimulate these cells to cause an immune reaction.
   d. Haptens are small molecules that can combine with larger ones, becoming antigenic.
2. Lymphocyte origins
   a. Lymphocytes originate in red bone marrow and are released into the blood before they differentiate.
   b. Some reach the thymus, where they mature into T cells.
   c. Others, the B cells, mature in the red bone marrow.
   d. Both T cells and B cells reside in lymphatic tissues and organs.
3. Lymphocyte functions
   a. Some T cells interact with antigen-bearing agents directly, providing the cellular immune response.
   b. T cells secrete cytokines, such as interleukins, that enhance cellular responses to antigens.
   c. T cells may also secrete substances that are toxic to their target cells.
   d. B cells interact with antigen-bearing agents indirectly, providing the humoral immune response.
   e. Varieties of T cells and B cells number in the millions.

f. The members of each variety respond only to a specific antigen.
  g. As a group, the members of each variety form a clone.
4. T cells and the cellular immune response
  a. T cells are activated when an antigen-presenting cell displays a foreign antigen.
  b. When a macrophage acts as an accessory cell, it phagocytizes an antigen-bearing agent, digests the agent, and displays the antigens on its cell membrane in association with certain MHC proteins.
  c. A helper T cell becomes activated when it encounters displayed antigens for which it is specialized to react.
  d. An activated helper T cell contacts a B cell that carries the foreign antigen the T cell previously encountered on an antigen-presenting cell.
  e. In response, the T cell secretes cytokines, stimulates B cell proliferation, and attracts macrophages.
  f. Cytotoxic T cells recognize foreign antigens on tumor cells and cells whose surfaces indicate that they are infected by viruses.
  g. Memory T cells respond quickly to subsequent antigen exposure.
5. B cells and the humoral immune response
  a. A B cell is activated when it encounters an antigen that fits its antigen receptors.
  b. An activated B cell proliferates (especially when stimulated by a T cell), enlarging its clone.
  c. Some activated B cells specialize into antibody-producing plasma cells.
  d. Antibodies react against the antigen-bearing agent that stimulated their production.
  e. An individual's diverse B cells defend against a very large number of pathogens.
6. Types of antibodies
  a. Antibodies are soluble proteins called immunoglobulins.
  b. The five major types of immunoglobulins are IgG, IgA, IgM, IgD, and IgE.
7. Antibody actions
  a. Antibodies directly attach to antigens, activate complement, or stimulate local tissue changes that are unfavorable to antigen-bearing agents.
  b. Direct attachment results in agglutination, precipitation, or neutralization.
  c. Activated proteins of complement attract phagocytes, alter cells so they become more susceptible to phagocytosis, and rupture foreign cell membranes (lysis).
8. Immune responses
  a. The first reaction to an antigen is called a primary immune response.
    (1) During this response, antibodies are produced for several weeks.
    (2) Some B cells remain dormant as memory cells.
  b. A secondary immune response occurs rapidly as a result of memory cell response if the same antigen is encountered later.
9. Practical classification of immunity
  a. Naturally acquired immunity arises in the course of natural events, whereas artificially acquired immunity is the consequence of a medical procedure.
  b. Active immunity lasts much longer than passive immunity.
  c. A person who encounters a pathogen and has a primary immune response develops naturally acquired active immunity.
  d. A person who receives a vaccine containing a dead or weakened pathogen, or part of one, develops artificially acquired active immunity.
  e. A person who receives an injection of antibodies has artificially acquired passive immunity.
  f. When antibodies pass through a placental membrane from a pregnant woman to her fetus, the fetus develops naturally acquired passive immunity.
10. Allergic reactions
  a. Allergic reactions are excessive and misdirected immune responses that may damage tissue.
  b. Delayed-reaction allergy, which can occur in anyone and inflame the skin, results from repeated exposure to antigens.
  c. Immediate-reaction allergy is an inborn ability to overproduce IgE.
    (1) Allergic reactions result from mast cells bursting and releasing allergy mediators such as histamine.
    (2) The released chemicals cause allergy symptoms such as hives, hay fever, asthma, eczema, or gastric disturbances.
11. Transplantation and tissue rejection
  a. A transplant recipient's immune system may react against the donated tissue, an event termed a tissue rejection reaction.
  b. Matching donor and recipient tissues and using immunosuppressive drugs can minimize tissue rejection.
  c. Immunosuppressive drugs may increase susceptibility to infection.
12. Autoimmunity
  a. In autoimmune disorders, the immune system manufactures autoantibodies that attack a person's own body tissues.
  b. Autoimmune disorders may result from a previous viral infection, faulty T cell development, or reaction to a nonself antigen that resembles a self antigen.

REVIEW EXERCISES

1. Explain how the lymphatic system is related to the cardiovascular system. (p. 367)
2. Trace the general pathway of lymph from the interstitial spaces to the bloodstream. (p. 367)
3. Describe the primary functions of lymph. (p. 369)
4. Explain why physical exercise promotes lymphatic circulation. (p. 369)
5. Describe the structure and functions of a lymph node. (p. 370)
6. Describe the structure and functions of the thymus. (p. 371)
7. Describe the structure and functions of the spleen. (p. 371)
8. Distinguish between specific and nonspecific body defenses against infection. (p. 373)
9. Explain *species resistance*. (p. 373)
10. Describe how enzymatic actions function as defense mechanisms. (p. 374)
11. Define *interferon*, and explain its action. (p. 374)
12. List the major effects of inflammation, and explain why each occurs. (p. 374)
13. Identify the major phagocytic cells in blood and other tissues. (p. 374)
14. Distinguish between an antigen and an antibody. (p. 375)
15. Define *hapten*. (p. 375)
16. Review the origin of T cells and B cells. (p. 375)
17. Explain *cellular immunity*. (p. 376)
18. Explain the function of *plasma cells*. (p. 376)
19. Explain *humoral immunity*. (p. 376)

20. Define *clone* of lymphocytes. (p. 376)
21. Describe how T cells become activated. (p. 376)
22. Explain the function of *memory cells*. (p. 377)
23. Explain how a B cell is activated. (p. 377)
24. List the major types of immunoglobulins, and describe where each occurs. (p. 378)
25. Explain two mechanisms by which antibodies directly attach to antigens. (p. 380)
26. Explain the function of complement. (p. 380)
27. Distinguish between a primary and a secondary immune response. (p. 380)
28. Distinguish between active and passive immunity. (p. 381)
29. Define *vaccine*. (p. 381)
30. Explain how a vaccine produces its effect. (p. 381)
31. Explain the relationship between an allergic reaction and an immune response. (p. 381)
32. Distinguish between an antigen and an allergen. (p. 381)
33. List the major events leading to a delayed-reaction allergic response. (p. 381)
34. Describe how an immediate-reaction allergic response may occur. (p. 382)
35. Explain the relationship between tissue rejection and an immune response. (p. 383)
36. Describe a method used to reduce the severity of a tissue rejection reaction. (p. 383)
37. Explain the relationship between autoimmunity and an immune response. (p. 383)

CRITICAL THINKING

1. The immune response is specific, diverse, and equipped with memory. Give examples of each of these characteristics.
2. On whom should experimental AIDS vaccines be evaluated?
3. There are more people needing transplants than there are organs available. Discuss the pros and cons of the following proposed rationing systems for determining who should receive transplants: (a) first come, first served; (b) people with the best tissue and blood type match; (c) patients whose need for an organ is caused by infection or disease, as opposed to those whose need for an organ was preventable, such as a lung destroyed by smoking; (d) the youngest people; (e) the wealthiest people; (f) the most important people.
4. Why is a transplant consisting of fetal tissue less likely to provoke an immune rejection response than tissue from an adult?
5. T cells "learn" to recognize self from nonself during prenatal development. How could this learning process be altered to prevent allergies? To enable a person to accept a transplant?
6. Some parents keep their preschoolers away from other children to prevent them from catching illnesses. How might these well-meaning parents actually be harming their children?
7. One out of every 310,000 children who receives the vaccine for pertussis (whooping cough) develops permanent brain damage. The risk of suffering such damage from pertussis is about 1 in 30,000. Some parents refuse to vaccinate their children because of the few reported cases of adverse reaction to the vaccine. What are the dangers, both to the individual and to the population, when parents refuse to allow their children to be vaccinated against pertussis?
8. A xenograft is tissue from a nonhuman animal used to a replace a body part in a human. For example, pigs are being developed to provide cardiovascular spare parts because their hearts and blood vessels are similar to ours. To increase the likelihood of such a xenotransplant working, researchers genetically engineer pigs to produce human antigens on their cell surfaces. How can this improve the chances of a human body not rejecting such a transplant?
9. How can the removal of enlarged lymph nodes for microscopic examination aid in diagnosing certain diseases?
10. Why is injecting a substance into the skin like injecting it into the lymphatic system?
11. Why does vaccination provide long-lasting protection against a disease, while gamma globulin (IgG) provides only short-term protection?
12. What functions of the lymphatic system would be affected by being born without a thymus?

WEB CONNECTIONS

Visit the website for additional study questions and more information about this chapter at:

http://www.mhhe.com/shieress8

Unit 5
Absorption and Excretion

chapter **15**

Digestion and Nutrition

PREVENTING VITAMIN D DEFICIENCY. In recent years, many people have been trying to limit their exposure to the sun because of links between excessive sunning and skin cancers. Too little sun, however, can lead to vitamin D deficiency, because exposure to the ultraviolet wavelengths in sunlight is needed to convert certain precursors in skin cells into vitamin D.

Vitamin D is necessary for intestinal absorption of phosphorus and calcium, minerals that are essential for the health of the skeletal system. A deficiency of the vitamin causes the parathyroid glands to become overactive, taking more calcium from bones and causing osteoporosis. Demineralization, which removes phosphorus from bones, causes the bone-softening condition osteomalacia. In youngsters, lack of vitamin D leads to the bone-weakening condition rickets.

Evidence for vitamin D deficiency related to sun avoidance has a long history. The link between lack of sunlight and development of rickets was noted in 1822, and a century later, researchers realized that sun exposure helps reverse the disease in children. Other evidence comes from diverse sources, such as women who wear veils and naval personnel serving 3-month tours of duty on submarines.

Because older people tend to be outdoors less than younger individuals, the Institute of Medicine suggests that daily vitamin D intake escalate with age (see the following table).

AGE RANGE (YEARS)	INTERNATIONAL UNITS OF VITAMIN D
<50	200
50–70	400
70+	600

In addition, brief exposure to the sun can do wonders for maintaining vitamin D levels, without raising the risk of developing skin cancer—just 5 minutes of sunshine two or three times a week should do the trick.

Photo:
Sunshine is necessary for an adequate supply of vitamin D.

Chapter Objectives

After studying this chapter, you should be able to do the following:

15.1 Introduction
1. Describe the general functions of the digestive system. (p. 393)
2. Name the major organs of the digestive system. (p. 393)

15.2 General Characteristics of the Alimentary Canal
3. Describe the structure of the wall of the alimentary canal. (p. 394)
4. Explain how the contents of the alimentary canal are mixed and moved. (p. 395)

15.3 Mouth
5. Name the structures of the mouth and describe their functions. (p. 396)
6. Describe how different types of teeth are adapted for different functions. (p. 398)
7. List the parts of a tooth. (p. 398)

15.4–15.10 Salivary Glands—Large Intestine
8. List the enzymes the digestive organs and glands secrete, and describe the function of each. (p. 399)
9. Describe how digestive secretions are regulated. (p. 399)
10. Describe the mechanism of swallowing. (p. 400)
11. Explain how the products of digestion are absorbed. (p. 404)
12. Describe the defecation reflex. (p. 419)

15.11 Nutrition and Nutrients
13. List the major sources of carbohydrates, lipids, and proteins. (p. 420)
14. Describe how cells utilize carbohydrates, lipids, and amino acids. (p. 420)
15. List the fat-soluble and water-soluble vitamins, and summarize the general functions of each vitamin. (p. 423)
16. List the major minerals and trace elements, and summarize the general functions of each. (p. 424)
17. Describe an adequate diet. (p. 426)

Aids to Understanding Words

aliment- [food] *aliment*ary canal: Tubelike portion of the digestive system.

chym- [juice] *chym*e: Semifluid paste of food particles and gastric juice formed in the stomach.

decidu- [falling off] *decidu*ous teeth: Teeth that are shed during childhood.

gastr- [stomach] *gastr*ic gland: Portion of the stomach that secretes gastric juice.

hepat- [liver] *hepat*ic duct: Duct that carries bile from the liver to the common bile duct.

lingu- [tongue] *lingu*al tonsil: Mass of lymphatic tissue at the root of the tongue.

nutri- [nourish] *nutri*ent: Substance needed to nourish body cells.

peri- [around] *peri*stalsis: Wavelike ring of contraction that moves material along the alimentary canal.

pyl- [gatekeeper] *pyl*oric sphincter: Muscle that serves as a valve between the stomach and small intestine.

vill- [hairy] *vill*i: Tiny projections of mucous membrane in the small intestine.

Key Terms

alimentary canal (al″ĭ-men′tar-e kah-nal′)
bile (bīl)
calorie (kal′o-re)
chyme (kīm)
feces (fe′sēz)
gastric juice (gas′trik jōōs)
intestinal villus (in-tes′tĭ-nal vil′us); plural: **villi** (vil′i)
intrinsic factor (in-trin′sik fak′tor)
malnutrition (mal″nu-trish′un)
mesentery (mes′en-ter″e)
mineral (min′er-al)
nutrient (nu′tre-ent)
pancreatic juice (pan″kre-at′ik jōōs)
peristalsis (per″ĭ-stal′sis)
vitamin (vi′tah-min)

15.1 Introduction

Digestion is the mechanical and chemical breakdown of foods and the absorption of the resulting nutrients by cells. The organs of the digestive system carry out these processes. The **digestive system** consists of the alimentary canal, which extends about 9 meters from the mouth to the anus, and several accessory organs, which secrete substances used in the process of digestion into the canal. The **alimentary canal** includes the mouth, pharynx, esophagus, stomach, small intestine, large intestine, and anal canal; the accessory organs include the salivary glands, liver, gallbladder, and pancreas (fig. 15.1; reference plates 4, 5, and 6, pp. 25–27). Overall,

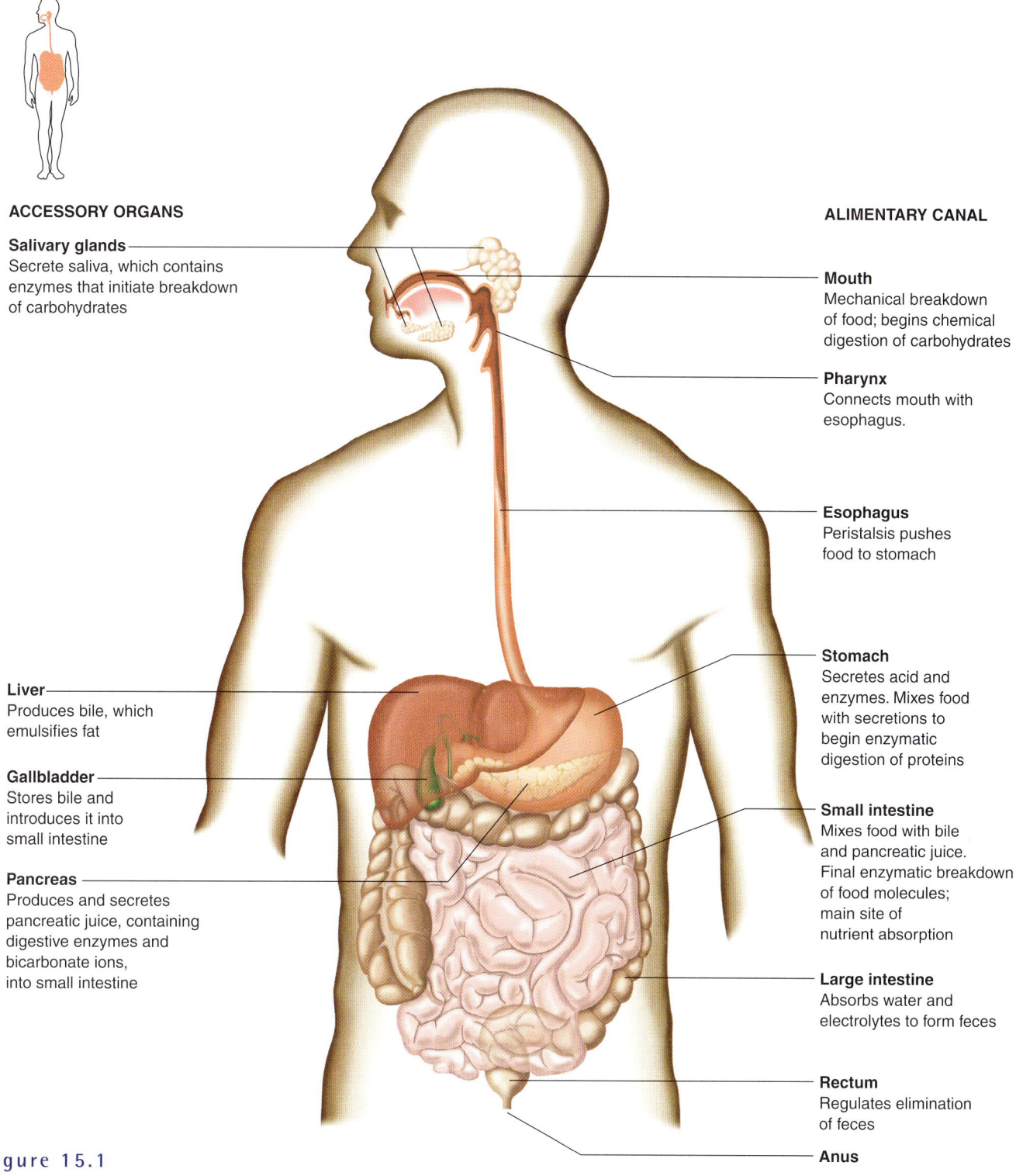

ACCESSORY ORGANS

Salivary glands
Secrete saliva, which contains enzymes that initiate breakdown of carbohydrates

Liver
Produces bile, which emulsifies fat

Gallbladder
Stores bile and introduces it into small intestine

Pancreas
Produces and secretes pancreatic juice, containing digestive enzymes and bicarbonate ions, into small intestine

ALIMENTARY CANAL

Mouth
Mechanical breakdown of food; begins chemical digestion of carbohydrates

Pharynx
Connects mouth with esophagus.

Esophagus
Peristalsis pushes food to stomach

Stomach
Secretes acid and enzymes. Mixes food with secretions to begin enzymatic digestion of proteins

Small intestine
Mixes food with bile and pancreatic juice. Final enzymatic breakdown of food molecules; main site of nutrient absorption

Large intestine
Absorbs water and electrolytes to form feces

Rectum
Regulates elimination of feces

Anus

Figure 15.1
Major organs of the digestive system.

the digestive system is a tube, open at both ends, that has a surface area of 186 square meters. It supplies body cells with nutrients.

15.2 General Characteristics of the Alimentary Canal

The alimentary canal is a muscular tube that passes through the body's ventral cavity. It is specialized in certain regions to carry on particular functions, but the structure of its wall, how it moves food, and its innervation are similar throughout its length (fig. 15.2).

Structure of the Wall

The wall of the alimentary canal consists of four distinct layers that are developed to different degrees from region to region. Beginning with the innermost tissues, these layers are (fig. 15.3):

1. **Mucosa,** or **mucous membrane** (mu´kus mem´brān) Surface epithelium, underlying connective tissue, and a small amount of smooth muscle form this layer. In some regions, the mucosa develops folds and tiny projections that extend into the passageway, or **lumen,** of the digestive tube and increase the mucosa's absorptive surface area. The mucosa may also contain glands that are tubular invaginations into which the lining cells secrete mucus and digestive enzymes. The mucosa protects the tissues beneath it and carries on secretion and absorption.

2. **Submucosa** The submucosa contains considerable loose connective tissue as well as glands, blood vessels, lymphatic vessels, and nerves organized into a network called a plexus. Its vessels nourish surrounding tissues and carry away absorbed materials.

3. **Muscular layer** This layer, which produces movements of the tube, consists of two coats of smooth muscle tissue and some nerves organized into a plexus. The fibers of the inner coat encircle the tube. When these *circular fibers* contract, the tube's diameter decreases. The fibers of the outer muscular coat run lengthwise. When these *longitudinal fibers* contract, the tube shortens.

4. **Serosa,** or **serous layer** (se´rus la´er) The *visceral peritoneum* comprises the serous layer, or outer covering, of the tube. The cells of the serosa protect underlying tissues and secrete serous fluid, which moistens and lubricates the tube's outer surface so that organs within the abdominal cavity slide freely against one another.

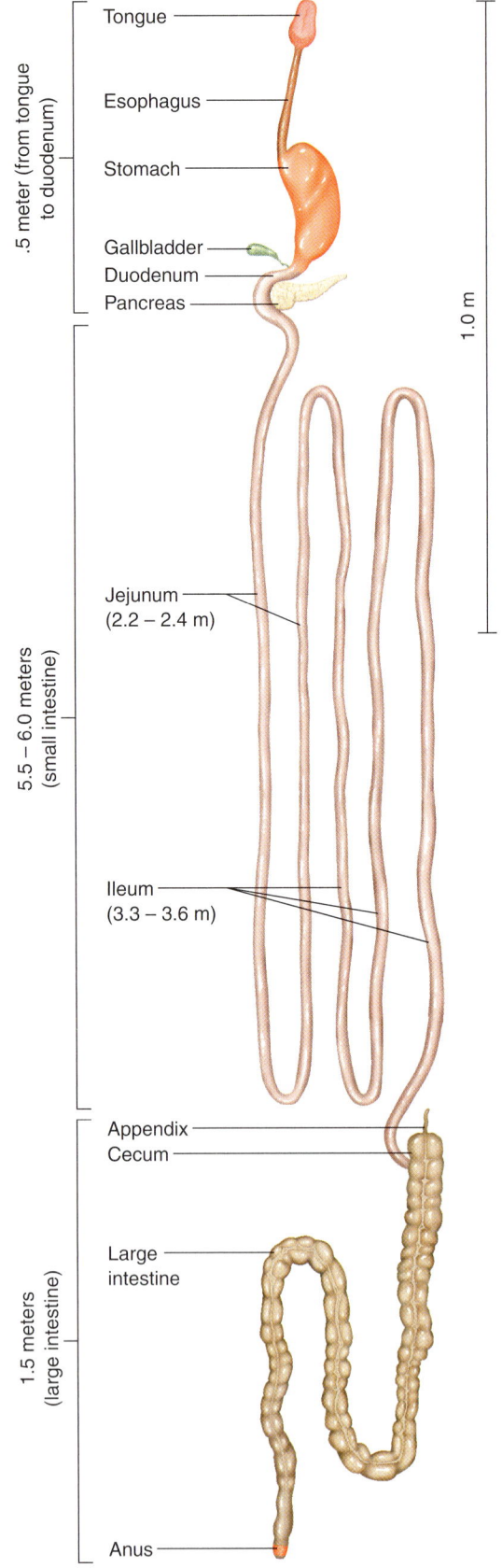

Figure 15.2
The alimentary canal is a muscular tube about 9 meters long.

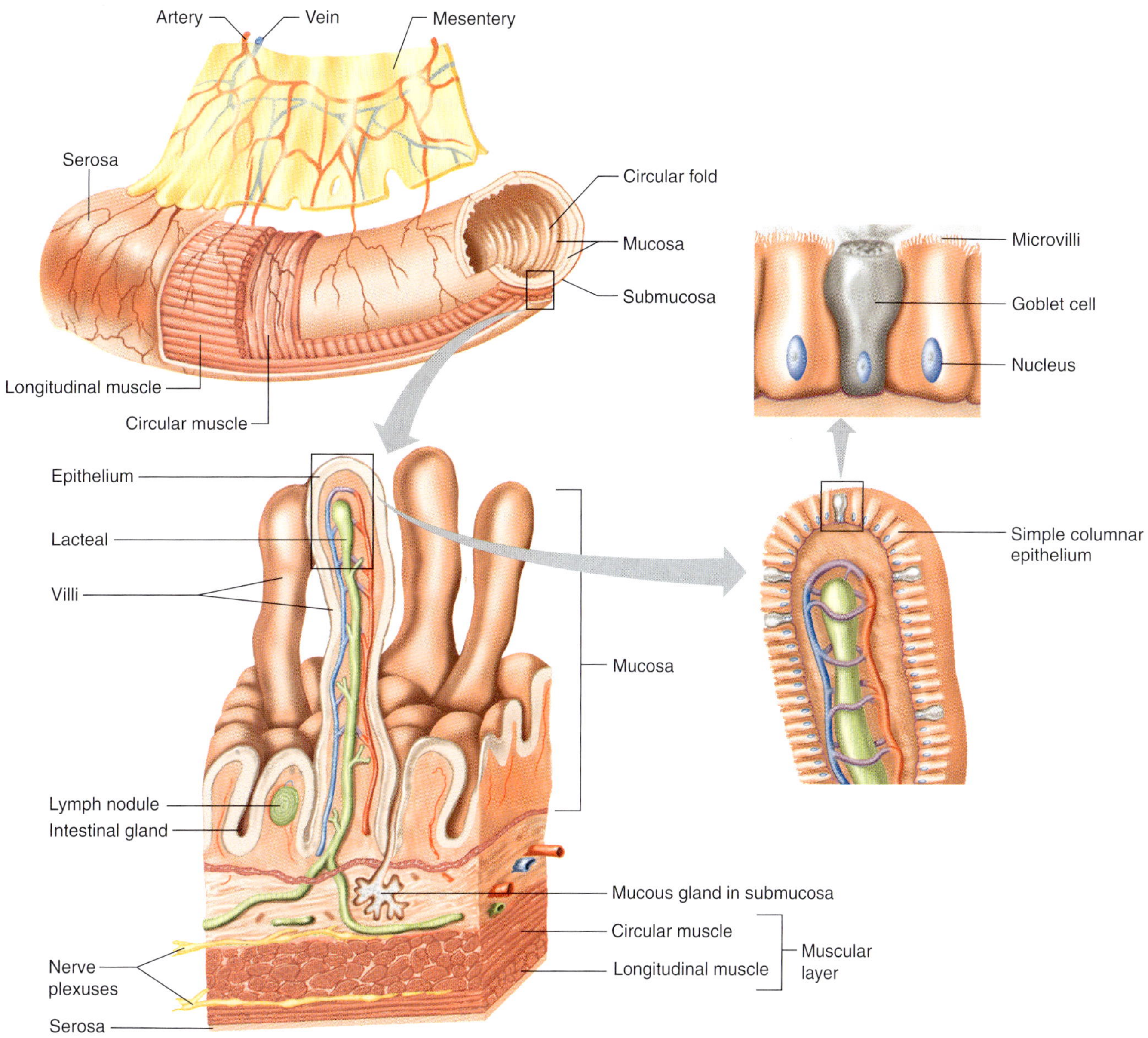

Figure 15.3
The wall of the small intestine, as in other portions of the alimentary canal, includes four layers: an inner mucosa, a submucosa, a muscular layer, and an outer serosa.

Movements of the Tube

The motor functions of the alimentary canal are of two basic types—*mixing movements* and *propelling movements*. Mixing occurs when smooth muscles in small segments of the tube contract rhythmically. For example, when the stomach is full, waves of muscular contractions move along its walls from one end to the other. These waves mix food with digestive juices that the mucosa secretes (fig. 15.4A).

Propelling movements include a wavelike motion called **peristalsis** (per´´ĭ-stal´sis). When peristalsis occurs, a ring of contraction appears in the wall of the tube. At the same time, the muscular wall just ahead of the ring relaxes. As the peristaltic wave moves along, it pushes the tubular contents ahead of it (fig. 15.4B).

CHECK YOUR RECALL

1. Which organs constitute the digestive system?
2. Describe the wall of the alimentary canal.
3. Name the two types of movements that occur in the alimentary canal.

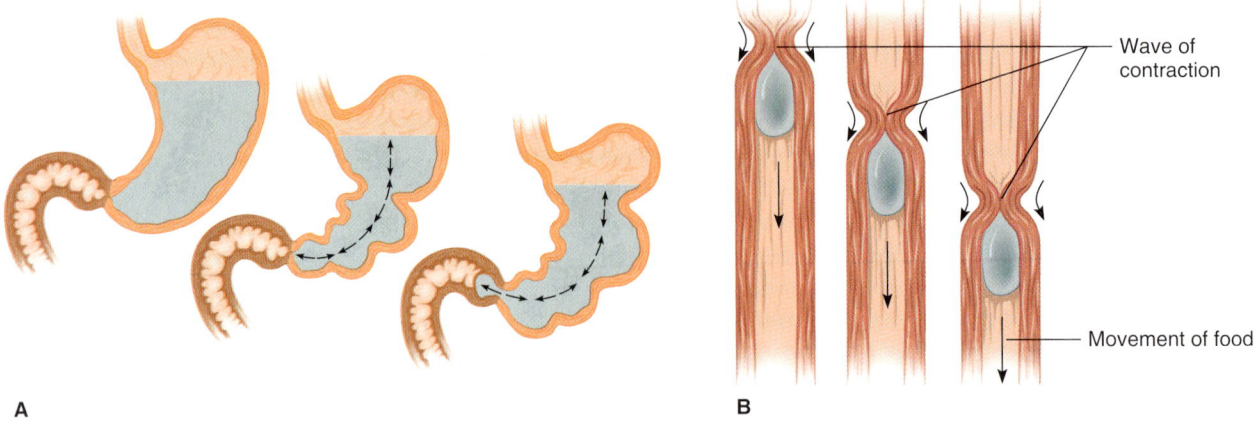

Figure 15.4
Movements through the alimentary canal. (*A*) Mixing movements occur when small segments of the muscular wall of the alimentary canal rhythmically contract. (*B*) Peristaltic waves move the contents along the canal.

15.3 Mouth

The **mouth** receives food and begins digestion by mechanically reducing the size of solid particles and mixing them with saliva. The lips, cheeks, tongue, and palate surround the mouth, which includes a chamber between the palate and tongue called the *oral cavity,* as well as a narrow space between the teeth, cheeks, and lips called the *vestibule* (fig. 15.5).

Cheeks and Lips

The **cheeks** consist of outer layers of skin, pads of subcutaneous fat, muscles associated with expression and chewing, and inner linings of moist, stratified squamous epithelium. The **lips** are highly mobile structures that surround the mouth opening. They contain skeletal muscles and sensory receptors useful in judging the temperature and texture of foods. Their normal reddish color is due to the many blood vessels near their surfaces.

Tongue

The **tongue** nearly fills the oral cavity when the mouth is closed. Mucous membrane covers the tongue, and a membranous fold called the **frenulum** connects the midline of the tongue to the floor of the mouth.

The *body* of the tongue is largely composed of skeletal muscle. These muscles mix food particles with saliva during chewing and move food toward the pharynx during swallowing. The tongue also helps move food underneath the teeth for chewing. Rough projections called **papillae** on the tongue surface provide friction, which helps handle food. These papillae also contain taste buds (see chapter 10, p. 258).

The posterior region, or *root,* of the tongue is anchored to the hyoid bone. It is covered with rounded masses of lymphatic tissue called **lingual tonsils** (fig. 15.6).

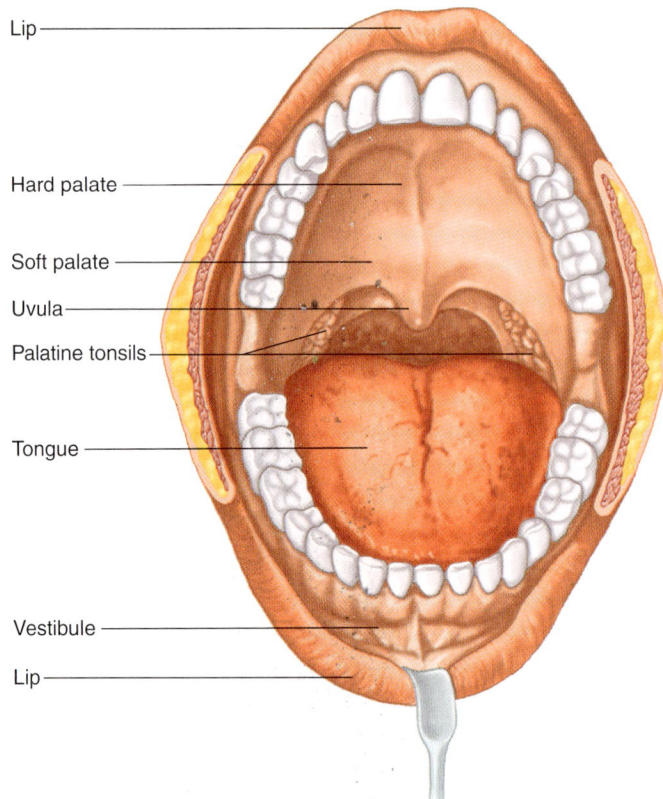

Figure 15.5
The mouth is adapted for ingesting food and preparing it for digestion.

Palate

The **palate** forms the roof of the oral cavity and consists of a hard anterior part (*hard palate*) and a soft posterior part (*soft palate*). The soft palate forms a muscular arch, which extends posteriorly and downward as a cone-shaped projection called the **uvula.**

During swallowing, muscles draw the soft palate and the uvula upward. This action closes the opening

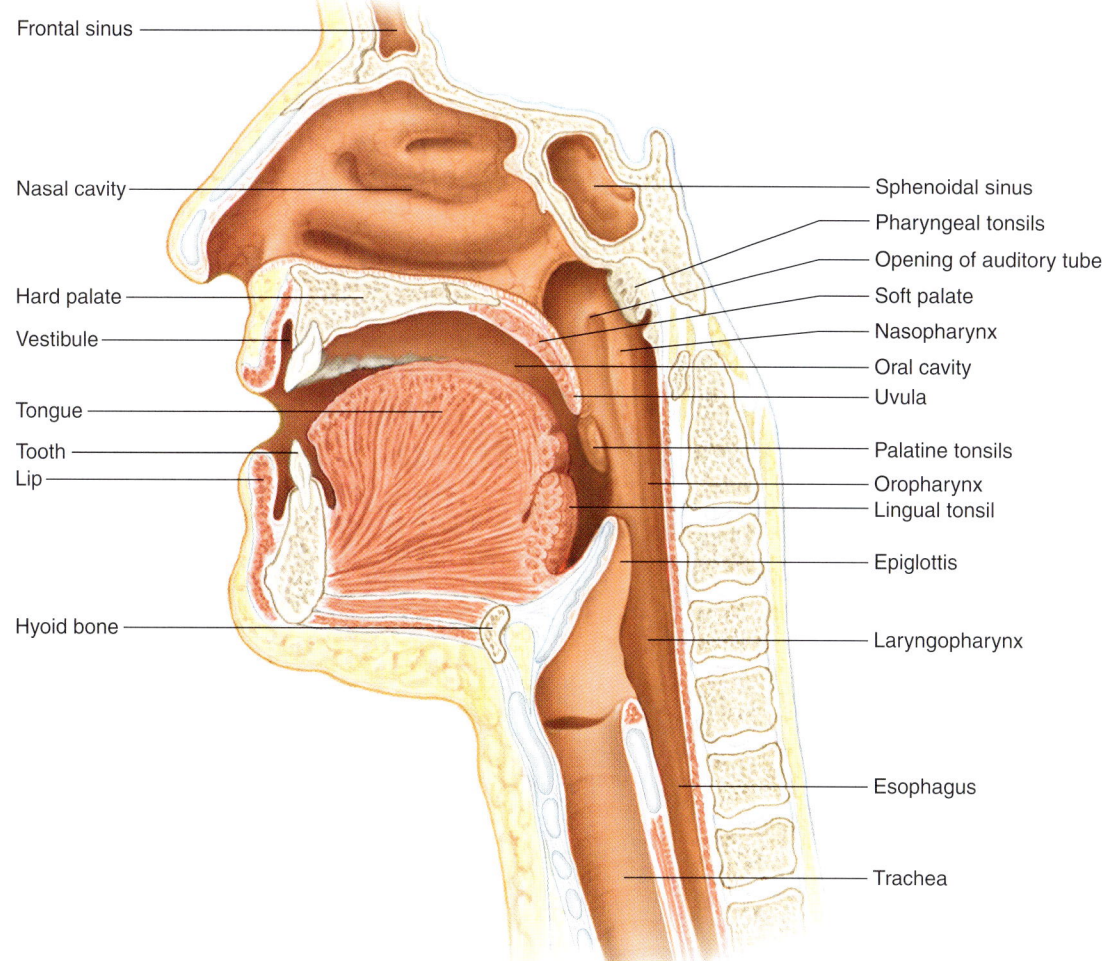

Figure 15.6
Sagittal section of the mouth, nasal cavity, and pharynx.

between the nasal cavity and the pharynx, preventing food from entering the nasal cavity.

In the back of the mouth, on either side of the tongue and closely associated with the palate, are masses of lymphatic tissue called **palatine tonsils** (see figs. 15.5 and 15.6). These structures lie beneath the epithelial lining of the mouth and, like other lymphatic tissues, help protect the body against infection.

The palatine tonsils are common sites of infection, and when inflamed, produce *tonsillitis*. Infected tonsils may swell so greatly that they block the passageways of the pharynx and interfere with breathing and swallowing. Because the mucous membranes of the pharynx, auditory tubes, and middle ears are continuous, such an infection can travel from the throat into the middle ears (*otitis media*).

When tonsillitis occurs repeatedly and does not respond to antibiotic treatment, the tonsils are sometimes surgically removed (*tonsillectomy*). However, tonsillectomies are done less often today than they were a generation ago because the tonsils' role in immunity is now recognized.

Other masses of lymphatic tissue, called **pharyngeal tonsils,** or *adenoids,* are on the posterior wall of the pharynx, above the border of the soft palate (fig. 15.6). If the adenoids enlarge and block the passage between the pharynx and the nasal cavity, they also may be surgically removed.

CHECK YOUR RECALL

1. How does the tongue function as part of the digestive system?
2. What is the role of the soft palate in swallowing?
3. Where are the palatine tonsils located?

Teeth

Two different sets of **teeth** form during development. The members of the first set, the *primary teeth* (deciduous teeth), usually erupt through the gums at regular intervals between the ages of six months and two to four years (fig. 15.7). There are twenty deciduous teeth—ten in each jaw.

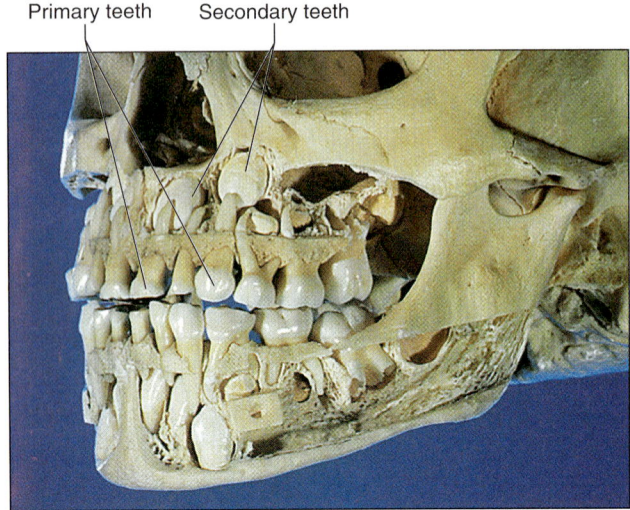

Figure 15.7
This partially dissected child's skull reveals primary and developing secondary teeth in the maxilla and mandible.

The primary teeth are usually shed in the same order they appeared. Before this happens, though, their roots are resorbed. Pressure from the developing *secondary teeth* (permanent teeth) then pushes the primary teeth out of their sockets. This secondary set consists of thirty-two teeth—sixteen in each jaw (fig. 15.8). The secondary teeth usually begin to appear at six years, but the set may not be complete until the third molars appear between seventeen and twenty-five years.

Teeth break pieces of food into smaller pieces. This action increases the surface area of food particles, allowing digestive enzymes to react more effectively with the food molecules.

Different teeth are adapted to handle food in different ways. The *incisors* are chisel-shaped, and their sharp edges bite off large pieces of food. The *cuspids* are cone-shaped, and they grasp and tear food. The *bicuspids* and *molars* have somewhat flattened surfaces and are specialized for grinding food particles (fig. 15.8). Table 15.1 summarizes the number and kinds of teeth that appear during development.

Each tooth consists of two main portions—the *crown*, which projects beyond the gum, and the *root*, which is anchored to the alveolar process of the jaw. Where these portions meet is called the *neck* of the tooth.

Glossy, white *enamel* covers the crown. Enamel mainly consists of calcium salts and is the hardest substance in the body. Unfortunately, if damaged by abrasive action or injury, enamel is not replaced.

The bulk of a tooth beneath the enamel is composed of *dentin*, a substance much like bone, but somewhat harder. Dentin surrounds the tooth's central cavity (pulp cavity), which contains a combination of blood vessels, nerves, and connective tissue called *pulp*. Blood vessels and nerves reach this cavity through tubular *root canals* extending into the root.

A thin layer of bonelike material called *cementum* encloses the root. A *periodontal ligament* surrounds the cementum. This ligament contains bundles of thick collagenous fibers, which pass between the cementum and the bone of the alveolar process, firmly attaching the tooth to the jaw. It also contains blood vessels and nerves (fig. 15.9).

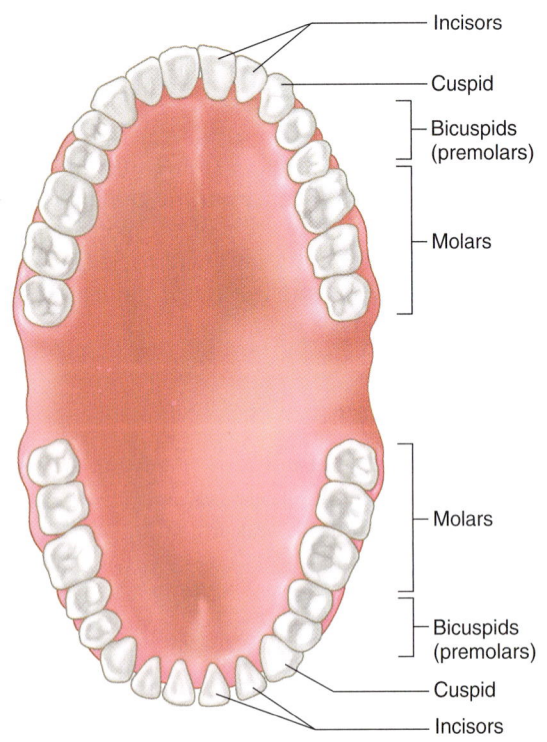

Figure 15.8
The secondary teeth of the upper and lower jaws.

TABLE 15.1 PRIMARY AND SECONDARY TEETH

PRIMARY TEETH (DECIDUOUS)		SECONDARY TEETH (PERMANENT)	
Type	Number	Type	Number
Incisor		Incisor	
Central	4	Central	4
Lateral	4	Lateral	4
Cuspid	4	Cuspid	4
		Bicuspid	
		First	4
		Second	4
Molar		Molar	
First	4	First	4
Second	4	Second	4
		Third	4
Total	20	Total	32

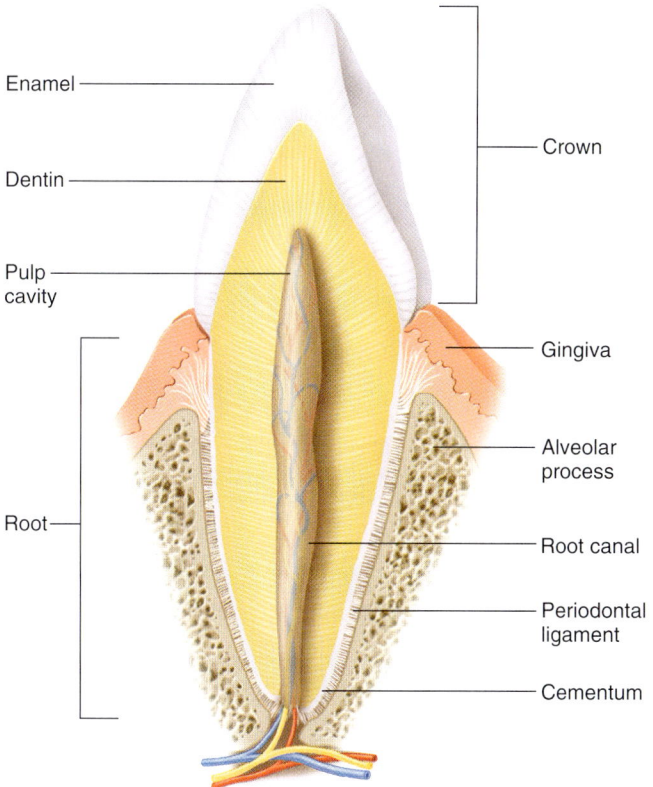

Figure 15.9
A section of a cuspid tooth.

CHECK YOUR RECALL

1. How do primary teeth differ from secondary teeth?
2. Describe the structure of a tooth.
3. Explain how a tooth is attached to the bone of the jaw.

15.4 Salivary Glands

The **salivary glands** secrete saliva. This fluid moistens food particles, helps bind them, and begins the chemical digestion of carbohydrates. Saliva is also a solvent, dissolving foods so that they can be tasted, and it helps cleanse the mouth and teeth.

Salivary Secretions

Within a salivary gland are two types of secretory cells—*serous cells* and *mucous cells.* These cells are present in varying proportions within different glands. Serous cells produce a watery fluid that contains the digestive enzyme **amylase.** This enzyme splits starch and glycogen molecules into disaccharides—the first step in the chemical digestion of carbohydrates. Mucous cells secrete a thick liquid called **mucus,** which binds food particles and lubricates during swallowing.

When a person sees, smells, tastes, or even thinks about pleasant food, parasympathetic nerve impulses elicit the secretion of a large volume of watery saliva. Conversely, food that looks, smells, or tastes unpleasant inhibits parasympathetic activity so that less saliva is produced, and swallowing may become difficult.

Major Salivary Glands

Three pairs of major salivary glands—the parotid, submandibular, and sublingual glands—and many minor ones are associated with the mucous membranes of the tongue, palate, and cheeks (fig. 15.10). The **parotid glands** are the largest of the major salivary glands. Each gland lies anterior and somewhat inferior to each ear, between the skin of the cheek and the masseter muscle. The parotid glands secrete a clear, watery fluid that is rich in amylase.

Topic of Interest — DENTAL CARIES

Sticky foods, such as caramel, lodge between the teeth and in the crevices of molars, feeding bacteria such as *Actinomyces, Streptococcus mutans,* and *Lactobacillus.* These microbes metabolize carbohydrates in the food, producing acid by-products that destroy tooth enamel and dentin. The bacteria also produce sticky substances that hold them in place.

If a person eats a candy bar, for example, but does not brush the teeth soon afterward, the actions of the acid-forming bacteria may produce decay in the tooth enamel, called *dental caries.* Unless a dentist cleans and fills the resulting cavity that forms where enamel is destroyed, the damage will spread to the underlying dentin. The tooth becomes very sensitive.

Dental caries can be prevented in several ways:

1. Brush and floss teeth regularly.
2. Have regular dental exams and cleanings.
3. Drink fluoridated water or receive a fluoride treatment. Fluoride is actually incorporated into the enamel's chemical structure, strengthening it.
4. Have the dentist apply a sealant to children's and adolescents' teeth where crevices might hold onto decay-causing bacteria. The sealant is a coating that keeps acids from eating away at tooth enamel.

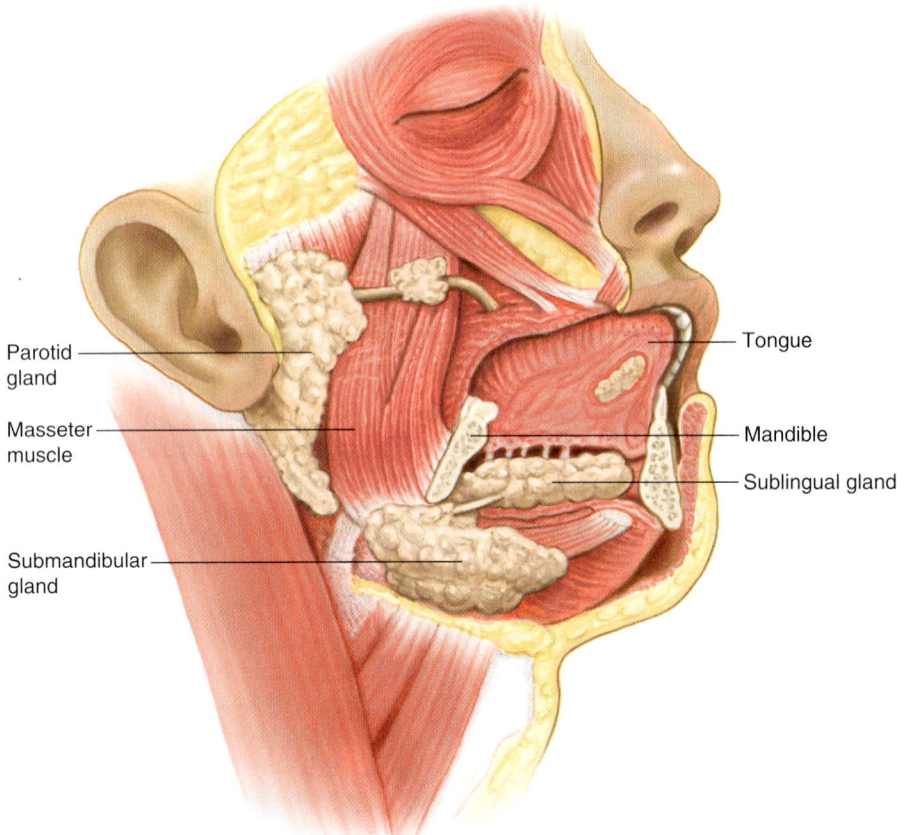

Figure 15.10
Locations of the major salivary glands.

The **submandibular glands** are located in the floor of the mouth on the inside surface of the lower jaw. The secretory cells of these glands are predominantly serous, with a few mucous cells. Consequently, the submandibular glands secrete a more viscous fluid than the parotid glands.

The **sublingual glands,** the smallest of the major salivary glands, are on the floor of the mouth inferior to the tongue. Their secretory cells are primarily the mucous type, making their secretions thick and stringy.

 CHECK YOUR RECALL

1. What is the function of saliva?
2. What stimulates salivary glands to secrete saliva?
3. Where are the major salivary glands located?

15.5 Pharynx and Esophagus

The pharynx is a cavity posterior to the mouth from which the tubular esophagus leads to the stomach (see fig. 15.1). The pharynx and the esophagus do not digest food, but both are important passageways whose muscular walls function in swallowing.

Structure of the Pharynx

The **pharynx** (far´inks) connects the nasal and oral cavities with the larynx and esophagus. It has three parts (see fig. 15.6):

1. The **nasopharynx** communicates with the nasal cavity and provides a passageway for air during breathing.
2. The **oropharynx** is posterior to the soft palate and inferior to the nasopharynx. It is a passageway for food moving downward from the mouth and for air moving to and from the nasal cavity.
3. The **laryngopharynx,** just inferior to the oropharynx, is a passageway to the esophagus.

Swallowing Mechanism

Swallowing has three stages. In the first stage, which is voluntary, food is chewed and mixed with saliva. Then the tongue rolls this mixture into a mass (bolus) and forces it into the pharynx.

Computer simulation experiments show that each food requires an optimum range of number of chews to form a bolus. Eating raw carrots, for example, requires 20 to 25 chews.

The second stage begins as food stimulates sensory receptors around the pharyngeal opening. This triggers the swallowing reflex, which includes the following actions:

1. The soft palate raises, preventing food from entering the nasal cavity.
2. The hyoid bone and the larynx are elevated. A flap-like structure attached to the larynx, called the *epiglottis,* closes off the top of the trachea so that food is less likely to enter.
3. The tongue is pressed against the soft palate, sealing off the oral cavity from the pharynx.
4. The longitudinal muscles in the pharyngeal wall contract, pulling the pharynx upward toward the food.
5. Muscles in the lower portion of the pharynx relax, opening the esophagus.
6. A peristaltic wave begins in the pharyngeal muscles and forces food into the esophagus.

The swallowing reflex momentarily inhibits breathing. Then, during the third stage of swallowing, peristalsis transports the food in the esophagus to the stomach.

Esophagus

The **esophagus** (ĕ-sof´ah-gus), a straight, collapsible tube about 25 centimeters long, is a food passageway from the pharynx to the stomach (see figs. 15.1 and 15.6). The esophagus begins at the base of the pharynx and descends posterior to the trachea, passing through the mediastinum. It penetrates the diaphragm through an opening, the *esophageal hiatus,* and is continuous with the stomach on the abdominal side of the diaphragm.

Mucous glands are scattered throughout the submucosa of the esophagus. Their secretions moisten and lubricate the tube's inner lining.

Just above where the esophagus joins the stomach, some circular smooth muscle fibers in the esophageal wall thicken, forming the *lower esophageal sphincter* (lo´er ĕ-sof´´ah-je´al sfing´ter), or cardiac sphincter (fig. 15.11). These fibers usually remain contracted, and they close the entrance to the stomach, preventing regurgitation of the stomach contents into the esophagus. When peristaltic waves reach the stomach, the muscle fibers temporarily relax and allow the swallowed food to enter.

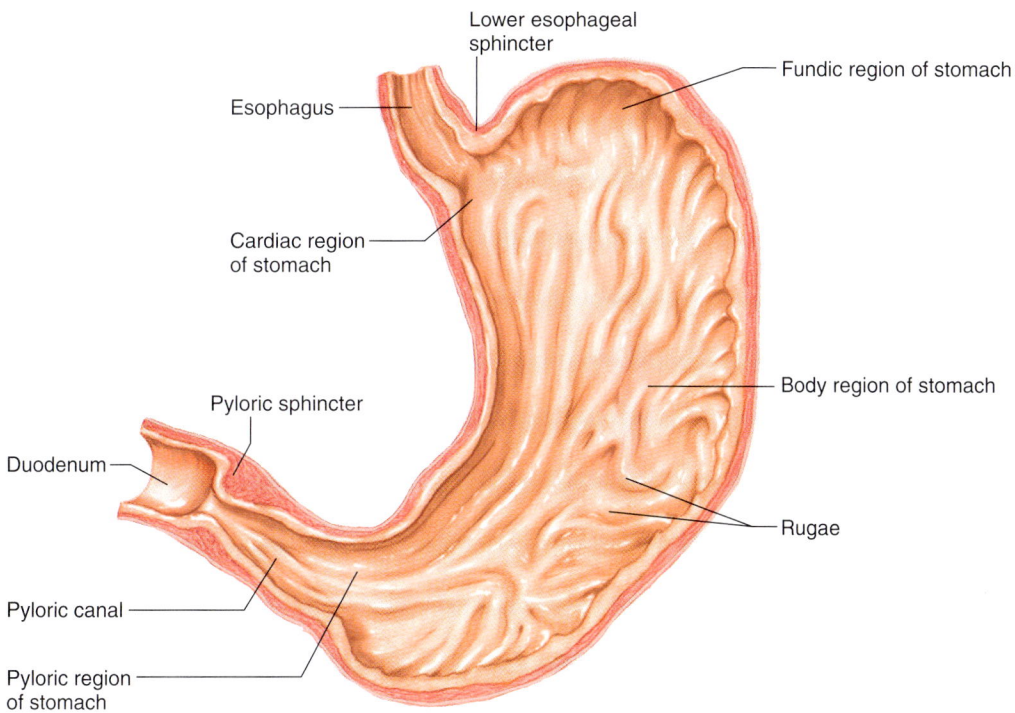

Figure 15.11
Major regions of the stomach.

In a *hiatal hernia,* a portion of the stomach protrudes through a weakened area of the diaphragm, through the esophageal hiatus, and into the thorax. As a result of a hiatal hernia, regurgitation (reflux) of gastric juice into the esophagus may inflame the esophageal mucosa, causing heartburn, difficulty in swallowing, or ulceration and blood loss. In response to the destructive action of gastric juice, columnar epithelium may replace the squamous epithelium that normally lines the esophagus. This condition, called *Barrett's esophagus,* increases the risk of developing esophageal cancer.

CHECK YOUR RECALL

1. Describe the regions of the pharynx.
2. List the major events that occur during swallowing.
3. What is the function of the esophagus?

15.6 Stomach

The **stomach** is a J-shaped, pouchlike organ that hangs inferior to the diaphragm in the upper left portion of the abdominal cavity and has a capacity of about 1 liter or more (figs. 15.1 and 15.11; reference plates 4 and 5, pp. 25–26). Thick folds (rugae) of mucosal and submucosal layers mark the stomach's inner lining and disappear when the stomach wall is distended. The stomach receives food from the esophagus, mixes the food with gastric juice, initiates protein digestion, carries on limited absorption, and moves food into the small intestine.

Parts of the Stomach

The stomach is divided into the cardiac, fundic, body, and pyloric regions (fig. 15.11). The *cardiac region* is a small area near the esophageal opening. The *fundic region,* which balloons superior to the cardiac portion, is a temporary storage area. The dilated *body region,* which is the main part of the stomach, lies between the fundic and pyloric portions. The *pyloric region* narrows and becomes the *pyloric canal* as it approaches the small intestine.

At the end of the pyloric canal, the muscular wall thickens, forming a powerful circular muscle, the **pyloric sphincter** (pylorus). This muscle is a valve that controls gastric emptying.

Gastric Secretions

The mucous membrane that forms the inner lining of the stomach is thick, its surface studded with many small openings called *gastric pits.* Gastric pits are at the ends of tubular **gastric glands** (fig. 15.12).

Gastric glands generally contain three types of secretory cells. *Mucous cells* (goblet cells) occur in the

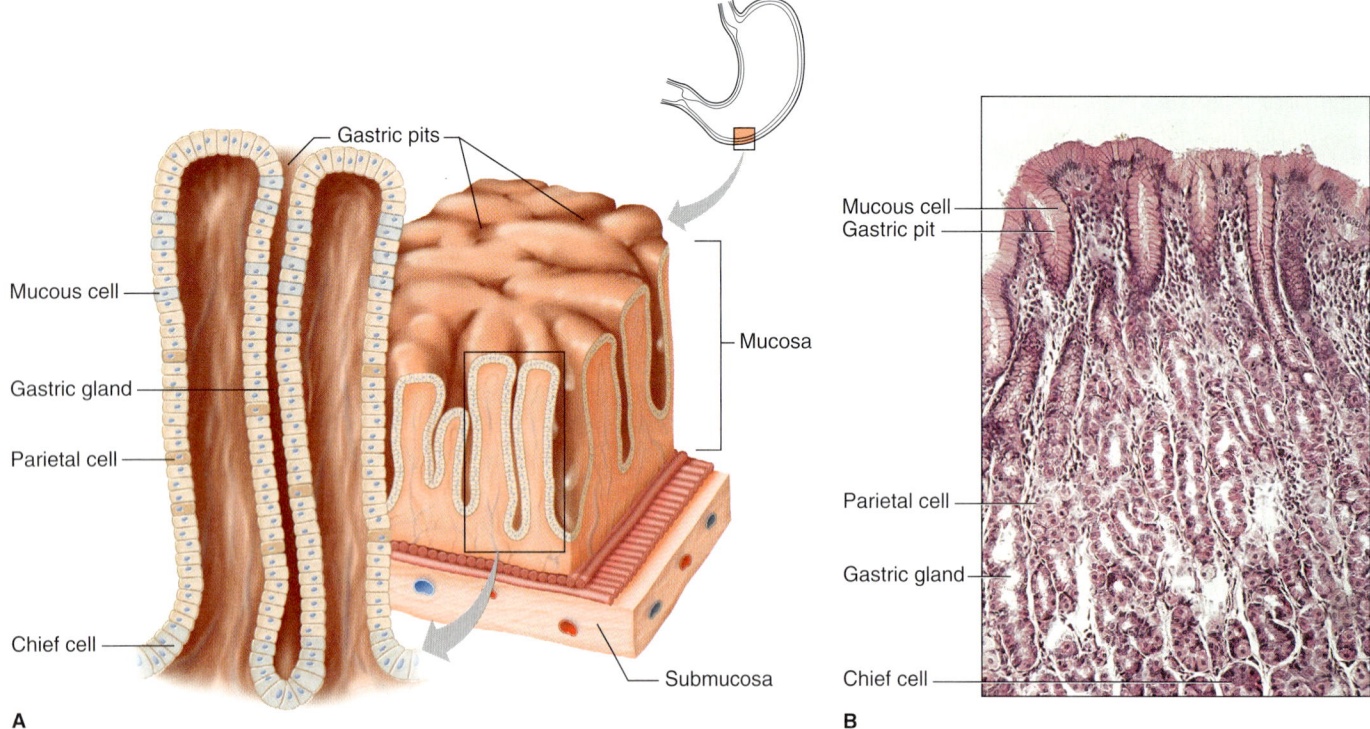

Figure 15.12
Lining of the stomach. (*A*) Gastric glands include mucous cells, parietal cells, and chief cells. (*B*) The mucosa of the stomach is studded with gastric pits that are the openings of the gastric glands.

necks of the glands, near the openings of the gastric pits. *Chief cells* and *parietal cells* are found in the deeper parts of the glands. The chief cells secrete digestive enzymes, and the parietal cells release hydrochloric acid. The products of the mucous cells, chief cells, and parietal cells together form **gastric juice** (gas´trik jōōs).

Of the several digestive enzymes in gastric juice, **pepsin** is by far the most important. The chief cells secrete it as the inactive enzyme precursor **pepsinogen.** However, when pepsinogen contacts the hydrochloric acid from the parietal cells, it is cut down rapidly, forming pepsin. Pepsin begins the digestion of nearly all types of dietary protein into polypeptide strands. This enzyme is most active in an acidic environment, provided by the hydrochloric acid in gastric juice.

The mucous cells of the gastric glands secrete large quantities of thin mucus. In addition, the cells of the mucous membrane associated with the stomach's inner lining and between the gastric glands release a more viscous and alkaline secretion, which coats the inside of the stomach wall. This coating normally prevents the stomach from digesting itself.

Another component of gastric juice is **intrinsic factor** (in-trin´sik fak´tor), which the parietal cells secrete. Intrinsic factor helps the small intestine absorb vitamin B_{12}. Table 15.2 summarizes the components of gastric juice.

CHECK YOUR RECALL

1. What are the secretions of the chief cells and parietal cells?
2. Which is the most important digestive enzyme in gastric juice?
3. Why doesn't the stomach digest itself?

 The 40 million cells that line the stomach's interior can secrete 2 to 3 quarts (about 2 to 3 liters) of gastric juice per day.

Regulation of Gastric Secretions

Gastric juice is produced continuously, but the rate varies considerably and is controlled both neurally and hormonally. When a person tastes, smells, or even sees appetizing food, or when food enters the stomach, parasympathetic impulses on the vagus nerves stimulate acetylcholine (Ach) release from nerve endings. This Ach stimulates gastric glands to secrete large amounts of gastric juice, which is rich in hydrochloric acid and pepsinogen. These parasympathetic impulses also stimulate certain stomach cells to release the peptide hormone **gastrin,** which increases the secretory activity of gastric glands (fig. 15.13).

As food moves into the upper part of the small intestine, acid triggers sympathetic nerve impulses that inhibit gastric juice secretion. At the same time, proteins and fats in this region of the intestine cause the intestinal wall to release the peptide hormone **cholecystokinin** (ko´´le-sis´´to-ki´nin). This hormonal action decreases gastric motility as the small intestine fills with food.

Gastrin stimulates cell growth in the mucosa of the stomach and intestines, except where gastrin is produced. This cell growth helps repair mucosal cells damaged by normal stomach function, disease, or medical treatments.

TABLE 15.2 — MAJOR COMPONENTS OF GASTRIC JUICE

COMPONENT	SOURCE	FUNCTION
Pepsinogen	Chief cells of the gastric glands	Inactive form of pepsin
Pepsin	Formed from pepsinogen in the presence of hydrochloric acid	A protein-splitting enzyme that digests nearly all types of dietary protein
Hydrochloric acid	Parietal cells of the gastric glands	Provides the acid environment needed for the conversion of pepsinogen into pepsin and for the action of pepsin
Mucus	Goblet cells and mucous glands	Provides a viscous, alkaline protective layer on the inside stomach wall
Intrinsic factor	Parietal cells of the gastric glands	Aids in vitamin B_{12} absorption

An *ulcer* is an open sore in the mucous membrane resulting from localized tissue breakdown. Gastric ulcers occur in the stomach and duodenal ulcers occur in the region of the small intestine nearest the stomach.

For many years, both types of ulcers were attributed to stress and treated with medications to decrease stomach acid secretion. In 1982, two Australian researchers boldly suggested that stomach infection by the bacterium *Helicobacter pylori* causes gastric ulcers.

When the medical community did not believe them, one of the researchers swallowed the bacteria to demonstrate their effect—and soon developed stomach pain. Still, it was twelve years before U.S. government physicians advised colleagues to treat gastric ulcers as an infection in people with evidence of *Helicobacter pylori*. Today, a short course of antibiotics, often combined with acid-lowering drugs, can cure a gastric ulcer.

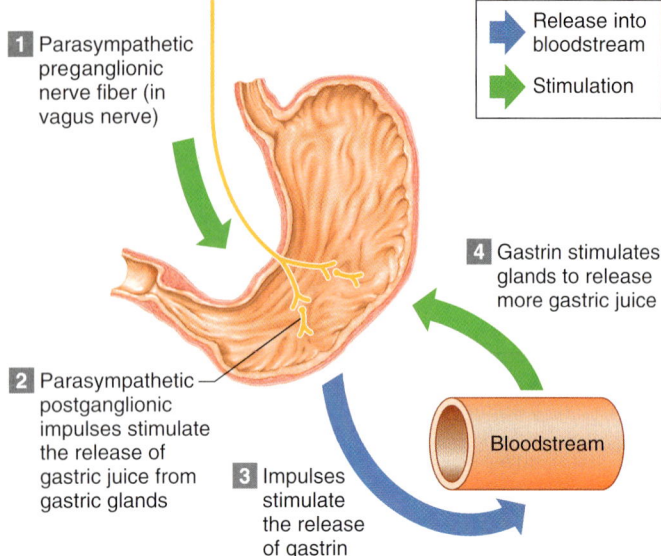

Figure 15.13
The secretion of gastric juice is regulated in part by parasympathetic nerve impulses that stimulate the release of gastric juice and gastrin.

Gastric Absorption

Gastric enzymes begin breaking down proteins, but the stomach wall is not well adapted to absorbing digestive products. The stomach absorbs only small quantities of water and certain salts as well as alcohol and some lipid-soluble drugs.

CHECK YOUR RECALL

1. What controls gastric juice secretion?
2. What is the function of cholecystokinin?
3. What substances can the stomach absorb?

Mixing and Emptying Actions

Following a meal, the mixing movements of the stomach wall aid in producing a semifluid paste of food particles and gastric juice called **chyme** (kīm). Peristaltic waves push the chyme toward the pyloric region of the stomach, and as chyme accumulates near the pyloric sphincter, this muscle begins to relax. Stomach contractions push chyme a little at a time into the small intestine.

The rate at which the stomach empties depends on the fluidity of the chyme and the type of food present. For example, liquids usually pass through the stomach quite rapidly, but solids remain until they are well mixed with gastric juice. Fatty foods may remain in the stomach 3–6 hours; foods high in proteins move through more quickly; and carbohydrates usually pass through faster than either fats or proteins.

As chyme enters the **duodenum** (the first portion of the small intestine), accessory organs—the pancreas, liver, and gallbladder—add their secretions.

*V*omiting results from a complex reflex that empties the stomach through the esophagus, pharynx, and mouth. Irritation or distension in the stomach or intestines can trigger vomiting. Sensory impulses travel from the site of stimulation to the *vomiting center* in the medulla oblongata, and several motor responses follow. These include taking a deep breath, raising the soft palate and thus closing the nasal cavity, closing the opening to the trachea (glottis), relaxing the circular muscle fibers at the base of the esophagus, contracting the diaphragm so it presses downward over the stomach, and contracting the abdominal wall muscles to increase pressure inside the abdominal cavity. As a result, the stomach is squeezed from all sides, forcing its contents upward and out.

CHECK YOUR RECALL

1. How is chyme produced?
2. What factors influence how quickly chyme leaves the stomach?

15.7 Pancreas

The **pancreas,** discussed as an endocrine gland in chapter 11 (p. 296), also has an exocrine function—secretion of a digestive juice called **pancreatic juice** (pan´kre-at´ik jōōs).

Structure of the Pancreas

The pancreas is closely associated with the small intestine. It extends horizontally across the posterior abdominal wall in the C-shaped curve of the duodenum (figs. 15.1 and 15.14).

The cells that produce pancreatic juice, called *pancreatic acinar cells,* make up the bulk of the pancreas. These cells cluster around tiny tubes, into which they release their secretions. The smaller tubes unite to form larger ones, which in turn give rise to a *pancreatic duct* extending the length of the pancreas. The pancreatic duct usually connects with the duodenum at the same place where the bile duct from the liver and gallbladder joins the duodenum, although other connections may be present (fig. 15.14). A hepatopancreatic sphincter controls the movement of pancreatic juices into the duodenum.

Pancreatic Juice

Pancreatic juice contains enzymes that digest carbohydrates, fats, nucleic acids, and proteins. The carbohydrate-digesting enzyme **pancreatic amylase** splits

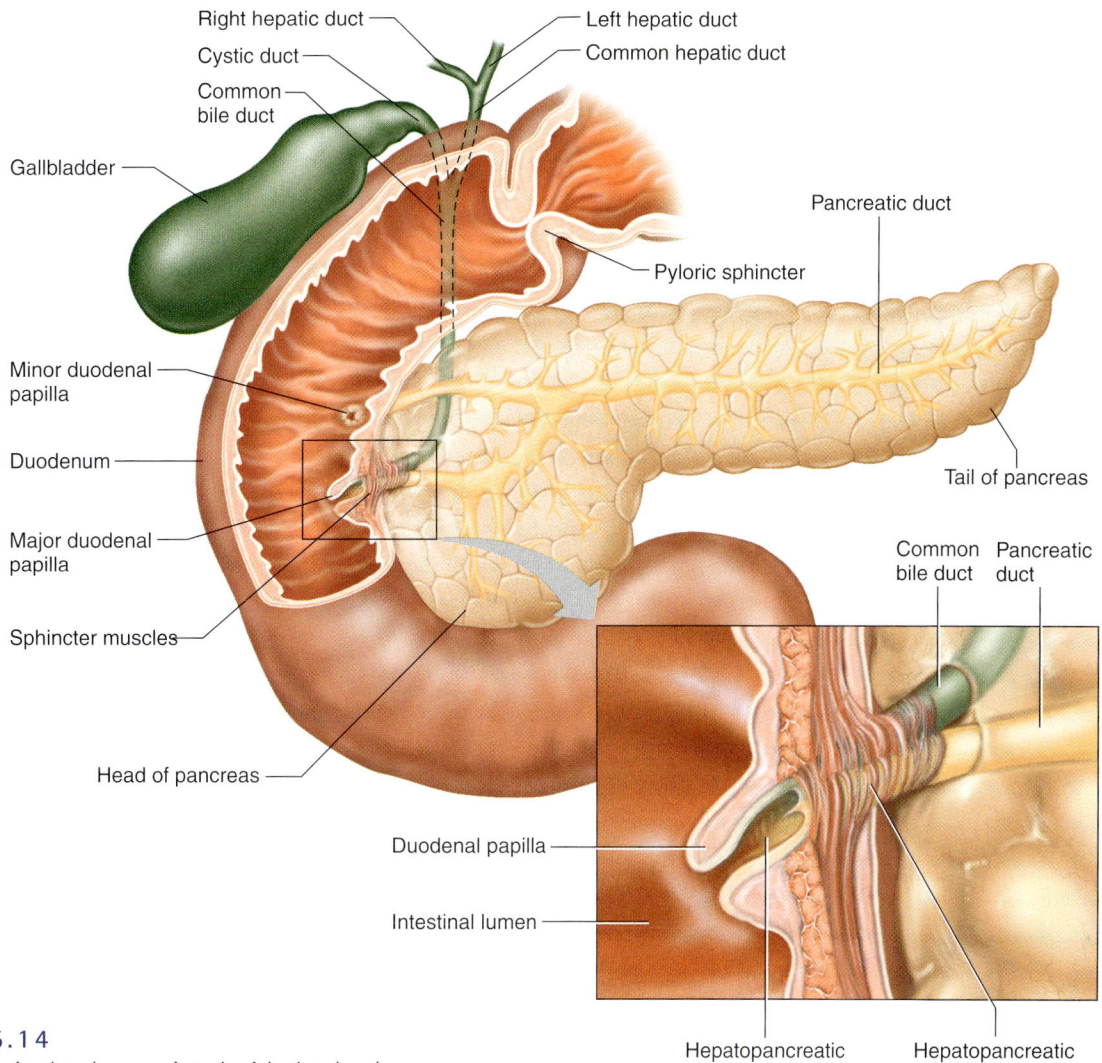

Figure 15.14
The pancreas is closely associated with the duodenum.

molecules of starch or glycogen into double sugars (disaccharides). The fat-digesting enzyme **pancreatic lipase** breaks triglyceride molecules into fatty acids and glycerol. Pancreatic juice also contains two **nucleases**, which are enzymes that break down nucleic acid molecules into nucleotides.

The protein-splitting (proteolytic) enzymes are **trypsin, chymotrypsin,** and **carboxypeptidase.** These enzymes split the bonds between particular combinations of amino acids in proteins. Because no single enzyme can split all the possible amino acid combinations, complete digestion of protein molecules requires several types of enzymes.

The proteolytic enzymes are stored within tiny structures in cells called *zymogen granules*. These enzymes, like gastric pepsin, are secreted in inactive forms. After they reach the small intestine, other enzymes activate them. For example, the pancreatic cells release inactive *trypsinogen,* which is activated to trypsin when it contacts the enzyme **enterokinase** secreted by the mucosa of the small intestine.

A painful condition called *acute pancreatitis* results from a blockage in the release of pancreatic juice. Trypsinogen, activated as pancreatic juice builds up, digests parts of the pancreas. Alcoholism, gallstones, certain infections, traumatic injuries, or the side effects of some drugs can cause pancreatitis.

Regulation of Pancreatic Secretion

As with gastric and small intestinal secretions, the nervous and endocrine systems regulate release of pancreatic juice. For example, when parasympathetic impulses stimulate gastric juice secretion, other parasympathetic impulses stimulate the pancreas to release digestive enzymes. Also, as acidic chyme enters the duodenum, the duodenal mucous membrane releases the peptide hormone **secretin** into the bloodstream (fig. 15.15). This hormone stimulates secretion of pancreatic juice that has a high concentration of bicarbonate ions. These ions neutralize the acid of chyme and provide a favorable environment for digestive enzymes in the intestine.

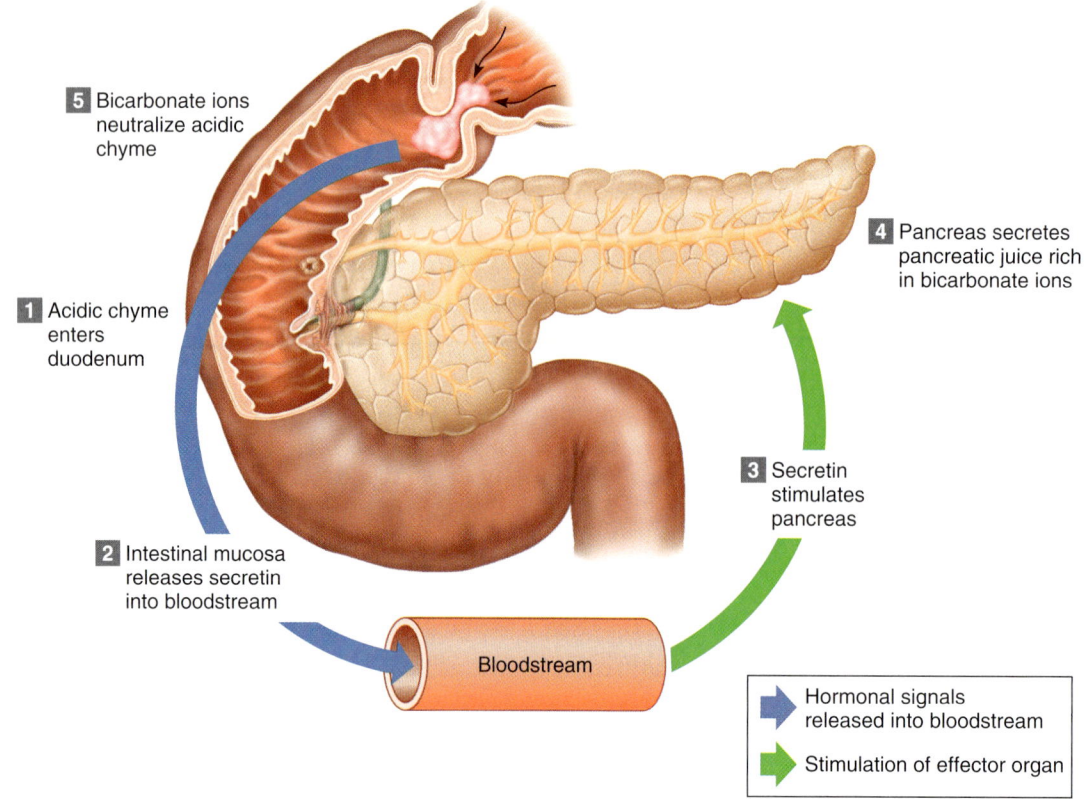

Figure 15.15
Acidic chyme entering the duodenum from the stomach stimulates the release of secretin, which in turn stimulates the release of pancreatic juice.

Proteins and fats in chyme within the duodenum also stimulate the intestinal wall to release the hormone **cholecystokinin.** Like secretin, cholecystokinin travels via the bloodstream to the pancreas. Pancreatic juice secreted in response to cholecystokinin has a high concentration of digestive enzymes.

In cystic fibrosis, abnormal chloride channels in cells in various tissues cause water to be drawn into the cells from interstitial spaces. This dries out secretions in the lungs and pancreas, leaving a very sticky mucus that impairs the functioning of these organs. When the pancreas is plugged with mucus, its secretions, containing digestive enzymes, cannot reach the duodenum. Individuals with cystic fibrosis must take digestive enzyme supplements—usually as a powder mixed with a soft food such as applesauce—to prevent malnutrition.

✓ CHECK YOUR RECALL

1. List the enzymes in pancreatic juice.
2. What are the functions of the enzymes in pancreatic juice?
3. What regulates secretion of pancreatic juice?

15.8 Liver

The **liver** is located in the upper right quadrant of the abdominal cavity, just inferior to the diaphragm. It is partially surrounded by the ribs, and extends from the level of the fifth intercostal space to the lower margin of the ribs. The reddish-brown liver is well supplied with blood vessels (see fig. 15.1 and reference plates 4 and 5, pp. 25–26).

 The average adult liver is the heaviest organ in the body at around 3 pounds.

Liver Structure

A fibrous capsule encloses the liver, and connective tissue divides the organ into a large *right lobe* and a smaller *left lobe* (fig. 15.16). Each lobe is separated into many tiny **hepatic lobules,** which are the liver's functional units (figs. 15.17 and 15.18). A lobule consists of many hepatic cells radiating outward from a *central vein*. Vascular channels called **hepatic sinusoids** separate platelike groups of these cells from each other.

CHAPTER 15 *Digestion and Nutrition* 407

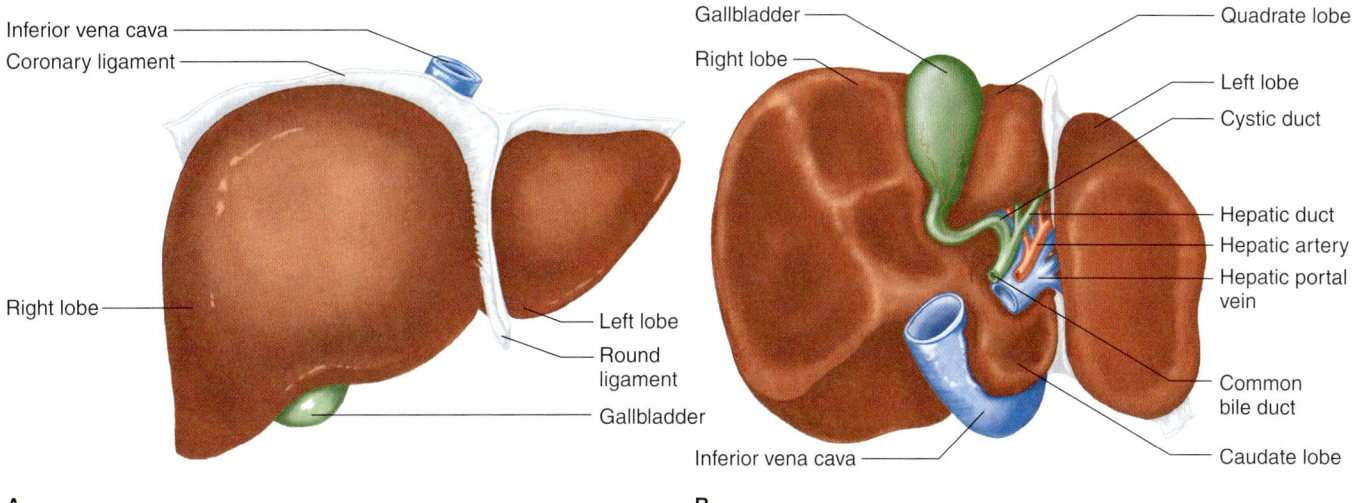

Figure 15.16
Lobes of the liver, viewed (*A*) anteriorly and (*B*) inferiorly.

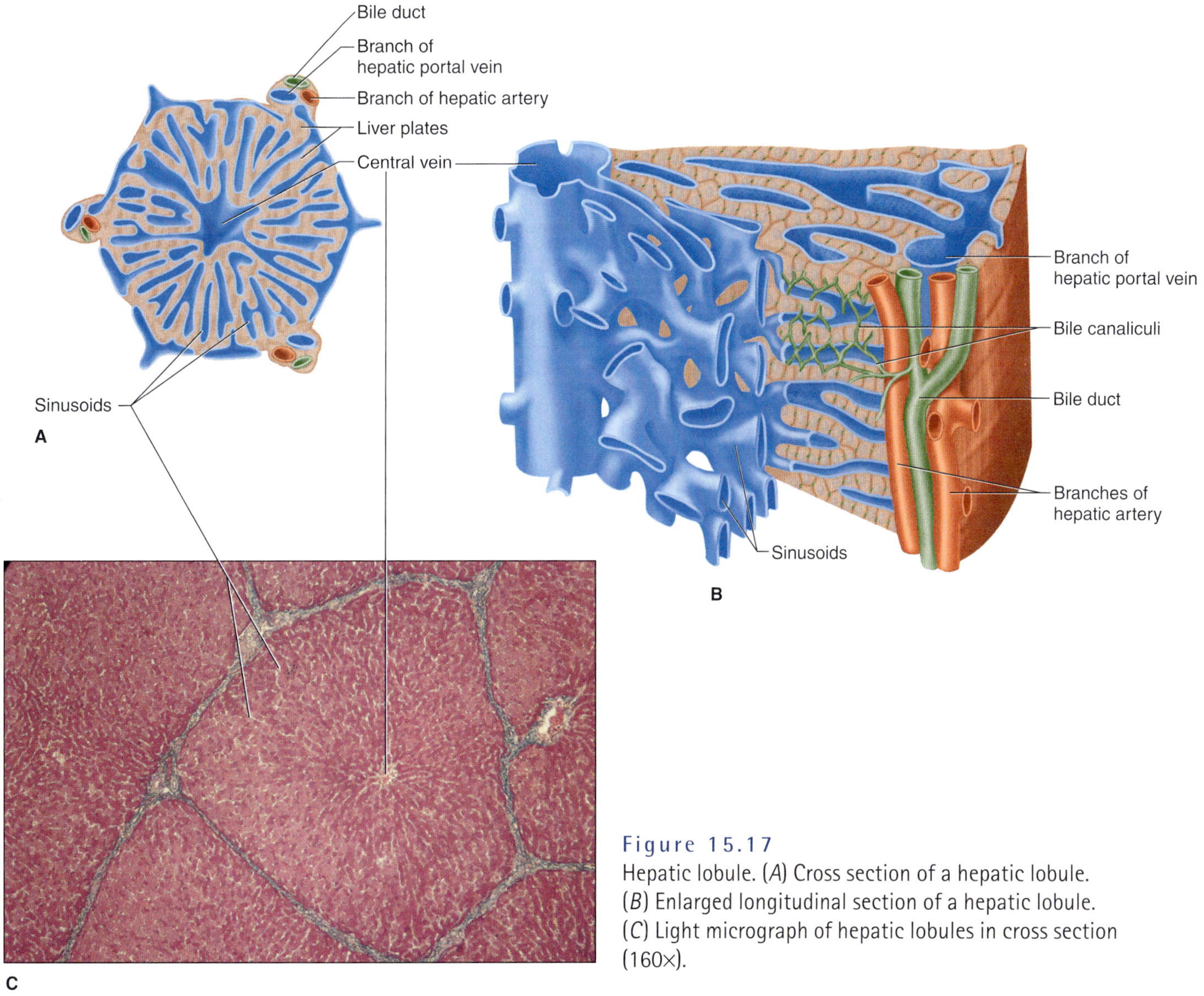

Figure 15.17
Hepatic lobule. (*A*) Cross section of a hepatic lobule.
(*B*) Enlarged longitudinal section of a hepatic lobule.
(*C*) Light micrograph of hepatic lobules in cross section
(160×).

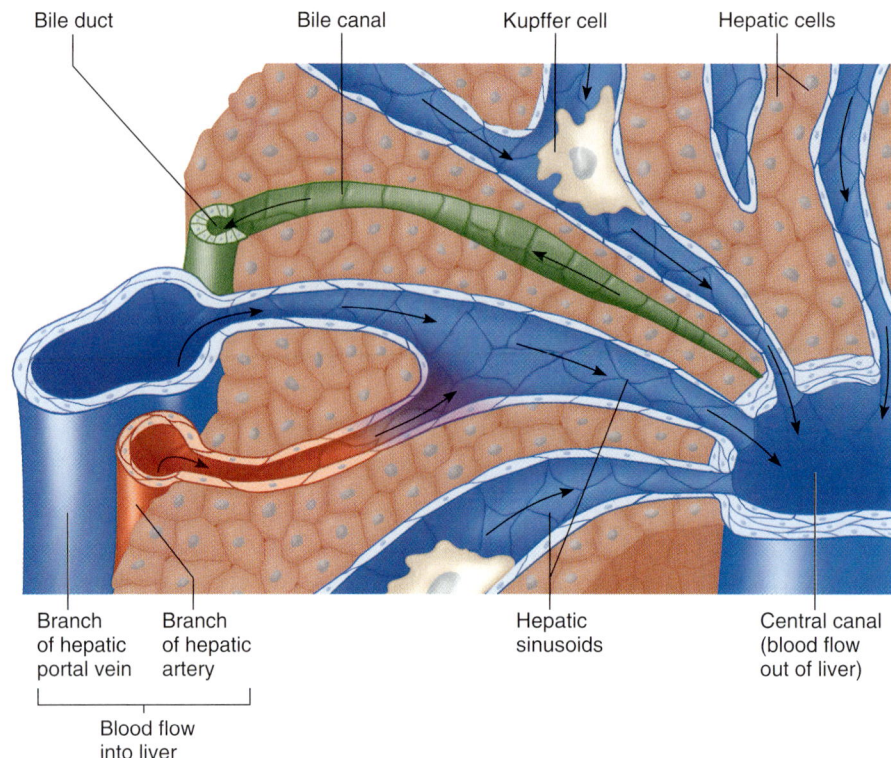

Figure 15.18
The paths of blood and bile within a hepatic lobule.

Blood from the digestive tract, which is carried in the *hepatic portal vein* (see chapter 13, p. 358), brings newly absorbed nutrients into the sinusoids and nourishes the hepatic cells.

Large phagocytic macrophages called *Kupffer cells* are fixed to the inner linings of the hepatic sinusoids. They remove bacteria or other foreign particles that enter the blood in the portal vein through the intestinal wall. Blood passes from these sinusoids into the central veins of the hepatic lobules and exits the liver.

Within the hepatic lobules are many fine *bile canals,* which receive secretions from hepatic cells. The canals of neighboring lobules unite to form larger ducts and then converge to become the **hepatic ducts.** These ducts merge, in turn, to form the **common hepatic duct.**

Liver Functions

The liver has many important metabolic activities. Recall from chapter 11 (p. 296) that the liver plays a key role in carbohydrate metabolism by helping maintain the normal concentration of blood glucose. Liver cells responding to hormones such as insulin and glucagon lower the blood glucose level by polymerizing glucose to glycogen, and raise the blood glucose level by breaking down glycogen to glucose or by converting noncarbohydrates into glucose.

The liver's effects on lipid metabolism include oxidizing fatty acids at an especially high rate (p. 421); synthesizing lipoproteins, phospholipids, and cholesterol; and converting portions of carbohydrate and protein molecules into fat molecules. The blood transports fats synthesized in the liver to adipose tissue for storage.

The most vital liver functions concern protein metabolism. They include deaminating amino acids; forming urea (p. 422); synthesizing plasma proteins, such as clotting factors (see chapter 12, p. 316); and converting certain amino acids to other amino acids.

The liver also stores many substances, including glycogen, iron, and vitamins A, D, and B_{12}. In addition, macrophages in the liver help destroy damaged red blood cells and phagocytize foreign antigens. The liver also removes toxic substances such as alcohol from blood (detoxification) and secretes bile.

Many of these liver functions are not directly related to the digestive system and are discussed in other chapters. Bile secretion, however, is important to digestion and is explained next in this chapter. Table 15.3 summarizes the major functions of the liver.

CHECK YOUR RECALL

1. Describe the location of the liver.
2. Describe a hepatic lobule.
3. Review liver functions.

Topic of Interest

HEPATITIS

Hepatitis is an inflammation of the liver. It has several causes, but the various types have similar symptoms.

For the first few days, hepatitis may resemble the flu, producing mild headache, low fever, fatigue, lack of appetite, nausea and vomiting, and sometimes stiff joints. By the end of the first week, more distinctive symptoms arise: a rash, pain in the upper right quadrant of the abdomen, dark and foamy urine, and pale feces. The skin and sclera of the eyes begin to turn yellow from accumulating bile pigments (jaundice). Great fatigue may continue for two or three weeks, and then gradually the person begins to feel better.

This is hepatitis in its most common, least dangerous acute guise. About half a million people develop hepatitis in the United States each year, and 6,000 die. In a rare form called *fulminant hepatitis,* symptoms occur suddenly and severely, along with altered behavior and personality. Medical attention is necessary to prevent kidney or liver failure or coma. Hepatitis that persists for more than six months is termed chronic. As many as 300 million people worldwide are hepatitis carriers. They do not have symptoms but can infect others. Five percent of carriers develop liver cancer.

Only rarely does hepatitis result from alcoholism, autoimmunity, or the use of certain drugs. Usually, one of several types of viruses can cause hepatitis. Viral types are distinguished by the route of infection and by biochemical differences, such as gene sequences and surface proteins. The viral types are classified as follows:

Hepatitis A spreads by contact with food or objects contaminated with virus-containing feces. In day-care centers, it spreads through diaper changing. An outbreak affecting children in several states was traced to contaminated strawberries in school lunches. The course of hepatitis A is short and mild.

Hepatitis B spreads by contact with virus-containing body fluids, such as blood, saliva, or semen. It may be transmitted by blood transfusions, hypodermic needles, or sexual activity.

Hepatitis C is believed to be responsible for about half of all cases of hepatitis. This virus is primarily transmitted in blood—by sharing razors or needles, from pregnant woman to fetus, or in blood transfusions or use of blood products. As many as 60% of individuals infected with the hepatitis C virus suffer chronic symptoms.

Hepatitis D occurs in people already infected with the hepatitis B virus. It is blood-borne and associated with blood transfusions and intravenous drug use. About 20% of individuals infected with this virus die.

Hepatitis E virus is usually transmitted in water contaminated with feces in developing nations—not to residents, who are immune, but most often to visitors.

Hepatitis F passes from feces and can infect other primates.

Hepatitis G accounts for many cases of fulminant hepatitis.

Because a virus usually causes hepatitis, antibiotic drugs, which are effective against bacteria, are not helpful. Usually, the person must just wait out the symptoms. Hepatitis C, however, sometimes responds to a form of interferon, an immune system biochemical given as a drug.

TABLE 15.3 — MAJOR FUNCTIONS OF THE LIVER

GENERAL FUNCTION	SPECIFIC FUNCTION
Carbohydrate metabolism	Polymerizes glucose to glycogen; breaks down glycogen to glucose; converts noncarbohydrates to glucose
Lipid metabolism	Oxidizes fatty acids; synthesizes lipoproteins, phospholipids, and cholesterol; converts portions of carbohydrate and protein molecules into fats
Protein metabolism	Deaminates amino acids; forms urea; synthesizes plasma proteins; converts certain amino acids to other amino acids
Storage	Stores glycogen, iron, and vitamins A, D, and B_{12}
Blood filtering	Removes damaged red blood cells and foreign substances by phagocytosis
Detoxification	Removes toxins from blood
Secretion	Secretes bile

Composition of Bile

Bile (bīl) is a yellowish-green liquid that hepatic cells continuously secrete. In addition to water, bile contains *bile salts, bile pigments* (bilirubin and biliverdin), *cholesterol,* and *electrolytes.* Of these, bile salts are the most abundant and are the only bile substances that have a digestive function. Bile pigments are breakdown products of hemoglobin from red blood cells and are normally excreted in the bile (see chapter 12, p. 309).

Jaundice turns the skin and whites of the eyes yellow. The distinctive skin color reflects buildup of bile pigments. The condition can have several causes. In *obstructive jaundice*, bile ducts are blocked. In *hepatocellular jaundice*, the liver is diseased. In *hemolytic jaundice*, red blood cells are destroyed too rapidly.

Gallbladder

The **gallbladder** is a pear-shaped sac located in a depression on the liver's inferior surface. It connects to the **cystic duct,** which in turn joins the common hepatic duct (see figs. 15.1 and 15.20). The gallbladder is lined with epithelial cells and has a strong, muscular layer in its wall. It stores bile between meals, reabsorbs water to concentrate bile, and releases bile into the small intestine.

The common hepatic and cystic ducts join to form the *common bile duct*. It leads to the duodenum (see fig. 15.14), where the *hepatopancreatic sphincter* guards its exit. This sphincter normally remains contracted, so that bile collects in the common bile duct and backs up into the cystic duct. When this happens, bile flows into the gallbladder, where it is stored.

Cholesterol in bile may precipitate and form crystals called *gallstones* under certain conditions (fig. 15.19). Gallstones entering the common bile duct may block bile flow into the small intestine and cause considerable pain. A surgical procedure called *cholecystectomy* removes the gallbladder when gallstones are obstructive. The surgery is often performed with a laser, which shortens recovery time.

Regulation of Bile Release

Normally, bile does not enter the duodenum until cholecystokinin stimulates the gallbladder to contract. The intestinal mucosa releases this hormone in response to proteins and fats in the small intestine. (Recall its action to stimulate pancreatic enzyme secretion.) The hepatopancreatic sphincter usually remains contracted until a peristaltic wave in the duodenal wall approaches it. Then the sphincter relaxes, and bile is squirted into the small intestine (fig. 15.20). Table 15.4

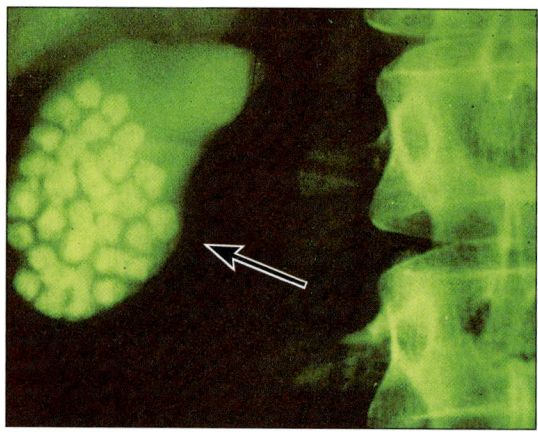

Figure 15.19
Radiograph of a gallbladder that contains gallstones (arrow).

summarizes the hormones that help control digestive functions.

Functions of Bile Salts

Bile salts aid digestive enzymes. Bile salts affect *fat globules* (clumped molecules of fats) much like a soap or detergent would affect them. That is, bile salts break fat globules into smaller droplets, an action called **emulsification** that greatly increases the total surface area of the fatty substance. The tiny fat droplets then mix with water. Fat-splitting enzymes (lipases) can then digest fat molecules more effectively.

Bile salts also enhance absorption of fatty acids, cholesterol, and the fat-soluble vitamins A, D, E, and K. Lack of bile salts results in poor lipid absorption and vitamin deficiencies.

 CHECK YOUR RECALL

1. Explain how bile originates.
2. Describe the function of the gallbladder.
3. How is secretion of bile regulated?
4. How do bile salts function in digestion?

TABLE 15.4 — HORMONES OF THE DIGESTIVE TRACT

HORMONE	SOURCE	FUNCTION
Gastrin	Gastric cells, in response to food	Causes gastric glands to increase their secretory activity
Cholecystokinin	Intestinal wall cells, in response to proteins and fats in the small intestine	Causes gastric glands to decrease their secretory activity and inhibits gastric motility; stimulates pancreas to secrete fluid with a high digestive enzyme concentration; stimulates gallbladder to contract and release bile
Secretin	Cells in the duodenal wall, in response to acidic chyme entering the small intestine	Stimulates pancreas to secrete fluid with a high bicarbonate ion concentration

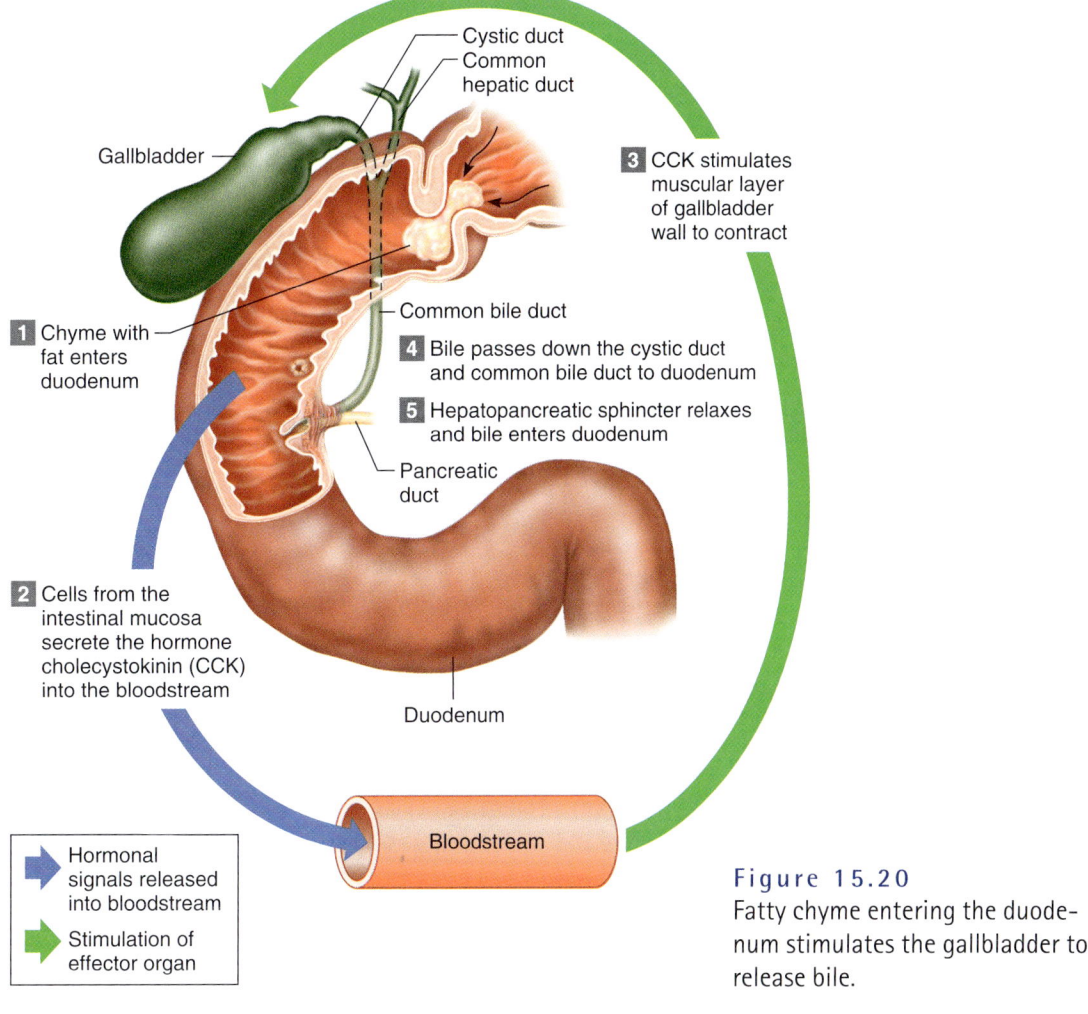

Figure 15.20
Fatty chyme entering the duodenum stimulates the gallbladder to release bile.

15.9 Small Intestine

The **small intestine** is a tubular organ that extends from the pyloric sphincter to the beginning of the large intestine. With its many loops and coils, it fills much of the abdominal cavity (see fig. 15.1 and reference plates 4 and 5, pp. 25–26).

The small intestine receives secretions from the pancreas and liver. It also completes digestion of the nutrients in chyme, absorbs the products of digestion, and transports the residues to the large intestine.

Parts of the Small Intestine

The small intestine consists of three portions: the duodenum, the jejunum, and the ileum (figs. 15.21 and 15.22). The **duodenum,** which is about 25 centimeters long and 5 centimeters in diameter, lies posterior to the parietal peritoneum and is the most fixed portion of the small intestine. It follows a C-shaped path as it passes anterior to the right kidney and the upper three lumbar vertebrae.

The remainder of the small intestine is mobile and lies free in the peritoneal cavity. The proximal two-fifths

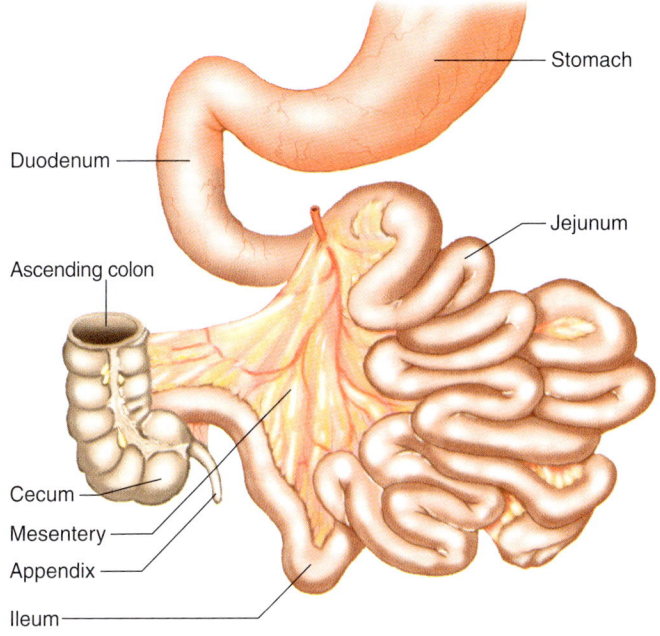

Figure 15.21
The three parts of the small intestine are the duodenum, the jejunum, and the ileum.

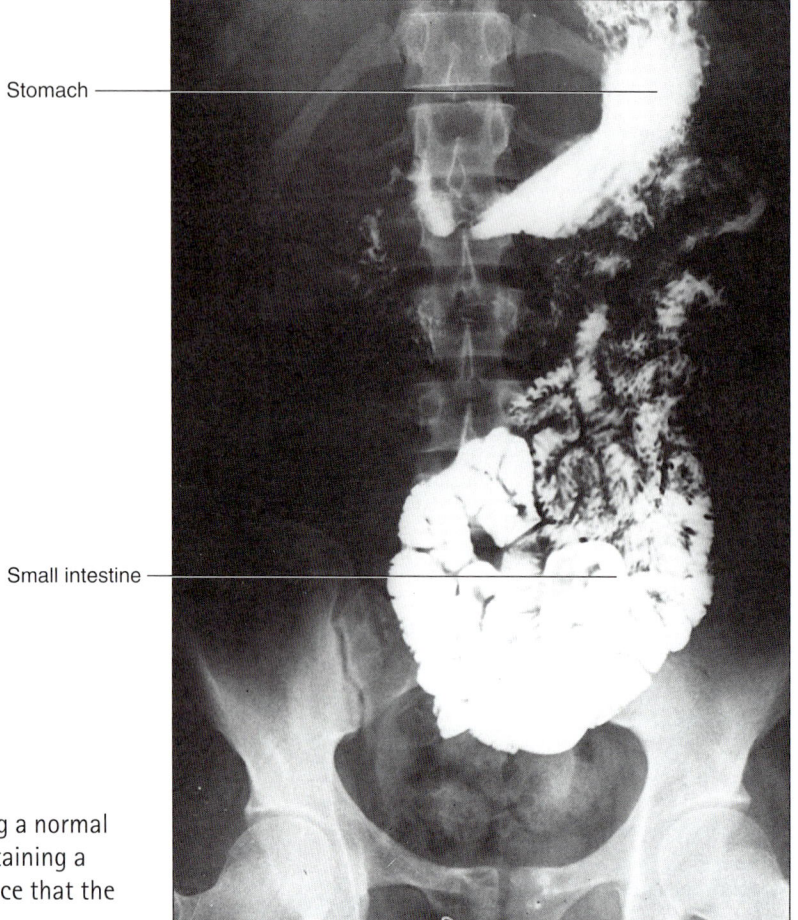

Figure 15.22
Radiograph showing a normal small intestine containing a radiopaque substance that the patient ingested.

of this portion is the **jejunum,** and the remainder is the **ileum.** A double-layered fold of peritoneal membrane called **mesentery** (mes´en-ter´´e) suspends these portions from the posterior abdominal wall (figs. 15.21 and 15.23). The mesentery supports the blood vessels, nerves, and lymphatic vessels that supply the intestinal wall. The jejunum and ileum are not distinctly separate parts; however, the diameter of the jejunum is greater and its wall thicker, more vascularized, and more active than that of the ileum.

A filmy, double fold of peritoneal membrane called the *greater omentum* drapes like an apron from the stomach over the transverse colon and the folds of the small intestine. If infections occur in the wall of the alimentary canal, cells from the omentum may adhere to the inflamed region and help wall it off so that the infection is less likely to enter the peritoneal cavity (fig. 15.23).

Structure of the Small Intestine Wall

Throughout its length, the inner wall of the small intestine appears velvety due to many tiny projections of mucous membrane called **intestinal villi** (in-tes´tĭ-nal vil´i) (figs. 15.24 and 15.25; see fig. 15.3). These structures are most numerous in the duodenum and the prox-

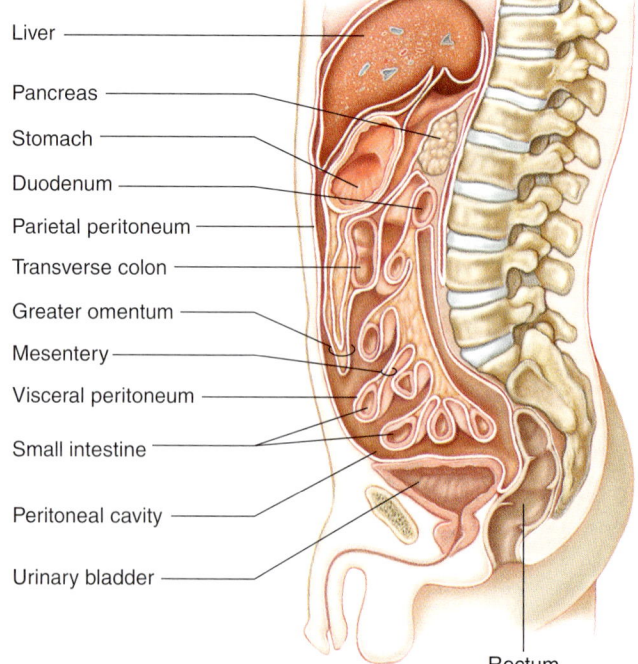

Figure 15.23
Mesentery formed by folds of the peritoneal membrane suspends portions of the small intestine from the posterior abdominal wall.

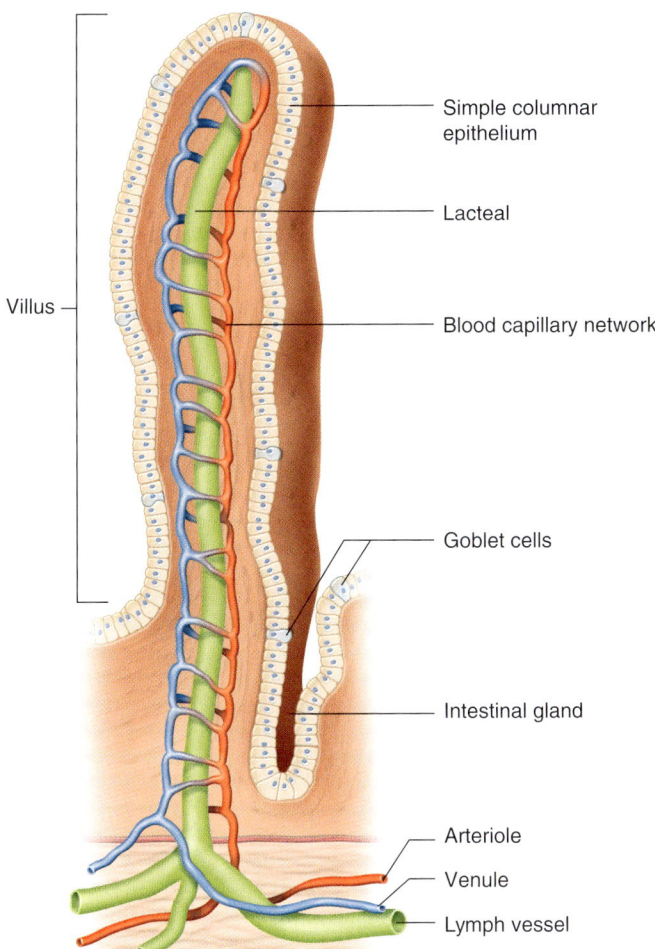

Figure 15.24
Structure of a single intestinal villus.

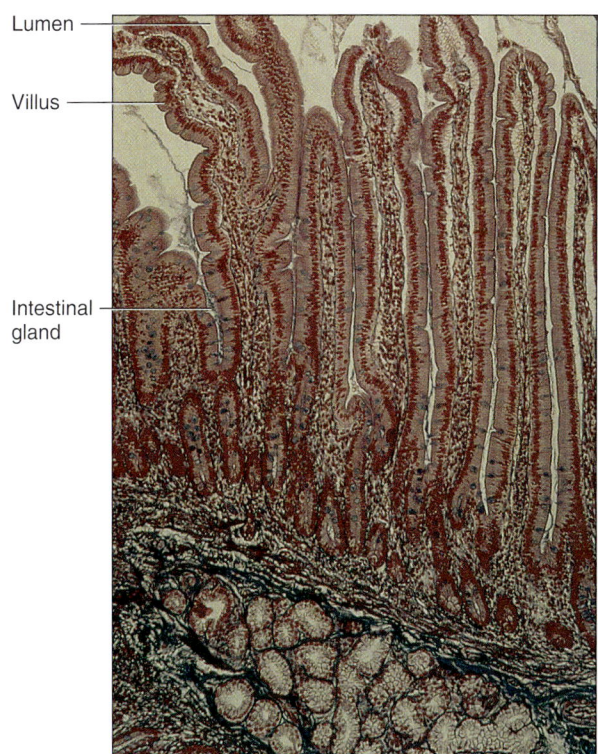

Figure 15.25
Light micrograph of intestinal villi from the wall of the duodenum (50×).

imal portion of the jejunum. They project into the lumen of the alimentary canal, contacting the intestinal contents. Villi greatly increase the surface area of the intestinal lining, aiding the absorption of digestive products.

Each villus consists of a layer of simple columnar epithelium and a core of connective tissue containing blood capillaries, a lymphatic capillary called a **lacteal,** and nerve fibers. Blood capillaries and lacteals carry away absorbed nutrients, and nerve fibers transmit impulses to stimulate or inhibit villus activities. Between the bases of adjacent villi are tubular **intestinal glands** that extend downward into the mucous membrane (figs. 15.24 and 15.25; see fig. 15.3).

> The epithelial cells that form the lining of the small intestine are continually replaced. New cells form within the intestinal glands by mitosis and migrate outward onto the villus surface. When the migrating cells reach the tip of the villus, they are shed. This *cellular turnover* renews the small intestine's epithelial lining every three to six days. As a result, nearly one-quarter of the bulk of feces consists of dead epithelial cells from the small intestine.

Secretions of the Small Intestine

Mucus-secreting goblet cells are abundant throughout the mucosa of the small intestine. In addition, many specialized *mucus-secreting glands* in the submucosa within the proximal portion of the duodenum secrete large quantities of thick, alkaline mucus in response to certain stimuli.

The intestinal glands at the bases of the villi secrete large amounts of a watery fluid. The villi rapidly reabsorb this fluid, which carries digestive products into the villi. The fluid the intestinal glands secrete has a nearly neutral pH (6.5–7.5), and it lacks digestive enzymes. However, the epithelial cells of the intestinal mucosa have digestive enzymes embedded in the membranes of their microvilli on their luminal surfaces. These enzymes break down food molecules just before absorption takes place. The enzymes include **peptidases,** which split peptides into their constituent amino acids; **sucrase, maltase,** and **lactase,** which split the double sugars (disaccharides) sucrose, maltose, and lactose into the simple sugars (monosaccharides) glucose, fructose, and galactose; and **intestinal lipase,** which splits fats into fatty acids and glycerol. Table 15.5 summarizes the sources and actions of the major digestive enzymes.

TABLE 15.5 SUMMARY OF THE MAJOR DIGESTIVE ENZYMES

ENZYME	SOURCE	DIGESTIVE ACTION
Salivary enzyme		
Amylase	Salivary glands	Begins carbohydrate digestion by breaking down starch and glycogen to disaccharides
Gastric enzyme		
Pepsin	Gastric glands	Begins protein digestion
Pancreatic enzymes		
Amylase	Pancreas	Breaks down starch and glycogen into disaccharides
Pancreatic lipase	Pancreas	Breaks down fats into fatty acids and glycerol
Proteolytic enzymes	Pancreas	Breaks down proteins or partially digested proteins into peptides
(a) Trypsin		
(b) Chymotrypsin		
(c) Carboxypeptidase		
Nucleases	Pancreas	Breaks down nucleic acids into nucleotides
Intestinal enzymes		
Peptidase	Mucosal cells	Breaks down peptides into amino acids
Sucrase, maltase, lactase	Mucosal cells	Breaks down disaccharides into monosaccharides
Intestinal lipase	Mucosal cells	Breaks down fats into fatty acids and glycerol
Enterokinase	Mucosal cells	Converts trypsinogen into trypsin

Many adults do not produce sufficient lactase to adequately digest lactose, or milk sugar. In this condition, called *lactose intolerance,* lactose remains undigested, increasing the osmotic pressure of the intestinal contents and drawing water into the intestines. At the same time, intestinal bacteria metabolize undigested sugar, producing organic acids and gases. The overall result of lactose intolerance is bloating, intestinal cramps, and diarrhea. To avoid these symptoms, people with lactose intolerance can take lactase in pill form before eating dairy products. Infants with lactose intolerance can drink formula based on soybeans rather than milk. Genetic evidence suggests that lactose intolerance may be the "normal" condition, with ability to digest lactose the result of a mutation that occurred recently in our evolutionary past. Because lactose intolerance most often affects adults, and is not seen in other primates, it may not have adversely affected prehistoric humans, who did not live long enough to experience symptoms.

Regulation of Small Intestine Secretions

Goblet cells and intestinal glands secrete their products when chyme provides both chemical and mechanical stimulation. Distension of the intestinal wall activates the nerve plexuses within the wall and stimulates parasympathetic reflexes that also trigger release of small intestine secretions.

CHECK YOUR RECALL

1. Describe the parts of the small intestine.
2. What is the function of an intestinal villus?
3. What is the function of the intestinal glands?
4. List the intestinal digestive enzymes.

Absorption in the Small Intestine

Villi greatly increase the surface area of the intestinal mucosa, making the small intestine the most important absorbing organ of the alimentary canal. So effective is the small intestine in absorbing digestive products, water, and electrolytes that very little absorbable material reaches its distal end.

Carbohydrate digestion begins in the mouth with the activity of salivary amylase, and enzymes from the intestinal mucosa and pancreas complete the process in the small intestine. Villi absorb the resulting monosaccharides, which enter blood capillaries. Simple sugars are absorbed by facilitated diffusion or active transport (see chapter 3, pp. 59 and 62).

Pepsin activity begins protein digestion in the stomach, and enzymes from the intestinal mucosa and the pancreas complete digestion in the small intestine. During this process, large protein molecules are broken down into amino acids, which are then actively transported into the villi and carried away by the blood.

Enzymes from the intestinal mucosa and pancreas digest fat molecules almost entirely. The resulting fatty acids and glycerol molecules diffuse into villi epithelial cells. The endoplasmic reticula of the cells use the fatty acids to resynthesize fat molecules similar to those digested. These fats are encased in protein to form *chylomicrons,* which make their way to the lacteals of the villi. Lymph in the lacteals and other lymphatic vessels carries chylomicrons to the bloodstream (see chapter 14, p. 367) (fig. 15.26). Some fatty acids with relatively short carbon chains may be absorbed directly into the blood capillary of a villus without being changed back into fat.

Chylomicrons transport dietary fats to muscle and adipose cells. Similarly, VLDL (very-low-density lipo-

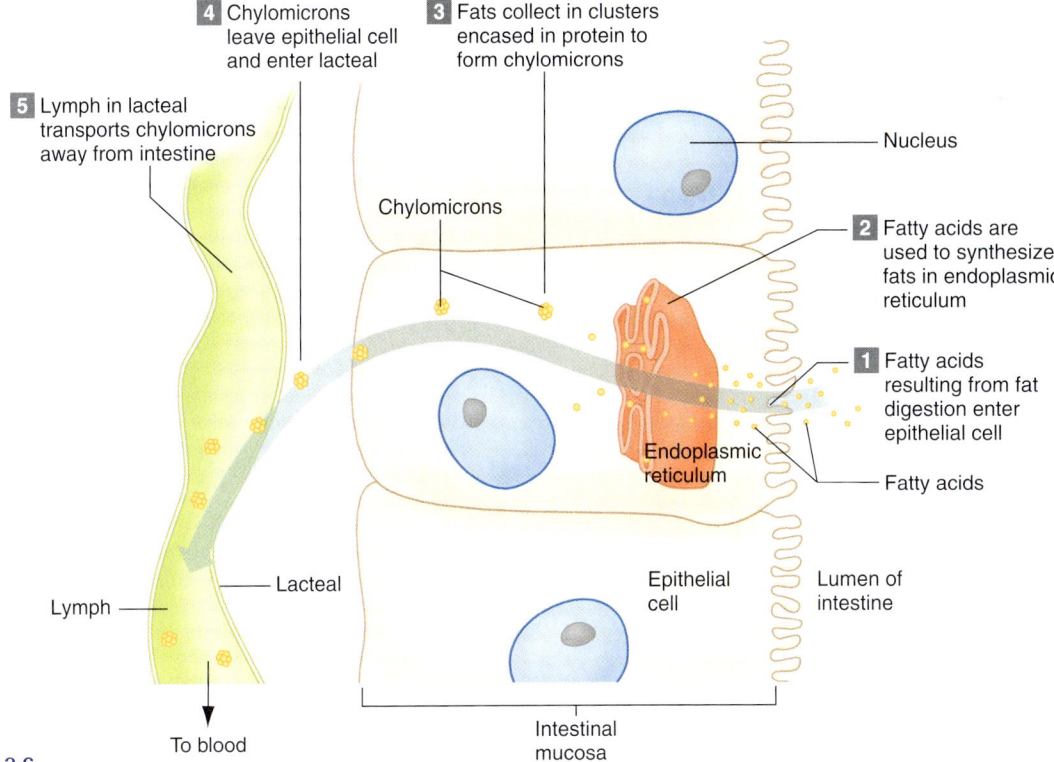

Figure 15.26
Fat absorption has several steps.

protein) molecules, produced in the liver, transport triglycerides synthesized from excess dietary carbohydrates. After VLDL molecules deliver their loads of triglycerides to adipose cells, an enzyme, *lipoprotein lipase,* converts their remnants to LDL (low-density lipoproteins). Because most of the triglycerides have been removed, LDL molecules have a higher cholesterol content than do the original VLDL molecules. Various cells, including liver cells, have surface receptors that combine with apoproteins associated with LDL molecules. These cells slowly remove LDL from plasma by receptor-mediated endocytosis, supplying cells with cholesterol (see chapter 3, p. 63).

After chylomicrons deliver their triglycerides to cells, their remnants are transferred to HDL (high-density lipoproteins) molecules. HDL molecules, which form in the liver and small intestine, transport chylomicron remnants to the liver, where they enter cells rapidly by receptor-mediated endocytosis. The liver disposes of the cholesterol it obtains in this manner by secreting it into bile or by using it to synthesize bile salts.

The intestine reabsorbs much of the cholesterol and bile salts in bile, which are then transported back to the liver, and the secretion-reabsorption cycle repeats. During each cycle, some of the cholesterol and bile salts escape reabsorption, reach the large intestine, and are eliminated with the feces.

The intestinal villi absorb electrolytes by active transport and water by osmosis in addition to the products of carbohydrate, protein, and fat digestion. Table 15.6 summarizes the absorption process.

TABLE 15.6 — INTESTINAL ABSORPTION OF NUTRIENTS

NUTRIENT	ABSORPTION MECHANISM	MEANS OF TRANSPORT
Monosaccharides	Facilitated diffusion and active transport	Blood in capillaries
Amino acids	Active transport	Blood in capillaries
Fatty acids and glycerol	Facilitated diffusion of glycerol; diffusion of fatty acids into cells	Lymph in lacteals
	(a) Most fatty acids are resynthesized into fats and incorporated into chylomicrons for transport	
	(b) Some fatty acids with relatively short carbon chains are transported without being changed back into fats	Blood in capillaries
Electrolytes	Diffusion and active transport	Blood in capillaries
Water	Osmosis	Blood in capillaries

CHECK YOUR RECALL

1. Which substances resulting from digestion of carbohydrate, protein, and fat molecules does the small intestine absorb?
2. Describe how fatty acids are absorbed and transported.

In *malabsorption*, the small intestine digests, but does not absorb, some nutrients. Causes of malabsorption include surgical removal of a portion of the small intestine, obstruction of lymphatic vessels due to a tumor, or interference with the production and release of bile as a result of liver disease.

Another cause of malabsorption is a reaction to *gluten*, found in certain grains, especially wheat and rye. This condition is called *celiac disease*. Microvilli are damaged, and in severe cases, villi may be destroyed. Both of these effects reduce the absorptive surface of the small intestine, preventing absorption of some nutrients. Symptoms of malabsorption include diarrhea, weight loss, weakness, vitamin deficiencies, anemia, and bone demineralization.

Movements of the Small Intestine

Like the stomach, the small intestine carries on mixing movements and peristalsis. The mixing movements include small, periodic, ringlike contractions that cut chyme into segments and move it back and forth.

Weak peristaltic waves propel chyme short distances through the small intestine. Consequently, chyme moves slowly through the small intestine, taking from 3 to 10 hours to travel its length.

If the small intestine wall becomes overdistended or irritated, a strong *peristaltic rush* may pass along the organ's entire length. This movement sweeps the contents of the small intestine into the large intestine so quickly that water, nutrients, and electrolytes that would normally be absorbed are not. The result is *diarrhea*, characterized by more frequent defecation and watery stools. Prolonged diarrhea causes imbalances in water and electrolyte concentrations.

At the distal end of the small intestine, the **ileocecal sphincter** joins the small intestine's ileum to the large intestine's cecum (fig. 15.27). Normally, this

Figure 15.27
Parts of the large intestine (anterior view).

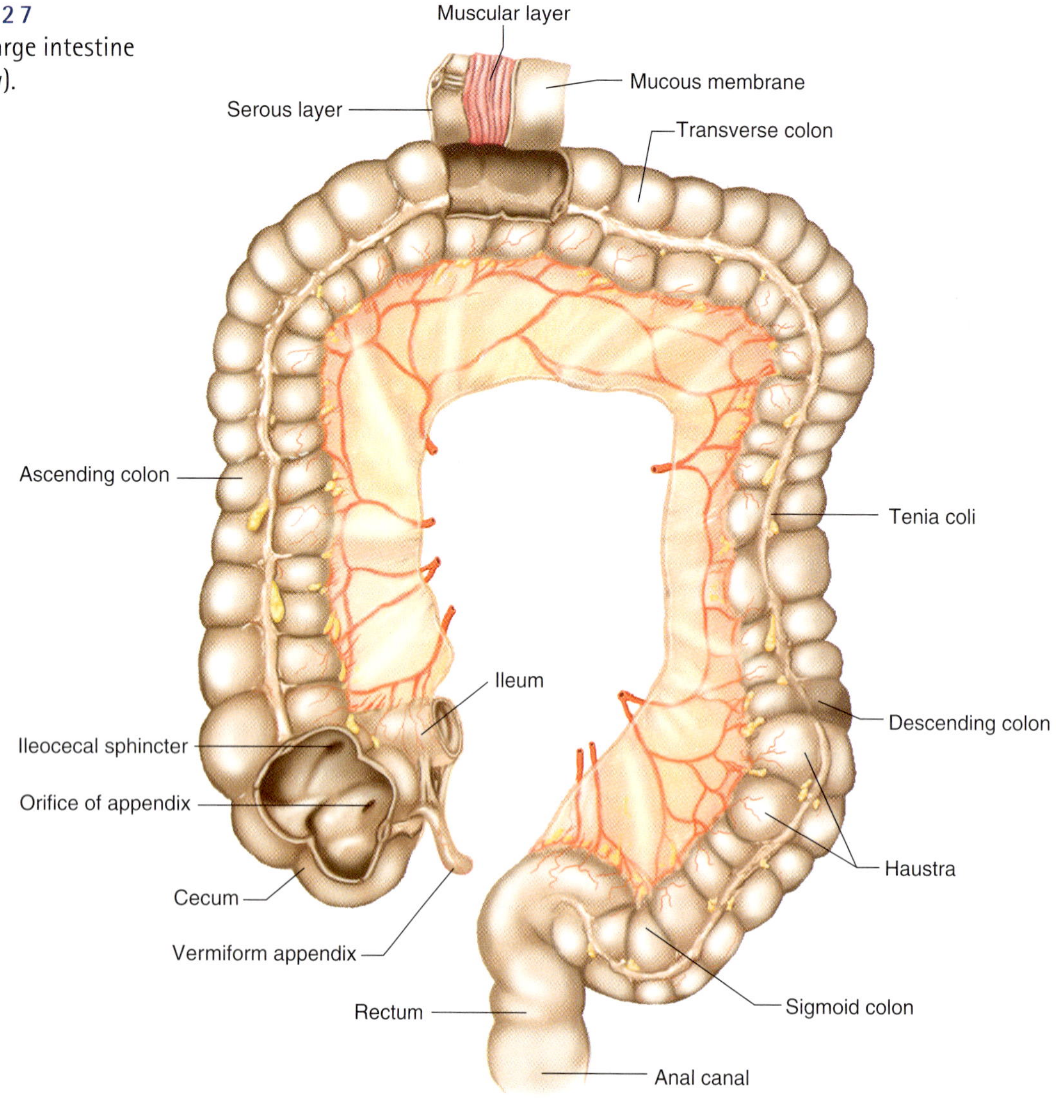

sphincter remains constricted, preventing the contents of the small intestine from entering the large intestine, and the contents of the large intestine from backing up into the ileum. However, after a meal, a gastroileal reflex increases peristalsis in the ileum and relaxes the sphincter, forcing some of the contents of the small intestine into the cecum.

CHECK YOUR RECALL

1. Describe the movements of the small intestine.
2. What stimulus relaxes the ileocecal sphincter?

15.10 Large Intestine

The **large intestine** is so named because its diameter is greater than that of the small intestine. This portion of the alimentary canal is about 1.5 meters long, and it begins in the lower right side of the abdominal cavity, where the ileum joins the cecum. From there, the large intestine ascends on the right side, crosses obliquely to the left, and descends into the pelvis. At its distal end, it opens to the outside of the body as the anus (see fig. 15.1).

The large intestine absorbs water and electrolytes from chyme remaining in the alimentary canal. It also forms and stores feces.

Parts of the Large Intestine

The large intestine consists of the cecum, colon, rectum, and anal canal (figs. 15.27 and 15.28; reference plates 4 and 5, pp. 25–26). The **cecum,** at the beginning of the large intestine, is a dilated, pouchlike structure that hangs slightly below the ileocecal opening. Projecting downward from it is a narrow tube with a closed end called the **vermiform appendix.** The human appendix has no known digestive function. However, it contains lymphatic tissue.

> In *appendicitis,* the appendix becomes inflamed and infected. Surgery is required to prevent the appendix from rupturing. If it does break open, the contents of the large intestine may enter the abdominal cavity and cause a serious infection of the peritoneum called *peritonitis.*

The **colon** is divided into four portions—the ascending, transverse, descending, and sigmoid colons. The **ascending colon** begins at the cecum and travels upward against the posterior abdominal wall to a point just inferior to the liver. There, it turns sharply to the left and becomes the **transverse colon.** The transverse colon is the longest and most movable part of the large

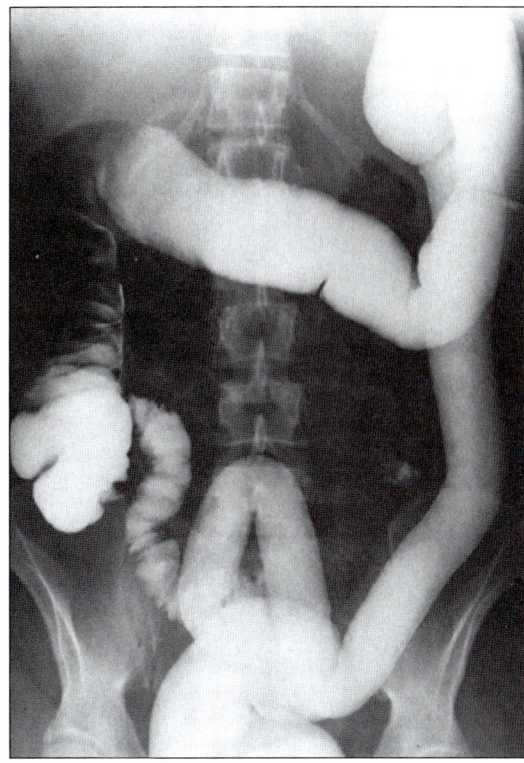

Figure 15.28
Radiograph of the large intestine containing a radiopaque substance that the patient ingested.

intestine. It is suspended by a fold of peritoneum and sags in the middle below the stomach. As the transverse colon approaches the spleen, it turns abruptly downward and becomes the **descending colon.** At the brim of the pelvis, the descending colon makes an S-shaped curve called the **sigmoid colon** and then becomes the rectum.

The **rectum** lies next to the sacrum and generally follows its curvature. The peritoneum firmly attaches the rectum to the sacrum, and the rectum ends about 5 centimeters below the tip of the coccyx, where it becomes the anal canal (see fig. 15.27).

The last 2.5–4.0 centimeters of the large intestine form the **anal canal** (fig. 15.29). The mucous membrane in the canal is folded into a series of six to eight longitudinal *anal columns*. At its distal end, the canal opens to the outside as the **anus.** Two sphincter muscles guard the anus—an *internal anal sphincter muscle* composed of smooth muscle under involuntary control and an *external anal sphincter muscle* composed of skeletal muscle under voluntary control.

CHECK YOUR RECALL

1. What is the general function of the large intestine?
2. Describe the parts of the large intestine.

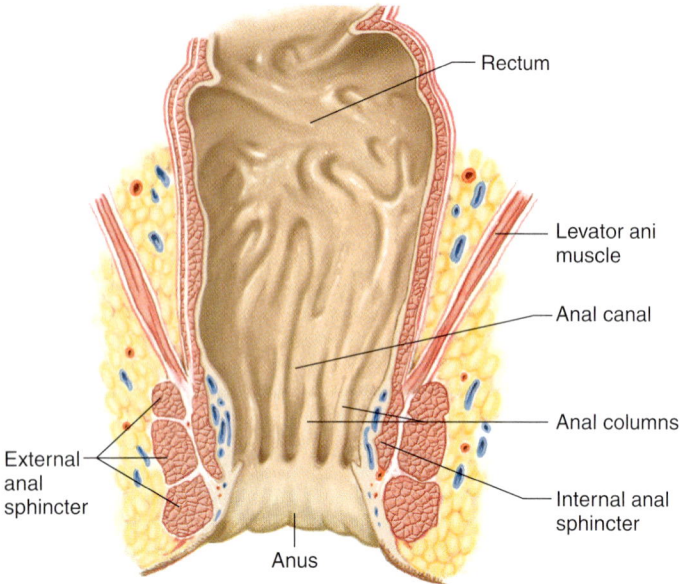

Figure 15.29
The rectum and the anal canal are at the distal end of the alimentary canal.

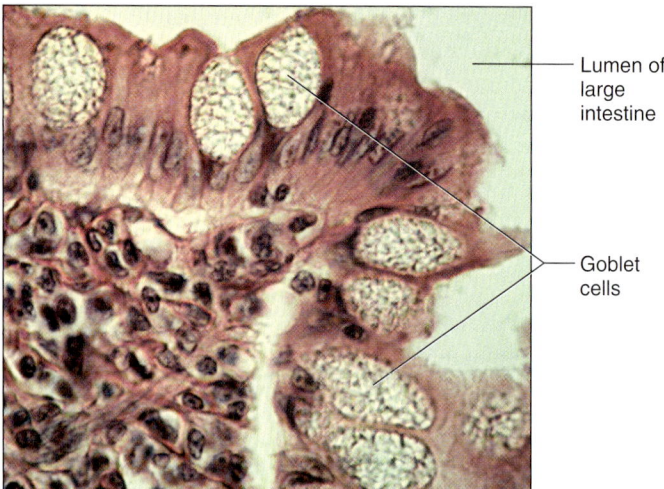

Figure 15.30
Light micrograph of the large intestinal mucosa (560×).

Hemorrhoids are, literally, a pain in the rear. Enlarged and inflamed branches of the rectal vein in the anal columns cause intense itching, sharp pain, and sometimes bright red bleeding. The hemorrhoids may be internal or bulge out of the anus. Causes of hemorrhoids include anything that puts prolonged pressure on the delicate rectal tissue, including obesity, pregnancy, constipation, diarrhea, and liver disease.

Eating more fiber-rich foods and drinking lots of water can usually prevent or cure hemorrhoids. Warm soaks in the tub, cold packs, and careful wiping of painful areas also help, as do external creams and ointments. Surgery—with a scalpel or a laser—can remove severe hemorrhoids.

Structure of the Large Intestine Wall

The wall of the large intestine is composed of the same types of tissues as other parts of the alimentary canal but has some unique features. The large intestine wall, for example, lacks the villi characteristic of the small intestine. Also, the layer of longitudinal muscle fibers does not uniformly cover the large intestine wall. Instead, the fibers form three distinct bands (teniae coli) that extend the entire length of the colon (see fig. 15.27). These bands exert tension lengthwise on the wall, creating a series of pouches (haustra).

Functions of the Large Intestine

Unlike the small intestine, which secretes digestive enzymes and absorbs digestive products, the large intestine has little or no digestive function. However, the mucous membrane that forms the large intestine's inner lining contains many tubular glands. Structurally, these glands are similar to those of the small intestine, but they are composed almost entirely of goblet cells (fig. 15.30). Consequently, mucus is the large intestine's only significant secretion.

Mucus secreted into the large intestine protects the intestinal wall against the abrasive action of the materials passing through it. Mucus also binds particles of fecal matter, and its alkalinity helps control the pH of the large intestine contents.

Chyme entering the large intestine contains materials that the small intestine did not digest or absorb. It also contains water, electrolytes, mucus, and bacteria. The large intestine normally absorbs water and electrolytes in the proximal half of the tube. Substances that remain in the tube become feces and are stored for a time in the distal portion of the large intestine.

The many bacteria that normally inhabit the large intestine, called *intestinal flora,* break down some of the molecules that escape the actions of human digestive enzymes. For instance, cellulose, a complex carbohydrate in food of plant origin, passes through the alimentary canal almost unchanged, but colon bacteria can break down cellulose and use it as an energy source. These bacteria, in turn, synthesize certain vitamins, such as K, B_{12}, thiamine, and riboflavin, which the intestinal mucosa absorbs. Bacterial actions in the large intestine may produce intestinal gas (flatus).

 The colon is home to 100 trillion bacteria.

 CHECK YOUR RECALL

1. How does the structure of the large intestine differ from that of the small intestine?
2. What substances does the large intestine absorb?

Topic of Interest

A BRIEF HISTORY OF CONSTIPATION

The idea that the normal contents of the large intestine can poison a person is not new. An Egyptian papyrus from the sixteenth century B.C. traces the origins of many diseases to various decomposing foods in the lower digestive tract. The discovery in the late eighteenth century that bacteria are normal residents of a human's intestines added to the concept of "intestinal autointoxication," a belief that bacteria mixed with leftovers from digestion could poison us from within. From then until the present time, many societies have attributed a variety of ills to constipation, which was thought to be a consequence of an increasingly urban lifestyle accompanied by less exercise and an unhealthy diet.

In the 1920s and 1930s, people widely feared that constipation would cause "sewer-like blood." Parents forced children to defecate daily, preferably in the morning so they wouldn't need to fret over a missed movement all day long. People discovered that eating bran helped them meet this daily requirement, and brands of bran flourished, one even called DinaMite. Gizmos and gadgets galore, various foods from yeast to yogurt, and many types of laxatives became staples in grocery stores. Some people even had parts of their large intestines removed to lessen the likelihood that the foul contents would kill them (fig. 15A).

In the second half of the twentieth century, after antibiotics had helped control many infectious diseases, attention turned to cancer. The dietary connection to constipation extended to cancer, and the idea that certain foods can either cause or prevent cancers of the large intestine or rectum (colorectal cancer) arose, based largely on studies of populations. People whose diets were low in meat and fat and high in fruits and vegetables tended to have a lower incidence of colorectal cancer than populations whose diets were fatty. In the 1970s, the "fiber hypothesis" gained favor, echoing the earlier popularity of bran cereal. However, two studies published in 2000 showed that, in more than 3,500 individuals, low-fat, high-fiber diets had no effect on the recurrence of intestinal polyps, which are growths that often precede development of cancer. These results were confusing, because epidemiological studies continue to show associations between high-fiber diets and lower incidence of colorectal cancer. Further studies are needed. It is possible that some other aspect of these cultures—such as exercise habits or the way meat is prepared—prevents colorectal cancer. Meanwhile, bran, fruits, and vegetables remain healthful foods that can prevent constipation.

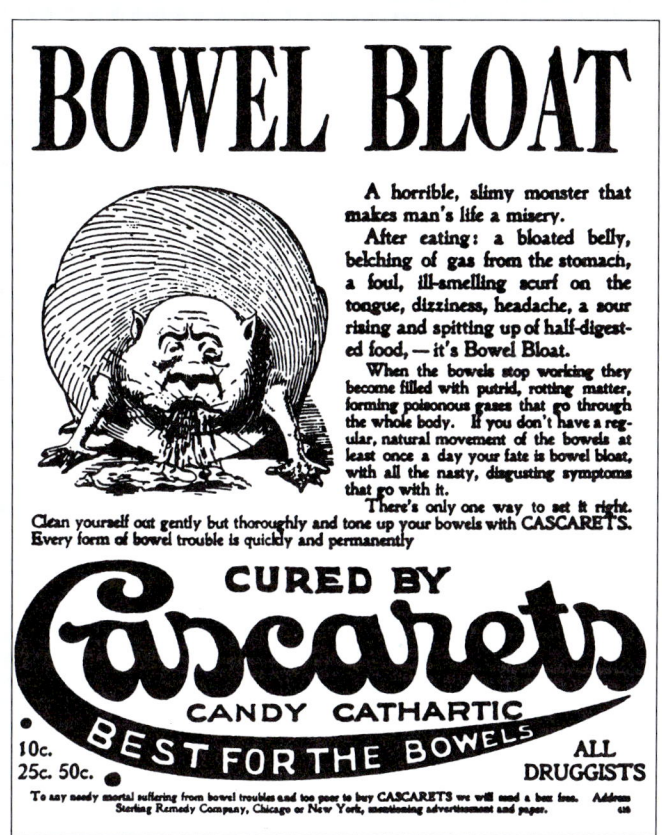

Figure 15A

In the first half of the twentieth century, an industry revolved around the perceived necessity of a daily bowel movement.

Movements of the Large Intestine

The movements of the large intestine—mixing and peristalsis—are similar to those of the small intestine, although usually slower. Also, peristaltic waves of the large intestine happen only two or three times each day. These waves produce *mass movements* in which a large section of the intestinal wall constricts vigorously, forcing the intestinal contents toward the rectum. Typically, mass movements follow a meal as a result of the gastrocolic reflex initiated in the small intestine. Irritations of the intestinal mucosa also can trigger such movements. For instance, a person with an inflamed colon (colitis) may experience frequent mass movements.

A person usually can initiate a *defecation reflex* by holding a deep breath and contracting the abdominal wall muscles. This action increases internal abdominal pressure and forces feces into the rectum. As the rectum fills, its wall distends, triggering the defecation reflex. This stimulates peristaltic waves in the descending colon, and the internal anal sphincter relaxes. At the same time, other reflexes involving the sacral region of the spinal cord strengthen the peristaltic waves, lower the diaphragm, close the glottis, and contract the abdominal wall muscles. These actions further increase internal abdominal pressure and squeeze the rectum. The external anal sphincter is signaled to relax, and the feces are forced to the outside. Contracting the external anal sphincter allows voluntary inhibition of defecation.

Feces

Feces (fē´sēz), include materials that were not digested or absorbed, plus water, electrolytes, mucus, shed intestinal cells, and bacteria. Usually, feces are about 75% water, and their color derives from bile pigments altered by bacterial action. Feces' pungent odor results from a variety of compounds that bacteria produce.

CHECK YOUR RECALL

1. How does peristalsis in the large intestine differ from peristalsis in the small intestine?
2. List the major events that occur during defecation.
3. Describe the composition of feces.

15.11 Nutrition and Nutrients

Nutrition is the study of nutrients and how the body utilizes them. **Nutrients** (nu´tre-enz) include carbohydrates, lipids, proteins, vitamins, minerals, and water (see chapter 2, p. 39, and chapter 4, p. 77). Foods provide these nutrients, and digestion breaks them down to sizes that can be absorbed and transported in the bloodstream. Nutrients that human cells cannot synthesize, such as certain amino acids, are called **essential nutrients.**

Carbohydrates

Carbohydrates are organic compounds used primarily to supply energy for cellular processes.

Carbohydrate Sources

Carbohydrates are ingested in a variety of forms, including starch from grains and vegetables; glycogen from meats; disaccharides from cane sugar, beet sugar, and molasses; and monosaccharides from honey and fruits. Digestion breaks down complex carbohydrates into monosaccharides, which are small enough to be absorbed.

Cellulose is a complex carbohydrate that is abundant in food—it gives celery its crunch and lettuce its crispness. Humans cannot digest cellulose, so the portion of it that is not broken down by intestinal flora passes through the alimentary canal largely unchanged. Thus, cellulose provides bulk (also called fiber or roughage) against which the muscular wall of the digestive system can push, facilitating the movement of food.

Carbohydrate Utilization

The monosaccharides absorbed from the digestive tract include *fructose, galactose,* and *glucose*. Liver enzymes convert fructose and galactose into glucose, which is the carbohydrate form most commonly oxidized for cellular fuel.

Some excess glucose is polymerized to form *glycogen* and stored in the liver and muscles. When required to supply energy, glucose can be mobilized rapidly from glycogen. However, only a certain amount of glycogen can be stored, and excess glucose is usually converted into fat and stored in adipose tissue.

Cells use carbohydrates as starting materials for the synthesis of such vital biochemicals as the five-carbon sugars *ribose* and *deoxyribose,* required for production of the nucleic acids RNA and DNA. Carbohydrates are also required to synthesize the disaccharide *lactose* (milk sugar) when the breasts are actively secreting milk.

Many cells obtain energy by oxidizing fatty acids. Some cells, however, such as neurons, normally require a continuous supply of glucose for survival. Even a temporary decrease in the glucose supply may seriously impair nervous system function. Consequently, the body requires a minimum amount of carbohydrates. If foods do not provide an adequate carbohydrate supply, the liver may convert some noncarbohydrates, such as amino acids from proteins, into glucose. Thus, the requirement for glucose has physiological priority over the requirement to synthesize proteins from available amino acids.

Carbohydrate Requirements

Because carbohydrates provide the primary fuel source for cellular processes, the requirement for carbohydrates varies with individual energy expenditure. Physically active individuals require more fuel than sedentary ones. The minimum carbohydrate requirement in the human diet is unknown, but a daily carbohydrate intake of at least 125–175 grams is necessary to spare protein (that is, to avoid protein breakdown) and to avoid metabolic disorders resulting from excess fat utilization. A rule of thumb suggests that carbohydrates should supply about 60% of a person's diet.

CHECK YOUR RECALL

1. List several common sources of carbohydrates.
2. Explain the importance of cellulose in the diet.
3. Explain why the requirement for glucose has priority over protein synthesis.

Lipids

Lipids are organic compounds that include fats, oils, and fatlike substances (see chapter 2, p. 39). They supply energy for cellular processes and for building structures such as cell membranes. Lipids include fats, phospholipids, and cholesterol. The most common dietary lipids are the fats called *triglycerides* (tri-glis´er-īdz).

Lipid Sources

Triglycerides are found in plant- and animal-based foods. Saturated fats (which should comprise no more than 10% of the diet) are found mainly in foods of animal origin, such as meats, eggs, milk, and lard, as well as in palm and coconut oils. Unsaturated fats are in seeds, nuts, and plant oils, such as those from corn, peanuts, and olives.

Cholesterol is abundant in liver and egg yolk and, to a lesser extent, in whole milk, butter, cheese, and meats. It is generally not present in foods of plant origin.

Lipid Utilization

Many foods contain lipids in the form of phospholipids, cholesterol, or the most common dietary lipids, triglycerides. Recall from chapter 2 (p. 40) that a triglyceride molecule consists of a glycerol and three fatty acids.

Lipids serve a variety of physiological functions, but the main one is to supply energy. Gram for gram, fats contain more than twice as much chemical energy as carbohydrates or proteins.

Before a triglyceride molecule can release energy, it must undergo hydrolysis. Digestion breaks triglycerides down into fatty acids and glycerol. After being absorbed, these products are transported in lymph and blood to tissues. Figure 15.31 shows that, upon hydrolysis, some of the resulting fatty acid portions can react to form molecules of acetyl coenzyme A by a series of reactions called **beta oxidation.** Excess acetyl coenzyme A can be converted into compounds called *ketone bodies,* such as acetone, which later may be changed back to acetyl coenzyme A. In either case, the resulting acetyl coenzyme A can be oxidized in the citric acid cycle. The glycerol portions of the triglyceride molecules can also enter metabolic pathways leading to the citric acid cycle, or they can be used to synthesize glucose. Fatty acid molecules released from fat hydrolysis can also combine to form fat molecules by anabolic processes and be stored in fat tissue.

The liver can convert fatty acids from one form to another, but it cannot synthesize certain fatty acids, called **essential fatty acids.** *Linoleic acid,* for example, is required for phospholipid synthesis, which in turn is necessary for cell membrane formation and the transport of circulating lipids. Good sources of linoleic acid

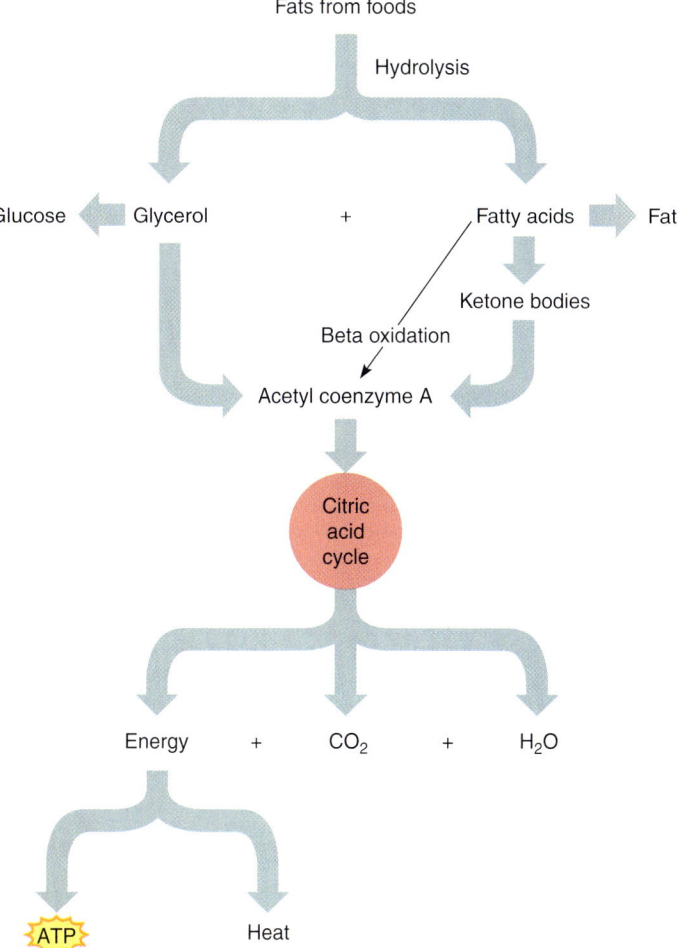

Figure 15.31
The body digests fats from foods into glycerol and fatty acids, which may enter catabolic pathways and be used as energy sources.

include corn, cottonseed, and soy oils. Other essential fatty acids are linolenic acid and arachidonic acid.

The liver uses free fatty acids to synthesize triglycerides, phospholipids, and lipoproteins that may then be released into the blood. Thus, the liver regulates circulating lipids. It also controls the total amount of cholesterol in the body by synthesizing cholesterol and releasing it into the bloodstream, or by removing cholesterol from the bloodstream and excreting it into bile. The liver also uses cholesterol to produce bile salts.

Cholesterol is not an energy source. It provides structural material for cell and organelle membranes and is a starting material for the synthesis of certain sex hormones and adrenal hormones.

Adipose tissue stores excess triglycerides. If the blood lipid concentration drops (in response to fasting, for example), some of these triglycerides are hydrolyzed into free fatty acids and glycerol, and then released into the bloodstream.

Lipid Requirements

The amounts and types of fats required for health are unknown. Because linoleic acid is an essential fatty acid, nutritionists recommend that formula-fed infants receive 3% of their energy intake in the form of linoleic acid to prevent deficiency conditions. With lipids supplying not more than 30% of daily food intake, a typical adult diet that includes a variety of foods usually provides adequate fats. Fat intake must also supply required amounts of fat-soluble vitamins.

Before 1993, "lite" and similar markings on prepared food packages could, and did, mean almost anything: A bottle of vegetable oil labeled "lite" referred to the product's color; a "lite" cheesecake referred to its texture! Now, "lite," or any similar spelling, has a distinct meaning: The product must have a third fewer calories than the "real thing" or half the fat calories.

Be wary of claims that a food product is "99% fat-free." This usually refers to percentage by weight—not calories, which is what counts. A creamy concoction that is 99% fat-free may be largely air and water, and in that form, fat comprises very little of it. But when the air is compressed and the water absorbed, as happens in the stomach, the fat percentage may skyrocket.

 CHECK YOUR RECALL

1. Which fatty acid is an essential nutrient?
2. What is the liver's role in the utilization of lipids?
3. What are the functions of cholesterol?

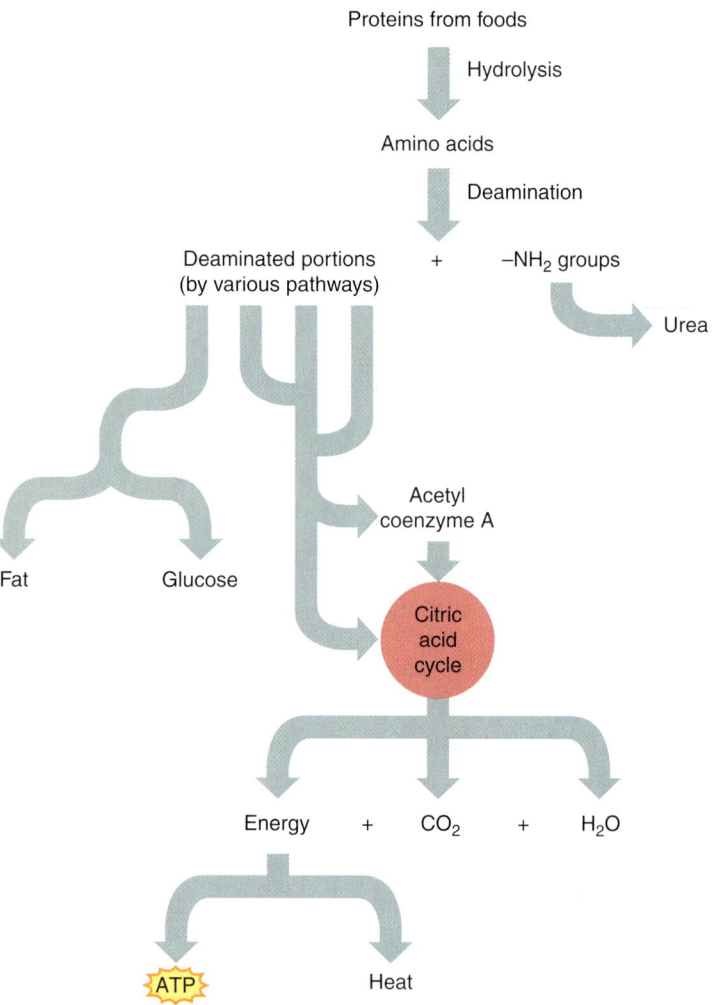

Figure 15.32
The body digests proteins from foods into amino acids but must deaminate these smaller molecules before they can be used as energy sources.

Proteins

Proteins are polymers of amino acids with a wide variety of functions. Proteins include enzymes that control metabolic rates, clotting factors, the keratin of skin and hair, elastin and collagen of connective tissue, plasma proteins that regulate water balance, the muscle components actin and myosin, certain hormones, and the antibodies that protect against infection. Proteins may also supply energy.

When protein molecules are used as energy sources, they are broken down into amino acids. Then the amino acids undergo **deamination,** a process that occurs in the liver and removes their nitrogen-containing portions (—NH_2 groups) (see fig. 2.15, p. 42). These —NH_2 groups then react to form the waste *urea,* which is excreted in urine.

Depending upon the particular amino acids involved, the remaining deaminated portions are decomposed in one of several pathways (fig. 15.32). Some of these pathways lead to formation of acetyl coenzyme A, and others lead more directly to the steps of the citric acid cycle. As energy is released from the cycle, some of it is captured in ATP molecules. If energy is not required immediately, the deaminated portions of the amino acids may react to form glucose or fat molecules in other metabolic pathways.

Protein Sources

Foods rich in proteins include meats, fish, poultry, cheese, nuts, milk, eggs, and cereals. Legumes, including beans and peas, contain lesser amounts. Digestion breaks down proteins into their component amino acids—smaller molecules that intestinal tissues absorb and blood transports.

The cells of an adult can synthesize all but eight required amino acids, and the cells of a child can produce all but ten. Amino acids that the body can synthe-

TABLE 15.7	AMINO ACIDS IN FOODS
Alanine	Leucine (e)
Arginine (ch)	Lysine (e)
Asparagine	Methionine (e)
Aspartic acid	Phenylalanine (e)
Cysteine	Proline
Glutamic acid	Serine
Glutamine	Threonine (e)
Glycine	Tryptophan (e)
Histidine (ch)	Tyrosine
Isoleucine (e)	Valine (e)

Eight essential amino acids (e) cannot be synthesized by human cells and must be provided in the diet. Two additional amino acids (ch) are essential in growing children.

size are termed nonessential; those that it cannot synthesize are **essential amino acids.** This term refers only to dietary intake since all amino acids are needed for normal protein synthesis.

All of the amino acids must be present in the body at the same time for growth and tissue repair to occur. In other words, if one essential amino acid is missing from the diet, normal protein synthesis cannot take place. This is because many proteins include all twenty types of amino acids. Table 15.7 lists the amino acids in foods and indicates those that are essential.

Proteins are classified as complete or incomplete on the basis of the amino acid types they provide. **Complete proteins,** which include those available in milk, meats, and eggs, contain adequate amounts of the essential amino acids. **Incomplete proteins,** such as *zein* in corn, which has too little of the essential amino acids tryptophan and lysine, are unable by themselves to maintain human tissues or to support normal growth and development. A protein called *gliadin* in wheat is an example of a **partially complete protein.** It does not contain enough lysine to promote growth, but it contains enough to maintain life.

Plant proteins typically contain too little of one or more essential amino acids to provide adequate nutrition for a person. Combining appropriate plant foods can provide a diversity of dietary amino acids. For example, beans are low in methionine but have enough lysine. Rice lacks lysine but has enough methionine. Thus, a meal of beans and rice provides enough of both types of amino acids.

CHECK YOUR RECALL

1. Which foods are rich sources of proteins?
2. Why are some amino acids called essential?
3. Distinguish between complete and incomplete proteins.

Protein Requirements

Proteins supply essential amino acids. They also provide nitrogen and other elements for the synthesis of nonessential amino acids and certain nonprotein nitrogenous substances. Consequently, the amount of protein individuals require varies according to body size, metabolic rate, and nitrogen requirements.

For an average adult, nutritionists recommend a daily protein intake of about 0.8 grams per kilogram of body weight, or 10% of a person's diet. For a pregnant woman, the recommendation increases an additional 30 grams of protein per day. Similarly, a nursing mother requires an extra 20 grams of protein per day to maintain a high level of milk production.

A food's potential energy can be expressed in calories, which are units of heat. A **calorie** (kal´o-re) is defined as the amount of heat required to raise the temperature of a gram of water by 1° Celsius. The calorie used to measure food energy, however, is 1,000 times greater. This larger calorie (Cal) is technically a kilocalorie, but nutritional studies commonly refer to it simply as a calorie. As a result of cellular oxidation, 1 gram of carbohydrate or 1 gram of protein yields about 4.1 calories, but 1 gram of fat yields 9.5 calories.

CHECK YOUR RECALL

1. What are the physiological functions of proteins?
2. How much protein is recommended for an adult diet?

Vitamins

Vitamins (vi´tah-minz) are organic compounds (other than carbohydrates, lipids, and proteins) that are required in small amounts for normal metabolic processes, but that body cells cannot synthesize in adequate amounts. Thus, they are essential nutrients that must come from foods.

Vitamins can be classified on the basis of solubility because some are soluble in fats (or fat solvents) and others are soluble in water. *Fat-soluble vitamins* include vitamins A, D, E, and K; *water-soluble vitamins* include the B vitamins and vitamin C.

Fat-Soluble Vitamins

Because fat-soluble vitamins dissolve in fats, they associate with lipids and are influenced by the same factors that affect lipid absorption. For example, bile salts in

TABLE 15.8 — FAT-SOLUBLE VITAMINS

VITAMIN	CHARACTERISTICS	FUNCTIONS	SOURCES AND RDA* FOR ADULTS
Vitamin A	Occurs in several forms; synthesized from carotenes; stored in liver; stable in heat, acids, and alkalis; unstable in light	Necessary for synthesis of visual pigments, mucoproteins, and mucopolysaccharides; for normal development of bones and teeth; and for maintenance of epithelial cells	Liver, fish, whole milk, butter, eggs, leafy green vegetables, yellow and orange vegetables and fruits; RDA = 4,000–5,000 IU†
Vitamin D	A group of steroids; resistant to heat, oxidation, acids, and alkalis; stored in liver, skin, brain, spleen, and bones	Promotes absorption of calcium and phosphorus; promotes development of teeth and bones	Produced in skin exposed to ultraviolet light; in milk, egg yolk, fish liver oils, fortified foods; RDA = 400 IU
Vitamin E	A group of compounds; resistant to heat and visible light; unstable in presence of oxygen and ultraviolet light; stored in muscles and adipose tissue	An antioxidant; prevents oxidation of vitamin A and polyunsaturated fatty acids; may help maintain stability of cell membranes	Oils from cereal seeds, salad oils, margarine, shortenings, fruits, nuts, and vegetables; RDA = 30 IU
Vitamin K	Occurs in several forms; resistant to heat, but destroyed by acids, alkalis, and light; stored in liver	Needed for synthesis of prothrombin, which functions in blood clotting	Leafy green vegetables, egg yolk, pork liver, soy oil, tomatoes, cauliflower; RDA = 55–70 µg

*RDA = recommended dietary allowance.
†IU = international unit.

the intestine promote absorption of these vitamins. As a group, fat-soluble vitamins are stored in moderate quantities within various tissues. They resist the effects of heat; therefore, cooking and food processing usually do not destroy them. Table 15.8 lists the fat-soluble vitamins and their characteristics, functions, sources, and recommended dietary allowances (RDA) for adults.

CHECK YOUR RECALL

1. What are vitamins?
2. How do bile salts affect absorption of fat-soluble vitamins?

Water-Soluble Vitamins

The water-soluble vitamins include the B vitamins and vitamin C. The **B vitamins** are several compounds that are essential for normal cellular metabolism. They help oxidize carbohydrates, lipids, and proteins. Since the B vitamins often occur together in foods, they are usually referred to as the *vitamin B complex*. Members of this group differ chemically and functionally. Cooking and food processing destroy some of them.

Vitamin C (ascorbic acid) is one of the least stable vitamins and is fairly widespread in plant foods. It is necessary for collagen production, the conversion of folacin to folinic acid, and the metabolism of certain amino acids. Vitamin C also promotes iron absorption and synthesis of certain hormones from cholesterol. Table 15.9 lists the water-soluble vitamins and their characteristics, functions, sources, and RDAs for adults.

 English ships carried limes to protect the sailors from scurvy. American ships carried cranberries.

CHECK YOUR RECALL

1. Name the water-soluble vitamins.
2. What is the vitamin B complex?
3. Distinguish between fat-soluble and water-soluble vitamins.

Minerals

Dietary **minerals** (min´er-alz) are elements other than carbon that are essential in human metabolism. Plants usually extract minerals from soil, and humans obtain minerals from plant foods or from animals that have eaten plants.

Characteristics of Minerals

Minerals are responsible for about 4% of body weight and are most concentrated in the bones and teeth. Minerals are usually incorporated into organic molecules. For example, phosphorus is found in phospholipids, iron in hemoglobin, and iodine in thyroxine. Some minerals are part of inorganic compounds, such as the calcium phosphate of bone. Other minerals are free ions, such as sodium, chloride, and calcium ions in blood.

Minerals compose parts of the structural materials in all body cells. They also constitute portions of enzyme

TABLE 15.9 — WATER-SOLUBLE VITAMINS

VITAMIN	CHARACTERISTICS	FUNCTIONS	SOURCES AND RDA* FOR ADULTS
Thiamine (vitamin B_1)	Destroyed by heat and oxygen, especially in alkaline environment	Part of coenzyme required to oxidize carbohydrates; coenzyme required for ribose synthesis	Lean meats, liver, eggs, whole-grain cereals, leafy green vegetables, legumes; RDA = 1.5 mg
Riboflavin (vitamin B_2)	Stable to heat, acids, and oxidation; destroyed by bases and ultraviolet light	Part of enzymes and coenzymes required to oxidize glucose and fatty acids and for cellular growth	Meats, dairy products, leafy green vegetables, whole-grain cereals; RDA = 1.7 mg
Niacin (nicotinic acid)	Stable to heat, acids, and alkalis; converted to niacinamide by cells; synthesized from tryptophan	Part of coenzymes required to oxidize glucose and to synthesize proteins, fats, and nucleic acids	Liver, lean meats, peanuts, legumes; RDA = 20 mg
Vitamin B_6	Group of three compounds; stable to heat and acids; destroyed by oxidation, bases, and ultraviolet light	Coenzyme required to synthesize proteins and certain amino acids, to convert tryptophan to niacin, to produce antibodies, and to synthesize nucleic acids	Liver, meats, bananas, avocados, beans, peanuts, whole-grain cereals, egg yolk; RDA = 2 mg
Pantothenic acid	Destroyed by heat, acids, and bases	Part of coenzyme A required to oxidize carbohydrates and fats	Meats, whole-grain cereals, legumes, milk, fruits, vegetables; RDA = 10 mg
Cyanocobalamin (vitamin B_{12})	Complex, cobalt-containing compound; stable to heat; inactivated by light, strong acids, and strong bases; absorption regulated by intrinsic factor from gastric glands; stored in liver	Part of coenzyme required to synthesize nucleic acids and to metabolize carbohydrates; plays role in myelin synthesis; needed for normal red blood cell production	Liver, meats, milk, cheese, eggs; RDA = 3–6 µg
Folacin (folic acid)	Occurs in several forms; destroyed by oxidation in acid environment or by heat in alkaline environment; stored in liver, where it is converted into folinic acid	Coenzyme required for metabolism of certain amino acids and for DNA synthesis; promotes red blood cell production	Liver, leafy green vegetables, whole-grain cereals, legumes; RDA = 0.4 mg
Biotin	Stable to heat, acids, and light; destroyed by oxidation and bases	Coenzyme required to metabolize amino acids and fatty acids, and to synthesize nucleic acids	Liver, egg yolk, nuts, legumes, mushrooms; RDA = 0.3 mg
Ascorbic acid (vitamin C)	Closely related to monosaccharides; stable in acids but destroyed by oxidation, heat, light, and bases	Required to produce collagen, to convert folacin to folinic acid, and to metabolize certain amino acids; promotes absorption of iron and synthesis of hormones from cholesterol	Citrus fruits, tomatoes, potatoes, leafy green vegetables; RDA = 60 mg

*RDA = recommended dietary allowance.

molecules, contribute to the osmotic pressure of body fluids, and play vital roles in nerve impulse conduction, muscle fiber contraction, blood coagulation, and maintaining the pH of body fluids.

Major Minerals

The minerals *calcium* and *phosphorus* account for nearly 75% by weight of the mineral elements in the body; thus, they are **major minerals.** Other major minerals, each of which accounts for 0.05% or more of the body weight, include potassium, sulfur, sodium, chlorine, and magnesium. Table 15.10 lists the distribution, functions, sources, and adult RDAs of major minerals.

 There is enough phosphorus in the human body to make two thousand match tips.

 CHECK YOUR RECALL

1. How are minerals obtained?
2. What are the major functions of minerals?

Trace Elements

Trace elements are essential minerals found in minute amounts, each making up less than 0.005% of adult body weight. They include iron, manganese, copper, iodine, cobalt, zinc, fluorine, selenium, and chromium. Table 15.11 lists the distribution, functions, sources, and adult RDAs of the trace elements.

 The iron in a human being could make a small nail.

TABLE 15.10 MAJOR MINERALS

MINERAL	DISTRIBUTION	FUNCTIONS	SOURCES AND RDA* FOR ADULTS
Calcium (Ca)	Mostly in the inorganic salts of bones and teeth	Structure of bones and teeth; essential for nerve impulse conduction, muscle fiber contraction, and blood coagulation; increases permeability of cell membranes; activates certain enzymes	Milk, milk products, leafy green vegetables; RDA = 800 mg
Phosphorus (P)	Mostly in the inorganic salts of bones and teeth	Structure of bones and teeth; component in nearly all metabolic reactions; constituent of nucleic acids, many proteins, some enzymes, and some vitamins; occurs in cell membrane, ATP, and phosphates of body fluids	Meats, cheese, nuts, whole-grain cereals, milk, legumes; RDA = 800 mg
Potassium (K)	Widely distributed; tends to be concentrated inside cells	Helps maintain intracellular osmotic pressure and regulate pH; promotes metabolism; needed for nerve impulse conduction and muscle fiber contraction	Avocados, dried apricots, meats, nuts, potatoes, bananas; RDA = 2,500 mg
Sulfur (S)	Widely distributed; abundant in skin, hair, and nails	Essential part of various amino acids, thiamine, insulin, biotin, and mucopolysaccharides	Meats, milk, eggs, legumes; no RDA established
Sodium (Na)	Widely distributed; large proportion occurs in extracellular fluids and bound to inorganic salts of bone	Helps maintain osmotic pressure of extracellular fluids and regulate water movement; needed for conduction of nerve impulses and contraction of muscle fibers; aids in regulation of pH and in transport of substances across cell membranes	Table salt, cured ham, sauerkraut, cheese, graham crackers; RDA = 2,500 mg
Chlorine (Cl)	Closely associated with sodium (as chloride); most highly concentrated in cerebrospinal fluid and gastric juice	Helps maintain osmotic pressure of extracellular fluids, regulate pH, and maintain electrolyte balance; essential in formation of hydrochloric acid; aids transport of carbon dioxide by red blood cells	Same as for sodium; no RDA established
Magnesium (Mg)	Abundant in bones	Needed in metabolic reactions that occur in mitochondria and are associated with ATP production; plays role in the breakdown of ATP to ADP	Milk, dairy products, legumes, nuts, leafy green vegetables; RDA = 300–350 mg

*RDA = recommended dietary allowance.

CHECK YOUR RECALL

1. Distinguish between a major mineral and a trace element.
2. Name the major minerals and trace elements.

Adequate Diets

An adequate diet provides sufficient energy, essential fatty acids, essential amino acids, vitamins, and minerals to support optimal growth and to maintain and repair body tissues. Because individual requirements for nutrients vary greatly with age, sex, growth rate, amount of physical activity, and level of stress, as well as with genetic and environmental factors, designing a diet that is adequate for everyone is impossible. However, nutrients are so widely distributed in foods, that satisfactory amounts and combinations can usually be obtained in spite of individual food preferences. Figure 15.33 depicts the **food pyramid** system for a healthy diet.

If the diet lacks essential nutrients or a person fails to use available foods to best advantage, **malnutrition** (mal´´nu-trish´un) results. This condition may be due to either *undernutrition*, producing the symptoms of deficiency diseases, or to *overnutrition*, arising from excess nutrient intake.

The factors leading to malnutrition vary. For example, a deficiency condition may stem from lack of food or poor-quality food. On the other hand, malnutrition may result from overeating or taking too many vitamin supplements.

TABLE 15.11 TRACE ELEMENTS

TRACE ELEMENT	DISTRIBUTION	FUNCTIONS	SOURCES AND RDA* FOR ADULTS
Iron (Fe)	Primarily in blood; stored in liver, spleen, and bone marrow	Part of hemoglobin molecule; catalyzes vitamin A formation; incorporated into a number of enzymes	Liver, lean meats, dried apricots, raisins, enriched whole-grain cereals, legumes, molasses; RDA = 10–18 mg
Manganese (Mn)	Most concentrated in liver, kidneys, and pancreas	Occurs in enzymes needed for fatty acid and cholesterol synthesis, urea formation, and normal functioning of the nervous system	Nuts, legumes, whole-grain cereals, leafy green vegetables, fruits; RDA = 2.5–5 mg
Copper (Cu)	Most highly concentrated in liver, heart, and brain	Essential for hemoglobin synthesis, bone development, melanin production, and myelin formation	Liver, oysters, crabmeat, nuts, whole-grain cereals, legumes; RDA = 2–3 mg
Iodine (I)	Concentrated in thyroid gland	Essential component for synthesis of thyroid hormones	Food content varies with soil content in different geographic regions; iodized table salt; RDA = 0.15 mg
Cobalt (Co)	Widely distributed	Component of cyanocobalamin; needed for synthesis of several enzymes	Liver, lean meats, milk; no RDA established
Zinc (Zn)	Most concentrated in liver, kidneys, and brain	Constituent of several enzymes involved in digestion, respiration, bone metabolism, liver metabolism; necessary for normal wound healing and maintaining skin integrity	Meats, cereals, legumes, nuts, vegetables; RDA = 15 mg
Fluorine (F)	Primarily in bones and teeth	Component of tooth structure	Fluoridated water; RDA = 1.5–4.0 mg
Selenium (Se)	Concentrated in liver and kidneys	Occurs in enzymes	Lean meats, fish, cereals; RDA = 0.05–2.00 mg
Chromium (Cr)	Widely distributed	Essential for use of carbohydrates	Liver, lean meats, wine; RDA = 0.05–2.00 mg

*RDA = recommended dietary allowance.

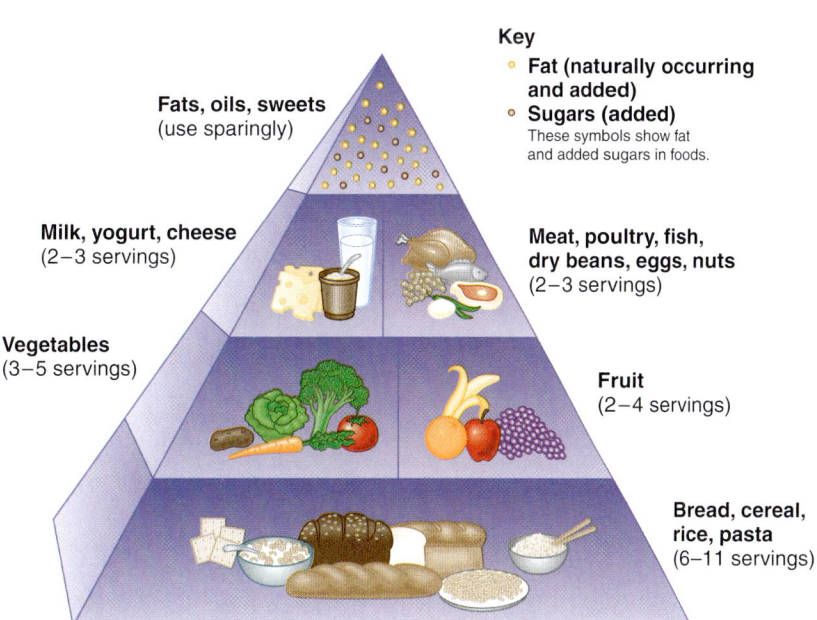

Figure 15.33
The U.S. Department of Agriculture introduced the food pyramid in 1992 as a guideline to healthy eating. In contrast to former food group plans, the pyramid gives an instant idea of which foods should make up the bulk of the diet—whole grains, fruits, and vegetables.

CHECK YOUR RECALL

1. What is an adequate diet?
2. What factors influence individual requirements for nutrients?
3. What causes malnutrition?

Clinical Terms Related to the Digestive System and Nutrition

achalasia (ak´´ah-la´ze-ah) Failure of the smooth muscle to relax at some junction in the digestive tube, such as between the esophagus and stomach.
achlorhydria (ah´klor-hi´dre-ah) Lack of hydrochloric acid in gastric secretions.
anorexia nervosa (ă-nah-rek´se-ah ner vo´sah) Self-starvation.
aphagia (ah-fa´je-ah) Inability to swallow.
cachexia (kah-kek´se-ah) State of chronic malnutrition and physical wasting.
celiac disease (se´le-ak di´zēz´) Inability to digest or use fats and carbohydrates.
cholecystitis (ko´´le-sis-ti´tis) Inflammation of the gallbladder.
cholelithiasis (ko´´le-li-thi´ah-sis) Stones in the gallbladder.
cholestasis (ko´´le-sta´sis) Blockage in bile flow from the gallbladder.
cirrhosis (si-ro´sis) Liver condition in which the hepatic cells degenerate and the surrounding connective tissues thicken.
diverticulitis (di´´ver-tik´´u-li´tis) Inflammation of small pouches (diverticula) that form in the lining and wall of the colon.
dumping syndrome (dum´ping sin´drōm) Symptoms, including diarrhea, that often occur following a gastrectomy.
dysentery (dis´en-ter´´e) Intestinal infection by viruses, bacteria, or protozoans that causes diarrhea and cramps.
dyspepsia (dis-pep´se-ah) Indigestion; difficulty in digesting a meal.
dysphagia (dis-fa´je-ah) Difficulty in swallowing.
enteritis (en´´tĕ-ri´tis) Inflammation of the intestine.
esophagitis (e-sof´´ah-ji´tis) Inflammation of the esophagus.
gastrectomy (gas-trek´to-me) Partial or complete removal of the stomach.
gastrostomy (gas-tros´to-me) Creation of an opening in the stomach wall through which food and liquids can be administered when swallowing is not possible.
glossitis (glŏs-si´tis) Inflammation of the tongue.
hyperalimentation (hi´´-per-al´´-ĭ-men-ta´shun) Long-term intravenous nutrition.
ileitis (il´´e-i´tis) Inflammation of the ileum.
pharyngitis (far´´in-ji´tis) Inflammation of the pharynx.
polyphagia (pol´´e-fa´je-ah) Overeating.
pyloric stenosis (pi-lor´ik stĕ-no´sis) Congenital obstruction at the pyloric sphincter due to an enlarged pyloric muscle.
pylorospasm (pi-lor´o-spazm) Spasm of the pyloric portion of the stomach or of the pyloric sphincter.
pyorrhea (pi´´o-re´ah) Inflammation of the dental periosteum with pus formation.
stomatitis (sto´´mah-ti´tis) Inflammation of the lining of the mouth.

Clinical Connection

Saliva is vital for tasting and processing food, so that it can be swallowed and digested. It also keeps teeth healthy by washing away bacteria and plaque. In a condition called xerostomia, or "dry mouth," saliva production is insufficient. As a result, chewed food does not soften enough and is difficult to swallow. Even licking an envelope can be impossible for a person with this condition.

Physicians today no longer accept xerostomia as a normal consequence of aging, but instead ask a patient which drugs he or she is taking. Often, dry mouth is a side effect of a medication. Hundreds of medications can cause xerostomia, including drugs that treat depression, hypertension, cancer, and allergies. Radiation to the head or neck to treat cancer can cause xerostomia. Infection of the salivary glands or Sjogren's syndrome can also cause dry mouth.

Sometimes identifying and avoiding a causative medication can relieve dry mouth. If this isn't possible, the Mayo Clinic Health Letter suggests the following strategies:

- Regularly sip water.
- Avoid mouth breathing.
- Suck on sugar-free hard candy or gum.
- Use room vaporizers to add moisture to the environment.

If these measures fail, saliva substitute sprays are available over-the-counter, or a physician can prescribe a medication that increases production of saliva, such as pilocarpine.

Organization

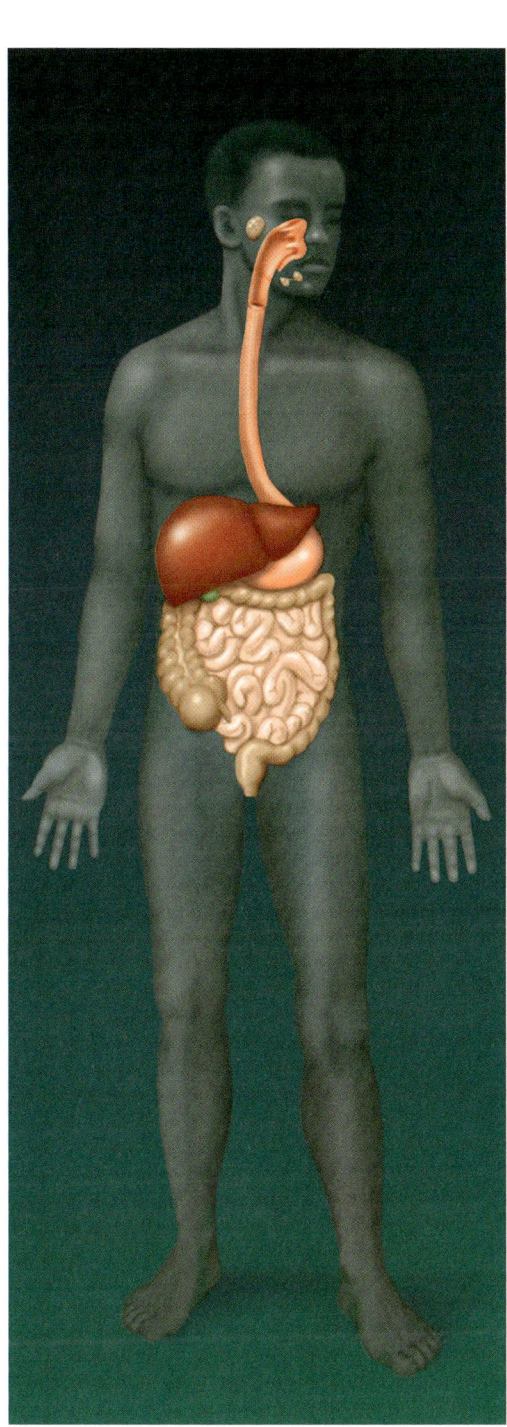

Digestive System
The digestive system ingests, digests, and absorbs nutrients for use by all body cells.

Integumentary System
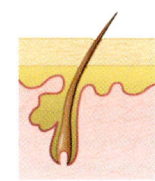
Vitamin D activated in the skin plays a role in absorption of calcium from the digestive tract.

Cardiovascular System
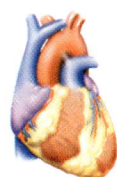
The bloodstream carries absorbed nutrients to all body cells.

Skeletal System

Bones are important in mastication. Calcium absorption is necessary to maintain bone matrix.

Lymphatic System

The lymphatic system plays a major role in the absorption of fats.

Muscular System
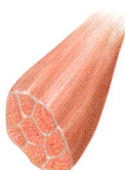
Muscles are important in mastication, swallowing, and the mixing and moving of digestion products through the gastrointestinal tract.

Respiratory System
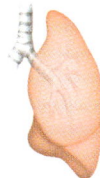
The digestive system and the respiratory system share common anatomical structures.

Nervous System
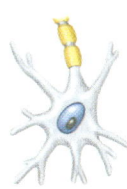
The nervous system can influence digestive system activity.

Urinary System

The kidneys and liver work together to activate vitamin D.

Endocrine System

Hormones can influence digestive system activity.

Reproductive System

In a woman, nutrition is essential for conception and normal development of an embryo and fetus.

SUMMARY OUTLINE

15.1 Introduction (p. 393)
Digestion is the process of mechanically and chemically breaking down foods and absorbing the breakdown products. The digestive system consists of an alimentary canal and several accessory organs.

15.2 General Characteristics of the Alimentary Canal (p. 394)
Regions of the alimentary canal perform specific functions.
1. Structure of the wall
   The wall consists of four layers—the mucosa, submucosa, muscular layer, and serosa.
2. Movements of the tube
   Motor functions include mixing and propelling movements.

15.3 Mouth (p. 396)
The mouth receives food and begins digestion.
1. Cheeks and lips
   a. Cheeks consist of outer layers of skin, pads of fat, muscles associated with expression and chewing, and inner linings of epithelium.
   b. Lips are highly mobile and contain sensory receptors.
2. Tongue
   a. The tongue's rough surface handles food and contains taste buds.
   b. Lingual tonsils are on the root of the tongue.
3. Palate
   a. The palate includes hard and soft portions.
   b. The soft palate closes the opening to the nasal cavity during swallowing.
   c. Palatine tonsils are located on either side of the tongue in the back of the mouth.
4. Teeth
   a. There are twenty primary and thirty-two secondary teeth.
   b. Teeth mechanically break food into smaller pieces, increasing the surface area exposed to digestive actions.
   c. Each tooth consists of a crown and root, and is composed of enamel, dentin, pulp, nerves, and blood vessels.
   d. A periodontal ligament attaches a tooth to the alveolar process.

15.4 Salivary Glands (p. 399)
Salivary glands secrete saliva, which moistens food, helps bind food particles, begins chemical digestion of carbohydrates, makes taste possible, and helps cleanse the mouth.
1. Salivary secretions
   Salivary glands include serous cells that secrete digestive enzymes and mucous cells that secrete mucus.
2. Major salivary glands
   a. The parotid glands secrete saliva rich in amylase.
   b. The submandibular glands produce viscous saliva.
   c. The sublingual glands primarily secrete mucus.

15.5 Pharynx and Esophagus (p. 400)
The pharynx and esophagus are important passageways.
1. Structure of the pharynx
   The pharynx is divided into a nasopharynx, oropharynx, and laryngopharynx.
2. Swallowing mechanism
   Swallowing occurs in three stages:
   a. Food is mixed with saliva and forced into the pharynx.
   b. Involuntary reflex actions move the food into the esophagus.
   c. Peristalsis transports food to the stomach.
3. Esophagus
   a. The esophagus passes through the diaphragm and joins the stomach.
   b. Circular muscle fibers at the distal end of the esophagus help prevent regurgitation of food from the stomach.

15.6 Stomach (p. 402)
The stomach receives food, mixes it with gastric juice, carries on a limited amount of absorption, and moves food into the small intestine.
1. Parts of the stomach
   a. The stomach is divided into cardiac, fundic, body, and pyloric regions.
   b. The pyloric sphincter is a valve between the stomach and small intestine.
2. Gastric secretions
   a. Gastric glands secrete gastric juice.
   b. Gastric juice contains pepsin, hydrochloric acid, and intrinsic factor.
3. Regulation of gastric secretions
   a. Parasympathetic impulses and the hormone gastrin enhance gastric secretion.
   b. Food in the small intestine reflexly inhibits gastric secretions.
4. Gastric absorption
   The stomach wall may absorb a few substances, such as water and other small molecules.
5. Mixing and emptying actions
   a. Mixing movements help produce chyme. Peristaltic waves move chyme into the pyloric region.
   b. The muscular wall of the pyloric region regulates chyme movement into the small intestine.
   c. The rate of emptying depends on the fluidity of chyme and the type of food present.

15.7 Pancreas (p. 404)
1. Structure of the pancreas
   a. The pancreas produces pancreatic juice that is secreted into a pancreatic duct.
   b. The pancreatic duct leads to the duodenum.
2. Pancreatic juice
   a. Pancreatic juice contains enzymes that can split carbohydrates, fats, nucleic acids, and proteins.
   b. Pancreatic juice has a high bicarbonate ion concentration that helps neutralize chyme and causes intestinal contents to be alkaline.
3. Hormones regulate pancreatic secretion
   a. Secretin stimulates the release of pancreatic juice with a high bicarbonate ion concentration.
   b. Cholecystokinin stimulates the release of pancreatic juice with a high concentration of digestive enzymes.

15.8 Liver (p. 406)
1. Liver structure
   a. The right and left lobes of the liver consist of hepatic lobules, the functional units of the gland.
   b. Bile canals carry bile from hepatic lobules to hepatic ducts.
2. Liver functions
   a. The liver metabolizes carbohydrates, lipids, and proteins; stores some substances; filters blood; destroys toxins; and secretes bile.
   b. Bile is the only liver secretion that directly affects digestion.
3. Composition of bile
   a. Bile contains bile salts, bile pigments, cholesterol, and electrolytes.
   b. Only the bile salts have digestive functions.
4. Gallbladder
   a. The gallbladder stores bile between meals.
   b. A sphincter muscle controls release of bile from the common bile duct.

5. Regulation of bile release
    a. Cholecystokinin from the small intestine stimulates bile release.
    b. The sphincter muscle at the base of the common bile duct relaxes as a peristaltic wave in the duodenal wall approaches.
6. Functions of bile salts
    Bile salts emulsify fats and aid in the absorption of fatty acids, cholesterol, and certain vitamins.

15.9 Small Intestine (p. 411)

The small intestine receives secretions from the pancreas and liver, completes nutrient digestion, absorbs the products of digestion, and transports the residues to the large intestine.

1. Parts of the small intestine
    The small intestine consists of the duodenum, jejunum, and ileum.
2. Structure of the small intestine wall
    a. The wall is lined with villi that greatly increase the surface area and aid in mixing and absorption.
    b. Intestinal glands are located between the villi.
3. Secretions of the small intestine
    a. Secretions include mucus and digestive enzymes.
    b. Digestive enzymes split molecules of sugars, proteins, and fats into simpler forms.
4. Regulation of small intestine secretions
    Gastric juice, chyme, and reflexes stimulated by distension of the small intestine wall stimulate small intestine secretions.
5. Absorption in the small intestine
    a. Enzymes on microvilli perform the final steps in digestion.
    b. Villi absorb monosaccharides, amino acids, fatty acids, and glycerol.
    c. Fat molecules with longer chains of carbon atoms enter the lacteals of the villi.
    d. Fatty acids with relatively short carbon chains enter blood capillaries of the villi.
6. Movements of the small intestine
    a. Movements include mixing and peristalsis.
    b. The ileocecal sphincter controls movement of the intestinal contents from the small intestine into the large intestine.

15.10 Large Intestine (p. 417)

The large intestine reabsorbs water and electrolytes, and forms and stores feces.

1. Parts of the large intestine
    a. The large intestine consists of the cecum, colon, rectum, and anal canal.
    b. The colon is divided into ascending, transverse, descending, and sigmoid portions.
2. Structure of the large intestine wall
    a. The large intestine wall resembles the wall in other parts of the alimentary canal.
    b. The large intestine wall has a unique layer of longitudinal muscle fibers arranged in distinct bands.
3. Functions of the large intestine
    a. The large intestine has little or no digestive function.
    b. It secretes mucus.
    c. The large intestine absorbs water and electrolytes.
    d. The large intestine forms and stores feces.
4. Movements of the large intestine
    a. Movements are similar to those in the small intestine.
    b. Mass movements occur two to three times each day.
    c. A defecation reflex stimulates defecation.
5. Feces
    a. Feces consist largely of water, undigested material, electrolytes, mucus, and bacteria.
    b. The color of feces is due to bile pigments that have been altered by bacterial actions.

15.11 Nutrition and Nutrients (p. 420)

Nutrition is the study of nutrients and how the body utilizes them.

1. Carbohydrates
    a. Carbohydrate sources
        (1) Starch, glycogen, disaccharides, and monosaccharides are carbohydrates.
        (2) Cellulose is a polysaccharide that human enzymes cannot digest.
    b. Carbohydrate utilization
        (1) Oxidation releases energy from glucose.
        (2) Excess glucose is stored as glycogen or converted to fat.
        (3) Most carbohydrates supply energy.
    c. Carbohydrate requirements
        (1) Humans survive with a wide range of carbohydrate intakes.
        (2) A rule of thumb suggests carbohydrates make up 60% of a person's diet.
2. Lipids
    a. Lipid sources
        (1) Foods of plant and animal origin provide triglycerides.
        (2) Foods of animal origin provide most cholesterol.
    b. Lipid utilization
        (1) The liver and adipose tissue control triglyceride metabolism.
        (2) Linoleic acid, linolenic acid, and arachidonic acid are essential fatty acids.
        (3) Lipids supply energy and are used to build cell structures.
    c. Lipid requirements
        (1) The amounts and types of fats required for health are unknown.
        (2) Fat intake should not exceed 30% of a person's diet and must be sufficient to carry fat-soluble vitamins.
3. Proteins
    a. Protein sources
        (1) Meats, dairy products, cereals, and legumes provide most proteins.
        (2) Complete proteins contain adequate amounts of all the essential amino acids.
        (3) Incomplete proteins lack adequate amounts of one or more essential amino acids.
    b. Protein utilization
        Proteins serve as structural materials, function as enzymes, and provide energy.
    c. Protein requirements
        Proteins and amino acids must supply essential amino acids and nitrogen for the synthesis of nitrogen-containing molecules. Proteins should make up 10% of a person's diet.
4. Vitamins
    a. Fat-soluble vitamins
        (1) These include vitamins A, D, E, and K.
        (2) They are carried in lipids and are influenced by the same factors that affect lipid absorption.
        (3) They resist the effects of heat; thus, cooking or food processing does not destroy them.
    b. Water-soluble vitamins
        (1) This group includes the B vitamins and vitamin C.
        (2) B vitamins make up a group (the vitamin B complex) and oxidize carbohydrates, lipids, and proteins.
        (3) Cooking or processing food destroys some water-soluble vitamins.

5. Minerals
    a. Characteristics of minerals
        (1) Most minerals are in the bones and teeth.
        (2) Minerals are usually incorporated into organic molecules; some occur in inorganic compounds or as free ions.
        (3) They serve as structural materials, function in enzymes, and play vital roles in metabolic processes.
    b. Major minerals include calcium, phosphorus, potassium, sulfur, sodium, chlorine, and magnesium.
    c. Trace elements include iron, manganese, copper, iodine, cobalt, zinc, fluorine, selenium, and chromium.
6. Adequate diets
    a. An adequate diet provides sufficient energy and essential nutrients to support optimal growth, maintenance, and repair of tissues.
    b. Individual requirements vary so greatly that designing a diet that is adequate for everyone is not possible.
    c. Malnutrition is poor nutrition due to lack of food or failure to make the best use of available food.

REVIEW EXERCISES

1. List and describe the locations of the major parts of the alimentary canal. (p. 393)
2. List and describe the locations of the accessory organs of the digestive system. (p. 393)
3. Name the four layers of the wall of the alimentary canal. (p. 394)
4. Distinguish between mixing movements and propelling movements. (p. 395)
5. Define *peristalsis*. (p. 395)
6. Discuss the functions of the mouth and its parts. (p. 396)
7. Distinguish among the lingual, palatine, and pharyngeal tonsils. (p. 396)
8. Compare the primary and secondary teeth. (p. 397)
9. Describe the structure of a tooth. (p. 398)
10. Explain how a tooth is anchored in its socket. (p. 398)
11. List and describe the locations of the major salivary glands. (p. 399)
12. Explain how the secretions of the salivary glands differ. (p. 399)
13. Discuss the digestive functions of saliva. (p. 399)
14. Explain the function of the esophagus. (p. 401)
15. Describe the structure of the stomach. (p. 402)
16. List the enzymes in gastric juice, and explain the function of each enzyme. (p. 403)
17. Explain how gastric secretions are regulated. (p. 403)
18. Define *cholecystokinin*. (p. 403)
19. Describe the location of the pancreas and the pancreatic duct. (p. 404)
20. List and explain the function of each enzyme found in pancreatic juice. (p. 404)
21. Explain how pancreatic secretions are regulated. (p. 405)
22. Describe the structure of the liver. (p. 406)
23. List the major functions of the liver. (p. 408)
24. Describe the composition of bile. (p. 409)
25. Explain the functions of bile salts. (p. 410)
26. List and describe the locations of the parts of the small intestine. (p. 411)
27. Describe the functions of intestinal villi. (p. 413)
28. Name and explain the function of each enzyme of the intestinal mucosa. (p. 413)
29. Explain how the secretions of the small intestine are regulated. (p. 414)
30. Summarize how each major type of digestive product is absorbed. (p. 414)
31. List and describe the locations of the parts of the large intestine. (p. 417)
32. Explain the general functions of the large intestine. (p. 418)
33. Describe the defecation reflex. (p. 419)
34. List some common sources of carbohydrates. (p. 420)
35. Summarize the importance of cellulose in the diet. (p. 420)
36. Explain why a temporary drop in the blood glucose concentration may impair nervous system functioning. (p. 420)
37. List some common sources of lipids. (p. 421)
38. Describe the liver's role in fat metabolism. (p. 421)
39. List some common sources of protein. (p. 422)
40. Distinguish between essential and nonessential amino acids. (p. 423)
41. Distinguish between complete and incomplete proteins. (p. 423)
42. Discuss the general characteristics of fat-soluble vitamins. (p. 423)
43. List the fat-soluble vitamins, and describe the major functions of each. (p. 424)
44. List the water-soluble vitamins, and describe the major functions of each. (p. 425)
45. Discuss the general characteristics of the mineral nutrients. (p. 424)
46. List the major minerals, and describe the major functions of each. (p. 426)
47. List the trace elements, and describe the major functions of each. (p. 427)
48. Define *adequate diet*. (p. 426)
49. Define *malnutrition*. (p. 426)

CRITICAL THINKING

1. How does mechanical digestion enhance chemical digestion?
2. How can too little fat in the diet lead to a vitamin deficiency, even if a person takes vitamin supplements?
3. Why are vitamins required only in small amounts?
4. How can people consume vastly different diets, yet all obtain adequate nourishment?
5. How would removal of 95% of the stomach (subtotal gastrectomy) to treat severe ulcers or cancer affect the digestion and absorption of foods? How would the patient's eating habits have to be altered? Why?
6. Why may a person with an inflammation of the gallbladder (cholecystitis) also develop an inflammation of the pancreas (pancreatitis)?
7. Why does the blood sugar concentration of a person whose diet is relatively low in carbohydrates remain stable?
8. Examine the label information on the packages of a variety of dry breakfast cereals. Which types of cereals provide adequate sources of vitamins and minerals? Which major nutrients are lacking in these cereals?

WEB CONNECTIONS

Visit the website for additional study questions and more information about this chapter at:

http://www.mhhe.com/shieress8

chapter **16**

Respiratory System

A CASE OF POISONING. A seventy-eight-year-old man was brought to the emergency room with chest pain, shortness of breath, fatigue, confusion, an irregular pulse, and abnormal heart rhythm. Just as physicians began treating him for suspected heart failure, his wife was brought in, with similar symptoms that had begun about 2 hours after her husband fell ill. The couple stayed in the hospital for observation for two days, then were released because their symptoms had abated. But after they had been home for two days, they returned, with the same symptoms.

Physicians suspected the couple had made errors in taking certain medications, but this wasn't the case. When nurses asked the couple about how and where they lived, the source of the problem became clear—kerosene space heaters in a small, unventilated apartment. The man and woman were suffering from carbon monoxide (CO) poisoning.

The normal percentage of hemoglobin molecules that bind CO in people who do not smoke is 2%. The man and woman's levels were 13–14% at the hospital and probably above 20% in their home. The increased CO binding prevented oxygen from binding, which starved the tissues of oxygen and led to the couple's symptoms. They were lucky that they developed symptoms—otherwise, they might have lost consciousness, gone into a coma, and died. A medical journal report describing the case led several physicians to begin measuring levels of hemoglobin-binding CO in patients with chest pain. This led to identification of several more cases of "the great imitator"—CO poisoning.

Photo:
Smog contains carbon monoxide, which in large amounts prevents oxygen from binding to hemoglobin. Indoors, carbon monoxide poisoning usually results from exposure to the gas from malfunctioning space heaters.

Chapter Objectives

After studying this chapter, you should be able to do the following:

16.1 Introduction
1. List the general functions of the respiratory system. (p. 435)

16.2 Organs of the Respiratory System
2. Name and describe the locations of the organs of the respiratory system. (p. 435)
3. Describe the functions of each organ of the respiratory system. (p. 435)

16.3 Breathing Mechanism
4. Explain the mechanisms of inspiration and expiration. (p. 441)
5. Name and define each of the lung volumes and respiratory capacities. (p. 446)

16.4 Control of Breathing
6. Locate the respiratory center, and explain how it controls normal breathing. (p. 447)
7. Discuss how various factors affect the respiratory center. (p. 448)

16.5 Alveolar Gas Exchanges
8. Describe the structure and function of the respiratory membrane. (p. 449)

16.6 Gas Transport
9. Explain how air and blood exchange gases and how blood transports these gases. (p. 450)

Aids to Understanding Words

alveol- [small cavity] *alveol*us: Microscopic air sac within a lung.
bronch- [windpipe] *bronch*us: Primary branch of the trachea.
cric- [ring] *cric*oid cartilage: Ring-shaped mass of cartilage at the base of the larynx.
epi- [upon] *epi*glottis: Flaplike structure that partially covers the opening into the larynx during swallowing.
hem- [blood] *hem*oglobin: Pigment in red blood cells that transports oxygen and carbon dioxide.

Key Terms

alveolus (al-ve′o-lus); plural: **alveoli** (al-ve′o-li)
bronchial tree (brong′ke-al tre)
carbaminohemoglobin (kar-bam′′ĭ-no-he′′mo-glo′bin)
carbonic anhydrase (kar-bon′ik an-hi′drās)
cellular respiration (sel′u-lar res′′pĭ-ra′shun)
expiration (ek′′spĭ-ra′shun)
glottis (glot′is)
hemoglobin (he′′mo-glo′bin)
hyperventilation (hi′′per-ven′′tĭ-la′shun)
inspiration (in′′spĭ-ra′shun)
oxyhemoglobin (ok′′se-he′′mo-glo′bin)
partial pressure (par′shil presh′ur)
pleural cavity (ploo′ral kav′ĭ-te)
respiratory capacities (re-spi′rah-to′′re kah-pas′ĭ-tēz)
respiratory center (re-spi′rah-to′′re sen′ter)
respiratory membrane (re-spi′rah-to′′re mem′brān)
respiratory volume (re-spi′rah-to′′re vol′ūm)
surfactant (ser-fak′tant)

16.1 Introduction

Cells require oxygen to break down nutrients, release energy, and produce ATP, and must excrete the carbon dioxide that results. Obtaining oxygen and removing carbon dioxide are the primary functions of the *respiratory system*. The respiratory system includes tubes that remove particles from (filter) incoming air and transport air into and out of the lungs. The system also includes many microscopic air sacs where gases are exchanged. The respiratory organs also trap particles from incoming air, help control the temperature and water content of incoming air, produce vocal sounds, and play important roles in the sense of smell and the regulation of blood pH.

The entire process of gas exchange between the atmosphere and body cells is called **respiration.** The events of respiration include: (1) movement of air into and out of the lungs—commonly called breathing or *ventilation;* (2) gas exchange between the blood and the air in the lungs (external respiration); (3) gas transport in blood between the lungs and body cells; and (4) gas exchange between the blood and the body cells (internal respiration). The process of oxygen utilization and carbon dioxide production at the cellular level is called **cellular respiration** (sel´u-lar res´´pĭ-ra´shun).

16.2 Organs of the Respiratory System

The organs of the respiratory system can be divided into two groups, or tracts. Those in the *upper respiratory tract* include the nose, nasal cavity, paranasal sinuses, and pharynx. Those in the *lower respiratory tract* include the larynx, trachea, bronchial tree, and lungs (fig. 16.1; reference plates 3, 4, 5, and 6, pp. 24–27).

Nose

Bone and cartilage support the **nose** internally. Its two *nostrils* are openings through which air can enter and leave the nasal cavity. Many internal hairs guard the nostrils, preventing entry of large particles carried in the air.

Nasal Cavity

The **nasal cavity** is a hollow space behind the nose (fig. 16.1). The **nasal septum,** composed of bone and cartilage, divides the nasal cavity into right and left portions. **Nasal conchae** are bones that curl out from the lateral walls of the nasal cavity on each side, dividing the cavity into passageways (fig. 16.2). Nasal conchae

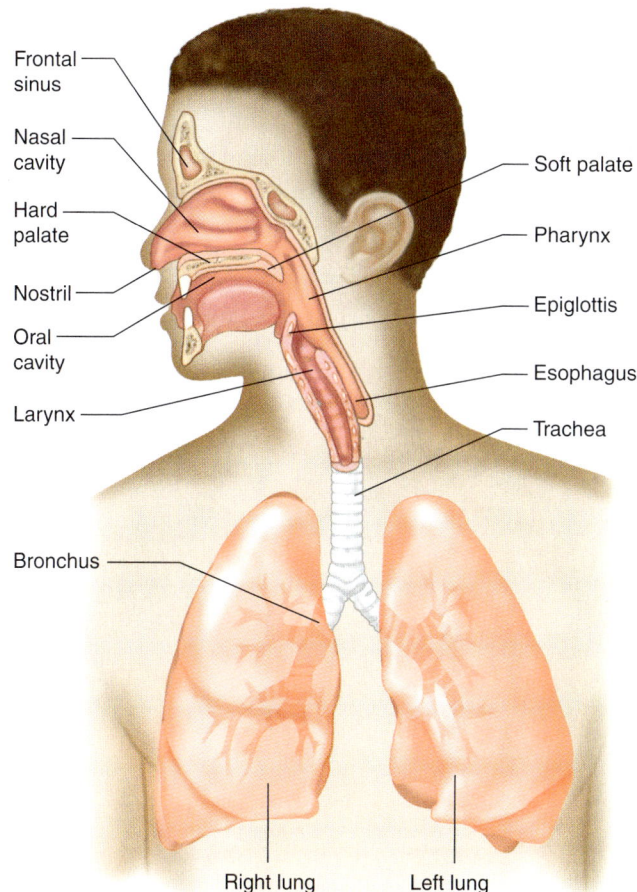

Figure 16.1
Organs of the respiratory system.

also support the mucous membrane that lines the nasal cavity and help increase its surface area.

The mucous membrane contains pseudostratified ciliated epithelium that is rich in mucus-secreting goblet cells (see chapter 5, p. 94). It also includes an extensive network of blood vessels, and as air passes over the membrane, heat leaves the blood and warms the air, adjusting the air's temperature to that of the body. In addition, incoming air is moistened as water evaporates from the mucous lining. The sticky mucus that the mucous membrane secretes entraps dust and other small particles entering with the air.

The nasal septum is usually straight at birth, although it sometimes bends as the result of a birth injury. It remains straight throughout early childhood, but as a person ages, the septum bends toward one side or the other. Such a *deviated septum* may obstruct the nasal cavity, making breathing difficult.

As the cilia of the epithelial lining move, they push a thin layer of mucus and entrapped particles toward

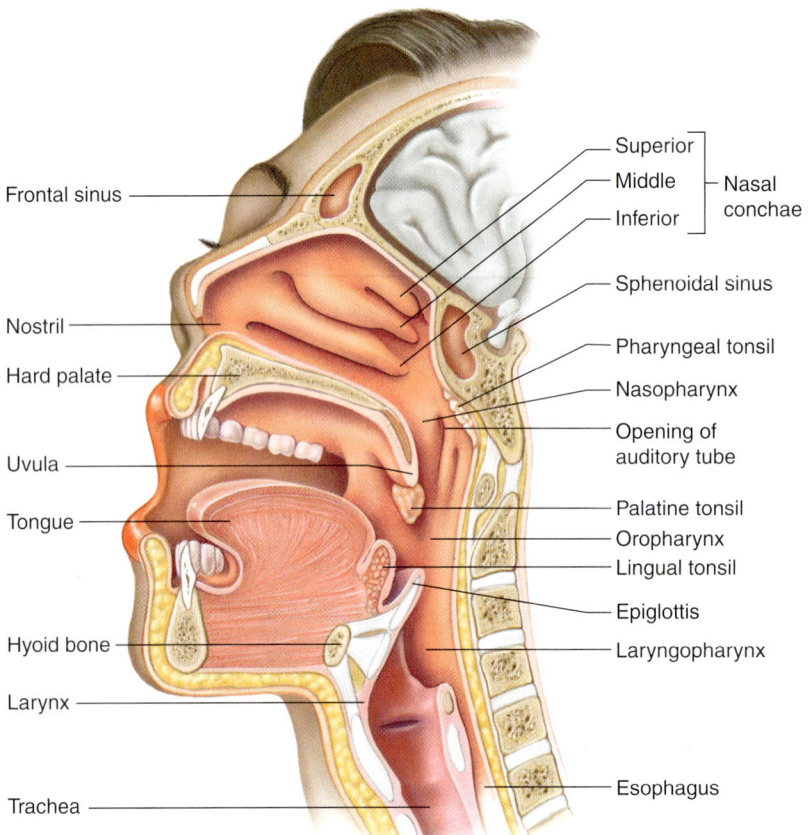

Figure 16.2
Major features of the upper respiratory tract.

the pharynx, where the mucus is swallowed (fig. 16.3). In the stomach, gastric juice destroys microorganisms in the mucus.

 A spore of the bacterium that causes anthrax is only half a micron wide. When spores are coated with powder to create a "bioweapon," they are still small enough to bypass the hairs and mucus in the nose, reaching the lungs, where they cause deadly inhalation anthrax. The bacteria release a toxin that causes death.

CHECK YOUR RECALL

1. What is respiration?
2. Which organs constitute the respiratory system?
3. What are the functions of the mucous membrane that lines the nasal cavity?

Paranasal Sinuses

Recall from chapter 7 (p. 136) that the **paranasal sinuses** are air-filled spaces located within the *maxillary, frontal, ethmoid,* and *sphenoid bones* of the skull and opening into the nasal cavity. Mucous membranes line the sinuses and are continuous with the lining of the nasal cavity. The paranasal sinuses reduce the weight of the skull and are resonant chambers that affect the quality of the voice.

 A painful sinus headache can result from blocked drainage caused by an infection or allergic reaction.

CHECK YOUR RECALL

1. Where are the paranasal sinuses located?
2. What are the functions of the paranasal sinuses?

Pharynx

The **pharynx** (far´inks) or throat is behind the oral cavity and between the nasal cavity and larynx (see fig. 16.1). It is a passageway for food traveling from the oral cavity to the esophagus and for air passing between the nasal cavity and larynx. It also helps produce the sounds of speech. Chapter 15 (p. 400) describes the subdivisions of the pharynx—nasopharynx, oropharynx, and laryngopharynx (see fig. 16.2).

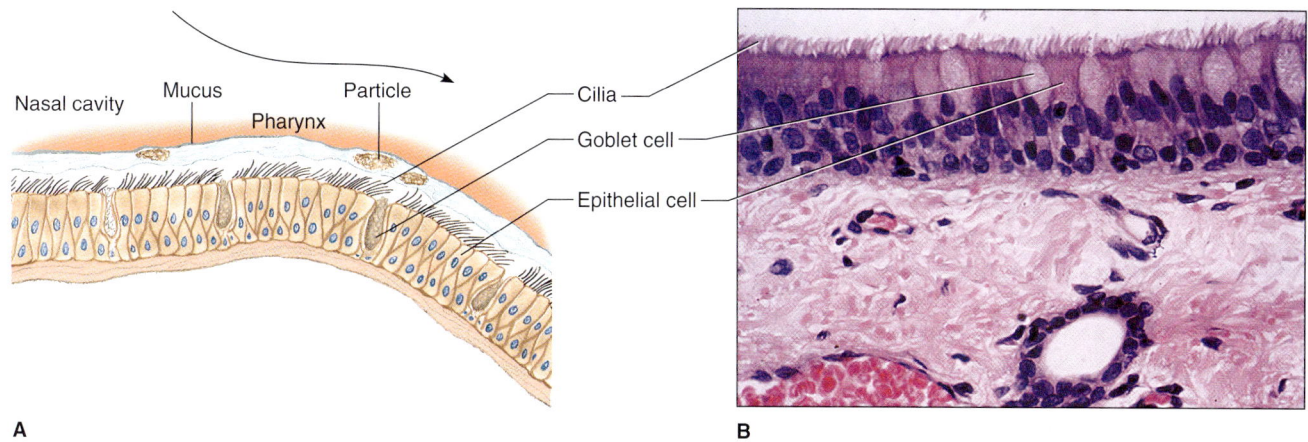

Figure 16.3
Mucus movement in the respiratory tract. (A) Cilia move mucus and trapped particles from the nasal cavity to the pharynx. (B) Micrograph of ciliated epithelium in the respiratory tract (275×).

Larynx

The **larynx** (lar′inks) is an enlargement in the airway at the top of the trachea and below the pharynx. It conducts air in and out of the trachea and prevents foreign objects from entering the trachea. It also houses the *vocal cords*.

The larynx is composed of a framework of muscles and cartilages bound by elastic tissue. The largest of the cartilages are the *thyroid* ("Adam's apple"), *cricoid,* and *epiglottic cartilages* (fig. 16.4).

Inside the larynx, two pairs of horizontal folds composed of muscle tissue and connective tissue with a covering of mucous membrane extend inward from the lateral walls. The upper folds are called *false vocal cords* because they do not produce sounds (fig. 16.5A). Muscle fibers within these folds help close the airway during swallowing.

The lower folds of muscle tissue and elastic fibers are the *true vocal cords*. Air forced between these vocal cords causes the cords to vibrate from side to side, which generates sound waves. Changing the shapes of the pharynx and oral cavity and using the tongue and lips transform these sound waves into words.

Contracting or relaxing muscles that alter the tension on the vocal cords controls the pitch (musical tone) of a sound. Increasing tension raises pitch, and decreasing tension lowers pitch. The intensity (loudness) of a sound reflects the force of air passing through the vocal folds. Stronger blasts of air produce louder sound; weaker blasts produce softer sound.

D amage to the nerves (recurrent laryngeal nerves) that supply the laryngeal muscles can alter the quality of a person's voice. These nerves pass through the neck as parts of the vagus nerves, and they can be injured by trauma or surgery to the neck or thorax. Nodules or other growths on the margins of the vocal folds that interfere with the free flow of air over the folds can also cause vocal problems. Surgery can remove such lesions.

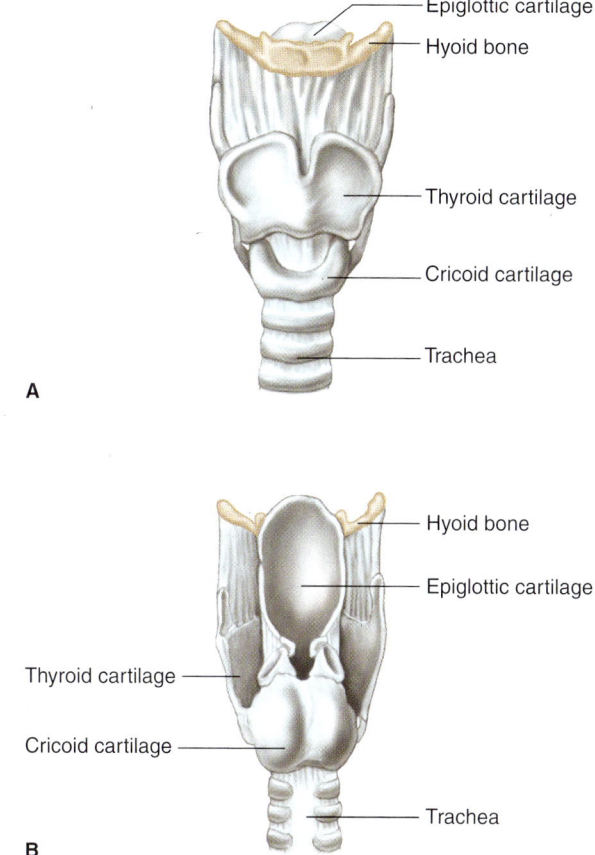

Figure 16.4
Larynx. (A) Anterior and (B) posterior views of the larynx.

During normal breathing, the vocal cords are relaxed, and the opening between them, called the **glottis** (glot′is), is a triangular slit (fig. 16.5). However, when food or liquid is swallowed, muscles within the false vocal cords close the glottis, which prevents food or liquid from entering the trachea.

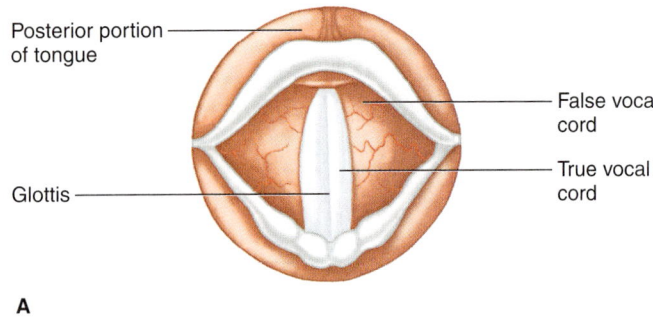

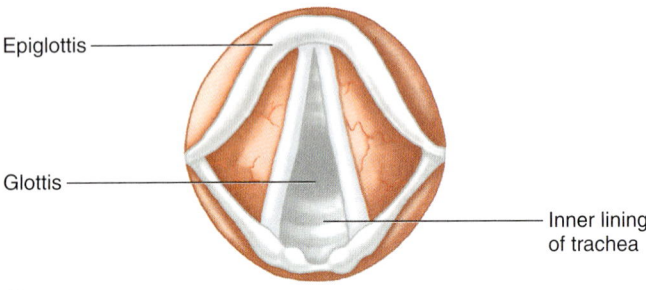

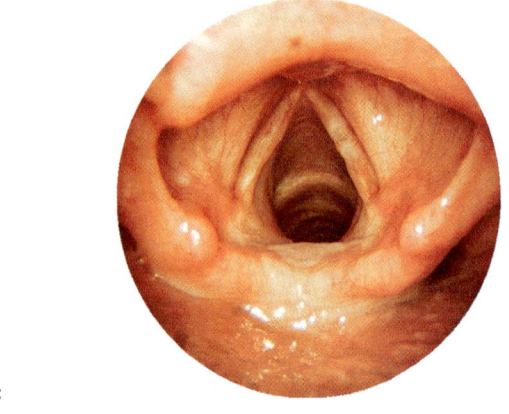

Figure 16.5
The vocal cords as viewed from above with the glottis (A) closed and (B) open. (C) Photograph of the glottis and vocal folds.

The epiglottic cartilage supports a flaplike structure called the **epiglottis.** This structure usually stands upright and allows air to enter the larynx. During swallowing, however, the larynx rises, and the epiglottis presses downward to partially cover the opening into the larynx. This helps prevent foods and liquids from entering the air passages (see chapter 15, p. 401).

L aryngitis—hoarseness or lack of voice—occurs when the mucous membrane of the larynx becomes inflamed and swollen due to an infection or an irritation from inhaled vapors, and prevents the vocal cords from vibrating as freely as before. Laryngitis is usually mild, but may be dangerous if swollen tissues obstruct the airway and interfere with breathing. Inserting a tube (endotracheal tube) into the trachea through the nose or mouth can restore the passageway until the inflammation subsides.

CHECK YOUR RECALL

1. Describe the structure of the larynx.
2. How do the vocal cords produce sounds?
3. What is the function of the glottis? The epiglottis?

Trachea

The **trachea** (tra´ke-ah) or windpipe is a flexible, cylindrical tube about 2.5 centimeters in diameter and 12.5 centimeters in length (fig. 16.6). It extends downward in front of the esophagus and into the thoracic cavity, where it splits into right and left bronchi.

A ciliated mucous membrane with many goblet cells lines the trachea's inner wall. This membrane filters incoming air and moves entrapped particles upward into the pharynx, where the mucus can be swallowed.

Within the tracheal wall are about twenty C-shaped pieces of hyaline cartilage, one above the other. The open ends of these incomplete rings are directed posteriorly, and smooth muscle and connective tissues fill the gaps between the ends. These cartilaginous rings prevent the trachea from collapsing and blocking the airway. The soft tissues that complete the rings in the back allow the nearby esophagus to expand as food moves through it to the stomach.

Bronchial Tree

The **bronchial tree** (brong´ke-al tre) consists of branched airways leading from the trachea to the microscopic air sacs in the lungs (fig. 16.7). Its branches begin with the right and left **primary bronchi,** which arise from the trachea at the level of the fifth thoracic vertebra.

A short distance from its origin, each primary bronchus divides into secondary bronchi, which in turn branch into finer and finer tubes. Among these smaller tubes are **bronchioles** that continue to divide, giving rise to very thin tubes called **alveolar ducts.** These ducts lead to thin-walled outpouchings called **alveolar sacs.** Alveolar sacs are clusters of smaller microscopic air sacs called **alveoli,** (al-ve´o-li; singular, al-ve´o-lus) which lie within capillary networks (figs. 16.8 and 16.9).

The structure of a bronchus is similar to that of the trachea, but as finer and finer branching tubes appear, the amount of cartilage in the walls decreases and finally disappears in the bronchioles. As the cartilage diminishes, a layer of smooth muscle surrounding the tube becomes more prominent. This muscular layer remains even in the smallest bronchioles, but only a few muscle fibers are in the alveolar ducts.

The branches of the bronchial tree are air passages whose mucous membranes filter incoming air and distribute the air to alveoli throughout the lungs. The alveoli provide a large surface area of thin simple squamous

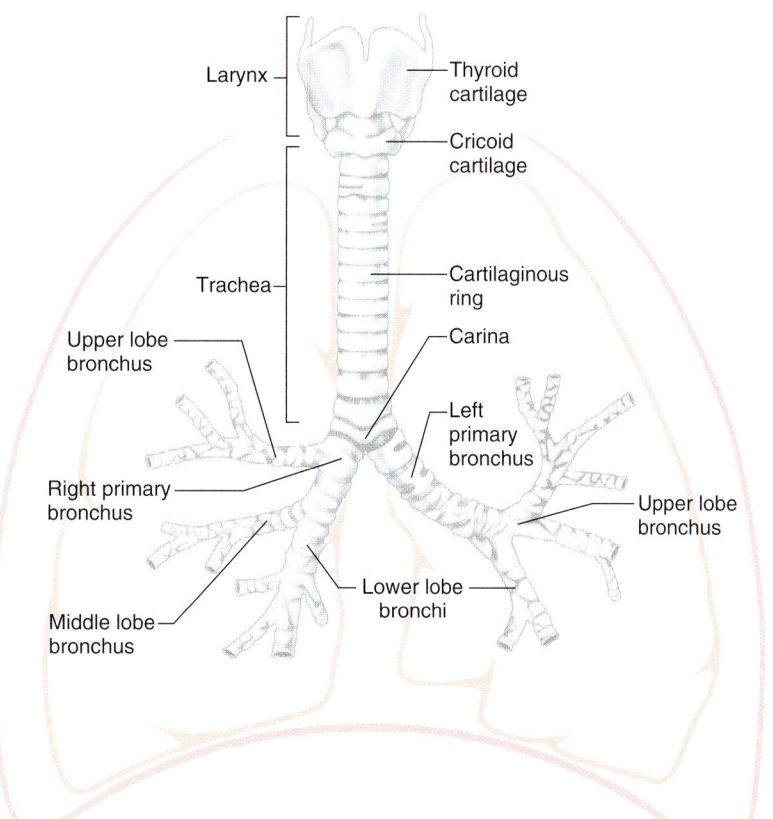

Figure 16.6
The trachea transports air between the larynx and the bronchi.

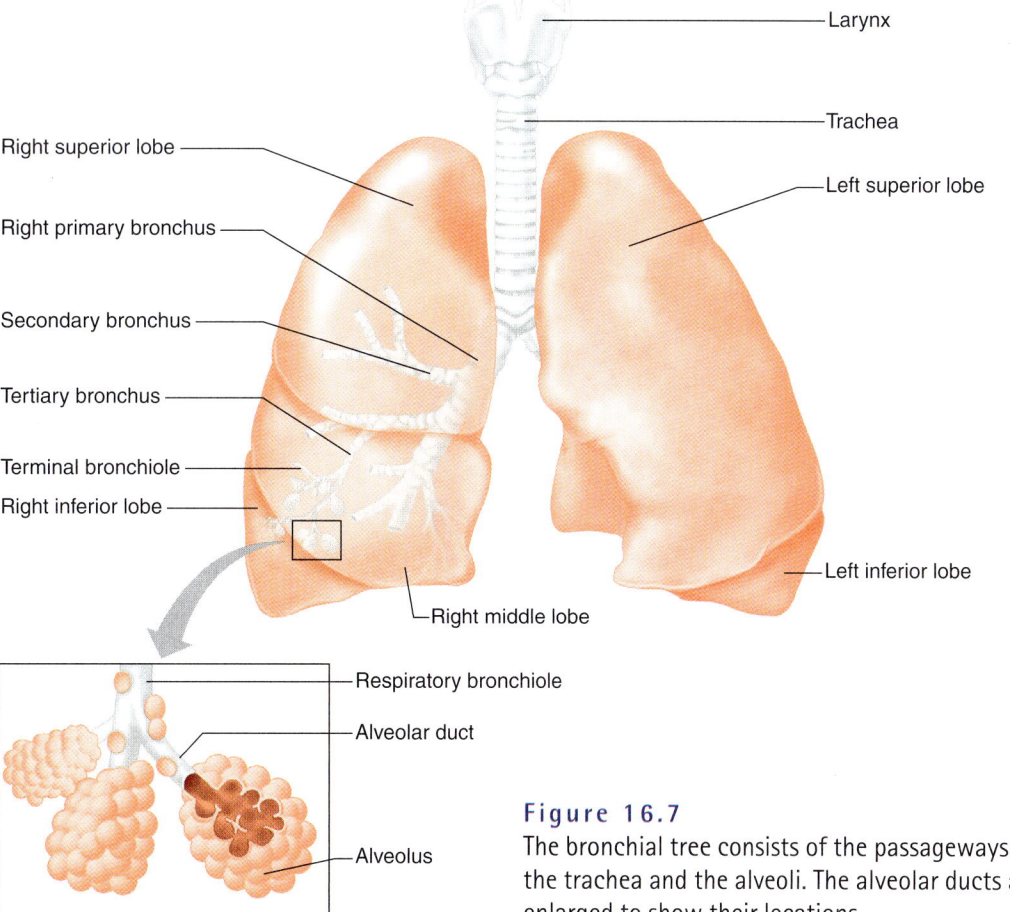

Figure 16.7
The bronchial tree consists of the passageways that connect the trachea and the alveoli. The alveolar ducts and alveoli are enlarged to show their locations.

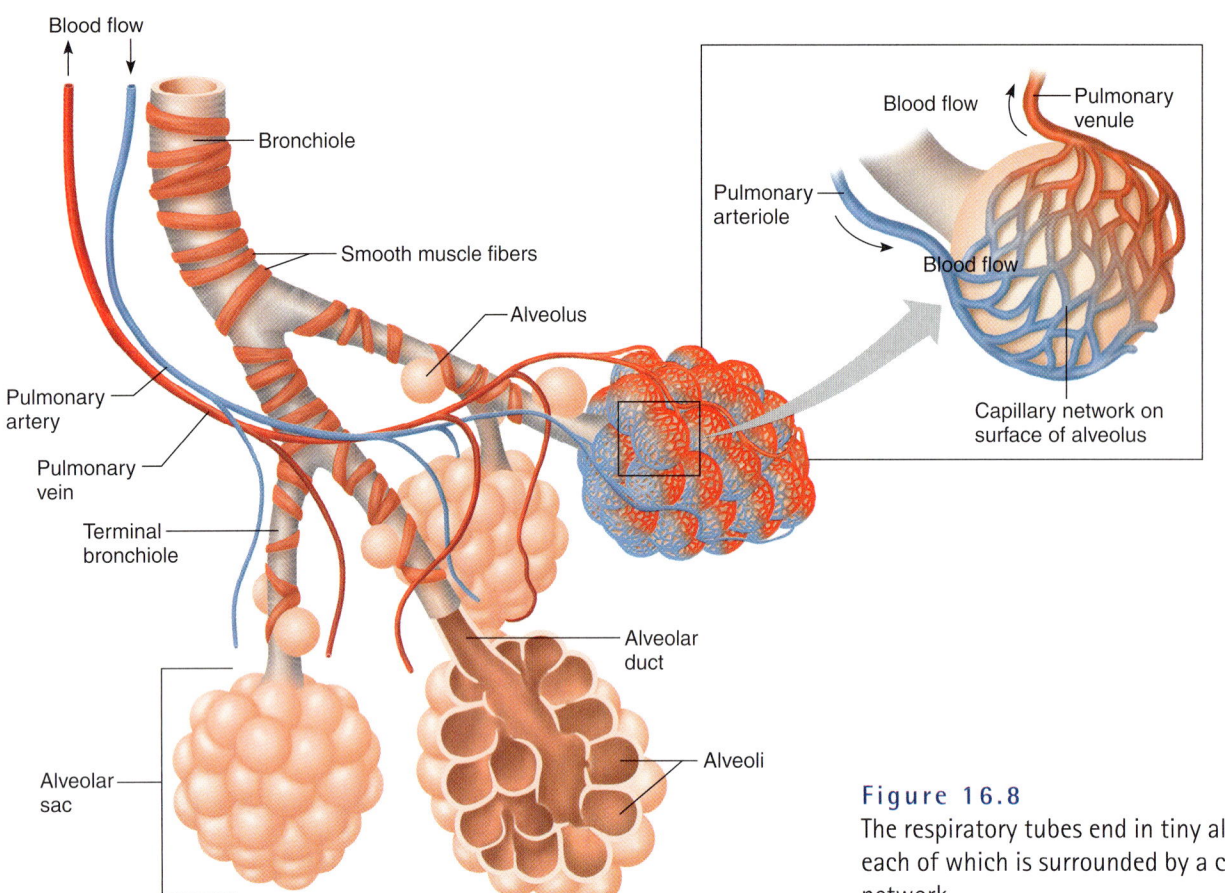

Figure 16.8
The respiratory tubes end in tiny alveoli, each of which is surrounded by a capillary network.

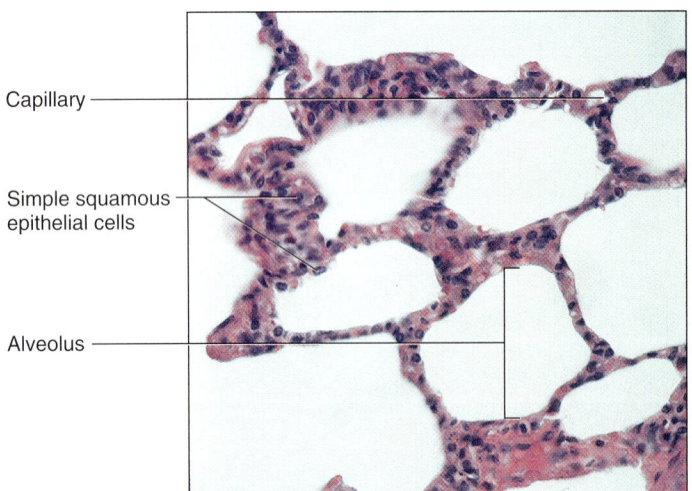

Figure 16.9
Light micrograph of alveoli (250×).

epithelial cells through which gases can easily be exchanged. Oxygen diffuses through alveolar walls and enters the blood in nearby capillaries, and carbon dioxide diffuses from the blood through the walls and enters alveoli (fig. 16.10).

An adult lung has about 300 million alveoli, providing a total surface area nearly half the size of a tennis court.

CHECK YOUR RECALL

1. What is the function of the cartilaginous rings in the tracheal wall?
2. Describe the bronchial tree.
3. Explain how gases are exchanged in the alveoli.

Lungs

The **lungs** are soft, spongy, cone-shaped organs in the thoracic cavity (see fig. 16.1 and reference plates 4 and 5, pp. 25–26). The mediastinum separates the right and left lungs medially, and the diaphragm and thoracic cage enclose them.

Each lung occupies most of the thoracic space on its side. A bronchus and some large blood vessels suspend each lung in the cavity. These tubular structures enter the lung on its medial surface. A layer of serous membrane, the **visceral pleura,** firmly attaches to each

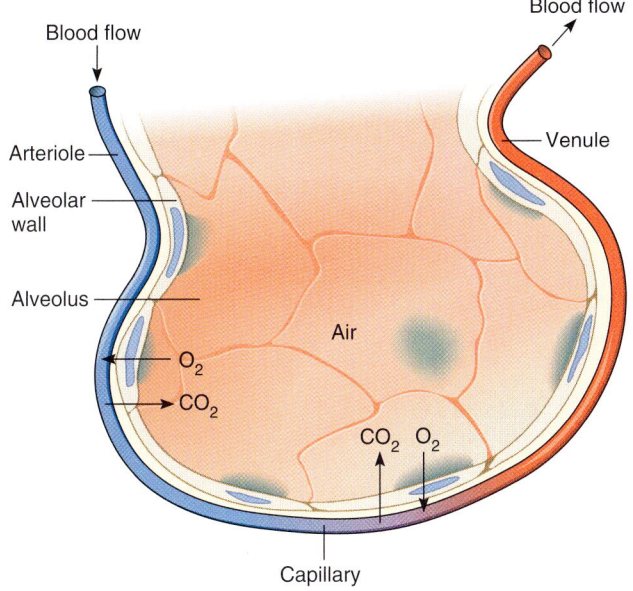

Figure 16.10
Oxygen (O_2) diffuses from air within the alveolus into the capillary, while carbon dioxide (CO_2) diffuses from blood within the capillary into the alveolus.

lung surface and folds back to become the **parietal pleura.** The parietal pleura, in turn, forms part of the mediastinum and lines the inner wall of the thoracic cavity (fig. 16.11).

No significant space exists between the visceral and parietal pleurae, but the potential space between them is called the **pleural cavity** (ploo´ral kav´ĭ-te). It contains a thin film of serous fluid that lubricates adjacent pleural surfaces, reducing friction as they move against one another during breathing. This fluid also helps hold the pleural membranes together, as explained in the next section.

The right lung is larger than the left one and is divided into three lobes. The left lung has two lobes (see fig. 16.1).

A major branch of the bronchial tree supplies each lobe. A lobe also has connections to blood and lymphatic vessels and lies within connective tissues. Thus, a lung includes air passages, alveoli, blood vessels, connective tissues, lymphatic vessels, and nerves. Table 16.1 summarizes the characteristics of the major parts of the respiratory system.

> **CHECK YOUR RECALL**
>
> 1. Where are the lungs located?
> 2. What is the function of serous fluid within the pleural cavity?
> 3. What kinds of structures make up a lung?

16.3 Breathing Mechanism

Breathing, or ventilation, is the movement of air from outside the body into and out of the bronchial tree and alveoli. The actions providing these air movements are termed **inspiration** (in´´spĭ-ra´shun), or inhalation, and **expiration** (ek´´spĭ-ra´shun), or exhalation.

Inspiration

Atmospheric pressure due to the weight of air is the force that moves air into the lungs. At sea level, this pressure is sufficient to support a column of mercury

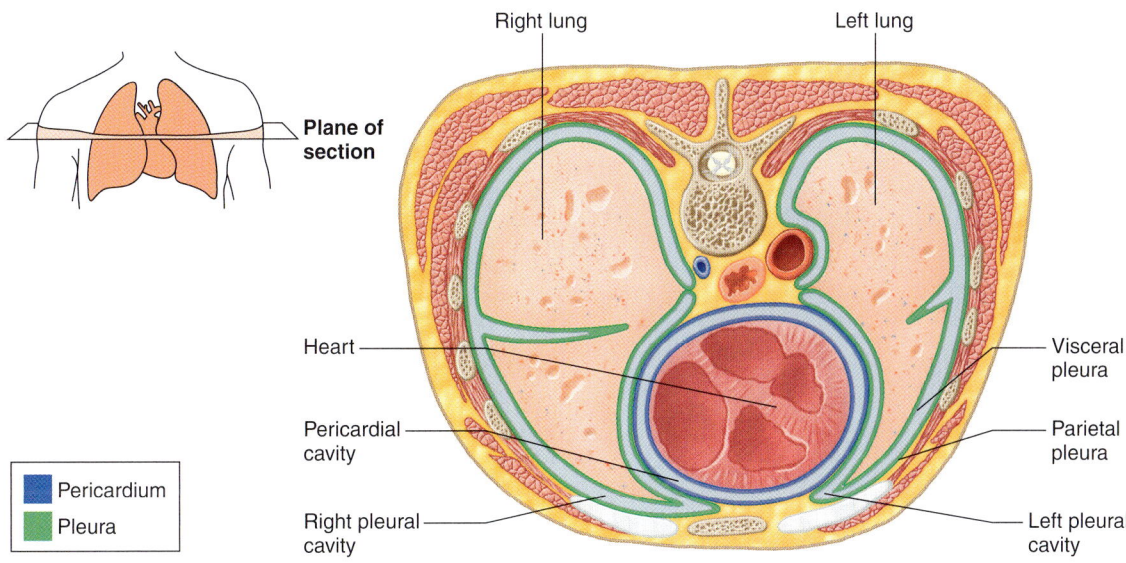

Figure 16.11
The potential spaces between the pleural membranes, called the left and right pleural cavities, are shown here as actual spaces.

TABLE 16.1 — PARTS OF THE RESPIRATORY SYSTEM

PART	DESCRIPTION	FUNCTION
Nose	Part of face centered above mouth, in and below space between eyes	Nostrils provide entrance to nasal cavity; internal hairs begin to filter incoming air
Nasal cavity	Hollow space behind nose	Conducts air to pharynx; mucous lining filters, warms, and moistens incoming air
Paranasal sinuses	Hollow spaces in certain skull bones	Reduce weight of skull; serve as resonant chambers
Pharynx	Chamber behind oral cavity and between nasal cavity and larynx	Passageway for air moving from nasal cavity to larynx and for food moving from oral cavity to esophagus
Larynx	Enlargement at top of trachea	Passageway for air; prevents foreign objects from entering trachea; houses vocal cords
Trachea	Flexible tube that connects larynx with bronchial tree	Passageway for air; mucous lining continues to filter particles from incoming air
Bronchial tree	Branched tubes that lead from trachea to alveoli	Conducts air from trachea to alveoli; mucous lining continues to filter incoming air
Lungs	Soft, cone-shaped organs that occupy a large portion of the thoracic cavity	Contain air passages, alveoli, blood vessels, connective tissues, lymphatic vessels, and nerves of the lower respiratory tract

about 760 millimeters (mm) high in a tube. Thus, normal air pressure is equal to 760 mm of mercury (Hg).

Air pressure is exerted on all surfaces in contact with the air, and because people breathe air, the inside surfaces of their lungs also are subjected to pressure. The pressures on the inside of the lungs and alveoli and on the outside of the thoracic wall are about the same.

If the pressure inside the lungs and alveoli decreases, atmospheric pressure will push outside air into the airways. That is what happens during normal inspiration. Impulses carried on the phrenic nerves, which are associated with the cervical plexuses (see chapter 9, p. 240), stimulate muscle fibers in the dome-shaped *diaphragm* below the lungs to contract. The diaphragm moves downward, the thoracic cavity enlarges, and the pressure within the alveoli falls to about 2 mm Hg below that of atmospheric pressure. In response, atmospheric pressure forces air into the airways (fig. 16.12).

Pressure and volume are related in an opposite or inverse way. For example, if we pull back on the plunger of a syringe, the volume inside the barrel increases, causing the air pressure inside to decrease. Outside air is then pushed into the syringe by atmo-

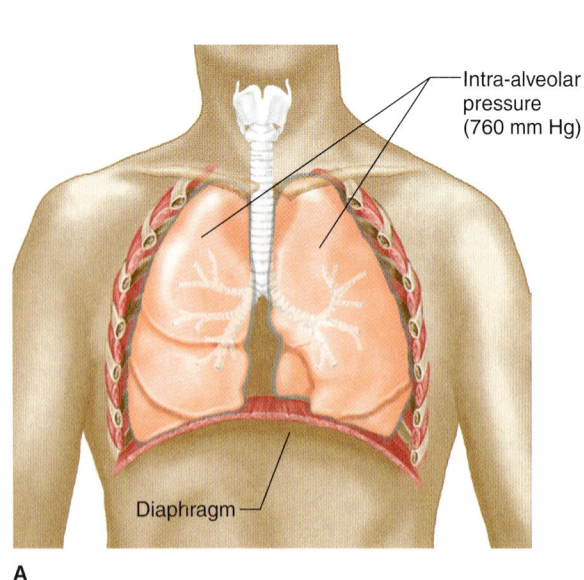
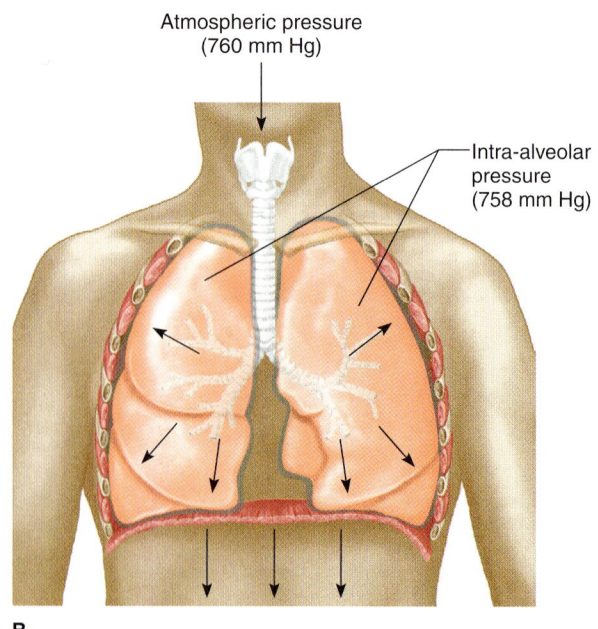

Figure 16.12

Normal inspiration. (*A*) Prior to inspiration, the intra-alveolar pressure is 760 mm Hg. (*B*) The intra-alveolar pressure decreases to about 758 mm Hg as the thoracic cavity enlarges, and atmospheric pressure forces air into the airways.

spheric pressure. In contrast, if we push on the plunger of a syringe, the volume inside the syringe is reduced, but the pressure inside increases, forcing air out into the atmosphere. The movement of air into and out of the lungs occurs in much the same way.

While the diaphragm is contracting and moving downward, the *external (inspiratory) intercostal muscles* between the ribs may be stimulated to contract. This raises the ribs and elevates the sternum, enlarging the thoracic cavity even more. As a result, the pressure inside is reduced further, and the relatively greater atmospheric pressure forces even more air into the airways.

Lung expansion in response to movements of the diaphragm and chest wall depends on movements of the pleural membranes. Any separation of the pleural membranes decreases pressure in the intrapleural space, tending to hold these membranes together. In addition, only a thin film of serous fluid separates the parietal pleura on the inner wall of the thoracic cavity from the visceral pleura attached to the surface of the lungs. The water molecules in this fluid greatly attract the pleural membranes and each other, helping to hold the moist surfaces of the pleural membranes tightly together, much as a wet coverslip sticks to a microscope slide. As a result of these factors, when the intercostal muscles move the thoracic wall upward and outward, the parietal pleura moves too, and the visceral pleura follows it. This helps expand the lung in all directions.

Although the moist pleural membranes play a role in expansion of the lungs, the moist inner surfaces of the alveoli have the opposite effect. Here, the attraction of water molecules creates a force called **surface tension** that makes it difficult to inflate the alveoli and may actually cause them to collapse. Certain alveolar cells, however, synthesize a mixture of lipoproteins called **surfactant** (ser-fak´tant). Surfactant, which is secreted continuously into alveolar air spaces, reduces the alveoli's tendency to collapse, especially when lung volumes are low, and makes it easier for inspiratory efforts to inflate the alveoli.

Surfactant is particularly important in the minutes after birth, when the newborn's lungs inflate for the first time. Premature infants often suffer respiratory distress syndrome because they do not produce sufficient surfactant. To help many of these newborns survive, physicians drip synthetic surfactant into the tiny lungs through an endotracheal tube. A ventilator machine especially geared to an infant's size assists breathing.

If a person needs to take a deeper than normal breath, the diaphragm and external intercostal muscles contract more forcefully. Additional muscles, such as the pectoralis minor and sternocleidomastoid, can also pull the thoracic cage farther upward and outward, enlarging the thoracic cavity and decreasing internal pressure (fig. 16.13).

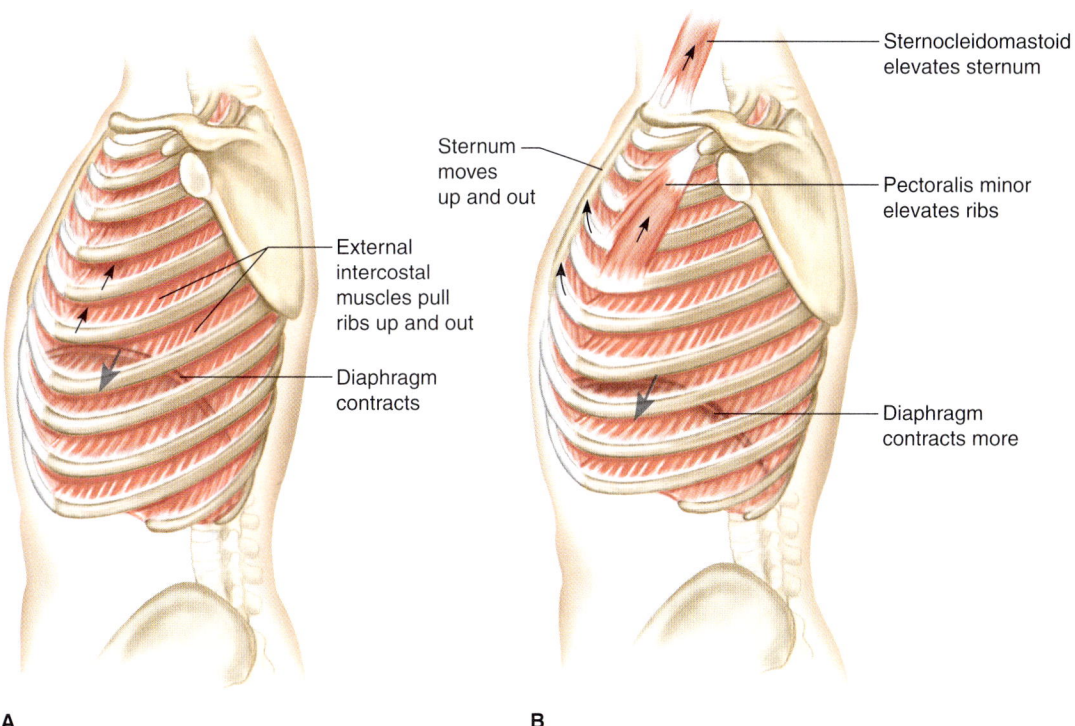

Figure 16.13
Maximal inspiration. (*A*) Shape of the thorax at the end of normal inspiration. (*B*) Shape of the thorax at the end of maximal inspiration, aided by contraction of the sternocleidomastoid and pectoralis minor muscles.

Topic of Interest

EMPHYSEMA AND LUNG CANCER

Emphysema is a progressive, degenerative disease that destroys alveolar walls. As a result, clusters of small air sacs merge to form larger chambers, which drastically decreases the surface area of the respiratory membrane and thereby reduces the volume of gases that can be exchanged through the membrane. Alveolar walls lose some of their elasticity, and capillary networks associated with the alveoli diminish (fig. 16A).

Loss of tissue elasticity makes it increasingly difficult for a person with emphysema to force air out of the lungs because normal expiration involves the passive elastic recoil of inflated tissues. Consequently, the person must exert abnormal muscular effort to exhale.

Emphysema may develop in response to prolonged exposure to respiratory irritants, such as those in tobacco smoke and polluted air. The disease may also result from an inherited enzyme deficiency.

Lung cancer, like other cancers, is the uncontrolled growth of abnormal cells that rob normal cells of nutrients and oxygen, eventually crowding them out. Some cancerous growths in the lungs result secondarily from cancer cells that have spread (metastasized) from other parts of the body, such as the breasts, intestines, liver, or kidneys. Cancers that begin in the lungs are called *primary pulmonary cancers*. These may arise from epithelia, connective tissue, or blood cells. The most common form originates from epithelium and is called *bronchogenic carcinoma*. This type of cancer is a response to irritation, such as prolonged exposure to tobacco smoke (fig. 16B). Susceptibility to primary pulmonary cancers may be inherited.

Once cancer cells appear, they grow into masses that obstruct air passages and reduce gas exchange. Furthermore, bronchogenic carcinoma can spread quickly and establish secondary cancers in the lymph nodes, liver, bones, brain, or kidneys.

Lung cancer is difficult to control. Usually, it is treated with surgery, ionizing radiation, and drugs (chemotherapy). Despite treatment, however, the survival rate among lung cancer patients remains low.

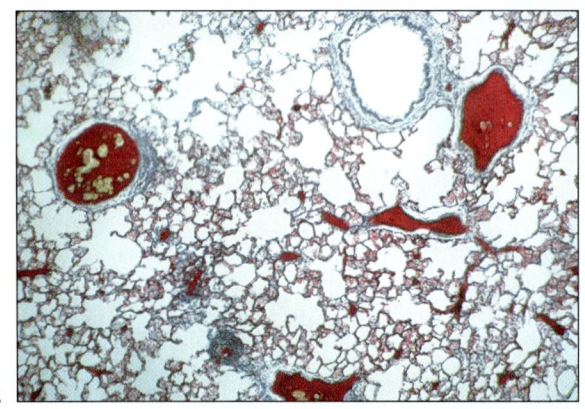
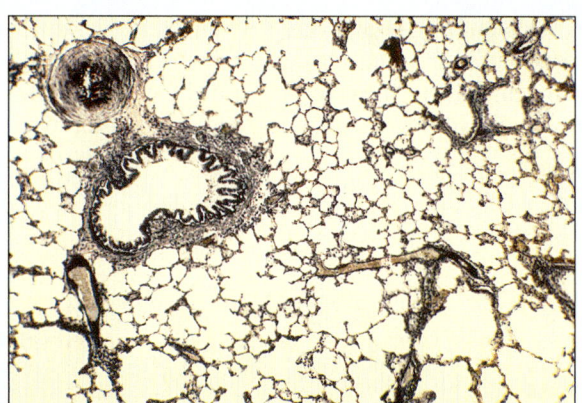

Figure 16A

Lung tissue. (A) Normal lung tissue. (B) As emphysema develops, the alveoli tend to merge, forming larger chambers.

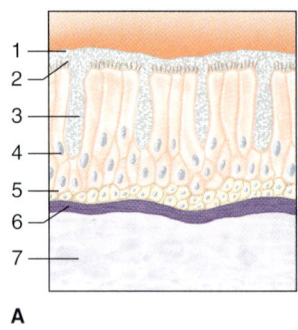
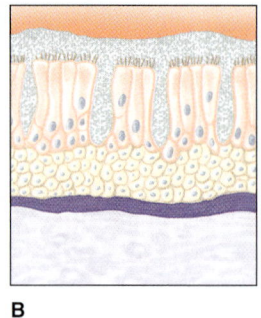
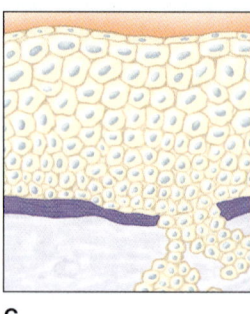

Figure 16B

Lung cancer usually starts in the lining (epithelium) of a bronchus. (A) The normal lining shows (4) columnar cells with (2) hairlike cilia, (3) goblet cells that secrete (1) mucus, and (5) basal cells from which new columnar cells arise. (6) A basement membrane separates the epithelial cells from (7) the underlying connective tissue. (B) In the first stage of lung cancer, the basal cells divide repeatedly. The goblet cells secrete excess mucus, and the cilia are less efficient in moving the heavy mucus secretion. (C) Continued division of basal cells displaces the columnar and goblet cells. The basal cells penetrate the basement membrane and invade the deeper connective tissue.

The first breath is the toughest. A newborn must use 20 times the energy to take the first breath as is necessary for subsequent breaths. This is because each of the millions of alveoli start out only partially inflated.

Expiration

The forces responsible for normal expiration come from the *elastic recoil* of tissues and from surface tension. The lungs and thoracic wall contain considerable elastic tissue, and lung expansion during inspiration stretches this tissue. Also, as the diaphragm lowers, it compresses the abdominal organs beneath it. As the diaphragm and external intercostal muscles relax following inspiration, the elastic tissues cause the lungs and thoracic cage to recoil and return to their original shapes. Similarly, the abdominal organs spring back into their previous shapes, pushing the diaphragm upward (fig. 16.14A). At the same time, the surface tension that develops between the moist surfaces of the alveolar linings decreases the diameters of the alveoli. Each of these factors increases alveolar pressure about 1 mm Hg above atmospheric pressure, so that the air inside the lungs is forced out through respiratory passages. Thus, normal resting expiration is a passive process.

Because low pressure and wet surfaces hold the visceral and parietal pleural membranes together, no actual space normally exists in the pleural cavity between them. A puncture in the thoracic wall, however, allows atmospheric air to enter the pleural cavity and create a real space between the membranes. This condition, called *pneumothorax,* may collapse the lung on the affected side because of the lung's elasticity. The condition of a collapsed lung is called *atelectasis.*

If a person needs to exhale more air than normal, the posterior *internal (expiratory) intercostal muscles* can be contracted (fig. 16.14B). These muscles pull the ribs and sternum downward and inward, increasing the pressure in the lungs. Also, the *abdominal wall muscles,* including the external and internal obliques, transversus abdominis, and rectus abdominis, can squeeze the abdominal organs inward (see fig. 8.19, p. 192). Thus, the abdominal wall muscles can increase pressure in the abdominal cavity and force the diaphragm still higher against the lungs. These actions squeeze additional air out of the lungs.

 **CHECK YOUR RECALL**

1. Describe the events in inspiration.
2. How does expansion of the chest wall expand the lungs during inspiration?
3. What forces cause normal expiration?

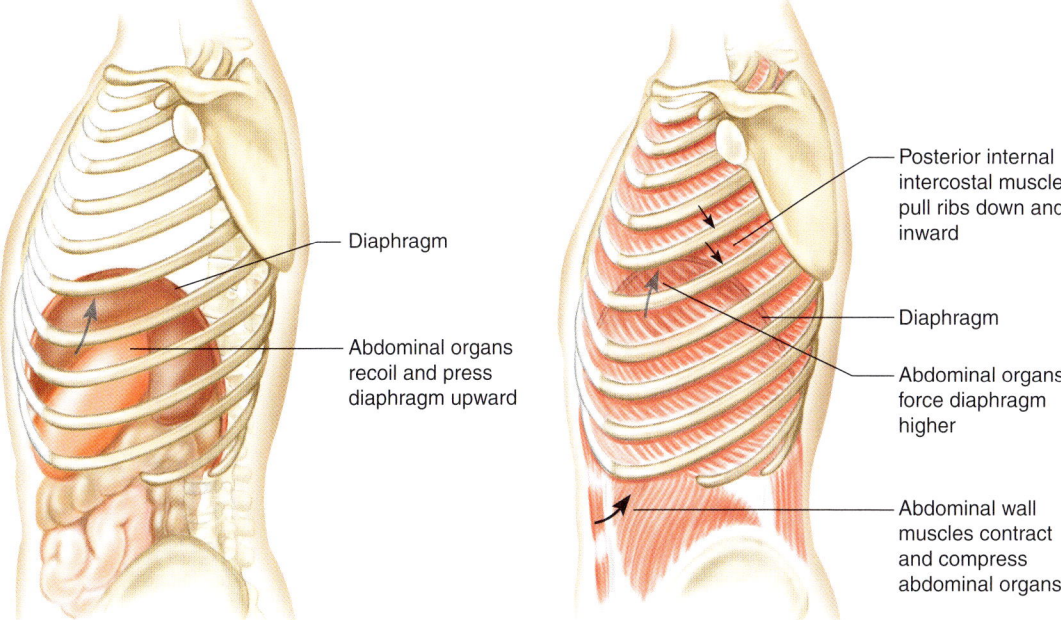

Figure 16.14
Expiration. (*A*) Normal resting expiration is due to elastic recoil of lung tissues, the thoracic wall, and the abdominal organs. (*B*) Contraction of abdominal wall muscles and posterior internal intercostal muscles aids maximal expiration.

Air movements that occur in addition to breathing are called *nonrespiratory movements*. They are used to clear air passages, as in coughing and sneezing, or to express emotion, as in laughing and crying.

Nonrespiratory movements usually result from *reflexes,* although sometimes they are initiated voluntarily. A *cough,* for example, can be produced through conscious effort or may be triggered by the presence of a foreign object in an air passage.

The act of coughing involves taking a deep breath, closing the glottis, and forcing air upward from the lungs against the closure. Then the glottis is suddenly opened, and a blast of air is forced upward from the lower respiratory tract. Usually, this rapid rush of air removes the substance that triggered the reflex.

A *sneeze* is much like a cough, but it clears the upper respiratory passages rather than the lower ones. This reflex is usually initiated by a mild irritation in the lining of the nasal cavity, and in response, a blast of air is forced up through the glottis. This time, the air is directed into the nasal passages by depressing the uvula, thus closing the opening between the pharynx and the oral cavity. A sneeze can propel a particle out of the nose at 200 miles per hour.

Laughing involves taking a breath and releasing it in a series of short expirations. *Crying* consists of very similar movements, and sometimes it is necessary to note a person's facial expression in order to distinguish laughing from crying.

A *hiccup* is caused by sudden inspiration due to a spasmodic contraction of the diaphragm while the glottis is closed. Air striking the vocal folds causes the sound of the hiccup. We do not know the function, if any, of hiccups.

Yawning may aid respiration by providing an occasional deep breath. During normal, quiet breathing, not all of the alveoli are ventilated, and some blood may pass through the lungs without becoming well oxygenated. This low blood oxygen concentration somehow triggers the yawn reflex, prompting a very deep breath that ventilates more of the alveoli.

Respiratory Air Volumes and Capacities

Different intensities in breathing move different volumes of air in or out of the lungs. *Spirometry* measures such air volumes, revealing four distinct **respiratory volumes** (re-spi´rah-to´´re vol´ūmz).

One inspiration plus the following expiration is called a **respiratory cycle.** The volume of air that enters (or leaves) during a single respiratory cycle is termed the **tidal volume.** About 500 milliliters (mL) of air enter during a normal, resting inspiration. Approximately the same volume leaves during a normal, resting expiration. Thus, the **resting tidal volume** is about 500 mL (fig. 16.15).

During forced inspiration, air in addition to the resting tidal volume enters the lungs. This extra volume is the **inspiratory reserve volume** (complemental air), and at maximum, it equals about 3,000 mL.

During forced expiration, the lungs can expel up to about 1,100 mL of air beyond the resting tidal volume. This quantity is called the **expiratory reserve volume** (supplemental air). Even after the most forceful expiration, however, about 1,200 mL of air remains in the lungs. This is called the **residual volume.**

Residual air remains in the lungs at all times, and consequently, newly inhaled air always mixes with air already in the lungs. This prevents the oxygen and carbon dioxide concentrations in the lungs from fluctuating greatly with each breath.

Combining two or more of the respiratory volumes yields four **respiratory capacities** (re-spi´rah-to´´re kah-pas´ĭ-tēz). Combining the inspiratory reserve volume (3,000 mL) with the tidal volume (500 mL) and the expiratory reserve volume (1,100 mL) gives the **vital capacity** (4,600 mL). This is the maximum amount of air a person can exhale after taking the deepest breath possible.

The tidal volume (500 mL) plus the inspiratory reserve volume (3,000 mL) gives the **inspiratory capacity** (3,500 mL), which is the maximum volume of air a

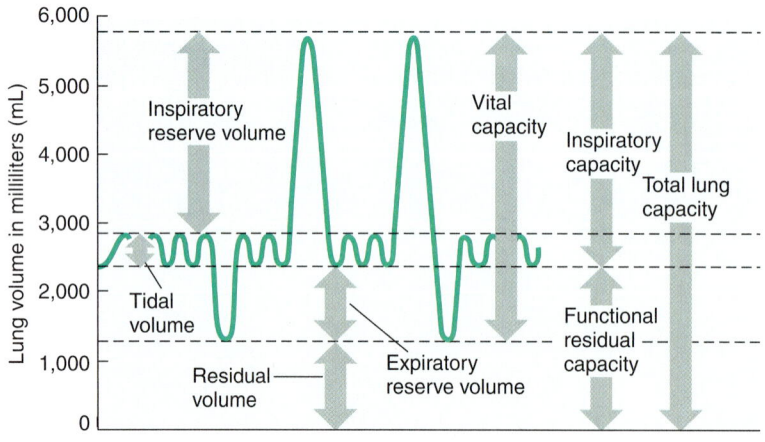

Figure 16.15
Respiratory volumes and capacities.

person can inhale following a resting expiration. Similarly, the expiratory reserve volume (1,100 mL) plus the residual volume (1,200 mL) equals the **functional residual capacity** (2,300 mL), which is the volume of air that remains in the lungs following a resting expiration.

The vital capacity plus the residual volume equals the **total lung capacity** (about 5,800 mL). This total varies with age, sex, and body size.

Some of the air that enters the respiratory tract during breathing does not reach the alveoli. This volume (about 150 mL) remains in the passageways of the trachea, bronchi, and bronchioles. Because gas is not exchanged through the walls of these passages, this air is said to occupy *anatomic dead space*. Table 16.2 summarizes the respiratory air volumes and capacities.

An instrument called a *spirometer* measures respiratory air volumes, except residual volume, which requires a special technique. Such measurements are used to evaluate the courses of emphysema, pneumonia, and lung cancer, conditions in which functional lung tissue is lost. These measurements may also track the progress of diseases such as bronchial asthma that obstruct air passages.

CHECK YOUR RECALL

1. What is tidal volume?
2. Distinguish between inspiratory and expiratory reserve volumes.
3. How is vital capacity determined?
4. How is total lung capacity calculated?

16.4 Control of Breathing

Normal breathing is a rhythmic, involuntary act that continues even when a person is unconscious. The respiratory muscles, however, are under voluntary control.

Respiratory Center

Groups of neurons that comprise the **respiratory center** (re-spi´rah-to´´re sen´ter) in the brain stem control both inspiration and expiration. These neurons are widely scattered throughout the pons and medulla oblongata (fig. 16.16). Two areas of the respiratory center are of special interest—the rhythmicity area of the medulla and the pneumotaxic area of the pons.

The **medullary rhythmicity area** includes two neuron groups that extend the length of the medulla oblongata. They are the dorsal respiratory group and the ventral respiratory group.

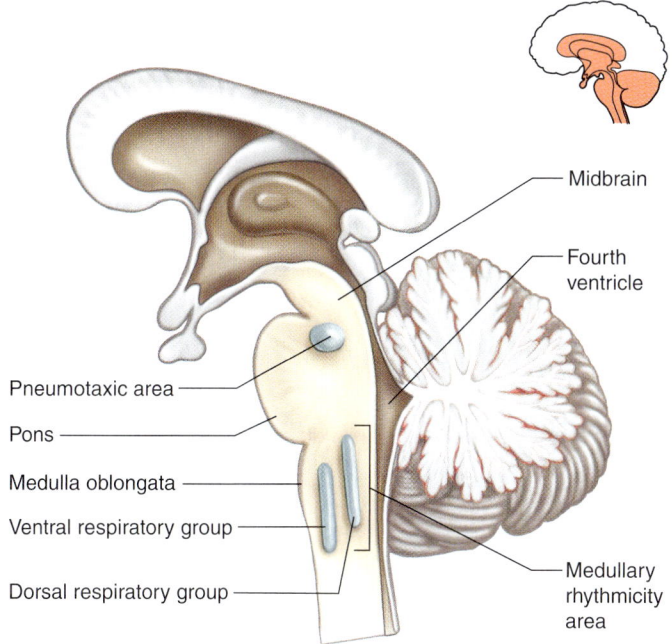

Figure 16.16
The respiratory center is in the pons and the medulla oblongata.

TABLE 16.2   RESPIRATORY AIR VOLUMES AND CAPACITIES

NAME	VOLUME*	DESCRIPTION
Tidal volume (TV)	500 mL	Volume moved in or out of lungs during respiratory cycle
Inspiratory reserve volume (IRV)	3,000 mL	Volume that can be inhaled during forced breathing in addition to tidal volume
Expiratory reserve volume (ERV)	1,100 mL	Volume that can be exhaled during forced breathing in addition to tidal volume
Residual volume (RV)	1,200 mL	Volume that remains in lungs even after maximal expiration
Inspiratory capacity (IC)	3,500 mL	Maximum volume of air that can be inhaled following exhalation of tidal volume: IC = TV + IRV
Functional residual capacity (FRC)	2,300 mL	Volume of air that remains in the lungs following exhalation of tidal volume: FRC = ERV + RV
Vital capacity (VC)	4,600 mL	Maximum volume of air that can be exhaled after taking the deepest breath possible: VC = TV + IRV + ERV
Total lung capacity (TLC)	5,800 mL	Total volume of air that the lungs can hold: TLC = VC + RV

*Values are typical for a tall, young adult.

The *dorsal respiratory group* controls the basic rhythm of inspiration. These neurons emit bursts of impulses that signal the diaphragm and other inspiratory muscles to contract. The impulses of each burst begin weakly, strengthen for about 2 seconds, and cease abruptly. The breathing muscles that contract in response to the impulses steadily increase the volume of air entering the lungs. The neurons remain inactive while expiration occurs passively, and then they emit another burst of inspiratory impulses, repeating the inspiration-expiration cycle.

The *ventral respiratory group* is quiet during normal breathing, but when more forceful breathing is required, some of these neurons generate impulses that increase inspiratory movements. Other neurons in the group activate muscles associated with forceful expiration (fig. 16.17).

The neurons in the **pneumotaxic area** of the pons continuously transmit impulses that inhibit the inspiratory bursts originating from the dorsal respiratory group. In this way, the pneumotaxic neurons control breathing rate. More specifically, when pneumotaxic inhibition is strong, the inspiratory bursts are shorter, and the breathing rate increases; when pneumotaxic inhibition is weak, the inspiratory bursts are longer, and the breathing rate decreases.

CHECK YOUR RECALL

1. Where is the respiratory center?
2. Describe how the respiratory center maintains a normal breathing pattern.
3. Explain how the breathing pattern may change.

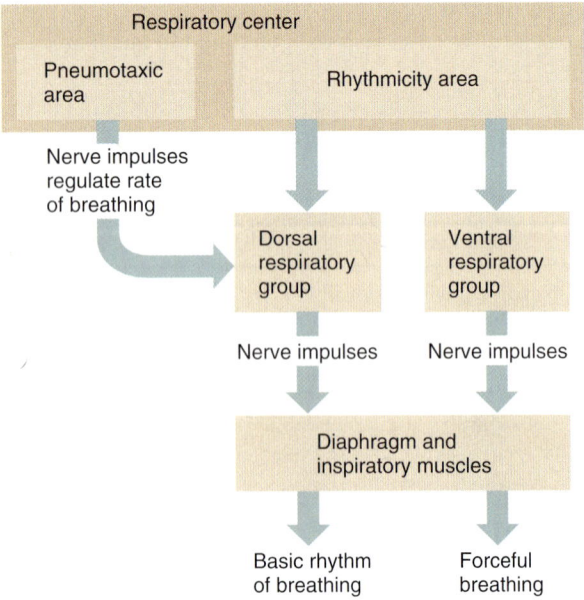

Figure 16.17
The medullary rhythmicity and pneumotaxic areas of the respiratory center control breathing.

Factors Affecting Breathing

The respiratory center affects breathing rate and depth, and so do certain chemicals in body fluids, the degree to which lung tissues stretch, and a person's emotional state. For example, *chemosensitive areas* (central chemoreceptors) within the respiratory center, located in the ventral portion of the medulla oblongata near the origins of the vagus nerves, sense changes in the cerebrospinal fluid (CSF) concentrations of carbon dioxide and hydrogen ions. If either of these concentrations rises, the central chemoreceptors signal the respiratory center, and respiratory rate and tidal volume increase. As a result of the increased ventilation, more carbon dioxide is exhaled, the blood and CSF concentrations of these chemicals fall, and breathing rate decreases.

Adding carbon dioxide to air can stimulate the rate and depth of breathing. Ordinary air is about 0.04% carbon dioxide. Inhaling air containing 4% carbon dioxide usually doubles the breathing rate.

Low blood oxygen has little direct effect on the central chemoreceptors associated with the respiratory center. Instead, *peripheral chemoreceptors* in specialized structures called the *carotid* and *aortic bodies* sense changes in blood oxygen concentration. Peripheral chemoreceptors are in the walls of certain large arteries (the carotid arteries and the aorta) in the neck and thorax (fig. 16.18). Stimulated peripheral chemoreceptors transmit impulses to the respiratory center, increasing breathing rate. However, blood oxygen concentration must be very low to trigger this mechanism. Thus, oxygen plays only a minor role in the control of normal respiration.

An *inflation reflex* helps regulate depth of breathing. This reflex occurs when stretched lung tissues stimulate stretch receptors in the visceral pleura, bronchioles, and alveoli. The sensory impulses of this reflex travel via the vagus nerves to the pneumotaxic area of the respiratory center and shorten the duration of inspiratory movements. This action prevents overinflation of the lungs during forceful breathing.

Emotional upset can alter the normal breathing pattern. Fear and pain typically increase the breathing rate. Conscious control of breathing is also possible because the respiratory muscles are voluntary. A person can stop breathing altogether for a very short time.

If breathing stops, blood concentrations of carbon dioxide and hydrogen ions rise, and oxygen concentration falls. These changes (primarily the increased carbon dioxide) stimulate the respiratory center, and soon the urge to inhale overpowers the desire to hold the breath. A person can increase breath-holding time by breathing rapidly and deeply in advance. This action, called **hyperventilation** (hi˝per-ven´tĭ-la´shun), lowers the blood carbon dioxide concentration. Following hyper-

16.5 Alveolar Gas Exchanges

The parts of the respiratory system discussed so far conduct air in and out of air passages. The alveoli carry on the vital process of exchanging gases between the air and the blood.

Alveoli

Alveoli are microscopic air sacs clustered at the distal ends of the narrowest respiratory tubes, the alveolar ducts (see fig. 16.8). Each alveolus consists of a tiny space within a thin wall that separates it from adjacent alveoli.

Respiratory Membrane

The wall of an alveolus consists of an inner lining of simple squamous epithelium and a dense network of capillaries, which are also lined with simple squamous epithelial cells. Thin, fused basement membranes separate the layers of these flattened cells, and in the spaces between the cells are elastic and collagenous fibers that support the alveolar wall. At least two thicknesses of epithelial cells and a layer of fused basement membranes separate the air in an alveolus from the blood in a capillary (fig. 16.19). These layers comprise the **respiratory membrane** (re-spi´rah-to´´re mem´brān) across which blood and alveolar air exchange gases.

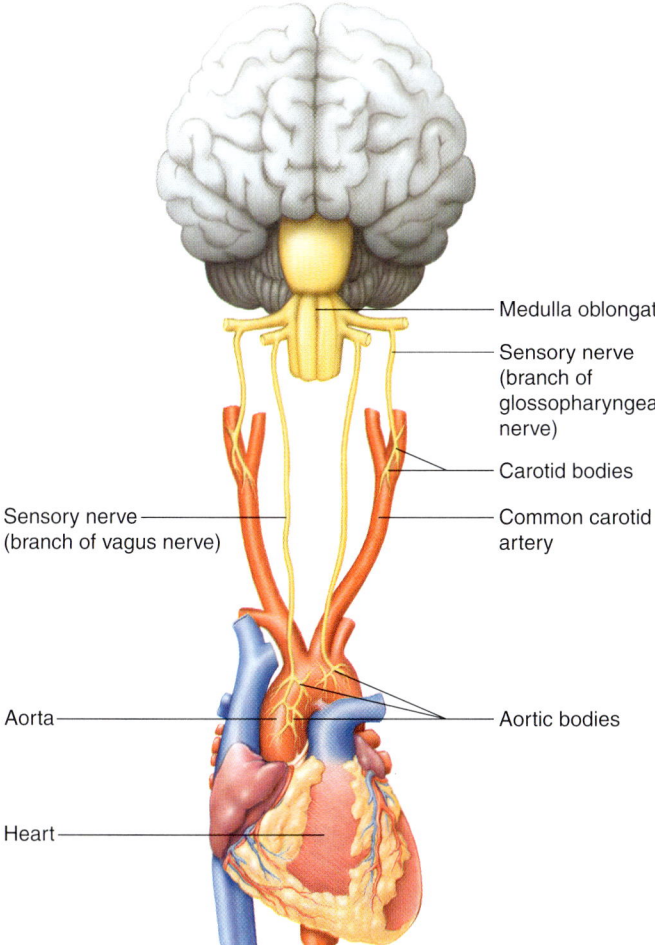

Figure 16.18
Decreased blood oxygen concentration stimulates chemoreceptors in the carotid and aortic bodies.

ventilation, it takes longer than usual for the carbon dioxide concentration to produce an overwhelming effect on the respiratory center. (*Note:* Prolonging breath-holding in this way can cause abnormally low blood oxygen levels. Hyperventilation should never be used to help hold the breath while swimming because the person may lose consciousness underwater and drown.)

Sometimes, a person who is emotionally upset may hyperventilate, become dizzy, and lose consciousness. This condition is due to a lowered carbon dioxide concentration followed by a rise in pH (respiratory alkalosis), a localized vasoconstriction of cerebral arterioles, and consequently, a decreased blood flow to nearby brain cells. Interference with the oxygen supply to the brain causes fainting.

✓ CHECK YOUR RECALL

1. What chemical factors affect breathing?
2. Describe the inflation reflex.
3. How does hyperventilation decrease the respiratory rate?

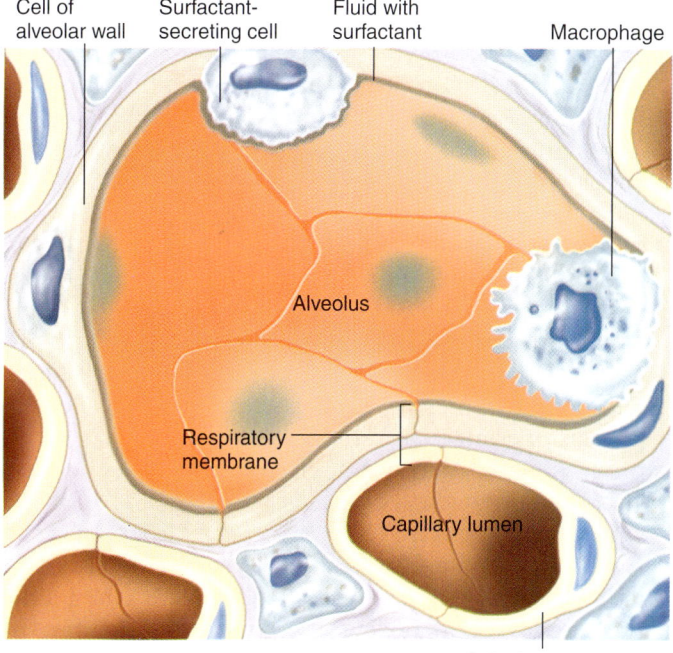

Figure 16.19
The respiratory membrane consists of the walls of the alveolus and the capillary.

Topic of Interest

EXERCISE AND BREATHING

Moderate to heavy physical exercise greatly increases the amount of oxygen the skeletal muscles use. For example, a young man at rest utilizes about 250 mL of oxygen per minute, but maximal exercise may require 3,600 mL of oxygen per minute.

As oxygen utilization increases, the volume of carbon dioxide produced also increases. Because decreased blood oxygen and increased blood carbon dioxide concentrations stimulate the respiratory center, exercise increases breathing rate. Studies reveal, however, that blood oxygen and carbon dioxide concentrations usually do not change during exercise. This reflects the respiratory system's effectiveness in obtaining oxygen and releasing carbon dioxide to the outside.

The cerebral cortex and sensory structures called *proprioceptors* that are associated with muscles and joints cause much of the increased breathing rate during vigorous exercise. Specifically, whenever the cerebral cortex signals skeletal muscles to contract, it also transmits stimulating impulses to the respiratory center. Muscular movements stimulate proprioceptors, triggering a *joint reflex* that travels to the respiratory center, increasing breathing rate.

When breathing rate increases during exercise, increased blood flow is also required to power skeletal muscles. Thus, physical exercise places demands on both the cardiovascular and respiratory systems. If either of these systems fails to keep up with cellular demands, the person begins to feel out of breath. This feeling usually reflects an inability of the heart and blood vessels to move enough blood between the lungs and cells, rather than the respiratory system's inability to provide enough air.

 If all of the capillaries that surround the alveoli were unwound and laid end to end, they would extend for about 620 miles.

Diffusion Across the Respiratory Membrane

Recall from chapter 3 (p. 58) that molecules diffuse from regions where they are in higher concentration toward regions where they are in lower concentration. For gases, it is more useful to think of diffusion occurring from regions of higher pressure toward regions of lower pressure. The pressure of a gas determines the rate at which it diffuses from one region to another.

Measured by volume, ordinary air is about 78% nitrogen, 21% oxygen, and 0.04% carbon dioxide. Air also contains small amounts of other gases that have little or no physiological importance.

In a mixture of gases such as air, each gas accounts for a portion of the total pressure the mixture produces. The amount of pressure each gas contributes is called the **partial pressure** (par´shil presh´ur) of that gas and is proportional to its concentration. For example, because air is 21% oxygen, oxygen accounts for 21% of the atmospheric pressure (21% of 760 mm Hg), or 160 mm Hg. Thus, the partial pressure of oxygen, symbolized P_{O_2}, in atmospheric air is 160 mm Hg. Similarly, the partial pressure of carbon dioxide (P_{CO_2}) in air is 0.3 mm Hg.

Gas molecules from the air may enter, or dissolve, in liquid. This is what happens when carbon dioxide is added to a carbonated beverage, or when inspired gases dissolve in the blood in the alveolar capillaries.

When a mixture of gases dissolves in blood, the resulting concentration of each gas is proportional to its partial pressure. Each gas diffuses between blood and its surroundings from areas of higher partial pressure to areas of lower partial pressure until the partial pressures in the two regions reach equilibrium. For example, the P_{CO_2} in capillary blood is 45 mm Hg, but the P_{CO_2} in alveolar air is 40 mm Hg. Because of the difference in these partial pressures, carbon dioxide diffuses from blood, where its partial pressure is higher, across the respiratory membrane and into alveolar air (fig. 16.20). When blood leaves the lungs, its P_{CO_2} is 40 mm Hg, which is the same as the P_{CO_2} of alveolar air. Similarly, the P_{O_2} of capillary blood is 40 mm Hg, but that of alveolar air is 104 mm Hg. Thus, oxygen diffuses from alveolar air into blood, and blood leaves the lungs with a P_{O_2} of 104 mm Hg. (Because of the large volume of air always in the lungs, as long as breathing continues, alveolar P_{O_2} stays relatively constant at 104 mm Hg.)

A number of factors affect diffusion across the respiratory membrane. More surface area, shorter distance, greater solubility of gases, and a steeper partial pressure gradient all favor increased diffusion. Thus, diseases that harm the respiratory membrane, such as pneumonia, or diseases that reduce the surface area for diffusion, such as emphysema, require increased P_{O_2} for treatment.

The respiratory membrane is normally so thin that certain soluble chemicals other than carbon dioxide may diffuse into alveolar air and be exhaled. This is why breath analysis can reveal alcohol in the blood or acetone on the breath of a person who has untreated diabetes mellitus. Breath analysis may also detect substances associated with kidney failure, certain digestive disturbances, and liver disease.

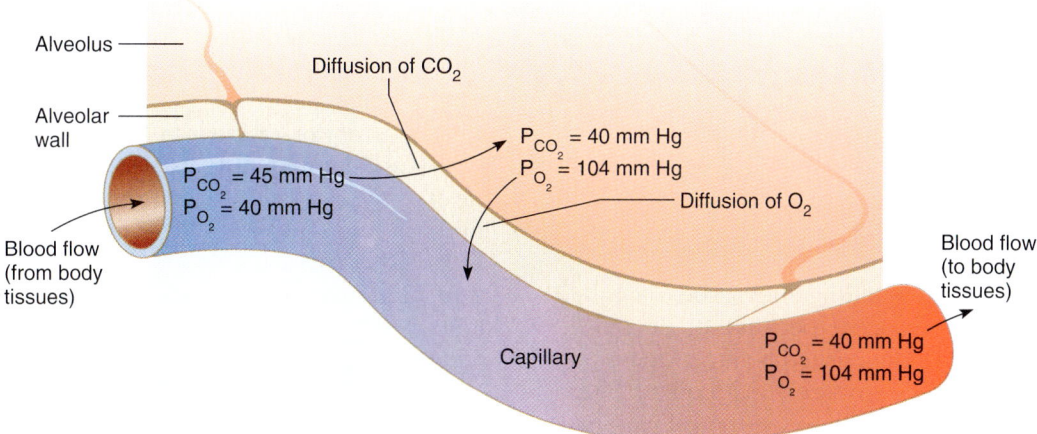

Figure 16.20
Gases are exchanged between alveolar air and capillary blood because of differences in partial pressures.

CHECK YOUR RECALL

1. Describe the structure of the respiratory membrane.
2. What is the partial pressure of a gas?
3. What force moves oxygen and carbon dioxide across the respiratory membrane?

16.6 Gas Transport

Blood transports oxygen and carbon dioxide between the lungs and cells. As these gases enter blood, they dissolve in the liquid portion (plasma) or combine chemically with blood components.

Oxygen Transport

Almost all the oxygen (over 98%) that blood transports combines with the iron-containing protein **hemoglobin** (he˝-mo-glo´bin) in red blood cells. The remainder of the oxygen dissolves in plasma.

In the lungs, where the P_{O_2} is relatively high, oxygen dissolves in blood and combines rapidly with the iron atoms of hemoglobin, forming **oxyhemoglobin** (ok˝-se-he˝-mo-glo´bin) (fig. 16.21). The chemical bonds between oxygen and hemoglobin molecules are unstable, and as the P_{O_2} decreases, oxyhemoglobin molecules release oxygen, which diffuses into nearby cells that have depleted their oxygen supplies in cellular respiration.

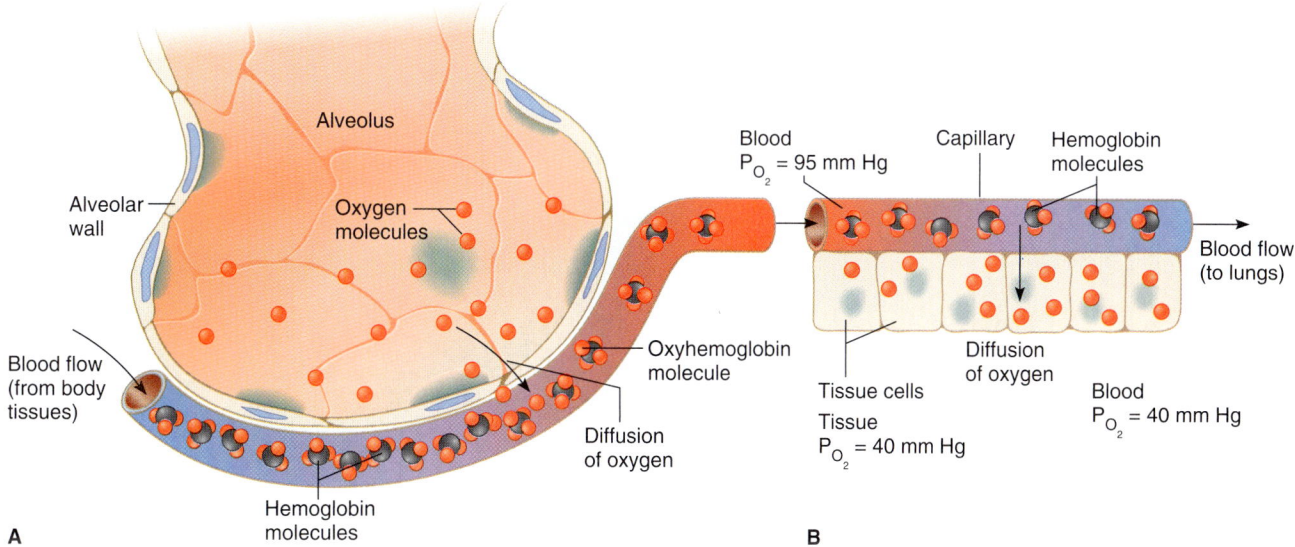

Figure 16.21
Blood transports oxygen. (*A*) Oxygen molecules, entering blood from the alveolus, bond to hemoglobin, forming oxyhemoglobin. (*B*) In the regions of the body cells, oxyhemoglobin releases oxygen. Note that much oxygen is still bound to hemoglobin at the P_{O_2} of systemic venous blood.

Several other factors affect the amount of oxygen that oxyhemoglobin releases. More oxygen is released as the blood concentration of carbon dioxide increases, as blood becomes more acidic, or as blood temperature increases. This explains why more oxygen is released to skeletal muscles during physical exercise. The increased muscular activity and oxygen utilization increase carbon dioxide concentration, decrease pH, and raise temperature. Less active cells receive proportionately less oxygen.

A deficiency of O_2 reaching the tissues is called **hypoxia**. Hypoxia may occur because of decreased arterial P_{O_2} (*hypoxemia*), diminished ability of the blood to transport O_2 (anemic hypoxia), inadequate blood flow (ischemic hypoxia), or a defect at the cellular level (histotoxic hypoxia) as in cyanide poisoning.

CHECK YOUR RECALL

1. How is oxygen transported from the lungs to cells?
2. What stimulates blood to release oxygen to tissues?

Carbon Dioxide Transport

Blood flowing through the capillaries of the tissues gains carbon dioxide because the tissues have a relatively high P_{CO_2}. Blood transports carbon dioxide to the lungs in one of three forms: as carbon dioxide dissolved in plasma, as part of a compound formed by bonding to hemoglobin, or in the form of a bicarbonate ion (fig. 16.22).

The amount of carbon dioxide that dissolves in plasma is determined by its partial pressure. The higher the P_{CO_2} of the tissues, the more carbon dioxide will go into solution. However, only about 7% of the carbon dioxide that blood transports is in this form.

Unlike oxygen, which combines with the iron atoms (part of the "heme" portion) of hemoglobin molecules, carbon dioxide bonds with the amino groups ($–NH_2$) of the "globin" or protein portion of these molecules. Consequently, oxygen and carbon dioxide do not compete for binding sites, and a hemoglobin molecule can transport both gases at the same time.

Carbon dioxide combining with hemoglobin forms a loosely bound compound called **carbaminohemoglobin** (kar-bam´´ĭ-no-he´´mo-glo´bin). This molecule decomposes readily in regions of low P_{CO_2}, releasing its carbon dioxide. This method of transporting carbon dioxide is theoretically quite effective, but carbaminohemoglobin forms slowly. Only about 23% of the carbon dioxide that blood transports is in this form.

The most important carbon dioxide transport mechanism involves the formation of **bicarbonate ions**

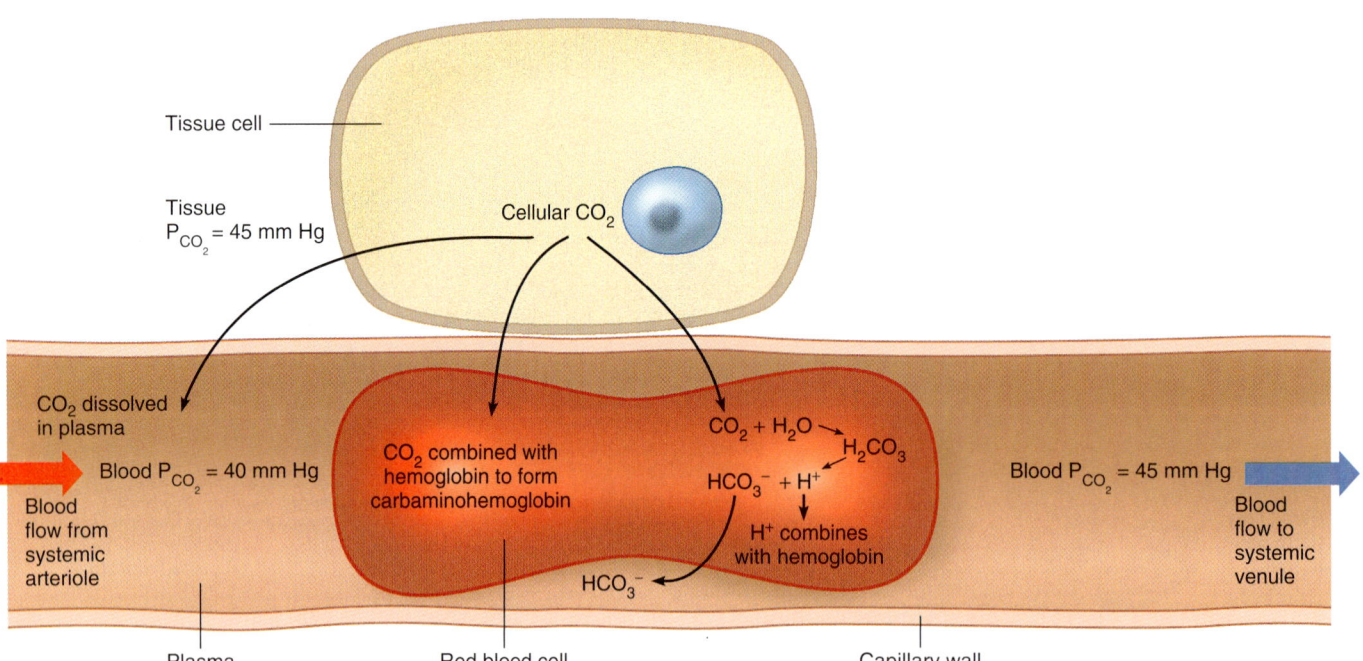

Figure 16.22
Carbon dioxide produced by tissue cells is transported in the blood plasma in a dissolved state, combined with hemoglobin, or in the form of bicarbonate ions (HCO_3^-).

(HCO_3^-). Carbon dioxide reacts with water to form carbonic acid (H_2CO_3):

$$CO_2 + H_2O \rightarrow H_2CO_3$$

This reaction occurs slowly in plasma, but much of the carbon dioxide diffuses into red blood cells. These cells contain the enzyme **carbonic anhydrase** (kar-bon´ik an-hi´drās), which speeds the reaction between carbon dioxide and water. The resulting carbonic acid then dissociates, releasing hydrogen ions (H^+) and bicarbonate ions (HCO_3^-):

$$H_2CO_3 \rightarrow H^+ + HCO_3^-$$

Most of the hydrogen ions combine quickly with hemoglobin molecules and, thus, do not accumulate and greatly change blood pH. The bicarbonate ions diffuse out of red blood cells and enter the plasma. Nearly 70% of the carbon dioxide that blood transports is in this form.

When blood passes through the capillaries of the lungs, its dissolved carbon dioxide diffuses into alveoli in response to the relatively low P_{CO_2} of alveolar air (fig. 16.23). At the same time, hydrogen ions and bicarbonate ions in red blood cells recombine to form carbonic acid, and under the influence of carbonic anhydrase, the carbonic acid quickly breaks down to yield carbon dioxide and water:

$$H^+ + HCO_3^- \rightarrow H_2CO_3 \rightarrow CO_2 + H_2O$$

Carbaminohemoglobin also releases its carbon dioxide, and carbon dioxide continues to diffuse out of blood until the P_{CO_2} of blood and of alveolar air are in equilibrium. Table 16.3 summarizes transport of blood gases.

CHECK YOUR RECALL

1. Describe three forms in which blood can transport carbon dioxide from cells to the lungs.
2. How can hemoglobin carry oxygen and carbon dioxide at the same time?
3. How is carbon dioxide released from blood into the lungs?

TABLE 16.3 GASES TRANSPORTED IN BLOOD

GAS	REACTION INVOLVED	SUBSTANCE TRANSPORTED
Oxygen	1–2% dissolves in plasma 98–99% combines with iron atoms of hemoglobin molecules	Oxyhemoglobin
Carbon dioxide	About 7% dissolves in plasma	Carbon dioxide
	About 23% combines with amino groups of hemoglobin molecules	Carbaminohemoglobin
	About 70% reacts with water to form carbonic acid; the carbonic acid then dissociates to release hydrogen ions and bicarbonate ions	Bicarbonate ions

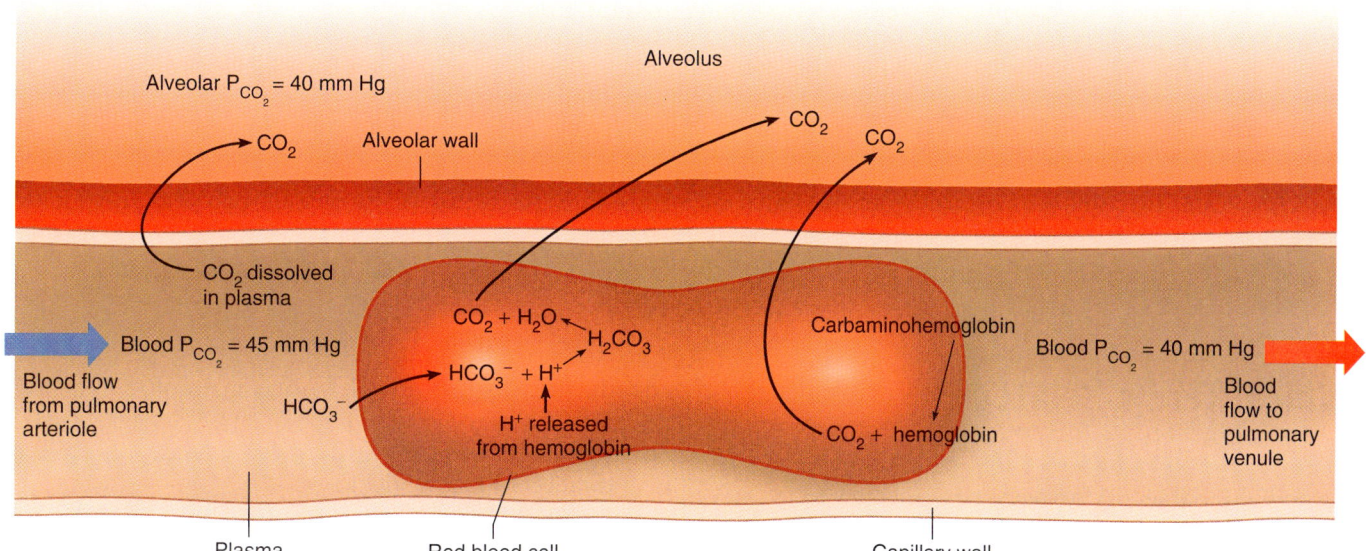

Figure 16.23
In the lungs, carbon dioxide diffuses from the blood into the alveoli.

Genetics Connection

CYSTIC FIBROSIS

"Woe to that child which when kissed on the forehead tastes salty. He is bewitched and soon must die." So went a seventeenth-century British saying about a child with cystic fibrosis (CF). But despite this early observation of a link between salty perspiration and poor health, researchers didn't unravel the cause of this common inherited disorder until recently.

History of an Illness

In 1938, physicians first described CF as a defect in channels leading from certain glands. This causes formation of extremely thick, sticky mucus, which encourages infections by microorganisms not otherwise common in the lungs. A clogged pancreas prevents digestive juices from reaching the intestines and thus impairs absorption of nutrients. Often the first sign of the illness is salty sweat, noticed when a parent kisses a child. Affected children are small and subject to frequent respiratory infections, and in the past were often diagnosed as "failing to thrive."

Until 1960, patients rarely lived past age ten, but then physicians developed ways to fight the symptoms. Antibiotics control the respiratory infections, and daily "postural drainage" exercises shake the stifling mucus free from the lungs—a parent catches the expelled mucus in a jar. Digestive enzymes mixed into soft foods can enhance nutrient absorption.

In 1989, geneticists discovered the gene and protein defect behind CF, allowing development of more targeted treatments. The gene encodes a protein called the "cystic fibrosis transmembrane regulator," or CFTR for short. It is an ion channel that controls chloride transport out of cells, but also controls flow of water, ATP, and sodium in ways that are still not well understood. In most people with CF, the chloride channel is missing one amino acid, and is so deformed that it never reaches the cell's surface. With chloride ions unable to leave the cell, water moves in by osmosis. This dries out the mucus, clogging the affected organs.

By the early 1990s, new drugs to treat CF had entered the arsenal. Some drugs allowed more sodium to enter or chloride to leave the lung lining cells. A natural enzyme, deoxyribonuclease, is used to degrade the DNA that accumulates in infected lungs as white blood cells cause inflammation. Several experimental gene therapies can introduce functional CFTR genes into affected cells, but so far these treatments work only for short times or restore function to only a small part of the respiratory tract.

Does Cystic Fibrosis Protect Against Diarrheal Disease?

CF is the most common inherited illness among Caucasians. It is inherited from two parents who are carriers, and the sexes are about equally affected. In some European populations, as many as 5% of the people may be carriers. Why is CF so common, if the symptoms so devastate health? The answer may lie in a genetic phenomenon in which being a carrier for an inherited disease protects against another disease, often an infectious one.

Cystic fibrosis carriers may be resistant to diarrheal diseases, which in many parts of the world cause much infant mortality. Part of the normal CFTR protein that sticks out of intestinal lining cells forms a receptor that *Salmonella typhi,* the bacterial cause of typhoid fever, latches onto. In CF carriers, the bacteria can't hold onto enough cells to cross into the digestive tract and cause typhoid fever. Mice with a version of CF show similar protection against cholera toxin, which causes severe diarrhea. Additional evidence supporting this connection comes from anecdotal reports and searches through records of CF patients, in which a carrier or affected individual who has had typhoid or cholera has never been documented.

If the CF-diarrhea hypothesis holds up, it is possible that, one day, better understanding of how cystic fibrosis develops may lead to treatments or even preventive strategies for both CF and diarrheal diseases.

Clinical Terms Related to the Respiratory System

anoxia (ah-nok´se-ah) Absence or deficiency of oxygen within tissues.

apnea (ap-ne´ah) Temporary cessation of breathing.

asphyxia (as-fik´se-ah) Oxygen deficiency and excess carbon dioxide in blood and tissues.

atelectasis (at´´e-lek´tah-sis) Collapse of a lung or part of a lung.

bradypnea (brad´´e-ne´ah) Abnormally slow breathing.

bronchiolectasis (brong´´ke-o-lek´tah-sis) Chronic dilation of the bronchioles.

bronchitis (brong-ki´tis) Inflammation of the bronchial lining.

Cheyne-Stokes respiration (chān stōks res´´pĭ-ra´shun) Irregular breathing consisting of a series of shallow breaths that increase in depth and rate, followed by breaths that decrease in depth and rate.

Organization

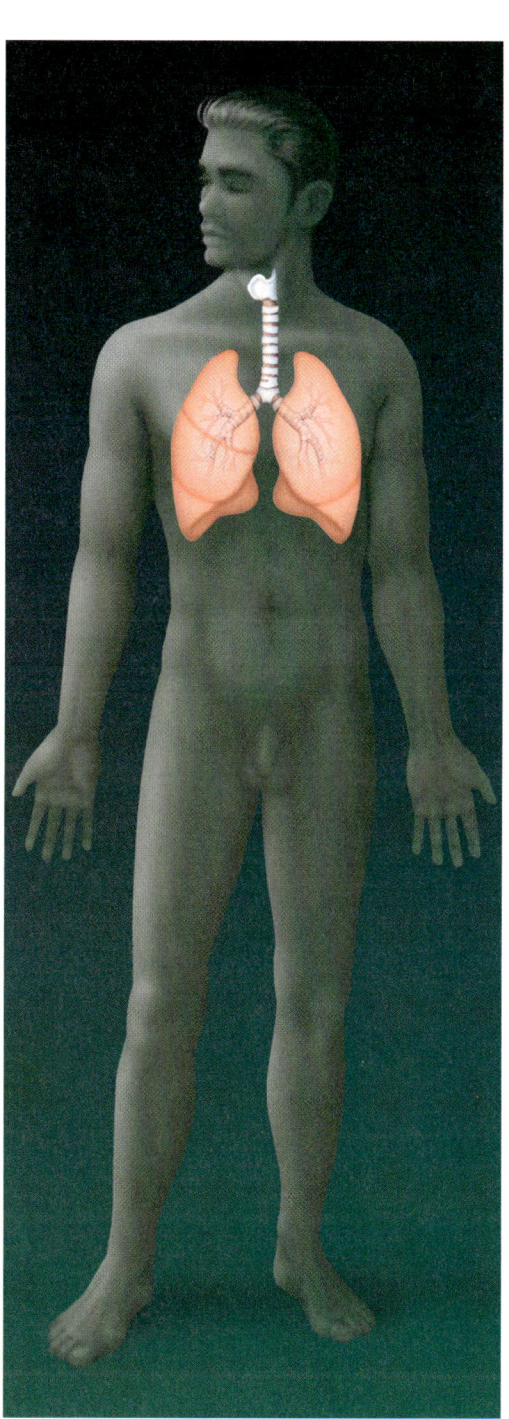

Respiratory System
The respiratory system provides oxygen for the internal environment and excretes carbon dioxide.

Integumentary System

Stimulation of skin receptors may alter respiratory rate.

Cardiovascular System
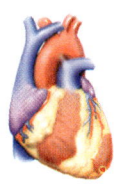
As the heart pumps blood through the lungs, the lungs oxygenate the blood and excrete carbon dioxide.

Skeletal System

Bones provide attachments for muscles involved in breathing.

Lymphatic System

Cells of the immune system patrol the lungs and defend against infection.

Muscular System
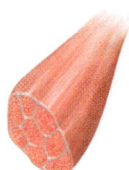
The respiratory system eliminates carbon dioxide produced by exercising muscles.

Digestive System
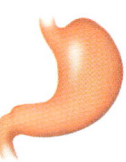
The digestive system and respiratory system share openings to the outside.

Nervous System
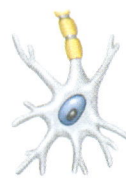
The brain controls the respiratory system. The respiratory system helps control pH of the internal environment.

Urinary System
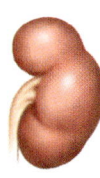
The kidneys and the respiratory system work together to maintain blood pH. The kidneys compensate for water lost through breathing.

Endocrine System

Hormonelike substances control the production of red blood cells that transport oxygen and carbon dioxide.

Reproductive System

Respiration increases during sexual activity. Fetal "respiration" begins before birth.

dyspnea (disp´ne-ah) Difficulty breathing.
eupnea (up-ne´ah) Normal breathing.
hemothorax (he´´mo-tho´raks) Blood in the pleural cavity.
hypercapnia (hi´´per-kap´ne-ah) Excess carbon dioxide in the blood.
hyperoxia (hi´´per-ok´se-ah) Excess oxygen in the blood.
hyperpnea (hi´´perp-ne´ah) Increase in the depth and rate of breathing.
hyperventilation (hi´´per-ven´´tĭ-la´shun) Prolonged, rapid, and deep breathing.
hypoxemia (hi´´pok-se´me-ah) Deficiency in blood oxygenation.
hypoxia (hi-pok´se-ah) Diminished availability of oxygen in tissues.
lobar pneumonia (lo´ber nu-mo´ne-ah) Pneumonia that affects an entire lung lobe.
pleurisy (ploo´rĭ-se) Inflammation of the pleural membranes.
pneumoconiosis (nu´´mo-ko´´ne-o´sis) Accumulation of particles from the environment in the lungs and the reaction of tissues to them.
pneumothorax (nu´´mo-tho´raks) Entrance of air into the space between the pleural membranes, followed by lung collapse.
rhinitis (ri-ni´tis) Inflammation of the nasal cavity lining.
sinusitis (si´´nŭ-si´tis) Inflammation of the sinus cavity lining.
tachypnea (tak´´ip-ne´ah) Rapid, shallow breathing.
tracheotomy (tra´´ke-ot´o-me) Incision in the trachea for exploration or for removal of a foreign object.

Clinical Connection

Emphysema can make routine tasks such as shopping or walking a dog impossible. One treatment approach is to remove the affected portions of the lungs. Supposedly, the remaining lung tissue regains some elastic recoil, and lung function improves. The technique, called "lung volume reduction surgery," was developed in the 1970s but was not widely done, then became popular again in the 1990s. But a recent study reveals that it is helpful for only some patients, and may actually harm certain others.

Of the 700 procedures performed in the mid 1990s, many patients never left the intensive care unit afterwards. So the National Institutes of Health and the Health Care Financing Administration called for controlled clinical trials of the procedure, comparing it to standard medical management for emphysema. Results were telling. For patients with emphysema widespread throughout the lungs, very low forced expiratory volume, or hampered ability to get rid of carbon monoxide—about 1 in 8 individuals requesting the procedure—lung volume reduction surgery causes more harm than good. The procedure is valuable only if physicians carefully screen candidates.

SUMMARY OUTLINE

16.1 Introduction (p. 435)
The respiratory system includes tubes that remove particles from incoming air and transport air to and from the lungs and the air sacs where gases are exchanged. Respiration is the entire process of gas exchange between the atmosphere and body cells.

16.2 Organs of the Respiratory System (p. 435)
The organs of the respiratory system can be divided into two groups. The upper respiratory tract includes the nose, nasal cavity, paranasal sinuses, and pharynx; the lower respiratory tract includes the larynx, trachea, bronchial tree, and lungs.

1. Nose
   a. Bone and cartilage support the nose.
   b. The nostrils are openings for air.
2. Nasal cavity
   a. Nasal conchae divide the nasal cavity into passageways and help increase the surface area of the mucous membrane.
   b. The mucous membrane filters, warms, and moistens incoming air.
   c. Ciliary action carries particles trapped in mucus to the pharynx, where they are swallowed.
3. Paranasal sinuses
   a. The paranasal sinuses are spaces in the bones of the skull that open into the nasal cavity.
   b. Mucous membrane lines sinuses.
4. Pharynx
   a. The pharynx is behind the oral cavity and between the nasal cavity and larynx.
   b. It is a passageway for air and food.
5. Larynx
   a. The larynx conducts air and helps prevent foreign objects from entering the trachea.
   b. It is composed of muscles and cartilages and is lined with mucous membrane.
   c. The larynx contains the vocal cords, which vibrate from side to side and produce sounds when air passes between them.
   d. The glottis and epiglottis help prevent foods and liquids from entering the trachea.
6. Trachea
   a. The trachea extends into the thoracic cavity in front of the esophagus.
   b. It divides into right and left bronchi.
7. Bronchial tree
   a. The bronchial tree consists of branched air passages that lead from the trachea to the air sacs.
   b. Alveoli are at the distal ends of the narrowest tubes, the alveolar ducts.
8. Lungs
   a. The mediastinum separates the left and right lungs, and the diaphragm and thoracic cage enclose them.

b. The visceral pleura attaches to the surface of the lungs. The parietal pleura lines the thoracic cavity.
   c. Each lobe of the lungs is composed of alveoli, blood vessels, and supporting tissues.

16.3 Breathing Mechanism (p. 441)
Changes in the size of the thoracic cavity accompany inspiration and expiration.
1. Inspiration
   a. Atmospheric pressure forces air into the lungs.
   b. Inspiration occurs when the pressure inside alveoli decreases.
   c. Pressure within alveoli decreases when the diaphragm moves downward and the thoracic cage moves upward and outward.
   d. Surface tension aids lung expansion.
2. Expiration
   a. Elastic recoil of tissues and surface tension within alveoli provide the forces of expiration.
   b. Thoracic and abdominal wall muscles aid expiration.
3. Respiratory air volumes and capacities
   a. One inspiration followed by one expiration is a respiratory cycle.
   b. The amount of air that moves in (or out) during a single respiratory cycle is the tidal volume.
   c. Additional air that can be inhaled is the inspiratory reserve volume. Additional air that can be exhaled is the expiratory reserve volume.
   d. Residual volume remains in the lungs after a maximal expiration.
   e. The vital capacity is the maximum amount of air a person can exhale after taking the deepest breath possible.
   f. The inspiratory capacity is the maximum volume of air a person can inhale following exhalation of the tidal volume.
   g. The functional residual capacity is the volume of air that remains in the lungs after a person exhales the tidal volume.
   h. The total lung capacity equals the vital capacity plus the residual volume.

16.4 Control of Breathing (p. 447)
Normal breathing is rhythmic and involuntary.
1. Respiratory center
   a. The respiratory center is in the brain stem and includes portions of the pons and medulla oblongata.
   b. The medullary rhythmicity area includes two groups of neurons.
      (1) The dorsal respiratory group controls the basic rhythm of breathing.
      (2) The ventral respiratory group increases inspiratory and expiratory movements during forceful breathing.
   c. The pneumotaxic area regulates breathing rate.
2. Factors affecting breathing
   a. Chemicals, stretching of lung tissues, and emotional states affect breathing.
   b. Chemosensitive areas (central chemoreceptors) are associated with the respiratory center.
      (1) Blood concentrations of carbon dioxide and hydrogen ions influence the central chemoreceptors.
      (2) Stimulation of these receptors increases breathing rate.
   c. Peripheral chemoreceptors are in the walls of certain large arteries.
      (1) These chemoreceptors sense low oxygen concentration.
      (2) When oxygen concentration is low, breathing rate increases.
   d. Overstretching lung tissues triggers an inflation reflex.
      (1) This reflex shortens the duration of inspiratory movements.
      (2) The inflation reflex prevents overinflation of the lungs during forceful breathing.
   e. Hyperventilation decreases blood carbon dioxide concentration, but *this is very dangerous when done before swimming underwater.*

16.5 Alveolar Gas Exchanges (p. 449)
Gas exchange between air and blood occurs in alveoli.
1. Alveoli
   Alveoli are tiny air sacs clustered at the distal ends of alveolar ducts.
2. Respiratory membrane
   a. This membrane consists of alveolar and capillary walls.
   b. Blood and alveolar air exchange gases across this membrane.
3. Diffusion across the respiratory membrane
   a. The partial pressure of a gas is proportional to the concentration of that gas in a mixture or the concentration dissolved in a liquid.
   b. Gases diffuse from regions of higher partial pressure toward regions of lower partial pressure.
   c. Oxygen diffuses from alveolar air into blood. Carbon dioxide diffuses from blood into alveolar air.

16.6 Gas Transport (p. 451)
Blood transports gases between the lungs and cells.
1. Oxygen transport
   a. Blood mainly transports oxygen in combination with hemoglobin molecules.
   b. The resulting oxyhemoglobin is unstable and releases its oxygen in regions where the P_{O_2} is low.
   c. More oxygen is released as the blood concentration of carbon dioxide increases, as blood becomes more acidic, and as blood temperature increases.
2. Carbon dioxide transport
   a. Carbon dioxide may be carried in solution, bound to hemoglobin, or as a bicarbonate ion.
   b. Most carbon dioxide is transported in the form of bicarbonate ions.
   c. The enzyme carbonic anhydrase speeds the reaction between carbon dioxide and water to form carbonic acid.
   d. Carbonic acid dissociates to release hydrogen ions and bicarbonate ions.

REVIEW EXERCISES

1. Describe the general functions of the respiratory system. (p. 435)
2. Distinguish between the upper and lower respiratory tracts. (p. 435)
3. Explain how the nose and nasal cavity filter incoming air. (p. 435)
4. Describe the locations of the major paranasal sinuses. (p. 436)
5. Distinguish between the pharynx and larynx. (p. 436)
6. Name and describe the locations of the larger cartilages of the larynx. (p. 437)
7. Distinguish between the false vocal cords and the true vocal cords. (p. 437)
8. Compare the structure of the trachea with that of the branches of the bronchial tree. (p. 438)
9. Distinguish between the visceral pleura and parietal pleura. (p. 440)
10. Explain normal inspiration and forced inspiration. (p. 441)
11. Define *surface tension*, and explain how it aids breathing. (p. 443)
12. Define *surfactant*, and explain its function. (p. 443)
13. Explain normal expiration and forced expiration. (p. 445)
14. Distinguish between the vital capacity and total lung capacity. (p. 446)

15. Describe the location of the respiratory center, and name its major components. (p. 447)
16. Describe the control of the basic rhythm of breathing. (p. 447)
17. Explain the function of the pneumotaxic area of the respiratory center. (p. 448)
18. Describe the function of the chemoreceptors in the carotid and aortic bodies. (p. 448)
19. Describe the inflation reflex. (p. 448)
20. Define *hyperventilation*, and explain how it affects the respiratory center. (p. 448)
21. Define *respiratory membrane*, and explain its function. (p. 449)
22. Explain the relationship between the partial pressure of a gas and its diffusion rate. (p. 450)
23. Summarize the gas exchanges across the respiratory membrane. (p. 450)
24. Describe how blood transports oxygen. (p. 451)
25. List three factors that increase release of blood oxygen. (p. 452)
26. Explain how the blood transports carbon dioxide. (p. 452)

CRITICAL THINKING

1. It is below 0°F outside, but the dedicated runner bundles up and hits the road anyway. "You're crazy," shouts a neighbor. "Your lungs will freeze." Why is the well-meaning neighbor wrong?
2. Why does breathing through the mouth dry out the throat?
3. George Washington went for a walk in the freezing rain on a bleak December day in 1799. The next day, he had trouble breathing and swallowing. A doctor suggested cutting a hole in the president's throat so he could breathe, but other doctors voted him down, instead bleeding the patient, plastering his throat with bran and honey, and placing beetles on his legs to produce blisters. Soon, Washington's voice became muffled, his breathing was more labored, and he grew restless. For a short time, he seemed euphoric; then he died. Washington had epiglottitis, in which the epiglottis swells to ten times its normal size. How does this diagnosis explain his symptoms? Which suggested treatment might have worked?
4. Why can you not commit suicide by holding your breath?
5. When a woman is very close to delivering a baby, she may hyperventilate. Breathing into a paper bag regulates her breathing. How does this action return her breathing to normal?
6. Why were the finishing times of endurance events rather slow at the 1968 Olympics, held in 2,200-meter-high Mexico City?
7. Why might it be dangerous for a heavy smoker to use a cough suppressant?
8. Emphysema reduces the lungs' capacity to recoil elastically. Which respiratory air volumes does emphysema affect?
9. What changes would you expect in the relative concentrations of blood oxygen in a patient who breathes rapidly and deeply for a prolonged time? Why?
10. If a person has stopped breathing and is receiving pulmonary resuscitation, would it be better to administer pure oxygen or a mixture of oxygen and carbon dioxide? Why?

WEB CONNECTIONS

Visit the website for additional study questions and more information about this chapter at:

http://www.mhhe.com/shieress8

chapter 17

Urinary System

HEMOLYTIC UREMIC SYNDROME. Felicia had looked forward to summer camp all year, especially the overnight hikes. A three-day expedition in July was wonderful, but five days after returning to camp, Felicia developed severe abdominal cramps. So did seventeen other campers and two counselors, some of whom had bloody diarrhea, too. Several of the stricken campers were hospitalized, Felicia among them. While the others improved in a few days and were released, Felicia suffered from a complication called hemolytic uremic syndrome (HUS). Her urine had turned bloody, and a blood analysis revealed severe anemia and lack of platelets.

Camp personnel reported the outbreak to public health officials, who quickly recognized the signs of food poisoning and traced the illness to hamburgers cooked outdoors on the camping trip. The burgers were not cooked long enough to kill *Escherichia coli*, bacteria which release a poison called shigatoxin.

Most people who eat meat tainted with the toxin become ill, but usually they just have cramps and diarrhea for several days. However, in about 6% of affected people, mostly children, HUS develops when the bloodstream transports the toxin to the kidneys, where it destroys cells of the microscopic capillaries that normally filter proteins and blood cells from forming urine. With the capillaries compromised, proteins and blood cells, as well as damaged kidney cells, appear in the urine.

HUS is a leading cause of acute renal (kidney) failure, killing 3–5% of affected children. Felicia was lucky. Blood clotted around the sites of her damaged kidney cells, and new cells formed. Three weeks after the bloody urine began, her urine was once again clear, and she was healthy.

Photo:
Undercooked beef tainted with a strain of *E. coli* that produces a powerful toxin is the source of many cases of hemolytic uremic syndrome, which damages the glomerular capillaries of nephrons within the kidneys. Changing the feed given to cattle can prevent colonization of these bacteria.

Chapter Objectives

After studying this chapter, you should be able to do the following:

17.1 Introduction
1. Name and list the general functions of the organs of the urinary system. (p. 461)

17.2 Kidneys
2. Describe the locations and structure of the kidneys. (p. 461)
3. List the functions of the kidneys. (p. 462)
4. Trace the pathway of blood through the major vessels within a kidney. (p. 462)
5. Describe a nephron, and explain the functions of its major parts. (p. 463)

17.3 Urine Formation
6. Explain how glomerular filtrate is produced, and describe its composition. (p. 466)
7. Explain the factors that affect the rate of glomerular filtration and how this rate is regulated. (p. 467)
8. Discuss the role of tubular reabsorption in urine formation. (p. 469)
9. Define *tubular secretion*, and explain its role in urine formation. (p. 471)

17.4 Urine Elimination
10. Describe the structure of the ureters, urinary bladder, and urethra. (p. 473)
11. Explain the process and control of micturition. (p. 475)

Aids to Understanding Words

calyc- [small cup] major *calyc*es: Cuplike divisions of the renal pelvis.
cort- [covering] renal *cort*ex: Shell of tissues surrounding the inner kidney.
detrus- [to force away] *detrus*or muscle: Muscle within the bladder wall that expels urine.
glom- [little ball] *glom*erulus: Cluster of capillaries within a renal corpuscle.
mict- [to pass urine] *mict*urition: Process of expelling urine from the bladder.
nephr- [pertaining to the kidney] *nephr*on: Functional unit of a kidney.
papill- [nipple] renal *papill*ae: Small elevations that project into a renal calyx.
trigon- [triangular shape] *trigon*e: Triangular area on the internal floor of the urinary bladder.

Key Terms

afferent arteriole (af′er-ent ar-te′re-ōl)
detrusor muscle (de-truz′or mus′l)
efferent arteriole (ef′er-ent ar-te′re-ōl)
glomerular capsule (glo-mer′u-lar kap′sul)
glomerulus (glo-mer′u-lus)
juxtaglomerular apparatus (juks″tah-glo-mer′u-lar ap″ah-ra′tus)
micturition (mik″tu-rish′un)
nephron (nef′ron)
peritubular capillary (per″ĭ-tu′bu-lar kap′ĭ-ler″e)
renal corpuscle (re′nal kor′pusl)
renal cortex (re′nal kor′teks)
renal medulla (re′nal mĕ-dul′ah)
renal tubule (re′nal tu′būl)
retroperitoneal (re″tro-per″ĭ-to-ne′al)

17.1 Introduction

Cells produce a variety of wastes which are toxic if they accumulate. Body fluids, such as blood and lymph, carry wastes from the tissues that produce them, and other structures remove wastes from blood and transport them to the outside. The respiratory system removes carbon dioxide from the blood, and the *urinary system* removes certain salts and nitrogenous wastes. The urinary system also helps maintain the normal concentrations of water and electrolytes within body fluids, regulates the pH and volume of body fluids, and helps control red blood cell production and blood pressure.

The urinary system consists of a pair of kidneys, which remove substances from blood, form urine, and help regulate certain metabolic processes; a pair of tubular ureters, which transport urine from the kidneys; a saclike urinary bladder, which stores urine; and a tubular urethra, which conveys urine to the outside of the body. Figure 17.1 and reference plate 6 (p. 27) show these organs.

17.2 Kidneys

A **kidney** is a reddish-brown, bean-shaped organ with a smooth surface. An adult kidney is about 12 centimeters long, 6 centimeters wide, and 3 centimeters thick, and is enclosed in a tough, fibrous capsule (fig. 17.2).

Location of the Kidneys

The kidneys lie on either side of the vertebral column in a depression high on the posterior wall of the abdominal cavity. The upper and lower borders of the kidneys are generally at the levels of the twelfth thoracic and third lumbar vertebrae, respectively. The left kidney is usually 1.5–2.0 centimeters higher than the right one.

The kidneys are positioned **retroperitoneally,** which means they are behind the parietal peritoneum and against the deep muscles of the back. Connective tissue and masses of adipose tissue surround the kidneys and hold them in position (see fig. 1.10, p. 11).

Kidney Structure

The lateral surface of each kidney is convex, but its medial side is deeply concave. The resulting medial depression leads into a hollow chamber called the **renal sinus.** The entrance to this sinus is termed the *hilum,* and through it pass blood vessels, nerves, lymphatic vessels, and the ureter (see fig. 17.1).

The superior end of the ureter expands to form a funnel-shaped sac called the **renal pelvis** inside the renal sinus. The pelvis subdivides into two or three tubes, called *major calyces* (singular, *calyx*), and these in turn subdivide into several *minor calyces* (fig. 17.2).

A series of small elevations called *renal papillae* project into the renal sinus from its wall. Tiny openings that lead into a minor calyx pierce each projection.

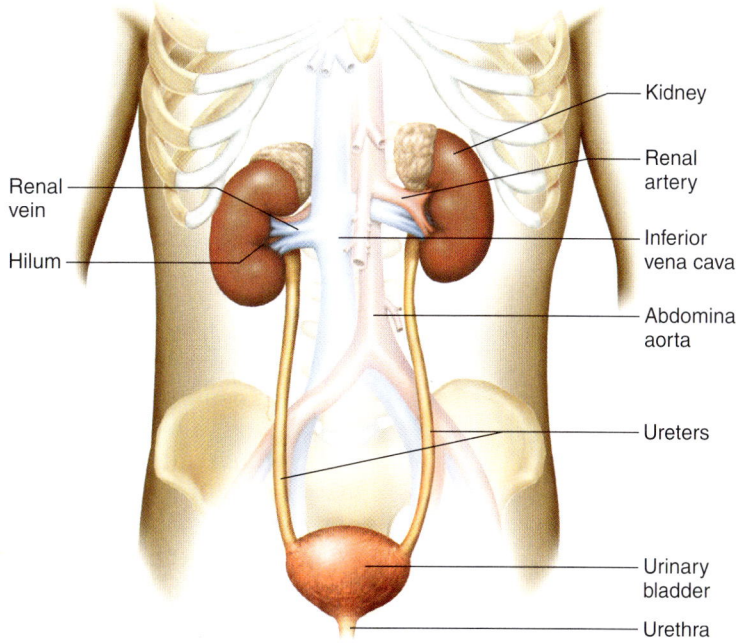

Figure 17.1

The urinary system includes the kidneys, ureters, urinary bladder, and urethra. Notice the relationship of these structures to the major blood vessels.

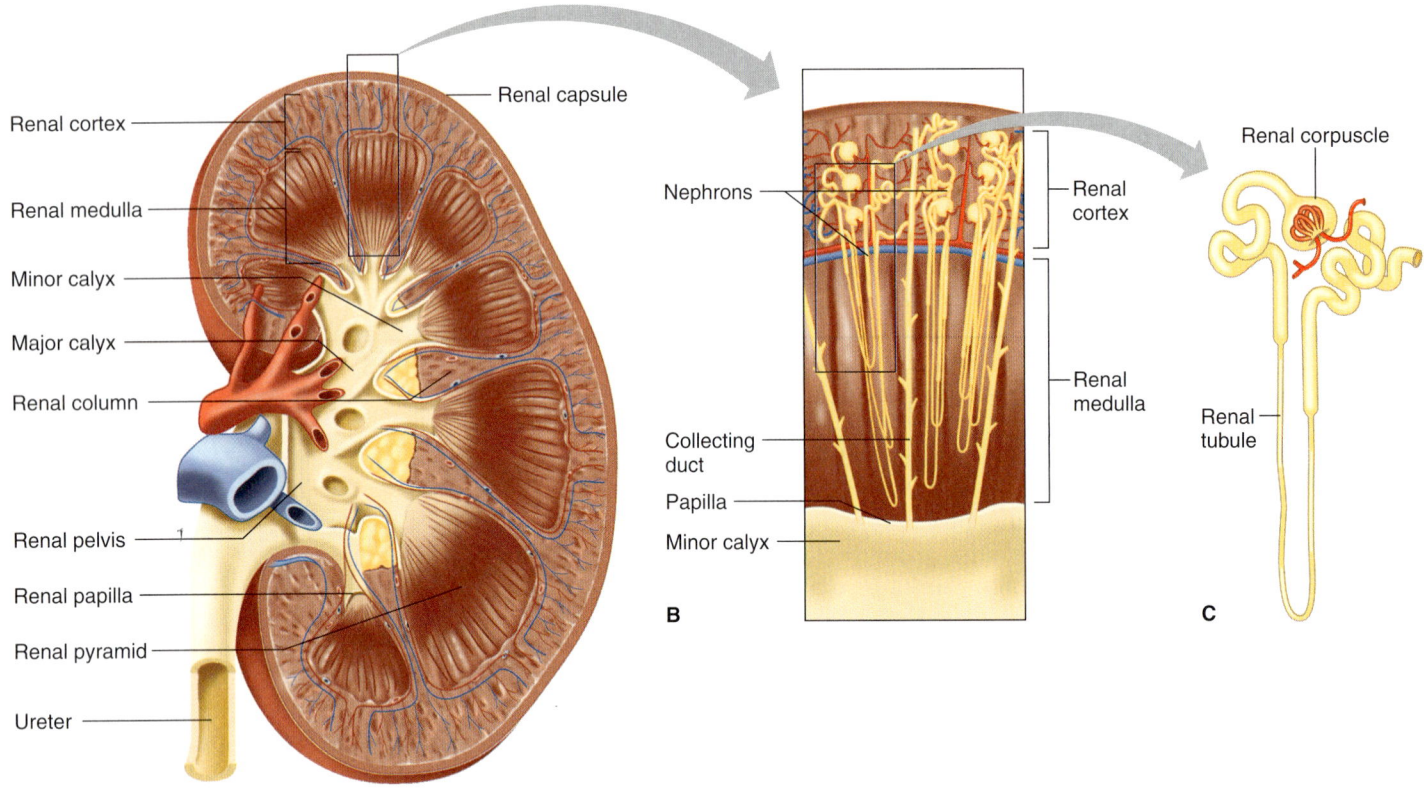

Figure 17.2
The kidney. (A) Longitudinal section of a kidney. (B) A renal pyramid containing nephrons. (C) A single nephron.

Each kidney has two distinct regions—an inner medulla and an outer cortex. The **renal medulla** is composed of conical masses of tissue called *renal pyramids* and appears striated. The **renal cortex** forms a shell around the medulla and dips into the medulla between renal pyramids, forming *renal columns*. The granular appearance of the cortex is due to the random organization of tiny tubules associated with the **nephrons** (nef´ronz), the kidney's functional units (fig. 17.2).

 CHECK YOUR RECALL

1. Where are the kidneys located?
2. Describe kidney structure.
3. Name the kidney's functional unit.

Kidney Functions

The kidneys help maintain homeostasis by regulating the composition and volume of the extracellular fluid. They accomplish this by removing metabolic wastes from the blood and combining them with excess water and electrolytes to form urine, which they then excrete.

The kidneys perform several other functions as well: They secrete the hormone erythropoietin (see chapter 12, p. 308) to help control the rate of red blood cell formation; play a role in the activation of vitamin D; and help maintain blood pressure by secreting the enzyme renin. Since blood pressure depends on adequate blood volume, kidneys also help control blood pressure by maintaining the volume of the extracellular fluid.

Renal Blood Vessels

The **renal arteries,** which arise from the abdominal aorta, supply blood to the kidneys. These arteries transport a large volume of blood. When a person is at rest, the renal arteries usually carry 15–30% of the total cardiac output into the kidneys.

A renal artery enters a kidney through the hilum and gives off several branches, called *interlobar arteries,* which pass between the renal pyramids. At the junction between the medulla and cortex, the interlobar arteries branch to form a series of incomplete arches, the *arcuate arteries,* which in turn give rise to *interlobular arteries*. The final branches of the interlobular arteries, called **afferent arterioles** (af´er-ent ar-te´re-ōlz), lead to the nephrons (figs. 17.3 and 17.4).

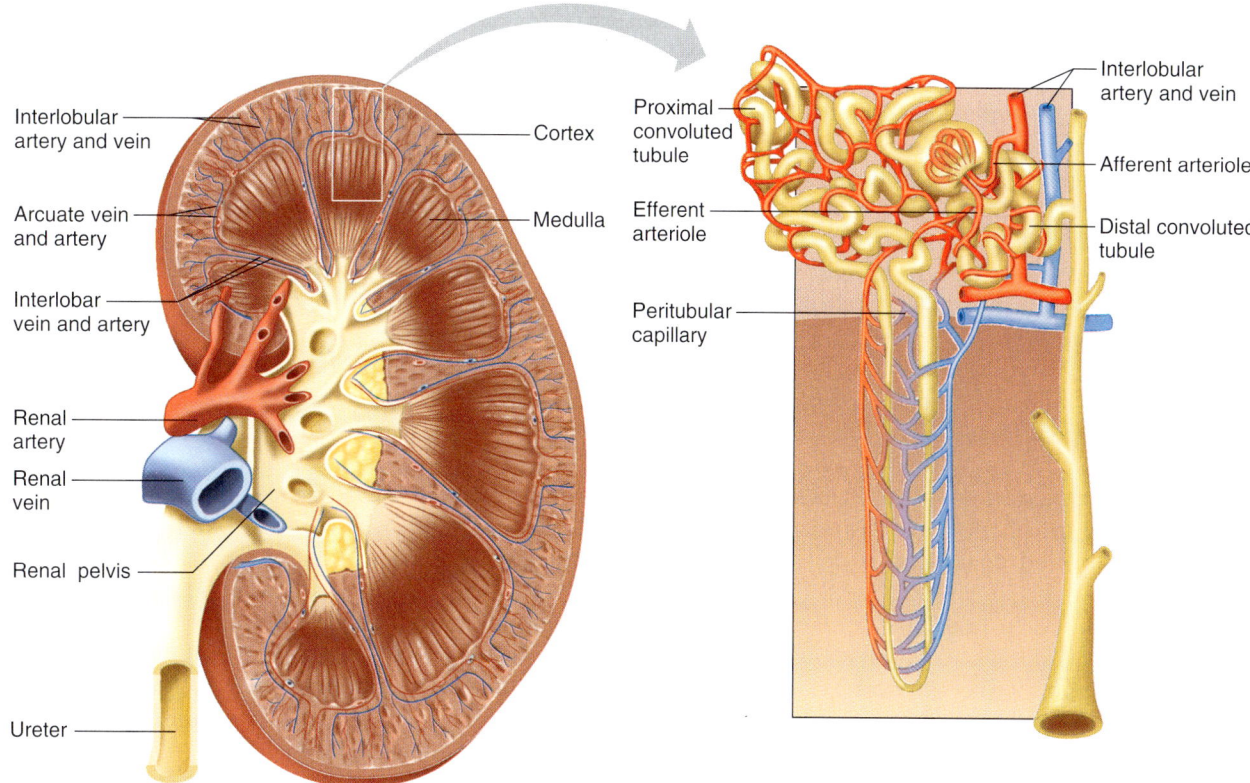

Figure 17.3
Main branches of the renal artery and renal vein.

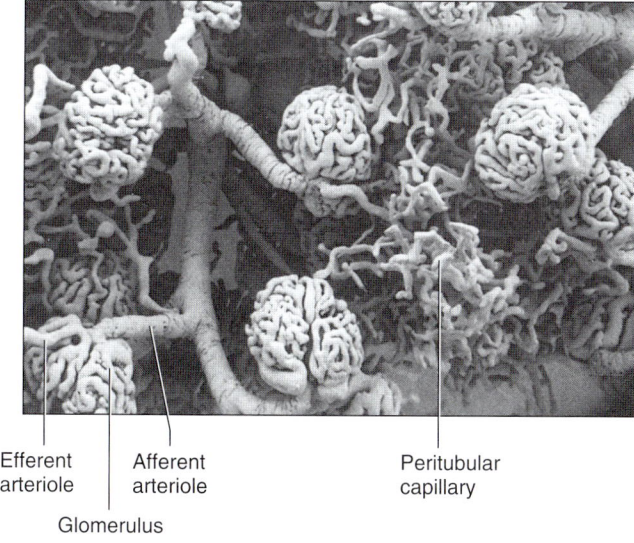

Figure 17.4
Scanning electron micrograph of a cast of the renal blood vessels associated with glomeruli (200×). From *Tissues and Organs: A Text-Atlas of Scanning Electron Microscopy,* by R. G. Kessel and R. H. Kardon, ©1979 W. H. Freeman and Company, all rights reserved.

Venous blood returns through a series of vessels that correspond generally to arterial pathways. The **renal vein** then joins the inferior vena cava as it courses through the abdominal cavity (see fig. 17.1).

A kidney transplant can help patients with end-stage renal disease. This procedure requires a kidney from a living or recently deceased donor whose tissues are antigenically similar (histocompatible) to those of the recipient. A surgeon places the kidney in the depression on the medial surface of the right or left ilium (iliac fossa). The surgeon then connects the renal artery and vein of the donor kidney to the recipient's iliac artery and vein, respectively, and the ureter of the donor kidney to the dome of the recipient's urinary bladder.

Nephrons

Nephron Structure

A kidney contains about 1 million nephrons. Each nephron consists of a **renal corpuscle** and a **renal tubule** (see fig. 17.2C). Fluid flows through renal tubules on its way out of the body.

A renal corpuscle is composed of a tangled cluster of blood capillaries called a **glomerulus** (glo-mer´u-lus). Glomerular capillaries filter fluid, the first step in urine formation. A thin-walled, saclike structure called a **glomerular capsule** (glo-mer´u-lar kap´sul) surrounds the glomerulus (fig. 17.5). The glomerular capsule, an expansion at the proximal end of a renal tubule, receives the fluid the glomerulus filters. The renal tubule leads away from the glomerular capsule and becomes highly coiled. This coiled portion is called the *proximal convoluted tubule*.

The proximal convoluted tubule dips toward the renal pelvis, becoming the *descending limb of the nephron loop* (loop of Henle). The tubule then curves back toward its renal corpuscle and forms the *ascending limb of the nephron loop*. The ascending limb returns to the region of the renal corpuscle, where it becomes highly coiled again and is called the *distal convoluted tubule*.

Distal convoluted tubules from several nephrons merge in the renal cortex to form a *collecting duct* (technically not part of the nephron), which in turn passes into the renal medulla and enlarges as other collecting ducts join it. The resulting tube empties into a minor calyx through an opening in a renal papilla. Figure 17.6 summarizes the structure of a nephron and its associated blood vessels.

CHECK YOUR RECALL

1. List the general functions of the kidneys.
2. Trace the blood supply to the kidney.
3. Name the parts of a nephron.

Blood Supply of a Nephron

The cluster of capillaries that forms a glomerulus arises from an afferent arteriole. After passing through the glomerular capillaries, blood (less any filtered fluid) enters an **efferent arteriole** (ef´er-ent ar-te´re-ōl), whose diameter is smaller than that of the afferent vessel (see fig. 17.4). This is instead of entering a venule, the usual circulatory route. The efferent arteriole resists blood flow to some extent. This backs up blood into the glomerulus, increasing pressure in the glomerular capillary.

The efferent arteriole branches into a complex, freely interconnecting network of capillaries, called the **peritubular capillary** (per´´ĭ-tu´bu-lar kap´ĭ-ler´´e) **system,** that surrounds the renal tubule (see figs. 17.4 and 17.6). Blood in the peritubular capillary system is under low pressure. After flowing through the capillary network, the blood rejoins blood from other branches of the peritubular capillary system and enters the venous system of the kidney.

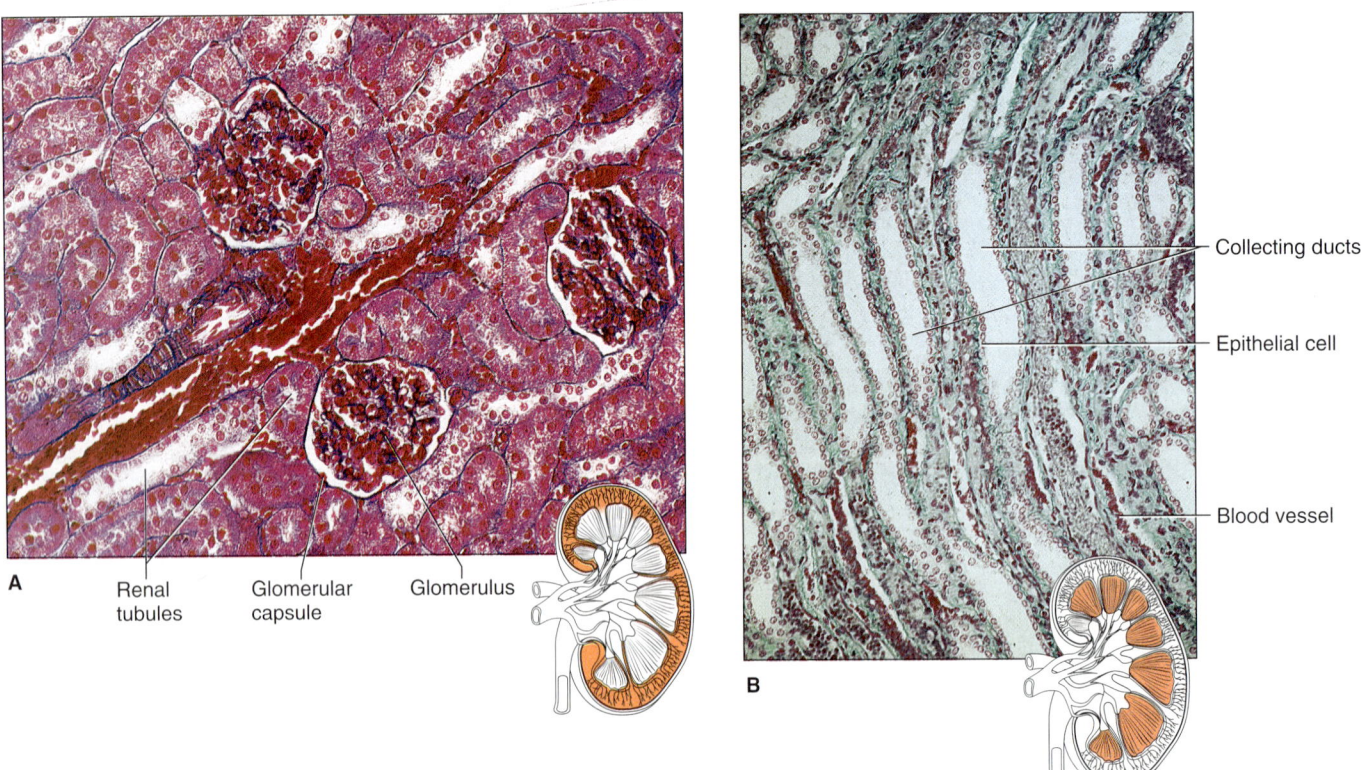

Figure 17.5
Microscopic view of the kidney. (*A*) Light micrograph of a section of the human renal cortex (220×). (*B*) Light micrograph of the renal medulla (80×).

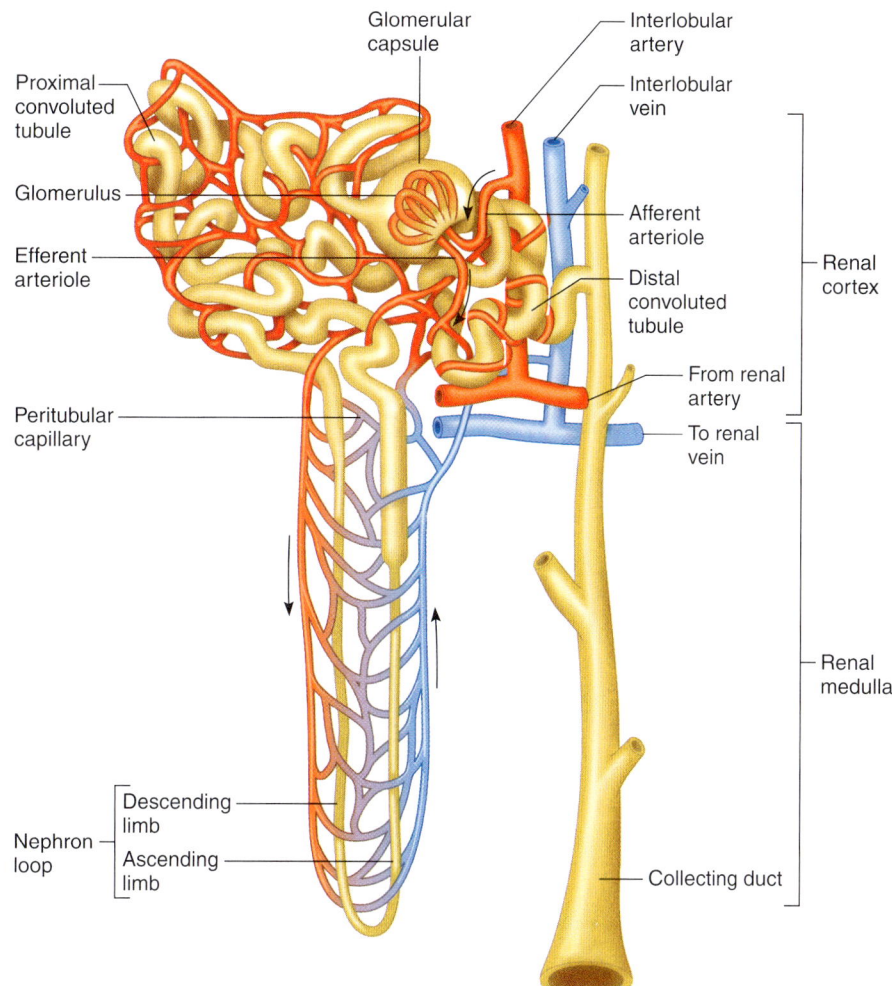

Figure 17.6
Structure of a nephron and the blood vessels associated with it. Arrows indicate the direction of blood flow.

Juxtaglomerular Apparatus

Near its origin, the distal convoluted tubule passes between and contacts afferent and efferent arterioles. At the point of contact, the epithelial cells of the distal tubule are quite narrow and densely packed. These cells comprise a structure called the *macula densa*.

Close by, in the walls of the arterioles near their attachments to the glomerulus, are some enlarged smooth muscle cells called *juxtaglomerular cells*. With cells of the macula densa, they constitute the **juxtaglomerular apparatus** (juks″tah-glo-mer´u-lar ap″ah-ra´tus), or juxtaglomerular complex (fig. 17.7). Its function in the control of renin secretion is described later in this chapter.

CHECK YOUR RECALL

1. Describe the system of blood vessels that supplies a nephron.
2. What structures form the juxtaglomerular apparatus?

17.3 Urine Formation

Urine formation begins with filtration of plasma by the glomerular capillaries, a process called **glomerular filtration.** However, glomerular filtration produces 180 liters of fluid, more than four times total body water, every 24 hours. Glomerular filtration could not continue for very long unless most of this filtered fluid were returned to the internal environment. The kidney accomplishes this by the process of **tubular reabsorption,** selectively reclaiming just the right amounts of substances, such as water, electrolytes and glucose, that the body needs. Waste products and excess substances are allowed out of the body. Some substances that the body must eliminate, such as hydrogen ions and certain toxins, are removed even faster than through filtration alone by the process of **tubular secretion.**

The final product of these three processes is **urine.** By selectively excreting waste products and excess materials in the urine, the kidneys contribute to homeostasis by maintaining the composition of the internal environment.

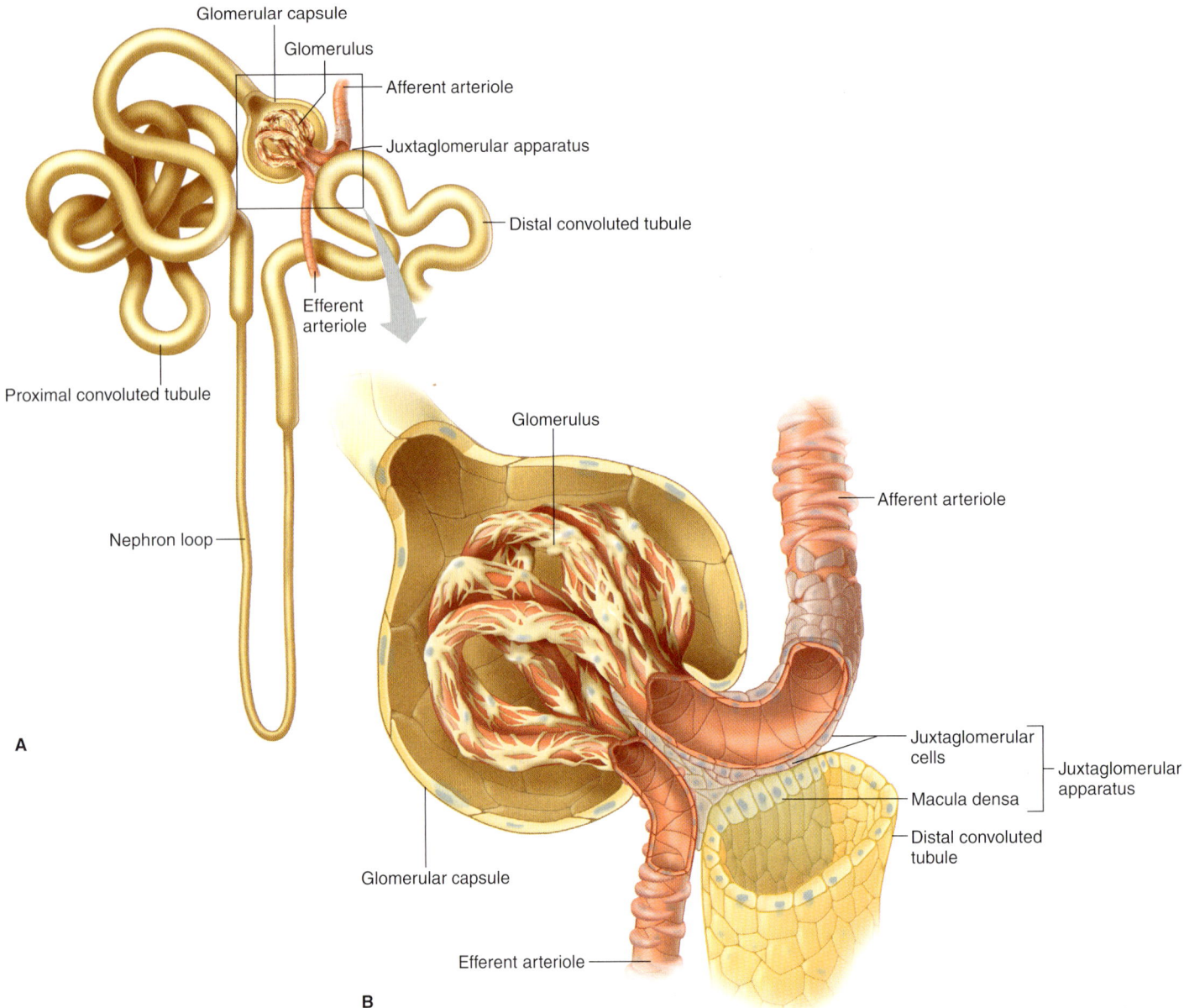

Figure 17.7
Juxtaglomerular apparatus. (*A*) Location of the juxtaglomerular apparatus. (*B*) Enlargement of a section of the juxtaglomerular apparatus, which consists of the macula densa and the juxtaglomerular cells.

Glomerular Filtration

Urine formation begins when water and certain dissolved substances are filtered out of glomerular capillaries and into glomerular capsules (fig. 17.8*A*). This filtration is similar to filtration at the arteriolar ends of other capillaries. However, many tiny openings (fenestrae) in glomerular capillary walls make glomerular capillaries much more permeable than capillaries in other tissues (fig. 17.8*B*).

The glomerular capsule receives the resulting **glomerular filtrate,** which is similar in composition to the filtrate that becomes tissue fluid elsewhere in the body. That is, glomerular filtrate is mostly water and the same components as blood plasma, except for the large protein molecules. Table 17.1 shows the relative concentrations of some substances in plasma, glomerular filtrate, and urine.

Filtration Pressure

As in other capillaries, the hydrostatic pressure of blood forces substances through the glomerular capillary wall. (Recall that glomerular capillary pressure is high compared to that of other capillaries.) The osmotic pressure of plasma in the glomerulus and the hydrostatic pres-

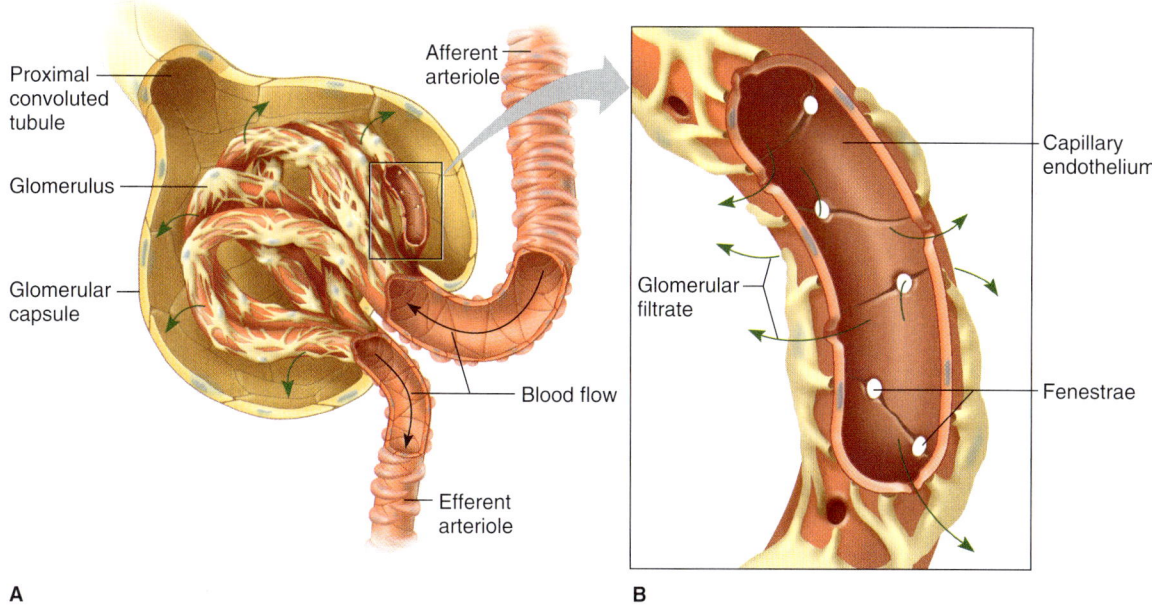

Figure 17.8
Glomerular filtration. (A) The first step in urine formation is filtration of substances out of glomerular capillaries and into the glomerular capsule. (B) Glomerular filtrate passes through the fenestrae of the capillary endothelium.

TABLE 17.1	RELATIVE CONCENTRATIONS OF SUBSTANCES IN THE PLASMA, GLOMERULAR FILTRATE, AND URINE		
	CONCENTRATIONS (mEq/L)		
SUBSTANCE	PLASMA	GLOMERULAR FILTRATE	URINE
Sodium (Na^+)	142	142	128
Potassium (K^+)	5	5	60
Calcium (Ca^{+2})	4	4	5
Magnesium (Mg^{+2})	3	3	15
Chloride (Cl^-)	103	103	134
Bicarbonate (HCO_3^-)	27	27	14
Sulfate (SO_4^{-2})	1	1	33
Phosphate (PO_4^{-3})	2	2	40
	CONCENTRATIONS (mg/100 mL)		
SUBSTANCE	PLASMA	GLOMERULAR FILTRATE	URINE
Glucose	100	100	0
Urea	26	26	1,820
Uric acid	4	4	53

Note: mEq/L = milliequivalents per liter.

sure inside the glomerular capsule also influence this movement. An increase in either of these pressures opposes movement out of the capillary and thus reduces filtration. The net pressure forcing substances out of the glomerulus is the **net filtration pressure,** and it is normally always positive, favoring filtration at the glomerulus.

If arterial blood pressure plummets, as can occur during *shock*, glomerular hydrostatic pressure may fall below the level required for filtration. At the same time, epithelial cells of the renal tubules may not receive sufficient nutrients to maintain their high metabolic rates. As a result, cells die (tubular necrosis), impairing renal functions. Such changes can cause chronic renal failure.

Filtration Rate

The glomerular filtration rate is directly proportional to net filtration pressure. Consequently, factors that affect glomerular hydrostatic pressure, glomerular plasma osmotic pressure, or hydrostatic pressure in the glomerular capsule also affect filtration rate. For example, any change in the diameters of the afferent and efferent arterioles changes glomerular hydrostatic pressure, also altering the glomerular filtration rate.

The afferent arteriole, through which blood enters the glomerulus, may constrict in response to sympathetic nerve impulses. Blood flow diminishes, filtration pressure decreases, and filtration rate drops. On the other hand, if the efferent arteriole (through which blood leaves the glomerulus) constricts, blood backs up into the glomerulus, net filtration pressure increases, and filtration rate rises. Vasodilation of these vessels causes opposite effects.

In capillaries, the plasma colloid osmotic pressure that attracts water inward (see chapter 12, p. 315) opposes the blood pressure that forces water and dissolved substances outward. During filtration through the capillary wall, proteins remaining in the plasma raise colloid osmotic pressure within the glomerular capillary. As this pressure rises, filtration decreases. Conversely, conditions that decrease plasma colloid osmotic pressure, such as a decrease in plasma protein concentration, increase the filtration rate.

In *glomerulonephritis,* the glomerular capillaries are inflamed and become more permeable to proteins, which appear in the glomerular filtrate and in urine (proteinuria). At the same time, the protein concentration in blood plasma decreases (hypoproteinemia), and this decreases blood colloid osmotic pressure. As a result, less tissue fluid moves into the capillaries, and edema develops.

The hydrostatic pressure in the glomerular capsule sometimes changes because of an obstruction, such as a stone in a ureter or an enlarged prostate gland pressing on the urethra. If this occurs, fluids back up into renal tubules and raise the hydrostatic pressure in the glomerular capsule. Because any increase in capsular pressure opposes glomerular filtration, the filtration rate may decrease significantly.

In an average adult, the glomerular filtration rate for the nephrons of both kidneys is about 125 milliliters per minute, or 180,000 milliliters (180 liters, or nearly 45 gallons) in 24 hours. Only a small fraction is excreted as urine. Instead, most fluid that passes through the renal tubules is reabsorbed and reenters the plasma.

CHECK YOUR RECALL

1. Which processes form urine?
2. Which forces affect net filtration pressure?
3. Which factors influence the rate of glomerular filtration?

Regulation of Filtration Rate

The glomerular filtration rate is usually relatively constant. To help maintain homeostasis, however, the glomerular filtration rate may increase when body fluids are in excess and decrease when the body must conserve fluid.

Sympathetic nervous system reflexes that respond to changes in blood pressure and blood volume can alter the glomerular filtration rate. If blood pressure or volume drops sufficiently, afferent arterioles vasoconstrict, decreasing the glomerular filtration rate. This helps ensure that less urine forms when the body must conserve water. Conversely, vasodilation of afferent arterioles increases the glomerular filtration rate to counter increased blood volume or blood pressure.

Another mechanism to control filtration rate uses the enzyme *renin.* Juxtaglomerular cells secrete renin in response to three types of stimuli: (1) whenever special cells in the afferent arteriole sense a drop in blood pressure; (2) in response to sympathetic stimulation; and (3) when the macula densa (see fig. 17.7) senses decreased amounts of chloride, potassium, and sodium ions reaching the distal tubule. Once in the bloodstream, renin reacts with the plasma protein *angiotensinogen* to form *angiotensin I.* A second enzyme (*angiotensin converting enzyme,* or ACE) in the lungs and in plasma quickly converts angiotensin I to *angiotensin II.*

Angiotensin II carries out a number of actions that help maintain sodium balance, water balance, and blood pressure (fig. 17.9). Angiotensin II vasoconstricts the efferent arteriole, which causes blood to back up into the glomerulus, thus raising glomerular capillary hydrostatic pressure. This important action helps minimize the decrease in glomerular filtration rate when systemic blood pressure is low. Angiotensin II has a major effect on the kidneys by stimulating secretion of the adrenal hormone aldosterone, which stimulates tubular reabsorption of sodium.

The heart secretes another hormone, atrial natriuretic peptide (ANP), when blood volume increases. ANP increases sodium excretion by a number of mechanisms, including increasing the glomerular filtration rate.

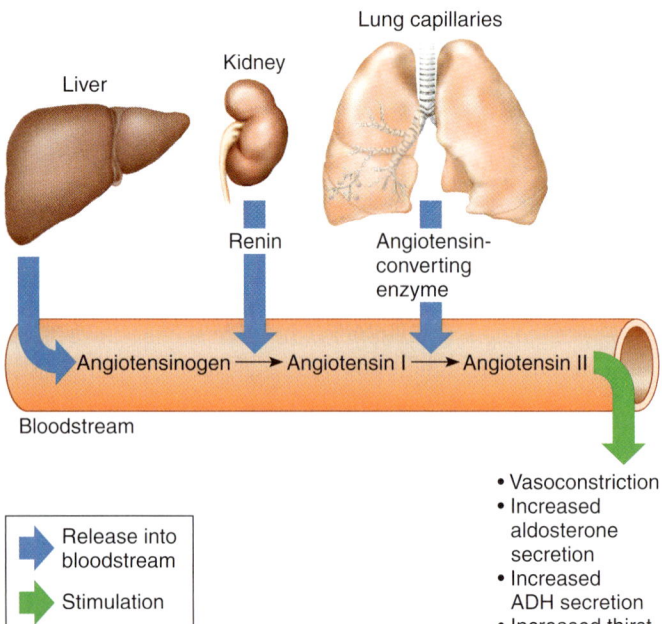

Figure 17.9

The formation of angiotensin II in the bloodstream involves several organs and includes multiple actions that conserve sodium and water.

Elevated blood pressure (hypertension) is sometimes associated with excess release of renin, followed by increased formation of the vasoconstrictor angiotensin II. Patients with this form of high blood pressure often take a drug that prevents the formation of angiotensin II by inhibiting the action of the enzyme that converts angiotensin I into angiotensin II. Such a drug is called an *angiotensin converting enzyme inhibitor* (ACE inhibitor).

CHECK YOUR RECALL

1. What is the function of the macula densa?
2. How does renin help regulate filtration rate?

Tubular Reabsorption

Comparing the composition of glomerular filtrate entering the renal tubule with that of urine leaving the tubule reveals that the fluid changes as it passes through the tubule (see table 17.1). For example, glucose is present in glomerular filtrate but absent in urine. In contrast, urea and uric acid are much more concentrated in urine than in glomerular filtrate. Such changes in fluid composition are largely the result of **tubular reabsorption** (tu´bu-lar re-ab-sorp´shun), the process by which substances are transported out of the tubular fluid, through the epithelium of the renal tubule, and into the interstitial fluid. These substances then diffuse into the peritubular capillaries (fig. 17.10).

Tubular reabsorption returns substances to the internal environment. The term *tubular* is used because this process is controlled by the epithelial cells that make up the renal tubules. In tubular reabsorption, substances must first cross the cell membrane facing the inside of the tubule and then cross the cell membrane facing the interstitial fluid.

The basic rules for movements across cell membranes apply to tubular reabsorption. Substances moving down a concentration gradient must be lipid-soluble, or there must be a carrier or channel for that substance. Active transport, requiring ATP, may move substances uphill against a concentration gradient.

Most of the plasma flowing through the kidney escapes filtration. Approximately 80% continues on through the peritubular capillaries. Peritubular capillary blood is under low pressure because it has already passed through two arterioles. Also, the wall of the peritubular capillary is more permeable than that of other capillaries. Both of these factors enhance the rate of fluid reabsorption from the renal tubule.

Tubular reabsorption occurs throughout the renal tubule, but most of it takes place in the proximal convoluted portion. The epithelial cells here have many microscopic projections called *microvilli* that form a "brush border" on their free surfaces. These tiny extensions greatly increase the surface area exposed to glomerular filtrate and enhance reabsorption.

Segments of the renal tubule are adapted to reabsorb specific substances, using particular modes of

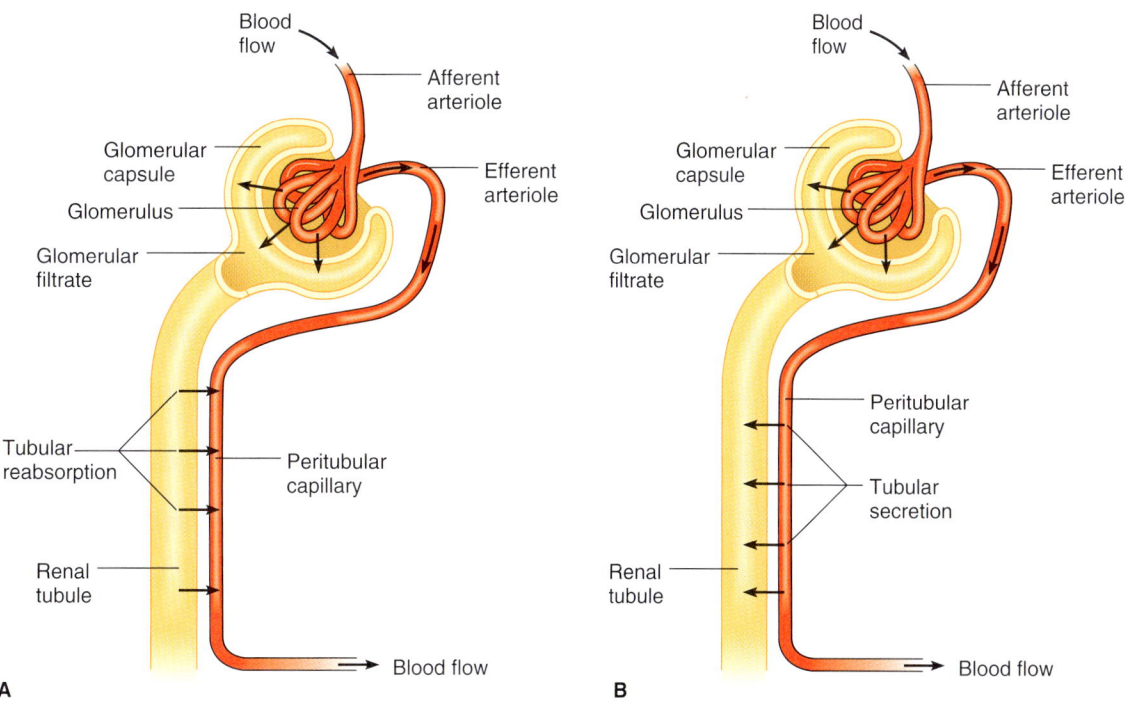

Figure 17.10

Two processes in addition to glomerular filtration contribute to urine formation. (*A*) Reabsorption is the process by which substances are transported from the glomerular filtrate into the blood within the peritubular capillary. (*B*) Secretion is the process by which substances are transported from the blood within the peritubular capillary into the renal tubule.

transport. Active transport, for example, reabsorbs glucose through the walls of the proximal tubule. Osmosis rapidly reabsorbs water through the epithelium of the proximal tubule. However, portions of the distal tubule and collecting duct are almost impermeable to water, a characteristic important in the regulation of urine concentration and volume, as described later in this chapter.

Active transport utilizes carrier molecules in cell membranes (see chapter 3, p. 62). These carriers transport certain molecules across the membrane, release them, and then repeat the process. Such a mechanism, however, has a *limited transport capacity;* that is, it can only transport a certain number of molecules in a given time because the number of carriers is limited.

Usually, carrier molecules are able to transport all of the glucose in glomerular filtrate. But when the plasma glucose concentration increases to a critical level, called the *renal plasma threshold,* more glucose molecules are in the filtrate than the active transport mechanism can handle. As a result, some glucose remains in the tubular fluid and is excreted in urine.

G lucose in urine, called *glucosuria* (or *glycosuria*), may occur following intravenous administration of glucose. It may also occur in a patient with diabetes mellitus whose blood glucose concentration rises abnormally (see chapter 11, p. 298).

Amino acids enter the glomerular filtrate and are reabsorbed in the proximal convoluted tubule. Three different active transport mechanisms reabsorb different groups of amino acids, whose members have similar structures. Normally, only a trace of amino acids remains in urine.

Glomerular filtrate is nearly free of protein except for traces of albumin, a small protein that is taken up by endocytosis through the brush border of epithelial cells lining the proximal convoluted tubule. Proteins inside epithelial cells are broken down to amino acids, which then move into the blood of the peritubular capillary.

The epithelium of the proximal convoluted tubule also reabsorbs other substances, including creatine; lactic, citric, uric, and ascorbic (vitamin C) acids; and phosphate, sulfate, calcium, potassium, and sodium ions. Active transport mechanisms with limited transport capacities reabsorb these chemicals. However, these substances usually do not appear in urine until glomerular filtrate concentration exceeds a particular substance's threshold.

Sodium and Water Reabsorption

Substances that remain in the renal tubule become more concentrated as water is reabsorbed from the filtrate. Water reabsorption occurs passively by osmosis, primarily in the proximal convoluted tubule, and is closely associated with the active reabsorption of sodium ions (fig. 17.11). If sodium reabsorption

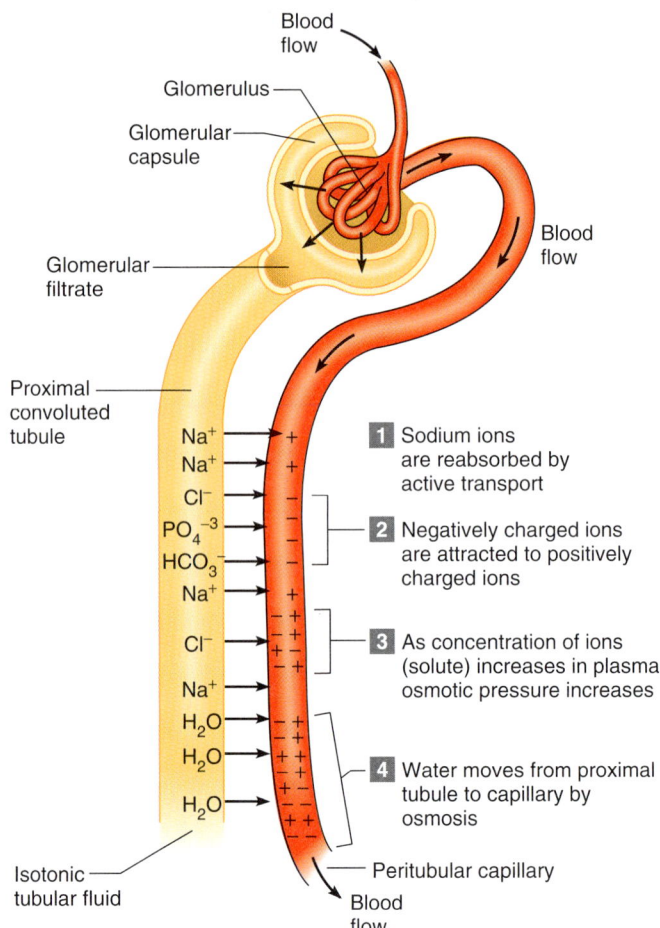

Figure 17.11
In the proximal portion of the renal tubule, osmosis reabsorbs water in response to active transport reabsorbing sodium and other solutes.

increases, water reabsorption increases; if sodium reabsorption decreases, water reabsorption decreases also.

Active transport (the sodium pump) reabsorbs about 70% of sodium ions in the proximal segment of the renal tubule. As these positively charged ions (Na^+) move through the tubular wall, negatively charged ions, including chloride ions (Cl^-), phosphate ions (PO_4^{-3}), and bicarbonate ions (HCO_3^-), accompany them. These negatively charged ions move because of the electrochemical attraction between particles of opposite charge. This *passive transport* does not require direct expenditure of cellular energy.

As active transport moves more sodium ions into the peritubular capillary, along with various negatively charged ions, the concentration of solutes within the peritubular blood increases. Since water moves across cell membranes from regions of lesser solute concentration (hypotonic) toward regions of greater solute concentration (hypertonic), water moves by osmosis from the renal tubule into the peritubular capillary. Movement of solutes and water into the peritubular capillary greatly reduces the fluid volume within the renal tubule. The end of the proximal convoluted tubule is in osmotic equilibrium, and the remaining tubular fluid is isotonic.

CHECK YOUR RECALL

1. Which chemicals are normally present in the glomerular filtrate but not in urine?
2. Which mechanisms reabsorb solutes from the glomerular filtrate?
3. Describe the role of passive transport in urine formation.

Regulation of Urine Concentration and Volume

Active transport continues to reabsorb sodium ions as the tubular fluid moves through the nephron loop, the distal convoluted tubule, and the collecting duct. Water is absorbed passively by osmosis in various segments of the renal tubule. As a result, almost all the sodium ions and water that enter the renal tubule as part of the glomerular filtrate are reabsorbed before urine is excreted.

The hormones aldosterone and ADH (antidiuretic hormone) may stimulate additional reabsorption of sodium and water, respectively. The changes in sodium and water excretion in response to these hormones are the final adjustments the kidney makes to maintain a constant internal environment.

As discussed in chapter 11 (p. 293), the adrenal gland secretes aldosterone in response to changes in the blood concentrations of sodium and potassium ions. Aldosterone stimulates the distal tubule to reabsorb sodium and secrete potassium. Angiotensin II is another important stimulator of aldosterone secretion.

Neurons in the hypothalamus produce ADH, which the posterior pituitary releases in response to a decreasing water concentration in blood or a decrease in blood volume. When ADH reaches the kidney, it increases the water permeability of the epithelial linings of the distal convoluted tubule and collecting duct, and water moves rapidly out of these segments by osmosis—that is, water is reabsorbed. Consequently, urine volume falls, and soluble wastes and other substances become more concentrated, which minimizes loss of body fluids when dehydration is likely.

If body fluids contain excess water, ADH secretion decreases. As blood levels of ADH drop, the epithelial linings of the distal segment and collecting duct become less permeable to water, less water is reabsorbed, and urine is more dilute, excreting the excess water. Table 17.2 summarizes the role of ADH in urine production.

Urea and Uric Acid Excretion

Urea is a by-product of amino acid catabolism. Consequently, its plasma concentration reflects the amount of protein in the diet. Urea enters the renal tubule by filtration. About 50% of it is reabsorbed, and the remainder is excreted in urine.

TABLE 17.2 ROLE OF ADH IN REGULATING URINE CONCENTRATION AND VOLUME

1. Concentration of water in blood decreases.
2. Increase in osmotic pressure of body fluids stimulates osmoreceptors in hypothalamus of brain.
3. Hypothalamus signals posterior pituitary to release ADH.
4. Blood carries ADH to kidneys.
5. ADH causes distal convoluted tubules and collecting ducts to increase water reabsorption by osmosis.
6. Urine becomes concentrated, and urine volume decreases.

Uric acid is a product of the metabolism of certain organic bases in nucleic acids. Active transport reabsorbs all the uric acid normally present in glomerular filtrate, but a small amount is secreted into the renal tubule for excretion in urine.

Elevated concentrations of uric acid in the plasma cause *gout*. Because uric acid is relatively insoluble, it precipitates when in excess. In gout, crystals of uric acid are deposited in joints and other tissues, causing inflammation and extreme pain. The joints of the great toes are most often affected, but other hand and feet joints may also be involved. Drugs that inhibit uric acid reabsorption, thus increasing its excretion, are used to treat gout. People once thought gluttony caused gout, but the condition may be inherited.

CHECK YOUR RECALL

1. How does the hypothalamus regulate urine concentration and volume?
2. Explain how urea and uric acid are excreted.

Tubular Secretion

In **tubular secretion** (tu´bu-lar se-kre´shun), certain substances move from the plasma of the peritubular capillary into the fluid of the renal tubule. As a result, the amount of a particular chemical excreted in the urine may exceed the amount filtered from the plasma in the glomerulus (fig. 17.10*B*). As in the case of tubular reabsorption, the term *tubular* is used because this process is controlled by the epithelial cells that make up the renal tubules.

Active transport mechanisms similar to those that function in reabsorption secrete some substances. Secretory mechanisms, however, transport substances in the opposite direction. For example, the epithelium of the proximal convoluted segment actively secretes certain organic compounds, including penicillin, creatinine, and histamine, into the tubular fluid.

Hydrogen ions are also actively secreted throughout the entire renal tubule. Secretion of hydrogen ions is important in regulating the pH of body fluids, as chapter 18 (p. 491) explains.

Most potassium ions in the glomerular filtrate are actively reabsorbed in the proximal convoluted tubule, but some may be secreted in the distal segment and collecting duct. During this process, active reabsorption of sodium ions from the tubular fluid produces a negative electrical charge within the tubule. Because positively charged potassium ions (K^+) and hydrogen ions (H^+) are attracted to negatively charged regions, these ions move passively through the tubular epithelium and enter the tubular fluid (fig. 17.12). Potassium ions are also secreted by active processes.

To summarize, urine forms as a result of the following:

- Glomerular filtration of materials from blood plasma.
- Reabsorption of substances, including glucose; water; urea; proteins; creatine; amino acids; lactic, citric, and uric acids; and phosphate, sulfate, calcium, potassium, and sodium ions.
- Secretion of substances, including penicillin, histamine, phenobarbital, hydrogen ions, ammonia, and potassium ions.

Table 17.3 summarizes some specific functions of the nephron segments and the collecting duct.

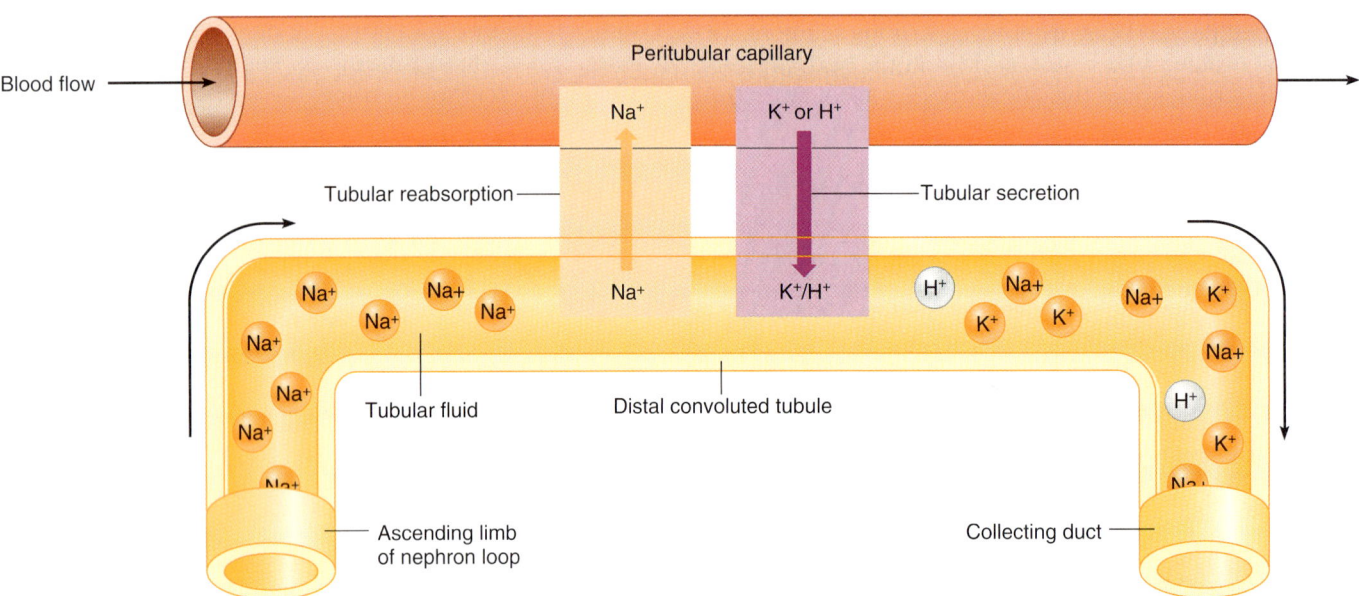

Figure 17.12
In the distal convoluted tubule, potassium ions (or hydrogen ions) may be passively secreted in response to the active reabsorption of sodium ions.

TABLE 17.3 — FUNCTIONS OF NEPHRON COMPONENTS

PART	FUNCTION
RENAL CORPUSCLE	
Glomerulus	Filtration of water and dissolved substances from plasma
Glomerular capsule	Receives glomerular filtrate
RENAL TUBULE	
Proximal convoluted tubule	Reabsorption of glucose; amino acids; creatine; lactic, uric, citric, and ascorbic acids; phosphate, sulfate, calcium, potassium, and sodium ions by active transport
	Reabsorption of water by osmosis
	Reabsorption of chloride ions and other negatively charged ions by electrochemical attraction
	Active secretion of substances such as penicillin, histamine, creatinine, and hydrogen ions
Descending limb of nephron loop	Reabsorption of water by osmosis
Ascending limb of nephron loop	Reabsorption of sodium, potassium, and chloride ions by active transport
Distal convoluted tubule	Reabsorption of sodium ions by active transport
	Reabsorption of water by osmosis
	Secretion of hydrogen and potassium ions both actively and passively by electrochemical attraction
Collecting duct	Reabsorption of water by osmosis

(Note: Although the collecting duct is not anatomically part of the nephron, it is included here because of its functional importance.)

CHECK YOUR RECALL

1. Define *tubular secretion*.
2. Which substances are actively secreted?
3. How does sodium reabsorption affect potassium secretion?

Urine Composition

Urine composition reflects the amounts of water and solutes that the kidneys must eliminate from the body or retain in the internal environment to maintain homeostasis. The composition differs considerably from time to time because of variations in dietary intake and physical activity. Urine is about 95% water, and usually contains urea and uric acid. It may also contain a trace of amino acids and a variety of electrolytes, whose concentrations vary directly with amounts in the diet (see table 17.1).

The volume of urine produced is usually between 0.6 and 2.5 liters per day, depending on fluid intake, environmental temperature, relative humidity of the surrounding air, and the person's emotional condition, respiratory rate, and body temperature. Urine output of 50–60 milliliters per hour is normal; output of less than 30 milliliters per hour may indicate kidney failure.

Glucose, proteins, hemoglobin, ketones, and blood cells are not normally in urine, but circumstances may explain their presence. Glucose in urine may follow a large intake of carbohydrates, proteins may appear following vigorous physical exercise, and ketones may appear after a prolonged fast. Pregnant women may have glucose in their urine as birth nears.

CHECK YOUR RECALL

1. List the normal constituents of urine.
2. What factors affect urine volume?

17.4 Urine Elimination

After urine forms in the nephrons, it passes from the collecting ducts through openings in the renal papillae and enters the calyces of the kidney (see fig. 17.2). From there, it passes through the renal pelvis, and a ureter conveys it to the urinary bladder (see fig. 17.1 and reference plate 6, p. 27). The urethra excretes urine to the outside.

Ureters

Each **ureter** is a tube about 25 centimeters long that begins as the funnel-shaped renal pelvis. It extends downward behind the parietal peritoneum and runs parallel to the vertebral column. Within the pelvic cavity, each ureter courses forward and medially, joining the urinary bladder from underneath.

The ureter wall has three layers. The inner layer, or *mucous coat,* is continuous with the linings of the renal tubules and the urinary bladder. The middle layer, or *muscular coat,* consists largely of smooth muscle fibers. The outer layer, or *fibrous coat,* is connective tissue (fig. 17.13).

The muscular walls of the ureters propel the urine. Muscular peristaltic waves, originating in the renal pelvis, force urine along the length of the ureter. When a peristaltic wave reaches the urinary bladder, a jet of urine spurts into the bladder. A flaplike fold of mucous membrane covers the opening through which urine enters the bladder. This fold acts as a valve, allowing urine to enter the bladder from the ureter but preventing it from backing up.

Kidney stones, which are usually composed of uric acid, calcium oxalate, calcium phosphate, or magnesium phosphate, can form in the collecting ducts and renal pelvis. Such a stone passing into a ureter causes severe pain that begins in the region of the kidney and radiates into the abdomen, pelvis, and lower limbs. It may also cause nausea and vomiting.

About 60% of kidney stones pass from the body on their own. Other stones were once removed surgically but are now shattered with intense sound waves. In this procedure, called *extracorporeal shock-wave lithotripsy* (*ESWL*), the patient is placed in a stainless steel tub filled with water. A spark-gap electrode produces shock waves underwater, and a reflector concentrates and focuses the shock-wave energy on the stones. The resulting sandlike fragments then leave in urine.

CHECK YOUR RECALL

1. Describe the structure of a ureter.
2. How is urine moved from the renal pelvis to the urinary bladder?
3. What prevents urine from backing up from the urinary bladder into the ureters?

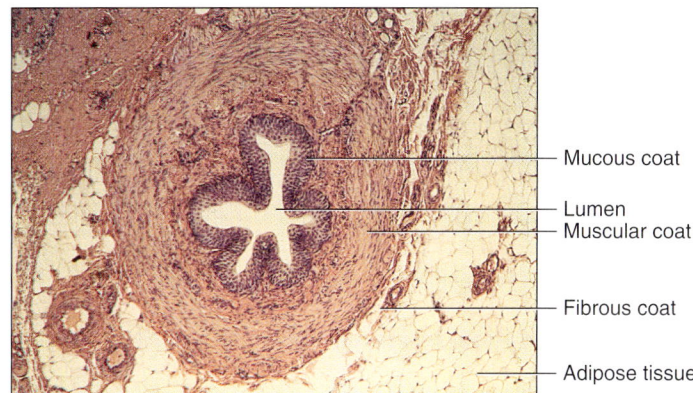

Figure 17.13
Cross section of a ureter (75×).

Urinary Bladder

The **urinary bladder** is a hollow, distensible, muscular organ that stores urine and forces it into the urethra (see fig. 17.1 and reference plate 6, p. 27). It is within the pelvic cavity, behind the symphysis pubis and beneath the parietal peritoneum.

The pressure of surrounding organs alters the shape of the somewhat spherical bladder. When empty, the inner wall of the bladder forms many folds, but as the bladder fills with urine, the wall becomes smoother. At the same time, the superior surface of the bladder expands upward into a dome.

The internal floor of the bladder includes a triangular area called the *trigone,* which has an opening at each of its three angles (fig. 17.14). Posteriorly, at the base of the trigone, the openings are those of the ureters. Anteriorly, at the apex of the trigone, a short, funnel-shaped extension, called the *neck* of the bladder, contains the opening into the urethra.

The wall of the urinary bladder has four layers. The inner layer, or *mucous coat,* includes several thicknesses of transitional epithelial cells. The thickness of this tissue changes as the bladder expands and contracts. Thus, during distension, the tissue may be only two or three cells thick; during contraction, it may be five or six cells thick (see chapter 5, p. 97).

The second layer of the bladder wall is the *submucous coat.* It consists of connective tissue and contains many elastic fibers.

The third layer of the bladder wall, or *muscular coat,* is composed primarily of coarse bundles of smooth muscle fibers. These bundles are interlaced in all directions and at all depths, and together they comprise the **detrusor muscle** (de-truz´or mus´l). The portion of the detrusor muscle that surrounds the neck of the bladder forms an *internal urethral sphincter.* Sustained contraction of this muscle prevents the bladder from emptying until pressure within the bladder increases to a certain level. The detrusor muscle is innervated with parasympathetic nerve fibers that function in the micturition reflex, discussed next.

The outer layer of the bladder wall, or *serous coat,* consists of the parietal peritoneum. This layer is only on the bladder's upper surface. Elsewhere, the outer coat is composed of connective tissue.

Because the linings of the ureters and the urinary bladder are continuous, infectious agents, such as bacteria, may ascend from the urinary bladder into the ureters. Inflammation of the urinary bladder, called *cystitis,* is more common in women than in men because the female urethral pathway is shorter. Inflammation of the ureter is called *ureteritis.*

CHECK YOUR RECALL

1. Describe the trigone of the urinary bladder.
2. Describe the structure of the bladder wall.
3. What kind of nerve fibers supply the detrusor muscle?

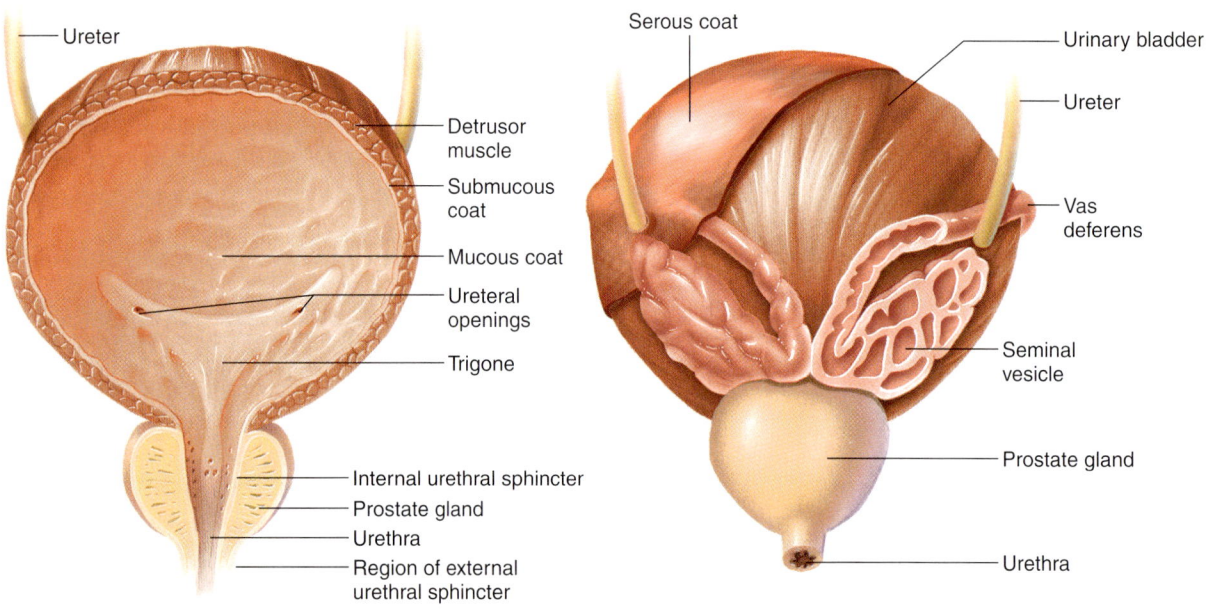

Figure 17.14
A male urinary bladder. (*A*) Longitudinal section. (*B*) Posterior view.

Micturition

Micturition (mik´´tu-rish´un), or urination, is the process that expels urine from the urinary bladder. In micturition, the detrusor muscle contracts, as do muscles in the abdominal wall and pelvic floor. At the same time, muscles in the thoracic wall and diaphragm do not contract. Micturition also requires relaxation of the *external urethral sphincter*. This muscle, which is part of the urogenital diaphragm described in chapter 8 (p. 194), surrounds the urethra about 3 centimeters from the bladder and is composed of voluntary skeletal muscle tissue.

Distension of the bladder wall as it fills with urine stimulates stretch receptors, triggering the micturition reflex. The *micturition reflex center* is in the spinal cord. When sensory impulses from the stretch receptors signal the reflex center, parasympathetic motor impulses travel to the detrusor muscle, which contracts rhythmically in response. A sensation of urgency accompanies this action.

The urinary bladder may hold as much as 600 milliliters of urine before stimulating pain receptors, but the urge to urinate usually begins when it contains about 150 milliliters. As urine volume increases to 300 milliliters or more, the sensation of fullness intensifies, and contractions of the bladder wall become more powerful. When these contractions are strong enough to force the internal urethral sphincter open, another reflex signals the external urethral sphincter to relax, and the bladder can empty.

Because the external urethral sphincter is composed of skeletal muscle, it is under conscious control. Thus, the sphincter muscle ordinarily remains contracted until a person decides to urinate. Nerve centers in the brain stem and cerebral cortex that can partially inhibit the micturition reflex aid this control. When a person decides to urinate, the external urethral sphincter relaxes, and the micturition reflex is no longer inhibited. Nerve centers within the pons and the hypothalamus of the brain heighten the micturition reflex. Consequently, the detrusor muscle contracts, and urine is excreted through the urethra. Within a few moments, the neurons of the micturition reflex fatigue, the detrusor muscle relaxes, and the bladder begins to fill with urine again.

> Damage to the spinal cord above the sacral region destroys voluntary control of urination. However, if the micturition reflex center and its sensory and motor fibers are uninjured, micturition may continue to occur reflexly. In this case, the bladder collects urine until its walls stretch enough to trigger a micturition reflex, and the detrusor muscle contracts in response. This condition is called an *automatic bladder*.

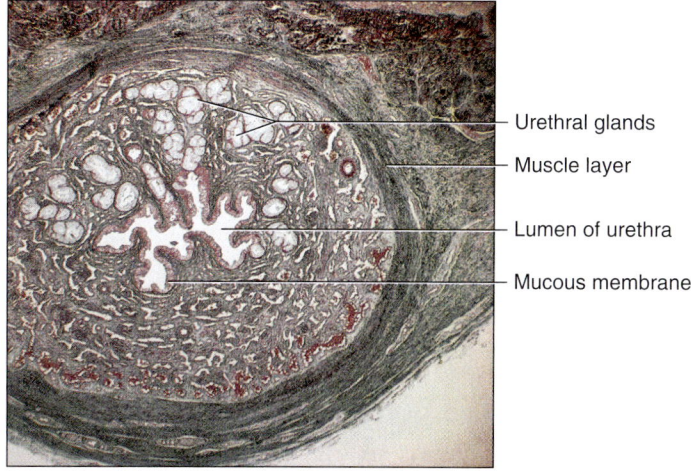

Figure 17.15
Cross section through the urethra (10×).

Urethra

The **urethra** is a tube that conveys urine from the urinary bladder to the outside (see fig. 17.1 and reference plate 7, p. 28). Its wall is lined with mucous membrane and contains a thick layer of smooth muscle tissue, whose fibers are generally directed longitudinally. The urethral wall also contains numerous mucous glands, called *urethral glands,* which secrete mucus into the urethral canal (fig. 17.15).

 CHECK YOUR RECALL

1. Describe micturition.
2. How is it possible to consciously inhibit the micturition reflex?
3. Describe the structure of the urethra.

Clinical Terms Related to the Urinary System

anuria (ah-nu´re-ah) Absence of urine due to failure of kidney function or to an obstruction in a urinary pathway.
bacteriuria (bak-te´´re-u´re-ah) Bacteria in urine.
cystectomy (sis-tek´to-me) Surgical removal of the urinary bladder.
cystitis (sis-ti´tis) Inflammation of the urinary bladder.
cystoscope (sis´to-skōp) Instrument used to visually examine the interior of the urinary bladder.
cystotomy (sis-tot´o-me) Incision of the urinary bladder wall.
diuresis (di´´u-re´sis) Increased urine excretion.
diuretic (di´´u-ret´ik) Substance that increases urine production.

Organization

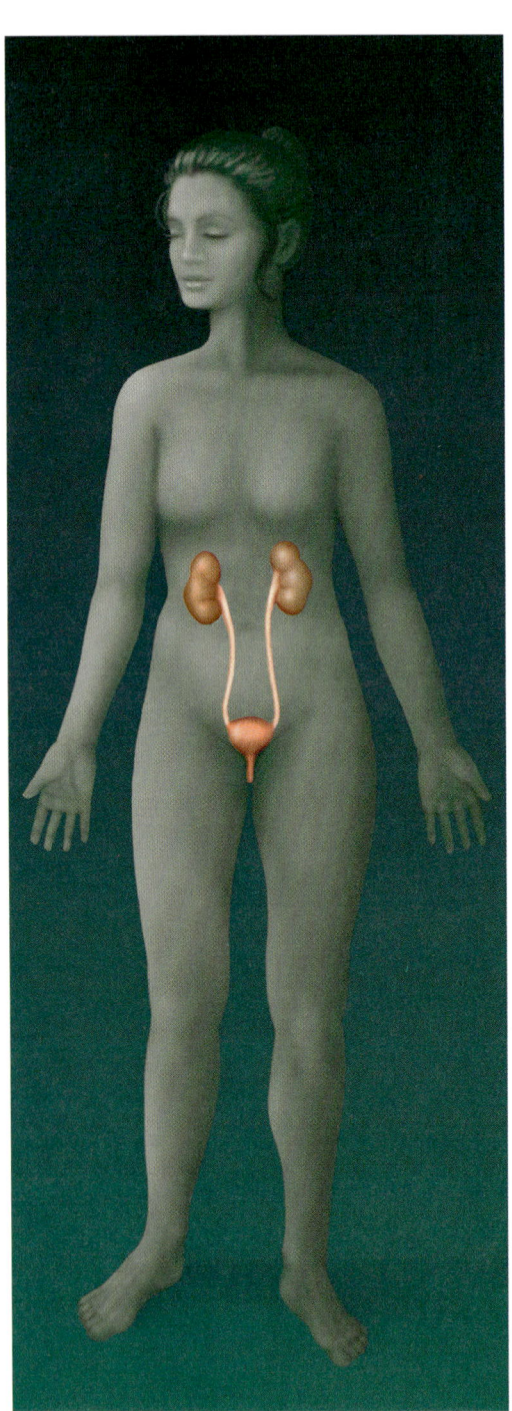

Urinary System
The urinary system controls the composition of the internal environment.

Integumentary System
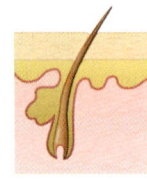
The urinary system compensates for water loss due to sweating. The kidneys and skin both play a role in vitamin D production.

Skeletal System

The kidneys and bone tissue work together to control plasma calcium levels.

Muscular System
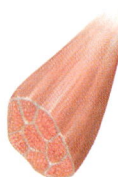
Muscle tissue controls urine elimination from the bladder.

Nervous System
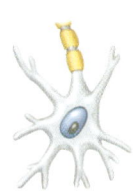
The nervous system influences urine production and elimination.

Endocrine System

The endocrine system influences urine production.

Cardiovascular System

The urinary system controls blood volume. Blood volume and blood pressure play a role in determining water and solute excretion.

Lymphatic System
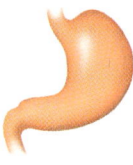
The kidneys control extracellular fluid volume and composition (including lymph).

Digestive System
The kidneys compensate for fluids lost by the digestive system.

Respiratory System
The kidneys and the lungs work together to control the pH of the internal environment.

Reproductive System

The urinary system in males shares organs with the reproductive system. The kidneys compensate for fluids lost from the male and female reproductive systems.

dysuria (dis-u´re-ah) Painful or difficult urination.
hematuria (hem´´ah-tu´re-ah) Blood in urine.
incontinence (in-kon´ti-nens) Inability to control urination and/or defecation reflexes.
nephrectomy (ně-frek´to-me) Surgical removal of a kidney.
nephrolithiasis (nef´´ro-lĭ-thi´ah-sis) Kidney stones.
nephroptosis (nef´´rop-to´sis) Movable or displaced kidney.
oliguria (ol´´ĭ-gu´re-ah) Scanty urine output.
polyuria (pol´´ě-u´re-ah) Excess urine output.
pyelolithotomy (pi´´ě-lo-lĭ-thot´o-me) Removal of a stone from the renal pelvis.
pyelonephritis (pi´´ě-lo-ne-fri´tis) Inflammation of the renal pelvis.
pyelotomy (pi´´ě-lot´o-me) Incision into the renal pelvis.
pyuria (pi-u´re-ah) Pus (white blood cells) in urine.
uremia (u-re´me-ah) Accumulation in blood of substances ordinarily excreted in urine.
ureteritis (u-re´´ter-i´tis) Inflammation of the ureter.
urethritis (u´´re-thri´tis) Inflammation of the urethra.

Clinical Connection

A 25-year-old man arrived at the hospital in acute renal failure following three days of malaise. He claimed to have taken only cough medicine, then slept 36 hours. A urinalysis revealed blood and protein, but it was the renal biopsy that alarmed the emergency room physician—the kidney tubules were damaged, with telltale crystals of calcium oxalate obstructing their lumens. The physician knew that these crystals result from drinking ethylene glycol, a component of antifreeze, but the patient denied such activity. He was sent home and continued to undergo hemodialysis as an outpatient. A week later, he returned with seizures and other signs of neurological impairment. Over the next two months, with aggressive treatment, his kidney function returned to normal. It was then that he admitted to having drunk the colorless, odorless, sweet-tasting antifreeze, in a suicide attempt. Because the dose that he took was not large enough to damage his kidneys to the extent seen, physicians suspected that an ingredient in the cough medicine increased the toxicity of the chemical.

SUMMARY OUTLINE

17.1 Introduction (p. 461)
The urinary system consists of the kidneys, ureters, urinary bladder, and urethra.

17.2 Kidneys (p. 461)
1. Location of the kidneys
   a. The kidneys are high on the posterior wall of the abdominal cavity.
   b. They are behind the parietal peritoneum.
2. Kidney structure
   a. A kidney contains a hollow renal sinus.
   b. The ureter expands into the renal pelvis.
   c. Renal papillae project into the renal sinus.
   d. Each kidney divides into a medulla and a cortex.
3. Kidney functions
   a. The kidneys maintain homeostasis by removing metabolic wastes from blood and excreting them.
   b. They also help regulate red blood cell production; blood pressure; and the volume, composition, and pH of body fluids.
4. Renal blood vessels
   a. Arterial blood flows through the renal artery, interlobar arteries, arcuate arteries, interlobular arteries, afferent arterioles, glomerular capillaries, efferent arterioles, and peritubular capillaries.
   b. Venous blood returns through a series of vessels that correspond to the arterial pathways.
5. Nephrons
   a. Nephron structure
      (1) A nephron is the functional unit of the kidney.
      (2) It consists of a renal corpuscle and a renal tubule.
         (a) The corpuscle consists of a glomerulus and glomerular capsule.
         (b) Segments of the renal tubule include the proximal convoluted tubule, nephron loop (ascending and descending limbs), and distal convoluted tubule, which empties into a collecting duct.
      (3) The collecting duct empties into the minor calyx of the renal pelvis.
   b. Blood supply of a nephron
      (1) The glomerular capillary receives blood from the afferent arteriole and passes it to the efferent arteriole.
      (2) The efferent arteriole gives rise to the peritubular capillary system, which surrounds the renal tubule.
   c. Juxtaglomerular apparatus
      (1) The juxtaglomerular apparatus is at the point of contact between the distal convoluted tubule and the afferent and efferent arterioles.
      (2) It consists of the macula densa and juxtaglomerular cells.

17.3 Urine Formation (p. 465)
Nephrons remove wastes from blood and regulate water and electrolyte concentrations. Urine is the end product.
1. Glomerular filtration
   a. Urine formation begins when water and dissolved materials filter out of glomerular capillaries.
   b. Glomerular capillaries are much more permeable than the capillaries in other tissues.
   c. The composition of the filtrate is similar to that of tissue fluid.
2. Filtration pressure
   a. Filtration is due mainly to hydrostatic pressure inside glomerular capillaries.
   b. The osmotic pressure of plasma and the hydrostatic pressure in the glomerular capsule also affect filtration.
   c. Filtration pressure is the net force moving material out of the glomerulus and into the glomerular capsule.

3. Filtration rate
   a. Rate of filtration varies with filtration pressure.
   b. Filtration pressure changes with the diameters of the afferent and efferent arterioles.
   c. As colloid osmotic pressure in the glomerulus increases, filtration rate decreases.
   d. As hydrostatic pressure in a glomerular capsule increases, filtration rate decreases.
   e. The kidneys produce about 125 milliliters of glomerular fluid per minute, most of which is reabsorbed.
4. Regulation of filtration rate
   a. Glomerular filtration rate remains relatively constant, but may increase or decrease as required.
   b. Increased sympathetic nerve activity can decrease glomerular filtration rate.
   c. When the macula densa senses decreased amounts of chloride, potassium, and sodium ions in the distal tubule, it causes juxtaglomerular cells to release renin.
   d. This triggers a series of changes leading to vasoconstriction of afferent and efferent arterioles, which may affect glomerular filtration rate, and aldosterone secretion, which stimulates tubular sodium reabsorption.
5. Tubular reabsorption
   a. Substances are selectively reabsorbed from glomerular filtrate.
   b. The peritubular capillary's permeability adapts it for reabsorption.
   c. Most reabsorption occurs in the proximal tubule, where epithelial cells have microvilli.
   d. Different modes of transport reabsorb various substances in particular segments of the renal tubule.
      (1) Active transport reabsorbs glucose and amino acids.
      (2) Osmosis reabsorbs water.
   e. Active transport mechanisms have limited transport capacities.
6. Sodium and water reabsorption
   a. Substances that remain in the filtrate are concentrated as water is reabsorbed.
   b. Active transport reabsorbs sodium ions.
   c. As positively charged sodium ions move out of the filtrate, negatively charged ions follow them.
   d. Water is passively reabsorbed by osmosis.
7. Regulation of urine concentration and volume
   a. Most sodium is reabsorbed before urine is excreted.
   b. Antidiuretic hormone increases the permeability of the distal convoluted tubule and collecting duct, promoting water reabsorption.
8. Urea and uric acid excretion
   a. Diffusion passively reabsorbs urea. About 50% of the urea is excreted in urine.
   b. Active transport reabsorbs uric acid. Some uric acid is secreted into the renal tubule.
9. Tubular secretion
   a. Secretion transports substances from plasma to the tubular fluid.
   b. Various organic compounds are secreted actively.
   c. Potassium and hydrogen ions are secreted both actively and passively.
10. Urine composition
    a. Urine is about 95% water, and it also usually contains urea and uric acid.
    b. Urine contains varying amounts of electrolytes and may contain a trace of amino acids.
    c. Urine volume varies with fluid intake and with certain environmental factors.

17.4 Urine Elimination (p. 473)
1. Ureters
   a. The ureter extends from the kidney to the urinary bladder.
   b. Peristaltic waves in the ureter force urine to the urinary bladder.
2. Urinary bladder
   a. The urinary bladder stores urine and forces it through the urethra during micturition.
   b. The openings for the ureters and urethra are at the three angles of the trigone.
   c. A portion of the detrusor muscle forms an internal urethral sphincter.
3. Micturition
   a. Micturition expels urine.
   b. Micturition contracts the detrusor muscle and relaxes the external urethral sphincter.
   c. Micturition reflex
      (1) Distension stimulates stretch receptors in the bladder wall.
      (2) The micturition reflex center in the spinal cord sends parasympathetic motor impulses to the detrusor muscle.
      (3) As the bladder fills, its internal pressure increases, forcing the internal urethral sphincter open.
      (4) A second reflex relaxes the external urethral sphincter unless voluntary control maintains its contraction.
      (5) Nerve centers in the cerebral cortex and brain stem aid control of urination.
4. Urethra
   The urethra conveys urine from the urinary bladder to the outside.

REVIEW EXERCISES

1. Name and list the general functions of the organs of the urinary system. (p. 461)
2. Describe the external and internal structure of a kidney. (p. 461)
3. List the functions of the kidneys. (p. 462)
4. Name the vessels through which blood passes as it travels from the renal artery to the renal vein. (p. 462)
5. Distinguish between a renal corpuscle and a renal tubule. (p. 464)
6. Name the parts through which fluid passes from the glomerulus to the collecting duct. (p. 464)
7. Describe the location and structure of the juxtaglomerular apparatus. (p. 465)
8. Define *filtration pressure*. (p. 466)
9. Compare the composition of glomerular filtrate with that of blood plasma. (p. 467)
10. Explain how the diameters of the afferent and efferent arterioles affect the rate of glomerular filtration. (p. 467)
11. Explain how changes in the osmotic pressure of blood plasma affect the glomerular filtration rate. (p. 468)
12. Explain how the hydrostatic pressure of a glomerular capsule affects the rate of glomerular filtration. (p. 468)
13. Describe two mechanisms by which the body regulates filtration rate. (p. 468)
14. Discuss how tubular reabsorption is selective. (p. 469)
15. Explain how the peritubular capillary is adapted for reabsorption. (p. 469)
16. Explain how epithelial cells of the proximal convoluted tubule are adapted for reabsorption. (p. 469)
17. Explain why active transport mechanisms have limited transport capacities. (p. 470)

18. Define *renal plasma threshold*. (p. 470)
19. Explain how amino acids and proteins are reabsorbed. (p. 470)
20. Describe the effect of sodium reabsorption on the reabsorption of negatively charged ions. (p. 470)
21. Explain how sodium reabsorption affects water reabsorption. (p. 470)
22. Describe the function of ADH. (p. 471)
23. Compare the processes that reabsorb urea and uric acid. (p. 471)
24. Explain how potassium ions may be secreted passively. (p. 472)
25. List the common constituents of urine and their sources. (p. 473)
26. List some of the factors that affect the urine volume produced daily. (p. 473)
27. Describe the structure and function of a ureter. (p. 473)
28. Explain how the muscular wall of the ureter helps move urine. (p. 473)
29. Describe the structure and location of the urinary bladder. (p. 474)
30. Define *detrusor muscle*. (p. 474)
31. Distinguish between the internal and external urethral sphincters. (p. 474)
32. Describe the micturition reflex. (p. 475)
33. Explain how the micturition reflex can be voluntarily controlled. (p. 475)

CRITICAL THINKING

1. Imagine you are adrift at sea. Why will you dehydrate more quickly if you drink seawater instead of fresh water to quench your thirst?
2. Urinary tract infections frequently accompany sexually transmitted diseases. Why?
3. Would an excess or deficiency of renin be likely to cause hypertension (high blood pressure)? Cite a reason for your answer.
4. Why is protein in the urine a sign of kidney damage? What structures in the kidney are probably affected?
5. Why are people following high-protein diets advised to drink large quantities of water?
6. How might very low blood pressure impair kidney function?
7. An infant is born with narrowed renal arteries. What effect will this condition have on urine volume?
8. If a patient who has had major abdominal surgery receives intravenous fluids equal to the blood volume lost during surgery, would you expect urine volume to be greater or less than normal? Why?
9. If blood pressure plummets in a patient in shock as a result of a severe injury, how would you expect urine volume to change? Why?

WEB CONNECTIONS

Visit the website for additional study questions and more information about this chapter at:

http://www.mhhe.com/shieress8

chapter 18

Water, Electrolyte, and Acid-Base Balance

HEATSTROKE CAN BE DEADLY. August 2, 2001, was another 90° high-humidity day at training camp for the Minnesota Vikings in Mankato. The day before, offensive tackle Korey Stringer hadn't been able to participate in afternoon practice, citing exhaustion—but he vowed to make it the next morning. He did, but did not feel well. After vomiting three times, he walked over to an air-conditioned shelter, dizzy and breathing heavily. Trainers recognized the signs of heat exhaustion and took Stringer to a nearby medical facility, but it was too late. On arrival, Stringer's body temperature was a life-threatening 108°F, and he soon lost consciousness. To the shock and dismay of his teammates, he died at 1:50 the next morning.

Korey Stringer died of heatstroke, which occurs rapidly when the body is exposed to a heat index (heat considering humidity) of more than 105°F and body temperature rises to above 106°F. On that August day, the heat index was 110°F. Under these conditions, the body stops sweating, so heat can no longer be dissipated, and the organs fail. The situation is worse if the individual is heavy or if the body is covered. Stringer weighed 335 pounds and was exercising in full football gear.

During the heat wave of 2001, several athletes in their teens also succumbed to heatstroke in the weeks following Stringer's death. According to the Centers for Disease Control and Prevention, more than 300 people die in the United States each year from this preventable condition, most of them either elderly people or infants, who may have poor temperature control. Despite knowing the symptoms, heatstroke is unpredictable, because people have different limits. In the wake of Stringer's death, many players remembered feeling dizzy or experiencing chills when the weather was hot, but continuing to exercise anyway. Athletic trainers typically weigh players twice a day and are alerted to possible heatstroke if an athlete suddenly loses 6 to 8 pounds. After Stringer's death, sports medicine specialists advised the National Football League to shorten or change the time of practices when heat and humidity become dangerous, to enforce water breaks, and to allow players at least a week to adjust to a different climate before wearing full gear. Stringer's experience may save others by calling attention to the danger of heatstroke. Following is a list of the symptoms of heatstroke:

Headache
Dizziness
Exhaustion
Profuse sweating, which then stops
Dry, hot, and red skin
Pulse elevated as high as 180 beats per minute
Increased respiratory rate
Disorientation
Losing consciousness or having a seizure
Rapid rise in body temperature

Photo:
Korey Stringer was an offensive tackle for the Minnesota Vikings who died of heatstroke.

Chapter Objectives

After studying this chapter, you should be able to do the following:

18.1 Introduction
1. Explain water and electrolyte balance, and discuss the importance of this balance. (p. 482)

18.2 Distribution of Body Fluids
2. Describe how the body fluids are distributed within compartments, how fluid composition differs between compartments, and how fluids move from one compartment to another. (p. 482)

18.3 Water Balance
3. List the routes by which water enters and leaves the body, and explain how water intake and output are regulated. (p. 484)

18.4 Electrolyte Balance
4. Explain how electrolytes enter and leave the body, and how the intake and output of electrolytes are regulated. (p. 487)

18.5 Acid-Base Balance
5. List the major sources of hydrogen ions in the body. (p. 488)
6. Distinguish between strong and weak acids and bases. (p. 489)
7. Explain how chemical buffer systems, the respiratory center, and the kidneys minimize changing pH values of the body fluids. (p. 489)

18.6 Acid-Base Imbalances
8. Describe the causes and consequences of elevation or decrease of body fluid pH. (p. 491)

Aids to Understanding Words

de- [separation from] *de*hydration: Removal of water from the cells or body fluids.
extra- [outside] *extra*cellular fluid: Fluid outside the body cells.
im- [not] *im*balance: Condition in which factors are not in equilibrium.
intra- [within] *intra*cellular fluid: Fluid within the body cells.
neutr- [neither one nor the other] *neutr*al: Solution that is neither acidic nor basic.

Key Terms

acid (as´id)
acid-base buffer system (as´id-bās buf´er sis´tem)
base (bās)
electrolyte balance (e-lek´tro-līt bal´ans)
extracellular (ek´´strah-sel´u-lar)
intracellular (in´´trah-sel´u-lar)
transcellular (trans-sel´u-lar)
water balance (wot´er bal´ans)

18.1 Introduction

Two types of substances that are important in maintaining homeostasis in the body are water and **electrolytes,** molecules that release ions in water. The quantities of water and electrolytes must be in *balance,* meaning that the amounts entering the body must equal the amounts leaving it. Thus, the body needs mechanisms to (1) replace lost water and electrolytes, and (2) excrete any excess.

Because electrolytes are dissolved in the water of body fluids, water balance and electrolyte balance are interdependent. Consequently, anything that alters electrolyte concentrations necessarily alters the water concentration by either adding or removing solutes. Likewise, anything that changes the water concentration changes electrolyte concentrations by either concentrating or diluting them.

 A human being is 60% water. Losing one-tenth of that amount can be fatal.

18.2 Distribution of Body Fluids

Body fluids are not uniformly distributed throughout tissues but occur in regions, or *compartments,* of different volumes that contain fluids of varying compositions. The movement of water and electrolytes between these compartments is regulated to stabilize both their distribution and the composition of body fluids.

Fluid Compartments

The body of an average adult female is about 52% water by weight, and that of an average male is about 63% water. These proportions differ because females generally have more adipose tissue, which contains little water, than do males. Water in the body (about 40 liters), together with its dissolved electrolytes, is distributed into two major compartments—an intracellular fluid compartment and an extracellular fluid compartment.

The **intracellular** (in´´trah-sel´u-lar) **fluid compartment** includes all the water and electrolytes that cell membranes enclose. In other words, intracellular fluid is fluid within cells, and in an adult, it represents about 63% by volume of total body water.

The **extracellular** (ek´´strah-sel´u-lar) **fluid compartment** includes all the fluid outside cells—within the tissue spaces (interstitial fluid), blood vessels (plasma), and lymphatic vessels (lymph). Epithelial layers separate a specialized fraction of extracellular fluid from other extracellular fluids. This **transcellular** (trans-sel´u-lar) **fluid** includes *cerebrospinal fluid* of the central nervous system, *aqueous* and *vitreous humors* of the eyes, *synovial fluid* of the joints, *serous fluid* within body cavities, and fluid *secretions* of the exocrine glands. The fluids of the extracellular compartment constitute about 37% by volume of total body water (fig. 18.1).

Body Fluid Composition

Extracellular fluids generally have similar compositions, including high concentrations of sodium, chloride, and bicarbonate ions. These fluids include a greater concentration of calcium ions, and lesser concentrations of potassium, magnesium, phosphate, and sulfate ions than does intracellular fluid. The blood plasma fraction of extracellular fluid contains considerably more protein than does either interstitial fluid or lymph.

Intracellular fluid contains high concentrations of potassium, phosphate, and magnesium ions. It includes a greater concentration of sulfate ions and lesser concentrations of sodium, chloride, and bicarbonate ions than do extracellular fluids. Intracellular fluid also has a greater protein concentration than does plasma. Figure 18.2 shows these relative concentrations.

 CHECK YOUR RECALL

1. How are water balance and electrolyte balance interdependent?
2. Describe the normal distribution of water within the body.
3. Which electrolytes are in higher concentrations in extracellular fluids? In intracellular fluid?
4. How does protein concentration vary in different body fluids?

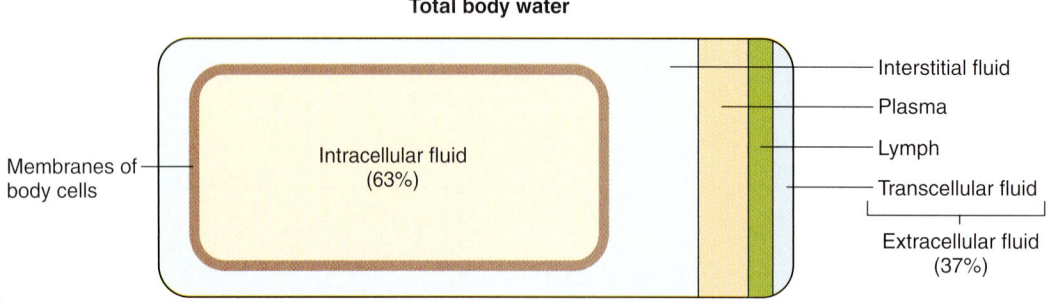

Figure 18.1
Cell membranes separate fluid in the intracellular compartment from fluid in the extracellular compartment. Approximately two-thirds of the water in the body is inside cells.

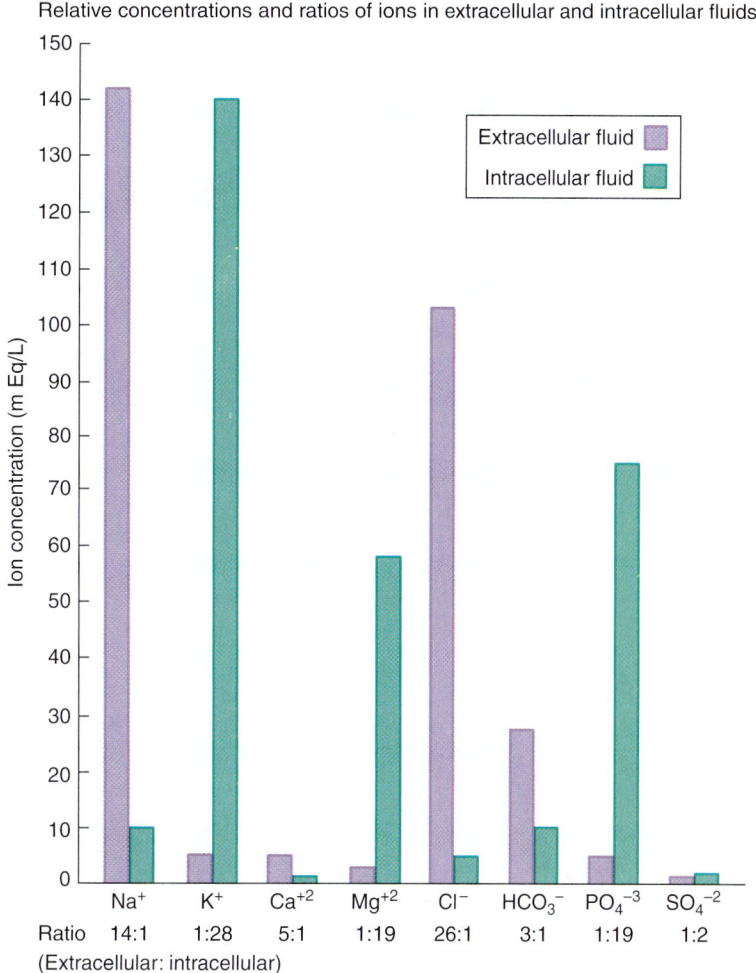

Figure 18.2
Extracellular fluids have relatively high concentrations of sodium (Na^+), calcium (Ca^{+2}), chloride (Cl^-), and bicarbonate (HCO_3^-) ions. Intracellular fluid has relatively high concentrations of potassium (K^+), magnesium (Mg^{+2}), phosphate (PO_4^{-3}), and sulfate (SO_4^{-2}) ions.

Movement of Fluid Between Compartments

Two major factors regulate the movement of water and electrolytes from one fluid compartment to another: *hydrostatic pressure* and *osmotic pressure* (fig. 18.3). As explained in chapter 13 (p. 345), fluid leaves the plasma at the arteriolar ends of capillaries and enters the interstitial spaces because of the net outward force of hydrostatic pressure (blood pressure). Fluid returns to the plasma from the interstitial spaces at the venular ends of capillaries because of the net inward force of *colloid osmotic pressure* due to the plasma proteins. Likewise, as mentioned in chapter 14 (p. 369), the hydrostatic pressure that develops within interstitial spaces forces the fluid in interstitial spaces into lymph capillaries. Lymph circulation returns interstitial fluid to the plasma.

Pressures similarly control fluid movement between the intracellular and extracellular compartments. Because hydrostatic pressure within the cells and surrounding

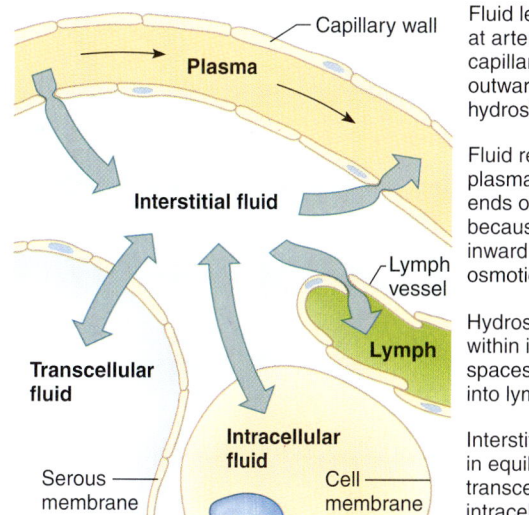

Figure 18.3
Net movements of fluids between compartments result from differences in hydrostatic and osmotic pressures.

interstitial fluid is ordinarily equal and stable, a change in osmotic pressure is the likely cause of any net fluid movement.

The sodium ion concentration in extracellular fluids is especially high. A decrease in this concentration causes net movement of water from the extracellular compartment into the intracellular compartment by osmosis. As a consequence, the cells swell. Conversely, if the sodium ion concentration in interstitial fluid increases, the net movement of water is outward from the intracellular compartment, and cells shrink as they lose water.

CHECK YOUR RECALL

1. Which factors control the movement of water and electrolytes from one fluid compartment to another?
2. How does the sodium ion concentration within body fluids affect the net movement of water between the compartments?

18.3 Water Balance

Water balance (wot´er bal´ans) exists when total water intake equals total water output. Homeostatic mechanisms maintain water balance.

Water Intake

The volume of water gained each day varies from individual to individual. An average adult living in a moderate environment takes in about 2,500 milliliters. Of this amount, drinking water or beverages supply probably 60%, while moist foods provide another 30%. The remaining 10% is a by-product of the oxidative metabolism of nutrients and is called **water of metabolism** (fig. 18.4A).

 The desert rodent known as the kangaroo rat does not have to drink water because it survives on the water of metabolism alone.

Regulation of Water Intake

The primary regulator of water intake is thirst. The intense feeling of thirst derives from the osmotic pressure of extracellular fluids and a *thirst center* in the hypothalamus. As the body loses water, the osmotic pressure of extracellular fluids increases. This stimulates *osmoreceptors* in the thirst center, which cause the person to feel thirsty and to seek water.

Thirst is a homeostatic mechanism, normally triggered whenever total body water decreases by as little as 1%. As a thirsty person drinks water, the act of drinking and the resulting stomach wall distension trigger nerve impulses that inhibit the thirst mechanism. Thus, drinking stops even before the swallowed water is absorbed.

CHECK YOUR RECALL

1. What is water balance?
2. Where is the thirst center located?
3. What stimulates fluid intake? What inhibits it?

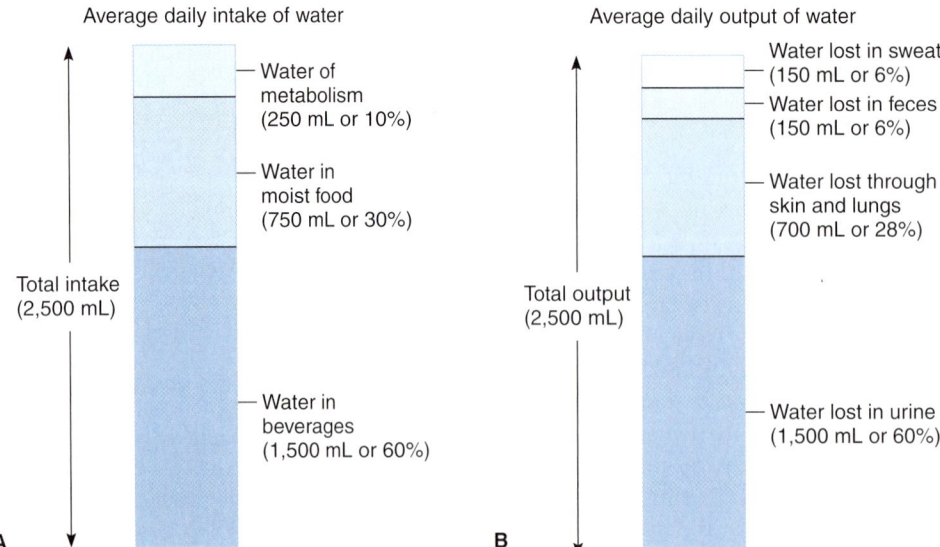

Figure 18.4

Water balance. (*A*) Major sources of body water. (*B*) Routes by which the body loses water. Urine production is most important in the regulation of water balance.

Topic of Interest

WATER BALANCE DISORDERS

Among the more common disorders that reflect an imbalance in the concentration of body fluids are dehydration, water intoxication, and edema.

Dehydration

In *dehydration,* water output exceeds water intake. This condition may develop following excess sweating or prolonged water deprivation accompanied by continued water output. In either case, as water is lost, the extracellular fluid becomes increasingly more concentrated, and water leaves cells by osmosis (fig. 18A). Dehydration may also accompany prolonged vomiting or diarrhea that depletes body fluids. During dehydration, the skin and mucous membranes of the mouth feel dry, and body weight drops. Hyperthermia may develop as the body's temperature-regulating mechanism becomes less effective due to lack of water for sweat.

Because infants' kidneys are less able to conserve water than those of adults, infants are more likely to become dehydrated. Elderly people are also especially susceptible to developing water imbalances because the sensitivity of their thirst mechanism decreases with age, and physical disabilities may make it difficult for them to obtain adequate fluids.

The treatment for dehydration is to replace the lost water and electrolytes. If only water is replaced, the extracellular fluid becomes more dilute than normal, producing a condition called water intoxication.

Water Intoxication

Babies rushed to emergency rooms because they are having seizures sometimes have drunk too much water, a rare condition called *water intoxication*. This can occur when a baby under six months of age is given several bottles of water a day or very dilute infant formula. The hungry infant drinks the water, and soon its tissues swell with the excess fluid. When the serum sodium level drops, the eyes begin to flutter, and a seizure occurs. As extracellular fluid becomes hypotonic, water enters the cells rapidly by osmosis (fig. 18B). Coma resulting from

continued

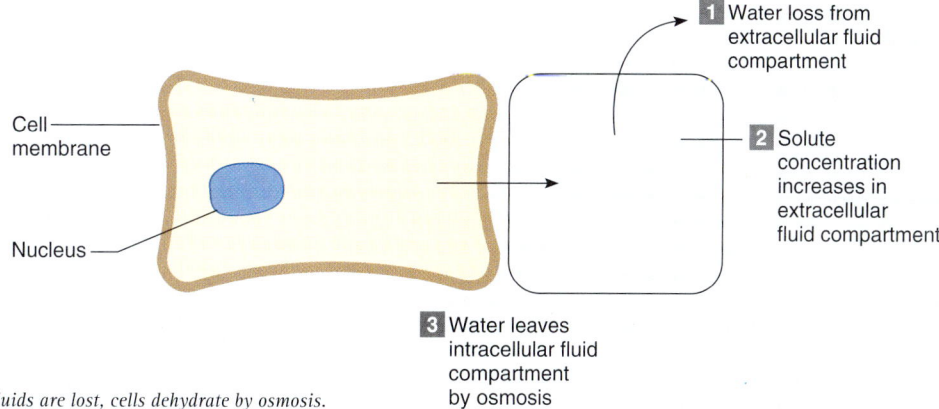

Figure 18A
If excess extracellular fluids are lost, cells dehydrate by osmosis.

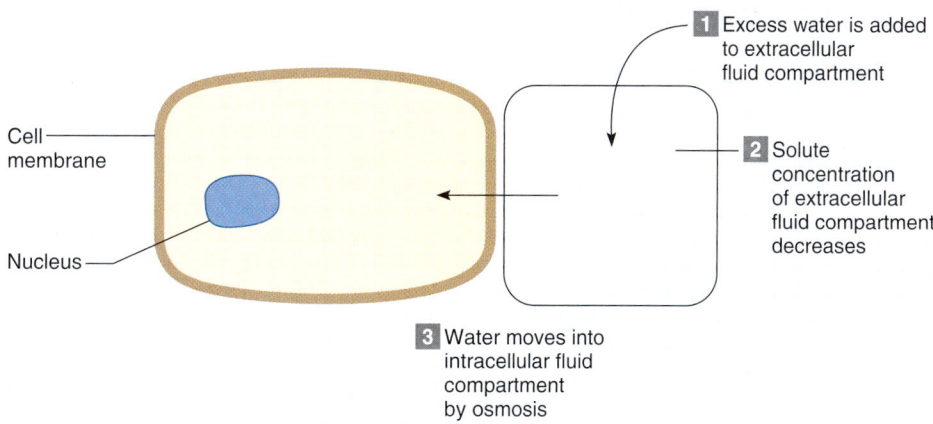

Figure 18B
If excess water is added to the extracellular fluid compartment, cells gain water by osmosis.

Topic of Interest

WATER BALANCE DISORDERS

Continued

swelling brain tissues may follow unless water intake is restricted and hypertonic salt solutions given. Usually, recovery is complete within a few days.

Edema

Edema is an abnormal accumulation of extracellular fluid within the interstitial spaces. A variety of factors can cause edema, including decrease in the plasma protein concentration (hypoproteinemia), obstruction of lymphatic vessels, increased venous pressure, and increased capillary permeability.

Hypoproteinemia may result from liver disease that hinders plasma protein synthesis; kidney disease (glomerulonephritis) that damages glomerular capillaries, allowing proteins to enter urine; or starvation, in which amino acid intake is insufficient to support synthesis of plasma proteins. In each of these instances, the plasma protein concentration decreases, which decreases plasma colloid osmotic pressure, reducing the normal return of tissue fluid to the venular ends of capillaries. Consequently, tissue fluid accumulates in the interstitial spaces.

As discussed in chapter 14 (p. 370), edema may result from *lymphatic obstructions* due to surgery or parasitic infections of lymphatic vessels. Back pressure develops in the lymphatic vessels, interfering with the normal movement of tissue fluid into them. At the same time, proteins that the lymphatic circulation ordinarily removes accumulate in interstitial spaces, raising the osmotic pressure of interstitial fluid. This effect draws still more fluid into the interstitial spaces.

If blood outflow from the liver into the inferior vena cava is blocked, venous pressure within the liver and portal blood vessels increases greatly. As a result, fluid with a high protein concentration is exuded from the surfaces of the liver and intestine into the peritoneal cavity. This increases the osmotic pressure of the abdominal fluid, which in turn attracts more water into the peritoneal cavity by osmosis. This condition, called *ascites*, distends the abdomen and is quite painful.

Edema may also result from increased capillary permeability accompanying *inflammation*. Recall that inflammation is a response to tissue damage and usually releases chemicals such as histamine from damaged cells. Histamine causes vasodilation and increased capillary permeability, so that excess fluid is filtered out of capillaries and enters interstitial spaces. Table 18A summarizes the factors that can cause edema.

TABLE 18A FACTORS ASSOCIATED WITH EDEMA

FACTOR	CAUSE	EFFECT
Low plasma protein concentration	Liver disease and failure to synthesize proteins; kidney disease and loss of proteins in urine; lack of proteins in diet due to starvation	Plasma colloid osmotic pressure decreases; less fluid enters venular ends of capillaries by osmosis
Obstruction of lymphatic vessels	Surgical removal of portions of lymphatic pathways; certain parasitic infections	Back pressure in lymphatic vessels interferes with movement of fluid from interstitial spaces into lymph capillaries
Increased venous pressure	Venous obstructions or faulty venous valves	Back pressure in veins interferes with reabsorption of fluid from interstitial spaces into venular ends of capillaries
Inflammation	Tissue damage	Capillaries become abnormally permeable; fluid leaks from plasma into interstitial spaces

Water Output

Water normally enters the body only through the mouth, but it can be lost by a variety of routes. These include obvious losses in urine, feces, and sweat (sensible perspiration), as well as less obvious losses, such as evaporation of water from the skin (insensible perspiration) and from the lungs during breathing.

If an average adult takes in 2,500 milliliters of water each day, then 2,500 milliliters must be eliminated to maintain water balance. Of this volume, perhaps 60% is lost in urine, 6% in feces, and 6% in sweat. About 28% is lost by evaporation from the skin and lungs (fig. 18.4B). These percentages vary with environmental temperature and relative humidity, as well as with physical exercise.

Regulation of Water Output

The primary means of regulating water output is urine production. The distal convoluted tubules and collecting ducts of the nephrons are most important in regulating the volume of water excreted in the urine. The epithelial linings of these segments of the renal tubule remain relatively impermeable to water unless antidiuretic hormone (ADH) is present. ADH increases the permeability of the distal tubule and collecting duct, thereby increasing water reabsorption and reducing urine production. In the absence of ADH, less water is reabsorbed, and more urine is produced (see chapter 17, p. 471).

CHECK YOUR RECALL

1. By what routes does the body lose water?
2. What role do renal tubules play in water balance regulation?

Diuretics are substances that promote urine production. A number of familiar chemicals, such as caffeine in coffee and tea, have diuretic effects, as do a variety of drugs used to reduce the volume of body fluids.

Diuretics produce their effects in different ways. Some, such as alcohol and certain narcotic drugs, promote urine formation by inhibiting ADH release. Other diuretics, such as caffeine, inhibit the reabsorption of sodium ions or other solutes in portions of renal tubules. As a consequence, the osmotic pressure of the tubular fluid increases, reducing osmotic reabsorption of water and increasing urine volume.

18.4 Electrolyte Balance

Electrolyte balance (e-lek´tro-līt bal´ans) exists when the quantities of electrolytes the body gains equal those lost. Homeostatic mechanisms maintain electrolyte balance.

Electrolyte Intake

The electrolytes most important to cellular functions release sodium, potassium, calcium, magnesium, chloride, sulfate, phosphate, bicarbonate, and hydrogen ions. Foods provide most of these electrolytes, but drinking water and other beverages are also sources. Some electrolytes are by-products of metabolic reactions.

Regulation of Electrolyte Intake

Ordinarily, responding to hunger and thirst provides sufficient electrolytes. A severe electrolyte deficiency may produce a *salt craving*, a strong desire to eat salty foods.

Electrolyte Output

The body loses electrolytes by perspiring, with more lost in sweat on warmer days and during strenuous exercise. Varying amounts of electrolytes are lost in the feces. However, kidney function and urine production vary electrolyte output to maintain balance.

CHECK YOUR RECALL

1. Which electrolytes are most important to cellular functions?
2. Which mechanisms ordinarily regulate electrolyte intake?
3. By what routes does the body lose electrolytes?

Regulation of Electrolyte Output

Precise concentrations of positively charged ions, such as sodium (Na^+), potassium (K^+), and calcium (Ca^{+2}), are required for nerve impulse conduction, muscle fiber contraction, and maintenance of cell membrane potential. *Sodium ions* account for nearly 90% of positively charged ions in extracellular fluids. The kidneys and the hormone aldosterone regulate these ions. Aldosterone, which the adrenal cortex secretes, increases sodium ion reabsorption in the distal convoluted tubules and collecting ducts of the kidneys' nephrons.

Aldosterone also regulates potassium ions. A rising potassium ion concentration directly stimulates the adrenal cortex to secrete aldosterone. This hormone enhances tubular reabsorption of sodium ions, and at the same time, causes tubular secretion of potassium ions (fig. 18.5).

Recall from chapter 11 (p. 291) that the parathyroid glands, as well as calcitonin from the thyroid gland, regulate the concentration of *calcium ions* in extracellular fluids. Calcium ion concentration dropping below normal directly stimulates the parathyroids to secrete parathyroid hormone. Parathyroid hormone increases the concentrations of calcium and phosphate ions in extracellular fluids.

Generally, the regulatory mechanisms that control positively charged ions secondarily control the concentrations of negatively charged ions. For example, renal tubules reabsorb chloride ions (Cl^-), the most abundant negatively charged ions in extracellular fluids, in response to active tubular reabsorption of

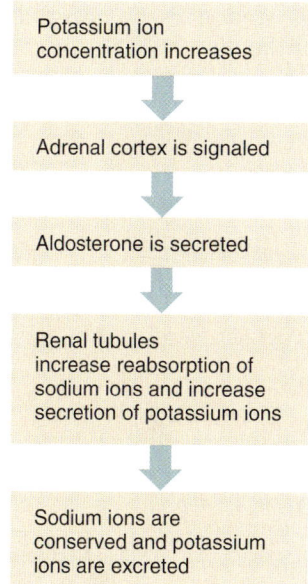

Figure 18.5

If the potassium ion concentration increases, the kidneys conserve sodium ions and excrete potassium ions.

Topic of Interest

SODIUM AND POTASSIUM IMBALANCES

Extracellular fluids usually have high sodium ion concentrations, and intracellular fluid usually has a high potassium ion concentration. Renal regulation of sodium is closely related to that of potassium because secretion (and excretion) of potassium accompanies active reabsorption of sodium (under the influence of aldosterone). Therefore, conditions resulting from sodium ion imbalance often also involve potassium ion imbalance.

Such disorders include:

1. *Low sodium concentration (hyponatremia)* Possible causes of sodium deficiencies include prolonged sweating, vomiting, or diarrhea; renal disease in which sodium is inadequately reabsorbed; adrenal cortex disorders in which aldosterone secretion is insufficient to promote sodium reabsorption (Addison disease); and drinking too much water. One possible effect of hyponatremia is the development of hypotonic extracellular fluid that promotes water movement into cells by osmosis, producing symptoms of water intoxication.
2. *High sodium concentration (hypernatremia)* Possible causes of elevated sodium concentration include excess water loss by evaporation (despite decreased sweating), as may occur during high fever, and increased water loss accompanying diabetes insipidus. In one form of diabetes insipidus, ADH secretion is insufficient for renal tubules to maintain water conservation. Hypernatremia may disturb the central nervous system, causing confusion, stupor, and coma.
3. *Low potassium concentration (hypokalemia)* Possible causes of potassium deficiency include the release of excess aldosterone by the adrenal cortex (Cushing syndrome), which increases renal excretion of potassium; use of diuretic drugs that promote potassium excretion; kidney disease; and prolonged vomiting or diarrhea. Possible effects of hypokalemia include muscular weakness or paralysis, respiratory difficulty, and severe cardiac disturbances, such as atrial or ventricular arrhythmias.
4. *High potassium concentration (hyperkalemia)* Possible causes of elevated potassium concentration include renal disease, which decreases potassium excretion; use of drugs that promote renal conservation of potassium; the release of insufficient aldosterone by the adrenal cortex (Addison disease); and a shift of potassium from intracellular to extracellular fluid, a change that accompanies an increase in plasma hydrogen ion concentration (acidosis). Possible effects of hyperkalemia include paralysis of the skeletal muscles and severe cardiac disturbances, such as cardiac arrest.

---

sodium ions. That is, negatively charged chloride ions are electrically attracted to positively charged sodium ions and accompany them as they are reabsorbed (see chapter 17, p. 470).

Active transport mechanisms with limited transport capacities partially regulate some negatively charged ions, such as phosphate ions (PO_4^{-3}) and sulfate ions (SO_4^{-2}). Thus, if extracellular phosphate ion concentration is low, renal tubules reabsorb phosphate ions. On the other hand, if the renal plasma threshold is exceeded, excess phosphate is excreted in urine.

✓ CHECK YOUR RECALL

1. How does aldosterone regulate sodium and potassium ion concentration?
2. How is calcium regulated?
3. What mechanism regulates the concentrations of most negatively charged ions?

18.5 Acid–Base Balance

Recall from chapter 2 (p. 37) that electrolytes that ionize in water and release hydrogen ions are called **acids** and that substances that combine with hydrogen ions are called **bases.** Maintenance of homeostasis depends on controlling the concentrations of acids and bases within body fluids.

Sources of Hydrogen Ions

Most of the hydrogen ions in body fluids originate as by-products of metabolic processes, although the digestive tract may directly absorb small quantities. The major metabolic sources of hydrogen ions include the following (fig. 18.6):

1. **Aerobic respiration of glucose** This process produces carbon dioxide and water. Carbon dioxide diffuses out of cells and reacts with the water in extracellular fluids to form *carbonic acid,* which then ionizes to release hydrogen ions and bicarbonate ions:

$$H_2CO_3 \rightarrow H^+ + HCO_3^-$$

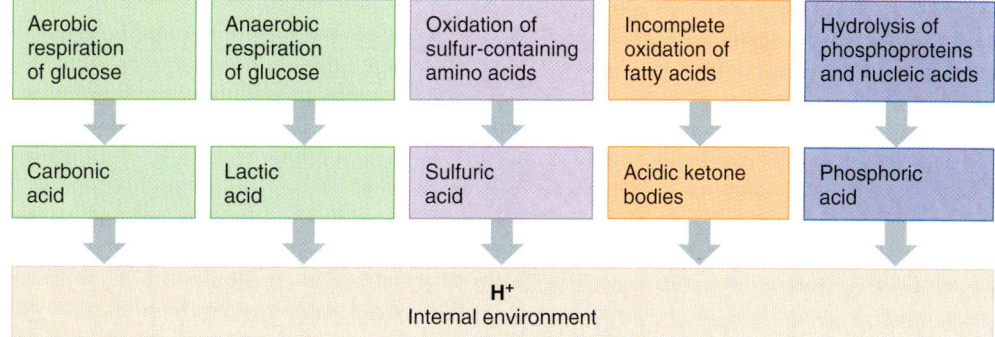

Figure 18.6
Some of the metabolic processes that provide hydrogen ions.

2. **Anaerobic respiration of glucose** Anaerobically metabolized glucose produces *lactic acid,* which adds hydrogen ions to body fluids.
3. **Incomplete oxidation of fatty acids** This process produces *acidic ketone bodies,* which increase hydrogen ion concentration.
4. **Oxidation of sulfur-containing amino acids** This process yields *sulfuric acid* (H_2SO_4), which ionizes to release hydrogen ions.
5. **Breakdown (hydrolysis) of phosphoproteins and nucleic acids** Phosphoproteins and nucleic acids contain phosphorus. Their oxidation produces *phosphoric acid* (H_3PO_4), which ionizes to release hydrogen ions.

The acids resulting from metabolism vary in strength. Thus, their effects on the hydrogen ion concentration of body fluids vary.

CHECK YOUR RECALL

1. Distinguish between an acid and a base.
2. What are the major sources of hydrogen ions in the body?

Strengths of Acids and Bases

Acids that ionize more completely are *strong acids,* and those that ionize less completely are *weak acids.* For example, the hydrochloric acid (HCl) of gastric juice is a strong acid, but the carbonic acid (H_2CO_3) produced when carbon dioxide reacts with water is weak.

Bases release ions, such as hydroxyl ions (OH^-), which can combine with hydrogen ions and thereby lower their concentration. Thus, sodium hydroxide (NaOH), which releases hydroxyl ions, and sodium bicarbonate ($NaHCO_3$), which releases bicarbonate ions (HCO_3^-), are bases. Strong bases dissociate to release more OH^- or its equivalent than do weak bases. Often, the negative ions themselves are called bases. For example, HCO_3^- acting as a base combines with H^+ from the strong acid HCl to form the weak acid carbonic acid (H_2CO_3).

Regulation of Hydrogen Ion Concentration

Acid-base buffer systems, the respiratory center in the brain stem, and the nephrons in the kidneys regulate hydrogen ion concentration, as measured by pH, in body fluids.

Acid-Base Buffer Systems

Acid-base buffer systems occur in all body fluids and consist of chemicals that combine with excess acids or bases. More specifically, the chemical components of a buffer system can combine with strong acids, which release many hydrogen ions, converting them into weak acids, which release fewer hydrogen ions. Likewise, these buffers can combine with strong bases to convert them into weak bases. Such activity helps minimize pH changes in body fluids. The three most important acid-base buffer systems in body fluids are:

1. **Bicarbonate buffer system** The bicarbonate buffer system, which is present in both intracellular and extracellular fluids, uses the bicarbonate ion (HCO_3^-), acting as a weak base, and carbonic acid (H_2CO_3), acting as a weak acid. In the presence of excess hydrogen ions, bicarbonate ions combine with hydrogen ions to form carbonic acid, thus minimizing any increase in the hydrogen ion concentration of the body fluids:

$$H^+ + HCO_3^- \rightarrow H_2CO_3$$

On the other hand, if conditions are basic or alkaline, carbonic acid dissociates to release bicarbonate ion and hydrogen ion:

$$H_2CO_3 \rightarrow H^+ + HCO_3^-$$

It is important to remember that, even though this reaction releases bicarbonate ion, the increase of free hydrogen ions at equilibrium is what minimizes the shift toward a more alkaline pH.

2. **Phosphate buffer system** The phosphate acid-base buffer system is also present in both intracellular and extracellular body fluids. It is particularly

important in the control of hydrogen ion concentrations in the tubular fluid of the nephrons and in urine. This buffer system consists of two phosphate ions—monohydrogen phosphate (HPO_4^{-2}) and dihydrogen phosphate ($H_2PO_4^-$). Under acidic conditions, monohydrogen phosphate ions react with hydrogen ions to produce dihydrogen phosphate:

$$H^+ + HPO_4^{-2} \rightarrow H_2PO_4^-$$

Under alkaline conditions, dihydrogen phosphate ions release hydrogen ions:

$$H_2PO_4^- \rightarrow H^+ + HPO_4^{-2}$$

3. **Protein buffer system** The protein acid-base buffer system consists of the plasma proteins, such as albumins, and certain proteins within the cells, including the hemoglobin of red blood cells. As described in chapter 2 (p. 43), proteins are chains of amino acids. Some of these amino acids have freely exposed groups of atoms, called *amino groups* ($-NH_2$). When the solution pH falls, these amino groups can accept hydrogen ions:

$$-NH_2 + H^+ \rightarrow -NH_3^+$$

Some amino acids within a protein molecule also contain freely exposed *carboxyl groups* ($-COOH$). When the solution pH rises, these carboxyl groups can ionize, releasing hydrogen ions:

$$-COOH \rightarrow -COO^- + H^+$$

Thus, protein molecules can function as bases by accepting hydrogen ions into their amino groups or as acids by releasing hydrogen ions from their carboxyl groups. This special property allows protein molecules to operate as an acid-base buffer system, minimizing changes in pH.

Table 18.1 summarizes the actions of the three major buffer systems.

Neurons are particularly sensitive to changes in the pH of body fluids. If the interstitial fluid becomes more alkaline than normal (alkalosis), neurons become more excitable, and seizures may result. Conversely, acidic conditions (acidosis) depress neuron activity, reducing the level of consciousness.

CHECK YOUR RECALL

1. What is the difference between a strong acid or base and a weak acid or base?
2. How does a chemical buffer system help regulate the pH of body fluids?
3. List the major buffer systems of the body.

TABLE 18.1 CHEMICAL ACID-BASE BUFFER SYSTEM

BUFFER SYSTEM	CONSTITUENTS	ACTIONS
Bicarbonate system	Bicarbonate ion (HCO_3^-)	Combines with hydrogen ions under acidic conditions
	Carbonic acid (H_2CO_3)	Releases hydrogen ions under alkaline conditions
Phosphate system	Monohydrogen phosphate (HPO_4^{-2})	Combines with hydrogen ions under acidic conditions
	Dihydrogen phosphate ($H_2PO_4^-$)	Releases hydrogen ions under alkaline conditions
Protein system (and amino acids)	$-NH_2$ group of an amino acid or protein	Accepts a hydrogen ion in the presence of excess acid
	$-COOH$ group of an amino acid or protein	Releases a hydrogen ion in the presence of excess base

The Respiratory Center

The respiratory center in the brain helps regulate the hydrogen ion concentration in the body fluids by controlling the rate and depth of breathing (see chapter 16, p. 448). Specifically, if the body cells increase their carbon dioxide production, as during physical exercise, carbonic acid production increases. As carbonic acid dissociates, the concentration of hydrogen ions increases, and the pH of the internal environment drops. Such an increasing concentration of carbon dioxide in the central nervous system and the subsequent increase in hydrogen ion concentration in the cerebrospinal fluid stimulate chemosensitive areas within the respiratory center.

The respiratory center responds by increasing the depth and rate of breathing, causing the lungs excrete more carbon dioxide. This causes the hydrogen ion concentration in body fluids to return toward normal because the released carbon dioxide comes from carbonic acid (fig. 18.7):

$$H_2CO_3 \rightarrow CO_2 + H_2O$$

Conversely, if body cells are less active, production of carbon dioxide and hydrogen ions in body fluids remains relatively low. As a result, breathing rate and depth stay closer to resting levels.

The Kidneys

Nephrons help regulate the hydrogen ion concentration of body fluids by excreting hydrogen ions in urine. Recall from chapter 17 (p. 471) that epithelial cells lining certain segments of the renal tubules secrete hydrogen ions into the tubular fluid.

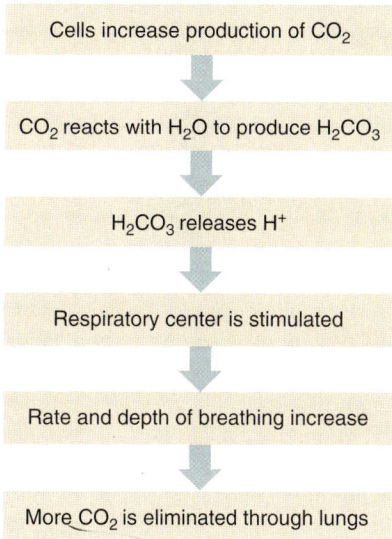

Figure 18.7
An increase in carbon dioxide production causes an increase in carbon dioxide elimination.

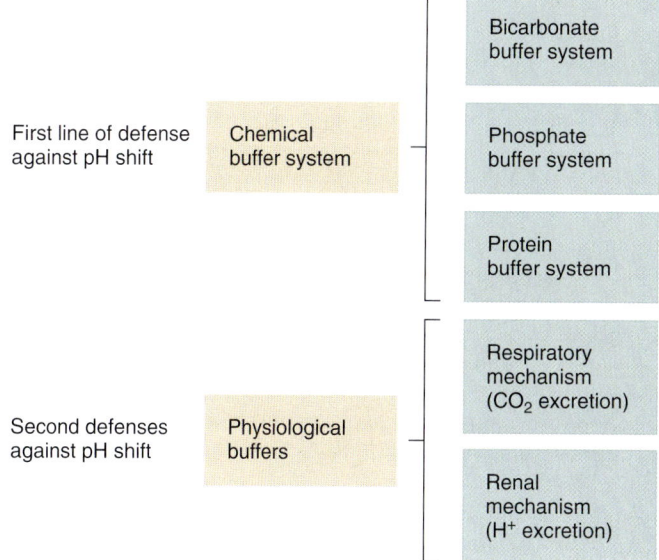

Figure 18.8
Chemical buffers act rapidly, while physiological buffers may require several minutes to several days to begin resisting a change in pH.

Rates of Regulation

The various regulators of hydrogen ion concentration operate at different rates. Acid-base buffers can convert strong acids or bases into weak acids or bases almost immediately. For this reason, these chemical buffer systems are sometimes called the body's *first line of defense* against shifts in pH.

Physiological buffer systems, such as the respiratory and renal mechanisms, function more slowly and constitute the body's *second line of defense*. The respiratory mechanism may require several minutes to begin resisting a change in pH, and the renal mechanism may require one to three days to regulate a changing hydrogen ion concentration. Figure 18.8 compares the actions of chemical buffers and physiological buffers.

 CHECK YOUR RECALL

1. How does the respiratory system help regulate acid-base balance?
2. How do the kidneys respond to excess hydrogen ions?
3. How do the rates of action differ between chemical and physiological buffer systems?

18.6 Acid–Base Imbalances

Chemical and physiological buffer systems generally maintain the hydrogen ion concentration of body fluids within very narrow pH ranges. The pH of arterial blood is normally 7.35–7.45. Abnormal conditions may disturb the acid-base balance. A pH value below 7.35 produces *acidosis*. A pH above 7.45 produces *alkalosis*. Such

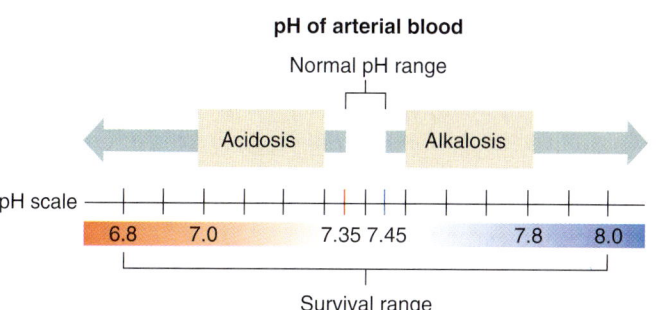

Figure 18.9
If the pH of arterial blood drops to 6.8 or rises to 8.0 for more than a few hours, the person usually cannot survive.

shifts in the pH of body fluids can be life-threatening. A person usually cannot survive if the pH of body fluids drops to 6.8 or rises to 8.0 for longer than a few hours (fig. 18.9).

Accumulation of acids or loss of bases, either of which increases the hydrogen ion concentrations of body fluids, causes acidosis. Conversely, loss of acids or accumulation of bases, and the consequent decreases in hydrogen ion concentrations, cause alkalosis (fig. 18.10).

Acidosis

The two major types of acidosis are respiratory acidosis and metabolic acidosis. Factors that increase carbon dioxide concentration, also increasing the concentration

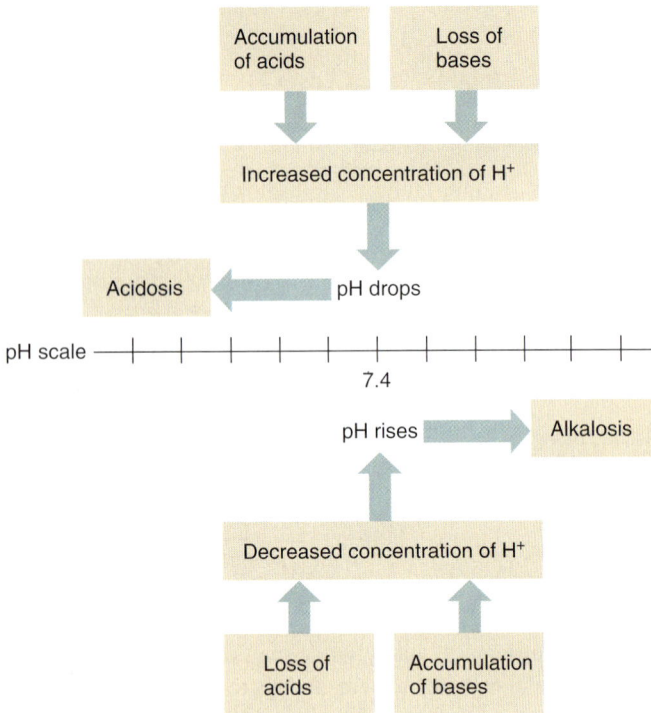

Figure 18.10
Acidosis results from accumulation of acids or loss of bases. Alkalosis results from loss of acids or accumulation of bases.

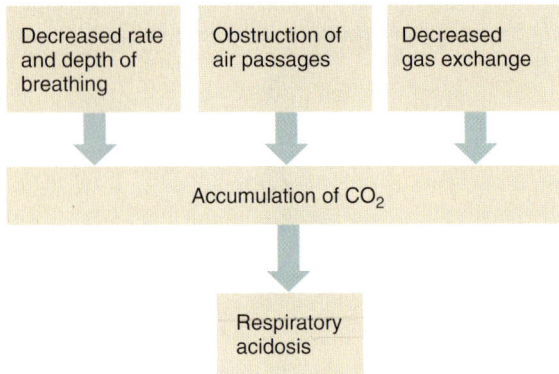

Figure 18.11
Some of the factors that lead to respiratory acidosis.

of carbonic acid (the respiratory acid), cause respiratory acidosis. Metabolic acidosis is due to accumulation of any other acids in the body fluids or to loss of bases, including bicarbonate ions.

Respiratory acidosis may be due to hindered pulmonary ventilation, which increases carbon dioxide concentration. This may result from the following conditions:

1. Injury to the respiratory center of the brain stem, decreasing rate and depth of breathing.
2. Obstructions in air passages that interfere with air movement into alveoli.
3. Diseases that decrease gas exchange, such as pneumonia, or reduce the surface area of the respiratory membrane, such as emphysema.

Figure 18.11 summarizes the factors that can lead to respiratory acidosis. Any of these conditions can increase the level of carbonic acid and hydrogen ions in body fluids, lowering pH. Chemical buffers, such as hemoglobin, may resist this shift in pH. At the same time, rising concentrations of carbon dioxide and hydrogen ions stimulate the respiratory center, increasing the breathing rate and depth and thereby lowering the carbon dioxide concentration. Also, the kidneys may begin to excrete more hydrogen ions. Eventually, these chemical and physiological buffers return the pH of the body fluids to normal. The acidosis is thus *compensated*.

The symptoms of respiratory acidosis result from depression of central nervous system function. They include drowsiness, disorientation, stupor, labored breathing, and cyanosis. In *uncompensated acidosis,* the person may become comatose and die.

Metabolic acidosis is due to accumulation of nonrespiratory acids or loss of bases. Factors that may lead to this condition include the following:

1. Kidney disease that reduces glomerular filtration so that the kidneys fail to excrete acids produced in metabolism (uremic acidosis).
2. Prolonged vomiting with loss of the alkaline contents of the upper intestine and the stomach contents. (Losing only the stomach contents produces metabolic alkalosis.) Vomiting can empty not only the stomach, but also the first foot or so of the intestine.
3. Prolonged diarrhea with loss of excess alkaline intestinal secretions (especially in infants).
4. In diabetes mellitus, metabolic reactions convert some fatty acids into ketone bodies, such as *acetoacetic acid, beta-hydroxybutyric acid,* and *acetone.* Normally, these molecules are rare, and cells oxidize them as energy sources. However, if fats are being utilized too quickly, as in diabetes mellitus, ketone bodies may accumulate faster than they can be oxidized and be excreted in urine (ketonuria). The lungs may excrete acetone, which is volatile and imparts a fruity odor to the breath. More seriously, acetoacetic acid and beta-hydroxybutyric acid may accumulate and lower pH (ketoacidosis), and also combine with bicarbonate ions in the renal tubules. As a result, excess bicarbonate ions are excreted in the urine, interfering with the function of the bicarbonate acid-base buffer system.

Figure 18.12 summarizes the factors leading to metabolic acidosis. In each case, pH is lowered. Countering this lower pH are chemical buffer systems,

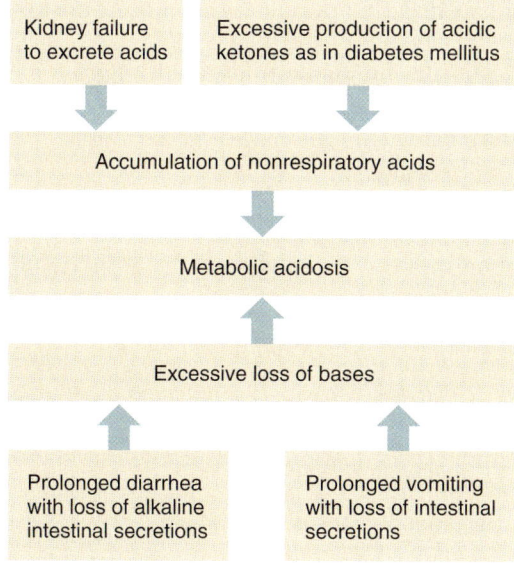

Figure 18.12
Some of the factors that lead to metabolic acidosis.

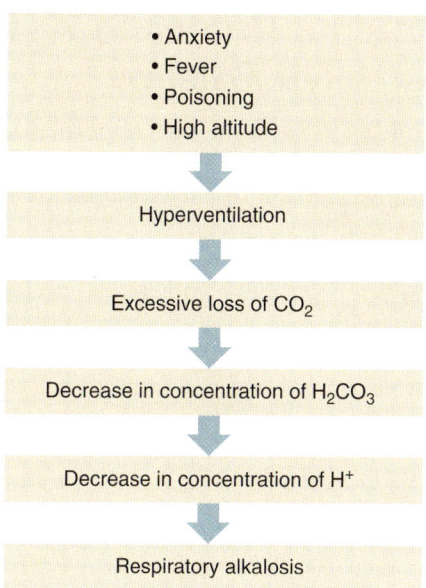

Figure 18.13
Some of the factors that lead to respiratory alkalosis.

which accept excess hydrogen ions; the respiratory center, which increases breathing rate and depth; and the kidneys, which excrete more hydrogen ions.

Alkalosis

The two major types of alkalosis are respiratory alkalosis and metabolic alkalosis. Respiratory alkalosis results from excessive loss of carbon dioxide and consequent loss of carbonic acid. Metabolic alkalosis is due to excessive loss of hydrogen ions or gain of bases.

Respiratory alkalosis develops as a result of *hyperventilation* (described in chapter 16, p. 448), losing too much carbon dioxide and consequently decreasing carbonic acid and hydrogen ion concentrations. Hyperventilation may occur in response to anxiety or may accompany fever or poisoning from salicylates, such as aspirin. At high altitudes, hyperventilation may be a response to low oxygen partial pressure. Musicians can hyperventilate when providing a large volume of air when playing sustained passages on wind instruments. In each case, rapid, deep breathing depletes carbon dioxide, and the pH of body fluids increases. Figure 18.13 illustrates the factors leading to respiratory alkalosis.

Chemical buffers, such as hemoglobin, that release hydrogen ions resist the increase in pH. The lower concentrations of carbon dioxide and hydrogen ions decrease stimulation of the respiratory center. This inhibits the hyperventilation, thus reducing further carbon dioxide loss. At the same time, the kidneys secrete fewer hydrogen ions, and the urine becomes alkaline as bases are excreted.

Symptoms of respiratory alkalosis include lightheadedness, agitation, dizziness, and tingling sensations. In severe cases, peripheral nerves may spontaneously trigger impulses, and muscles may respond with tetanic contractions (see chapter 8, p. 182).

Metabolic alkalosis results from a great loss of hydrogen ions or from a gain in bases, both of which increase blood pH (alkalemia). This condition may follow gastric drainage (lavage), prolonged vomiting of stomach contents, or use of certain diuretic drugs. Because gastric juice is very acidic, its loss leaves body fluids more basic. Metabolic alkalosis may also develop from ingesting too much antacid, such as sodium bicarbonate. Symptoms of metabolic alkalosis include decreased breathing rate and depth, which in turn increases the blood carbon dioxide concentration. Figure 18.14 illustrates the factors leading to metabolic alkalosis.

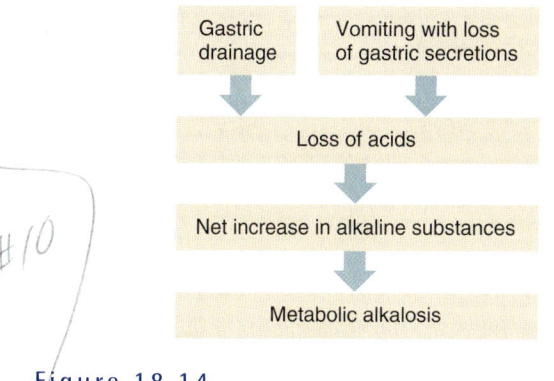

Figure 18.14
Some of the factors that lead to metabolic alkalosis.

Clinical Terms Related to Water and Electrolyte Balance

acetonemia (as˝ĕ-to-ne´me-ah) Abnormal amounts of acetone in blood.
acetonuria (as˝ĕ-to-nu´re-ah) Abnormal amounts of acetone in urine.
albuminuria (al-bu´´mĭ-nu´re-ah) Albumin in urine.
anasarca (an´´ah-sar´kah) Widespread accumulation of tissue fluid.
antacid (ant-as´id) Substance that neutralizes an acid.
anuria (ah-nu´re-ah) Absence of urine excretion.
azotemia (az´´o-te´me-ah) Accumulation of nitrogenous wastes in blood.
diuresis (di´´u-re´sis) Increased urine production.
glucosuria (glu´´ko-su´re-ah) Excess sugar in urine.
hyperglycemia (hi´´per-gli-se´me-ah) Abnormally high blood sugar level.
hyperkalemia (hi´´per-kah-le´me-ah) Excess potassium in the blood.
hypernatremia (hi´´per-na-tre´me-ah) Excess sodium in the blood.
hyperuricemia (hi´´per-u´´rĭ-se´me-ah) Excess uric acid in the blood.
hypoglycemia (hi´´po-gli-se´me-ah) Abnormally low blood sugar level.
ketonuria (ke´´to-nu´re-ah) Ketone bodies in the urine.
ketosis (ke´to´sis) Acidosis due to excess ketone bodies in body fluids.
proteinuria (pro´´te-ĭ-nu´re-ah) Protein in the urine.
uremia (u-re´me-ah) Toxic condition resulting from nitrogenous wastes in the blood.

SUMMARY OUTLINE

18.1 Introduction (p. 482)
Maintenance of water and electrolyte balance requires that equal quantities of these substances enter and leave the body. Altering the water balance affects the electrolyte balance.

18.2 Distribution of Body Fluids (p. 482)
1. Fluid compartments
   a. The intracellular fluid compartment includes the fluids and electrolytes that are enclosed by cell membranes.
   b. The extracellular fluid compartment includes all the fluids and electrolytes outside the cell membranes.
2. Body fluid composition
   a. Extracellular fluids have high concentrations of sodium, chloride, and bicarbonate ions. These fluids include a greater concentration of calcium ions with less potassium, magnesium, phosphate, and sulfate ions than does intracellular fluid. Plasma contains more protein than does either interstitial fluid or lymph.
   b. Intracellular fluid contains high concentrations of potassium, phosphate, and magnesium ions. It also has a greater concentration of sulfate ions and lesser concentrations of sodium, chloride, calcium, and bicarbonate ions than does extracellular fluid.
3. Movement of fluid between compartments
   a. Hydrostatic and colloid osmotic pressure regulate fluid movements.
      (1) Hydrostatic pressure forces fluid out of plasma, and osmotic pressure returns fluid to plasma.
      (2) Hydrostatic pressure drives fluid into lymph vessels.
      (3) Colloid osmotic pressure regulates fluid movement in and out of cells.
   b. Sodium ion concentrations are especially important in regulating fluid movement.

18.3 Water Balance (p. 484)
1. Water intake
   a. Most water comes from consuming liquids or moist foods.
   b. Oxidative metabolism produces some water.
2. Regulation of water intake
   a. Thirst is the primary regulator of water intake.
   b. Drinking and the resulting stomach distension inhibit thirst.
3. Water output
   Water is lost in urine, feces, and sweat, and by evaporation from the skin and lungs.
4. Regulation of water output
   The distal convoluted tubules and collecting ducts of the nephrons regulate water output.

18.4 Electrolyte Balance (p. 487)
1. Electrolyte intake
   a. The most important electrolytes in body fluids release ions of sodium, potassium, calcium, magnesium, chloride, sulfate, phosphate, bicarbonate, and hydrogen.
   b. These ions are obtained in foods and beverages or as by-products of metabolic processes.
2. Regulation of electrolyte intake
   a. Food and drink usually provide sufficient quantities of electrolytes.
   b. A severe electrolyte deficiency may produce a salt craving.
3. Electrolyte output
   a. Electrolytes are lost through perspiration, feces, and urine.
   b. Quantities lost vary with temperature and physical exercise.
   c. Most electrolytes are lost through the kidneys.
4. Regulation of electrolyte output
   a. Concentrations of sodium, potassium, and calcium ions in body fluids are particularly important.
   b. The adrenal cortex secretes aldosterone to regulate sodium and potassium ions.
   c. Parathyroid hormone and calcitonin regulate calcium ions.
   d. The mechanisms that control positively charged ions secondarily regulate negatively charged ions.

18.5 Acid-Base Balance (p. 488)
Acids are electrolytes that release hydrogen ions. Bases combine with hydrogen ions. Body fluid pH must remain within a certain range.
1. Sources of hydrogen ions
   a. Aerobic respiration of glucose produces carbonic acid.
   b. Anaerobic respiration of glucose produces lactic acid.
   c. Incomplete oxidation of fatty acids releases acidic ketone bodies.
   d. Oxidation of sulfur-containing amino acids produces sulfuric acid.

e. Hydrolysis of phosphoproteins and nucleic acids produces phosphoric acid.
  2. Strengths of acids and bases
     a. Acids vary in ionization extent.
        (1) Strong acids, such as hydrochloric acid, ionize more completely.
        (2) Weak acids, such as carbonic acid, ionize less completely.
     b. Bases vary in strength also.
  3. Regulation of hydrogen ion concentration
     a. Acid-base buffer systems
        (1) Buffer systems convert strong acids into weaker acids or strong bases into weaker bases.
        (2) They include the bicarbonate buffer system, phosphate buffer system, and protein buffer system.
        (3) Buffer systems minimize pH changes.
     b. The respiratory center controls the rate and depth of breathing to regulate pH.
     c. Kidney nephrons secrete hydrogen ions to regulate pH.
     d. Chemical buffers act more rapidly. Physiological buffers act less rapidly.

18.6 Acid–Base Imbalances (p. 491)
  1. Acidosis
     a. Respiratory acidosis results from increases in concentration of carbon dioxide and carbonic acid.
     b. Metabolic acidosis results from accumulation of other acids or loss of bases.
  2. Alkalosis
     a. Respiratory alkalosis results from loss of carbon dioxide and carbonic acid.
     b. Metabolic alkalosis results from loss of hydrogen ions or gain of bases.

REVIEW EXERCISES

1. Explain how water balance and electrolyte balance are interdependent. (p. 482)
2. Name the body fluid compartments, and describe their locations. (p. 482)
3. Explain how extracellular and intracellular fluids differ in composition. (p. 482)
4. Describe the control of fluid movements between body fluid compartments. (p. 483)
5. Prepare a list of sources of normal water gain and loss to illustrate how water intake equals water output. (p. 484)
6. Define *water of metabolism*. (p. 484)
7. Explain how water intake is regulated. (p. 484)
8. Explain how nephrons regulate water output. (p. 486)
9. List the most important electrolytes in body fluids. (p. 487)
10. Explain how electrolyte intake is regulated. (p. 487)
11. List the routes by which electrolytes leave the body. (p. 487)
12. Explain how the adrenal cortex regulates electrolyte output. (p. 487)
13. Describe the role of the parathyroid glands in regulating electrolyte balance. (p. 487)
14. Distinguish between an acid and a base. (p. 488)
15. List five sources of hydrogen ions in body fluids, and name an acid that originates from each source. (p. 489)
16. Distinguish between a strong acid and a weak acid, and name an example of each. (p. 489)
17. Distinguish between a strong base and a weak base, and name an example of each. (p. 489)
18. Explain how an acid-base buffer system functions. (p. 489)
19. Describe how the bicarbonate buffer system resists changes in pH. (p. 489)
20. Explain why a protein has acidic as well as basic properties. (p. 490)
21. Explain how the respiratory center functions in the regulation of the acid-base balance. (p. 490)
22. Explain how the kidneys function in the regulation of the acid-base balance. (p. 491)
23. Distinguish between a chemical buffer system and a physiological buffer system. (p. 491)
24. Distinguish between respiratory and metabolic acid-base disturbances. (p. 492)
25. Describe how the body compensates for acid-base disturbances. (p. 492)

CRITICAL THINKING

1. After eating an undercooked hamburger, a twenty-five-year-old male developed diarrhea due to infection with a strain of *Escherichia coli* that produces a shigatoxin. How would this affect his blood pH, urine pH, and respiratory rate?
2. A student hyperventilates and is disoriented just before an exam. Is this student likely to be experiencing acidosis or alkalosis? How will the body compensate in an effort to maintain homeostasis?
3. A ten-year-old female is rescued from a swimming pool after several minutes of floundering in the water. What is (are) the cause(s) of the girl's acidosis? What treatment(s) will bring the body back to homeostasis?
4. A thirty-eight-year-old woman contracted *Mycoplasma* pneumonia and ran a temperature of 104°F for five days. Even though the woman drank copious amounts of liquid, her blood pressure dropped to 70/50, indicating dehydration. Should the woman receive intravenous hypertonic glucose or normal saline? Why?
5. Some time ago, several newborn infants died due to an error in which sodium chloride was substituted for sugar in their formula. What symptoms would this produce? Why are infants more prone to the hazard of excess salt intake than adults?
6. An elderly, semiconscious patient is tentatively diagnosed as having acidosis. What components of the arterial blood will be most valuable in determining if the acidosis is of respiratory origin?
7. Radiation therapy may damage the mucosa of the stomach and intestines. What effect might this have on the patient's electrolyte balance?
8. If the right ventricle of a patient's heart is failing so that venous pressure is increased, what changes might occur in the patient's extracellular fluid compartments?

WEB CONNECTIONS

Visit the website for additional study questions and more information about this chapter at:

http://www.mhhe.com/shieress8

chapter 19

Unit 6
The Human Life Cycle

Reproductive Systems

TREATING ERECTILE DYSFUNCTION. Erectile dysfunction (impotence), in which the penis cannot become erect or sustain an erection, was until recently not often talked about. Then, in the spring of 1998, Viagra® (sildenafil) became available, a drug that enables about half of all men who take it to produce and maintain erections. The drug was originally developed to treat chest pain. Its effects on the penis were noted when participants in the clinical trials reported improved sex lives and refused to return extra pills! In most cases, Viagra appears to be safe and effective for men.

Viagra works by interfering with certain signals to the penis. The process of erection depends upon a very small molecule, nitric oxide (NO). The penis consists of two chambers of spongy tissue that surround blood vessels. When the vessels fill with blood, as they do following sexual stimulation, the organ engorges and stiffens. The stimulation causes neurons as well as the endothelial cells that line the interiors of the blood vessels to release NO. The NO then enters the muscle cells that form the middle layers of the blood vessels, relaxing them by activating a series of other chemicals. The vessels dilate and fill with blood, and the penis becomes erect. One of the chemicals, cGMP, must be present for awhile for an erection to persist. Viagra blocks the enzyme that normally breaks down cGMP, thereby sustaining the erection.

Photo:
Even with normal penile function, having too many abnormally shaped sperm cells can impair a man's fertility.

Chapter Objectives

After studying this chapter, you should be able to do the following:

19.1 Introduction
1. State the general functions of the male and female reproductive systems. (p. 498)

19.2 Organs of the Male Reproductive System
2. Name the parts of the male reproductive system, and describe the general functions of each part. (p. 498)
3. Outline the process of spermatogenesis. (p. 500)
4. Trace the path of sperm cells from their site of formation to the outside. (p. 500)

19.3 Hormonal Control of Male Reproductive Functions
5. Explain how hormones control the activities of male reproductive organs and the development of male secondary sex characteristics. (p. 505)

19.4 Organs of the Female Reproductive System
6. Name the parts of the female reproductive system, and describe the general functions of each part. (p. 507)
7. Outline the process of oogenesis. (p. 508)

19.5 Hormonal Control of Female Reproductive Functions
8. Describe how hormones control the activities of female reproductive organs and the development of female secondary sex characteristics. (p. 513)
9. Describe the major events that occur during a menstrual cycle. (p. 513)

19.6 Mammary Glands
10. Review the structure of the mammary glands. (p. 515)

19.7 Birth Control
11. List several methods of birth control, and describe the relative effectiveness of each method. (p. 516)

19.8 Sexually Transmitted Diseases
12. List the general symptoms of sexually transmitted diseases. (p. 520)

Aids to Understanding Words

andr- [man] *andr*ogens: Male sex hormones.
ejacul- [to shoot forth] *ejacul*ation: Process of expelling semen from the male reproductive tract.
fimb- [fringe] *fimb*riae: Irregular extensions on the margin of the infundibulum of the uterine tube.
follic- [small bag] *follic*le: Ovarian structure that contains an egg.
genesis- [origin] spermato*genesis*: Process by which sperm cells are formed.
germ- [to bud or sprout] *germ*inal epithelium: Tissue that gives rise to sex cells by special cell division.
labi- [lip] *labi*a minora: Flattened, longitudinal folds that extend along the margins of the female vestibule.
mens- [month] *mens*trual cycle: Monthly female reproductive cycle.
mons- [an eminence] *mons* pubis: Rounded elevation overlying the pubic symphysis in a female.
puber- [adult] *puber*ty: Time when a person becomes able to reproduce.

Key Terms

androgen (an´dro-jen)
contraception (kon´´trah-sep´shun)
ejaculation (e-jak´´u-la´shun)
emission (e-mish´un)
estrogen (es´tro-jen)
gonadotropin (go-nad´´o-trōp´in)
meiosis (mi-o´sis)
menopause (men´o-pawz)
menstrual cycle (men´stroo-al si´kl)
oogenesis (ō´´o-jen´ĕ-sis)
orgasm (or´gazm)
ovulation (o´´vu-la´shun)
primary follicle (pri´ma-re fol´ĭ-kl)
progesterone (pro-jes´tĕ-rōn)
puberty (pu´ber-te)
semen (se´men)
spermatogenesis (sper´´mah-to-jen´ĕ-sis)
testosterone (tes-tos´tĕ-rōn)
zygote (zi´gōt)

19.1 Introduction

The male and female reproductive systems are a connected series of organs and glands that produce and nurture sex cells and transport them to sites of fertilization. Sex cells have one set of genetic instructions, compared to two in other cells, so that when sex cells join at fertilization, the right amount of genetic information, held in 46 chromosomes, is restored. Some of the reproductive organs secrete hormones vital in the development and maintenance of secondary sex characteristics and the regulation of reproductive physiology.

19.2 Organs of the Male Reproductive System

Organs of the male reproductive system produce and maintain male sex cells, or *sperm cells;* transport these cells and supporting fluids to the outside; and secrete male sex hormones. A male's *primary sex organs* (gonads) are the two testes in which sperm cells and male sex hormones form. The *accessory sex organs* of the male reproductive system are the internal and external reproductive organs (fig. 19.1; reference plates 3 and 4, pp. 24–25).

Testes

The **testes** (singular, *testis*) are ovoid structures about 5 centimeters in length and 3 centimeters in diameter. Both testes are within the cavity of the saclike *scrotum*.

Structure of the Testes

A tough, white, fibrous capsule encloses each testis. Along the capsule's posterior border, the connective tissue thickens and extends into the testis, forming thin septa that subdivide the testis into about 250 *lobules*.

Each lobule contains one to four highly coiled, convoluted **seminiferous tubules,** each approximately 70 centimeters long uncoiled. These tubules course posteriorly and unite to form a complex network of channels. These channels give rise to several ducts that join a tube called the *epididymis*. The epididymis coils on the outer

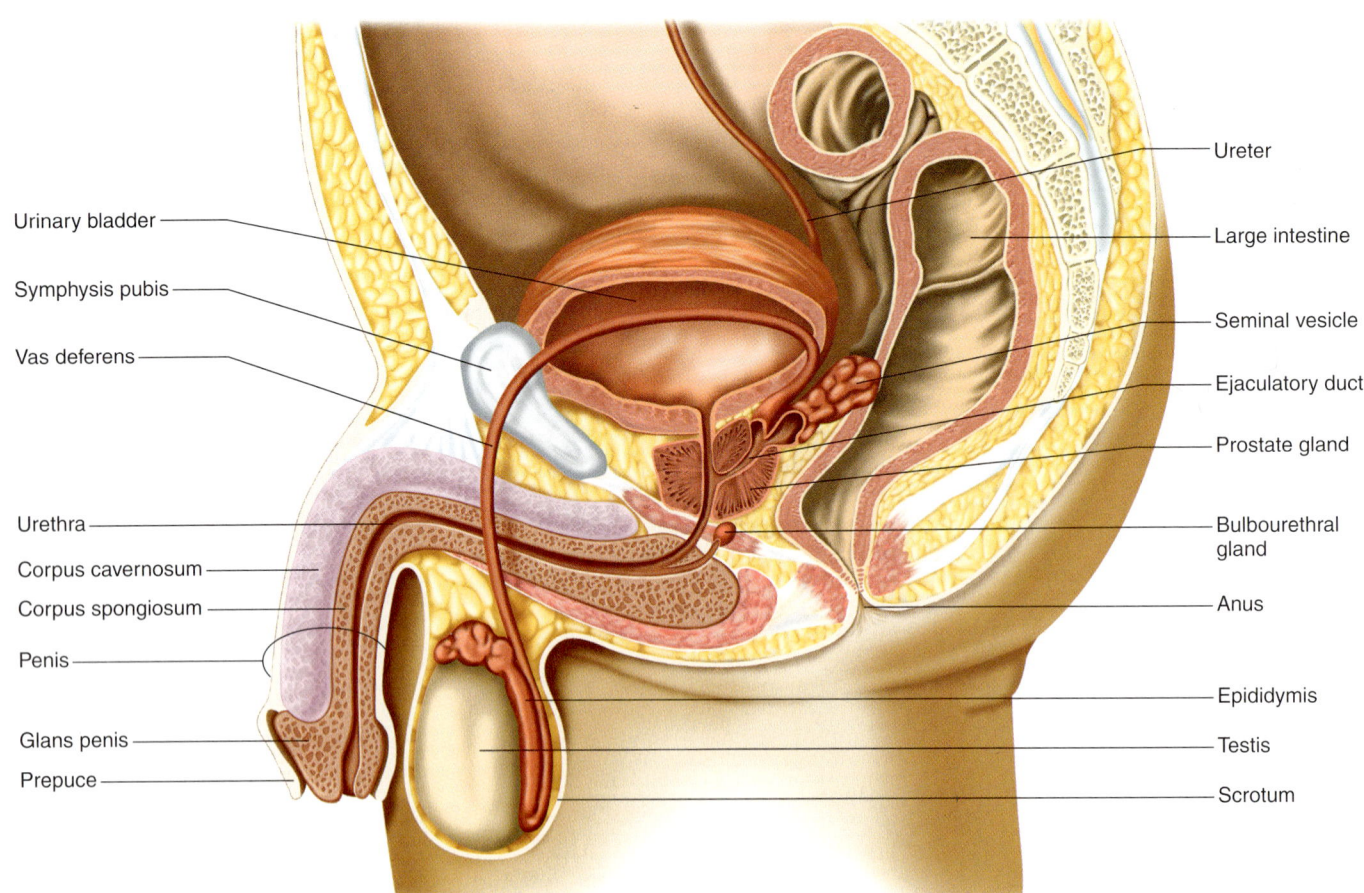

Figure 19.1
Male reproductive organs (sagittal view). The paired testes are the primary sex organs, and the other structures, both internal and external, are accessory sex organs.

surface of the testis and continues to become the *vas deferens* (fig. 19.2A).

A specialized stratified epithelium with **spermatogenic cells,** which give rise to sperm cells, lines the seminiferous tubules. Other specialized cells, called **interstitial cells** (cells of Leydig), lie in the spaces between the seminiferous tubules (fig. 19.2B, C). Interstitial cells produce and secrete male sex hormones.

The epithelial cells of the seminiferous tubules can give rise to *testicular cancer,* a common cancer in young men. In most cases, the first sign is a painless testis enlargement or a scrotal mass that attaches to a testis.

If a biopsy (tissue sample) reveals cancer cells, surgery can remove the affected testis (orchiectomy). Radiation and/or chemotherapy often prevent(s) the cancer from recurring.

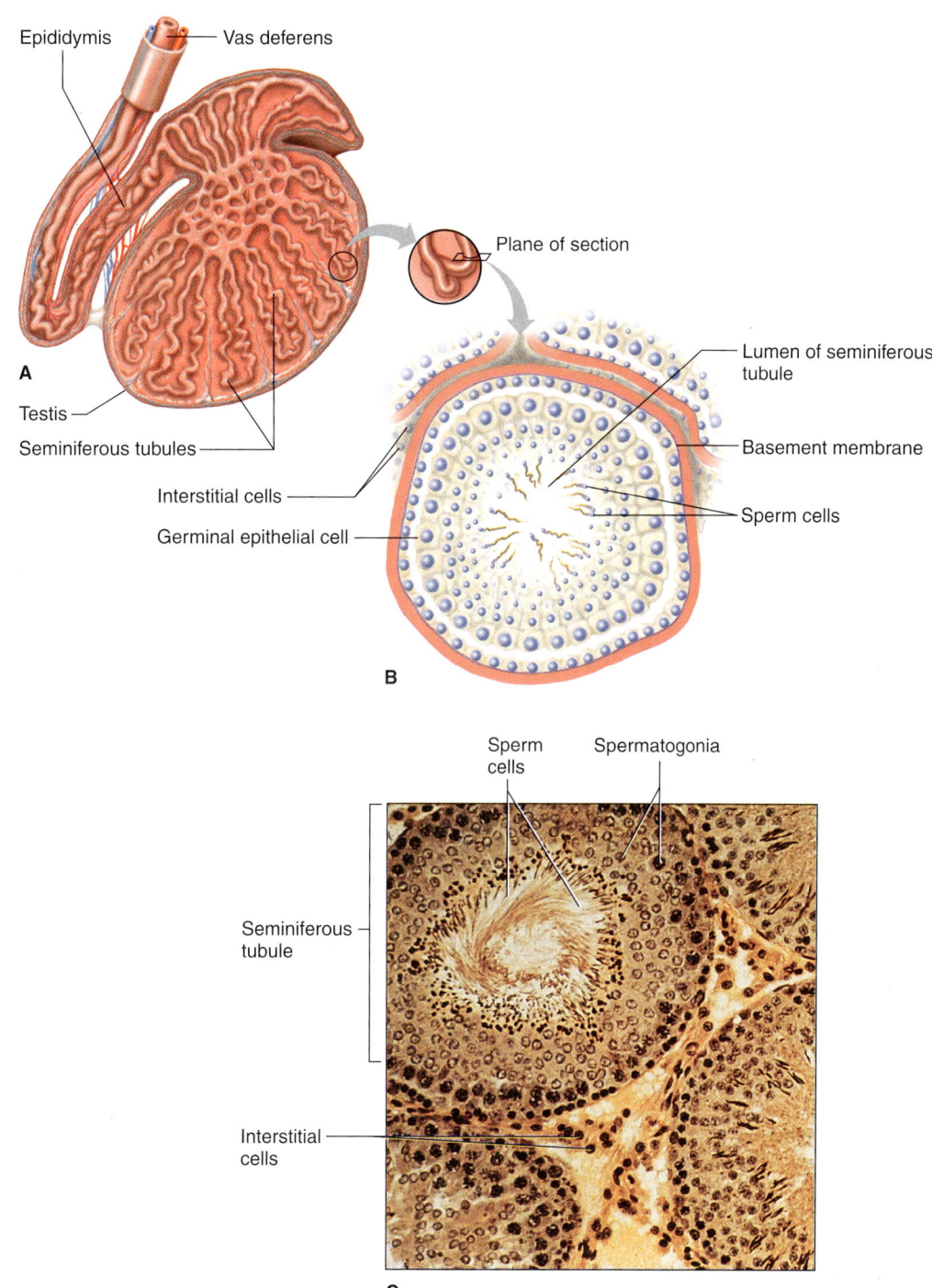

Figure 19.2
Structure of the testis. (*A*) Sagittal section of a testis. (*B*) Cross section of a seminiferous tubule. (*C*) Light micrograph of a seminiferous tubule (200×).

> **CHECK YOUR RECALL**
> 1. Describe the structure of a testis.
> 2. Where are the sperm cells produced within the testes?
> 3. Which cells produce male sex hormones?

Formation of Sperm Cells

The epithelium of the seminiferous tubules consists of supporting cells and spermatogenic cells. Supporting cells provide a scaffolding for the spermatogenic cells, and also nourish and regulate them.

Males produce sperm cells continually throughout their reproductive lives. Sperm cells collect in the lumen of each seminiferous tubule, then pass to the epididymis, where they accumulate and mature.

A mature sperm cell is a tiny, tadpole-shaped structure about 0.06 millimeters long. It consists of a flattened head, a cylindrical midpiece (body), and an elongated tail (fig. 19.3; see fig. 3.9B, p. 57).

The oval *head* of a sperm cell is composed primarily of a nucleus and contains highly compacted chromatin consisting of 23 chromosomes. A small protrusion at its anterior end, called the *acrosome,* contains enzymes that help the sperm cell penetrate an egg cell during fertilization. (Chapter 20, p. 528, describes this process.)

The *midpiece* of a sperm cell has a central, filamentous core and many mitochondria in a spiral. The *tail* (flagellum) consists of several microtubules enclosed in an extension of the cell membrane. The mitochondria provide ATP for the tail's lashing movement, which propels the sperm cell through fluid.

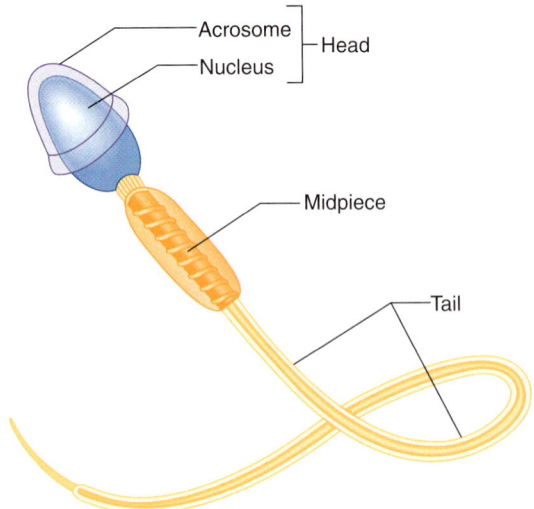

Figure 19.3
Parts of a mature sperm cell.

Spermatogenesis

Sperm cells form in a process called **spermatogenesis** (sper″mah-to-jen′ĕ-sis). In the male embryo, spermatogenic cells are undifferentiated and are called *spermatogonia* (fig. 19.4). Each spermatogonium contains 46 chromosomes in its nucleus, the usual number for human cells. Beginning during embryonic development, hormones stimulate spermatogonia to undergo mitosis (see chapter 3, p. 65), and some of them enlarge to become *primary spermatocytes. Supporting cells* help sustain the developing sperm cells.

At puberty, the primary spermatocytes then reproduce by a special type of cell division called **meiosis** (mi-o′sis). Each primary spermatocyte divides to form two *secondary spermatocytes.* Each of these cells, in turn, divides to form two *spermatids,* which mature into sperm cells. Meiosis also reduces the number of chromosomes in each cell by one-half. Consequently, for each primary spermatocyte that undergoes meiosis, four sperm cells, with 23 chromosomes in each of their nuclei, form. Figure 19.5 depicts spermatogenesis.

> **CHECK YOUR RECALL**
> 1. Explain the function of supporting cells in the seminiferous tubules.
> 2. Describe the structure of a sperm cell.
> 3. Review the events of spermatogenesis.

Male Internal Accessory Organs

The internal accessory organs of the male reproductive system are specialized to nurture and transport sperm cells. These structures include the epididymides, vasa deferentia, ejaculatory ducts, and urethra, as well as the seminal vesicles, prostate gland, and bulbourethral glands.

Epididymis

Each **epididymis** (ep″ĭ-did′ĭ-mis; plural, *epididymides*) is a tightly coiled, threadlike tube about 6 meters long (see figs. 19.1 and 19.2). The epididymis is connected to ducts within a testis. It emerges from the top of the testis, descends along the posterior surface of the testis, and then courses upward to become the vas deferens.

Immature sperm cells reaching the epididymis are nonmotile. However, as rhythmic peristaltic contractions help move these cells through the epididymis, they mature. Following this aging process, sperm cells have the potential to move independently and fertilize egg cells. But they usually do not "swim" until after ejaculation.

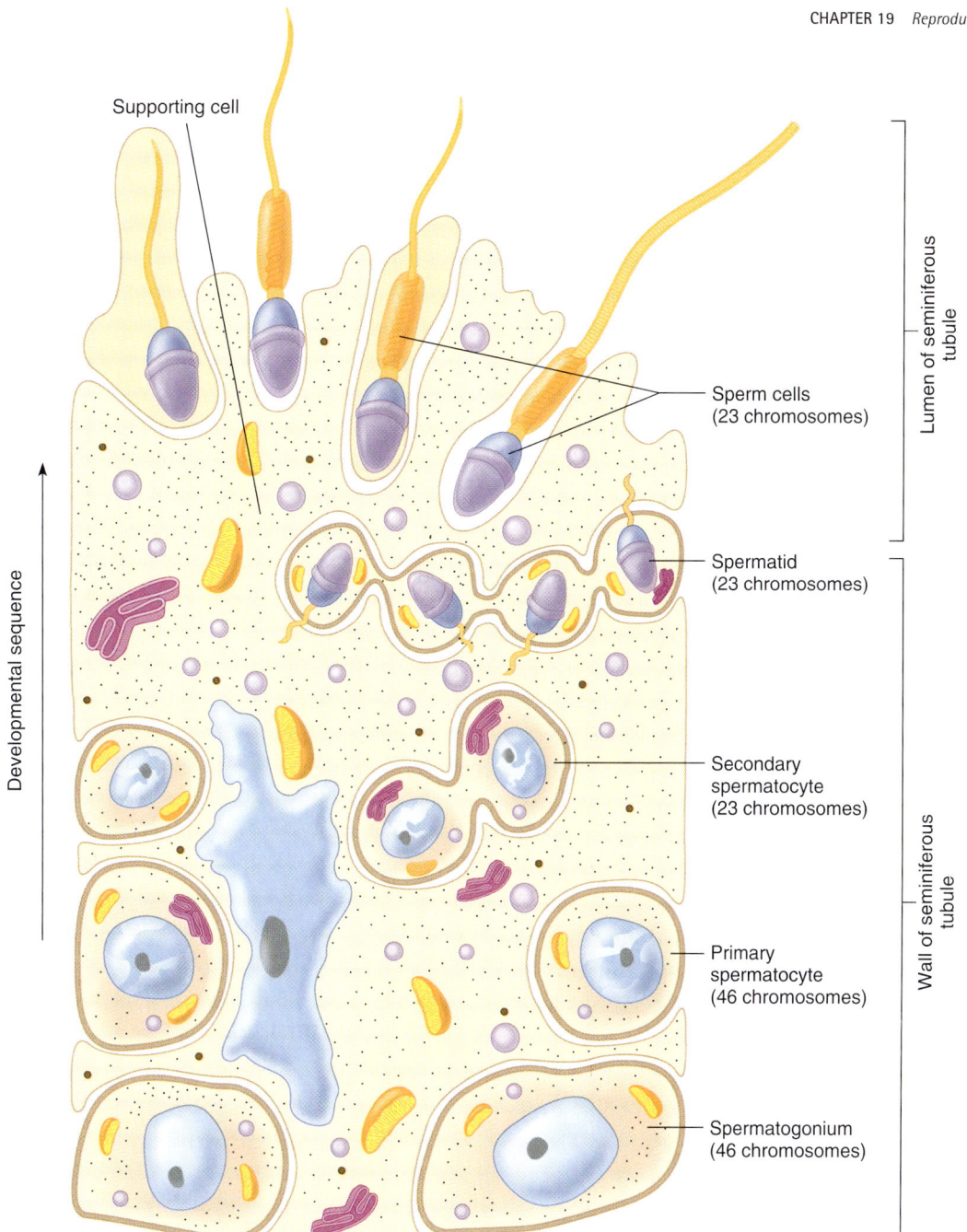

Figure 19.4
Spermatogonia give rise to primary spermatocytes by mitosis. The spermatocytes, in turn, give rise to sperm cells by meiosis. Note that as the cells approach the lumen, they mature.

Vas Deferens

Each **vas deferens** (vas def´er-enz; plural, *vasa deferentia*), also called a ductus deferens, is a muscular tube about 45 centimeters long (see fig. 19.1). It passes upward along the medial side of a testis and through a passage in the lower abdominal wall (inguinal canal), enters the abdominal cavity, and ends behind the urinary bladder. Just outside the prostate gland, the vas deferens unites with the duct of a seminal vesicle to form an **ejaculatory duct,** which passes through the prostate gland and empties into the urethra.

Seminal Vesicle

A **seminal vesicle** is a convoluted, saclike structure about 5 centimeters long that is attached to the vas deferens near the base of the urinary bladder (see fig. 19.1). The glandular tissue lining the inner wall of a seminal vesicle secretes a slightly alkaline fluid. This

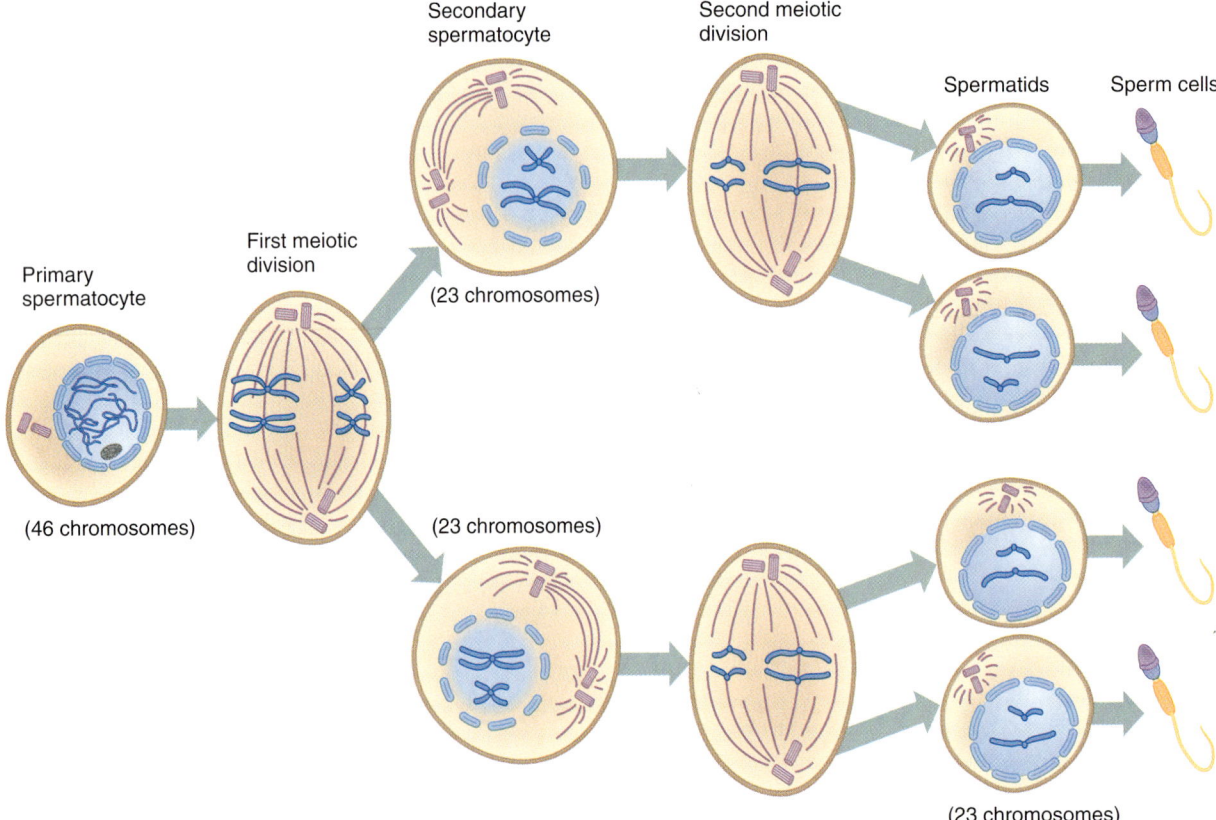

Figure 19.5
Spermatogenesis involves two successive meiotic divisions.

fluid helps regulate the pH of the tubular contents as sperm cells travel to the outside. Seminal vesicle secretions also contain *fructose,* a monosaccharide that provides energy to sperm cells, and *prostaglandins* (see chapter 11, p. 284), which stimulate muscular contractions within the female reproductive organs, aiding the movement of sperm cells toward the egg cell.

CHECK YOUR RECALL

1. Describe the structure of the epididymis.
2. Trace the path of the vas deferens.
3. What is the function of a seminal vesicle?

Prostate Gland

The **prostate gland** is a chestnut-shaped structure about 4 centimeters across and 3 centimeters thick that surrounds the proximal portion of the urethra, just inferior to the urinary bladder (see fig. 19.1). It is enclosed in connective tissue and composed of many branched tubular glands, whose ducts open into the urethra.

The prostate gland secretes a thin, milky fluid with an alkaline pH. This secretion neutralizes the fluid containing sperm cells, which is acidic due to the presence of metabolic wastes that stored sperm cells produce. Prostatic fluid also enhances the motility of sperm cells and helps neutralize the acidic secretions of the vagina.

Bulbourethral Glands

The two **bulbourethral glands** (Cowper's glands) are each about a centimeter in diameter and are inferior to the prostate gland within muscle fibers of the external urethral sphincter (see fig. 19.1). Bulbourethral glands have many tubes whose epithelial linings secrete a mucuslike fluid in response to sexual stimulation. This fluid lubricates the end of the penis in preparation for sexual intercourse. However, females secrete most of the lubricating fluid for sexual intercourse.

Semen

Semen (se′men) is the fluid the male urethra conveys to the outside during ejaculation. It consists of sperm cells from the testes and secretions of the seminal vesicles, prostate gland, and bulbourethral glands. Semen is slightly alkaline (pH about 7.5), and its contents include prostaglandins and nutrients.

The volume of semen released at one time varies from 2 to 5 milliliters. The average number of sperm cells in the fluid is about 120 million per milliliter.

Topic of Interest

PROSTATE ENLARGEMENT

The prostate gland is relatively small in boys, begins to grow in early adolescence, and reaches adult size several years later. Usually, the gland does not grow again until age fifty, when in about half of all men, it enlarges enough to press on the urethra. This produces a feeling of pressure because the bladder cannot empty completely and the man feels the need to urinate frequently. Retained urine can lead to infection and inflammation, bladder stones, or kidney disease.

Medical researchers do not know what causes prostate enlargement. Risk factors include a fatty diet, having had a vasectomy, and possibly occupational exposure to batteries or to the metal cadmium. The enlargement may be benign or cancerous. Because prostate cancer is highly treatable if detected early, men should have their prostates examined regularly.

Diagnostic tests include a rectal exam as well as a blood test to detect prostate-specific antigen (PSA), a cell surface protein normally found on prostate cells. Elevated PSA levels can indicate an enlarged prostate, possibly from a benign or cancerous growth. Ultrasound may provide further information. Table 19A summarizes the treatments for an enlarged prostate.

TABLE 19A   MEDICAL TREATMENTS FOR AN ENLARGED PROSTATE GLAND

Surgical removal of prostate
Radiation
Drugs to block testosterone's growth-stimulating effect on prostate
Microwave energy delivered through a probe inserted into urethra or rectum
Balloon inserted into urethra and inflated with liquid
Tumor frozen with liquid nitrogen delivered by probe through skin
Device (stent) inserted between lobes of the prostate to relieve pressure on the urethra

---

Sperm cells are nonmotile while in the ducts of the testis and epididymis, but begin to swim as they mix with accessory gland secretions. Sperm cells cannot naturally fertilize an egg cell, however, until they enter the female reproductive tract. Acquiring the ability to fertilize an egg cell is called *capacitation,* and it reflects weakening of the sperm cells' acrosomal membranes.

CHECK YOUR RECALL

1. Where is the prostate gland located?
2. What are the functions of the prostate gland's secretion?
3. What are the components of semen?

Male External Reproductive Organs

The male external reproductive organs are the scrotum, which encloses the testes, and the penis. The urethra passes through the penis.

Scrotum

The **scrotum** is a pouch of skin and subcutaneous tissue that hangs from the lower abdominal region posterior to the penis (see fig. 19.1). A medial septum subdivides the scrotum into two chambers, each of which encloses a testis. Each chamber also contains a serous membrane, which covers the testis and helps ensure that it moves smoothly within the scrotum. The scrotum protects and aids in temperature regulation of the testes, important to sex cell production.

Penis

The **penis** is a cylindrical organ that conveys urine and semen through the urethra to the outside (see fig. 19.1). During erection, it enlarges and stiffens so that it can be inserted into the vagina during sexual intercourse.

The *body,* or shaft, of the penis has three columns of erectile tissue—a pair of dorsally located *corpora cavernosa* and a single, ventral *corpus spongiosum*. A tough capsule of dense connective tissue surrounds each column. Skin, a thin layer of subcutaneous tissue, and a layer of connective tissue enclose the penis.

The corpus spongiosum, through which the urethra extends, enlarges at its distal end to form a sensitive, cone-shaped **glans penis.** The glans covers the ends of the corpora cavernosa and bears the urethral opening (external urethral orifice). The skin of the glans is very thin and hairless, and contains sensory receptors for sexual stimulation. A loose fold of skin called the *prepuce* (foreskin) begins just posterior to the glans and extends anteriorly to cover the glans as a sheath. The prepuce is sometimes removed by a surgical procedure called *circumcision.*

CHECK YOUR RECALL

1. Describe the structure of the penis.
2. What is circumcision?

Topic of Interest

MALE INFERTILITY

Male infertility—the inability of sperm cells to fertilize an egg cell—has several causes. If, during fetal development, the testes do not descend into the scrotum, the higher temperature of the abdominal cavity or inguinal canal destroys any sperm cells developing in the seminiferous tubules. Certain diseases, such as mumps, may inflame the testes (orchitis) and cause infertility by destroying cells in the seminiferous tubules.

Both the quality and quantity of sperm cells are essential factors in a man's ability to father a child. If a sperm head is misshapen, if a sperm cell cannot swim, or if sperm cells are too few, completing the journey to the egg cell may be impossible.

Computer-aided sperm analysis (CASA) uses criteria for normalcy in human male seminal fluid and the sperm cells it contains. For this analysis, a man abstains from intercourse for two to three days and then provides a sperm sample, which must be examined within the hour. The man also provides information about his reproductive history and possible exposure to toxins. The sperm sample is placed on a slide under a microscope, and a video camera sends an image to a videocassette recorder, which projects a live or digitized image. The camera also sends the image to a computer, which traces sperm cell trajectories and displays them on a monitor. Figure 19A shows a CASA of normal sperm cells, depicting different swimming patterns as they travel. Table 19B lists the components of a semen analysis.

TABLE 19B	SEMEN ANALYSIS
CHARACTERISTIC	NORMAL VALUE
Volume	2–5 milliliters/ejaculate
Sperm cell density	120 million cells/milliliter
Percent motile	> 40%
Motile sperm cell density	> 8 million/milliliter
Average velocity	> 20 micrometers/second
Motility	> 8 micrometers/second
Percent abnormal morphology	< 40%
White blood cells	> 5 million/milliliter

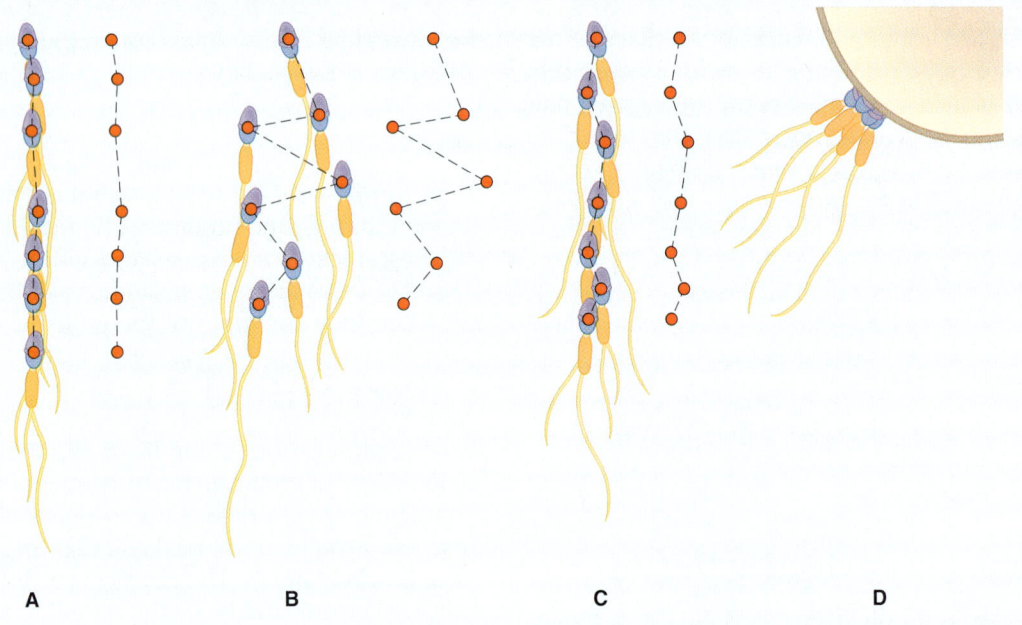

A   B   C   D

Figure 19A
A computer tracks sperm cell movements. In semen, sperm cells swim in a straight line (A), but as they are activated by biochemicals normally found in the woman's body, their trajectories widen (B). The sperm cells in (C) are in the mucus of a woman's cervix, and the sperm cells in (D) are attempting to digest through the structures surrounding an egg cell.

Erection, Orgasm, and Ejaculation

During sexual stimulation, parasympathetic nerve impulses from the sacral portion of the spinal cord release the vasodilator nitric oxide, causing the arteries leading into the penis to dilate. At the same time, the increasing pressure of arterial blood entering the vascular spaces of erectile tissue compresses the veins of the penis, reducing the flow of venous blood away from the penis. Consequently, blood accumulates in erectile tissues, and the penis swells and elongates, producing an **erection.**

The culmination of sexual stimulation is **orgasm** (or´gazm), a pleasurable feeling of physiological and psychological release. Emission and ejaculation accompany male orgasm.

Emission (e-mish´un) is the movement of sperm cells from the testes and secretions from the prostate gland and seminal vesicles into the urethra, where they mix to form semen. Emission occurs in response to sympathetic nerve impulses from the spinal cord, which stimulate peristaltic contractions in smooth muscle within the walls of the testicular ducts, epididymides, vasa deferentia, and ejaculatory ducts. At the same time, other sympathetic impulses stimulate rhythmic contractions of the seminal vesicles and prostate gland.

As the urethra fills with semen, sensory impulses are stimulated and pass into the sacral portion of the spinal cord. In response, motor impulses are transmitted from the cord to certain skeletal muscles at the base of the penile erectile columns, causing them to contract rhythmically. This increases the pressure within the erectile tissues and helps force semen through the urethra to the outside, a process called **ejaculation** (e-jak´´u-la´shun).

The sequence of events during emission and ejaculation is coordinated so that fluid from the bulbourethral glands is expelled first. This is followed by the release of fluid from the prostate gland, the passage of sperm cells, and finally the ejection of fluid from the seminal vesicles.

Immediately after ejaculation, sympathetic impulses constrict the arteries that supply the erectile tissue, reducing blood inflow. Smooth muscles within the walls of the vascular spaces partially contract again, and the veins of the penis carry excess blood out of these spaces. The penis gradually returns to its flaccid state.

Table 19.1 summarizes the functions of the male reproductive organs.

Spontaneous emissions and ejaculations commonly occur in adolescent males during sleep and thus are called *nocturnal emissions.* Changes in hormonal concentrations that accompany adolescent development and sexual maturation cause these emissions.

CHECK YOUR RECALL

1. What controls blood flow into penile erectile tissues?
2. Distinguish among orgasm, emission, and ejaculation.
3. Review the events associated with emission and ejaculation.

19.3 Hormonal Control of Male Reproductive Functions

The hypothalamus, anterior pituitary gland, and testes secrete hormones that control male reproductive functions. These hormones initiate and maintain sperm cell production and oversee the development and maintenance of male secondary sex characteristics.

Hypothalamic and Pituitary Hormones

Prior to ten years of age, the male body is reproductively immature. It is childlike, with spermatogenic cells undifferentiated. Then, a series of changes leads to development of a reproductively functional adult. The hypothalamus controls many of these changes.

Recall from chapter 11 (p. 286) that the hypothalamus secretes gonadotropin-releasing hormone (GnRH), which enters blood vessels leading to the anterior pituitary gland. In response, the anterior pituitary secretes the **gonadotropins** (go-nad´´o-trōp´inz), called *luteinizing hormone (LH)* and *follicle-stimulating hormone (FSH).* LH, which in males is also called interstitial cell-stimulating hormone (ICSH), promotes development of testicular interstitial cells, and they in turn secrete male sex hormones. FSH stimulates the supporting cells of the seminiferous tubules to respond to the effects of the

TABLE 19.1   FUNCTIONS OF THE MALE REPRODUCTIVE ORGANS

ORGAN	FUNCTION
Testis	
Seminiferous tubules	Produce sperm cells
Interstitial cells	Produce and secrete male sex hormones
Epididymis	Stores sperm cells undergoing maturation; conveys sperm cells to vas deferens
Vas deferens	Conveys sperm cells to ejaculatory duct
Seminal vesicle	Secretes an alkaline fluid containing nutrients and prostaglandins that helps neutralize the acidic components of semen
Prostate gland	Secretes an alkaline fluid that helps neutralize semen's acidity and enhances sperm cell motility
Bulbourethral gland	Secretes fluid that lubricates end of penis
Scrotum	Encloses, protects, and regulates temperature of testes
Penis	Conveys urine and semen to outside of body; inserted into vagina during sexual intercourse; glans penis is richly supplied with sensory nerve endings associated with feelings of pleasure during sexual stimulation

male sex hormone *testosterone*. Then, in the presence of FSH and testosterone, these supporting cells stimulate spermatogenic cells to undergo spermatogenesis, giving rise to sperm cells (fig. 19.6). The supporting cells also secrete a hormone called *inhibin,* which inhibits the anterior pituitary gland by negative feedback, and thus prevents oversecretion of FSH.

Male Sex Hormones

Male sex hormones are termed **androgens** (an´dro-jenz). Testicular interstitial cells produce most of them, but the adrenal cortex synthesizes small amounts (see chapter 11, p. 295). **Testosterone** (tes-tos´te-rōn) is the most abundant androgen. It loosely attaches to plasma proteins for secretion and transport in blood.

Testosterone secretion begins during fetal development and continues for several weeks following birth; then it nearly ceases during childhood. Between the ages of thirteen and fifteen, a young man's androgen production usually increases rapidly. This phase in development, when an individual becomes reproductively functional, is **puberty** (pu´ber-te). After puberty, testosterone secretion continues throughout the life of a male.

Actions of Testosterone

During puberty, testosterone stimulates enlargement of the testes (the *male primary sex organs*) and accessory organs of the reproductive system, as well as development of *male secondary sex characteristics,* which are special features associated with the adult male body. Secondary sex characteristics in the male include:

1. Increased growth of body hair, particularly on the face, chest, axillary region, and pubic region. Sometimes, hair growth on the scalp slows.
2. Enlargement of the larynx and thickening of the vocal folds, with lowering of the pitch of the voice.
3. Thickening of the skin.
4. Increased muscular growth, broadening of the shoulders, and narrowing of the waist.
5. Thickening and strengthening of the bones.

Testosterone also increases the rate of cellular metabolism and red blood cell production, so that the average number of red blood cells in a cubic millimeter of blood is usually greater in males than in females. Testosterone stimulates sexual activity by affecting certain portions of the brain.

Regulation of Male Sex Hormones

The extent to which male secondary sex characteristics develop is directly related to the amount of testosterone that interstitial cells secrete. A negative feedback system involving the hypothalamus regulates testosterone output (see fig. 19.6).

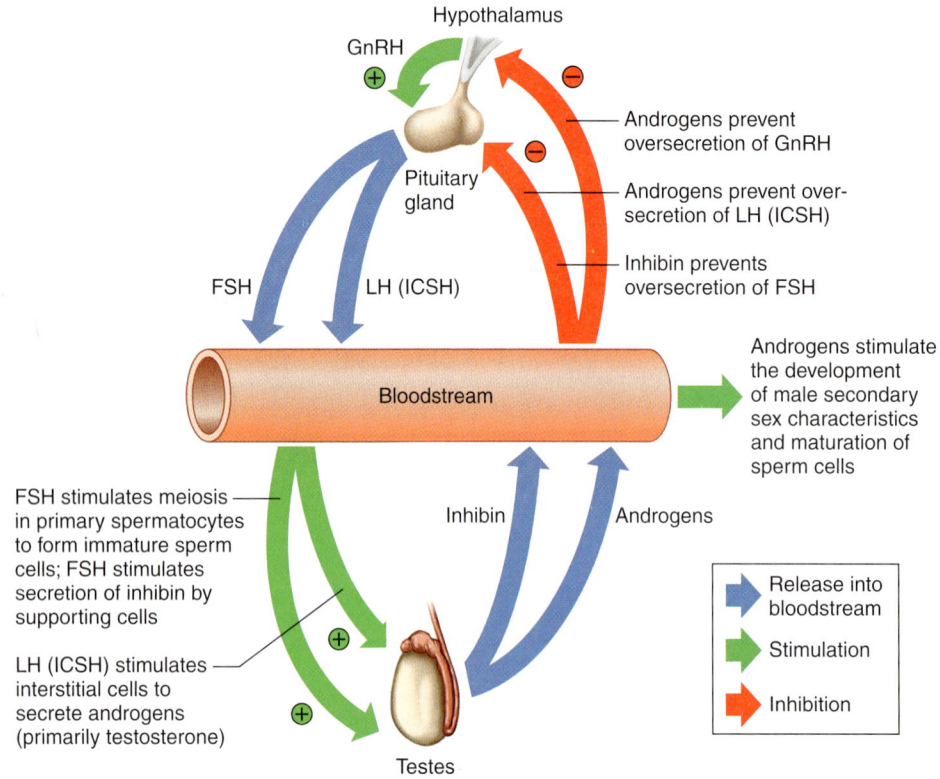

Figure 19.6
The hypothalamus controls maturation of sperm cells and development of male secondary sex characteristics. A negative feedback mechanism operating between the hypothalamus, the anterior lobe of the pituitary gland, and the testes controls the concentration of testosterone.

An increasing blood testosterone concentration inhibits the hypothalamus, and its stimulation of the anterior pituitary gland by GnRH decreases. As the pituitary's secretion of LH (ICSH) falls in response, the amount of testosterone the interstitial cells release decreases.

As the blood testosterone concentration drops, the hypothalamus becomes less inhibited, and it once again stimulates the anterior pituitary to release LH. Increasing LH secretion then causes interstitial cells to release more testosterone, and the blood testosterone concentration increases.

Testosterone level decreases somewhat during and after the *male climacteric,* a decline in sexual function associated with aging. At any given age, the testosterone concentration in the male body is regulated to remain relatively constant.

CHECK YOUR RECALL

1. Which hormone initiates the changes associated with male sexual maturity?
2. Describe several male secondary sex characteristics.
3. List the functions of testosterone.
4. Explain how the secretion of male sex hormones is regulated.

19.4 Organs of the Female Reproductive System

The organs of the female reproductive system produce and maintain the female sex cells, or egg cells (ova); transport these cells to the site of fertilization; provide a favorable environment for a developing offspring; move the offspring to the outside; and produce female sex hormones. A female's *primary sex organs* (gonads) are the two ovaries, which produce the female sex cells and sex hormones. The *accessory sex organs* of the female reproductive system are the internal and external reproductive organs (fig. 19.7; reference plates 5 and 6, pp. 26–27).

Ovaries

The two **ovaries** are solid, ovoid structures, each about 3.5 centimeters long, 2 centimeters wide, and 1 centimeter thick. The ovaries lie in shallow depressions in the lateral wall of the pelvic cavity (fig. 19.7).

Ovary Structure

Ovarian tissues are subdivided into two indistinct regions—an inner *medulla* and an outer *cortex.* The ovarian medulla is composed of loose connective tissue

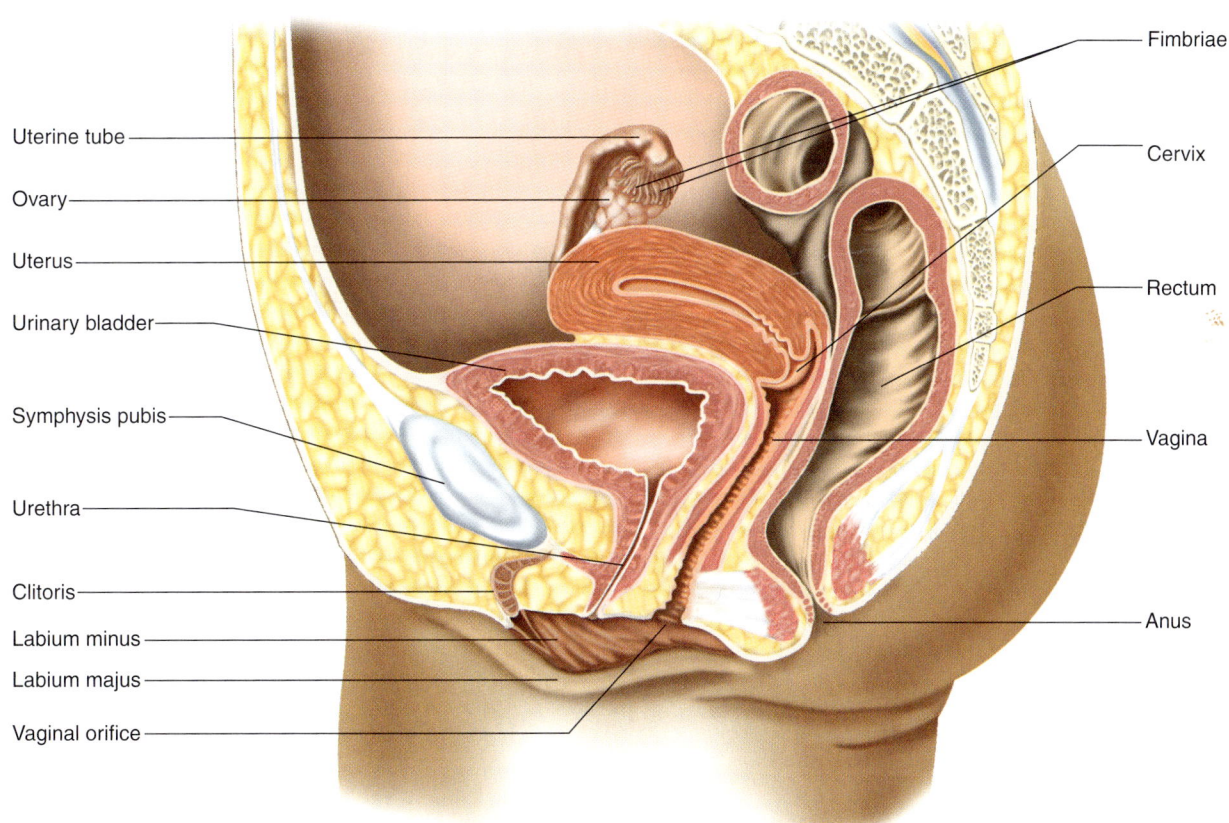

Figure 19.7
Female reproductive organs (sagittal view). The paired ovaries are the primary sex organs, and the other structures, both internal and external, are accessory sex organs.

and contains many blood vessels, lymphatic vessels, and nerve fibers. The ovarian cortex consists of more compact tissue and has a granular appearance due to tiny masses of cells called *ovarian follicles*.

A layer of cuboidal epithelium covers the ovary's free surface. Just beneath this epithelium is a layer of dense connective tissue.

CHECK YOUR RECALL

1. What are the primary sex organs of the female?
2. Describe the structure of an ovary.

Primordial Follicles

During prenatal (before birth) development of a human female, small groups of cells in the outer region of the ovarian cortex form several million **primordial follicles.** Each of these structures consists of a single, large cell, called a *primary oocyte*, which is closely surrounded by epithelial cells called *follicular cells*.

Early in development, primary oocytes begin to undergo meiosis, but the process soon halts and does not continue until the individual reaches puberty. Once the primordial follicles appear, no new ones form. Instead, the number of oocytes in the ovary steadily declines as many of the oocytes degenerate. Of the several million oocytes formed originally, only a million or so remain at birth, and perhaps 400,000 are present at puberty. The ovary releases fewer than 400 or 500 oocytes during a female's reproductive life.

Oogenesis

Oogenesis (o˝o-jen´ĕ-sis) is the process of egg cell formation. Beginning at puberty, some primary oocytes are stimulated to continue meiosis. As in the case of sperm cells, the resulting cells have one-half as many chromosomes (23) in their nuclei as their parent cells.

When a primary oocyte divides, the distribution of the cytoplasm is unequal. One of the resulting cells, called a *secondary oocyte,* is large, and the other, called the *first polar body,* is small (fig. 19.8).

The large secondary oocyte can be fertilized by a sperm cell, at which point it is considered an egg. Upon fertilization, the oocyte divides unequally to produce a tiny *second polar body* and a large fertilized egg cell, or **zygote** (zi´gōt).

The polar bodies have no further function and soon degenerate. Their role in reproduction is to allow the egg cell to accumulate the large amount of cytoplasm necessary to nurture a zygote and early embryo.

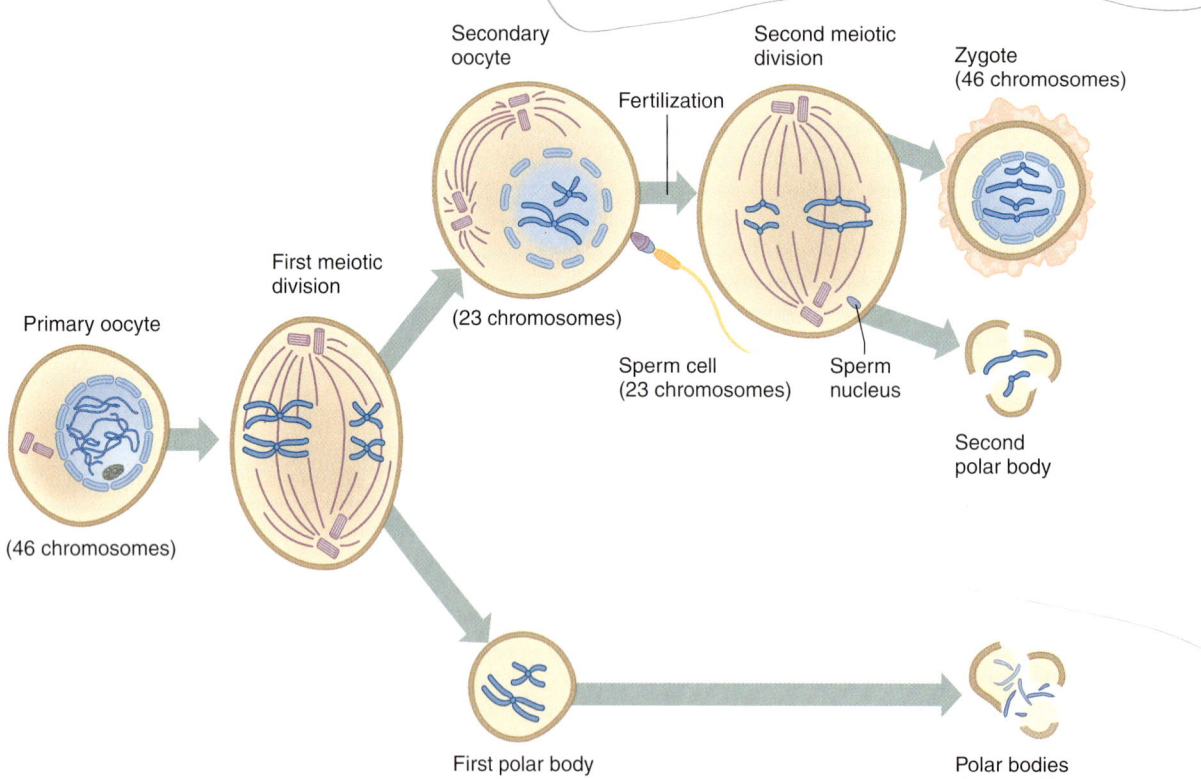

Figure 19.8
During oogenesis, a single egg cell (secondary oocyte) results from meiosis in a primary oocyte. If the egg cell is fertilized, it generates a second polar body and becomes a zygote. (Note: The second meiotic division does not occur in the egg cell if it is not fertilized by a sperm cell.)

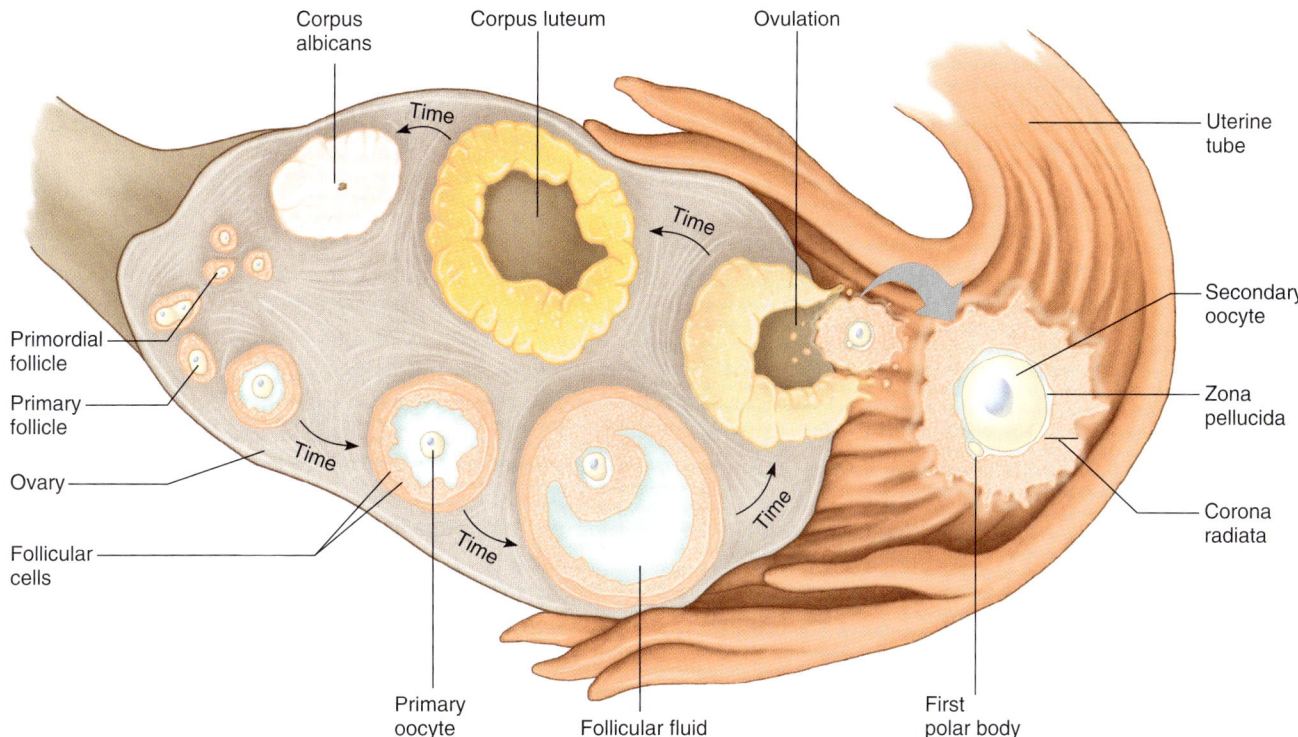

Figure 19.9
Within an ovary, as a follicle matures, a developing oocyte enlarges and becomes surrounded by follicular cells and fluid. Eventually, the mature follicle ruptures, releasing the secondary oocyte.

 The largest cell in the human body is the ovum. The smallest cell is the sperm.

 CHECK YOUR RECALL

1. How does the timing of egg cell production differ from that of sperm cells?
2. Describe the major events of oogenesis.

Follicle Maturation

At puberty, the anterior pituitary gland secretes increased amounts of FSH, and the ovaries enlarge in response. At the same time, some of the primordial follicles mature into **primary follicles** (pri´ma-re fol´ĭ-klz). Figure 19.9 traces the maturation of a follicle within an ovary.

During maturation, a primary oocyte enlarges, and surrounding follicular cells proliferate by mitosis. These follicular cells become organized into layers, and soon a cavity (*antrum*) appears in the cellular mass. A clear *follicular fluid* fills the cavity and bathes the oocyte. The enlarging fluid-filled cavity presses the oocyte to one side. In time, the follicle reaches a diameter of 10 millimeters or more and bulges outward on the ovary surface like a blister.

The secondary oocyte is a large, spherical cell, surrounded by a membrane (zona pellucida) and attached to a mantle of follicular cells (corona radiata) (fig. 19.10). Processes from the follicular cells extend through the zona pellucida and supply the oocyte with nutrients.

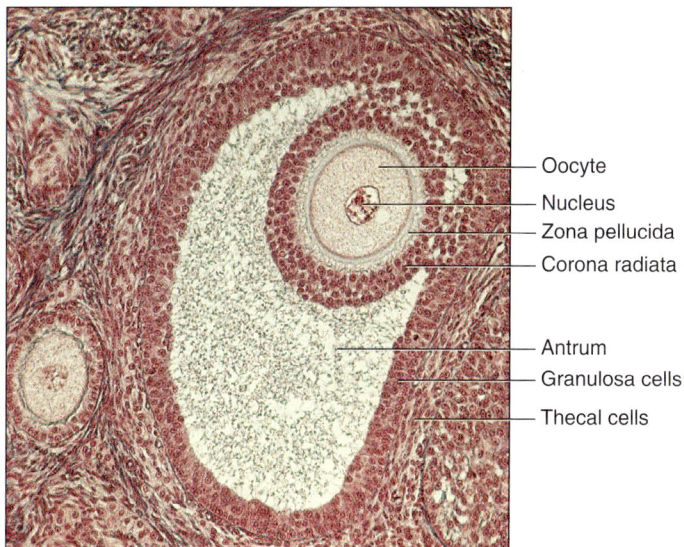

Figure 19.10
Light micrograph of a maturing follicle (200×).

As many as twenty primary follicles may begin maturing at any one time, but one follicle usually outgrows the others. Typically, only the dominant follicle fully develops, and the other follicles degenerate.

Ovulation

As a follicle matures, its primary oocyte undergoes oogenesis, giving rise to a secondary oocyte and a first polar body. The process called **ovulation** (o˝vu-la´shun) releases these cells from the follicle.

Hormones from the anterior pituitary gland trigger ovulation, causing the mature follicle to swell rapidly and its wall to weaken. Eventually, the wall ruptures, and follicular fluid and the secondary oocyte ooze from the ovary's surface (see fig. 19.9).

After ovulation, the secondary oocyte and one or two layers of follicular cells surrounding it are usually propelled to the opening of a nearby uterine tube (fig. 19.11). If the oocyte is not fertilized within a relatively short time, it degenerates.

CHECK YOUR RECALL

1. What changes occur in a follicle and its oocyte during maturation?
2. What causes ovulation?
3. What happens to an oocyte following ovulation?

Female Internal Accessory Organs

The internal accessory organs of the female reproductive system include a pair of uterine tubes, a uterus, and a vagina.

Uterine Tubes

The **uterine tubes** (fallopian tubes or oviducts) open near the ovaries (fig. 19.11). Each tube is about 10 centimeters long and passes medially to the uterus, penetrates its wall, and opens into the uterine cavity.

Near each ovary, a uterine tube expands to form a funnel-shaped *infundibulum* (in˝fun-dib´u-lum), which partially encircles the ovary medially. Fingerlike extensions called *fimbriae* (fim´bre) fringe the infundibulum margin. Although the infundibulum generally does not touch the ovary, one of the larger extensions connects directly to the ovary.

Simple columnar epithelial cells, some of which are *ciliated,* line the uterine tube. The epithelium secretes mucus, and the cilia beat toward the uterus. These actions help draw the secondary oocyte and expelled follicular fluid into the infundibulum following ovulation. Ciliary action and peristaltic contractions of the uterine tube's muscular layer help transport the secondary oocyte down the uterine tube.

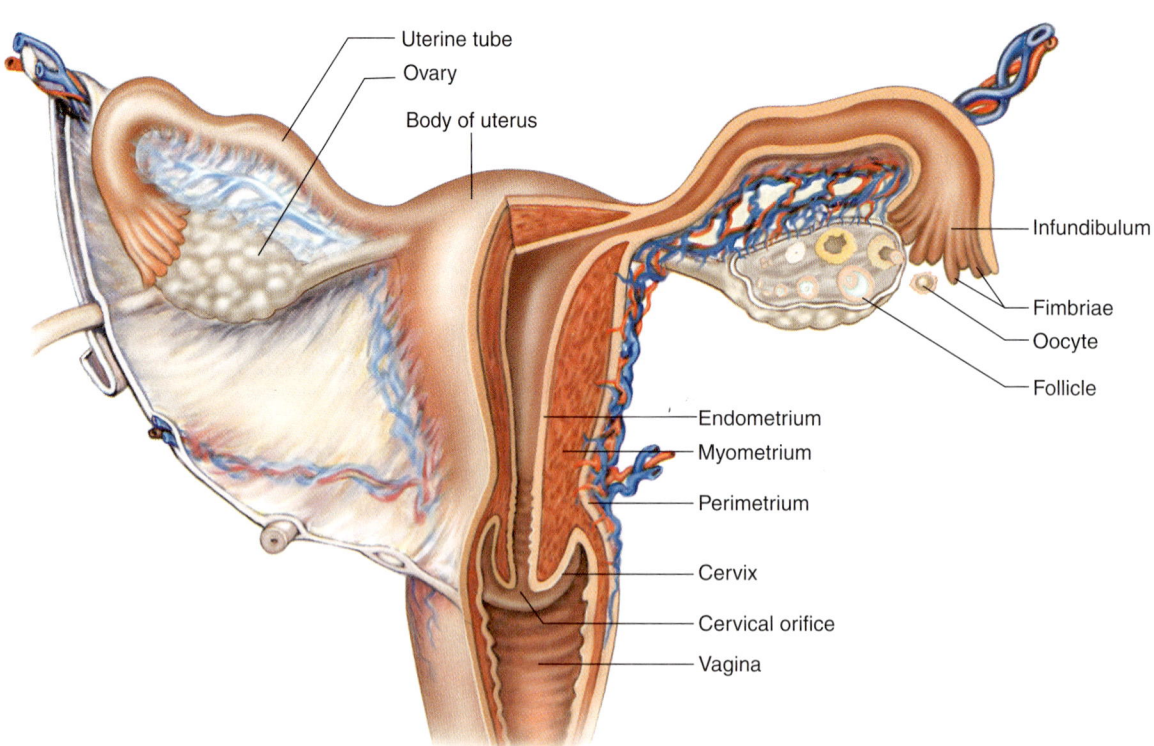

Figure 19.11
The funnel-shaped infundibulum of the uterine tube partially encircles the ovary (posterior view).

Uterus

If the secondary oocyte is fertilized in the uterine tube, becoming an ovum or egg cell, the **uterus** receives the embryo and sustains its development. The uterus is a hollow, muscular organ shaped somewhat like an inverted pear.

The size of the uterus changes greatly during pregnancy. In its nonpregnant, adult state, the uterus is about 7 centimeters long, 5 centimeters wide (at its broadest point), and 2.5 centimeters in diameter. It is located medially within the anterior portion of the pelvic cavity, superior to the vagina, and usually bends forward over the urinary bladder.

The upper two-thirds, or *body*, of the uterus has a dome-shaped top (fig. 19.11). The uterine tubes enter the top of the uterus at its broadest part. The lower third of the uterus is called the **cervix.** This tubular part extends downward into the upper portion of the vagina. The cervix surrounds the opening called the *cervical orifice*, through which the uterus opens to the **vagina.**

The uterine wall is thick and has three layers (fig. 19.12). The **endometrium,** the inner mucosal layer, is covered with columnar epithelium and contains abundant tubular glands. The **myometrium,** a thick, middle, muscular layer, consists largely of bundles of smooth muscle fibers. During the monthly female reproductive cycles and during pregnancy, the endometrium and myometrium change extensively. The **perimetrium** is an outer serosal layer that covers the body of the uterus and part of the cervix.

 During pregnancy, the uterus expands to 500 times its normal size.

A procedure called the *Pap (Papanicolaou) smear test* can usually detect cancer of the cervix. A sample of cervical tissue is smeared on a glass slide, stained, and sent to a laboratory, where computer image recognition software is used to identify cancer cells. Cervical cancer detected and treated early has a high cure rate.

Vagina

The vagina is a fibromuscular tube, about 9 centimeters long, extending from the uterus to the outside (see fig. 19.7). It conveys uterine secretions, receives the erect penis during sexual intercourse, and provides an open channel for offspring during birth.

The vagina extends upward and back into the pelvic cavity. It is posterior to the urinary bladder and urethra, anterior to the rectum, and attached to these structures by connective tissues.

A thin membrane of connective tissue and stratified squamous epithelium called the **hymen** partially closes the *vaginal orifice*. A central opening of varying size allows uterine and vaginal secretions to pass to the outside.

The vaginal wall has three layers. The inner *mucosal layer* is stratified squamous epithelium. This layer lacks mucous glands; the mucus in the lumen of the vagina comes from uterine glands and from vestibular glands at the mouth of the vagina.

The middle *muscular layer* consists mainly of smooth muscle fibers. A thin band of striated muscle at the lower end of the vagina helps close the vaginal opening. However, a voluntary muscle (bulbospongiosus) is primarily responsible for closing this orifice.

The outer *fibrous layer* consists of dense connective tissue interlaced with elastic fibers. It attaches the vagina to surrounding organs.

 CHECK YOUR RECALL

1. How is a secondary oocyte moved along a uterine tube?
2. Describe the structure of the uterus.
3. Describe the structure of the vagina.

Female External Reproductive Organs

The external accessory organs of the female reproductive system include the labia majora, labia minora, clitoris, and vestibular glands. These structures surround

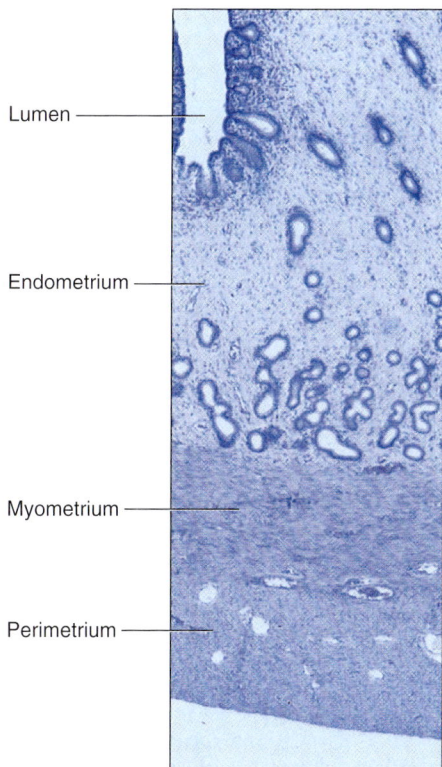

Figure 19.12
Light micrograph of the uterine wall (10.5×).

the openings of the urethra and vagina, and compose the **vulva** (see fig. 19.7).

Labia Majora

The **labia majora** (singular, *labium majus*) enclose and protect the other external reproductive organs. They correspond to the scrotum in males and are composed of rounded folds of adipose tissue and a thin layer of smooth muscle, covered by skin.

The labia majora lie close together. A cleft that includes the urethral and vaginal openings separates the labia longitudinally. At their anterior ends, the labia merge to form a medial, rounded elevation of adipose tissue called the *mons pubis,* which overlies the symphysis pubis (see fig. 19.7).

Labia Minora

The **labia minora** (singular, *labium minus*) are flattened, longitudinal folds between the labia majora (see fig. 19.7). They are composed of connective tissue richly supplied with blood vessels, giving a pinkish appearance. Posteriorly, the labia minora merge with the labia majora, while anteriorly, they converge to form a hoodlike covering around the clitoris.

Clitoris

The **clitoris** (klit´o-ris) is a small projection at the anterior end of the vulva between the labia minora (see fig. 19.7). It is usually about 2 centimeters long and 0.5 centimeters in diameter, including a portion embedded in surrounding tissues. The clitoris corresponds to the penis in males and has a similar structure. It is composed of two columns of erectile tissue called *corpora cavernosa.* At its anterior end, a small mass of erectile tissue forms a **glans,** which is richly supplied with sensory nerve fibers.

Vestibule

The labia minora enclose a space called the **vestibule.** The vagina opens into the posterior portion of the vestibule, and the urethra opens in the midline, just anterior to the vagina and about 2.5 centimeters posterior to the glans of the clitoris.

A pair of **vestibular glands,** corresponding to the bulbourethral glands in males, lie one on either side of the vaginal opening. Beneath the mucosa of the vestibule on either side is a mass of vascular erectile tissue called the **vestibular bulb.**

CHECK YOUR RECALL

1. What is the male counterpart of the labia majora? Of the clitoris?
2. Which structures are within the vestibule?

Erection, Lubrication, and Orgasm

Erectile tissues in the clitoris and around the vaginal entrance respond to sexual stimulation. Following such stimulation, parasympathetic nerve impulses from the sacral portion of the spinal cord release the vasodilator nitric oxide, causing the arteries associated with the erectile tissues to dilate. As a result, blood inflow increases, and the erectile tissues swell. At the same time, the vagina expands and elongates.

If sexual stimulation is sufficiently intense, parasympathetic impulses stimulate the vestibular glands to secrete mucus into the vestibule. This moistens and lubricates the tissues surrounding the vestibule and the lower end of the vagina, facilitating insertion of the penis into the vagina.

The clitoris is abundantly supplied with sensory nerve fibers, which are especially sensitive to local stimulation. The culmination of such stimulation is orgasm.

Just prior to orgasm, the tissues of the outer third of the vagina engorge with blood and swell. This increases the friction on the penis during intercourse. Orgasm initiates a series of reflexes involving the sacral and lumbar portions of the spinal cord. In response to these reflexes, the muscles of the perineum and the walls of the uterus and uterine tubes contract rhythmically. These contractions help transport sperm cells through the female reproductive tract toward the upper ends of the uterine tubes. Table 19.2 summarizes the functions of the female reproductive organs.

TABLE 19.2	FUNCTIONS OF THE FEMALE REPRODUCTIVE ORGANS
ORGAN	FUNCTION
Ovary	Produces oocytes and female sex hormones
Uterine tube	Conveys secondary oocyte toward uterus; site of fertilization; conveys developing embryo to uterus
Uterus	Protects and sustains embryo during pregnancy
Vagina	Conveys uterine secretions to outside of body; receives erect penis during sexual intercourse; provides open channel for offspring during birth process
Labia majora	Enclose and protect other external reproductive organs
Labia minora	Form margins of vestibule; protect openings of vagina and urethra
Clitoris	Produces feelings of pleasure during sexual stimulation due to abundant sensory nerve endings in glans
Vestibule	Space between labia minora that contains vaginal and urethral openings
Vestibular glands	Secrete fluid that moistens and lubricates vestibule

The clitoris contains nearly 8,000 nerve fibers, the densest collection of any body part. The corresponding part of the penis has only 4,000 nerve fibers.

CHECK YOUR RECALL

1. What events result from parasympathetic stimulation of the female reproductive organs?
2. What changes occur in the vagina just prior to and during orgasm?

19.5 Hormonal Control of Female Reproductive Functions

The hypothalamus, anterior pituitary gland, and ovaries secrete hormones that control the development and maintenance of female secondary sex characteristics, maturation of female sex cells, and changes that occur during the monthly reproductive cycle.

Female Sex Hormones

The female body is reproductively immature until about age ten. Then the hypothalamus begins to secrete increasing amounts of GnRH. GnRH, in turn, stimulates the anterior pituitary to release the gonadotropins FSH and LH. These hormones play primary roles in controlling female sex cell maturation and in producing female sex hormones.

Several tissues, including the ovaries, the adrenal cortices, and the placenta (during pregnancy), secrete female sex hormones belonging to two major groups—**estrogens** (es´tro-jenz) and **progesterone** (pro-jes´tĭ-rōn). *Estradiol* is the most abundant of the estrogens, which also include *estrone* and *estriol*.

The ovaries are the primary source of estrogens (in a nonpregnant female). At puberty, under the influence of the anterior pituitary, the ovaries secrete increasing amounts of estrogens. Estrogens stimulate enlargement of accessory organs, including the vagina, uterus, uterine tubes, ovaries, and external reproductive structures. Estrogens also develop and maintain the *female secondary sex characteristics,* special features associated with the adult female body, which include:

1. Development of the breasts and the ductile system of the mammary glands within the breasts.
2. Increased deposition of adipose tissue in the subcutaneous layer generally and in the breasts, thighs, and buttocks particularly.
3. Increased vascularization of the skin.

The ovaries are also the primary source of progesterone (in a nonpregnant female). This hormone promotes changes in the uterus during the female reproductive cycle, affects the mammary glands, and helps regulate the secretion of gonadotropins from the anterior pituitary.

Androgen (male sex hormone) concentrations produce certain other changes in females at puberty. For example, increased hair growth in the pubic and axillary regions is due to androgen secreted by the adrenal cortices. Conversely, development of the female skeletal configuration, which includes narrow shoulders and broad hips, is a response to a low androgen concentration.

CHECK YOUR RECALL

1. What stimulates sexual maturation in a female?
2. What is the function of estrogens?
3. What is the function of androgen in a female?

Female Reproductive Cycle

The female reproductive cycle, or **menstrual cycle** (men´stroo-al si´kl), consists of regular, recurring changes in the uterine lining, which culminate in menstrual bleeding (menses). At the same time, changes in the ovary constitute the **ovarian cycle.** Such cycles usually begin around age thirteen and continue into middle age, then cease.

> Women athletes may have disturbed menstrual cycles, ranging from diminished menstrual flow (oligomenorrhea) to complete stoppage (amenorrhea). The more active an athlete, the more likely it is that she will have menstrual problems. This effect results from a loss of adipose tissue and a consequent decline in estrogens, which adipose tissue synthesizes from adrenal androgen.

A female's first menstrual cycle, called **menarche** (mĕ-nar´ke), occurs after the ovaries and other organs of the reproductive control system have matured and begun responding to certain hormones. Then, hypothalamic secretion of GnRH stimulates the anterior pituitary to release threshold levels of FSH and LH. FSH stimulates maturation of an ovarian follicle. The follicular cells produce increasing amounts of estrogens and some progesterone. LH stimulates certain ovarian cells to secrete precursor molecules (testosterone), which are also used to produce estrogens.

In a young female, estrogens stimulate the development of secondary sex characteristics. Estrogens secreted during subsequent menstrual cycles continue the development and maintenance of these characteristics.

As shown in the diagram of the menstrual cycles in figure 19.13, increasing concentration of estrogens during the first week or so of a menstrual cycle changes the uterine lining, thickening the glandular endometrium

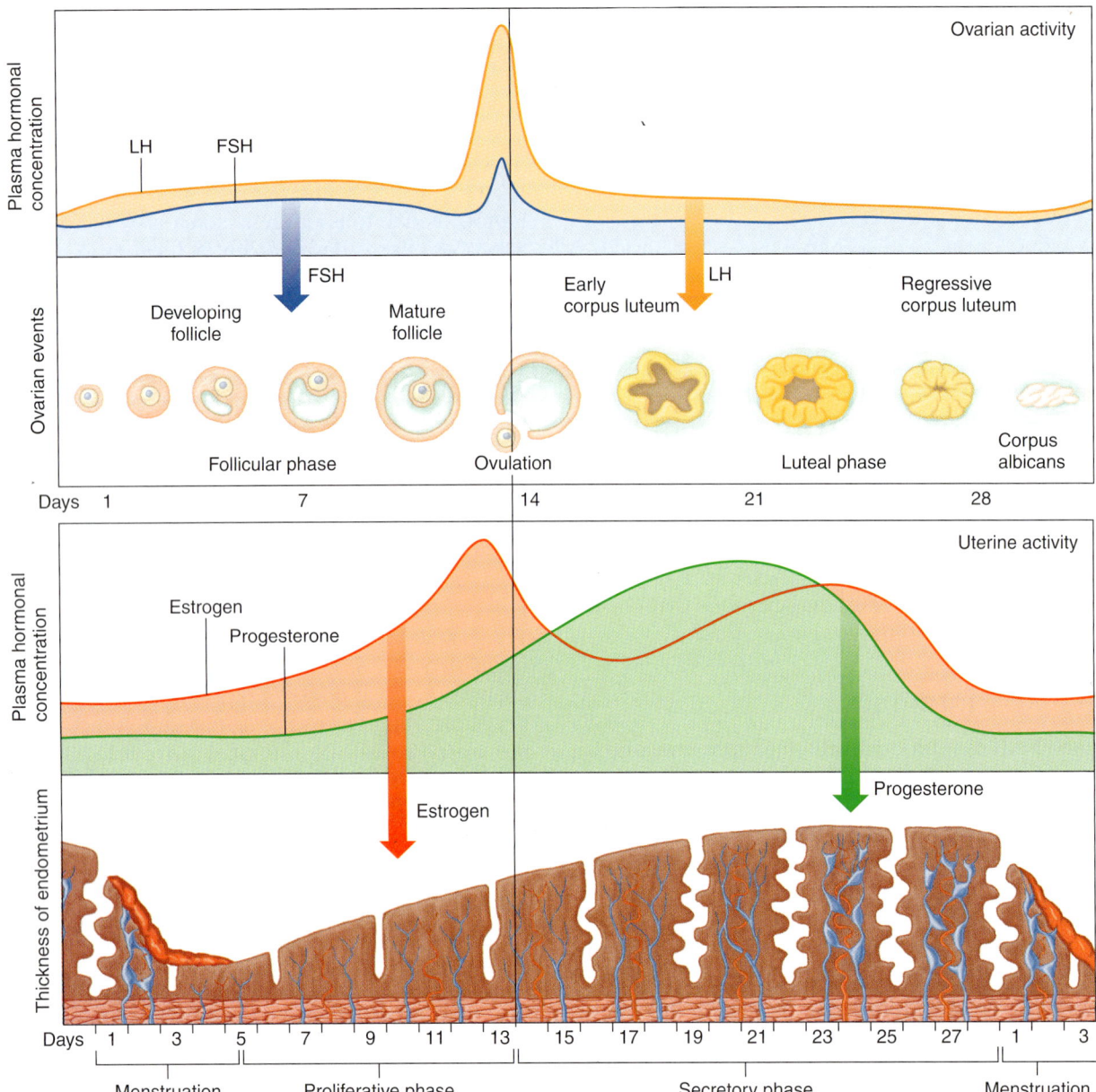

Figure 19.13
Major events in the female menstrual cycle.

(proliferative phase). Meanwhile, the developing follicle completes maturation, and by the fourteenth day of the cycle, the follicle appears on the ovary surface as a blisterlike bulge. Within the follicle, the follicular cells, which surround and connect the oocyte to the inner wall, loosen. Follicular fluid accumulates rapidly.

While the follicle matured, estrogens that it secreted inhibited the anterior pituitary's release of LH, but allowed LH to be stored in the gland. Estrogens also made anterior pituitary cells more sensitive to the action of GnRH, which the hypothalamus released in rhythmic pulses about 90 minutes apart.

Near the fourteenth day of follicular development, anterior pituitary cells finally release the stored LH in response to the GnRH pulses. The resulting surge in LH concentration, which lasts about 36 hours, weakens and ruptures the bulging follicular wall, which sends the oocyte and follicular fluid from the ovary (ovulation).

Following ovulation, the remnants of the follicle within the ovary change rapidly. The space that the follicular fluid occupied fills with blood, which soon clots, and under the influence of LH, follicular cells enlarge to form a temporary glandular structure called a **corpus luteum** ("yellow body").

Follicular cells secrete progesterone during the first part of the menstrual cycle. Corpus luteum cells secrete abundant progesterone and estrogens during the last half of the cycle. Consequently, as a corpus luteum forms, blood progesterone concentration increases sharply.

Progesterone causes the endometrium to become more vascular and glandular. It also stimulates uterine glands to secrete more glycogen and lipids (secretory phase). As a result, endometrial tissues fill with fluids containing nutrients and electrolytes, providing a favorable environment for embryo development.

Estrogens and progesterone inhibit the anterior pituitary's release of LH and FSH. Consequently, no other follicles are stimulated to develop when the corpus luteum is active. However, if a sperm cell does not fertilize the oocyte released at ovulation, the corpus luteum begins to degenerate on about the twenty-fourth day of the cycle. Eventually, connective tissue replaces it. The remnant of such a corpus luteum is called a *corpus albicans*.

When the corpus luteum ceases to function, concentrations of estrogens and progesterone decline rapidly, and blood vessels in the endometrium constrict in response. This reduces the supply of oxygen and nutrients to the thickened uterine lining, and these lining tissues soon disintegrate and slough off. At the same time, blood escapes from damaged capillaries, creating a flow of blood and cellular debris that passes through the vagina as the *menstrual flow* (menses). This flow usually begins about the twenty-eighth day of the cycle and continues for three to five days while the concentrations of estrogens are relatively low. The menstrual flow marks the end of a menstrual cycle and the beginning of a new cycle. Table 19.3 summarizes the menstrual cycle.

Low blood concentrations of estrogens and progesterone at the beginning of the menstrual cycle mean that the hypothalamus and anterior pituitary are no longer inhibited. Consequently, FSH and LH concentrations soon increase, stimulating a new follicle to mature. As this follicle secretes estrogens, the uterine lining undergoes repair, and the endometrium begins to thicken again.

Menopause

After puberty, menstrual cycles continue at regular intervals into the late forties or early fifties, when they become increasingly irregular. Then, in a few months or years, the cycles cease altogether. This period in life is called **menopause** (men´o-pawz), or the female climacteric.

Aging of the ovaries causes menopause. After about thirty-five years of cycling, few primary follicles remain to respond to pituitary gonadotropins. Consequently, the follicles no longer mature, ovulation does not occur, and blood concentration of estrogens plummets.

Reduced concentrations of estrogens and lack of progesterone may change the female secondary sex characteristics. The breasts, vagina, uterus, and uterine tubes may shrink, and the pubic and axillary hair may thin.

CHECK YOUR RECALL

1. Trace the events of the menstrual cycle.
2. What causes menstrual flow?
3. What are some changes that may occur at menopause?

19.6 Mammary Glands

The **mammary glands** are accessory organs of the female reproductive system that are specialized to secrete milk following pregnancy (fig. 19.14; reference plate 1, p. 22). They are in the subcutaneous tissue of the anterior thorax within elevations called *breasts*. The breasts overlie the *pectoralis major* muscles and extend from the second to the sixth ribs and from the sternum to the axillae.

A *nipple* is located near the tip of each breast at about the level of the fourth intercostal space. A circular area of pigmented skin, called the *areola*, surrounds each nipple.

TABLE 19.3   MAJOR EVENTS IN A MENSTRUAL CYCLE

1. Anterior pituitary gland secretes follicle-stimulating hormone (FSH) and luteinizing hormone (LH).
2. FSH stimulates maturation of a follicle.
3. Follicular cells produce and secrete estrogens.
    a. Estrogens maintain secondary sex traits.
    b. Estrogens cause uterine lining to thicken.
4. Anterior pituitary releases a surge of LH, which stimulates ovulation.
5. Follicular cells become corpus luteum cells, which secrete estrogens and progesterone.
    a. Estrogens continue to stimulate uterine wall development.
    b. Progesterone stimulates the uterine lining to become more glandular and vascular.
    c. Estrogens and progesterone inhibit anterior pituitary from secreting LH and FSH.
6. If the egg cell is not fertilized, the corpus luteum degenerates and no longer secretes estrogens and progesterone.
7. As concentrations of estrogens and progesterone decline, blood vessels in the uterine lining constrict.
8. Uterine lining disintegrates and sloughs off, producing menstrual flow.
9. Anterior pituitary is no longer inhibited and again secretes FSH and LH.
10. The menstrual cycle repeats.

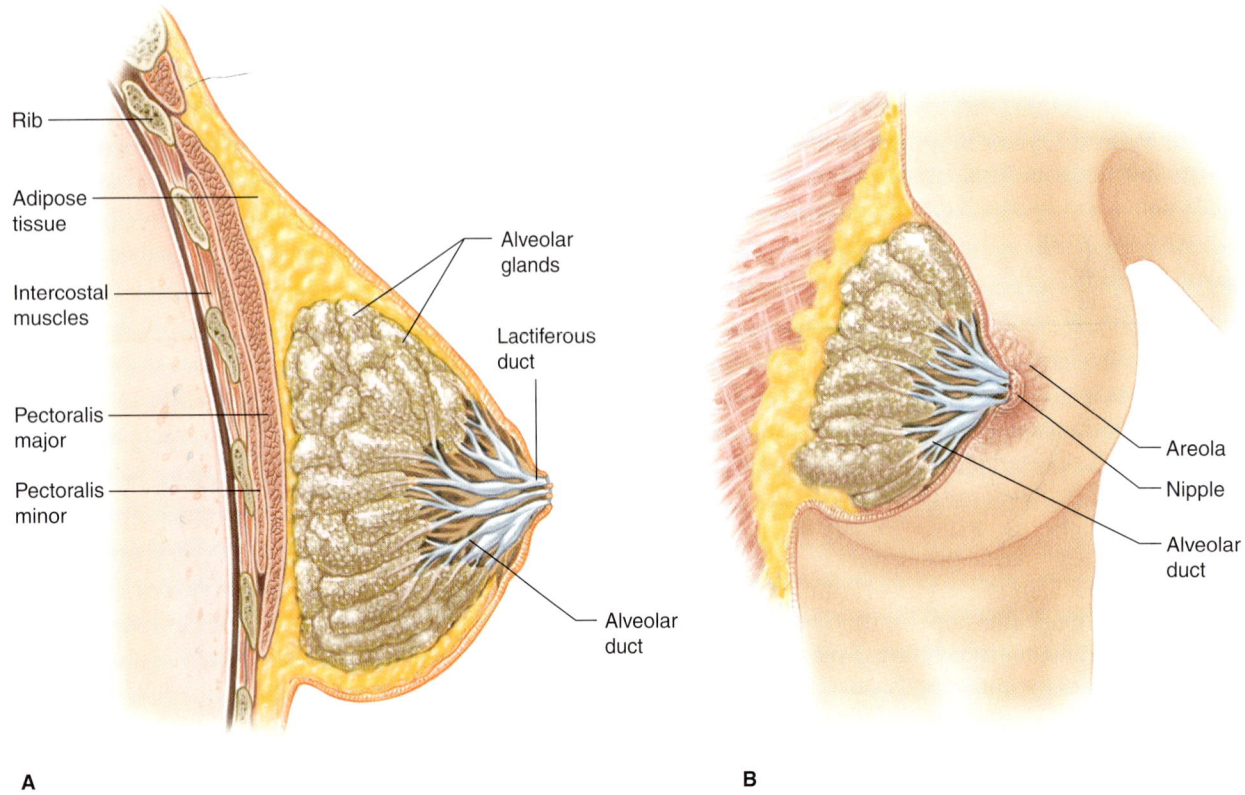

Figure 19.14
Structure of the female breast and mammary glands. (*A*) Sagittal section. (*B*) Anterior view.

A mammary gland is composed of fifteen to twenty lobes. Each lobe contains glands (alveolar glands) and an alveolar duct that leads to a lactiferous duct that leads to the nipple and opens to the outside. Dense connective and adipose tissues separate the lobes. These tissues also support the glands and attach them to the fascia of the underlying pectoral muscles. Other connective tissue, which forms dense strands called *suspensory ligaments,* extends inward from the dermis of the breast to the fascia, helping to support the breast's weight.

The mammary glands of males and females are similar. As children reach puberty, the glands in males do not develop, whereas in females, ovarian hormones stimulate development of the glands. As a result, the alveolar glands and ducts enlarge, and fat forms deposits around and within the breasts. Chapter 20 (p. 532) describes the hormonal mechanism that stimulates mammary glands to produce and secrete milk.

> **CHECK YOUR RECALL**
> 1. Describe the structure of a mammary gland.
> 2. What changes do ovarian hormones cause in mammary glands?

19.7 Birth Control

Birth control is the voluntary regulation of the number of offspring produced and the time at which they are conceived. This control requires a method of **contraception** (kon´´trah-sep´shun) to avoid fertilization of a secondary oocyte following sexual intercourse (coitus) or to prevent the hollow ball of cells (a blastocyst) that will develop into an embryo from implanting in the uterine wall.

Coitus Interruptus

Coitus interruptus is withdrawal of the penis from the vagina before ejaculation, which prevents entry of sperm cells into the female reproductive tract. This method can still result in pregnancy because a male may find it difficult to withdraw just prior to ejaculation. Also, small quantities of semen containing sperm cells may reach the vagina before ejaculation occurs.

Rhythm Method

The *rhythm method* (also called timed coitus or natural family planning) requires abstinence from sexual intercourse a few days before and a few days after ovulation. The rhythm method results in a relatively high rate of pregnancy because accurately identifying infertile times to have intercourse is difficult. Another disadvantage of the rhythm method is that it restricts spontaneity in sexual activity.

Mechanical Barriers

Mechanical barrier contraceptives prevent sperm cells from entering the female reproductive tract during sexual intercourse.

The *male condom* consists of a thin latex or natural membrane sheath placed over the erect penis before intercourse to prevent semen from entering the vagina upon ejaculation (fig. 19.15A). A *female condom* resembles a small plastic bag. A woman inserts it into her vagina prior to intercourse. The device blocks sperm cells from reaching the cervix.

Some men feel that a condom decreases the sensitivity of the penis during intercourse, and its use may interrupt spontaneity of the sex act. However, condoms are inexpensive and may also protect against sexually transmitted diseases.

Another mechanical barrier is the *diaphragm*, a cup-shaped device with a flexible ring forming the rim. A woman inserts the diaphragm into the vagina so that it covers the cervix, preventing entry of sperm cells into the uterus (fig. 19.15B). To be effective, a diaphragm must be fitted for size by a physician, inserted properly, and used with a chemical spermicide applied to the surface adjacent to the cervix and to the rim of the diaphragm. The device must be left in position for several hours following sexual intercourse. A diaphragm can be inserted up to 6 hours prior to sexual contact.

Similar to but smaller than the diaphragm is the *cervical cap*, which adheres to the cervix by suction. A woman inserts it with her fingers before intercourse. For centuries, different societies have used cervical caps made of such varied substances as beeswax, lemon halves, paper, and opium poppy fibers.

Chemical Barriers

Chemical barrier contraceptives include creams, foams, and jellies with spermicidal properties (fig. 19.15C).

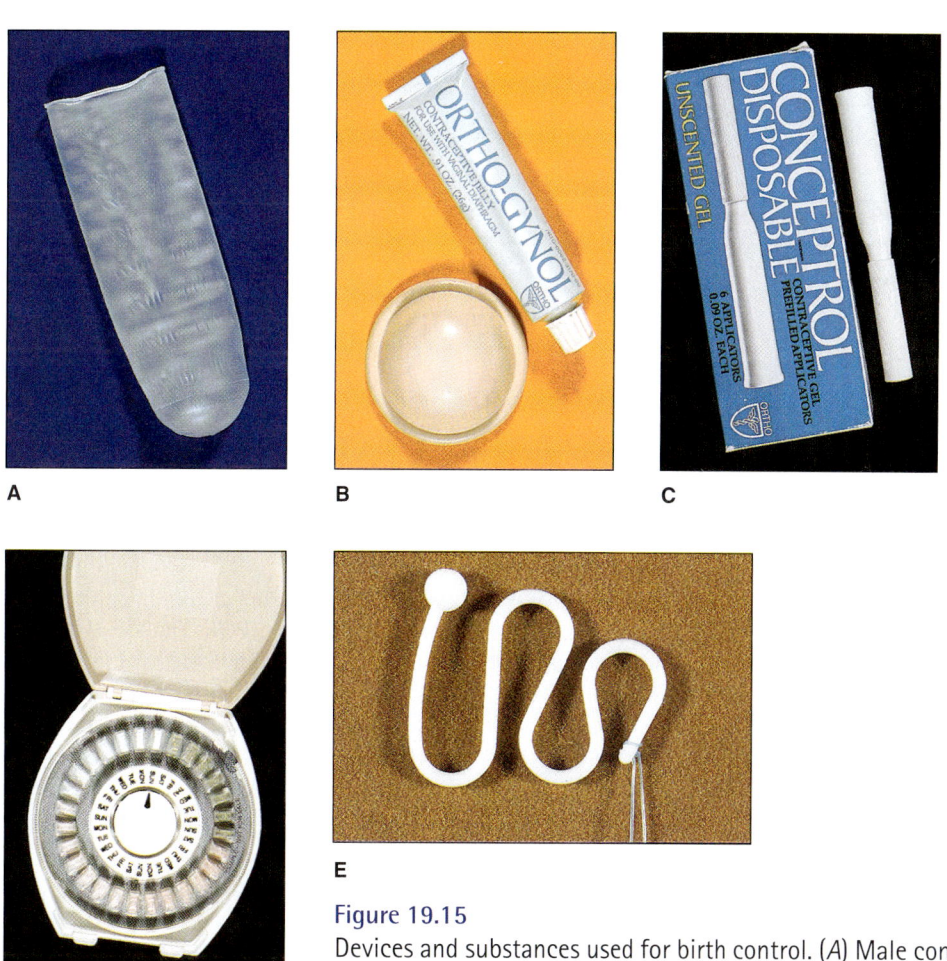

Figure 19.15
Devices and substances used for birth control. (A) Male condom. (B) Diaphragm and spermicide. (C) Spermicidal gel. (D) Oral contraceptives. (E) IUD. (Photographs are not to scale.)

Topic of Interest

TREATING BREAST CANCER

One in eight women will develop breast cancer at some point in her life (table 19C). Breast cancer is really several illnesses. As medical research reveals the cellular and molecular characteristics that distinguish subtypes of the disease, treatments old and new are being increasingly tailored to individuals. This "rational" approach may delay progression of the disease and increase the survival rate.

Warning Signs

Changes that could signal breast cancer include a small area of thickened tissue, a dimple, a change in contour, or a nipple that is flattened, points in an unusual direction, or produces a discharge. A woman can note these changes by performing a monthly "breast self-exam," in which she lies flat on her back with one arm raised behind her head and systematically feels all parts of each breast. But sometimes breast cancer gives no warning at all—early signs of fatigue and feeling ill may not occur until the disease has spread beyond the breast.

After finding a lump, the next step is a physical exam, in which a health-care provider palpates the breast and does a mammogram, an X-ray scan that can pinpoint the location and approximate extent of abnormal tissue (fig. 19B). An ultrasound scan can distinguish between a cyst (a fluid-filled sac of glandular tissue) and a tumor (a solid mass). If an area is suspicious, a thin needle is used to take a biopsy (sample) of the tissue, whose cells will be scrutinized for the telltale characteristics of cancer.

Eighty percent of the time, a breast lump is a sign of fibrocystic breast disease, which is benign (noncancer-

TABLE 19C		BREAST CANCER RISK	
BY AGE	ODDS	BY AGE	ODDS
25	1 in 19,608	60	1 in 24
30	1 in 2,525	65	1 in 17
35	1 in 622	70	1 in 14
40	1 in 217	75	1 in 11
45	1 in 93	80	1 in 10
50	1 in 50	85	1 in 9
55	1 in 33	95 or older	1 in 8

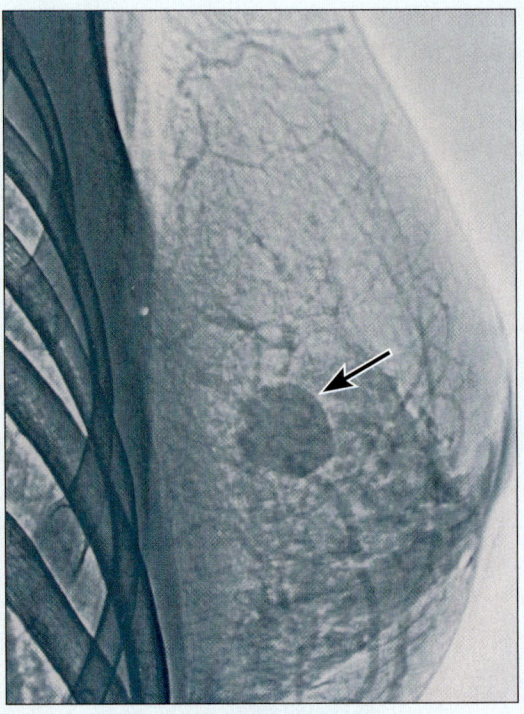

Figure 19B
Mammogram of a breast with a tumor (arrow).

Within the vagina, such chemicals create an unfavorable environment for sperm cells.

Chemical barrier contraceptives are fairly easy to use but have a high failure rate when used alone. They are most effective when used with a condom or diaphragm.

Oral Contraceptives

An *oral contraceptive*, or birth control pill, contains synthetic estrogen-like and progesterone-like chemicals (fig. 19.15D). In women, these drugs disrupt the normal pattern of gonadotropin secretion and prevent the LH surge that triggers ovulation. They also interfere with buildup of the uterine lining necessary for implantation.

Oral contraceptives, if used correctly, prevent pregnancy nearly 100% of the time. However, they may cause nausea, retention of body fluids, increased skin pigmentation, and breast tenderness. Also, some women, particularly those over age thirty-five who smoke, may develop intravascular blood clots, liver disorders, or high blood pressure when using certain types of oral contraceptives.

Injectable Contraception

An intramuscular injection of Depo-Provera (medroxyprogesterone acetate) protects against pregnancy for three months by preventing the maturation and release of a secondary oocyte. It also alters the uterine lining,

ous). The lump may be a cyst or a solid, fibrous mass of connective tissue called a fibroadenoma. Treatment for fibrocystic breast disease includes taking vitamin E or synthetic androgens under a doctor's care, lowering caffeine intake, and examining unusual lumps further.

Surgery, Radiation, and Chemotherapies

If biopsied breast cells are cancerous, treatment usually begins with surgery. A lumpectomy removes a small tumor and some surrounding tissue; a simple mastectomy removes a breast; and a modified radical mastectomy removes the breast and surrounding lymph nodes, but preserves the pectoral muscles. Radical mastectomies, which remove the muscles too, are rarely done anymore. In addition, a few lymph nodes are typically examined, which allows a physician to identify the ones that are affected and must be removed.

Most breast cancers are then treated with radiation and combinations of chemotherapeutic drugs, plus newer drugs that are sometimes targeted to certain types of breast cancer. Standard chemotherapies kill all rapidly dividing cells, and those used for breast cancer include fluorouracil, doxorubicin, cyclophosphamide, and methotrexate. A newer chemotherapeutic agent is paclitaxol, which was originally derived from the bark of yew trees.

Drugs called selective estrogen receptor modulators (SERMs) are used for women whose cancer cells have receptors for estrogens. These drugs include tamoxifen, which has been used for more than 20 years, and a newer drug called raloxifene. SERMs block the receptors so that estrogens cannot bind and trigger division of cancer cells. In contrast to standard chemotherapies, which are given for weeks or months, SERMs are taken for many years. Tamoxifen may also be able to prevent cancer in certain women who are at very high risk due to family history.

Another new breast cancer drug, Herceptin, can help women whose cancer cells bear many receptors that bind a particular growth factor. Herceptin is a type of immune system biochemical called a monoclonal antibody. It prevents the growth factor from stimulating cell division.

Prevention Strategies

Many health-care providers advise women to have baseline mammograms by the age of forty and yearly mammograms after that, or beginning at age fifty, depending upon individual medical and family histories. Although a mammogram can detect a tumor up to two years before it can be felt, it can also miss some tumors. Thus, breast self-exam is also important in early detection.

Genetic tests are becoming available that can identify women who have inherited certain variants of genes—such as BRCA1, BRCA2, p53, and her-2/neu—that place them at very high risk for developing breast cancer. Women at high risk can be tested more frequently, and some have even had their breasts removed because they have inherited a gene variant that, in their families, predicts a very high risk of developing breast cancer. In one family, a genetic test told a woman whose two sisters and mother had inherited breast cancer that she had escaped their fate, and she canceled the scheduled surgery. Yet her young cousin, who thought she was free of the gene because it was inherited through her father, found by genetic testing that she would likely develop breast cancer. A subsequent mammogram revealed that the disease had already begun.

Only 5 to 10% of all breast cancers arise from an inherited tendency. Much current research seeks to identify the environmental triggers that cause the majority of cases.

making it less hospitable for a blastocyst. Use of Depo-Provera requires a doctor's care because potential side effects make it risky for women with certain medical conditions.

Contraceptive Implants

A *contraceptive implant* is a set of small progesterone-containing capsules or rods inserted surgically under the skin of a woman's arm or scapular region. The progesterone, which the implant releases slowly, prevents ovulation in much the same way as do oral contraceptives. A contraceptive implant is effective for up to five years, and removal of the implant reverses its contraceptive action.

Intrauterine Devices

An *intrauterine device (IUD)* is a small, solid object that a physician places within the uterine cavity (fig. 19.15E). An IUD interferes with implantation of a blastocyst, perhaps by inflaming the uterine tissues.

The uterus may spontaneously expel the IUD, or the IUD may produce abdominal pain or excessive menstrual bleeding. It may also injure the uterus or produce other serious health problems. A physician should regularly check IUD placement.

Surgical Methods

Surgical methods of contraception sterilize the male or female. In the male, a physician performs a *vasectomy,*

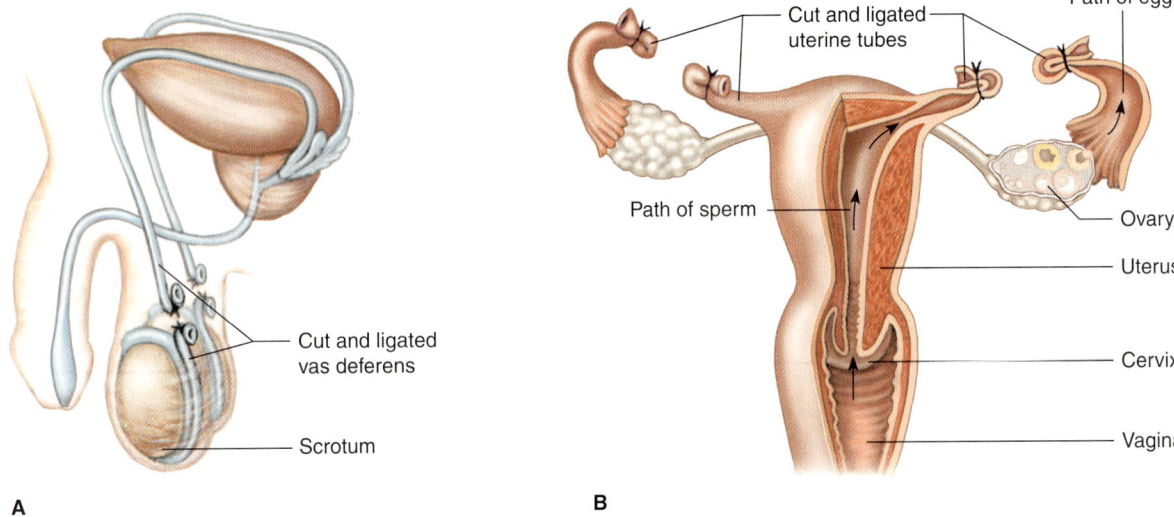

Figure 19.16
Surgical methods of birth control. (*A*) In a vasectomy, each vas deferens is cut and ligated. (*B*) In a tubal ligation, each uterine tube is cut and ligated.

removing a small section of each vas deferens near the epididymis and tying (ligating) the cut ends of the ducts. A vasectomy is a simple operation with few side effects, although it may cause some pain for a week or two.

After a vasectomy, sperm cells cannot leave the epididymis; thus, they are not included in the semen. However, sperm cells may already be present in portions of the ducts distal to the cuts. Consequently, the sperm count may not reach zero for several weeks.

The corresponding procedure in the female is *tubal ligation*. The uterine tubes are cut and ligated so that sperm cells cannot reach an egg cell.

Neither a vasectomy nor a tubal ligation changes hormonal concentrations or sex drives. These procedures, shown in figure 19.16, are the most reliable forms of contraception. Reversing them requires microsurgery.

CHECK YOUR RECALL

1. What factors make the rhythm method less reliable than some other methods of contraception?
2. Describe two methods of contraception that use mechanical barriers.
3. How do oral contraceptives, injectable contraceptives, and contraceptive implants prevent pregnancy?

19.8 Sexually Transmitted Diseases

The twenty recognized **sexually transmitted diseases (STDs)** are often called "silent infections" because the early stages may not produce symptoms, especially in women. Table 19.4 describes the six most prevalent STDs. By the time symptoms appear, it is often too late to prevent complications or spread of the infection to sex-

TABLE 19.4 SOME SEXUALLY TRANSMITTED DISEASES

DISEASE	CAUSE	SYMPTOMS	TREATMENT
Acquired immune deficiency syndrome	Human immunodeficiency virus	Fever, weakness, infections, cancer	Drugs to treat or delay symptoms; no cure
Chlamydia infection	Bacteria of genus *Chlamydia*	Painful urination and intercourse, mucous discharge from penis or vagina	Antibiotics
Genital herpes	Herpes virus 2	Genital sores, fever	Antiviral drug (acyclovir)
Genital warts	Human papilloma virus	Warts on genitals	Chemical or surgical removal
Gonorrhea	*Neisseria gonorrhoeae* bacteria	In women, usually none; in men, painful urination	Antibiotics
Syphilis	*Treponema pallidum* bacteria	Initial chancre sore usually on genitals or mouth; rash 6 months later; several years with no symptoms as infection spreads; finally damage to heart, liver, nerves, brain	Antibiotics

ual partners. Many STDs have similar symptoms, some of which are also seen in diseases or allergies that are not sexually related. A physician should be consulted if one or a combination of the following symptoms appears:

1. Burning sensation during urination.
2. Pain in the lower abdomen.
3. Fever or swollen glands in the neck.
4. Discharge from the vagina or penis.
5. Pain, itching, or inflammation in the genital or anal area.
6. Pain during intercourse.
7. Sores, blisters, bumps, or a rash anywhere on the body, particularly the mouth or genitals.
8. Itchy, runny eyes.

One possible complication of the STDs gonorrhea and chlamydia is **pelvic inflammatory disease,** in which bacteria enter the vagina and spread throughout the reproductive organs. The disease begins with intermittent cramps, followed by sudden fever, chills, weakness, and severe cramps. Hospitalization and intravenous antibiotics can stop the infection. The uterus and uterine tubes are often scarred.

Acquired immune deficiency syndrome (AIDS) is an STD that destroys the immune system. Infections and often cancer, diseases that the immune system usually conquers, overrun the body. The AIDS virus (human immunodeficiency virus, or HIV) passes from one person to another in body fluids such as semen and blood. Unprotected intercourse and using a needle containing contaminated blood are the most frequent routes of transmission in the U.S.

CHECK YOUR RECALL

1. Why are sexually transmitted diseases often called "silent infections"?
2. What are some common symptoms of sexually transmitted diseases?

Clinical Terms Related to the Reproductive Systems

amenorrhea (a-men´´o-re´ah) Absence of menstrual flow, usually due to a disturbance in hormonal concentrations.
conization (ko´´nĭ-za´shun) Surgical removal of a cone of tissue from the cervix for examination.
curettage (ku´´rĕ-tahzh´) Surgical procedure in which the cervix is dilated and the endometrium of the uterus is scraped (commonly called D and C, for dilation and curettage).
dysmenorrhea (dis´´men-ŏ-re´ah) Painful menstruation.
endometriosis (en´´do-me´´tre-o´sis) Tissue similar to the inner lining of the uterus occurring within the pelvic cavity.
endometritis (en´´do-mĕ-tri´tis) Inflammation of the uterine lining.
epididymitis (ep´ĭ-did´´ĭ-mi´tis) Inflammation of the epididymis.
hematometra (hem´´ah-to-me´trah) Accumulation of menstrual blood within the uterine cavity.
hysterectomy (his´´tĕ-rek´to-me) Surgical removal of the uterus.
mastitis (mas´´ti´tis) Inflammation of a mammary gland.
oophorectomy (o´´of-o-rek´to-me) Surgical removal of an ovary.
oophoritis (o´´of-o-ri´tis) Inflammation of an ovary.
orchiectomy (or´´ke-ĕk´to-me) Surgical removal of a testis.
orchitis (or-ki´tis) Inflammation of a testis.
prostatectomy (pros´´tah-tek´to-me) Surgical removal of a portion or all of the prostate gland.
prostatitis (pros´´tah-ti´tis) Inflammation of the prostate gland.
salpingectomy (sal´´pin-jek´to-me) Surgical removal of a uterine tube.
vaginitis (vaj´´ĭ-ni´tis) Inflammation of the vaginal lining.
varicocele (var´ĭ-ko-sēl´´) Distension of the veins within the spermatic cord.

Clinical Connection

Bruce Reimer was born in 1965. At age 8 months, most of his penis was accidentally burned off during a circumcision procedure. Physicians and psychologists advised the parents to "reassign" the child's gender as female. At 22 months of age, corrective surgery created Brenda from Bruce. But Brenda continually fought attempts to raise her as a girl and at age 14, threatened suicide unless allowed to live as a male. He took the name David Reimer, and eventually married, adopted his wife's children, and is now a young grandfather. Apparently, surgery could not silence David's XY chromosome constitution—that of a male. Since then, several studies of infants born with very small penises and reared as girls overwhelmingly confirm Reimer's experience that nature has a greater effect on gender identity than nurture. In the past, physicians based the decision to remove a small or damaged penis and reassign sex as female on a yardstick of sorts. If a newborn's stretched organ exceeded an inch, he was deemed a he. If the protrusion was under 3/8 of an inch, she was deemed a she. Organs that fell in between were shortened into a clitoris during the first week of life, and girlhood officially began. Today, such decisions rest more on an individual's chromosomal sex, and in some cases, surgery is delayed until a person can decide for him- or herself.

Organization

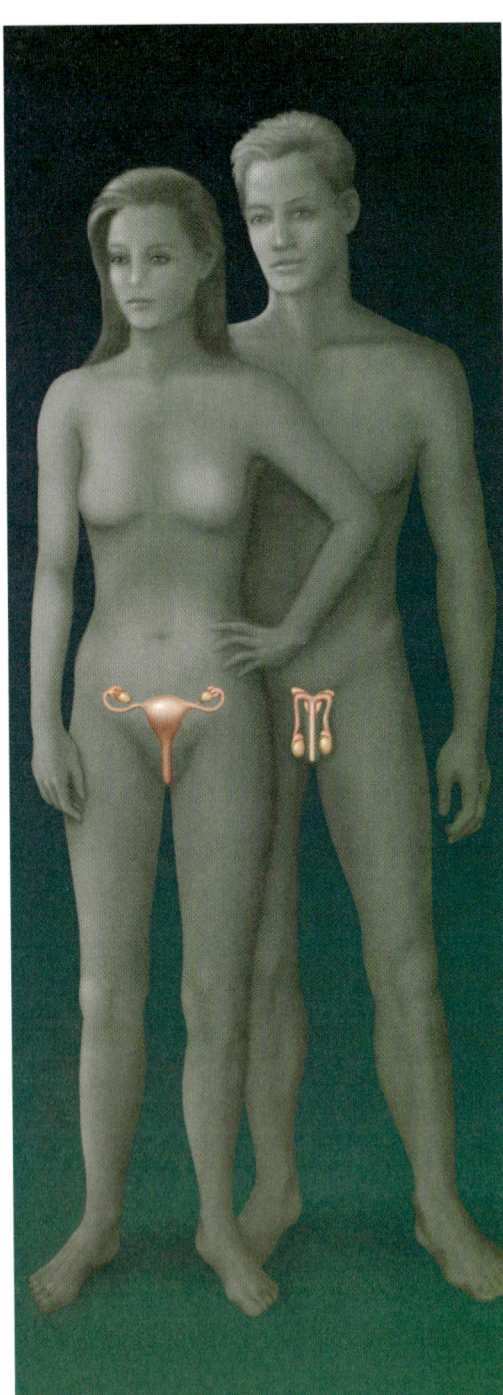

Reproductive Systems
Gamete production, fertilization, fetal development, and childbirth are essential for survival of the species.

Integumentary System
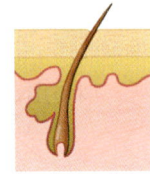
Skin sensory receptors play a role in sexual pleasure.

Skeletal System

Bones can be a temporary source of calcium during lactation.

Muscular System
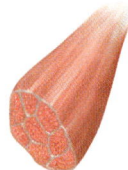
Skeletal, cardiac, and smooth muscles all play a role in reproductive processes and sexual activity.

Nervous System
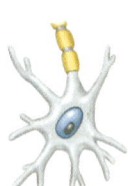
The nervous system plays a major role in sexual activity and sexual pleasure.

Endocrine System

Hormones control the production of ova in the female and sperm in the male.

Cardiovascular System
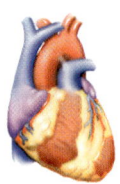
Blood pressure is necessary for the normal function of erectile tissue in the male and female.

Lymphatic System

Special mechanisms inhibit the female immune system from attacking sperm as foreign invaders.

Digestive System
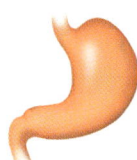
Proper nutrition is essential for the formation of normal gametes and for normal fetal development during pregnancy.

Respiratory System
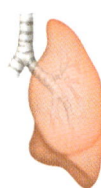
During pregnancy, the placenta provides oxygen to the fetus and removes carbon dioxide.

Urinary System

Male urinary and reproductive systems share common structures. Kidneys compensate for fluid loss from the reproductive systems. Pregnancy may cause fluid retention.

SUMMARY OUTLINE

19.1 Introduction (p. 498)
Reproductive organs produce sex cells and sex hormones, sustain these cells, or transport them from place to place.

19.2 Organs of the Male Reproductive System (p. 498)
The primary male sex organs are the testes, which produce sperm cells and male sex hormones. Accessory organs include the internal and external reproductive organs.

1. Testes
   a. Structure of the testes
      (1) The testes are composed of lobules separated by connective tissue and filled with seminiferous tubules.
      (2) The epithelium lining the seminiferous tubules produces sperm cells.
      (3) The interstitial cells produce male sex hormones.
   b. Formation of sperm cells
      (1) The epithelium lining the seminiferous tubules includes supporting cells and spermatogenic cells.
         (a) Supporting cells support and nourish spermatogenic cells.
         (b) Spermatogenic cells give rise to sperm cells.
      (2) A sperm cell consists of a head, midpiece, and tail.
   c. Spermatogenesis
      (1) Spermatogonia give rise to sperm cells.
      (2) Meiosis reduces the number of chromosomes in sperm cells by one-half (from 46 to 23).
      (3) Spermatogenesis produces four sperm cells from each primary spermatocyte.
2. Male internal accessory organs
   a. Epididymis
      (1) The epididymis is a tightly coiled tube that leads into the vas deferens.
      (2) It stores and nourishes immature sperm cells and promotes their maturation.
   b. Vas deferens
      (1) The vas deferens is a muscular tube.
      (2) It passes through the inguinal canal, enters the abdominal cavity, courses medially into the pelvic cavity, and ends behind the urinary bladder.
      (3) It fuses with the duct from the seminal vesicle to form the ejaculatory duct.
   c. Seminal vesicle
      (1) The seminal vesicle is a saclike structure attached to the vas deferens.
      (2) It secretes an alkaline fluid that contains nutrients, such as fructose, and prostaglandins.
   d. Prostate gland
      (1) The prostate gland surrounds the urethra just inferior to the urinary bladder.
      (2) It secretes a thin, milky fluid that neutralizes the pH of semen and the acidic secretions of the vagina.
   e. Bulbourethral glands
      (1) The bulbourethral glands are two small structures inferior to the prostate gland.
      (2) They secrete a fluid that lubricates the penis in preparation for sexual intercourse.
   f. Semen
      (1) Semen consists of sperm cells and secretions of the seminal vesicles, prostate gland, and bulbourethral glands.
      (2) This fluid is slightly alkaline and contains nutrients and prostaglandins.
      (3) Sperm cells in semen begin to swim, but these sperm cells are unable to fertilize egg cells until they enter the female reproductive tract.
3. Male external reproductive organs
   a. Scrotum
      The scrotum is a pouch of skin and subcutaneous tissue that encloses the testes for protection and temperature regulation.
   b. Penis
      (1) The penis is specialized to become erect for insertion into the vagina during sexual intercourse.
      (2) Its body is composed of three columns of erectile tissue.
4. Erection, orgasm, and ejaculation
   a. During erection, the vascular spaces within the erectile tissue engorge with blood.
   b. Orgasm is the culmination of sexual stimulation. Emission and ejaculation accompany male orgasm.
   c. Semen moves along the reproductive tract as smooth muscle in the walls of the tubular structures contracts by reflex.

19.3 Hormonal Control of Male Reproductive Functions (p. 505)

1. Hypothalamic and pituitary hormones
   a. The male body remains reproductively immature until the hypothalamus releases gonadotropin-releasing hormone (GnRH), which stimulates the anterior pituitary gland to release gonadotropins.
   b. Follicle-stimulating hormone (FSH) stimulates spermatogenesis.
   c. Luteinizing hormone (LH), known in males as interstitial cell-stimulating hormone (ICSH), stimulates interstitial cells to produce male sex hormones.
2. Male sex hormones
   a. Male sex hormones are called androgens, with testosterone the most important.
   b. Androgen production increases rapidly at puberty.
   c. Actions of testosterone
      (1) Testosterone stimulates development of the male reproductive organs.
      (2) It also develops and maintains male secondary sex characteristics.
   d. Regulation of male sex hormones
      (1) A negative feedback mechanism regulates testosterone concentration.
         (a) A rising testosterone concentration inhibits the hypothalamus and reduces the anterior pituitary's secretion of gonadotropins.
         (b) As testosterone concentration falls, the hypothalamus signals the anterior pituitary to secrete gonadotropins.
      (2) The testosterone concentration remains relatively stable from day to day.

19.4 Organs of the Female Reproductive System (p. 507)
The primary female sex organs are the ovaries, which produce female sex cells and sex hormones. Accessory organs are internal and external.

1. Ovaries
   a. Ovary structure
      (1) Each ovary is subdivided into a medulla and a cortex.
      (2) The medulla is composed of connective tissue, blood vessels, lymphatic vessels, and nerves.
      (3) The cortex contains ovarian follicles and is covered by cuboidal epithelium.
   b. Primordial follicles
      (1) During prenatal development, groups of cells in the ovarian cortex form millions of primordial follicles.

(2) Each primordial follicle contains a primary oocyte and a layer of follicular cells.
(3) The primary oocyte begins meiosis, but the process halts until puberty.
(4) The number of oocytes steadily declines throughout a female's life.
  c. Oogenesis
(1) Beginning at puberty, some oocytes are stimulated to continue meiosis.
(2) When a primary oocyte undergoes oogenesis, it gives rise to a secondary oocyte in which the original chromosome number is reduced by one-half (from 46 to 23).
(3) Fertilization of a secondary oocyte produces a zygote.
  d. Follicle maturation
(1) At puberty, FSH initiates follicle maturation.
(2) During maturation, the oocyte enlarges, the follicular cells multiply, and a fluid-filled cavity forms.
(3) Usually, only one follicle at a time fully develops.
  e. Ovulation
(1) Ovulation is the release of an oocyte from an ovary.
(2) A rupturing follicle releases the oocyte.
(3) After ovulation, the oocyte is drawn into the opening of the uterine tube.
2. Female internal accessory organs
  a. Uterine tubes
(1) The end of each uterine tube expands, and its margin bears irregular extensions.
(2) Ciliated cells that line the tube and peristaltic contractions in the wall of the tube help transport the oocyte down the uterine tube. Fertilization may occur.
  b. Uterus
(1) The uterus receives the embryo and sustains it during development.
(2) The uterine wall includes the endometrium, myometrium, and perimetrium.
  c. Vagina
(1) The vagina receives the erect penis, conveys uterine secretions to the outside, and provides an open channel for the fetus during birth.
(2) Its wall consists of mucosal, muscular, and fibrous layers.
3. Female external reproductive organs
  a. Labia majora
(1) The labia majora are rounded folds of adipose tissue and skin.
(2) The upper ends form a rounded elevation over the symphysis pubis.
  b. Labia minora
(1) The labia minora are flattened, longitudinal folds between the labia majora.
(2) They are well supplied with blood vessels.
  c. Clitoris
(1) The clitoris is a small projection at the anterior end of the vulva. It corresponds to the male penis.
(2) It is composed of two columns of erectile tissue.
  d. Vestibule
(1) The vestibule is the space between the labia minora.
(2) The vestibular glands secrete mucus into the vestibule during sexual stimulation.
4. Erection, lubrication, and orgasm
  a. During periods of sexual stimulation, the erectile tissues of the clitoris and vestibular bulbs engorge with blood and swell.
  b. The vestibular glands secrete mucus into the vestibule and vagina.
  c. During orgasm, the muscles of the perineum, uterine wall, and uterine tubes contract rhythmically.

19.5 Hormonal Control of Female Reproductive Functions (p. 513)

The hypothalamus, anterior pituitary gland, and ovaries secrete hormones that control sex cell maturation and the development and maintenance of female secondary sex characteristics.

1. Female sex hormones
  a. A female body remains reproductively immature until about ten years of age, when gonadotropin secretion increases.
  b. The most important female sex hormones are estrogens and progesterone.
(1) Estrogens develop and maintain most female secondary sex characteristics.
(2) Progesterone changes the uterus.
2. Female reproductive cycle
  a. FSH initiates a menstrual cycle by stimulating follicle maturation.
  b. Maturing follicular cells secrete estrogens, which maintain the secondary sex traits and thicken the uterine lining.
  c. Secretion of a relatively large amount of LH by the anterior pituitary triggers ovulation.
  d. Following ovulation, follicular cells give rise to the corpus luteum.
(1) The corpus luteum secretes progesterone, which causes the uterine lining to become more vascular and glandular.
(2) If an oocyte is not fertilized, the corpus luteum begins to degenerate.
(3) As concentrations of estrogens and progesterone decline, the uterine lining disintegrates, causing menstrual flow.
  e. During this cycle, estrogens and progesterone inhibit the release of LH and FSH. As concentrations of estrogens and progesterone decline, the anterior pituitary secretes FSH and LH again, stimulating a new menstrual cycle.
3. Menopause
  a. Menopause is termination of the menstrual cycle due to aging of the ovaries.
  b. Reduced concentrations of estrogens and lack of progesterone may cause regressive changes in female secondary sex characteristics.

19.6 Mammary Glands (p. 515)

1. The mammary glands are in the subcutaneous tissue of the anterior thorax.
2. They are composed of lobes that contain glands and a duct.
3. Dense connective and adipose tissues separate the lobes.
4. Ovarian hormones stimulate female breast development.
  a. Alveolar glands and ducts enlarge.
  b. Fat is deposited around and within the breasts.

19.7 Birth Control (p. 516)

Birth control is voluntary regulation of how many children are produced and when they are conceived. It usually involves some method of contraception.

1. Coitus interruptus is withdrawal of the penis from the vagina before ejaculation.
2. Rhythm method is abstinence from sexual intercourse for several days before and after ovulation.
3. Mechanical barriers
  a. Males and females can use condoms.
  b. Females can also use diaphragms and cervical caps.

4. Chemical barriers
   Spermicidal creams, foams, and jellies provide an unfavorable environment in the vagina for sperm survival.
5. Oral contraceptives
   Birth control pills contain synthetic estrogen-like and progesterone-like substances that disrupt a female's normal pattern of gonadotropin secretion and prevent ovulation and the normal buildup of the uterine lining.
6. Injectable contraception
   Intramuscular injection with medroxyprogesterone acetate every three months acts similarly to oral contraceptives to prevent pregnancy.
7. Contraceptive implants
   a. A contraceptive implant is a set of progesterone-containing capsules or rods inserted under a woman's skin.
   b. Progesterone released from the implant prevents ovulation.
8. Intrauterine devices (IUD)
   An IUD is a solid object inserted in the uterine cavity that prevents pregnancy by interfering with implantation of a blastocyst.
9. Surgical methods
   Vasectomies in males and tubal ligations in females are surgical sterilization procedures.

19.8 Sexually Transmitted Diseases (p. 520)

1. Sexually transmitted diseases (STDs) are passed during sexual contact and may go undetected for years.
2. The twenty recognized STDs share similar symptoms.

REVIEW EXERCISES

1. List the general functions of the male reproductive system. (p. 498)
2. Distinguish between the primary and accessory male reproductive organs. (p. 498)
3. Describe the structure of a testis. (p. 498)
4. Review the process of meiosis. (p. 500)
5. Describe the epididymis, and explain its function. (p. 500)
6. Trace the path of the vas deferens from the epididymis to the ejaculatory duct. (p. 501)
7. On a diagram, locate the seminal vesicles, prostate gland, and bulbourethral glands, and describe the composition of their secretions. (p. 501)
8. Describe the composition of semen. (p. 502)
9. Describe the structure of the penis. (p. 503)
10. Explain the mechanism that produces penile erection. (p. 504)
11. Distinguish between emission and ejaculation. (p. 505)
12. Explain the mechanism of ejaculation. (p. 505)
13. Explain the role of gonadotropin-releasing hormone (GnRH) in the control of male reproductive functions. (p. 505)
14. List several male secondary sex characteristics. (p. 506)
15. Explain the regulation of testosterone concentration. (p. 506)
16. List the general functions of the female reproductive system. (p. 507)
17. Describe the structure of an ovary. (p. 507)
18. Describe how a follicle matures. (p. 509)
19. On a diagram, locate the uterine tubes, and explain their function. (p. 510)
20. Describe the structure of the uterus. (p. 511)
21. On a diagram, locate the clitoris, and describe its structure. (p. 512)
22. Explain the role of gonadotropin-releasing hormone (GnRH) in regulating female reproductive functions. (p. 513)
23. List several female secondary sex characteristics. (p. 513)
24. Define *menstrual cycle*. (p. 513)
25. Summarize the major events in a menstrual cycle. (p. 513)
26. Describe the structure of a mammary gland. (p. 515)
27. Define *contraception*. (p. 516)
28. List several methods of contraception, and explain how each prevents pregnancy. (p. 516)
29. List several symptoms of sexually transmitted diseases. (p. 520)

CRITICAL THINKING

1. How are the human male and female reproductive tracts similar? How are the structures of the testis and ovary similar?
2. Why must the chromosome number be halved in sperm cells and oocytes?
3. Some men are unable to become fathers because their spermatids do not mature into sperm. Injection of their spermatids into their partner's secondary oocytes sometimes results in conception. A few men have fathered healthy babies this way. Why would this procedure work with spermatids, but not with primary spermatocytes?
4. *Contraception* literally means "against conception." According to this definition, is an intrauterine device a contraceptive? Why or why not?
5. Understanding the causes of infertility can be valuable in developing new birth control methods. Cite a type of contraceptive based on each of the following causes of infertility: (a) failure to ovulate due to a hormonal imbalance; (b) a large fibroid tumor that disturbs the uterine lining; (c) endometrial tissue blocking uterine tubes; (d) low sperm count (too few sperm per ejaculate).
6. How can a couple use "fertility awareness" methods to conceive a child or to prevent pregnancy?
7. Sometimes a sperm cell fertilizes a polar body rather than an oocyte. An embryo does not develop, and the fertilized polar body degenerates. Why is a polar body unable to support development of an embryo?
8. What changes, if any, would a male who has had one testis removed experience? A female who has had one ovary removed?
9. Does a tubal ligation cause a woman to enter menopause prematurely? Why or why not?

WEB CONNECTIONS

Visit the website for additional study questions and more information about this chapter at:

http://www.mhhe.com/shieress8

chapter 20

Pregnancy, Growth, and Development

THE JOY AND SORROW OF MULTIPLE BIRTHS. A human uterus can best accommodate one fetus, and this is why most births are "singletons." About one in eighty pregnancies produces twins, and although these babies are often smaller and born earlier than singletons, most fare quite well. The picture isn't as bright as the number of fetuses increases. An Iowa couple became the parents of healthy septuplets in late 1997, and two of the children have lingering medical problems. A year later, a couple in Texas had octuplets, seven of whom survived. These families are two relative success stories.

After the McCaughey septuplets were born, Mario and Jane Simeone, of Tucson, Arizona, decided to tell the public that multiple pregnancies do not always have such happy endings. The Simeones learned this from their own tragedy. After six years of undergoing treatment for infertility, Jane delivered triplets on June 21, 1997, two girls and a boy, fifteen weeks premature. Within three weeks, both girls had died, and the boy, Mario Jr., remained hospitalized, gaining strength and weight. Although Mario Jr. came home by summer's end and has been healthy, his parents cannot forget his two sisters. One in ten "multiples" does not survive to see a first birthday. Those that do are more likely to have seizures, blindness, cerebral palsy, and mental retardation than singletons. Many multiple conceptions and pregnancies end before survival is possible.

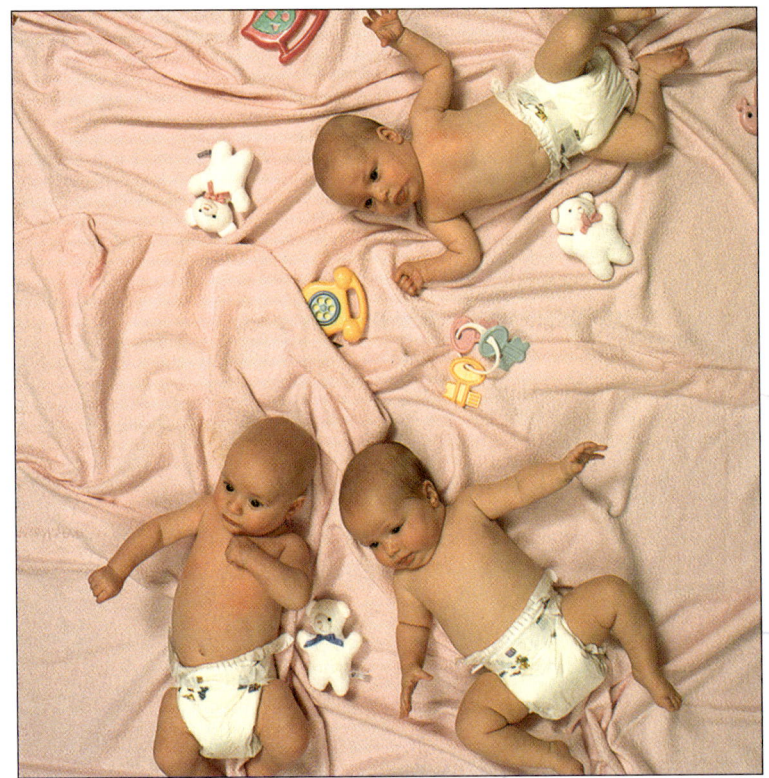

Photo:
Multiples such as triplets and quadruplets are more likely to be born with health problems than "singletons."

Chapter Objectives

After studying this chapter, you should be able to do the following:

20.1 Introduction
1. Distinguish between growth and development. (p. 528)
2. Distinguish between the prenatal and the postnatal periods. (p. 528)

20.2 Pregnancy
3. Define *pregnancy,* and describe the process of fertilization. (p. 528)

20.3 Prenatal Period
4. Describe the major events of cleavage. (p. 529)
5. Distinguish between an embryo and a fetus. (p. 530)
6. Describe the formation and function of the placenta. (p. 530)
7. Describe the hormonal changes in the maternal body during pregnancy. (p. 531)
8. Explain how the primary germ layers originate, and list the structures each layer produces. (p. 533)
9. Describe the major events of the embryonic stage of development. (p. 534)
10. Describe the major events of the fetal stage of development. (p. 537)
11. Trace the general path of blood through the fetal cardiovascular system. (p. 538)
12. Describe the birth process, and explain the role of hormones in this process. (p. 542)

20.4 Postnatal Period
13. Describe the major cardiovascular and physiological adjustments required of the newborn. (p. 545)

Aids to Understanding Words

allant- [sausage-shaped] *allant*ois: Tubelike structure extending from the yolk sac into the connecting stalk of the embryo.

chorio- [skin] *chorio*n: Outermost membrane surrounding the fetus and other fetal membranes.

cleav- [to divide] *cleav*age: Period of development characterized by division of the zygote into smaller and smaller cells.

lacun- [pool] *lacun*a: Space between the chorionic villi that fills with maternal blood.

morul- [mulberry] *morul*a: Embryonic structure consisting of a solid ball of about sixteen cells that resembles a mulberry.

nat- [to be born] pre*nat*al: Period of development before birth.

troph- [nurture] *troph*oblast: Cellular layer that surrounds the inner cell mass and helps nourish it.

umbil- [navel] *umbil*ical cord: Structure attached to the fetal naval (umbilicus) that connects the fetus to the placenta.

Key Terms

amnion (am´ne-on)
chorion (ko´re-on)
cleavage (klēv´ij)
embryo (em´bre-o)
fertilization (fer´´tĭ-lĭ-za´shun)
fetus (fe´tus)
gastrula (gas´troo-lah)
neonatal period (ne´´o-na´tal)
placenta (plah-sen´tah)
postnatal period (pōst-na´tal)
prenatal period (pre-na´tal)
primary germ layers (pri´mar-e jerm la´erz)
umbilical cord (um-bil´ĭ-kal kord)
zygote (zi´gōt)

20.1 Introduction

A sperm cell and an oocyte unite, forming a zygote, and the journey of prenatal development begins. Following thirty-eight weeks of cell division, growth, and specialization into distinctive tissues and organs, a new human being enters the world.

Before birth, an individual grows and develops. Growth is an increase in size. In humans and other many-celled organisms, growth entails an increase in cell numbers, followed by enlargement of the newly formed cells. Development, which includes growth, is the continuous process by which an individual changes from one life phase to another. These life phases include a **prenatal period** (pre-na´tal pe´re-od), which begins with fertilization and ends at birth, and a **postnatal** (pōst-na´tal) **period,** which begins at birth and ends at death.

20.2 Pregnancy

Pregnancy (preg´nan-se) is the presence of a developing offspring in the uterus. It consists of three three-month periods called trimesters. Pregnancy results from the union of a secondary oocyte and a sperm cell, an event called **fertilization** (fer´´tĭ-lĭ-za´shun).

Transport of Sex Cells

Prior to fertilization, a female ovulates an egg cell (secondary oocyte), which enters a uterine tube. During sexual intercourse, the male deposits semen containing sperm cells in the vagina near the cervix. To reach the oocyte, the sperm cells must move upward through the uterus and uterine tube. Prostaglandins in the semen stimulate lashing of sperm tails and muscular contractions within the walls of the uterus and uterine tube, which help sperm cells move. Also, high concentrations of estrogens during the first part of the menstrual cycle stimulate the uterus and cervix to secrete a thin, watery fluid that promotes sperm transport and survival. Conversely, during the latter part of the cycle, when progesterone concentration is high, the female reproductive tract secretes a viscous fluid that hampers sperm transport and survival. These changes in the penetrability of the cervical mucus increase the chance that sperm will reach the oocyte when a woman is most fertile.

Sperm cells reach the upper portions of the uterine tube within an hour following sexual intercourse. Many sperm cells may reach the egg cell, but only one actually fertilizes it (fig. 20.1).

Fertilization

A sperm cell that reaches the oocyte invades the follicular cells that adhere to the oocyte's surface (corona radi-

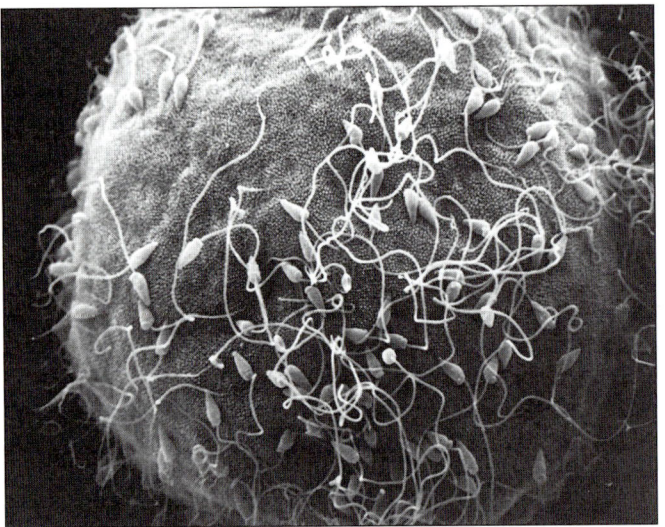

Figure 20.1
Scanning electron micrograph of sperm cells on the surface of an egg cell (1,100×). Only one sperm cell actually fertilizes the egg.

ata) and binds to the *zona pellucida* surrounding the oocyte cell membrane. The acrosome of the sperm cell releases enzymes (including hyaluronidase) that aid penetration by digesting proteins in the zona pellucida (fig. 20.2). However, at least several hundred sperm cells must be present to produce enough enzymes to enable one to penetrate. This is why males with very low sperm counts are said to be subfertile.

Union of the oocyte and sperm cell, rendering the structure a "fertilized egg," triggers lysosome-like vesicles just beneath the egg cell membrane to release enzymes that harden the zona pellucida. This reduces the chance that more than one sperm cell will penetrate.

Once a sperm cell enters the egg cell's cytoplasm, the nucleus in the sperm cell's head swells (fig. 20.3). The approaching nuclei from the two sex cells are called pronuclei, until they join. The egg cell then divides unequally to form a large cell, which becomes the fertilized egg, and a tiny second polar body, which is later expelled. Meiosis ends. Next, the nuclei of the egg cell and sperm cell unite. Their nuclear membranes fall apart, and their chromosomes mingle, completing fertilization.

Because the sperm cell and the egg cell each provides 23 chromosomes, the product of fertilization is a cell with 46 chromosomes—the usual number in a human body cell (somatic cell). This cell, called a **zygote** (zi´gōt), is the first cell of the future offspring.

 Every human being once spent about half an hour as a single cell.

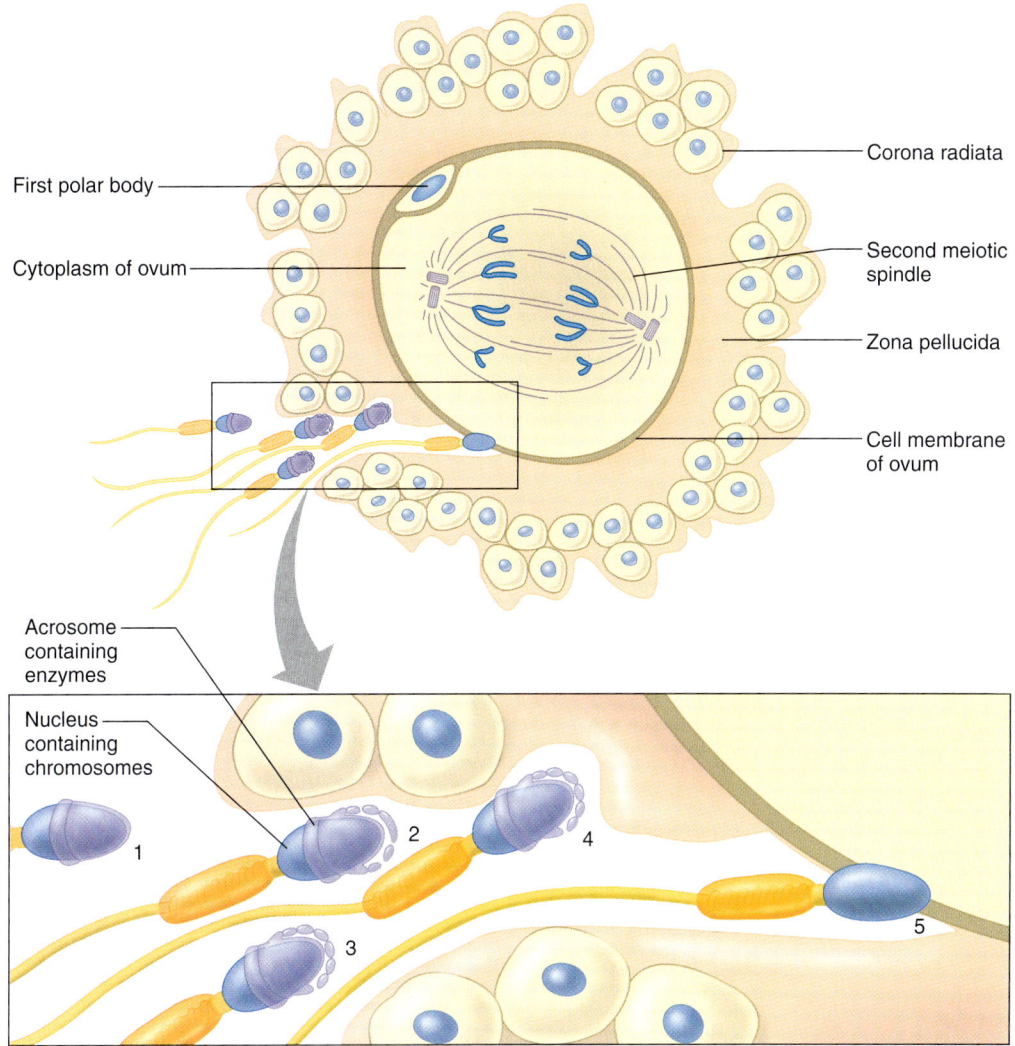

Figure 20.2
Steps in fertilization: (*1*) The sperm cell reaches the corona radiata surrounding egg cell. (*2*) The acrosome of the sperm cell releases a protein-digesting enzyme. (*3*, and *4*) The sperm cell penetrates the zona pellucida surrounding the egg cell. (*5*) The sperm cell's membrane fuses with the egg cell's membrane.

The approximate time of fertilization is fourteen days before the expected onset of the next menstrual period. The estimated time of birth is 266 days from fertilization. Most women give birth within ten to fifteen days of this calculated time.

CHECK YOUR RECALL

1. What factors aid the movements of the egg and sperm cells through the female reproductive tract?
2. Where in the female reproductive system does fertilization normally take place?
3. List the events of fertilization.

20.3 Prenatal Period

Early Embryonic Development

About 30 hours after forming, the zygote undergoes *mitosis,* giving rise to two new cells (blastomeres) (fig. 20.4*B*). These cells, in turn, divide into four cells, which divide into eight cells, and so forth. These divisions occur rapidly, with little time for growth. Thus, each division yields smaller cells. This phase of early rapid cell division is termed **cleavage** (klēv´ij) (see fig. 20.3).

During cleavage, the tiny mass of cells moves through the uterine tube to the uterine cavity. This trip takes about three days, and by then the structure consists of a solid ball of about sixteen cells. The ball is

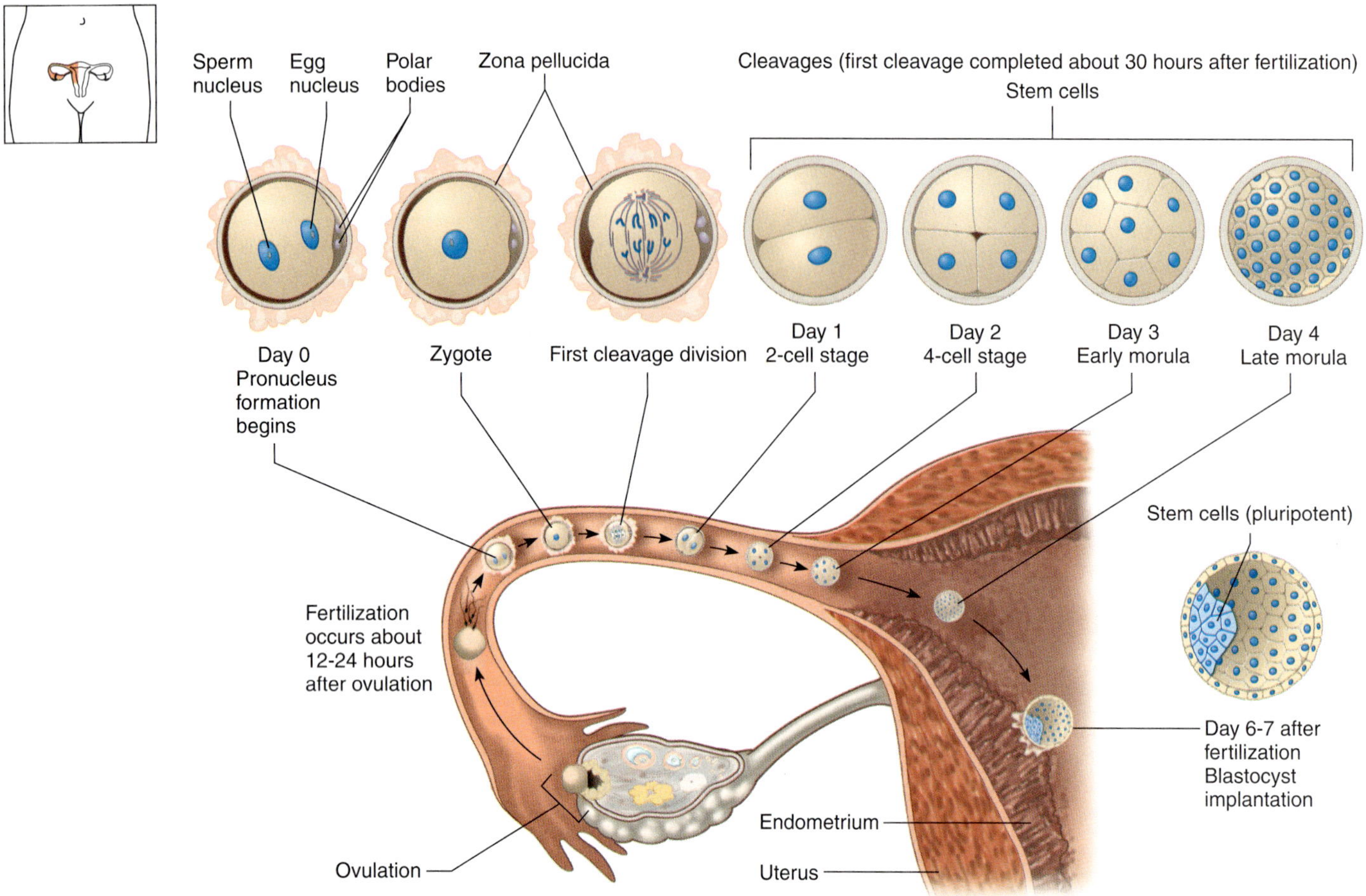

Figure 20.3
Stages of early human development.

called a *morula,* Latin for mulberry, which it resembles (fig. 20.4C).

The morula remains free within the uterine cavity for about three days. During this stage, the zona pellucida of the original egg cell degenerates. Then the morula hollows out, forming a *blastocyst,* which begins to attach to the endometrium. By the end of the first week of development, the blastocyst superficially implants in the endometrium (fig. 20.5). Up until this point, the cells that will become the developing offspring are pluripotent stem cells, which means that they can give rise to several specialized types of daughter cells, as well as yield additional stem cells.

About the time of implantation, certain cells on the inner face of the blastocyst organize into a group (inner cell mass) that will give rise to the offspring. This marks the beginning of the **embryonic stage** of development. The offspring is termed an **embryo** (em′bre-o) until the end of the eighth week, when the basic structural form of the human body is recognizable. After the eighth week and until birth, the offspring is called a **fetus** (fe′tus). Rudiments of all organs are present by the end of embryonic development. These organs and other structures enlarge and specialize during fetal development.

Eventually, the cells surrounding the embryo, with cells of the endometrium, form a complex vascular structure called the **placenta** (plah-sen′tah). This organ attaches the embryo to the uterine wall and exchanges nutrients, gases, and wastes between maternal blood and the embryo's blood. The placenta also secretes hormones.

CHECK YOUR RECALL

1. What is cleavage?
2. What is implantation?
3. How do an embryo and a fetus differ?

Sometimes two ovarian follicles release egg cells simultaneously, and if both are fertilized, the resulting zygotes develop into fraternal (dizygotic) twins. Such twins are no more alike genetically than siblings. Twins may also develop from a single fertilized egg (monozygotic twins). This may happen if two inner cell masses form within a blastocyst and each produces an embryo. Monozygotic twins usually share a single placenta and are genetically identical. Thus, they are always the same sex and are very similar in appearance.

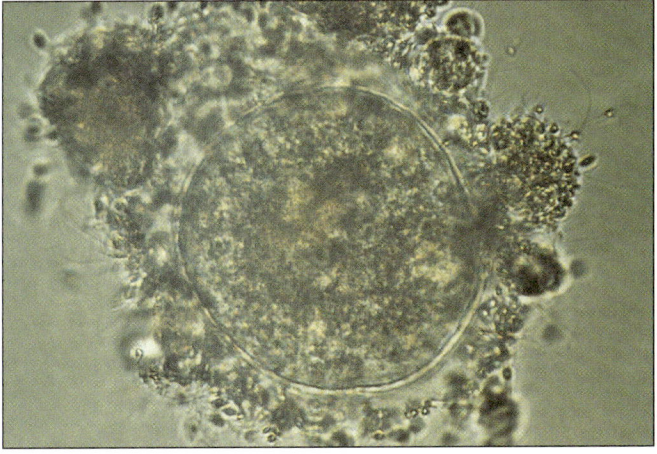

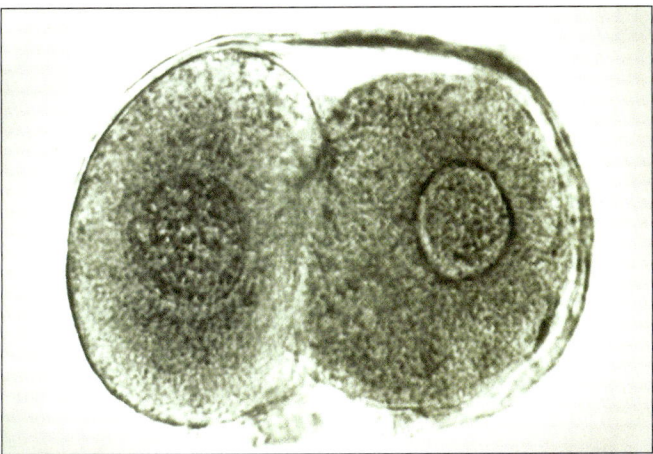

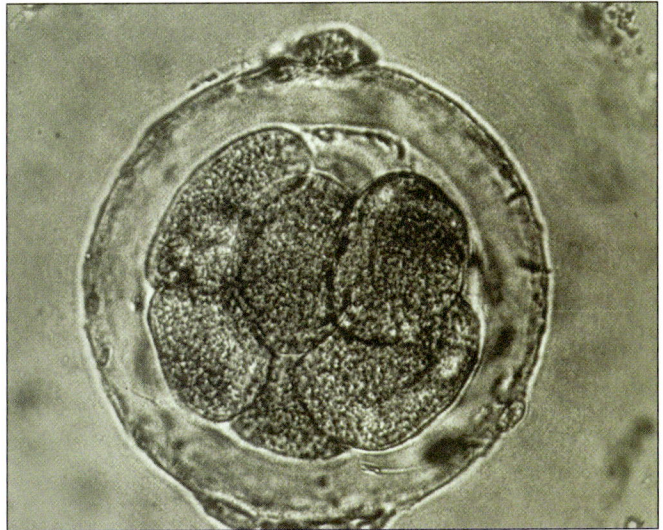

Figure 20.4
Light micrographs of (*A*) a human egg surrounded by follicular cells and sperm cells (250×), (*B*) the two-cell stage (600×), and (*C*) a morula (500×).

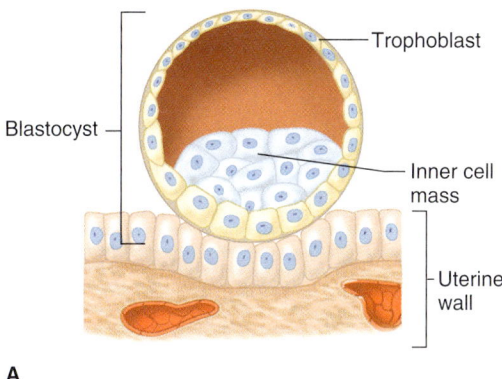

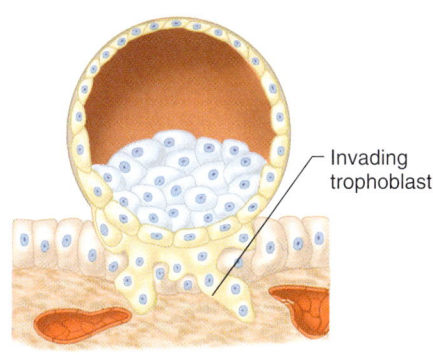

Figure 20.5
About the sixth day of development, the blastocyst (*A*) contacts the uterine wall and (*B*) begins to implant. The trophoblast, which will help form the placenta, secretes hCG, a hormone that maintains the pregnancy.

Hormonal Changes During Pregnancy

During a typical menstrual cycle, the corpus luteum degenerates about two weeks after ovulation. Consequently, concentrations of estrogens and progesterone decline rapidly, the uterine lining breaks down, and the endometrium sloughs away as menstrual flow. If this occurs following implantation, the embryo is lost (spontaneously aborted).

The hormone **human chorionic gonadotropin (hCG)** normally prevents spontaneous abortion. Cells from the outer blastocyst form a layer of embryonic cells called the trophoblast, which surrounds the developing embryo and later helps form the placenta (fig. 20.5). The trophoblast secretes hCG. This hormone, similar in function to luteinizing hormone (LH), maintains the corpus luteum, which continues secreting estrogens and progesterone, stimulating the uterine wall to grow and develop. At the same time, hCG inhibits the anterior pituitary's release of follicle-stimulating hormone (FSH) and LH, halting the normal menstrual cycle.

Topic of Interest

FEMALE INFERTILITY

Infertility is the inability to conceive after a year of trying. In 90% of cases, infertility has a physical cause, and 60% of the time, the abnormality is in the female's reproductive system.

A common cause of female infertility is insufficient secretion (hyposecretion) of gonadotropic hormones by the anterior pituitary, resulting in absence of ovulation (anovulation). Testing the urine for *pregnanediol,* a product of progesterone metabolism, detects an anovulatory cycle. Because progesterone concentration normally rises after ovulation, no increase in pregnanediol in the urine during the latter part of the menstrual cycle suggests lack of ovulation.

Fertility specialists can treat anovulation due to hyposecretion of gonadotropic hormones by administering human chorionic gonadotropin (hCG) obtained from human placentas. Another ovulation-stimulating biochemical, human menopausal gonadotropin (hMG), contains luteinizing hormone (LH) and follicle-stimulating hormone (FSH) and is obtained from the urine of postmenopausal women. However, either hCG or hMG may overstimulate the ovaries and cause many follicles to release secondary oocytes simultaneously, which may result in multiple births if several are fertilized.

Another cause of female infertility is *endometriosis,* in which small pieces of the inner uterine lining (endometrium) move up through the uterine tubes during menstruation and implant in the abdominal cavity. Here, the tissue changes in a similar way to the uterine lining during the menstrual cycle. The abnormally located tissue breaks down at the end of the cycle but cannot be expelled. Instead, it remains in the abdominal cavity, irritating the lining (peritoneum) and causing considerable pain. This tissue also stimulates formation of fibrous tissue (fibrosis), which may encase the ovary, preventing ovulation or obstructing the uterine tubes.

Sexually transmitted diseases (STDs), such as gonorrhea, cause some women to become infertile. These infections can inflame and obstruct the uterine tubes or stimulate production of viscous mucus that plugs the cervix and prevents sperm entry.

Women become infertile if their ovaries must be removed, such as to treat cancer. In an experimental procedure, healthy ovarian tissue can be implanted in her upper arm, and healthy oocytes removed later. The oocytes can be fertilized *in vitro* (see Topic of Interest on page 535).

Finding the right treatment for a particular patient requires determining the infertility's cause. Table 20A describes diagnostic tests for female infertility.

TABLE 20A   TESTS TO ASSESS FEMALE INFERTILITY

TEST	WHAT IT CHECKS
Hormone levels	Whether ovulation occurs
Ultrasound	Placement and appearance of reproductive organs and structures
Postcoital test	Cervix examined soon after unprotected intercourse to see if mucus is thin enough to allow sperm through
Endometrial biopsy	Small piece of uterine lining sampled and viewed under microscope to see if it can support an embryo
Hysterosalpingogram	Dye injected into uterine tube and followed with scanner shows if tube is clear or blocked
Laparoscopy	Small, lit optical device inserted near navel to detect scar tissue blocking tubes, which ultrasound may miss

---

Secretion of hCG continues at a high level for about two months, then declines by the end of four months. Detecting this hormone in urine or blood is the basis of pregnancy tests. The corpus luteum persists throughout pregnancy, but its function as a hormone source becomes less important after the first three months (first trimester), when the placenta secretes sufficient estrogens and progesterone (fig. 20.6).

Placental estrogens and *placental progesterone* maintain the uterine wall during the second and third trimesters of pregnancy. The placenta also secretes a hormone called **placental lactogen** that, with placental estrogens and progesterone, stimulates breast development and prepares the mammary glands for milk secretion. Placental progesterone and a polypeptide hormone called *relaxin* from the corpus luteum inhibit

Very early in pregnancy, while vast hormonal changes sweep a woman's body and the embryo rapidly increases in size and complexity, the woman may not yet realize what is happening. Early signs of pregnancy resemble those of approaching menstruation, such as bloating and irritable mood. As the pregnancy continues, the woman's blood volume increases by one-third, and her bones may weaken if she does not receive adequate dietary calcium. Muscle spasms may occur in response to rapid weight gain. In the later months, the fetus pushing against the woman's internal organs can produce heartburn, shortness of breath, and frequent urination. Fetal movements become noticeable by the fourth or fifth month, first as slight flutterings, then as jabs, kicks, and squirming movements.

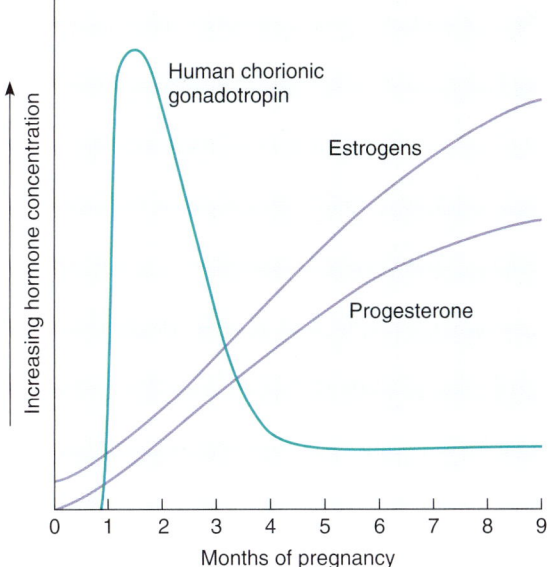

Figure 20.6
Relative concentrations of three hormones in maternal blood during pregnancy.

TABLE 20.1 HORMONAL CHANGES DURING PREGNANCY

1. Following implantation, cells of the embryo begin to secrete human chorionic gonadotropin.
2. Human chorionic gonadotropin maintains the corpus luteum, which continues to secrete estrogens and progesterone.
3. The developing placenta secretes large quantities of estrogens and progesterone.
4. Placental estrogens and progesterone:
   a. stimulate the uterine lining to continue development.
   b. maintain the uterine lining.
   c. inhibit the anterior pituitary's secretion of follicle-stimulating hormone (FSH) and luteinizing hormone (LH).
   d. stimulate development of mammary glands.
   e. inhibit uterine contractions (progesterone).
   f. enlarge the reproductive organs (estrogens).
5. Relaxin from the corpus luteum also inhibits uterine contractions and relaxes the pelvic ligaments.
6. The placenta secretes placental lactogen that stimulates breast development.
7. Aldosterone from the adrenal cortex promotes renal reabsorption of sodium.
8. Parathyroid hormone from the parathyroid glands helps maintain a high concentration of maternal blood calcium.

the smooth muscles in the myometrium, suppressing uterine contractions until the birth process begins.

The high concentration of placental estrogens during pregnancy enlarges the vagina and external reproductive organs. Also, relaxin relaxes the ligaments joining the symphysis pubis and sacroiliac joints during the last week of pregnancy, allowing greater movement at these joints and aiding the passage of the fetus through the birth canal.

Other hormonal changes of pregnancy include increased adrenal secretion of aldosterone, which promotes renal reabsorption of sodium and leads to fluid retention. The parathyroid glands secrete parathyroid hormone, which helps maintain a high concentration of maternal blood calcium (see chapter 11, p. 291). Table 20.1 summarizes the hormonal changes of pregnancy.

CHECK YOUR RECALL

1. Which hormone normally prevents spontaneous abortion?
2. What is the source of the hormones that sustain the uterine wall during pregnancy?
3. What other hormonal changes occur during pregnancy?

Embryonic Stage

The embryonic stage extends until the eighth week of prenatal development. During this time, the placenta forms, the main internal organs develop, and the major external body structures appear.

Early in the embryonic stage, the cells of the inner cell mass organize into a flattened **embryonic disc** with two distinct layers—an outer *ectoderm* and an inner *endoderm*. A short time later, the ectoderm and endoderm fold, and a third layer of cells, the *mesoderm*, forms between them. All organs form from these three cell layers, called the **primary germ layers** (pri´mar-e jerm la´erz) (fig. 20.7). A *connecting stalk* attaches the embryonic disc to the developing placenta. The two-week embryo, with its three primary germ layers, is called a **gastrula** (gas´troo-lah). Table 20.2 summarizes the stages of early embryonic development.

Ectodermal cells give rise to the nervous system, portions of special sensory organs, the epidermis, hair,

TABLE 20.2 STAGES AND EVENTS OF EARLY HUMAN PRENATAL DEVELOPMENT

STAGE	TIME PERIOD	PRINCIPAL EVENTS
Fertilized ovum	12–24 hours following ovulation	Oocyte fertilized; zygote has 23 pairs of chromosomes and is genetically distinct
Cleavage	30 hours to third day	Mitosis increases cell number
Morula	Third to fourth day	Solid ball of cells
Blastocyst	Fifth day through second week	Hollowed ball forms trophoblast (outside) and inner cell mass, which implants and flattens to form embryonic disc
Gastrula	End of second week	Primary germ layers form

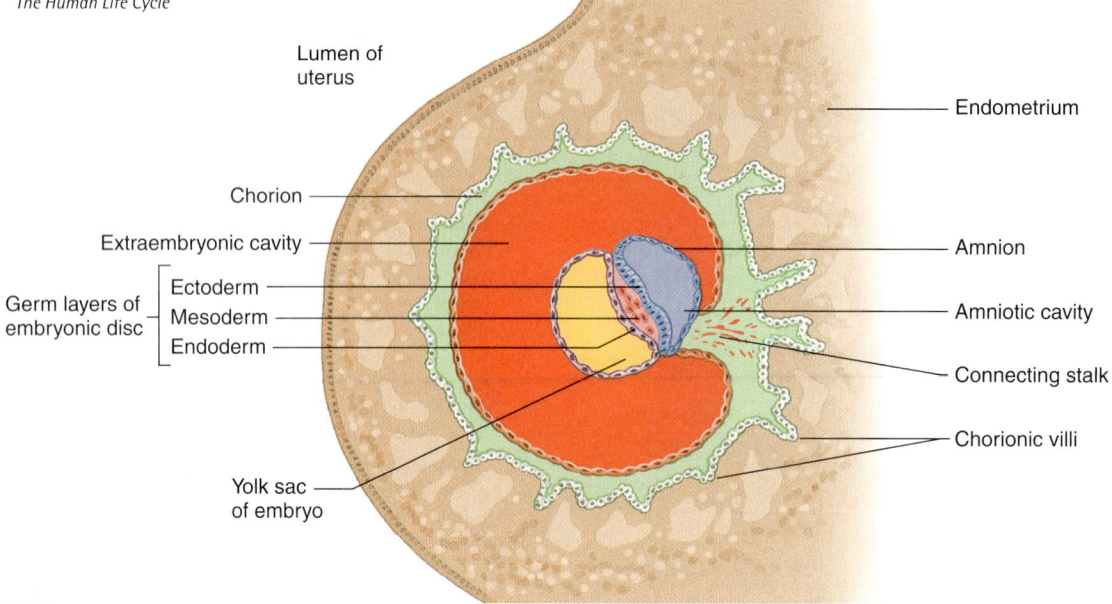

Figure 20.7
Early in the embryonic stage of development, the three primary germ layers form.

nails, glands of the skin, and linings of the mouth and anal canal. Mesodermal cells form all types of muscle tissue, bone tissue, bone marrow, blood, blood vessels, lymphatic vessels, connective tissues, internal reproductive organs, kidneys, and the epithelial linings of the body cavities. Endodermal cells produce the epithelial linings of the digestive tract, respiratory tract, urinary bladder, and urethra.

As the embryo implants in the uterus, proteolytic enzymes from the trophoblast break down endometrial tissue, providing nutrients for the developing embryo. A second layer of cells begins to line the trophoblast, and together these two layers form a structure called the **chorion** (ko´re-on). Soon, slender projections grow out from the trophoblast, including the new cell layer, eroding their way into the surrounding endometrium by continuing to secrete proteolytic enzymes. These projections become increasingly complex and form the highly branched **chorionic villi,** which are well established by the end of the fourth week.

As the chorionic villi develop, embryonic blood vessels appear within them and are continuous with those passing through the connecting stalk to the body of the embryo. At the same time, irregular spaces called **lacunae** form around and between the villi. These spaces fill with maternal blood that escapes from eroded endometrial blood vessels.

During the fourth week of development, the flat embryonic disc is transformed into a cylindrical structure. The head and jaws develop, the heart beats and forces blood through the blood vessels, and tiny buds, which will give rise to the upper and lower limbs, form (fig. 20.8).

During the fifth through the seventh weeks, as figure 20.8 shows, the head grows rapidly and becomes rounded and erect. The face, with developing eyes, nose, and mouth, becomes more humanlike. The upper and lower limbs elongate, and fingers and toes appear

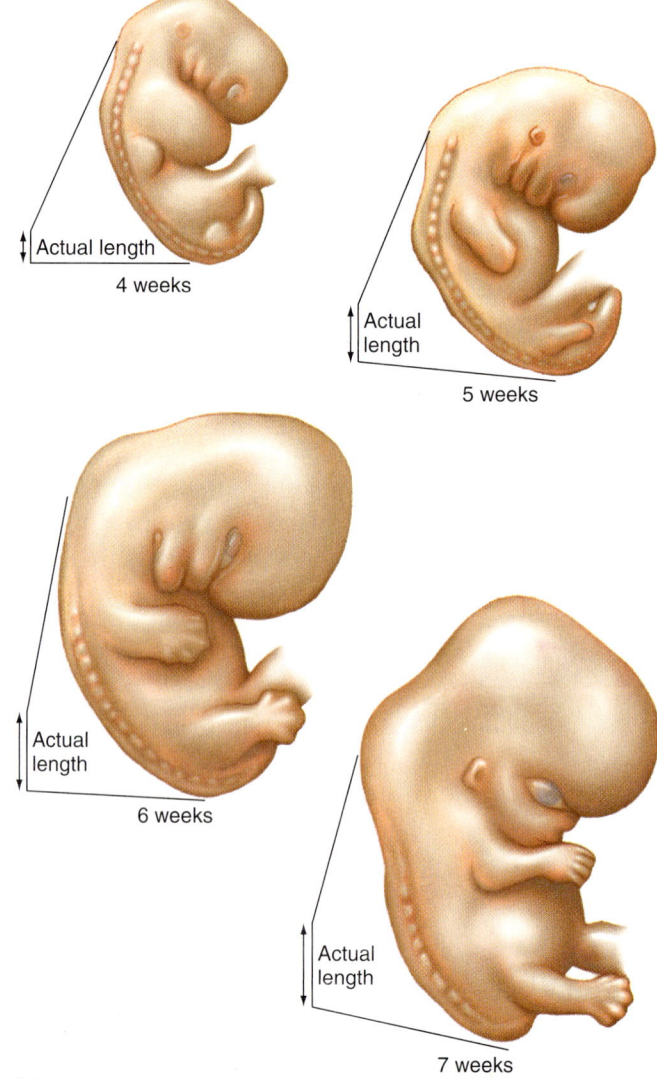

Figure 20.8
In the fifth through the seventh weeks of development, the embryonic body and face develop a humanlike appearance.

Topic of Interest — ASSISTED REPRODUCTIVE TECHNOLOGIES

Michele and Ray L'Esperance wanted children badly, but Michele's uterine tubes had been removed due to scarring. A procedure called *in vitro fertilization (IVF)* enabled the couple to have children.

First, Michele received human menopausal gonadotropin to stimulate development of ovarian follicles. When an ultrasound scan showed that the follicles had grown to a certain diameter, she received human chorionic gonadotropin (which acts like LH) to induce ovulation. Then, Michele's physician used an optical instrument called a laparoscope to examine the interior of her abdomen and take the largest oocytes from an ovary. The oocytes were incubated at 37°C in a medium buffered at pH 7.4. When the oocytes matured, they were mixed in a laboratory dish with Ray's sperm cells, which had been washed to remove inhibitory factors. Secretions from Michele's reproductive tract were added to activate the sperm.

Next, fertilized egg cells were selected and incubated in a special medium for about 60 hours. At this stage, five of the eight- to sixteen-cell balls of cells were transferred through Michele's cervix and into her uterus to increase the chances that one or two would complete development. (Today, fewer fertilized eggs are transferred because of medical problems associated with multiple births, as the chapter opener describes.) The L'Esperances beat the odds—they had healthy quintuplets (fig. 20A)!

Success rates for IVF vary from clinic to clinic, ranging from 0 to 40%, with the average about 17%. Pregnancy via IVF is expensive, costing thousands of dollars. Table 20B describes other assisted reproductive technologies.

Figure 20A
In vitro fertilization worked for Michele and Ray L'Esperance. Five fertilized ova implanted in Michele's uterus are now Erica, Alexandria, Veronica, Danielle, and Raymond. But many couples are disappointed with the high failure rate of the technology.

TABLE 20B — ASSISTED REPRODUCTIVE TECHNOLOGIES

TECHNOLOGY	PROCEDURE	CONDITION IT TREATS
Artificial insemination	Donated sperm cells or pooled specimens are placed near a woman's cervix.	Male infertility—lack of sperm cells or low sperm count
Surrogate mother	An oocyte fertilized in vitro is implanted in a woman other than the one who donated the oocyte. The surrogate, or "gestational mother," gives the newborn to the "genetic mother" and her partner, the sperm donor.	Female infertility—lack of a uterus
Gamete intrafallopian transfer (GIFT)	Oocytes are removed from a woman's ovary, then placed along with donated sperm cells into a uterine tube.	Female infertility—bypasses blocked uterine tube
Zygote intrafallopian transfer (ZIFT)	An oocyte fertilized in vitro is placed in a uterine tube. It travels to the uterus on its own.	Female infertility—bypasses blocked uterine tube
Embryo adoption	A woman is artificially inseminated with sperm cells from a man whose partner cannot ovulate healthy oocytes. If the woman conceives, the morula is flushed from her uterus and implanted in the uterus of the sperm donor's partner.	Female infertility—a woman has nonfunctional ovaries, but a healthy uterus

(fig. 20.9). By the end of the seventh week, all the main internal organs are present, and as these structures enlarge, the body takes on a humanlike appearance.

Until about the end of the eighth week, the chorionic villi cover the entire surface of the former trophoblast. However, as the embryo and the chorion surrounding it enlarge, only those villi that remain in contact with the endometrium endure. The others degenerate, and the portions of the chorion to which they were attached become smooth. Thus, the region of the chorion still in contact with the uterine wall is restricted to a disc-shaped area that becomes the placenta.

A thin **placental membrane** separates embryonic blood within the capillary of a chorionic villus from

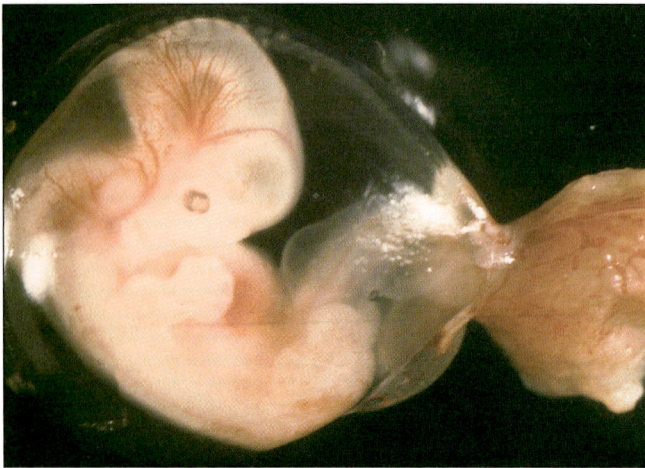

Figure 20.9
Human embryo after about six weeks of development (6.5×).

maternal blood in a lacuna. Across this membrane, which is composed of the epithelium of the chorionic villus and the epithelial wall of the capillary inside the villus, maternal and embryonic blood exchange substances (fig. 20.10). Oxygen and nutrients diffuse from the maternal blood into the embryo's blood, and carbon dioxide and other wastes diffuse from the embryo's blood into the maternal blood. Various substances also cross the placental membrane by active transport and pinocytosis.

> If a pregnant woman takes an addictive substance, her newborn may suffer from withdrawal symptoms when amounts of the chemical it is accustomed to receiving suddenly plummet. Newborn addiction occurs with certain drugs of abuse, such as heroin, and with some prescription drugs used to treat anxiety. It also occurs with very large doses of vitamin C. Although vitamin C is not addictive, if a fetus is accustomed to megadoses, the sudden drop in vitamin C level after birth may bring on symptoms of deficiency.

CHECK YOUR RECALL

1. Describe the major events of the embryonic stage of development.
2. Which tissues and structures develop from ectoderm? From mesoderm? From endoderm?
3. How do embryonic and maternal blood exchange substances?
4. Describe how the placenta forms.

The embryonic portion of the placenta is the chorion and its villi; the maternal portion is the area of the uterine wall where the villi attach (fig. 20.11). When fully formed, the placenta is a reddish-brown disc about 20 centimeters long and 2.5 centimeters thick, and weighing about 0.5 kilogram.

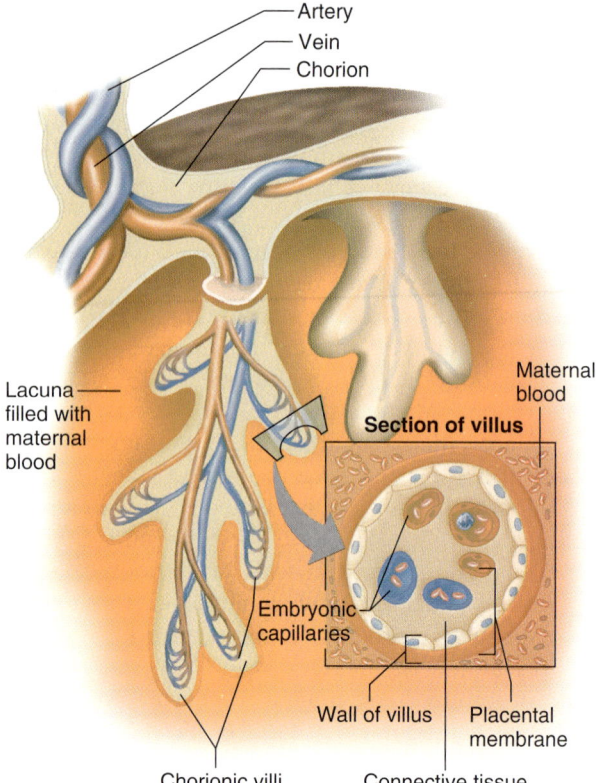

Figure 20.10
As illustrated in the section of villus (lower part of figure), the placental membrane consists of the epithelial wall of an embryonic capillary and the epithelial wall of a chorionic villus.

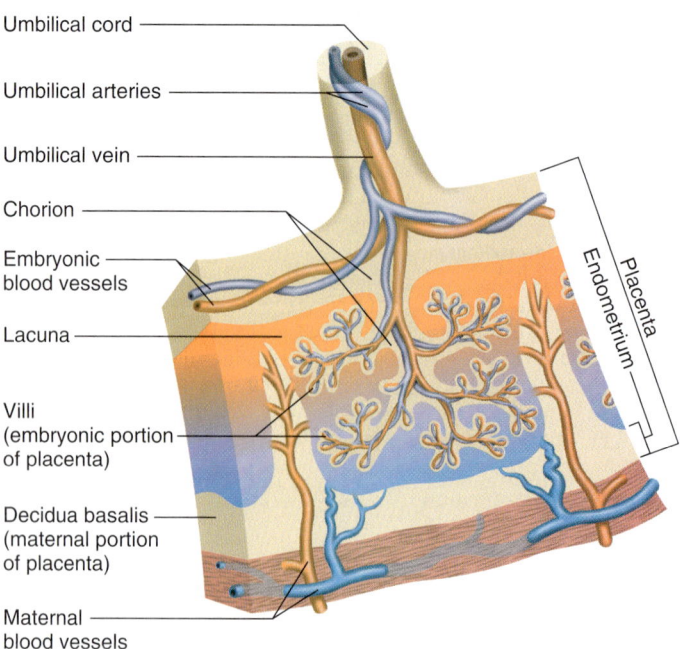

Figure 20.11
The placenta consists of an embryonic portion and a maternal portion.

While the placenta forms, another membrane, called the **amnion** (am´ne-on), develops around the embryo during the second week. Its margin attaches around the edge of the embryonic disc, and **amniotic fluid** fills the space between the amnion and the embryonic disc.

As the embryo becomes more cylindrical, the amnion margins fold, enclosing the embryo in the amnion and amniotic fluid. The amnion envelops the tissues on the underside of the embryo, by which the embryo attaches to the chorion and the developing placenta. In this manner, the **umbilical cord** (um-bil´ĭ-kal kord) forms (fig. 20.12).

The umbilical cord contains three blood vessels—two *umbilical arteries* and one *umbilical vein*—that transport blood between the embryo and the placenta (see fig. 20.11). The umbilical cord suspends the embryo in the *amniotic cavity*. The amniotic fluid allows the embryo to grow freely without compression from surrounding tissues and also protects the embryo from jarring movements of the woman's body.

Two other embryonic membranes form during development—the yolk sac and the allantois (fig. 20.12). The **yolk sac** forms during the second week and attaches to the underside of the embryonic disc. It forms blood cells in the early stages of development and gives rise to the cells that later become sex cells.

The **allantois** (ah-lan´to-is) forms during the third week as a tube extending from the early yolk sac into the connecting stalk of the embryo. It, too, forms blood cells and gives rise to the umbilical arteries and vein.

By the beginning of the eighth week, the embryo is usually 30 millimeters long and weighs less than 5 grams. It is recognizable as human (fig. 20.13).

CHECK YOUR RECALL

1. What is the function of amniotic fluid?
2. Which blood vessels are in the umbilical cord?
3. What is the significance of the yolk sac?

Fetal Stage

The **fetal stage** begins at the end of the eighth week of development and lasts until birth. During this period, growth is rapid, and body proportions change considerably. At the beginning of the fetal stage, the head is disproportionately large, and the lower limbs are short. Gradually, the proportions become more like those of a child.

During the third month, growth in body length accelerates, but head growth slows. The upper limbs

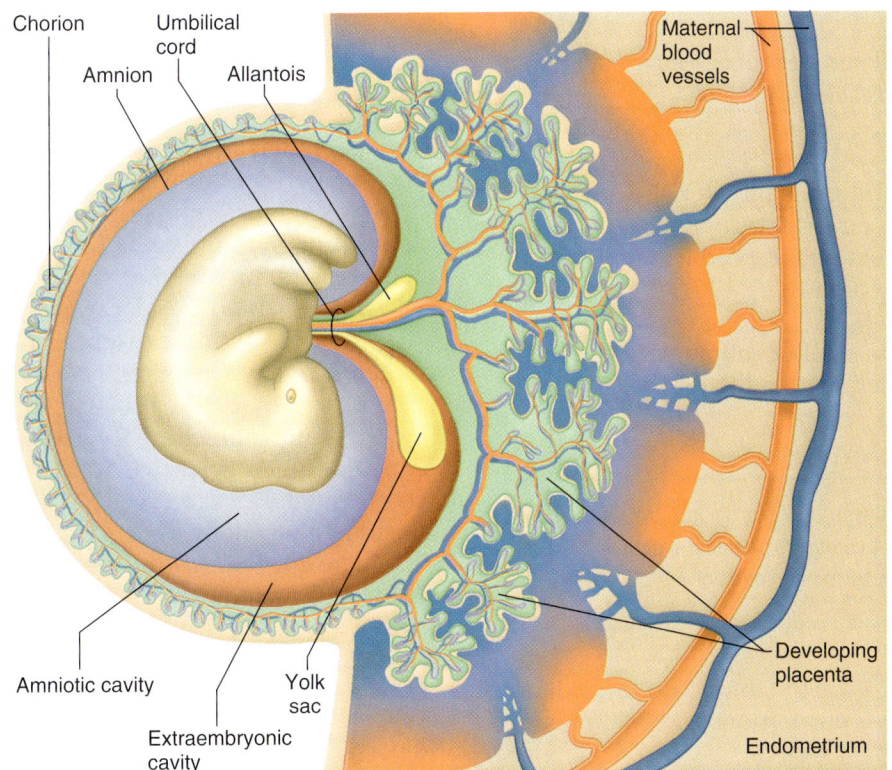

Figure 20.12
As the amnion develops, it surrounds the embryo, and the umbilical cord begins to form from structures in the connecting stalk.

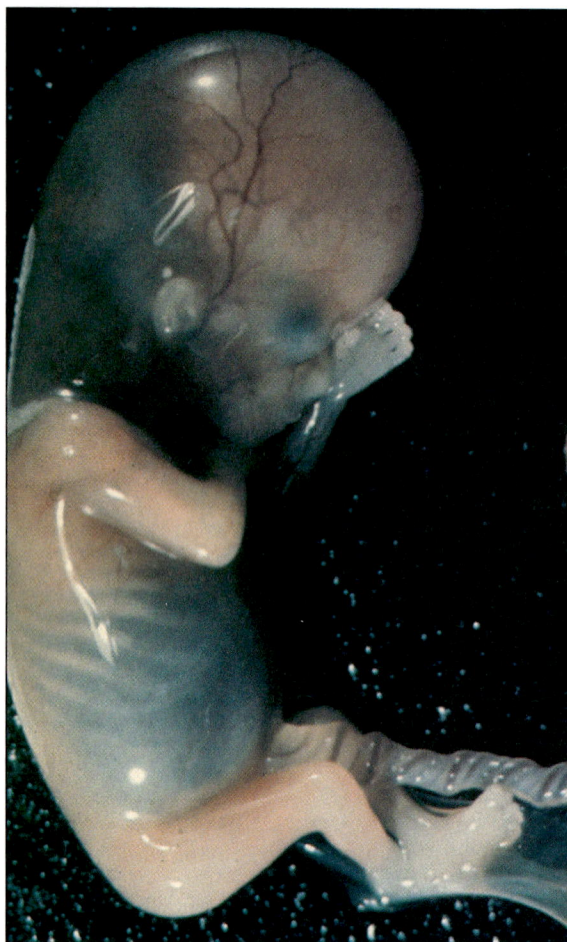

Figure 20.13
By the beginning of the eighth week of development, the embryonic body is recognizable as human (6×).

During the sixth month, the fetus gains substantial weight. Eyebrows and eyelashes appear. The skin is quite wrinkled and translucent, and blood vessels in the skin give the fetus a reddish appearance.

In the seventh month, fat is deposited in subcutaneous tissues, making the skin smoother. The eyelids, which fused during the third month, reopen. At the end of this month, the fetus is about 40 centimeters long.

In the final trimester, fetal brain cells rapidly form networks, as organs specialize and grow. A layer of fat is laid down beneath the skin. In the male, the testes descend from regions near the developing kidneys, through the inguinal canal, and into the scrotum. The digestive and respiratory systems mature last, which is why premature infants often have difficulty digesting milk and breathing.

At the end of the ninth month (on average, 266 days), the fetus is *full-term*. It is about 50 centimeters long and weighs 2.7–3.6 kilograms. The skin has lost its downy hair, but sebum and dead epidermal cells still coat it. Hair usually covers the scalp. The fingers and toes have well-developed nails. The skull bones are largely ossified. As figure 20.14 shows, the fetus is usually positioned upside down, with its head toward the cervix.

CHECK YOUR RECALL

1. What major changes occur during the fetal stage of development?
2. Describe a full-term fetus.

reach the length they will maintain throughout development, and ossification centers appear in most bones. By the twelfth week, the external reproductive organs are distinguishable as male or female.

In the fourth month, the body grows rapidly and reaches a length of up to 20 centimeters. The lower limbs lengthen considerably, and the skeleton continues to ossify.

 A four-month-old fetus will startle and turn away if a bright light is flashed on the pregnant woman's belly. Fetuses also react to sudden loud noises.

In the fifth month, growth slows. The lower limbs reach their final relative proportions. Skeletal muscles contract, and the pregnant woman may feel fetal movements. Hair appears on the head. Fine, downy hair and a cheesy mixture of dead epidermal cells and sebum from the sebaceous glands cover the skin.

Fetal Blood and Circulation

Throughout fetal development, the maternal blood supplies oxygen and nutrients and carries away wastes. These substances diffuse between maternal and fetal blood through the placental membrane, and umbilical blood vessels carry them to and from the fetal body.

The fetal blood and cardiovascular system must adapt to intrauterine existence. The concentration of oxygen-carrying hemoglobin in fetal blood is about 50% greater than in maternal blood, and fetal hemoglobin has a greater attraction for oxygen than does adult hemoglobin. At a particular oxygen partial pressure, fetal hemoglobin can carry 20–30% more oxygen than can adult hemoglobin.

Figure 20.15 shows the path of blood in the fetal cardiovascular system. The umbilical vein transports blood rich in oxygen and nutrients from the placenta to the fetus. This vein enters the body and extends along the anterior abdominal wall to the liver. About half the blood it carries passes into the liver, and the rest enters a vessel called the **ductus venosus,** which bypasses the liver.

The ductus venosus extends a short distance and joins the inferior vena cava. There, oxygenated blood from the placenta mixes with deoxygenated blood from

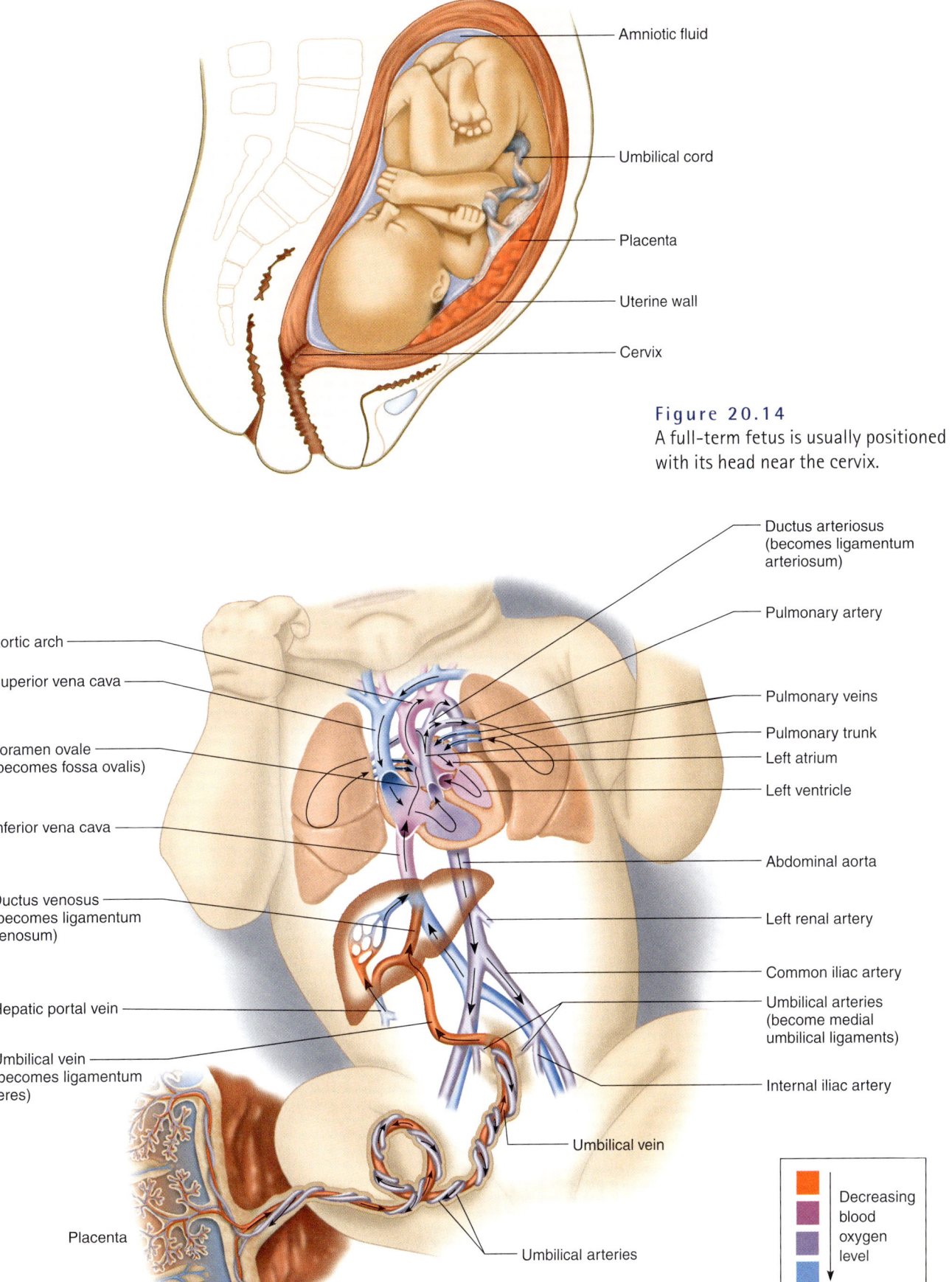

Figure 20.14
A full-term fetus is usually positioned with its head near the cervix.

Figure 20.15
The general pattern of fetal circulation.

Topic of Interest

SOME CAUSES OF BIRTH DEFECTS

Thalidomide

The idea that the placenta always protects the embryo and fetus from harmful substances was tragically disproven between 1957 and 1961, when 10,000 children in Europe were born with flippers in place of limbs. Doctors soon determined that the teratogen (an agent that causes a birth defect) was the mild tranquilizer *thalidomide,* which all of the mothers of deformed infants had taken early in pregnancy, the time when limbs form. The United States was spared a thalidomide disaster because an astute government physician noted the drug's adverse effects on experimental monkeys and halted testing.

Rubella

At about the same time as the thalidomide crisis, another teratogen, a virus, was sweeping the United States. In the early 1960s, a *rubella* (German measles) epidemic caused 20,000 birth defects and 30,000 stillbirths. Successful vaccination programs have since greatly lowered the incidence of "congenital rubella syndrome" in many countries.

Alcohol

A pregnant woman who has as few as one or two alcoholic drinks a day, or perhaps a large amount at a crucial time in prenatal development, risks *fetal alcohol syndrome (FAS)* in her unborn child. Animal studies show that even small amounts of alcohol can alter fetal brain chemistry. Thus, it is best to avoid drinking alcohol entirely when pregnant or when trying to become pregnant.

A child with FAS has a small head, misshapen eyes, and a flat face and nose (fig. 20B). He or she grows slowly before and after birth. Intellect is impaired, ranging from minor learning disabilities to mental retardation. Teens and young adults with FAS are short and have small heads. Many remain at an early grade-school level of development, and they often lack social and communication skills.

In the United States today, FAS is the third most common cause of mental retardation in newborns. One to three of every 1,000 infants has the syndrome, and more than 40,000 of these children are born each year.

Cigarettes

Chemicals in cigarette smoke stress a fetus. Carbon monoxide crosses the placenta and plugs sites on the fetus's hemoglobin molecules that bind oxygen. Other chemicals in smoke prevent nutrients from reaching the fetus. Studies comparing the placentas of smokers and nonsmokers show that smoke-exposed placentas lack important growth factors. The result of these assaults is poor growth before and after birth. Cigarette smoking during pregnancy is linked to spontaneous abortion, stillbirth, prematurity, and low birth weight.

Nutrients

Certain nutrients in large amounts, particularly vitamins, act in the body as drugs. The acne medication *isotretinoin* (Accutane) is a derivative of vitamin A that causes spontaneous abortions and defects of the heart, nervous system, and face. A vitamin A-based drug used to treat psoriasis, as well as excesses of vitamin A itself, also cause birth defects because some forms of the vitamin are stored in body fat for up to three years after ingestion.

Malnutrition in a pregnant woman threatens the fetus. Obstetric records of pregnant women before, during, and

---

the lower parts of the fetal body. This mixture continues through the vena cava to the right atrium.

In an adult heart, blood from the right atrium enters the right ventricle and is pumped through the pulmonary trunk and arteries to the lungs (see chapter 13, p. 333). The fetal lungs, however, are nonfunctional, and blood largely bypasses them. Much of the blood from the inferior vena cava that enters the fetal right atrium is shunted directly into the left atrium through an opening in the atrial septum called the **foramen ovale.** Blood passes through the foramen ovale because blood pressure is somewhat greater in the right atrium than in the left atrium. Furthermore, a small valve on the left side of the atrial septum overlies the foramen ovale and helps prevent blood from moving in the reverse direction.

The rest of the fetal blood entering the right atrium, including a large proportion of the deoxygenated blood entering from the superior vena cava, passes into the right ventricle and out through the pulmonary trunk. Only a small volume of blood enters the pulmonary circuit because the lungs are collapsed, and their blood vessels have a high resistance to blood flow. However, enough blood does reach lung tissues to sustain them.

Most of the blood in the pulmonary trunk bypasses the lungs by entering a fetal vessel called the **ductus arteriosus,** which connects the pulmonary trunk to the descending portion of the aortic arch. As a result of this connection, blood with a relatively low oxygen concentration, which is returning to the heart through the superior vena cava, bypasses the lungs. At the same time, it is prevented from entering the portion of the aorta that branches to the heart and brain.

The more highly oxygenated blood that enters the left atrium through the foramen ovale mixes with a small amount of deoxygenated blood returning from the pulmonary veins. This mixture moves into the left ventricle

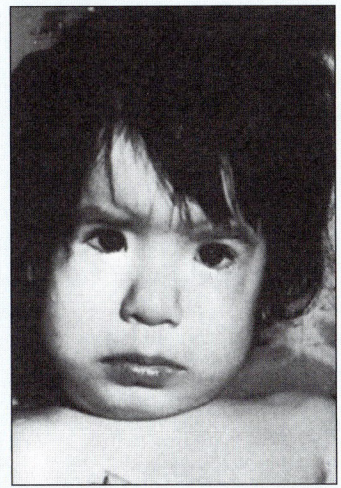 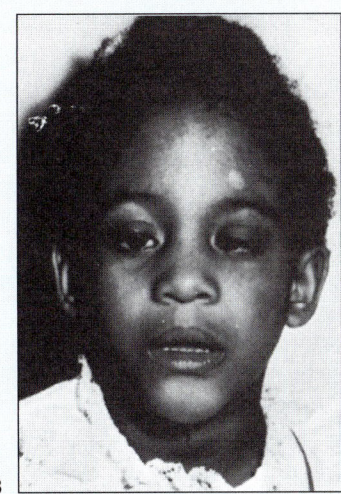 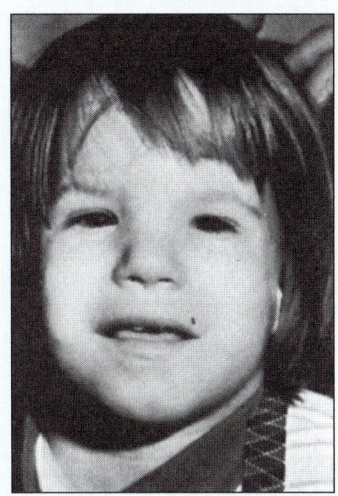

Figure 20B
Fetal alcohol syndrome. Some children whose mothers drank alcohol during pregnancy have characteristic flat faces that are strikingly similar in children of different races. Women who drink excessively while pregnant have a 30–45% chance of having a child affected to some degree by prenatal exposure to alcohol. However, only 6% of exposed offspring have full-blown fetal alcohol syndrome.

after World War II link inadequate nutrition early in pregnancy to an increase in spontaneous abortions. The aborted fetuses had very little brain tissue. More recent studies reveal that malnutrition or starvation before birth causes shifts in metabolism to make the most of calories from food. This protective action, however, sets the stage for developing obesity and associated disorders, such as type II diabetes and cardiovascular disease, in adulthood. Poor nutrition later in pregnancy affects placenta development. The infant has a low birth weight and is at high risk for short stature, tooth decay, delayed sexual development, learning disabilities, and possibly mental retardation.

Occupational Hazards

The workplace can be a source of teratogens. Women who work with textile dyes, lead, certain photographic chemicals, semiconductor materials, mercury, and cadmium have increased rates of spontaneous abortion and delivering children with birth defects. The male's role in environmentally caused birth defects is not well understood. However, men whose jobs expose them to sustained heat, such as smelter workers, glass manufacturers, and bakers, may produce sperm that can fertilize an oocyte but possibly lead to spontaneous abortion or a birth defect. A virus or a toxic chemical carried in semen may also cause a birth defect.

and is pumped into the aorta. Some of it reaches the myocardium through the coronary arteries, and some reaches the brain tissues through the carotid arteries.

Blood carried by the descending aorta is partially oxygenated and partially deoxygenated. Some of it is carried into the branches of the aorta that lead to the lower regions of the body. The rest passes into the umbilical arteries, which branch from the internal iliac arteries and lead to the placenta. There, the blood is reoxygenated (fig. 20.15).

Table 20.3 summarizes the major features of fetal circulation. At birth, the fetal cardiovascular system

TABLE 20.3   FETAL CARDIOVASCULAR ADAPTATIONS

ADAPTATION	FUNCTION
Fetal blood	Has greater oxygen-carrying capacity than blood in an adult
Umbilical vein	Carries oxygenated blood from placenta to fetus
Ductus venosus	Conducts about half the blood from umbilical vein directly to inferior vena cava, bypassing liver
Foramen ovale	Conveys much blood entering right atrium from inferior vena cava, through atrial septum, and into left atrium, bypassing lungs
Ductus arteriosus	Conducts some blood from pulmonary trunk to aorta, bypassing lungs
Umbilical arteries	Carry blood from internal iliac arteries to placenta

Genetics Connection

FETAL CHROMOSOME CHECKS

The chromosomes in a cell's nucleus provide clues to the individual's health. A chromosome number other than 46 signals a serious medical condition, as do chromosomes that have missing or extra material. Sampling fetal cells and preparing charts of the chromosomes can help prenatally diagnose these conditions. Certain tests can also be applied to sampled cells to detect disorders caused by abnormal or missing single genes.

Ultrasound, in which sound waves bounced off a fetus are converted into an image, can detect large-scale structural anomalies that are part of certain chromosomal syndromes. Also, blood tests (maternal serum marker tests) performed on a pregnant woman at fifteen weeks measure hormones (alpha fetoprotein, human chorionic gonadotropin, and a form of estrogen). Abnormal levels indicate that fetal cells may have an extra chromosome. Doctors follow up questionable ultrasound or blood test results with one of the following procedures that examines fetal chromosomes (fig. 20C).

Chorionic Villus Sampling

Chorionic villus sampling (CVS) examines the chromosomes in chorionic villus cells, which are genetically identical to fetal cells because they are derived from the same fertilized egg. The test carries a risk of causing miscarriage. Thus, only women who have previously had a child with a detectable chromosome abnormality usually have the test. CVS is performed at or after the tenth week of gestation.

Amniocentesis

Amniocentesis is performed after the fourteenth week of gestation. A physician uses ultrasound to guide a needle into the amniotic sac and withdraws about 5 milliliters of fluid. Fetal fibroblasts in the fluid are cultured and their chromosomes checked. It takes about a week to grow these cells. A faster technique uses DNA probes to highlight specific chromosomes.

Amniocentesis carries about a 0.5% chance of causing miscarriage. Only women whose risk of having a fetus with a chromosomal anomaly equals or exceeds the risk of the procedure are offered amniocentesis. This includes women of any age who have had a child with a detectable chromosomal abnormality and women over age thirty-five. (Older women are more likely to produce oocytes that have extra or missing chromosomes, which can lead to abnormal fetuses if fertilized.)

Fetal Cell Sorting

A new way to check fetal chromosomes is fetal cell sorting, which separates rare fetal cells from a pregnant woman's bloodstream. A device called a fluorescence-activated cell sorter can pull out the fetal cells. The technique is safer than CVS or amniocentesis because the fetus and its membranes are not touched.

Fetal cell sorting traces its roots to 1957, when an autopsy on a pregnant woman revealed cells from a very early embryo lodged in a blood vessel in her lung. Researchers were able to tell that the cells were from an embryo only because the cells had Y chromosomes, which female cells lack. Since then, researchers have found that fetal cells enter the maternal circulation in up to 70% of all pregnancies, and may remain for decades in the woman's body, sometimes triggering an immune attack years later. Fetal cell sorting is still experimental.

---

must adjust when the placenta ceases to function and the newborn begins to breathe.

The umbilical cord usually contains two arteries and one vein. A small percentage of newborns have only one umbilical artery. Since this condition is often associated with other cardiovascular disorders, the vessels within the severed cord are routinely counted following birth.

CHECK YOUR RECALL

1. Which umbilical vessel carries oxygen-rich blood to the fetus?
2. What is the function of the ductus venosus?
3. How does fetal circulation allow blood to bypass the lungs?

Birth Process

Pregnancy usually continues for thirty-eight weeks from conception, which is forty weeks from the woman's last menstrual period. Pregnancy ends with the *birth process*. A period of rapid changes and intense physical demands on the pregnant woman begins hours or days before the birth.

A declining progesterone concentration plays a major role in initiating birth. During pregnancy, progesterone suppresses uterine contractions. As the placenta ages, the progesterone concentration within the uterus declines, which stimulates synthesis of a prostaglandin that promotes uterine contractions. At the same time, the cervix begins to thin and then open. Changes in the cervix may begin a week or two before other signs of labor occur.

Another stimulant of the birth process is stretching of the uterine and vaginal tissues late in pregnancy.

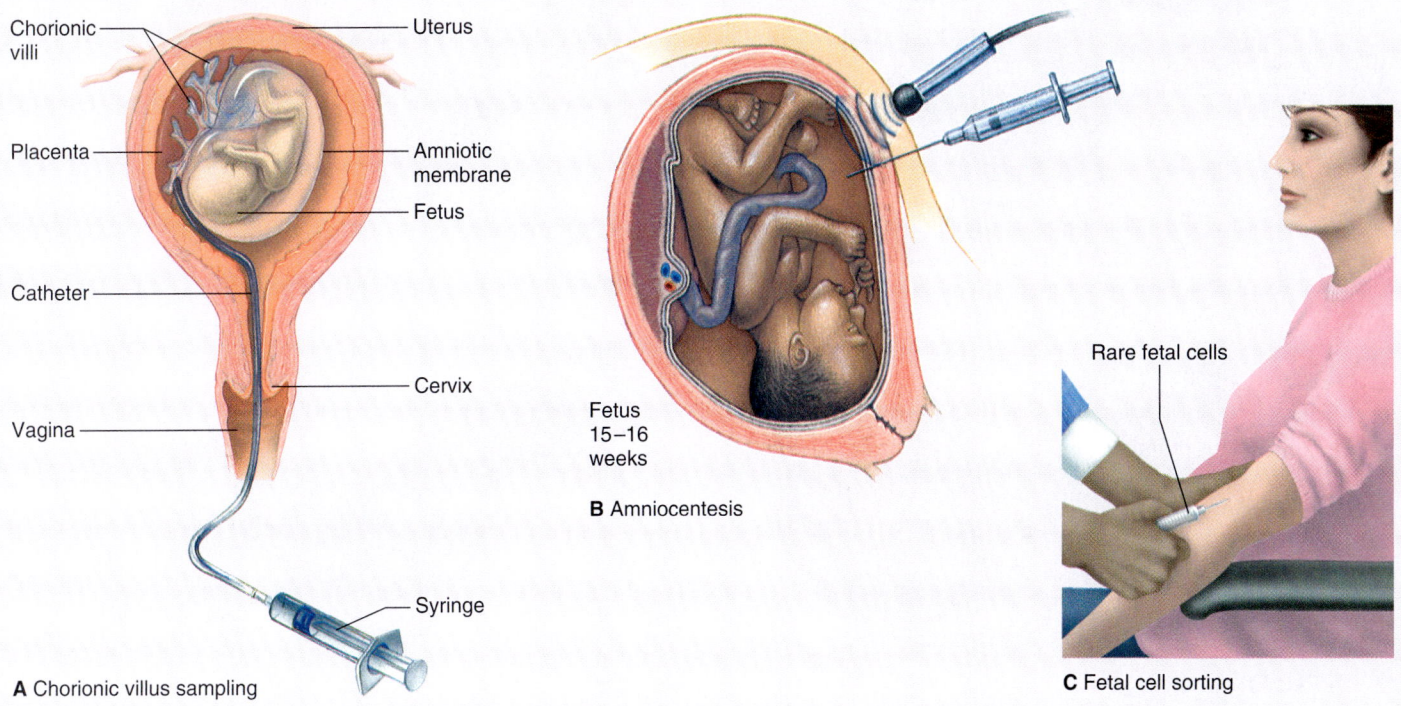

Figure 20C
Three ways to check a fetus's chromosomes. (A) Chorionic villus sampling (CVS) removes cells of the chorionic villi, whose chromosomes match those of the fetus because they all descend from the fertilized ovum. CVS is usually performed earlier than amniocentesis. (B) In amniocentesis, a needle is inserted into the uterus to collect a sample of amniotic fluid, which contains fetal cells. The cells are grown in the laboratory and then dropped onto a microscope slide to spread the chromosomes. The chromosomes are then stained and arranged into a chromosome chart (karyotype). Amniocentesis is performed after the fifteenth week of gestation. (C) Fetal cell sorting separates fetal cells in the woman's circulation. A genetic counselor interprets the results of these tests for patients.

This initiates nerve impulses to the hypothalamus, which in turn signals the posterior pituitary gland to release the hormone **oxytocin** (see chapter 11, p. 288). Oxytocin stimulates powerful uterine contractions. Combined with the greater excitability of the myometrium due to the decline in progesterone secretion, oxytocin aids labor in its later stages.

During labor, rhythmic muscular contractions begin at the top of the uterus and extend down its length. Since the fetus is usually positioned head downward, labor contractions force the head against the cervix (fig. 20.16). This action stretches the cervix, which elicits a reflex that stimulates still stronger labor contractions. Thus, a *positive feedback system* operates, in which uterine contractions produce more intense uterine contractions. At the same time, continuing cervix dilation reflexly stimulates the posterior pituitary to increase oxytocin release. As labor continues, positive feedback stimulates abdominal wall muscles to contract, which also helps force the fetus through the cervix and vagina to the outside.

> An infant passing through the birth canal can tear the delicate tissues between the vulva and anus (perineum). To avoid a ragged tear, a physician makes an *episiotomy,* a clean cut in the perineal tissues.

Following birth of the fetus, the placenta separates from the uterine wall, and uterine contractions expel it through the birth canal. Bleeding accompanies the expelled placenta, termed the *afterbirth,* because the separation damages vascular tissues. However, oxytocin stimulates continued uterine contraction, which compresses the bleeding vessels and minimizes blood loss.

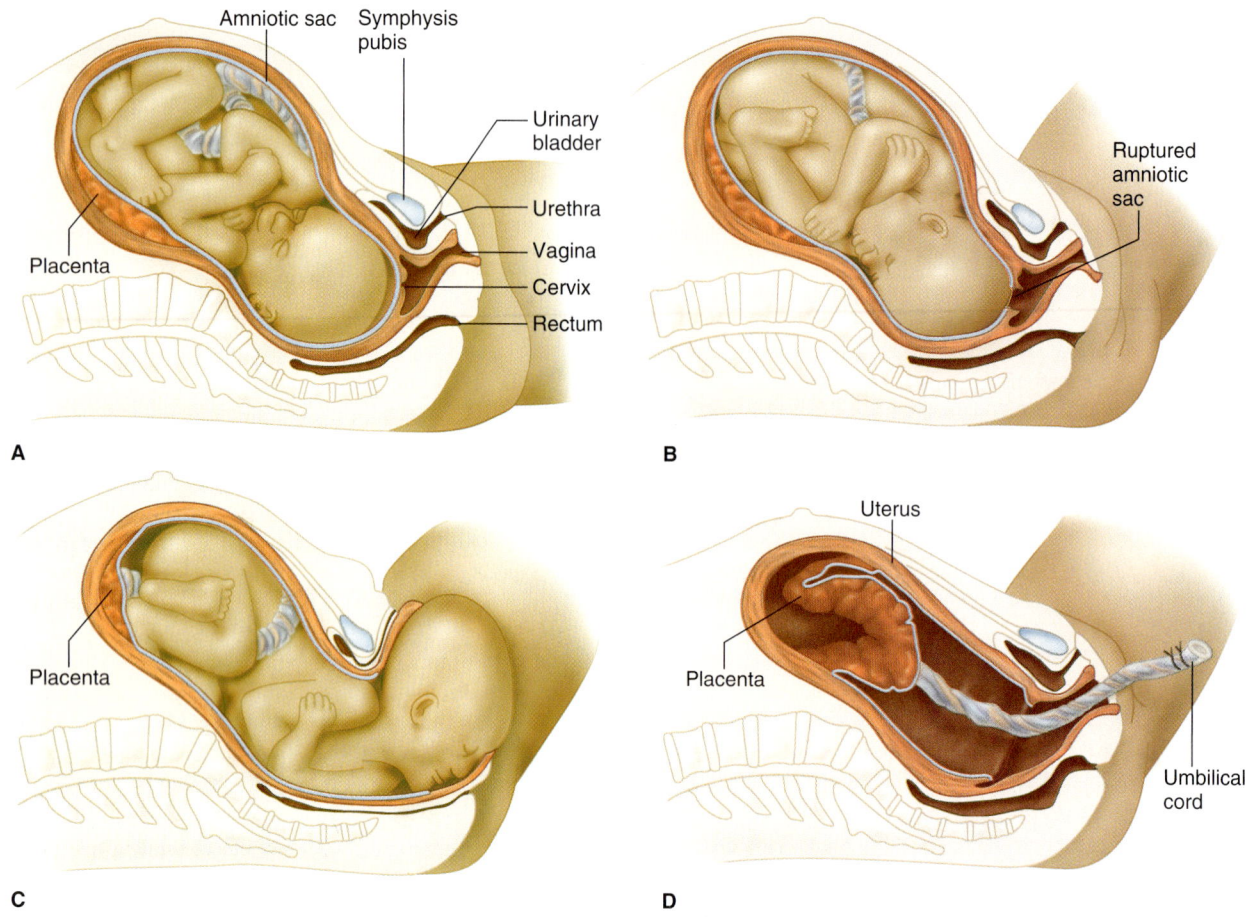

Figure 20.16
Stages in birth. (A) Fetal position before labor, (B) dilation of the cervix, (C) expulsion of the fetus, (D) expulsion of the placenta.

Breast-feeding also contributes to returning the uterus to its original, prepregnancy size, as the suckling of the newborn stimulates the release of oxytocin from the posterior pituitary.

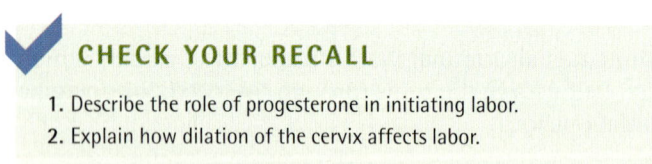

CHECK YOUR RECALL
1. Describe the role of progesterone in initiating labor.
2. Explain how dilation of the cervix affects labor.

20.4 Postnatal Period

Following birth, both mother and newborn experience physiological and structural changes.

Milk Production and Secretion

During pregnancy, placental estrogens and progesterone stimulate further development of the mammary glands. Estrogens cause the ductile systems to grow and branch and deposit abundant fat around them. Progesterone stimulates development of the alveolar glands at the ends of the ducts. Placental lactogen also promotes these changes.

Hormonal activity doubles breast size during pregnancy, and the mammary glands become capable of secreting milk. However, milk is not secreted because placental progesterone inhibits milk production, and placental lactogen blocks the action of *prolactin* (see chapter 11, p. 287).

Following childbirth and the expulsion of the placenta, maternal blood concentrations of placental hormones decline rapidly. In two or three days, prolactin, which is no longer inhibited, stimulates the mammary glands to secrete milk. Meanwhile, the glands secrete a thin, watery fluid called *colostrum*. Colostrum is rich in proteins, but its carbohydrate and fat concentrations are lower than those of milk. Colostrum contains antibodies from the mother's immune system that protect the newborn from certain infections.

Milk ejection requires contraction of specialized *myoepithelial cells* surrounding the alveolar glands (fig. 20.17). Suckling or mechanical stimulation of the nipple or areola elicits the reflex action that controls this process. Impulses from sensory receptors within the breasts go to the hypothalamus, which signals the poste-

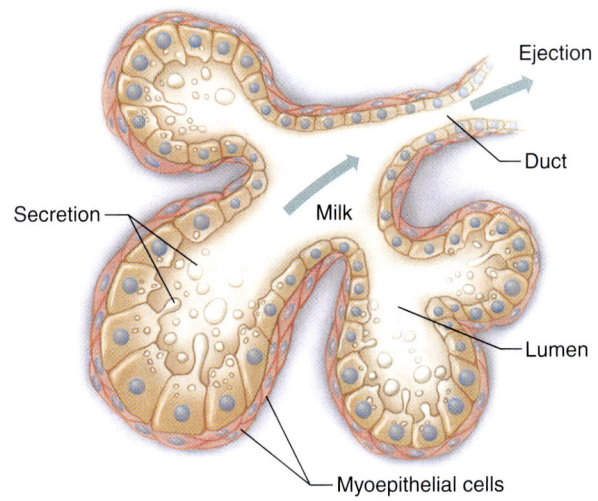

Figure 20.17
Myoepithelial cells eject milk from an alveolar gland.

rior pituitary gland to release oxytocin. Oxytocin travels in the bloodstream to the breasts and stimulates myoepithelial cells to contract. As a result, milk is ejected into a suckling infant's mouth in about 30 seconds.

As long as milk is removed from the breasts, release of prolactin and oxytocin continues, and the mammary glands produce milk. If milk is not removed regularly, the hypothalamus inhibits prolactin secretion, and within about one week, the mammary glands stop producing milk.

Human milk is the best possible food for human babies. The milk of other animals contains different concentrations of nutrients than human milk.

CHECK YOUR RECALL

1. How does pregnancy affect the mammary glands?
2. What stimulates the mammary glands to produce milk?
3. What causes milk to flow into the ductile system of a mammary gland?

Neonatal Period

The **neonatal** (ne″o-na′tal) **period** begins abruptly at birth and extends to the end of the first four weeks. At birth, the newborn must make quick physiological adjustments to become self-reliant. It must respire, obtain and digest nutrients, excrete wastes, and regulate body temperature.

 The largest newborn of recent times was a 24-pound 4-ounce baby boy born in Turkey.

A newborn's most immediate requirement is to obtain oxygen and excrete carbon dioxide. The first breath must be particularly forceful because the newborn's lungs are collapsed, and its small airways offer considerable resistance to air movement. Also, surface tension tends to hold the moist membranes of the lungs together. However, the lungs of a full-term fetus continuously secrete *surfactant* (see chapter 16, p. 443), which reduces surface tension. After the first powerful breath begins to expand the lungs, breathing eases.

> Premature infants' survival chances increase directly with age and weight. Survival is more likely if the lungs are sufficiently developed with the thin respiratory membranes necessary for rapid exchange of oxygen and carbon dioxide, and if the lungs produce enough surfactant to reduce alveolar surface tension. A fetus of less than twenty-four weeks or weighing less than 600 grams at birth seldom survives, even with intensive medical care. *Neonatology* is the medical field that deals with premature and ill newborns.

The newborn has a high metabolic rate, and its immature liver may be unable to supply enough glucose to support its metabolic requirements. Consequently, the newborn typically utilizes stored fat for energy.

A newborn's kidneys are usually unable to produce concentrated urine, so they excrete a dilute fluid. For this reason, the newborn may become dehydrated and develop a water and electrolyte imbalance. Also, some of the newborn's homeostatic control mechanisms may not function adequately. For example, the temperature-regulating system may be unable to maintain a constant body temperature.

When the placenta ceases to function and breathing begins, the newborn's cardiovascular system also changes. Following birth, the umbilical vessels constrict. The umbilical arteries close first, and if the umbilical cord is not clamped or severed for a minute or so, blood continues to flow from the placenta to the newborn through the umbilical vein, adding to the newborn's blood volume. Similarly, the ductus venosus constricts shortly after birth and appears in the adult as a fibrous cord (ligamentum venosum) superficially embedded in the wall of the liver.

The foramen ovale closes as a result of blood pressure changes in the right and left atria as fetal vessels constrict. As blood ceases to flow from the umbilical vein into the inferior vena cava, the blood pressure in the right atrium falls. Also, as the lungs expand with the first breathing movements, resistance to blood flow through the pulmonary circuit decreases, more blood enters the left atrium through the pulmonary veins, and the blood pressure in the left atrium increases.

As the blood pressure in the left atrium rises and that in the right atrium falls, the valve on the left side of the atrial septum closes the foramen ovale. In most individuals, this valve gradually fuses with the tissues along the margin of the foramen. In an adult, a depression called the *fossa ovalis* marks the site of the previous opening.

The ductus arteriosus, like the other fetal vessels, constricts after birth. After the ductus arteriosus closes, blood can no longer bypass the lungs by moving from the pulmonary trunk directly into the aorta. In an adult, a cord called the *ligamentum arteriosum* represents the ductus arteriosus.

> In *patent ductus arteriosus (PDA)*, the ductus arteriosus fails to close completely. This condition is common in newborns whose mothers were infected with rubella virus (German measles) during the first three months of pregnancy.
>
> After birth, the metabolic rate and oxygen consumption in neonatal tissues increase, in large part to maintain body temperature. If the ductus arteriosus remains open, the neonate's blood oxygen concentration may be too low to adequately supply tissues, including the myocardium. If PDA is not corrected surgically, the heart may fail, even though the myocardium is normal.

Changes in the newborn's cardiovascular system are gradual. Constriction of the ductus arteriosus may be functionally complete within 15 minutes, but the permanent closure of the foramen ovale may take up to a year.

Recall that fetal hemoglobin is slightly different and has a greater affinity for oxygen than the adult type. Fetal hemoglobin production falls after birth, and by the time an infant is four months old, most of the circulating hemoglobin is the adult type. Figure 20.18 illustrates cardiovascular changes in the newborn.

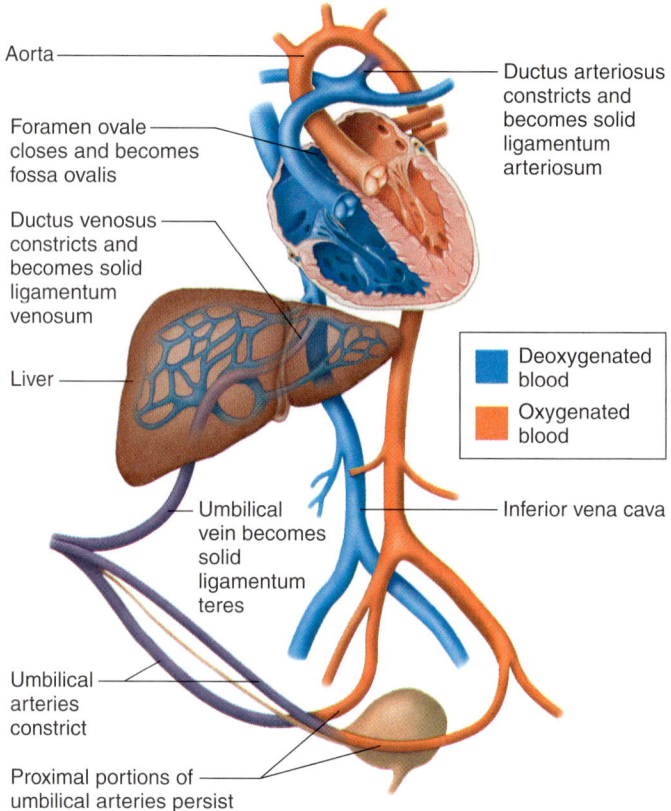

Figure 20.18
Major changes in the newborn's cardiovascular system.

CHECK YOUR RECALL

1. Why must a newborn's first breath be particularly forced?
2. What does a newborn use for energy during its first few days of life?
3. How do the kidneys of a newborn differ from those of an adult?
4. What changes occur in the newborn's cardiovascular system?

Table 20.4 summarizes the major events during the neonatal period as well as those of the later stages of human development. Table 20.5 outlines aging-related changes.

Clinical Terms Related to Pregnancy, Growth, and Development

abruptio placentae (ab-rup´she-o plah-cen´tā) Premature separation of the placenta from the uterine wall.

dizygotic twins (di´´zi-got´ik twinz) Twins resulting from two sperm cells fertilizing two egg cells.

hydatidiform mole (hi´´dah-tid´ĭ-form mōl) Abnormal pregnancy resulting from a pathologic ovum; a mass of cysts.

hydramnios (hi-dram´ne-os) Excess amniotic fluid.

intrauterine transfusion (in´´trah-u´ter-in transfu´zhun) Transfusion administered by injecting blood into the fetal peritoneal cavity before birth.

lochia (lo´ke-ah) Vaginal discharge following childbirth.

meconium (mĕ-ko´ne-um) Anal discharge from the digestive tract of a full-term fetus or a newborn.

monozygotic twins (mon´´o-zi-got´ik twinz) Twins resulting from one sperm cell fertilizing one egg cell, which then splits.

perinatology (per´´ĭ-na-tol´o-je) Branch of medicine concerned with the fetus after twenty-five weeks of development and with the newborn for the first four weeks after birth.

postpartum (pōst-par´tum) Occurring after birth.

teratology (ter´´ah-tol´o-je) Study of substances that cause abnormal development and congenital malformations.

trimester (tri-mes´ter) Each third of the total period of pregnancy.

ultrasonography (ul´´trah-son-og´rah-fe) Technique used to visualize the size and position of fetal structures from patterns of deflected ultrasonic waves.

TABLE 20.4 — STAGES IN POSTNATAL DEVELOPMENT

STAGE	TIME PERIOD	MAJOR EVENTS
Neonatal period	Birth to end of fourth week	Newborn begins to respire, eat, digest nutrients, excrete wastes, regulate body temperature, and make cardiovascular adjustments
Infancy	End of fourth week to one year	Growth rate is high; teeth begin to erupt; muscular and nervous systems mature so that coordinated activities are possible; communication begins
Childhood	One year to puberty	Growth rate is high; deciduous teeth erupt and are replaced by permanent teeth; high degree of muscular control is achieved; bladder and bowel controls are established; intellectual abilities mature
Adolescence	Puberty to adulthood	Person becomes reproductively functional and emotionally more mature; growth spurts occur in skeletal and muscular systems; high levels of motor skills are developed; intellectual abilities increase
Adulthood	Adolescence to old age	Person remains relatively unchanged anatomically and physiologically; degenerative changes begin to occur
Senescence	Old age to death	Degenerative changes continue; body becomes less and less able to cope with demands; death usually results from mechanical disturbances in the cardiovascular system or from disease processes that affect vital organs

TABLE 20.5 — AGING-RELATED CHANGES

ORGAN SYSTEM	AGING-RELATED CHANGES
Integumentary system	Degenerative loss of collagenous and elastic fibers in dermis; decreased production of pigment in hair follicles; reduced activity of sweat and sebaceous glands Skin thins, wrinkles, and dries out; hair turns gray and then white
Skeletal system	Degenerative loss of bone matrix Bones become thinner, less dense, and more likely to fracture; stature may shorten due to compression of intervertebral discs and vertebrae
Muscular system	Loss of skeletal muscle fibers; degenerative changes in neuromuscular junctions Loss of muscular strength
Nervous system	Degenerative changes in neurons; loss of dendrites and synaptic connections; accumulation of lipofuscin in neurons; decreases in sensation Decreasing efficiency in processing and recalling information; decreasing ability to communicate; diminished sense of smell and taste; loss of elasticity of lenses and consequent loss of ability to accommodate for close vision
Endocrine system	Reduced hormonal secretions Decreased metabolic rate; reduced ability to cope with stress; reduced ability to maintain homeostasis
Cardiovascular system	Degenerative changes in cardiac muscle; decrease in lumen diameters of arteries and arterioles Decreased cardiac output; increased resistance to blood flow; increased blood pressure
Lymphatic system	Decrease in efficiency of immune system Increased incidence of infections and neoplastic diseases; increased incidence of autoimmune diseases
Digestive system	Decreased motility in gastrointestinal tract; reduced secretion of digestive juices Reduced efficiency of digestion
Respiratory system	Degenerative loss of elastic fibers in lungs; fewer alveoli Reduced vital capacity; increase in dead air space; reduced ability to clear airways by coughing
Urinary system	Degnerative changes in kidneys; fewer functional nephrons Reductions in filtration rate, tubular secretion, and reabsorption
Reproductive systems	
Male	Reduced secretion of sex hormones; enlargement of prostate gland; decrease in sexual energy
Female	Degenerative changes in ovaries; decrease in secretion of sex hormones; menopause; regression of secondary sex characteristics

Clinical Connection

Preeclampsia, also called "toxemia of pregnancy," is a complication that produces dangerously high blood pressure in a pregnant woman. Evidence suggests that preeclampsia may be passed on through the male. For many years, obstetricians routinely asked their patients if their mothers had preeclampsia, because it has a tendency to occur in women whose mothers were affected. However, in 1998 a study of 1.7 million pregnancies in Norway revealed that if a man's first wife had preeclampsia, his second wife had double the average risk of developing the condition too. Another study in 2001 on 298 men and 237 women in Utah found that women whose mothers-in-law had experienced preeclampsia when pregnant with the women's husbands faced approximately twice the risk of developing the condition themselves. Somehow, a gene from the male must affect the placenta in a way that elevates the pregnant woman's blood pressure.

SUMMARY OUTLINE

20.1 Introduction (p. 528)
Growth is an increase in size. Development is the process of changing from one life phase to another.

20.2 Pregnancy (p. 528)
Pregnancy is the presence of a developing offspring in the uterus.
1. Transport of sex cells
    a. A male deposits semen in the vagina during sexual intercourse.
    b. A sperm cell lashes its tail to move, and is aided by muscular contractions in the female reproductive tract.
2. Fertilization
    a. An enzyme helps a sperm cell penetrate the zona pellucida.
    b. When a sperm cell penetrates an egg cell membrane, changes in the membrane and the zona pellucida prevent entry of additional sperm cells.
    c. Fusion of the nuclei of a sperm cell and an egg cell completes fertilization.
    d. The product of fertilization is a zygote with 46 chromosomes.

20.3 Prenatal Period (p. 529)
1. Early embryonic development
    a. Cells undergo mitosis, giving rise to smaller and smaller cells during cleavage.
    b. The developing offspring moves down the uterine tube to the uterus, where it implants in the endometrium.
    c. The offspring is called an embryo through the eighth week of development. Thereafter, it is a fetus.
    d. Eventually, embryonic and maternal cells form a placenta.
2. Hormonal changes during pregnancy
    a. Embryonic cells produce human chorionic gonadotropin (hCG), which maintains the corpus luteum.
    b. Placental tissue produces high concentrations of estrogens and progesterone.
        (1) Estrogens and progesterone maintain the uterine wall and inhibit secretion of follicle-stimulating hormone (FSH) and luteinizing hormone (LH).
        (2) Progesterone and relaxin inhibit contraction of uterine muscles.
        (3) Estrogens cause enlargement of the vagina.
        (4) Relaxin helps relax the ligaments of the pelvic joints.
    c. Placental lactogen stimulates development of the breasts and mammary glands.
    d. During pregnancy, increased aldosterone secretion promotes retention of sodium and body fluid. Increased secretion of parathyroid hormone helps maintain a high concentration of maternal blood calcium.
3. Embryonic stage
    a. The embryonic stage extends from the beginning of the second week through the eighth week of development.
    b. During this stage, the placenta and main internal and external body structures develop.
    c. The cells of the inner cell mass organize into primary germ layers.
    d. The embryonic disc becomes cylindrical and attaches to the developing placenta.
    e. The placental membrane consists of the epithelium of the chorionic villi and the epithelium of the capillaries inside the villi.
        (1) Oxygen and nutrients diffuse from maternal blood across the placental membrane and into fetal blood.
        (2) Carbon dioxide and other wastes diffuse from fetal blood across the placental membrane and into maternal blood.
    f. A fluid-filled amnion develops around the embryo.
    g. The umbilical cord forms as the amnion envelops the tissues attached to the underside of the embryo.
    h. The yolk sac forms on the underside of the embryonic disc.
    i. The allantois extends from the yolk sac into the connecting stalk.
    j. By the beginning of the eighth week, the embryo is recognizable as human.
4. Fetal stage
    a. The fetal stage extends from the end of the eighth week of development until birth.
    b. Existing structures grow and mature. Only a few new parts appear.
    c. The fetus is full-term at the end of thirty-eight weeks.
5. Fetal blood and circulation
    a. Umbilical vessels carry blood between the placenta and the fetus.
    b. Fetal blood carries a greater concentration of oxygen than does maternal blood.
    c. Blood enters the fetus through the umbilical vein and partially bypasses the liver through the ductus venosus.
    d. Blood enters the right atrium and partially bypasses the lungs through the foramen ovale.
    e. Blood entering the pulmonary trunk partially bypasses the lungs through the ductus arteriosus.
    f. Blood enters the umbilical arteries from the internal iliac arteries.
6. Birth process
    a. During pregnancy, placental progesterone inhibits uterine contractions.
    b. A variety of factors promote birth.
        (1) A decreasing progesterone concentration and the release of a prostaglandin initiate the birth process.
        (2) The posterior pituitary gland releases oxytocin.
        (3) Oxytocin stimulates uterine muscles to contract, and labor begins.
    c. Following birth, placental tissues are expelled.

20.4 Postnatal Period (p. 544)
1. Milk production and secretion
    a. Following childbirth, concentrations of placental hormones decline, the action of prolactin is no longer blocked, and the mammary glands begin to secrete milk.
    b. A reflex response to mechanical stimulation of the nipple stimulates the posterior pituitary to release oxytocin, which causes the alveolar ducts to eject milk.
2. Neonatal period
    a. The neonatal period extends from birth to the end of the first four weeks.
    b. The newborn must begin to respire, obtain nutrients, excrete wastes, and regulate body temperature.
    c. The first breath must be powerful to expand the lungs.
    d. The liver is immature and unable to supply sufficient glucose, so the newborn depends primarily on stored fat for energy.
    e. A newborn's immature kidneys cannot concentrate urine well.
    f. A newborn's homeostatic mechanisms may function imperfectly, and body temperature may be unstable.
    g. The cardiovascular system changes when placental circulation ceases.
        (1) Umbilical vessels constrict.
        (2) The ductus venosus constricts.
        (3) A valve closes the foramen ovale as blood pressure in the right atrium falls and pressure in the left atrium rises.
        (4) The ductus arteriosus constricts.

REVIEW EXERCISES

1. Define *growth* and *development*. (p. 528)
2. Define *pregnancy*. (p. 528)
3. Describe how sperm cells move within the female reproductive tract. (p. 528)
4. Describe the process of fertilization. (p. 528)
5. Describe the process of cleavage. (p. 529)
6. Describe the process of implantation. (p. 530)
7. Define *embryo*. (p. 530)
8. Define *fetus*. (p. 530)
9. Describe the formation of the placenta, and explain its functions. (p. 530)
10. Explain the major hormonal changes in the maternal body during pregnancy. (p. 531)
11. Explain how the primary germ layers form. (p. 533)
12. List the major body parts derived from ectoderm, mesoderm, and endoderm. (p. 533)
13. Define *placental membrane*. (p. 535)
14. Distinguish between the chorion and the amnion. (p. 537)
15. Explain the function of amniotic fluid. (p. 537)
16. Describe the formation of the umbilical cord. (p. 537)
17. Explain how the yolk sac and allantois form. (p. 537)
18. List the major changes in the fetal stage of development. (p. 537)
19. Describe a full-term fetus. (p. 538)
20. Compare the properties of fetal hemoglobin with those of adult hemoglobin. (p. 538)
21. Trace the pathway of blood from the placenta to the fetus and back to the placenta. (p. 538)
22. Discuss the events that occur during the birth process. (p. 542)
23. Explain the roles of prolactin and oxytocin in milk production and secretion. (p. 544)
24. Explain why a newborn's first breath must be particularly forceful. (p. 545)
25. Explain why newborns tend to develop water and electrolyte imbalances. (p. 545)
26. Describe the changes in the newborn's cardiovascular system. (p. 545)

CRITICAL THINKING

1. Why can twins resulting from a single fertilized egg exchange blood or receive organ transplants from each other without rejection, while twins resulting from two fertilized eggs sometimes cannot?
2. What symptoms may appear if a newborn's ductus arteriosus fails to close?
3. What kinds of studies and information are required to determine whether a man's exposure to a potential teratogen can cause birth defects years later? How would such analysis differ if a woman were exposed?
4. In Aldous Huxley's book *Brave New World*, egg cells are fertilized *in vitro* and develop assembly-line style. To render some of the embryos less intelligent, lab workers give them alcohol. What medical condition does this scenario invoke?
5. What technology would enable a fetus born in the fourth month to survive in a laboratory setting? (This is not yet possible.)
6. Toxins usually cause more severe medical problems if exposure is during the first eight weeks of pregnancy rather than during the later weeks. Why?
7. Milk from a cow has a higher percentage of protein and a lower percentage of fat than does human milk. Why do you think this is so?

WEB CONNECTIONS

Visit the website for additional study questions and more information about this chapter at:

http://www.mhhe.com/shieress8

Appendix

AIDS TO UNDERSTANDING WORDS

acetabul-, vinegar cup: *acetabul*um
adip-, fat: *adip*ose tissue
agglutin-, to glue together: *agglutin*ation
aliment-, food: *aliment*ary canal
allant-, sausage-shaped: *allant*ois
alveol-, small cavity: *alveol*us
an-, without: *an*aerobic respiration
ana-, up: *ana*bolic
andr-, man: *andr*ogens
append-, to hang something: *append*icular
ax-, axis: *ax*ial skeleton
bil-, bile: *bil*irubin
-blast, budding: osteo*blast*
brady-, slow: *brady*cardia
bronch-, windpipe: *bronch*us
calat-, something inserted: inter*calat*ed disc
calyc-, small cup: *calyc*es
cardi-, heart: peri*cardi*um
carp-, wrist: *carp*als
cata-, down: *cata*bolic
chondr-, cartilage: *chondr*ocyte
chorio-, skin: *chorio*n
choroid, skinlike: *choroid* plexus
chym-, juice: *chym*e
-clast, broken: osteo*clast*
cleav-, to divide: *cleav*age
cochlea, snail: *cochlea*
condyl-, knob: *condyl*e
corac-, beaklike: *corac*oid process
cort-, covering: *cort*ex
cribr-, sievelike: *cribr*iform plate
cric-, ring: *cric*oid cartilage
crin-, to secrete: endo*crin*e
crist-, ridge: *crist*a galli
cut-, skin: sub*cut*aneous
cyt-, cell: *cyt*oplasm
de-, undoing: *de*amination
decidu-, falling off: *decidu*ous
dendr-, tree: *dendr*ite
derm-, skin: *derm*is
detrus-, to force away: *detrus*or muscle
di-, two: *di*saccharide
diastol-, dilation: *diastol*e
diuret-, to pass urine: *diuret*ic
dors-, back: *dors*al
ejacul-, to shoot forth: *ejacul*ation
embol-, stopper: *embol*us
endo-, within: *endo*plasmic reticulum
epi-, upon: *epi*thelial tissue
erg-, work: syn*erg*ist
erythr-, red: *erythr*ocyte
exo-, outside: *exo*crine gland

extra-, outside: *extra*cellular
fimb-, fringe: *fimb*riae
follic-, small bag: hair *follic*le
fov-, pit: *fov*ea
funi-, small cord or fiber: *funi*culus
gangli-, a swelling: *gangli*on
gastr-, stomach: *gastr*ic gland
-gen, to be produced: aller*gen*
-genesis, origin: spermato*genesis*
germ-, to bud or sprout: *germ*inal
glen-, joint socket: *glen*oid cavity
-glia, glue: neuro*glia*
glom-, little ball: *glom*erulus
glyc-, sweet: *glyc*ogen
-gram, something written: electrocardio*gram*
hema-, blood: *hema*toma
hemo-, blood: *hemo*globin
hepat-, liver: *hepat*ic duct
homeo-, same: *homeo*stasis
humor-, fluid: *humor*al
hyper-, above: *hyper*tonic
hypo-, below: *hypo*tonic
im-, (or in-), not: *im*balance
immun-, free: *immun*ity
inflamm-, to set on fire: *inflamm*ation
inter-, between: *inter*phase
intra-, inside: *intra*membranous
iris, rainbow: *iris*
iso-, equal: *iso*tonic
kerat-, horn: *kerat*in
labi-, lip: *labi*a
labyrinth, maze: *labyrinth*
lacri-, tears: *lacri*mal gland
lacun-, pool: *lacun*a
laten-, hidden: *laten*t
-lemm, rind or peel: neuri*lemma*
leuko-, white: *leuko*cyte
lingu-, tongue: *lingu*al tonsil
lip-, fat: *lip*ids
-logy, study of: physio*logy*
-lyte, dissolvable: electro*lyte*
macro-, large: *macro*phage
macula, spot: *macula* lutea
meat-, passage: auditory *meat*us
melan-, black: *melan*in
mening-, membrane: *mening*es
mens-, month: *mens*trual cycle
meta-, change: *meta*bolism
mict-, to pass urine: *mict*urition
mit-, thread: *mit*osis
mono-, one: *mono*saccharide
mons-, mountain: *mons* pubis
morul-, mulberry: *morul*a

moto-, moving: *moto*r
mut-, change: *mut*ation
myo-, muscle: *myo*fibril
nat-, to be born: pre*nat*al
nephr-, kidney: *nephr*on
neutr-, neither one nor the other: *neutr*al
nod-, knot: *nod*ule
nutri-, nourish: *nutri*ent
odont-, tooth: *odont*oid process
olfact-, to smell: *olfact*ory
-osis, abnormal increase in production: leukocyt*osis*
oss-, bone: *oss*eous tissue
papill-, nipple: *papill*ary muscle
para-, beside: *para*thyroid glands
pariet-, wall: *pariet*al membrane
patho-, disease: *patho*gen
pelv-, basin: *pelv*ic cavity
peri-, around: *peri*cardial membrane
phag-, to eat: *phag*ocytosis
pino-, to drink: *pino*cytosis
pleur-, rib: *pleur*al membrane
plex-, interweaving: choroid *plex*us
poie-, to make: hemoto*poie*sis
poly-, many: *poly*unsaturated
pseudo-, false: *pseudo*stratified epithelium
puber-, adult: *puber*ty
pylor-, gatekeeper: *pylor*ic sphincter
sacchar-, sugar: mono*sacchar*ide
sarco-, flesh: *sarco*plasm
scler-, hard: *scler*a
seb-, grease: *seb*aceous gland
sens-, feeling: *sens*ory neuron
-som, body: ribo*som*e
squam-, scale: *squam*ous epithelium
stasis-, standing still: homeo*stasis*
strat-, layer: *strat*ified
syn-, together: *syn*thesis
systol-, contraction: *systol*e
tachy-, rapid: *tachy*cardia
tetan-, stiff: *tetan*ic
thromb-, clot: *thromb*ocyte
toc-, birth: oxy*toc*in
-tomy, cutting: ana*tomy*
trigon-, triangle: *trigon*e
troph-, well fed: muscular hyper*troph*y
-tropic, influencing: adrenocortico*tropic*
tympan-, drum: *tympan*ic membrane
umbil-, navel: *umbil*ical cord
ventr-, belly or stomach: *ventr*icle
vill-, hair: *vill*i
vitre-, glass: *vitre*ous humor
zym-, ferment: en*zym*e

Glossary

A phonetic guide to pronunciation follows each glossary word. Any unmarked vowel that ends a syllable or stands alone as a syllable has the long sound. Thus, the word *play* is phonetically spelled pla. Any unmarked vowel followed by a consonant has the short sound. The word *tough*, for instance, is phonetically spelled tuf. If a long vowel appears in the middle of a syllable (followed by a consonant), it is marked with a macron (¯), the sign for a long vowel. Thus, the word *plate* is phonetically spelled plāt. Similarly, if a vowel stands alone or ends a syllable, but has a short sound, it is marked with a breve (˘).

A

abdomen (ab-do'men) Portion of body between diaphragm and pelvis. p. 16

abdominopelvic cavity (ab-dom″ĭ-no-pel'vik kav'ĭ-te) The space between the diaphragm and the lower portion of the trunk of the body. p. 8

abduction (ab-duk'shun) Movement of a body part away from the midline. p. 159

absorption (ab-sorp'shun) The taking in of substances by cells or membranes. p. 4

accessory organ (ak-ses'o-re or'gan) Organ that supplements the functions of other organs. p. 267

accommodation (ah-kom″o-da'shun) Adjustment of lens of eye for close or distant vision. p. 271

acetylcholine (as″ĕ-til-ko'lēn) A type of neurotransmitter, which is a biochemical secreted at axon ends of many neurons; transmits nerve messages across synapses. p. 176

acetylcholinesterase (as″ĕ-til-ko″lin-es'ter-ās) An enzyme that catalyzes breakdown of acetylcholine. p. 176

acetyl coenzyme A (as'ĕ-til ko-en'zīm) An intermediate compound produced during oxidation of carbohydrates and fats. p. 79

acid (as'id) A substance that ionizes in water to release hydrogen ions. p. 37

acidosis (as″ĭ-do'sis) A relative increase in acidity of body fluids. p. 37

ACTH Adrenocorticotropic hormone. p. 288

actin (ak'tin) A protein that, with myosin, contracts muscle fibers. p. 173

action potential (ak'shun po-ten'shal) The sequence of electrical changes when a nerve cell membrane is exposed to a stimulus that exceeds its threshold. p. 213

active transport (ak'tiv trans'port) Process that requires an expenditure of energy to move a substance across a cell membrane, usually moved against the concentration gradient. p. 62

adduction (ah-duk'shun) Movement of body part toward the midline. p. 159

adenosine diphosphate (ah-den'o-sēn di-fos'fāt) **(ADP)** Molecule produced when adenosine triphosphate loses a terminal phosphate. p. 78

adenosine triphosphate (ah-den'o-sēn tri-fos'fāt) **(ATP)** Organic molecule that stores and releases energy for use in cellular processes. p. 78

adenylate cyclase (ah-den'i-lāt si'klās) An enzyme activated when certain hormones combine with receptors on cell membranes circularizing ATP to cyclic AMP. p. 283

ADH Antidiuretic hormone. p. 288

adipose tissue (ad'ĭ-pōs tish'u) Fat-storing tissue. p. 101

ADP Adenosine diphosphate. p. 78

adrenal cortex (ah-dre'nal kor'teks) The outer portion of the adrenal gland. p. 292

adrenal gland (ah-dre'nal gland) Endocrine gland located on the superior portion of each kidney. p. 292

adrenal medulla (ah-dre'nal me-dul'ah) The inner portion of the adrenal gland. p. 292

adrenergic fiber (ad″ren-er'jik fi'ber) A nerve fiber that secretes norepinephrine at the axon terminal. p. 244

adrenocorticotropic hormone (ad-re″no-kor″te-ko-trōp'ik hor'mōn) **(ACTH)** Hormone that the anterior pituitary secretes to stimulate activity in adrenal cortex. p. 288

aerobic respiration (a″er-o'bik res″pĭ-ra'shun) The complete, energy-releasing, breakdown of glucose to carbon dioxide and water, in the presence of oxygen. p. 78

afferent arteriole (af'er-ent ar-te're-ōl) Vessel that conveys blood to glomerulus of nephron within kidneys. p. 462

agglutination (ah-gloo″tĭ-na'shun) Clumping of blood cells in response to a reaction between an antibody and an antigen. p. 320

agranulocyte (a-gran'u-lo-sīt) A nongranular leukocyte. p. 311

albumin (al-bu'min) A plasma protein that helps regulate osmotic concentration of blood. p. 314

aldosterone (al dos'ter-ōn″) A hormone that the adrenal cortex secretes to regulate sodium and potassium ion concentrations and fluid volume. p. 293

alimentary canal (al'ĭ-men'tar-e kah-nal') The tubular portion of digestive tract that leads from the mouth to the anus. p. 393

alkaline (al'kah-līn) Pertaining to or having properties of a base or alkali; basic. p. 37

alkalosis (al″kah-lo'sis) A relative increase in alkalinity of body fluids. p. 37

allantois (ah-lan'to-is) An embryonic structure that forms umbilical cord blood vessels. p. 537

allergen (al'er-jen) A foreign substance capable of stimulating an allergic reaction. p. 381

all-or-none response (al'or-nun' re-spons') Phenomenon in which a muscle fiber completely contracts when it is exposed to a stimulus of threshold strength. p. 182

alveolar duct (al-ve'o-lar dukt') A fine tube that carries air to an air sac of the lungs. p. 438

alveolus (al-ve'o-lus) (plural, *alveoli*) An air sac of a lung; saclike structure. p. 438

amino acid (ah-me'no as'id) A small organic compound that contains an amino group (–NH$_2$) and a carboxyl group (–COOH); structural unit of a protein molecule. p. 43

amnion (am'ne-on) An extraembryonic membrane that encircles a developing fetus and contains amniotic fluid. p. 537

amniotic fluid (am″ne-ot'ik floo'id) Fluid within the amniotic cavity that surrounds the developing fetus. p. 537

ampulla (am-pul′ah) An expansion at the end of each semicircular canal that contains a crista ampullaris. p. 266

amylase (am′ĭ-lās) An enzyme that hydrolyzes starch. p. 399

anabolism (ah-nab′o-liz″em) Metabolic process by which larger molecules are synthesized from smaller ones; anabolic metabolism. p. 74

anaerobic respiration (an-a″er-o′bik res″pĭ-ra′shun) The energy-releasing breakdown of glucose to lactic acid, in the absence of oxygen. p. 179

anaphase (an′ah-fāz) Stage in mitosis when duplicate chromosomes move to opposite poles of cell. p. 65

anatomy (ah-nat′o-me) Branch of science dealing with the form and structure of body parts. p. 3

androgen (an′dro-jen) A male sex hormone such as testosterone. p. 506

antagonist (an-tag′o-nist) A muscle that acts in opposition to a prime mover. p. 187

antebrachial (an″te-bra′ke-al) Pertaining to the forearm. p. 16

antecubital (an″te-ku′bĭ-tal) The region in front of the elbow joint. p. 16

anterior (an-te′re-or) Pertaining to the front. p. 13

anterior pituitary (an-te′re-or pĭ-tu′ĭ-tār″e) The front lobe of pituitary gland. p. 286

antibody (an′tĭ-bod″e) A protein (immunoglobulin) that B cells of the immune system produce in response to the presence of a nonself antigen; it reacts with the antigen. p. 376

anticodon (an″tĭ-ko′don) Three contiguous nucleotides of a transfer RNA molecule that are complementary to a specific mRNA codon. p. 83

antidiuretic hormone (an″tĭ-di″u-ret′ik hor′mōn) **(ADH)** Hormone that the posterior pituitary lobe releases to enhance water conservation by kidneys. p. 288

antigen (an′tĭ-jen) A chemical that stimulates cells to produce antibodies. p. 375

aorta (a-or′tah) Major systemic artery that receives blood from left ventricle. p. 332

aortic sinus (a-or′tik si′nus) Swelling in aortic wall behind each cusp of the semilunar valve. p. 351

aortic valve (a-or′tik valv) Flaplike structures in wall of aorta near its origin that prevent blood from returning to left ventricle of heart. p. 332

apocrine gland (ap′o-krin gland) A type of gland whose secretions contain parts of secretory cells. p. 97

aponeurosis (ap″o-nu-ro′sis) A sheet of connective tissue by which certain muscles are attached to adjacent muscles. p. 172

appendicular (ap″en-dik′u-lar) Pertaining to upper or lower limbs. p. 8

aqueous humor (a′kwe-us hu′mor) Watery fluid that fills the anterior cavity of the eye. p. 271

arachnoid mater (ah-rak′noid ma′ter) Delicate, weblike middle layer of meninges. p. 222

arrector pili muscle (ah-rek′tor pil′i mus′l) Smooth muscle in skin associated with a hair follicle. p. 117

arrhythmia (ah-rith′me-ah) Abnormal heart action characterized by a loss of rhythm. p. 341

arteriole (ar-te′re-ōl) A small branch of an artery that communicates with a capillary network. p. 342

artery (ar′ter-e) A vessel that transports blood away from heart. p. 342

articular cartilage (ar-tik′u-lar kar′tĭ-lij) Hyaline cartilage that covers ends of bones in synovial joints. p. 128

articulation (ar-tik″u-la′shun) Joining of structures at a joint. p. 156

ascending tract (ah-send′ing trakt) Group of nerve fibers in the spinal cord that transmits sensory impulses upward to the brain. p. 224

ascorbic acid (as-kor′bik as′id) Vitamin C; a water-soluble vitamin. p. 425

assimilation (ah-sim″ĭ-la′shun) The action of chemically changing absorbed substances. p. 4

association area (ah-so″se-a′shun a′re-ah) Region of the cerebral cortex controlling memory, reasoning, judgment, and emotions. p. 229

astrocyte (as′tro-sīt) A type of neuroglial cell that connects neurons to blood vessels. p. 209

atherosclerosis (ath″er-o-sklĕ-ro′sis) Condition in which fatty substances accumulate on inner linings of arteries. p. 318

atmospheric pressure (at″mos-fēr′ik presh′ur) Pressure exerted by the weight of air; about 760 millimeters of mercury at sea level. p. 441

atom (at′om) Smallest particle of an element that has the properties of that element. p. 7

atomic number (ah-tom′ik num′ber) Number of protons in an atom of an element. p. 32

atomic weight (ah-tom′ik wāt) Number approximately equal to number of protons plus number of neutrons in an atom of an element. p. 32

ATP Adenosine triphosphate. p. 78

ATPase Enzyme that causes ATP molecules to release the energy stored in their terminal phosphate bonds. p. 176

atrioventricular bundle (a″tre-o-ven-trik′u-lar bun′dl) **(A-V bundle)** Group of specialized fibers that conduct impulses from the atrioventricular node to the ventricular muscle of heart. p. 338

atrioventricular node (a″tre-o-ven-trik′u-lar nōd) **(A-V node)** Specialized mass of cardiac muscle fibers in the interatrial septum of heart; transmits cardiac impulses from sinoatrial node to the A-V bundle. p. 338

atrium (a′tre-um) (plural, *atria*) Chamber of heart that receives blood from veins. p. 331

auditory (aw′di-to″re) Pertaining to the ear, or sense of hearing. p. 260

auditory ossicle (aw′di-to″re os′i-kl) A bone of the middle ear. p. 261

auditory tube (aw′di-to″re tūb) The tube that connects the middle ear cavity to the pharynx; eustachian tube. p. 261

auricle (aw′ri-kl) An earlike structure; the portion of the heart that forms the wall of an atrium. p. 331

autonomic nervous system (aw″to-nom′ik ner′vus sis′tem) Portion of nervous system that controls the actions of the viscera. p. 235

axial (ak′se-al) Pertaining to the head, neck, and trunk. p. 8

axillary (ak′sĭ-ler″e) Pertaining to the armpit. p. 16

axon (ak′son) A nerve fiber that conducts a nerve impulse away from a neuron cell body. p. 207

B

baroreceptor (bar″o-re-sep′tor) Sensory receptor in the blood vessel wall stimulated by changes in pressure. p. 340

basal ganglion (ba′sal gang′gle-on) Mass of gray matter deep within a cerebral hemisphere of brain. p. 229

base (bās) Substance that ionizes in water to release hydroxide ions (OH⁻) or other ions that combine with hydrogen ions. p. 37

basement membrane (bās′ment mem′brān) A layer of nonliving material that anchors epithelial tissue to underlying connective tissue. p. 92

basophil (ba′so-fil) White blood cell containing cytoplasmic granules that stain with basic dye. p. 312

B cell (sel) Lymphocyte that produces and secretes antibodies to fight foreign substances in body; B lymphocyte. p. 375

beta oxidation (ba′tah ok″sĭ-da′shun) Chemical process by which fatty acids are broken down into molecules of acetyl coenzyme A. p. 421

bicuspid valve (bi-kus′pid valv) Heart valve between left atrium and left ventricle; mitral valve. p. 332

bile (bīl) Fluid secreted by liver and stored in gallbladder. p. 409

bilirubin (bil″ĭ-roo′bin) A bile pigment produced from hemoglobin breakdown. p. 309

biliverdin (bil″ĭ-ver′din) A bile pigment produced from hemoglobin breakdown. p. 309

biotin (bi′o-tin) A water-soluble vitamin; member of the vitamin B complex. p. 425

blastocyst (blas′to-sist) An early stage of prenatal development that consists of a hollow ball of cells. p. 530

brachial (bra′ke-al) Pertaining to the arm. p. 16

brain stem (brān stem) Portion of brain that includes midbrain, pons, and medulla oblongata. p. 233

Broca's area (bro′kahz a′re-ah) Region of frontal lobe that coordinates complex muscular actions of mouth, tongue, and larynx, making speech possible. p. 229

bronchial tree (brong′ke-al tre) The bronchi and their branches that carry air from the trachea to the alveoli of the lungs. p. 438

bronchiole (brong′ke-ōl) A small branch of a bronchus within the lung. p. 438

bronchus (brong′kus) (plural, *bronchi*) A branch of the trachea that leads to the lung. p. 438

buccal (buk′al) Pertaining to the mouth and inner lining of the cheeks. p. 16

buffer (buf′er) A substance that can react with a strong acid or base to form a weaker acid or base and thus resist change in pH. p. 489

buffer system (buf′er sis′tem) Sets of chemical reactions that occur in body fluids to maintain a particular pH. p. 489

bulbourethral gland (bul″bo-u-re′thral gland) Gland that secretes a viscous fluid into the male urethra during sexual excitement; Cowper's gland. p. 502

bursa (bur′sah) A saclike, fluid-filled structure, lined with synovial membrane, near a joint. p. 157

C

calcitonin (kal″sĭ-to′nin) Hormone the thyroid gland secretes to help regulate blood calcium concentration. p. 290

calorie (kal′o-re) A unit used to measure heat energy and the energy content of foods. p. 423

canaliculus (kan″ah-lik′u-lus) Microscopic canal that connects lacunae of bone tissue. p. 103

cancellous bone (kan′sĕ-lus bōn) Bone tissue with a latticework structure; spongy bone. p. 128

capacitation (kah-pas″i-ta′shun) Activation of sperm cell to fertilize an egg cell. p. 503

capillary (kap′ĭ-ler″e) A small blood vessel that connects an arteriole and a venule. p. 343

carbohydrate (kar″bo-hi′drāt) An organic compound that contains carbon, hydrogen, and oxygen, in a 1:2:1 ratio. p. 39

carbonic anhydrase (kar-bon′ik an-hi′drās) Enzyme that catalyzes the reaction between carbon dioxide and water to form carbonic acid. p. 453

carbaminohemoglobin (kar-bam″ĭ-no-he″mo-glo′bin) Compound formed by the union of carbon dioxide and hemoglobin. p. 452

carboxypeptidase (kar-bok″se-pep′tĭ-dās) A protein-splitting enzyme in pancreatic juice. p. 405

cardiac conduction system (kar′de-ak kon-duk′shun sis′tem) System of specialized cardiac muscle fibers that conducts cardiac impulses from the S-A node into the myocardium. p. 338

cardiac cycle (kar′de-ak si′kl) A series of myocardial contractions and relaxations that constitutes a complete heartbeat. p. 336

cardiac muscle (kar′de-ak mus′l) Specialized muscle tissue found only in the heart. p. 106

cardiac output (kar′de-ak owt′poot) The volume of blood per minute pumped by the heart (calculated by multiplying stroke volume in milliliters by heart rate in beats per minute). p. 348

carpal (kar′pal) A bone of the wrist. p. 136

carpus (kar′pus) Wrist; wrist bones as a group. p. 151

cartilage (kar′tĭ-lij) Type of connective tissue in which cells are located within lacunae and are separated by a semisolid matrix. p. 102

cartilaginous joint (kar-tĭ-laj′ĭ-nus joint) Two or more bones joined by cartilage. p. 157

catabolism (kă-tab′o-lizm) Metabolic process that breaks down large molecules into smaller ones; catabolic metabolism. p. 74

catalyst (kat′ah-list) A chemical that increases the rate of a chemical reaction but is not permanently altered by the reaction. p. 36

celiac (se′le-ak) Pertaining to the abdomen. p. 16

cell (sel) The structural and functional unit of an organism. p. 7

cell body (sel bod′e) Portion of nerve cell that includes a cytoplasmic mass and a nucleus, and from which nerve fibers extend. p. 207

cell membrane (sel mem′brān) The selectively permeable outer boundary of a cell consisting of a phospholipid bilayer embedded with proteins; plasma membrane or cytoplasmic membrane. p. 49

cellular immune response (sel′u-lar i-mūn′ ri-spons′) The attack of T cells and their secreted products on foreign cells. p. 376.

cellular respiration (sel′u-lar res″pĭ-ra′shun) Cellular process that releases energy from organic compounds. p. 77

cellulose (sel′u-lōs) Polysaccharide abundant in plant tissues that human enzymes cannot break down. p. 420

cementum (se-men′tum) Bonelike material that fastens the root of a tooth into its bony socket. p. 398

central canal (sen′tral kah-nal′) Tiny channel in bone tissue that contains a blood vessel; tube within spinal cord that is continuous with brain ventricles and contains cerebrospinal fluid. pp. 129, 224

central nervous system (sen′tral ner′vus sis′tem) **(CNS)** Portion of nervous system that consists of the brain and spinal cord. p. 207

centriole (sen′tre-ōl) A cellular organelle built of microtubules that organizes the mitotic spindle. p. 56

centromere (sen′tro-mēr) Portion of chromosome to which spindle fibers attach during mitosis. p. 65

centrosome (sen′tro-sōm) Cellular organelle consisting of two centrioles. p. 56

cephalic (sĕ-fal′ik) Pertaining to the head. p. 16

cerebellar cortex (ser″ĕ-bel′ar kor′teks) The outer layer of the cerebellum. p. 234

cerebellum (ser″ĕ-bel′um) Portion of brain that coordinates skeletal muscle movement. p. 234

cerebral cortex (ser′ĕ-bral kor′teks) The outer layer of the cerebrum. p. 228

cerebral hemisphere (ser′ĕ-bral hem′ĭ-sfēr) One of the large, paired structures that constitute the cerebrum. p. 226

cerebrospinal fluid (ser″ĕ-bro-spi′nal floo′id) Fluid occupying the ventricles of the brain, subarachnoid space of the meninges, and the central canal of the spinal cord. p. 222

cerebrum (ser′ĕ-brum) Portion of the brain that occupies the upper part of cranial cavity and provides higher mental functions. p. 225

cervical (ser′vĭ-kal) Pertaining to the neck. p. 16

cervix (ser′viks) Narrow, inferior end of the uterus that leads into vagina. p. 511

chemoreceptor (ke″mo-re-sep′tor) A receptor that is stimulated by the binding of certain chemicals. p. 253

chief cell (chēf sel) Cell of gastric gland that secretes various digestive enzymes, including pepsinogen. p. 403

cholecystokinin (ko″le-sis″to-ki′nin) Hormone the small intestine secretes to stimulate release of pancreatic juice from the pancreas and bile from the gallbladder. p. 403

cholesterol (ko-les′ter-ol) A lipid produced by body cells used to synthesize steroid hormones; excreted into bile. p. 421

cholinergic fiber (ko″lin-er′jik fi′ber) A nerve fiber that secretes acetylcholine at the axon terminal. p. 244

chondrocyte (kon′dro-sīt) A cartilage cell. p. 102

chorion (ko′re-on) Extraembryonic membrane that forms the outermost covering around a fetus and contributes to formation of the placenta. p. 534

chorionic villi (ko″re-on′ik vil′i) Projections that extend from the outer surface of the chorion and help attach an embryo to the uterine wall. p. 534

choroid coat (ko′roid kōt) The vascular, pigmented middle layer of the wall of the eye. p. 270

choroid plexus (ko′roid plek′sus) Mass of specialized capillaries that secretes cerebrospinal fluid into a ventricle of brain. p. 230

chromatid (kro′mah-tid) One-half of a replicated chromosome or a single unreplicated chromosome. p. 65

chromatin (kro′mah-tin) DNA and complexed protein that condenses to form chromosomes during mitosis. p. 57

chromatophilic substance (kro″mah-to-fil′ik sub′stans) **(Nissl bodies)** Membranous sacs within cytoplasm of nerve cells that have ribosomes attached to their surfaces. p. 209

chromosome (kro′mo-sōm) Rodlike structure that condenses from chromatin in a cell's nucleus during mitosis. p. 57

chylomicron (ki″lo-mi′kron) A microscopic droplet of fat in the blood following fat digestion. p. 414

chyme (kīm) Semifluid mass of partially digested food that passes from the stomach to the small intestine. p. 404

chymotrypsin (ki″mo-trip′sin) A protein-splitting enzyme in pancreatic juice. p. 405

cilia (sil′e-ah) Microscopic, hairlike processes on the exposed surfaces of certain epithelial cells. p. 56

ciliary body (sil′e-er″e bod′e) Structure associated with the choroid layer of the eye that secretes aqueous humor and contains the ciliary muscle. p. 270

circadian rhythm (ser″kah-de′an rithm) Pattern of repeated behavior associated with cycles of night and day. p. 297

circle of Willis (ser′kl uv wil′is) An arterial ring on the ventral surface of the brain. p. 353

circular muscles (ser′ku-lar mus′lz) Muscles whose fibers are organized in circular patterns, usually around an opening or in the wall of a tube; sphincter muscles. p. 271

circulation (ser-ku-la′shun) Path of fluid through a system of vessels. p. 4

circumduction (ser″kum-duk′shun) Movement of a body part, such as a limb, so that the end follows a circular path. p. 159

cisternae (sis-ter′ne) Enlarged portions of sarcoplasmic reticulum near the actin and myosin filaments of a muscle fiber. p. 174

citric acid cycle (sit′rik as′id si′kl) Series of chemical reactions that oxidizes certain molecules, releasing energy; Krebs cycle. p. 78

cleavage (klēv′ij) The early successive divisions of blastocyst cells into smaller and smaller cells. p. 529

clitoris (kli′to-ris) Small, erectile organ located in the anterior portion of the vulva; corresponding to the penis. p. 512

clone (klōn) A group of cells that originate from a single cell and are therefore genetically identical. p. 376

CNS Central nervous system. p. 207

coagulation (ko-ag″u-la′shun) Blood clotting. p. 333

cochlea (kok′le-ah) Portion of inner ear that contains hearing receptors. p. 262

codon (ko′don) Set of three nucleotides of a messenger RNA molecule corresponding to a particular amino acid. p. 81

coenzyme (ko-en′zīm) A nonprotein organic molecule required for the activity of a particular enzyme. p. 76

cofactor (ko′fak-tor) A small molecule or ion that must combine with an enzyme for activity. p. 76

collagen (kol′ah-jen) Protein in white fibers of connective tissues and in bone matrix. p. 100

common bile duct (kom′mon bīl dukt) Tube that transports bile from the cystic duct to the duodenum. p. 410

compact bone (kom-pakt′ bōn) Dense tissue in which cells are organized in osteons (Haversian systems) with no spaces. p. 128

complement (kom′plĕ-ment) A group of enzymes activated by the combination of antibody with antigen; enhances reaction against foreign substances within body. p. 380

complete protein (kom-plēt′ pro′te-in) A protein that contains adequate amounts of the essential amino acids to maintain body tissues and to promote normal growth and development. p. 423

compound (kom′pownd) A substance composed of two or more chemically bonded elements. p. 35

condyle (kon′dīl) A rounded process of a bone, usually forming a joint. p. 137

cone (kōn) Color receptor in the retina of the eye. p. 273

conjunctiva (kon″junk-ti′vah) Membranous covering on the anterior surface of the eye. p. 267

connective tissue (kŏ-nek′tiv tish′u) One of the basic types of tissue that includes bone, cartilage, and loose connective tissue. p. 98

convergence (kon-ver′jens) Nerve impulses arriving at the same neuron. p. 219

convolution (kon″vo-lu′shun) Elevation on a structure's surface caused by infolding of structure upon itself. p. 226

cornea (kor′ne-ah) Transparent anterior portion of the outer layer of the eye wall. p. 269

coronary artery (kor′o-na″re ar′ter-e) An artery that supplies blood to wall of heart. p. 333

coronary sinus (kor'o-na"re si'nus) A large vessel on the posterior surface of the heart into which cardiac veins drain. p. 336

corpus albicans (kor'pus al'bĭ-kanz) Remnant of corpus luteum in ovary; composed of fibrous connective tissue. p. 515

corpus callosum (kor'pus kah-lo'sum) A mass of white matter within the brain; composed of nerve fibers connecting the right and left cerebral hemispheres. p. 226

corpus luteum (kor'pus loot'e-um) Structure that forms from the tissues of a ruptured ovarian follicle and secretes female hormones. p. 514

cortisol (kor'tĭ-sol) A glucocorticoid that the adrenal cortex secretes. p. 294

costal (kos'tal) Pertaining to the ribs. p. 16

covalent bond (ko'va-lent bond) Chemical bond formed by electron sharing between atoms. p. 35

cranial (kra'ne-al) Pertaining to the cranium. p. 8

cranial nerve (kra'ne-al nerv) Nerve that arises from the brain. p. 236

creatine phosphate (kre'ah-tin fos'fāt) A muscle biochemical that stores energy. p. 179

crest (krest) A ridgelike projection of a bone. p. 137

cricoid cartilage (kri'koid kar'tĭ-lij) A ringlike cartilage that forms the lower end of the larynx. p. 437

crista ampullaris (kris'tah am-pul'ar-is) Sensory organ within a semicircular canal that functions in the sense of dynamic equilibrium. p. 266

cubital (ku'bi-tal) Pertaining to the elbow. p. 16

cutaneous (ku-ta'ne-us) Pertaining to the skin. p. 113

cystic duct (sis'tik dukt) The tube that connects the gallbladder to the common bile duct. p. 410

cytocrine secretion (si'to-krin se-kre'shun) Transfer of melanin granules from melanocytes into adjacent epithelial cells. p. 116

cytoplasm (si'to-plazm) The contents of a cell, excluding the nucleus and cell membrane. p. 49

D

deamination (de-am"ĭ-na'shun) Removing amino groups (–NH₂) from amino acids. p. 422

deciduous teeth (de-sid'u-us tēth) Teeth that are shed and replaced by permanent teeth; primary teeth. p. 397

decomposition (de-kom"po-zish'un) The breakdown of molecules into simpler compounds. p. 36

defecation (def"ĕ-ka'shun) The discharge of feces from the rectum through the anus. p. 419

dehydration synthesis (de"hi-dra'shun sin'thĕ-sis) Anabolic process that joins small molecules by releasing the equivalent of a water molecule; synthesis. p. 74

dendrite (den'drīt) Nerve fiber that transmits impulses toward a neuron cell body. p. 207

dentin (den'tin) Bonelike substance that forms the bulk of a tooth. p. 398

deoxyhemoglobin (de-ok"se-he"mo-glo'bin) Hemoglobin to which oxygen is not bound. p. 307

deoxyribonucleic acid (de-ok'si-ri"bo-nu-kle"ik as'id) **(DNA)** The genetic material; a double-stranded polymer of nucleotides, each containing a phosphate group, a nitrogenous base (adenine, thymine, guanine, or cytosine), and the sugar deoxyribose. p. 44

depolarization (de-po"lar-ĭ-za'shun) The loss of an electrical charge on the surface of a cell membrane. p. 213

dermis (der'mis) The thick layer of the skin beneath the epidermis. p. 113

descending tract (de-send'ing trakt) Group of nerve fibers that carries nerve impulses from the brain through the spinal cord. p. 224

detrusor muscle (de-trūz'or mus'l) Muscular wall of the urinary bladder. p. 474

diapedesis (di"ah-pĕ-de'sis) Squeezing of leukocytes between the cells of blood vessel walls. p. 313

diaphragm (di'ah-fram) A sheetlike structure largely composed of skeletal muscle and connective tissue that separates thoracic and abdominal cavities; also, a caplike contraceptive device inserted in the vagina. pp. 8, 517

diaphysis (di-af'ĭ-sis) The shaft of a long bone. p. 128

diastole (di-as'tol-le) Phase of the cardiac cycle when a heart chamber wall relaxes. p. 336

diastolic pressure (di-a-stol'ik presh'ur) Lowest arterial blood pressure reached during diastolic phase of cardiac cycle. p. 347

diencephalon (di"en-sef'ah-lon) Portion of the brain in the region of the third ventricle that includes the thalamus and hypothalamus. p. 232

differentiation (dif"er-en"she-a'shun) Cell specialization. p. 67

diffusion (dĭ-fu'zhun) Random movement of molecules from a region of higher concentration toward one of lower concentration. p. 58

digestion (di-jes'chun) Breaking down of large nutrient molecules into smaller molecules that can be absorbed; hydrolysis. p. 4

dipeptide (di-pep'tīd) A molecule composed of two joined amino acids. p. 74

disaccharide (di-sak'ah-rīd) A sugar produced by the union of two monosaccharides. p. 39

distal (dis'tal) Further from the midline or origin; opposite of *proximal*. p. 13

divergence (di-ver'jens) A spreading apart. p. 219

DNA Deoxyribonucleic acid. p. 44

dorsal root (dor'sal root) The sensory branch of a spinal nerve by which it joins the spinal cord. p. 238

dorsal root ganglion (dor'sal root gang'gle-on) Mass of sensory neuron cell bodies in the dorsal root of a spinal nerve. p. 238

dorsal (dors'al) Pertaining to the back surface of a body part. p. 16

ductus arteriosus (duk'tus ar-te"re-o'sus) Blood vessel that connects the pulmonary artery and the aorta in a fetus. p. 540

ductus venosus (duk'tus ven-o'sus) Blood vessel that connects the umbilical vein and the inferior vena cava in a fetus. p. 538

dura mater (du'rah ma'ter) Tough outer layer of the meninges. p. 222

dynamic equilibrium (di-nam'ik e"kwĭ-lib're-um) Maintenance of balance when the head and body are suddenly moved or rotated. p. 265

E

eccrine gland (ek'rin gland) Sweat gland that maintains body temperature. p. 119

ECG Electrocardiogram. p. 339

ectoderm (ek'to-derm) The outermost primary germ layer. p. 533

edema (ĕ-de'mah) Fluid accumulation within the tissue spaces. p. 61

effector (ĕ-fek'tor) A muscle or gland that responds to stimulation. p. 207

efferent arteriole (ef'er-ent ar-te're-ol) The vessel that conducts blood away from the glomerulus of a kidney nephron. p. 464

ejaculation (e-jak″u-la′shun) Discharge of sperm-containing semen from the male urethra. p. 505

elastin (e-las′tin) Protein that comprises the yellow, elastic fibers of connective tissue. p. 100

electrocardiogram (el-lek″tro-kar′de-o-gram″) (**ECG** or **EKG**) A recording of the electrical activity associated with the heartbeat. p. 339

electrolyte (e-lek′tro-līt) A substance that ionizes in water solution. p. 37

electrolyte balance (e-lek′tro-līt bal′ans) Condition when the quantities of electrolytes entering the body equal those leaving it. p. 487

electron (e-lek′tron) A small, negatively charged particle that revolves around the nucleus of an atom. p. 31

element (el′ĕ-ment) A chemical substance with only one type of atom. p. 31

embolus (em′bo-lus) A blood clot or gas bubble that obstructs a blood vessel. p. 318

embryo (em′bre-o) A prenatal stage of development after germ layers form but before rudiments of all organs are present. p. 530

emission (e-mish′un) The movement of sperm cells from the vas deferens into the ejaculatory duct and urethra. p. 505

emulsification (e-mul″sĭ-fĭ-ka′shun) Breaking up of fat globules into smaller droplets by the action of bile salts. p. 410

enamel (e-nam′el) Hard covering on the exposed surface of a tooth. p. 398

endocardium (en″do-kar′de-um) Inner lining of the heart chambers. p. 331

endochondral bone (en″do-kon′dral bōn) Bone that begins as hyaline cartilage that is subsequently replaced by bone tissue. p. 130

endocrine gland (en′do-krin gland) A gland that secretes hormones directly into the blood or body fluids. p. 97

endocytosis (en″do-si-to′sis) Process by which a cell membrane envelops a substance and draws it into the cell in a vesicle. p. 62

endoderm (en′do-derm) The innermost primary germ layer. p. 533

endolymph (en′do-limf) Fluid contained within the membranous labyrinth of the inner ear. p. 262

endometrium (en″do-me′tre-um) The inner lining of uterus. p. 511

endomysium (en″do-mis′e-um) The sheath of connective tissue surrounding each skeletal muscle fiber. p. 173

endoplasmic reticulum (en-do-plaz′mic rĕ-tik′-u-lum) Organelle composed of a system of connected membranous tubules and vesicles along which protein is synthesized. p. 52

endosteum (en-dos′te-um) Tissue lining the medullary cavity within a bone. p. 129

endothelium (en″do-the′le-um) The layer of epithelial cells that forms the inner lining of blood vessels and heart chambers. p. 342

energy (en′er-je) An ability to cause something to move and thus do work. p. 76

enzyme (en′zīm) A protein that catalyzes a specific biochemical reaction. p. 41

eosinophil (e″o-sin′o-fil) White blood cell containing cytoplasmic granules that stain with acidic dye. p. 312

ependyma (ĕ-pen′dĭ-ma) A membrane composed of neuroglial cells that lines the ventricles of the brain. p. 209

epicardium (ep″ĭ-kar′de-um) The visceral portion of the pericardium on the surface of heart. p. 330

epicondyle (ep″ĭ-kon′dīl) A projection of bone above a condyle. p. 137

epidermis (ep″ĭ-der′mis) The outer epithelial layer of the skin. p. 113

epididymis (ep″ĭ-did′ĭ-mis) Highly coiled tubule that leads from the seminiferous tubules of the testis to the vas deferens. p. 500

epidural space (ep″ĭ-du′ral spās) The space between the dural sheath of the spinal cord and the bone of the vertebral canal. p. 222

epigastric region (ep″ĭ-gas′trik re′jun) The upper middle portion of the abdomen. p. 14

epiglottis (ep″ĭ-glot′is) Flaplike, cartilaginous structure at the back of the tongue near the entrance to the trachea. p. 438

epimysium (ep″i-mis′e-um) The outer sheath of connective tissue surrounding a skeletal muscle. p. 172

epinephrine (ep″ĭ-nef′rin) A hormone the adrenal medulla secretes during times of stress. p. 292

epiphyseal plate (ep″ĭ-fiz′e-al plāt) Cartilaginous layer within the long bone epiphysis that grows. p. 130

epiphysis (ĕ-pif′ĭ-sis) The end of a long bone. p. 128

epithelium (ep″ĭ-the′le-um) Tissue type that covers all free body surfaces. p. 92

equilibrium (e″kwĭ-lib′re-um) A state of balance between two opposing forces. p. 58

erythroblast (ĕ-rith′ro-blast) An immature red blood cell. p. 310

erythrocyte (ĕ-rith′ro-sīt) A red blood cell. p. 307

erythropoietin (ĕ-rith″ro-poi′ĕ-tin) A kidney hormone that promotes red blood cell formation. p. 308

esophagus (ĕ-sof′ah-gus) Tubular portion of the digestive tract that leads from the pharynx to the stomach. p. 401

essential amino acid (ĕ-sen′shal ah-me′no as′id) Amino acid required for health that body cells cannot synthesize in adequate amounts. p. 423

essential fatty acid (ĕ-sen′shal fat′e as′id) Fatty acid required for health that body cells cannot synthesize in adequate amounts. p. 421

estrogens (es′tro-jenz) A group of hormones that stimulates the development of female secondary sex characteristics. p. 513

evaporation (e″vap′o-ra-shun) Changing a liquid into a gas. p. 120

eversion (e-ver′zhun) Outward turning movement of the sole of the foot. p. 159

exchange reaction (eks-chānj re-ak′shun) A chemical reaction in which parts of two kinds of molecules trade positions. p. 36

excretion (ek-skre′shun) Elimination of metabolic wastes. p. 4

exocrine gland (ek′so-krin gland) A gland that secretes its products into a duct or onto a body surface. p. 97

exocytosis (eks-o-si-to′sis) Transport of substances out of a cell in vesicles. p. 62

expiration (ek″spĭ-ra′shun) Expulsion of air from the lungs. p. 441

extension (ek-sten′shun) Movement increasing the angle between parts at a joint. p. 159

extracellular (ek″strah-sel′u-lar) Outside of cells; refers to the internal environment, body fluids outside individual cells. p. 482

extrapyramidal tract (ek″strah-pĭ-ram′ĭ-dal trakt) Nerve tracts, other than the corticospinal tracts, that transmit impulses from the cerebral cortex to the spinal cord. p. 224

F

facet (fas′et) A small, flattened surface of a bone. p. 137

facilitated diffusion (fah-sil″ĭ-tāt′ed dĭ-fu′zhun) Diffusion in which carrier molecules transport substances across membranes from a region of higher concentration to a region of lower concentration. p. 59

facilitation (fah-sil″ĭ-ta′shun) Subthreshold stimulation of a neuron makes it more responsive to further stimulation. p. 219

fascia (fash′e-ah) A sheet of fibrous connective tissue that encloses a muscle. p. 172

fat (fat) Adipose tissue; an organic molecule containing glycerol and fatty acids. p. 40

fatty acid (fat′e as′id) A building block of a fat molecule. p. 40

feces (fe′sēz) Material expelled from the digestive tract during defecation. p. 420

fertilization (fer″tĭ-lĭ-za′shun) The union of an egg cell and a sperm cell. p. 528

fetus (fe′tus) A human embryo after eight weeks of development. p. 530

fibrin (fi′brin) Insoluble, fibrous protein formed from fibrinogen during blood coagulation. p. 318

fibrinogen (fi-brin′o-jen) Plasma protein converted into fibrin during blood coagulation. p. 318

fibroblast (fi′bro-blast) Cell that produces fibers in connective tissues. p. 99

fibrous joint (fi′brus joint) Two or more bones joined by connective tissue containing many fibers. p. 156

filtration (fil-tra′shun) Movement of material through a membrane as a result of hydrostatic pressure. p. 61

fissure (fish′ur) A narrow cleft that separates parts, such as the lobes of the cerebrum. p. 226

flexion (flek′shun) Bending at a joint to decrease the angle between bones. p. 159

follicle (fol′ĭ-kl) A pouchlike depression or cavity. p. 508

follicle-stimulating hormone (fol′ĭ-kl stim′u-la″ting hor′mōn) **(FSH)** A substance that the anterior pituitary secretes to stimulate follicular development in a female or sperm cell production in a male. p. 288

follicular cells (fŏ-lik′u-lar selz) Ovarian cells that surround a developing egg cell and secrete female sex hormones. p. 508

fontanel (fon″tah-nel′) Membranous region between certain cranial bones in the skull of a fetus or infant. p. 137

foramen (fo-ra′men) (plural, *foramina*) An opening, usually in a bone or membrane. p. 137

foramen magnum (fo-ra′men mag′num) Opening in the occipital bone of the skull through which the spinal cord passes. p. 136

foramen ovale (fo-ra′men o-val′e) Opening in the interatrial septum of the fetal heart. p. 540

fossa (fos′ah) A depression in a bone or other part. p. 137

fovea (fo′ve-ah) A tiny pit or depression. p. 137

fovea centralis (fo′ve-ah sen-tral′is) Region of the retina, consisting of densely packed cones, that provides the greatest visual acuity. p. 272

frenulum (fren′u-lum) A fold of tissue that anchors and limits movement of a body part. p. 396

frontal (frun′tal) Pertaining to the forehead. p. 17

FSH Follicle-stimulating hormone. p. 288

functional syncytium (funk′shun-al sin-sish′e-um) Merging cells performing as a unit; those of the heart are joined electrically. p. 338

G

gallbladder (gawl′blad-er) Saclike organ associated with the liver that stores and concentrates bile. p. 410

ganglion (gang′gle-on) (plural, *ganglia*) A mass of neuron cell bodies, usually outside the central nervous system. p. 211

gastric gland (gas′trik gland) Gland within the stomach wall that secretes gastric juice. p. 402

gastric juice (gas′trik jōōs) Secretion of the gastric glands within the stomach. p. 403

gastrin (gas′trin) Hormone that the stomach lining secretes to stimulate gastric juice secretion. p. 403

gene (jēn) Portion of a DNA molecule that encodes information to synthesize a protein, a control sequence, or tRNA or rRNA; the unit of inheritance. p. 79

genetic code (jĕ-net′ik kōd) Information for synthesizing proteins that is enclosed in the nucleotide sequence of DNA molecules. p. 81

GH Growth hormone. p. 287

gland (gland) A group of cells that secrete a product. p. 97

globin (glo′bin) The protein portion of a hemoglobin molecule. p. 309

globulin (glob′u-lin) A type of protein in blood plasma. p. 316

glomerular capsule (glo-mer′u-lar kap′sūl) Proximal portion of a renal tubule that encloses the glomerulus of a nephron; Bowman's capsule. p. 464

glomerulus (glo-mer′u-lus) A capillary tuft within the glomerular capsule of a nephron. p. 464

glottis (glot′is) Slitlike opening between the true vocal folds or vocal cords. p. 437

glucagon (gloo′kah-gon) Hormone that pancreatic islets of Langerhans secrete to release glucose from storage. p. 296

glucocorticoid (gloo″ko-kor′tĭ-koid) Any one of a group of hormones that the adrenal cortex secretes to influence carbohydrate, fat, and protein metabolism. p. 294

glucose (gloo′kōs) A monosaccharide in blood that is the primary source of cellular energy. p. 39

gluteal (gloo′te-al) Pertaining to the buttocks. p. 17

glycerol (glis′er-ol) An organic compound that is a building block for fat molecules. p. 40

glycogen (gli′ko-jen) A polysaccharide that stores glucose in the liver and muscles. p. 39

glycolysis (gli-kol′ĭ-sis) The energy-releasing breakdown of glucose to produce 2 pyruvic acid molecules. p. 78

glycoprotein (gli″ko-pro′te-in) A compound composed of a carbohydrate combined with a protein. p. 41

goblet cell (gob′let sel) An epithelial cell specialized to secrete mucus. p. 93

Golgi apparatus (gol′je ap″ah-ra′tus) An organelle that prepares cellular products for secretion. p. 54

gonadotropin (go-nad″o-trōp′in) A hormone that stimulates activity in the gonads. p. 505

granulocyte (gran′u-lo-sīt) A leukocyte with granules in its cytoplasm. p. 311

gray matter (gra mat′er) Region of the central nervous system that generally lacks myelin and thus appears gray. p. 209

groin (groin) Region of the body between the abdomen and thighs. p. 17

growth (grōth) Process by which a structure enlarges. p. 4

growth hormone (grōth hor′mōn) **(GH)** A hormone that the anterior pituitary secretes to promote growth of the organism; somatotropin. p. 287

H

hair follicle (hār fol′ĭ-kl) Tubelike depression in the skin in which a hair develops. p. 117

hapten (hap′ten) A small molecule that combines with a larger one; forms an antigenic substance. p. 375

hematocrit (he-mat′o-krit) The volume percentage of red blood cells within a sample of whole blood. p. 307

hematoma (he″mah-to′mah) A mass of coagulated blood within tissues or a body cavity. p. 319

heme (hēm) The iron-containing portion of a hemoglobin molecule. p. 309

hemocytoblast (he″mo-si′to-blast) A cell that gives rise to blood cells. p. 308

hemoglobin (he″mo-glo′bin) Pigment of red blood cells that transports oxygen. p. 133

hemopoiesis (he″mo-poi-e′sis) The production of blood cells. p. 133

hemorrhage (hem′o-rij) Loss of blood from the cardiovascular system; bleeding. p. 319

hemostasis (he″mo-sta′sis) The stoppage of bleeding. p. 317

hepatic (hĕ-pat′ik) Pertaining to the liver. p. 406

hepatic lobule (hĕ-pat′ik lob′ul) A functional unit of liver. p. 406

holocrine gland (ho′lo-krin gland) A gland whose secretion contains entire secretory cells. p. 97

homeostasis (ho″me-o-sta′sis) A state of balance in which the body's internal environment remains in the normal range. p. 5

hormone (hor′mōn) A substance that an endocrine gland secretes and that the blood or body fluids transport. p. 281

humoral immune response (hu′mor-al i-mūn′ ri-spons′) Circulating antibodies' destruction of cells bearing foreign (nonself) antigens. p. 376

hydrogen bond (hi′dro-jen bond) A weak chemical bond between a hydrogen atom and an atom of oxygen or nitrogen. p. 35

hydrolysis (hi-drol′ĭ-sis) Enzymatically adding a water molecule to split a molecule into smaller portions. p. 75

hydrostatic pressure (hy″dro-stat′ik presh′ur) Pressure exerted by fluids, such as blood pressure. p. 61

hydroxide ion (hi-drok′sīd i′on) OH^-. p. 37

hymen (hi′men) A membranous fold of tissue that partially covers the vaginal opening. p. 511

hyperplasia (hi″per-pla′ze-ah) The increased production and growth of new cells. p. 67

hypertonic (hi″per-ton′ik) A solution with a greater osmotic pressure than the solution with which it is compared. p. 60

hyperventilation (hi″per-ven″tĭ-la′shun) Abnormally deep and prolonged breathing. p. 448

hypochondriac region (hi″po-kon′dre-ak re′jun) The portion of the abdomen on either side of the middle or epigastric region. p. 14

hypogastric region (hi″po-gas′trik re′jun) The lower middle portion of the abdomen. p. 14

hypothalamus (hi″po-thal′ah-mus) A portion of the brain below the thalamus and forming the floor of the third ventricle. p. 232

hypotonic (hi″po-ton′ik) A solution with a lower osmotic pressure than the solution (usually body fluids) to which it is compared. p. 61

I

iliac region (il′e-ak re′jun) Portion of the abdomen on either side of the lower middle or hypogastric region. p. 14

ilium (il′e-um) One of the bones of a coxal bone or hipbone. p. 152

immunity (ĭ-mu′nĭ-te) Resistance to the effects of specific disease-causing agents. p. 373

immunoglobulin (im″u-no-glob′u-lin) Globular plasma proteins that function as antibodies. p. 378

implantation (im″plan-ta′shun) The embedding of an embryo in the lining of the uterus. p. 530

impulse (im′puls) A wave of depolarization conducted along a nerve or muscle fiber. p. 215

incomplete protein (in″kom-plēt′ pro′te-in) A protein that lacks adequate amounts of essential amino acids. p. 423

inferior (in-fer′e-or) Situated below something else; pertaining to the lower surface of a part. p. 13

inflammation (in″flah-ma′shun) A tissue response to stress that causes blood vessel dilation and fluid accumulation in the affected region. p. 120

inguinal (ing′gwĭ-nal) Pertaining to the groin region. p. 17

inorganic (in″or-gan′ik) Chemical substances that lack carbon and hydrogen. p. 37

insertion (in-ser′shun) The end of a muscle attached to a movable part. p. 159

inspiration (in″spĭ-ra′shun) Breathing in; inhalation. p. 441

insula (in′su-lah) A cerebral lobe deep within the lateral sulcus. p. 228

insulin (in′su-lin) A hormone that the pancreatic islets of Langerhans secrete to control carbohydrate metabolism. p. 296

integumentary (in-teg-u-men′tar-e) Pertaining to the skin and its accessory organs. p. 12

intercalated disc (in-ter″kah-lāt′ed disk) Membranous boundary between adjacent cardiac muscle cells. p. 106

intercellular fluid (in″ter-sel′u-lar floo′id) Tissue fluid between cells other than blood cells. p. 482

interneuron (in″ter-nu′ron) A neuron between a sensory neuron and a motor neuron; internuncial or association neuron. p. 212

interphase (in′ter-fāz) Period between two cell divisions when a cell is carrying on its normal functions and prepares for division. p. 65

interstitial cell (in″ter-stish′al sel) A hormone-secreting cell between seminiferous tubules of the testis. p. 499

interstitial fluid (in″ter-stish′al floo′id) Same as intercellular fluid. p. 482

intervertebral disc (in″ter-ver′tĕ-bral disk) A layer of fibrocartilage between bodies of adjacent vertebrae. p. 134

intestinal gland (in-tes′tĭ-nal gland) Tubular gland at the base of a villus within the intestinal wall. p. 413

intestinal juice (in-tes′tĭ-nal jōōs) The fluid that intestinal glands secrete, containing digestive enzymes. p. 413

intracellular (in″trah-sel′u-lar) Within cells. p. 482

intracellular fluid (in″trah-sel′u-lar floo′id) Fluid within cells. p. 482

intramembranous bone (in″trah-mem′brah-nus bōn) Bone that forms from membranelike layers of primitive connective tissue. p. 130

intrinsic factor (in-trin′sik fak′tor) A substance that the gastric glands produce to promote absorption of vitamin B_{12}. p. 403

inversion (in-ver′zhun) Inward turning movement of the sole of the foot. p. 159

involuntary (in-vol′un-tar″e) Not consciously controlled; functions automatically. p. 106

ion (i′on) An atom or molecule with an electrical charge. p. 32

ionic bond (i-on′ik bond) A chemical bond formed between two ions by transfer of electrons. p. 34

iris (i′ris) Colored, muscular portion of the eye that surrounds the pupil and regulates its size. p. 271

ischemia (is-ke′me-ah) Deficiency of blood in a body part. p. 254

isotonic (i″so-ton′ik) A solution with the same osmotic pressure as the solution (usually body fluids) with which it is compared. p. 60

isotope (i′so-tōp) An atom that has the same number of protons as other atoms of an element but has a different number of neutrons. p. 33

J

joint (joint) The union of two or more bones; articulation. p. 156

juxtaglomerular apparatus (juks″tah-glo-mer′u-lar ap″ah-ra′tus) Structure located in the arteriolar walls near the glomerulus that regulates renal blood flow. p. 465

K

keratin (ker′ah-tin) Protein in epidermis, hair, and nails. p. 94

keratinization (ker″ah-tin″ĭ-za′shun) The process by which cells form fibrils of keratin and harden. p. 113

ketone body (ke′tōn bod′e) Type of compound produced during fat catabolism, including acetone, acetoacetic acid, and betahydroxybuteric acid. p. 421

kinase (ki′nās) An enzyme that activates a precursor form of another enzyme by adding a phosphate group. p. 283

Kupffer cell (koop′fer sel) Large, fixed phagocyte in the liver that removes bacterial cells from the blood. p. 408

L

labor (la′bor) The process of childbirth. p. 542

labyrinth (lab′ĭ-rinth) The system of connecting tubes within the inner ear, including the cochlea, vestibule, and semicircular canals. p. 261

lacrimal gland (lak′rĭ-mal gland) Tear-secreting gland. p. 268

lactase (lak′tās) Enzyme that catalyzes breakdown of lactose into glucose and galactose. p. 413

lactation (lak-ta′shun) Production of milk by the mammary glands. p. 299

lacteal (lak′te-al) A lymphatic capillary associated with a villus of the small intestine. p. 413

lactic acid (lak′tik as′id) An organic compound formed from pyruvic acid during anaerobic respiration. p. 179

lacuna (lah-ku′nah) A hollow cavity. p. 103

lamella (lah-mel′ah) A layer of matrix in bone tissue. p. 103

laryngopharynx (lah-ring″go-far′ingks) The lower portion of the pharynx near opening to the larynx. p. 400

larynx (lar′inks) Structure located between the pharynx and trachea that houses the vocal cords. p. 437

latent period (la′tent pe′re-od) Time between application of a stimulus and the beginning of a response in a muscle fiber. p. 182

lateral (lat′er-al) Pertaining to the side. p. 13

leukocyte (lu′ko-sīt) A white blood cell. p. 311

lever (lev′er) A simple mechanical device consisting of a rod, fulcrum, weight, and a source of energy that is applied to some point on the rod. p. 133

LH Luteinizing hormone. p. 288

ligament (lig′ah-ment) A cord or sheet of connective tissue binding two or more bones at a joint. p. 100

limbic system (lim′bik sis′tem) A group of connected structures within the brain that produces emotional feelings. p. 233

lingual (ling′gwal) Pertaining to the tongue. p. 396

lipase (li′pās) A fat-digesting enzyme. p. 405

lipid (lip′id) A fat, oil, or fatlike compound that usually has fatty acids in its molecular structure. p. 39

lipoprotein (lip-o-pro′te-in) A complex of lipid and protein. p. 316

lumbar (lum′bar) Pertaining to the region of the loins. p. 17

lumen (lu′men) Space within a tubular structure, such as a blood vessel or intestine. p. 394

luteinizing hormone (lu′te-in-īz″ing hor′mōn) (**LH; ICSH** in males) A hormone that the anterior pituitary secretes that controls formation of the corpus luteum in females and testosterone secretion in males. p. 288

lymph (limf) Fluid that the lymphatic vessels carry. p. 368

lymphatic pathway (lim-fat′ik path′wa) A pattern of vessels that transport lymph. p. 367

lymph node (limf nōd) A mass of lymphoid tissue. p. 368

lymphocyte (lim′fo-sīt) A type of white blood cell that provides immunity; B cell or T cell. p. 312

lysosome (li′so-sōm) Organelle that contains digestive enzymes. p. 55

M

macrophage (mak′ro-fāj) A large phagocytic cell. p. 99

macula lutea (mak′u-lah lu′te-ah) A yellowish depression in the retina of the eye that is associated with acute vision. p. 272

malignant (mah-lig′nant) The power to threaten life; cancerous. p. 119

malnutrition (mal″nu-trish′un) Physical symptoms resulting from lack of specific nutrients. p. 426

mammary (mam′ar-e) Pertaining to the breast. p. 17

marrow (mar′o) Connective tissue in spaces within bones that includes blood-forming stem cells. p. 129

mast cell (mast sel) A cell to which antibodies, formed in response to allergens, attach, bursting the cell and releasing allergy mediators, which cause symptoms. p. 99

mastication (mas″tĭ-ka′shun) Chewing movements. p. 190

matrix (ma′triks) The intercellular material of connective tissue. p. 98

matter (mat′er) Anything that has weight and occupies space. p. 31

meatus (me-a′tus) A passageway or channel, or the external opening of a passageway. p. 137

mechanoreceptor (mek″ah-no-re-sep′tor) A sensory receptor that senses mechanical stimulation, such as changes in pressure or tension. p. 253

medial (me′de-al) Toward or near the midline. p. 13

mediastinum (me″de-ah-sti′num) Tissues and organs of the thoracic cavity that form a septum between the lungs. p. 8

medulla oblongata (mĕ-dul′ah ob″long-gah′tah) Portion of the brain stem between the pons and the spinal cord. p. 233

medullary cavity (med′u-lār″e kav′ĭ-te) Cavity containing marrow within the diaphysis of long bone. p. 129

megakaryocyte (meg″ah-kar′e-o-sīt) A large bone marrow cell that gives rise to blood platelets. p. 314

meiosis (mi-o′sis) Cell division that halves the genetic material, resulting in egg and sperm cells (gametes). p. 500

melanin (mel′ah-nin) Dark pigment normally found in skin and hair. p. 114

melanocyte (mel′ah-no-sīt) Melanin-producing cell. p. 114

melatonin (mel″ah-to′nin) A hormone that the pineal gland secretes. p. 297

memory cell (mem′o-re sel) B lymphocyte or T lymphocyte produced in the primary immune response that can be activated rapidly if the same antigen is encountered in the future. p. 377

meninx (me′ninks) (plural, *meninges*) A membrane that covers the brain and spinal cord. p. 222

meniscus (men-is′kus) (plural, *menisci*) Fibrocartilage that separates the articulating surfaces of bones in the knee. p. 157

menopause (men′o-pawz) Termination of the menstrual cycle. p. 515

menstrual cycle (men′stroo-al si′kl) Reoccurring changes in the uterine lining of a woman of reproductive age due to cycling hormones. p. 513

menstruation (men″stroo-a′shun) Shedding of blood and other tissue from the uterine lining at the end of a female reproductive cycle. p. 515

merocrine gland (mer′o-krin gland) A gland whose cells secrete a fluid without losing cytoplasm. p. 97

mesentery (mes′en-ter″e) A fold of peritoneal membrane that attaches an abdominal organ to the abdominal wall. p. 412

mesoderm (mez′o-derm) The middle primary germ layer. p. 533

messenger RNA (mes′in-jer) RNA that transmits information for a protein's amino acid sequence from the cell nucleus to the cytoplasm. p. 81

metabolism (mĕ-tab′o-lizm) All chemical changes within cells considered together. p. 4

metacarpal (met″ah-kar′pal) Bone of the hand between the wrist and the finger bones. p. 136

metaphase (met′ah-fāz) Stage in mitosis when chromosomes align in the middle of the cell. p. 65

metatarsal (met″ah-tar′sal) Bone of the foot between the ankle and the toe bones. p. 136

microfilament (mi″kro-fil′ah-ment) A tiny rod of actin protein in cytoplasm that provides structural support or movement. p. 55

microglial cell (mi-krog′le-al sel) A neuroglial cell that supports neurons and phagocytizes. p. 208

microtubule (mi″kro-tu′būl) A minute, hollow rod of the protein tubulin. p. 55

microvillus (mi″kro-vil′us) (plural, *microvilli*) A tiny, cylindrical process that extends from some epithelial cell membranes and increases membrane surface area. p. 93

micturition (mik″tu-rish′un) Urination. p. 475

midbrain (mid′brān) A small region of the brain stem between the diencephalon and the pons. p. 233

mineral (min′er-al) Element not found in organic compounds that is essential in human metabolism. p. 424

mineralocorticoid (min″er-al-o-kor′tĭ-koid) A hormone that the adrenal cortex secretes to influence electrolyte concentrations in body fluids. p. 293

mitochondrion (mi″to-kon′dre-on) (plural, *mitochondria*) Organelle housing enzymes that catalyze reactions of aerobic respiration. p. 54

mitosis (mi-to′sis) Division of a somatic cell to form two genetically identical cells. p. 65

mixed nerve (mikst nerv) Nerve that includes both sensory and motor neuron fibers. p. 220

molecular formula (mo-lek′u-lar fōr′mu-lah) An abbreviation for the number of atoms of each element in a compound. p. 35

molecule (mol′ĕ-kūl) A particle composed of two or more joined atoms. p. 7

monocyte (mon′o-sīt) A type of white blood cell that is a phagocyte. p. 312

monosaccharide (mon″o-sak′ah-rīd) A simple sugar, such as glucose or fructose. p. 39

motor area (mo′tor a′re-ah) The region of the brain from which impulses to muscles or glands originate. p. 228

motor end plate (mo′tor end plāt) Specialized portion of muscle fiber membrane at a neuromuscular junction. p. 175

motor nerve (mo′tor nerv) A nerve that consists of motor neuron fibers. p. 220

motor neuron (mo′tor nu′ron) A neuron that transmits impulses from the central nervous system to an effector. p. 212

motor unit (mo′tor u′nit) A motor neuron and its associated muscle fibers. p. 175

mucosa (mu-ko′sah) The membrane that lines tubes and body cavities that open to the outside of body; mucous membrane. p. 394

mucous cell (mu′kus sel) Glandular cell that secretes mucus. p. 98

mucous membrane (mu′kus mem′brān) Mucosa. p. 113

mucus (mu′kus) Fluid secretion of the mucous cells. p. 98

muscle impulse (mus′el im′puls) Impulse that travels along the sarcolemma to the transverse tubules. p. 176

muscle tissue (mus′el tish′u) Contractile tissue consisting of filaments of actin and myosin, which slide past each other, shortening cells. p. 105

mutagen (mu′tah-jen) An agent that can cause mutations. p. 87

mutation (mu-ta′shun) A change in a gene. p. 87

myelin (mi′ĕ-lin) Fatty material that forms a sheathlike covering around some nerve fibers. p. 209

myocardium (mi″o-kar′de-um) Muscle tissue of the heart. p. 331

myofibril (mi″o-fi′bril) Contractile fibers within muscle cells. p. 173

myoglobin (mi″o-glo′bin) A pigmented compound in muscle tissue that stores oxygen. p. 179

myogram (mi′o-gram) A recording of a muscular contraction. p. 182

myometrium (mi″o-me′tre-um) The layer of smooth muscle tissue within the uterine wall. p. 511

myosin (mi′o-sin) A protein that, with actin, contracts and relaxes muscle fibers. p. 173

N

nasal cavity (na′zal kav′ĭ-te) Space within the nose. p. 435

nasal concha (na′zal kong′kah) Shell-like bone extending out from the wall of the nasal cavity; a turbinate bone. p. 435

nasal septum (na′zal sep′tum) A wall of bone and cartilage that separates the nasal cavity into two portions. p. 435

nasopharynx (na″zo-far′ingks) Portion of the pharynx associated with the nasal cavity. p. 400

negative feedback (neg′ah-tiv fēd′bak) A mechanism activated by an imbalance that corrects it. p. 5

neonatal (ne″o-na′tal) The first four weeks of life. p. 545

nephron (nef′ron) The functional unit of a kidney, consisting of a renal corpuscle and a renal tubule. p. 462

nerve (nerv) A bundle of nerve fibers. p. 207

nerve impulse (nerv im′puls) The electrochemical process of depolarization and repolarization along a nerve fiber. p. 215

nervous tissue (ner′vus tish′u) Neurons and neuroglia. p. 107

neurilemma (nu″rĭ-lem′ah) Sheath formed from Schwann cells on the outside of some nerve fibers. p. 209

neurofibril (nu″ro-fi′bril) A fine, cytoplasmic thread that extends from the cell body into the process of a neuron. p. 209

neuroglial cell (nu-rog′le-ahl sel) Specialized cell of the nervous system that produces myelin, communicates between cells, and maintains the ionic environment. p. 107

neuromuscular junction (nu″ro-mus′ku-lar jungk′shun) Point of contact between a nerve cell and muscle cell. p. 175

neuron (nu′ron) A nerve cell. p. 107

neurotransmitter (nu″ro-trans′mit-er) Chemical that an axon end secretes to stimulate a muscle fiber to contract or a neuron to fire an impulse. p. 217

neutral (nu′tral) Neither acidic nor alkaline. p. 37

neutron (nu′tron) An electrically neutral subatomic particle. p. 31

neutrophil (nu′tro-fil) A type of phagocytic leukocyte. p. 311

niacin (ni′ah-sin) A vitamin of the B-complex group; nicotinic acid. p. 425

nonelectrolyte (non″e-lek′tro-līt) A substance that does not dissociate into ions when dissolved. p. 38

nonprotein nitrogenous substance (non-pro′te-in ni-troj′ĕ-nus sub′stans) A substance, such as urea or uric acid, that contains nitrogen but is not a protein. p. 316

norepinephrine (nor″ep-ĭ-nef′rin) A neurotransmitter released from the axons of some nerve fibers. p. 217

nuclease (nu′kle-ās) An enzyme that catalyzes decomposition of nucleic acids. p. 405

nucleic acid (nu-kle′ik as′id) A substance composed of bonded nucleotides; RNA or DNA. p. 43

nucleolus (nu-kle′o-lus) (plural, *nucleoli*) A small structure within the cell nucleus that contains RNA and proteins. p. 57

nucleotide (nu′kle-o-tīd″) A building block of a nucleic acid molecule, consisting of a sugar, a nitrogenous base, and a phosphate group. p. 43

nucleus (nu′kle-us) (plural, *nuclei*) A cellular organelle enclosed by a double-layered, porous membrane and containing DNA; the dense core of an atom composed of protons and neutrons. pp. 49, 31

nutrient (nu′tre-ent) A chemical that the body requires from the environment. p. 420

O

occipital (ok-sip′ĭ-tal) Pertaining to the lower, back portion of the head. p. 17

olfactory (ol-fak′to-re) Pertaining to the sense of smell. p. 257

olfactory nerves (ol-fak′to-re nervz) First pair of cranial nerves, which conduct impulses associated with the sense of smell. p. 236

oligodendrocyte (ol″ĭ-go-den′dro-sīt) A type of neuroglial cell that forms myelin. p. 208

oocyte (o′o-sīt) An immature egg cell. p. 508

oogenesis (o″o-jen′ĕ-sis) Differentiation of an egg cell. p. 508

optic (op′tik) Pertaining to the eye. p. 269

optic chiasma (op′tik ki-az′mah) X-shaped structure on the underside of the brain; formed by a partial crossing over of optic nerve fibers. p. 232

optic disc (op′tik disk) Region in the retina of the eye where nerve fibers exit to become part of the optic nerve. p. 272

oral (o′ral) Pertaining to the mouth. p. 17

organ (or′gan) A structure consisting of a group of tissues with a specialized function. p. 7

organelle (or″gah-nel′) A part of a cell that performs a specialized function. p. 7

organic (or-gan′ik) Carbon-containing molecules. p. 37

organism (or′gah-nizm) An individual living thing. p. 7

organ system (or′gan sis′tem) A group of organs coordinated to carry on a specialized function. p. 7

orgasm (or′gaz-um) The culmination of sexual excitement. p. 505

orifice (or′ĭ-fis) An opening. p. 533

origin (or′ĭ-jin) End of a muscle that attaches to a relatively immovable part. p. 159

oropharynx (o″ro-far′inks) Portion of the pharynx in the posterior part of the oral cavity. p. 400

osmosis (oz-mo′sis) Diffusion of water through a selectively permeable membrane in response to a concentration gradient. p. 60

osmotic pressure (oz-mot′ik presh′ur) The amount of pressure needed to stop osmosis; a solution's potential pressure caused by nondiffusible solute particles in the solution. p. 60

ossification (os″ĭ-fĭ-ka′shun) The formation of bone tissue. p. 130

osteoblast (os′te-o-blast″) A bone-forming cell. p. 130

osteoclast (os′te-o-klast″) A cell that erodes bone. p. 130

osteocyte (os′te-o-sīt) A mature bone cell. p. 129

osteon (os′te-on) A cylinder-shaped unit containing bone cells that surround a central canal; Haversian system. p. 103

otolith (o′to-lith) A small particle of calcium carbonate associated with the receptors of equilibrium. p. 265

oval window (o′val win′do) Opening between the stapes and the inner ear. p. 261

ovarian (o-va′re-an) Pertaining to the ovary. p. 508

ovary (o′vah-re) The primary female reproductive organ; an egg-cell-producing organ. p. 507

oviduct (o′vĭ-dukt) A tube that leads from the ovary to the uterus; uterine tube or fallopian tube. p. 510

ovulation (o″vu-la′shun) The release of an egg cell from a mature ovarian follicle. p. 510

oxidation (ok″sĭ-da′shun) Process by which oxygen combines with another chemical; the removal of hydrogen or the loss of electrons; opposite of *reduction*. p. 77

oxygen debt (ok′sĭ-jen det) The amount of oxygen required after physical exercise to convert accumulated lactic acid to glucose. p. 179

oxyhemoglobin (ok″sĭ-he″mo-glo′bin) Compound formed when oxygen combines with hemoglobin. p. 451

oxytocin (ok″sĭ-to′sin) A hormone that the posterior pituitary releases to contract smooth muscles in the uterus and mammary glands. p. 288

P

pacemaker (pās′māk-er) Mass of specialized muscle tissue that controls the rhythm of the heartbeat; sinoatrial node. p. 338

pain receptor (pān re″sep′tor) Sensory nerve ending associated with the feeling of pain. p. 253

palate (pal′at) The roof of the mouth. p. 396

palatine (pal′ah-tīn) Pertaining to the palate. p. 141

palmar (pahl′mar) Pertaining to the palm of the hand. p. 17

pancreas (pan′kre-as) Glandular organ in the abdominal cavity that secretes hormones and digestive enzymes. pp. 295, 423

pancreatic (pan″kre-at′ik) Pertaining to the pancreas. p. 404

pancreatic juice (pan″kre-at′ik jōōs) Digestive secretions of the pancreas. p. 404

pantothenic acid (pan″to-then′ik as′id) A vitamin of the B-complex group. p. 425

papilla (pah-pil′ah) Tiny, nipplelike projection. p. 258

papillary muscle (pap′ĭ-ler″e mus′l) Muscle that extends inward from the ventricular walls of the heart and to which the chordae tendineae attach. p. 331

paranasal sinus (par″ah-na′zal si-nus) An air-filled cavity in a cranial bone; lined with mucous membrane and connected to the nasal cavity. p. 436

parasympathetic nervous system (par″ah-sim″pah-thet′ik ner′vus sis′tem) Portion of the autonomic nervous system that arises from the brain and sacral region of the spinal cord. p. 240

parathyroid gland (par″ah-thi′roid gland) One of four small endocrine glands embedded in the posterior portion of the thyroid gland. p. 291

parathyroid hormone (par″ah-thi′roid hor′mōn) **(PTH)** Hormone that the parathyroid glands secrete to help regulate the level of blood calcium and phosphate ions. p. 291

parietal (pah-ri′ĕ-tal) Pertaining to the wall of a cavity. p. 10

parietal cell (pah-ri′ĕ-tal sel) Cell of a gastric gland that secretes hydrochloric acid and intrinsic factor. p. 403

parietal pleura (pah-ri′ĕ-tal ploo′rah) Membrane that lines the inner wall of the thoracic cavity. p. 10

parotid glands (pah-rot′id glandz) Large salivary glands on the sides of the face just in front and below the ears. p. 399

partial pressure (par′shal presh′ur) The pressure one gas produces in a mixture of gases. p. 450

pathogen (path′o-jen) A disease-causing agent. p. 373

pectoral (pek′tor-al) Pertaining to the chest. p. 17

pectoral girdle (pek′tor-al ger′dl) Portion of the skeleton that supports and attaches the upper limbs. p. 147

pelvic (pel′vik) Pertaining to the pelvis. p. 8

pelvic girdle (pel′vik ger′dl) Portion of skeleton to which the lower limbs attach. p. 151

pelvis (pel′vis) Bony ring formed by the sacrum and coxal bones. p. 152

penis (pe′nis) Male external reproductive organ through which the urethra passes. p. 503

pepsin (pep′sin) Protein-splitting enzyme that the stomach gastric glands secrete. p. 403

pepsinogen (pep-sin′o-jen) Inactive form of pepsin. p. 403

peptide bond (pep′tīd bond) Bond between the carboxyl group of one amino acid and the amino group of another. p. 74

pericardial (per″ĭ-kar′de-al) Pertaining to the pericardium. p. 10

pericardium (per″ĭ-kar′de-um) Serous membrane that surrounds the heart. p. 329

perichondrium (per″ĭ-kon′dre-um) Layer of fibrous connective tissue that encloses cartilaginous structures. p. 102

perilymph (per′ĭ-limf) Fluid in the space between the membranous and osseous labyrinths of the inner ear. p. 262

perimetrium (per-ĭ-me′tre-um) The outer serosal layer of the uterine wall. p. 511

perimysium (per″i-mis′e-um) Sheath of connective tissue that encloses a bundle of skeletal muscle fibers. p. 172

perineal (per″ĭ-ne′al) Pertaining to the perineum. p. 17

perineum (per″ĭ-ne′um) Body region between the scrotum or urethral opening and the anus. p. 17

periodontal ligament (per″e-o-don′tal lig′ah-ment) Fibrous membrane that surrounds a tooth and attaches it to the jawbone. p. 398

periosteum (per″e-os′te-um) Fibrous connective tissue covering on the surface of a bone. p. 128

peripheral (pĕ-rif′er-al) Pertaining to parts near the surface or toward the outside. p. 13

peripheral nervous system (PNS) (pĕ-rif′er-al ner′vus sis′tem) The portions of the nervous system outside the central nervous system. p. 207

peripheral resistance (pĕ-rif′er-al re-zis′tans) Resistance to blood flow due to friction between the blood and the walls of the blood vessels. p. 348

peristalsis (per″ĭ-stal′sis) Rhythmic waves of muscular contraction in the walls of certain tubular organs. p. 184

peritoneal (per″ĭ-to-ne′al) Pertaining to the peritoneum. p. 10

peritoneal cavity (per″i-to-ne′al kav′ĭ-te) Potential space between the parietal and visceral peritoneal membranes. p. 10

peritoneum (per″ĭ-to-ne′um) A serous membrane that lines the abdominal cavity and encloses the abdominal viscera. p. 10

peritubular capillary (per″ĭ-tu′bu-lar kap′ĭ-ler″e) Capillary that surrounds a renal tubule and functions in reabsorption and secretion during urine formation. p. 464

pH scale (pH scāl) The negative logarithm of the hydrogen ion concentration used to indicate the acidic or alkaline condition of a solution; values range from 0 to 14. p. 37

phagocytosis (fag″o-si-to′sis) Process by which a cell engulfs and digests solid substances. p. 62

phalanx (fa′langks) (plural, *phalanges*) Bone of a finger or toe. p. 151

pharynx (far′inks) Portion of the digestive tube between the mouth and the esophagus. p. 400

phospholipid (fos″fo-lip′id) A lipid that contains two fatty acid molecules and a phosphate group combined with a glycerol molecule. p. 40

photoreceptor (fo″to-re-sep′tor) A sensory receptor sensitive to light energy; rods and cones. p. 253

physiology (fiz″e-ol′o-je) The study of body functions. p. 3

pia mater (pi′ah ma′ter) Inner layer of meninges that encloses the brain and spinal cord. p. 222

pineal gland (pin′e-al gland) A small structure located in the central part of the brain that secretes the hormone melatonin, which controls certain biological rhythms. p. 297

pinocytosis (pin″o-si-to′sis) Process by which a cell engulfs droplets of fluid from its surroundings. p. 62

pituitary gland (pĭ-tu′ĭ-tār″e gland) Endocrine gland attached to the base of the brain consisting of anterior and posterior lobes. p. 286

placenta (plah-sen′tah) Structure that attaches the fetus to the uterine wall, providing for delivery of nutrients to and removal of wastes from the fetus. p. 530

plantar (plan′tar) Pertaining to the sole of the foot. p. 17

plasma (plaz′mah) Fluid portion of circulating blood. p. 307

plasma cell (plaz′mah sel) Antibody-producing cell that forms when activated B cells proliferate. p. 377

plasma protein (plaz′mah pro′te-in) Proteins dissolved in blood plasma. p. 314

platelet (plāt′let) Cytoplasmic fragment formed in bone marrow that helps blood clot. p. 314

pleural (ploo′ral) Pertaining to pleura or membranes surrounding the lungs. p. 10

pleural cavity (ploo′ral kav′ĭ-te) Potential space between pleural membranes. p. 10

pleural membranes (ploo′ral mem′brānz) Serous membranes that enclose the lungs and line the chest wall. p. 10

plexus (plek′sus) A network of interlaced nerves or blood vessels. p. 239

polar body (po′lar bod′e) Small, nonfunctional cell that is a product of meiosis in the female. p. 508

polarization (po″lar-ĭ-za′shun) An electrical charge on a cell membrane surface due to an unequal distribution of positive and negative ions on either side of the membrane. p. 212

polypeptide (pol″e-pep′tīd) A compound formed by the union of many amino acid molecules. p. 75

polysaccharide (pol″e-sak′ah-rīd) A carbohydrate composed of many joined monosaccharides. p. 39

polyunsaturated fatty acid (pol″e-un-sach′ĕ-ra-ted fat′e as′id) A fatty acid containing one or more double bonds in its carbon atom chain. p. 40

pons (ponz) A portion of the brain stem above the medulla oblongata and below the midbrain. p. 233

popliteal (pop″lĭ-te′al) Pertaining to region behind the knee. p. 17

positive feedback system (poz′ĭ-tiv fēd′bak sis′tem) Process by which changes cause additional similar changes, producing unstable conditions. p. 318

posterior (pos-tēr′e-or) Toward the back; the opposite of *anterior*. p. 13

posterior pituitary (pos-tēr′e-or pĭ-tu′ĭ-tār″e) The lobe of pituitary gland that secretes oxytocin and antidiuretic hormone (vasopressin). p. 286

postganglionic fiber (pōst″gang-gle-on′ik fi′ber) Autonomic nerve fiber on the distal side of a ganglion. p. 241

postnatal (pōst-na′tal) After birth. p. 528

preganglionic fiber (pre″gang-gle-on′ik fi′ber) Autonomic nerve fiber on the proximal side of a ganglion. p. 241

pregnancy (preg′nan-se) The condition in which a female has a developing offspring in her uterus. p. 528

prenatal (pre-na′tal) Before birth. p. 528

primary follicle (pri′ma-re fol′ĭ-kl) Primordial follicle that begins to mature in response to hormonal changes in a female at puberty. p. 509

primary germ layers (pri′ma-re jerm la′ers) Three layers of embryonic cells that develop into specific tissues and organs; ectoderm, mesoderm, and endoderm. p. 533

primary sex organs (pri′ma-re seks or′ganz) Sex-cell-producing parts; testes in males and ovaries in females. pp. 498, 507

prime mover (prīm moov′er) Muscle responsible for a particular body movement. p. 187

PRL Prolactin. p. 287

process (pros′es) A prominent bone projection. p. 137

progesterone (pro-jes′tĕ-rōn) A female hormone that the corpus luteum and placenta secrete. p. 513

projection (pro-jek′shun) Process by which the brain causes a sensation to seem to come from the region of the body being stimulated. p. 253

prolactin (pro-lak′tin) **(PRL)** A hormone that the anterior pituitary secretes to stimulate milk production in the mammary glands. p. 287

pronation (pro-na′shun) Downward or backward rotation of palm of hand. p. 159

prophase (pro′fāz) Stage of mitosis when chromosomes become visible. p. 65

proprioceptor (pro″pre-o-sep′tor) A nerve ending that senses changes in muscle or tendon tension. p. 450

prostaglandins (pros″tah-glan′dins) A group of compounds with powerful, hormonelike effects. p. 284

prostate gland (pros′tāt gland) Gland surrounding the male urethra below the urinary bladder that adds its secretion to semen just prior to ejaculation. p. 502

protein (pro′te-in) Nitrogen-containing organic compound of joined amino acid molecules. p. 41

prothrombin (pro-throm′bin) Plasma protein that functions in the formation of blood clots. p. 318

proton (pro′ton) A positively charged particle in an atomic nucleus. p. 31

protraction (pro-trak′shun) A forward movement of a body part. p. 159

proximal (prok′sĭ-mal) Closer to the midline or origin; opposite of *distal*. p. 13

pseudostratified (soo″do-strat′ĭ-fīd) A single layer of cells appearing as more than one layer because the nuclei occupy different positions in the cells. p. 94

PTH Parathyroid hormone. p. 291

puberty (pu′ber-te) Stage of development in which the reproductive organs become functional. p. 506

pulmonary (pul′mo-ner″e) Pertaining to the lungs. p. 351

pulmonary circuit (pul′mo-ner″e ser′kit) System of blood vessels that carries blood between the heart and the lungs. p. 351

pulse (puls) The surge of blood felt through the walls of arteries due to the contraction of the heart ventricles. p. 347

pupil (pu′pil) Opening in iris through which light enters the eye. p. 271

Purkinje fibers (pur-kin′je fi′berz) Specialized muscle fibers that conduct cardiac impulses from the A-V bundle into the ventricular walls. p. 339

pyramidal cell (pĭ-ram′ĭ-dal sel) A large, pyramid-shaped neuron in the cerebral cortex. p. 228

pyruvic acid (pi-roo′vik as′id) An intermediate product of carbohydrate oxidation. p. 78

R

radioactive (ra″de-o-ak′tiv) An atom that releases energy at a constant rate. p. 33

rate-limiting enzyme (rāt lim′i-ting en′zīm) An enzyme, usually present in small amounts, that controls the rate of a metabolic pathway by regulating one of its steps. p. 79

recruitment (re-kroōt′ment) Increase in number of motor units activated as stimulation intensity increases. p. 183

red marrow (red mar′o) Blood-cell-forming tissue in spaces within bones. p. 133

referred pain (re-ferd′ pān) Pain that feels as if it is originating from a part other than the site being stimulated. p. 255

reflex (re′fleks) A rapid, automatic response to a stimulus. p. 220

reflex arc (re′fleks ark) A nerve pathway, consisting of a sensory neuron, interneuron, and motor neuron, that forms the structural and functional bases for a reflex. p. 220

refraction (re-frak′shun) A bending of light as it passes between media of different densities. p. 273

relaxin (re-lak′sin) Hormone from the corpus luteum that inhibits uterine contractions during pregnancy. p. 532

renal (re′nal) Pertaining to the kidney. p. 461

renal corpuscle (re′nal kor′pusl) Part of a nephron that consists of a glomerulus and a glomerular capsule. p. 463

renal cortex (re′nal kor′teks) The outer portion of a kidney. p. 462

renal medulla (re′nal mĕ-dul′ah) The inner portion of a kidney. p. 462

renal pelvis (re′nal pel′vis) The cavity in a kidney that channels urine to the ureter. p. 461

renal tubule (re′nal tu′būl) Portion of a nephron that extends from the renal corpuscle to the collecting duct. p. 463

renin (re′nin) Enzyme that kidneys release to maintain blood pressure and blood volume. p. 468

replication (rep″lĭ-ka′shun) Production of an exact copy of a DNA molecule. p. 86

reproduction (re″pro-duk′shun) Offspring formation. p. 4

respiration (res″pĭ-ra′shun) Cellular process that releases energy from nutrients; breathing. p. 4

respiratory capacities (re-spi′rah-to″re kah-pas′ĭ-tēz) Value obtained by adding two or more respiratory volumes. p. 446

respiratory center (re-spi′rah-to″re sen′ter) Portion of the brain stem that controls breathing depth and rate. p. 447

respiratory membrane (re-spi′rah-to″re mem′brān) Membrane composed of a capillary wall and an alveolar wall through which blood and inspired air exchange gases. p. 449

response (re-spons′) The action resulting from a stimulus. p. 216

resting potential (res′ting po-ten′shal) The difference in electrical charge between the inside and the outside of an undisturbed nerve cell membrane. p. 213

reticular formation (rĕ-tik′u-lar for-ma′shun) A complex network of nerve fibers within the brain stem that arouses the cerebrum. p. 233

retina (ret′ĭ-nah) Inner layer of the eye wall that contains the visual receptors. p. 272

retinal (ret′ĭ-nal) A form of vitamin A; retinene. p. 274

retinene (ret′ĭ-nēn) Chemical precursor of rhodopsin, a visual pigment. p. 274

retraction (rĕ-trak′shun) Movement of a part toward the back. p. 159

retroperitoneal (ret″ro-per″i-to-ne′al) Located behind the peritoneum. p. 461

reversible reaction (re-ver′sĭ-bl re-ak′shun) Chemical reaction in which the end products can change back into the reactants. p. 36

rhodopsin (ro-dop′sin) Light-sensitive pigment in the rods of the retina; visual purple. p. 274

riboflavin (ri″bo-fla′vin) A vitamin of the B-complex group; vitamin B_2. p. 425

ribonucleic acid (ri″bo-nu-kle′ik as′id) **(RNA)** Nucleic acid whose nucleotides each contain the sugar ribose, a phosphate group, and a nitrogenous base (adenine, uracil, guanine, or cytosine). p. 43

ribose (ri′bōs) A five-carbon sugar in RNA. p. 81

ribosome (ri′bo-sōm) Organelle composed of RNA and protein that is a structural support for protein synthesis. p. 53

RNA Ribonucleic acid. p. 43

rod (rod) A type of receptor that provides colorless vision. p. 273

rotation (ro-ta′shun) Movement turning a body part on its longitudinal axis. p. 159

round window (rownd win′do) A membrane-covered opening between the inner ear and the middle ear. p. 262

S

sagittal (saj′ĭ-tal) A plane or section that divides a structure into right and left portions. p. 14

S-A node (nōd) Sinoatrial node. p. 338

sarcomere (sar′ko-mēr) The structural and functional unit of a myofibril. p. 174

sarcoplasmic reticulum (sar″ko-plaz′mik rĕ-tik′u-lum) Membranous network of channels and tubules within a muscle fiber, corresponding to the endoplasmic reticulum of other cells. p. 174

saturated fatty acid (sat′u-rāt″ed fat′e as′id) Fatty acid molecule that contains as many hydrogen as possible, and therefore has no double bonds between its carbon atoms. p. 40

Schwann cell (shwahn sel) A type of neuroglial cell that surrounds a fiber of a peripheral neuron, forming the neurilemmal sheath and myelin. p. 209

sclera (skle′rah) White, fibrous outer layer of the eyeball. p. 269

scrotum (skro′tum) A pouch of skin that encloses the testes. p. 503

sebaceous gland (se-ba′shus gland) Skin gland that secretes sebum. p. 118

sebum (se′bum) Oily secretion of sebaceous glands. p. 118

secretin (se-kre′tin) Hormone that the small intestine secretes to stimulate the pancreas to release pancreatic juice. p. 405

selectively permeable (se-lek′tiv-le per′me-ah-bl) A membrane that allows some molecules through but not others; semipermeable. p. 51

semen (se′men) Fluid containing sperm cells and secretions discharged from male reproductive tract at ejaculation. p. 502

semicircular canal (sem″ĭ-ser′ku-lar kah-nal′) Tubular structure within the inner ear that contains receptors providing the sense of dynamic equilibrium. p. 266

seminiferous tubule (sem″ĭ-nif′er-us tu′būl) Tubule within the testes where sperm cells form. p. 498

sensation (sen-sa′shun) A feeling resulting from the brain's interpretation of sensory nerve impulses. p. 253

sensory adaptation (sen′so-re ad″ap-ta′shun) Sensory receptors becoming less responsive after constant repeated stimulation. p. 253

sensory area (sen′so-re a′re-ah) A portion of the cerebral cortex that receives and interprets sensory nerve impulses. p. 229

sensory nerve (sen′so-re nerv) A nerve composed of sensory nerve fibers. p. 220

sensory neuron (sen′so-re nu′ron) A neuron that transmits an impulse from a receptor to the central nervous system. p. 212

sensory receptor (sen′so-re re″sep′tor) A specialized structure associated with the peripheral end of a sensory neuron specific to detecting a particular sensation and triggering a nerve impulse in response, which is transmitted to the central nervous system. p. 220

serotonin (se″ro-to′nin) A vasoconstrictor that blood platelets release when blood vessels break, controlling bleeding. Also a neurotransmitter. pp. 317, 217

serous cell (ser′us sel) Glandular cell that secretes a watery fluid with high enzyme content. p. 98

serous fluid (ser′us floo′id) The secretion from a serous cell. p. 98

serous membrane (ser′us mem′brān) Membrane that lines a cavity without an opening to the outside of the body. p. 113

simple sugar (sim′pl shoog′ar) Monosaccharide. p. 39

sinoatrial node (si″no-a′tre-al nōd) **(S-A node)** Specialized tissue in the wall of the right atrium that initiates cardiac cycles; pacemaker. p. 338

sinus (si′nus) A cavity or space in a bone or other body part. p. 137

skeletal muscle (skel′ĭ-tal mus′l) Type of voluntary muscle tissue in muscles attached to bones. p. 105

smooth muscle (smooth mus′l) Type of involuntary muscle tissue in the walls of hollow viscera. p. 105

solute (sol′ūt) The substance dissolved in a solution. p. 60

solution (so-lu′shun) Homogenous mixture of substances (solutes) within a dissolving medium (solvent). p. 60

solvent (sol′vent) The liquid portion of a solution in which a solute is dissolved. p. 60

somatic nervous system (so-mat′ik ner′vus sis′tem) Motor pathways of peripheral nervous system that lead to the skin and skeletal muscles. p. 235

special sense (spesh′al sens) Sense that involves receptors associated with specialized sensory organs, such as the eyes and ears. p. 253

spermatid (sper′mah-tid) An intermediate stage in sperm cell formation. p. 500

spermatocyte (sper-mat′o-sīt) An early stage in sperm cell formation. p. 500

spermatogenesis (sper″mah-to-jen′ĕ-sis) Sperm cell production. p. 500

spermatogonium (sper″mah-to-go′ne-um) Undifferentiated spermatogenic cell in the germinal epithelium of a seminiferous tubule. p. 500

sphincter (sfingk′ter) A circular muscle that closes an opening or the lumen of a tubular structure. p. 402

spinal cord (spi′nal kord) Portion of the central nervous system extending from the brain stem through the vertebral canal. p. 224

spinal nerve (spi′nal nerv) Nerve that arises from the spinal cord. p. 224

spleen (splēn) A large organ in the upper left region of the abdomen. p. 371

spongy bone (spunj′e bōn) Bone that consists of bars and plates separated by irregular spaces; cancellous bone. p. 128

squamous (skwa′mus) Flat or platelike. p. 92

starch (starch) A polysaccharide common in foods of plant origin. p. 39

static equilibrium (stat′ik e″kwĭ-lib′re-um) The maintenance of balance when the head and body are motionless. p. 265

steroid (ste′roid) A type of organic molecule including complex rings of carbon and hydrogen atoms. p. 40

stimulus (stim′u-lus) (plural, *stimuli*) A change in environment that triggers a response from an organism or a cell. p. 213

stomach (stum′ak) Digestive organ between the esophagus and small intestine that stores food. p. 402

stratified (strat′ĭ-fīd) Arranged in layers. p. 94

stratum basale (strat′tum ba′sal-e) The deepest layer of the epidermis, in which cells divide; stratum germinativum. p. 113

stratum corneum (stra′tum kor′ne-um) Outer, horny layer of the epidermis. p. 114

stressor (stres′or) A factor capable of stimulating a stress response. p. 299

stroke volume (strōk vol′ūm) The amount of blood the ventricle discharges with each heartbeat. p. 348

structural formula (struk′cher-al for′mu-lah) A representation of the way atoms bond to form a molecule, using symbols for each element and lines to indicate chemical bonds. p. 36

subarachnoid space (sub″ah-rak′noid spās) The space within the meninges between the arachnoid mater and the pia mater. p. 222

subcutaneous (sub″ku-ta′ne-us) Beneath the skin. p. 113

sublingual (sub-ling′gwal) Beneath the tongue. p. 400

submucosa (sub″mu-ko′sah) Layer of connective tissue underneath a mucous membrane. p. 394

substrate (sub′strāt) The substance upon which an enzyme acts. p. 75

sucrase (su′krās) Digestive enzyme that catalyzes the breakdown of sucrose. p. 413

sucrose (soo′krōs) A disaccharide; table sugar. p. 39

sulcus (sul′kus) (plural, *sulci*) A shallow groove, such as that between the convolutions on the brain surface. p. 226

summation (sum-ma′shun) Increased force of contraction by a skeletal muscle fiber when twitches occur so rapidly that the next twitch occurs before the previous twitch relaxes. p. 182

superficial (soo″per-fish′al) Near the surface. p. 13

superior (soo-pe′re-or) Structure higher than another structure. p. 13

supination (soo″pĭ-na′shun) Forearm rotation so palm faces upward when the arm is outstretched. p. 159

surface tension (sur′fis ten′shun) The force that holds moist membranes together due to the attraction of water molecules. p. 443

surfactant (ser-fak′tant) Substance produced by the lungs that reduces the surface tension within the alveoli. p. 443

suture (soo′cher) An immovable joint, such as that between flat bones of skull. p. 137

sweat gland (swet gland) Exocrine gland in skin that secretes a mixture of water, salt, urea, and other bodily wastes. p. 119

sympathetic nervous system (sim″pah-thet′ik ner′vus sis′tem) Portion of the autonomic nervous system that arises from the thoracic and lumbar regions of the spinal cord. p. 240

symphysis (sim′fĭ-sis) A slightly movable joint between bones separated by a pad of fibrocartilage. p. 157

synapse (sin′aps) The junction between the axon of one neuron and the dendrite or cell body of another neuron. p. 216

synaptic knob (sĭ-nap′tik nob) Tiny enlargement at the end of an axon that secretes a neurotransmitter. p. 217

syncytium (sin-sish′e-um) A mass of merging cells. p. 338

synergist (sin′er-jist) A muscle that assists the action of a prime mover. p. 187

synovial fluid (sĭ-no′ve-al floo′id) Fluid that the synovial membrane secretes. p. 157

synovial joint (sĭ-no′ve-al joint) A freely movable joint. p. 157

synovial membrane (sĭ-no′ve-al mem′brān) Membrane that forms the inner lining of the capsule of a freely movable joint. p. 113

synthesis (sin′thĕ-sis) Building large molecules from smaller ones that join. p. 36

systemic circuit (sis-tem′ik ser′kit) The vessels that conduct blood between the heart and all body tissues except the lungs. p. 351

systole (sis′to-le) Phase of the cardiac cycle when a heart chamber wall contracts. p. 336

systolic pressure (sis-tol′ik presh′ur) Arterial blood pressure during the systolic phase of the cardiac cycle. p. 347

T

target cell (tar′get sel) A specific cell on which a hormone exerts its effect. p. 281

tarsus (tar′sus) The ankle bones. p. 155

taste bud (tāst bud) Organ containing receptors associated with the sense of taste. p. 258

T cell (sel) A type of lymphocyte that interacts directly with antigen-bearing particles, producing the cellular immune response. p. 371

telophase (tel′o-fāz) Stage in mitosis when newly formed cells separate. p. 65

tendon (ten′don) A cordlike or bandlike mass of white fibrous connective tissue that connects a muscle to a bone. p. 100

testis (tes′tis) (plural, *testes*) Primary male reproductive organ; sperm-cell-producing organ. p. 498

testosterone (tes-tos′tĕ-rōn) Male sex hormone that the interstitial cells of the testes secrete. p. 506

tetanic contraction (tet′ah-nik kon-trak′shun) Continuous, forceful muscular contraction without relaxation. p. 182

thalamus (thal′ah-mus) A mass of gray matter at base of the cerebrum in the wall of the third ventricle. p. 232

thermoreceptor (ther″mo-re-sep′tor) A sensory receptor sensitive to temperature changes; heat and cold receptors. p. 253

thiamine (thi′ah-min) Vitamin B_1. p. 425

thoracic (tho-ras′ik) Pertaining to the chest. p. 8

threshold stimulus (thresh′old stim′u-lus) Stimulation level that must be exceeded to elicit a nerve impulse or a muscle contraction. p. 182

thrombus (throm′bus) A blood clot that remains at its formation site in a blood vessel. p. 318

thymosins (thi′mo-sins) A group of peptides that the thymus gland secretes to increase production of certain types of white blood cells. p. 297

thymus (thi′mus) A glandular organ in the mediastinum behind the sternum and between the lungs. p. 297

thyroid gland (thi′roid gland) Endocrine gland located just below the larynx and in front of the trachea that secretes thyroid hormones. p. 289

thyroxine (thi-rok′sin) One hormone that the thyroid gland secretes. p. 289

tissue (tish′u) A group of similar cells that performs a specialized function. p. 7

trachea (tra′ke-ah) Tubular organ that leads from the larynx to the bronchi. p. 438

transcellular fluid (trans″sel′u-lar floo′id) A portion of the extracellular fluid, including the fluid within special body cavities. p. 482

transcription (trans-krip′shun) Manufacturing a complementary RNA from DNA. p. 81

transfer RNA (trans′fer) RNA molecule that carries an amino acid to a ribosome in protein synthesis. p. 83

translation (trans-la′shun) Assembly of an amino acid chain according to the sequence of base triplets in an mRNA molecule. p. 83

transverse tubule (trans-vers′ tu′būl) Membranous channel that extends inward from a muscle fiber membrane and passes through the fiber. p. 174

tricuspid valve (tri-kus′pid valv) Heart valve between the right atrium and the right ventricle. p. 331

triglyceride (tri-glis′er-īd) A lipid composed of three fatty acids combined with a glycerol molecule. p. 40

triiodothyronine (tri″i-o″do-thi′ro-nēn) A type of thyroid hormone. p. 289

trochanter (tro-kan′ter) A broad process on a bone. p. 137

trypsin (trip′sin) An enzyme in pancreatic juice that breaks down protein molecules. p. 405

tubercle (tu′ber-kl) A small, rounded process on a bone. p. 137

tuberosity (tu″bĕ-ros′ĭ-te) An elevation or protuberance on a bone. p. 137

twitch (twich) A brief muscular contraction followed by relaxation. p. 182

tympanic membrane (tim-pan′ik mem′brān) A thin membrane that covers the auditory canal and separates the external ear from the middle ear; the eardrum. p. 261

U

umbilical cord (um-bil′ĭ-kal kord) Cordlike structure that connects the fetus to the placenta. p. 537

umbilical region (um-bil′ĭ-kal re′jun) The central portion of the abdomen. p. 14

unsaturated fatty acid (un-sat′u-rāt″ed fat′e as′id) Fatty acid molecule with one or more double bonds between the atoms of its carbon chain. p. 40

urea (u-re′ah) A nonprotein nitrogenous substance resulting from protein metabolism. p. 422

ureter (u-re′ter) A muscular tube that carries urine from the kidney to the urinary bladder. p. 473

urethra (u-re′thrah) Tube leading from the urinary bladder to the outside of body. p. 475

urine (u′rin) Wastes and excess water removed from the blood and excreted by the kidneys into the ureters, to the urinary bladder, and out of the body through the urethra. p. 465

uterine (u′ter-in) Pertaining to the uterus. p. 510

uterine tube (u′ter-in tūb) Tube that extends from the uterus on each side toward an ovary and transports sex cells; fallopian tube or oviduct. p. 510

uterus (u′ter-us) Hollow, muscular organ within the female pelvis in which a fetus develops. p. 511

utricle (u′trĭ-kl) An enlarged portion of the membranous labyrinth of the inner ear. p. 265

uvula (u′vu-lah) A fleshy portion of the soft palate that hangs down above the root of the tongue. p. 396

V

vaccine (vak′sēn) A substance that contains antigens used to stimulate an immune response. p. 381

vagina (vah-ji′nah) Tubular organ that leads from the uterus to the vestibule of the female reproductive tract. p. 511

vascular (vas′ku-lar) Pertaining to blood vessels. p. 329

vas deferens (vas def′er-ens) (plural, *vasa deferentia*) Tube that leads from the epididymis to the urethra of the male reproductive tract. p. 501

vasoconstriction (vas″o-kon-strik′shun) A decrease in the diameter of a blood vessel. p. 342

vasodilation (vas″o-di-la′shun) An increase in the diameter of a blood vessel. p. 342

vein (vān) A vessel that carries blood toward the heart. p. 346

vena cava (ve′nah kav′ah) One of two large veins (superior and inferior) that convey deoxygenated blood to the right atrium of the heart. p. 331

ventral root (ven′tral root) Motor branch of a spinal nerve by which it attaches to the spinal cord. p. 239

ventricle (ven′trĭ-kl) A cavity, such as brain ventricles filled with cerebrospinal fluid, or heart ventricles that contain blood. pp. 230, 331, 347

venule (ven′ūl) A vessel that carries blood from capillaries to a vein. p. 346

vesicle (ves'ĭ-kl) Membranous cytoplasmic sac formed by infolding of the cell membrane. p. 56

villus (vil'us) (plural, *villi*) Tiny, fingerlike projection that extends outward from the lining of the small intestine. p. 412

visceral (vis'er-al) Pertaining to the contents of a body cavity. p. 10

visceral peritoneum (vis'er-al per'ĭ-to-ne'um) Membrane that covers organ surfaces within the abdominal cavity. p. 10

visceral pleura (vis'er-al ploo'rah) Membrane that covers the surfaces of the lungs. p. 10

viscosity (vis-kos'ĭ-te) The tendency for a fluid to resist flowing due to the internal friction of its molecules. p. 348

vitamin (vi'tah-min) An organic compound, other than a carbohydrate, lipid, or protein, required for normal metabolism but that the body cannot synthesize in adequate amounts. p. 423

vitreous humor (vit're-us hu'mor) Substance between the lens and the retina of the eye. p. 272

vocal cords (vo'kal kordz) Folds of tissue within the larynx that vibrate and produce sounds. p. 437

voluntary (vol'un-tār"e) Capable of being consciously controlled. p. 105

vulva (vul'vah) The external female reproductive parts that surround the vaginal opening. p. 512

W

water balance (wot'er bal'ans) When the volume of water entering the body is equal to the volume leaving it. p. 484

water of metabolism (wot'er uv mĕ-tab'o-lizm) Water produced as a by-product of metabolic processes. p. 484

Y

yellow marrow (yel'o mar'o) Fat storage tissue found in certain bone cavities. p. 133

Z

zygote (zi'got) Cell produced when an egg and a sperm fuse; fertilized egg cell. p. 508

zymogen granule (zi-mo'jen gran'ūl) A cellular structure that stores inactive forms of protein-splitting enzymes in a pancreatic cell. p. 405

Credits

PHOTOS

Chapter 1
Opener: Andreas G. Nerlich; **1Aa:** © SPL / Photo Researchers; **1Ab:** © CNRI / SPL / Photo Researchers.

Chapter 2
Opener: © P.M. Motta & S. Correr / Science Photo Library / Photo Researchers; **2A:** © Mark Antman / The Image Works; **2Ba:** © SIU / Peter Arnold; **2.8A,B, 2.17:** Courtesy of John W. Hole, Jr.

Chapter 3
Opener: © AP / Wide World Photos; **3.4A:** © Dr. Don Fawcett / Photo Researchers; **3.5A:** © Gordon Leedale / Biophoto Associates; **3.6A:** © Bill Longcore / Photo Researchers; **3.8A:** © Don W. Fawcett / Visuals Unlimited; **3.9A:** © Oliver Meckes / Photo Researchers; **3.9B:** © Manfred Kage / Peter Arnold; **3.10B:** © Stephen L. Wolfe; **3.15A–C:** © David M. Phillips / Visuals Unlimited; **3.21A–E:** © Ed Reschke; **3A:** © Dr. Tony Brain / SPL / Photo Researchers.

Chapter 4
Opener: © Superstock.

Chapter 5
Opener: © Federation of American Societies for Experimental Biology / Photo Courtesy of Niclas L'Heureux; **5.3b:** Victor Eroschenko; **5.4b:** © Ed Reschke; **5.5B:** © G.W. Willis M.D., / Visuals Unlimited; **5.6B:** Victor Eroschenko; **5.7B:** © Richard Kessel / Visuals Unlimited; **5.8B, D:** © Ed Reschke; **5.10:** © David M. Phillips / Visuals Unlimited; **5.11:** © Manfred Kage / Peter Arnold, Inc.; **5.12:** © Veronica Burmeistre / Visuals Unlimited; **5.13B – 5.15B:** © Ed Reschke; **5.16B:** © Fred Hossler / Visuals Unlimited; **5.17B:** © Ed Reschke / Peter Arnold; **5.18B:** © R. Calentine / Visuals Unlimited; **5.19B:** © Victor B. Eichler, Ph.D.; **5.19C:** © Biophoto Associates / Photo Researchers; **5.20B, 5.21B, 5.22B:** © Ed Reschke; **5.23B:** © Manfred Kage / Peter Arnold; **5.24B:** © Ed Reschke.

Chapter 6
Opener: © Owen Franken / Stock Boston; **6.1B:** Victor Eroschenko; **6.2:** © Victor B. Eichler, Ph.D.; **6.3A:** © M. Schliwa / Visuals Unlimited; **6.4B:** © Victor B. Eichler, Ph.D.; **6.5:** © CNRI / SPL / Photo Researchers; **6.6:** © Per H. Kjeldsen; **6A(a–c):** © SPL / Photo Researchers.

Chapter 7
Opener: © John Reader / SPL / Photo Researchers; **7.2:** © Ed Reschke; **7.4:** © Biophoto Associates / Photo Researchers; **7.21B:** Courtesy, Eastman Kodak; **7.26B:** © Martin M. Rotker.

Reference Plates
Plate 8: © McGraw-Hill Higher Education/Photo by Jim Womack; **Plate 9:** © McGraw-Hill Higher Education/Photo by Jim Womack; **Plate 10:** Courtesy of John W. Hole, Jr.; **Plate 11:** McGraw-Hill Higher Education/Photo by Jim Womack.

Chapter 8
Opener: OSF/PLUMTRE, A. / Animals Animals; **8.3A, 8.9B:** © H.E. Huxley; **8A:** © Gianni Giansanti / Corbis; **8B:** © Muscular Dystrophy Association; **Page 204 (both):** © John Carl Mese.

Chapter 9
Opener: Walt Disney (Photo courtesy Kobal Collection); **9.1:** © Ed Reschke; **9.13:** © Don Fawcett / Photo Researchers; **9.21B:** © Per H. Kjeldsen / University of Michigan; **9.24:** © Martin M. Rotker / Photo Researchers, Inc.

Chapter 10
Opener: Reprinted with Permission from the Democrat and Chronicle; **10.7:** Dave Roberts / Science Photo Library / Photo Researchers, Inc.; **10.10:** © Fred Hossler / Visuals Unlimited; **10.22:** © Per H. Kjeldsen / University of Michigan; **10.23A:** © Carroll Weiss / Camera M.D. Studios; **10.25C:** © Frank S. Werblin.

Chapter 11
Opener: © Superstock; **11.15:** © Ed Reschke; **11A:** Courtesy, Eli Lilly and Company Archives.

Chapter 12
Opener: © Kathy Guise, Courtesy Apex Bioscience; **12.2B:** © Bill Longcore / Photo Researchers; **12.6–12.10:** © Ed Reschke; **12A(b):** Andrew Syred / Science Photo Library / Photo Researchers, Inc.; **12A(a):** © Joaquin Carrillo-Farga / Photo Researchers, Inc.; **12.13:** © SPL / Photo Researchers; **12.14A:** Peter Arnold, Inc. / Ed Reschke; **12.14B:** © Camera M.D. Studios; **12.17C:** © G. W. Willis, MD / Visuals Unlimited; **12.17D:** © George W. Wilder / Visuals Unlimited.

Chapter 13
Opener: © UPI / Corbis-Bettmann; **13.5:** © The McGraw-Hill Companies, Inc. / University of Michigan Biomedical Communications; **13.21:** © Don W. Fawcett, from Bloom, W. and Fawcett, D.W., TEXTBOOK OF HISTOLOGY, 10th edition, W.B. Saunders Co., 1975 photo by T. Kuwabara; **13A:** © Ed Reschke / Peter Arnold; **13B:** © J & L Weber / Peter Arnold; **13C:** © Alfred Pasieka / Peter Arnold; **13.23:** © Carolina Biological / Visuals Unlimited.

Chapter 14
Opener: © Jim Olive / Peter Arnold; **14.3:** © Ed Reschke; **14.7:** © Dr. Kent M. Van De Graaff; **14.9B:** © John Cunningham / Visuals Unlimited; **14.10B:** © Biophoto Associates / Photo Researchers; **14.12:** Courtesy of Sloan-Kettering; **14A:** Institute for Cancer Research / Etienne de Harven, M.D., and Nina Lampen / © Corbis; **14B:** © The Courier-Journal; **14C:** © 1995 Jessica Boyatt; **14D:** © Los Angeles Times Photo by Tammy Lechner.

Chapter 15

Opener: PhotoDisc #LS014016; **15.7:** © VideoSurgery / Photo Researchers, Inc.; **15.12B:** © Ed Reschke; **15.17C:** © Victor B. Eichler, Ph.D.; **15.19:** © Carroll Weiss / Camera M.D. Studios; **15.22:** © Armed Forces Institute of Pathology; **15.25:** © Manfred Kage / Peter Arnold; **15.28:** © James L. Shaffer; **15.30:** © Ed Reschke; **15A:** SPECIAL COLLECTIONS DIVISION UNIVERSITY OF WASHINGTON LIBRARIES Neg. No. UW 17931.

Chapter 16

Opener: © John Lawlor; **16.3B:** © Biophoto Associates / Photo Researchers, Inc., **16.5C:** © CNRI / Phototake; **16.9:** © McGraw-Hill Higher Education / Bob Coyle, photographer; **16A(a, b):** © Victor B. Eichler, Ph.D.

Chapter 17

Opener: © Gerald Zanetti / CorbisStockMarket.com; **17.4:** Copyright by R.G. Kessel and R.H. Kardon, *Tissues and Organs: A Text-Atlas of Scanning Electron Microscopy*. W.H. Freeman, 1979, all rights reserved; **17.5A:** © Biophoto Associates / Photo Researchers; **17.5B:** © Manfred Kage / Peter Arnold; **17.13:** © Per H. Kjeldsen; **17.15:** © Ed Reschke.

Chapter 18

Opener: AP Photo / File, Tom Olmscheid.

Chapter 19

Opener: © Dr. Tony Brain / SPL / Photo Researchers; **19.2C:** © Biophoto Associates / Photo Researchers; **19.10:** © Ed Reschke / Peter Arnold; **19.12:** © The McGraw-Hill Companies, Inc. / Carol D. Jacobson, Ph.D., Department of Veterinary Anatomy, Iowa State University; **19.15A–E:** © McGraw-Hill Higher Education / Bob Coyle, photographer; **19B:** Courtesy, Southern Illinois University School of Medicine.

Chapter 20

Opener: © Sean O'Brien / Custom Medical Stock Photo; **20.1:** From M. Tegner and D. Epel, "Sea Urchin Sperm: Eggs Interaction Studied with the Scanning Electron Microscope," *Science* 179: 685–688. © 1973 by the AAAS; **20.4A:** © A. Tsiaras / Photo Researchers; **20.4B:** © Omikron / Photo Researchers; **20.4C:** © Petit Format / Nestle / Photo Researchers; **20A:** © *People Weekly* © Taro Yamasaki; **20.9, 20.13:** © Donald Yaeger / Camera M.D. Studios; **20Ba–c:** From Streissguth, A.P., Landesman-Dwyer, S., Martin, J.C., & Smith, D.W. (1980). "Teratogenic effects of alcohol in human and laboratory animals." *Science* 209 (18): 353–361. © 1980 American Association for the Advancement of Science.

Index

APPLICATION INDEX

BOXED READINGS

A Brief History of Constipation, 419
A New Muscle Discovered?, 190
Assisted Reproductive Technologies, 535
Atherosclerosis, 344
Biological Rhythms, 299
Cancer, 67
Coagulation Disorders, 319
Conquering Inherited Immune Deficiency—Children Who Made Medical History, 384–85
Cystic Fibrosis, 454
Dental Caries, 399
Diabetes Mellitus, 297
Drug Abuse, 235
Elevated Body Temperature, 122
Emphysema and Lung Cancer, 444
Exercise and Breathing, 450
Exercise and the Cardiovascular System, 350
Factors Affecting Synaptic Transmission, 218
Faulty Ion Channels Cause Inherited Disease, 52
Female Infertility, 532
Fetal Chromosome Checks, 542
Headache, 258
Hepatitis, 409
Immunity Breakdown: AIDS, 382–83
Inherited Diseases of Muscle, 186
Leukemia, 315
Male Infertility, 504
Mutations, 87
Of Genomes, Chips and SNPs, 84
Osteoporosis, 135
Prostate Enlargement, 503
Radioactive Isotopes, 33
Repair of a Bone Fracture, 132
Skin Cancer, 119
Sodium and Potassium Imbalances, 488
Some Causes of Birth Defects, 540–41
Steroids and Athletes—An Unhealthy Combination, 181
Treating Breast Cancer, 518–19
Use and Disuse of Skeletal Muscles, 183
Water Balance Disorders, 485–86

CLINICAL APPLICATIONS

Acidosis, 37
Acne, 118
Addison Disease, 295
Adrenal Tumor, 293
Alkalosis, 37
Amblyopia, 268
Amenorrhea, 513
Anabolic Steroids, 75
Anaphylactic Shock, 382
Anesthetics, 215
Angina Pectoris, 333
Angiotensin Converting Enzyme Inhibitors, 469
Anosmia, 258
Appendicitis, 417
Arthroscopy, 161
Atherosclerosis, 40
Automatic Bladder, 475
Blood-Brain Barrier, 345
Bone Marrow Transplant, 134
Botulism, 178
Breath Analysis, 450
Calories, 423
Carbon Dioxide, 448
Carcinomas, 97
Cataract, 270
Cell Proliferation, 107
Cerebral Cortex Injury, 229
Cerebral Vascular Accident, 349
Cleft Palate, 141
Collagen Injections, 100
Computerized Tomography, 15
Corneal Transplant, 270
Coronary Thrombosis, 349
Cranial Nerve Injury, 237
Cushing Syndrome, 295
Cyanosis, 307
Cystic Fibrosis, 406
Cystitis, 474
Deafness, 264
Deviated Septum, 435
Diabetes Insipidus, 288
Diplopia, 268
Diuretics, 487
DNA Fingerprinting, 81
Dwarfism, 287
Ear Infections, 261
Edema, 316
Endocrine Paraneoplastic Syndrome, 296
Endosymbiont Theory, 54
Epiphyseal Injury, 131
Episiotomy, 543
Erythroblastosis Fetalis, 323
Fat-Free Foods, 422
Fertilization, 529
Floaters, 272
Gastric Ulcer, 403
Gastrin, 403
Glaucoma, 271
Glomerulonephritis, 468
Glucosuria, 470, 473
Gout, 471
Heart Sounds, 338
Heart Transplant, 336
Hemodialysis, 59
Hemorrhoids, 418
Herniated Disc, 145
Hiatal Hernia, 402
Hip Fracture, 154
Huntington Disease, 230
Hypertension, 349, 469
Hyperthyroidism, 290, 292
Hypertonic Solutions, 61
Hyperventilation, 449
Hypothyroidism, 290, 292
Hypotonic Solutions, 61
Hypoxemia, 212
Immunity, 373
Injections, 113
Intermediate Pituitary Lobe, 286
Intestinal Cellular Turnover, 413
Ischemia, 212
Jaundice, 410
Kidney Stones, 473
Kidney Transplant, 463
Labor Induction, 288
Lactose Intolerance, 414
Laryngitis, 438
Lumbar Puncture, 230
Lymphatic System, 369
Lymph Node Infection, 371
Malabsorption, 416
Maternal Antibodies, 379
Mitral Valve Prolapse, 332
Motion Sickness, 267
Murmur, 338
Muscle Pull, 183
Muscle Strain, 175
Myelination, 209
Myocardial Infarction, 333
Neural Stem Cells, 209
Newborn Addiction, 536
Night Blindness, 274
Nocturnal Emissions, 505
Nonrespiratory Air Movements, 446
Obesity, 101
Oligomenorrhea, 513
Osteoarthritis, 158
Pain, 254
Pancreatitis, 405
Pap Smear, 511
Parkinson Disease, 230
Patent Ductus Arteriosus, 546
Pericarditis, 330
Physiologic Jaundice, 311
Plant Proteins, 423
Pneumothorax, 445
Pregnancy, 532
Premature Infants, 545
Pressure Ulcers, 116
Prostate Cancer, 131
Prothrombin Time, 318
Reflex Function, 221
Respiratory Membrane, 450
Rheumatoid Arthritis, 158
Rigor Mortis, 180
Seizures, 490
Shock, 467
Sickle Cell Disease, 309
Sinus Headache, 436
Sound Intensity, 264
Spina Bifida, 142
Spinal Nerve Injury, 240
Spirometry, 447
Stem Cells, 118
Subdural Hematoma, 223
Superantigens, 380
Surfactant, 445
Temporomandibular Joint Syndrome, 190
Tendinitis, 173
Tenosynovitis, 173
Testicular Cancer, 499
Thromboplastin Time, 318
Tonsillitis, 397
Transfusion Reaction, 320
Twins, 530
Umbilical Cord, 542
Ureteritis, 474
Vaccines, 381
Varicose Veins, 359
Venipuncture, 358
Vital Signs, 5
Vitamin D, 116
Vocal Fold Nodules, 437
Vomiting, 404
Whiplash, 240
Young Transient Amplifying Cells, 118

TABLES

A Comparison Between the Nervous System and the Endocrine System, 281
A Comparison of DNA and RNA Molecules, 86
A Comparison of T Cells and B Cells, 376
Aging-Related Changes, 547
Amino Acids in Foods, 423

Antigens and Antibodies of the ABO Blood Group, 321
Assisted Reproductive Technologies, 535
Atomic Structure of Elements 1 Through 12, 32
Bones of the Adult Skeleton, 137
Breast Cancer Risk, 518
Cellular Components of Blood, 314
Characteristics of Blood Vessels, 347
Characteristics of Life, 4
Chemical Acid-Base Buffer System, 490
Comparative Effects of Epinephrine and Norepinephrine, 293
Connective Tissues, 105
Differences Between the Female and Male Skeleton, 154
Effects of Neurotransmitter Substances on Visceral Effectors or Actions, 245
Elements in the Human Body, 31
Epithelial Tissues, 99
Events Leading to the Conduction of a Nerve Impulse, 216
Events Leading to the Release of a Neurotransmitter, 219
Factors Associated with Edema, 486
Fat-Soluble Vitamins, 424
Features of Vertebrae, 145
Fetal Cardiovascular Adaptations, 541
Functions of Cranial Nerves, 238
Functions of Nephron Components, 472
Functions of the Female Reproductive Organs, 512
Functions of the Male Reproductive Organs, 505
Gases Transported in Blood, 453
Heart Valves, 332
HIV Transmission, 383
Hormonal Changes During Pregnancy, 533
Hormones of the Adrenal Gland, 294
Hormones of the Digestive Tract, 410
Hormones of the Pituitary Gland, 289
Hormones of the Thyroid Gland, 290
Important Groups of Lipids, 41
Inflammation, 121
Inorganic Substances Common in Cells, 38
Intestinal Absorption of Nutrients, 415
Major Branches of the External and Internal Carotid Arteries, 354
Major Components of Gastric Juice, 403
Major Events in a Menstrual Cycle, 515
Major Events of Muscle Contraction and Relaxation, 178
Major Functions of the Liver, 409
Major Minerals, 426
Medical Treatments for an Enlarged Prostate Gland, 503
Movements Through Cell Membranes, 63
Muscle and Nervous Tissues, 107
Muscles Associated with the Eyelids and Eyes, 269
Muscles of Facial Expression, 188
Muscles of Mastication, 190
Muscles of the Abdominal Wall, 195
Muscles of the Pelvic Outlet, 196
Muscles That Move the Ankle, Foot, and Toes, 199
Muscles That Move the Arm, 193
Muscles That Move the Forearm, 194
Muscles That Move the Head, 190
Muscles That Move the Leg, 198
Muscles That Move the Pectoral Girdle, 191
Muscles That Move the Thigh, 197
Muscles That Move the Wrist, Hand, and Fingers, 195
Organic Compounds of Cells, 44
Parts of a Reflex Arc, 222
Parts of the Respiratory System, 442
Plasma Lipoproteins, 317
Plasma Proteins, 316
Practical Classification of Immunity, 381
Preferred and Permissible Blood Types for Transfusions, 321
Primary and Secondary Teeth, 398
Principal Branches of the Aorta, 352
Protein Synthesis, 86
Relative Concentrations of Substances in the Plasma, Glomerular Filtrate, and Urine, 467
Respiratory Air Volumes and Capacities, 447
Role of ADH in Regulating Urine Concentration and Volume, 471
Semen Analysis, 504
Some Neurotransmitters and Representative Actions, 218
Some Nucleotide Sequences of the Genetic Code, 82
Some Particles of Matter, 35
Some Sexually Transmitted Diseases, 520
Stages and Events of Early Human Prenatal Development, 533
Stages in Postnatal Development, 547
Steps in Antibody Production, 377
Steps in the Generation of Sensory Impulses from the Ear, 264
Structures and Functions of Cytoplasmic Organelles, 58
Subdivisions of the Nervous System, 236
Summary of the Major Digestive Enzymes, 414
Terms Used to Describe Skeletal Structures, 137
Tests to Assess Female Infertility, 532
Tissues, 92
Trace Elements, 427
Types of Glandular Secretions, 98
Types of Hormones, 282
Types of Joints, 159
Types of Muscle Tissue, 185
Water-Soluble Vitamins, 425

Index

SUBJECT INDEX

Page numbers in *italics* refer to figures; page numbers followed by a "t" designate tables.

A

A band, *173, 174*
Abdomen
   CT scan of, *15*
   subdivisions of abdominal area, 14, *15*
Abdominal, definition of, 16, *16*
Abdominal cavity, 8, *9, 28*
Abdominal viscera, veins from, 358–59, *358*
Abdominal wall
   arteries to, 355
   muscles of, 194, 195t, 445
   veins from, 358
Abdominopelvic cavity, 8, *9*
Abdominopelvic membranes, 10, *11*
Abducens nerve (VI), *236,* 237, 238t
Abduction, 159, *160*
ABO blood group, 320–23, *321,* 321t, *322*
Abortion, spontaneous, 531, 540–41
Abruptio placentae, 546
Absorption, 4t, 12–13
Accessory nerve (XI), *236,* 237, 238t
   cranial branch of, 237, 238t
   spinal branch of, 237, 238t
Accessory organs
   of digestion, 393, *393*
   sex organs
      female, 507
      male, 498
   visual, 267–68, *268*
Accommodation, *270,* 271
ACE. *See* Angiotensin converting enzyme
Acetabulum, 152, *152*
Acetaminophen, 122
Acetoacetic acid, 492
Acetone, 492
Acetonemia, 494
Acetonuria, 494
N-Acetylaspartate, 108
Acetylcholine, 176–77, 184, 217–18, 218t, 244, *244,* 340, 403
Acetylcholinesterase, 176–77, 218

Acetyl coenzyme A, 77, 421–22, *421, 422*
Achalasia, 428
Achlorhydria, 428
Acid, 36–37
   definition of, 488
   strength of, 489
Acid-base balance, 488–91
   disorders of, 491–93
Acid-base buffer systems, 489–91, 490t, *491*
Acidic solution, 37
Acidosis, 37, 297, 490–92, *491, 492*
   compensated, 492
   metabolic, 492, *493*
   respiratory, 492, *492*
   uncompensated, 492
Acne, 118, 121
Acquired immune deficiency syndrome. *See* AIDS
Acquired immunity, 373, 381, 381t
Acromegaly, 161, 286
Acromial, definition of, 16, *16*
Acromion process, 148, *148, 149,* 157
Acrosome, 500, *500,* 528, *529*
ACTH. *See* Adrenocorticotropic hormone
Actin, 55, *55,* 173–74, *173, 174,* 184
   in cardiac muscle, 186
   in muscle contraction, 176, *176, 177, 178*
Action potential, 213–15, *215*
Active immunity
   artificially acquired, 381, 381t
   naturally acquired, 381, 381t
Active site, 76, *76*
Active transport, 62, *62,* 63t, 414, 470–71
Activity site, on receptor, 283
Acute leukemia, 315
Acute pain fiber, 255–56
Acute paralytic poliomyelitis, 201
Adam's apple. *See* Thyroid cartilage
Adaptive immunity. *See* Acquired immunity
Addiction, 235
Addison disease, 295, 488
Adduction, 159, *160*
Adductor longus, *25, 26, 27, 28,* 188, 196, *196,* 197t
Adductor magnus, *28,* 188, 196, *196,* 197, 197t
Adenine, 79, *80,* 81

Adenoids. *See* Pharyngeal tonsils
Adenosine deaminase deficiency, 384–85, *384, 385*
Adenylate cyclase, 283, *284*
ADH. *See* Antidiuretic hormone
Adipocytes, 101
Adipose tissue, 101, *101,* 105t, 117, *117*
Adolescence, 547t
ADP (adenosine diphosphate), 78, *78*
Adrenal cortex, 287, 292, *292,* 513
   hormones of, 293–94, 294t
Adrenalectomy, 300
Adrenal gland, *27,* 282, 292–95, *292*
Adrenalin. *See* Epinephrine
Adrenal medulla, 292, *292*
   hormones of, 292–93, 293t
   tumor in, 293
Adrenergic fiber, 244, *244*
Adrenocorticotropic hormone (ACTH), 287, 289t, 294–96, *294,* 298, *299*
Adrenogenital syndrome, 300
Adulthood, 547t
Aerobic respiration, 78, 488–89, *489*
Afferent arteriole, 462, *463, 464, 465, 466, 467, 467, 469*
Afferent neuron. *See* Sensory neuron
Afterbirth, 543
Age-related changes, 547t
Agglutination, 320, 380
Agonist (muscle), 187
Agranulocytes, *310,* 311–12, 314t
AIDS, 313–14, 318, 382–83, 520t, 521
Ajulemic acid, 245
Albumins, 314–16, 316t, 470, 490
Albuminuria, 494
Alcohol/alcoholism, 405
   fetal alcohol syndrome, 540, *541*
Aldosterone, 282t, 293, 294t, 468, *468,* 471, 487–88, *487,* 533, 533t
Alimentary canal, 393, *393*
   length of, 394, *394*
   movements of tube, 395, *396*
   wall of, 394, *395*
Alkaline taste, 259–60
Alkalosis, 37, 490–91, *491, 492,* 493, *493*
   metabolic, 493, *493*
   respiratory, 493, *493*

Allantois, 537, *537*
Allergen, 381
Allergic reaction, 313, 379, 381–82
Allograft, 386
All-or-none response
   muscle contraction, 182
   nerve impulse, 216
Alopecia, 121
Alpha cells, 295–96
Alpha fetoprotein, 542
Alpha globulin, 316, 316t
Alpha radiation, 33
Alveolar arch, 141–42
Alveolar duct, 438, *439,* 440, *440,* 516, *516*
Alveolar gas exchange, 449–51, *449, 451*
Alveolar gland, 516, *516,* 544, *545*
Alveolar pressure, 445
Alveolar process, 141, *141,* 399
Alveolar sac, 438, *440*
Alveolus, *334, 439, 440,* 443–44, *444, 449, 449*
Alzheimer disease, 44
Amblyopia, 268, 275
Amenorrhea, 513, 521
Amine hormones, 282t, 283
Amino acids, 43, 422–23
   absorption in small intestine, 414, *415,* 415t
   in blood, 316, 359
   dietary sources of, 422–23, 423t
   essential, 423, 423t
   neurotransmitters, 217, 218t
   in protein synthesis, 81–83, *82, 83*
   structure of, *42*
   sulfur-containing, 489, *489*
Amino group, *42,* 43, 490
Amniocentesis, 542, *543*
Amnion, *534,* 537, *537*
Amniotic cavity, *534,* 537, *537*
Amniotic fluid, 537, *539*
Amoeboid motion, 313
Amphetamines, 235
Ampulla (ear), *262, 266, 266*
Amylase
   pancreatic, 404–5, 414t
   salivary, 399, 414t
Anabolic steroids, 75, 181, *181,* 235
Anabolism, 74–75, *74*
Anaerobic respiration, 78, 179, 489, *489*
Anal canal, 393, *416,* 417, *418*

Anal column, 417, *418*
Analgesia, 235, 247
Analgesic, 247
Anal sphincter
    external, *195*, 417, *418*, 420
    internal, 417, *418*
Anandamide, 245
Anaphase, *64*, 65, *66*
Anaphylactic shock, 382
Anasarca, 494
Anastomosis, 359
Anatomical neck, of humerus, *149*, 150
Anatomical position, 13
Anatomical terminology, 13–17
Anatomic dead space, 447
Anatomy, 3
Androgens, 506–7, *506*, 513, 519
    adrenal, 294, 294t
Anemia, 29, 309, 315, 452
Anesthesia, 235, 247
Anesthetic drug, 215, 221
Aneurysm, 327
Angina pectoris, 333
Angiogenesis, in cancer, 67
Angioplasty, 344
Angiospasm, 359
Angiotensin I, 468, *468*
Angiotensin II, 468–69, *468*, 471
Angiotensin converting enzyme (ACE), 468, *468*
Angiotensin converting enzyme (ACE) inhibitor, 469
Angiotensinogen, 468, *468*
Anisocytosis, 323
Ankle, 158
    bones of, 155–56, *155*, *156*
    muscles that move, 198, 199t
Ankylosis, 161
Annulus fibrosus, 157
Anopia, 275
Anorexia, 245
Anorexia nervosa, 428
Anosmia, 258
Anovulation, 532
Anoxia, 454
Antacid, 493–94
Antagonist (muscle), 187
Antebrachial, definition of, 16, *16*
Antecubital, definition of, 16, *16*
Anterior (ventral), definition of, 13
Anterior chamber of eye, *269*, 271
Anterior fontanel, *143*
Anterior funiculus, 224, *225*
Anterior horn, 224, *225*
Anterior median fissure, 224
Anthrax, 436
Anti-A antibody, 320–23, *321*, 321t, *322*
Anti-B antibody, 320–23, *321*, 321t, *322*
Antibiotic, 454
Antibody, 41, 313, 316, 320, 376–77
    actions of, 379–80, *380*
    in colostrum, 379
    types of, 378–79
Anticodon, 83, *83*
Antidepressant, 218
Antidiuretic, 288
Antidiuretic hormone (ADH), 282t, 288, 289t, 296, 299–300, 471, 471t, 486–88
Antigen, 320, 375
Antigen D, 322–23, *323*
Antigen-presenting cells, 376
Antihemophilic plasma, 323
Antiserum, 381
Antrum, *509*
Anuria, 475, 494
Anus, *195*, *393*, *394*, 417, *498*, *507*
Aorta, *11*, *329*, *330*, *331*, 332–33, *334*, *335*, *341*, 351, *352*, 352t, *356*, 448, *449*, *541*, *546*
    abdominal, *28*, 351, *352*, 352t, *461*, *539*
    ascending, 351, 352t
    descending, *27*, *28*, 351, 352t
    thoracic, 351, 352t
Aortic arch, *25*, *26*, *27*, *28*, 351, *352*, 352t, *539*
Aortic body, 448, *449*
Aortic sinus, 351
Aortic valve, *331*, 332–33, *332*, 332t, *333*, *334*, *336*, *336*
Apex, of heart, *329*, *335*
Aphagia, 428
Aphasia, 247
Apnea, 454
Apocrine gland, *97*, *98*, 98t, 120
Aponeurosis, 172
Apoprotein, 316
Apoptosis, 68–69
Appendicitis, 417
Appendicular portion, of body, 8
Appendicular skeleton, 136, 137t
Appendix, *25*, *26*, *394*, *411*, *416*, 417
Aqueous humor, *269*, 271, 482
Arachidonic acid, 421
Arachnoid mater, 222, *223*, 231
Arches of foot, 156
Arcuate artery, 462, *463*
Arcuate vein, *463*
Areola, *22*, 515, *516*
Areolar tissue. *See* Loose connective tissue
Arm. *See* Upper limb
Arousal, 234
Arrector pili muscle, *114*, 117, *117*
Arrhythmia, 341, 488
Arsenicosis, 72
Arsenic poisoning, 72, *72*
Arterial blood pressure. *See* Blood pressure
Arteriography, 359
Arteriole, 342, *343*, 347t
Arteriosclerosis, 344, 349
Artery(ies), *329*, 342, 347t, 351–55
    cross section of, *319*
    to neck, head, and brain, 353–54, *353*, *354*
    to pelvis and lower limb, 355
    to shoulder and upper limb, 354–55, *355*
    smooth muscle of, 342
    to thoracic and abdominal walls, 355
    wall of, 342, *342*
Arthralgia, 161
Arthritis, 158
Arthrocentesis, 161
Arthrodesis, 161
Arthroplasty, 161
Arthroscopy, 161
Articular cartilage, 128, *128*, 131, *131*, 157, *157*, *158*
Articulating process
    inferior, 142
    superior, 142
Articulation. *See* Joint
Artificial heart, 362
Artificial insemination, 535t
Artificially acquired immunity
    active, 381, 381t
    passive, 381, 381t
Ascending limb of nephron loop, 464, *465*, 472, 472t
Ascending tract, 224
Ascites, 486
Ascorbic acid. *See* Vitamin C
Aspartic acid, 217
Aspartocyclase deficiency, 108
Asphyxia, 454
Aspirin, 122, 493
Asplenia, 386
Assimilation, 4t
Assisted reproductive technology, 535, *535*, 535t
Association area, 229
Association neuron. *See* Interneuron
Association study, 84–85
Aster, *66*
Asthma, 382, 447
Astrocytes, *208*, 209
Asystole, 359
Ataxia, 247
Atelectasis, 445, 454
Atherosclerosis, 40, 300, 318, *319*, 344, *344*, 349
Athletes. *See also* Exercise
    anabolic steroid use by, 75, 181, *181*
    cardiovascular system of, 350
    heatstroke in, 480, *480*
    menstrual problems in, 513
Athlete's foot, 121
Atlas, 143, *146*
Atmospheric pressure, 5, 441–42, *442*
Atom, 7, *8*, 31, 35t
    bonding of, 32–35, *32*, *34*
Atomic number, 32, 32t
Atomic radiation, 33
Atomic structure, 31–32, *31*
Atomic weight, 32, 32t
ATP (adenosine triphosphate), 55
    production of
        in catabolism, *80*
        in cellular respiration, 77–78, *77*
    structure of, 78, *78*
    uses of, 78
        in active transport, 62
        in muscle contraction, 176, *177*, 179, *179*, 184
ATPase, 176
Atrial diastole, 336, *336*, *337*
Atrial natriuretic peptide, 298, 468
Atrial syncytium, 338, *339*
Atrial systole, 336, *336*, *337*
Atrioventricular bundle (A-V bundle), 338–40, *338*, *339*, *340*
Atrioventricular node (A-V node), 338, *338*, *339*, 340, *341*
Atrioventricular valve (A-V valve), 331
Atrium, 331
    left, *26*, *329*, *331*, 332–33, *334*, *335*, 351, *539*, *545*
    right, *26*, *329*, *330*, *331*, *331*, *333*, *334*, *335*, 351, *540*, *545*
Atrophy, muscle, 183
Audiometry, 275
Auditory area, *228*, 229
Auditory nerve, 251
Auditory nerve pathways, 264
Auditory ossicles, *260*, 261, *261*
Auditory tube, 261, *397*, *436*
Auricle (ear), *260*, 261
Auricle (heart), *330*, 331, *335*
Auricular artery, 353, 354t
Australopithecus afarensis, 126
Australopithecus africanus, 126, *126*
Autoantibody, 383
Autocrine secretion, 281
Autograft, 386
Autoimmune disorder, 158, 297
Autoimmunity, 383–86
Automatic bladder, 475
Autonomic nerve fiber, 240–41, *241*
Autonomic nervous system, 207, 235, 240–45
    characteristics of, 240
    control of autonomic activity, 244–45
    neurotransmitters of, 244, *244*, 245t
    parasympathetic division of, 240–41, *243*, 245t
    sympathetic division of, 240–41, *242*, 245t
A-V bundle. *See* Atrioventricular bundle
A-V node. *See* Atrioventricular node
A-V valve. *See* Atrioventricular valve
Axial portion, of body, 8
Axial skeleton, 134, 137t
Axillary, definition of, 16, *16*
Axillary artery, *25*, 354, *355*, *356*
Axillary nerve, *239*, 240
Axillary vein, *25*, 357–58, *357*, *360*
Axis, 143, *146*
Axon, 207, *207*, *209*, *210*, 216

Axonal hillock, 209, *210*
Azotemia, 494
Azygos vein, 358, *360*

B

Back pain, 170
Bacteria, colon, 418, 420
Bacteriuria, 475
Balance. *See* Equilibrium
Ball-and-socket joint, 158, *158,* 159t
Band cells, *310*
Barbiturate, 235
Baroreceptor, 340–41, *341, 349, 349*
Barrett's esophagus, 402
Basal cell carcinoma, 119, *119*
Basal metabolic rate (BMR), 290
Basal nuclei, 229–30, *230,* 233
Base, 36–37
    definition of, 488
    strength of, 489
Base of heart, 329
Basement membrane, 67, 92, *93, 94, 95, 96, 97,* 113, *114, 115, 117*
Basic solution, 37
Basilar artery, 353, *353, 354*
Basilar membrane, 262, *263*
Basilic vein, 357–58, *357, 360*
Basophils, *310,* 311–13, *312,* 314t, 382
B cells, *310,* 376t, 377t, *379,* 382
    activation of, 377, *378*
    functions of, 376
    humoral immune response and, 377
    memory, 377, *379,* 380
    origin of, 375, *375, 376*
Becker muscular dystrophy, 186
Benzodiazepine, 235
Beryllium, 32t
Beta cells, 295–97
Beta globulin, 316, 316t
Beta-hydroxybutyric acid, 492
Beta oxidation, 421, *421*
Beta radiation, 33
Bicarbonate, 38, 38t, 405, *406,* 452–53, *452,* 488–89, 490t, 492
    in blood, 317, 467t
    in body fluids, 482, *483*
    electrolyte balance, 487–88
    in glomerular filtrate, 467t
    reabsorption in renal tubule, 470, *470*
    in urine, 467t
Bicarbonate buffer system, 489, 490t, *491*
Biceps brachii, *23,* 133, 185, *185, 187, 188, 192,* 193, *193,* 194t
Biceps femoris, *188,* 196, *197,* 198t, *199*
Bicuspid (tooth), 398, *398,* 398t
Bicuspid valve, 331–33, *331, 332,* 332t, *333, 334,* 336, *336*
    mitral valve prolapse, 332

Bile, *393,* 408, *411*
    composition of, 409
    release of, 410
Bile canal, 408, *408*
Bile canaliculi, *407*
Bile duct, *407, 408*
    common, *295, 405,* 410, *411*
Bile pigments, 309, 409
Bile salts, 409–10, 415, 421, 423
Bili lights, 311
Bilirubin, 116, 309, 311, *311,* 409
Biliverdin, 309, 409
Binding site, on receptor, 283
Biological clock, 298–99
Biological rhythm, 299
Biopsy, 499, 518
Biotin, 425t
Bioweapon, 381, 436
Bipolar neuron, 211, *211, 272*
Birth control, 516–20
Birth control pill, *517, 518*
Birth date, 529
Birth defects, 540–41, *541*
Birthmark, 121
Birth process, 288, 542–44, *544*
Bitter taste, 259–60
Bladder. *See* Urinary bladder
Blastocyst, 530, *530, 531,* 533t
Blastomere, 529
Blepharitis, 275
Blindness, 268
Blind spot, 272
Blood, 38, 103, *104,* 105t, 305–24
    composition of, 307
    fetal, 538–42, 541t
    gas transport in, 451–53, *451, 452, 453,* 453t
    oxygen-carrying capacity of, 308–9
    pH of, 37, 491–93, *491*
Blood-brain barrier, 345
Blood cells, 307, 314t, 348
    formation of, 133
    in urine, 473
Blood clot, 121, 132, 314, 318, 374
Blood coagulation, 316, 318–20, *318, 320*
    disorders of, 319
Blood gases, 316, 450
Blood group, 320–23
Blood pressure, 232–33, 293, 293t, 295, 299–300, 316, 341–42, 345–51, *347,* 462, 468–69, 483
    control of, 6, 348–50, *349*
    factors that influence, 348, *348*
    filtration and, 61, *61*
    in pregnancy, 547
Blood substitute, 305, *305*
Blood supply, to heart, 333–36, *334, 335*
Blood vessels, 341–46, 347t
    dermal, 116, *117*
    lab-built, 90, *90*
    spasm of, 317, *320*
Blood viscosity, 348

Blood volume, 293, 300, 307, 309, 316, 346, 351, 468
    blood pressure and, 348, *348*
BMR. *See* Basal metabolic rate
Body
    of penis, 503
    of sternum, 147, *147*
    of tongue, 396
    of uterus, *510,* 511
    of vertebrae, 142, 145t
Body cavities, 8–10, *9*
Body coverings, 12
Body fluids
    compartments of, 482, *482*
    composition of, 482, *483*
    movement between compartments, 483–84, *483*
Body position, 234, 267
Body region, of stomach, *401,* 402
Body regions, 14–17, *16*
Body sections, 13–14
Body weight, 232
Boil, 121
Bonaparte, Napoleon, 72, *72*
Bone, 103, *104,* 105t
    cancer of, 131
    development and growth of, 130–31, *130, 131*
    fracture of, 132, *132,* 135
    functions of, 131–34
    homeostasis of bone tissue, 131
    microscopic structure of, 129, *129*
    parts of long bone, 128–29, *128*
    remodeling of, 131
    resorption and deposition of, 131–32, *134,* 291–92
    tumor of, 162
Bone cells, 67
Bone marrow. *See* Marrow
Bone marrow transplant, 134, 309, 315
Bone mass, 135
Bony callus, 132, *132*
Bony labyrinth, *262*
Boren, Laura Cay, 384, *384*
Boron, 32t
Botulinum toxin, 178
Botulism, 178
Bovine spongiform encephalopathy, 44
Brachial, definition of, 16, *16*
Brachial artery, *25,* 354, *356*
    deep, 354, *355, 356*
Brachialis, *188,* 193, *193,* 194t
Brachial plexus, *25,* 239, *240*
Brachial vein, 357–58, *357, 360*
Brachiocephalic artery, *26, 28,* 351, *352,* 352t, *353, 356*
Brachiocephalic vein, *25, 27,* 357, *357, 360*
Brachioradialis, *188,* 193, *193, 194,* 194t
Bradypnea, 454
Brain, 225–34, *227*
    arteries to, 353–54, *354*
    veins from, 357, *357*
Brain scan, 276

Brain stem, 225, *226, 227, 228, 230, 232,* 233, *234*
Brain waves, 5
Breast, *22,* 513, 515, 544–45
    cancer of, 84, 87, 518–19, *518,* 518t
    lymph drainage of, *368,* 370
    self-examination of, 518
Breastbone. *See* Sternum
Breast-feeding, 544
Breath analysis, 450
Breath-holding, 448
Breathing, 435
    control of, 233, 447–49, *447, 448, 449*
    exercise and, 450
    factors affecting, 448–49, *449*
    first breath, 443, 545
    lymph circulation and, 369
    mechanism of, 441–47
    rate and depth of, 490–92, *491*
Breathing rate, 293, 293t, 450
Broca's area. *See* Motor speech area
Bronchial artery, 352t
Bronchial tree, 438–40, *439, 440,* 442t
Bronchiole, 438, *439, 440*
Bronchiolectasis, 454
Bronchitis, 454
Bronchogenic carcinoma, 444
Bronchus, *27, 435,* 438, *439*
Buccal, definition of, 16, *16*
Buccinator, 187, 188t, *189*
Buffer systems, 489–90, 490t
Buffy coat, 307
Bulbospongiosus, 194, *195,* 196t, 511
Bulbourethral gland, *498,* 502, 505t
Bundle branch, *338, 339*
Burning, 77
Burn patient, 118
Bursae, 157, *157*
B vitamins, 424
Bypass graft surgery, 344

C

Cachexia, 428
Caffeine, 218, 487, 519
Calcaneal tendon, *188, 199*
Calcaneus, 155, *155, 156, 199*
Calcification, 130
Calcitonin, 133, *134,* 290–91, 290t, 487
Calcium, 31t, 38t, 424–25
    in blood, 134, 290–91, *291,* 317, 467t
    in body fluids, 482, *483*
    in bone, 133–34, *134*
    in coagulation, 318, *320*
    dietary, 135
    electrolyte balance, 487–88
    functions and distribution in body, 426t
    in glomerular filtrate, 467t
    in heart action, 341

intestinal absorption of, 391
in muscle contraction, 176, *177*, 184
in nerve impulse transmission, 217
requirements and dietary sources, 426t
in urine, 467t
Calcium carbonate, 103, 265
Calcium channel, 52
Calcium oxalate, 477
Calcium phosphate, 103, 424
Callus, in fracture repair
bony, 132, *132*
cartilaginous, 132
Callus, skin of palms or soles, 114
Calorie, 423
CAM. *See* Cellular adhesion molecule
Canaliculi, 103, *104*, 129, *129*
Canal of Schlemm. *See* Scleral venous sinus
Canavan disease, 107–8
Cancer, 67, *67*. *See also specific sites and types*
diagnosis and monitoring of, 84
endocrine paraneoplastic syndrome, 296
Cancer vaccine, 119
Cannabinoid, 235
Capillary, 329, 343–46, *343*, *345*, 347t, *367*
exchanges in, 345–46, *345*
Capitate, *151*
Capitulum, *149*, 150
Capsule, of lymph node, 370, *370*
Carbaminohemoglobin, 452–53, *453*
Carbohydrates, 39, *39*, 44t
absorption in small intestine, 414, *415*, 415t
catabolism of, 75, *80*
dietary sources of, 420
digestion of, 399, 405, 413, 414t
metabolism in liver, 408
requirements for, 420
utilization of, 420
Carbon, 31t, 32t
covalent bond formation, 35
Carbonate, 38
in bone, 134
Carbon dioxide, 38, 38t
from aerobic respiration, *77*, 78
in air, 450
alveolar gas exchange, 449–51, *449*, *451*
in cerebrospinal fluid, 448
diffusion from blood, 440, *441*
diffusion out of cells, 59, *59*
respiratory rate and, 448
transport in blood, *329*, 452–53, *452*, 453t
Carbonic acid, 453, 488–93, *489*, 490t, *491*
Carbonic anhydrase, 453
Carbon monoxide poisoning, 433, *433*, 540
Carboxyl group, *42*, 43, 490

Carboxypeptidase, 405, 414t
Carbuncle, 121
Carcinoma, 97
Cardiac arrest, 488
Cardiac center, 233
Cardiac control center, 340–41, *341*
Cardiac cycle, 336, *336*, *337*
electrocardiogram, 339–40, *340*
regulation of, 340–41, *341*
Cardiac muscle, 184–85, 185t, 331
Cardiac muscle fiber, 338, *339*
Cardiac muscle tissue, 105–6, *106*, 107t
Cardiac output, 348–49
Cardiac plexus, *243*
Cardiac region, of stomach, *401*, 402
Cardiac sphincter. *See* Lower esophageal sphincter
Cardiac tamponade, 359
Cardiac vein, 335–36, *335*
Cardioaccelerator reflex, 349
Cardioinhibitor reflex, 349, *349*
Cardiology, 17
Cardiomyopathy, 186
Cardiovascular system, 12, 327–64
aging-related changes in, 547t
exercise and, 350
Carina, *439*
Carotene, 116
Carotid artery, 448
common, *23*, *24*, *26*, *27*, *28*, *341*, 351, *352*, 352t, 353–54, *353*, *355*, *356*, *449*
external, 353–54, *353*, 354t, *356*
internal, 353–54, *353*, *354*, 354t, *356*
Carotid body, 448, *449*
Carotid sinus, *341*, *353*, 354
Carpal, definition of, 16, *16*
Carpals, *135*, 136, 137t, 151, *151*
Carrier protein, 59–60, *60*, 62, *62*
Cartilage, 102–3, *102*, *103*
elastic, 102, *102*, 105t
fibrocartilage, 103, *103*, 105t, 132, *132*, 157
hyaline, 102, *102*, 105t, 130–31, *131*
Cartilaginous callus, 132
Cartilaginous joint, 157, 159t
Catabolism, 75
Catalase, 75
Catalysis, 75
Catalyst, 36, 75
Cataract, 270
Cauda equina, 238, *239*
Caudate nucleus, 229, *230*
Causalgia, 275
Cecum, *25*, *26*, *394*, 411, 416, 417
Celiac, definition of, 16
Celiac artery, *27*, 351, *352*, 352t, *356*
Celiac disease, 416, 428
Celiac ganglion, *242*
Celiac plexus, *242*, 243
Cell(s), 7, *8*
chemical constituents of, 37–44
structure of, 49–58

Cell body, of neuron, 207, *207*, 209, *210*
Cell cycle, 64–69, *64*
checkpoints in, 64, *64*, *66*
regulation of, 64
Cell death, 68–69
Cell differentiation, 67, *68*
Cell division, 107
Cell lysis, 380, *380*
Cell membrane, 49–52, *50*, 58t
characteristics of, 50–51
membrane potential, 212–15
movements through, 58–63
permeability of, 51, 58, 60, 213
structure of, 51–52, *51*
Cell motility, 55
Cell proliferation, 107
Cellular adhesion molecule (CAM), 52
Cellular immune response, 376–77
Cellular respiration, 77–78, *77*, 435
in muscle, 179
Cellular turnover, 413
Cellulose, 420
Cementum, 398, *399*
Central canal, 103, *104*, 129, *129*, 209, 224, *225*, 408
Central chemoreceptor, 448
Central nervous system (CNS), 207, *212*, 236t
Central nervous system (CNS) depressant, 235
Central nervous system (CNS) stimulant, 235
Central sulcus, 227, *228*
Central vein, 406, *407*
Centriole, *50*, 56, *56*, 65, *66*
Centromere, 65, *66*
Centrosome, 56, *56*, 58t, 65
Cephalic, definition of, 16, *16*
Cephalic vein, *357*, 358, *360*
Cerebellar cortex, 234
Cerebellar peduncle, *232*, 234, *234*
Cerebellum, *223*, 225, *227*, *228*, *230*, 234, *234*
Cerebral aqueduct, 230, *231*
Cerebral arterial circle, 353, *354*
Cerebral artery, *354*, 354t
posterior, 353
Cerebral cortex, 228, 256, *286*
functional regions of, 228–29, *228*
injury to, 229
Cerebral hemispheres, 226
hemisphere dominance, 229, *230*
Cerebral palsy, 247
Cerebral thrombosis, 319
Cerebral vascular accident. *See* Stroke
Cerebrospinal fluid (CSF), 222, 230, *231*, 482
composition of, 448
pH of, 490
secretion of, 230
Cerebrum, *223*, 225, *227*
functions of, 228–30, *228*, *230*
structure of, 226–28, *228*

Ceruminous gland, 98t, 120
Cervical, definition of, 16, *16*
Cervical cap, 517
Cervical enlargement, 224, *224*
Cervical nerves, 238
Cervical orifice, *510*, 511
Cervical plexus, *239*, 240
Cervical vertebrae, 137t, 143, *144*, *145*, 145t
Cervix, *507*, *510*, 511, *539*
in birth process, 542–43, *544*
cancer of, 511
CFTR. *See* Cystic fibrosis transmembrane regulator
Charcot-Marie-Tooth disease, 186
Checkpoints, cell cycle, 64, *64*, *66*
Cheeks, 396
Chemical barrier(s), against pathogens, 374
Chemical barrier contraceptive, 517–18, *517*
Chemical energy, 76–77
Chemical reaction, 36
Chemistry, 31
Chemoreceptor, 253
central, 448
peripheral, 448
Chemosensitive area, 448
Chemotaxis, 374, 380, *380*
Chemotherapy, 519
Chest pain, 433
Chest wall, 443
Chewing, 396, 400–401
Cheyne-Stokes respiration, 454
Chief cells, *402*, 403, 403t
Childbirth. *See* Birth process
Childhood, 547t
Chlamydia infection, 520t, 521
Chloride, 34, *34*, 38t, 424–25
in blood, 317, 467t
in body fluids, 482, *483*
electrolyte balance, 487–88
in glomerular filtrate, 467t
reabsorption in renal tubule, 470, *470*
in urine, 467t
Chloride channel, 406, 454
faulty, 52
Chlorine, 31t, 426t
Chlorolabe, 275
Cholecystectomy, 410
Cholecystitis, 428
Cholecystokinin, 403, 406, 410t, *411*
Cholelithiasis, 428
Cholera, 454
Cholestasis, 428
Cholesterol, 40, *41*
in bile, 409–10
in blood, 415, 421
dietary, 421
disposal of, 415
entry into cells, 63
functions of, 421
membrane, 51, *51*
transport on lipoproteins, 316, 317t, 414–15

Cholinergic fiber, 244, *244*
Chondrocytes, 102, *102, 103*
Chordae tendineae, 331–32, *331*
Chorion, 534–35, *534, 536, 537*
Chorionic villi, 534–36, *534, 536*
Chorionic villus sampling, 542, *543*
Choroid artery, *353,* 354t
Choroid coat, *269,* 270–71, *270, 272*
Choroid plexus, 209, 230, *231*
Chromatid, 65
Chromatin, *50,* 57, *57,* 58t, *66*
Chromatophilic substance, 209, *210*
Chromium, 31t, 425, 427t
Chromosome, 57. *See also* Meiosis; Mitosis
Chronic leukemia, 315
Chronic pain fiber, 255–56
Chronobiology, 298
Chylomicron, 316, 317t, 414, *415*
Chyme, 404, *406,* 411
Chymotrypsin, 405, 414t
Cilia, *50,* 56, *57,* 58t, 93–94, *95,* 257, *437,* 510
Ciliary body, *269,* 270, *270*
Ciliary ganglion, *243*
Ciliary muscles, 270–71, *270*
Ciliary process, 270, *270*
Circadian rhythm, 298–99
Circle of Willis. *See* Cerebral arterial circle
Circular muscle fibers, of alimentary tract, 394, *395*
Circular sulcus, 228
Circulation, 4t, *329*
  fetal, 538–42, *539,* 541t
  newborn, 545–46, *546*
  paths of, 351
Circumcision, 503
Circumduction, 159, *160*
Circumflex artery, *355*
Cirrhosis, 428
Cisterna chyli, *368*
Cisternae, of sarcoplasmic reticulum, 174, *174*
Citrated whole blood, 323
Citric acid cycle, 77–78, *77, 80,* 421–22
Clavicle, *22, 23, 135,* 136, 137t, 147–48, *148, 157, 192*
Cleavage, 529, *530,* 533t
Cleft palate, 141
Clitoris, *195, 507,* 512–13, 512t, 521
Clone, 376
Cloning, 47, *47*
Clotting factors, 318, 374
CNS. *See* Central nervous system
Coagulation. *See* Blood coagulation
Cobalt, 31t, 425, 427t
Cobalt-60, 33
Cocaine, 235
Coccygeal nerves, 238
Coccyx, *135,* 136, 137t, *144,* 145, 145t, *146,* 152, 154t

Cochlea, *260,* 262–63, *262*
Cochlear branch, of vestibulocochlear nerve, 237, 238t, 264
Cochlear duct, 262, *262*
Cochlear implant, 251, *251*
Cochlear nerve, *262, 263*
Codeine, 235
Codon, 81, 83, *83*
Coenzyme, 76
Coenzyme A, *80*
Cofactor, 76
Coitus interruptus, 516
Cold receptor, 254
Colitis, 419
Collagen, *43,* 100, 129
Collagen injections, 100
Collagenous fiber, 100–102, *100, 101, 102, 103,* 105t
Collarbone. *See* Clavicle
Collecting duct, 368–69, *368, 369,* 462, 464, *464, 465, 472,* 472t, 486
Colles fracture, 161
Colloid, 288, *289*
Colloid osmotic pressure, 315–16, *345,* 345t, 369, 468, 483
Colon, 417
  ascending, *25, 26,* 411, *416,* 417
  descending, *25, 26, 27, 416,* 417
  sigmoid, *26, 27, 416,* 417
  transverse, *25, 26,* 412, *416,* 417
Colony-stimulating factor (CSF), 311, 376
Color blindness, 275, 284
Colorectal cancer, 419
Color vision, 274–75
Colostrum, 379, 544
Columnar epithelium, 92
  pseudostratified, 94, *95,* 99t
  simple, 93, *94,* 99t
  stratified, 96, *96,* 99t
Coma, 247
Comatose state, 234
Comedone, 118
Compact bone, 128–31, *128, 129, 131, 132*
Complement, 380, *380*
Complementary base pairing, 79, *80, 81, 82, 83, 86, 86*
Complete protein, 423
Complex carbohydrate, 39
Compound, 35
Computer-aided sperm analysis, 504, *504*
Computerized tomography (CT), 15, *15*
Concentration gradient, 58–60, *59, 60,* 345
Condom
  female, 517
  male, 517, *517*
Conductive deafness, 264
Condyle, 137t
  of humerus, 150

Condyloid joint, 158, 159t
Cone(s), *272,* 273–75, *274*
Cone, David, 327, *327*
Conformation, of proteins, 43–44, *43*
Congenital rubella syndrome, 540, 546
Congestive heart failure, 359
Conization, 521
Conjunctiva, 267, *267*
Conjunctivitis, 275
Connecting stalk, 533, *534, 537*
Connective tissue, 92, 92t, 98–105, 105t
  cell types in, 99, *99*
  characteristics of, 98
  coverings around skeletal muscle, 172–73, *172*
  specialized, 100
Connective tissue cells, *68*
Connective tissue fibers, 99–100, *100*
Connective tissue proper, 100
Constipation, 419, *419*
Contraception, 516–20
Contraceptive implant, 519
Contracture, 201
Convergence, 219, *219*
Convex lens, 273, *273*
Convolution, 226, *227*
Convulsion, 201
Coordination, 12
Copper, 31t, 425, 427t
Coracobrachialis, *23, 24, 25,* 191, *192,* 193t
Coracoid process, 148, *148, 149*
Cordotomy, 247
Corn (skin), 114
Cornea, *267, 269, 269*
Corneal transplant, 270
Coronal (frontal) plane, 14, *14*
Coronal suture, 136, *138, 139, 141, 143, 167, 169*
Corona radiata, 509, *509,* 529
Coronary artery, *333, 334,* 335, *335, 352,* 352t, *356*
Coronary embolism, 349
Coronary ligament, *407*
Coronary sinus, 331, *331, 333, 335,* 336
Coronary thrombosis, 319, 333, 349
Coronoid fossa, *149,* 150
Coronoid process, *139, 142, 150,* 151, *158*
Corpora cavernosa, 503, 512
Cor pulmonale, 359
Corpus albicans, *509,* 514, 515
Corpus callosum, 226, *227, 229, 234*
Corpus cavernosa, *498*
Corpus luteum, *509,* 514–15, *514,* 531–32
Corpus spongiosum, *498,* 503
Cortex, of ovary, 507–8
Cortical bone. *See* Compact bone
Corticospinal tract, 224, *226, 228,* 233

Corticotropin-releasing hormone (CRH), 287, 293–94, *294,* 298, *299*
Cortisol, 282t, 293–94, 294t, 298–300, *299*
Costal, definition of, 16
Costal cartilage, *147,* 148
Coughing, 233, 446
Covalent bond, *34,* 35
  double, 35
  single, 35
Cowper's gland. *See* Bulbourethral gland
Coxae, 135, 151–52, *152, 158*
Coxal, definition of, 16, *16*
Coxal bone, 137t
Cramp, muscle, 180, 254
Cranial branch, of accessory nerve, 237, 238t
Cranial capacity, 126
Cranial cavity, 8, *9*
Cranial nerves, 236–37, *236,* 238t
  injury to, 237
Craniotomy, 247
Cranium, 134, *135,* 136–40, *223*
Creatine phosphate, 179, *179*
Creatine phosphokinase, 179
Crest of bone, 137t
Cretinism, 290
Creutzfeldt-Jakob disease, variant, 44, 69
CRH. *See* Corticotropin-releasing hormone
Cribriform plate, 140, *140, 168,* 257
Cricoid cartilage, 437, *437, 439*
Crista ampullaris, 266, *266*
Cristae, 54, *55*
Crista galli, 140, *140, 141, 168*
Cross-bridges, myosin, 176, *176*
Cross section, 14, *14*
Crown, of tooth, 398, *399*
Crural, definition of, 16, *16*
Crying, 446
CSF. *See* Cerebrospinal fluid; Colony-stimulating factor
CT. *See* Computerized tomography
Cubital, definition of, 16, *16*
Cuboid, *156*
Cuboidal epithelium, 92
  simple, 93, *94,* 99t
  stratified, 96, *96,* 99t
Cuneiform
  intermediate, *156*
  lateral, *156*
  medial, 155, *156*
Cupula, 266, *266*
Curettage, 521
Cushing syndrome, 295, 488
Cusp(s), of tricuspid valve, 331
Cuspid, 398, *398,* 398t, *399*
Cutaneous carcinoma, 119
Cutaneous membrane. *See* Skin
Cutshall, Cynthia, 385
Cyanide poisoning, 76, 452
Cyanocobalamin. *See* Vitamin B_{12}
Cyanolabe, 275

Cyanosis, 116, 307
Cyclic AMP (cAMP), 283–84, *284*
Cyclophosphamide, 519
Cyst, 121
Cystectomy, 475
Cystic duct, *26, 405, 407*, 410, *411*
Cystic fibrosis, 52, 406, 454
  diarrheal disease and, 454
Cystic fibrosis transmembrane regulator (CFTR), 454
Cystitis, 474–75
Cystoscope, 475
Cystotomy, 475
Cytocrine secretion, 116
Cytokine, 376–77, *378, 379*
Cytokinesis, *64*, 65–67, *66*
Cytology, 17
Cytoplasm, 49, 52–56
Cytosine, 79, *80*, 81
Cytoskeleton, 52, 55
Cytosol, 49, 52
Cytotoxic T cells, 376–77, 383

D

DaSilva, Ashanti, 385, *385*
Deamination, 422, *422*
Death, 5
Decibel, 264
Decidua basalis, *536*
Deciduous teeth. *See* Primary teeth
Decomposition reaction, 36
Dedifferentiation, 67
Deep, definition of, 13
Defecation reflex, 419
De Humani Corporis Fabrica, 3
Dehydration, 288, 348, 485, *485*
Dehydration synthesis, 74–75, *74, 75*
Delayed-reaction allergy, 381–82
Deltoid, *22, 23*, 187, *188*, 191, *191, 192*, 193t
Deltoid tuberosity, *149*, 150
Denaturation, of proteins, 43
Dendrite, 207, *207*, 209, *210*, 216
Dens, 143, 145t
Dense connective tissue, 100–102, *101*, 105t
Dental caries, 399
Dentin, 398–99, *399*
Deoxyhemoglobin, 307
Deoxyribose, 420
Depolarization, 213–14
Depo-Provera, 518–19
Depression (joint movement), 159, *161*
Depression (mental illness), 218
Dermal papilla, *114*, 116
Dermatitis, 121
Dermatology, 17
Dermis, 113, *114*, 116, *117*
Descending limb of nephron loop, *464, 465*, 472t
Descending tract, 224
Detoxification, 408
Detrusor muscle, 474–75, *474*
Development, of bone, 130–31, *130, 131*

Deviated septum, 435
Diabetes insipidus, 288, 300, 488
Diabetes mellitus, 84, 297, *297*, 300, 470, 492, *493*
  type 1 (insulin-dependent), 297, 386
  type 2 (noninsulin-dependent), 297
Diacylglycerol, 284
Dialysis, 59
Diapedesis, 313, *313*, 374
Diaphoresis, 333
Diaphragm (contraceptive), 517, *517*
Diaphragm (muscle), 8, *9*, 24, *25, 26, 27, 28, 330*, 401, 440, 442–43, *442, 443*, 445, *445*
Diaphysis, 128–31, *128*
Diarrhea, 488, 492
  cystic fibrosis and diarrheal disease, 454
Diastole, 336, *336*
Diastolic pressure, 347, *347*
Diencephalon, 225, *227*, 232–33, *232*
Diet, adequate, 426, *427*
Differential white blood cell count, 313–14
Differentiation, 67, *68*
Diffusion, 58–59, *59*, 62t
  across respiratory membrane, 450
  exchanges in capillaries, 345
  facilitated, 59–60, *60*, 63t, 414
Digestion, 4t, 75, 391–420
  definition of, 393
  hormonal control of, 410t
Digestive enzymes, 403, 403t, 406, 413, 414t, 454
Digestive glands, 298
Digestive system, 12, 393, *393*
  aging-related changes in, 547t
Digital, definition of, 16, *16*
Digital artery, *355*
Dihydrogen phosphate, 490, 490t
Dilantin, 218
Dilated cardiomyopathy, hereditary idiopathic, 186
Dipeptide, 74–75, *75*
Diplopia, 268, 275
Disaccharide, 39, *39*, 74, *74*, 420
Distal, definition of, 13
Distal convoluted tubule, *463, 464, 465, 466*, 472, 472t, 486
Diuresis, 475, 494
Diuretic, 288, 349, 475, 487–88, 493
Divergence, 219, *219*
Diverticulitis, 428
Dizygotic twins, 530, 546
DNA, 44, *44*, 79
  comparison to RNA, 86t
  double helix, 79, *80*
  repair of, 87
  replication of, 65, 86, *86*
  errors in, 87

  sequencing of, 84
  structure of, 79, *80*
  transcription of, 81, *82*
DNA chip, 84
DNA fingerprinting, 81
Dolly (cloned sheep), 47, *47*
Dominant hemisphere, 229
Dopamine, 217, 218t, 229
Dorsal, definition of, 16
Dorsal arch vein, *357*
Dorsal cavity, 8, *9*
Dorsalis pedis artery, 355, *356*
Dorsal respiratory group, *447*, 448, *448*
Dorsal root, *223, 225*, 238
Dorsal root ganglion, *223, 225*, 238, *241*
Dorsiflexion, 159, *160*
Dorsum, definition of, *16*
Double covalent bond, 35
Double helix, 79, *80*
Double sugar, 39
Doxorubicin, 519
Dried plasma, 323
Drug abuse, 235
Drug tolerance, 235
Dry mouth. *See* Xerostomia
Duchenne muscular dystrophy, 186, *186*
Ductus arteriosus, *539*, 540, 541t, 546, *546*
Ductus deferens. *See* Vas deferens
Ductus venosus, 538–40, *539*, 541t, 545, *546*
Dumping syndrome, 428
Duodenal ulcer, 403
Duodenum, *26, 27*, 394, 404, *405, 406*, 411, *411, 412*
Duplicate gene, 186
Dural sinus, *223*
Dural space, *231*
Dura mater, 222, *223*, 231
Dwarfism, 287
Dynamic equilibrium, 265–67, *266*
Dysentery, 428
Dysmenorrhea, 521
Dyspepsia, 428
Dysphagia, 428
Dyspnea, 333, 456
Dystrophin, 173, 186
Dystrophin-associated glycoprotein, 186
Dysuria, 477

E

Ear, 103, 260–64
Eardrum. *See* Tympanic membrane
Ear infection, 261
Earwax, 120
Eccrine gland, 111, *117*, 119–20
ECG. *See* Electrocardiogram
Ectoderm, 533–34, *534*
Eczema, 121, 382
Edema, 61, 316, 346, 370, 374, 468, 486, 486t

Effector, 5, *5, 6, 7*, 207, *221, 222*t
Efferent arteriole, *463, 464, 465, 466, 467*, 467, 469
Efferent neuron. *See* Motor neuron
Egg cells, 65, 528, *528, 529*
Ejaculation, 502, 504–5
Ejaculatory duct, *498*, 501
Elastic cartilage, 102, *102*, 105t
Elastic fiber, 100–101, *100, 101, 102*
Elastic recoil, 445, 456
Elastin, 100
Elbow joint, 133, *133*, 148, 157–58, *158*
Electrocardiogram (ECG), *337*, 339–40, *340*, 350
Electroencephalogram, 247
Electrolyte(s), 37, 482
  absorption in small intestine, 415, 415t
  in bile, 409
  in plasma, 317
Electrolyte balance, 232, 295, 487–88, *487*
Electromyography, 186, 201
Electron, 31–32, *31*, 32t, 35t
Electron shell, 32, *32*
Electron transport chain, 77–78, *77, 80*
Elements, 31
  in human body, 31t
Elevation, 159, *161*
Embolectomy, 359
Embolus, 318, 333, 344
Embryo, 530
Embryo adoption, 535t
Embryonic development, 529–30, *530, 541*
Embryonic disc, 533–34
Embryonic stage, 530, 533–37, *534, 536, 538*
Emission, 505
Emmetropia, 275
Emotional response, 229, 233
Emphysema, 444, *444*, 447, 450, 456, 492
Emulsification, 410
Enamel, 398–99, *399*
Encephalitis, 247
Encephalogram, 5
Endarterectomy, 359
Endocarditis, 332, 338
Endocardium, *330*, 331
Endochondral bone, 130–31, *130*
Endocrine gland, 97
Endocrine paraneoplastic syndrome, 296
Endocrine system, 12, 279–301. *See also* Hormones
  aging-related changes in, 547t
  characteristics of, 281, *282*
  compared to nervous system, 281, 281t
  control of hormone secretion, 285, *285, 286*
Endocrinology, 17

Endocytosis, 62–63, 63t, 69
  receptor-mediated, 63, 63t, *64*, 415
Endoderm, 533–34, *534*
Endolymph, 262, *262, 266*
Endometriosis, 521, 532
Endometritis, 521
Endometrium, *510*, 511, *511*, 513–15, *534, 536*
Endomysium, *172*, 173
Endoneurium, *220*
Endoplasmic reticulum, 50, 52–53, *53*, 58t
Endorphin, 218t, 257
Endosteum, 129, *129*
Endosymbiont theory, 54
Endothelial cells, 90
Endothelium, 342–43, *342, 343*
Endotracheal tube, 438
Energy, for metabolic reactions, 76–78
Enkephalin, 218t, 256–57
Enteritis, 428
Enterokinase, 405, 414t
Enucleation, 275
Enzyme, 41, 74
  action of, 75–76, *76*
  active site of, 76, *76*
  factors that alter, 76
  rate-limiting, 79
  regulatory, 79
  specificity of, 76
Eosinophils, *310*, 311–13, *312*, 314t
Ependymal cells, *208*, 209
Epicardium, 330–31, *330*
Epicondyle, 137t
  of humerus, *149*, 150
Epicranial aponeurosis, *189*
Epicranius, 187, 188t, *189*
Epidemiology, 17
Epidermal ridge, 116
Epidermis, 94–96, 113–16, *114, 117*
Epididymis, *25*, 498–500, *498, 499*, 505t
Epididymitis, 521
Epidural space, 222, *223*
Epigastric region, 14, *15*
Epiglottic cartilage, 437, *437*
Epiglottis, *397*, 401, *435, 436*, 438
Epilepsy, 247
Epileptic seizure, 218
Epimysium, 172, *172*, 220
Epinephrine, 217, 282t, 292–93, 293t, 298, *299*, 350, 382
Epineurium, 220, *220*
Epiphyseal plate, *128*, 130–31, *131*
Epiphysiolysis, 161
Epiphysis, 128, *128*, 130–31
  injury to, 131
Episiotomy, 543
Epithelial cells, 49, *49*, 56
Epithelium, 92–98, 92t, 98t
  columnar, 92
    pseudostratified, 94, *95*, 99t
    simple, 93, *94*, 99t
    stratified, 96, *96*, 99t

  cuboidal, 92
    simple, 93, *94*, 99t
    stratified, 96, *96*, 99t
  general characteristics of, 92
  glandular, 97–98, *98*, 99t
  squamous, 92
    simple, 92–93, *93*, 99t
    stratified, 94–96, *95*, 99t
    transitional, 96–97, *97*, 99t
Equilibrium (balance), 253, 264–67, *265, 266*
  dynamic, 265–67, *266*
  static, 264–65, *265*
Equilibrium (chemical), 58
Erectile dysfunction, 496
Erection
  female, 512
  male, 496, 503–5
Erythema, 122
Erythroblast(s), *310*
Erythroblastosis fetalis, 323
Erythrocytes. *See* Red blood cell(s)
Erythrolabe, 275
Erythropoietin, 298, 308, *309*, 462
Escherichia coli, 459, *459*
Esophageal artery, 352t
Esophageal hiatus, 401–2
Esophagitis, 428
Esophagus, *11, 27, 28*, 393, *393, 394, 397*, 400–402, *401, 435, 436*
  cancer of, 402
Essential amino acids, 423, 423t
Essential fatty acids, 421–22
Essential hypertension, 349
Essential nutrient, 420
Estradiol, 513
Estriol, 513
Estrogen(s), 135, 282t, 294, 298, 513–15, *514*, 528, 544
  placental, 532–33, *533*, 533t
Estrogen receptor, 519
Estrogen replacement therapy, 135
Estrone, 513
Ethmoidal sinus, *138*, 140, 436
Ethmoid bone, 137t, *138, 139, 140, 141, 167, 168, 169*
Ethylene glycol poisoning, 477
Euphoria, 235
Eupnea, 456
Eustachian tube. *See* Auditory tube
Eversion, 159, *161*
Exchange reaction, 36
Excretion, 4t, 12–13
Exercise
  blood flow and, 343
  breathing and, 450
  cardiovascular system and, 350
  muscle response to, 183
  osteoporosis and, 135
  oxygen debt in muscle, 179, *180*
  respiratory activity and, 6
Exocrine gland, 97, 281
Exocytosis, 54, 62–63, 63t, 69
Exophthalmos, 275, 290, 300
Expanding gene, 186

Expiration, 441, 444–45, *445*
Expiratory reserve volume, *446*, 447t
Extension, 159, *160*, 187
Extensor carpi radialis brevis, 193, *194*, 195t
Extensor carpi radialis longus, 193, *193, 194*, 195t
Extensor carpi ulnaris, *194*, 195t
Extensor digitorum, 187, *194*, 195t
Extensor digitorum longus, *188*, 198, *198, 199*, 199t
External auditory meatus, *139, 140, 169*, 260–61, *260*
External ear, 260–61, *260*
External oblique, *22, 23, 24*, 187, *188, 192*, 194, 195t, 445
Extracellular fluid, 60, 482–83, *482, 483*, 487
Extracorporeal shock-wave lithotripsy, 473
Extrapyramidal tract, 224
Extrinsic muscles, of eye, 268, *268*, 269t
Eye, 237, *267*
  movement of, 237, 268, *268*, 269t
  structure of, 269–73, *269, 270, 272, 273*
Eyelash, 267
Eyelid, 267, *267*, 269t

F

Facet, 137t, 143
  of rib, 147
Facial artery, *353*, 354t
Facial bones, 134, *135*
Facial expression, 237
  muscles of, 187, 188t, *189*
Facial nerve (VII), 236, 237, 238t, *243*
Facial skeleton, 141–42
Facial vein, *357, 360*
Facilitated diffusion, 59–60, *60*, 63t, 414
Facilitation, 219
Factor VIII, 318
Fainting, 449
Falciform ligament, *24*
Fallopian tube. *See* Uterine tube
False ribs, 147, *147*
False vocal cords, 437, *438*
Falx cerebelli, 234
Falx cerebri, 226
Fascia, 172
Fascicle, 172–73, *172*, 220, *220*
Fasting, 101
Fast muscle fiber, 183
Fast twitch muscle fiber, 182
Fat(s), 40, *41*. *See also* Lipid(s)
  catabolism of, 80
  dietary, 40
Fat droplets, 101, *101*
Fat-free food, 422
Fat globule, 410
Fat-soluble vitamins, 410, 423–24, 424t

Fatty acids, 40, *40*, 421
  absorption in small intestine, 414, *415*, 415t
  essential, 421–22
  metabolism of, 293
  oxidation of, 489, *489*
  polyunsaturated, 40
  unsaturated, 40, *40*
Feces, 418, 486–87
Female reproductive cycle, 513–15, *514*, 515t
Female reproductive system, 13
  aging-related changes in, 547t
  external reproductive organs, 511–12
  hormonal control of, 513–15
  internal accessory organs, 510–11, *510*
  organs of, 507–13, *507*, 512t
Femoral, definition of, 16, *17*
Femoral artery, *23, 24, 26*, 355, *356*
Femoral nerve, *23, 24*, 25, *239*, 240
Femoral vein, *22, 23, 24, 26*, 359, *360*
Femur, *28*, 128, *128, 135*, 136, 137t, *152*, 154, *154, 157, 158*
  fracture of, 154
Fenestrae, 467
Fertilization, 508, *508*, 528–29, *528, 529*
Fetal alcohol syndrome, 540, *541*
Fetal cell(s), in maternal circulation, 386, 542
Fetal cell sorting, 542, *543*
Fetal stage, 537–38
Fetus, 530
  blood of, 538–42, 541t
  checking chromosomes of, 542–43, *543*
  circulation in, 538–42, *539*, 541t
  movements of, 532, 538
Fever, 122, 374, 488
Fiber, dietary, 418–20
Fibrillation, 201
Fibrin, 43, 318, *318*, 320
Fibrinogen, 314, 316, 316t, 318, *320*, 374
Fibroblast(s), 90, 99, *99, 100, 101*, 107, 121, 318, 374
Fibroblast growth factor, 344
Fibrocartilage, 103, *103*, 105t, 132, *132*, 157
Fibrocystic breast disease, 518
Fibrosis, 201
Fibrositis, 201
Fibrous joint, 156, *156*, 159t
Fibrous layer
  of ureter, 473, *473*
  of uterus, 511
Fibrous pericardium, 329, *330*
Fibula, *135*, 136, 137t, 155, *155, 199*
Fight or flight response, 293, 298, *299*

Filtration, 369
    across cell membrane, 61, *61*, 63t
    exchanges in capillaries, 345
Fimbriae, of uterine tubes, *507*, 510, *510*
Finger(s), muscles that move, 193–94, 195t
Fingerprints, 116
First breath, 443, 545
First messenger, 283
First ventricle, 230
Fissure, 226
Fixed cells, 99
Flagella, *50*, 56, *57*, 58t
Flatus, 418
Flexion, 159, *160*, 187
Flexor carpi radialis, 193, *193*, 195t
Flexor carpi ulnaris, 193, *193*, *194*, 195t
Flexor digitorum longus, 198, *199*, 199t
Flexor digitorum profundus, 193, 195t
Floaters, 272
Floating ribs, 147, *147*
Fluorine, 31t, 32t, 399, 425, 427t
Fluorouracil, 519
Folacin, 309, *311*, 425t
Folic acid. *See* Folacin
Follicle-stimulating hormone (FSH), 282t, 287, 289t, 505–6, *506*, 513, *514*, 515, 532
Follicular cells
    ovarian, 508–9, *509*
    thyroid, 288–90, *289*
Follicular fluid, 509–10, *509*, 514
Fontanel, 137t, 142, *143*
Food
    fats in, 40
    requirement for, 5
Food poisoning, 178, 380, 459, *459*
Food pyramid, 426, *427*
Food selection, 257
Foot
    bones of, 155–56, *155*, *156*
    muscles that move, 198, 199t
Foramen, 137t
Foramen magnum, 136, *139*, *140*, *141*, *168*, *224*, 353
Foramen ovale, *539*, 540, 541t, 545, *546*
Forearm. *See* Upper limb
Foreskin. *See* Prepuce
Formed elements, in blood, 103, 307, *307*
Formula, 35–36, *36*
    molecular, 35–36, *36*
    structural, 36, *36*
Fossa, 137t
Fossa ovalis, *539*, 546, *546*
Fourth ventricle, 230, *231*, *232*, 447
Fovea, 137t
Fovea capitis, 154, *154*
Fovea centralis, *269*, 272, *273*, 274

Fracture, 132, *132*, 135
    of femur, 154
Freely movable joint, 156
Free nerve ending, 254
Frenulum, 396
Frontal, definition of, *16*, 17
Frontal bone, 136, 137t, *138*, *139*, *140*, *141*, 142, *167*, *168*, 169
Frontal eye field, 228, *228*
Frontalis, 187, *188*, 189
Frontal lobe, 227, *228*
Frontal sinus, *10*, 136, *138*, *141*, *397*, *435*, *436*, *436*
Frontal suture, *143*
Fructose, 39, 75, 420, 502
FSH. *See* Follicle-stimulating hormone
Full-term infant, 538
Fulminant hepatitis, 409
Functional residual capacity, *446*, 447, 447t
Functional syncytium, 338
Fundic region, of stomach, *401*, 402
Fusion inhibitor, 383

G

GABA. *See* Gamma-aminobutyric acid
Galactose, 39, 420
Gallbladder, *11*, *24*, *25*, *26*, 295, 393, *394*, *405*, *407*, 410, *410*, 411
Gallium-67, 33
Gallstones, 405, 410, *410*
Gamete intrafallopian transfer (GIFT), 535t
Gametogenesis, 65
Gamma-aminobutyric acid (GABA), 217, 218t, 235
Gamma globulin, 316, 316t, 378, 381
Gamma radiation, 33
Ganglia, 211, 236, 240
Ganglion cells, *272*
Gas exchange, 435
    alveolar, 449–51, *449*, *451*
Gas transport, in blood, 451–53, *451*, *452*, *453*, 453t
Gastrectomy, 428
Gastric artery, 351, *352*
Gastric gland, 402, *402*
Gastric juice, 374, 402–4, *404*
Gastric pit, 402, *402*
Gastric ulcer, 403
Gastric vein, *358*, 359
Gastrin, 403, *404*, 410t
Gastrocnemius, *188*, *197*, 198, *198*, *199*, 199t
Gastrocolic reflex, 419
Gastroenterology, 17
Gastroileal reflex, 417
Gastrointestinal ulcer, 300
Gastrostomy, 428
Gastrula, 533t
Gender identity, 521

Gene, 79
Gene expression, 83–84
    correlation to clinical signs and symptoms, 84
    in different tissues, *85*
General interpretative area, 228, *229*
General stress syndrome, 298–300, *299*
Gene therapy, 108, 119, 385, 454
Genetic code, 79, 81–83, 82t
Genetic information, 79
Genital, definition of, *16*, 17
Genital herpes, 520t
Genital warts, 520t
Genome, human, 3, 79, 84–85, *85*
Geriatrics, 17
Germinal center, *370*
German measles. *See* Rubella
Gerontology, 17
GIFT. *See* Gamete intrafallopian transfer
Gigantism, 287
Gingiva, *399*
Gland, 97, *98*, 98t
Gland cells, 68
Glandular epithelium, 97–98, *98*, 99t
Glans penis, *498*, 503
Glaucoma, 271
Gleevec, 315
Glenoid cavity, 148, *149*
Gliding joint, 158, 159t
Globin, 309
Globulins, 314, 316, 316t
Globus pallidus, 229, *230*
Glomerular capsule, 464, *464*, *465*, *466*, *467*, 468, *469*, *470*, 472t
Glomerular filtrate, 466, *467*, 467t, *469*, *470*, *470*
Glomerular filtration, 465–66, *467*
    filtration pressure, 466–67
Glomerular filtration rate, 467–68
    regulation of, 468, *468*
Glomerulonephritis, 468
Glomerulus, *463*, 464, *464*, *465*, *466*, *467*, *469*, *470*, 472t
Glossitis, 428
Glossopharyngeal nerve (IX), 236, 237, 238t, *243*, 449
Glottis, 437, *438*
Glucagon, 285, 295–96, *296*, 300, 408
Glucocorticoid, 293
Glucose, 36, *36*, 38–39, *39*, 75, 420
    in blood, 285, 293, 293t, 295–96, *296*, 316, 359, 408, 467t, 470
    energy from, 77
    entry into cells, 59, 296
    in glomerular filtrate, 467t
    renal plasma threshold, 470
    in urine, 467t, 470, 473
Glucosuria, 470, 494
Glutamic acid, 217, 218t
Gluteal, definition of, *16*, 17
Gluten, 416

Gluteus maximus, 187, *188*, *195*, 196, *197*, 197t
Gluteus medius, *28*, *188*, 196, *197*, 197t
Gluteus minimus, 196, 197t
Glycerol, 40, *40*, 421
Glycine, 217
Glycogen, 39, 74, 293, 296, 408, 420
Glycolysis, 77–78, *77*, *80*
Glycoproteins, 54, 282t
Glycosuria, 297
GnRH. *See* Gonadotropin-releasing hormone
Gobea, Andrew, 385, *385*
Goblet cells, 93–94, *94*, *95*, 113, *395*, 402–3, *402*, 403t, 413–14, *413*, 418, *418*, 435, *437*, 438
Goiter, 290, 300
Golgi apparatus, *50*, 53–54, *54*, 58t
Gonadal artery, 351, 352t, *356*
Gonadal vein, *360*
Gonadotropin(s), 287, 298, 505–6, *506*, 532
Gonadotropin-releasing hormone (GnRH), 505, *506*, 513–14
Gonorrhea, 520t, 521, 532
Gooseflesh, 117
Gout, 471
G_1 phase, *64*, 65, 66
G_2 phase, *64*, 65, 66
G protein, 283, *284*
    abnormal, 284
Gracilis, *25*, *26*, *27*, *28*, *188*, 196, *196*, *197*, 197t, *199*
Graft-versus-host disease, 134, 386
Granulation, 121
Granulation tissue, 132
Granulocytes, *310*, 311, 314t
Granulosa cells, *509*
Gray commissure, 224, *225*
Gray hair, 117
Gray matter, 209, *223*, 224, *225*
Greater omentum, 412, *412*
Greater trochanter, 154, *154*
Greater tubercle, *149*, 150
Great saphenous vein, *22*, *23*, *26*
Ground substance, 98, *100*, 102
Growth, 4t
    of bone, 130–31, *130*, *131*
Growth hormone, 282t, 286–87, 289t, 299–300
Growth hormone release-inhibiting hormone, 287
Growth hormone releasing hormone, 287
Guanine, 79, *80*, 81
Gustatory cells. *See* Taste cells
Gustatory cortex, 260
Gynecology, 17

H

Hair, 117–18, *117*, *118*, 374
    color of, 117
Hair cells, 263, *263*, 265–66, *265*, *266*

Hair follicle, *114*, 116–18, *117*, *118*
Hair root, 117, *117*
Hair shaft, *114*, 117, *117*
Half-life, 33
Hallucination, 235
Hamate, *151*
Hamstring group, 198t
Hand
  bones of, 151, *151*
  functions of, *4*
  muscles that move, 193–94, 195t
Hangover, 258
Hapten, 375
Hard palate, 141, 396, *396*, *397*, *435*, *436*
Hashish, 235
Haustra, *416*, 418
Haversian canal. *See* Central canal
Hay fever, 382
hCG. *See* Human chorionic gonadotropin
HCT. *See* Hematocrit
HDL. *See* High-density lipoprotein
Head
  arteries to, 353–54, *353*, *354*
  cavities within, 10, *10*
  CT scan of, *15*
  injury to, 223
  muscles that move, *189*, 190, 190t
  veins from, 357, *357*
Headache, 258
  migraine, 258
  sinus, 436
  tension, 3
Head movement, 265–66, *266*
Head of bone, 137t
  of femur, 152, 154, *154*
  of fibula, 155, *155*
  of humerus, 148–50, *148*, *149*
  of radius, 150, *150*
  of rib, 147
  of ulna, 151
Head of sperm, 500, *500*
Head position, 264–65, *265*
Hearing, 253, 260–64, 264t
Hearing loss, 251, 264
Hearing receptor cells, 263–64
Heart, *4*, *11*, *25*, *441*, *449*
  actions of, 336–41
    blood pressure and, 348, *348*
  artificial, 362
  blood supply to, 333–36, *334*, *335*
  chambers of, 331–32, *331*, *332*, 332t
  conduction system of, 338–39, *338*, *339*
  coverings of, 329–30
  location of, 329
  path of blood through, 333, *334*
  size of, 329
  skeleton of, 333, *333*
  structure of, 329–36

  valves of, 331–32, *331*, *332*, 332t
  wall of, 330–31, *330*
Heart attack, 255, 299, 319, 333, 350
Heartburn, 402
Heart failure, 350, 359, 362
Heart rate, 232–33, 293, 293t, 340–41, 348
  theoretical maximum, 350
Heart sounds, 337–38, *337*
Heart transplant, 336, 365, *365*
Heat
  production in muscle, 180
  requirement for, 5
Heatstroke, 480
Helicobacter pylori, 403
Helium, 32t
Helper T cells, 377, *378*, *379*
Hemarthrosis, 161
Hematocrit (HCT), 307
Hematology, 17
Hematoma, 132, *132*, 162, 318
Hematometra, 521
Hematuria, 477
Heme, 309
Hemianopsia, 275
Hemiazygos vein, 358
Hemiplegia, 247
Hemisphere dominance, 229, *230*
Hemochromatosis, 29
Hemocytoblasts, 308, *310*, 311, 314
Hemodialysis, 59
Hemoglobin, 29, *29*, 43, 116, 133, 179, 305, *305*, 307, 309, 433, 451–52, *451*, *452*, 453t, 490
  fetal, 538, 546
Hemolysin, 323
Hemolytic disease of the newborn. *See* Erythroblastosis fetalis
Hemolytic jaundice, 410
Hemolytic uremic syndrome, 459, *459*
Hemophilia, 319
Hemopoiesis, 133
Hemorrhage, 346, 348
Hemorrhagic telangiectasia, 323
Hemorrhoids, 418
Hemostasis, 317–20, *320*
Hemothorax, 456
Heparin, 99, 313
Heparinized whole blood, 323
Hepatic artery, 351, *352*, *407*, *408*
Hepatic duct, *405*, *407*, 408
  common, 408
Hepatic portal system, 359
Hepatic portal vein, *358*, 359, *407*, 408, *408*, *539*
Hepatic sinusoid, 359, 406, *407*, 408, *408*
Hepatic vein, 359, *360*
Hepatitis A, 409
Hepatitis B, 409
Hepatitis C, 409
Hepatitis D, 409
Hepatitis E, 409

Hepatitis F, 409
Hepatitis G, 409
Hepatocellular jaundice, 410
Hepatopancreatic sphincter, 404, *405*, 410, *411*
Hereditary idiopathic dilated cardiomyopathy, 186
Herniated disc, 145
Heroin, 235
Herpes, 122
Hexose, 39
Hiatal hernia, 402
Hiccup, 446
High altitude adaptation, 308, 493
High-density lipoprotein (HDL), 316, 317t, 415
Hilum
  of kidney, 461, *461*
  of lymph node, 370, *370*
Hinge joint, 158, *158*, 159t
Hipbone. *See* Coxae
Hip fracture, 154
Hip joint, 158, *158*
Hirsutism, 300
Histamine, 99, 218t, 313, 346, 382, 486
Histology, 17
Histotoxic hypoxia, 452
HIV. *See* Human immunodeficiency virus
Hives, 382
hMG. *See* Human menopausal gonadotropin
Holocrine gland, 97–98, *98*, 98t, 118
Homeostasis, 5–7, *5*, *6*, 120, *121*
  of bone tissue, 131
  endocrine system in, 279–301
Homeostatic mechanisms, 5, *5*, *6*
Homo sapiens, 126
Hormones, 12, 279, *279*, 281
  action of, 281–85, *283*, *284*
  changes during pregnancy, 531–33, *531*, *533*, 533t
  control of reproductive system
    female, 513–15
    male, 505–7, *506*
  control of secretion of, 285, *285*, *286*
  of digestive tract, 410t
  local, 281
  types of, 282t
Human chorionic gonadotropin (hCG), 531–32, *531*, *533*, 533t, 535, 542
Human immunodeficiency virus (HIV), 382–83
  drugs targeting, 383
  entry into bloodstream, 69
  long-term nonprogressors, 383
  progression of infection to AIDS, 383
  resistance to HIV infection, 87, 383
  transmission of, 382, 383t, 521
Human menopausal gonadotropin (hMG), 531, 535

Human remains, identification of, 81
Humerus, *25*, *135*, 136, 137t, 148–50, *148*, *149*, *157*, *158*
Humoral immune response, 376–77
Hunger, 232, 487
Hunter, C.J., 181
Huntington disease, 87, 230, 247
Hyaline cartilage, 102, *102*, 105t, 130–31, 157
Hybrid fixator, 132
Hydatidiform mole, 546
Hydramnios, 546
Hydrochloric acid, 37, 403, 403t, 489
Hydrocortisone. *See* Cortisol
Hydrogen, 31t, 32t
  atoms of, *32*
  molecule of, *34*
Hydrogen bond, 35
Hydrogen ions, 37, *37*
  in cerebrospinal fluid, 448
  electrolyte balance, 487–88
  secretion from renal tubule, 471–72, *472*
  sources of, 488–89, *489*
Hydrogen peroxide, 75
Hydrolysis, 75
Hydromorphone, 235
Hydrostatic pressure, 5, 483, *483*
Hydroxyl ions, 37, *37*
Hydroxyurea, 309
Hymen, 511
Hyoid bone, 134, *135*, 137t, *397*, 401, *436*, *437*
Hyperalgesia, 275
Hyperalimentation, 428
Hypercalcemia, 300, 341
Hypercapnia, 456
Hyperextension, 159, *160*
Hyperglycemia, 297, 300, 494
Hyperhidrosis, 111
Hyperkalemia, 341, 488, 494
Hyperkalemic periodic paralysis, 52
Hypernatremia, 488, 494
Hyperoxia, 456
Hyperparathyroidism, 292
Hyperplasia, 67
Hyperpnea, 456
Hypertension, 349, 469
Hyperthermia, 122, 485
Hyperthyroidism, 290
Hypertonic solution, 60–61, *61*
Hypertrophy, muscle, 183
Hyperuricemia, 494
Hyperventilation, 448–49, 456, 493
Hypocalcemia, 300, 341
Hypochondriac region, 14, *15*
Hypodermis. *See* Subcutaneous layer
Hypogastric plexus, *243*
Hypogastric region, 14, *15*
Hypoglossal nerve (XII), *236*, 237, 238t
Hypoglycemia, 295–96, 300, 494

Hypokalemia, 341, 488
Hyponatremia, 488
Hypoparathyroidism, 292
Hypophyseal portal vein, 286, *287*
Hypophysectomy, 300
Hypoproteinemia, 468, 486, 486t
Hypothalamus, *230*, 232–33, *232*, 244, *282*, 285–86, *286*, *287*, *288*, *288*, 299, *299*, 341
  control of male reproductive system, 505–6, *506*
Hypothyroidism, 290
Hypotonic solution, 61, *61*
Hypoxemia, 212, 452, 456
Hypoxia, 307, 452, 456
Hysterectomy, 521
Hysterosalpingogram, 532t
H zone, *173*, *174*

I

I band, 173–74, *173*, *174*
Ibuprofen, 122
Ig. *See* Immunoglobulin
Ileitis, 428
Ileocecal sphincter, 416–17, *416*
Ileum, *26*, *394*, *411*, 412, *416*
Iliac artery, *463*
  common, *26*, *27*, 351, *352*, 352t, 355, *356*
  external, 355, *356*
  internal, 355, *356*, *539*
Iliac crest, *28*, *152*, 152
Iliac region, 14, *15*
Iliac spine, anterior superior, *22*, *24*, *28*, *152*, 153
Iliacus, 196, *196*, 197t
Iliac vein, *463*
  common, 359, *360*
  external, 359, *360*
  internal, 359, *360*
Iliotibial band, *197*
Ilium, 152, *152*, 153
Illusion, 235
Immediate-reaction allergy, 382
Immovable joint, 156
Immune response
  cellular, 376–77
  humoral, 376–77
  primary, 380, *380*
  secondary, 380, *380*
  in small intestine, 69
Immune surveillance, 370
Immune system, in HIV infection, 382–83
Immunity, 313, 365–87
  acquired, 373, 381, 381t
  active. *See* Active immunity
  adaptive. *See* Immunity, acquired
  innate, 373
  passive. *See* Passive immunity
  practical classification of, 381, 381t
Immunocompetence, 386
Immunodeficiency, 382–86
Immunoglobulin (Ig), 376, 378–79

Immunoglobulin A (IgA), 378
Immunoglobulin D (IgD), 378
Immunoglobulin E (IgE), 379, 382
Immunoglobulin G (IgG), 378, 380
Immunoglobulin M (IgM), 378, 380
Immunology, 17
Immunosuppressive drug, 383
Impetigo, 122
Implantation, 530, *531*
Impotence, 496
Incisor, 398, *398*, 398t
Incomplete protein, 423
Incontinence, 477
Incus, 137t, *260*, 261, *261*
Infant, 547t. *See also* Neonate
  skull of, 142, *143*
Infarction, 319
Infection, 122, 313, 373
  body defenses against, 373
  nonspecific defenses against, 373–74
  specific defenses against. *See* Immunity
Inferior, definition of, 13
Inferior oblique, *268*, 269t
Inferior rectus, *268*, 269t
Inferior vena cava, *27*, *28*, 331, *331*, *334*, *335*, 359, *360*, *407*, *461*, *539*
Infertility
  female, 532, 532t, 535, 535t
  male, 87, 496, *496*, 504, *504*, 535, 535t
Inflammation, 120, 121t, 313, 374, 380, 486, 486t
Inflation response, 448
Infradian rhythm, 299
Infraorbital foramen, *138*, *167*
Infrapatellar bursa, *157*
Infraspinatus, *188*, 191, *191*, *192*, 193t
Infundibulum, 232, 286, 510, *510*
Inguinal, definition of, 16, *17*
Inherited disease, 87
  faulty ion channels causing, 52
  of muscle, 186, *186*
Inhibin, 506
Injectable contraception, 518–19
Injections, 113
Innate immunity, 373
Inner cell mass, 530
Inner ear, 261–64, *262*, *263*, *264*
Inner tunic, 269, 272, *272*, *273*
Inorganic salts, in bone, 133–34, *134*
Inorganic substances, 37–38, 38t
Inositol triphosphate, 284
Insertion, of skeletal muscle, 159, 185–87, *185*
Insomnia, 300
Inspiration, 441–44, *442*, *443*
Inspiratory capacity, 446–47, *446*, 447t
Inspiratory reserve volume, 446, *446*, 447t
Insula, 228
Insulin, 59, 285, 295–96, *296*, 408. *See also* Diabetes mellitus

Insulin receptor, 296
Integration, 12
Integrative function, of nervous system, 207
Integrin, 124
Integumentary system, 12, 111–24
  aging-related changes in, 547t
Intellectual processing, 229
Intercalated disc, 106, *106*, 184–85
Intercostal artery, *356*
  anterior, 355
  posterior, 352t, 355
Intercostal muscle, *516*
  external (inspiratory), *23*, *24*, *28*, 443, *443*, 445
  internal (expiratory), *24*, *28*, *192*, 445, *445*
Intercostal nerve, *239*, 240
Intercostal vein, 358
Interferon, 374, 376, 409
Interleukin, 311
Interleukin-1, 122, 376
Interleukin-2, 376
Interlobar artery, 462, *463*
Interlobar vein, *463*
Interlobular artery, 462, *463*, *465*
Interlobular vein, *463*, *465*
Intermediate lobe, 286
Internal environment, 5
Internal oblique, *23*, *24*, *192*, 194, 195t, 445
Interneuron, 212, *212*, 220, 222t, 224
Internuncial neuron. *See* Interneuron
Interphase, *64*, 65, *66*
Interstitial cells, 499, 505t
Interstitial cell stimulating hormone. *See* Luteinizing hormone
Interstitial fluid. *See* Tissue fluid
Interstitial space, 483
Intertubercular groove, *149*, 150
Interventricular foramen, *231*
Intervertebral disc, *28*, 103, 134, 142, *144*, 157
  herniated, 145
Intervertebral foramen, 142–43, *144*, 239
Intestinal autointoxication, 419
Intestinal flora, 418
Intestinal gas, 418
Intestinal gland, *395*, 413–14, *413*
Intestine. *See* Large intestine; Small intestine
Intra-alveolar pressure, 442
Intracellular fluid, 60, 482–83, *482*, *483*
Intracranial pressure, 230, 258
Intradermal injection, 113
Intramembranous bone, 130, *130*
Intramuscular injection, 113
Intrauterine device (IUD), 517, 519
Intrauterine transfusion, 546
Intrinsic factor, 403, 403t
Invasiveness, of cancer cells, 67

Inversion, 159, *161*
In vitro fertilization (IVF), 535
Involuntary muscle, 106
Iodine, 31t, 290, 424–25, 427t
Iodine-131, 33, *33*
Ion(s), 32–34, 35t
  membrane potential and, 212–13, *213*
Ion channel, 51–52, 59
  faulty, 52
Ionic bond, 34, *34*
Iridectomy, 275
Iris, *269*, 271, *271*
Iritis, 275
Iron, 31t, 309, 374, 408, 424–25, 427t
  imbalances in, 29
Iron-deficiency anemia, 29
Iron lung, 201
Ischemia, 212, 254, 340, 344, 452
Ischial spine, *152*, 153
Ischial tuberosity, *152*, 153
Ischiocavernosus, 194, *195*, 196t
Ischium, 152–53, *152*
Islets of Langerhans, 295–96, *295*
Isotonic solution, 60, *61*
Isotope, radioactive, 33, *33*
Isotretinoin, 123
IUD. *See* Intrauterine device
IVF. *See* In vitro fertilization

J

Jaundice, 311, 409–10
Jejunum, *26*, *394*, *411*, 412
Jet lag, 300
Johnson, Ben, 181, *181*
Joint, 156–61
  movements of, 159, *160*, *161*
Joint capsule, 157, *157*, *158*
Joint cavity, *158*
Joint reflex, 450
Judgment, 229
Jugular vein
  external, *24*, 357, *357*, *360*
  internal, *23*, *24*, *27*, 357, *357*, *360*, *368*
Junctional fiber, 338, *338*, *339*
Juxtaglomerular apparatus, 465, *466*
Juxtaglomerular cells, 465, *466*, 468

K

Kangaroo rat, 484
Karyotype, fetal, 542–43, *543*
Keloid, 122
Keratin, 43, 72, 94–96, 113, 124
Keratinization, 94–96, 113
Keratinocytes, 116
Keratitis, 275
Ketoacidosis, 492
Ketone body, 421, *421*, 492
  acidic, 489, *489*
Ketonuria, 473, 492, 494
Ketosis, 494

Kidney, *11, 27, 282,* 461–65, *461, 462,* 487
   in acid-base balance, 490–91, *491*
   artificial, 59
   blood vessels of, 462–63, *462*
   functions of, 462
   location of, 461
   structure of, 461–62
Kidney failure, 473
   acute, 477
   chronic, 467
Kidney stone, 292, 473
Kidney transplant, 463
Kilocalorie, 423
Kneecap. *See* Patella
Knee-jerk reflex, 220–21, *221*
Knee joint, 157, *157*
Kupffer cells, 359, 408, *408*
Kuru, 69

L

Labium major, *507,* 512, 512t
Labium minus, *507,* 512, 512t
Labor, 543, *544*
   induction of, 288
Labyrinth, 261–62
Labyrinthectomy, 275
Labyrinthitis, 275
Lacrimal apparatus, 268, *268*
Lacrimal bone, 137t, *138, 139,* 142, *167, 169*
Lacrimal gland, 267–68, *268*
Lacrimal sac, 268, *268*
Lactase, 413–14, 414t
Lactation, 288, 516, 544–45, *545*
Lacteal, 365, *395,* 413–14, *413,* 415
Lactic acid, 489, *489*
   in muscle, 179–80, *180*
Lactiferous duct, 516, *516*
Lactose, 39, 420
Lactose intolerance, 414
Lacunae
   of cartilage, 102, *102, 104*
   around chorionic villi, 534, 536, *536*
   of bone, *104,* 129, *129*
Lambdoidal suture, 136, *139, 141, 169*
Lamellae, of bone, 103, *104*
Lamina, of vertebrae, 142, *145,* 145t
Laminectomy, 161, 247
Laparoscopy, 532t, 535
Large intestine, *11, 393, 393, 394,* 416, 417–20, 498
   functions of, 418
   intestinal flora, 418, 420
   movements of, 419–20
   parts of, 417, *417, 418*
   wall of, *416,* 418
Laryngitis, 438
Laryngopharynx, *397,* 400, *436, 436*
Larynx, *23, 26, 372,* 401, 435, *436,* 437–38, *437, 439,* 442t

Laser angioplasty, 344
Latent period, muscle, 182, *182*
Lateral, definition of, 13
Lateral condyle
   of femur, 154, *154*
   of tibia, 154
Lateral epicondyle
   of femur, *154*
   of humerus, *149*
Lateral funiculus, 224, *225*
Lateral geniculate body, 275
Lateral horn, 224, *225*
Lateral malleolus, 155, *155*
Lateral rectus, *268, 269,* 269t
Lateral sulcus, 227, *228*
Lateral ventricle, 230, *231*
Latissimus dorsi, *23, 24,* 188, 191, *191,* 193t
Laughing, 446
LDL. *See* Low-density lipoprotein
Leg. *See* Lower limb
Leg ulcer, 124
Lens, *269,* 270–71, *270*
Lens fiber, 270
L'Esperance family, 535, *535*
Lesser trochanter, 154, *154*
Lesser tubercle, *149,* 150
Leukemia, 134, 315, *315*
Leukocytes. *See* White blood cell(s)
Leukocytosis, 313
Leukopenia, 313
Leukotriene, 382
Levator ani, 194, *195,* 196t, *418*
Levator palpebrae superioris, 267, *267, 268,* 269t
Levator scapulae, 191, *191,* 191t, *192*
Lever, 133, *133*
LH. *See* Luteinizing hormone
Licorice, 348
Life
   characteristics of, 4, 4t
   maintenance of, 4–7
Ligament, 100–101, 128
Ligamentum arteriosum, *539, 546, 546*
Ligamentum capitis, 154, *158*
Ligamentum teres, *539*
Ligamentum venosum, *539,* 545, *546*
Light refraction, 273, *273*
Limb-girdle muscular dystrophy, 186
Limbic system, 233, 235
Linea alba, *23,* 192, 194
Linea terminalis. *See* Pelvic brim
Lingual artery, *353,* 354t
Lingual tonsils, 396, *397, 436*
Linoleic acid, 421–22
Linolenic acid, 421
Lipase, 410
   intestinal, 413, 414t
   pancreatic, 405
Lipid(s), 39–41, *40, 41,* 41t, 44t
   absorption in small intestine, 414, *415,* 415t
   in blood, 316, 359

   catabolism of, 75
   dietary sources of, 421
   digestion of, 405, 410, 413, 414t
   metabolism in liver, 408
   requirements for, 422
   utilization of, 421, *421*
Lipid bilayer, 51, *51*
Lipoprotein, 316, 317t
Lipoprotein lipase, 415
Lips, 396, *396, 397*
"Lite" foods, 422
Lithium, 32t
Liver, *11, 24, 25, 26,* 359, 393, *393,* 406–10
   cancer of, 409
   functions of, 408, 409t
   structure of, 406–8, *407, 408*
Lobar pneumonia, 456
Lobes
   of liver, 406, *407*
   of lungs, *439,* 441
Lobules
   of liver, 406, *407*
   of spleen, 371
   of testes, 498
   of thymus, 371, *372*
Local current, 215
Local hormone, 281
Lochia, 546
Long bone, 128–29, *128*
Longitudinal fissure, 226, *230, 234*
Longitudinal muscle fibers, of alimentary tract, 394, *395*
Longitudinal section, 14, *14*
Long-QT syndrome, 52
Loose connective tissue, 100, *100,* 105t
Low-density lipoprotein (LDL), 316, 317t, 415
Lower esophageal sphincter, 401, *401*
Lower limb
   arteries to, 355
   bones of, 136, 137t, 154–56, *154, 155, 156*
   muscles that move ankle, foot, and toes, 198, 199t
   muscles that move leg, 196, *198,* 198t, *199*
   veins from, 359
Lower respiratory tract, 435
LSD, 235
Lubrication, for sexual intercourse, 512
Lumbago, 161
Lumbar, definition of, *16, 17*
Lumbar artery, *352,* 352t, 355, *356*
Lumbar enlargement, 224
Lumbar nerves, 238
Lumbar puncture, 230, 258
Lumbar region, 14, *15*
Lumbar vein, 358, *360*
Lumbar vertebrae, 137t, 144, *144, 145,* 145t
Lumbosacral plexus, *239,* 240
Lumen, of alimentary tract, 394

Lumpectomy, 519
Lunate, *151*
Lungs, *11, 24, 25, 26, 330,* 435, *439,* 440–41, *441,* 442t
   cancer of, 444, *444,* 447
   fetal, 540
   of newborn, 443, 445, 545
Lung volume reduction surgery, 456
Lunula, 118, *118*
Luteinizing hormone (LH), 282t, 287, 289t, 505–7, *506,* 513–15, *514,* 518, 532
Lymph, 368–69, *368,* 482, *482, 483, 483*
   formation of, 369
   functions of, 369
   movement of, 369–70
Lymphadenectomy, 386
Lymphadenitis, 371
Lymphadenopathy, 386
Lymphadenotomy, 386
Lymphangitis, 371
Lymphatic capillary, 367–69, *367, 369*
Lymphatic duct, 368–69, *368*
Lymphatic pathways, 367–69, *367*
Lymphatic system, 12, 365–87
   aging-related changes in, 547t
   functions of, 365
Lymphatic trunk, 368–69, *368, 369*
Lymphatic vessels, 346, 367–69, *367, 368, 369, 370, 413, 414*
   obstruction of, 486, *486*
Lymph node, *367, 368,* 368, *369,* 370–71, *370, 371,* 519
   functions of, 370–71
   location of, 370, *371*
   structure of, 370, *370*
Lymph nodule, 370, *370*
Lymphocytes, 312–13, *313,* 314t, 370
   functions of, 376
   origins of, 375, *375, 376*
   production of, 371
Lymphocytopenia, 386
Lymphocytosis, 386
Lymphoid leukemia, 315
Lymphoma, 386
Lymphosarcoma, 386
Lymph sinus, 370, *370*
Lysine, 423, 423t
Lysosomes, *50,* 55, 58t, 63, 313, 376
Lysozyme, 268, 374

M

Macrocytosis, 323
Macromolecule, 7, *8*
Macrophages, 99, *99,* 309, *310,* 312, 370–71, 373–74, 376, 382, 449
Macula (ear), *262,* 265, *265*
Macula densa, 465, *466,* 468

Macula lutea, 272, *273*
Mad cow disease, 44
Mad Hatter, 205, *205*
Magnesium, 31t, 32, 38, 38t, 134, 425
  in blood, 317, 467t
  in body fluids, 482, *483*
  electrolyte balance, 487–88
  functions and distribution in body, 426t
  in glomerular filtrate, 467t
  requirements and dietary sources, 426t
  in urine, 467t
Major calyx, 461, *462*
Major histocompatibility complex (MHC), 376–77
Major minerals, 425, 426t
Malabsorption, 416
Male climacteric, 507
Male reproductive system, 13
  aging-related changes in, 547t
  external reproductive organs, 503–5
  hormonal control of, 505–7, *506*
  internal accessory organs, 500–503
  organs of, 498–505, *498*
Malleus, 137t, *260*, 261, *261*
Malnutrition, 426
  in pregnancy, 540–41
Maltase, 413, 414t
Mammary, definition of, *16, 17*
Mammary gland, *22*, 98t, 120, 515–16, *516*, 532, 544–45
Mammillary body, 232, *232*
Mammogram, 518, *518*
Mandible, 136, 137t, *138, 139, 141*, 142, *143, 167*, 169, 400
Mandibular condyle, *139*, 142, *169*
Mandibular division, of trigeminal nerve, 237, 238t
Mandibular fossa, *139*, 140, *168*
Manganese, 31t, 425, 427t
Manometer, 230
Manubrium, 147, *147*
Marijuana, 235, 245
Marinol, 245
Marrow, 129, 131, 133–34
Masseter, *188, 189*, 190, 190t, *399*
Mass movements, 419
Mast cells, 99, *99*, 382
Mastectomy, 519
Mastication, muscles of, *189*, 190, 190t
Mastitis, 521
Mastoid fontanel, *143*
Mastoid process, *139*, 140, *169*
Maternal antibody, 379
Maternal serum marker test, 542
Mate selection, 279, *279*
Matrix, intercellular, 92, 98
Matter
  definition of, 31
  particles of, 35t
  structure of, 31–37

Maxilla, 137t, *138, 139*, 141, *141, 143, 167, 168, 169*
Maxillary artery, *353*, 354t
Maxillary division, of trigeminal nerve, 237, 238t
Maxillary sinus, *138*, 141, 436
M cells, 69
Meatus, 137t
Mechanical barrier(s), against pathogens, 374
Mechanical barrier contraceptives, 517, *517*
Mechanoreceptor, 253–55
Meconium, 546
Medial, definition of, 13
Medial condyle
  of femur, 154, *154*
  of tibia, 154
Medial epicondyle, of femur, *154*
Medial malleolus, 155, *155*
Medial rectus, *268, 269*, 269t
Median cubital vein, *357, 358, 360*
Median nerve, *239*, 240
Mediastinal artery, 352t
Mediastinum, 8–9, 440–41
Medulla, of ovary, 507–8
Medulla oblongata, *226, 227, 232*, 233, *447, 449*
Medullary cavity, *128*, 129, 131, *131, 132*, 133
Medullary rhythmicity area, 447–48, *447*
Megakaryoblasts, *310*
Megakaryocytes, *310*, 314
Meiosis, 65, 500, *501, 502, 508, 508*
Meissner's corpuscle, 253, *254*
Melanin, 114–17, *115*, 119, 270, 286
Melanocytes, 114–17, *115*, 119, 270
Melanocyte-stimulating hormone (MSH), 286
Melanoma, 119, *119*
Melatonin, 297–98, 300
Membrane(s), types of, 113
Membrane potential, 212–15. *See also* Action potential
  ion distribution and, 212–13, *213*
  potential changes, 213
  resting potential, 213, *214*
Membranous labyrinth, 262, *262*
Memory, 207, *228*, 229
Memory B cells, 377, *379*, 380
Memory T cells, 377
Menarche, 513
Ménière's disease, 275
Meninges, 222–23, *223, 227*
Meniscus, 157, *157*
Menopause, 515
Menstrual cycle, 298, 513–15, *514*, 515t
Menstrual flow, 515, 531
Mental, definition of, *16, 17*
Mental foramen, *138, 167*
Meperidine, 235

Mercury poisoning, 205, *205*
Merocrine gland, 97–98, *98*, 98t
Mesenteric artery
  inferior, 27, 351, *352*, 352t, *356*
  superior, 27, 351, *352*, 352t, *356*
Mesenteric ganglion, *242*
Mesenteric vein
  inferior, *358*, 359
  superior, 27, *358*, 359
Mesentery, *24, 26*, 395, *411, 412*, *412*
Mesoderm, 533–34, *534*
Messenger RNA (mRNA), 81–83, *82, 83*, 282, *283*
Metabolic acidosis, 492, *493*
Metabolic alkalosis, 493, *493*
Metabolic pathway, 79, *80*
Metabolic rate, 293t
Metabolic reactions, 74–75
  control of, 75–76
  energy for, 76–78
Metabolism, 4, 72–87
Metacarpals, *135*, 136, 137t, 151, *151*
Metallic taste, 259–60
Metaphase, 64, 65, 66
Metaphase plate, 66
Metastasis, 67
Metatarsals, *135*, 136, 137t, 155, 156, *156*
Metatarsus, 155
Methadone, 235
Methamphetamine, 235
Methionine, 423, 423t
Methotrexate, 519
MHC. *See* Major histocompatibility complex
Microchimerism, 386
Microcytosis, 323
Microfilaments, 55, *55*, 58t
Microglial cells, 208, *208*
Microtubules, *50*, 55–56, *55*, 58t, 65, *66*
Microvilli, *50*, 93
  of renal tubule, 469
  of small intestine, *395*, 416
Micturition, 475
Micturition reflex center, 475
Midbrain, *226, 227*, 233, *447*
Middle ear, 261, *261*
Middle ear cavity, *10, 10*
Middle tunic, 269–70, *270*
Midpiece, of sperm, 500, *500*
Migraine, 258
Milk, production and secretion of, 544–45, *545*
Mineral(s), 424–25
  major, 425, 426t
  trace elements, 425, 427t
Mineralocorticoid, 293
Minor calyx, 461, *462*
Mitochondria, *50*, 54–55, *55*, 58t, 78
Mitosis, *64*, 65, *66*, 529
Mitral valve. *See* Bicuspid valve
Mitral valve prolapse, 332
Mixed nerve, 220

Mixing movement, of alimentary canal, 395, *396*, 416–17, 419
M line, *173*, 174
Molar, 398, *398*, 398t
Mole (skin), 119, 122
Molecular formula, 35–36, *36*
Molecule, 7, *8*, 35, 35t
Monoamine(s), 217–18, 218t
Monoamine oxidase, 218
Monoblasts, *310*
Monocytes, *310*, 312–13, *312*, 314t, 374
Monohydrogen phosphate, 490, 490t
Mononuclear phagocytic system, 374
Monoplegia, 247
Monosaccharide, 39, *39*, 74, *74*, 414, 420
Monozygotic twins, 530, 546
Mons pubis, *22*, 512
Morula, 530, *530, 531*, 533t
Motion sickness, 267
Motor area, of cerebral cortex, 228
Motor cortex, *226*
Motor end plate, 175, *175*
Motor function, of nervous system, 207
Motor nerve, 220
Motor neuron, 175–76, *175*, 212, *212*, 220, *221, 222*, 222t
Motor speech area, 228, *228*
Motor unit, 175, *175*
  recruitment of, 183
Mouth, 4, *393, 393*, 396–400, *397, 400*
Movement, 4t, 12, 133, *133*
  of joints, 159, *160, 161*
mRNA. *See* Messenger RNA
MSH. *See* Melanocyte-stimulating hormone
Mucin, 98
Mucosa
  of alimentary canal, 394, *395*
  of stomach, *402*
  of urethra, 475, *475*
  of vaginal wall, 511
Mucous cells, 98, 399. *See also* Goblet cells
Mucous coat
  of ureter, 473, *473*
  of urinary bladder, 474, *474*
Mucous membrane, 113, 374, 435. *See also* Mucosa
Mucus, 93–94, *94*, 98, 374, 399, 403t
  in large intestine, 418
  in respiratory tract, 435–36, *437*, 438
Mucus-secreting glands, 413
Multiple births, 526, *526*, 532, 535
Multiple sclerosis, 47, 247
Multipolar neuron, *211*, 212
Multiunit smooth muscle, 184
Mummy, 1, *1*
Murmur, 338
Muscle cells, 49, *49*, 55, 67, 68

Muscle contraction, 176–80, 178t
  all-or-none response, 182
  energy sources for, 179, *179*
  heat production, 180
  myosin and actin in, 176, *176, 177, 178*
  oxygen debt, 179, *180*
  oxygen supply for, 179
  recording of, 182, *182*
  recruitment of motor units, 183
  skeletal muscle, 176–80
  smooth muscle, 184
  stimulus for, 176–77
  strength of, 183
  summation, 182, *182*
  sustained, 183
  threshold stimulus for, 180–81
Muscle fatigue, 179–80, 183
Muscle fiber, 105, *105*
Muscle impulse, 176
Muscle pull, 183
Muscle relaxation, 177, 178t
Muscle strain, 175
Muscle tissue, 92, 92t, 105–6, 107t. *See also* Cardiac muscle tissue; Skeletal muscle tissue; Smooth muscle tissue
Muscle tone, 183
Muscular coat
  of ureter, 473, *473*
  of urinary bladder, 474
Muscular dystrophy, 173, 186, *186*, 201
Muscular layer
  of alimentary tract, 394, *395*
  of large intestine, *416*
  of urethra, 475, *475*
  of vaginal wall, 511
Muscular rheumatism, 201
Muscular system, 12, 170–201. *See also* Cardiac muscle; Skeletal muscle; Smooth muscle
  aging-related changes in, 547t
  inherited diseases of, 186, *186*
Musculocutaneous nerve, 25, *239*, 240
Mutagen, 87
Mutation, 87
  cancer and, 67
Myalgia, 201
Myasthenia gravis, 201
Myelin, 209, *210*
  disorders of, 108
Myelinated fiber, 209–10, 216
Myelin sheath, 209, *211*, 216
Myeloblasts, *310*
Myelocytes, *310*
Myeloid leukemia, 315, *315*
Myocardial infarction. *See* Heart attack
Myocardium, *330*, 331
Myoepithelial cells, 544–45, *545*
Myofibril, 55, *172*, 173, *173, 174*
Myoglobin, 43, 179
Myogram, 182, *182*
Myokymia, 201
Myology, 201

Myoma, 201
Myometrium, *510*, 511, *511*
Myopathy, 201
Myosin, 173–74, *173, 174*, 184
  in muscle contraction, 176, *176, 177, 178*
Myositis, 201
Myotomy, 201
Myotonia, 186, 201
Myotonic dystrophy, 186

N

Nail(s), 118, *118*
Nail bed, 118, *118*
Nail plate, 118, *118*
Nandrolone, 181
Nasal, definition of, *16*, 17
Nasal bone, 137t, *138, 139, 141, 142, 143, 167, 169*
Nasal cavity, 10, *10*, 257, 268, *397*, 435, *435, 437*, 442t
Nasal conchae
  inferior, 137t, *138, 141*, 142, *167*, 435, *436*
  middle, *138*, 140, *167*, 435, *436*
  superior, 140, 257, *257*, 435, *436*
Nasal septum, 257, 435
  deviated septum, 435
Nasolacrimal duct, 268, *268*
Nasopharynx, *397*, 400, 436, *436*
Naturally acquired immunity
  active, 381, 381t
  passive, 381, 381t
Navicular, *155, 156*
Neck
  arteries to, 353–54, *353*
  veins from, 357, *357*
Neck of femur, 154, *154*
Neck of tooth, 398
Neck of urinary bladder, 474
Necrosis, 116, 344
Negative feedback, 5–6, *6*, 285, *285, 286*, 293, *294*, 296, *296*, 308, 506, *506*
Neon, 32t
Neonatal period, 545–46, *546*, 547t
Neonate
  addiction in, 536
  circulation in, 545–46, *546*
  first breath, 443, 545
  immunity in, 379, 381
Neonatology, 17, 545
Nephrectomy, 477
Nephrolithiasis, 477
Nephrology, 17
Nephron, 462, *462*, 472t
  in acid-base balance, 490–91, *491*
  blood supply to, 464, *465*
  structure of, 463–64, *464*
Nephroptosis, 477
Nerve, 207, *220*
  types of, 220, *220*
Nerve cells. *See* Neuron
Nerve conduction velocity, 186

Nerve impulse, 12, 107, 175, 207, 215–16, *215*, 216t
  all-or-none response, 216
  conduction of, 216
  processing of, 219, *219*
Nerve pathway, 216, 220–21
Nerve tract, 224
Nervous system, 12, 205–47
  aging-related changes in, 547t
  compared to endocrine system, 281, 281t
  functions of, 207–8
  stimulation of glands, 285, *286*
  subdivisions of, 236t
Nervous tissue, 92, 92t, 107, *107*, 107t
Neuralgia, 247, 275
Neural stem cells, 47, 209
Neurilemma, 209–10, *211*
Neuritis, 247, 276
Neurofibril, 209, *210*
Neuroglial cells, 108, 205, 207–9, *207, 208*
Neurology, 17
Neuromuscular junction, 175, *175*
Neuron, 49, *49*, 67, 68, 107, 205, 207, *207*, 490
  classification of, 211–12, *211*
  membrane potential. *See* Membrane potential
  structure of, 209–10, *210*
Neuronal pools, 219, *219*
Neuropeptide, 217, 218t, 256–57
Neurotransmitter, 175–76, 184, 217–18, *217*, 218t, 219t
  of autonomic nervous system, 244, *244*, 245t
  excitatory actions of, 217, 218t
  inhibitory actions of, 217, 218t
Neutral solution, 37
Neutron, 31–32, *31*, 32t, 35t
Neutrophil(s), *310*, 311–13, *312*, 314t, 374
Neutrophilia, 324
Newborn. *See* Neonate
Niacin, 425t
Nicotinic acid. *See* Niacin
Night blindness, 274
Nipple, *22*, 515, *516*
Nissl body. *See* Chromatophilic substance
Nitric oxide, 218t, 342, 496, 504, 512
Nitrogen, 32t
  in air, 450
  in human body, 31t
Nocturnal emissions, 505
Node of Ranvier, 209, *210, 211*, 216
Nonelectrolytes, 38
Nonprotein nitrogenous substances, 316
Nonrespiratory air movements, 446
Nonspecific defense, against infection, 373–74
Nonsteroid hormones, 283–84, *284*

Norepinephrine, 184, 217, 218t, 244, *244*, 282t, 292–93, 293t, *299*, 340, 350
Normal range, 7
Normoblasts, *310*
Nose, 103, 435, 442t
Nostril, 435, *435, 436*
Nuclear envelope, 56, *57, 58*t, 65, *66*
Nuclear membrane, 50
Nuclear pore, 56, *57*
Nuclease, 405, 414t
Nucleic acids, 43–44, *44*, 44t, 79
  catabolism of, 489, *489*
Nucleolus, 50, 57, *57*, 58t, 65
Nucleotide, 43–44, *43*, 79, *80*
Nucleus, atomic, 31, *31*
Nucleus, cell, 49, *50*, 56–57, *57*
Nucleus, gray matter, 233
Nucleus pulposus, 157
Nutrients, 420–27
  in plasma, 316
Nutrition, 420–27
  definition of, 420
  in pregnancy, 540–41
Nystagmus, 276

O

Obesity, 101
Oblique section, 14, *14*
Obstetrics, 17
Obstructive jaundice, 410
Obturator foramen, *152*, 153
Obturator nerve, *239*, 240
Occipital, definition of, *16*, 17
Occipital artery, *353*, 354t
Occipital bone, 136–40, 137t, *139, 140, 141, 143, 168, 169*
Occipital condyle, *139*, 140, *168*
Occipitalis, 187, *188, 189*
Occipital lobe, 227, *228*
Occupational teratogens, 541
Oculomotor nerve (III), *236*, 237, 238t, *243*
Odorant molecules, 257–58
Olecranon fossa, *149*, 150
Olecranon process, 151, 157, *158*
Olfactory bulb, 236, *236*, 257
Olfactory code, 258
Olfactory complex, 258
Olfactory nerve (I), 236, *236*, 238t
Olfactory nerve pathway, 257–58
Olfactory organs, 257
Olfactory receptor, 257, *257*
Olfactory receptor cells, 236, 257–58, *257*
Olfactory stimulation, 258
Olfactory tract, 236, *236*, 257–58, *257*
Oligodendrocytes, 208–10, *208*
Oligomenorrhea, 513
Oliguria, 477
Oncogene, 67
Oncology, 17
Oocyte
  primary, 508–10, *509*
  secondary, 508–10, *508, 509*

Oogenesis, 508, *508*
Oophorectomy, 521
Oophoritis, 521
Ophthalmic artery, 354t
Ophthalmic division, of trigeminal nerve, 237, 238t
Ophthalmic vein, *357*
Ophthalmology, 17
Opiate, 235, 245
Opioid receptor, 235
Opportunistic infection, 382
Opsin, 274
Opsonization, 380, *380*
Optic chiasma, 232, *232*, 275, *275, 286*
Optic disc, *269*, 272, *273*
Optic foramen, 236
Optic nerve (II), *232*, 236–37, *236*, 238t, 269, *269*, 272, *272*, 275, *275, 286*
Optic radiations, 275, *275*
Optic tract, 232, *232*, 236, 275, *275*
Oral, definition of, *16, 17*
Oral cavity, 10, *10*, 396, *397, 435*
Oral contraceptives, *517*, 518
Orbicularis oculi, 187, *188*, 188t, *189*, 267, *267*, 269t
Orbicularis oris, 187, *188*, 188t, *189*
Orbit, 267
Orbital, definition of, *16, 17*
Orbital cavity, 10, *10*
Orchiectomy, 499, 521
Orchitis, 504, 521
Organ, 7, *8*
Organelles, 7, *8*, 49, *50*, 58t
Organic substances, 37, 39–44, 44t
Organisms, 7, *8*
  requirements of, 4–5
Organization
  of human body, 8–13
  levels of, 7, *8*
Organ of Corti, 262–63, *263, 264*
Organ system, 7, *8*, 12–13
Organ transplant, 365. *See also specific organs*
  rejection of, 383
Orgasm
  female, 512
  male, 504–5
Origin, of skeletal muscle, 159, 185–87, *185*
Oropharynx, *397*, 400, 436, *436*
Orthopedics, 17, 161
Osmoreceptor, 288, 484
Osmosis, 60–61, *60*, 63t, 470, 485, *485*
  exchanges in capillaries, 345, *345*
Osmotic pressure, 60, 315, 483, *483*
Osseous labyrinth, 261–62
Ossification center
  primary, 130, *131*
  secondary, 130, *131*
Ostealgia, 161
Ostectomy, 161

Osteoarthritis, 158, 170
Osteoblasts, 130–33, 291
Osteochondritis, 161
Osteochondroma, 162
Osteoclasts, 130–33, *132*, 290–91
Osteocytes, 103, *104*, 129–30, *129*, 291
Osteogenesis, 161
Osteogenesis imperfecta, 161
Osteoma, 161
Osteomalacia, 161, 391
Osteomyelitis, 162
Osteon, 103, 129, *129*
Osteonecrosis, 162
Osteopathology, 162
Osteoporosis, 135
Osteotomy, 162
Otic, definition of, *16, 17*
Otic ganglion, *243*
Otitis media, 276, 397
Otolaryngology, 17
Otolith, 265, *265*
Otosclerosis, 276
Outer tunic, 239
Oval window, *260*, 261, *262*
Ovarian cycle, 513–15, *514*, 515t
Ovarian follicle, 508
Ovary, *26, 27, 282,* 287, 298, 507–10, *507, 509, 510,* 512t, 513
Overnutrition, 426
Overweight, 101
Oviduct. *See* Uterine tube
Ovulation, *509*, 510, *510*, 514, *514,* 528
Oxidation, 77
Oxidative phosphorylation, 78
Oxycodone, 235
Oxygen, 31t, 32t, 38, 38t
  in air, 450
  alveolar gas exchange, 449–51, *449, 451*
  diffusion into body cells, 59, *59*
  diffusion through alveolar walls, 440, *441*
  for muscle contraction, 179
  requirement for, 5
  transport in blood, *329*, 451–52, *451*, 453t
Oxygen-carrying capacity, of blood, 308–9
Oxygen debt, 179, *180*
Oxyhemoglobin, 307, 451–52, *451*
Oxytocin, 282t, 288, 289t, 543, 545

P

Pacemaker, 338
Pacinian corpuscle, 253, *254*
Packed red cells, 324
Pain, 221, 232, 253–54, 374
  referred, 255, *255, 256*
  regulation of pain impulses, 256–57
  visceral, 255, *255, 256*
Pain nerve fiber, 255–56

Pain receptor, *222*, 253–54, 260
Palate, 396–97, *396*
Palatine bone, 137t, *139*, 141, *141, 168*
Palatine process, 141
Palatine tonsils, *396,* 397, *397, 436*
Palmar, definition of, *16, 17*
Palmaris longus, 193, *193*, 195t
Palpitation, 359
Pancreas, *11, 27*, 98t, 285, 295–96, *295, 296*, 393, *393, 394*, 404–6, *405, 406*, 412
  secretions of, 405–6, *406*
  structure of, 295, *295*, 404, *405*
Pancreatic acinar cells, 404
Pancreatic duct, *295*, 404, *405, 411*
Pancreatic juice, *393*, 405–6, *406*
Pancreatitis, acute, 405
Pancytopenia, 324
Pantothenic acid, 425t
Papillae
  dermal, *114,* 116
  renal, 461, *462*
  on tongue, 258, *259*, 396
Papillary muscles, 331–32, *331*
Pap smear, 511
Paracrine secretion, 281
Paralysis, 201
Paranasal sinus, 436, 442t
Paraplegia, 247
Parasympathetic division, of autonomic nervous system, 240–41, *243*, 245t
Parathyroidectomy, 300
Parathyroid gland, *282,* 290–92, *291*
Parathyroid hormone (PTH), 133, *134,* 282t, 290–92, *291*, 296, 487, 533, 533t
Paravertebral ganglia, 241
Paresis, 201
Parietal bone, 136, 137t, *138, 139, 140, 141, 143, 167, 168, 169*
Parietal cells, *402*, 403, 403t
Parietal lobe, 227, *228*
Parietal pericardium, 10, *11*, 329, *330*
Parietal peritoneum, 10, *11, 412*
Parietal pleura, 10, *11,* 441, *441,* 443
Parkinson disease, 47, 230
Parotid gland, 399, *400*
Partially complete protein, 423
Partial pressure, 450, *451,* 452
Passive immunity
  artificially acquired, 381, 381t
  naturally acquired, 381, 381t
Passive transport, 470
Patella, *135,* 136, 137t, 154, 157, *157, 196, 198*
Patellar, definition of, *16, 17*
Patellar ligament, 155, *196, 198,* 221, *221*
Patellar tendon reflex. *See* Knee-jerk reflex

Patent ductus arteriosus, 546
Pathogen, 373. *See also* Infection
Pathology, 17
Paxil, 218
PCP, 235
Pectoral, definition of, *16, 17*
Pectoral girdle, 136, 137t, 147–48, *148*
  muscles that move, 190–91, 191t, *192*
Pectoralis major, *22, 23,* 187, *188,* 191, *192,* 193t, 515, *516*
Pectoralis minor, *23,* 191, 191t, *192,* 443, *443,* 516
Pedal, definition of, *16, 17*
Pediatrics, 17
Pedicle, 142, *145,* 145t
Pediculosis, 122
Pelvic, definition of, *17*
Pelvic brim, 153, *153*
Pelvic cavity, 8, *9*
  male versus female, 154t
Pelvic diaphragm, 194
Pelvic girdle, 136, 137t, 151–53, *152*
  male versus female, 154t
Pelvic inflammatory disease, 521
Pelvic outlet, muscles of, 194, *195,* 195t
Pelvis, 136, 152
  arteries to, 355
  female, *153*
  male, *153*
  veins from, 359
Penicillin, 375, 382
Penis, *24, 25,* 195, 498, 503, 505t, 521
Pepsin, 374, 403, 403t, 414, 414t
Pepsinogen, 403, 403t
Peptidase, 413, 414t
Peptide bond, 74–75, *75,* 83, *83*
Peptide hormones, 283
Percutaneous transluminal angioplasty, 344
Perfluorocarbons, 305
Perforating canal, 129, *129*
Pericardial artery, 352t
Pericardial cavity, *9,* 10, *27,* 330, *330, 441*
Pericardial membranes, 10
Pericardial sac, *24*
Pericardiectomy, 362
Pericarditis, 330
Pericardium
  fibrous, 329, *330*
  parietal, 10, *11,* 329, *330*
  visceral, 10, *11,* 329, *330*
Perichondrium, 102
Perilymph, 262, *262*
Perimetrium, *510,* 511, *511*
Perimysium, 172, *172*
Perinatology, 546
Perineal, definition of, *16, 17*
Perineurium, *220*
Periodontal ligament, 398, *399*
Periosteum, 128, *128, 129,* 130, *132,* 172
Peripheral, definition of, 13

Peripheral chemoreceptor, 448
Peripheral membrane protein, 51
Peripheral nervous system (PNS), 207, *212*, 235–40, 236t. *See also* Autonomic nervous system
  cranial nerves, 236–37, *236*, 238t
  spinal nerves, 238–40, *239*
Peripheral resistance, 348–49, *348*
Peristalsis, 184, *393*, 395, *396*, 401, 404, 416–17, 419, 473
Peristaltic rush, 416
Peritoneal cavity, 10, *11, 412*
Peritoneal membranes, 10
Peritoneum
  parietal, 10, *11, 412*
  visceral, 10, *11*, 394, *412*
Peritonitis, 417
Peritubular capillary, *463*, 464, *465*, 469–70, *469, 470, 472*
Permanent teeth. *See* Secondary teeth
Peroneal artery, *356*
Peroneus longus, *188*, 198, *198, 199*, 199t
Peroneus tertius, 198, *199*, 199t
Peroxisomes, 55, 58t
Perpendicular plate, 140, *141*
PET. *See* Positron emission tomography
Peyer's patches, 370
pH, 37
  of blood, 37, 491–93, *491*
  of gastric juice, 374
  pH scale, 37, *37*
Phagocytes, 62, 313, 374
Phagocytosis, 62–63, *63*, 63t, 374
Phalanges
  of fingers, *135*, 136, 137t, 151, *151*
  of toes, *135*, 136, 137t, *155*, 156, *156*
Pharmacology, 17
Pharyngeal fossa, *436*
Pharyngeal tonsils, *397, 397*
Pharyngitis, 428
Pharynx, 393, *393*, 400–402, 435, *436, 437*, 442t
Pheochromocytoma, 300
Pheromones, 279
Phlebitis, 362
Phlebotomy, 362
Phosphate, 38, 38t, 78
  in blood, 290–91, 317, 467t
  in body fluids, 482, *483*
  electrolyte balance, 487–88
  in glomerular filtrate, 467t
  reabsorption in renal tubule, 470, *470*
  in urine, 467t
Phosphate buffer system, 489–90, 490t, *491*
Phosphate group, 43
Phosphodiesterase, 284
Phospholipids, 40, *41*, 41t
  membrane, 51, *51*
Phosphoprotein, 489, *489*

Phosphoric acid, 489, *489*
Phosphorus, 31t, 424–25
  functions and distribution in body, 426t
  intestinal absorption of, 391
  requirements and dietary sources, 426t
Phosphorylation, of proteins, 283, *284*
Photoreceptor, 253, 272
Phrenic artery, 352t, 355
Phrenic nerve, *239*, 240, 442
Physiologic jaundice, 311
Physiology, 3
Pia mater, 222, *223*, 231
Pigmented epithelium, 273–74, *274*
Pilocarpine, 428
Pimples, 118
Pineal gland, 232, *232*, 282, 297–98
Pinocytosis, 62–63, 63t
Pisiform, *151*
Pituitary gland, *232*, 282, 286–88, *286*
  anterior, 285–87, *286, 287, 288*, 289t
  control of male reproductive system, 505–6, *506*
  intermediate lobe, 286
  posterior, 232, 286, *286, 287, 288*, 289t
Pivot joint, 158, 159t
Placenta, 298, 513, 530, *531*, 533, 533t, 535–36, *536, 537, 539*, 540, 543, *544*
Placental lactogen, 532, 533t, 544
Placental membrane, 535–36, *536*
Plantar, definition of, *16*, 17
Plantar flexion, 159, *160*
Plant proteins, 423
Plaque, atherosclerotic, 344, *344*
Plasma, 103, *104*, 307, *307*, 314–17, 369, 482–83, *482*
  composition of, 467t
  nutrients of, 316
Plasma cells, *310*, 376–77, *379*
Plasma membrane. *See* Cell membrane
Plasmaphoresis, 324
Plasma proteins, 314–16, 316t, 345, 348, 486, 486t
Platelet(s), 103, *104*, 307, *307, 310*, 314, 314t, 324
  formation of, 133
Platelet count, 314
Platelet plug, 317, *317, 320*
Platysma, 187, 188t, *189*
Pleura
  parietal, 10, *11*, 441, *441, 443*
  visceral, 10, *11*, 440–41, *441, 443*
Pleural cavity, *9*, 10, *11, 27*, 441
Pleural membranes, 10, 443
Pleurisy, 456
Plexus, 239
Pneumoconiosis, 456
Pneumonia, 447, 450, 492

Pneumotaxic area, *447*, 448
Pneumothorax, 445
PNS. *See* Peripheral nervous system
Podiatry, 17
Poikilocytosis, 324
Poison, 76
Polar body, *530*
  first, *508, 508, 509*, 510, *529*
  second, 508, *508*, 528
Polarization, 212
Poliomyelitis, 201
Polydipsia, 297
Polymorphism, 84
Polypeptide, 75
Polyphagia, 300, 428
Polysaccharide, 39, *39*
Polyunsaturated fatty acid, 40
Polyuria, 477
Pons, *226, 227, 232, 233, 234*
Popliteal, definition of, *16*, 17
Popliteal artery, 355, *356*
Popliteal vein, 359, *360*
Pore, *117*, 120
Positive feedback, 318, 543
Positron emission tomography (PET), 276
Postcoital test, 532t
Posterior (dorsal), definition of, 13
Posterior cavity of eye, *269*, 272
Posterior chamber of eye, *269*, 271
Posterior communicating artery, *354*
Posterior fontanel, *143*
Posterior funiculus, 224, *225*
Posterior horn, 224, *225*
Posterior median sulcus, 224, *225*
Postganglionic fiber, 241, *241*
Postganglionic neuron, *242*, *243*, 292
Postnatal period, 528, 544–46, 547t
Postpartum period, 546
Postpolio syndrome, 201
Postsynaptic neuron, 216–17, *216, 217*
Postural drainage, 454
Posture, 183, 234, 264
Potassium, 31t, 38, 38t, 425
  in action potential, 213–15, *215*
  in blood, 293, 295, 317, 467t, 471
  in body fluids, 482, *483*
  in bone, 134
  electrolyte balance, 487–88, *487*
  functions and distribution in body, 426t
  in glomerular filtrate, 467t
  in heart action, 341
  imbalances of, 488
  in membrane potential, 213, *214*
  requirements and dietary sources, 426t
  secretion from renal tubule, 472, *472*
  in urine, 467t

Potassium channel, 52
  faulty, 52
Potential difference, 213
P-Q interval, 340
Precapillary sphincter, 343, *343*
Precipitation reaction, 380
Precocious puberty, 284
Preeclampsia, 547
Preganglionic fiber, 241, *241*
Preganglionic neuron, *242, 243*
Pregnancy, 511, 528–42
  drug abuse during, 536
  fetal cells in maternal circulation, 386, 542
  hormonal changes during, 531–33, *531, 533*, 533t
  preeclampsia in, 547
  Rh blood group and, *323, 323*
Pregnancy test, 532
Pregnanediol, 532
Premature infant, 538, 545
Prenatal development, 286, 528–42, 533t
  of bones, 130–31, *130, 131*
Prepatellar bursa, *157*
Prepuce, *498*, 503
Pressure, requirement for, 5
Pressure sense, 253, *254*
Pressure ulcer, 116
Presynaptic neuron, 216, *216, 217*
Primary bronchus, 438, *439*
Primary follicle, *509*
Primary germ layers, 533, *534*
Primary immune response, 380, *380*
Primary sex organs
  female, 507
  male, 498, 506
Primary teeth, 397–98, 398t
Prime mover, 187
Primordial follicle, 508–9, *509*
Prion protein, 44, 69
Problem solving, *228*
Process, bony, 128, 137t
Proerythroblasts, *310*
Progesterone, 298, 513–15, *514*, 519, 528, 532, 542, 544
  placental, 532, *533*, 533t
Progranulocytes, *310*
Projection, 253
Prolactin, 282t, 287, 289t, 544–45
Pronation, 159, *160*
Pronator quadratus, 193, *193*, 194t
Pronator teres, 193, *193*, 194t
Pronucleus, 528, *530*
Propelling movements, of alimentary canal, 395, *396*
Prophase, *64, 65, 66*
Proprioceptor, 267, 450
Prostaglandin, 502, 528, 542
Prostaglandin D_2, 382
Prostatectomy, 521
Prostate gland, *474, 498*, 502, 505t
  cancer of, 131, 503
  enlargement of, 503, 503t

Prostate-specific antigen (PSA), 503
Protease inhibitor, 383
Protective function, of bone, 131–32
Protein(s), 41–43, *42, 43,* 44t, 422–23
　catabolism of, 75, *80*
　complete, 423
　conformation of, 43–44, *43*
　denaturation of, 43
　dietary sources of, 422–23, 423t
　digestion of, 405, 413, 414t
　incomplete, 423
　membrane, 51, *51*
　metabolism in liver, 408
　partially complete, 423
　phosphorylation of, 283, *284*
　plant, 423
　requirements for, 423
　structure of, *42, 43*
　synthesis of, 53, 74, 75, 81–83, *82, 83,* 293
　in urine, 473
Protein buffer system, 490, 490t, *491*
Protein hormones, 282t, 283
Protein kinase, 283, *284*
Proteinuria, 468, 494
Proteomics, 84
Prothrombin, 318, *320*
Prothrombin activator, 318, *320*
Prothrombin time (PT), 318
Proton, 31–32, *31,* 32t, 35t
Protraction, 159, *161*
Proximal, definition of, 13
Proximal convoluted tubule, *463, 464,* 465, *466, 467,* 469–70, *470,* 472t
Prozac, 218
Pruritus, 122
PSA. *See* Prostate-specific antigen
Pseudostratified columnar epithelium, 94, *95,* 99t
Psoas major, 196, *196,* 197t
Psoriasis, 122
Psychiatry, 17
Psychological stress, 298
PT. *See* Prothrombin time
Pterygium, 276
PTH. *See* Parathyroid hormone
PTT. *See* Thromboplastin time
Puberty, 298
　female, 508–9, 513, 516
　male, 500, 506
　precocious, 284
Pubic arch, 153, *153*
Pubis, 152–53, *152*
Pulmonary artery, *26, 329, 331, 332–33, 334, 335,* 351, *440,* 539
Pulmonary cancer, primary, 444
Pulmonary circuit, *329,* 351
Pulmonary embolism, 319
Pulmonary plexus, *242, 243*
Pulmonary trunk, *25, 26, 330, 331,* 332–33, *335,* 351, *539,* 540

Pulmonary valve, 332–33, *332,* 332t, *333, 336, 336*
Pulmonary vein, *26, 329, 331,* 332–33, *334, 335,* 351, *440, 539*
Pulp, of tooth, 398
Pulp cavity, 398, *399*
Pulse, 299, 347, 355
"Pump," 62
Pupil, *269,* 271
　dilatation of, 271, *271*
Purkinje fibers, 331, *338, 339, 339*
Purpura, 324
Pus, 374
Pustule, 118, 122
Putamen, 229, *230*
P wave, 339, *340*
Pyelolithotomy, 477
Pyelonephritis, 477
Pyelotomy, 477
Pyloric canal, *401,* 402
Pyloric region, of stomach, *401,* 402
Pyloric sphincter, *401,* 402, 405
Pyloric stenosis, 428
Pylorospasm, 428
Pyorrhea, 428
Pyramidal cells, 228
Pyramidal tract, 224, *232*
Pyramiding (anabolic steroids), 181
Pyruvic acid, *77,* 179
Pyuria, 477

Q

QRS complex, 339–40, *340*
Quadriceps femoris muscle group, 196, 198t, 221
Quadriplegia, 247
Q wave, 339

R

Radial artery, 354–55, *355, 356*
Radial nerve, *239,* 240
Radial notch, 150, *150*
Radial recurrent artery, *355*
Radial tuberosity, 150, *150*
Radial vein, 357, *357, 360*
Radiation therapy, 519
Radioactive isotopes, 33, *33*
Radiology, 17
Radiotherapy, 33
Radius, 133, *135,* 136, 137t, *148,* 150, *150, 151, 158*
Raloxifene, 519
Rate-limiting enzyme, 79
Reactive oxygen species, 87
Reasoning, 229
Receptor, 5, *5, 6, 7,* 41, 51, *221,* 222t, 253
　types of, 253
Receptor cells, 212
Receptor end, 212
Receptor-mediated endocytosis, 63, 63t, *64,* 415
Recruitment, of motor units, 183
Rectal vein, 418

Rectum, *26, 28, 393,* 412, *416, 417, 418,* 507
Rectus abdominis, *22, 23,* 188, 192, 194, 195t, 445
Rectus femoris, *23, 25, 26, 27,* 188, 196, *196, 197,* 198t
Recurrent laryngeal nerve, 437
Red blood cell(s), *29,* 68, 103, *104,* 307, *307, 308,* 314t, 451–53, *452, 453,* 506
　antigens of, 320–23
　destruction of, 309, *311*
　hemolysis of, 61
　life span of, 308
　production of, 133, 308–9, *309, 310, 311,* 462
　shrinkage of, 61
Red blood cell count, 308
Red blood cell substitute, 305
Red marrow, 103, *128,* 133, 307–8, *310, 311,* 312, 315, 375, *375*
Red pulp, of spleen, 371, *373*
Reduction division, 65
Referred pain, 255, *255, 256*
Reflex, 220–21, *221,* 446
Reflex arc, 220, 222t
Reflex center, 220–21, 233–34
Refraction, light, 273, *273*
Regenerative medicine, 134
Regurgitation, 402
Reissner's membrane. *See* Vestibular membrane
Relative position, 13
Relaxin, 532–33, 533t
Releasing hormones, 286, *287*
Renal artery, 351, *352,* 352t, *356,* 461, 462–63, *463, 539*
Renal capsule, *462*
Renal column, *462, 462*
Renal corpuscle, 463–64
Renal cortex, 462, *462, 464, 465*
Renal failure. *See* Kidney failure
Renal medulla, 462, *462, 464, 465*
Renal pelvis, 461, *461, 463*
Renal plasma threshold, of glucose, 470
Renal sinus, 461
Renal tubule, *462,* 463–64, *464*
Renal vein, *360,* 461, 463, *463*
Renin, 462, 468–69, *468*
Replication of DNA, 65, 86, *86*
　errors in, 87
Repolarization, 214
Reproduction, 4t, 13
Reproductive system, 496–522
　female. *See* Female reproductive system
　male. *See* Male reproductive system
Residual volume, 446, *446,* 447t
Respiration, 4t, 435
　aerobic, 78, 488–89, *489*
　cellular, 77–78, *77,* 179, 435
　external, 435
　internal, 435
Respiratory acidosis, 492, *492*

Respiratory activity
　exercise and, 6
　venous blood flow and, 350
Respiratory air volumes, 446–47, *446,* 447t
Respiratory alkalosis, 449, *493,* 493
Respiratory bronchiole, *439*
Respiratory capacities, 446–47, *446,* 447t
Respiratory center, 233, 447–48, *447, 448,* 490–93, *491*
Respiratory cycle, 446
Respiratory distress syndrome, 445
Respiratory membrane, 449–50, *449*
Respiratory system, 13, 433–56, 442t
　aging-related changes in, 547t
　functions of, 435
　organs of, 435–41, *435*
Responsiveness, 4t
Resting neuron, 212
Resting potential, 213, *214*
Reticular fiber, 100
Reticular formation, 233–35
Reticulocytes, *310*
Reticuloendothelial system. *See* Mononuclear phagocytic system
Retina, *269, 270,* 272–73, *272, 273*
Retinal, 274
Retinitis pigmentosa, 276, 284
Retinoblastoma, 276
Retraction, 159, *161*
Retroperitoneal region, 461
Reverse transcriptase, 382
Reversible reaction, 36
R group, *42, 43*
Rh blood group, 320, 322–23, *323*
　Rh-negative blood, 322–23, *323*
　Rh-positive blood, 322–23, *323*
Rheumatoid arthritis, 158, 386
Rhinitis, 456
Rhodopsin, 274
RhoGAM, 323
Rhomboideus, *188*
Rhomboideus major, 191, *191,* 191t
Rhythmicity, of smooth muscle, 184
Rhythmicity area, medullary, 447–48
Rhythm method, 516–17
Riboflavin, 418, 425t
Ribose, 420
Ribosome, 50, 53, *53,* 81–83, *82, 83*
　formation of, 57
Ribs, *11, 28, 135,* 136, 137t, 144, 147, *147, 148,* 443, *443*
　first rib, 157
Rickets, 391
Rigor mortis, 180
RNA, 43–44, *44,* 81
　comparison to DNA, 86t
RNA polymerase, 81

Rods, *272,* 273–74, *274*
Root canal, 398, *399*
Root of tongue, 396
Root of tooth, 398
Rotation, 159, *160*
Rough endoplasmic reticulum, *50,* 53, *53*
Round ligament of liver, 407
Round ligament of uterus, *26*
Round window, *260, 262, 262*
Rubella, 540, 546
Rugae, *401,* 402
R wave, 339

S

Saccule, *262,* 265
Sacral, definition of, *16, 17*
Sacral artery, *352,* 352t
Sacral canal, 145, *146*
Sacral foramen
　anterior, *28*
　dorsal, 145, *146*
　pelvic, 145, *146*
Sacral hiatus, 145, *146*
Sacral nerves, 238
Sacral promontory, *153*
Sacroiliac joint, 153, 158, 533
Sacrum, *28, 135,* 136, 137t, *144,* 145, 145t, *146,* 151, *152,* 154t
Saddle joint, 158, 159t
Sagittal plane, 14, *14*
Sagittal suture, 136, *143*
Saliva, *393,* 396, 428
Salivary gland, 98t, *393, 393,* 399–400, *400,* 428
Saliva substitute, 428
Salpingectomy, 521
Salt(s), 38
Saltatory conduction, 216
Salt craving, 487
Salty taste, 259–60
S-A node. See Sinoatrial node
Saphenous vein
　great, 359, *360*
　small, 359, *360*
Sarcolemma, *172, 173, 174, 175*
Sarcomere, *173, 174, 174*
Sarcoplasm, 173, *174*
Sarcoplasmic reticulum, *172, 173,* 174, *176*
Sartorius, *22, 23, 24, 26, 27,* 187, *188,* 196, *196, 197,* 198t, *199*
Scab, 121
Scabies, 122
Scala media, *263*
Scala tympani, 262, *263*
Scala vestibuli, 262, *263*
Scalp, *223*
Scaphoid, *151*
Scapula, *135,* 136, 137t, 147–48, *148, 149, 157, 192*
Scar, 107, 121
Schwann cells, 209, *210, 211*
Sciatic nerve, *239,* 240
SCID. See Severe combined immune deficiency
Scintillation counter, 33, *33*

Sclera, 269, *269, 270, 272*
Scleral venous sinus, 271
Scleroderma, 386
Scrotum, *25, 195, 498,* 503, 505t
Scurvy, 424
Sealant, for teeth, 399
Sebaceous gland, 98t, *114,* 116, *117,* 118, *118*
Seborrhea, 122
Sebum, 118
Secondary bronchus, *439*
Secondary immune response, 380, *380*
Secondary sex characteristics
　female, 513, 515
　male, 506
Secondary teeth, 398, *398,* 398t
Second messenger, 283–84
Second ventricle, 230
Secretin, 405, *406,* 410t
Secretion, 54, *54,* 482
Secretory vesicle, *50*
Sedation, 235
Seizure, 218, 490
Selective estrogen receptor modulator, 519
Selectively permeable membrane, 51
Selective serotonin reuptake inhibitor, 218
Selenium, 425, 427t
Sella turcica, 140, *140, 141, 168,* 286
Semen, 502
Semen analysis, 504, *504*
Semicircular canals, 262, *262,* 265–67, *266*
Semilunar valve, 332
Semimembranosus, *188,* 196, *197, 198t, 199*
Seminal vesicle, *474, 498,* 501–2, 505t
Seminiferous tubules, 498, *499,* 500
Semispinalis capitis, *189,* 190, 190t
Semitendinosus, *188,* 196, *197,* 198t, *199*
Senescence, 547t
Sensation, 253
Sensorineural deafness, 264
Sensory adaptation, 253, 258, 260
Sensory area, of cerebral cortex, 229
Sensory function, of nervous system, 207
Sensory nerve, 220
Sensory nerve fiber, 220, 253, *254*
Sensory neuron, 212, *212, 221, 222,* 222t
Sensory receptor, 116, 207, 253
Septicemia, 324
Septum, of heart, 331, *331*
Serotonin, 217, 218t, 256–57, 317
Serous cells, 98, 399
Serous fluid, 10, 98, 113, 482
Serous layer
　of alimentary tract, 394, *395*
　of large intestine, *416*

　of urinary bladder, 474, *474*
Serous membrane, 113
Serratus anterior, *22, 23,* 188, 191, 191t, *192*
Serum, 318
Set point, 5, *5, 6,* 7
　of body temperature, 120, *121,* 122
Severe combined immune deficiency (SCID), 384–85, *384, 385*
Sex assignment, 521
Sex hormones
　adrenal, 294–95, 295t
　female, 513
　male, 506–7, *506*
Sexually transmitted disease (STD), 520–21, 520t, 532
Shigatoxin, 459
Shin bone. See Tibia
Shin splint, 201
Shivering, 6, 120
Shock, 467
Shoulder, *157,* 158
　arteries to, 354–55, *355*
　veins from, 357–58, *357*
Shoulder blade. See Scapula
Sickle cell disease, 134, 309
Sight, 267–75
Signal transduction, 50–51
Simple columnar epithelium, 93, *94,* 99t
Simple cuboidal epithelium, 93, *94,* 99t
Simple diffusion, 58–59, *59*
Simple squamous epithelium, 92–93, *93,* 99t
Simple sugar, 39
Single covalent bond, 35
Single nucleotide polymorphism (SNP), 84
Sinoatrial node (S-A node), 338, *338, 339,* 340, *341,* 349
Sinus, 10, *10,* 136, 137t, *138*
Sinus headache, 436
Sinusitis, 456
Sinus rhythm, 362
Sjogren's syndrome, 428
Skeletal muscle. *See also specific muscles*
　of abdominal wall, 194, 195t
　actions of, 185–87, *185*
　atrophy of, 183
　characteristics of, 185t
　connective tissue coverings of, 172–73, *172*
　contraction of. See Muscle contraction
　discovery of new muscle, 190
　of eyelids and eye, 269t
　of facial expression, 187, 188t, *189*
　hypertrophy of, 183
　insertion of, 159, 185–87, *185*
　interactions of muscles, 187
　lymph movement and, 369
　major muscles, 187–98

　of mastication, *189,* 190, 190t
　origin of, 159, 185–87, *185*
　of pelvic outlet, 194, *195,* 196t
　structure of, 172–75, *172*
　that move ankle, foot, and toes, 198, 199t
　that move arm, 191–93, *192,* 193t
　that move forearm, 193, *193, 194,* 194t
　that move head, *189,* 190, 190t
　that move leg, 196, *198,* 198t, *199*
　that move pectoral girdle, 190–91, 191t, *192*
　that move thigh, 196, *196, 197,* 197t
　that move wrist, hand, and fingers, 193–94, 195t
　venous blood flow and, 350
Skeletal muscle fiber, *172,* 173–74, *173, 174*
Skeletal muscle tissue, 105, *105,* 107t
Skeletal system, 12, 126–62, 166
　aging-related changes in, 547t
Skeleton
　female, 154t
　male, 154t
　organization of, 134–36, *135*
　weight of, 136
Skeleton of the heart, 333, *333*
Skin, 92, 94–96, 111–24
　accessory organs of, 117–20
　cancer of, 119, *119*
　color of, 116
　stem cells of, 124
　tissues of, 113–17, *114, 115*
Skin cells, *68*
Skull, 134, *135,* 136–42
　anterior view of, *138, 167*
　floor of cranial cavity, *168*
　infantile, 142, *143*
　inferior view of, *139*
　lateral view of, *139, 169*
　male versus female, 154t
　of prehistoric humans, 126, *126*
　sagittal section of, *141*
Sleep, 233–34
　position of body during, 170, *170*
Sleep-wake cycle, 299–300
Sliding filament model, 176, *177*
Slightly movable joint, 156
Slow muscle fiber, 183
Slow twitch muscle fiber, 182
Small intestine, *11, 24, 25, 393, 393, 394, 395,* 411–17, *412*
　absorption in, 414–16, *415,* 415t
　cellular turnover in, 413
　movements of, 416–17
　parts of, 411–12, *411, 412*
　secretions of, 413–14, 414t
　wall of, 412–13, *413*
Small intestine transplant, 365
Smallpox, 381

Smell, sense of, 236, 253, 257–58, *257*, 279
Smog, *433*
Smoking, in pregnancy, 540
Smooth endoplasmic reticulum, 50, 53, *53*
Smooth muscle, 184, 185t
   of blood vessel walls, 317, 342
   contraction of, 184
   lymph movement and, 369
   multiunit, 184
   rhythmicity of, 184
   visceral, 184
Smooth muscle cells, 90, *90*
Smooth muscle fiber, 184
Smooth muscle tissue, 105–6, *106*, 107t
Sneezing, 233, 446
SNP. *See* Single nucleotide polymorphism
Sodium, 31t, 32t, 38, 38t, 424–25
   in action potential, 213–15, *215*
   active transport of, 62
   atoms of, 34, *34*
   in blood, 293, 295, 317, 467t, 471
   in body fluids, 482, *483*
   in bone, 134
   dietary, 349
   electrolyte balance, 487–88, *487*
   in extracellular fluid, 484
   functions and distribution in body, 426t
   in glomerular filtrate, 467t
   imbalances of, 488
   in membrane potential, 213, *214*
   reabsorption in renal tubule, 470–71, *470*
   requirements and dietary sources, 426t
   in urine, 467t
Sodium channel, 52
   faulty, 52
Sodium chloride, 34, *34*, 36
Sodium hydroxide, 37, 489
Sodium-potassium pump, 213, *214*
Soft palate, 396, *396*, *397*, 401, *435*
Soft spot. *See* Fontanel
Soleus, *188*, 198, *198*, *199*, 199t
Solute, 38
Solvent, 38
Somatic nervous system, 207, 235
Somatic senses, 253–57
Sound intensity, 264
Sour taste, 259–60
Specialized cells, 67
Specialized connective tissue, 100
Special senses, 257–75
Species resistance, 373–74
Specific defenses, against pathogens. *See* Immunity
Speech, 228–29, *228*, 237, 436–37
   vocal problems, 437
Spermatid, 500, *501*, *502*

Spermatocyte
   primary, 500, *501*, *502*
   secondary, 500, *501*, *502*
Spermatogenesis, 500, *502*
Spermatogenic cells, 499–500
Spermatogonia, *499*, 500, *501*
Sperm cells, 56, *57*, 65, 87, 498, *499*, 500, *500*, *501*, *502*, 508, 528, 529, 530
   abnormal, 504, *504*
   formation of, 500
   transport of, 528
Spermicide, 517–18, *517*
S phase, 64, 65, *66*
Sphenoidal sinus, *10*, *138*, 140, *141*, *286*, *397*, 436, *436*
Sphenoid bone, 137t, *138*, *139*, 140, *140*, *143*, *167*, *168*, *169*, *286*
Sphenoid fontanel, *143*
Sphenomandibularis, 190
Sphenopalatine ganglion, *243*
Spherocytosis, 324
Spina bifida, 142
Spinal branch, of accessory nerve, 237, 238t
Spinal cord, *11*, 223, *224*, *232*
   functions of, 224–25, *226*
   injury to, 47
   structure of, 224, *224*, *225*
Spinal nerves, 223, *224*, *225*
   injury to, 240
Spinal reflex, 225
Spindle fibers, 65, *66*
Spine (projection of bone), 137t
   of scapula, 148, *149*
Spinothalamic tract, *224*, *226*
Spinous process, 142–44, *145*, 145t
Spirometry, 446–47
Spleen, *11*, *25*, *26*, *27*, 371–73, *372*, *373*
Splenectomy, 386
Splenic artery, 351, *352*, *373*
Splenic vein, *358*, 359, *373*
Splenitis, 386
Splenius capitis, *189*, 190, 190t
Splenomegaly, 386
Splenotomy, 386
Spongy bone, 128–29, *128*, *129*, 130, *131*, 132, *132*
Squamosal suture, *138*, *139*, 140, *141*, *169*
Squamous cell carcinoma, 119, *119*
Squamous epithelium, 92
   simple, 92–93, *93*, 99t
   stratified, 94–96, *95*, 99t
Stacking (anabolic steroids), 181
Stanozolol, 181
Stapedius, 187
Stapes, 137t, *260*, 261, *261*
Staphylococcus aureus, 380
Starch, 39, 420
Starling's law of the heart, 349
Static equilibrium, 264–65, *265*
STD. *See* Sexually transmitted disease

Stem cell(s), 67, 118, 124, 362. *See also* Hemocytoblasts
   embryonic, 530, *530*
   from umbilical cord, 134
Stem cell island, 124
Stem cell technology, 47, *47*
Stem cell transplant, 315
Sterilization (contraception), 519–20
Sternal, definition of, 16, *17*
Sternocleidomastoid, *22*, *23*, 187, *188*, *189*, 190, 190t, *192*, 443, *443*
Sternum, *11*, *24*, *135*, 136, 137t, 147, *147*, *148*
Steroid, 40, *41*, 41t
   anabolic, 75, 181, *181*, 235
Steroid hormone, 282–83, 282t, *283*
   adrenal, 293–94, 294t
Stomach, *11*, *24*, *25*, *26*, 393, *393*, *394*, 401, 402–4, *402*, *411*, *412*
   absorption in, 404
   mixing and emptying actions of, 404
   parts of, *401*, *402*
   secretions of, 402–4, *402*, 403t, *404*
Stomatitis, 428
Stratified columnar epithelium, 96, *96*, 99t
Stratified cuboidal epithelium, 96, *96*, 99t
Stratified squamous epithelium, 94–96, *95*, 99t
Stratum basale, 113–14, *114*, *115*
Stratum corneum, 114, *114*, *115*
Stratum granulosum, 114, *115*
Stratum lucidum, 114, *115*
Stratum spinosum, 114, *115*
Strep throat, 122
Streptococcal infection, 332
Stress
   health and, 298–300, *299*
   responses to, 298–300, *299*
   types of, 298
Stressor, 298
Stress test, 350
Striations, muscle, 105, *105*, *106*, 173–74, *173*, *174*
Stringer, Korey, 480, *480*
Stroke, 299, 319, 349
Stroke volume, 348
Strong acid, 489
Strong base, 489
Structural formula, 36, *36*
Styloid process, *139*, 140, *141*, 150–51, *150*
Subarachnoid space, 222, *223*, 230, *231*
Subchondral plate, *157*
Subclavian artery, *25*, *26*, *27*, *28*, 327, 351, *352*, 352t, 353, *353*, *355*, *356*
Subclavian vein, *24*, *25*, *27*, 357, *357*, *358*, *360*, 368, 369, *369*
Subcutaneous injection, 113

Subcutaneous layer, 113, *114*, 116–17
Subdeltoid bursa, *157*
Subdural hematoma, 223
Sublingual gland, 400, *400*
Submandibular ganglion, *243*
Submandibular gland, 400, *400*
Submucosa
   of alimentary tract, 394, *395*
   of stomach, *402*
   of urinary bladder, 474, *474*
Subpatellar fat, *157*
Subscapularis, *24*, 191, *192*, 193t
Substance P, 218t
Substrate, 75–76, *76*
Suckling, 544
Sucrase, 413, 414t
Sucrose, 35, 39, 75
Sudoriferous gland. *See* Sweat gland
Sugar, 39
Sulcus, 226, *227*
Sulfate, 38, 38t
   in blood, 317, 467t
   in body fluids, 482, *483*
   electrolyte balance, 487–88
   in glomerular filtrate, 467t
   in urine, 467t
Sulfur, 31t, 425, 426t
Sulfuric acid, 489, *489*
Summation
   membrane potential, 213
   muscle contraction, 182, *182*
Sunlight exposure, 119, 391, *391*
Sunscreen, 119
Superantigen, 380
Superficial, definition of, 13
Superficial transversus perinei, 194, *195*, 196t
Superior, definition of, 13
Superior oblique, *268*, 269t
Superior orbital fissure, *167*
Superior rectus, *267*, *268*, 269t
Superior vena cava, *26*, *27*, *330*, 331, *331*, *334*, *335*, *357*, *360*, 539
Supination, 159, *160*
Supinator, 193, *193*, 194t
Support, 12
Support function, of bone, 131–32
Supporting cells
   nervous tissue, 107, *107*
   in spermatogenesis, 500
Suprachiasmatic nucleus, 298
Supraorbital foramen, 136, *138*
Supraorbital notch, 136, *167*
Suprapatellar bursa, *157*
Suprarenal artery, 351, 352t, *356*
Supraspinatus, 191, *191*, *192*, 193t
Surface tension, 443, 445
Surfactant, 443, 445, *449*, 545
Surgical neck, of humerus, *149*, 150
Surrogate mother, 535t
Suspensory ligament (breast), 516
Suspensory ligament (eye), *269*, 270–71, *270*
Sustained contraction, 183

Suture, 136, 137t, 156, *156*
Swallowing, 233, 237, 396–97, 437–38
　mechanism of, 400–401
S wave, 339
Sweat, 111, 120, 281, 374, 487
　salty, 454
Sweat gland, 98t, *114*, 116, 119–20
Sweat gland duct, *114*
Sweat gland pore, *114*
Sweating, 6, 111, *111*, 120, 480, 485–86, 488
Sweet taste, 259–60
Sympathetic chain ganglia, *242*
Sympathetic division, of autonomic nervous system, 240–41, *242*, 245t
Symphysis pubis, *27, 28*, 152, 153, *153*, 157, *498, 507*, 533
Synapse, 216–18, *216, 217*
Synaptic cleft, 175–76, *175*, 216, *216, 217*, 218
Synaptic knob, *210*, 216–18, *217*
Synaptic transmission, 216–18, *216, 217*
Synaptic vesicle, *175*, 217, *217*
Synergist (muscles), 187
Synesthesia, 235, 276
Synovectomy, 162
Synovial fluid, 113, 157, *157*, 482
Synovial joint, 113, 157–58, *157*, 159t
Synovial membrane, 113, 157, *157, 158*
Synthesis reaction, 36
Syphilis, 520t
Systemic circuit, *329*, 351
Systemic lupus erythematosus, 386
Systole, 336, *336*
Systolic pressure, 347, *347*

T

T₃. *See* Triiodothyronine
T₄. *See* Thyroxine
Tail, of sperm, 500, *500*
Talus, 155, *155, 156*
Tamoxifen, 519
Target cells, 12, 281
Tarsal, definition of, *16, 17*
Tarsals, *135*, 136, 137t, 155, *156*
Tarsus, 155, *155*
Taste, 229, 253, 258–60
Taste bud, 258–60, *259*, 396
Taste cells, 259–60, *259*
Taste hair, 259, *259*
Taste nerve pathway, 260
Taste pore, 259, *259*
Taste receptor, 237, 259, *259*
T cells, *310*, 371, 376t, 377t, 382
　activation of, 376
　cellular immune response and, 376–77
　cytotoxic, 376–77, 383
　functions of, 376
　helper, 377, *378, 379*
　in HIV infection, 382–83
　memory, 377
　origin of, 375, *375, 376*
Tears, 268, 374
Tectorial membrane, 263, *263*
Teeth, 397–99, *397, 398*, 398t, *399*
　dental caries, 399
　grinding of, 190
　primary, 397–98, 398t
　secondary, 398, *398*, 398t
Telomerase, 67
Telomere, 65
Telophase, *64*, 65, *66*
Temperature, body, 232, 341, 435, 480
　biological rhythms and, 299
　elevated, 122
　regulation of, 6, 7, 120, *121*
　set point of, 120, *121*, 122
Temperature sense, 232, 253–54
Temporal artery, *353*, 354t, *356*
Temporal bone, 137t, *138, 139*, 140, *140, 141, 143, 167, 168, 169*
Temporalis, *188, 189*, 190, 190t
Temporal lobe, 227, *228*
Temporal process, 142
Temporal vein, *360*
Temporomandibular joint syndrome, 190
Tendinitis, 173
Tendon, 100–101, 128, 172–73, *172*
Teniae coli, *416*, 418
Tenosynovitis, 173
Tenosynovium, 173
Tension headache, 258
Tensor fasciae latae, *23, 26, 27*, 188, 196, *196, 197*, 197t
Tentorium cerebelli, 227
Teratogen, 540
Teratology, 546
Teres major, *24*, 188, 191, *191, 192*, 193t
Teres minor, 188, 191, *191, 192*, 193t
Terminal bronchiole, *439*, 440
Tertiary bronchus, *439*
Testis, *25*, *282*, 287, 298, 498–500, *498, 499*, 505t
　cancer of, 499
　undescended, 504
Testosterone, 181, 282t, 298, 506–7, *506*, 513
Tetanic contraction, 182, *182*, 292
Thalamus, *226*, 230, 232–33, *232, 234*
Thalassemia, 324
Thalidomide, 540
Thallium-201, 33
Thecal cells, *509*
Thermoreceptor, 253
Thiamine, 418, 425t
Thick filament, *172*, 173–74, *173, 176, 176, 177*
Thigh, muscles that move, *196, 196, 197*, 197t
Thin filament, *172*, 173–74, *173, 174, 176, 176, 177*
Third ventricle, 230, *231, 232*
Thirst, 6, *468*, 484–85, 487
Thirst center, 484
Thoracic artery, internal, 355
Thoracic cage, 136, 137t, 147, *147*, 440
Thoracic cavity, *8, 9, 28*, 441, 443
Thoracic duct, 368, *368*
Thoracic membranes, 10, *11*
Thoracic nerves, 238
Thoracic vertebrae, 137t, 144, *144, 145*, 145t, 147, *147, 223*
Thoracic wall
　arteries to, 355
　puncture in, 445
　veins from, 358
Thought processing, 229
Threshold potential, 213
Threshold stimulus, for muscle contraction, 180–81
Thrombin, 318, *320*
Thrombocytes. *See* Platelet(s)
Thrombocytopenia, 315
Thrombophlebitis, 362
Thromboplastin time (PTT), 318
Thrombopoietin, 314
Thrombotic thrombocytopenic purpura, 324
Thrombus, 318, 333, 344
Thymectomy, 300, 386
Thymine, 79, *80*, 81
Thymitis, 386
Thymocytes, 371
Thymosin, 298, 371
Thymus gland, *282*, 298, 371, *372*, 375
Thyrocervical artery, 354
Thyrocervical axis, *353*
Thyroid artery, *353*, 354t
Thyroid cartilage, *24, 25*, 437, *437, 439*
Thyroidectomy, 300
Thyroid gland, *23, 24, 25*, 33, *33*, *282*, 284, 288–90, *289, 291*, 372
　hormones of, 289–90, *289*, 290t
　structure of, 288–89, *289*
Thyroiditis, 300
Thyroid-stimulating hormone (TSH), 282t, 287, *288*, 289t
Thyrotropin-releasing hormone (TRH), 282t, 287, *288*
Thyroxine, 289–90, 290t
Tibia, *135*, 136, 137t, 154–55, *155, 157, 198*
Tibial artery
　anterior, 355, *356*
　posterior, 355, *356*
Tibialis anterior, *188*, 198, *198, 199*, 199t
Tibialis posterior, 198, 199t
Tibial tuberosity, 154–55, *155*
Tibial vein
　anterior, 359, *360*
　posterior, 359, *360*
Tidal volume, 446, *446*, 447t
Tinea pedis, 121
Tinnitus, 276
Tissue, 7, *8*, 90–108
Tissue engineering, 90, *90*
Tissue fluid, 369, 482, *482, 483*
Tissue plasminogen activator, 319
Tissue rejection, 383
Tissue thromboplastin, 318
Toes
　mummy's prosthetic, 1, *1*
　muscles that move, 198, 199t
Tongue, 258–60, *259*, 394, 396, *396, 397, 400*, 436
Tonometer, 271
Tonsillectomy, 397
Tonsillitis, 397
Tonsils, 370, 396–97, *396, 397*
Tools, Bob, 362
Tooth. *See* Teeth
Torso
　female, *22, 26, 27, 28*
　male, *23, 24, 25*
Torticollis, 201
Total body water, 482
Total lung capacity, 446, 447, 447t
Touch receptor, *114*, 116
Touch sense, 232, 253, *254*
Toxemia of pregnancy, 547
Toxicology, 17
Toxic shock syndrome, 380
Toxoid, 381
Trace elements, 425, 427t
Trachea, *24, 26, 27*, 372, 397, *435, 436, 437, 438, 439*, 442t
Tracheotomy, 382
Trachoma, 276
Training, 179
Transcellular fluid, 482, *482, 483*
Transcription, 81, *82*, 86t, 282, *283*
Transcytosis, 69
Transdermal patch, 113
Transfer RNA (tRNA), *82*, 83, *83*
Transfusion, 305, 320–23, 348
Transfusion reaction, 320–23
Transitional epithelium, 96–97, *97*, 99t
Translation, 81–83, *82, 83*, 86t
Transmembrane protein, 51
Transmissible spongiform encephalopathy, 44
Transplant. *See specific organs*
Transport throughout body, 12
Transport vesicles, 54, *54*
Transverse fissure, 227, *227*
Transverse foramen, 142, *145*, 145t
Transverse (horizontal) plane, 14, *14*
Transverse process, 142–43, *145*, 145t
Transverse tubules, 174, *174*, 176, 184
Transversus abdominis, *23, 28*, 192, 194, 195t, 445
Trapezium, *151*
Trapezius, *22*, 188, 191, *191*, 191t, *192*
Trapezoid, *151*
Trepanation, 126

TRH. *See* Thyrotropin-releasing hormone
Triceps brachii, 133, *133*, 188, *192*, 193, *194*, 194t
Trichosiderin, 117
Tricuspid valve, 331–33, *331*, *332*, 332t, *333*, *334*, 336, *336*
Trigeminal nerve (V), *236*, 237, 238t
   mandibular division of, 237, 238t
   maxillary division of, 237, 238t
   ophthalmic division of, 237, 238t
Triglycerides, 40, *40*, 41t, 74, *74*
Trigone, 474, *474*
Triiodothyronine, 289–90, 290t
Trimester, 546
Triplets, 526
Triquetum, *151*
tRNA. *See* Transfer RNA
Trochanter, 137t
Trochlea, *149*, 150, *158*
Trochlear nerve (IV), *236*, 237, 238t
Trochlear notch, *150*, 151
Trophoblast, 531, *531*, 534
Tropomyosin, 176, *176*, *177*
Troponin, 176, *176*, *177*
True ribs, 147, *147*
True vocal cords, 437, *438*
Trypsin, 405, 414t
Trypsinogen, 405
TSH. *See* Thyroid-stimulating hormone
T tubules. *See* Transverse tubules
Tubal ligation, 520, *520*
Tubercle, 137t, 145
   of rib, 147
Tuberosity, 137t
Tubular necrosis, 467
Tubular reabsorption, 465, 469–70, *469*, *470*, *472*, 487–88, *487*
Tubular secretion, 465, *469*, 471–72, *472*, 487–88, *487*
Tubulin, 55, *55*, 65
Tumor suppressor gene, 67
Tunica externa, 342, *342*
Tunica interna, 342, *342*
Tunica media, 342, *342*
T wave, 339–40, *340*
Twins, 526
   dizygotic, 530, 546
   monozygotic, 530, 546
Twitch, 182–83
Tympanic cavity, *260*, 261
Tympanic membrane, *260*, 261
Tympanoplasty, 276
Type O blood, 320–23, *321*, 321t
Typhoid fever, 454

U

Ulcer
   definition of, 122
   gastrointestinal, 403
   leg, 124
   pressure, 116

Ulna, *135*, 136, 137t, *148*, 150, 151, *151*, *158*
Ulnar artery, 354, *355*, *356*
Ulnar nerve, *239*, 240
Ulnar notch, 151
Ulnar recurrent artery, *355*
Ulnar vein, 357, *357*, *360*
Ultradian rhythm, 299
Ultrasound, of fetus, 542, 546
Umami, 259–60
Umbilical, definition of, *16*, 17
Umbilical artery, *536*, 537, *539*, 541, 541t, 545, *546*
Umbilical cord, *536*, 537, *537*, *539*, 542, 545
Umbilical cord stem cells, 134, 315
Umbilical ligament, *539*
Umbilical region, 14, *15*
Umbilical vein, *536*, 537–38, *539*, 545
Umbilicus, 22
Undernutrition, 426
Unipolar neuron, 211, *211*
Universal donor, 321
Universal recipient, 321
Unmyelinated fiber, 209, 216
Unsaturated fatty acid, 40, *40*
Upper limb
   arteries to, 354–55, *355*
   bones of, 136, 137t, 148–51, *149*, *150*, *151*
   muscles that move arm, 191–93, *192*, 193t
   muscles that move forearm, 193, *193*, *194*, 194t
   veins from, 357–58, *357*
Upper respiratory tract, 435, *436*
Uracil, 81
Urea, 316, 422, *422*, 467t, 469, 471, 473
Uremia, 494
Uremic acidosis, 492
Ureter, *26*, *27*, *461*, 462, 473, *473*, *498*
Ureteritis, 474, 477
Urethra, *28*, *461*, 474, 475, *475*, *498*, *507*
Urethral gland, 475, *475*
Urethral orifice, *195*
   external, 503
Urethral sphincter
   external, 474, 475
   internal, 474–75, *474*
Urethritis, 477
Uric acid, 316, 467t, 469, 471, 473
Urinary bladder, *24*, *25*, *26*, *27*, *412*, *461*, 474, *474*, *498*, *507*
Urinary system, 13, 459–76, *461*
   aging-related changes in, 547t
   functions of, 461
Urine, 465, 487, 490
   composition of, 467t, 473
   concentration of, 471
   elimination of, 473–75
   formation of, 465–73
   volume of, 471, 473, 475, 487
   water output via, 485

Urine test, for anabolic steroids, 181
Urogenital diaphragm, 194
Urology, 17
Urticaria, 122
Uterine contractions, 288, 542–44
Uterine tube, *26*, *507*, *509*, 510, *510*, 512t
Uterus, *26*, *27*, *507*, 511, *511*, 512t. *See also* Pregnancy
   menstrual cycle, 513–15, *514*, 515t
Utricle, *262*, 265
Uveitis, 276
Uvula, 396, *396*, *397*, 436

V

Vaccine, 381
Vagina, *28*, *507*, *510*, 511, 512t
Vaginal orifice, *195*, 511
Vaginitis, 521
Vagotomy, 247
Vagus nerve (X), *236*, 237, 238t, *243*, 341, *341*, *404*, 449
Valve
   heart, 331–32, *331*, *332*, 332t
   lymphatic vessels, 368–69, *368*
   venous, *342*, 346, *346*
Valvotomy, 362
Varicocele, 521
Varicose veins, 359
Vascular headache, 258
Vas deferens, *25*, *474*, *498*, 499, *499*, 501, 505t, 520
Vasectomy, 519–20, *520*
Vasoconstriction, 233, 258, 317, 342–43, 349–50, 449, 468
Vasodilation, 233, 258, 342–43, 349, 468
Vasomotor center, 233, 349
Vasomotor fiber, 342
Vasospasm, 317, *320*
Vastus intermedius, *27*, 196, 198t
Vastus lateralis, *25*, *26*, *27*, 188, 196, *196*, *197*, 198t, *199*
Vastus medialis, *25*, *26*, 188, 196, *196*, 198t
Vein(s), 329, 346, 347t, 355–59
   from abdominal and thoracic walls, 358
   from abdominal viscera, 358–59, *358*
   from brain, head, and neck, 357, *357*
   from lower limb and pelvis, 359
   from upper limb and shoulder, 357–58, *357*
   venous pathways, 355–56
   wall of, *342*
Vena cava, *329*, 333
Venipuncture, 358
Venoconstriction, 350–51
Venography, 362
Venous blood flow, 350–51
Venous pressure, increased, 486, 486t
Venous sinus, *357*, 373

Ventilation. *See* Breathing
Ventral cavity, 8, *9*
Ventral respiratory group, *447*, 448, *448*
Ventral root, *223*, *225*, 238–39
Ventricle (brain), 209, 230, *231*
Ventricle (heart), 331
   left, *26*, *329*, *330*, *331*, 332–33, *334*, *335*, 351, *539*
   right, *26*, *329*, *330*, 331–33, *331*, *334*, *335*, 351, 540
Ventricular diastole, 336, *336*, *337*, 347
Ventricular syncytium, 338, *339*
Ventricular systole, 336, *336*, *337*, 347
Venule, 329, 346, 347t, *367*
Vermiform appendix. *See* Appendix
Vermis, 234
Vertebrae, *11*, *28*, 142–46, 145t, *223*
Vertebral, definition of, *16*, 17
Vertebral arch, 142
Vertebral artery, 353, *353*, *354*, *356*
Vertebral canal, 8, *9*, 142, *224*
Vertebral column, 134–36, *135*, 137t, 142–46
Vertebral foramen, 142, *145*
Vertebral vein, *357*
Vertebra prominens, *144*
Vertebrochondral ribs. *See* False ribs
Vertebrosternal ribs. *See* True ribs
Vertigo, 276
Very-low-density lipoprotein (VLDL), 316, 317t, 414–15
Vesalius, Andreas, *3*
Vesicles, 53, 56, 58t
Vestibular branch, of vestibulocochlear nerve, 237, 238t, 265
Vestibular bulb, 512
Vestibular gland, 512, 512t
Vestibular membrane, 262
Vestibular nerve, *262*
Vestibule
   of ear, 265
   female reproductive organ, 512, 512t
   of mouth, 396, *396*, *397*
Vestibulocochlear nerve (VIII), *236*, 237, 238t, *260*
   cochlear branch of, 237, 238t, 264
   vestibular branch of, 237, 238t, 265
Vetter, David, 384, *384*
Viagra, 496
Victoria, Queen of England, 318
Villi, intestinal, 412, *413*, 414
Virilism, 300
Viscera, 8
Visceral pain, 255, *255*, 256
Visceral pericardium, 10, *11*, *329*, *330*